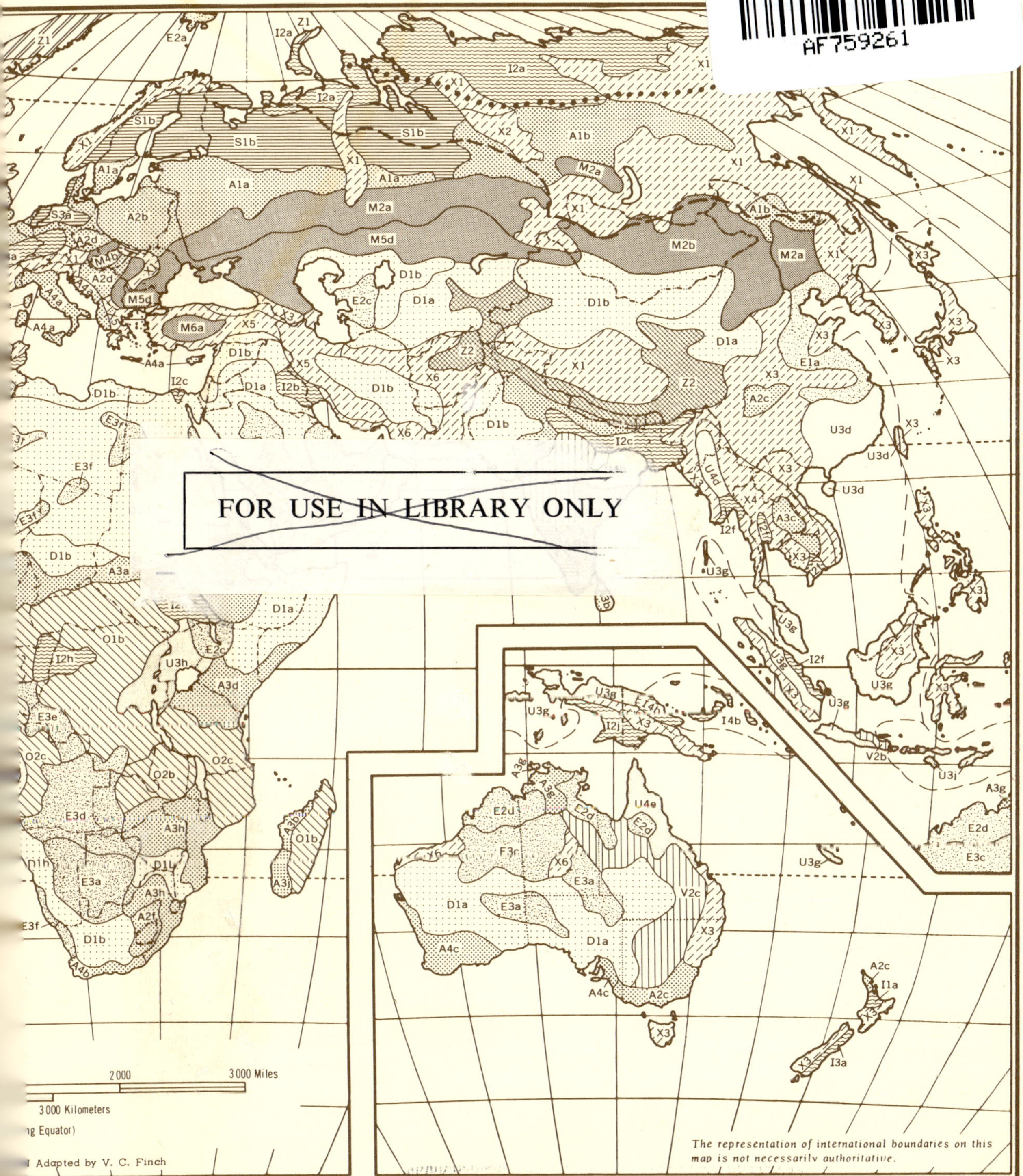

MAY 1972

USDA-SCS-HYATTSVILLE MD 1972

The

ENCYCLOPEDIA

of

SOIL SCIENCE, PART 1

ENCYCLOPEDIA OF EARTH SCIENCES SERIES

Volume

I THE ENCYCLOPEDIA OF OCEANOGRAPHY/*Rhodes W. Fairbridge*

II THE ENCYCLOPEDIA OF ATMOSPHERIC SCIENCES AND ASTROGEOLOGY/*Rhodes W. Fairbridge*

III THE ENCYCLOPEDIA OF GEOMORPHOLOGY/*Rhodes W. Fairbridge*

IVA THE ENCYCLOPEDIA OF GEOCHEMISTRY AND ENVIRONMENTAL SCIENCES/*Rhodes W. Fairbridge*

VI THE ENCYCLOPEDIA OF SEDIMENTOLOGY/*Rhodes W. Fairbridge and Joanne Bourgeois*

VII THE ENCYCLOPEDIA OF PALEONTOLOGY/*Rhodes W. Fairbridge and David Jablonski*

VIII THE ENCYCLOPEDIA OF WORLD REGIONAL GEOLOGY, PART 1: Western Hemisphere (Including Antarctica and Australia)/*Rhodes W. Fairbridge*

XII THE ENCYCLOPEDIA OF SOIL SCIENCE, PART 1: Physics, Chemistry, Biology, Fertility, and Technology/*Rhodes W. Fairbridge and Charles W. Finkl, Jnr.*

ENCYCLOPEDIA OF EARTH SCIENCES, VOLUME XII

The ENCYCLOPEDIA *of* SOIL SCIENCE PART 1

Physics, Chemistry, Biology, Fertility, and Technology

EDITED BY

Rhodes W. Fairbridge

Columbia University

Charles W. Finkl, Jnr.

Ocean Sciences Center
Nova University

Dowden, Hutchinson & Ross, Inc.
Stroudsburg Pennsylvania

Library of Congress Catalog Card Number: 78-31233
ISBN: 0-87933-176-3

83 82 81 80 79 5 4 3 2 1
Manufactured in the United States of America.

Library of Congress Cataloging in Publication Data
Main entry under title:
The Encyclopedia of soil science.
(Encyclopedia of earth sciences; no. 12)
Bibliography: p.
Includes index.
CONTENTS: pt. 1. Physics, chemistry, biology, fertility, and technology.
1. Soil science–Dictionaries. I. Fairbridge, Rhodes Whitmore, 1914- II. Finkl, Charles W., 1941- III. Series: Encyclopedia of earth sciences series; v. 12.
S592.E52 631.4'03 78-31233
ISBN 0-87933-176-3

Distributed world wide by Academic Press,
a subsidiary of Harcourt Brace Jovanovich,
Publishers.

FOREWORD

The soil is the link between the rock core of the earth and the life on its surface. Despite its importance, that link is a negligible fraction of the earth's mass, being no more than the outermost thin rind of the surface crust, itself a minor part of the mass of the planet. Yet the soil provides the foothold for the land plants on which mankind relies for food, fiber, and shelter.

Occurring generally on the land surface, soils have been commonplace throughout the existence of mankind. Because they are so commonplace and usually lack striking features, soils have long been taken for granted. About 2,000 years ago Columella wrote that the earth was ". . . the common mother of all things —because she has always brought forth all things and is destined to bring them forth continuously. . . ."

Recognition of the soil as a natural object worthy of study has occurred gradually over the last 150 years, chiefly within the last century. Interest in soils seems to have been stimulated initially by a widespread belief, stated eloquently by Malthus, that population growth would soon outstrip the food supply. Ways and means of increasing food production thus seemed vital; many people searched for methods of increasing agricultural production. The science of chemistry was advancing at the time and chemists applied their techniques to the study of soils. Their early investigations consisted largely of analyses of samples. Actually, a surface layer of the soil a few centimeters or few inches thick was considered the entire soil. Examination of anything below the surface layer was not believed to be necessary.

In reality, the soil mantle on the land surface is commonly a meter or so (a few feet) deep but ranges from a few millimeters (a fraction of an inch) at the one extreme to about three meters (ten feet) at the other. The soil profile (a vertical slice down through the mantle), consisting of related and either contrasting or similar horizontal layers (horizons), forms the basis for the study and understanding of the nature and origin of soils.

From its modest beginnings, the study of the soil mantle has grown until it is now known that the soil is as complex as it is commonplace, that there is a large variety of kinds of soils both locally and regionally over the world, and that many approaches and techniques are needed in order to understand the nature and properties of soils.

There is still much concern about the capacity of the earth to provide food and fiber for a growing population. Current knowledge of the soil resources of the world as a whole indicates, however, that they could support at least a 30-fold increase in production. To achieve marked increases in total output, the management of soils would have to be greatly improved. To this end, better understanding of soils is needed. Moreover, sustained production of food and fiber at high levels would be greatly facilitated by better understanding of soils.

This volume is therefore an attempt to provide information on the nature and properties of soils as a contribution toward that understanding. The emphasis is on soil physics, chemistry, biology, and fertility. The articles cover topics such as soil water, mineralogy of soil materials, particle sizes, the composition of organic matter, and microbial life in soils and their capacity to provide nutrient elements for plant growth. This list of topics is illustrative rather than comprehensive. It is hoped that information in this volume will make it easier to provide a "greater measure of care to our soils, this slight superficial and inconstant covering of the earth. . . ."

ROY W. SIMONSON

PREFACE

Soil science is both a pure and applied science that systematizes what we know about soil. For a long time the study of soil was regarded as an applied science, i.e., a branch of agriculture that dealt with soil-related problems. Modern interests have, however, enlarged the original concept to include the study of soil as a natural body; the importance of soil resources in reclamation, recreation, urban planning, pollution abatement, and other nonagricultural activities are also recognized as integral to the discipline. Soil science thus embraces that body of knowledge that pertains to soil systems, i.e., the study of soil as independent natural bodies at the surface of the earth, as a surficial mantle of weathered rock, as a medium for plant growth, and as a constituent of the world ecological resource base. Soil is a living entity, and it is, indeed, the viable nature of the biologically active outer crust of the earth (*pedosphere*) that distinguishes soil science from geology, which generally treats the inactive (non-living) parts of the earth (e.g., the lithosphere).

Because the scope of soil science is broad and involves the application of other sciences—especially biology, botany, ecology, chemistry, physics, and geology—the discipline is subdivided in many different ways. The International Society of Soil Science, for example, lists the following seven commissions: (I) soil physics; (II) soil chemistry; (III) soil biology; (IV) soil fertility and plant nutrition; (V) soil genesis, classification and cartography, (VI) soil technology; and (VII) soil mineralogy. We used this breakdown as the basis for grouping the various subfields of soil science into two companion volumes (see below).

In the Soviet Union and a number of other countries, the terms *pedology* and *soil science* are considered synonymous. However, some scientists (especially those in the United States) tend to equate pedology with soil genesis, classification, and cartography. Because of the different connotations associated with the concept, scope, and definition of pedology, we have used "soil science" as a general term with distinctive subtitles to distinguish the material comprehended.

The Encyclopedia of Soil Science was originally conceived as a single, comprehensive volume, but, as a result of the wide scope of the discipline and rising production costs of large volumes, the initial plan had to be modified. Two volumes will now appear under the general egis of "soil science." This volume, *Soil Science, Part 1,* is number XII in the *Encyclopedia of Earth Sciences;* it deals with salient aspects of soil physics, chemistry, biology, fertility, and technology. Facets of soil morphology, genesis, classification, and geography will be considered in *Part 2,* a companion volume scheduled for publication in the near future.

For several decades now, scientists of the world, and most conspicuously those of the United States, have made great strides in the procurement of knowledge. The development of scientific pursuits has fostered new fields of study, which in turn have created a demand for new books and periodicals. As in other sciences, no less impressive is the rise in prestige and dignity of soil science as an independent natural science. With the need for increased world food production, soil scientists are endeavoring to meet the challenge for a more rational and intensive use of world soil resources.

This volume gives a comprehensive, alphabetical treatment of basic soil science. Professional soil scientists, agronomists, geologists, and others will no doubt find this book a useful reference volume that summarizes basic facts and concepts. Graduate students, reference librarians, and others who deal with the nature, properties, and attributes of soil will have access to essential information and be directed to pertinent specialized literature. By now your volume editor should indeed be exceedingly well informed having edited some 1850 pages of typewritten manuscript and authored several entries himself, all in the compilation of this single work. But, things being such as they are and his memory being somewhat less than perfect, alas, he will return to the volume again and again. For others wishing to check a fine point or to just confirm something they already "knew," this reference volume should be a helpful reminder.

Numerous terms are defined in this encyclopedia; it is emphasized, however, that this book is *not* a dictionary but rather a compendium of knowledge. We have tried to present up-to-date information that is relevant to the central concepts of the various sub-disciplines of soil science. Controversial issues will not be avoided, but new and untested ideas are not emphasized. A specific entry may, however, reflect the individual outlook of a particular author. The interested reader will hopefully appreciate the personal touch in the many articles prepared by a broad international spectrum of contributors. In addition, with more than a hundred authors from approximately twenty countries and nearly 150 major entries, some minor overlap in coverage is inevitable. However, the extensive cross-referencing should allow the reader to note differences in opinion or approach in related topics.

Publications and Standard References

The growth of soil science, especially in recent decades, is reflected by the publication of numerous new journals, texts, and specialized volumes. Even though this encyclopedia does not cover the field in depth, it does give an overview and directs the reader to relevant publications in various branches of soil science and related fields. A selection of recent texts as well as older reference books that complement this volume are listed below. Those books marked with an asterisk (*) contain rather extensive bibliographies.

General References

Brady, N. C., 1974. *The Nature and Properties of Soils.* New York: Macmillan, 639p.

De Sigmond, A. A. J., 1938. *The Principles of Soil Science.* London: Thomas Murby, 362p.

Donahue, R. L.; Miller, R. W.; and Schickluna, J. C., 1977. *Soils: An Introduction to Soils and Plant Growth.* Englewood Cliffs, New Jersey: Prentice-Hall, 612p.

Duchaufour, P. 1965. *Précis de Pédologie.* Paris: Masson, 481p.

Foth, H. D., 1978. *Fundamentals of Soil Science.* New York: Wiley, 436p.

Gaucher, G., 1968. *Traité de Pédologie: Agricole Le Sol.* Paris: Dunod, 578p.

Kellogg, C. E., 1966. *The Soils That Support Us.* New York: Macmillan, 370p.

Leeper, G. W., 1964. *Introduction to Soil Science.* Parkville, Victoria: Melbourne University Press, 253p.

*Scheffer, F.; and Schachtschabel, P., 1970. *Lehrbuch der Bodenkunde.* Stuttgart: Enke, 349p.

Tamhane, R. V.; Motiramani, D. P.; Bali, Y. P.; and Donahue, R. L., 1966. *Soils: Their Chemistry and Fertility in Tropical Asia.* New Delhi: Prentice-Hall of India Private Limited, 475p.

*U.S. Department of Agriculture Staff, 1938. *Soils and Men, Yearbook of Agriculture 1938.* Washington, D.C.: U.S. Government Printing Office, 1232p.

U.S. Department of Agriculture Staff, 1957. *Soil, Yearbook of Agriculture 1957.* Washington, D.C.: U.S. Government Printing Office, 784p.

Soil Physics

*Baver, L. D.; Gardner, W. H.; and Gardner, W. R., 1972. *Soil Physics.* New York: Wiley, 498p.

*Childs, E. C., 1969. *An Introduction to the Physical Basis of Soil Water Phenomena.* New York: Interscience, 493p.

Hillel, D., 1971. *Soil and Water: Physical Principles and Processes.* New York: Academic, 288p.

Ingles, O. G., and Metcalf, J. B., 1973. *Soil Stabilization.* New York: Wiley, 374p.

*Kirkham, D., and Powers, W. C., 1972. *Advanced Soil Physics.* New York: Interscience, 534p.

*Moldenhauer, W. C., ed., 1975. *Soil Conditioners.* Madison, Wisconsin: Soil Science Society of America, 186p.

Yong, R. N., and Warkentin, B. P., 1975. *Soil Properties and Behaviour.* Amsterdam: Elsevier, 449p.

Soil Chemistry

*Bear, F. E., 1964. *Chemistry of the Soil.* New York: Reinhold, 515p.

*Bolt, G. H., and Bruggenwert, M. G. M., 1976. *Soil Chemistry, A. Basic Elements.* Amsterdam: Elsevier, 281p.

*McLaren, A. D., General Editor, 1967 et seq. *Soil Biochemistry.* New York: Marcel Dekker, 4 Vols.: (1) McLaren, A. D., and Peterson, G. H., 1967, 528p.; (2) McLaren, A. D., and Skujinš, J., 1971, 544p.; (3) Paul, E. A., and McLaren, A. D., 1975, 352p.; and (4) Paul, E. A., and McLaren, A. D., 1975, 296p.

Marshall, C. E., 1977. *The Physical Chemistry and Mineralogy of Soils; Vol. 2, Soils in Place.* New York: Wiley, 313p.

van Olphen, H., 1977. *An Introduction to Clay Colloid Chemistry.* New York: Wiley, 318p.

Soil Biology

*Alexander, M., 1961. *Introduction to Soil Microbiology.* New York: Wiley, 472p.

Andrews, W. A., ed., 1972. *A Guide to the*

Study of Soil Ecology. Toronto: Prentice-Hall, 198p.

*Brauns, A., 1968. *Praktische Bodenbiologie.* Stuttgart: Gustav Fischer, 470p.

*Clarke, F. E., 1949. *Ecological Associations Among Soil Microorganisms.* Paris: UNESCO, 350p.

Gray, T. R. G., and Williams, S. T., 1971. *Soil Microorganisms.* Edinburgh: Oliver and Boyd, 240p.

*Kühnelt, W., 1961. *Soil Biology: With Special Reference to the Animal Kingdom.* London: Faber and Faber, 397p. (Transl. by N. Walker)

*Müller, G., 1965. *Bodenbiologie.* Jena: Gustav Fischer, 889p.

*Pochon, J., and de Barjac, H., 1958. *Traité de Microbiologie des Sols.* Paris: Dunod, 685p.

Soil Fertility

*Black, C. A., 1968. *Soil-Plant Relationships.* New York: Wiley, 792p.

Finck, A., 1969. *Pflanzernernährung in Stichworten,* Kiel: Hirt, 200p.

*Martin, J. H., and Leonard, W. N., 1967. *Principles of Field Crop Production.* New York: Macmillan, 1044p.

*Russell, E. W., 1973. *Soil Conditions and Plant Growth.* London: Longmans, 849p.

*Sprague, H. B., ed., 1964. *Hunger Signs in Crops: A Symposium.* New York: McKay, 461p.

Tisdale, S. L., and Nelson, W. L., 1975. *Soil Fertility and Fertilizers.* New York: Macmillan, 694p.

Soil Mechanics

*Jumakis, A. R., 1962. *Soil Mechanics.* New York: Van Nostrand, 791p.

Lambe, T. W., and Whitman, R. V., 1969. *Soil Mechanics.* New York: Wiley, 553p.

*Lee, I. K., 1968. *Soil Mechanics.* Sydney: Butterworths, 625p.

Sowers, G. B., and Sowers, G. F., 1970. *Introductory Soil Mechanics and Foundations.* New York: Macmillan, 556p.

Terzaghi, K., 1943. *Theoretical Soil Mechanics.* New York: Wiley, 510p.

Soil Mineralogy

*Brown, G., 1961. *The X-Ray Identification and Crystal Structure of Clay Minerals.* London: Mineralogical Society, 544p.

*Dixon, J. B.; Weed, S. B.; et al, eds., 1977. *Minerals in Soil Environments.* Madison, Wisconsin: Soil Science Society of America, 958p.

Farmer, V. C., ed., 1974. *The Infrared Spectra of Minerals.* London: Mineralogical Society, 539p.

Grim, R. E., 1968. *Clay Mineralogy.* New York: McGraw-Hill, 596p.

*Millott, G., 1970. *Geology of Clays.* New York: Springer-Verlag, 429p.

Soil Technology

Anon., 1966. *Testing Soils for Engineering Purposes.* Sydney: Standards Association of Australia, 121p.

Anon., 1970. *Special Procedures for Testing Soil and Rock for Engineering Purposes.* Philadelphia: American Society for Testing and Materials, 630p.

*Black, C. A., ed., 1965. *Methods of Soil Analysis.* Madison, Wisconsin: American Society of Agronomy, 1572p.

Jackson, M. L., 1960. *Soil Chemical Analysis.* Englewood Cliffs, New Jersey: Prentice-Hall, 498p.

Jones, J. B., ed., 1974. *Handbook on Reference Methods for Soil Testing.* Athens, Georgia: University of Georgia Press, 101p.

Piper, C. S., 1950. *Soil and Plant Analysis.* Adelaide, South Australia: University of Adelaide Press, 368p.

Walsh, L. K., ed., 1971. *Instrumental Methods for Analysis of Soils and Plant Tissue.* Madison, Wisconsin: Soil Science Society of America, 222p.

The study of physical, chemical, and biological soil systems requires investigation from diverse points of view. Thus in addition to the mainline journals that deal exclusively with the multifaceted nature of soils, there are other periodicals that emphasize certain aspects of the soil environment or individual soil components, e.g., silicate clay minerals. The following list, which is not intended to be comprehensive (for a more complete review of source materials, see the article *Journals, Soil Science*), features some standard serial publications.

Agronomy Journal (American Society of Agronomy, 677 South Segoe Road, Madison, Wisconsin, USA 53711)

Association Francaise Pour l'Etude du Sol Bulletin (C.N.R.A., Route de Saint-Cyr, Versailles 78000, France)

Australian Journal of Soil Research (Editorial and Publications Service, C.S.I.R.O., 314 Albert Street, East Melbourne, Victoria, Australia 3002)

Canadian Journal of Soil Science (Agricultural Institute of Canada, 151 Slater Street,

Suite 907, Ottawa, Ontario, Canada K1P 5H4)

Catena (Catena Verlag, Brockenblick 8, D 3302 Cremlingen-Destedt, West Germany)

Egyptian Journal of Soil Science (National Information and Documentation Centre, Dokki, Cairo, Egypt, Egyptian Arab Republic)

Geoderma (Elsevier Scientific Publishing Company, Post Office Box 330, 1000 AH Amsterdam, The Netherlands)

Geotechnique (Thomas Telford Limited, Publications Division, 26–34 Old Street, London, England EC1V 9AD)

Journal of Environmental Quality (American Society of Agronomy, Crop Science Society of America, and Soil Science Society of America, 677 South Segoe Road, Madison, Wisconsin, USA 53711)

Journal of Soil Science (Oxford University Press, Press Road, Neasden, London, England NW10 ODD)

Moscow University Soil Science (English translation of *Moscow Universitet Vestnik Pochvovedenie*) (Allerton Press, Inc., 150 Fifth Avenue, New York, New York, USA 10011)

Pedobiologia (VEB Gustav Fischer Verlag, Villengang 2, Postfach 176, 69 Jena, East Germany)

Pédologie (Société Belge de Pédologie, Rozier 44, 9000 Ghent, Belgium)

Philippine Journal of Soils (Department of Agriculture and Natural Resources, Agricultural Information Division, Diliman, Quezon City, Philippines)

Plant and Soil (Martinus Nijhoff, Lange Voorhout 9–11, Box 269, The Hague, The Netherlands)

Revue d'Ecologie et de Biologie du Sol (Centre des Revues Dunod Gauthier Villars, B.P. 119, 93104 Montreuil Cedex, France)

Soil Biology and Biochemistry (Pergamon Press Limited, Headington Hill Hall, Oxford, England OX3 OBW)

Soil Conservation (U.S. Department of Agriculture, Washington, D.C., USA 20250)

Soil Science (The Williams and Wilkins Company, 428 East Preston Street, Baltimore, Maryland, USA 21202)

Soil Science and Plant Nutrition (c/o Japan Academic Societies Center, 2-4-16 Yayoi, Bunkyo-ku, Tokyo 113, Japan)

Soil Science Society of America Journal (Soil Science Society of America, 677 South Segoe Road, Madison, Wisconsin, USA 53711)

Sols Africans (Organization of African Unity, Inter-African Bureau for Soils, B.P. 1352, Bangui, Central African Republic)

Soviet Soil Science (English translation of *Pochvovedeniye* and other Russian periodicals) (Scripta Publishing Corporation, 1511 K Street NW, Washington, D.C., USA 20005)

Zeitschrift für Pflanzenernährung und Bodenkunde (Verlag Chemie Gmbh., Poppelallee 3, Postfach 1260, 6940 Weinheim, West Germany)

In addition to primary soil science journals, review journals, such as *Advances in Agronomy* (Academic Press) and *Earth Science Reviews* (Elsevier), regularly canvass specific soil subjects. Also available are several abstracting services that publish special sections on soils: *Soil and Fertilizers* (Commonwealth Bureau of Soils, Rothamsted, Harpenden, England), *Chemical Abstracts, Bibliography and Index of Geology*, and *Biological Abstracts*. The International Society of Soil Science (c/o FAO, Via delle Terme di Caracalla, Rome, Italy 00153) publishes the *Transactions* of its international congresses and also a quarterly *Bulletin* that lists new texts and reports. A particularly useful reference is *Geo Abstracts A: Landforms and the Quaternary* (Geo Abstracts Ltd., University of East Anglia, Norwich, England NR4 7TJ), which lists a wide range of topics in soil science.

Information of interest to soils specialists is often issued on an occasional basis by various governmental agencies in many countries. Additional primary sources for further information about soils in the United States, for example, are the federal and state departments of agriculture which publish leaflets and

TABLE 1. Technical Service Centers of the United States Soil Conservation Service.

NEBRASKA 134 South 12th Street Lincoln, NE 68508 (402) 475–3371
OREGON Federal Building, Rm. 209 511 N.W. Broadway Portland, OR 97209 (503) 221–2824
PENNSYLVANIA 7600 West Chester Pike Upper Darby, PA 19082 (215) 352–5800, Ext. 333
TEXAS Fort Worth Federal Center P.O. Box 11222 Fort Worth, TX 76110 (817) 334–5456

technical bulletins. Queries may be sent to the state officers, or alternatively one may write to the Director of Information, U.S. Department of Agriculture, Washington, D.C. 20250. Another important source of soils information is the Soil Conservation Service, a branch of the U.S. Department of Agriculture. Specific requests indicating what information is required should be directed to regional Soil Conservation Service facilities. Table 1 gives the location of regional centers that provide technical services. Other government agencies such as the Commonwealth scientific and industrial research organizations (e.g., C.S.I.R.O.: Australia; C.S.I.R.: New Zealand, South Africa) and local departments of agriculture also produce noteworthy reports that deal with various aspects of soil science.

How to Use This Encyclopedia

The alphabetical organization of topics provides easy access to the mainstream articles. Because there is such a wide range of subjects, many specific concepts or applications have been grouped under general headings. For example, to look up "cation exchange," simply turn to the Cs where the topic is listed. The main in-text cross-reference, however, directs the reader to the more general topic of *Exchange Phenomena,* which considers both anion and cation exchange. Should a subject not appear as a main entry or as an in-text cross-reference, the comprehensive subject index may be consulted at the back of the volume. An author citation index is also provided.

Most articles are extensively cross-referenced, both in the body of the text and at the end of the reference list for each entry. For example, the abbreviation "q.v." (*quode vide*) in the context, "carbon cycle (q.v.)" means that an entry by that title appears in the volume. Cross-references to other volumes in the series are also included. The following is a list of published and planned volumes:

I:	Oceanography
II:	Atmospheric Sciences and Astrogeology
III:	Geomorphology
IVA:	Geochemistry and Environmental Sciences
IVB:	Mineralogy
V:	Petrology
VI:	Sedimentology
VII:	Paleontology
VIII:	World Regional Geology, Part 1: Western Hemisphere
VIII:	World Regional Geology, Part 2: Europe and Asia
VIII:	World Regional Geology, Part 3: Africa and Middle East
IX:	Stratigraphy
X:	Structural Geology
XII:	Soil Science, Part 1: Physics, Chemistry, Biology, Fertility, and Technology
XII:	Soil Science, Part 2: Morphology, Genesis, Classification, and Geography
XIII:	Applied Geology
XIV:	Petroleum Geology
XV:	Beaches and Coastal Environments
XVI:	Natural Resources and Energy Conservation
XVII:	Snow, Ice, and Glaciology
XVIII:	Geohydrology and Water Supply
XIX:	World Ore Deposits
XX:	Ore Genesis and Metallogeny
XXI:	Mining and Mineral Resources
XXII:	Volcanoes and Volcanology
XXIII:	Geophysics
XXIV:	History of Geology

Each entry is followed by a list of references. Some references are not included in the text; they were listed, however, because they may lead the reader to additional, more detailed information or to summaries that contain extensive bibliographies (indicated by an asterisk). A concerted effort was made to provide complete bibliographic information so that materials could be easily retrieved in libraries or through reference services. Journal titles are generally abbreviated according to the style recommended by *Serial Publications Commonly Cited in Technical Bibliographies of the United States Geological Survey*, or according to the format suggested in the *Handbook and Style Manual for ASA, CSSA, and SSSA Publications.*

Units of Measure

Standard abbreviations are used for measures of area, length, volume, mass, pressure, and temperature. The Systéme International d'Unités (SI), as endorsed by the International Organization for Standardization in 1960, is used throughout the volume. Because some tables, figures, and maps retain units of the English system (Customary Units) or because it is convenient to retain older units for practical purposes [e.g., pounds per acre (kg 10^{-4} m^{-2}), hundred weight per acre (1.26 kg 10^{-2} m^{-2}), tons per acre (25.14 kg 10^{-2} m^{-2}); exchange capacity is still expressed in the practical unit of mequiv./100g], equivalent units are usually given in parentheses. With such a wide range of topics it is impossible to list here all the units used in this volume. All units are, however, defined locally as they are used in each article. For those readers not especially familiar with SI units or for those who may wish to brush up on the refinements of the traditional metric system, the following tables (2, 3, and 4) may be useful. Because some SI units are of inconvenient size, certain prefixes may be used to indicate fractions or multiples of the basic or derived units, as indicated in Table 5. Also there are a number of familiar terms that differ from the corresponding SI units only by a power of ten. Even though these units are not part of the SI and their use is to be discouraged, they will probably be used for some time. Some of the units commonly applied by soils specialists are listed in Table 6. Some SI units and customary equivalents, and conversion factors are given in Tables 7 and 8, respectively. Still other symbols or letter combinations, some of which are peculiar to soil science, are listed in Table 9. The FAO Soils Bulletin 28 (1975), *S.I. Units and Nomenclature in Soil Science,* coordinates many special quantities used in soil science and as such is a valuable reference.

TABLE 2. Basic and Supplementary SI Units.

Quantity	Unit Name	Unit Symbol
Length	Meter	m
Mass	Kilogram	kg
Time	Second	s
Electric current	Ampere	A
Thermodynamic temperature	Degree Kelvin	°K
Luminous intensity	Candela	cd
Amount of substance	Mole	mol
Plane angle	Radian	rad
Solid angle	Steradian	sr

Source: Reprinted from *Australian Jour. Soil Research* 9 (1971).

TABLE 3. Derived SI Units with Special Names

Physical Quantity	Unit Name	Unit Symbol	Unit Definition
Energy	Joule	J	$kg\ m^2\ s^{-2}$
Force	Newton	N	$kg\ m\ s^{-2} = J\ m^{-1}$
Power	Watt	W	$kg\ m^2\ s^{-3} = J\ s^{-1}$
Electric charge	Coulomb	C	A s
Electric potential difference	Volt	V	$kg\ m^2\ s^{-3}\ A^{-1} = J\ A^{-1}\ s^{-1}$
Electric resistance	Ohm	Ω	$kg\ m^2\ s^{-3}\ A^{-2} = V\ A^{-1}$
Electric capacitance	Farad	F	$A^2\ s^4\ kg^{-1}\ m^{-2}\ A\ s\ V^{-1}$
Magnetic flux	Weber	Wb	$kg\ m^2\ s^{-2}\ A^{-1} = V\ s$
Luminous flux	Lumen	lm	cd sr
Illumination	Lux	lx	$cd\ sr\ m^{-2}$
Frequency	Hertz	Hz	s^{-1}

Source: Reprinted from *Australian Jour. Soil Research* 9 (1971).

TABLE 4. Other Derived SI Units.

Physical Quantity	SI Unit	Symbol
Area	Square meter	m^2
Volume	Cubic meter	m^3
Density	Kilogram per cubic meter	$kg\ m^{-3}$
Velocity	Meter per second	$m\ s^{-1}$
Angular velocity	Radian per second	$rad\ s^{-1}$
Acceleration	Meter per second squared	$m\ s^{-2}$
Pressure	Newton per square meter	$N\ m^{-2}$
Kinematic viscosity, diffusion coefficient	Square meter per second	$m^2\ s^{-1}$
Dynamic viscosity	Newton second per square meter	$N\ s\ m^{-2}$
Electric field strength	Volt per meter	$V\ m^{-1}$
Magnetic field strength	Ampere per meter	$A\ m^{-1}$
Luminance	Candela per square meter	$cd\ m^{-2}$
Heat capacity (specific)	Joule per kilogram degree Kelvin	$J\ K^{-1}$
Thermal conductivity	Watt per meter degree Kelvin	$W\ m^{-1}\ K^{-1}$
	Joule per meter second degree Kelvin	$J\ s^{-1}\ m^{-1}\ K^{-1}$
Surface tension	Newton per meter	$N\ m^{-1}$
	Joule per square meter	$J\ m^{-2}$

Source: Reprinted from *Australian Jour. Soil Research* 9 (1971).

TABLE 5. Prefixes for SI Units.

Fraction	Prefix	Symbol	Multiple	Prefix	Symbol
10^{-1}	deci	d	10	deka	da
10^{-2}	centi	c	10^2	hecto	h
10^{-3}	milli	m	10^3	kilo	k
10^{-6}	micro	μ	10^6	mega	M
10^{-9}	nano	n	10^9	giga	G
10^{-12}	pico	p	10^{12}	tera	T
10^{-15}	femto	f			
10^{-18}	atto	a			

Source: Reprinted from *Australian Jour. Soil Research* 9 (1971).

TABLE 6. Named Units that Are Decimal Fractions and Multiples of SI Units.

Physical Quantity	Name	Symbol	Definition
Length	Angstrom	A	10^{-10} m = 10^{-1} nm
Length	Micron	μm	10^{-6} m
Area	Hectare	ha	10^4 m^2
Volume	Liter	l	10^{-3} m^3
Mass	Ton	t	10^3 kg
Mass	Gram	g	10^{-3} kg
Force	Dyne	dyn	10^{-5} N
Pressure	Bar	bar	10^5 N m^{-2}
Energy	Erg	erg	10^{-7} J
Kinematic viscosity, diffusion coefficient	Stokes	St	10^{-4} m^2 s^{-1}
Dynamic viscosity	Poise	P	10^{-1} kg m^{-1} s^{-1}

Source: Reprinted from *Australian Jour. Soil Research* 9 (1971).

TABLE 7. SI Units and Customary Equivalents. (The values given here have been rounded to four significant digits, except for 0.01 bar which is the exact equivalent of 1 kPa. Hectare (ha), an alternate name for square hectometer (hm^2) is restricted to the measurement of land or water areas. Liter (l) is a special name for cubic decimeter (dm^3) and its use is restricted to the measurement of liquids and gases. Metric ton (t), an alternate name for megagram (Mg), is normally restricted to commercial usage.)

SI Unit		U.S. Customary Equivalent	
Length			
millimeter (mm)	=	0.039 37	inch (in)
meter (m)	=	3.281	feet (ft)
	=	1.094	yards (yd)
kilometer (km)	=	0.621 4	mile (mi)
	=	0.540 0	mile, nautical (nmi)
Area			
centimeter^2 (cm^2)	=	0.155 0	inch^2 (in^2)
meter^2 (m^2)	=	10.76	feet^2 (ft^2)
	=	1.196	yards^2 (yd^2)
	=	0.000 247 1	acre
hectometer^2 (hm^2)	=	2.471	acres
	=	0.003 861	section (640 acres or 1 mi^2)
kilometer^2 (km^2)	=	0.386 1	mile^2 (mi^2)
Volume			
centimeter^3 (cm^3)	=	0.061 02	inch^3 (in^3)
decimeter^3 (dm^3)	=	61.02	inches^3 (in^3)
(i.e. 1 liter)	=	2.113	pints (pt)
	=	1.057	quarts (qt)
	=	0.264 2	gallon (gal)
	=	0.035 31	foot^3 (ft^3)
meter^3 (m^3)	=	35.31	feet^3 (ft^3)
(i.e. 1000 liters)	=	1.308	yards^3 (yd^3)
	=	264.2	gallons (gal)
	=	0.000 810 7	acre-foot (acre-ft)
hectometer^3 (hm^3)	=	810.7	acre-feet (acre-ft)
kilometer^3 (km^3)	=	0.239 9	mile^3 (mi^3)

TABLE 7 continued

SI Unit			U.S. Customary Equivalent
Volume per unit time (flow)			
decimeter^3 per second (dm^3/s)	=	0.035 31	foot^3 per second (ft^3/s)
	=	2.119	feet^3 per minute (ft^3/min)
decimeter^3 per second (dm^3/s)	=	15.85	gallons per minute (gal/min)
meter^3 per second (m^3/s)	=	35.31	feet^3 per second (ft^3/s)
	=	15 850	gallons per minute (gal/min)
Mass			
gram (g)	=	0.035 27	ounce avoirdupois (oz avdp)
kilogram (kg)	=	2.205	pounds avoirdupois (lb avdp)
megagram (mg)	=	1.102	tons, short (2 000 lb)
	=	0.984 2	ton, long (2 240 lb)
Mass per unit volume (density)			
kilogram per meter^3 (kg/m^3)	=	0.062 43	pound per foot^3 (lb/ft^3)
Pressure			
kilopascal (kPa)	=	0.145 0	pound-force per inch^2 (lbf/in^2)
	=	0.009 869	atmosphere, standard (atm)
	=	0.01	bar
	=	0.296 1	inch of mercury at 60°F (in Hg)
megapascal (MPa)	=	10	bars
Temperature			

temp kelvin (K) = [temp deg Fahrenheit (°F) + 459.67]/1.8
temp deg Celsius (°C) = [temp deg Fahrenheit (°F) - 32]/1.8

Source: Reprinted from *U.S. Geol. Survey Jour. Research* endcover.

TABLE 8. Conversion Factors for U.S. and Metric Units.

To Convert Column 1 into Column 2, Multiply By	Column 1	Column 2	To Convert Column 2 into Column 1, Multiply By
Length			
0.621	kilometer (km)	mile (mi)	1.609
1.094	meter (m)	yard (yd)	0.914
0.394	centimeter (cm)	inch (in)	2.54

TABLE 8 continued

To Convert Column 1 into Column 2, Multiply By	Column 1	Column 2	To Convert Column 2 into Column 1, Multiply By
	Area		
0.386	kilometer^2 (km^2)	mile^2 (mi^2)	2.590
247.1	kilometer^2 (km^2)	acre (acre)	0.00405
2.471	hectare (ha)	acre (acre)	0.405
	Volume		
0.00973	meter^3 (m^3)	acre-inch	102.8
3.532	hectoliter (hl)	cubic foot (ft^3)	0.2832
2.838	hectoliter (hl)	bushel (bu)	0.352
0.0284	liter (l)	bushel (bu)	35.24
1.057	liter (l)	quart (liquid) (qt)	0.946
	Mass		
1.102	ton (metric)	ton (U.S.)	0.9072
2.205	quintal (q)	hundredweight, cwt (short)	0.454
2.205	kilogram (kg)	pound (lb)	0.454
0.035	gram (g)	ounce (avdp) (oz)	28.35
	Pressure		
14.50	bar	lb/inch^2 (psi)	0.06895
0.9869	bar	atmosphere (atm)	1.013
0.9678	kg (weight)/cm^2	atmosphere (atm)	1.033
14.22	kg (weight)/cm^2	lb/inch^2 (psi)	0.07031
14.70	atmosphere (atm)	lb/inch^2 (psi)	0.06805
	Yield or Rate		
0.446	ton (metric)/hectare	ton (U.S.)/acre	2.240
0.892	kg/ha	lb/acre	1.12
0.892	quintal/hectare	hundredweight/acre	1.12
	Temperature		
$\left(\frac{9}{5}{}^\circ C\right) + 32$	Celsius	Fahrenheit	$\frac{5}{9}$ (°F − 32)
	−17.8C	0F	
	0C	32F	
	20C	68F	
	100C	212F	
	Water Measurements		
8.108	hectare-meters (ha-m)	acre-feet	0.1233
97.29	hectare-meters (ha-m)	acre-inches	0.01028
0.08108	hectare-centimeters (ha-cm)	acre-feet	12.33
0.973	hectare-centimeters (ha-cm)	acre-inches	1.028
0.00973	meters^3 (m^3)	acre-inches	102.8
0.981	hectare-centimeters/hour (ha-cm/hour)	feet^3/sec	1.0194
440.3	hectare-centimeters/hour (ha-cm/hour)	U.S. gallons/min	0.00227
0.00981	meters^3/hour (m^3/hour)	feet^3/sec	101.94
4.403	meters^3/hour (m^3/hour)	U.S. gallons/min	0.227

Source: Reprinted from *Soil Sci. Soc. America Jour.* endcover by permission of the publisher, the Soil Science Society of America.

TABLE 9. Other Principal Symbols or Letter Combinations.

Symbol	Description	Unit
General Abbreviations		
A, B	Chemical species	
(*A*)	Activity of species *A*	
A	Concentration of species *A*	
log	Logarithm to base 10	
ln	Logatharim to base *e*	
Letter Combinations		
AEC	Anion exchange capacity	mequiv/100 g
CEC	Cation exchange capacity	mequiv/100 g
DDL	Diffuse double layer	-
EC	Electrical conductivity	mmho/cm
EC_e	Electrical conductivity of the saturation extract	mmho/cm
EC_{FC}	Electrical conductivity of the soil solution at field capacity	mmho/cm
EC_{iw}, EC_{dw}	Electrical conductivity of the irrigation water, drainage water	mmho/cm
EMF	Electromotive force	Volt
ESP	Exchangeable sodium percentage	-
FC	Soil moisture content at field capacity	cm^3/100 g
HHB	Hydrophilic-hydrophoic balance	-
LR	Leaching requirement	-
RSC	Residual sodium concentration	eq/m^3 = mequiv/l
SAR	Sodium adsorption ratio	$(mmol/l)^{1/2}$
SP	Soil moisture content at saturation	cm^3/100 g
ZPC	Zero point of charge	-
Subscripts		
+	Monovalent cation species, positive charge	
2+	Divalent cation species, positive charge	
-	Monovalent anion species, negative charge	

N.B. Volume II of this series (*Encyclopedia of Atmospheric Sciences and Astrogeology*) contains a very long entry devoted to *Units, Numbers, Constants, and Symbols;* this entry contains conversion tables that may be useful to soil scientists and other readers.

Acknowledgements

A work of this magnitude necessarily involves the generous cooperation and coordination of numerous people. Though there were some headaches and tense moments trying to meet deadlines, the responses of our contributors were gratifying. We are especially indebted to our foreign contributors for their patience and perseverance during the many rewrites of papers that had to be translated into English. Some topics seemed to defy an author; thus, it was expedient for us to prepare the entry. In these instances we are most grateful to our colleagues who read the articles and made suggestions for improvement.

To those contributors who volunteered to write additional articles, and to other helpers who indicated topics or concepts that should be incorporated into the volume, and to those who suggested the names of other experts as potential contributors, a special word of thanks is due: I. P. Abrol (India); G. Aubert (France); Bryon W. Bache (Scotland); J. Dan (Israel); R. Dudal (Italy); Walter R. Gardner (USA); I. P. Gerasimov (USSR); Leland H. Gile (USA); F. D. Hole (USA); G. D. Hubble (Australia); R. J. Huggett (USA); Victor J. Kilmer (USA); V. A. Kovda (USSR); R. F. Loxton (South Africa); W. M. McArthur (Australia); M. A. Mustafa (Zambia); L. J. Pons (The Netherlands); J. P. Quirk (Australia); Mathias Stelly (USA); Cliff D. Ollier (Australia); D. S. Orlov (USSR); Pedro Sanchez (USA); Hubert J. Morel-Seytoux (USA); E. L. Skidmore (USA); F. J. Stevenson (USA); L. van der Plas (The Netherlands); D. C. Van Dijk (Australia); R. Tavernier (Belgium); John M. de Villiers (Rhodesia); R. Webster (Great Britain); E. P. Whiteside (USA); and Dan H. Yaalon (Israel).

Roy W. Simonson (formerly Director of Soil Classification and Correlation, U.S. Soil Conservation Service) has advised us throughout the project and has helped proof read many manuscripts; he has lent the fine point of his felt-tip pen to hone some seemingly intractable prose, and made valuable suggestions as to potential contributors for troublesome topics. His longtime association with soil science and field work in many countries made his tutelage in the preparation of this volume particularly valuable. To our able and respected colleague, a special word of gratitude is extended.

For their patience and determined effort to clear up nagging problems in the final stages of production, we thank the production editor, Bernice Wisniewski, and our copyeditor, Marci Nugent. Both managed to ease the pains of nearly three years' hard work to get the job done.

CHARLES W. FINKL, JNR.
RHODES W. FAIRBRIDGE

CONTRIBUTORS

I. P. ABROL, Division of Soils and Agronomy, Central Soil Salinity Research Institute, (I.C.A.R.) Karnal, 132001 (Haryama), India. *Gypsum Requirement; Salt Leaching.*

RICHARD W. ARNOLD, Dept. of Agronomy, Cornell University, Ithaca, New York 14853. *Soil Science and Society.*

BRYON W. BACHE, The Macauly Institute for Soil Research, Craigiebuckler, Aberdeen, Scotland AB9 2QJ. *Activity Ratio; Base Saturation; Chemical Composition; Soil Reaction; Soil Solution.*

DALE E. BAKER, Dept. of Agronomy, Pennsylvania State University, University Park, Pennsylvania 16802. *Electrochemistry; Ionic Activities.*

ROBERT H. BECK, Dept. of Agronomy, Purdue University, West Lafayette, Indiana 47907. *Spectral Characteristics.*

LEON BERNSTEIN, 1795 New Hampshire Drive, Costa Mesa, California 92626. *Irrigation.*

D. F. BEZDICEK, Dept. of Agronomy and Soils, Washington State University, Pullman, Washington 99163. *Nitrogen Fixation.*

GEORGE R. BLAKE, Dept. of Soil Science, University of Minnesota, St. Paul, Minnesota 55101. *Particle Density.*

G. H. BOLT, Dept. of Soil and Fertilizers, Agricultural University, Wageningen, The Netherlands. *Pollution.*

JOHANNES BOUMA, Soil Survey Institute, Box 98, Wageningen, The Netherlands. *Water Movement.*

HERMAN BOUWER, U.S. Dept. of Agriculture, Federal Research, Western Region, U.S. Water Conservation Laboratory, 4331 East Broadway, Phoenix, Arizona 85040. *Conductivity, Hydraulic; Pore Space, Drainable.*

ROBERT BRINKMAN, Dept. of Soil Science and Geology, Agricultural University, Wageningen, The Netherlands. *Clay Mineral Decomposition.*

K. W. BROWN, Dept. of Soil and Crop Sciences, Texas A & M University, College Station, Texas 77843. *Lysimeters.*

GAYLON S. CAMPBELL, Dept. of Agronomy and Soils, Washington State University, Pullman, Washington 99163. *Energy Balance.*

JOHN W. CARY, U.S. Dept. of Agriculture, Agricultural Research, Western Region, Snake River Conservation Research Center, Route 1, Box 186, Kimberly, Idaho 83341. *Pore-Size Distribution.*

C. D. CARSON, Dept. of Agriculture, Southwest Texas State University, San Marcos, Texas 78666. *Acidity.*

D. K. CASSEL, Dept. of Soil Science, North Carolina State University, Raleigh, North Carolina 27607. *Profiles, Physical Modification.*

ROBERT B. CATE, USAID, Lockheed Electronics Company, Inc., 1830 NASA Road 1, Houston, Texas 77058. *Law of the Minimum.*

MARTIN VEALE CHESHIRE, The Macaulay Institute of Soil Research, Craigiebuckler, Aberdeen, Scotland AB9 2QJ. *Polysaccharide.*

FRANCIS E. CLARK, U.S. Dept. of Agriculture, Agricultural Research Service, Fort Collins, Colorado 80521. *Soil Microbiology.*

J. D. COLWELL, Division of Soils, C.S.I.R.O., P.O. Box 639, Canberra City, A.C.T., Australia 2601. *Statistical-Chemical Characterizations.*

JOHN C. COREY, Savannah River Laboratory, E. I. du Pont de Nemours and Co., Aiken, South Carolina 29801. *Radioisotopes.*

RICHARD B. COREY, Dept. of Soil Science, University of Wisconsin, Madison, Wisconsin 53706. *Soil Chemistry.*

M. E. J. CURZON, Dept. of Caries Research, Eastman Dental Center, 800 Main Street, Rochester, New York 14603. *Dental Caries and Soils.*

MURRAY D. DAWSON, Dept. of Soil Science, Oregon State University, Corvallis, Oregon 97331. *Edaphology; Training in Soil Science.*

F. A. M. DE HAAN, Dept of Soils and Fertilizers, Agricultural University, Wageningen, The Netherlands. *Pollution.*

E. DE JONG, Saskatchewan Institute of Pedology, Saskatoon, Saskatchewan, Canada S7N 0W0. *Aeration, Respiration, and Atmosphere.*

C. C. DELWICHE, Dept. of Land, Air, and Water Resources, University of California, Davis, California 95616. *Carbon Cycle.*

J. B. DIXON, Dept. of Soil and Crop Sciences, Texas A & M University, College Station, Texas 77843. *Acidity.*

JAMES DRAGUN, Dept. of Agronomy, Pennsylvania State University, University Park, Pennsylvania 16802. *Electrochemistry.*

R. DUDAL, Land and Water Development Division, FAO/UNESCO, Via delle Terme di Caracalla, Rome, Italy 00100. *International Society of Soil Science.*

J. DUMANSKI, Land Resource Research Institute, Agriculture Canada, Central Experimental Farm, Ottawa, Ontario, Canada K1A 0C6. *Information Systems.*

S. Al. EL-SWAIFY, Dept. of Agronomy and Soil Science, University of Hawaii, Honolulu, Hawaii 96822. *Irrigation Water.*

W. W. EMERSON, Division of Soils, C.S.I.R.O., Private Bag No. 2, Glen Osmond, South Australia, Australia 5064. *Aggregate Stability.*

RICHARD M. ESHELMAN, Dept. of Agronomy, Pennsylvania State University, University Park, Pennsylvania 16802. *Ionic Activities.*

CHARLES W. FINKL, JNR., Ocean Sciences Center, Nova University, Port Everglades, Florida 33004. *Aeragronomy; Conductivity, Electrical; Leaching; Macronutrients; Micronutrients; Organic Matter; Phosphorus Cycle; Soil; Soil Conditioners; Soil Pores; Thixotrophy, Thixotropism.*

JOHN E. FOSS, Dept. of Agronomy, University of Maryland, College Park, Maryland 20742. *Soil Science.*

J. R. FRENEY, Division of Plant Industry, C.S.I.R.O., P.O. Box 1600, Canberra City, A.C.T., Australia 2601. *Sulfur Transformations.*

WALTER H. GARDNER, Dept. of Agronomy and Soils, Washington State University, Pullman, Washington 99164. *Soil Water; Water Content and Retention.*

W. R. GARDNER, Dept. of Soil Science, University of Wisconsin, Madison, Wisconsin 53706. *Field Water Cycle.*

ROBERT G. GAST, Dept. of Soil Science, University of Minnesota, St. Paul, Minnesota 55101. *Exchange Phenomena.*

CARLO GESSA, Agricultural Chemistry Institute, University of Sassari, Via Enrico De Nicola, 07100, Sassari, Italy. *Buffers, Buffering.*

JOHN E. GIESEKING, Dept. of Agronomy, University of Illinois, Urbana, Illinois 61801. *Soil Components, Inorganic; Soil Components, Organic.*

R. J. GILKES, Dept. of Soil Science and Plant Nutrition, University of Western Australia, Nedlands, Western Australia, Australia 6009. *Clay Mineralogy; Hydrous Oxides; Mineral Analysis.*

WILLIAM R. GILL, National Tillage Machinery Laboratory, Agricultural Research, Science and Education Administration, U.S. Department of Agriculture, P.O. Box 792, Auburn, Alabama 36830. *Tillage.*

J. E. GILLOTT, Dept. of Civil Engineering, University of Calgary, Calgary, Alberta, Canada T2N 1N4. *Clays, Physical Chemistry.*

P. H. GROENEVELT, Dept. of Land Resource Science, University of Guelph, Guelph, Ontario, Canada. *Envelope-Pressure Potential; Transport Processes.*

AMOS HADAS, Institute of Soils and Water, The Volcani Center, Bet Dagan, Israel. *Conductivity, Thermal; Heat Capacity; Thermal Regimes.*

R. J. HANKS, Dept. of Soil Science and Biometeorology, Utah State University, Logan, Utah 84322. *Evaporation.*

STUART A. HARRIS, Dept. of Geography, University of Calgary, Calgary, Alberta, Canada T2N 1N4. *Humic Acid-Fulvic Acid Ratio.*

ROGER HARTMANN, Faculty of Agricultural Sciences, State University of Ghent, Coupure Linka 533, Ghent, Belgium B-9000. *Aggregation.*

TSUTOMU HATTORI, Institute for Agricultural Research, Tohoku University, Sendai, Japan. *Microhabitats.*

MICHAEL HERLIHY, An Foras Talúntais, The Agricultural Institute, Johnstown Castle Research Centre, Wexford, Ireland. *Enzyme Activity; Fertilizers, Organic; Rhizosphere.*

ED HIGHLEY, Division of Entomology, C.S.I.R.O., P.O. Box 1700, Canberra City, A.C.T. 2601, Australia. *Soil Physics.*

RICHARD M. HIGHSMITH, JR., Dept. of Geography, Oregon State University, Corvallis, Oregon 97331. *Soil Conservation Service.*

DAVID E. HILL, Dept. of Soil and Water, Connecticut Agricultural Experiment Station, 123 Huntington Street, Box 1106, New Haven, Connecticut 06504. *Percolation.*

S. B. HILL, Dept. of Entomology, Macdonald Campus, McGill University, Ste. Anne de Bellevue, Quebec, Canada HOA 1CD. *Fauna.*

RONALD C. HIRSCHFELD, Geotechnical Engineers, Inc., 1017 Main Street, Winchester, Massachusetts 01890. *Soil Mechanics.*

M. M. HOLLERBACH, Botanical Institute of USSR, Leningrad 197022, USSR. *Algae.*

M. L. JACKSON, Dept. of Soil Science, University of Wisconsin, Madison, Wisconsin 53706. *Soil Mineralogy.*

RAY D. JACKSON, U.S. Dept. of Agriculture, Federal Research, Western Region, U.S. Water Conservation Laboratory, 4331 East Broadway, Phoenix, Arizona 85040. *Water Fluxes.*

J. BENTON JONES, JR., Dept. of Horticulture, University of Georgia, Athens, Georgia 30602. *Soil Tests.*

DENNIS R. KEENEY, Dept. of Soil Science, University of Wisconsin, Madison, Wisconsin 53706. *Adsorption Phenomena.*

VICTOR J. KILMER, 302 Nottingham Road, Florence, Alabama 35630. *Nutrient Losses.*

M. B. KIRKHAM, Dept. of Agronomy, Oklahoma State University, Stillwater, Oklahoma 74074. *Sludge Disposal; Trace Elements.*

B. KLOOSTERMAN, Soil Research Institute, Central Experimental Farm, Ottawa, Ontario, Canada K1A 0C6. *Information Systems.*

F. KUNC, Institute of Microbiology, Czechoslovak Academy of Science, 142 20 Praha 4-KRC, Budejovicka 1083, Czechoslovakia. *Metabolic Activity; Microbial Ecology and Clay Minerals.*

KERMIT STILLMAN LaFLEUR, Dept. of Agronony and Soils, Clemson University, Clemson, South Carolina 29631. *Soil-Solvent Interactions.*

R. LAL, International Institute of Tropical Agriculture, Oyo Road, P.M.B. 5320, Ibadan, Nigeria. *Mulch; Zero-Tillage.*

DAVID T. LEWIS, Dept. of Agronomy, University of Nebraska, Lincoln, Nebraska 68583. *Bulk Density.*

A. J. LOW, 50 Sewell Avenue, Wokingham, Berks, England. *Soil Structure.*

D. KEITH McE. KEVAN, Dept. of Entomology, Macdonald Campus, McGill University, Ste. Anne de Bellevue, Quebec, Canada HOA 1CD. *Fauna.*

E. O. McLEAN, Agronomy Dept., Ohio State University, Columbus, Ohio 43210. *Lime, Liming; pH, Field Determination.*

L. DONALD MEYER, U.S.D.A. Sedimentation Laboratory, Oxford, Mississippi 38658. *Water Erosion.*

FRED P. MILLER, Dept. of Agronomy, University of Maryland, College Park, Maryland 20742. *Soils, Nonagricultural Uses.*

H. J. MOREL-SEYTOUX, Engineering Research Center, Colorado State University, Fort Collins, Colorado 80523. *Capillary Pressure; Imbibition; Infiltration; Permeability; Wetting Front.*

NORMAN R. MORROW, Petroleum Recovery Research Center, New Mexico Institute of Mining and Technology, Socorro, New Mexico 87801. *Thermodynamics, Soil Moisture.*

Y. MUALEM, Engineering Research Center, Colorado State University, Fort Collins, Colorado 80523. *Capillary Pressure; Permeability.*

M. A. MUSTAFA, Faculty of Agriculture, University of Khartoum, Shambat, Sudan. *Dispersion Phenomena.*

H. NISHITA, Laboratory of Nuclear Medicine and Radiation Biology, 900 Veteran Avenue, University of California, Los Angeles, California 90024. *Thermoluminescence.*

P. H. NYE, Dept. of Agricultural Science, University of Oxford, Parks Road, Oxford, England OX1 3PF. *Diffusion Processes.*

J. J. OERTLI, Botanical Institute, University of Basel, Basel, Switzerland. *Fertilizers, Inorganic; Plant Nutrients; Soil Fertility.*

D. S. ORLOV, Dept. Soil Science, Moscow State University, 177234 Moscow, USSR. *Physical Chemistry.*

BERNARD L. PARSONS, Dept. of Agricultural Engineering, Virginia Polytechnic Institute and State University, Blacksburg, Virginia 20461. *Conservation.*

E. A. PAUL, Saskatchewan Institute of Pedology, Saskatoon, Saskatchewan, Canada S7N 0W0. *Aeration, Respiration, and Atmosphere.*

W. J. PAYNE, Dept. of Biochemistry, University College, Cardiff, England CF1 1XL. *Denitrification.*

G. A. PETERSON, Dept. of Agronomy, University of Nebraska, Lincoln, Nebraska 68508. *Management of Soils.*

CHARLES W. RALSTON, School of Forestry and Environmental Studies, Duke University, Durham, North Carolina 27706. *Redox Reactions.*

W. K. ROBERTSON, Dept. of Soil Science, University of Florida, Gainesville, Florida 32611. *Profiles, Physical Modification.*

PEDRO A. SÁNCHEZ, Centro Internacional de Agricultura Tropical, Apartado Aéreo 6713, Cali, Colombia. *Puddling.*

G. O. SCHWAB, Dept. of Agricultural Engineering, Ohio State University, Columbus, Ohio 43210. *Soil Drainage.*

H. D. SCOTT, Dept. of Agronomy, University of Arkansas, Fayetteville, Arkansas 72701. *Herbicide Diffusion.*

H. M. SELIM, Agronomy Dept., Louisiana State University, Baton Rouge, Louisiana 70803. *Flow Theory.*

J. C. SHEPPARD, Dept. of Chemical and Nuclear Engineering, Washington State University, Pullman, Washington 99163. *Radiocarbon Dating.*

E. A. SHTINA, Agricultural Institute, Octjabrask Prospect 131, Kirov 610039, USSR. *Algae.*

ROY W. SIMONSON, P.O. Box 427, College Park, Maryland 20740. *Phosphorus Cycle.*

ARIEH SINGER, Dept. of Soil and Water Science, Hebrew University of Jerusalem, Rehovot, Israel. *Clay Mineral Formation.*

EDWARD L. SKIDMORE, Dept. of Agronomy, Kansas State University, Manhattan, Kansas 66506. *Surface Soil Water Content; Wind Erosion.*

PETER SMART, Civil Engineering Dept., University of Glasgow, Glasgow, Scotland G12 8LT. *Microstructure Manipulation.*

MILTON A. SPRAGUE, Dept. of Soils and Crops, Rutgers University, New Brunswick, New Jersey 08903. *Agronomy:*

GARY C. STEINHARDT, Dept. of Agronomy, Purdue University, West Lafayette, Indiana 47907. *Particle-Size Distribution.*

F. J. STEVENSON, Dept. of Agronomy, University of Illinois, Urbana, Illinois 61801. *Carbon-Nitrogen Ratio; Humus; Nitrogen Cycle.*

O. TALIBUDEEN, Rothamstead Experimental Station, Harpenden, Hertfordshire, England AL5 2JQ. *Nutrient Potential.*

B. K. G. THENG, Soil Bureau, D.S.I.R., Lower Hutt, New Zealand. *Clay-Organic Reactions.*

M. TSCHAPEK, Institute of Edaphology and Hydrology, National University, Bahia Blanca, Argentina. *Hydrophility, Hydrophobicity.*

PAUL W. UNGER, Southwestern Great Plains Research Center, U.S. Dept. of Agriculture, Bushland, Texas 79012. *Moisture Management.*

CHARLES E. WEAVER, School of Geophysical Sciences, Georgia Institute of Technology, Atlanta, Georgia 30332. *Clay Minerals, Silicates.*

W. O. WILLIAMSON, College of Earth and Mineral Sciences, Pennsylvania State University, University Park, Pennsylvania 16802. *Flocculation.*

HANS F. WINTERKORN, Dept. of Civil Engineering, Princeton University, Princeton, New Jersey 08540. *Electroosmosis; Hygroscopicity, Hygroscopic Constant; Soil Stabilization.*

A. G. WOLLUM, II, Dept. of Soil Science, North Carolina State University, Raleigh, North Carolina 27607. *Microorganism Transport.*

D. H. YAALON, Dept. of Geology/Pedology, Hebrew University of Jerusalem, Jerusalem, Israel. *Journals, Soil Science.*

HUGO ZUNINO, Dept. of Inorganic and Analytical Chemistry, University of Chile, Casilla 233, Santiago, Chile. *Metal-Complexing Macromolecules.*

A

ACIDITY

Acid soils are by definition soils with a pH below 7 on a scale of 0 to 14. More importantly, acid soils may have exchangeable aluminum in addition to plant nutrient cations (Ca^{2+}, Mg^{2+}, K^+, etc.) on the cation exchange sites. Acid soils are sometimes considered synonymous with infertile or poor soils, which indeed they are in some cases. An acid soil may, however, be an asset in growing coniferous and many shrubby plants that grow poorly or not at all in alkaline soils (soils with a pH above 7).

As acidity becomes intense, more cation exchange sites are saturated with aluminum, and fewer nutrient ions are available to plants. When the cation exchange sites become about 60% saturated with aluminum, appreciable aluminum enters the soil solution (Kamprath, 1970). Many plants can tolerate only traces of aluminum in the soil solution (q.v.). Thus the major effect of soil acidity on plants is the increased solubility of ions such as aluminum and manganese, which can be present in toxic quantities.

Occurrence of Acid Soils

Acid soils are most extensive in the eastern half of the contiguous United States. The distribution of acid soils in the world as a whole is illustrated by the Ultisols, Oxisols, and Spodosols (outlined in the endpapers).

Acidity is a prominent feature of Ultisols, which have 35% or less base saturation (that is, 35% of the exchangeable cations are Ca^{2+}, Mg^{2+}, K^+, and Na^+). In addition to those Ultisols in the E United States, they also occur in W Africa, EC Africa, Brazil, Paraguay, NE India, Bangladesh, E China, and NE Australia. Oxisols, which are also typically acid, occur in large areas of NC South America and SC Africa.

Spodosols are characteristically acid. Mobile aluminum and iron have played roles in their genesis. Spodosols occur in the NE and SE United States, Canada, N Europe, and Russia. They usually form in cooler climates in more sandy parent materials than Ultisols. Acidity often limits crop production on Ultisols, Oxisols, and Spodosols, but calcium carbonate is commonly added to alleviate the acid problem (see *Lime, Liming*).

Role of Hydrogen and Aluminum

Exchangeable hydrogen and aluminum cations are largely responsible for soil acidity. Although pH indicates activity of hydrogen ions in solution, the source of hydrogen ions is frequently the hydrolysis of aluminum. Hydrogen and aluminum ions are in equilibrium in acid soils. Hydrogen ions readily attack clay mineral structures releasing trivalent aluminum, which may replace ions on exchange sites (see *Exchange Phenomena*).

In very acid soil, aluminum is soluble, and much of it occurs as hydrated exchangeable ions. Exchangeable aluminum ions are in equilibrium with aluminum ions in the soil solution (q.v.). As the aluminum in solution hydrolyzes, an increase in acidity occurs:

$$Al^{3+} + 2H_2O \leftrightarrows Al(OH)^{2+} + H_3O^+ \qquad (1)$$

As the pH of a soil-extracting solution is lowered (Table 1), more Al^{3+} forms from hydroxy aluminum forms (equation 1) or from dissolution of silicate minerals (McLean et al., 1958). The hydroxy aluminum forms may be monoaluminum compounds or polynuclear complexes (Figure 1). Hydronium ions also replace Al^{3+} present on cation-exchange sites, although the extent of this reaction may be small.

TABLE 1. Aluminum Extracted from a Cambridge Soil Horizon with 1*N* Ammonium Acetate at Various pH Levels.

pH	Aluminum (meq/100 g)
5.5	7.0
5.2	9.3
5.0	11.9
4.8	17.2
4.5	22.6
4.2	27.7
4.0	30.2
3.0	32.1

After McLean et al., 1958.

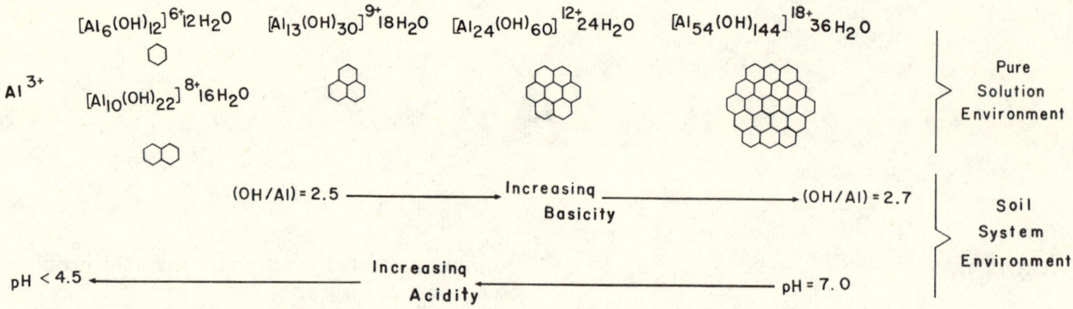

FIGURE 1. Proposed aluminum structures from a pure solution system compared to measured parameters in a soil clay system.

The species of aluminum ions present are variable with pH. The forms of aluminum vary from mostly exchangeable Al^{3+} under very acid conditions (pH< 4.5) to aluminum-hydroxy ions at higher pH (4.5 to 6.5). In general, the net positive charge of the hydroxy aluminum species becomes less as the pH increases and then becomes negative in the alkaline pH range (Figure 1). Acidity from hydroxy aluminum ion species would be produced as the following hydrolysis examples indicate.

$$Al(OH)^{2+} + 2H_2O \rightleftharpoons Al(OH)_2^+ + H_3O^+ \quad (2)$$

$$Al(OH)_2^+ + 2H_2O \rightleftharpoons Al(OH)_3 + H_3O^+ \quad (3)$$

Apparently, much of the hydroxy aluminum material in soils is present as *polynuclear complexes* (Figure 1). These hydroxy materials are often "fixed" in soil clay mineral interlayers. Evidence suggests that this "fixed," or nonexchangeable, aluminum occurs over an OH/Al molar ratio of about 2.5 to 2.7. In a system in which the OH/Al molar ratio is less than 2.5, the aluminum may be Al^{3+} or one of the polymers.

Hsu and Bates (1964) have proposed that hydroxy aluminum polymer structures in a pure solution system are comparable to OH/Al molar ratios in clay or soil environments in which hydroxy aluminum is often fixed (Figure 1). An equilibrium is established with the soil solution. Thus the pH and pH-dependent charge of the soil system can be greatly affected by these aluminum-hydroxy polymers (see *Lime, Liming*).

Soil Materials and Acidity

Organic Matter. The humic fraction of soil organic matter (q.v.) is most important in attracting and dissociating hydrogen ions. Nonhumic material includes those compounds that can be easily attacked by *microorganisms.* The variability of acidity produced by humic constituents is due to their diverse nature. The three main humic fractions (see *Humus*) are *humic acid, fulvic acid,* and *humin* (Schnitzer and Khan, 1972). Usually carboxyl and hydroxyl groups are the most prevalent in the fulvic-acid fraction, which constitutes 25% to 75% of the total organic matter in most soils. These oxygen-containing functional groups furnish acidity by dissociating hydrogen.

Another important aspect of the role of organic matter in soil acidity is its complexing of metal ions such as iron and aluminum (see *Metal-complexing Macromolecules*). Hydrolysis of these organometallic complexes contributes to acidity. Also of great importance is the influence of organic matter on the availability of aluminum. Complexing of aluminum in a soil with high organic matter content reduces solution and exchangeable aluminum (Figure 2) (Pionke and Corey, 1967; Thomas, 1975).

As mineralization of organic matter occurs, the carbon dioxide produced combines with water to give carbonic acid (see *Soil Reaction*). Also as organic matter decomposes, acid-forming ions (hydrogen, aluminum, iron, and sulfur) are liberated into solution and increase soil acidity.

Aluminosilicate Clay Minerals. Soil clay silicates constitute one of the most important sources of acidity. Adsorption of hydrogen or aluminum is governed to a large extent by the availability of clay mineral-exchange sites of both permanent and pH-dependent charges. Minerals differ in their susceptibility to attack during chemical weathering (q.v. in vol. XII). Hydrogen attacks the montmorillonite structure and liberates ions such as aluminum, magnesium, and iron. The soluble aluminum, magnesium, and iron then compete for remaining cation-exchange sites. Drying promotes hydroxylation of aluminum on exchange sites yielding hydrogen by hydrolysis (Barshad, 1960).

The solubilities of various soil minerals

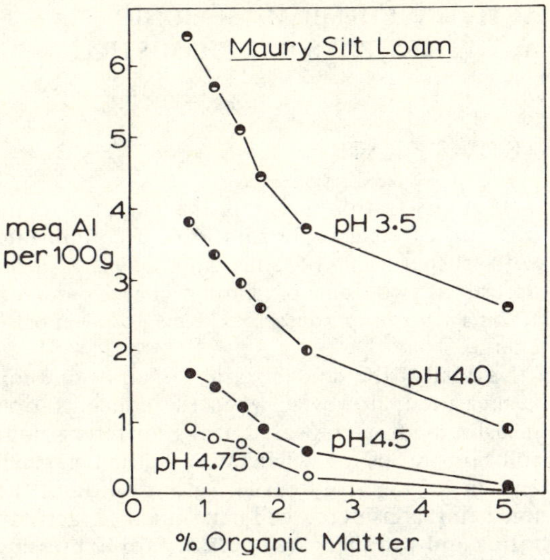

FIGURE 2. A relation between percent organic matter and exchangeable aluminum at several pH levels. (From Thomas, 1975)

definitely (q.v. in vol. IVB) affect soil acidity. The buffering intensity—that is, the amount of strong acid or base required to change the pH of a system by a given amount—of minerals depends on their respective solubilities at different pH levels (van Breemen and Wielemaker, 1974). The clay minerals montmorillonite and kaolinite are strong buffers in the pH range of 3 to 4 due to their relative solubilities at this level. At pH 4.5 to 6.0, minerals such as microcline feldspar, gibbsite, and clay micas are strong buffers (see *Buffers, Buffering*).

An important aspect of the weathering of soil clay minerals and of acidity is the presence of aluminum- and iron-hydroxy interlayers. The hydroxy aluminum present contributes to the amount of pH-dependent charge. This hydroxy material greatly affects the availability of solution aluminum, especially in the pH range of 4.5 to 6.5. Hydroxy iron and aluminum in interlayers in clays give a buffering action to the hydrogen and aluminum-iron system (Coleman and Thomas, 1964). Hydroxy aluminum interlayers occur in soils under the following conditions: (1) moderately active weathering, which furnishes aluminum ions; (2) moderately acid pH (about 5.0); (3) low organic matter content; and (4) frequent wetting and drying (Rich, 1968). These conditions are common in soils classified as "Ultisols" (see *Hydrous Oxides*).

Hydroxy Compounds and Iron and Aluminum Oxides. In addition to their occurrence in clay interlayers, hydroxy forms of aluminum and iron occur on external surfaces of silicate clays and in separate phases. The acidity associated with these materials may be produced by hydrolysis during their formation from iron and aluminum in solution. The occurrence as amorphous coatings on silicate particles can reduce exchange acidity by blocking exchange sites or increase acidity by dissolving to be further hydrolyzed.

Other Acid-forming Substances. Many materials leave an acid residue in the soil. The acid residue is produced primarily by the nitrification of ammonium ions. When oxygen is present, ammonium ions can form acidity as shown by the following equation:

$$NH_4^+ + 2O_2 \rightarrow 2H^+ + NO_3^- + H_2O \qquad (4)$$

Ammonium ions can also be adsorbed on the exchange complex of colloids, which may cause an eventual increase in acidity by displacing base-forming cations (Ca^{2+}, Mg^{2+}, K^+, etc.) into groundwater to be leached away. Sources of ammonium ions may be fertilizers (q.v.) such as anhydrous ammonia, ammonium nitrate, and ammonium sulfate or hydrolysis products of urea or other organic-nitrogen compounds.

Sulfur minerals are less common in soils than the aluminosilicate minerals discussed previously, but sulfide minerals react faster when weathering occurs under oxidizing conditions and have a more pronounced effect on soil acidity. Sulfur liberated as a decomposition product can be oxidized and hydrated to form sulfuric acid. This chemical and biological process depends on the availability of oxygen and the mineral's resistance to attack (see *Sulfur Transformations*).

Summary

Soil acidity arises from several different sources. One source is the dissociation of functional groups of humus, and another is the microbiologic decomposition of organic matter. Other sources of soil acidity are silicate clay minerals and hydrous oxides of iron and aluminum. Acid salts such as fertilizer salts that undergo either hydrolysis or nitrification also contribute acidity to soil. Sulfide compounds commonly found in coastal marshes contribute acidity to soils when the marshes are drained and oxidation occurs. Each source contributes acidity in a particular soil pH range and in different amounts. The quantity and intensity of acidity produced depend in some part on the equilibrium that exists between hydrogen and aluminum ions in the soil.

C. D. CARSON
J. B. DIXON

References

Barshad, I., 1960. The effect of the total chemical composition and crystal structure of soil minerals on the nature of the exchange cation in acidified clays in naturally occurring acid soil, *7th Internat. Congr. Soil Sci. Trans.* (Madison),2, 435–444.

Bear, F. E., 1965. *Chemistry of the Soil.* New York: Reinhold, 515p.

Brady, N. C., 1974. *The Nature and Properties of Soils.* New York: Macmillan 639p.

Breemen, N. van, and Wielemaker, W. G., 1974. Buffer intensities and equilibrium pH of minerals and soils. II. Theoretical and actual pH of minerals and soils, *Soil Sci. Soc. America Proc.*, **38**, 61–66.

Coleman, N. T., and Thomas, G. W., 1964. Buffer curves of acid clays as affected by the presence of ferric iron and aluminum, *Soil Sci. Soc. America Proc.*, **28**, 187–190.

Coleman, N. T., and Thomas, G. W., 1967. The basic chemistry of soil acidity, *in* R. W. Pearson and F. Adams, eds., *Soil Acidity and Liming.* Madison: American Society of Agronomy, 1–41.

Hsu, Pa Ho, and Bates, T. R., 1964. Formation of x-ray amorphous and crystalline aluminum hydroxides, *Mineralog. Mag.*, **33**, 749–768.

Kamprath, E. J., 1970. Exchangeable aluminum as a criterion for liming leached mineral soils, *Soil Sci. Soc. America Proc.*, **34**, 252–254.

Kellogg, C. E., 1975. *Agricultural Development: Soil, Food, People, Work.* Madison: Soil Science Society of America, 233p.

McLean, E. O.; Heddleson, M. R.; Bartlett, R. J.; and Holowaychuk, N., 1958. Aluminum in soils: I. Extraction methods and magnitudes in clays and Ohio soils, *Soil Sci. Soc. America Proc.*, **22**, 383–387.

Peionke, H. B., and Corey, R. B., 1967. Relations between acidic aluminum and soil pH, clay, and organic matter, *Soil Sci. Soc. America Proc.*, **31**, 749–752.

Rich, C. I., 1968. Hydroxy interlayers in expansible layer silicates, *Clays and Clay Minerals,* **16**, 15–30.

Schnitzer, M., and Khan, S. U., 1972. *Humic substances in the environment.* New York: Marcel Dekker, 327p.

Thomas, G. W., 1975. The relationship between organic matter content and exchangeable aluminum in acid soil, *Soil Sci. Soc. America Proc.*, **39**, 591.

Cross-references: *Base Saturation; Buffers, Buffering; Exchange Phenomena; Lime, Liming; pH, Field Determination; Soil Reaction.*

ACTINOMYCETES—See SOIL MICROBIOLOGY

ACTIVITY, BIOLOGICAL—See ENZYME ACTIVITY; METABOLIC ACTIVITY

ACTIVITY, CHEMICAL—See IONIC ACTIVITIES; PHYSICAL CHEMISTRY

ACTIVITY RATIO

"Activity ratios" express the chemical composition of electrolyte solutions in equilibrium with suspensions of soils and clays. A *ratio* describes two ions of similar charge, and an *activity product* describes ions of opposite charge.

Normally, the concentration of only one ion is measured; however, when the concentration measurement is repeated using different soil: solution ratios or when after adding a small amount of electrolyte, the concentration of the ion being considered will have changed, activity ratios and products are necessary for expressing the relationship. The concentration of the ion changes after the addition of electrolyte because the surfaces of the clay and humus colloids of the soil contain negative electrically charged sites that are neutralized by adsorbed cations. The factors mentioned alter the distribution of cations and anions between the charged surfaces and the solution.

Theory and Definitions

Chemical Potential, Activity, and Concentration. When considering chemical reactions, the term of the *Gibbs free-energy function* that relates to the chemical composition of the system is the *chemical potential,* μ. For a given component, μ is defined in terms of the *activity, a,* by the relationship:

$$\mu = \mu^{\circ} + RT \ln a \qquad (1)$$

where R is the gas constant, T is the temperature °K, and μ° is the *standard chemical potential,* an arbitrary reference point given by the value of μ at unit activity.

The activity can be qualitatively described as "effective concentration," and it is related to the actual concentration, m mol/l, by the activity coefficient, γ, that is, $a = m\,\gamma$. The activity coefficient has a value close to unity for very dilute solutions ($<10^{-4}$ mol/l); but as the concentration increases, the activity coefficient at first decreases (more so for ions of higher charge) and then increases again as concentrations approach 1.0 mol/l. It can be evaluated theoretically or determined experimentally by a variety of techniques (Robinson and Stokes, 1965).

The Electrical Double Layer. When an electrically charged surface is in equilibrium with a solution, electrical neutrality is maintained by adsorbed counterions, which produces an *electrical double layer.* The structure of such a double layer is shown in Figure 1. The diagrammatic representation [Figure 1(a)] of the distribution of cations adjacent to the negatively charged surface shows the *Stern layer,* a localized layer of cations near the surface, and the *Gouy, or diffuse, layer,* occurring farther from the surface. An exponential drop in electrical potential to zero in the external solution is associated with the electrical double layer [Figure 1(b)]. The excess of cations over anions in the solution phase balances the negative charge of the surface. The concentration of cations and anions in a given volume element is shown as a function of distance from the surface in Figure 1(c).

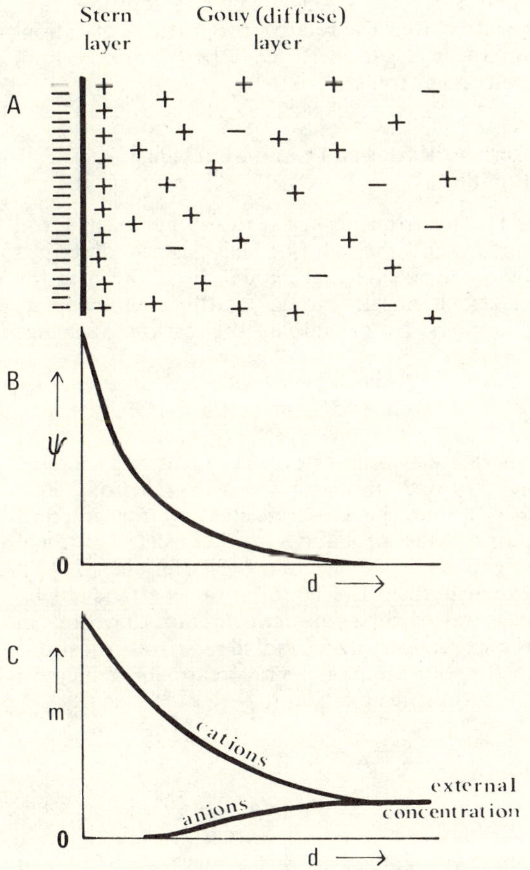

FIGURE 1. The electrical double layer. (a) diagrammatic representation of the distribution of counterions adjacent to a negatively charged surface. (b) Electrical potential ψ at a point in the solution as a function of distance d from the charged surface. (c) Concentration m of cations and anions in solution as a function of distance from the surface.

The "thickness" of the double layer is thought of as the distance from the surface of a plane of charge with electrical capacity equal to that of the double layer; thickness decreases as the valency and concentration of the electrolyte increase. For typical soil systems the double-layer thickness measures between 1 and 10 nm, but the distance from the surface at which the charge is totally neutralized by the diffuse layer will be several times greater than this. This topic is discussed in detail by van Olphen (1977).

The Electrochemical Potential. In the electrically charged system described previously, the electrochemical potential of an ion replaces the chemical potential as the appropriate term of the Gibbs free-energy function (Guggenheim, 1929). For a given ion species, this function is defined by:

$$\bar{\mu} = \bar{\mu}^{\circ} + zF\Psi + RT\ln a \quad (2)$$

where z is the valency of the ion, F is Faraday's constant, and Ψ is the electrical potential in solution. At equilibrium, $\bar{\mu}$ must remain constant throughout the system, which requires the activity a to vary with distance from the surface because of the variation of Ψ. This variation depends on both the concentration of external electrolyte and the charge of the ions in the system.

Consider the electrochemical potential of two ions of differing charge, say, K^+ and Ca^{2+}, in a plane of constant potential Ψ. The electrochemical potentials are given by:

$$\bar{\mu}_K = \bar{\mu}_K^{\circ} + F\Psi + RT\ln a_K \quad (3)$$

and

$$\bar{\mu}_{Ca} = \bar{\mu}_{Ca}^{\circ} + 2F\Psi + RT\ln a_{Ca} \quad (4)$$

Subtracting (4) from 2 × (3), we get:

$$2\bar{\mu}_K - \bar{\mu}_{Ca} = 2\bar{\mu}_K^{\circ} - \bar{\mu}_{Ca}^{\circ} + RT\ln \frac{(a_K)^2}{(a_{Ca})} \quad (5)$$

which may be simplified to:

$$f(\Delta\bar{\mu}) = \log \frac{(a_K)^2}{(a_{Ca})} \quad (6)$$

Activity Ratios. Equation 6 shows that a function of the difference in electrochemical potential is directly related to the logarithm of a ratio of the two ion activities. The difference in valency of the ions is also considered. Since $f(\Delta\bar{\mu})$ is a constant throughout the system, the activity ratio must also be constant, and it defines the chemical potential of one ion in solution with reference to that of another ion.

Similar reasoning may be applied to any ion pair, taking account of the sign and charge of the ions. In many natural soils, Ca^{2+} is the dominant cation, and it is therefore frequently used as the ion to which activities of other ions are related, producing activity ratios and products such as a_{Mg}/a_{Ca}; $(a_H)^2/a_{Ca}$; $(a_{Al})^2/(a_{Ca})^3$; $a_{Ca} \times (a_{H_2PO_4})^2$; $a_{Ca} \times a_{HPO_4}$. The activity ratios for cation pairs are often expressed in such forms as AR_{K-Ca} or AR_{Mg-Ca}. It is sometimes convenient to express the "reduced" activity ratio, that is, $a_K/(a_{Ca})^{1/2}$, on a negative logarithmic scale, which yields functions such as pK − ½pCa, pH − ½pCa, and pH_2PO_4 + ½pCa, which are frequently, but inaccurately, referred to as "potassium potentials," "phosphate potentials," etc.

The Ratio Law

Schofield (1947) proposed a "ratio law" governing the distribution of soil cations. He expressed it as follows: "When cations in a solution are in equilibrium with a larger number of exchangeable ions, a change in the concentration of the solution will not disturb the equilibrium if the concentrations of all the monovalent ions are changed in one ratio, those of all the divalent ions in the square of that ratio and those of all the trivalent ions in the cube of that ratio."

It appears that the ratio Schofield referred to in this statement was that in which the concentration of ions was changed. However, it is clear from Schofield's paper that his emphasis when formulating this law was on the index of this ratio to be used for cations of different valency. The statement introduces another factor, namely, a change in the concentration of the solution. This change may be caused either by diluting the whole system (altering the soil:solution ratio) or by adding electrolyte. The ratio law defines the situations under which these changes will not disturb the equilibrium. In these situations the ψ terms will also cancel out in equations similar to equations 2–5, provided the ion distribution at the surface is not altered. Thus the activity ratio should remain constant in spite of these changes.

The ratio law has subsequently been interpreted rather differently since it has been found experimentally that for a number of ion pairs and a variety of soils (Beckett 1971), activity ratios and products remain approximately constant while solution concentrations vary up to 0.02 mol/l of divalent cations, even when these variations are not strictly in accord with the requirements of Schofield's statement. The activity ratio has therefore become accepted as a relatively stable characteristic of a soil sample in that it changes little with changing solution concentration.

There are, however, many soils for which the ratio law does not hold or holds only at lower electrolyte concentrations. A fundamental condition for conformity to the ratio law is the effective exclusion of anions from the sites where the exchangeable cations are situated (i.e., the inner part of the double layer in the model illustrated in Figure 1). This condition occurs only if there is a high potential drop between the charged surfaces and the solution. In practical terms, this drop requires that the external electrolyte concentration be low and that the soils have a high surface negative charge. At high electrolyte concentration for soils with a low negative charge and for soils with measurable amounts of positive charge on their surfaces, small changes in solution composition alter the relative proportions of cations on the soil surfaces, and activity ratios do not remain constant.

Activity Ratios and Cation-Exchange Equilibria

The distribution of cations between a soil surface and the solution may also be described by a cation-exchange model, in addition to the electrochemical model outlined previously. Consider, for example, the cation exchange reaction:

$$\text{Soil Ca} + 2\text{KCl} \rightleftharpoons \text{Soil K}_2 + \text{CaCl}_2$$

There are a number of ways of formulating the equilibrium quotient for this reaction (Bolt 1967), but the most satisfactory for practical purposes is probably the "corrected rational selectivity coefficient" (Helfferich, 1962), which defines the equilibrium coefficient K^K_{Ca} in terms of the *equivalent fraction* of cations in the exchanger phase and the *activity* of cations in the solution phase. This treatment reduces to the following equation:

$$K^K_{Ca} = \frac{(q_K)^2}{q_{Ca}\, q_0} \times \frac{a_{Ca}}{(a_K)^2} \qquad (7)$$

where q_K and q_{Ca} are the amounts of K^+ and Ca^{2+} adsorbed at the surface, and q_0 is the total charge (cation exchange capacity) of the exchanger (that is, $q_0 = q_K + q_{Ca}$, measured in milli-equivalents per kilogram soil). This selectivity coefficient is not necessarily constant for all conditions, but it can be related to the thermodynamic equilibrium constant **K** by introducing terms for the activity coefficients of ions in the adsorbed phase (Bolt, 1967), provided that solvent and electrolyte sorption effects are disregarded.

According to equation 7, the *activity ratio in solution* relates to two important properties of the soil: (1) the composition of the exchange phase, i.e., the relative proportions of K^+ and Ca^{2+} adsorbed at the negatively charged surface, and (2) the relative affinity of these cations for the soil surface as reflected by the selectivity coefficient K^{K}_{Ca}. Reasoning similar to that used here for potassium and calcium can be used for any cation pair.

Measurement and Use of Activity Ratio

The determination of the equilibrium activity ratio AR^e of an ion pair for a soil sample serves to characterize the sample with respect to the free-energy difference of the two ions whether the ratio law holds or not. To calculate an activity ratio rigorously requires a knowledge of the chemical composition of the soil solution, but two other approximate methods are frequently used:

(1) A soil sample is equilibrated with a dilute electrolyte, say, 5 mmol/l $CaCl_2$ if Ca^{2+} is the dominant cation of the soil, at a 1:1 soil:solution ratio. This ratio provides sufficient solution for analysis with minimum alteration of the amounts of exchangeable cations.

(2) An *adsorption-desorption isotherm* is constructed by shaking a number of samples of the soil at a wider soil:solution ratio, say, 1:10, with incremental additions of the ion under investigation such as K^+ but at a constant concentration of the reference cation such as 5 or 10 mmol/l $CaCl_2$. Chemical determination of the concentrations of K^+ allows the calculation of the amount desorbed or adsorbed (ΔK, mmol/kg soil) at each point. A typical plot resulting from such an experiment is shown in Figure 2, and the equilibrium activity ratio, AR^{e}_{K-Ca} for the soil is obtained from the *null* point, $\Delta K = 0$, where potassium is neither lost to nor gained from the solution.

In addition to demonstrating the equilibrium activity ratio, the sorption isotherm shown in Figure 2 also shows how the activity can be expected to change when the ion is removed from a field soil by leaching or crop uptake or when it is added in fertilizers. The gradient of the isotherm ($\delta\Delta K/\delta AR$) defines the *buffer capacity* of the soil for the ion, but because the isotherms are rarely linear, a comparison of buffer capacities of different soils must be made at some standard AR value. The two linear regions of the curve illustrated in Figure 2 reflect potassium-calcium exchange on two different types of surface sites, and the two intercepts K_1 and K_2 give the capacity of these sites for retaining potassium.

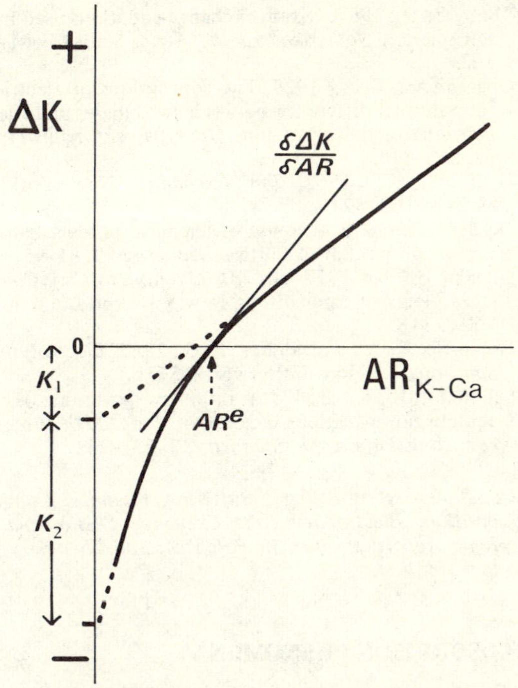

FIGURE 2. Sorption isotherm for potassium, with calcium as the reference cation. The diagram illustrates the estimation of the equilibrium activity ratio (AR^e), the buffer capacity ($\delta\Delta K/\delta AR$), and adsorption capacities (K_1 and K_2) of two different types of potassium-adsorbing sites.

Plant growth is complex and is controlled by a network of nutrient interactions so that the application of soil cation-activity ratios to growth studies is by no means straightforward (Beckett, 1972), and the use of free energy concepts in relation to plant nutrition has recently been strongly challenged (Oertli, 1973). However, the use of activity ratios and products is the only method of defining the chemical potential of ions in a charged system such as soil. The use of individual ion concentrations may be perfectly satisfactory in practical situations in which the incidental factors mentioned at the beginning of this article do not vary, but only because this restriction maintains both the reference ion concentration and the activity-coefficients constant.

BRYON W. BACHE

References

*Beckett, P. H. T., 1971. Potassium potentials—a review, *Potash Review,* Subject 5, 30th Suite, 1–41.
*Beckett, P. H. T., 1972. Critical cation activity ratios, *Adv. Agron.,* **24**, 379–412.

*Bolt, G. H., 1967. Cation exchange equations used in soil science, *Netherlands Jour. Agric. Sci.*, **15**, 81–103.

Guggenheim, E. A., 1929. The conceptions of electrical potential difference between two phases and the individual activities of ions, *Jour. Phys. Chemistry*, **33**, 842–849.

Helfferich, F., 1962. *Ion Exchange.* New York: McGraw-Hill, 624p.

Oertli, J. J., 1973. The use of chemical potentials to express nutrient availabilities, *Geoderma*, **9**, 81–95.

Olphen, H., van 1977. *An Introduction to Clay Colloid Chemistry.* 2nd edition. New York and London: Wiley, 318.

Robinson, R. A., and Stokes, R. S., 1965. *Electrolyte Solutions.* London: Butterworth, 571p.

Schofield, R. K., 1947. A ratio law governing the equilibrium of cations in solution, *Proc. 11th Internat. Cong. Pure Appl. Chemistry*, **3**, 257–261.

Cross-references: *Buffers, Buffering; Chemical Composition; Electrochemistry; Exchange Phenomena; Ionic Activities; Nutrient Potential; Soil Chemistry; Soil Solution.*

ADSORPTION PHENOMENA

Adsorption is the adhesion of ions or molecules, in an extremely thin layer, to the surface of soil solids with which they are in contact (Ellis and Knezek, 1972). In more practical terms, "adsorption" refers to the existence of a higher concentration of a particular component at the surface of a solid phase than is present in bulk solution of soils and sediments. The more general term "sorption" is often used, particularly when the adsorption mechanisms are not known. Adsorption is the major general retention mechanism for many organic and inorganic compounds and metals, particularly the metals in the transition series. Thus the adsorption-desorption phenomena play an important role in regulating the availability of several plant nutrients (q.v.) and the rate of leaching (q.v.), volatilization, or breakdown, of organic pesticides. The adsorption-desorption phenomena also protect water supplies by retaining numerous potential pollutants including nutrients, heavy metals, pesticides, and pathogens. Compounds or ions adsorbed on a soil surface are in equilibrium with the soil solution (q.v.) and are capable of *desorption* (Ellis and Knezek, 1972; Jenne, 1968; McLaren and Peterson, 1965).

Precipitation Reactions

In the past the sorption of inorganic ions by soils was often thought to be due largely to *precipitation reactions,* i.e., the formation of a sparingly soluble solid phase. However, careful studies have shown that the solubility product principle will not account for the extremely low concentrations of phosphorus and many of the metals in the bulk solution of an aerated soil (Lindsay, 1972). These studies have brought about the realization that precipitation generally dominates only at relatively high concentrations of the reactants.

Ion Exchange

Cation exchange, the interchange between a cation in solution and another cation on the surface of any surface-active material, is probably the most important adsorption mechanism for the cationic plant nutrients (potassium, calcium, and magnesium) in soils. A similar type of *anion exchange* may be involved in the retention of nitrate and chloride in acid, highly weathered soils carrying a net positive charge (see *Exchange Phenomena*).

Adsorption

Physical sorption involves the attachment of the sorbent and sorbate through weak atomic and molecular interaction forces (*van der Waal forces*) that operate when the electron clouds of the atoms do not overlap sufficiently to cause strong attractive forces. The activation energy for this type of attraction is characteristically low, much lower than is normally observed for ions (e.g., orthophosphate) which are bound more strongly.

Most anions, as well as several heavy metals, exhibit a property termed *specific adsorption.* This property involves the exchange of the ion with surface ligands to form partly covalent bonds with lattice ions. The net result is that the amount of ion adsorbed is far greater than would be expected for nonspecifically adsorbed species that are adsorbed according to their relative abundance (Mott, 1970).

The magnitude of specific adsorption for most species depends on pH. The specific adsorption is predictable because the materials in most soils active in sorption are believed to be the hydrous oxides of aluminum, iron, and manganese. These materials are *amphoteric,* exhibiting a zero point of charge. Thus, as a general rule, anions are more tightly sorbed at acid than at alkaline pHs, while the reverse is true for cations.

Two equations are most often used to quantitatively describe sorption of ions from solution onto solids, namely, the *Freundlich* and the *Langmuir isotherms.* The Freundlich isotherm has the form:

*Additional bibliographic references may be found in this work.

$$x = kc^n \qquad \text{or} \qquad \log x = \log k + n \log c$$

where x is the amount sorbed per unit sorbent at concentration c of sorbate, and k and n are constants. The Langmuir equation has the form:

$$\frac{M}{x/m} = \frac{1}{Kb} + \frac{M}{b}$$

where M is the activity (moles per liter) of the ion; x/m is the amount of ion M sorbed per unit of sorbate (usually meq/100 g); K is a constant related to the bonding energy; and b is the maximum amount of the ion that will be sorbed by a given sorbate.

The disadvantages of these equations are that they provide little information on mechanisms of sorption and assume uniformity of sorption sites and noninteraction laterally of sorbed molecules. Further, sorption data often deviate from these models at low and high ion concentrations, likely due to different types of sorption sites or mechanisms. However, the literature contains numerous examples of experimental data that do agree with these isotherms.

Organic Sorption

Organic compounds, including materials such as proteins, enzymes, viruses, pesticides, and bacteria can be sorbed to varying degrees by soils (McLaren and Peterson, 1965; Marshall, 1971; Green, 1974; Weed and Weber, 1974). As with inorganic species, the mechanisms are extremely complex and are even more difficult to categorize because of differing chemical and physical characteristics of natural and synthesized organic materials.

In general, the retention of pesticides by soils relates more to the organic matter content than any other single property (Weed and Weber, 1974), while considerable evidence exists for formation of relatively stable clay-organic complexes (McLaren and Peterson, 1965). Bacterium-soil particle interactions appear very complex (Marshall, 1971).

DENNIS R. KEENEY

References

Ellis, B. G., and Knezek, B. D., 1972. Adsorption reactions of micronutrients in soils, *in* J. J. Mortvedt, P. M. Giordano, and W. L. Lindsay, eds, *Micronutrients in Agriculture.* Madison: Soil Science Society of America, 59–78.

Green, R. E., 1974. Pesticide-clay-water interactions, *in* W. D. Guenzi, J. E. Ahlrichs, M. E. Bloodworth, G. Chesters, and R. G. Nash, eds., *Pesticides in Soil and Water.* Madison: Soil Science Society of America 3–38.

Jenne, E. A., 1968. Controls on Mn, Fe, Co, Ni, Cu and Zn concentrations in soils and water: The significant role of hydrous Mn and Fe oxides, *Adv. Chem. Ser.,* 73, 337–387.

Keeney, D. R., and Wildung, R. E. 1977. Chemical properties of soils, *in* F. J. Stevenson and L. F. Elliot, eds., *Soils for Management of Organic Wastes and Waste Waters.* Madison: American Society of Agronomy, 379–402.

Lindsay, W. L., 1972. Inorganic phase equilibria of micronutrients in soils, *in* J. J. Mortvedt, P. M. Geordano, and W. L. Lindsay, eds., *Micronutrients in Agriculture.* Madison: Soil Science Society of America, 41–57.

Marshall, K. C., 1971. Sorptive interactions between soil particles and microorganisms, *in* A. D. McLaren and J. Skujins, eds., *Soil Biochemistry,* New York: Marcel Dekker, 409–445.

McLaren, A. D., and Peterson, G. H., 1965. Physical chemistry and biological chemistry of clay mineral-organic nitrogen complexes, *in* W. V. Bartholomew and F. E. Clark, eds., *Soil Nitrogen. Agronomy,* **10,** 261–286.

Mott, C. B., 1970. Sorption of anions by soils, *Soc. Chem. Ind. Monogr. (London),* 37, 40–42.

Weed, S. B., and Weber, J. B., 1974. Pesticide-organic matter interactions, *in* W. D. Guenzi, J. E. Ahlrichs, M. E. Bloodworth, G. Chesters, and R. G. Nash, eds., *Pesticides in Soil and Water.* Madison: Soil Science Society of America, 39–66.

Cross-references: *Activity Ratio; Clay-organic Reactions; Electrochemistry; Exchange Phenomena; Herbicide Diffusion; Hydrous Oxides; Leaching; Soil Solution; Soil-solvent Reactions. Vol. XVI: Trace Element Contamination.*

AERAGRONOMY

Aeragronomy, a term derived from the combination of "aeronautics" and "agronomy" (q.v.), has given soil technology new dimensions in space and time. It has eased several agronomic operations to provide more effective land management in some localities. Aeragronomy has provided the foundations for soil-conservation farming systems (Campbell, 1955), especially on unstable hill slopes; made possible the mechanization of hill country farming; and revolutionized fertility building and pasture management in New Zealand.

The use of aircraft in aerial firefighting and crop dusting (Barrett, 1975; Baxter, 1975) was first developed in the United States, but it has been used worldwide in a wide variety of new applications. Recent developments include control of noxious animals by dropping poisoned baits from planes, distribution of weedicides and pesticides (Larson, 1975), and delivery of fencing materials to overcome transport problems in remote intractable areas. In New Zealand, for example, topdressing rough hill country pasture and seeding clover or grass from aircraft has been integrated into soil-conservation practices in an effort to control water movement and soil erosion (Campbell, 1968). By maintaining adequate vegetation cover on hill slopes through the use of these techniques, it has been possible to reduce sheet wash, scree development, and gully erosion hazards.

CHARLES W. FINKL, JNR.

References

Aviation Weekly, 1975. U.S. agricultural aircraft, **102**, 107+.

Barrett, P., 1975. Crop dusters, *Mech. Illus.*, **71**, 58–59.

Baxter, G., 1975. Dust thou must, *Flying*, **96**, 83–86.

Campbell, D. A., 1955. Giving wings to soil conservation, *Bull. Soil Conserv. Rivers Control Coun. N.Z.*, 1–72.

Campbell, D. A., 1968. Aeragronomy–The key to soil conservation in New Zealand, *9th Internat. Congr. Soil Sci. Trans. (Adelaide)*, **3**, 395–405.

Larson, G. C., 1975. Let us spray, *Flying*, **96**, 26–27.

Cross-references: *Agronomy; Conservation; Management of Soils.*

AERATION, RESPIRATION, AND ATMOSPHERE

The *soil atmosphere* is that portion of the soil system composed of gases occurring between the soil liquids and solids. Its composition at any one time and depth depends on the balance between the rate of production and use of the various gases in the soil (primarily *respiration*) and the rate of exchange between the soil atmosphere and atmospheric air (*soil aeration*). The first processes tend to accentuate the differences between soil air and atmospheric air, whereas the second has the opposite effect.

Soil Aeration

Gas exchange between the respiring sites and the atmosphere may be visualized as a two-step process: first, the movement between the atmosphere and the air-filled soil pores and, second, between the air-filled pores and the center of biological activity. The center of activity may be a root or microbial growth on or in an aggregate. Since living tissues and soil particles are hydrated, the second step must occur through water films of varying thickness.

Movement of gases between the atmosphere and the soil air occurs as mass flow and diffusion. *Mass flow* results from differences in total air pressure arising from meteorological events such as wind or rain and changes in barometric pressure or temperature. *Diffusion* is attributable to a difference in partial pressure, or concentration, created by soil respiration and results in the movement of individual components of the soil air.

Mass Flow. Mass flow is described by *Darcy's law* provided the pressure gradients are not large. For practical purposes, the effect of gravity can be disregarded; thus

$$v = -\frac{k^1}{\eta}\frac{dp}{dx} \qquad (1)$$

where

v = volume flow rate (cm s^{-1})
k^1 = permeability of the soil (cm^2)
η = viscosity of the air
P = air pressure (dyn cm^{-2})
x = distance (cm)

The permeability coefficient depends on the number and size of the air-filled pores. Evans and Kirkham (1949) report in situ values for air permeability of 1 to 40 × 10^{-8} cm^2 for medium-textured soils at various moisture contents; data for disturbed soils or coarser materials tend to be consiberably higher (e.g., see Kimball and Lemon, 1971).

Changes in temperature cause the soil air to contract or expand, and atmospheric air to be sucked in or soil air expelled. The effect is not large since the coefficient of expansion is 1/273 per degree Celsius. The effect of the daily temperature wave should be confined to the top few centimeters of the soil, and its contribution to the total gas exchange is often estimated at less than 0.1%. Baver et al (1972) attribute less than 0.1% of the gas exchange to wind action. Fluctuations in air pressure at the soil surface result in the mixing of air in the topsoil, which enhances aeration, especially on coarse soils (Kimball and Lemon, 1971). Rainfall displaces soil air as it infiltrates into the soil and carries dissolved O_2 with it. As the water drains or evaporates, air is sucked in at the surface. In many cases the initial displacement of soil air is not complete as evidenced by increased air pressures below the wetting front (Garner, Donaldson, and Taylor, 1969). The contribution of rain to gas exchange depends on its frequency and intensity but is estimated at less than 10%. The preceding discussion indicates that mass flow contributes to soil aeration in only a minor degree and that diffusion must constitute the major process.

Diffusion Through Air Spaces. As soil air contains more CO_2 and less O_2 than the atmosphere, CO_2 diffuses from the soil to the atmosphere while O_2 diffuses in the opposite direction. This diffusion is described by *Fick's law:*

$$q_i = -D_{sa,i}\frac{dc_i}{dz} \qquad (2)$$

where

q_i = diffusion rate of gas i (g cm^{-2} s^{-1})
$D_{sa,i}$ = diffusion constant in the soil (cm^2 s^{-1})
c_i = concentration in soil air (g cm^{-3})
z = depth (cm)

In soil aeration the equation of conservation of matter must include a CO_2 source or an O_2 sink due to respiration:

$$\frac{dG_i}{dt} = -\frac{dq_i}{dz} + a_i \qquad (3)$$

where

$G_i = c_iV_A$ = amount of gas per unit volume of soil (g cm^{-3})
V_A = air-filled pore space (cm^3 cm^{-3})
t = time (s)
a = respiration rate per unit volume of soil (g cm^{-3} s^{-1})

The respiration rate (or activity) is positive when the gas is evolved (CO_2) and negative when the gas is used (O_2). Combining equations 2 and 3 gives:

$$\frac{dG_i}{dt} = \frac{d}{dz}\left(D_{sa,i}\frac{dc_i}{dz}\right) + a_i \qquad (4)$$

Usually several simplifying assumptions are made before attempting to solve equation 4. For example:

1. The moisture content is steady, implying that neither the air-filled pore space nor the diffusion constant changes with time; thus

$$\frac{dG_i}{dt} = V_A\frac{dc_i}{dt}$$

2. The concentration of the gases is steady, hence

$$\frac{dc_i}{dt} = 0$$

The resulting equation:

$$\frac{d}{dz}\left(D_{sa,i}\frac{dc_i}{dz}\right) + a_i = 0 \qquad (5)$$

has been solved subject to the boundary conditions that at the soil surface the concentration is equal to that in the air and that there is no diffusion of the gas over the lower boundary of the active layer (that is, $dc_i/dz = 0$). Solutions for equation 5 have been obtained when the diffusion constant and soil activity did not change with depth and when both changed with depth (Wesseling 1962). Figure 1 shows a typical example of the results of such calculations.

The diffusion constant of a gas in soil is less than in air due to the presence of soil particles and water films. The most general relationship between the two is:

$$D_{sa}/D_a = kV_A{}^n \qquad (6)$$

where

D_a = diffusion constant in air (cm^2 s^{-1})
k = proportionality constant
n = constant

The ratio D_{sa}/D_a is sometimes called the *relative diffusivity*. The simplest equation for the relative diffusivity was proposed by Penman (1940):

$$D_{sa}/D_a = 0.66\ V_A \qquad (7)$$

Often diffusion ceases before the air-filled pore space equals zero; presumably the remaining air-filled pores are blocked by water films. To take this into account, the following equation is sometimes used:

$$D_{sa}/D_a = a(V_A - b) \qquad (8)$$

where a and b are constants. de Vries (1950) has introduced a particle shape factor in the equation for D_{sa}/D_a to account for the shape of the crumb and the shape of the particles forming the crumb. The results show that the relationship between D_{sa}/D_a and V_A becomes more curvilinear as the particle shape deviates from spherical.

Bakker and Hidding (1970) have listed several of the proposed equations for D_{sa}/D_a. In disturbed samples, the relationship between D_{sa}/D_a and V_A is relatively similar, but for undisturbed samples quite large differences are noted (Ayres, Button, and deJong, 1972). Unfortunately, very few measurements of the relative diffusivity are available at low values for V_A, the range where aeration is most critical. Puddling of surface soils reduces the diffusion constant at low values for air-filled porosity; at 20% air-filled porosity or greater, the differences disappear (Bakker and Hidding, 1970).

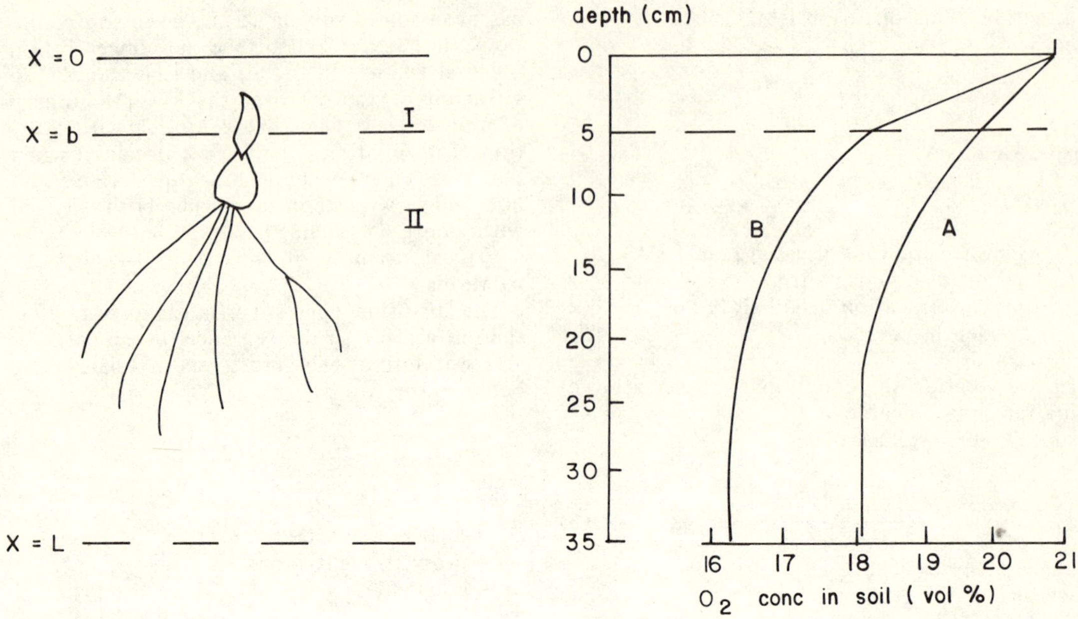

FIGURE 1. Computed oxygen concentration in soils with a nonpuddled (curve *A*) and puddled topsoil (curve *B*). V_A of layers I and II = 0.10; α of layer I = 0; α of layer II = 0.132×10^{-9} g O_2 $cm^{-3}s^{-1}$; L = 35 cm, and b = 5 cm. (From Bakker and Hidding, 1970)

The diffusion constant increases with temperature and decreases with barometric pressure, but the effects are usually small:

$$D_{a(T,\,P)} = D_{a(T_0,\,P_0)}\left(\frac{T}{T_0}\right)^{m}\frac{P_0}{P} \qquad (9)$$

where

$D_{a(T,\,P_0)}$ = diffusion constant at temperature T_0 and pressure P_0 in air

T = actual temperature (°K)
P = actual pressure (atm)
m = varies from 1.5 to 2.0 (for diffusion in air, m is 1.75 for O_2 and 2.00 for CO_2)

The diffusion constant depends also on the composition of the air, but the difference in composition between soil air and atmospheric air is usually disregarded.

The thermal gradients in the soil may require the addition of another term to the diffusion equation (Richter, 1972):

$$q_i = -D_a\frac{dc_i}{dx} - c_i D_T\frac{dT}{dx} \qquad (10)$$

where D_T is the thermal diffusion constant ($cm^2\ s^{-1}\ deg^{-1}$). As D_T is only 1/100 to 1/1000 of D_a, the second term is normally not considered important, but the evidence on thermal diffusion is inconclusive (de Jong, 1973).

Diffusion Through Water Films. Fick's law also describes the diffusion of a gas through water, but the diffusion constants in water are approximately 1/10,000 of the values in air (Table 1). In soil water the effective diffusion constant is further reduced due to the presence of solid particles and the tortuosity of the water films. The effective diffusion constant through soil water (D_{sw}) is related to the diffusion constant in pure water (D_w) by the equation:

TABLE 1. Diffusion Constants of CO_2, O_2, and N_2 in Air and Water and their Solubility in Water at 20°C.

	Diffusion Constant[a] (cm^2/s)		Solubility[b] (Absorption Coefficient)
	Air	Water	
CO_2	0.161	0.177×10^{-4}	0.878
O_2	0.205	0.180×10^{-4}	0.031
N_2	0.205	0.164×10^{-4}	0.015

[a]Dolz, R. E., and Luve, G. T. (1970), D_a corrected using equation 9.
[b]Volume of gas (reduced to 0°C and 1 atm) absorbed per unit volume of water when the pressure of the gas is 1 atm (Hodgman, Weast, and Selby, 1958).

$$D_{sw} = D_w \Gamma \lambda V_w \qquad (11)$$

where

V_w = moisture content ($cm^3\ cm^{-3}$)
Γ = tortuosity factor
γ = factor accounting for the interaction between the gas and the soil (usually assumed to be 1)

The driving force for the diffusion is the concentration gradient in the liquid, which, through the solubility constants (Table 1), is related to the concentrations in the gas phase. As a result, the same concentration difference between two gas phases separated by a water film will move about 16 times as much CO_2 as O_2.

Aeration of Microsites

Roots. Models for O_2 distribution around and in plant roots consider the root as a cylindrical sink for O_2 surrounded by a liquid-solid shell. Lemon (1962) derived the following equations to describe the O_2 concentration.

Within the root:

$$c_i = c_R + \frac{\alpha}{4D_i}(R^2 - r_i^2) \qquad (12)$$

At the root surface:

$$c_R = c_P - \frac{\alpha R^2}{2D_{sw}} \ln \frac{R}{r_e} \qquad (13)$$

where

c_i, c_R, c_P = O_2 concentration ($g\ cm^{-3}$) in water inside the root at radius r_i, at the root surface with radius R, and adjacent to the air-filled pores at radius r_e, respectively
α = O_2 uptake per unit volume of root ($g\ cm^{-3}\ s^{-1}$)
D_i = diffusion constant inside the root ($cm^2\ s^{-1}$)
D_{sw} = diffusion constant in the liquid-solid shell ($cm^2\ s^{-1}$)

The equations indicate that the thickness of the solid-liquid shell is an important factor determining the O_2 concentration at the root surface. For the same values of c_R and α, the O_2 concentration in the center of thicker roots will be less than in the center of thinner roots.

Luxmoore et al (1970*a,b,c*) elaborated on Lemon's model by including a longitudinal O_2 flux in the plant tissue (Figure 2) and by making the O_2 uptake of the roots dependent on the oxygen concentration. Their analysis showed that the O_2 content of plant roots is primarily controlled by the thickness of the solid-liquid shell and the gas-filled porosity of the roots; root respiration and root permeability have only minor effects. Roots with a high porosity such as rice had a higher O_2 concentration within the root and were less dependent on soil aeration than low-porosity maize roots.

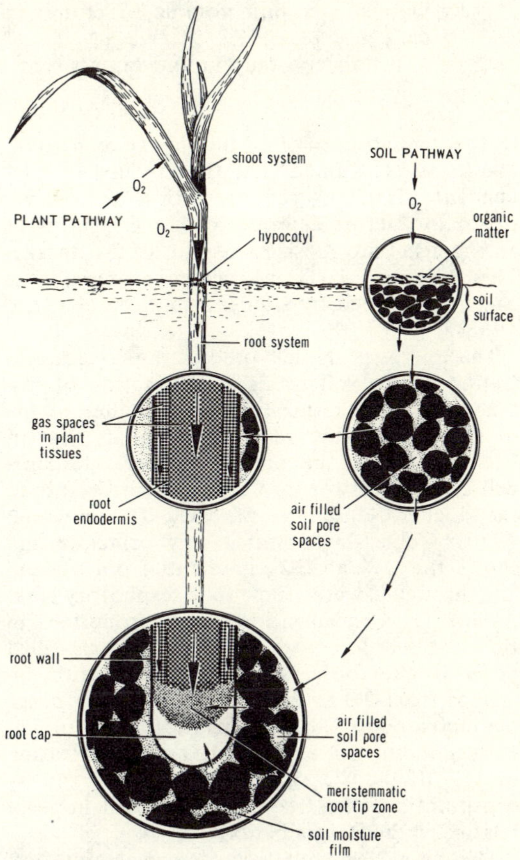

FIGURE 2. Soil- and plant-aeration pathways. (From Luxmoore et al., 1970*a*)

Aggregates. The O_2 concentration in a water-saturated, spherical aggregate with an active microbial population is given by:

$$c_r = c_P + \frac{\alpha(R^2 - r^2)}{6D_{sw}} \qquad (14)$$

where

c_r, c_P = O_2 concentration ($g\ cm^{-3}$) of water inside the crumb at radius r and at the outside of the crumb at radius R, respectively

α = O_2 use per unit volume of crumb (g cm^{-3} s^{-1})

D_{sw} = diffusion constant in the crumb (cm^2 s^{-1})

In the derivation of this equation (Greenwood, 1967), it is assumed that the O_2 use is independent of the O_2 concentration. Obviously, the major factors determining the O_2 status of an aggregate are its size and rate of O_2 uptake. The diffusion of O_2 into single bacterial cells can be described by similar equations (Griffin, 1968).

The microsite models predict the O_2 concentration in the soil water as a function of distance from the respiratory sink. If the volume of O_2 used in the respiration equals that of CO_2 produced, the partial pressure gradients will be much lower for CO_2 than for O_2 due to the higher solubility of CO_2 (Greenwood, 1970). Table 2 illustrates this principle and shows the O_2 and CO_2 gradients from the soil air, through a water film, to a respiratory sink. As the O_2 concentration drops from 19% in the air-filled soil pore to 2% in the gas-filled pores in the root, the CO_2 concentration increases from 2% to 2.6%. The maximum possible increase in CO_2 concentration is about 1%, corresponding to a drop in O_2 concentration of 21%. If the CO_2 production exceeds the O_2 consumption, the CO_2 gradients will increase relative to the O_2 gradients.

The microsite models are an oversimplification of the real geometry. The perfect coaxial solid-liquid cylindrical shell around a root or the perfect spherical solid-liquid crumb just do not occur in nature. Nonetheless, the models provide a working approximation that is useful in understanding the real system.

Characterizing Soil Aeration

The potential aeration of a soil can be inferred from its pore size distribution. If a soil has a large number of macro (noncapillary) pores, aeration is not likely a problem. A minimum air-filled pore space of 10% is often considered necessary for adequate aeration. The actual aeration can be assessed by measuring the air-filled pore space, the composition of the soil air, the oxygen-diffusion rate to a platinum electrode, or the redox potential.

TABLE 2. O_2 and CO_2 Concentrations (1g cm^{-3}) from Soil Air, through a Water Film, to a Respiratory Sink.

	Air-filled Space in Root	Water Film		Air-filled Pore
O_2	27(= 2%)	0.9	8.4	253(= 19%)
CO_2	48(= 2.6%)	44	34	37(= 2%)

The composition of the soil air is determined on as small a sample as possible. Using gas chromatography, samples of 1 ml can be analyzed accurately; in older techniques larger samples were needed, and contamination with atmospheric air was often a problem. Sampling techniques are described by van Bavel (1965) who also discusses the determination of O_2 and CO_2 with portable analyzers. Electrodes have been described that can be used for continuous in situ monitoring of CO_2 (Jensen, van Gundy, and Stolzy, 1965) and O_2 (Willey and Tanner 1963).

The platinum-microelectrode technique, which measures oxygen diffusion, does not give intrinsic values (McIntyre, 1970), but it may be useful for comparative purposes. Similarly, the usefulness of redox potentials depends on the extent to which they provide information about the biological or chemical conditions that are of interest.

Dasberg and Bakker (1970) measured soil air composition (O_2 and CO_2), oxygen-diffusion rate to platinum electrodes, redox potentials, and air-filled pore space in a pot experiment. They concluded that a continuous record of soil air content is a better index of aeration under fluctuating soil moisture conditions than either the O_2 content or oxygen-diffusion rate. No consistent relationship between redox potential and air-filled pore space or O_2 content was apparent, probably because none of the O_2 values were low enough to cause measurable reduction (the lowest O_2 content was about 1%). It has been suggested (Greenwood, 1968) that an assessment of the extent of anaerobic zones could be a useful measure of soil aeration. Such an assessment could be made using ferrous iron or metabolic products as indicators.

Soil Respiration

Biological Reactions. The literature contains two definitions of *respiration:* (1) the uptake and use of gaseous oxygen as a hydrogen acceptor and (2) an energy-yielding metabolic process in which either organic compounds or reduced inorganic compounds serve as electron donors. Life is dependent on the process of oxidation or *dehydrogenation,* which means the removal of hydrogen (usually as 2H, or in the form of H^+ plus an electron *e*) by means of the electron-transfer system of cells. In all aerobic organisms, whether eukaryotic or prokaryotic, oxygen is utilized as the final electron acceptor. In microorganisms, especially bacteria, alternate electron acceptors are available. These include $SO_4^=$, NO_3^-, CO_2, and a number of their partially reduced intermediate products. Oxidized manganese and iron and a number of other possible compounds such as iodine, mercury, and possibly even phosphate can also be used. Where alternate electron acceptors are used, oxidation does not depend on O_2, yet the other processes of life are the same. This process is termed *anaerobic respiration* (Hattori 1973).

In addition to respiration, *fermentation* and *photosynthesis* yield energy and result in the generation of ATP and reducing power, which are the bases of all life. In fermentation, organic compounds serve as both electron donors and acceptors, producing a mixture of by-products, some of which are more oxidized, others more reduced, thus maintaining a strict oxidation-reduction balance. Most fermentation is restricted to anaerobic or facultative anaerobic bacteria. Photosynthesis is carried out in the upper layers of soil or sediment systems by a number of algae and photosynthetic bacteria and reverses the normal process of CO_2 evolution in the soil.

Depending on the availability of energy-yielding substrates and the presence of a sink for electron transfer, microorganisms can carry out a number of reactions that affect the overall gas status of soil or sediment systems. Autotrophic organisms such as the nitrifiers and sulfur oxidizers use CO_2 during the oxidation of inorganic compounds. The oxidized inorganic compound can in turn act as an electron acceptor under O_2-free conditions in the presence of an available energy supply. *Thiobacillus denitrificans* can oxidize H_2S or S by using NO_3 as an electron acceptor according to the following equation:

$$5S + 6NO_3^- + 2H_2O \rightarrow 5SO_4^{=} + 3N_2 + 4H^+$$

CO_2 is used as a carbon source, and in the process CO_2 is reduced to methane. Although H_2 is a common by-product of anaerobic metabolism, evolution of H_2 from flooded soils is small because CH_4-producing bacteria use H_2 as a source of energy for growth (Yoshida, 1975):

$$CO_2 + 4H_2 \rightarrow CH_4 + 2H_2O$$

The preceding discussion indicates that no one term can describe the gas exchange during energy-yielding reactions attributable to biological activity in the soil system. However, respiration, anaerobic respiration fermentation, and photosynthesis can cover the reactions when using the definition of respiration as an energy-yielding metabolic process in which either organic compounds or reduced inorganic compounds serve as electron donors.

Lundegardh (1927) defined soil respiration as the uptake of O_2 and the evolution of CO_2. Many soil reactions that result in O_2 uptake or CO_2 evolution do not necessarily coincide in space or time, nor are they necessarily the best indicators of biological activity. In the remainder of this review, we will utilize the term "soil respiration" as defined by Lundegardh. No single better term is presently available.

In aerobic respiration about one-half of the carbon in organic residues is released as CO_2. The remainder is used for the formation of new cell material (Forest and Walker, 1971). Under these conditions root respiration and microbial decomposition of organic residues can be described by the equation:

$$CH_2O + O_2 \rightarrow CO_2 + H_2O + \text{cell material}$$

In fermentation, most of the organic compounds involved are already at the oxidation-reduction state of cell material. Little energy is required to transform the substrate into cell material; however, many partially reduced intermediates are produced, and the efficiency of substrate incorporation is low. In well-drained soils, the fermentation products produced in anaerobic microsites would be rapidly reoxidized by other organisms having access to O_2. As a consequence, the overall process in the soil could be said to be "aerobic" although a number of organisms and microsites may be involved.

The respiration quotient (RQ = volume of CO_2 produced/volume of O_2 used) has been used to characterize the soil respiration. Values of RQ in excess of one generally indicate anaerobic decomposition. The previous discussion on the large number of reactions involved in biological oxidation of organic and inorganic substrates and the occurrence of nonbiological processes using or producing O_2 and CO_2 indicate that RQ values must be interpreted with caution.

Partitioning Soil Respiration. The largest part of the CO_2 evolution from soil comes from microbial activity, and a ratio of 2:1 for microbial:root respiration is often quoted. Table 3 shows that root respiration contributes 10% to 55% of the CO_2 production. The lower values were obtained on bare roots removed from the soil system. Roots contain a rhizosphere (q.v.) and mycorrhyzal flora, and roughly half the respiration attributable to roots in the soil has been said to come from these organisms (Clark and Kemper, 1967; Paul and Warembourg, 1973). Under "forest," the contribution of the litter layer to total soil respiration can be separated from the mineral horizons (Table 4).

TABLE 3. Contribution of Roots to Total Soil Respiration.

Crop	Root Respiration (% Total)	Method	Reference
Field Crops			
Oats	30	Cropped compared with bare soil	Lundegardh (1927)
Wheat	20	$^{14}CO_2$	Johnen (1974)
Potatoes, kale, and tobacco	30–50	Cropped compared with bare soil	Brown, Fountaine, and Holden (1965)
Grass			
Short-grass prairie	19	$^{14}CO_2$ in conjunction with CO_2 flux	Paul and Warembourg (1973)
Short-grass prairie	15	CO_2 evolution on roots separated from soil	Singh and Coleman (in press)
Tall-grass prairie	40	Correlation between CO_2 evolution and root biomass	Kucera and Kirkham (1971)
Dry meadow	55	Vegetated area compared with clipped	Kosonen (1968)
Alluvial meadow	29–38	Determined on bare roots after removal of soil	Yastrebov (1958)
Broom sedge old field	<17	CO_2 evolution on roots separated from soil	Coleman 1973*b*)
Drained peat bog	33–50	Vegetated area compared with bare soil	Makarov (1960)
Forest[a]			
Pine over dogwood	<50	Soil contributed less than 56% (rest from litter layer)	Witkamp and Frank (1969)
Southern red oak	<11	Determined on roots separated from soil	Coleman (1973*b*)
Oak	40 (Winter 25, summer 55)	O_2 uptake, roots separated from soil	de Boois (1974)
Temperate deciduous	48	Extrapolated respiration data of different sized roots	Reichele et al (1973)

[a]Total includes contribution from litter.

Buried litter usually decomposes more rapidly than litter on the soil surface, and meteorological conditions greatly affect the rate of decomposition.

Measurement of Soil Respiration. The most common methods of measuring soil respiration involve placing a canopy on the soil surface. Care must be taken that no photosynthesis or shoot respiration occurs underneath the canopy and that the microclimate is disturbed as little as possible. The amount of CO_2 that evolves is measured by "static," or "dynamic," methods. Using *static* methods, the increase in CO_2 concentration is measured as a function of

TABLE 4. Contribution of the Litter Layer to Soil Respiration.

	Litter Respiration (% of Total)		Reference
Forest			
Pine over dogwood	28		Witkamp and Frank (1969)
Southern red oak	37–39		Coleman (1973*b*)
Oak (*L* + *F* layer)	50	*L* = 23, *F* = 28	de Boois (1974)
Temperate deciduous ($O_1 + O_2$)	45	O_1 = 30, O_2 = 15	Reichele et al. (1973)
Temperate deciduous	21		Edwards (1975)
Grass			
Broom sedge old field	6–16		Coleman (1973*b*)

time, or the evolved CO_2 is absorbed in alkali. Using the *dynamic* approach, soil respiration is calculated from the increased CO_2 concentration in a stream of air being pulled through the canopy. The static alkali-absorption method has been reported to overestimate (Witkamp and Frank, 1969) or to underestimate soil respiration (Coleman, 1973*a*). With the dynamic methods, the measured soil respiration depends very much on the air pressure in the box, (Kanemasu, Powers, and Sij, 1974).

The aerodynamic and the "soil-CO_2-profile" methods cause minimal disturbance. In the aerodynamic method, the CO_2 flux from the soil is estimated from the CO_2 gradient above the soil surface and the aerodynamic-transport coefficients. Problems arise from the effects of photosynthesis and respiration of the shoots on the CO_2 flux and the estimation of the transport coefficient within the plant canopy. The aerodynamic method includes the contribution of the litter layer to soil respiration. This is not the case for the soil-profile method, which calculates soil respiration from the CO_2 profile in the soil and the estimated diffusion constant of the soil (de Jong and Schappert, 1972). This soil-profile method neglects the possible contributions of CO_2 mass flow to soil respiration.

The respiration of specific soil layers' activity is usually determined in the laboratory with Warburg respirometers or similar equipment (Stotsky, 1965). Either disturbed (Lundegardh, 1927; Yastrebov, 1958; de Boois, 1974) or undisturbed (Penningsfeld, 1950) samples taken from specific soil horizons have been used. Witkamp and Frank (1969) have determined activity in situ by progressively removing layers from the top of the profile and measuring the changes in CO_2 evolution. The soil-CO_2-profile method of measuring soil respiration can be used to calculate activity at various depths.

Most investigators have measured CO_2 evolution rather than O_2 because it is technically easier and more sensitive. Unfortunately, CO_2 is involved in chemical reactions such as the formation or breakdown of carbonates, and it is highly soluble in water. Thus it evolves when water evaporates. Ahrens and Thalmann (1970) found that abiotic CO_2 evolution occurred from calcareous and calcium-free soils but that the main part of the abiotically produced CO_2 probably originated from chemical decarboxylation of the organic matter. Calculations by Ballard (1968) show that the loss of CO_2 in drainage water is not significant under most conditions. It has been suggested that O_2 is a better measure of soil respiration than CO_2, but O_2 may be used in chemical oxidation in the soil and does not participate in fermentation processes.

Factors Influencing Soil Respiration

Biomass. The soil organisms are one of the two major components affecting soil respiration, and they include diverse types ranging from fauna to fungi and bacteria. The other major component, roots, tends to be restricted to fewer species. All components are influenced in many similar ways by abiotic parameters (Bunnell and Tait, 1974). Examples of both a good correlation between CO_2 production and microbial numbers and of poor or no correlation can be found in the literature (Russell, 1973). Soil microorganisms are generally limited by available substrate, and any increase in the supply leads immediately to an increase in the numbers of organisms (Russell, 1973). Clark and Paul (1970) calculated that in a grassland soil the available substrate was sufficient only for the total microflora to reproduce a few times per year, and microsite activity must predominate. In an aerobic soil a large active population can result in an immediate CO_2 evolution without a large increase in biomass (Shields et al, 1973). On the other hand, a small population would require a lag period for development. As microorganisms have a high reproductive potential, the initial population size would not restrict the CO_2 evolution over the long-term period.

Soil Moisture. Field experiments have shown general relationships between soil moisture content and soil respiration (Krzysch 1965; de Jong, Schappert, and MacDonald, 1974); but in many cases the correlation coefficients have been low. The respiration of forest litter has also shown to be highly dependent on moisture content (de Boois, 1974).

In droughty soil the activity of the microorganisms is restricted by the high energy level of the soil water. The high energy level restricts the availability of soil water to microbes, the mobility of dissolved nutrients, and the mobility of the microorganisms in the water films (Russell, 1973). Excess water restricts microbial activity by its interference with aeration and the requirement for fermentation or anaerobic respiration to occur. Molecular oxygen is usually required prior to the degradation of aromatic rings predominant in many ligniferous components of plant cell material and soil humus. Therefore, soil carbon can accumulate under continually waterlogged conditions such as in peat formation.

The availability of water is usually expressed in terms of its activity (a_w) or soil moisture tension (SMT):

$$a_w = \frac{P_s}{P_w} = \frac{\text{RH}}{100}$$

SMT ≈ 3000 log a_w (at room temperature) atm

where

P_s = vapor pressure of the solution
P_w = vapor pressure of pure water at the same temperature
RH = relative humidity

The lower limits in a_w for pure cultures in solutions range from 0.99 to 0.93 for bacteria. The minimal a_w for some filamentous fungi ranges from 0.70 to 0.65 (Griffin, 1972). In soil, decomposition practically comes to a standstill as the soil moisture content drops below the permanent wilting point ($a_w \approx 0.989$).

Temperature. The effects of temperature are dependent to some extent on the moisture content and on the site characteristics. Areas that contain psycrophilic bacteria such as the tundra and ocean sediments show respiration below the freezing point of pure water. Tamm and Krzysch (1965) show that soil respiration ceased at -8°C; at -5°C soil respiration became measurable and increased as moisture and substrate conditions became more favorable (Figure 3). Similar data have been obtained for the tundra where the interaction between moisture content and temperature has been noted (Bunnell and Tait, 1974).

The relationship between temperature and soil respiration has a typical bell shape of which Figure 4 shows only the ascending half. The response surface can be described by a normal curve

$$Y = Y_{\max} \exp\left[-(M - X)^2/B^2\right]$$

(Russell, 1975). In the laboratory a second maximum is sometimes measured at higher temperatures due to the development of a thermophilic microbial population (Novak, 1973). The relationship between temperature and CO_2 respiration has been demonstrated in arable land (Krzysch, 1965; Tamm and Krzysch, 1965; Brown, Fountaine, and Holden, 1965); grassland (Makarov, 1960; Kucera and Kirkham, 1971); and forest (Edwards, 1975; Witkamp and Frank, 1969; de Boois, 1974).

Soil moisture often interacts with the effects of soil temperature on respiration and vice versa (Figure 4). In the field, temperature often controls soil respiration in early spring, fall, and winter, whereas in summer the soil moisture content is more critical and overrides the effects of temperature.

Disturbance. *Wetting-drying cycles,* or freezing and thawing, are known to stimulate soil respiration. Freezing and drying result in autolyses of microbial cells and disruption of soil aggregates. After restoration of favorable conditions, the surviving microorganisms use the supply of easily decomposable substrate leading to enhanced CO_2 production (Birch, 1958; Shields, Paul, and Lowe, 1974). The magnitude of the CO_2 flush depends on the decomposable organic material in the soil (especially the biomass) and the severity of the treatment.

The significance of disturbance (Russell, 1973; Greenwood, 1967) is documented, but the specific quantitative effects are not well known (de Boois, 1974; Coleman, 1973*b*; Lundegardh, 1927). Martel and Paul (1974) attributed the large and consistent decrease in soil organic matter during 70 years of culti-

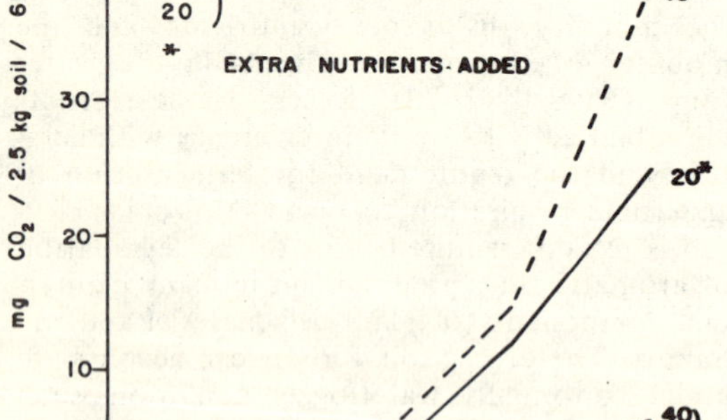

FIGURE 3. CO_2 production of a "Braunerde" as a function of temperature, moisture content, and substrate availability. (Plotted from Tamm and Krzysch, 1965)

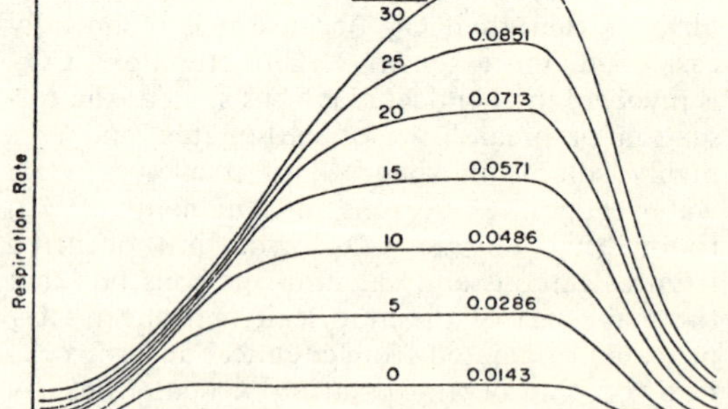

FIGURE 4. Semiperspective plot of computed respiration rate at different temperature and moisture levels. All curves actually begin at origin. (From Bunnell and Tait, 1974; copyright © 1974 by the IBP Tundra Biome Steering Committee)

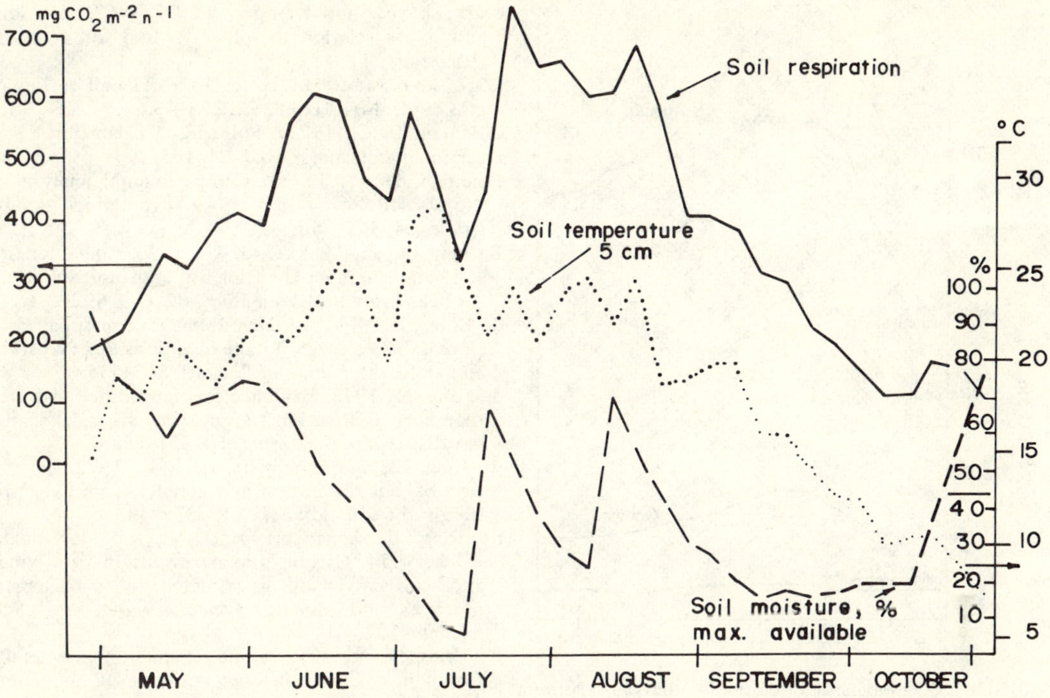

FIGURE 5. Variation in soil respiration, soil temperature, and available soil moisture under fodder beets during 1959. (Redrawn from Krzysch, 1965)

vation largely to a destruction of aggregates, which makes the soil organic matter available to microbial attack.

Vegetation. Vegetation affects soil respiration directly through root respiration and indirectly through its effects on soil moisture and (to a lesser extent) soil temperature. Vegetation also has a major effect on substrate availability. Maximum soil respiration of grassland seems to coincide with maximum vegetative growth (Makarov, 1960; Yastrebov, 1958; de Jong, Schappert, and MacDonald, 1974), whereas for cereals maximum soil respiration seems to coincide with the tillering stage (Vugakov and Popova, 1968). Changes in substrate availability are the major reasons for soil respiration decreases with depth.

Annual and Daily Pattern of Soil Respiration

Seasonal changes in soil respiration integrate the effects of soil moisture and temperature and substrate availability. In forest the variation in litter composition is especially important. Figure 5 shows the variation in soil respiration during the growing season in a cultivated field. As noted earlier, temperature appears to control soil respiration in the spring and fall, whereas soil moisture is the important controlling factor during the summer. de Jong, Schappert, and MacDonald (1974) observed that under semiarid conditions soil respiration was less highly correlated with soil moisture than with rainfall or number of days with rain in excess of 5 mm. They concluded that the number of wetting and drying cycles was mainly responsible for the large annual variations in soil respiration. Changes in soil temperture are largely responsible for the diurnal variation in soil temperature; Figure 6 shows typical examples of this diurnal cycle.

In natural ecosystems that have reached an equilibrium state, soil respiration should reflect productivity. In the absence of significant above-ground decomposing organisms, soil respiration is nearly equivalent to the total production in grassland (Coleman, 1973*a*; de Jong, Schappert, and MacDonald, 1974) when measurements are carried out over a number of years. The soil-respiration measurements must be representative for the various periods of the year, and short-term fluctuations in CO_2 production must be avoided. If these conditions are met, soil respiration can indicate climatic trends and primary productivity of various plant communities (Leith and Ouellette, 1962).

E. DE JONG
E. A. PAUL

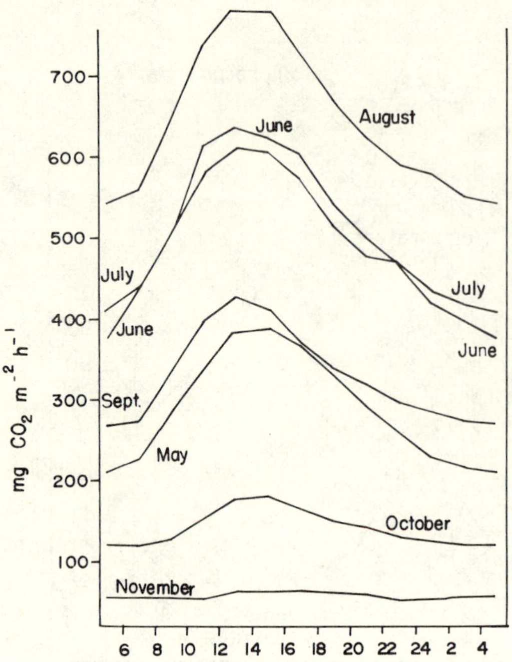

FIGURE 6. Average daily course of the soil respiration for different months of the year in the same field as Figure 5.

References

Ahrens, E., and Thalmann, A., 1970. Zur Frage der Abiotischen Co_2-Abspaltung aus karbonathaltigen Böden, *Z. Pfl. Ernähr. Bodenk.* **127**. 143–154.

Ayres, K. W., Button, R. G., and de Jong, E., 1972. Soil morphology and soil physical properties. I. Soil Aeration, *Canadian Jour. Soil Sci.*, **52**. 311–321.

Bakker, J. W., and Hidding, A. P., 1970. The influence of soil structure and air content on gas diffusion in soils, *Netherlands Jour. Agric. Sci.*, **18**, 37–48.

Ballard, T. M., 1968. Carbon dioxide production and diffusion in forest floor material–A study of gas exchange in biologically active, porous media. Ph.D. Thesis, University of Washington, 120p.

Bavel, C. H. M. van, 1951. A soil aeration theory based on diffusion, *Soil Sci.*, **72**, 33–46.

Bavel, C. H. M. van, 1965. Composition of soil atmosphere, *Agronomy*, **9**, 315–318.

Baver, L. D.; Gardner, W. H.; and Gardner, W. R., 1972. *Soil Physics.* New York: Wiley, 498p.

Birch, H. F., 1958. Pattern of humus decomposition in East African soils, *Nature*, **181**. 788.

Brown, N. J.; Fountaine, E. R.; and Holden, M. R., 1965. The oxygen requirement of crop roots and soils under near field conditions, *Jour. Agric. Sci.*, **64**, 195–203.

Bunnell, F. L., and Tait, D. E. N., 1974. Mathematical simulation models of decomposition processes, *in* A. J. Holding et al, eds., *Soil Organisms and Decomposition in Tundra.* Stockholm: Tundra Biome Steering Committee, 207–224.

Clark, F. E., and Kemper, W. D., 1967. Microbial activity in relation to soil water and soil aeration, *Agronomy*, **11**, 472–480.

Clark, F. E., and Paul, E. A., 1970. The microflora of grassland, *Adv. Agron.*, **22**, 375–435.

Coleman, D. C., 1973*a*. Soil carbon balance in a successional grassland, *Oikos*, **24**, 195–199.

Coleman, D. C., 1973*b*. Compartmental analysis of "total soil respiration": An exploratory study, *Oikos*, **24**, 361–366.

Dasberg, S., and Bakker, J. W., 1970. Characterizing soil aeration under changing soil moisture conditions for bean growth, *Agron. Jour.*, **62**, 689–692.

de Boois, H. M., 1974. Measurement of seasonal variations in the oxygen uptake of various litter layers of an oak forest, *Plant and Soil*, **40**, 545–555.

de Jong, E., 1973. Evidence for significance of other-than-normal diffusion transport in soil gas exchange –A discussion, *Geoderma*, **10**, 251–253.

de Jong, E., and Schappert, H. J. V., 1972. Calculation of soil respiration and activity from CO_2 profiles in the soil, *Soil Sci.*, **113**, 328–333.

de Jong, E., Schappert, H. J. V., and MacDonald, K. B., 1974. Carbon dioxide evolution from virgin and cultivated soil as affected by management practices and climate, *Canadian Jour. Soil Sci.*, **54**, 299–307.

de Vries, D. A., 1950. Some remarks on gaseous diffusion in soils, *4th Internat. Congr. Soil Sci. Trans.* **4**, 41–43.

Dolz, R. E., and Luve, G. T., eds., 1970. *Handbook of Tables for Applied Engineering Science.* Cleveland: The Chemical Rubber Company, pp. 453–454.

Edwards, N. T., 1975. Effects of temperature and moisture on carbon dioxide evolution in a mixed deciduous forest floor, *Soil Sci. Soc. America Proc.*, **39**, 361–365.

Enoch, H., and Dasberg, S., 1971. The occurrence of high CO_2 concentrations in soil air, *Geoderma*, **6**, 17–21.

Evans, D. D., and Kirkham, D., 1949. Measurement of the air permeability of soil *in situ*, *Soil Sci. Soc. America Proc.*, **14**, 65–73.

Forest, W. W., and Walker, D. J., 1971. The generation and utilization of energy during growth, *Adv. Microbiol. Phys.*, **5**, 213.

Garner, D. E., Donaldson, J. K., and Taylor, G. S., 1969. Entrapped air in a field site, *Soil Sci. Soc. America Proc.*, **33**, 634–635.

Greenwood, D. J., 1967. Measurement of microbial metabolism in soil, *in* T. R. G. Gray and D. Parkinson, eds., *The Ecology of Soil Bacteria.* Toronto: University of Toronto Press, 138–157.

Greenwood, D. J., 1968. Root growth and oxygen distribution in soil, *9th Internat. Congr. Soil Sci. Trans.*, **1**, 823–831.

Greenwood, D. J., 1970. Distribution of carbon dioxide in the aqueous phase of aerobic soils, *Jour. Soil Sci.*, **21**, 314–329.

Griffin, D. M., 1968. A theoretical study relating the concentration and diffusion of oxygen to the biology of organisms in the soil, *New Phytol.*, **67**, 561–577.

Griffin, D. M., 1972. *Ecology of Soil Fungi.* Syracuse: Syracuse University Press, 193p.

Hattori, T., 1973. *Microbial Life in the Soil–An Introduction.* New York: Marcel Dekker, 427p.

Hodgmen, C. D.; Weast, R. C.; and Selby, S. M., eds., 1958. *Handbook of Chemistry and Physics,* 39th ed. Cleveland: The Chemical Rubber Company, 1606.

Jensen, C. R.; van Gundy, S. D.; and Stolzy, L. H., 1965. Recording CO_2 in soil-root systems with a potentiometric membrane electrode, *Soil Sci. Soc. America Proc.,* **29**, 631–633.

Johnen, B. G., 1974. *Bildung, Menge and Umsetzung von Pflanzenwurzeln im Boden.* Bonn: Agrikulturchemischen Institut der Rheinishchen Friedrich-Wilhelms-Universitat Bonn, 125p.

Kanemasu, E. T.; Powers, W. L.; and Sij, T. W., 1974. Field chamber measurements of CO_2 flux from soil surface, *Soil Sci.,* **118**, 233–237.

Kimball, B. A., and Lemon, E. R., 1971. Air turbulence effects upon soil gas exchange, *Soil Sci. Soc. America Proc.,* **35**, 16–21.

Kosonen, M., 1968. The relation between the carbon dioxide production in the soil and the vegetation of a dry meadow, *Oikos,* **19**, 242–249.

Krzysch, G., 1965. Zur der Dynamik Bodenatmung während der Vegetations Zeit, *Z. Ack. Pfl. Bau.,* **122**, 108–140.

Kucera, C. L., and Kirkham, D. R., 1971. Soil respiration studies in tall-grass prairie in Missouri, *Ecology,* **52**, 912–915.

Leith, H., and Ouellette, R., 1962. Studies on the vegetation of the Gaspé Peninsula. II. The soil respiration of some plant communities, *Canadian Jour. Botany,* **40**, 127–140.

Lemon, E. R., 1962. Soil aeration and plant root relations. I. Theory, *Agron. Jour.,* **54**, 167–170.

Lundegardh, H., 1927. Carbon dioxide evolution of soil and crop growth, *Soil Sci.,* **23**, 417–453.

Luxmoore, R. J.; Stolzy, L. H.; and Letey, J., 1970*a*. Oxygen diffusion in the soil-plant system. I. A model, *Agron. Jour.,* **62**, 317–322.

Luxmoore, R. J., Stolzy, L. H., and Letey, J., 1970*b*. Oxygen diffusion in the soil-plant system. III. Oxygen concentration profiles, respiration rates, and the significance of plant aeration predicted for maize roots, *Agron. Jour.,* **62**, 325–329.

Luxmoore, R. J., Stolzy, L. H., and Letey, J., 1970*c*. Oxygen diffusion in the soil-plant system. IV. Oxygen concentration profiles, respiration rates, and radial oxygen losses predicted for rice roots, *Agron. Jour.,* **62**, 329–332.

McIntyre, D. S., 1970. The platinum microelectrode method for soil aeration measurement, *Adv. Agron.,* **22**, 235–283.

Makarov, B. N., 1958. Diurnal variation in soil respiration and in the carbon dioxide content of the layer of air next to the soil, *Akad. Nauk Azerbaydzhan. SSR Doklady,* **118**, 389–391.

Makarov, B. N., 1960. Respiration of soil and composition of soil air on drained peat-bog soils, *Sov. Soil Sci.,* **1960**, 154–160.

Martel, Y. A., and Paul, E. A., 1974. Effects of cultivation on the organic matter of grassland soils as determined by fractionation and radiocarbon dating, *Canadian Jour. Soil Sci.,* **54**, 419–426.

Novak, B., 1973. Ausnutzung biochemischer Teste in der Bodenmikrobiologie. II. Bestimmung der Abbaugeschwindigkeit organische Stoffe im Boden, *Zbl. Bakt. Abt.,* II, Bd., **128**, 316–335.

Paul, E. A., and Warembourg, F. L., 1973. Separation of microbial and root respiration, *Bull. Ecol. Res. Comm.* (Stockholm), **17**, 274.

Penman, H. L., 1940. Gas and vapour movements in soil (I and II). *Jour. Agric. Sci.,* **30**, 447–462, 570–581.

Penningsfeld, F., 1950. Uber Atmungsuntersuchungen an einigen Bodenprofilen, *Z. Pfl. Ernähr Bodenk.,* **50**, 135–164.

Reichele, D. E.; Dinger, B. E.; Edwards, N. T.; Harris, W. F.; and Sollins, P., 1973. Carbon flow and storage in a forest ecosystem, *in* G. M. Woodwell and E. V. Pecan, eds., *Proc. of the Symposium Carbon and the Biosphere.* New York: 24th Brookhaven Symposium in Biology, Upton, New York, May 16–18, 1972, 392p.

Richter, J., 1972. Evidence for significance of other-than-normal diffusion transport in soil gas exchange, *Geoderma,* **8**, 95–101.

Russell, E. W., 1973. *Soil conditions and plant growth.* London: Longman, 849p.

Russell, J. S., 1975. Systems analysis of soil ecosystems, *in* E. A. Paul and A. D. McLaren, eds., *Soil Biochemistry,* Vol. 3. New York: Marcel Dekker, 37–82.

Shields, J. A.; Paul, E. A.; and Lowe, W. E., 1974. Factors influencing the stability of labelled microbial materials in soils, *Soil Biol. Biochem.,* **6**, 31–37.

Shields, J. A.; Paul, E. A.; Lowe, W. E.; and Parkinson, D., 1973. Turnover of microbial tissue in soil under field conditions, *Soil Biol. Biochem.,* **5**, 753–764.

Singh, J. S., and Coleman, D. C. (1978). Evaluation of functional root biomass and translocation of photoassimilated carbon-14 in a short grass prairie ecosystem. J. Marshall, ed. *The Below Ground Ecosystem, Range Science Series (Colo. State University),* **26**, *123-132.*

Stotsky, G., 1965. Microbial respiration, *Agronomy,* **9**, 1550–1572.

Tamm, E., and Krzysch, G., 1965. Zur Dynamik der Bodenatmung und des CO_2-Gehaltes der bodennahen Luftschicht während der Vegetationsruhe, *Z. Ack. Pfl. Bau.,* **122**, 209–224.

Vugakov, P. S., and Popova, Y. P., 1968. Carbon dioxide regime in soils of the Krasnoyarsk forest steppe, *Sov. Soil Sci.,* **1968**, 795–801.

Wessling, J., 1962. Some solutions of the steady state diffusion of carbon dioxide through soils, *Netherlands J. Agric. Sci.,* **10**, 109–117.

Willey, C. R., and Tanner, C. B., 1963. Membrane-covered electrode for measurement of oxygen concentration in soil, *Soil Sci. Soc. America Proc.,* **27**, 511–515.

Witkamp, M., and Frank, M. L., 1969. Evolution of CO_2, from litter, humus, and subsoil of a pine stand, *Pedobiologia,* **9**, 358–365.

Yastrebov, M. T., 1958. Effect of the major biological factors on air composition of alluvial soils in the Klyas'ma river flood plain, *Sov. Soil Sci.,* **1958**, 1155–1162.

Yoshida, T., 1975. Microbial metabolism of flooded soils, *in* E. A. Paul and A. D. McLaren, eds., *Soil Biochemistry,* Vol. 4. New York: Marcel Dekker, 83–122.

Cross-references: *Diffusion Processes; Metabolic Activity; Nitrogen Cycle; Rhizosphere; Soil Pores; Soil Structure; Soil Water; Thermal Regimes.*

AGGREGATE STABILITY

The ability of a soil aggregate to withstand the stresses caused by immersion in water is known as *aggregate stability*. The term *aggregate* refers here to a portion of a soil sufficiently large so that the distribution of primary particles smaller than 2 mm in diameter within it is not greatly different from that of the whole soil. The stresses on an aggregate may arise solely from the wetting process itself. Additional stresses may be applied externally—for example, during prior cultivation—or for surface soil aggregates by the mechanical action of raindrops.

Breakup of Aggregates on Wetting

Slaking. The initial breakup of aggregates when immersed in water [Figures 1(a), (b)] is caused by differential swelling and entrapped air (Henin, 1938). The relative importance of the two factors depends on the surface area of the clay present and the amount of clay. The magnitude of the stresses induced decreases with increasing water content of the aggregates prior to immersion.

The faster an aggregate is wetted, the greater its tendency to slake. Its rate of wetting depends on both the precise method of wetting used and the permeability of the aggregate itself. Organic matter reduces slaking by enclosing the packets of clay particles present in an aggregate rather like a string bag, the packets of clay being free to swell inside. Another effect of organic matter, if sufficient quantity is present, is to increase the soil-water contact angle so that the rate of wetting is reduced. Slaking of aggregates can also be prevented by engulfing the clay particles in a solid matrix either naturally—by the deposition of carbonate, for example—or artificially by adding lime.

Dispersion. After immersion in water and any slaking, soluble salts present in an aggregate begin to diffuse out into the surrounding water. If the electrolyte content of the water is sufficiently low, then clay particles may gradually separate from the aggregate [Figures 1(c), (d)]. Exchangeable sodium is normally the main cause of dispersion. Its effect is enhanced by the presence of exchangeable magnesium rather than calcium. Exchangeable aluminum has the opposite effect. In the neutral to slightly acid soils used for agriculture, the aluminum ions are not normally present as simple ions but rather as polymers. On a larger scale very fine aluminum and iron oxides can prevent dispersion by acting as cationic bridges between clay particles (El-Swaify and Emerson, 1975). By linking particles together, as already indicated, organic matter can prevent dispersion despite the presence of appreciable exchangeable sodium. The presence of a source of divalent cations such as carbonate reduces dispersion by increasing the electrolyte content of the water.

Dispersion of Wet, Sheared Aggregates. An aggregate in which there is no exchangeable sodium present does not normally disperse in water unless it has been sheared first when wet. During shearing, water initially present in pores tends to become associated with the clay particles so that the average distance between these particles is increased. If the increase is sufficient, then the particles may disperse when the aggregate is immersed in water. The same factors that control the ease of dispersion of unsheared aggregates affect in a similar way the value of the minimum average distance between clay particles, i.e., minimum water content of the aggregate, before dispersion occurs, with one exception: organic matter, which can act in two ways. First, if there are sufficient organic links between the clay packets, shearing is unable to

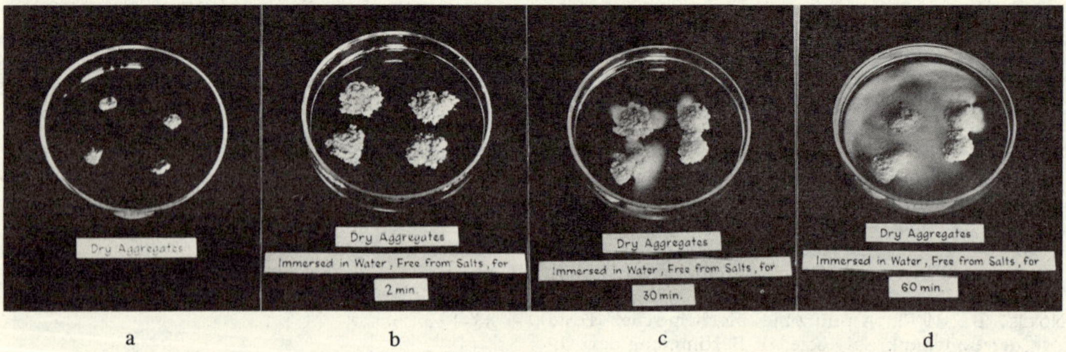

FIGURE 1. The breakup of aggregates on wetting where dry aggregates (a) at first rapidly slake in water (b) and then subsequently release clay at a slow rate by dispersion (c, d). (After Emerson, 1967)

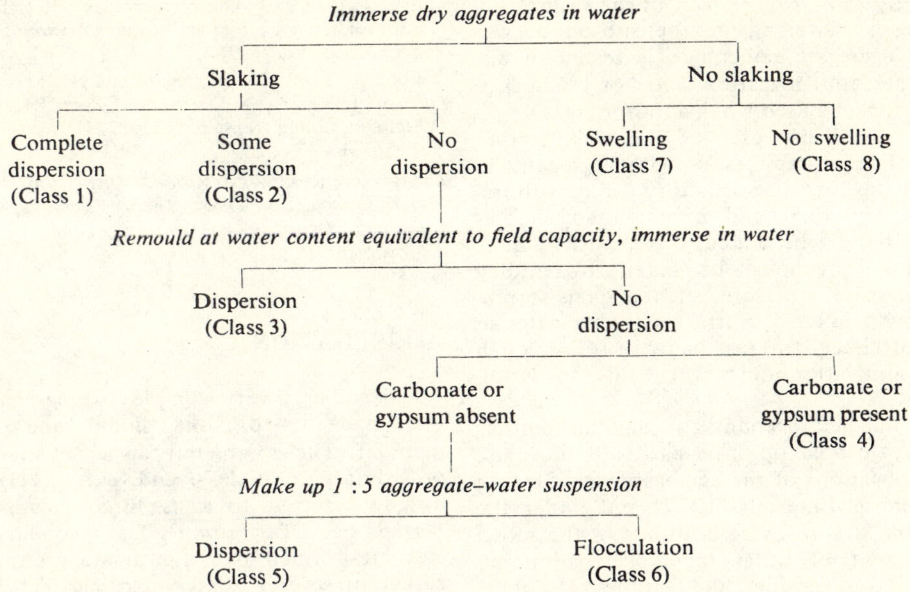

FIGURE 2. Scheme for determining class numbers of aggregates.

break the links, and there is reduced dispersion (as for unsheared aggregates). Second, if there are only a few links, then when these are broken by shearing, the individual polymer chains act as peptizing agents. Thus for some soils the surface aggregates may be more easily dispersed by shearing than the subsoil aggregates with similar clay content.

Measurement of Aggregate Stability

Water-stable Aggregation. This has been measured in mainly two ways: macroaggregation, started by Tiulin (1928), and microaggregation, introduced by Middleton (1930). Using macroaggregation, aggregates were first wetted, then immersed in water and the fragments gently sieved to determine their size distribution. Since the initial success with the method, the details of the prior wetting and the mechanics of sieving have been standardized (Kemper, 1965; Kemper and Chepil, 1965). Using microaggregation, wetted aggregates were immersed in water contained in a cylinder, which was then shaken end over end by hand. The degree of aggregation was measured by the proportion of particles smaller than 50 μ in suspension compared with the proportion of primary particles smaller than 50 μ present in the aggregates. Later the shaking was mechanized, and it and the method of wetting was partially standardized (Quirk, 1950). Microaggregation combines breakup of aggregates due to slaking and aggregates due to subsequent attrition (see *Microagregation*).

Aggregate Coherence in Water. Soil aggregates have been divided into eight stability classes based on the way the aggregates break up in water (Figure 2). The boundaries of the classes were chosen for practical reasons. For example, for class 3, a water content equivalent to "field capacity" was selected as the highest water content at which a soil was likely to be sheared by machinery or by stock. The equivalent water content taken is that of smaller than 2 mm aggregates wet at 100 cm suction. Loveday and Pyle (1973) have developed a *dispersion index* for aggregates in classes 1–3 by which 16 levels of stability are distinguished instead of the 3 here.

Application in the Field

Measurements of water-stable aggregation, as outlined here, are helpful for giving quantitative differences in the stability of aggregates from the same soil type due to different levels of organic matter, e.g., after different rotations (Clarke and Marshall, 1947). The methods are too arbitrary to be generally useful for elucidating causes of aggregate instability. For this, observation of the coherence of aggregates in water is a simple but powerful tool (Greenland, Rimmer, and Payne, 1975). Dispersion was chosen as the main diagnostic criterion, first, because it is easy to see. Second, dispersion causes the greatest changes in soil physical properties such as water transmission. Last, most agricultural crops are not very sensitive to aggregate stability so that crop growth is satisfactory unless dispersion of the surface soil or excessive swelling of the subsoil occurs. Usually aggregate coherence is tested in distilled water, but irrigation water or a sequence of waters can be used when appropriate.

There are many uses of the classification system. For example, soils with aggregates in classes 1 or 2 are unsatisfactory for agricultural crops (except rice), and an ameliorant such as gypsum has to be added. Also, with class 1 aggregates, care must be taken in building water-retention structures such as dams to prevent piping failures. Surface-soil aggregates in class 3 are likely to crust badly or form plough pans if cultivation under wet conditions is followed by rain.

More detailed accounts of the mechanisms of aggregate breakup, inter-particle bonds, and field applications of the aggregate classification are given in Emerson (1977) and under the entry *Soil Structure.* Descriptions of the actual measurement of water-stable aggregation are given in Baver, Gardner, and Gardner (1972).

W. W. EMERSON

References

Baver, L. D.; Gardner, W. H.; and Gardner, W. R., 1972. *Soil Physics.* New York: Wiley, 498p.

Clarke, G. B., and Marshall, T. J., 1947. The influence of cultivation on soil structure and its assessment in soils of variable mechanical composition, *Jour. CSIR,* **20**, 162–175.

El-Swaify, S. A., and Emerson, W. W., 1975. Changes in the physical properties of soil clays due to precipitated aluminum and iron hydroxides, *Soil Sci. Soc. America Proc.,* **39**, 1056–1063.

Emerson, W. W., 1967. A classification of soil aggregates based on their coherence in water, *Australian Jour. Soil Res.,* **5**, 47–57.

Emerson, W. W., 1977. Physical properties and structure, *in* J. S. Russell, and E. L. Greacen, eds., *Soil Factors in Crop Production in a Semi-arid Environment.* St. Lucia: Queensland University Press, 78–104.

Greenland, D. J.; Rimmer, D.; and Payne, D., 1975. Determination of the structural stability class of English and Welsh soils, using a water coherence test, *Jour. Soil Sci.,* **26**, 294–303.

Kemper, W. D., 1965. Aggregate stability, *in* C. A. Black, ed., *Methods of Soil Analysis.* Madison: American Society of Agronomy, 511–519.

Kemper, W. D., and Chepil, W. S., 1965. Size disbution of aggregates, *in* C. A. Black, ed., *Methods of Soil Analysis.* Madison: American Society of Agronomy, 499–510.

Loveday, J., and Pyle, J., 1973. The Emerson Dispersion test and its relationship to hydraulic conductivity, *Div. Soils Tech. Paper 15.* Melbourne: CSIRO, 7p.

Middleton, H. E., 1930. Properties of soils which influence erosion. *U.S. Dept. Agriculture Tech. Bull.,* 178.

Quirk, J. P., 1950. The measurement of stability of soil microaggregates in water, *Australian Jour. Agric. Res.,* **1**, 276–284.

Tiulin, A. F., 1928. Aggregate analysis as a method for determining soil structure, *Perm. Agric. Exp. Sta. Div. Chem. Report,* **2**, 77–122.

Cross-references: *Aggregation; Dispersion Phenomena; Flocculation; Flow Theory; Microaggregation; Particle Density; Pore Space, Drainable; Water Movement.*

AGGREGATION

The soil is a very complex, three-phase system made up of solid, liquid, and gaseous material. The completely dispersed individual, or primary, particles (sand, silt, or clay) are usually referred to as *textural separates* (see *Particle-size Distribution*). The secondary particles are formed by an intimate grouping together of several primary particles. These new units are the *structural units,* or *aggregates.* The individual natural soil aggregate, also called *ped* (Soil Survey Staff, 1951), can be defined as a naturally occurring cluster, or group, of soil particles in which the forces holding the particles together are much stronger than the forces between adjacent aggregates (Martin et al, 1955). The size distribution of the aggregates, their stability, and the amount and size distribution of the pore space both between and within the aggregates characterize the soil structure (q.v.). The terms "aggregation" and "soil structure" should not, however, be used interchangeably. "Aggregation" represents only one aspect of soil-structure formation while "soil structure" can result either from the building up of aggregates from dispersed materials or from the breaking down of larger coherent masses into favorably sized aggregates.

From an agricultural point of view, the binding of soil particles into aggregates is essential for the production of optimum *soil tilth.* Soil tilth is a general term, which signifies the ability of the aggregates to withstand destruction by the impact of implements, raindrops, or running water so that the air-moisture regime is maintained at a favorable level for plant growth and microbiological activity (see *Tillage*).

Mechanisms of Aggregate Formation

The diversity of soil constituents implicated as aggregating agents has produced many theories explaining the processes involved in soil aggregation. Martin et al (1955) proposed that clay is the predominant binding agent in soil aggregation. They suggested that organic materials do not act primarily to hold clay, silt, and sand grains together, but rather their chief role may be to modify the forces by which clay particles per se are attracted to one another. According to this view, cohesiveness between clay particles is the binding force in aggregation rather than the cementing action of organic molecules. Other investigators have stressed the importance of organic substances, produced largely through microbial activity, to soil aggregation. From the point of view of soil structure, aggregation requires a cementation, or binding together, of flocculated particles. Nevertheless, flocculation and aggregation are not synonymous. The former is primarily electrokinetic, although flocculation (*q.v.*) can occur as a result of electrostatic attraction between the positive edges and negative forces of clay minerals (Schofield and Samson, 1954). Thus flocculation may *aid* the aggregation process, but is is not aggregation in itself.

As the formation of aggregates comprises a complex interrelationship of physical, biological, and chemical reactions, different kinds of mechanisms have been proposed to explain the formation of aggregation in the soil (Martin et al, 1955; Harris, Chesters, and Allen, 1966).

Clay Particles. The clay particles themselves cohere and thus entrap or bridge between larger sand and soil grains. The cohesive forces between the particles arise from physicochemical mechanisms (Seed, Mitchell, and Chan, 1960), which may be:

1. The *van der Waals forces,* which vary inversely as the cube of the distance between particles.
2. The electrostatic attraction between negatively charged clay surfaces and positively charged clay edges.
3. The linking of the particles together through cationic bridges. There is a linking system between particles consisting of the negatively charged surface of a particle and oriented water molecules.
4. The cementation effects of organic matter, aluminum, and iron oxides. The iron oxides, in particular, may serve a dual purpose in aggregation. That part in solution may act as a flocculating agent, and the other part, which is more gelatinous in nature, may exert a cementation action.
5. The surface tension of the curved menisci at the air-water interfaces that are mostly present in the soil. Cohesion in wet soils takes place between the molecules of the liquid phase as bridges or films between adjacent particles.

Organic Matter. The organic colloidal material, produced largely through microbial activity, together with clay is responsible for the major portion of soil aggregation. Moreover, it is more effective than clay in causing the formation of stable aggregates with sand. The beneficial effects on soil aggregation originate from the integrated activity of microorganisms, fauna, and vegetation. Changes in aggregation following addition of organic material to soil indicate that aggregating effectiveness relates directly to microbial decomposition.

Organic residues and other products of microbial decomposition in the soil are thought to surround the soil particles and thus hold them together through cementing or encapsulating action. Organic material itself without biological transformation has little effect on soil structure. The conditions that favor high microbial activity result usually in early increased rates of aggregate formation due to fungal mycelia. However, high microbial activity also results in an increased rate of aggregate degradation due to an accelerated decomposition of the organic aggregating substances. Diverse bacteria, fungi, and actinomycetes are able to synthesize soil-binding substances. The major mechanism by which microorganisms effect soil aggregation involves the production of polysaccharides (*q.v.*) during microbial metabolism of organic matter in soil (Greenland, 1965). The formation of aggregates by microorganisms may be explained by adsorption phenomena, physical entanglement, and cementation by microbial mucilages (Harris, Chesters, and Allen, 1966).

The action of organic compounds in effecting aggregation may be very complex and at best is little understood. Many have postulated that soil particles are held together in aggregates by the organic compounds. There is, however, considerable evidence that another important role of these compounds may be in modifying the expression of cohesive forces between clay particles through adsorption on the surfaces of the clay. Thus active organic materials may be thought of as acting both to hold soil particles together and in other cases to hold clay particles apart.

As quartz, clay, and organic matter are the main constituents of the soil aggregate or crumb, the various possibilities of bonding arrangements are summarized in Figure 1. The clay particles are shown as *domains* consisting of several clay particles held together face to face. The domains may be held edge to face through aluminum bonds or edge to face, edge to edge, or face to face through organic colloids. These colloids can also bond clay particles to the surface of siliceous silt or sand particles.

Aggregate Formation

In light of the previous discussion, the following seems to best explain how aggregates are formed in agricultural soils. Aggregates result primarily from the action of natural agencies or any process by which parts of the soil are caused to clump together and separate from adjacent masses of soil. If soils are initially dispersed, flocculation is essential for aggregate

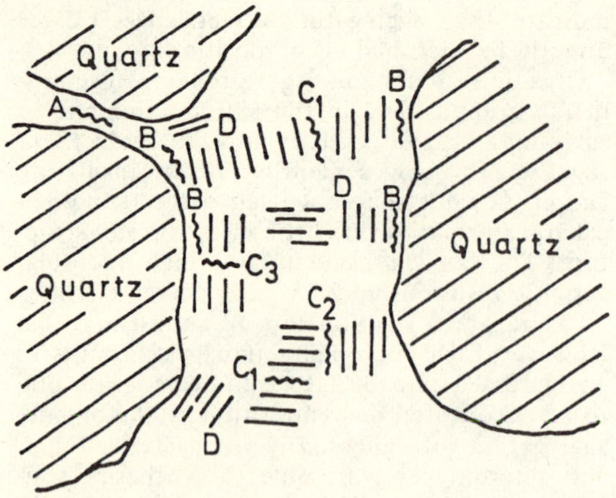

FIGURE 1. Suggested arrangements of quartz particles, clay domains, and organic matters in a soil aggregate. (*A*) Quartz-organic colloid-quartz; (*B*) quartz-organic colloid-clay domain; (*C*) clay domain-organic colloid-clay domain; (C_1) face-face; (C_2) edge-face; (C_3) edge-edge; (*D*) clay domain edge-clay domain face. (After Emerson, 1959)

formation; if they are partially puddled (see *Puddling*) or solid, fragmention into smaller units is the first essential. Thus there are two kinds of processes involved. The first concerns the building up of aggregates from dispersed materials; the second involves the breaking down of larger coherent masses into favorably sized aggregates. Since most soils become more dense and compact with continued farming, the second case is of greater interest.

Separation of parts of the soil mass may result from several factors such as the action of small animals, particularly earthworms as they ingest and intermix soil with partially decomposed organic matter and excrete the ingested material as surface casts or subsurface deposits, and the effect of tillage (*q.v.*), which is a function of the soil moisture content at the time of tillage.

Climate-dependent environmental variables include the effect of freezing and thawing where the aggregating action of frosts relates inversely to the rate of freezing and depends on the manner in which soil moisture crystallizes. Slow freezing results in the formation of few large ice crystals and large aggregates of soil particles. Since there is about a 9% increase in volume when water changes to ice, the pores are enlarged, which loosens the soil. Water is drawn from around the clay particles to the ice crystals during this process, creating a dehydration effect. The combination of ice-crystal pressure and dehydration causes aggregation. On the other hand, large numbers of small crystals are formed if the cooling is rapid and the integrated effect of many expansions is the breakdown of soil aggregates.

The effect of *wetting and drying* on aggregation is a function of the type of soil, the aggregate moisture at the time of wetting, and the intensity of the wetting treatment. The dehydration of a soil mass cannot be uniform, especially if the drying process is rapid. Consequently, unequal strains arise throughout the mass that tend to form *clods.* At least two processes operate to cause disruption of the clod into smaller units when the dried clod is wetted. The rapid intake of water causes unequal swelling throughout the clod, which produces fracturation and fragmentation along the cleavage planes. Moreover, the sorption of water into the capillaries results first in a compression of the occluded air and finally in a virtual explosion within the clod, as the pressure of the entrapped air exceeds the cohesion of the particles.

The *plant root system* is tremendously important, acting to separate and compress small clumps of soil, causing shrinkage and cracking due to desiccation, and making conditions favorable for the activity of microorganisms at the surface of these units. The exact mechanism of aggregate formation by plant root systems has not been established. The earliest explanation was based on pressures exerted by growing roots, which causes a separation to the soil particles adjacent to the root and a pressing together of these units into aggregates. In other words, each root hair that penetrates into large aggregates introduces a point of weakness. The penetration of sufficient root hairs produces granules. However, attempts to form aggregates by applying artificial pressure have failed to duplicate the water stability of natural field aggregates (Rogowski and Kirkham, 1962). This failure suggests that mechanical pressures of roots or soil fauna may not be basic aggregating mechanisms. Another possible aggregating factor is the dehydration of the soil in the vicinity of the root system as water is absorbed by the plant. This dehydration produces localized shrinkage and formation of fracture surfaces.

ROGER HARTMANN

References

Coughlan, K. J.; Fox, W. E.; and Hughes, J. D., 1973. Aggregation in swelling clay soils, *Aust. Jour. Soil Res.*, **11** (2), 133–142.

Emerson, W. W., 1959. The structure of soil crumbs. *Jour. Soil Sci.*, **10**, 235–244.

Greenland, D. J., 1965. Interaction between clays and organic compounds in soils. Part 1. Mechanisms of interaction between clays and defined organic compounds, *Soils Fert.*, **28**, 415–425.

Harris, R. F.; Chesters, G.; and Allen, O. N., 1966. Dynamics of soil aggregation, *Adv. Agron.*, **18**, 107–169.

Martin, J. P.; Martin, W. P.; Page, J. P.; Raney, W. A.; and Dement, J. D., 1955. Soil aggregation, *Adv. Agron.*, **7**, 1–37.

Rogowski, A. S., and Kirkham, D., 1962. Moisture, pressure and formation of water-stable aggregates, *Soil Sci. Soc. America Proc.*, **26**, 213–216.

Schofield, R. K., and Samson, H. R., 1954. Flocculation of kaolinite due to the attraction of oppositely charged crystal faces, *Discussions Faraday Soc.*, **18**, 135–145.

Seed, H. B.; Mitchell, J. K.; and Chan, C. K., 1960. The strength of compacted soils, *Am. Soc. Civil Engineers Res. Conf. on Shear Strength of Cohesive Soils*, 877–964.

Soil Survey Staff, 1951. *Soil Survey Manual.* U.S. Department of Agriculture Handbook No. 18, 225p.

Cross-references: *Aggregate Stability; Adsorption Phenomena; Clays, Physical Chemistry; Flocculation; Humus; Microhabitats; Microstructure, Manipulation; Organic Matter; Polysaccharide; Soil Components, Organic; Soil Structure; Tillage.*

AGRICULTURAL MINERALS—See FERTILIZERS, INORGANIC; FERTILIZERS, ORGANIC; LIME, LIMING. Vol. XVI: FERTILIZER MINERALS

AGROCHEMICALS—See HERBICIDE DIFFUSION; POLLUTION, SOIL CONDITIONERS; SOIL STABILIZATION

AGRONOMY

Agronomy is the science and practice of field crop production and soil management (*q.v.*). The term, derived from the Greek words *agros* ("field") and *nomos* ("to manage"), denotes the cultural management of the crop and the soil as it relates to the whole environment in the field. The term connotes group management of a large field planted with a single crop as opposed to the culture and care of individual specimens on a smaller scale as in horticulture. Agronomy normally excludes *silvaculture,* the management of forest species, but it encompasses essentially all other crops grown on a large field scale such as cereal and feed grains, fiber, sugar, oil crops, and forages.

The art of field crop production predates recorded history, but the science of agronomy was first recognized about the time of the first agricultural experiment station in Alsace in 1834. Agronomy became a designated branch of agricultural science in the United States just prior to 1900. As such, it was recognized as the coordinated use of the natural and physical sciences—including biology, botany, genetics, geology, chemistry, mathematics, and physics—in conducting controlled experiments with soils and crop species for the purpose of improved production.

The essential features of *agronomic science* encompass an evaluation of the whole environment for its influence on the amount and quality of crop growth. The physiological growth response of the crop is related to such aspects and mechanisms as climate, soil physics (*q.v.*), soil chemistry (*q.v.*), availability of essential minerals, water, light, and gaseous content of the atmosphere. Potential performance of the plants is determined by genetic makeup.

Production

Production measurements require a common denominator for comparative evaluation. *Yield* is exhibited as production per unit of land and input investment. *Quality* is identified as composition of desirable components in the harvested product. *Performance* of crops is most often represented per unit area of soil surface, but it is sometimes represented as return per leaf area, quantity of fertilizer or water applied, or quantity of labor and capital investment.

Crops

By adjusting cultural management of field crops, agronomists, more than any other group, face the challenge of providing an ever-increasing demand for agricultural production by a growing population. Agronomic crops include *foods* for direct human consumption, *feeds* for consumption by livestock, and *fibers,* certain plant parts of which are used by manufacturers for their strength characteristics.

Of prime importance among the food crops are the cereal and feed grains, corn (maize), wheat, oats, barley, rice, and rye, plus an assortment of legume seeds (Table 1). Feed crops include several of the same species of plants used for food plus forage and pasture crops and in many instances food and fiber crop residues following initial processing. Forages preserved by anaerobic fermentation are termed *silage crops,* and those dried are called *hay crops.* The herbage of a *soiling crop* is cut and fed green to livestock.

Root and tuber crops (named for those morphological structures of primary value) are often used interchangeably for food and feed.

TABLE 1. Harvested Area and World Production of Major Agronomic Crops

	Harvested Area (Million Hectares)			Production (Million Metric Tons)		
	1970	1971	1972	1970	1971	1972
Wheat (*Triticum*)	204	210	208	291	323	308
Corn (*Zea*)	109	111	105	325	291	283
Rice (*Oryza*)	130	132	128	291	299	287
Barley (*Hordeum*)	70	72	76	120	130	130
Oats (*Avena*)	32	30	29	53	55	49
Rye (*Secale*)	19	19	17	26	30	27
Soybeans (*Glycine*)	29	30	32	42	44	48
Peanuts (*Arachis*)	18	19	19	17	18	16
Flax seed (*Linum*)	7.2	6.0	5.6	4.1	3.7	2.6
Edible dried beans (*Phaseolus*)	8.4	8.4	8.0	125*	135*	127*
Cotton (*Gossypium*)	31	32	33	52**	57**	59**
Sugar crops (*Cane and beet*)	–	–	–	70	71	74

Source: Agricultural Statistics 1973, U.S. Department of Agriculture, Washington, D.C.
*Million cwt.
**Million bales of 480 lb, net.

The *sugar crops* (sugar beet and sugar cane) are grown for the sweet juice from which sucrose is extracted. *Oil seed crops* include soybeans, peanuts, rape, flax, sunflower, castor bean, and sesame. The oil is of value for both food and industrial uses, and the residue after extraction (with a few exceptions) is a primary source of protein in livestock feed.

Fiber crops are represented by cotton, flax, ramie, jute, kanef, hemp, and sisal where the length, diameter, and strength of their cellulose fibers are of special value. *Specialty crops* include seeds grown for planting plus others such as turf, tobacco, mint, and pyrethrum. *Cover crops* are grown for soil and water conservation purposes, *green manure crops* for soil improvement, and *trap crops* for insect control.

Choice of the crop to be grown requires a keen knowledge of its genetic adaptability to the particular environment of the region or field for which it is intended. Widely differing types and subspecies often exist that permit some species to be grown throughout much of the world. For example, making use of five different types, new cultivars of wheat may be grown from the tropics almost to the arctic regions and from low to high elevations. Sophisticated plant-breeding programs have produced genetically different cultivars of all the major crops, each well adapted to particular environments.

Management

The essential features of cultural management are those basic to the natural development of seed-bearing plants. They include (1) selection, preservation, and treatment of the seed; (2) destruction of unwanted vegetation or other plant pests; (3) treating, tilling, and smoothing the soil in preparing a seedbed conducive to good germination, emergence, and seedling growth; (4) proper placement of the seed in the soil with respect to rate of planting, depth, moisture, and anticipated seasonal environment; (5) corrective measures to the soil, including adjustments in drainage, acidity, water and mineral additives for good growth and development; (6) protection of the crop from weeds and other natural enemies; and (7) harvesting, processing, and storing the products of yield until they can be marketed.

Critical to the successful management of the crop are the determination of the need and manner of use to which the product will be put and the evaluation of the whole natural environment including soil structure and composition. In addition, the many features of climate must be considered, as well as the genetic potential of the crop cultivar with respect to yield, survival, and quality and the appraisal of the extent to which changes in the environment are economically practical to meet the requirements for plant growth and development. With this appraisal the agronomist can choose the crops best adapted to the environment, select the soil and climate most conducive to good performance of the crop, and change the environment, within economic limits, to favor efficient production and high quality.

Yield potential nearly always exceeds that obtained in practice, due to incomplete changes in the environment or control of weeds and parasitic pests. Climatic adaptation with respect to day length, humidity, and temperature extremes is of major importance. Mineral nutrition and available soil moisture are frequently the most limiting environmental factors in the production of adapted crop cultivars.

The addition of mineral elements required for growth has been facilitated in some parts of the world during recent years by the development of the fertilizer industry (see *Fertilizers*). The importance of soil amendments has increased as the demand for production has become greater since the natural supply of minerals in the soil is inadequate to meet crop needs at high production levels. Those elements that crop plants require from the soil in largest amounts are nitrogen, phosphorous, potassium, and calcium (see *Macronutrients*). As many as ten minor elements are also required but in much smaller amounts (see *Micronutrients*). Corrective measures of soil improvement constitute one of the greatest management opportunities to improve yield, quality, and efficiency of production.

Most crops experience only a short period of high quality during their growth cycle. Harvest management consists of claiming and preserving the product at this peak of optimum quality. Because each species varies in its morphological growth characteristics as well as its physiological pattern of development, in different environments specific harvest methods must be devised to best handle each crop.

The harvest of cereal grains, oil crops, and fibers is best as the plant reaches maturity and before natural shedding, disease or weather damage ensures. Natural dessication associated with maturity in dry climates or artificial drying to 15% moisture content permits safe storage of the grains and forage, keeping the products free of spoilage due to bacteria or mold growth. Prevention of insect and rodent damage during storage requires special precautions.

Forage crops are best harvested at an early stage of maturity for their high protein content, digestibility, and palatability characteristics. These high moisture content crops (if not consumed immediately) must be either dried as hay or fermented anaerobically as silage to prevent spoilage and loss of feed constituents. Cold or controlled atmosphere storage of low value per pound field crops is possible but generally an impractical investment.

MILTON A. SPRAGUE

References

Brady, N. C., 1974. *The Nature and Properties of Soils.* New York: MacMillan, 639p.

Heath, M. E.; Metcalf, D. S.; and Barnes, R. F., 1973. *Forages: The Science of Grassland Agriculture.* Ames: Iowa State College Press, 755p.

Martin, J. H., and Leonard, W. N., 1967. *Principles of Field Crop Production,* New York: MacMillan, 1044p.

Myers, W. M., 1958. Unfinished business, *Agron. Jour.*, **50**, 719–721.

Peterson, J. B., 1957. Examples of influence of soil science on agronomic practice, *Agron. Jour.*, 49, 651–659.

Watson, S. J., and Nash, M. J., 1960. *The Conservation of Grass and Forage Crops.* London: Oliver & Boyd, 758p.

Cross-references: *Aeragronomy; Edaphology; Fertilizers, Inorganic and Organic; Management of Soils; Plant Nutrients.*

AIR–See AERATION, RESPIRATION, AND ATMOSPHERE

ALGAE

Soil algae are lower chlorophyllic autotrophic plants that grow on or in the soil. They comprise different ecological algal groups such as *terrestrial algae* that grow as crusts or films at the soil surface, *aquatic-terrestrial* forms that grow at the surface of permanently damp soil, and *soil algae proper* that live below the surface between soil particles. *Soil algology* (soil phycology) is the science that deals with soil algae.

Development of Soil Algology

Soil algology originated with the merging of algology, phytocoenology, and soil microbiology (q.v.). Botanists have traditionally considered algal growth on the soil surface as simply one part of the whole vegetation complex. The introduction of microbiological principles by Bristol-Roach (1919, cited in Lund, 1967), however, helped to establish the composition and quantity of algae in soils and gave soil algology an independent status as a part of soil biology. Scientists of many different countries have played an important role in the development of soil algology. The most extensive and comprehensive investigations have, however, been conducted in Russia.

The development of soil algology has followed several principal trends. The *floristic-systematic* approach, for example, involves the determination of different species of algae in soils and the detailed study of separate groups of soil algae. The *ecologic-coenological* approach studies algal communities of different soils and their interrelationships with different environmental conditions. The *physiological* approach involves the investigation of nutrition, nitrogen fixation (q.v.), and tolerances of soil algae. Modern investigations are directed toward determining the role of soil algae in soil formation and toward assessing their importance in the turnover of energy and matter in soils (see *Organic Matter, Humus*) (Shtina, 1974).

Methods of Research

Algal species composition is commonly determined by cultural methods–i.e., in water, soil, and agar cultures–by using mineral nutritive media and their incubation in daylight. The abundance of algae is determined most often by one of the two following methods. The number of microscopic algae growing among soil particles is counted (1) by direct observation of soil suspensions, sometimes with the aid of luminescence microscopy by which the living cells are defined by the natural fluorescence of chlorophyll, or (2) by dilution cultures. In these cases the *biomass* (standing crop) is calculated on the basis of the cell volume.

The mass of terrestrial algal crusts and films is estimated by direct weighing of colonies, crusts, or algal felt taken from a certain area, usually 1 cm^2 or 1 dm^2, and by spectrophotometric methods that are based on extracting the chlorophyll and then measuring its concentration in a volume of soil.

Soil Algae

Algae are widespread in soils and are represented by a large variety of species (about 2000) such as those of the Chlorophyta (> 700), Cyanophyta (> 600), Bacillariophyta (~ 300), and Xanthophyta (~ 200). Some species of Euglenophyta, Pyrrophyta, Cryptophyta, and Rhodophyta have also been found.

Besides the aquatic algae, which are abundant in ponds, some species are dominantly found in soils. Many species of blue-green and mostly green and yellow-green algae were, for example, first discovered in soils.

Biological Peculiarities. Soil algae are microbial forms that depend on light for their development. Most soil algae seem to be completely autotrophic because they are capable of manufacturing all substances essential for their metabolic requirements. Soil algae continue photosynthesis even under conditions of low light intensity (0.04–0.1% of full daylight). Many species are facultative heterotrophs capable of using organic compounds in the light or in the dark; heterotrophic growth in the dark, however, proceeds at a slower rate than phototrophic growth in the light. Soil algae do not differ from other green plants in their nutritional requirements.

Even though the moisture optimal for the intensive growth of soil algae equals about 60–80% of the total soil water capacity, many algal species resist drought and remain alive in dry soils for long periods of time. Viable cells of microalgae are present in the atmosphere and are probably carried throughout the world by winds. Some surface species of soil algae, especially those in arid soils, can exist on condensed water. This resistance to drought is conditioned by: (1) the formation of slimy sheaths and covers that preserve waters; (2) the relatively small size of the organisms; (3) physiological properties of protoplasts; and (4) the ephemeral nature of the vegetation, i.e., the ability to quickly change from a state of rest to active growth and vice versa. Resistance to desiccation and wide range of tolerance to extreme temperatures and high salt concentrations also contribute to their hardiness. Cyanophyta and certain unicellular Chlorophyta are good examples of algae able to resist extreme conditions.

Most soil algae are found over a fairly wide pH range. Nearly neutral or slightly alkaline soils contain the most species but not necessarily the largest numbers of algae. Most Cyanophyta, for example, prefer neutral or alkaline soils. Chlorophyta may abound at almost any natural pH, but only certain species predominate in highly acid soils in which Cyanophyta and diatoms are usually rare or absent.

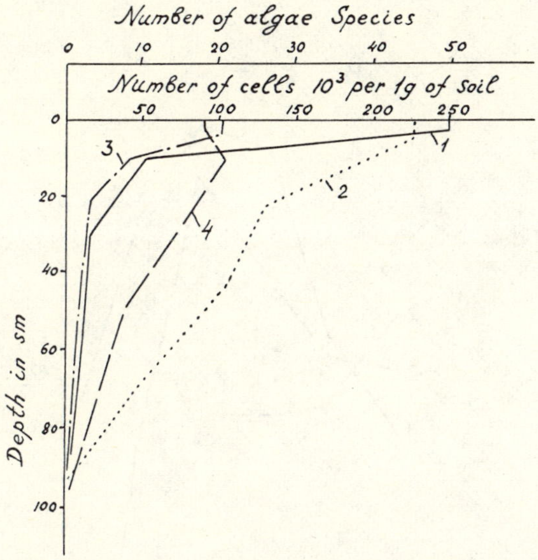

FIGURE 1. Vertical distribution of algae in virgin Renzina soil: (1) number of cells in meadow; (2) number of species in meadow; (3) number of cells in forest; and (4) number of species in forest. (After Shtina and Hollerbach, 1976)

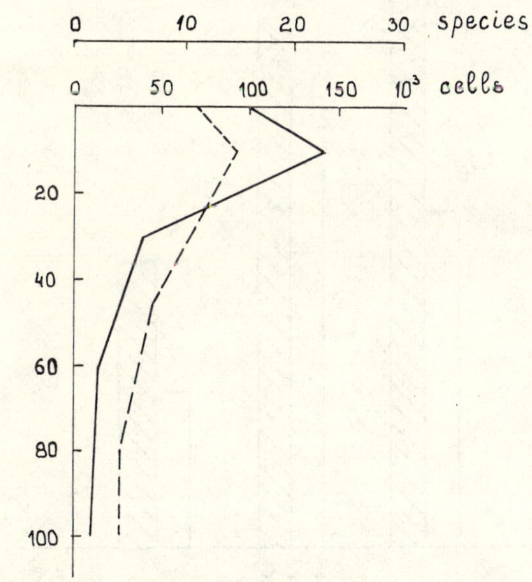

FIGURE 2. Vertical distribution of algae in ploughed Sod-Podzolic soil: —— number of cells; – – – number of species. (After Hollerbach and Shtina, 1969)

Numbers of Soil Algae. The actual number of algae in soil depends on the environmental conditions and is expressed in terms of 10^2–10^7 cells per gram of soil. As phototrophic organisms, soil algae normally develop within the limits of light penetration. The greatest variety and largest numbers of algae occur in association with optimal moisture conditions in the upper layers of the soil (Figure 1). The number of algae normally decreases sharply with depth except in ploughed fields in which soil algae may become abundant at about 10 cm depth (Figure 2).

According to data obtained by direct counting, the number of soil algae in virgin soils ranges from 500 to 300,000 cells per gram of soil at 0–10 cm below the surface, to 8 million cells per gram at 0–2 cm. In ploughed soils, however, the number of cells often reaches 2 million per gram from 0–10 cm and 35 million in a layer 0–1 cm below the surface. The numbers of algae vary during the year with changes in soil moisture and temperature (Figure 3).

Distribution of Soil Algae. Algal distribution in soil depends on the ecological peculiarities of the soil and associated vegetation, especially the soil water regime (q.v.), pH, and density of the plant cover.

Investigations in western Europe (Bristol-Roach, 1927; Petersen, 1935), USSR (Shtina, 1960; Hollerbach and Shtina, 1969), United States (Cameron, 1964; Shields and Durrell, 1964), and Australia (Shtina, 1975) have shown that algal associations change from ephemeral surface films to diffusive distributions in soil. Algae quickly colonize unvegitated soil surfaces where they grow quickly and intensively during occasional moist periods.

In arctic deserts and tundra regions especially, algal growths are formed by green, yellow-green, and blue-green algae, which are important pioneers in bare-ground colonization.

In tundra soils, green, blue-green, and diatoms develop only in the very top layer of mossy-sod, litter fall, or peat horizon.

Under forest vegetation in Podzolic, Grey Forest, and Grey-Brown Forest soils, algae develop mainly in the upper layer of soil, but they also develop on the forest floor. Algal associations in forest soils are rather similar throughout the whole zone with green and yellow-green algae, especially unicellular species (*Chlamydomonas, Coccomyxa, Chlorococcum, Chlorella, Pleurochloris*) and filamentous Ulotrichales (*Stichococcus and Hormidium*) prevailing there. They are accompanied by filamentous yellow-green algae and bryophilic diatoms. Blue-green algae (*Nostoc*), on the other hand, occur in deciduous forests only.

The most diffusive and rich associations of

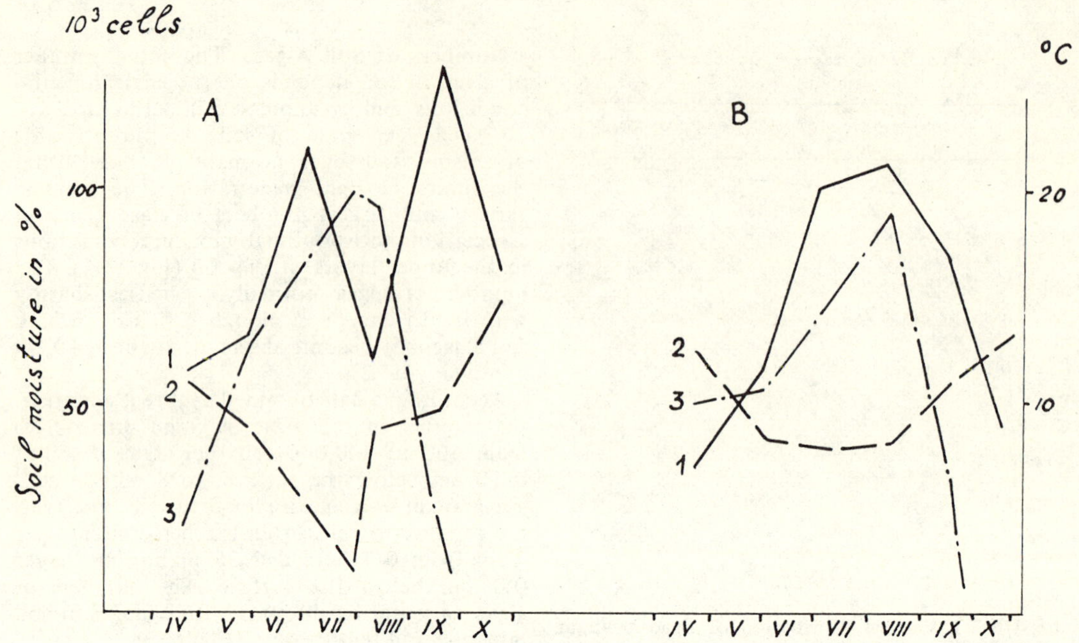

FIGURE 3. Seasonal changes of algae number in Sod-Podzolic soils: (A) clover fields; (B) fallow fields., (1) number of cells in layer 0–1 cm; (2) soil moisture, percent of total water capacity; soil temperature. (After Hollerbach and Shtina, 1969)

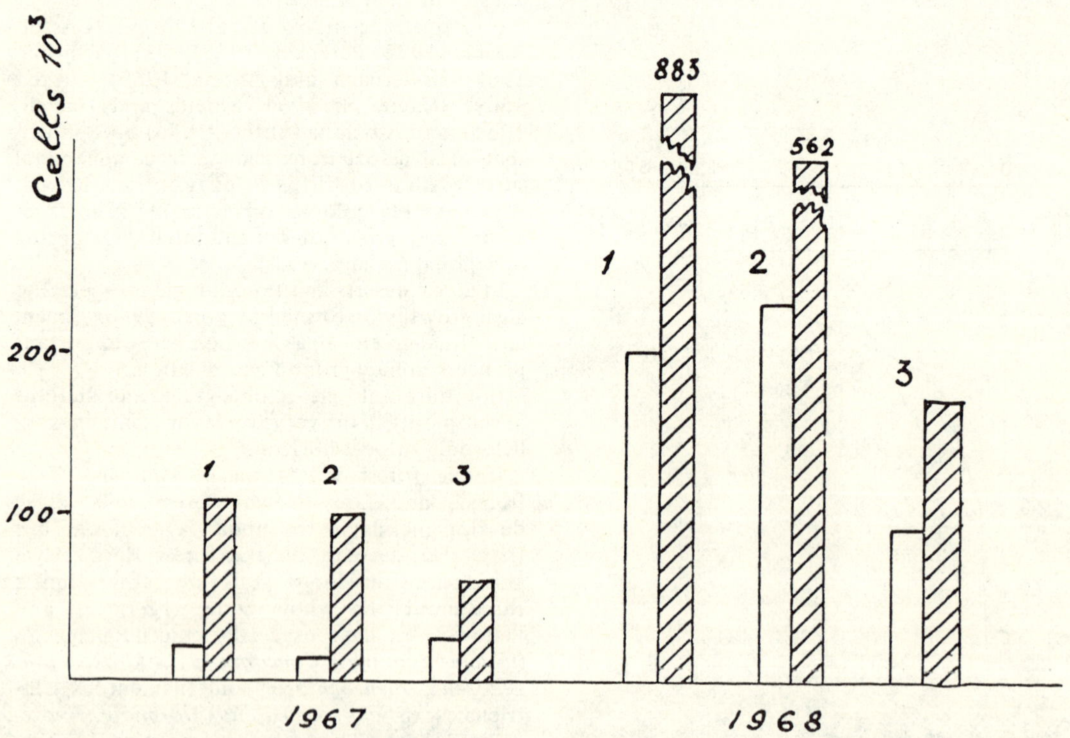

FIGURE 4. Effect of mineral fertilizers on algae number in Sod-Podzolic soils: (A) ordinary doses of fertilizers; (B) increasing doses; (1) rye field; (2) clover field; and (3) potato field. (After Shtina and Hollerbach, 1976)

soil algae develop under a herbaceous vegetation in Sod-Podzolic, Yellow-Podzolic, and Red-Podzolic soils and Chernozems of the meadows and steppes of Europe and Asia, in the prairies of United States, and in the tussock steppes of Australia. Green, blue-green, and yellow-green algae and diatoms dominate the algae population.

Many species grow on the surface, forming ephemeral films. Ephemeral surface films, which are formed by Cyanophyta species occur on the semiarid and arid soils of all continents. In semidesert phytocoenoses (dry steppe, semidesert, savanna), *Phormidium, Microcoleus, Schizothrix, Scytonema,* and *Nostoc* species are most abundant. In solonetzes they are accompanied by numerous diatoms, especially *Hantzschia amphioxys* and *Pinnularia borealis.*

The dominance of blue green-algae is even more pronounced in desert phytocoenoses in which they form films and crusts of only one or two species over vast areas of clay desert. In contrast, stony deserts are inhabited by lithophylic algae.

Cultivation of soil alters its physioal and chemical responses (see *Tillage*) as well as the character of soil algae associations. Here the role of Nostocales, small Bacillariophyta, increases, and the variety of Chlorophyta and Xanthophyta becomes greater. Both the quantity and composition of algae in cultivated soils are more dynamic than the initial virgin soils. Algal development is especially stimulated by the addition of mineral fertilizers (Figure 4) and the liming of acid soils.

Role of Algae in Soil Formation. The main processes conducted by algae in soil include the accumulation of organic substances, nitrogen fixation (q.v.), the alteration of physical properties of soil, and the fixation of mineral fertilizers.

At primary stages of soil formation, algae take part in the weathering of rock and in the creation of primary humus (q.v.). They constitute an initial successional stage on substrates that are poor in plant nutrients such as recent volcanic deposits and lifeless industrial dumps.

The accumulation of organic matter (q.v.) is most intensive in the algal development on the soil surface. The biomass of algae, expressed as *wet weight,* frequently reaches 600 kg/ha in a 0–10 cm layer and 1.5g/dm^2 at the surface (Tables 1 and 2). On the desert takyrs, algae are the main source of organic matter, forming 500 or even 1400 kg of dry mass per hectare. The algal biomass, however, changes in response to the cover vegetation. Thus for a Soddy-Podzolic soil kept in *pure fallow* (without plants), the monthly production was defined by summing the incremental increases of biomass

TABLE 1. Numbers and Biomass of Algae in the Surface Growth on Virgin Soils

Soils	Predominating Groups of Algae	Numbers of Cells (0000/cm^2)	Biomass (kg/ha) (wet weight)
Sod-Podzolic	Blue-green	11,800	177
Flood plain	Blue-green	15,780	
Flood plain	*Nostoc* commune		29–40*
Sandy steppe	Blue-green	22,560	
Crust solonetz	Blue-green	11,900	467
Meadow solonetz	Blue-green, diatoms	988	625
Solonetzes	*Nostoc* commune		72–217*
South chernozem	Blue-green	14,970	
Dark-chestnut	Blue-green, diatoms	5,103	511
Chestnut	Blue-green	7,654	181
Takyr-like	Blue-green		24*
Takyrs	Blue-green		500–1400*
Mountain deserts	Different groups		1500**

(After Shtina and Hollerbach, 1976).
*Dry algal weight.
**Algal crust with admixture of soil particles.

(Figure 5). This production was three times larger than the biomass ("standing crop"), i.e., the algal mass regenerated itself at least three times every month. During the five-month vegetative period, algal production reached 45,000,000 cells per gram with a biomass of about 500 kg/ha. Although algal productivity in terrestrial situations is much inferior to that of higher plants, the importance of the algal organic matter lies in its exceptional mobility, whereby it is involved in many soil processes.

The organic matter of algae greatly affects soil microflora and microfauna, as well as physicochemical properties of soil. Living algal secretions and their weak cells become food for many bacteria and fungi. Mucous covers of algae are inhabited by different bacteria, sometimes by fungi and even by unicellular algae that find a proper moisture medium there. Algal cells and talloms in soil thus act as stable centers of microbial associations.

When algae die, the organic matter from their cells enters the trophic chains of destruction of organic substances and prompts local outbursts of heterotrophic microflora. The application of algae to soil is normally accompanied by the activization of the soil microflora, which includes increases in oligonitrophils, ammonifiers, and a number of other microorganisms.

A considerable and evidently main part of the organic matter of algae becomes the food of various soil animals, including different species of *Protozoa, Nematodae, Acarina, Enchitreidae, Collembolae, Diplopodae,* earthworms, and even larvae of some insects. Some species of microfauna also graze specific "pastures" on the films of the algae. Consumption of algae by soil animals, phycofages, is one of the important factors governing changes in composition, numbers, and mass of algae. Animals also take part in the distribution of algae in soil.

TABLE 2. Algal Biomass in Surface Growth on Ploughed Soils.

Soils	Field Cultures	Predominate Algal Groups	Biomass (kg/ha) (wet weight)
Sod-Podzolic	Rye	Green	Up to 260
Peat-boggy	Rye	Green	Up to 1500
Peat-boggy	Spring crops	Green, yellow green	Up to 750
Peat-boggy	Grass	Green, yellow green	200
Serosem	Cultivated pasture	*Nostoc*	112–120*
DDR (soil unidentified)	Spring crops	Blue-green	1–34.5*
Rice fields (India)	Rice	Blue-green	2200*
Rice field	Rice	Blue-green	4500*

(After Shtina and Hollerbach, 1976).
*Dry algal weight.

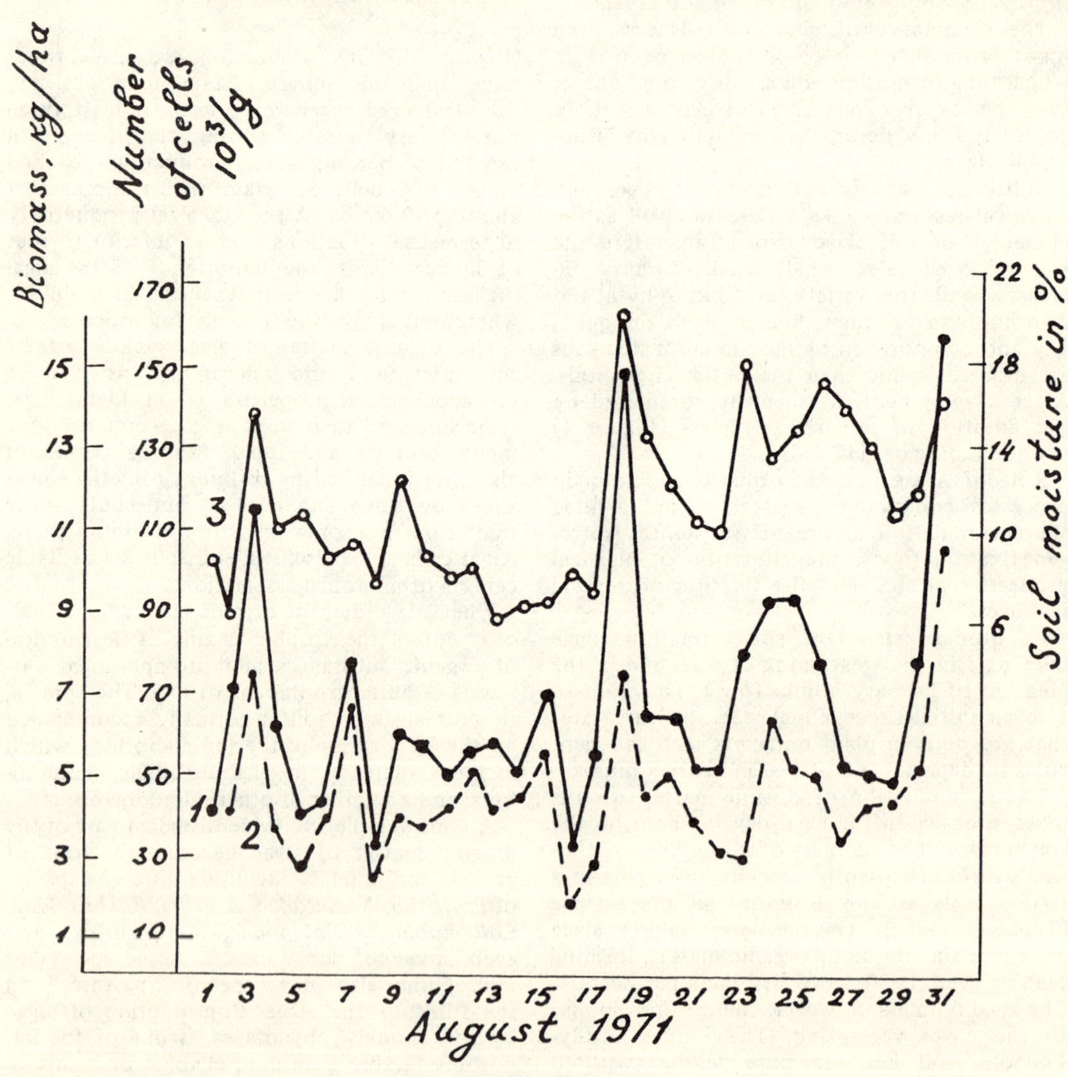

FIGURE 5. Dynamics of algae number and biomass in Sod-Podzolic soil (August 1971): (1) number of cells; (2) biomass; (3) soil moisture, percent soil weight. (After Shtina and Hollerbach, 1976)

Algae participate in nitrogen enrichment of soil either directly or indirectly. Many blue green algae, at least all heterocystous species, are, for example, capable of molecular nitrogen fixation. Many soils contain more than one hundred species of Cyanophyta, which are nitrogen fixers. Nitrogen accumulation by algae averages about 0.3–20 kg/ha in virgin soils of the temperate zone (Shtina and Hollerbach, 1976) but approaches 90 kg/ha on irrigated fields in the tropics (Singh, 1961). Nitrogen fixed by algae may also be absorbed by other algae, fungi, and higher plants. Algae can also stimulate the activity of some nitrogen-fixing bacteria, particularly *Azotobacter* and *Rhizobium*.

Influence on Soil Properties. Biological absorption of soluble mineral salts takes place during algal growth. Thus algal films with a biomass of 1.5g/dm^2 may contain up to 6 kg of nitrogen. Nevertheless, it is impossible for algae to compete with higher plants in absorption of mineral salts as the algal cells gradually liberate absorbed substances.

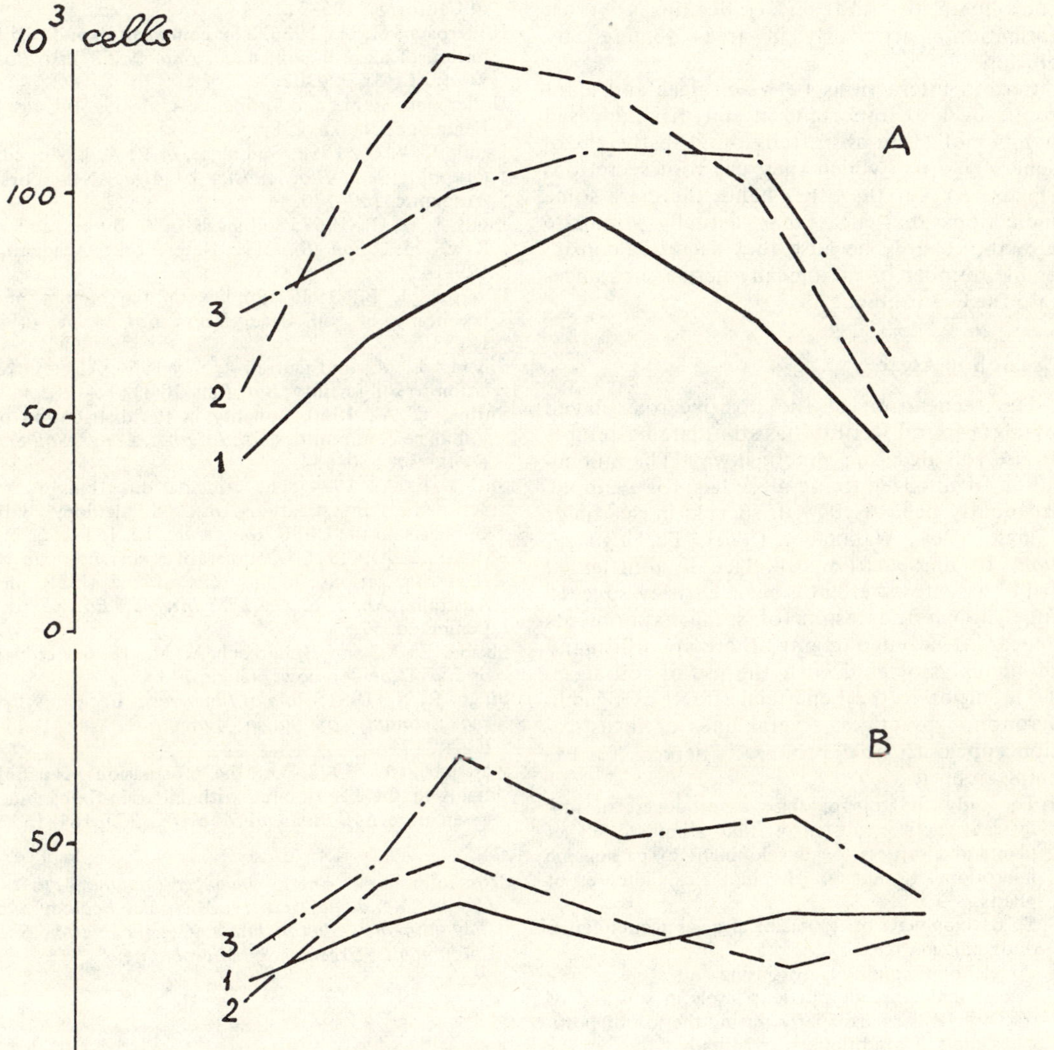

FIGURE 6. Dynamics of algae number in rhizospheres of clover and loose-bunch grass, two years old: (A) green algae; (b) diatoms algae (1) in control soil (soil free of roots); (2) in rhizosphere of clover; (3) in rhizosphere of grass. (After Hollerbach and Shtina, 1969)

Surface algal films can inhibit erosion and influence the water regime of soils. Filamentous algae with mucous cell walls (*Hormidium* spp.) or with a mucous cover (blue-green algae) are of a special importance in this regard because they connect solid particles and bind them together. In surface sandy soils, for example, proliferations of filamentous algae such as *Hormidium* and *Schizothrix* result in a total filament length of 22–65 m/cm^2 (Shtina and Hollerbach, 1976). Studies by Shields and Durrell (1964) and Lund (1967) indicate that soil algae may function as water-conserving stratum in semidesert habitats because the algae consolidate the soil surface leading to crust formation, particularly in areas denuded by erosion.

Certain interactions between algae and plant roots are also important in soil. An increased number of algae are often found in the rhizosphere (q.v.) in which they use root secretions (Figure 6). On the other hand, there are some indications that algae may actually stimulate growth, possibly because they secrete a considerable number of biologically active substances into the environment.

Use of Soil Algae

The recognition of the positive role played by algae in soil fertility has stimulated attempts to use soil algae in a practical way. The innoculation of nitrogen-fixing algae has, for example, frequently been tried with success in rice fields (Singh, 1961; Watanabe, 1973). The *algalization*, or innoculation, of algae in unirrigated fields has, however, not been entirely successful. Though the results of some experiments appear to be encouraging, there are still many problems associated with the use of soil algae. To incorporate the beneficial effects of algae in agronomic practices, several lines of investigation appear to hold promise. There is, for example, a need:

1. To study the role of algae as producers of biologically active substances that stimulate higher plants and supress the development of pathogenic microflora to increase the biological defenses of plants,
2. To develop ways of regulating algae as accumulators of organic matter,
3. To elaborate methods for using soil algae as biological indicators of nutrient levels in soil and of toxicity levels, particularly during the use of pesticides and soil conditioners, and finally,
4. To study the role of algae in the transformations of mineral fertilizers added to soils.

E. A. SHTINA
M. M. HOLLERBACH

References

Bristol-Roach, B. M., 1927. On the algae of some normal English soils, *Jour. Agric. Sci.*, **17**(4), 563–588.

Cameron, R. E., 1964. Terrestrial algae of southern Arizona, *Am. Microscop. Soc. Trans.*, **83**(2), 212–218.

Draganov, S. J., 1972. *Bibliography on Soil Algae 1895–1965*, Sofia: Publishing House of the Bulgarian Academy of Sciences, 92p.

Friedmann, E. I., and Galun, M., 1974. Desert algae, lichens and fungi, *in* G. W. Brown, Jr., ed., *Desert Biology*, **II**. New York, San Francisco, London, 165–212.

Fritsch, F. E., 1936. The role of terrestrial algae in nature, *Essays in Geobotany*. Berkeley: University of California, 195–217.

Hollerbach, M. M., 1936. The composition and distribution of algae in soil, *Akad. Nauk SSSR Inst. Bot. Trudy*, **11**(3), 99–302.

Hollerbach, M. M., and Shtina, E. A., 1969. *Soil Algae*. Leningrad: Nauka, 228p.

Lund, J. W. G., 1962. Soil algae, *in* R. A. Levin, ed., *Physiology and Biochemistry of Algae*. New York: Academic, 759–770.

Lund, J. W. G., 1967. Soil algae, *in* A. Burges, and F. Raw, eds., *Soil Biology*, New York: Academic, 129–147.

Petersen, J. B., 1935. Studies on the biology and taxonomy of soil algae, *Dansk Bot. Arkiv*, **8**(9), 1–183.

Shields, L. M., and Durrell, L. W., 1964. Algae in relation to soil fertility, *Bot. Rev.* **30**(1), 92–128.

Shtina, E. A., 1960. Zonality in the distribution of soil algae communities, *7th Internat. Congr. Soil Sci. Trans.*, **11**, 630–634.

Shtina, E. A., 1974. The principal directions of experimental investigations in soil algology with emphasis on the USSR, *Geoderma*, **12**, 151–156.

Shtina, E. A., 1975. Comparative characteristic of algal associations in the soils of the USSR and Australia, Abstr. of the *12th Internat. Bot. Congr.*, Leningrad, 57.

Shtina, E. A., and Hollerbach, M. M., 1976. *Ecology of Soil Algae*. Moscow: Nauka, 143p.

Singh, R. N., 1961. *Role of Blue-green Algae in Nitrogen Economy of Indian Agriculture*. New Delhi: ICAR, 173p.

Watanabe, A., 1973. On the inoculation of paddy fields in the Pacific area with nitrogen-fixing blue-green algae, *Soil Biol. and Biochem.*, **5**(I), 161–162.

Cross-references: *Energy Balance; Enzyme Activity; Humus; Metabolic Activity; Microbial Ecology and Clay Minerals; Microhabitats; Organic Matter; Soil Components, Organic; Soil Microbiology.*

ALKALINITY—See BASE SATURATION; SOIL REACTION. Vol. IVA: SOIL SALINITY AND ALKALINITY

B

BACTERIA–See CARBON CYCLE; DENITRIFICATION; METABOLIC ACTIVITY; NITROGEN CYCLE; NITROGEN FIXATION; SOIL MICROBIOLOGY; SULFUR TRANSFORMATIONS

BASE EXCHANGE–See EXCHANGE PHENOMENA

BASE SATURATION

Base saturation indicates the balance between acid and base cations adsorbed by the cation exchange complex (CEC) of a soil. The term is a partial misnomer because a base is a chemical compound that can react with an acid to form a salt; calcium hydroxide, $Ca(OH)_2$, is an appropriate example. In the present context, however, it is now understood to mean the *cation* of the base, that is, Ca^{2+}, as distinct from the cations H_3O^+ and $[Al(H_2O)_6]^{3+}$, which are acids.

If a neutral solution (that is, at pH 7) of a salt such as ammonium chloride percolates through a neutral soil, the leachate will contain an amount of cations equivalent to that in the initial solution, although the composition of the leachate will be different because of cation exchange; in particular, the leachate will be enriched in Ca^{2+}. If, on the other hand, the neutral salt solution passes through an acid soil, the leachate will be acid, and it will contain less base cations than the added solution because the base cations have been adsorbed by the soil to the extent that it was unsaturated.

Hissink (1925) introduced the concept of *degree of base saturation* and defined it as the ratio of the quantity of adsorbed base actually present in the soil to the total quantity the soil can adsorb. Hissink experienced difficulty in measuring total base adsorption because it increases almost indefinitely as a base is added to the soil; however, he was able to at least approximate the degree of base unsaturation from the amounts of $Ba(OH)_2$ that remained unreacted after addition to the soil.

Understanding of both cation exchange and soil acidity (q.v.) has improved considerably in recent years, and because these topics are intimately linked with base saturation, they will be discussed first.

Cation Exchange

The origin of cation exchange lies in the negative electric charges on the colloidal clay and humus particles of the soil matrix. There are two main components of negative charge: a constant, or permanent, charge and a variable, pH-dependent charge.

The *permanent charge* is generated by partial isomorphous substitution within the lattices of clay-size layer silicates, and in particular, the substitution of Al^{3+} for Si^{4+} in the tetrahedral sheet and of Mg^{2+} or Fe^{2+} for Al^{3+} in the octahedral sheet. The consequent deficiency in positive valency results in a crystal with an excess negative charge.

The *variable charge* is generated by the pH-dependent dissociation of hydroxyl groups, which may occur in (a) silanol groups located at the surfaces of aluminosilicate gels or the edges of layer-silicate crystals, (b) phenols, or (c) in carboxylic acids. Where appreciable amounts of organic matter are present such as in surface soils, the major part of the variable charge arises from proton dissociation by carboxyl and phenolic groups. Another mechanism contributing to pH-dependent charge is the "blocking" of negative charge by strongly adsorbed hydroxy aluminum cations. As the pH rises, these cations precipitate as aluminum hydroxide, and the negatively charged sites are then free to participate in cation-exchange reactions.

The adsorption of positively charged counterions (mainly Ca^{2+}, but also K^+, Na^+, NH_4^+, Mg^{2+}, Mn^{2+}, and Al^{3+}, depending on circumstances) by soil particles ensures their overall electrical neutrality. When a soil is suspended in or leached by a salt solution, the cations of the salt displace the freely diffusible counterions of the soil, producing cation-exchange phenomenon (See *Exchange Phenomena*).

Acid Soils

The nature of soil acidity was reviewed by Coleman and Thomas (1967). Soil acidity is a complicated field, but the main features can be illustrated by considering the reactions that occur when a base such as $Ca(OH)_2$ reacts with an acid soil.

1. Free hydrogen ions (i.e., hydrated protons, H_3O^+) in solution are neutralized by the base:

$$H_3O^+ + OH^- \rightarrow 2H_2O$$

It is important to realize that acid soils normally contain very little free hydrogen ion, except in very acid peat or muck soils, and that this reaction therefore accounts for very little of the base consumed.

2. Aluminum ions, which balance in part the surface negative charges of acid soils, are desorbed by the cations of the base and are then precipitated as aluminum hydroxide by hydroxyl ions:

$$2 \text{ soil Al} + 3Ca^{2+} \rightarrow 3 \text{ Soil Ca} + 2Al^{3+}$$

$$Al^{3+} + 3OH \rightarrow Al(OH)_3$$

The amount of surface charge is unaffected by these reactions. The components of acidity described under (1) and (2) are described as *exchange acidity* because they are acid cations displaced when the soil is leached by an unbuffered salt solution; they can be estimated by a direct titration of the leachate with a base.

3. Acidic hydroxyl groups, which occur in a number of soil components as mentioned above, ionize and are neutralized by the base:

$$H_2O + R\text{-}OH \rightarrow RO^- + H_3O^+;$$
$$H_3O^+ + OH^- \rightarrow 2\ H_2O$$

This reaction is analogous to the neutralization of a simple weak acid, but the acid groups involved occur on soil surfaces rather than in solution. This component is described as *hydrolytic acidity*.

An important effect of reaction 3 is to increase the negative electric charge, which is then balanced by adsorption of cations from the base. This increase is the pH-dependent charge referred to above, and it is of considerable magnitude in surface soils because of the humus they contain.

The total acidity of the soil is the sum of the exchange acidity and hydrolytic acidity as defined above; i.e., it is the total base consumed in bringing the soil from its natural pH to the final pH attained after reaction with the base. The total acidity has also been called *titratable acidity*, and some authors have referred to it as *exchange acidity* (e.g., Peech, 1965).

Cation Exchange Capacity (CEC)

CEC measures the ability of a soil to adsorb cations in exchangeable forms. It corresponds to the negative charge of the soil and is expressed in milli-equivalents per unit mass of soil (mequiv/kg).

The negative charge, and hence the CEC, increases as pH rises. This was explained previously and is illustrated in Figure 1. The capacity of the soil to adsorb exchangeable cations cannot therefore be defined unless a standard pH for its measurement is agreed upon. Bradfield and Allison (1933) proposed that "a soil saturated with bases is one which has reached equilibrium with a surplus of $CaCO_3$ at the partial pressure of CO_2 existing in the atmosphere." This corresponds to a pH of 8.2–8.3 and provides a rational basis for the determination of both CEC and total acidity. A buffer (see *Buffers, Buffering*) system containing triethanolamine and triethanolamine hydrochloride at pH 8.2, in 0.25 mol/l $BaCl_2$ solution has frequently been used for these determinations (Chapman, 1965; Peech, 1965).

The partial pressure of CO_2 in soils is generally greater than that in the atmosphere because of the respiration of soil fauna and flora so that

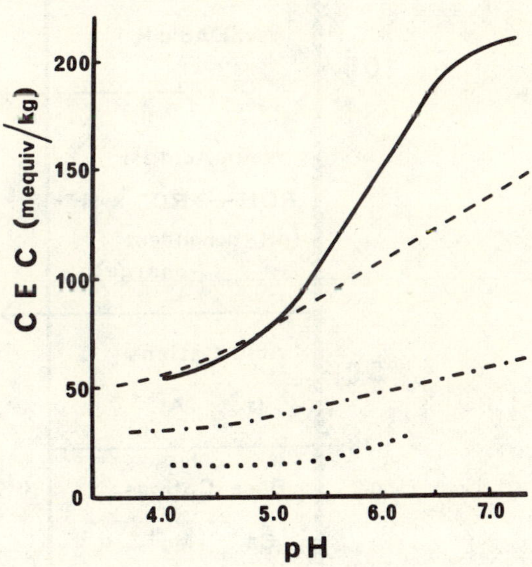

FIGURE 1. Effect of soil reaction (pH in 0.01 M $CaCl_2$ suspension) on CEC, for some contrasting soils adjusted to different reactions in the field. The effects are related mainly to the organic matter contents of the soils: —— 3.1% C; - - - 1.7% C; - • - • 0.5% C; • • • • 0.3% C (subsoil)

the pH of a soil in equilibrium with $CaCO_3$ is lower than 8.2. This is one reason for the more common practice of measuring CEC at the neutral point, pH 7, using a 1.0 mol/1 ammonium-acetate buffer solution.

A third approach in determining CEC is to leach soil with unbuffered salt solutions (1.0 mol/1 KC1 or NH_4Cl) that rapidly adjust to the natural reaction (pH) of the soil and displace only the exchangeable cations without generating extra pH-dependent charge. In this case CEC equals the sum of cations displaced from the soil, including Al^{3+} from acid soils of pH less than 5.2, as well as the base cation suite referred to previously.

Degree of Base Saturation

The degree of base saturation is defined simply as (ΣM)/CEC, or when expressed as a percentage, $V = 100(\Sigma M)$/CEC. Here ΣM refers to the sum of milli-equivalents per kilogram of exchangeable cations Ca^{2+}, Mg^{2+}, K^+, Na^+, NH_4^+, and so on. It is clear from the previous discussion that when defining the degree of base saturation, the pH at which the CEC is measured must be specified. In the majority of soils literature, the CEC measured at pH 7 in ammonium acetate (NH_4OAc) solution is taken as the standard for base saturation, although the other methods mentioned previously are perfectly valid for different purposes.

The components that contribute toward CEC and that account for base saturation and unsaturation are related as follows:

$$\text{CEC} = \begin{matrix}\text{exchangeable}\\ \text{base cations}\end{matrix} + \begin{matrix}\text{exchange acidity}\\ \text{(acid cations)}\end{matrix} + \begin{matrix}\text{hydrolytic}\\ \text{acidity}\end{matrix}$$

The effect of different methods of expressing CEC on the apparent degree of base saturation is illustrated diagrammatically in Figure 2.

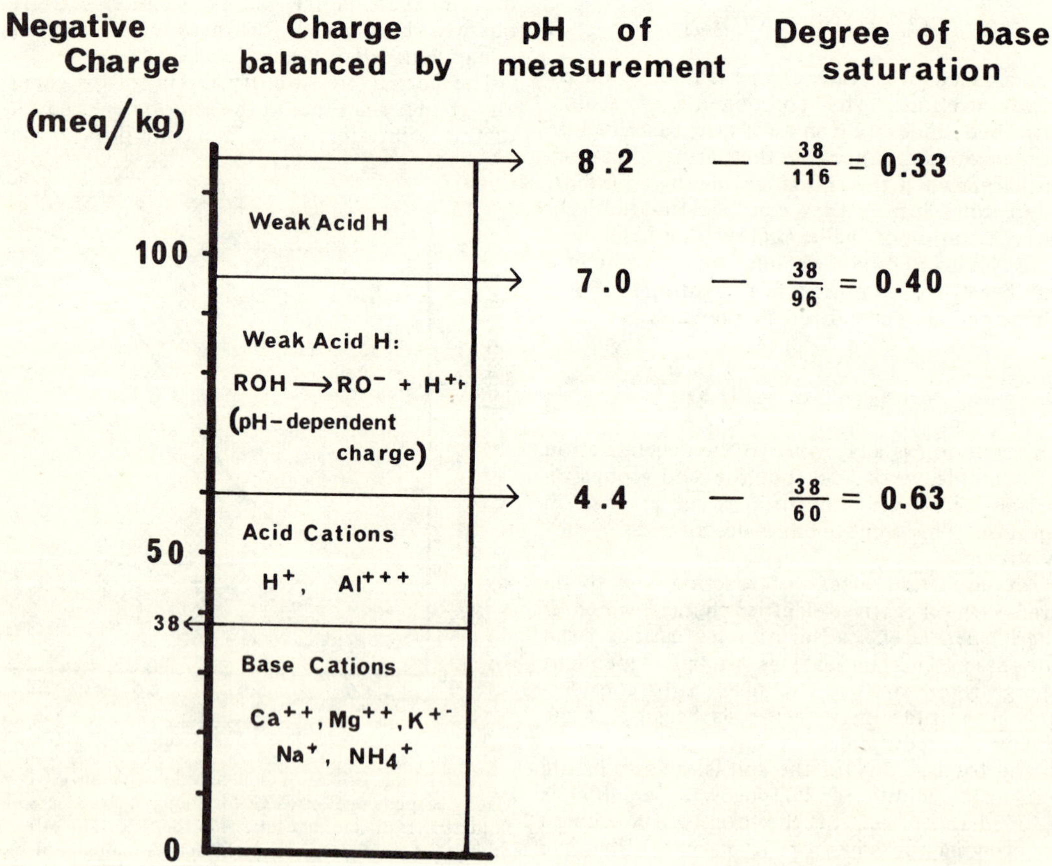

FIGURE 2. Diagrammatic illustration of components of negative charge and their relationship to degree of base saturation.

Base Saturation and Soil pH

Base-saturated soils are neutral in reaction while base-unsaturated soils are acid. Some relationship would therefore be expected between pH and the degree of base saturation. A number of workers have sought such a relationship, but all have found that for different soils there is no simple direct relationship, although soils with more organic matter generally give higher pH values at a given saturation (based on CEC at pH 7) than similar soils with less organic matter. This is illustrated in Figure 3 by statistical regression lines for the contributions of the inorganic and organic fractions of some German soils.

Good regression lines for pH against base saturation were obtained by Blosser and Jenny (1971) when they separated soils from the same parent material according to vegetational type (considered to be a function of climate) and according to *A* and *B* horizons (reflecting differences in organic matter content). Limiting the study to the permanent charge components of CEC and expressing base saturation in terms of the balance between exchangeable calcium and aluminum, still gave considerable variation in the relationship with pH for contrasting soils (Turner and Clark, 1966; Bache, 1974).

Early workers in this field (Pierre and Scarseth, 1931) considered that differences between soils in this relationship originated in the amount and nature of the soil colloids and in the strength of the soil acids. These factors are

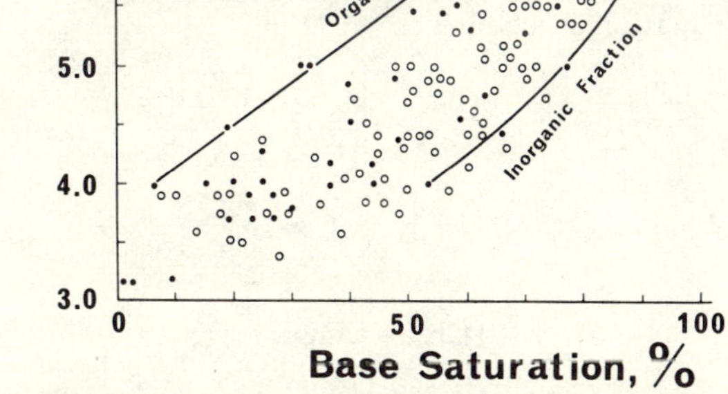

FIGURE 3. Relationships between soil pH (measured in 0.1 M KCl suspension) and degree of base saturation of some German soils, showing differing effects of organic and inorganic exchange sites. (From Schachtschabel and Renger, 1966) • Marsh soils ○ Parabrown earths

expressed in modern terminology by the amounts and the dissociation constants of the humus acids, the amounts of clay, and the differences between different clay minerals in their selectivity coefficients for exchange of the cations H_3O^+, Al^{3+} and (Ca, Mg)$^{2+}$.

Base Saturation and Soil Formation

Soil formation in a leaching environment is accompanied by a gradual loss of base cations, especially Ca^{2+}, from the cation-exchange complex. The soils thus become progressively unsaturated and increasingly acid. The hydrogen ions in solution, which cause this, originate either from the solubility of carbon dioxide:

$$2H_2O + CO_2 \rightarrow HCO_3^- + H_3O^+$$

or from soluble acids that form during the decomposition of humus in the surface horizons of the soil. Soluble acids are particularly important where the vegetation is coniferous forest, and these acids may remove Ca^{2+} by the formation of chelate complexes as well as by hydrogen-ion exchange.

Cations are lost initially from the pH-dependent exchange sites, which acquire protons, the reverse of the reactions described previously:

$$(R\text{–}O)_2Ca + 2H_3O^+ \rightarrow 2ROH + Ca^{2+} + 2H_2O$$

As the pH drops below 5.2, in the more advanced stages of development, Al^{3+} ions begin to be released from the surfaces of soil minerals and are strongly adsorbed onto the exchange sites, displacing more Ca^{2+} into solution. The cations in solution are removed from the soil by leaching, either as bicarbonate or as organic-acid salts.

There appear to be few studies that relate modern concepts of base unsaturation to soil genesis and systematics, although Turner and Clark (1966) have done so for the Podzolic and Brunisolic orders of the Canadian soil-classification system. Degree of base saturation, based on CEC determined by ammonium acetate at pH 7, is used in modern classification systems to distinguish between mollic and umbric epipedons, and between eutric and dystric *B* horizons; in each case the second member of these taxonomic terms has less than 50% base saturation.

BRYON W. BACHE

REFERENCES

Bache, B. W., 1974. Soluble aluminium and calcium-aluminium exchange in relation to the pH of acid soils, *Jour. Soil Sci.*, 25, 320–332.

Blosser, D. L., and Jenny, H., 1971. Correlations of soil pH and per cent base saturation as influenced by soil forming factors, *Soil Sci. Soc. America Proc.*, 35, 1017–1018.
Bradfield, R., and Allison, W. H., 1933. Criteria of base saturation in soils, *Internat. Soc. Soil. Sci. Trans. Comm. II*, A, 63–79.
Chapman, H. D., 1965. Cation-exchange capacity, *in* C. A. Black, ed., *Methods of Soil Analysis,* 2. Madison: American Society of Agronomy. 891–901.
*Coleman, N. T., and Thomas, G. W., 1967. The basic chemistry of soil acidity, *in* R. W. Pearson and F. Adams, eds., *Soil Acidity and Liming.* Madison: American Society of Agronomy, 1–41.
Hissink, D. J., 1925. Base exchange in soils, *Faraday Soc. Trans.*, 20, 551–556.
Peech, M., 1965. Exchange acidity, *in* C. A. Black, ed., *Methods of Soil Analysis,* 2. Madison: American Society of Agronomy, 905–913.
Pierre, W. H., and Scarseth, G. D., 1931. Determination of the percentage base saturation of soils and its value in different soils at definite pH values, *Soil Sci.*, 31, 99–114.
Schachtschabel, P., and Renger, M., 1966. Beziehung-zwischen V- und pH-Wert von Boden, *Z. Pfl. Dung. Bodenk.*, 112, 238–248.
Turner, R. C., and Clark, J. S., 1966. Lime potential in acid clay and soil suspensions, *Internat. Soc. Soil Sci. (Aberdeen) Trans. Comm. II and IV,* 207–215.

Cross-references: *Acidity; Buffers, Buffering; Chemical Composition; Electrochemistry; Exchange Phenomena; Leaching; Soil Reaction.*

BIOLOGY—See ALGAE; ENZYME ACTIVITY; FAUNA; METABOLIC ACTIVITY; MICROBIAL ECOLOGY AND CLAY MINERALS: SOIL MICROBIOLOGY

BONDING—See Vol. IVA: CRYSTAL CHEMISTRY

BUFFERS, BUFFERING

In the chemical sense of the word, *buffers* are those systems that tend to maintain a constant pH level despite the addition (within certain limits) of an acid or base or despite dilution. They usually consist of mixtures of either a weak acid and a salt formed from the acid and a strong base, or they consist of a weak base and a salt formed from the base and a strong acid.

Many chemical reactions and all biological processes take place only at specific pH levels. If there is no buffer action, hydrogen-ion concentration may alter and cause the reaction to slow down until it stops, or produce products different from those required.

*Additional bibliographic references may be found in this article.

Theory

Using the case of an aqueous solution of acetic acid and sodium acetate—i.e., a mixture of a weak acid and its salt formed from a strong base—the dissociation equilibrium of the acid and the relative constant can be represented by

$$CH_3-COOH + H_2O \rightleftharpoons CH_3^-COO^- + H_3O^+$$

$$K_a = \frac{[CH_3-COO^-]\,[H_3O^+]}{[CH_3-COOH]} \qquad (1)$$

The hydrolysis equilibrium of the salt and the relative constant can be represented by

$$CH_3-COO^- + H_2O \rightleftharpoons CH_3-COOH + OH^-$$

$$K_i = \frac{[CH_3-COOH]\,[OH^-]}{[CH_3-COO^-]} \qquad (2)$$

Considering that the concentrations $[CH_3-COO^-]$ and $[CH_3-COOH]$ indicated in (1) and (2) are the same and that the values $[H_3O^+]$ and $[OH^-]$ are interdependent, because $[OH^-][H_3O^+] = 10^{-14}$ at 25°C = K_w, either equation 1 or 2 can be rewritten in the form

$$[H_3O^+] = K_a \frac{[CH_3-COOH]}{[CH_3-COO^-]} \qquad (3)$$

Equation 3 shows that the concentration of hydrogen ions depends on the dissociation constant K_a and the ratio of the concentration of undissociated acetic acid to that of acetate ions.

If the value of $[H_3O^+]$ is expressed as pH, equation 3 becomes

$$pH = -\log K_a - \log \frac{[CH_3-COOH]}{[CH_3-COO^-]} \qquad (4)$$

When

$$\frac{[CH_3-COOH]}{[CH_3-COO^-]}$$

equals 1, equation 4 can be simplified into the form

$$pH = -\log K_a = pK_a \qquad (5)$$

Since the concentrations of undissociated acid and its anion can be identified by the analytical concentrations of the acid C_A and its salt C_S, the small fractions of dissociated acid and hydrolyzed salt can be ignored; and equation 4 becomes

$$pH = -\log K_a - \log \frac{C_A}{C_S} \quad (6)$$

that is,

$$pH = pK_a + \log \frac{C_S}{C_A} \quad (7)$$

Similar reasoning can be applied to solutions of a weak base and its salt from a strong acid ($NH_4OH + NH_4Cl$), whereupon the following analogous expressions are obtained:

$$[OH^-] = K_B \frac{C_B}{C_S} \quad (8)$$

$$pOH = pK_B + \log \frac{C_S}{C_B} \quad (9)$$

Equations 3, 7, 8 and 9 are the basic relations that govern a solution buffer mechanism, and they are known as the *Henderson and Hasselbach equations.*

Mechanism and Efficiency of Buffer Solutions

The buffer action a system produces whenever an increase or decrease in hydrogen-ion concentration occurs results from acid-base equilibrium. If a small amount of strong acid (HCl) is added to an equimolecular solution of acetic acid and sodium acetate, some of the acetate ion is converted into acetic acid; if a strong base (NaOH) is added, some of the acetic acid is converted into acetate ion. In other words, the strong acidity of the HCl is buffered by being converted into weak acidity, just as the strong base NaOH is buffered through the formation of its salt. Table 1 gives an idea of the effect obtained. The Henderson and Hasselbach equations show that the greatest buffering effect is obtained when the pH of the buffer equals pK_a (or pK_b). In fact, under these conditions, variations undergone by the ratio C_S/C_A (or C_S/C_B) are minimal when acid or base are added; and so must be variations in pH. Further, the amounts of acid or base that can be buffered are restricted to the limits within which the ratio C_S/C_A may be considered practically constant. It can be calculated that the maximum amount of acid or base that can be added without overcoming the buffer action varies between 1/50 and 1/100 of the buffer components. This information is important in practice because it means that it is necessary to maintain fairly high total concentrations of acid (or base) and its salt to ensure a good buffer efficiency.

Thus one can say that when effecting a buffered reaction, the choice of buffer system components will be made according to the following criteria: (1) use a weak acid whose pK is as near as possible to the pH at which the reaction is to take place; (2) calculate the amount of conjugate base to mix with the acid from the opposite Henderson and Hasselbach equation; (3) the ratio C_S/C_A must be contained within 0.1 and 10 to guarantee good efficiency.

It follows that the pH range within which the buffer action is manifest has the limits of $pK + 1$ and $pK - 1$.

The Salts of Multi-proton-donating Acids as Buffers

Mixtures comprising the several salts of a weak multi-proton-donating acid likewise constitute buffer systems. For example, consider the neutralization reaction of H_3PO_4. Initially the monosubstituted salt is formed, then the bisubstituted one, and finally the trisubstituted

TABLE 1. pH Variations When HCl or NaOH Are Added to a Buffered and Nonbuffered System.

	Composition	pH	Additions In Equivalents: HCl	Additions In Equivalents: NaOH	Resulting pH: With HCl	Resulting pH: With NaOH
Buffered system	1000 ml CH_3-COOH 0.1 N $CH_3-COONa$ 0.1 N	4.74	10^{-3}	10^{-3}	4.73	4.75
Nonbuffered system	1000 ml H_2O	7.00	10^{-3}	10^{-3}	3.00	11.00

$$H_3PO_4 + NaOH \rightarrow NaH_2PO_4 \qquad pK_{a_1} \cong 2.00$$

$$NaH_2PO_4 + NaOH \rightarrow Na_2HPO_4 \qquad pK_{a_2} \cong 7.00$$

$$Na_2HPO_4 + NaOH \rightarrow Na_3PO_4 \qquad pK_{a_3} \cong 11.00$$

When the pH of the system equals pK_{a_2}, the solution consists of an equimolecular mixture of mono and bisubstituted phosphates. This mixture constitutes a buffer system in which the acid-base equilibriums that typify it are conceptually similar to those in the $CH_3-COOH/CH_3-COONa$ system. In fact, the mono and bisubstituted salts are completely dissociated, and the $H_2PO_4^-$ and HPO_4^{2-} ions are in equilibrium:

$$H_2PO_4^- + H_2O \rightleftharpoons HPO_4^{2-} + H_3O^+$$

The $H_2PO_4^-$ ion is the acid, and the HPO_4^{2-} ion is the conjugate base; thus equation 7 can be applied

$$pH = pK_{a_2} + \log \frac{C_{HPO_4^{2-}}}{C_{H_2PO_4^-}} \qquad (10)$$

Similarly, it can be shown that the mixture of bi and trisubstituted phosphates is a buffer solution with maximum buffer action when the pH equals pK_{a_3}.

Buffering in Soil

Buffering in soil is defined as the resistance of the soil to variations in pH and is chiefly due to the colloidal humus and clay fractions. In soils containing appreciable amounts of phosphates and carbonates, buffering is also effected by these salts, as illustrated in the preceding section.

As early as 1922 Arrenhius sustained that soils with strong buffer effects were very fertile and advised organic fertilization to increase the buffer action. That a soil does not undergo marked fluctuations in pH following additions of acids or bases or as a result of dilution is of considerable practical importance if the organisms living in it, particularly microorganisms and the higher plants, are to continue their various activities under those particular pedological conditions. A large change in pH means a radical modification in soil environment, which severely affects the availability of plant nutrients and alters the pedogenetic processes. Compounds normally held in the soil solution (q.v.) may precipitate out, while others normally insoluble may be caused to dissolve. This whole question is of interest not only in relation to plant mineral nutrition but also to the pedological characteristics and evolution of the soil.

Factors Causing Alteration in Soil pH

Factors tending to modify soil pH can be divided into two groups:

1. Those that bring about an increase in hydrogen-ion concentration tending to lower the pH; and
2. Those that produce an increased adsorption of exchangeable bases tending to raise the pH.

Acidity (q.v.) is principally encouraged by the decomposition of organic matter and the application of certain fertilizers. Both inorganic (H_2CO_3, HNO_3, H_2SO_4) and organic acids are found among the products of organic molecular decay, while nitric and sulphuric acids are formed by microbial action on fertilizers such as ammonium sulphate. The use of fertilizers like K_2SO_4 or KCl will also tend to lower the pH.

Alkalinity is usually encountered in arid regions where the colloidal complex of the soil is generally highly saturated with exchangeable bases, the degree of this alkalinity depending on the proportion of sodium. In these regions the weathering processes that liberate the alkaline cations from the soil minerals are helped by the extremes of temperature and by the fact that precipitation is limited to certain periods of the year. The low rainfall, moreover, means that these exchangeable bases are not leached away. Other factors conducive to alkalinity can be such agricultural processes as irrigation (q.v.), liming (see *Lime, Liming*), or manuring [with superphosphates, Thomas meal, or $Ca(NO_3)_2$].

In theory, soil pH should not be affected by a greater or lesser water content because variations dependent on the degree of dissociation of the various acids should be compensated by the variations in the degree of hydrolysis of the "salts." In practice, however, the soil pH does vary seasonally, but this is due not so much to dilution effects as to variations in biotic activities. In summer the fall in pH would seem to correspond to a greater microorganic activity with the contributing factor of an increased acid-exudate production by the plant roots; in winter a rise in pH value is often seen, probably due to the reduction in biotic activity.

The Inorganic Colloidal Fraction as a Buffer System

The colloidal soil complex functions as a slightly ionized acid or a slightly ionized salt of a weak acid. The acid character of clay was reported by Bradfield (1923, 1925). He showed that the titration of acid clays with a base produced pH and conductivity variations, which can be interpreted in terms of a weak-acid model. From the first, soil acidity was thought to be caused by exchangeable aluminum. For a long time this theory was disputed, but it was finally accepted when Coleman and Harward (1953) demonstrated that hydrogen clay, prepared for treatment with HCl, tends to be spontaneously transformed into aluminum clay. Later, based on results reported by Coleman et al (1959), Thomas (1960), Mehlich, (1941, 1960) and Yuan (1963), the conclusion was formed that acid soils are generally aluminum saturated and have the characteristics of a weak acid resulting from the hydrolysis of the adsorbed $Al(OH_2)_6{}^{3+}$ ion:

$$Al(OH_2)_6{}^{3+} + H_2O \rightleftharpoons Al(OH_2)_5\,(OH^-)^{2+} + H_3O^+$$

But there are other, weaker acid groups present in the inorganic colloidal fraction, and these create buffering at pH values higher than 5.5. They belong to two different types of "structural form": the first has been identified as the aluminum interlayer, a hydroxy aluminum polymer, postulated as a hexagon or multiple-ring structure with a composition that may range from $Al_6(OH)_{12}{}^{6+}$ to $Al_n(OH)_{3n}$; the second is considered to be of lattice origin and consists of silanol groups (–SiOH).

Figure 1 shows the potentiometric-titration curves of a H-Al-bentonite from Sardinia in the absence and presence of electrolyte. The first buffer range (1) is due to the neutralization of the adsorbed H_3O^+ ions; the second (II) to the deprotonization of the Al^{3+} ions; and the third (III), which appears only in the presence of electrolyte, is due to the –OH lattice groups.

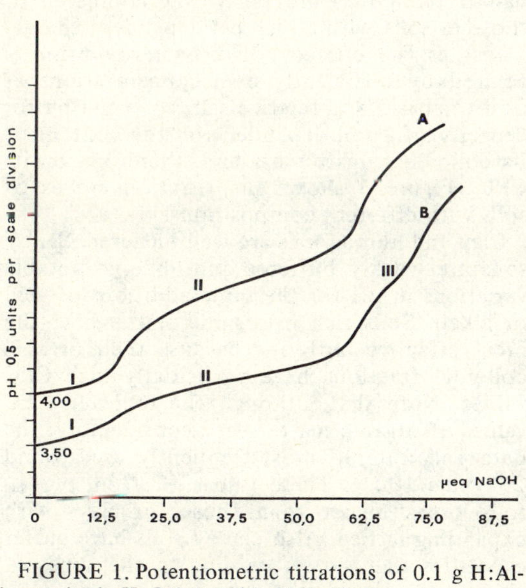

FIGURE 1. Potentiometric titrations of 0.1 g H:Al-bentonite in H_2O (*A*) and in N NaCl (*B*).

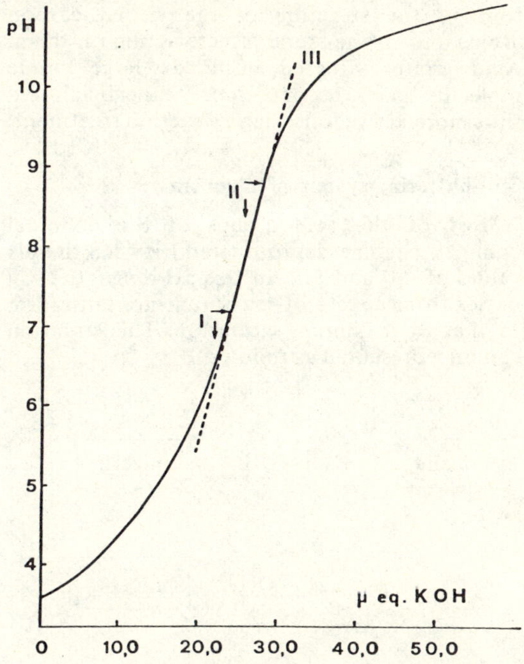

FIGURE 2. Potentiometric titration of 7.24 mg humic acid in 0.1 *N* KCl. (Adapted from Posner, 1964)

The Organic Colloidal Fraction as a Buffer System

On the basis of current knowledge of humic-acid structure, it can be held that the -COOH and phenolic -OH groups are chiefly responsible for acidity in the organic fraction. Figure 2 reproduces the titration curve of a humic acid extracted from a red brown earth (Posner, 1964). After premising that "one of the major difficulties in the interpretation of titration curves of this type is the choice of the end point," the author attributes the acidity between pH values 3.0 and 7.0 (zone I) to carboxylic groups; between 8.0 and 12.0 (zone III) to phenolic –OH groups; and between 7.0 and 8.0 (zone II) to the overlap of the other two groups as well as α-NH_2 groups. This interpretation is justified by the fact that the carboxylic acids and phenolic –OH groups have acidity constants around 10^{-5} and 10^{-10}, respectively. Subsequently Posner (1964) confirmed the validity of his attribution "in absence of evidence to the contrary" and showed that the CEC, average p*K* values, and range of p*K* values determined from the titration curves of humic acids vary with the method of extraction used. This observation is perfectly comprehensible since it is known that the acidity constant of a function is in strict relation to the structure of the vicinal carbon atoms and to the steric effects acting on them. Acid groups situated inside a "large humic molecule" can become more superficial and thus more active following extraction treatment.

Soil-buffering-system Mechanism

Most of the acid groups of the colloidal complex become deprotonated between the pH values of 4.0 and 8.5. In this pH range, the soil passes from a state of maximum nonsaturation to that of maximum saturation. This situation can be represented as follows:

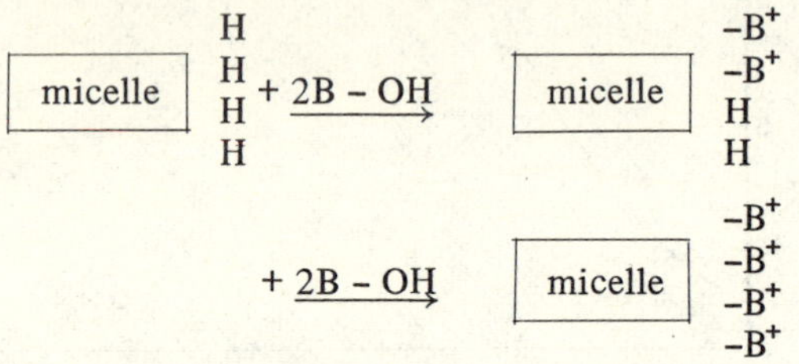

The acid form represented here is obviously a simplification; however, the hydrogen atoms shown should be considered as protons, which are freed whatever the structural form they belong to since, as Jackson (1963) has observed, "fundamentally, all acidity groups depend on proton dissociation, that is 'exchangeable H,' but differ only in the acidity strength of the functional groups." The weak-acid groups of the colloidal complex liberate protons into the soil solution; the state of equilibrium between undissociated and dissociated forms can be illustrated by:

$$\boxed{\text{micelle}}\,H + H_2O \rightleftharpoons \boxed{\text{micelle}}^{-} + H_3O^+$$

The addition of a base neutralizes the H_3O^+ ions and results in a further liberation of protons from the colloidal complex, leaving the pH more or less unvaried. When the acid function is partially salified, the colloidal complex can be compared to a mixture of a weak acid and its salt formed from a strong base, constituting exactly the form needed for acting as a buffer. In fact, buffering with respect to bases takes place through neutralization of the residual acid functions, while buffering with respect to acids is determined by the readsorption of protons into the colloidal micelle following a shift in the equilibrium of dissociation of the "weak acid" toward the undissociated form, with a consequent transfer of adsorbed exchangeable bases into the soil solution.

Buffer Capacity of Soil

The Henderson and Hasselbach equation applicable to soil could be expressed in the following way:

$$\text{pH} = \text{p}K + \log \frac{\alpha}{1 - \alpha} \qquad (11)$$

in which α is the degree of saturation, and pK assumes the meaning of apparent pK value corresponding to the pH value of the soil at 50% base saturation. Theoretically, the maximum buffering capacity of the soil should be manifest when the colloidal complex is 50% saturated. In effect, Peech (1941) obtained results, regarding a large number of Florida soils, that can be interpreted according to this equation. He set pH values against base-saturation percentages of the soils and traced a theoretical titration curve for a monobasic acid having a pK of 5.52, corresponding to the average apparent pK value of all the examined samples. His curve is not valid in respect to individual samples since their specific curves will deviate more or less widely according to the composition of their colloidal complexes. In other words, the pH at which a soil exerts its maximum buffer capacity varies from soil to soil according to its apparent pK. Mehlich (1941, 1960) showed that pH values of kaolinitic and montmorillonitic soils at 50% neutralization were significantly different. For the former, he determined apparent pK values of 6.5, and for the latter 4.5 to 5.0. Generalizing, it can be said that in soils where the permanent charge is a high proportion of the total charge, the apparent pK values are fairly low compared to those in soils with a high pH dependent charge.

Just as the efficacy of a buffer solution is assured by sufficiently high concentrations of acid (or base) and relative salt, so is the buffer capacity of a soil dependent on the content of its colloidal constituents and, therefore, on its CEC. Figure 3 shows the titration curves of soils with different compositions.

Clay and humus soils are well buffered. Sandy soils are weakly buffered and undergo notable variations in pH for the same additions of acid or alkali. Soils rich in organic matter have the most effective buffering because their organic colloidal fractions have particularly high CEC values. Note that although the reported CEC values of organic matter vary considerably, the data encountered most frequently are around 200 meq/100 g. These values are from two to four times higher than those for clays with expanding lattice. Also clayey soils have buffer capacities that vary according to the types of

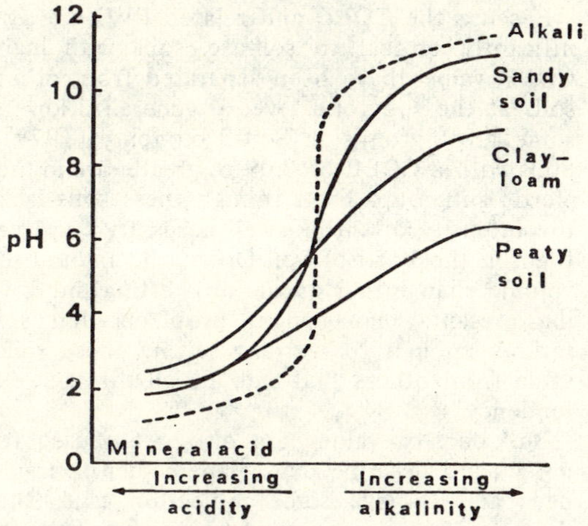

FIGURE 3. Buffering curves of different soils.

clay present; montmorillonitic soils, having higher CECs than kaolinitic soils, are more efficiently buffered.

CARLO GESSA

References

Arrenhius, O., 1922a. The potential acidity of soils, *Soil Sci.*, **14**, 223–232.

Bradfield, R., 1923. The nature of the acidity of colloidal clay of acid soils, *Jour. Am. Chem. Soc.*, **45**, 2669–2678.

Bradfield, R., 1925. The chemical nature of colloidal clay, *Jour. Am. Soc. Agron.*, **17**, 253–270.

Coleman, N. T., and Harward, M. E., 1953. The heats of neutralization of acid clays and cation exchange resins, *Jour. Am. Chem. Soc.*, **75**, 6045–6046.

Coleman, N. T.; Weed, S. B.; and McCracken, R. J., 1959. Cation exchange capacity and exchangeable cations in Piedmont soils of North Carolina, *Soil Sci. Soc. America Proc.*, **23**, 146–149.

Gessa, C., 1973. Interrelationships of a third buffer range with pH-dependent and permanent charges in bentonite, *Geoderma*, **10**, 299–306.

Gessa, C.; Melis, P.; Bellu, G.; and Testini, C., 1978. Inactivation of clay pH-dependent charge in organo-mineral complexes, *Jour. Soil Sci.*, **29**, 58–64.

Harward, M. E.; Coleman, N. T., 1954. Some properties of H- and Al-clays and exchange resins, *Soil Sci.*, 78, 181–188.

Hsu, P. H., and Bates, T. F., 1964. Fixation of hydroxy-aluminum polymers by vermiculite, *Soil Sci. America Soc. Proc.*, 28, 763–769.

Jackson, M. L., 1963. Aluminum bonding in soils; a unifying principle in soil science, *Soil Sci. America Soc. Proc.*, 27, 1–10.

Kappen, H., 1929. *Die Bodenaziditat*. Berlin: Springer-Verlag, 64–88.

Low, P. F., 1955. The role of aluminium in the titration of bentonite, *Soil Sci. Soc. America Proc.*, **19**, 135–139.

Martin, A. E., and Reeve, R., 1958. Chemical studies of Podzolic illuvial horizons. III. Titration curves of organic matter suspensions, *Jour. Soil Sci.*, 9, 89–100.

Mehlich, A., 1941. Base unsaturation and pH in relation to soil type, *Soil Sci. America Soc. Proc.*, 6, 150–156.

Mehlich, A., 1960. Charge characterization of soils, *7th Internat. Congr. Soil Sci. Trans.*, 2, 292–302.

Peech, M., 1941. Availability of ions in Linght Sandy soils as affected by soil reaction, *Soil Sci.*, **51**, 473–486.

Pommer, A. M., and Breger, I. A., 1960. Potenziometric titration and equivalent weight of humic acid, *Geochim. et Cosmochim. Acta*, **20**, 30–44.

Posner, A. M., 1964. Titration curves of humic acid, *8th Internat. Congr. Soil Sci. Trans.*, **3**, 161–174.

Posner, A. M., 1966. The humic acids extracted by various reagents from a soil, Part I. Yield, inorganic components, and titration curves, *Jour. Soil Sci.*, **17**, 65–78.

Rich, C. I., and Obenshain, S. S., 1955. Chemical and clay mineral properties of a red yellow Podzolic soil derived from muscovite schist, *Soil Sci. Soc. America Proc.*, **19**, 334–339.

Rich, C. I., 1960. Aluminium in interlayers of vermiculite, *Soil Sci. Soc. America Proc.*, **24**, 26–32.

Schwertmann, U., and Jackson, M. L., 1963. Hydrogen-aluminum clays: A third buffer range appearing in potenziometric titration, *Science*, **139**, 1052–1054.

Thomas, G. W., 1960. Forms of aluminum in cation exchangers, *7th Internat. Congr. Soil Sci. Trans.*, 2, 364–369.

Yuan, T. L., 1963. Some relationships among hydrogen, aluminum and pH in solution and soil systems, *Soil Sci.*, 95, 155–163.

Cross-references: *Acidity; Base Saturation; Exchange Phenomena; Fertilizers, Inorganic; Humus; Metabolic Activity; Soil Reaction; Soil Solution.*

BULK DENSITY

Brady (1974) defines *bulk density* as the mass (weight) of a unit volume of dry soil. Blake's (1965) definition states that this soil parameter is the ratio of the mass to the bulk, or microscopic volume of soil particles plus pore spaces in the sample. Therefore, the voume with which bulk density is concerned is that of the solid particles and pore space. Bulk density differs from *particle density* (q.v.) in that the latter concerns the relationship between the mass and volume of only the solid particles in the soil excluding the volume of pore space. Bulk-density and particle-density values are used to compute *pore space* (see *Soil Pores*) in a soil using the following relationship:

$$\% \text{ pore space} = 100 - \frac{Db}{Dp} \times 100$$

where

Db = bulk density, oven dry

Dp = particle density

Bulk density is expressed as a ratio of grams per cubic centimeter (g/cc). Since the mass of a cubic centimeter of water is about 1 g, the bulk density of water is said to be 1 g/1 cc or 1. Soil usually has a bulk density greater than water. For topsoils a value of 1.1–1.3 g/cc is about normal. For subsoils bulk-density values are usually greater than those for topsoils and range from 1.3–1.7 g/cc. The bulk density of the substratum may be greater or less than that of the subsoil. These figures are necessarily approximate for the bulk-density value of a soil depends on many soil conditions such as the way the soil is managed, or cultivated. Compaction, puddling (q.v.), and a weakening grade of structure all relate to high bulk-density values. Any management practice that encourages these conditions to exist increases bulk density, thereby decreasing pore space and restricting gas diffusion and water movement in the soil. These conditions create soil zones where roots cannot penetrate and thus cannot derive water and nutrients.

Another use of bulk-density values in computing soil parameters is that of determining the *COLE* of the soil. COLE is the "coefficient of linear extensibility" and is therefore a measure of the potential volume change (PVC) of a soil as wetting or drying takes place and swelling or shrinking goes on (Buol, Hole, and McCracken, 1973). COLE is an expression of the shrink-swell tendency of the soil and is directly related to the change that occurs in bulk density as wetting or drying takes place. This relationship is shown by the following expression:

$$\mathrm{COLE} = \sqrt[3]{\frac{Db_d}{Db_m}} - 1$$

where

Db_d = bulk density, dry

Db_m = bulk density, moist

Because the COLE and related PVC are significantly critical to soil-use, soils with high COLE values have been separated from other soils at the subgroup level of generalization in some great groups in Soil Taxonomy (1977). Soils with a COLE of 0.09 or greater are in the Vertic subgroup. Even though these soils lack the intensity of shrink-swell necessary to place them in the Vertisol Soil Order, the amount of volume change that occurs on wetting and drying presents management problems that are serious enough to warrant setting these soils aside from others that lack a high shrink-swell tendency.

Bulk-density values are also often used to detect and describe dense layers such as fragipans and to give some indication as to the degree of development of these pans (q.v. in Vol. XII, Pt. 2, *Soil Pans*). In addition, they are used to detect appreciable amounts of volcanic ash, to determine the degree of alteration of *C* horizons formed in saprolite, and to evaluate gains within and losses from various soil horizons if parent-material uniformity can be established (Buol, Hole, and McCracken, 1973).

DAVID T. LEWIS

References

Blake, G. R., 1965. Bulk density, *in* C. A. Black, ed., *Methods of Soil Analysis* **1**. Madison: American Society of Agronomy, 374–390.

Brady, N. C., 1974. *The Nature and Properties of Soils.* Toronto: Collier-MacMillan, 639p.

Buol, S. W.; Hole, F. D.; and McCracken, R. J., 1973. *Soil Genesis and Classification.* Ames: Iowa State University Press, 360p.

Grossman, R. B.; Brasher, B. R.; Franzmeier, D. P.; and Walker, J. L., 1968. Linear extensibility as calculated from natural clod bulk density measurements, *Soil. Sci. Soc. America Proc.*, **32**, 570–573.

Soil Survey Staff, 1977. Soil taxonomy: A basic system of soil classification for making and interpreting soil surveys, *U.S. Department of Agriculture Handbook No. 436*, 754p.

Cross-references: *Aggregation; Particle Density; Pore-size Distribution; Soil Mechanics; Soil Pores; Soil Structure.*

C

CALCIUM–See LIME, LIMING

CAPILLARY PRESSURE

Two fluids are *immiscible* if they retain a distinct phase identity when they are mixed. As a consequence of the different degrees of attraction between molecules of a different nature a tension exists at the interface, which is called *surface tension.* The surface tension σ has the dimension of a force per unit length of the interface edge (Figure 1).

If the interface between two fluids, e.g., water and air, in equilibrium is flat, the pressures (in the water and in the air) are equal. In contrast, if the interface is curved, as shown in Figure 2, there exists a pressure difference between the two phases. At equilibrium for a spherical bubble of air in water, it can be shown (e.g., Morel-Seytoux, 1969, p. 472) from a consideration of balance of forces on a small element of interface (Figure 2) that

$$p_a - p_w = \frac{2\sigma}{R} \qquad (1)$$

where p_a is air pressure, p_w is water pressure, and R is the radius of the bubble. The difference in pressure between the fluid at the higher pressure and the fluid at the lower pressure is called the *capillary pressure* and is usually denoted by the symbol p_c. If the interface is not spherical in shape (see Figure 2), equation 1 generalizes to a form, known as *Laplace's or Plateau's equation:*

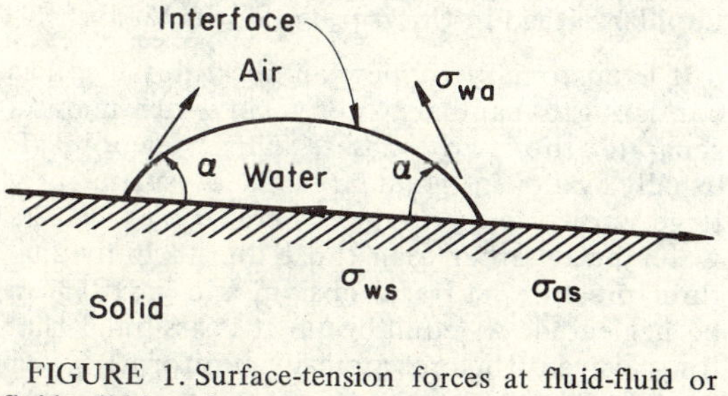

FIGURE 1. Surface-tension forces at fluid-fluid or fluid-solid interfaces.

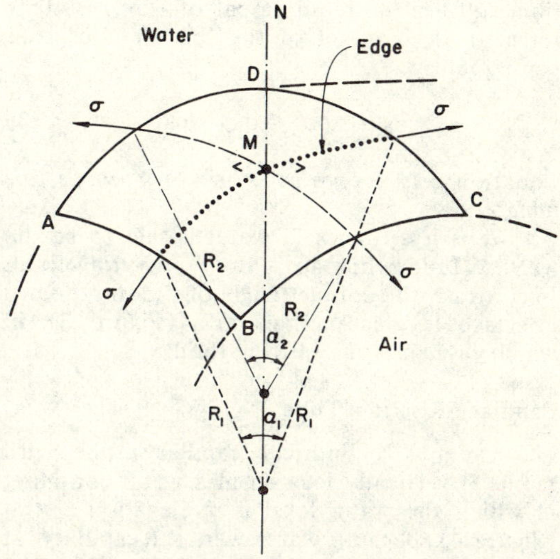

FIGURE 2. Capillary force equilibrium at an interface between two immiscible fluids.

$$p_c = \sigma\left(\frac{1}{R_1} + \frac{1}{R_2}\right) \qquad (2)$$

where $(1/R_1) + (1/R_2)$ is the total curvature of the interface at point m, and R_1 and R_2 are the radii of the curvature of interface lines in any two mutually orthogonal normal planes at M (see Figure 2). The radii of curvatures R_1 and/or R_2 are algebraic quantities, which are positive if the center of the curvature lies within the phase at higher pressure and negative otherwise.

By definition, the capillary pressure is always a *positive* (or zero) quantity. The capillary pressure is often expressed as an equivalent water height, denoted h_c and called *capillary-pressure head.* The term "capillary pressure" is misleading because the definition of capillary pressure assumes in no way the presence of a solid wall, as introduced by a capillary. It does, however, assume the presence of two immiscible fluids. However, the presence of a solid wall may have a bearing on the *shape* of the interface and consequently a bearing on the *magnitude* of the capillary pressure.

Wettability

Just as there exists a surface tension between immiscible fluids, there exists a surface tension between a fluid and a solid. The surface tension between water and air, σ_{wa}, differs from that between water and solid material, σ_{ws}. A water drop on a glass plate tends to spread as shown in Figure 1. The contact angle α between the water-air interface with the solid at equilibrium fulfilles the requirement of zero resultant force at the contact of the three phases, and consequently,

$$\cos \alpha = \left(\sigma_{as} - \sigma_{wa}\right) / \sigma_{wa} \qquad (3)$$

Equation 3 is known as *Dupré or Young's formula.*

As α is less than $\pi/2$, water is said to be the *wetting* fluid while air is the *nonwetting* fluid. In contrast, the contact angle of the air-mercury interface is greater than $\pi/2$, (Figure 3), in which case air is the wetting fluid.

Capillary Rise in a Tube

Consider a cylindrical capillary tube with radius R vertically located in a water container in which the water level is at elevation $z = z_0$ (Figure 4). Because water wets the capillary, at equilibrium the air-water interface has the shape of a spherical meniscus concave toward the air, of total curvature $2(\cos \alpha/R)$, where α is the contact angle. According to equation 1 since air pressure is atmospheric, the water pressure is

$$p_w = p_A - \frac{2\sigma \cos \alpha}{R}$$

where p_A is atmospheric pressure. If atmospheric pressure is used as a reference pressure, then the water (gauge) pressure at the interface is negative, i.e., less than atmospheric; in other words, the water is under *suction.* As a result of this imbalance, the water rises in the capillary tube up to an equilibrium level $h_c + z_0$. At hydrostatic equilibrium the total head of water must be constant; thus

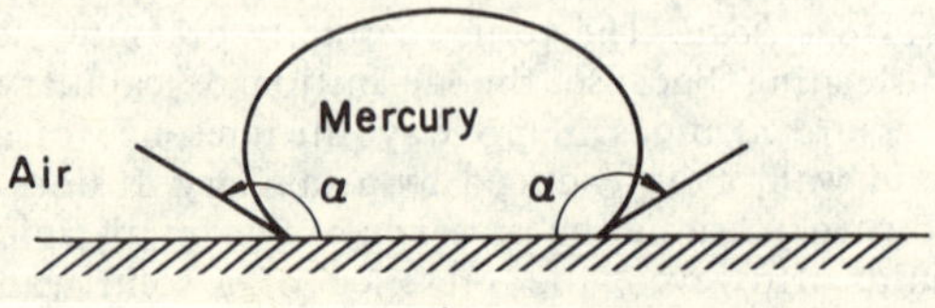

FIGURE 3. Contact angle of air-mercury interface.

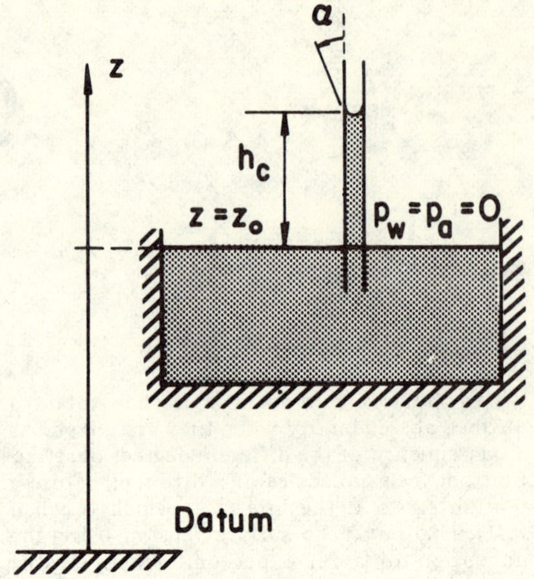

FIGURE 4. Capillary rise.

$$\phi = \frac{p_w}{\rho_w g} + z$$

where ρ_w is specific mass of water, and g is the acceleration of gravity. As the water (gauge) pressure is zero at $z = z_0$, it follows that $\Phi = z_0$ everywhere in the water phase. Consequently, at the meniscus level, one must have:

$$\frac{p_w}{\rho_w g} + z = -\frac{p_c}{\rho_w g} + h_c + z_0 = z_0$$

or

$$h_c = \frac{p_c}{\rho_w g} = \frac{2\sigma \cos \alpha}{R \gamma_w} \qquad (4)$$

where $\gamma_w = \rho_w g$ is the specific weight of water. Equation 4 is usually referred to as the *capillary rise equation.*

Capillary Head in Unsaturated Porus Media

It is customary to view an unsaturated soil as consisting of capillary "pores" in which menisci separate the two phases. This viewpoint is usually valid in granular soils at sufficiently large moisture contents where forces in the water body other than those due to individual fluid pressure, surface tension, and gravity can be neglected. At equilibrium it is assumed that for a given (macroscopically uniform) water content, the air-water interfaces have the same constant total curvature throughout the porous medium. Soil scientists traditionally define this state by the *capillary head*, $\psi = -P_c/\gamma_w$ which is the *negative* of the *capillary pressure head* defined previously. The total (hydraulic) head is thus

$$\Phi = \psi + z \quad (5)$$

Clearly at low water contents and with soils of fine texture such as clays, the major fraction of the water content in the medium is distributed in thin films covering the solid matrix. In such cases the interaction between the water-air interface and the water-solid interface must be considered, and the simple *mechanical* model for capillary pressure is no longer valid. This latter conclusion can be understood visually from a description of the microscopic and macroscopic geometry (Figures 5 and 6). Consider two wetted soil particles covered with thin, yet observable, films of water (Figure 5). So-called pendular (Stallman, 1964) water exists between the two soils particles. In one region (point *a* of Figure 5), the total curvature is positive; hence p_c is positive. In another region (point *b*), the overall curvature is negative, yielding a negative value of p_c. At equilibrium the air pressure is

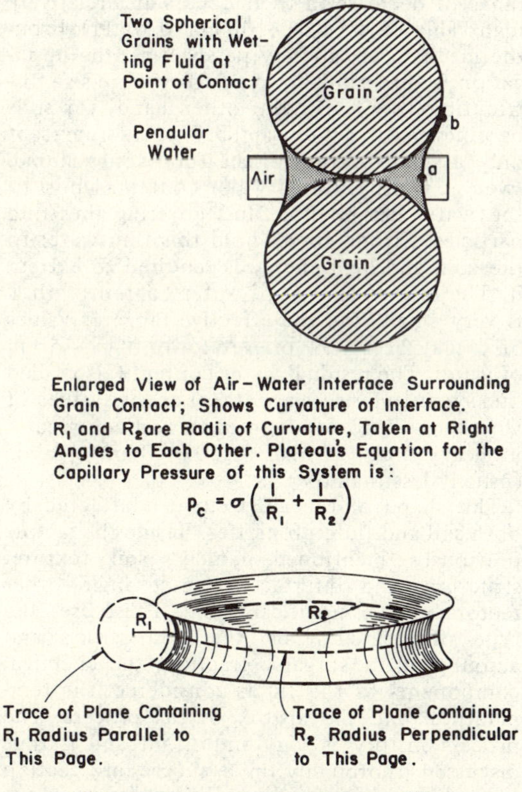

FIGURE 5. Geometry of capillary equilibrium.

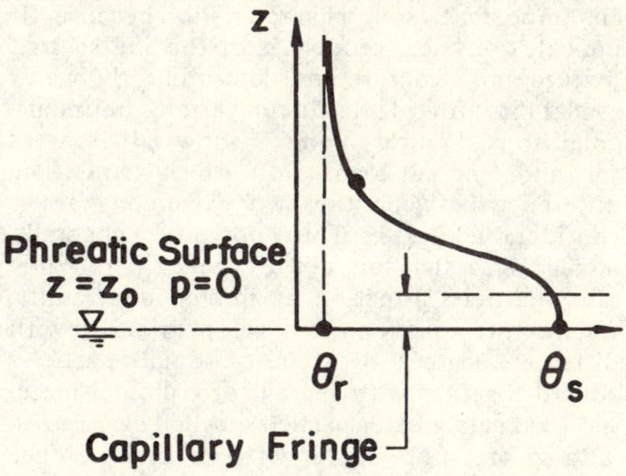

FIGURE 6. Capillary pressure versus water content.

constant, and based on the capillary geometrical mechanical model, the water pressure would be of different signs around each soil particle. As the water phase is continuous and in equilibrium, it is clear that *other* forces must compensate for this pressure difference in the water and maintain the equilibrium.

Figure 6 displays a profile of water content θ above a water-saturated medium at rest. Above the water table (*phreatic* surface) at $z = z_0$, the water content gradually decreases with elevation. Immediately above the water table the rate of change of water content with elevation, $d\theta/dz$, is negligible. As elevation rises, $d\theta/dz$ gradually decreases to an inflection point and then increases again toward zero as the water content approaches θ_r, the *residual* water content. This low residual value of water content may be observed at more than several hundred meters above the phreatic surface. As water is in equilibrium throughout the soil, the total head Φ is also constant and equal to z_0. If equation 2 were generally valid, then the pressure of water at the low water contents would be less than 10 atm. Actually, suction in water cannot exceed 0.7 atm, which shows that the mechanical model of capillary pressure is not generally valid. It is possible, however, to apply equation 5 for the saturated-unsaturated soil profile with ψ representing the cumulative effect of pressure and other *unspecified* (though effective) factors. The value of ψ for a given equilibrium water content is determined from the total head equation, for Φ equals constant, yielding

$$\psi = -(z - z_0) \quad (6)$$

where z is the elevation of the given water content. For convenience, soil scientists continue to refer to ψ as the "capillary head."

The dependence of ψ on the water content—sometimes called the *water-retention* curve—is an important soil characteristic because it indicates physical properties of the soil such as texture and structure and determines the water availability for plants under various boundary conditions. Figures 7 and 8 show $\psi(\theta)$ curves for sands and silt loam and for heavy soils. The difference between the two extreme types is considerable: (1) The porosity generally increases as the soil texture gets finer. While sand particles function as independent units, fine-texture soils tend to develop structure with different degrees of stability. Small particles adhere together with the aid of colloidal material to create greater particles, called *aggregates,* with greater pore space between them. Hence the observed water content at $\psi = 0$ in heavy soils is greater than in sandy soils. (2) Sands usually have a minimum critical value of ψ (in absolute value), which must be exceeded before a reduction in water content can be observed. Thus for $\psi_{cr} \leqslant \psi \leqslant 0$, θ remains essentially constant at the saturated value, θ_s. At this value of $\psi = \psi_{cr}$, air starts to penetrate into the soil, and for this reason it is also called the *air-entry value.* Soils of very fine texture have no clear air-entry value because reduction of the water content occurs first as a result of compression of the soil while it is still saturated with water. (3) The $\psi(\theta)$ curve of granular soils displays an abrupt drop in water content within a small range of decrease in ψ. It occurs at a relatively high value of ψ ($\psi > -100$ cm of water). From the mechanical capillary perspective, the implication is that sands have a narrow pore-size distribution (q.v.). On the other hand, for soils of fine texture the capillary pores represent only a small portion of the total void volume. Even at relatively high water contents, most of the water lies in thin films covering the solid particles. Strong forces hold this film water to the soil, and high energy is required to extract it. The rate of change of water content with ψ is very small, and the effective range of values of ψ may be as low or even lower than −150 m of water. The residual water content (also called the *irreducible water content*) at low values of ψ, where $d\theta/d\psi$ tends to zero, is much greater in heavy soils (as high as 20%) than in sands (usually less than 6%).

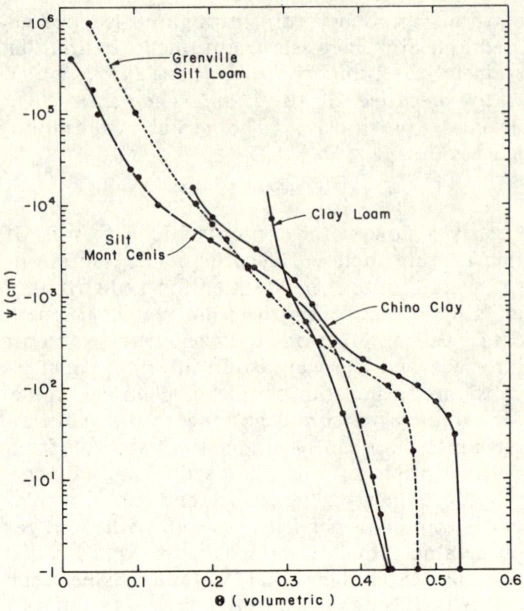

FIGURE 8. Water-retention curves for heavy soils: Grenville silt loam (Staple, 1965); Silt Mont Cenis (Vachaud, 1966), Clay loam (Bruce, 1972) and Chino clay (Gardner, 1959).

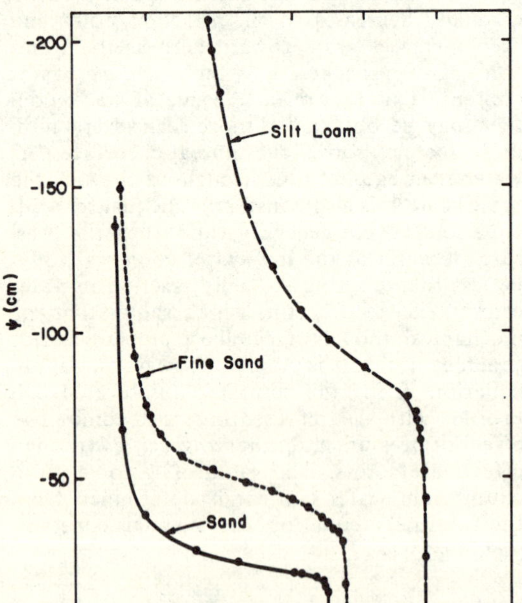

FIGURE 7. Water-retention curve for sand, fine sand, and silt loam. (After Brooks and Corey, 1966)

The shape of the $\psi(\theta)$ curve is affected by both soil and fluid properties. Beside the factors previously mentioned—namely, soil texture, structure, and interface contact angle—other factors play a significant role. These are: the type of soil minerals, the composition and amount of the organic fraction, the chemical components of the fluids considered, the temperature, and the history of the process. The process history is very important and is thus discussed thoroughly in a succeeding section headed "capillary hysteresis."

The water-retention curves $\theta(\psi)$ are determined experimentally. No theory has yet been developed that is accurate or simple enough to derive analytically the $\theta(\psi)$ curve. There is, however, value in fitting the observed data by analytical expressions. Such analytical yet empirical expressions serve two purposes: (1) They permit solutions to unsaturated-flow problems in closed form, and (2) they eliminate the truncation errors introduced in the estimation of ψ and $d\psi/d\theta$ in numerical solutions. A variety of expressions have been proposed in the literature. The simplest one is the power function:

$$\psi = a\theta^{-n} \tag{7}$$

where a and n are adjustable parameters. Equation 7 can be modified to account for air-entry value and residual water content in the form

$$\begin{aligned} \theta(\psi) &= \theta_s \quad \text{for} \quad \psi > \psi_{cr} \\ \psi &= S_c^{-n} \quad \text{for} \quad \psi < \psi_{cr} \end{aligned} \tag{8}$$

where

$$S_e = \frac{\theta - \theta_r}{\theta_s - \theta_r}$$

is called the *effective saturation.* For $\theta(\psi)$ curves with monotonically decreasing slope, a similar expression is used:

$$S_e = \frac{a}{a + \psi^n} \tag{9}$$

Two other expressions are respectively of exponential nature, namely;

$$\begin{aligned} \theta &= \theta_s \quad \psi \geqslant \psi_{cr} \\ S_e &= e^{\alpha(\psi - \psi_{cr})} \quad \psi < \psi_{cr} \end{aligned} \tag{10}$$

or logarithmic, namely;

$$\begin{aligned} \theta &= \theta_s \quad \psi \geqslant \psi_{cr} \\ S_e &= 1 - \frac{\ln(\psi - \psi_{cr} + 1)}{\ln(\psi_{15} - \psi_{cr} + 1)} \quad \psi < \psi_{cr} \end{aligned} \tag{11}$$

Equation 11 mixes dimensional and dimensionless terms and as a result can be used only if ψ is expressed in centimeters of water.

None of these analytical curves except that defined by equation 9 has an inflection point and can be accurately adjusted only to the experimental curves for a limited range of ψ or θ values. More flexible and realistic expressions can be derived such as

$$\psi = a\theta^n(\theta_s - \theta)^m \tag{12}$$

$$S_e = a\,\frac{\cos h\left[\left(\dfrac{\psi}{\psi_0}\right)^n + b\right] - c}{\cos h\left[\left(\dfrac{\psi}{\psi_0}\right)^n + b\right] + c} \tag{13}$$

where a, b, c, n, and ψ_0 are adjustable parameters (King, 1964), or any combination of the more elementary functions mentioned previously. These latter expressions are of limited value, however, as the amount of work required for adjustment is considerable, and they are too complicated to be used in the derivation of closed analytical solutions. For a detailed review and references, see Mualem and Dagan (1976).

Experimental Determination of Capillary Head

There are a variety of ways to measure the capillary head in a soil. One method determines directly the pressure difference between air and water and the corresponding water content in the soil. Suppose a rock (or soil) sample, completely saturated with water at atmospheric pressure, is placed in contact along a fraction of its surface with air. Pressure in the air phase is increased and remains constant. A certain volume of air penetrates the core and expels a corresponding amount of water, which is measured.

The air is retained in the porous medium by a semipermeable membrane that transmits the displaced water but not the air. Ultimately a no-flow condition is established. Due to the expulsion of a certain volume of water, the remaining water fills up only a fraction of the total pore space. (The fraction of the total pore volume occupied by water is called the *water saturation.*) Pressure in the air is increased again. When equilibrium is reached, a new and lower equilibrium water saturation prevails in the core. Ultimately, repeating the operation successively (and patiently), a curve of capillary head versus water saturation or water content is obtained (Figure 9). Because water was drained from the core in the procedure, the curve so obtained is labeled "drainage" or "drying curve." Because the core was initially *fully* saturated (no air present at all), the curve is labeled "first draining curve."

It is found experimentally that a point is reached when even a tremendous increase in capillary pressure no longer induces a saturation change. The water saturation is said to have

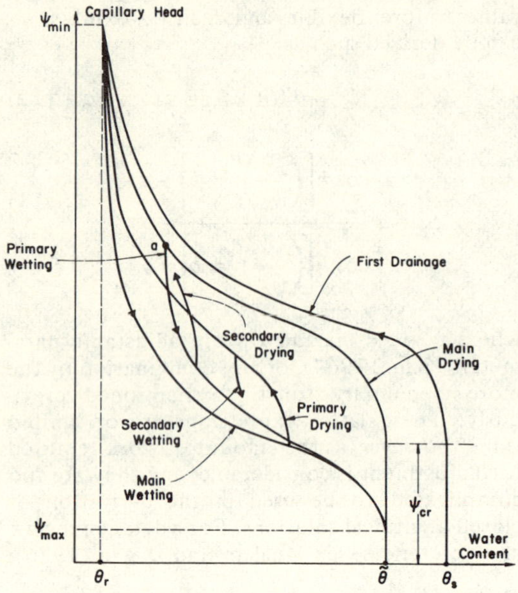

FIGURE 9. Nomenclature of capillary hysteresis.

reached its irreducible, or residual, value. The concept of irreducibility is relative to the process. For example, the residual water saturation could be reduced further by placing the core in an oven at an elevated temperature.

Capillary Hysteresis

Once residual water content has been established in the core by displacement by air, the *imbibition*, or *wetting*, curve can be obtained by letting the pressure drop stepwise and water imbibe back. Imbibition (q.v.), or wetting, are the terms used if the water content increases with time. However, a different curve is obtained (Figure 9), which implies that given a water content, several equilibrium states are possible depending on previous history. The capillary-pressure curve is said to display *hysteresis*. By the time the capillary pressure has returned to the zero value, all the air has not been expelled. A fraction of it has been trapped, the residual air content. This imbibition curve is referred to as the "main wetting curve." If the soil is drained again, the main drying curve is described. If the process is reversed before the capillary head has reached the value ψ_{min} (point *a* on Figure 9), a primary wetting curve is described. But if the process is reversed only after the value ψ_{min} has been reached, then the main wetting curve is described again. The main drying and main wetting curves form the *main hysteresis loop*. Depending on the location of the reversal point in the process of successive wetting and drying, the described curves are referred to as "primary," "secondary," "tertiary," etc.

The observed capillary hysteresis results from the cumulative effects of several factors. Probably the most important one is the geometry of the porous system. Let us adopt the mechanical capillary-pressure model discussed earlier. Under these conditions it is possible to depict different porous-matrix geometries that lead to hysteresis. They have in common the existence of cavities with converging and diverging walls. The isolated pores so constructed are connected by narrow channels, which permit *different* configurations of the interface at equilibrium for the *same* value of ψ. Figure 10 displays a pore with two degrees of filling, yet with the same curvature radius for the interface and consequently the same capillary head. Figure 11 displays another type of geometry that causes hysteresis (the "ink-bottle" effect) as the same curvature can exist with various degrees of filling of the void space.

A second effect is the hysteresis of the contact angle α. As stated previously in the discussion of the capillary rise in a capillary tube, the capillary pressure depends on the contact angle according to the formula

$$p_c = \frac{2\sigma}{r} \cos \alpha \tag{14}$$

In general, the wetting angle is not constant. It reaches its maximum value when water (or more general a liquid) moves toward a dry surface and takes it minimum value when it recedes. This phenomenon can be observed visually in the process of filling and emptying a capillary tube (Figure 12). As a result of this wetting angle hysteresis, a row consisting alter-

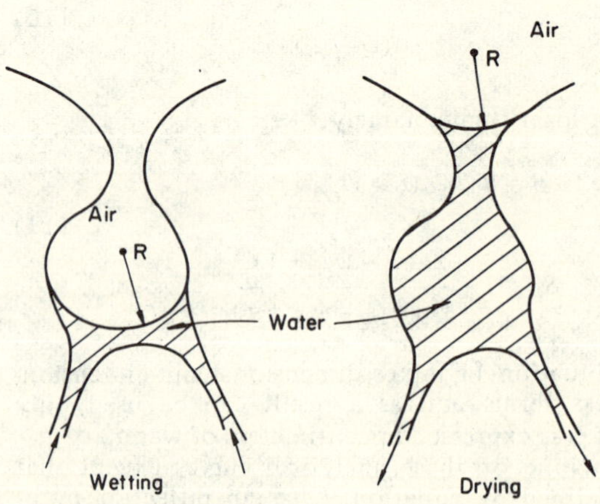

FIGURE 10. Different degrees of saturation for the same capillary pressure.

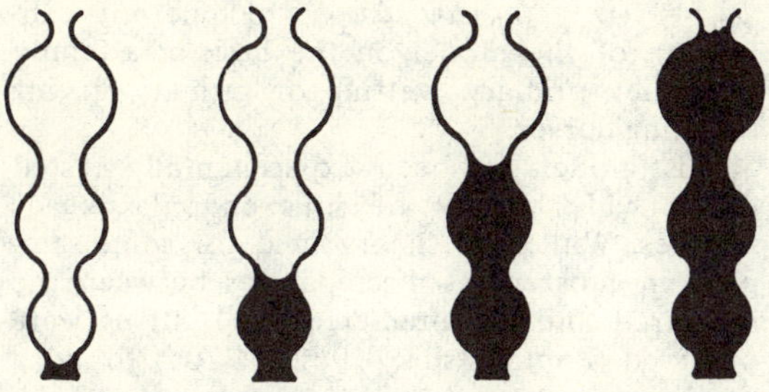

FIGURE 11. The "ink-bottle" effect.

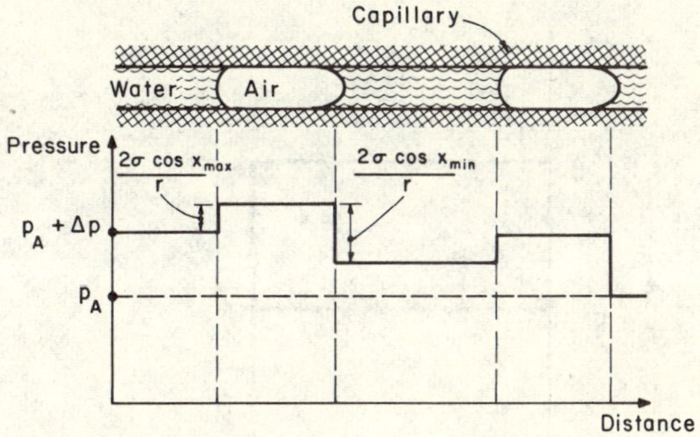

FIGURE 13. Equilibrium induced by a series of wetting-angle hysteresis.

nately of air bubbles and of water drops can maintain rest against a significant pressure drop between the two ends of a tube (Figure 13).

Entrapment of air during the imbibition (or wetting) process is another important factor. The appreciable difference between the first drainage (or drying) curve (abbreviated F.D. curve) and the main drying (M.D.) curve displayed on Figure 9 is the direct result of this air entrapment. This entrapment may be explained in a simple way by the closing of narrow entrances of pores or groups of pores by the wetting fluid in a slow wetting process. The air content in the sample varies with time as air dissolves in water and moves away by diffusion. It is a known fact that if ψ is kept constant for a long time, the water content increases as air disappears from the soil (Bloomsburg and Corey, 1964).

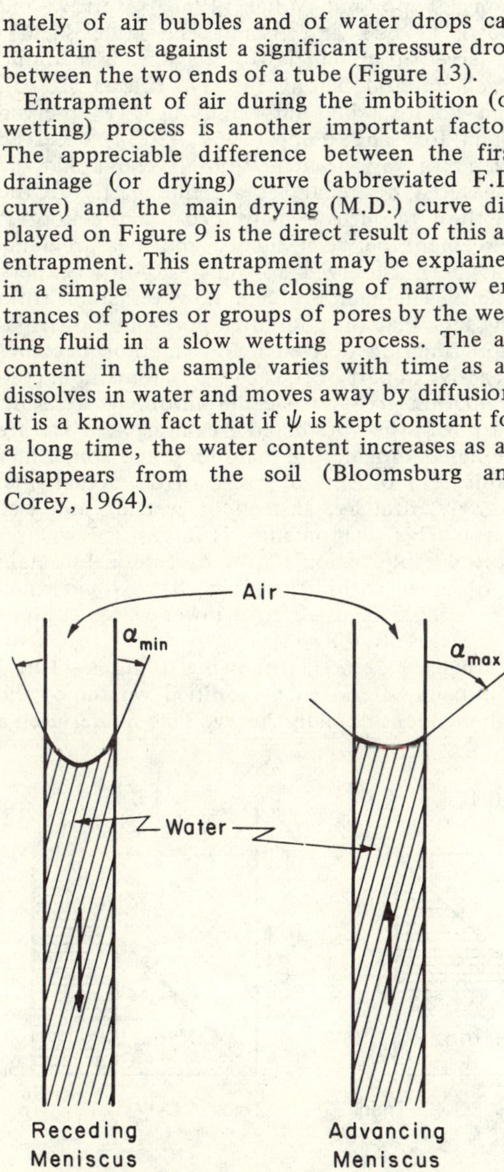

FIGURE 12. Hysteresis of wetting angle.

Comparisons of numerical studies of water redistribution in soils, including or excluding the hysteretic effects, have shown that the profiles differ substantially. In the presence of hysteresis higher water contents prevail in the upper part of the soil, and wetting of the lower part is delayed. This delay in the drainage rate enhances the evaporation from the soil surface. For a detailed review and references, see Mualem and Dagan (1972).

Theory of Capillary Hysteresis

The fundamental theory of hysteresis based on the "independent domains" concept was initiated by Preisach (1935) and Néel (1942–43) and thoroughly analyzed by Everett and his coworkers (1952–1955) and Enderby (1955). According to this theory a porous medium is viewed as a system consisting of independent elementary pore domains. Each domain is characterized by two length scales, ρ and r, which can be interpreted geometrically as the radius of the pore and the radius of its constricted connection with other pores, respectively. Using the capillary law ($\psi \sim a/R$), the variables ρ and r can be uniquely related to the wetting and drying capillary head ψ_w and ψ_d, respectively. The pore domain has therefore only two stable states, either empty or full (Figure 14). In a wetting process, the pore is empty until ψ reaches the value ψ_w at which time it flips over to a filled state. There is no change in water content of the pore when ψ is increased further. In drainage the pore remains water filled until ψ takes to value of ψ_d. At this point the pore is totally drained. The elementary pore domain displays hysteresis. It is assumed that these properties are independent of the states of the neighboring pores. Denoting ΔV as the pore volume and taking ψ_w and ψ_d as independent variables continuously distribu-

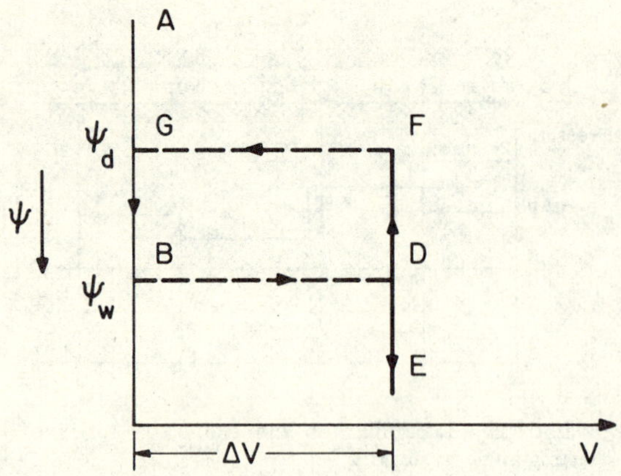

FIGURE 14. Hysteretical behavior of the isolated pore domain.

ted between ψ_{min} and ψ_{max}, it is possible to define a pore-water density distribution function

$$f(\psi_d, \psi_w) = \frac{\Delta V(\psi_d, \psi_w)}{V} \tag{15}$$

where V is the total volume of the sample. Now $f(\psi_w, \psi_d) d\psi_w d\psi_d$ represents the relative pore volume which is filled at ψ_w and drained at ψ_d due to differential changes $d\psi_w d\psi_d$. Essentially, $f(\psi_w, \psi_d)$ is positive with $\psi_w \geqslant \psi_d$. Figure 15(b) and 15(c) show the diagram of the filled-pore domains, "Néel diagrams," at a given θ for the main wetting and the main drying processes.

Assuming that the independent-domains theory applies conceptually, it is possible to relate the measurements of ψ and θ to the pore-water density function f. It is customary to operate with finite differences and to define approximately the integral function $F(\psi_{wi}, \psi_{dj}) = f(\psi_{wi}, \psi_{dj})\Delta\psi_w\Delta\psi_d$, which permits calibration of the model on the basis of a family of either primary wetting or primary drying scanning curves.

This model has been experimentally tested for a wide variety of soils and glass-beads samples. While the theory held for some samples, in most cases discrepancies between the measured and the predicted $\theta(\psi)$ curves were observed. Poulovassilis (1962, 1970a) found a good agreement between the predicted and the measured scanning curves for glass beads and sand. Other investigators reported considerable discrepancies between the theory and observations; Topp and Miller (1966), Morrow and Harris (1965), and Bomba and Miller (1967) observed discrepancies for glass-beads medium; Vachaud and Thony (1971), Talsma (1970), and Poulovassilis and Childs (1971) observed them for sand; Topp (1969, 1971*a*) observed them for various loam soils. It seems that the measured discrepancies are due partly to the method of measurement because the poorest agreement between theory and measurements has been observed for unsteady-flow experiments. The main problem, however, stems from the fact that the computed pore-water distribution function derived in this method becomes negative in part of the (ψ_w, ψ_d) domain, in contradiction to its basic definition. As a result, part of the predicted scanning curves are depicted outside the main loop and sometimes indicate a process opposite to the actual one, namely, drainage instead of wetting and vice versa. This phenomenon is intensified when a considerable region of the hysteretical domain is observed within the range $(0, \psi_{cr})$ and when the sample is dried to a lower water content θ_{min}, close to θ_r.

Mualem (1973), following Philip's (1964) approach, suggested a modified variant of the independent-domain theory. This model used a

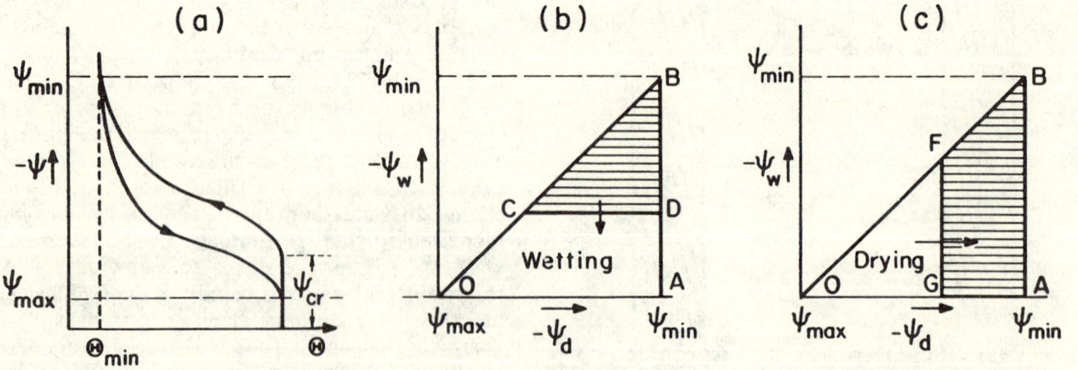

FIGURE 15. (a) Schematical representation of capillary hysteresis. (b) Filled-pore diagram (shadowed region) for the main wetting process. (c) Filled-pore diagram for the main drying process. (After Mualem, 1973)

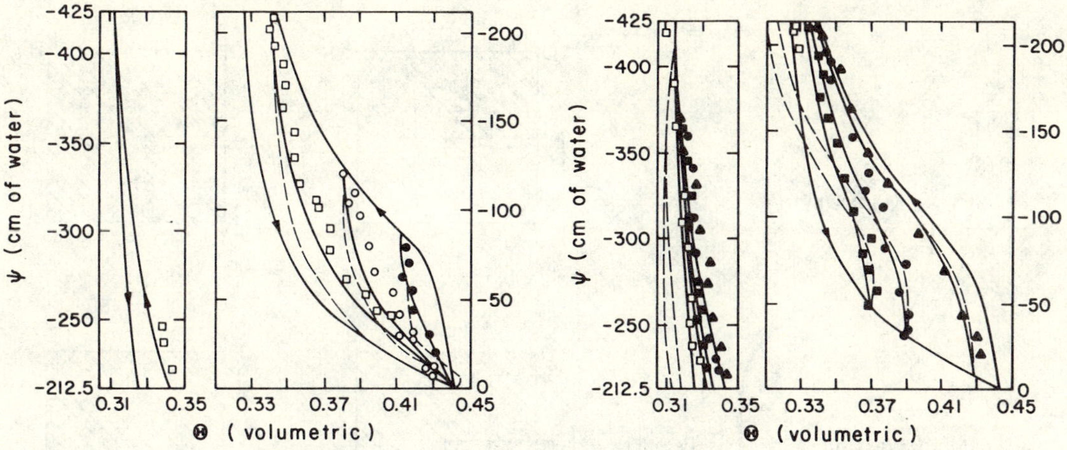

FIGURE 16. Scanning curves predicted by the usual Néel-Everett model (dashed line), by model I (solid line), and the measured data for Caribou silt loam (points). (After Mualem, 1973)

similarity hypothesis, according to which the pore-water distribution function $f(\psi_w, \psi_d)$ is represented as a product of two independent distribution functions, $h(\psi_w)\ell(\psi_d)$. This hypothesis considerably simplifies the computational procedures. Only the measured boundary curves are required for the calibration of the model, and the scanning curves are expressed analytically. Moreover, the modified model is better adapted to the independent-domain principles (Figure 16) as the computed pore-distribution function is always positive. As a result, the predicted $\theta(\psi)$ curves are in better agreement with observations than those derived by using the original Néel-Everett model.

A more flexible independent-domain theory, which accounts also for a reversible contribution of the pore domains to the wetting and drying processes, has been suggested by Mualem (1974). This model is underlain by three basic assumptions:

1. The two parameters r and ρ, characterizing a pore volume of the porous medium, vary in the whole range between R_{min} and R_{max} (which correspond to ψ_{min} and ψ_{max}, respectively, by the capillary law $R \sim 1/|\psi|$). After the normalization, that is,

$$\bar{r} = \frac{r - R_{min}}{R_{max} - R_{min}} \qquad \bar{\rho} = \frac{\rho - R_{min}}{R_{max} - R_{min}} \tag{16}$$

the pore-volume distribution function $f(\bar{r},\bar{\rho})$ is defined in the square $OABC$ of Figure 17.

2. In a wetting process under a change from $\psi(\bar{R})$ to $\psi(\bar{R} + d\bar{R})$, pores characterized by $\bar{R} \leqslant \bar{\rho} \leqslant \bar{R} + d\bar{R}$ are filled, while in drainage when ψ diminishes from $\psi(\bar{R})$ to $\psi(\bar{R} - d\bar{R})$, only the pores characterized by $\bar{R} - d\bar{R} \leqslant \bar{\rho} \leqslant \bar{R}$ and $\bar{R} \leqslant \bar{r} \leqslant 1$ are drained. Diagrams of the filled-pore domains in the $\bar{r}$, $\bar{\rho}$ plane (corresponding to four different processes) are given in Figure 17.

3. The simplifying similarity hypothesis

$$f(\bar{r},\bar{\rho}) = h(\bar{r})\ell(\bar{\rho}) \tag{17}$$

is adopted.

The effective water content θ is subsequently defined as

$$\theta = \Theta - \Theta_{min} \tag{18}$$

where Θ and Θ_{min} are the actual and the residual water content; and two integral distribution functions are defined as

$$L(\bar{R}) = \int_0^{\bar{R}} \ell(\bar{\rho})d\bar{\rho} \qquad H(\bar{R}) = \int_0^{\bar{R}} h(\bar{r})d\bar{r} \tag{19}$$

The effective water content in the main wetting proces $\theta_w(\bar{R})$ is determined by the integration of $f(\bar{r}, \bar{\rho})$ over the corresponding region of the filled pores [Figure 17(b)]:

$$\theta_w(R) = L(\bar{R})H(1) \tag{20}$$

and similarly, in the main drying process (Figure 17)

$$\theta_d(\bar{R}) = L(\bar{R}) + [L(1) - L(\bar{R})]H(\bar{R}) \tag{21}$$

Choosing $H(1) = 1$ and realizing that there is a unique relationship between $(\bar{R})$ and ψ, we may

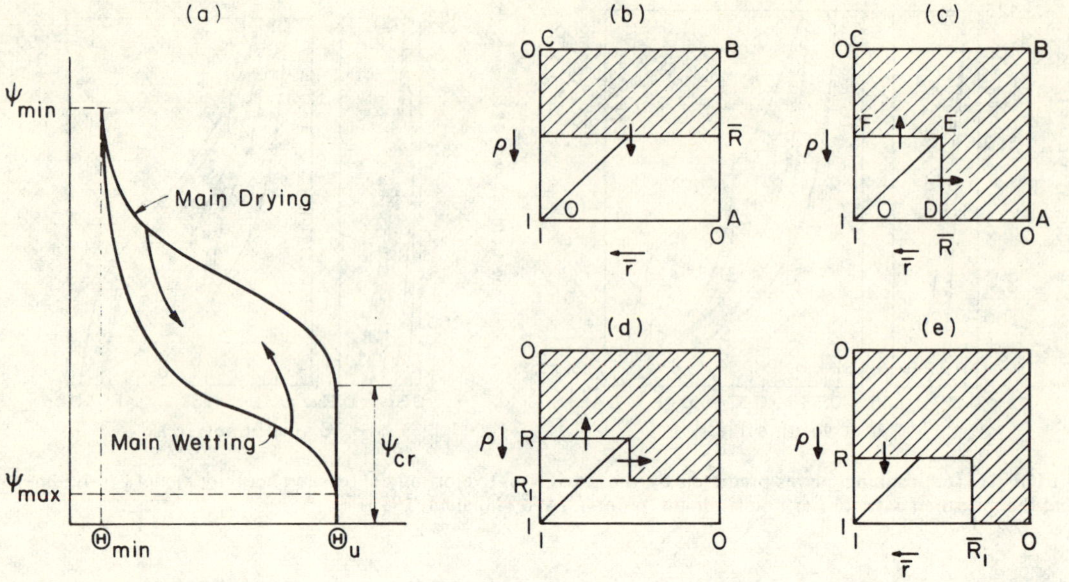

FIGURE 17. (a) Schematic representation of capillary hysteresis and the filled-pore diagrams in the $\bar{r}\rho$ plane (the shaded domans) for (b) the main wetting process, (c) the main drying process, (d) the primary drying process, and (e) the primary wetting process. (After Mualem, 1974)

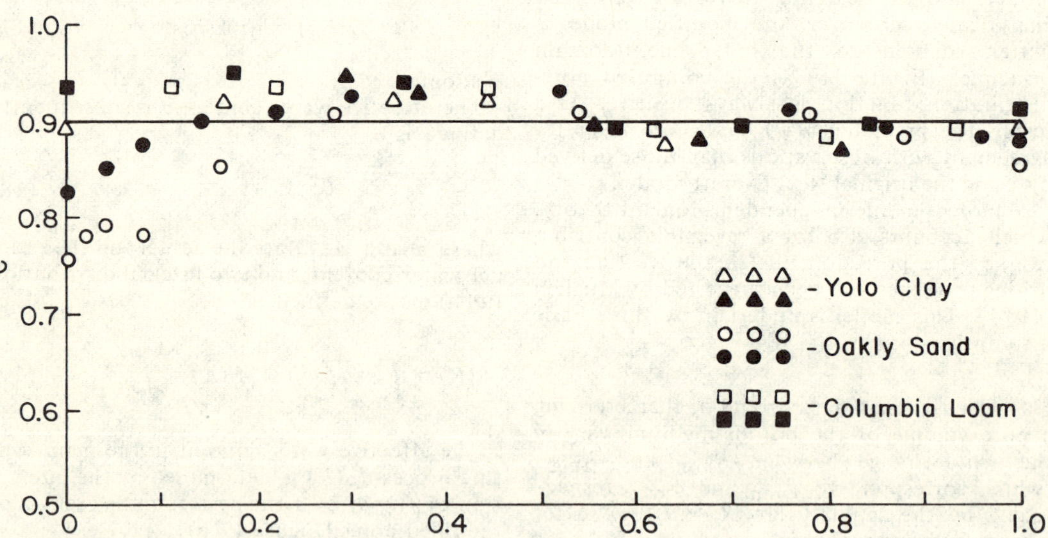

FIGURE 18. The function $C(S_e)$ for three soils measured after several cycles of imbibition and drying. The hollow symbols denote results in the wetting process, and the solid symbols denote measurements in the drying process. (After Mualem, 1974)

express H and L as functions depending directly on ψ by

$$L(\psi) = \theta_w(\psi) \quad (22)$$

$$H(\psi) = [\theta_d(\psi) - \theta_w(\psi)]/[\theta_u - \theta_w(\psi)] \quad (23)$$

Once $L(\psi)$ and $H(\psi)$ are determined from the two measured boundary curves of a given soil, any hysteretic path is described afterwards in a simple way with the aid of these two basic functions. For instance, the water content along a primary drying scanning curve [obtained by integrating $f(\bar{r}, \bar{\rho})$ over the region of filled pores of Figure 17(d) is given by

$$\theta\begin{pmatrix} & \psi_1 \\ \psi_{\min} & \psi \end{pmatrix} = \theta_w(\psi) + H(\psi)\,[L(\psi_1) - L(\psi)] \quad (24)$$

while the primary wetting scanning curve [Figure 17(e)] is given by

$$\theta\begin{pmatrix} \psi_{\max} & \psi \\ & \psi_1 \end{pmatrix} = \theta_w(\psi) + [\theta_u - L(\psi)]H(\psi_1) \quad (25)$$

This new model considerably facilitates the computational work as $H(\psi)$ and $L(\psi)$ are very simply related to $\theta(\psi)$ on the main boundary curve. Theory has been extended further to account for the entrapment of air. Using Cary's (1967) experiments, it was found that the entrapped air content constitutes a constant fraction of the water content. Figure 18 shows that the factor $c = \theta_w/(\theta_w + \theta_a)$, where θ_a is entrapped air content, is approximately 0.9. On this basis it was assumed that by rewetting a sample that was previously drained starting from saturation, a well-defined quantity of air is trapped. Moreover, it is assumed that subsequent drying and wetting cycles do not cause additional entrapment. This hypothesis permits predicting the first drying curve and any other scanning curves branching from it (Figure 19). The agreement between the forecasted $\theta(\psi)$ relationships and observations is quite satisfactory. It should be noted that this theory permits using the first drying curve and the main wetting curve, solely, for calibration of the model. This way, less experimental work is required while theory applies to the whole hysteretical domain.

In some cases, for soils having a narrow pore-size distribution with a major portion of the hysteretic loop in the range of air-entry value or when the soil is drained to a very low water content, the independent-domain models are not really satisfactory. In these cases the pore-interaction effects are significant. Everett

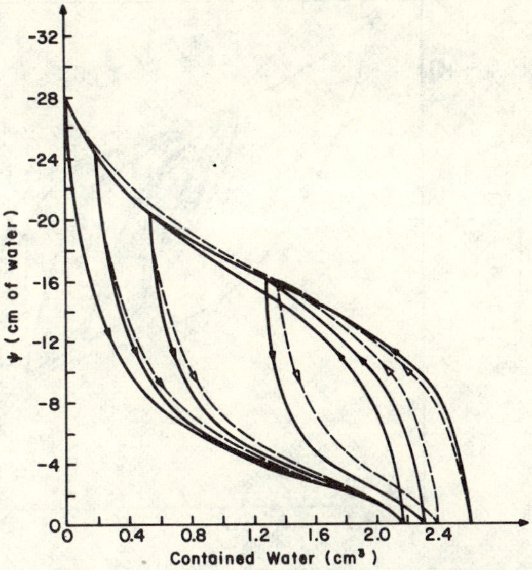

FIGURE 19. Measured hysteretic curves outside and inside the main loop (solid lines) and the predicted curves using model II (dashed lines) for glass-beads sample. (After Mualem, 1974)

(1967) and Topp (1971*b*) suggested a dependent-domain theory, which accounts for pore blockage against air entry near saturation and air blockage against water entry at low water content. While this theory leads to better agreement with observations, Topp's procedure is laborious and requires a large amount of measured data, comprising sets of both drying and wetting families of scanning curves. Poulovassilis and Childs (1971) suggested another dependent-domain model. This model, however, requires even more experimental data.

Recently Mualem (1974, 1976) and Mualem and Dagan (1975) have generalized Mualem's model previously discussed to incorporate the fundamental ideas of Everett and Topp. The computational procedure requires one primary drying scanning curve and one primary wetting scanning curve in addition to the boundary loop to determine the two functions corresponding to water and air blockage. They concluded that in most cases pore blockage against air entry is of lesser significance and can be disregarded to simplify computation. The generalized theory, however, significantly improves the prediction when the effect of the pore blockage against air entry is considerable (Figure 20). Thus the new models attain a high degree of generality and capability of representing observed phenomena more accurately.

Hysteretical phenomenon is observed also in

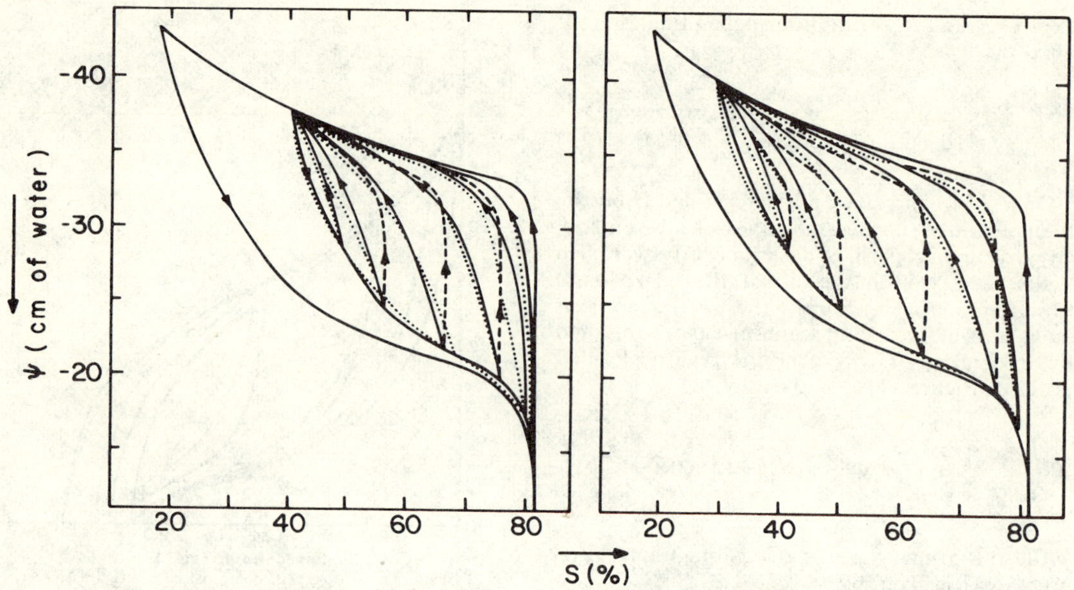

FIGURE 20. Two families of secondary drying scanning curves predicted with Model III (solid lines), computed with Model II (dashed lines), and measured (dotted lines) for the glass bead sample (after Mualem and Dagan, 1975).

the $K(\theta)$ and the $K(\psi)$ relationship. The subject is discussed under the title *Permeability*.

Y. MUALEM
H. J. MOREL-SEYTOUX

References

Bloomsburg, G. L., and Corey, A. T., 1964. Diffusion of entrapped air from porous media, *Hydrology Paper 5,* Colorado State University, Fort Collins.

Bomba, S. J., and Miller, E. E., 1967. *Secondary-scan Hysteresis in Glass-bead Media.* Preprint for presentation in the discussion section of the 1967 Meeting of the Soil Science Society of America, 62p.

Brooks, R. H., and Corey, A. T., 1964. Hydraulic properties of porous media, *Hydrology Paper 3,* Colorado State University, Fort Collins, 27p.

Bruce, R. R., 1972. Hydraulic conductivity evaluation of the soil profile from soil water retensions, *Soil Sci. Soc. America Proc.,* **36,** 555–560.

Cary, J. W., 1967. Experimental measurements of soil moisture hysteresis and entrapped air, *Soil Sci.,* **104,** 174–180.

Enderby, J. A., 1955. The domain model of hysteresis. 1. Independent domains, *Faraday Soc. Trans.,* **51,** 835–848.

Everett, D. H., 1954. A General approach to hysteresis. 3. A Formal treatment of the independent domain model of hysteresis. *Faraday Soc. Trans.,* **50,** 1077–1096.

Everett, D. H., 1955. A general approach to hysteresis. 4. An alternative formulation of the domain model, *Faraday Soc. Trans.,* **51,** 1551–1557.

Everett, D. H., 1967. Adsorption hysteresis, *in* E. A. Flood, ed., *Solid Gas Interface.* New York: Marcel Dekker, 1055–1113.

Everett, D. H., and Whitton, W. I., 1952. A general approach to hysteresis. 1, *Faraday Soc. Trans.,* **48,** 749–752.

Everett, D. H., and Smith, F. W., 1954. A general approach to hysteresis. 2. Development of the domain theory. *Faraday Soc. Trans.,* **50,** 187–197.

Gardner, W. R., 1959. Mathematics of isothermal water conduction in unsaturated soils, *Highway Res. Board, Nat. Res. Council, Rep. 40,* 78–82.

King, L. G., 1964. *Imbibition of fluids by porous solids.* Ph.D. Thesis, Colorado State University, Fort Collins, 231p.

Morel-Seytoux, H. J., 1969. Introduction to flow of immiscible liquids in porous media, *in* R. deWiest, ed., *Flow Through Porous Media.* New York: Academia, 455–516.

Morrow, N. R., and Harris, C. C., 1965. Capillary equilibrium in porous materials, *Soc. Petroleum Engineers Jour.,* **5,** 12–14.

Mualem, Y., 1973. Modified approach to capillary hysteresis based on a similarity hypothesis, *Water Resources Research,* **9**(5), 1324–1331.

Mualem, Y., 1974. A conceptual model of hysteresis, *Water Resources Research,* **10**(3), 514–520.

Mualem, Y., and Dagan, G., 1972. *Hysteresis in Unsaturated Porous Media: A Critical Review and New Simplified Approach, Rep. 4, Proj. A10-SWC-177.* Haifa: Technion-Israel Institute of Technology, 93p.

Mualem, Y., and Dagan, G., 1975. A dependent domain model of capillary hysteresis, *Water Resources Research,* **11**(3), 452–460.

Mualem, Y., 1976. Hysteretical models for prediction of the hydraulic conductivity of unsaturated porous media, *Water Resources Research,* **12**(6), 1248–1254.

Mualem, Y., and Dagan, G., 1976. Methods of predicting the hydraulic conductivity of unsaturated soils,

Project 442. Haifa: Technion–Israel Institute of Technology, 78p.
Néel, L., 1942. Théories des lois d'aimantation de Lord Rayleigh, 1, *Cahiers Phys.*, **12**, 1–20.
Néel, L., 1943. Théories des lois d'aimantation de Lord Rayleigh. 2, *Cahiers Phys.*, **13**, 19–30.
Philip, J. R., 1964. Similarity hypothesis for capillary hysteresis in porous materials, *Jour. Geophys. Research*, **69**(8), 1553–1562.
Poulovassilis, A., 1962. Hysteresis of pore water, an application of the concept of independent domains, *Soil Sci.*, **93**, 405–412.
Poulovassilis, A., 1970a. Hysteresis of pore water in granular porous bodies, *Soil Sci.*, **109**, 5–12.
Poulovassilis, A., 1970b. The effect of the entrapped air on the hysteresis curves of a porous body and its hydraulic conductivity, *Soil Sci.*, **109**, 154–162.
Poulovassilis, A., and Childs, E. E., 1971. The hystere- of pore water: The nonindependence of domain, *Soil Sci.*, **112**, 301–312.
Preisach, F., 1935. Uber die magnetische Nachwirkung, *Zeitschr. Physik*, **94**, 277–302.
Stallman, R. W., 1969. Multi-phase fluids in porous media–A review of theories pertinent to hydrological studies. *U.S. Geological Survey Prof. Paper 411-E*, 49p.
Staple, W. J., 1964. Moisture tension, diffusivity and conductivity of a loam soil during wetting and drying, *Canadian Jour. Soil Sci.*, **45**, 78–86.
Talsma, T., 1970. Hysteresis in two sands and the independent domain model, *Water Resources Research*, **6**(3), 964–970.
Topp, G. C., 1969. Soil water hysteresis measured in a sandy loam compared with the hysteretic domain model, *Soil Sci. Soc. America Proc.*, **33**, 645–651.
Topp, G. C., 1971a. Soil water hysteresis in silt loam and clay loam soils, *Water Resources Research*, **7**(4), 914–920.
Topp, G. C. 1971b. Soil-water hysteresis: The domain model theory extended to pore interaction conditions, *Soil Sci. Soc. America Proc.*, **35**, 219–225.
Topp, G. C., and Miller, E. E., 1966. Hysteretic moisture characteristics and hydraulic conductivities for glass-bead media, *Soil Sci. Soc. America Proc.*, **30**, 156–162.
Vachaud, G., 1966. Verification de la loi de Darcy generalisee et determination de la conductivite capilaire a partir d'une infiltration horizontal. Wageningen: AIHS-Symposium, 277–292.
Vachaud, G., and Thony, J. L., 1971. Hysteresis during infiltration and redistribution in a soil column at different initial water contents, *Water Resources Research*, **7**(1), 111–127.

Cross-references: *Conductivity, Hydraulic; Flow Theory; Imbibition; Infiltration; Percolation; Permeability; Pore Space, Drainable; Soil Structure; Water Content and Retention; Water Movement; Wetting Front.*

CARBON CYCLE

Carbon is the basic building block of all life on the planet Earth. The carbon cycle is a dominant feature of the biological world, and life processes are in turn the prominent agents of the geologic reactions involving carbon.

Overall Features

As with all elements, particularly those that are essential constituents of plants and animals, carbon is cycled through the various "compartments" of the earth in which it is distributed approximately as shown in Table 1. Excluding from consideration carbon in the unweathered crust of the earth, essentially all the carbon that has passed through the chemically and biologically active earth surface and atmosphere is now in sediments. The carbon cycle can therefore be further subdivided into the long-term processes that include sedimentary rocks and the short-term processes involving atmospheric carbon dioxide and the organic and inorganic carbon of the soil and oceans as mediated by biological and nonbiological reactions. Only that portion of the latter category of processes that involve soil carbon will be considered here.

Figure 1 outlines the soil carbon processes. Soil organic matter constitutes about four times the carbon content of the atmosphere reservoir and five or six times the land biomass (living plants and animals). The annual photosynthetic input of carbon to the soil is approximately 30 billion metric tons or about 4% of the atmosphere carbon dioxide. An equivalent quantity of carbon dioxide is delivered to the atmosphere from the soil by decay and other oxidative processes (not including an approximate 4.7 billion metric tons per year generated by fossil-fuel burning. A small quantity (about 2% of the photosynthetic input) finds its way to the sea, which in turn delivers carbon dioxide to the atmosphere or carbon to sediments. Thus the

TABLE 1. Distribution of Carbon in Some of the Main Compartments Involved in the Soil Carbon Cycle.

Compartment	Billions of Metric Tons of Carbon*
Atmosphere	700
Soil organic matter (nonliving)	2,500
Ocean organic matter (nonliving)	3,000
Land life forms	480
Ocean life forms	50
Dissolved carbonate-bicarbonate in oceans	38,400
Coal and petroleum	10,000
Sediments	60,000,000

*"Best" estimate based on the calculations of Bolin; Garrels, Mackenzie, and Hunt; and the author. Some of these are very uncertain as, for example, the estimate of biomass in the oceans, which is largely microflora in the ocean oozes and estimated with difficulty with little data.

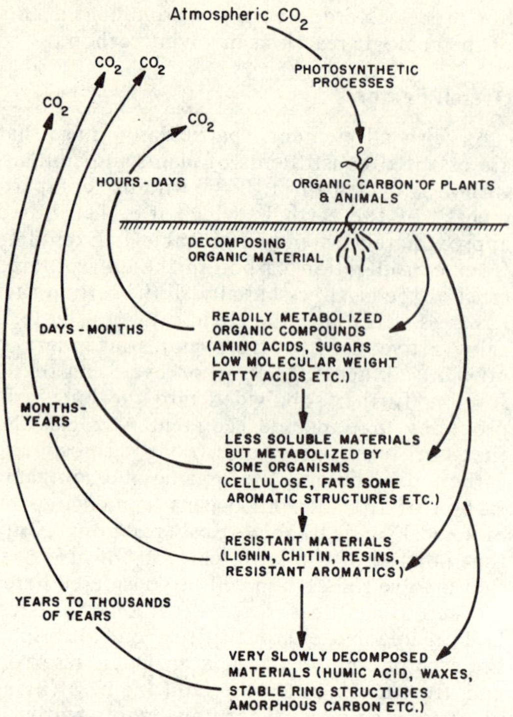

FIGURE 1. Schematic representation of overall carbon cycle. The vertical representation of compounds of increasing resistance to decomposition shows the nature of the degradation sequence and the tendency to enrich for resistant compounds but does not necessarily reflect the location of the various fractions in the soil profile.

organic constituents of the soil represent an approximate steady state, each constituent having a pool size reflecting an inverse relationship to the ease of its decomposition relative to its input rate.

The organic carbon content of the soil is largely a function of temperature and oxygen availability, ranging from soils that are predominantly carboniferous like the peat soils of waterlogged deltas or cool climates to those that are almost devoid of carbon in well-aerated tropical environments. In tropical environments, the carbon fixed by plants in the photosynthetic process, as it is returned to soil either as plant or animal residues, is rapidly reoxidized to carbon dioxide and returned to the atmosphere. In carboniferous soils decomposition is slow, and carbon-containing compounds become significant and sometimes dominant constituents of the soil. Decomposition is mainly microbial, and compounds such as readily soluble carbohydrates and amino acids are rapidly metabolized by the microflora of the soil whereas others such as lignin, keratin, highly saturated compounds, and some proteins or their derivatives tend to accumulate. These products of organic degradation all fall under the general heading of humus (q.v.), which cannot be defined in rigid chemical terms. From this group can be separated a few "nonhumic" materials, that is, materials that can be defined chemically (Figure 2) and "humic" substances, which are polymers of varied composition depending on their sources and formation conditions.

Humic Fractions

The definition and classification of the *humic fraction* of soil organic matter (q.v.), which is frequently the major portion thereof, is in a totally unsatisfactory state of disarray. This problem of definition and classification is not from lack of competent professional attention but rather the result of a remarkable complexity and variability. Classically, nomenclature and description have been based on arbitrary chemical extraction and precipitation procedures, one example of which is shown in Figure 3. No generalization can be made concerning the relative proportions of these fractions because they vary greatly from one soil to another. Spodosols, for example, have a larger extractable organic fraction than do soils containing *A*-horizon organic material.

Laboratory degradation of these biologically resistant materials yields considerable quantities of amino acids, amino sugars, phenolic compounds, and related compounds polymerized to more complex structures or combined with inorganic metals or compounds.

Influence on Soil Properties

These varied carbon compounds influence the association of clay minerals, the movement of ions through the soil, water-holding capacity, ion-exchange capacity, and other variables. Their relative concentration reflects a steady state in the decomposition process, but eventually all of them are subject to decomposition and thus the return of carbon dioxide to the atmosphere for further photosynthesis and further metabolic sequences.

Some of the more soluble intermediates of organic degradation tend to solubilize or chelate metallic ions (see *Metal-Complexing Macromolecules*) such as iron and accelerate their downward movement in the soil. As the organic material becomes oxidized, these chelated ions are precipitated so that in older soil profiles the zone of precipitation of iron may be characterized by a hardpan including ferric hydrate, clay particles, and some resistant organic materials. Frequently this layer is less permeable to water movement. It can be at varying depths from a meter or more below the soil surface.

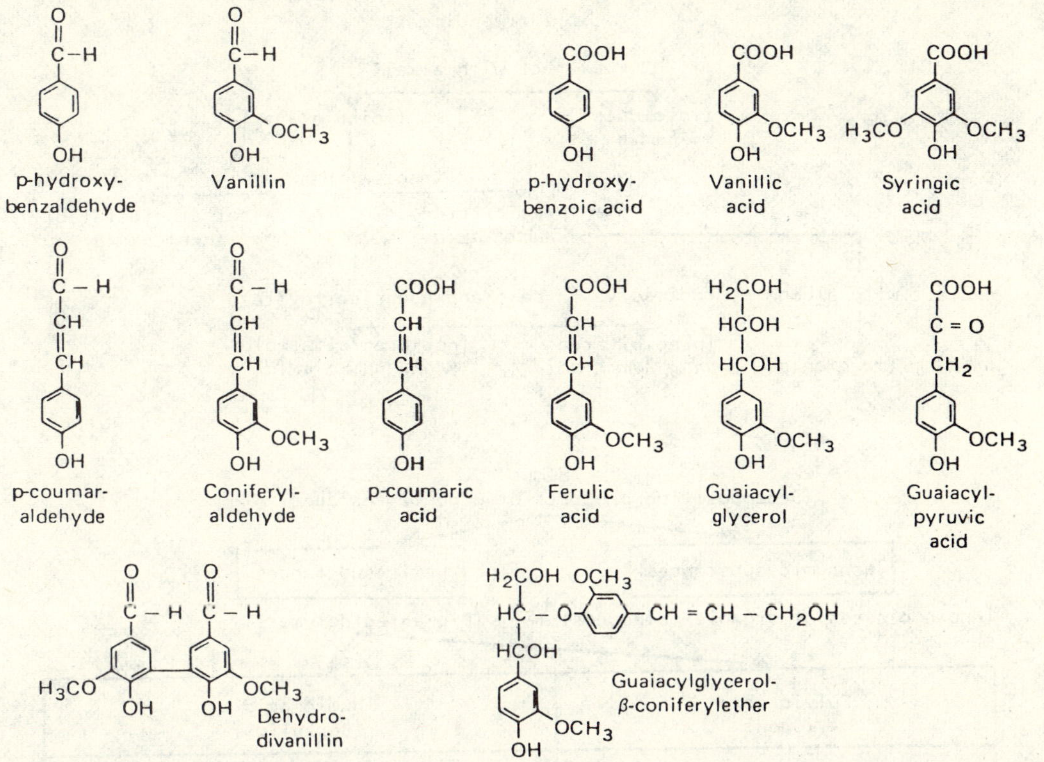

FIGURE 2. Examples of identifiable organic compounds found in the "humus" of the soil. (After Gieseking, 1975)

Climatic influences on rates of decomposition of organic matter result in a wide range of soil characteristics starting with essentially similar parent rocks. Many prairie soils (Mollisols) owe their good tillage characteristics and much of their moisture- and ion-retention capacities to an abundance of organic matter (6–8% in upper horizons decreasing to 1% or less at a depth of 1 m). Soils formed on similar parent material with similar rainfall in subarctic environments will have lower decomposition rates and may show strong leaching of the upper (*A*) horizon and an accumulation of organic materials in lower horizons (Spodosols). These are usually under forested areas, and vegetation type (also a function of climate) is partially responsible for these differences.

Peat Soils

Peat soils are an interesting extreme of the wide range of organic content of soils with the organic fraction sometimes constituting 90% or more of the soil mass. They are characteristic of flood plains or poorly drained basins where, because of poor aeration, decomposition has been anaerobic and has not kept pace with annual photosynthetic input. Their depths may range to 20 m or more, and they frequently provide excellent agricultural land. Their agricultural use requires drainage that increases aeration and accelerates microbial degradation. This releases mineral constituents of the parent organic materials, particularly nitrogen, contributing to soil fertility (q.v.). This slow microbial degradation results in a subsidence of the soil surface, increasing problems of drainage, and in the extreme forcing abandonment of the soil for agricultural use.

Although the degradation rate of soil organic materials varies greatly for different constituents, the process is essentially complete so that soils of great age, thousands or even hundreds of thousands of years, which may have experienced an input of several tons of organic material per square meter may contain only a few kilograms per square meter of organic carbon, this itself undergoing what Robert Frost has called "the slow smokeless burning of decay." Even amorphous carbon, although highly insoluble, does not persist indefinitely.

In any event, the organic composition of the soil influences the characteristics of that soil and in turn is influenced by other mineral and climatic variables. In the final analysis, the cycling of carbon from the atmosphere to the

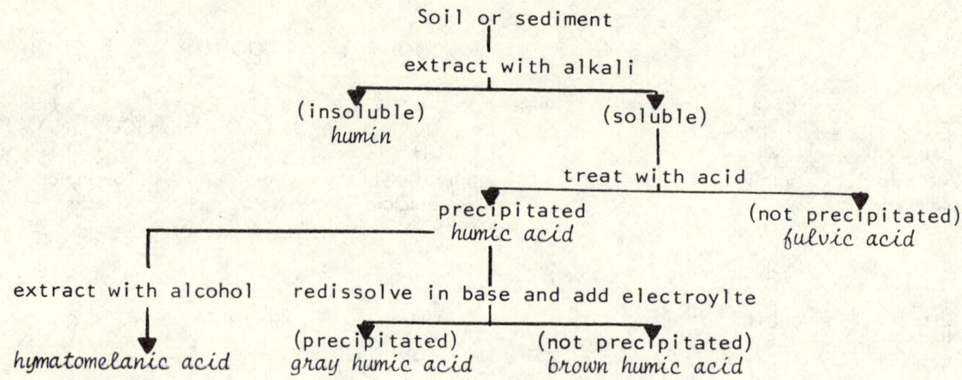

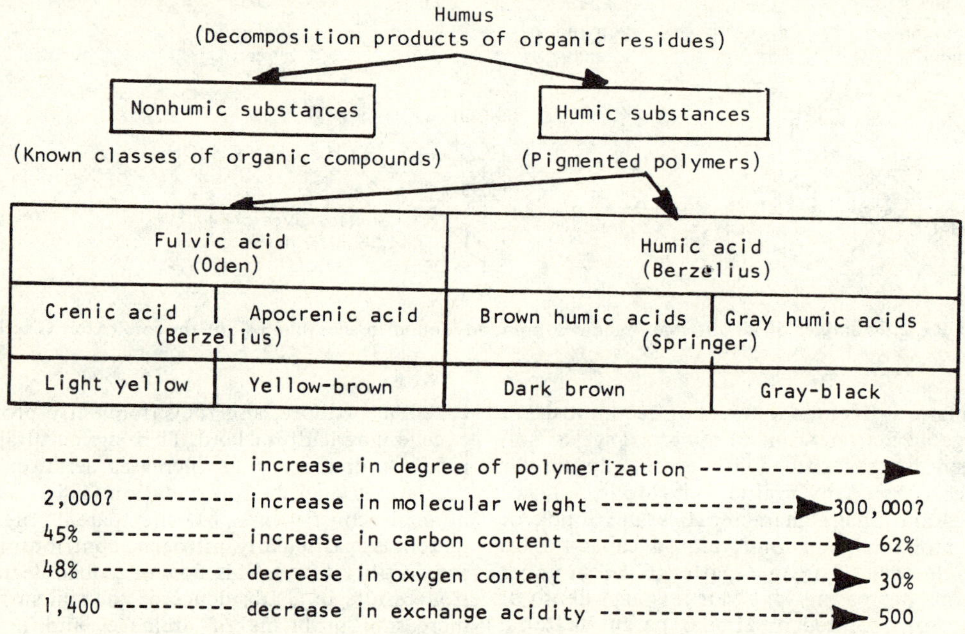

FIGURE 3. A classification of humic substances. (After Eglinton and Murphy, 1969)

soil and back is an almost steady-state situation with the residence time for a particular carbon atom in the soil ranging from a period of hours to thousands or tens of thousands of years for the carbon atom in more slowly decomposed organic structures. The mean time scale for the cycling of organic carbon, about 25 years, is more a function of temperature than any other variable.

C. C. DELWICHE

References

Berger, I. A., 1951. Chemical and structural relationship of lignin to humic substances, *Fuel*, **30**, 204–208.

Bolin, B., 1970. The carbon cycle, *Sci. Am.*, **223**(2), 124–132.

Eglington, G., and Murphy, M. T. J., eds., 1969. *Organic Geochemistry*. New York: Springer-Verlag, 828p.

Garrels, R. M.; Mackenzie, F. T.; and Hunt, C., 1975. *Chemical Cycles and the Global Environment*. Los Altos: Kaufmann, 206p.

Gieseking, J. E., ed., 1975. *Soil Components, Volume 1, Organic Components*. New York: Springer-Verlag, 534p.

Gray, T. R. G., and Parkinson, D., eds., 1968. *The Ecology of Soil Bacteria*. Toronto: University of Toronto Press, 681p.

Cross-references: *Exchange Phenomena; Humus; Metabolic Activity; Nitrogen Cycle; Organic Matter; Phosphorus Cycle; Soil Microbiology; Sulfur Transformations. vol. IVA, Carbon Cycle; Vol. XII, Weathering, Organic.*

CARBON-NITROGEN RATIO

A distinguishing characteristic of soils, sediments, and sedimentary rocks is the constancy of the *carbon-nitrogen* (C/N) *ratio.* For surface soils, and for the top layer of lake and marine sediments, the ratio generally falls within well-defined limits, usually from about 10 to 12 (Leighty and Shorey, 1930; Stevenson, 1959; Waksman, 1933). In most soils the C/N ratio decreases with increasing depth, often attaining values less than 5.

Decay of Organic Residues

The use of organic residues by soil organisms leads to incorporation of part of the carbon into microbial tissue, with the remainder being liberated as CO_2. As a general rule, about one-third of the applied carbon in fresh residues will remain in the soil after the first few months of decomposition. The decay process is accompanied by conversion of the organic form of nitrogen to ammonia and nitrate, and soil microorganisms use some of this nitrogen for synthesizing new cells. The nitrogen content of microbial tissue varies widely but will be of the order of 6 to 13% for bacteria and 3 to 6% for fungi. Thus the gradual transformation of plant raw material into stable organic matter (humus) leads to the establishment of a reasonably consistent relationship between carbon and nitrogen. Other factors that may be involved in the narrowing of the C/N ratio include chemical fixation of ammonia or amines by lignin-like substances (Mattson and Koutler-Andersson, 1943).

The C/N ratio of virgin soils formed under grass vegetation is normally lower than for soils formed under forest vegetation, and for the latter the C/N ratio of the humus layers is usually higher than for the mineral soil proper. Also the C/N ratio of a well-decomposed muck soil is lower than for a fibrous peat. As a general rule, conditions that encourage decomposition of organic matter result in the narrowing of the C/N ratio.

Fixed Ammonium

Traditionally, the C/N ratio has been assumed to be a characteristic of the indigenous organic matter. This concept is not strictly valid because a significant amount of the nitrogen in many fine-grained mineral samples occurs in an inorganic form, namely, as ammonium that is held within the lattice structures of clay minerals. Clays such as montmorillonite, illite, and vermiculite can hold considerable amounts of ammonium between the hexagonal layers formed by linking Si-O tetrahedra; the ammonium neutralizes negative charges arising from the substitution of aluminum for silica at the silica-oxygen tetrahedra. The unusually low C/N ratios reported for many subsoils now appears to be due to ammonium held by clays.

Table 1 gives the relationship between clay-bound (fixed) ammonium and the C/N ratio of several soils examined by Stevenson (1959). In the subsoils (*B* and *C* horizons) where the C/N ratio is low, rather high amounts of the nitrogen occur as fixed ammonium. An interesting point brought out by the data is that the *C/organic N ratio* also narrowed with increasing depth, a finding in agreement with Young's (1962) results. Possible reasons for apparent narrowing of the C/organic N ratio with depth include the following:

1. Underestimation of the amount of fixed ammonium in the samples. Major problems are encountered in estimating fixed ammonium in soil, and widely different values have been obtained when the various methods have been compared on a similar group of soils (e.g., see Bremner, Nelson, and Silva, 1967). Many of the earlier methods for determining fixed ammonium were found to give low results.
2. An increase in the relative amount of the soil organic matter that occurs in the form of amino acids. From time to time the suggestion has been made that the narrowing of the C/N ratio with depth is due to amino acids, but this hypothesis has not been supported by quantitative measurements suggesting that the percentage of the organic nitrogen as amino acid-N decreases rather than increaes with depth (Stevenson, 1959).
3. An increase with depth in the percentage of the organic nitrogen as amino sugars. Estimates for amino sugar-N have shown that from 5 to 10% of the nitrogen in the surface soil occurs in this form but that the percentage often increases with depth, occasionally reaching values as high as 20% in the *B* horizon (Sowden, 1959; Stevenson, 1957).
4. Eluviation of nitrogen-rich, humic-acid complexes. Mattson and Koutler-Andersson (1943) postulated that there was a relationship between the C/N ratio of the soil and the extent to which humic acids have been oxidized and enriched with nitrogen through autooxidative fixation of ammonia or amines. The enriched complexes were believed to be subject to leaching, thereby resulting in the narrowing of the ratio with depth. This theory has yet to be confirmed.
5. Occurrence of greater quantities of undecomposed plant residues (including roots) in the uppermost part of the solum. This explanation can be dismissed on the grounds that the apparent narrowing of the ratio with depth is due to organic substances associated with the clay fraction of the soil (Stevenson, 1959).

The C/N ratio of the soil usually decreases when soils are placed under cultivation, a result that can be partially explained on the basis that the percentage of the soil nitrogen as fixed ammonium increases slightly when soils are cropped.

TABLE 1. Carbon-Nitrogen Relationships in Some Terrestrial Soils.

Soil	Organic C (%)	Total N (%)	NH_4^+ - N ppm[a]	NH_4^+ - N % of N	$\frac{C}{N}$	$\frac{C}{\text{Organic N}}$
Elliott silt loam						
A_1	3.01	0.290	136	4.7	10.4	10.9
B_2	0.66	0.095	166	17.5	6.9	8.4
B_3	0.56	0.081	177	21.9	6.2	7.9
C_2	0.14	0.036	148	41.1	3.9	6.6
Blount silt loam						
A_1	2.28	0.205	123	6.0	11.1	11.8
B_1	0.47	0.071	148	20.9	6.6	8.4
C_1	0.29	0.053	186	35.1	5.5	8.4
Hagerstown silt loam						
A_1	1.16	0.122	113	9.3	9.5	10.5
B_1	0.47	0.067	124	18.5	7.0	8.6
B_3	0.35	0.062	108	17.4	5.6	6.8
Cisne silt loam						
A_1	1.42	0.163	66	4.1	8.7	9.1
B_2	0.21	0.031	58	18.7	6.8	8.3
B_3	0.38	0.064	91	14.2	5.9	6.9
C	0.19	0.033	93	28.2	5.8	8.0

After Stevenson, 1959.
[a]Includes trace amounts of nitrogen as "exchangeable" NH_4^+.

Marine and Lake Sediments

Deep marine and lake sediments often have very low ratios. In an early study Hall and Miller (1908) determined the C/N ratios of some deep clays and marls and reported that several specimens had C/N ratios less than 5.5; in one case a C/N ratio of 2.1 was reported. Mohamed (1949) found that the C/N ratios of some sediments from the Red Sea varied from 2.0 to 8.2 with a mean of 5.2. Arrhenius (1950) compiled existing information on the carbon and nitrogen contents of subaquatic sediments and found that in shelf sediments the C/N ratio attained values between 4.3 and 12.7. Waksman (1933) summarized information on the C/N ratios of marine sediments and concluded that the ratios ranged from 5.3 to 13.3.

In view of the preceding discussion on the C/N ratio of subsurface soils, it seems likely that the low ratios for marine and lake sediments are due to fixed ammonium. Obviously the C/N ratio cannot be considered a feature of the indigenous organic matter unless corrected for any fixed ammonium present.

Sediments and Sedimentary Rocks

The C/N ratios of sedimentary rocks vary considerably; both very high (>40) and very low (<5) values having been reported. Table 2 reproduces the results Stevenson (1962) obtained for six Paleozoic shales. For the three noncarbonaceous shales (those low in organic carbon), from one-half to two-thirds of the nitrogen occurred in the ammonium form. These shales had rather low C/N ratios (< 13). A considerably lower proportion of the nitrogen in the three shales high in organic matter occurred as ammonium, and they had somewhat higher C/N ratios (> 30). When the nitrogen values were corrected for the amounts of ammonium present, the resulting C/organic N ratios of the six shales were remarkably similar (34.9 to 43.5).

Attempts have been made from time to time to relate the C/N ratio of sediments and sedimentary rocks to such factors as source of organic matter, biological activity, depositional environment, and diagenesis processes. In as much as some of the nitrogen regarded as organic may exist as ammonium, these relationships require reexamination. For example, the low C/N ratios of certain sedimentary rocks have been attributed to selective preservation of nitrogen-rich organic compounds, whereas they could be a reflection of the dilution of the organic matter with clays containing high amounts of ammonium.

Summary

Notwithstanding the difficulties involved in interpreting published data on C/N ratios, certain conclusions seem warranted. The transformation of plant remains in soils and sediments leads to selective loss of carbon relative to nitrogen, and the residual material (humus) has a C/N ratio that varies somewhat but which is a distinctive environmental characteristic. The C/N ratio of the surface layers of soils

TABLE 2. Carbon-Nitrogen Relationships in Some Paleozoic Shales.

Sample No.	Organic C (%)	Total N (%)	NH_4^+ - N ppm[a]	NH_4^+ - N % of N	C/N	C/Organic N
1	0.63	0.061	499	73.1	9.4	36.8
2	0.64	0.051	344	67.5	12.6	38.6
3	1.05	0.081	521	64.3	13.0	36.3
4	5.31	0.173	236	13.6	30.7	35.5
5	13.84	0.372	539	14.5	37.2	43.5
6	12.08	0.403	564	14.0	30.0	34.9

After Stevenson, 1962.
[a]Includes trace amounts of nitrogen present "exchangeable" NH_4^+.

and sediments is to some extent a property of the so-called humic and fulvic acids (see *Humus*); for the lower horizons, the ratio may well reflect preservation of nitrogen-containing biochemical compounds that have been trapped or preserved within the expandable layers of clay minerals.

The C/N ratio of crop residues has long been of interest to soil scientists from the standpoint of the availability of soil and fertilizer nitrogen to plants. This aspect of the C/N ratio is covered in the section on the *Nitrogen Cycle*.

F. J. STEVENSON

References

Arrhenius, G., 1950. Carbon and nitrogen in subaquatic sediments, *Geochim. et Cosmochim. Acta*, **1**, 15–21.

Bremner, J. M.; Nelson, D. W.; and Silva. J. A., 1967. Comparison and evaluation of methods for determining fixed ammonium in soil, *Soil Sci. Soc. America Proc.*, **31**, 466–472.

Hall, D. A., and Miller, N. H. J., 1908. The nitrogen compounds of the fundamental rocks, *Jour. Agric. Sci.*, **2**, 343–345.

Leighty, W. R., and Shorey, E. C., 1930. Some carbon-nitrogen relations in soils, *Soil Sci.*, **30**, 257–266.

Mattson, S., and Koutler-Andersson, E., 1943. The acid-base condition in vegetation, litter, and humus: VIII, *Lantbruks Högskol. Ann.*, **12**, 70–100.

Mohamed, A. F., 1949. The distribution of organic matter in sediments from the northern Red Sea, *Am. Jour. Sci.*, **247**, 116–127.

Sowden, F. J., 1959. Investigations of the amounts of hexosamines found in various soils and methods for their determination, *Soil Sci.*, **88**, 138–143.

Stevenson, F. J., 1957. Investigations of aminopolysaccharides in soils, 2, *Soil Sci.*, **84**, 99–106.

Stevenson, F. J. 1959. Carbon-nitrogen relationships in soil, *Soil Sci.*, **88**, 201–208.

Stevenson, F. J., 1962. Chemical state of the nitrogen in rocks, *Geochim. et Cosmochim. Acta.*, **26**, 797–809.

Young, J. L., 1962. Inorganic soil nitrogen and carbon: Nitrogen ratios of some Pacific northwest soils, *Soil Sci.*, **93**, 397–404.

Waksman, S. A., 1933. On the distribution of organic matter in the sea bottom, and the chemical nature and origin of marine humus, *Soil Sci.*, **36**, 125–147.

Cross-references: *Carbon Cycle; Clay-Organic Reactions; Denitrification; Humus; Nitrogen Cycle; Organic Matter; Soil Components, Organic; Vol. IVA: Carbon/Nitrogen Ratio.*

CATION EXCHANGE—See EXCHANGE PHENOMENA

CATION EXCHANGE CAPACITY—See BASE SATURATION; EXCHANGE PHENOMENA

CHELATES, CHELATION—See CLAY-ORGANIC REACTIONS; METAL-COMPLEXING MACROMOLECULES. Vol. IVA: CHELATION

CHEMICAL COMPOSITION

As soil is a complex mixture of many components (See *Soil Components*) there is usually little value in determining the amount of a chemical element present without any indication of the fraction of the soil in which it occurs and its form of combination. Indeed, fractions that comprise only a small proportion of the total soil mass are often very important in determining its behavior. The following is a convenient classification of *soil fractions:*

1. The Organic Fraction
2. The Mineral Fraction
   a. Soluble in water: Simple inorganic ions
   b. Soluble in dilute acids: Carbonates
   c. Insoluble in dilute acids: Primary minerals, mainly occurring in sand and silt fractions; secondary minerals, usually occurring in the clay fraction (<2 μm diameter) but in strongly weathered soils forming larger, nondispersible aggregates and concretions.

Organic Fraction

Organic materials are added to soils as dead plant and animal remains. They are decomposed by the microflora and microfauna to form *humus* (q.v.), an amorphous material distinct from undecomposed litter (see *Organic Matter*). Well-humified organic matter contains about 58% carbon, so the amount of the soil organic fraction is usually specified by determining the organic carbon content and multiplying it by 1.73. Organic contents range from zero in some mineral subsoils, through 1 to 10% in arable topsoils, to nearly 100% (of the dry weight) in some peat and muck soils. The amounts in surface soils depend on the balance between accumulation and decomposition, and these processes in turn are influenced by temperature and moisture content: High temperatures and good aeration favor rapid decomposition, while low temperatures and waterlogging favor accumulation.

Apart from carbon, hydrogen, and oxygen, the organic fraction contains nitrogen, sulfur, and phosphorus. The proportions of these elements are often expressed as ratios compared to nitrogen taken as 10, and typical values are C:N = 80–150:10, S:N = 1.2–1.5:10, and P:N = 0.2–3.0:10. Metals such as aluminum, iron, manganese, and copper are also found in small amounts in humic complexes.

The organic compounds in humus are very different. The main portion appears to consist of polymers, some of which are formed by random condensation of phenols, amino acids, and other related microbial degradation products. A large number of compounds have been isolated from humus extracts, but many of these must be artifacts. Of particular interest, apart from the polyphenols, are amino acids (implying that humus contains protein), sugars (indicating carbohydrate fractions), and amino sugars. The sulfur seems to be part of the main humus fraction, probably as sulfur-containing amino acids and organic sulphates. In some soils much of the organic phosphorus is present as inositol polyphosphates, which appear not to be an integral part of the humus.

Water-soluble Components

The soluble-salt content of most soils is low so that the soil solution (q.v.) typically contains between 5 and 25 mmolc/l of calcium and magnesium salts, mainly as nitrate. In saline soils, however, the salt content is of the order of 100 mmole/l, and although still less than 1% of the soil mass, the soluble salts dominate the behavior of the soil and include also Na^+, Cl^-, HCO_3^-, and $SO_4^=$ ions.

The salt content is normally determined in a *saturation extract* prepared by wetting the soil until it is just saturated with water and filtering off the extract under reduced pressure. The filtrate may be analyzed chemically, but a rapid indication of the degree of salinity is given by measuring its *electrical conductivity* (q.v.). Conductivity values above 4 milliSiemens (mS), i.e. > 4 mmol/cm, indicate that crop production may be reduced by salt damage, while above 20 mS only salt-tolerant species can survive. The approximate conductivity at 25°C of a 100 mmol/l solution referred to above is 8–10 mS.

The "reaction" of soil (see *Soil Reaction*) is one of its most important diagnostic parameters. It is given by a pH measurement on the saturation extract or on a suspension of soil in water or in a dilute electrolyte solution. Strongly acid soils may have pH values down to 3.5, and strongly alkaline soils as high as 9.5, but more typical pH values of soils range from 5 to 8.

Carbonates

In soils formed from limestone rocks or other carbonate-containing sediments, carbonates occur mainly as calcite ($CaCO_3$) but sometimes as dolomite $(Ca,Mg)CO_3$. They are important in the buffer system (see *Buffers, Buffering*) that controls the pH and cation balance of soil, and also for their reactions with anions, particularly phosphate. In their reactions with anions, the particle size and surface area of the soil carbonates are more important than the amount.

Amounts of soil carbonate are estimated from the carbon dioxide evolved when the soil is treated with dilute acid

$$(CaCO_3 + 2H^+ \rightarrow Ca^{2+} + H_2O + CO_2)$$

the results being expressed as a percentage by weight of the soil.

In a leaching environment, soil carbonate is gradually removed by solution in carbonated water $[CaCO_3 + H_2O + CO_2 \rightarrow Ca(HCO_3)_2]$ so that topsoils contain less carbonate than subsoils or the parent material. The leached carbonate may be concentrated by chemical precipitation at depth in the soil profile.

Primary Minerals

Soil characterization includes analyses of particle-size distribution (q.v.), that is, the separation and determination of sand, silt, and clay fractions by sieving and sedimentation. Although this is not strictly chemical, some chemical properties of soil depend on the proportions of these size fractions. Further details of the topics mentioned here can be found in the relevant specialist articles (see *Mineral Analysis, Soil Mineralogy*), and in Bear (1964). Methods of determination are given in standard works (Jackson, 1958; Black, 1965; Hesse, 1971)

The mineral matter of soils is directly inherited from the parent material, although its composition is usually different depending on the age of the soil and the resistance of minerals to weathering. The minerals in the sand and silt fractions are mainly quartz and feldspars, plus a host of accessory minerals. Only the most resistant primary minerals remain in advanced stages of soil development, i.e., quartz (SiO_2) as the major component, with smaller amounts of heavy metal oxides such as hematite (Fe_2O_3), magnetite (Fe_3O_4), and rutile (TiO_2).

Secondary Minerals

The clay-sized (<2 μm effective diameter) fractions of many soils contain microcrystalline aluminosilicate layer minerals (see *Clay Minerals, Silicates*), and smaller amounts of hydrous oxides (q.v.) such as goethite, FeO.OH, and gibbsite, $Al(OH)_3$. The latter are more abundant in soils formed from basic rocks than from acidic rocks and are often amorphous; but in advanced stages of weathering, they crystallize and may also form concretions. The clay fractions of some soils such as those derived from volcanic ash consist almost wholly of amorphous aluminosilicates of variable composition (allophane), and similar material is associated with the surfaces of many crystalline particles. These amorphous fractions are particularly important in reactions with anions and humus polymers.

The nature of the clay fraction is reflected only partially in its chemical composition. The theoretical mole ratio SiO_2/Al_2O_3 of the type 1:1 Si:Al layer mineral kaolinite is 2, and of the type 2:1 mineral pyrophyllite is 4. But iron may proxy for aluminum in octahedral layers so that the ratio SiO_2/R_2O_3 (where R = Al + Fe) better reflects the mineral type. These ratios are also altered by aluminum substitution for silicon in the tetrahedral layer and by the presence of free aluminum and iron oxides/hydroxides, free silica, and allophane. However, the higher SiO_2/R_2O_3 ratios are commonly found for smectitic and illitic clays in weakly weathered environments, and the lower ratios for kaolinitic clays and hydroxides in strongly weathered environments. These generalizations are reflected in the composition of some subsoil clay fractions given in Table 1. Mollisols and Alfisols contain 2:1 Si:Al layer minerals, but with free sesquioxide (R_2O_3) in the Alfisol *B* horizon. Ultisols, Oxisols, and Spodosols are likely to contain 1:1 layer minerals plus increasing amounts of free sesquioxides. Inceptisols contain mainly allophane. Physical methods of analysis such as X-ray diffraction, thermal analysis, and electronoptical methods are necessary to identify the minerals present (see *Mineral Analysis*).

Selective dissolution techniques are used to estimate free aluminum and iron oxides/hydroxides, that is, those not associated with silicate mineral lattices. The iron compounds are dissolved effectively by the reducing action of sodium dithionite solution, and both aluminum and iron by acid ammonium oxalate. Pyrophosphate dissolves organically bound iron. An ill-defined "active" aluminum fraction

TABLE 1. Chemical Composition of Clay from *B* Horizons (Main Constituents Only)

Soil Type	Barnes Loam	Miami Silt Loam	Cecil Sandy Clay Loam	Columbiana Clay	Au Train Sand	Choyo Soil
Soil Order	Mollisol	Alfisol	Ultisol	Oxisol	Spodosol	Inceptisol
Percent (of ignited residue)						
SiO_2	55.43	48.5	39.7	32.2	25.1	30.2
Al_2O_3	21.9	27.6	38.3	44.3	60.8	45.8
Fe_2O_3	11.0	13.6	19.8	18.8	9.5	17.4
CaO	5.7	1.1	0.2	0.2	0.8	0.2
MgO	3.0	2.5	0.1	0.9	0.5	0.8
K_2O	1.6	5.2	0.3	0.1	1.0	0.6
Mole ratio						
SiO_2/Al_2O_3	4.29	3.41	1.76	1.24	0.88	1.12
SiO_2/R_2O_3	3.25	2.60	1.32	0.97	0.78	0.90

(Adapted from M. L. Jackson as cited in Bear, 1964)

can be removed by reagents such as salicylic acid. Amorphous aluminosilicate (allophane) is dissolved by sodium carbonate solution, and this fraction may also be detected by the release of hydroxyl when flouride reacts with the soil or clay.

Trace Elements

Some trace elements (q.v.) are biologically important, particularly manganese, cobalt, nickel, copper, zinc, molybdenum, boron, selenium, and iodine (see also *Micronutrients*). Iron is a trace element biologically, but it occurs in soils in major amounts (1–5%). Esimates of reactive fractions of these elements may be obtained by extracting the soil with reagents such as acetic acid, ammonium acetate, or ethylenediamine tetraacetic acid. The amounts of trace elements in these solutions are frequently determined by instrumental methods such as atomic-absorption spectroscopy, flame-emission spectroscopy, and solution electrochemical methods.

The total amounts of trace elements are mainly of interest geochemically and for mineral prospecting (see *Micronutrients*). A large number of elements can be determined simultaneously in one soil or rock sample using a multichannel-emission spectrometer. The amounts are much more variable than for the more abundant elements, and typical values for a number of trace elements are illustrated in Figure 1.

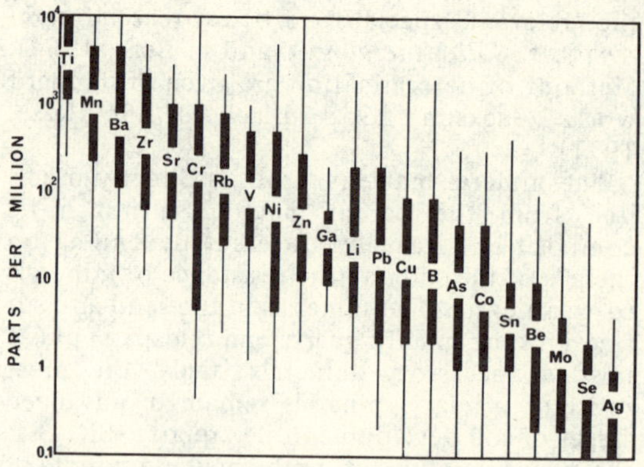

FIGURE 1. Range of contents of some trace elements commonly found in mineral soils. Thin lines indicate more unusual values; certain extremely high values from locations influenced by ore deposits have been ignored. (After R. L. Mitchell as cited in Bear, 1964)

Fertility Analysis

The main usefulness of soil is for growing crops, and the estimation of fertility is the underlying reason for much soil chemical analysis. Plants absorb nutrients in ionic form from soil solution so that the soil solution concentrations should give a good indications of nutrient supply (see *Plant Nutrients*). However, this is not necessarily so for nutrients present in low concentrations that are replenished by desorption from the solid matrix of the soil, for example, potassium, phosphate, and trace elements.

For cation nutrients (calcium, magnesium, and potassium), the exchangeable fraction gives a good indication of the nutrient potential (q.v.). Many empirical extractants have been used to dissolve phosphate from soils, and isotope-dilution methods have been used to measure labile surface fractions of phosphate and other nutrients. Recent work has shown the importance of the *adsorption isotherm* (i.e., the curve connecting the amount of labile solid phase nutrient with its chemical potential in solution) in dtermining nutrient supply to crops, and methods that bear some relationship to this are likely to be superior to those that do not.

Nitrate is a special case in that the whole of it is in solution, but its concentration fluctuates rapidly owing to absorption by roots and leaching. Humus provides a natural reserve because its slow decomposition releases ammonia, which is then oxidized to nitrate. Estimation of the longer term nitrate supply must therefore simulate these natural processes by incubation methods or by suitable chemical treatments.

BRYON W. BACHE

References

Bolt, G. H., and Bruggenwert, M. G. M., eds., 1976. *Soil Chemistry.* Amsterdam: Elsevier, 281p.

Bear. F. E., ed., 1964. *Chemistry of the Soil,* 2nd ed., ACS Monograph Series, vol. 160. Washington, D.C.: American Chemical Society, 515p.

Black, C. A., ed. 1965. *Methods of Soil Analysis.* Madison: American Society of Agronomy, 1580p.

Hesse, P. R., 1971. *A Textbook of Soil Chemical Analysis.* London: Murray, 520p.

Jackson, M. L., 1958. *Soil Chemical Analysis.* Englewood Cliffs, N.J.: Prentice-Hall, 498p.

Cross-references: *Buffers, Buffering: Carbon Cycle; Electrochemistry; Humus, Macronutrients; Micronutrients; Nitrogen Cycle; Phosphorus Cycle; Plant Nutrients; Polysaccharide; Radioisotopes; Soil Chemistry; Soil Components; Soil Solution; Sulfur Transformations; Trace Elements; Vol. IVB: Soil Minerals.*

CHEMISTRY OF SOILS—See ELECTROCHEMISTRY; PHYSICAL CHEMISTRY; SOIL CHEMISTRY

CLAY MINERAL DECOMPOSITION

Clay minerals in soils may be derived (inherited) from parent materials (i.e., they are "allochtonous" or "detritic") or they may be formed within the soil profile (see *Clay Mineral Formation*) by transformation (alteration) and neoformation (synthesis or authigenesis). Most mineral soils, however, contain inherited soil minerals (q.v. in Vol. IVB). The silicate clays, which form under conditions prevailing at the surface of the earth, may have originated under climatic conditions similar to or much different from those at the present. When environmental conditions differ from those of the original formation (due to climatic change, modification of soil drainage and aeration regimes, etc.), the clay minerals may become unstable and decompose.

Clay minerals commonly become unstable and hence decompose after a change that lowers or, less commonly, raises the activities of solutes in soils. The decomposition may be *incongruent dissolution,* i.e., breakdown combined with the formation of other crystalline or amorphous solids. The decomposition may *also* be *congruent dissolution,* i.e., breakdown and disappearance without leaving residues. The pathways of decomposition and the nature of residues depend on weathering conditions, including the milieu for reactions provided by the soils in which the minerals occur.

Processes of Clay Decomposition

Three major and two minor pathways of clay decomposition occur in soils but not all in the same soil simultaneously. The major pathways are: (a) *hydrolysis* by water containing carbon dioxide (Ollier, 1969); (b) *cheluviation,* or the dissolution of minerals and removal of aluminum and iron by chelating or complexing organic acids (Swindale and Jackson, 1956); and (c) *ferrolysis,* or the dissolution by a cyclic process depending on the alternate reduction of ferric oxides and oxidation of ferrous ions (Brinkman, 1970). The minor pathways are (a) dissolution by strong acids and (b) dissolution by strong alkalies. These are "minor" in that they affect a small proportion of the soils of the world. The modifier *strong* indicates soil reactions appreciably more acid or basic than they are for the major pathways. Neither heat nor oxidants or other strong reagents are considered because they rarely affect soils under natural conditions.

Clay decomposition may follow one or more major pathways in the same soil. Also, one pathway may be so dominant that the other pathways in the clay decomposition are of little consequence. Hydrolysis is the dominant process in well-drained soils of the humid tropics. It is also dominant in some situations in humid, temperate regions. Cheluviation is restricted to humid regions where the organic residues added to soils decay slowly enough to permit formation and persistence of organic acids. The pH is normally low as are the activities of calcium and magnesium in solution. Ferrolysis is restricted to seasonally wet soils in both temperate and tropical zones. This pathway requires that at least the surface horizon (or horizons) be subject to alternate oxidation and reduction in combination with some leaching.

Clay decomposition by one of the minor pathways occurs in soils of limited geographic areas. Dissolution by strong acids occurs most commonly in acid-sulfate soils, a large majority of which are derived from early Holocene sediments along seacoasts. After the soils are drained, substances such as pyrites oxidize to form the acids that attack clays. Strong acids are also added to soils in rain in some highly industrial regions and by local volcanic activity. Dissolution by strong alkalies occurs in sodic soils of arid regions. Some of these sodic soils are derived from sediments deposited in or near natron lakes.

Stability Diagram for Neutral to Acid Conditions

The clay decomposition processes under neutral and acid conditions can be summarized with the aid of a stability diagram (Figure 1). Experimental data do not yet cover all aspects of the model represented by the diagram; for example, the nature of aluminum chelates or of polynuclear aluminum ions in soils is poorly known. Figure 1 shows log (H_4SiO_4) (parentheses represent activities) on the abscissa, and log (Al^{3+}) + 3pH on the ordinate. The lower left-hand part represents soil solutions undersaturated with respect to the different minerals indicated; the lines represent solutions in equilibrium with a mineral; and their intersections represent solutions in equilibrium with two minerals. Regions above and to the right of the lines denote supersaturation with respect to one or more minerals. Lines continuing into such regions denote metastable equilibrium, which may govern clay decomposition where the approach to stable equilibrium is very slow.

The stability lines for most clay minerals differ with the activities of their constituent ions other than silicon or aluminum, as illustrated by the several parallel lines in the diagram for montmorillonite at different magnesium activities. The stability lines for well-crystallized

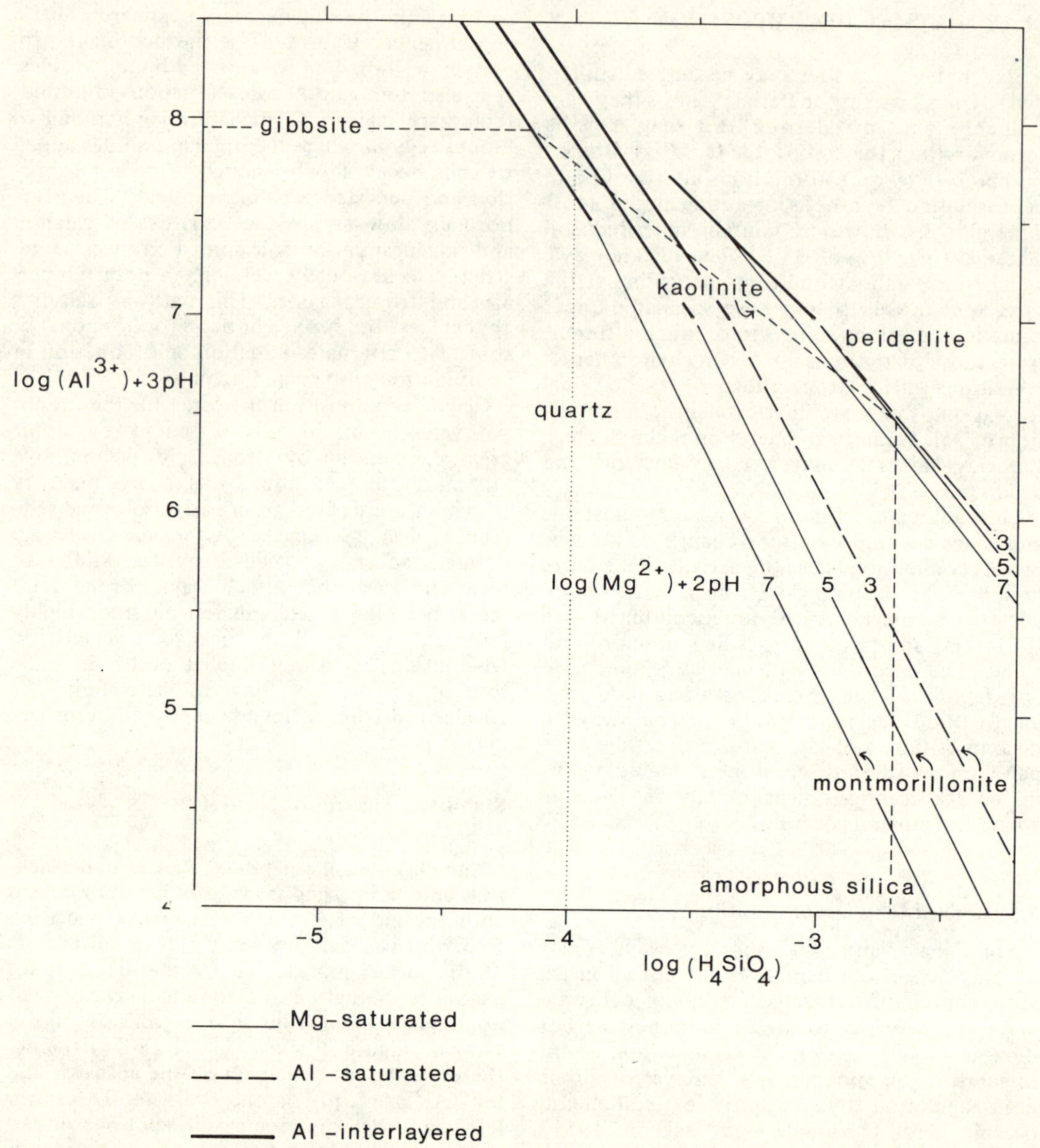

FIGURE 1. Stability diagram at 25°C and 1 bar total pressure. Equations for stability lines of gibbsite, kaolinite, beidellite, and quartz from Helgeson (1969); of amorphous silica from Stumm and Morgan (1970); and of montmorillonite from van Breemen, pers. comm. [Recalculated from Kittrick (1971) assuming the Belle Fourche montmorillonite was aluminum saturated at pH 2.7 to 3; log (Mg^{2+}) - 0.67 log (Al^{3+}) = 0.26 for aluminum-magnesium exchange; and log (Al^{3+}) + 3 pH = 7 for equilibrium between aluminum-saturated clay and aluminum-interlayered clay with interlayer composition $Al_2(OH)_5^+$]

gibbsite, kaolinite, and amorphous silica are constant.

The montmorillonite shown in the diagram is only an example. Stability lines for other smectites or for illites could have been shown, each with its own slope and with locations depending on its composition and on the activities of different combinations of ions such as Mg^{2+}, Fe^{2+}, or K^+. The beidellite described by Helgeson (1969), for example, has only a small range of stability, close to the intersection of the lines for kaolinite and amorphous silica (Figure 1). The slopes of the stability lines for most clay minerals are reasonably accurate, but their locations have various degrees of uncertainty.

Hydrolysis by Water and CO_2

Water in equilibrium with a CO_2 pressure of $10^{-3.5}$ bar (the normal atmospheric pressure of CO_2) has a pH near 5.6. Clay decomposition through hydrolysis depends on the nature of the minerals present, the pH, and the rates of reaction products' removal. Initial hydrolysis by such water of a magnesium-saturated smectite, for example, proceeds congruently and without producing a residue until enough aluminum has been dissolved to reach the solubility product of gibbsite, that is, log Al^{3+} + 3 pH (usually about 8). This represents an extremely low activity of Al^{3+}, on the order of 10^{-9} mol/l. If hydrolysis continues after that concentration is reached, further decomposition of smectite is accompanied by the formation of gibbsite. Activities of the silica and divalent cations dissolved from the smectite continue to rise in the soil solution. Essentially all iron released through hydrolysis remains as solid goethite or hematite. If leaching of the soil solution is rapid enough at this stage to prevent the activity of silica from reaching $10^{-4.15}$ mol/l (approximately 4 ppm of SiO_2), smectite and most other clay minerals would continue to decompose and leave residues enriched in gibbsite and iron oxides. Relative proportions of oxides of aluminum and iron would depend on the composition of the original clay minerals. Under these conditions, even quartz is unstable and may dissolve slowly. Clay decomposition of this general nature can occur in well-drained soils under perhumid and humid climates where the bulk of the plant residues decays on the soil surface and releases CO_2 to the atmosphere directly rather than within the soil.

If leaching is not rapid and minerals such as smectites, illites, and aluminum chlorites are present, silica activity may rise above $10^{-4.15}$ mol/l. Kaolinite then becomes stable over gibbsite, and the residue becomes the former mineral. Many soils in the humid tropics contain kaolinitic or ferruginous kaolinitic minerals of pedogenic origin. As the smectites or illites continue to decompose, silica activity will increase. An aluminum-interlayered smectite (an aluminum chlorite) will then become stable over kaolinite. Smectite will continue to decompose and aluminum chlorite to form until the first mineral has disappeared. So long as both minerals are present, the soil solution containing chiefly soluble SiO_2, some bases, and HCO_3 is near equilibrium with the two minerals. Pedogenic aluminum chlorites are common in many subtropical and tropical soils.

If kaolinite is present alone under the conditions described previously, it dissolves congruently, i.e., without leaving a residue. If silica activity exceeds 10^{-4} mol/l, quartz becomes a stable mineral and presumably could be formed in soils. However, quartz forms very slowly in most cases. Consequently, its presence has little effect on silica activity. That activity is controlled by the clay minerals (Marion et al, 1976) and amorphous silica if present.

Conditions of hydrolysis are different in well-drained soils of humid, temperate regions if a large proportion of the organic matter added as plant residues decays within the profile. The CO_2 pressure is increased to values of 10^{-2} or even 10^{-1} bar, and the soil solution becomes more acid (pH between 4.4 and 5). Initial dissolution of most clay minerals then proceeds congruently without leaving any residues. As the process of dissolution continues, the solubility product of kaolinite [log (Al^{3+}) + 3pH + log (H_4SiO_4) or about 3.8] is reached, and that mineral begins to form. Gibbsite is not formed under these conditions even if the removal of soil solution by leaching is rapid.

Total aluminum concentrations in solution may be considerably higher than Al^{3+} concentrations, particularly at high values for log (Al^{3+}) + 3pH near the gibbsite stability boundary. This condition may reflect equilibrium between Al^{3+} and soluble polymeric species as suggested by Richburg and Adams (1970), or it may be due to the presence of metastable polymeric species with lifetimes of several months such as those found in aqueous solutions by Turner (1968) and Smith (1971). Whatever the actual circumstances, they limit the formation of gibbsite from other clay minerals to CO_2 pressures of less than 10^{-2} bar.

Cheluviation

In solutions of chelating or complexing acids such as fulvic acids, salicylic acid, and p-hydroxy benzoic acid, and at a pH near 4, virtually all trivalent iron and aluminum are in chelated or complexed form. Concentrations of the two elements in those forms are several orders of magnitude higher than those of Al^{3+} and Fe^{3+}. Hence decomposition of kaolinite or gibbsite by a chelating acid at a pH about 4 is roughly equivalent to that of a nonchelating acid at pH 3.

Chelation of potassium and divalent cations such as magnesium is negligible at low pH values. Smectites and illite contain relatively less aluminum and more silicon, magnesium, and potassium than kaolinite. As a result, stabilities of smectites and illite are affected less by chelating anions than the stability of kaolinite. Small quantities of smectite are in fact present in the eluvial but not the deeper horizons of many podzols. Aluminum chlorite, kaolinite, illite, and other clay minerals are common in the deeper horizons.

Chelated metals normally move short distances in soils. They are generally precipitated in an accumulation horizon where part of the organic acids are oxidized by microbial action. This results in locally high Al^{3+} activity (still at relatively low pH) and stability of aluminum chlorite or aluminum-interlayered smectite. Such minerals as aluminum and, locally, iron oxides are common in the *B* horizons of podzols.

Ferrolysis

Clay decomposition by ferrolysis consists of two alternating phases: one occurring under reducing conditions with a slowly rising pH and the other under oxidizing conditions at low pH. During the latter phase, exchangeable ferrous iron is oxidized, producing ferric oxides and exchangeable hydrogen. This attacks the clay minerals as do strong acids, but the aluminum and other cations in the clay structure become exchangeable rather than soluble, while the corresponding silica may remain in the form of unsupported lattice edges of clay minerals. During the reducing phase, ferric oxides are reduced to ferrous ions, which displace exchangeable aluminum and other cations into the soil solution. Part of the silica in unsupported edges dissolves as well. The dissolved silica, some ferrous iron, displaced basic cations, and part of the displaced aluminum together with dissolved HCO_3^- may be removed by leaching in this phase.

Nonequilibrium concentrations of aluminum may be removed in solution under conditions of rising pH in the form of low-molecular-weight polynuclear ions with a lifetime of several weeks or months (Smith, 1971). Part of the aluminum is not leached out but is polymerized in the interlayer spaces of swelling clay minerals, forming incomplete octahedral sheets. Thus the originally swelling minerals are brought to a fixed 1.4 nm basal spacing, forming soil chlorites. Unlike aluminum chlorites formed in well-drained soils, those formed by ferrolysis contain various amounts of trapped, originally exchangeable, ferrous ions in their interlayers, which might increase their stability as compared to kaolinite.

Decomposition by Strong Acids

Decomposition of clay minerals by strong acids at pH values as low as 2.5 to 3.5 is initially congruent without leaving any residue. Dissolution continues to be congruent until the activity of dissolved silica reaches $10^{-2.7}$ mol/l (about 120 mg SiO_2/l). If leaching is rapid enough to keep activity below that level, dissolution continues to be congruent. On the other hand, if dissolution continues and activity exceeds that level, the residue first formed is amorphous silica in contrast to the gibbsite or kaolinite (and iron oxides) formed through hydrolysis. If leaching is very slow, the activities of aluminum and other cations in solution continue to rise until equilibrium is reached with the clay mineral, at which point dissolution stops. Where sulfuric acid is the agent of decomposition, the formation of $AlOHSO_4$ may limit the activity of aluminum (van Breemen, 1973). For example, log (Al^{3+}) + 3pH is limited to 6.3 when the activity of $SO_4{}^{2-}$ is about $10^{-2.5}$ mol/l.

Under very acid conditions the relative stabilities of clay minerals depend on activities of divalent cations and potassium. Several smectites and illites are stable over kaolinite at high activities and less stable at low activities of those basic cations. There is consequently no single weathering sequence for clay minerals under very acid soil conditions.

Decomposition by Strong Alkalies

Dissolution of clay minerals at high pH values and high activities of Na^+ and HCO_3^- seems to result in high concentrations of SiO_2 and aluminum in solution. The main aluminum species in solution is tetrahedral $Al(OH)^-_4$ in contrast to the octahedral (hexacoordinated) aluminum in clay minerals and in soil solutions of lower pH. Becuase of the strongly alkaline conditions and the presence of much sodium in solution, the common clay minerals are unstable and tend to dissolve with the formation of analcime ($NaAlSi_2O_6 \cdot H_2O$), a sodium zeolite (Schultz, Overstreet, and Barshad, 1964; Baldar and Whittig, 1968). Calculations based on data from Thompson (1973) suggest that analcime is more stable over kaolinite in the presence of amorphous silica when log (NA^+) + pH exceeds 8.25.

Decomposition Sequences

The list of clay minerals stable at intervals between near-neutral and extremely acid conditions (Figure 1) differs with changes in activities of divalent cations (especially magnesium) and potassium. With low activities of these ions, i.e., low-activity products of metal ions and OH^-, expressed as log (Mg^{2+}) + 2pH and log (K^+) + pH, the list from extremely acid to near-neutral, or from high to low log (H_4SiO_4), and concurrently from low to high log (Al^{3+}) + 3pH, may read as follows: amorphous silica, aluminum-saturated beidellite-type smectite, aluminum-interlayered smectite or soil chlorite (aluminum chlorite), kaolinite, gibbsite. With higher potassium and calcium activities, this list may change to: amorphous silica, calcium-saturated illite, aluminum-saturated illite, possibly aluminum-interlayered vermiculite, kaolinite, gibbsite. With high magnesium and calcium activities, the list may be: amorphous silica, calcium-saturated montmorillonite, possibly aluminum-saturated montmorillonite, aluminum-interlayered montmorillonite or soil chlorite (aluminum chlorite), kaolinite, gibbsite. The ranges of log (Al^{3+}) + 3pH and log (H_4SiO_4) over which each of these minerals is stable (or dissolves congruently) differ with the activities of basic cations in solution and with the composition of the clay minerals.

Clay decomposition in soils under conditions greatly different from those under which a given mineral is stable appears to proceed generally by a series of near-equilibrium, incongruent dissolution steps. The mineral dissolves with production of the adjacent mineral in the stability diagram, while the soil solution remains near equilibrium between the two. After completion of this dissolution, the composition of the percolating soil solution changes to near-equilibrium between the mineral formed, which is now slowly dissolved, and the next mineral. Decomposition would thus proceed until a mineral is formed that is stable over all others under the circumstances. This may then dissolve congruently (extremely slowly in many cases) with further leaching. A calcium-smectite under extremely acid conditions might produce the following decomposition sequence: caldium-smectite → aluminum-smectite → amorphous silica. Under near-neutral conditions the sequence might be: calcium smectite → aluminum-interlayered smectite → kaolinite → gibbsite.

Inasmuch as a soil generally contains pores of various diameters, leaching rates may differ over short distances (millimeters). One soil sample may thus contain minerals from several steps of a decomposition sequence.

The concepts of "weathering sequence" and "weathering stage" (Jackson and Sherman, 1953) originate from mineral sequences observed in nature mainly under one decomposition process: hydrolysis with water at low activities of carbonic acid. The fact that a decomposition sequence may differ with weathering conditions, e.g., the presence of dilute or strong acids or of chelating acids, makes it desirable to specify details of the weathering system, as, for example, hydrolysis or acidolysis (Hénin, Pedro, and Robert, 1968; Pedro, Jamagne, and Begon, 1969) as well as the soil mineralogy.

Rates of Clay Decomposition

The rates of clay decomposition in soil appear to be generally dependent on the rate of leaching, or solution renewal, and on the concentrations in the soil solution in equilibrium with dissolving and newly formed (clay) minerals rather than on dissolution kinetics. These may be first order in many cases and diffusion controlled through a layer of newly formed material in some incongruent dissolutions. Even at leaching rates of some 2000 mm/year, contact times of the soil solution with the clay minerals are of the order of a week per 10 cm thickness of soil material, and of one or two months over the thickness of a soil profile. These rates are sufficient to achieve near equilibrium in water or organic acids at the high solid-solution ratios prevailing in soils. Huang and Keller (1971, 1973), for example, estimated dissolution rates of the order of 10^{-3} to 10^{-2}% per day for clay minerals in organic acids and of the order of 10^{-3}% per day for silica from clay minerals in water.

Hydrolysis with production of residual gibbsite or kaolinite typically results in soil solutions with concentrations of the order of 1 to 5 mg SiO_2/l, roughly equivalent to a loss of 4 mm soil material per 1000 years under 2000 mm annual percolation, derived from decomposition of 1 to 10 mg clay mineral per liter of percolating water. Concentrations of aluminum and iron in solution are some orders of magnitude smaller than 1 mg/l. Similar concentrations, resulting from the decomposition of 1 to 5 mg clay per liter of percolating water, were derived by Pedro (1970) and Pedro et al (1970) from laboratory experiments at 65°C with strongly accelerated leaching.

Decomposition of clays by chelating acids may give rise to soil solutions containing between 5 and 100 or more mg SiO_2/l and similar total concentrations of various species of aluminum and iron (Huang and Keller, 1971). This would be roughly equivalent to the loss of 10 to 200 mm of soil material per 1000 years with 2000 mm of annual percolation. In strong mineral acids, silica concentrations are about 100 ppm. Concentrations of aluminum of all species are usually similar although they may be up to an order of magnitude higher (van Breemen, 1973). When decomposition of clay is due to strong mineral or chelating acids, some SiO_2 remains as a residue, but removal of silica is still greater than that during hydrolysis. The supply of reacting acid and the leaching rate appear to be important to the rate of clay decomposition in these cases.

In ferrolysis, the leaching rate and either the quantity of iron reduced in the wet season or the cation exchange capacity seem to determine the rate of clay decomposition because the amount of exchangeable hydrogen formed by oxidation is equivalent to the exchangeable ferrous iron minus the bicarbonate not yet leached out. This crude estimate leads to maximum possible rates of clay decomposition on the order of 0.1% per year, roughly equivalent to loss of 200 mm soil material per 1000 years if the layer subject to ferrolysis is 20 cm thick. Actual weathering rates may well be about an order of magnitude smaller than the indicated maxima for several reasons. Annual leaching rates are more commonly several hundred mm rather than 2000 mm; water normally does not move uniformly through the soil mass; cation exchange does not proceed with 100% efficiency; and the cation-exchange capacity decreases as ferrolysis progresses.

ROBERT BRINKMAN
(manuscript received 1976)

References

Baldar, N. A., and Whittig, L. D., 1968. Occurrence and synthesis of soil zeolites, *Soil Sci. Soc. America Proc.*, **32**, 235–238.

Breemen, N. van, 1973. Dissolved aluminum in acid sulfate soils and in acid mine waters, *Soil Sci. Soc. America Proc.*, **37**, 694–697.

Brinkman, R., 1970. Ferrolysis, a hydromorphic soil forming process, *Geoderma*, **3**, 199–206.

Helgeson, H. C., 1969. Thermodynamics of hydrothermal systems at elevated temperatures and pressures, *Am. Jour. Sci.*, **267**, 729–804.

Hénin, S.; Pedro, G.; and Robert, M., 1968. Considerations on the ideas of stability and instability of minerals as a function of environmental conditions; tentative classification of "agression systems," *9th Internat. Congr. Soil Sci. Trans. (Adelaide)*, **3**, 79–90. (in French)

Huang, W. H., and Keller, W. D., 1971. Dissolution of clay minerals in dilute organic acids at room temperature, *Am. Mineralogist*, **56**, 1083–1095.

Huang, W. H., and Keller, W. D., 1973. Kinetics and mechanism of dissolution of Fithian illite in two complexing organic acids, *Proc. Internat. Clay Conf. (Madrid)*, 321–331.

Jackson, M. L., and Sherman, G. D., 1953. Chemical weathering of minerals in soils, *Adv. Agron.*, **5**, 219–318.

Kittrick, J. A., 1971. Stability of montmorillonites: I. Belle Fourche and Clay Spur montmorillonites, *Soil Sci. Soc. America Proc.*, **35**, 140–145.

Marion, G. M.; Hendricks, D. M.; Dutt, G. R.; and Fuller, W. H., 1976. Aluminum and silica solubility in soils, *Soil Sci.*, **121**, 76–85.

Ollier, C., 1969. *Weathering.* Edinburgh: Oliver & Boyd, 304p.

Pedro, G., 1970. Alteration of calcareous materials in "laterizing" conditions: Experimental study of the evolution of an illitic marl, *C. R. Acad. Sci. Paris, Serie D*, **270**, 36–38. (in French)

Pedro, G.; Berrier, J.; and Tessier, D., 1970. Experiments on "allitic" weathering of kaolinitic and illitic dioctahedral clays, *Bull. Groupe Français Argiles*, **22**, 29–50. (in French)

Pedro, G.; Jamagne, M.; and Begon, J. C., 1969. Mineral interactions and transformations in relation to pedogenesis during the Quaternary, *Soil Sci.*, **107**, 462–469.

Richburg, J. S., and Adams, F., 1970. Solubility and hydrolysis of aluminum in soil solutions and saturated-paste extracts, *Soil Sci. Soc. America Proc.*, **34**, 728–734.

Schultz, R. K.; Overstreet, R.; and Barshad, I., 1964. Some unusual ionic exchange properties of sodium in certain salt-affected soils, *Soil Sci.*, **99**, 161–165.

Smith, R. W., 1971. Relations among equilibrium and nonequilibrium aqueous species of aluminum hydroxy complexes, *in Non-equilibrium Systems in Water Chemistry.* American Chemical Society Advanced Chemistry Series, **106**, 250–279.

Stumm, W., and Morgan, J. J., 1970. *Aquatic Chemistry.* New York: Wiley, 583p.

Swindale, L. D., and Jackson, M. L., 1956. Genetic processes in some residual podzolized soils of New Zealand, *6th Internat. Congr. Soil Sci. Trans. (Paris)*, **E**, 233–239.

Thompson, A. B., 1973. Analcime: free energy from hydrothermal data. Implications for phase equilibria and thermodynamic quantities for phases in $NaAlO_2$-SiO_2-H_2O. *Amer. Mineral.*, **58**, 277–286.

Turner, R. C., 1968. Conditions in solution during the formation of gibbsite in dilute Al salt solutions: 2. Effect of length of time of reaction on the formation of polynuclear hydroxyaluminum cations, the substitutions of other cations for OH^- in amorphous $Al(OH)_3$, and the crystallization of gibbsite, *Soil Sci.*, **106**, 338–344.

Cross-references: *Clay Mineral Formation; Clay Mineralogy; Clay Minerals, Silicates; Clays, Physical Chemistry; Exchange Phenomena; Ionic Activities; Leaching; Metal-complexing Macromolecules; Redox Reactions; Vol. IVA: Ion Exchange; Weathering, Chemical; Vol. IVB: Clays and Clay Minerals; Vol. VIA: Clay: Diagenesis; Vol. XII: Weathering and Soil Formation.*

CLAY MINERAL FORMATION

Clay is that size fraction of the soil that consists of particles of less then 2 μm equivalent spherical diameter. The minerals that usually predominate in the clay fraction are termed *clay minerals* (see *Clay Minerals, Silicates*). Clay minerals are hydrous silicates or alumosilicates, generally secondary, and they commonly form in nature by the alteration or weathering of primary minerals or crystallization from solutions. The occurrence of clay minerals in soils is due to one of the following three major processes: *inheritance, transformation,* or *neoformation.*

Inheritance

Clay minerals that form part of the parent materials of soils are passed over to the soil. If this transition does not involve transformations in their structure, the clay minerals are said to have been "inherited" (or to be "allochtonous," or "detritic"). Clay minerals present in soils formed on argillaceous parent materials like alluvial or loessial deposits, marls, shales, and clay stone are frequently inherited. Inheritance is also possible when only a fraction of the parent material consists of clay minerals like limestone. Inheritance is less frequent when the parent materials are igneous rocks. In tropical regions where a thick weathering zone (saprolite) on igneous rocks is frequent, soils may inherit the clay minerals present in that zone.

Inheritance depends on the relative stability in the *pedosphere* of the clay minerals that were passed over. Unstability results in transformation of the clay minerals into a new phase, more at equilibrium with the new conditions, or even their total disintegration. Strictly speaking, in this sense there is no real inheritance, and all clay minerals have been somewhat transformed during their transition from parent material to solum. For simplicity sake, however, minor changes are disregarded. In well-drained environments of the subhumid Mediterranean zone, for example, the smectite-kaolinite ratio in the insoluble residue of limestone is maintained in the clay fractions inherited by soils formed on that parent material. Palygorskite, on the other hand, which is much less stable and which is present in some of these limestones, disappears from the clay fractions of the soils formed, possibly having been transformed into some other clay mineral (Yaalon et al, 1966).

Transformation

Transformations are the changes that modify clay minerals or related alumosilicates, without altering the basic layer structure, and that occur during the transition parent material-soil. Alterations involving the transition of *diphormic* (1:1) into *triphormic* (2:1) *minerals,* or vice versa—for example, kaolinite ⇆ smectite transitions—could be included in this group, provided it is accepted that whole octahedral or tetrahedral layers are used in the change. If, on the contrary, it is believed that dissolved phases are indispensable in their alteration, they also have to be classified with the neoformations.

One distinguishes between *degradational transformations,* in the course of which the alteration involves loss of substances, and *aggradational transformations,* during which substances are added (Millot, 1964). Some examples for triphormic transformations include the following:

degradation →

illite ⇌ interstratified I-V ⇌ vermiculite
⇌ interstratified V-S ⇌ smectite
biotite ⇌ hydrobiotite ⇌ vermiculite (tri)
⇌ smectite (tri)
biotite ⇌ chlorite ⇌ interstratified C-V
⇌ vermiculite (di) ⇌ (smectite
(di)

← aggradation

Degradation. Degradation is the most common transformation in soils. The transformation illite → interstratified → vermiculite → smectite, involves opening of the structure, increases in specific surface, increases in exchange capacity and water content, and decreases in potassium content. It has been described by Jackson (1968) for many soils in which illite alteration occurs. The transformation biotite → interstratified → vermiculite → smectite is accompanied, in addition, by the oxidation of Fe^{++} to Fe^{3+} and by the migration of octahedral Mg^{++} into interlayer positions. Although some of these transformations involve more deep-seated alterations in the lattice composition, they can probably proceed without the intervention of a solution stage. Researches carried out on Scottish soils, for example, followed the full transformation sequence of biotite from various igneous rocks, through hydrobiotite interstages into vermiculite and occasionally also into smectite in the soils (Walker, MacEwan, and Wilson, cited by Norrish, 1972).

Aggradation. Aggradational transformations, characterized by the addition of elements to the mineral structure, are common in sedimentary environments. In the pedosphere they are rarer. These transformations are produced in a confined environment where leaching is not pronounced. Aggradational transformations of smectite into illite have been observed in loess-derived soils in Germany and Prairie soils in Canada (St. Arnaud and Mortland, cited by Niederbudde, 1975). The additional potassium required for their transformation is provided by release of potassium from degradational transformations (that take place simultaneously) by phytocycling and by tropospheric dust.

Neoformation

Neoformation (synthesis or authigenesis) occurs when clay minerals are formed from other clay minerals of a different structure, or from parent substances that were not clay minerals such as framework silicates, volcanic glass, ions in solution, and colloidal gels of appropriate composition. The clay minerals form because they represent the mineral phase possessing the lowest level of free energy attainable in the environment at the time of their formation.

Neoformation can be attained by direct crystallization from solution, replacement, modification of existing clay minerals, and crystallization from a colloidal gel.

Direct Crystallization from Solution. Evidence for clay formation from solution is indirect and is usually implied from mode of occurrence. Loose, strikingly well-formed crystals that partially or completely fill voids, chemical purity, and perfect microcrystallinity indicate formation from solution. Kaolinite, for example, presumably crystallized from solution, occurs in quartz geodes near Keokuk, Iowa, and in vugs in Mississippian dolomite (Keller, 1970). Various clay minerals, mainly from the smectite group, frequently found lining basalt amygdales, are assumed to have been crystallized from solution. Deposition of crystalline forms on coarse-grained, nonclay minerals is also taken to indicate neoformation from solution. Dioctahedral K-mica or illite, shown in electron micrographs to occur as stars of mica needles on quartz grains of sandstone, presumably crystallized from solution (Rex, cited by Keller, 1970). Pedogenic neoformation of palygorskite from solution was inferred from its occurrence as coating on soil peds (that did not contain palygorskite) and from its deposition on quartz grains in semiarid and arid soils of Australia (Figure 1) (Singer and Norrish, 1974). Neoformation from solution is easier to prove in sedimentary rocks than in soils.

Replacement. Pseudormorphic replacement of primary alumosilicates is a common formation process of clay minerals in the weathering zone and in soils. Decomposition of the host mineral and neoformation of the clay mineral out of the solubilized decomposition products take place along microfissures of the host material. Decomposition and neoformation take place simultaneously, and thus the external morphology of the host mineral is preserved up to a very advanced stage in its decomposition.

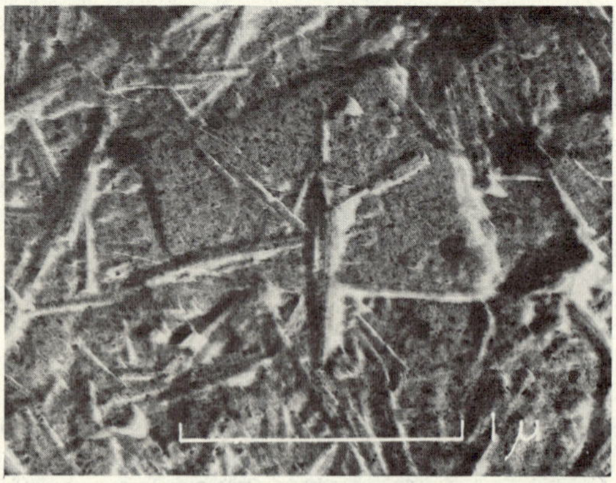

FIGURE 1. Neoformation of palygorskite fibers on top of a quartz skeleton grain in a semiarid soil from Australia. (After Singer and Norrish, 1974)

One of the most common alterations of this type is the sericitization of feldspars. The neoformed mica-structure appears to be favored by the presence of alkaline earth cations. Under mild to moderate leaching conditions, feldspars are then altered into illite, while plagioclase preferentially alters into halloysite:

$$\underset{\text{(sanidine)}}{3KAlSi_3O_8} + 8H_2O \rightarrow \underset{\text{(illite)}}{KAl_2(Si_3Al)O_{10}(OH)_2} + 6H_2SiO_3 + KOH$$

$$\underset{\text{(albite)}}{2NaAlSi_3O_8} + 9H_2O \rightarrow \underset{\text{(halloysite)}}{Al_2Si_2O_5(OH)_4 2H_2O} + 2H_2SiO_3 + 2NaOH$$

In the Hawaiian Islands, for example, plagioclase of basalt alters to halloysite. Simultaneously olivine is converted to serpentine or a trioctahedral smectite and various ferric hydrates and gels (Figure 2). Olivine may also alter directly to nontronite or magnesium-rich smectites (Loughnan, 1969). Other ferromagnesian minerals alter into trioctahedral smectites, chlorite, or vermiculite.

Under extreme leaching conditions, the rapid removal of alkalis and alkaline earth cations and also of silicon result in the pseudomorphic replacement of plagioclase by gibbsite:

$$NaAlSi_3O_{8(s)} + 8\ H_2O \rightarrow \underset{\text{(gibbsite)}}{Al(OH)_{3(s)}} + Na^+ + OH^- + 3H_4SiO_4$$

This replacement has frequently been observed in the soils and weathering zones of the tropics and subtropics (Harrison, Alexander et al, Cady, cited by Alexander and Cady, 1962).

Structural Modification of Existing Clay Minerals. The transition from triphormic (2:1) into diphormic (1:1) clay minerals, or vice versa, in the course of weathering is assumed to involve too many structural changes to be accepted as a transformation. Dissolved, or at least amorphous, phases seem to take part in these alterations, and they are therefore classified as neoformations.

Smectite, which has a large specific edge surface, is attacked by weathering acids with the subsequent decomposition of the layer lattice and release of aluminum. This aluminum tends to be deposited as an aluminohydroxy-hydronium interlayer betwen the 2:1 lattices

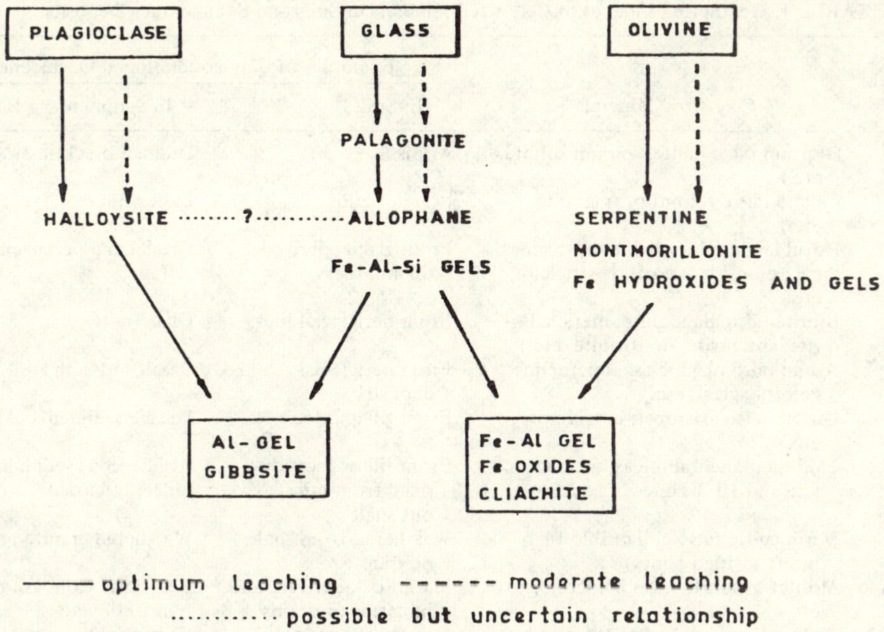

FIGURE 2. Weathering sequence of basaltic lavas, Hawaiian Islands. (After Bates, 1962, cited by Loughnan, 1969)

of the smectite. The deposition of gibbsite-like crystals in interlayer positions, which transforms the 2:1 lattices into 2:2 intergrades, continues as long as there are expansible-layer silicates in the weathering zone and is termed the *antigibbsite effect* (Jackson, 1968). Under extensive weathering and good drainage conditions, the 2:2 end member of the series, chlorite, can be obtained. A subsequent weathering stage, involving silica tetrahedra inversion, could give kaolinite. Kaolinite formation thus may be represented as a 2:1 → 2:2 → 1:1 alteration sequence. Only after further intensive leaching of the soil and a rise of the pH above the isoelectric pH of the hydroxyaluminum, a separate, crystalline gibbsite phase appears, as the hydroxyaluminum adsorbed on clay surfaces becomes negative and thus repelled and released.

These transitions are accompanied by the nearly total loss of alkaline earth cations and considerable losses in silica and are thus a function of weathering intensity. The persistence of clay mineral species in soils is controlled by weathering intensity and also the relative ease with which the mineral can weather. Different combinations of weathering intensities, mineral stabilities, and time of weathering may produce a given degree of weathering, allowing the arrangement of clay minerals and related minerals into a weathering sequence (Table 1). The stability index numbers given to the minerals in that sequence indicate increasing relative resistance to weathering.

Crystallization from a Colloidal Gel. The decomposition of primary alumosilicates under intense leaching conditions results frequently in the formation of colloidal gels of aluminum and iron hydroxides. These age and crystallize into crystalline iron and aluminum oxides and hydroxides upon changes in the environment such as the transition from permanently wet into alternately wet and dry conditions. These neoformations are particularly frequent during laterite (plinthite) horizon formation in the tropics and subtropics:

$$2Fe(OH)_{3\,(a,\,s)} + \underset{\text{(pseudoboehmite)}}{Al(OOH)_{(s)}} \rightarrow \underset{\text{(hematite)}}{Fe_2O_{3\,(s)}} + \underset{\text{(gibbsite)}}{Al(OH)_{3\,(s)}} + 2H_2O$$

The alteration of volcanic class particularly under strong leaching conditions, commonly yields amorphous, gel-like alumosilicates, of which the principal is allophane.

$$\underset{\text{(basic volcanic glass)}}{2NaCaAl_3Si_5O_{16\,(s)}} + 17H_2O + 6H^+ \rightarrow \underset{\text{(allophane)}}{3Al_2SiO_4(OH)_2 \cdot H_2O_{(s)}} + 2Na^+ + 2Ca^{++} + 7H_4SiO_4$$

TABLE 1. Weathering Indexes of Clay-sized Minerals in Soils and Sedimentary Deposits.

Weathering Index and Symbol	Clay-sized Minerals	Examples of Clay-sized Mineral Occurrences	
		In Soils	In Sedimentary Deposits
1 Gyp	Gypsum (also halite, sodium sulfate, etc.)	Aridisols	Drained brackish swamps
2 Clt	Calcite (also dolomite, aragonite, etc.)	C_{Ca} horizons	Limestones
3 Horn	Hornblende (also olivine, pyroxenes, e.g., diopside, anorthite, analcite, etc.)	From glacial-derived till and loess	Glacial and periglacial rock flour
4 Biot	Biotite (also glauconite, mafic chlorite, antigorite, nontronite, etc.)	From periglacial loess	Glauconite
5 Alb	Albite (also plagioclase, microcline, volcanic glass, etc.)	From periglacial deposits	Basalt outwash
6 Qtz	Quartz (also cristobalite, tridimite, etc.)	From till and loess	Pelagic sediments
7 Mi-d	1Md dioctahedral micas, (also muscovite and 10 Å zones of sericite, etc.)	From till or loess derived from micaceous shale	Argillaceous sediments, deeply buried
8 Verm	Vermiculite (also collapsible 14 Å interstratified zones)	Weathered from shale or granite	Weathered granite outwash
9 Mont	Montmorillonite (also beidellite, etc.)	Mafic rock-derived, and in basins or drains	Bentonite from ash deposited in fresh water
9 Al-Chl	Pedogenic dioctahedral chlorite (including interstratified 2:2 zones)	Commonly pedogenic in expansible-layer silicates	Commonly associated with montmorillonite
10 Allo	Allophane (sesquioxic, halloysitic, etc.)	Ando soils; Ultisols	Allophanized ash
10 Kaol	Kaolinite (also halloysite, etc.)	Ultisols	White clay deposits
11 Gib	Gibbsite (also boehmite, etc.)	Oxisols	Bauxite
12 Hem	Hematite (also goethite, limonite, lepidocrocite, magnetite, etc.)	Oxisols	Iron ore
13 Ana	Anatase (also rutile, ilmenite, leucoxene, zircon, corundum, etc.)	Oxisols	Titanium ore

(After Jackson, 1968).

Allophane forms by the isoelectric precipitation of silica and alumina gels, which are obtained by the primary alteration of glass. Allophane is typical for andosols and is widely distributed in soils formed from volcanic rocks in high rainfall areas. Ageing and dehydration of allophane can lead to halloysite formation. Further desilication of allophane produces gibbsite.

$$\text{Basic volcanic glass} \rightarrow \text{mixed gels of } SiO_2 + Al_2O_3 \rightarrow \text{allophane} \begin{matrix} \nearrow \text{gibbsite} \\ \searrow \text{halloysite} \end{matrix}$$

Factors Affecting Clay Mineral Formation in Soils

The stability diagrams in Figure 3 suggest that the nature of the clay minerals that form in a soil are determined by the chemical environment at the time of their formation. If the H_4SiO_4 activity is kept low in that environment, under conditions of rapid permeability and perfect drainage, the stable end product of weathering, by whatever process of alteration, is gibbsite. At higher H_4SiO_4 activities and in neutral or alkaline media [high (K/H) ratios], under conditions of slow permeability and poor drainage, mica-type minerals are formed. Under similar conditions, but with (K/H) low and (Na/H) variable, and a supply of soluble calcium and magnesium, smectite-type minerals form. Illite and smectite may represent, under continuing conditions of poor drainage, stable end products of weathering. They may, however, also merely represent transitory stages in the weathering sequence of clay minerals (Figure 2). Thus with progressive removal of alkalis, alkaline earth metals, and silica from the soil solution and its acidization, both illite and smectite can alter into kaolinite.

The principal factor that affects the chemistry of the weathering environment is climate. Rainfall controls the supply of moisture needed for

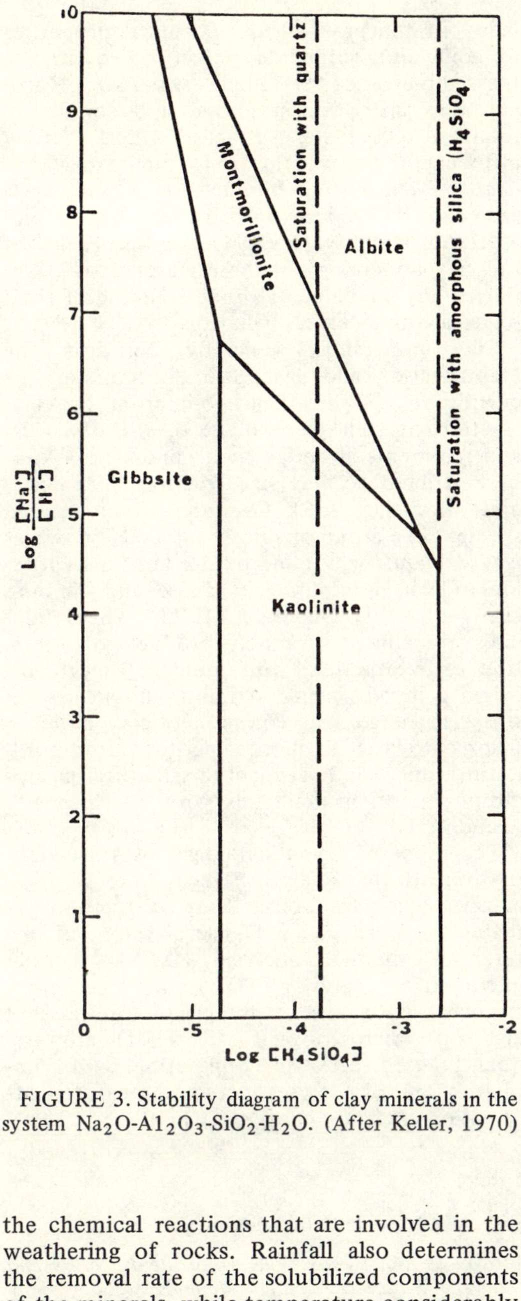

FIGURE 3. Stability diagram of clay minerals in the system Na_2O-Al_2O_3-SiO_2-H_2O. (After Keller, 1970)

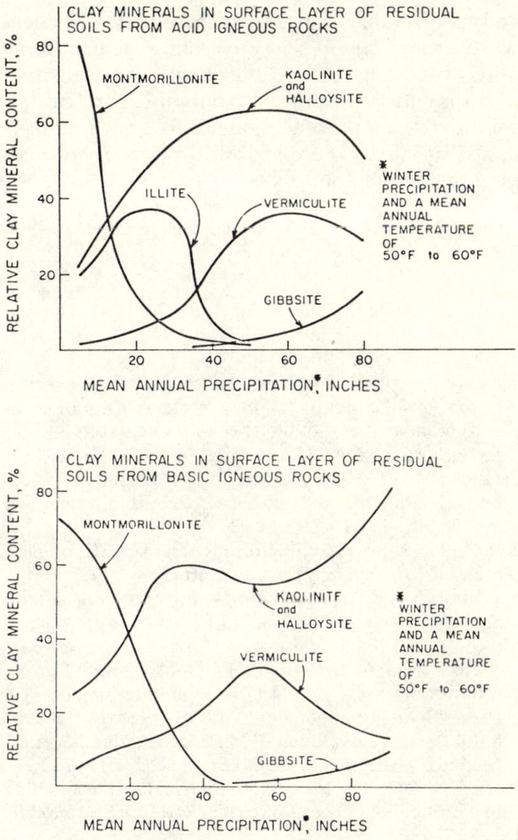

FIGURE 4. The effect of precipitation on the frequency distribution of clay minerals in the surface layer of residual soils in California: (a) Soils formed from acid rocks; (b) soils formed from basic rocks. (After Barshad, 1966)

the chemical reactions that are involved in the weathering of rocks. Rainfall also determines the removal rate of the solubilized components of the minerals, while temperature considerably influences the rate of these reactions. Thus in a study of several hundred Californian soils developed either from acid or basic igneous rocks, smectite was found to be predominant in soils at precipitations below 500 mm/ann, whereas kaolinite-halloysite is predominant at precipitations greater than 1000 mm/ann (Figure 4). The frequency distribution patterns of clay minerals as a function of rainfall are not affected markedly by the nature of the parent material except with respect to the occurrence of illite in the acid and not in the basic rock-derived soils.

Topography affects clay mineral neoformation by modifying the moisture regime of the soils. Only with topography that allows maximum water penetration and drainage can the full precipitation potential of the area become effective in clay mineral genesis. Poor drainage conditions inhibit the formation of clay minerals that are at the end of the weathering sequence even in high rainfall areas. Therefore the catenary-differentiation concept expresses itself also by the clay mineral distribution pattern (Yaalon et al, 1966).

The biosphere influences clay mineral genesis mainly by modifying the chemical environment. Organic decomposition products, primarily organic acids, speed up chemical weathering and clay mineral production. Chelation by organic components increases the solubility and mobility of inorganic substances, mainly iron and aluminum, that are released by chemical weathering. Finally, phytocycling is an important process in the mobilization and transport of inorganic substances from one part of the solum, or weathering, zone into another, thus contributing to the distribution pattern of clay minerals in the solum.

ARIEH SINGER

References

Alexander, Lyle T., and Cady, John G., 1962. Genesis and hardening of laterite in soils, *U.S. Dept. Agric. Tech. Bull. 1286*, 90p.

Barshad, J., 1966. The effect of a variation of precipitation on the nature of clay mineral formation in soils from acid and basic igneous rocks, *Proc. Internat. Clay Conf. (Jerusalem)*, **1**, 167–173.

Jackson, M. L., 1968. Weathering of primary and secondary minerals in soils, *9th Internat. Congr. Soil Sci. Trans.*, **4**, 281–292.

Keller, W. D., 1970. Environmental aspects of clay minerals, *Jour. Sed. Petrology*, **40**, 788–813.

Loughnan, F. C., 1969. *Chemical Weathering of the Silicate Minerals.* New York: Elsevier, 154p.

Millot, G., 1964. *Géologie des Argiles, Altérations, Sédimentologie, Géochimie.* Paris: Masson, 499p.

Niederbudde, E. A., 1975. Veränderungen von Dreischicht-Tonmineralien durch natives K in holozänen Lössboden Mitteldeutschlands und Niederbayerns, *Z. Pfl. Dung. Bodenk*, **2**, 217–234.

Norrish, K., 1972. Factors in the weathering of mica to vermiculite, *Proc. Internat. Clay Conf. (Madrid)*, **1**, 417–432.

Singer, A., and Norrish, K., 1974. Pedogenic palygorskite occurrences in Australia, *Am. Mineralogist*, **59**, 508–517.

Yaalon, D. H.; Nathan, Y.; Koyumdjisky, H.; and Dan, J., 1966. Weathering and catenary differentiation of clay minerals in soils on various parent materials in Israel, *Proc. Internat. Clay Conf. (Jerusalem)*, **1**, 187–198.

Cross-references: *Clay Mineral Decomposition; Clay Mineralogy; Clay Minerals, Silicates; Clays, Physical Chemistry; Ionic Activities; Leaching; Metal-complexing Macromolecules; Vol. IVA: Weathering, Chemical; Vol. IVB: Clays and Clay Minerals, Vol. VIA: Clay: Genesis; Clay: Diagenesis, Vol. XII, Pt. 2: Weathering and Soil Formation.*

CLAY MINERALOGY

The science of *clay mineralogy*, as it is currently practiced, dates from about 1930 when the newly developed X-ray-diffraction technique showed that the smallest particles in soils and sediments were usually crystalline. Prior to 1930 most scientists believed that amorphous materials comprised a significant portion of the finest-sized fractions (see *Particle-size Distribution*). It subsequently became apparent that many of the physical and chemical properties of soils and sediments reflect the nature of these minerals of very small crystal size. Nutrient ions are often adsorbed and sometimes retained on crystal surfaces, which thereby influence their availability to plants (see *Exchange Phenomena*). Interactions between crystals can determine soil structure (q.v.) and aggregate stability (q.v.). Thus a knowledge of the clay mineralogy of these materials is essential to any fundamental understanding of their physical and chemical behavior.

Clay mineralogy generally concerns the identification and description of minerals occurring in the less-than-two-micron, or clay, fraction of soils (see vol. IVB, *Soil Minerals*) and sediments although such minerals may not be restricted to this size fraction (Mackenzie and Mitchell, 1966). Commonly encountered species are a group of silicate minerals known as *clay minerals* (q.v.) and oxides and/or hydroxides of iron, aluminum, manganese, and titanium (Brown, 1961; Gieseking, 1975). These minerals are almost invariably products of alteration or weathering, with minerals of primary origin generally being of minor importance in this size range. The commonest clay minerals belong to the kaolin, smectite (montmorillonite), mica, and vermiculite structural groups although other minerals are occasionally encountered.

The modern clay mineralogist uses a variety of sophisticated analytical techniques to characterize clay fractions. X-ray diffraction remains the most generally useful tool, but differential thermal analysis (DTA), thermal gravimetric analysis (TGA), infrared spectrometry (IR), and scanning and transmission electron microscopy (SEM, TEM) are frequently used, often in conjunction with complete chemical analyses including measurement of cation-exchange capacity (CEC).

R. J. GILKES

References

Brown, G., ed., 1961. *The X-ray Identification and Crystal Structures of Clay Minerals.* London: Mineralogical Society, 544p.

Caillère, S., and Hénin, S., 1963. *Mineralogie des Argiles.* Paris: Masson et cie, 355p.

Farmer, V. C., ed., 1974. *The Infrared Spectra of Minerals.* London: Mineralogical Society, 539p.

Gard, J. A., ed., 1971. *The Electron-optical Investigation of Clays.* London: Mineralogical Society, 383p.

Gieseking, J. E., ed., 1975. *Soil Components: Inorganic Components.* New York: Springer-Verlag, 684p.

Grim, R. E., 1953. *Clay Mineralogy.* New York: McGraw-Hill, 384p.

Joint Committee on Powder Diffraction Standards, 1974. *Selected Powder Diffraction Data for Minerals, Publication DBM-1-23.* JCPDS, Swarthmore, Pa., 262p.
Mackenzie, R. C., ed., 1969. *Differential Thermal Analysis, Vol. 1: Fundamental Aspects.* London: Academic, 776p.
Mackenzie, R. C., and Mitchell, B. D., 1966. Clay mineralogy, *Earth Sci. Rev.*, **2**, 47–91.
Thorez, J., 1975. *Phyllosilicates and Clay Minerals: A Laboratory Handbook for their X-ray Diffraction Analysis.* Dison, Belgium: G. Lelotte, 579p.
Weaver, C. E., and Pollard, L. D., 1973. *The Chemistry of Clay Minerals.* Amsterdam: Elsevier, 213p.

Cross-references: *Clay Minerals, Silicates; Clays, Physical Chemistry; Exchange Phenomena; Hydrous Oxides; Mineral Analysis; Particle-size Distribution; Pedosesquioxides; Soil Mineralogy; Vol. IVB; Soil Minerals.*

CLAY MINERALS, SILICATES

Hydrous clay minerals are layer-lattice silicates that comprise tetrahedrally (Si, Al, Fe^{3+}) and octahedrally (Al, Fe^{3+}, Fe^{2+}, Mg) coordinated cations that form either sheets or chains (Brown, 1961). The basic structural units in layer silicates are silica sheets and brucite ($MgOH_2$) or gibbsite ($AlOH_3$) sheets. Silica sheets consist of SiO_4^{2-} tetrahedra connected at three corners in the same plane forming a hexagonal network with the tips of the tetrahedra all pointing in the same direction. This unit is called the *tetrahedral sheet.* The brucite or gibbsite sheet consists of two planes of hydroxyl ions which straddle a plane of magnesium or aluminum ions, which is octahedrally coordinated by the hydroxyls. These sheets, known as *octahedral sheets,* are combined so that the oxygens at the tips of the tetrahedra project into a plane of hydroxyls in the octahedral sheet and replace two-thirds of the hydroxyls. This combination of sheets forms a *layer.*

Classification of clay minerals is based on the different arrangements of tetrahedral and octahedral sheets (Gieseking, 1975). Table 1 gives a simplified classification scheme, and Figure 1 shows diagrammatic sketches of the structures of the most common clay minerals.

TABLE 1. Simplified Classification of Clay Minerals.

	Dioctahedral	Trioctahedral
1:1	Kaolinite (Al)	Serpentine (Mg)
2:1 Nonexpanded	Illite (Al) Glauconite (Fe)	"Biotite" (Fe, Mg)
2:1 (Low charge) Expanded Smectites	Montmorillonite (Si) Beidellite (Al) Nontronite (Fe)	Saponite (Mg) Hectorite (Li)
2:1 (High charge)	Vermiculite (Al)	Vermiculite (Mg, Fe)
2:1:1	Chlorites (Al)	Chlorites (Mg, Fe)
Mixed layer	Intergrowth of any two or three of the clay minerals listed above.	
Chain structure	Palygorskite (Al, Mg)	Sepiolite (Mg)

1:1 Clay Minerals

The 1:1 clay mineral type consists of one tetrahedral sheet and one octahedral sheet. This two-sheet type is divided into kaolinite (dioctahedral) and serpentine (trioctahedral) groups. The kaolinite minerals are nearly pure hydrous-aluminum silicates. The serpentine minerals, of which chrysotile and antigorite are the most common, consist of a tetrahedral sheet and an octahedral sheet containing magnesium with only minor amounts of aluminum. Other minerals in this group vary in composition as aluminum, iron, manganese, nickel, and chromium can substitute for magnesium in the octahedral sheet, and aluminum, ferric iron, and germanium can substitute for silicon in the tetrahedral sheet. Chamosite is the most common mineral in this group.

2:1 Clay Minerals

The three-sheet, or 2:1 layer-lattice, silicates consist of two silica tetrahedral sheets between which is an octahedral sheet. They include the mica and smectite groups that are by far the most abundant of the clay minerals. The pure end members of this type are talc, a hydrous-magnesium silicate; pyrophyllite, a hydrous-aluminum silicate; and minnesotaite, a hydrous-iron silicate.

The 2:1 structural unit of the micas is similar to that for talc, except that between the 2:1 units, there lies a plane of large interlayer cations, of which potassium is the most common, though sodium and calcium also occur. These interlayer cations, which fit into the hexagonal ring formed by the tetrahedral oxygen ions and bond adjacent 2:1 units, balance the charge due to the substitution of cations of lesser charge in the tetrahedral and/or octahedral sheet.

The mica group is subdivided on the basis of whether the species are dioctahedral (muscovite type) or trioctahedral (biotite type). Aluminum, and less commonly ferric iron, substitute for the silicon. The micas are further categorized according to the wide range of cations and the combination of cations that occur in the octahedral sheet. Aluminum (muscovite) and mag-

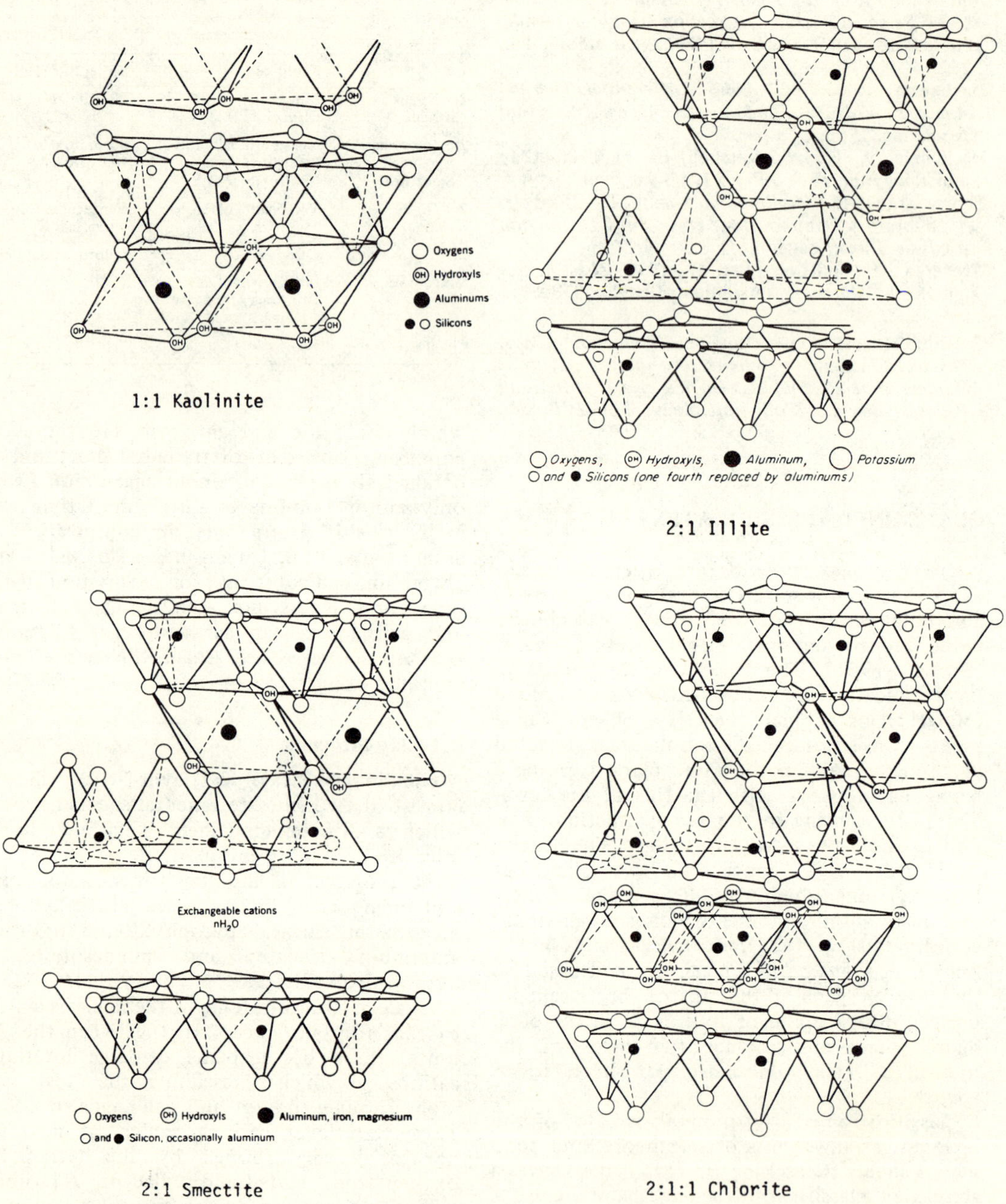

FIGURE 1. Diagrammatic sketch of the structures of the major clay minerals. (After Grim, 1968)

nesium (phlogopite) are the only two that occur alone in the octahedral sheets. Most micas have two or more cations in the octahedral sheets; aluminum, magnesium, and iron occur in several combinations in the octahedral sheets. Manganese, vanadium, chromium, lithium, titanium, and a variety of other cations can occur in varying amounts. The substitution of a cation of a lower charge for a cation of a higher charge in both the octahedral (for example, Mg^{2+} replacing Al^{3+}) and tetrahedral (for example, Al^{3+} replacing Si^{4+}) sheets gives the 2:1 layer a net negative charge that is balanced by the interlayer cations.

The fine-grained micas belong to the illite family, with the dioctahedral illites predominating over the trioctahedral types. The most common illite mineral is dioctahedral, which has approximately half as much aluminum substituting for silicon in the tetrahedral sheet as does muscovite. Approximately three-fourths of the octahedral cations are aluminum; minor amounts of ferric iron are present, and approximately one-eighth of the cations are divalent (magnesium and ferrous iron). This gives a total negative charge of approximately 0.75 as compared with a value of 1 for muscovite [per $O_{10}(OH)_2$].

Glauconite and celadonite are dioctahedral iron illites. In glauconites more than half of the octahedral positions are filled with iron, the more abundant being ferric iron. The aluminum content of the tetrahedral sheet is usually less than that of the aluminum illites, and the magnesium content of the octahedral sheet greater. Celadonite has more octahedral magnesium and less tetrahedral aluminum than glauconite. Trioctahedral illites are relatively rare as pure mineral concentrates. Because the interlayer cations are relatively easily leached from biotite, the clay-sized minerals are usually mixed-layer biotite-vermiculite (expanded biotite).

The expanded or expandable 2:1 clay minerals vary widely in chemical composition and in layer charge. These minerals are characterized by the presence of loosely bound cations and layers of water or polar organic molecules between the silica sheets. The interlayer width is reversibly variable. The interlayer water can be driven off at temperatures between 100° and 200°C. Sodium, calcium, hydrogen, magnesium, iron, and aluminum are the most common naturally occurring interlayer cations.

The dioctahedral subgroup is by far the most abundant. The layer charge on the expanded clays ranges from 0.3 to 0.8 per $O_{10}(OH)_2$ unit of structure. The low-charge (0.2–0.6) dioctahedral minerals, which have most of their charge originating in the octahedral sheet, are called *dioctahedral smectites,* or *montmorillonites.* Expanded-dioctahedral smectites, which have a relatively high tetrahedral charge content (0.4), are called *beidellites.* These clays commonly have a total charge of 0.7 or higher. Identical minerals, derived by leaching potassium from illite or muscovite, are referred to as *dioctahedral vermiculite.* There is apparently a complete gradation in composition and charge between the species of montmorillonite and beidellite. The ferric iron-rich variety is called *nontronite.*

Although they are relatively rare in soils, there is a wide spectrum of trioctahedral-expanded clays. The most common in the low-charge range (0.3–0.5) are hectorite, which contains magnesium and lithium in the octahedral sheet, and saponite, which has considerable magnesium in the octahedral sheet and some aluminum substitution in the tetrahedral sheet. Trioctahedral-expanded 2:1 minerals with a layer charge of 0.6–0.8 are called *vermiculites.* These minerals are usually coarser grained and have better crystal organization than most expanded clays. Though magnesium is the predominant ion in the octahedral sheet, sufficient amounts of aluminum and iron are present to give the sheet a positive charge. Slightly more than one out of three tetrahedral positions is filled with aluminum. The resulting layer charge can approach 1.0. Dioctahedral vermiculites are relatively rare.

2:1:1 Clay minerals

Chlorite can occur as a clay-sized mineral. Most consist of a 2:1 talc layer plus a brucite (magnesium) sheet. Most chlorites are trioctahedral, though dioctahedral chlorites are common. Some chlorites have both dioctahedral and trioctahedral sheets. Because substitution can occur in both the 2:1 layers and in the brucite sheet, the chlorites have a wide range of compositions.

There is a large number of clays that are not pure mineral types but consist of interstratified units of different chemical composition. These are called mixed-layer clays and have compositions intermediate between those of the two or more mineral components.

Chain-structure Clay Minerals

Palygorskite and sepiolite are clay minerals with a chain structure. Palygorskite has five octahedral positions, and sepiolite either eight or nine; both have relatively little tetrahedral substitution. The octahedral positions in sepiolite are filled largely with magnesium, and those in palygorskite with approximately half magnesium and half aluminum.

Chemical Analyses

Table 2 has average or typical chemical analyses of the major clay minerals (Weaver and Pollard, 1973). Table 3 shows the same chemical data in the form of structural formulas. Structural formulas are calculated assuming the anions in the 2:1 clays consist of $O_{10}(OH)_2$ and that there are four cations in tetrahedral coordination. The anion population of the 1:1 clays is assumed to be $O_5(OH)_4$ and the tetrahedral layer is believed to contain two cations.

CHARLES E. WEAVER

TABLE 2. Chemical Analyses of Clay Minerals.

	Illite*	Glauconite*	Montmorillonite*	Beidellite	Nontronite	Palygorskite	Sepiolite	Kaolinite	Chamosite
SiO_2	49.780	49.215	59.488	59.30	43.51	53.64	52.50	44.59	22.03
Al_2O_3	26.346	9.148	21.934	36.11	2.94	8.76	0.60	38.12	22.91
Fe_2O_3	4.304	17.951	3.770	0.50	28.62	3.36	2.90	1.43	0.49
FeO	0.611	3.427	0.197		0.99	0.23	0.70		36.68
MgO	2.753	3.584	3.548	0.10	0.05	9.05	21.31	0.06	1.91
CaO	0.321	0.644	1.176	0.02	2.22	2.02	0.47	0.12	0.07
Na_2O	0.245	0.460	0.824	3.98				0.08	0.08
K_2O	7.019	6.884	0.342	0.11					0.03
TiO_2	0.424		0.250			0.60			3.63
H_2O^+	7.120	8.482	8.380		6.62	10.89	9.21	13.91	10.65
H_2O^-	1.481				14.05	9.12	12.06	0.71	0.63

(From Weaver and Pollard, 1973)
*Average analysis.

TABLE 3. Structural Formula of Clay Minerals.

	Illite*	Glauconite*	Montmorillonite*	Beidellite	Nontronite	Vermiculite	Palygorskite	Sepiolite	Kaolinite	Chamosite
Octahedral										
Al	1.531	0.445	1.492	1.98	0.06	1.44	1.29		1.958	0.76
Fe^{3+}	0.225	1.010	0.187	0.02	1.86	0.16	0.37	0.47	0.039	0.02
Fe^{2+}	0.035	0.203	0.007		0.07		0.03	0.13		1.73
Mg	0.283	0.389	0.354	0.01	0.01	0.27	1.96	7.14	0.003	0.16
Tetrahedral										
Al	0.600	0.348	0.158	0.52	0.24	1.10	0.21	0.16	0.018	0.76
Si	3.405	3.649	3.837	3.48	3.76	2.90	7.79	11.80	1.982	1.24
Interlayer										
Ca	0.046	0.065	0.007				0.32	0.11	0.002	
Na	0.032	0.059	0.010			0.43			0.008	
K	0.615	0.657	0.004			0.37				

(From Weaver and Pollard, 1973).
*Average analysis.

References

Beutelspacher, H., and Van Der Marel, H. W., 1968. *Atlas of Electron Microscopy of Clay Minerals and Their Admixtures.* Amsterdam: Elsevier, 333p.

Brown, G., 1961. *The X-Ray Identification and Crystal Structures of Clay Minerals.* London: Mineralogical Society, 544p.

Caillère, S., and Hénin, S., 1963. *Mineralogie des Argilles.* Paris: Masson, 355p.

Gieseking, J. E., ed., 1975. *Soil Components; Vol. 2, Inorganic Components.* New York: Springer-Verlag, 684p.

Grim, R. E., 1968. *Clay Mineralogy.* New York: McGraw-Hill, 596p.

Grimshaw, R. W., 1971. *The Chemistry and Physics of Clays and Allied Ceramic Materials.* London: Benn, 1024p.

Loughnan, F. C., 1969. *Chemical Weathering of the Silicate Minerals.* Amsterdam: Elsevier, 154p.

Mackenzie, R. C., 1959. The classification and nomenclature of clay-minerals, *Clay Min. Bull.*, **4**, 52–66.

Marshall, C. E., 1964. *The Physical Chemistry and Mineralogy of Soils; Vol. 1, Soil Materials.* New York: Wiley, 388p.

Millot, G., 1970. *Geology of Clays.* Paris: Masson, 429p.

Warshaw, C. M., and Roy, R., 1961. Classification and a scheme for the identification of layer silicates, *Geol. Soc. America Bull.*, **72**, 1455–1492.

Weaver, C. E., and Pollard, L. D., 1973. *The Chemistry of Clay Minerals.* Amsterdam: Elsevier, 213p.

Cross-references: *Clay Mineralogy; Clays, Physical Chemistry; Clay-Organic Reactions; Exchange Phenomena; Flocculation; Microbial Ecology and Clay Minerals; Mineral Analysis; Soil Mineralogy, Vol. IVA: Colloids; Vol. IVB: Soil Minerals.*

CLAY-ORGANIC REACTIONS

The scientific study of clay-organic reactions is a relatively recent development, dating from the early 1940s since it was only in the preceding decade that the crystallinity and the elucidation of the structures of the main groups of clay minerals (q.v.) had gained general acceptance. This is not to say, however, that the chemical association between clays and organic substances, *complex formation,* was unknown before this time. An interaction between the clay fraction and the organic constituents of soils, for example, was postulated by the early soil chemists (e.g., Demolon and Barbier, 1929) to account for the stability and resistance to microbial decomposition of humus.

Despite its late beginnings, rapid advances have been made in understanding the underlying reaction mechanisms, and a self-consistent picture has emerged from the mass of experimental data. In particular, the application of infrared spectroscopy, in conjunction with classical x-ray diffractometry and electron microscopy, to the study of clay-organic reactions has provided valuable information on the bonding of organic molecules to clay mineral surfaces.

As the literature on the subject accumulated over the years, attempts were made to review existing knowledge. The earlier accounts by MacEwan (1962), Calvet (1963), and Greenland (1965) are of a general nature. Later attempts dealt with specific aspects, emphasizing formation mechanisms (Bailey and White, 1970; Brindley, 1970; Calvet and Chassin, 1973; Cloos, 1972; Farmer, 1971; Greenland, 1971; Mortland, 1970; Solomon, 1968; Theng, 1971; Theng and Walker, 1970; Weber, 1972; Weiss, 1969). Other reviews such as those by Jordan (1963), Nahin (1963), and Theng (1970) concern the practical applications of the clay-organic reaction. For a critical assessment of the literature up to 1972, the reader is referred to Theng (1974).

Clay Mineral Structures

Details of the structures of clay minerals and their identification have been given by Brown (1961) and by Grim (1953). Only the basic structural features of clays together with their architectural and organizational peculiarities that influence their behavior toward organic compounds are discussed here.

Clay minerals belong to the class of *layer silicates* (phyllosilicates) because their crystals are essentially made up of layers formed by condensation of *sheets* of $Si(O, OH)_4$ tetrahedra with those of $Al(OH)_6$ octahedra [Figure 1(a)]. Condensation in a 1:1 proportion gives rise to the two-sheet (dimorphic) type minerals, exemplified by kaolinite [Figure 1(b)]. The three-sheet (trimorphic) clays are formed by a 2:1 condensation, the octahedral $[Al(OH)_6]$ sheet being sandwiched between two tetrahedral sheets, as in pyrophyllite [Figure 1(c)]. A *clay crystal* is composed of di- or trimorphic layers continuous in the horizontal (*a b*) plane and stacked in a more or less regular array along the vertical (*c*) axis. There is scope in these structures for *isomorphous replacement,* the substitution of Si^{4+} and/or Al^{3+} for cations of similar size but of different (usually lower) charge. Thus if $3Mg^{2+}$ substitute for $2Al^{3+}$ in the octahedral sheet of pyrophyllite, the mineral talc results. In kaolinite, pyrophyllite, and talc the layers are essentially electrically neutral, which makes such minerals relatively unreactive toward organic compounds. In most instances, however, isomorphous replacement produces a structure that carries a permanent negative charge. This (positive) charge deficiency is balanced by *adsorption* (q.v.), the surface uptake of extraneous cations such as Na^+, K^+, Ca^{2+}, and Mg^{2+}. Apart from those associated with the external crystal surface, these cations occupy sites in the interlayer space and are often *exchangeable* (see *Exchange Phenomena*). The amount of substitution can be expressed as a charge per formula unit χ which represents half a unit layer [cf. Figure 1(c)]. Besides being a measure of the exchange capacity of the mineral, the magnitude of χ influences the cation-layer attraction and, more importantly, the behavior of the clay toward water (a polar liquid) and polar-organic compounds. This is so because such substances interact more with the exchangeable (saturating) inorganic cations than with the mineral surface. Table 1 illustrates these points. (See page 90.)

In *kaolinite* each layer is about 0.72 nm thick. Since no extraneous species are present in the interlayer space, this thickness also represents the repeat distance along the *c* axis of a crystal, referred to as the *basal or d(001) spacing* [Figure 1(b)]. The superposition of oxygen and hydroxyl planes of successive layers within a crystal produces O and OH pairing and strong O . . . OH hydrogen bonding. Such interlayer H bonds and the nonspecific, intermolecular (van der Waals) attractive forces must be overcome if *intercalation*—interlayer uptake of organic compounds—is to occur. For this reason, both the range and type of organics that may be intercalated are rather limited.

Many organic complexes of *halloysite* have been prepared by MacEwan as early as 1948. Penetration of the interlayer space by the organic species here proceeds with little diffi-

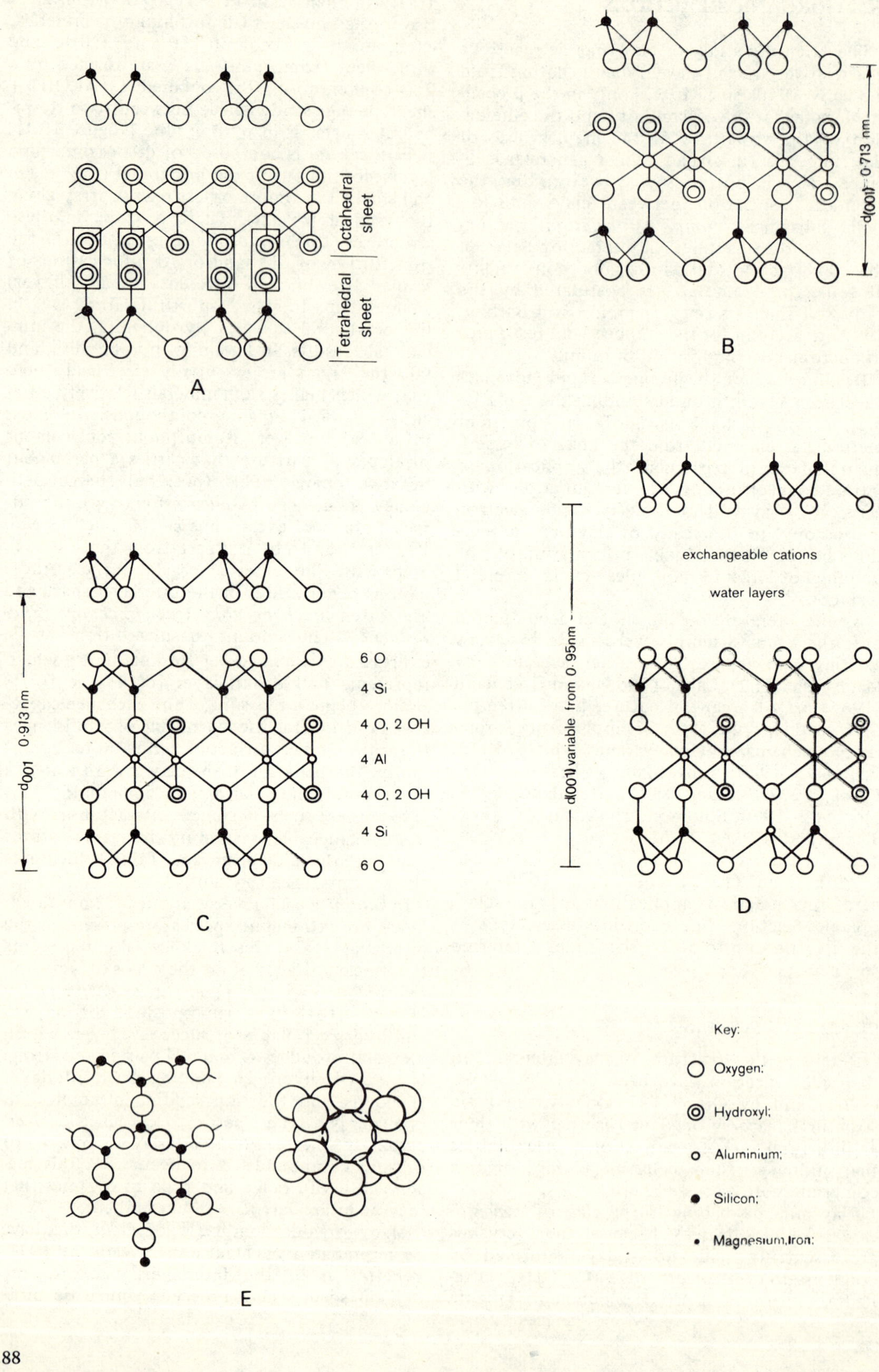
Octahedral sheet
Tetrahedral sheet
A
d(001) 0·713 nm
B
d001 0·913 nm
6 O
4 Si
4 O, 2 OH
4 Al
4 O, 2 OH
4 Si
6 O
C
d(001) variable from 0·95nm
exchangeable cations
water layers
D
E
Key:
Oxygen;
Hydroxyl;
Aluminium;
Silicon;
Magnesium,Iron;

culty because the (disordered) kaolinite-like layers within a crystal are already separated by a single layer of water molecules. The presence of this water in (hydrated) halloysite increases the basal spacing to ~ 1.01 nm, collapsing to ~ 0.72 nm on dehydration.

Of the trimorphic-layer silicates, *montmorillonite* has held the limelight for many years [Figure 1(d)] because it combines a relatively high exchange capacity (~ 1 meq g^{-1}) with a large surface area (~ 750 $m^2 g^{-1}$) that is accessible to a wide variety of positively charged and polar nonionic organic compounds (cf. Table 1). The charge per formula unit is such that the forces holding together successive layers within a crystal are rather weak. For this reason, montmorillonite almost invariably contains interlayer water at ambient temperature and humidity. By analogy with halloysite, this water can wholly or partially be replaced by polar-organic molecules. Similarly, the interlayer (inorganic) cations can be exchanged by positively charged organics. The basal spacing of montmorillonite varies depending on the nature of the exchangeable cations and on the water content of the system (Norrish, 1954). The minimum d(001) value, corresponding to a fully dehydrated (collapsed) state, is ~ 0.95 nm, which is generally taken as the thickness of an individual layer.

Vermiculite has a similar layer structure to montmorillonite but a higher (layer) charge (Table 1). The intercalation of extraneous species here is more difficult to achieve because more energy is required to separate the individual silicate layers, as compared with the lower charged montmorillonite.

Mica and *brittle mica* have a still higher χ value. Minerals of this group do not show interlayer expansion (swelling) in water and organic liquids. Except for some long-chain alkylammonium ions, the interlayer K^+ and Ca^{2+} ions cannot exchange for other cations in solution, or do so only with great difficulty (Table 1).

Formation of Complexes

Most clay-organic studies have been conducted with montmorillonite, and unless otherwise specified, the following discussion refers to complexes with this mineral species. The majority of complexes with small, well-defined organic compounds, fall essentially into two classes: those in which the adsorbed species exist (*a*) as a polar, uncharged compound, or (*b*) as a positively charged (cationic) molecule. The *a* class of complex usually arises from replacement of the interlayer water, which is associated with the exchangeable, inorganic cations or, in some instances, of a previously intercalated organic species. Class *b* complexes may be formed either by exchanging the inorganic cations for organic cations or by protonation at the clay surface of (already) adsorbed organic bases. Little, if any, adsorption occurs with negatively charged (anionic) organics because the clay itself acts as a (rigid) polyanion. Some anionic herbicides, however, may be *negatively adsorbed,* i.e., repelled at neutral and alkaline pH (Frissel and Bolt, 1962). *Anionic organics,* however, may be adsorbed under acid conditions (low pH) when the edges of clay crystals acquire positive charges by accepting H^+ ions (protons) from the bulk solution. Polyvalent cations occupying exchange sites at the clay surface or water molecules coordinated to such cations may act as a bridge between the anionic group of the organic and the clay and so give rise to appreciable adsorption. Some nonpolar hydrocarbons, both aliphatic and aromatic, also form interlayer complexes. The underlying mechanisms are as yet incompletely understood, but dipole-induced

FIGURE 1. Idealized representation of the structures of some clay minerals viewed along the *a* axis. Basal, or d(001), spacings, given in nm units, are indicated.

a. Condensation of $Si(O, OH)_4$ tetrahedral sheet with $Al(OH)_6$ octahedral sheet in a 1:1 proportion to yield the two-sheet (dimorphic) type minerals such as kaolinite.

b. The kaolinite layer structure showing characteristic d(001) spacing, which in halloysite increases to ~1.01 nm due to the presence of a monolayer of water molecules in the interlayer space.

c. The pyrophyllite layer structure showing d(001) spacing and the number of oxygens, silicons, hydroxyls, and aluminums making up the electrically neutral layer unit.

d. The montmorillonite layer structure showing variable d(001) spacing depending on the nature of the interlayer (exchangeable) cation and the hydration state of the crystal. The occurrence of isomorphous replacement in the tetrahedral and octahedral sheets, which produces a permanent negative layer charge, is also indicated.

e. Silica tetrahedral sheet projected on the *a b*, or basal, plane showing hexagonal depression into which an adsorbed molecule (broken circle) may "key." In reality, there is a departure from ideal hexagonal symmetry of six-membered rings due to a misfit between the tetrahedral and octahedral sheets. As a consequence, the symmetry is ditrigonal rather than hexagonal.

TABLE 1. Surface-Charge Density and Intercalating Ability of Some Layer-Silicate Minerals.

Type	Charge per Formula Unit (χ)	Group	Representative Species	Main Interlayer Cation	Organic Complexes	Interlayer Swelling in Water
1:1	∼0	Kaolinite-serpentine	Kaolinite halloysite	None None (?)	Highly polar organics variety of polar organics	No swelling, one layer of water
2:1	∼0	Pyrophyllite-talc	Pyrophyllite talc	None None	No complex No complex	No swelling No swelling
	∼0.25–0.6	Montmorillonite-saponite (or smectite)	Montmorillonite, beidellite, hectorite, saponite	Na^{+}, Ca^{2+}, etc.	Wide variety of cationic and uncharged organics, simple and polymeric	Variable, extensive with some monovalent cations
	∼0.6–0.9	Vermiculite	Vermiculite	Mg^{2+}	Charged and uncharged organics	Limited, may be extensive with special treatment
	∼1	Mica	Muscovite, biotite	K^{+} (Na^{+}, Li^{+})	Long-chain mono- and dialkylammonium ions	May swell slightly with special treatment
	∼2	Brittle mica	Margarite	Ca^{2+}	Long-chain alkylammonium ions	No swelling

dipole interactions may be involved (Eltantawy and Arnold, 1972). The rate of intercalation is slow, and the nature of the exchangeable interlayer cation, so crucial in the adsorption of polar molecules, is apparently of little importance here.

Complexes with Polar, Uncharged Species

There is persuasive evidence that polar organics are held to clay mineral surfaces through the exchangeable inorganic cations by *ion-dipole interactions.* This type of bonding is more important than the *mineral-to-organic* or *organic-to-organic interactions* (Theng, 1974).

With highly polarizing cations such as Mg^{2+} and Al^{3+} the coordinated water molecules in the *primary hydration shell* (closest to the cation) may not be replaceable by the organic compounds. Bonding of the adsorbed organic to the cation is then effected through a *water bridge.* As an example, the interaction of benzonitrile with an air-dry film of Mg^{2+} – montmorillonite can be represented by (Serratosa, 1968):

$$Mg^{2+}\,\overset{\displaystyle H}{\overset{|}{O}}\text{–H}\ldots N\equiv C\text{–}$$

This view is supported by the observation that the infrared frequency of the O-H stretching band of water decreases whereas that of the deformation band increases from the respective values given by the parent clay sample. In addition, the C ≡ N stretching frequency increases from 2228 cm^{-1} in liquid benzonitrile to 2240 cm^{-1} in the adsorbed state. For weakly polarizing cations such as Na^{+} and Ba^{2+} or where special treatments are used to dehydrate the cation, direct ion-dipole links may be established. Thus in the case of benzonitrile the C ≡ N group interacts directly with the cation on evacuating the system. The amount of displacement of the C ≡ N stretching frequency from that shown by the free liquid depends on the nature of the cation, being greater for the more highly polarizing type: 12 cm^{-1} for Ba^{2+}, 21 cm^{-1} for Ca^{2+}, and 33 cm^{-1} for Mg^{2+}. Similar findings have been reported for numerous organic compounds.

Because polar organics compete with water for essentially the same sites around the cation, little or no adsorption occurs from dilute aqueous solutions of molecules containing less than 5 carbon atoms. If, however, the molecule is sufficiently large to be adsorbed, its chemical character influences the adsorption process (Hoffmann and Brindley, 1960). A measure of this character is its *CH-activity,* which refers to the activation of methylene (CH_2) groups by neighboring electron-withdrawing groups. In their classical papers on clay-organic reactions, Bradley (1945) and MacEwan (1948) suggested that CH-activity was related to the ability of the molecule to form C-H . . . O-Si bonds with the silicate surface. Subsequent studies, however, have cast doubt on the importance of this mode of interaction (Tensmeyer, Hoffmann, and Brindley, 1960).

Intercalated molecules with strongly polar groups tend to adopt an α_{II} orientation in which the carbon zigzag plane lies parallel to the silicate surface whereas compounds without such groups prefer an α_{I} orientation in which the zigzag plane is perpendicular to the surface (Brindley and Hoffmann, 1962; Greene-Kelly, 1955, 1956). In many cases, α_{I} and α_{II} complexes give Δ *values* (MacEwan, 1948), the difference between the observed basal spacing of the complex and the assumed thickness (0.95 nm) of an individual montmorillonite layer, which are shorter by about 0.06–0.08 nm and by 0.04–0.06 nm, respectively, than the minimum (van der Waals) thickness of the molecules. This apparent contraction of contact distances is of general applicability to clay-organic systems and can be accounted for by a recession, or keying, of the molecules into the ditrigonal depressions of the silicate surface [cf. Figure 1e)] without having to invoke any specific chemical interaction such as C-H . . . O hydrogen bonding. At sufficiently high surface concentrations, up to 3 layers of molecules may be intercalated, or the molecules may adopt an upright, tilted orientation.

Complexes with Organic Cations

Exchange Reactions. Under suitable conditions most organic cations are capable of replacing the (interlayer) inorganic cations from their exchange sites in montmorillonite and vermiculite (Theng, 1974). Of the complexes formed in this way, those involving *n*-alkylammonium ions ($C_nH_{2n-1}\,NH_3^{+}$ or RNH_3^{+}) have received the greatest attention. Except for some bulky cations, the exchange reaction is stoichiometric until it reaches a maximum at or close to the exchange capacity of the clay. With long-chain cations ($>C_8$), adsorption may exceed this capacity; the excess is usually present as the amine salt (RNH_3 halide), an equivalent amount of the halide being taken up.

The affinity of the organic cation for the clay increases regularly with molecular weight—that is, the larger the cation or the longer the alkyl (R) chain, the stronger its adsorption. The incremental increase in the free energy (ΔG) of the reaction per carbon atom is generally ascribed to the increased contribution of the (additive) *van der Waals forces* to the adsorption energy as cation size increases; entropy and hydration effects, however, may also be involved.

Basal-spacing measurements by X-ray diffractometry indicate that the organic cation is adsorbed with its shortest axis perpendicular to the silicate surface, either in an α_I or α_{II} orientation. Again the observed apparent shortening of contact distances can be attributed to keying effects. Single-layer complexes of flatly oriented cations (basal spacing ~ 1.25–1.35 nm) are formed up to C_{10}, when the area of the cation is less than the area per exchange site. This is so because the organic ions adsorbed on one basal surface can fit into the gaps between those attached to the opposing surface in the interlayer region. Beyond C_{10}, however, the cation area exceeds the area per exchange site. The *lock-and-key arrangement* is no longer possible, and double-layer complexes (basal spacing ~ 1.65–1.75 nm) are obtained. At sufficiently high surface concentrations, long-chain alkylammonium ions ($>C_{10}$) may assume an "end-on" orientation in which the alkyl chain makes a high angle with the silicate surface. This tilted conformation is generally preferred in vermiculite and obtains even with short-chain derivatives because the distance separating the negative charges in the layer is less than in montmorillonite. The basal spacing of vermiculite complexes therefore tends to show a linear, rather than a stepwise, increase with the number of C atoms in the alkyl chain.

Protonation Reactions. Adsorbed organic bases may convert into their corresponding cations by accepting protons derived from essentially three types of sources:

a. Those occupying exchange sites as in acid-washed clays:

$$RNH_2 + H^{\pm}\text{ clay} \rightleftharpoons RNH_3^+ - \text{clay}$$

b. Dissociation of residual water molecules in the primary hydration shell of highly polarizing (inorganic) cations occupying exchange sites at the clay surface. This type of acidity, besides being dependent on the water content and the nature of the cation, M, is also influenced by the basicity of the adsorbed base, B. The reaction may be represented by:

$$[M(H_2O)_x]^{n+} + B \rightleftharpoons BH^+ + [M(OH)(H_2O)_{x-1}]^{(n-1)+}$$

c. Protonated species already present at the surface, that is, by proton transfer from the cationic form of one base, B_1H^+ to another adsorbed base, B_2,

$$B_1H^+ + B_2 \rightleftharpoons B_2H^+ + B_1$$

In which direction this reaction will move, for any given system, clearly depends on the relative basicity of B_1 and B_2 as well as on the concentration of the reactants and products. Related to this is the formation of a *hemisalt* complex, which may arise when the amount adsorbed of a base B exceeds that of protons capable of protonating B. The protonated species BH^+ then shares its proton with the neutral (excess) B to yield a hemisalt, a cation of the type $(B \ldots H \ldots B)^+$.

Complexes with Pesticides and Bioorganic Compounds

Organic compounds of known pesticidal activity may be uncharged, cationic, or anionic. The mechanisms controlling the interactions of clays with pesticides therefore vary according to the charge characteristics of the particular compound (Bailey and White, 1970; Cloos, 1972; Weber, 1972; Theng, 1974).

Cationic pesticides interact by an ion-exchange mechanism. The adsorption is often so strong that the compounds become virtually immobilized and lose their bioactivity on contact with soil-clay particles. Intercalation of the pesticide may also lead to conformational changes in the molecule, rendering the compound resistant to microbial degradation. Under conditions when (positive) adsorption does take place, the clay-organic link is usually weak indicating that negatively charged pesticides are readily desorbed and so remain mobile and active for relatively long periods after application. Basic, uncharged pesticides are adsorbed primarily by ion-dipole interactions, either directly or through a water bridge; they may also be protonated under acid conditions. Though many pesticides are only sparingly soluble in water, limiting their uptake by the mineral fraction of soils, appreciable amounts can be adsorbed by the top layers of the soil where they interact with the organic constituents. This is why the bioactivity of pesticides in soils is often related more to the amount of organic matter than to that of the clay minerals present (Weber, 1972).

Because of their importance to organic matter transformations in soils, the interactions of clays with amino acids, purines, pyrimidines, and nucleosides have received much attention (Theng, 1974). Amino acids exist as cations under acid conditions, and hence their adsorption by clays occurs largely through an ion-exchange process. At or close to the isoelectric pH, these compounds exist as dipolar ions and adsorb by ion-dipole and H-bonding type interactions. Very little adsorption occurs at alkaline pH when amino acids are negatively charged. Adsorption may be enhanced in the presence of exchangeable transition metal cations that are capable of forming stable coordination (chelation) complexes involving the amino and/or the carboxyl groups of the amino acid. Similarly, the interactions of purines, pyrimidines, and nucleosides with clays is pH-dependent, and steric factors related to molecular geometry and aromaticity are also important.

Complexes with Organic Polymers

Because of the large size of many polymers, van der Waals type interactions are most important. Adsorption may also be favored by a gain in entropy of the system if the attachment of one polymer molecule results in the displacement of many water molecules from the clay surface. This entropy effect plays an important part in the adsorption of some uncharged, linear polymers, especially those that tend to uncoil on contact with the mineral (Greenland, 1965, 1972).

Uncharged polymers such as the polyvinyl alcohols are strongly and practically irreversibly adsorbed. Ion-dipole interactions, O-H. . . O-Si bonding, and the entropy effect are involved in the process. Although individual polymer segments may be relatively only weakly held, the chain as a whole is strongly adsorbed because it is attached to the clay by numerous points of contact. Polycations are adsorbed primarily by an ion-exchange reaction, and hence the segment-surface interaction is relatively strong. Like uncharged polymers, polycations can penetrate the clay interlayers where they exist in a largely collapsed conformation.

Little uptake occurs with polyanions, and negative adsorption has been reported under some conditions. Positive uptake, however, can take place in the presence of polyvalent cations acting as a bridge between the anionic groups of the polymer and the surface or when the electrolyte concentration in the bulk solution is raised. Adsorption is normally confined to the external surface of the clay. At low pH the crystal edges, by acquiring positive charges, are important in providing adsorption sites for polyanions. Under these conditions, however, the negative charges on the polymer will also be mostly neutralized, and some intercalation may occur. Thus at pH $\sim$ 2.5 fulvic and humic acids that make up the bulk of soil organic matter can form interlayer complexes with montmorillonite whereas no intercalation occurs above pH 6 (Schnitzer and Khan, 1972). The uptake of humic and fulvic acids by montmorillonite at pH $\sim$ 7 increases linearly with concentration. The strength of adsorption depends on the ionic potential of the exchangeable cations, that is, on the ratio of ionic charge and size. The polymer is apparently capable of penetrating the pore space between oriented clay crystals where it is held by H bonding between the anionic groups of the molecule and water molecules in the primary hydration shell of polyvalent cations (Theng and Scharpenseel, 1975).

Proteins and nucleic acids are rapidly adsorbed by clays, the amount taken up decreasing with an increase in pH. Adsorption, especially when it occurs on interlayer surfaces of the clay, may partly protect these compounds from microbial attack. This may be ascribed to changes in molecular conformation of the adsorbed species and/or to the inaccessibility of specific groups in the macromolecule to microbial enzymes.

Clay-polymer systems are of great practical importance. One area of application is the improvement of the structure and foundation engineering properties of soils. To be an effective aggregate stabilizer, the polymer must be able to penetrate the pore space of the aggregate as well as be capable of forming strong interparticle bonds. At first sight, it might appear that polycations would be suited for this purpose, but this does not accord with practice. Apparently the rate of collapse of the polycation onto the negatively charged clay surface is so great that only very limited interparticle bridging can be achieved. Under certain pH conditions it is possible to stretch a polyanion chain through intersegment-charge repulsion and so optimize its interparticle bridging efficiency; under such conditions, however, the polyanion is also largely repelled by the clay surface. The number of segments taking part in interparticle bonding is therefore not likely to be large, and polyanions are generally not greatly more superior to polycations in stabilizing clay aggregates. On the other hand, the water-soluble, uncharged polyvinyl alcohol (PVA) when introduced into a natural soil or clay aggregate can bring about a large increase in the mechanical strength and water stability of the aggregate. Although in aqueous solution PVA exists as a random coil, the chain apparently uncoils after entering the large pores and slowly spreads like a "coat of paint" over the pore surface. This process binds together regions of oriented clay crystals (domains) that constitute the walls of the pores. On wetting, water moves slowly and gradually from these pores to the remainder of the pore space and aggregate slaking, caused by the sudden expulsion of entrapped air, is prevented (Greenland, 1972).

B. K. G. THENG

References

Bailey, G. W., and White, J. L., 1970. Review of adsorption and desorption of organic pesticides by soil colloids, with implications concerning pesticide bioactivity, *Jour. Agric. Food Chem.*, **12,** 324–332.

Bradley, W. F., 1945. Molecular associations between montmorillonite and some polyfunctional organic liquids, *Jour. Am. Chem. Soc.*, **67,** 975–981.

Brindley, G. W., 1970. Organic complexes of silicates, *Reunion Hispano-Belga de Minerales de la Arcilla (Madrid)*, 55–66.

Brindley, G. W., and Hoffmann, R. W., 1962. Orientation and packing of aliphatic chain molecules on montmorillonite, *Clays Clay Minerals,* **9,** 546–556.

Brown, G., ed., 1961. *The X-Ray Identification and Crystal Structures of Clay Minerals.* London: Mineralogical Society, 544p.

Calvet, R., 1963. Mise au point bibliographique. Les complexes organiques des argiles, *Ann. Agron.*, **14,** 31–117.

Calvet, R., and Chassin, 1973. Complexes organiques des argiles: Mécanismes de formation, méthodes d'étude et applications, *Bull. Groupe Fr. Argiles,* **25,** 87–113.

Cloos, P., 1972. Interactions entre les pesticides et la fraction minérale du sol, *Pédologie,* **22,** 148–173.

Demolon, A., and Barbier, G., 1929. Conditions de formation et constitution du complexe argilo-humique des sols, *Acad. Sci. Comptes Rendus* (Paris), **188,** 654–656.

Eltantawy, I. M., and Arnold, P. W., 1972. Adsorption of *n*-alkanes by Wyoming montmorillonite, *Nature Phys. Sci.*, **237,** 123–125.

Farmer, V. C., 1971. The characterization of adsorption bonds in clays by infrared spectroscopy, *Soil Sci.*, **112,** 62–68.

Frissel, M. J., and Bolt, G. H., 1962. Interaction between certain ionizable organic compounds (herbicides) and clay minerals, *Soil Sci.*, **94,** 284–291.

Greene-Kelly, R., 1955. Sorption of aromatic organic compounds by montmorillonite. I. Orientation studies, *Faraday Soc. Trans.*, **51,** 412–424.

Greene-Kelly, R., 1956. Sorption of saturated organic compounds by montmorillonite, *Faraday Soc. Trans.*, **52,** 1281–1286.

Greenland, D. J., 1965. Interaction between clays and organic compounds in soils, *Soils Fertilizers,* **28,** 415–425, 521–532.

Greenland, D. J. 1971. Interactions between humic and fulvic acids and clays, *Soil Sci.*, **111,** 34–41.

Greenland, D. J., 1972. Interactions between organic polymers and inorganic soil particles, in M. De Boodt, ed., *Proc. Symp. Fundamentals of Soil Conditioning (Ghent),* 37(3), 897–914.

Grim, R. E., 1953. *Clay Mineralogy.* New York: McGraw-Hill, 384p.

Hoffmann, R. W., and Brindley, G. W., 1960. Adsorption of non-ionic aliphatic molecules from aqueous solutions on montmorillonite. Clay-organic studies II, *Geochim. Cosmochim. Acta,* **20,** 15–29.

Jordan, J. W., 1963. Organophilic clay-base thickeners, *Clays Clay Minerals,* **10,** 299–308.

MacEwan, D. M. C., 1948. Complexes of clay with organic compounds. I. Complex formation between montmorillonite and halloysite and certain organic liquids, *Faraday Soc. Trans.*, **44,** 349–367.

MacEwan, D. M. C., 1962. Interlamellar reactions of clays and other substances, *Clays Clay Minerals,* **9,** 431–443.

Mortland, M. M., 1970. Clay-organic complexes and interactions, *Adv. Agron.*, **22,** 75–117.

Nahin, P. G., 1963. Perspectives in applied organo-clay chemistry, *Clays Clay Minerals,* **10,** 257–271.

Norrish, K., 1954. The swelling of montmorillonite, *Disc. Faraday Soc.*, **18,** 120–134.

Schnitzner, M., and Khan, S. U., 1972. *Humic Substances in the Environment,* New York: Marcel Dekker, 327p.

Serratosa, J. M., 1968. Infrared study of benzonitrile-montmorillonite complexes, *Am. Mineralogist,* **53,** 1244–1251.

Solomon, D. H., 1968. Clay minerals as electron acceptors and/or donors in organic reactions, *Clays Clay Minerals,* **16,** 31–39.

Tensmeyer, L. G.; Hoffmann, R. W.; and Brindley, G. W., 1960. Infrared studies of some complexes between ketones and calcium montmorillonite, *Jour. Phys. Chem.*, **64,** 1655–1662.

Theng, B. K. G., 1970. Interactions of clay minerals with organic polymers. Some practical applications, *Clays Clay Minerals,* **18,** 357–362.

Theng, B. K. G., 1971. Mechanisms of formation of colored clay-organic complexes. A review, *Clays Clay Minerals,* **19,** 383–390.

Theng, B. K. G., 1974. *The Chemistry of Clay-Organic Reactions.* London: Adam Hilger, 343p.

Theng, B. K. G., and Scharpenseel, H. W., 1975. Adsorption of ^{14}C-labelled humic acid by montmorillonite, *Proc. Internat. Clay Conf. (Mexico City),* 643–653.

Theng, B. K. G., and Walker, G. F., 1970. Interactions of clay minerals with organic monomers, *Israel Jour. Chem.*, **8,** 417–424.

Weber, J. B., 1972. Interaction of organic pesticides with particulate matter in aquatic and soil systems, *in* R. F. Gould, ed., *Fate of Organic Pesticides in the Aquatic Environment,* Advances in Chemistry Series 111 (Washington, D.C.), A.C.S., 55–120.

Weiss, A., 1969, Organic derivatives of clay minerals, zeolites, and related minerals, *in* G. Eglinton, and M. J. T. Murphy, eds., *Organic Geochemistry.* Berlin: Springer-Verlag, 737–781.

Cross-references: *Adsorption Phenomena; Aggregate Stability; Clay Minerals: Exchange Phenomena; Herbicide Diffusion; Humus; Microbial Ecology and Clay Minerals; Soil-solvent Reactions.*

CLAYS, PHYSICAL CHEMISTRY

The physicochemical behavior of clays is strongly influenced by their high specific surface area and by the net negative charge carried by many of the clay minerals.

Nature of Surface

The atoms at the surfaces are attracted most strongly on the side facing inward toward the solid. They have unsatisfied valencies on their outer sides, and the residual bonds are believed to be responsible for adsorption. The unsymmetrical distribution of forces causes strain, and surfaces often show high chemical reactivity.

Clays are composed of particles smaller than 2 μm (equivalent spherical diameter) and have properties of the colloidal state (Banin, 1976). Such small particles have a high specific surface area, which is made relatively larger in the clay minerals themselves because of their platy, or elongate, morphology (see *Clay Minerals*). In phyllosilicates the basal surface is much larger than the edge surface.

In 1:1-layer-structure minerals, one basal surface consists of oxygen ions linked to tetrahedrally coordinated silicons (*siloxane surface*). The opposite surface is composed of hydroxyl ions bonded to octahedrally coordinated cations. In the 2:1-layer-structure minerals, both outermost sheets are siloxane surfaces. Edges are different, being composed of oxygen or hydroxyl ions linked to alternating rows of silicon and either aluminum in dioctahedral minerals or divalent ions such as magnesium in trioctahedral minerals.

Many of the clay minerals carry a permanent negative charge due to unbalanced ionic substitutions in the structure, a rather unusual feature among hydrophobic colloids. The radius ratio of aluminum to oxygen is so very close to the critical value for transition from 4- to 6-fold coordination that aluminum is able to occupy both tetrahedral and octahedral sites. Replacement of Si^{4+} by Al^{3+} in tetrahedral positions imparts a unit negative charge to the structure. There is a considerable range in the nature of the ions that may be substituted within the structure of the clay minerals, and these ions may be distributed with varying degrees of order and disorder (Besson et al., 1974). The net negative charge is balanced by external cations, which are exchangeable in varying proportions depending on the mineral.

Clay soils contain various amounts of primary minerals reduced by attrition to the clay-sized range. The proportion and type of such minerals depends on the intensity and duration of chemical weathering. Fine quartz, however, is a very common constituent. Crystals of this mineral are generally surrounded by a disordered skin of the order of 0.5 μm thick, which constitutes a significant fraction of a particle in the less than 2 μm size range. In neutral or alkaline solution, adsorbed hydroxyl ions (a *silanol surface*) impart a negative charge to colloidal silica, but the charge is lost in acid solution. Reaction with aluminate ion, $Al(OH)_4{}^{1-}$, preserves the negative charge to lower pH (Iler, 1973).

Clays in Liquid Media

The principles worked out for other colloids apply to clays, but clay systems are unusually complex. A stable suspension is formed when there is a net repulsive force between the particles. The repulsive force originates in the electric charges that the particles carry. In clays there are three principal causes for this charge. First, there is the permanent negative charge previously mentioned; second, ions of the same type as those in the structure may be adsorbed; and third, ions different from those in the structure may be attached to active sites on the surface. The type of ion adsorbed depends on the composition, concentration, and pH of the solution; sometimes the charge carried by colloidal particles is reversible.

There is considerable evidence to support the suggestion made by Van Olphen (1977) that in layer-structure silicates, the negative charge is associated with the platy surfaces while a positive charge may develop on the edges. This phenomenon is attributed to the aluminum ions that occupy positions in both the octahedral and tetrahedral layers. In acid solution, aluminum hydroxide acquires a positive charge by adsorption of Al^{3+} as potential determining ions, and in alkaline solution the charge reverses to negative due to adsorption of hydroxyl or aluminate ions. Correspondingly, kaolinite in acid solution shows a capacity for anion adsorption, which it loses in alkaline solution. Negatively charged colloidal gold and negatively charged particles of montmorillonite have been shown by electron microscopy to be adsorbed at edges of kaolinite crystals.

A clay suspension does not have a net electric charge as the negative charge on the clay mineral particles is balanced by an equivalent positive charge carried by counterions in the solution; the particle charge and the charge carried by the counterions is referred to as the *electrical double layer.* The ionic concentration is higher within the double layer than in the bulk solution and falls off exponentially with distance from the particle surface. The nature and extent of the double layer depends on the composition and concentration of the solution, the charge density of the particle, and the solution temperature. There is a continuous interchange between ions within the diffuse double layer and ions in the solution, and since the distribution is statistical, there are ions present with the same polarity as that of the particle. It is considered that there is a linear drop in potential within an inner layer adjacent to the particle (sometimes called the *Stern layer*), the position of which depends on ionic size.

Chemical, electrostatic, and van der Waals bonds are involved in the interactions between the clay minerals themselves and between the clay minerals and ions in solution.

Some clay minerals, notably the smectites, vermiculite, and swelling chlorite, take up varying proportions of water and other liquids between the layers. Differences in the extent of swelling have been related to the nature of the interlayer cation and to the charge density. Estimates of this property have been based on chemical analysis and on measurement of the change in basal spacing of the complex formed between swelling clay mineral and alkylammonium cations of variable chain length (Lagaly and Weiss, 1969).

Brownian movement causes colloidal particles to be in a state of continuous erratic motion when in suspension. If there are net attractive forces between particles, chance approaches will often lead to collision and adhesion as the size of the units increases. The agglomerates settle out under gravity, and the suspension is said to *flocculate* (see *Flocculation*). In a stably dispersed colloidal system, the particles repel one another so the chances of collision and adhesion are small.

The *stability* of a colloidal suspension depends on the balance of attractive and repulsive forces when particles interact. If the diffuse double layers are extensive, it will be through these layers that initial contact will be made between approaching particles. Hence ions with like charges will encounter one another, and repulsive forces will rapidly increase. This is the condition in a stably dispersed colloidal suspension. If the double layers are compressed, their extent may be exceeded by the range of the van der Waals forces of attraction so when particles approach one another, there is a good probability of adhesion. "Adhesion" is the condition when a colloidal suspension is flocculated.

The *extent* of the double layer depends on the nature and concentration of the ions in solution. Ions of high valency and opposite charge to that on the colloidal particle are effective flocculating agents since they compress the double layer. Organic compounds such as alcohol that lower the dielectric constant of the medium and thereby compress the double-layer function in a similar manner.

While these principles apply to clays, the strong probability that edges may, under certain circumstances, acquire a positive charge in addition to the negative charge associated with the faces of the layer-structure silicates makes their behavior unusually complex. Clay minerals may associate in an edge-to-edge, edge-to-face, or face-to-face mode depending on the balance and distribution of attractive and repulsive forces. Explanations for rheological behavior, permeability of argillaceous sediments, and fabric of clay soils have been proposed on the basis of these ideas.

Surface Reactions

The ions that form the double layer are in dynamic equilibrium with ions in the remainder of the solution, and if the composition of the solution changes, a corresponding change takes place in the ionic population of the double layer. Such exchange reactions are governed by the *law of mass action* and may involve either cations or anions (see *Exchange Phenomena*). Cation-exchange measurement is commonly based on the determination of desorbed ammonium ions with which the clay has been previously saturated; other ions and a variety of analytical procedures have been used (Chhabra et al., 1976; Black, 1965). Quantitative values of exchange capacity are commonly given in mequiv/100 g (milliequivalents per 100 grams) of dry solid for neutral solutions. The exchange capacity of soils is derived from the clay minerals to a significant degree, but other minerals such as zeolites, noncrystalline constituents, and organic materials also make a contribution. Clay-organic interactions (q.v.), selective sorption and fixation, uptake of organic compounds, and catalytic syntheses are also important reactions at the surfaces of clay minerals.

J. E. GILLOTT

Financial support from the National Research Council of Canada is acknowledged.

References

Banin, A., 1976. Recent advances in research on the colloidal properties of clay systems, *Proc. Internat. Clay Conf.* Wilmette, Ill.: Applied Publishing, 311–317.

Besson, G.; Miljud, A.; Tchoubar, C.; and Mering, J., 1974. Order and disorder relations in the distribution of the substituents in smectites, illites and vermiculites, *Clays Clay Minerals,* **22,** 379–384.

Black, C. A., ed., 1965. *Methods of Soil Analysis,* Agronomy Series No. 9, vols. I and II. Madison: American Society Agronomy, 1572.

Chhabra, R.; Pleysier, J.; and Cremers, A., 1976. The measurement of the cation exchange capacity and exchangeable cations in soils: A new method, *Proc. Internat. Clay Conf.* Wilmette, Ill.: Applied Publishing, 439–449.

Edwards, A. P., and Bremner, J. M., 1967. Microaggregates in soils, *Jour. Soil Sci.,* **18,** 64–73.

Gillott, J. E., 1968. *Clay in Engineering Geology.* New York: Elsevier, 296p.

Greenland, C. J., 1965. Interaction between clays and organic compounds in soils, *Soils Fertilizers,* **28**(6), 521-532.

Iler, R. K., 1973. Colloidal silica, *in* E. Matijevic, ed., *Surface and Colloid Science.* New York: Wiley, **6**, 1-100.

Kidder, G., and Reed, L. W., 1972. Swelling characteristics of hydroxy-aluminum interlayered clays, *Clays Clay Minerals,* **20**, 13-20.

Lagaly, G., and Weiss, A., 1969. Determination of the layer charge in mica-type layer silicates. *Proc. Internat. Clay Conf. (Tokyo),* **I**, 61-80.

Sawhney, B. L., 1972. Selective sorption and fixation of cations by clay minerals: A review, *Clays Clay Minerals,* **20**, 93-100.

van Olphen, J., 1977. *An Introduction to Clay Colloid Chemistry,* 2nd ed. New York: Interscience, 318p.

Cross-references: *Adsorption Phenomena; Clay Minerals, Silicates; Clay-organic Reactions; Exchange Phenomena; Flocculation; Physical Chemistry; Redox Reactions; Soil Chemistry; Vol. IVA; Clay Minerals–Base Exchange; Colloids; Ion Exchange.*

COLLOIDS—See CLAY MINERALS, SILICATES

CONDUCTIVITY, ELECTRICAL

Electrical conductivity is the ratio of the electric current density to the electric field in a material. Electrical conductivity measurements are applied in studies of both agricultural and engineering problems. Soil scientists use such measurements mainly to characterize the salt contents of soil, irrigation or drainage waters, and occasionally as an indirect way of determining soil water content.

The electrical conductivity, or specific electrical conductance, of a solution is the reciprocal value of its electrical resistivity. Electrical resistance R (ohms) is directly proportional to the distance L (cm) between two parallel electrodes immersed in a solution or soil extract and inversely proportional to the cross-sectional area A (cm^2) of the material (solution) between the electrodes. Thus $R = rL/A$, where r is a proportionality constant known as the electrical resistivity. The units for electrical resistivity (r) are ohm-cm. The electrical resistance (R) is expressed in ohms; its reciprocal value ($1/R$) ohm^{-1} (or mho) is the electrical conductance, also referred to as siemens (S) in the S.I. system. Similarly, the reciprocal of the electrical resistivity ($1/r$) is the electrical conductivity (EC). The units of $1/r$ or EC are 1/ohm-cm or mho/cm, or in S.I. units S cm^{-1} (= kg^{-1} m^{-3} s^{-3} A^2). This is the standard unit for conductivity (Bower and Wilcox, 1965).

Because most soil extracts have an EC value that is usually less than a thousandth of a unit (mho/cm), it is customary to choose a smaller subunit. The expression "millimho per centimeter" or "mmho/cm (EC × 10^3)," for example, has been found to be a practical conductivity unit for most soil salinity. Because the electrical conductivity of aqueous salt solutions increases by about 2% per centigrade degree, the standard temperature for reporting EC measurements is 25°C (Bolt and Bruggenwert, 1976).

EC Measurements in Soil

The electrical conductivity should ideally be measured in an extract from a soil sample with a moisture content between field capacity and the permanent wilting point. Because such an extract is very difficult to obtain and because the concentration of soluble salts in the soil solution (q.v.) is influenced by the moisture content, variable water-soil ratios have been used, especially those of the ratios 1:1, 2:1, 5:1, and 10:1 (Carpena et al., 1968).

Soil salinity is normally characterized according to the electrical conductivity of the saturation extract (ECe), which is obtained by suction-filtration of a soil-water paste. The electrical conductivity of the extract is determined according to the relation

$$\text{ECe, mho/cm at } 25^\circ\text{C} = \frac{0.0014118 \times R_{std}}{R_{ext}}$$

where the value 0.0014118 is the electrical conductivity of a 0.01*N* KCl solution, R_{std} the resistance of the conductivity cell filled with its 0.01*N* KCl solution, and R_{ext} the resistance of the cell containing the extract (Bower and Wilcox, 1965).

The electrical conductivity, the total ionic concentration, and the osmotic pressure of the soil-water extracts are positively correlated. Due to the somewhat poorer correlation between electrical conductivity and the salt concentration on a weight basis, the following conversions are frequently used:

Salt concentration (mg/1)
= 640 × EC(mmho/cm)

Total cation concentration (meq/1)
= 10 × EC (mmho/cm)

Osmotic pressure (atm)
= 0.30 × EC (mmho/cm)

EC of Irrigation Waters

The concentration of total soluble salts is an important criterion of quality of irrigation water (q.v.) because its salinity level can produce yield-limiting soil-salinity values under general field conditions (Branson et al, 1975). The concentration levels of total ionizable salts in irrigation waters are commonly inferred from electrical conductivity values. Commercial instruments are available for measuring the salinity of irrigation waters by a rapid conductivity measurement with an accuracy of ±0.33 parts per thousand over the range 0.2–50 parts per thousand. The unit commonly used for expressing irrigation water electrical conductivity (ECiw) is micromhos per centimeter (μmho/cm). This is frequently used as an index of total soluble salt concentration calibrated to parts per million (ppm) or milliequivalents per liter (meq/1).

EC Measurements and Soil Water Content

Soil water content and matric potential may be determined indirectly from measurements of electrical conductivity (Gardner, 1965; Marshall, 1963). A common technique employs the use of buried porous gypsum blocks that contain electrodes. Their water content varies according to changes in the suction of soil water (q.v.), i.e., the water content of the blocks comes to equilibrium with that of the soil. Estimates of soil water content may thus be obtained by calibrating readings of electrical conductivity from the gypsum blocks. The so-called gypsum block method is sometimes used for measuring water content of soils in the field (see *Water Content and Retention*).

Interpretation of Conductivity Readings

The level of salt concentration in the soil has a direct bearing on crop growth. The harmful effects of excess soluble salts on plant growth have been thoroughly investigated by the U.S. Salinity Laboratory Staff (1954), reported by Bernstein (1970) and reviewed by Russell (1973). The expected effects on yields of salts have been determined for a wide range of plants. The U.S. Salinity Laboratory Staff (1954) has also developed a classification of saline soils that anticipates expected damages to crops (Table 1). If, for example, the ECe is less than 4 mmho/cm, no crop is likely to suffer from general salt injury, but when the conductivity exceeds 8 mmho/cm, only tolerant crops will yield satisfactorily. When the ECe exceeds 16 mmho/cm, only a few very tolerant crops are likely to give economic yields.

General groups of saline soils are recognized on the basis of salt contents. The salt-affected soils (q.v. in Vol. XII) listed in Table 1 are separated by limiting values chosen for their general significance to plant responses. This division into saline and nonsaline soils is generally taken at ECe 4 mmho/cm. Thus when the ECe exceeds this critical value the concentration of neutral soluble salts is sufficient to interfere seriously with growth of most plants.

CHARLES W. FINKL, JNR.

References

Bernstein, L., 1970. Salt tolerance of plants, *U.S. Dept. Agric. Inform. Bull. No. 283,* 23p.

Bolt, G. H., and Bruggenwert, M. G. M., eds., 1976. *Soil Chemistry.* Amsterdam: Elsevier, 281p.

Bower, C. A., and Wilcox, L. V., 1965. Soluble salts, *in* C. A. Black, ed., *Methods of Soil Analysis.* Madison: American Society of Agronomy, 933–951.

TABLE 1. Expected Effects of Salt on Crop Growth.

Conductivity of Saturation Extract (mmho/cm)*	Total Salt Content (%)	Crop Reaction	Classification
< 2	< 0.1	Salinity effects mostly negligible	Nonsaline
2–4	0.1–0.15	Yields of very sensitive crops may be restricted	Slightly saline
4–8	0.15–0.35	Yields of many crops restricted	Moderately saline
8–16	0.35–0.70	Only tolerant crops yield satisfactorily	Very saline
> 16	> 0.70	Only a few very tolerant crops yield satisfactorily	Extremely saline

After U.S. Soil Salinity Laboratory Staff, 1954.
*mS cm^{-1} in S.I.

Branson, R. L.; Pratt, P. F.; Rhoades, J. D.; and Oster, J. D., 1975. Water quality in irrigated watersheds, *Jour. Environ. Qual.*, 4(1), 33–40.

Carpena, O.; Guillen, M. G.; Fernandez, F. G.; and Caro, M., 1968. Saline soil classification using the 5:1 aqueous extract, *9th Internat. Congr. Soil Sci. Trans. (Adelaide)*, 483–490.

Gardner, W. H., 1965. Water content, *in* C. A. Black, ed., *Methods of Soil Analysis*. Madison: American Society of Agronomy, 82–127.

Marshall, T. J., 1963. Relations between water and soil, *Commonwealth Bur. Soils Tech. Commun. No. 50*, 91p.

Russell, E. W., 1973. *Soil Conditions and Plant Growth*. London: Longmans, 849p.

U.S. Salinity Laboratory Staff, 1954. Diagnosis and improvement of saline and alkali soils, *in* L. A. Richards, ed., *U.S. Department Agriculture Handbook No. 956*, 160p.

Wilcox, L. V., 1950. Electrical conductivity, *Am. Water Works Assoc. Jour.*, **42**, 775–776.

Cross-references: *Electrochemistry; Irrigation Water; Salt-Leaching; Soil Solution; Water Content and Retention; Vol. IVA: Soil Salinity and Alkalinity; Vol. XII, Pt. 2: Salt-affected Soils.*

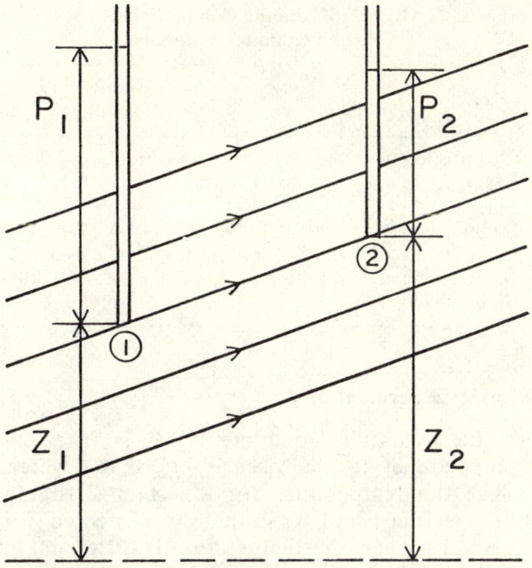

FIGURE 1. Pressure head and elevation head in a system of parallel streamlines.

CONDUCTIVITY, HYDRAULIC

Hydraulic conductivity K is the proportionality factor in Darcy's equation (Darcy, 1856), which relates the velocity v of water in a soil or other porous medium to the hydraulic gradient i as follows:

$$v = Ki$$

The velocity is the gross velocity, or *Darcy velocity* which is the velocity as if the water were also moving through the solid particles of the medium. Thus the volume rate of flow through a certain cross-sectional area normal to the flow direction is calculated as vA, where A is the gross area of the cross-section (solids as well as pores). The actual, or macroscopic, velocity of water in the pores is about equal to v/n, where n is the porosity of the medium. The *hydraulic gradient* i is the rate of decrease in the total head (sum of pressure head and elevation head) along a streamline in the direction of flow. If the streamlines are parallel (Figure 1), i is calculated as

$$i = \frac{(p_1 + z_1) - (p_2 + z_2)}{\ell}$$

where

p = pressure head (height to which water will rise above bottom of vertical, open-ended tube inserted to point where pressure head is to be measured)

z = elevation head (height of point above arbitrary horizontal reference plane)

ℓ = distance between two points on streamline where p is measured

The subscripts 1 and 2 refer to p and z at points 1 and 2, respectively (Figure 1). Since i is dimensionless, K has the dimension of velocity (length/time).

Units and Orders of Magnitude

Customary units for K are m/day for practical and field work and cm/s for theoretical and laboratory work. Orders of magnitude for K of granular materials are given in Table 1.

English units of hydraulic conductivity include inches per hour or per day, feet per day, and U.S. gallons per day through a cross-sectional area of 1 ft^2/unit hydraulic gradient. The latter unit, standardized to 15.56°C (60°F) is called the *meinzer* and is used in groundwater hydraulics. The various units are converted to m/day as follows:

1 cm/sec	= 864 m/day
1 inch/day	= 0.0254 m/day
1 foot/day	= 0.3048 m/day
1 meinzer	= 0.04075 m/day (at 15.56°C)

TABLE 1. Hydraulic Conductivity of Selected Granular Materials.

Material	Conductivity (m/day)
Clay (dispersed)	Essentially 0–0.1
Loam	0.1–1
Fine sand	1–2
Medium sand	2–5
Coarse sand	5–20
Natural sand and gravel mixes	10–100
Clean gravels	50–1000

Intrinsic permeability

The hydraulic conductivity K is inversely proportional to the viscosity μ of the water. Thus the temperature for a certain K should be specified, and K should be corrected for viscosity when the temperature is different. In practice, these specifications are seldom necessary because K cannot be evaluated sufficiently accurately to warrant such correction. The constancy of the product $K\mu$ is the basis for the *intrinsic permeability*, which expresses the permeability of a soil or other porous medium as a property of the medium only, independent of density and viscosity of the fluid. The intrinsic permeability k is defined as

$$k = \frac{K\mu}{\rho g}$$

where

K = hydraulic conductivity (length/time)
μ = viscosity of fluid (centipoises)
ρ = density of fluid
g = acceleration due to gravity

The dimension of k is length^2, and the customary unit is the *darcy*, which is 0.987×10^{-8} cm^2. For water at 20°C, the darcy corresponds to a hydraulic-conductivity value of 0.831 m/day.

Factors Affecting Hydraulic Conductivity

The hydraulic conductivity concept and Darcy's equation are based on laminar flow. Subsurface water flow is usually laminar. Turbulent flow may occur in media with large pores and high hydraulic gradients. Turbulent flow in porous media has been observed to start at Reynolds numbers of 60 to 700 (Hubbert, 1956). The *Reynolds number* for porous media is based on the Darcy velocity and the average particle diameter. For most subsurface flow systems, the Reynolds number is less than 1, well below the range of turbulent flow.

The hydraulic conductivity of a porous medium depends on the sizes and configurations of the pores. Anything affecting the pore configuration will affect K. If some of the pores are filled with air, the pore area available for water flow is reduced, and hence K is reduced. Such air may be entrapped, which can occur below a water table after it has been rising into unsaturated material. The K value of soil after rapid rewetting may be only one-half the K value of the soil at complete saturation (Bouwer, 1966). Entrapped air below the water table may also be formed by biological activity or by air getting out of solution as water enters groundwater of a higher temperature, which sometimes happens in groundwater recharge with wells and is called "air binding." Because of air in pores, K of unsaturated soil (*vadose zones*) is much lower than K at saturation. As the water content of the material decreases, the unsaturated hydraulic conductivity decreases. The unsaturated hydraulic conductivity must therefore be expressed in relation to the water content or the water tension (negative pressure head) of the material.

The concentration and cationic composition of the soil water determines whether clay particles in a soil are flocculated or dispersed, which in turn determines the structure of the soil and hence K. If water with a high sodium content enters the soil, clay particles may become dispersed, which results in a reduction in K. Conversely, water with divalent or trivalent ions or water with a high salt concentration (seawater) will cause clay particles to flocculate. When such water enters a soil, the structure can be improved and hence K increases (McNeal, 1968).

In dense clays, water moves through the very fine pores between clay particles. Surfaces of clay particles are negatively charged, causing adsorption of cations. Polar molecules, like water, are also influenced by the clay-surface charges, giving water near the clay particles a crystalline or ice-like structure. This water is not as mobile as free water, and pore water may not move at all at small hydraulic gradients. Thus a certain "threshold" gradient may be required to initiate water movement through such fine pores (Swartzendruber, 1969). Once water moves through the pores, the amount of water molecules participating in the flow then depends on the gradient. These phenomena give the net effect of a gradient-dependent K value. Additional phenomena such as the flow of water changing the orientation of clay particles and electrokinetic effects in dense clays also cause K to be affected by the hydraulic gradient (Elnaggar, Karadi, and Krizek, 1974).

Like the imbrication of gravel in alluvial materials, soil particles also tend to be oriented, particularly when the soil has been deposited in water. This orientation causes the pore configuration in the vertical direction to differ from that in the horizontal direction, which in turn causes a lower K for vertical flow than for horizontal flow, even in seemingly uniform soil. Directional hydraulic conductivity, called *anisotropy*, is usually considered in two dimensions only (vertical and horizontal). For a large aquifer, or groundwater, basin, however, directional K may also be distinguished in the horizontal plane. The K value will usually be greater parallel to the direction of the stream at the time the alluvial materials were deposited than normal thereto because coarse-textured layers tend to be more continuous in the direction of old streambeds.

Most alluvial soil consists of horizontal layers of materials with different textures. The layers may be thin (several millimeters or less) or thick (several centimeters of more). When a layered soil is treated as one medium, it should be considered as an anisotropic soil with K greater in horizontal than in vertical direction. This is because the average vertical K—calculated as the harmonic mean of the K values of the individual layers,

$$\frac{n}{\left(\frac{1}{K_1} + \ldots + \frac{1}{K_n}\right)}$$

if the layers are equally thick is lower than average K for horizontal direction, which is calculated as the arithmetic mean

$$\frac{(K_1 + \ldots + K_n)}{n}$$

for equally thick layers (Bouwer, 1969).

Measurement of Hydraulic Conductivity

Since K is controlled by the pore configuration, it seems logical to relate K to the pore-size distribution of the medium. Because there are so many variations in pore size, shape, and arrangement, this relationship is, however, only applicable to uniform materials consisting of spherical particles. Such materials are seldom, if ever, encountered in practice. Thus direct measurement of K for the soil material(s) in question is preferred. The simplest technique is to take an "undisturbed" cylindrical sample of the soil, saturate it if necessary, and let water flow through it in the laboratory (Bouwer and Jackson, 1974). From the velocity and the hydraulic gradient observed on the sample, K can be calculated with Darcy's equation. Because truly undisturbed samples are difficult to obtain and the sample size is relatively small, laboratory methods have limited usefulness and direct measurement of K in the field is usually preferred.

The field methods for measuring K are divided into those where K is measured below the water table and those where K is measured in initially unsaturated soil (above the water table or where there is no water table at all). Most methods for measuring K below the water table are pumped-hole techniques, where a vertical cylindrical hole is dug, and the water level in the hole is allowed to reach equilibrium with the water table. A volume of water is then suddenly removed from the hole and the subsequent rate of rise of the water level in the hole is measured. The K value for the soil or aquifer material surrounding the hole is then calculated from this rate of rise and the geometry of the system. These techniques include the piezometer, tube, auger hole, and well methods (listed in order of increasing volume of soil or aquifer "sampled" for K) (Bouwer and Jackson, 1974).

With the *piezometer method*, a small-diameter tube (usually 10 cm or less in diameter) is vertically installed to the depth at which K is to be measured. A small cavity is augered below the bottom of the tube. After equilibrating and then lowering the water level in the tube, water flows into the cavity causing the water level in the tube to rise. If the height of the cavity is large as compared with the diameter, the piezometer yields K in mainly horizontal direction. As the height of the cavity decreases, the measured K reflects more and more the hydraulic conductivity in vertical direction. If the cavity height is zero, the method is called the "tube method," which is used for cased auger holes.

With the *auger-hole technique*, an auger hole is dug into the soil below the water table. The hole might range from 1 to several m deep and 10 to 20 cm in diameter. The K value, calculated from the rise of the water level in the auger hole after sudden removal of a quantity of water, primarily reflects K in horizontal direction. When the auger-hole principle is applied to groundwater wells, it is usually called the "slug test" (Lohman, 1972). A modification of the auger-hole method is the two- or four-hole technique where water is pumped from one auger hole into another a few meters away. This creates a steady-state flow between the holes, and K is evaluated from the water level difference between the holes.

Another technique is the traditional *well-pumping test,* where water is continually pumped from a well, and K (or rather the product of K and the thickness of the aquifer) for horizontal flow is evaluated from the pumping rate and the draw-down of the water table, as measured with observation wells located 10 to 100 m or more away from the well (Lohman, 1972).

When the soil for which K needs to be determined is not saturated at the time of the test (e.g., above a water table), a part of the soil is artificially wetted, and K is determined from a flow system created in that wetted zone. The methods for measuring K in initially unsaturated soil include the shallow-well pump-in, the infiltration-gradient, and double-tube methods (Bouwer and Jackson, 1974).

The *shallow-well, pump-in method* consists of applying water to an auger hole and maintaining a constant water level in the hole until the infiltration rate has become constant, which may take several days. The value of K of the wetted zone around the hole is then calculated from the final infiltration rate and the system geometry. The resulting value primarily reflects K in horizontal direction.

With the *infiltration-gradient techniques,* water is applied to a cylindrical device that may be installed in the surface of the soil, like the infiltrometer and air-entry permeameter techniques, or to two concentric tubes in an auger hole. The vertical hydraulic gradient in the (upper) portion of the wetted zone is measured with tensiometers or piezometers. The gradient can also be determined from the pressure drop in the water above the ground if the infiltrometer is a closed cylinder and the supply of water is stopped, like the air-entry permeameter. The infiltration rate is also measured so that K of the wetted soil can be calculated with Darcy's equation.

With the *double-tube method,* two concentric tubes are placed in an auger hole with the center tube penetrating the bottom a few centimeters. Water is applied to both tubes to wet the soil below the tubes. The K value of the wetted zone below the tubes is determined from a flow system created in this zone by letting the water levels in the tubes differ. Field measurements consist of determining the rate of water-level drop in the center tube with the water level in the outer tube first kept constant and then allowed to fall at the same rate as that in the inner tube. The resulting K value reflects primarily K in vertical direction but is also affected by K in horizontal direction. Combining the double-tube K with a measurement of the vertical K by using the infiltration-gradient technique on the same test installation enables calculation of K in the horizontal direction.

In selecting a method for K measurement, care should be taken that the flow direction and the soil region on which K is determined approach those in the system to be analyzed or predicted as much as possible. Where the soil region sampled for K is small in relation to that in the flow system, several K measurements are needed (Bouwer and Jackson, 1974). Sometimes K can be evaluated from existing flow systems, e.g., flow to underground drains, drawdown to pumped wells, and groundwater mounds below recharge facilities. Soils and aquifers are inherently nonuniform, and adequate characterization of K in the field often is one of the main difficulties in predicting or analyzing underground flow systems.

HERMAN BOUWER

References

Bonell, M., 1976. Some comments on the association between saturated hydraulic conductivity and texture of holderness boulder clay, *Catena,* **3,** 77–90.

Bouwer, H., 1966. Rapid field measurement of air entry value and hydraulic conductivity of soil as significant parameters in flow system analysis, *Water Resources Res.,* **2,** 729–738.

Bouwer, H., 1969. Planning and interpreting soil permeability measurements, *Jour. Irrigation and Drainage Div., Am. Soc. Civil Engrs.,* 95, 391–402.

Bouwer, H., and Jackson, R. D., 1974. Determining soil properties, *in* J. van Schilfgaarde, ed., *Drainage for Agriculture.* Madison: American Society of Agronomy, 611–672.

Darcy, H., 1856. *Les Fontaines Publiques de la Ville de Dijon.* Paris: V. Dalmont, 570–590–594.

Elnaggar, H. A.; Karadi, G. M.; and Krizek, R. J., 1974. Non-Darcian flow in clay soils, *in* R. B. Dowdell, and H. W. Stoll, eds., *Flow: Its Measurement and Control in Science and Industry,* vol. 1. Pittsburgh: Instrument Society of America, 53–61.

Hubbert, M. K., 1956. Darcy's law and the field equations of underground fluids, *Trans. Am. Inst. Mining and Metallurgy Engrs.,* **27,** 222–239.

Lohman, S. W., 1972. Ground-water hydraulics, *U.S. Geol. Surv. Prof. Paper 708,* 70p.

McNeal, B. L., 1968. Prediction of the effect of mixed-salt solutions on the soil hydraulic conductivity, *Soil Sci. Soc. America Proc.,* **32,** 190–193.

Swartzendruber, D., 1969. The flow of water in unsaturated soils, *in* R. J. M. De Wiest, ed., *Flow Through Porous Media.* New York: Academic Press, 215–292.

Cross-references: *Capillary Pressure; Field Water Cycle; Flow Theory; Infiltration; Percolation; Permeability; Pore Space, Drainable; Soil Water; Water Content and Retention; Water Movement.*

CONDUCTIVITY, THERMAL

Thermal conductivity is the time rate of heat transfer by conductance through unit thickness across a unit area due to a unit temperature dif-

TABLE 1. Thermal Conductivity of Soils and Soil Constituents.

Material	Thermal Conductivity [λ (Cal $cm^{-1}s^{-1}$ °C^{-1})]	Temperature (°C)	S (% Saturation)	Source
Quartz	21.4×10^{-3}	20	–	de Vries (1963)
	19.0×10^{-3}	40	–	de Vries (1963)
	17.8×10^{-3}	60	–	de Vries (1963)
Sand dry	3.7×10^{-3}	–	100	van Duin (1963)
Sand wet	0.37×10^{-3}	–	0	van Duin (1963)
Clay minerals	7.0×10^{-3}	–	0	van Duin (1963)
Clay soil wet	3.18×10^{-3}	–	100	van Duin (1963)
Clay soil dry	0.37×10^{-3}	–	0	van Duin (1963)
Water	1.32×10^{-3}	0	–	de Vries (1963)
Water	1.5×10^{-3}	40	–	de Vries (1963)
Vapor saturation	0.059×10^{-3}	5	100	de Vries (1963)
Vapor saturation	0.55×10^{-3}	40	100	de Vries (1963)
Organic matter	0.6×10^{-3}	–	–	de Vries (1963)
Air	0.05×10^{-3}	–	–	Rosenberg (1974)
Air dry	0.06×10^{-3}	20	–	de Vries (1963)
Air dry	0.07×10^{-3}	60	–	de Vries (1963)

ference; the unit used is Cal cm^{-1} s^{-1} °C^{-1}. Since the soil is made up of granulated solids, between which are voids filled with water, air, or both, its thermal conductivity depends on the volume fraction and thermal conductivity of each of its constituents. Moreover, the size, shape, mode of the solid particles arrangement (see *Soil Structure*), and consequently the solid-water interfacial relationship affect the soil thermal conductivity. Table 1 gives values of thermal conductivity for the various soil constituents and soils.

It is obvious from Table 1 that the ratios between the thermal conductivities of quartz, clay, and water to air (dry)–340, 117, and 24 to 1, respectively–are very great. Thus the thermal conductivity of the soil depends quite heavily on the bulk density (q.v.), closeness of the contacts between the solid particles, and the degree of water saturation.

The data in Figure 1 show the dependence of the thermal conductivity of different soils on their solids and water volume fractions. Yet various soils having the same mineral constituents and bulk density but a different texture may differ in their thermal conductivities due to different grain-size distributions.

Factors Affecting Thermal Properties

Soil Type and Porosity. The thermal conductivity of a given granulated material diminishes with decreasing particle size due to the reduction of surface contact between the particles, even though the total porosity and volume fractions of solids and air are the same. Increasing the bulk density of a granulated material improves the interparticle contacts due to increased closeness of their packing, thus improving the thermal conductivity (see Figure 1).

Soil Water Content. From the data shown in Figure 1, it is obvious that the increase in thermal conductivity due to increased bulk density is small at low volumetric water contents in comparison to changes caused by variation in

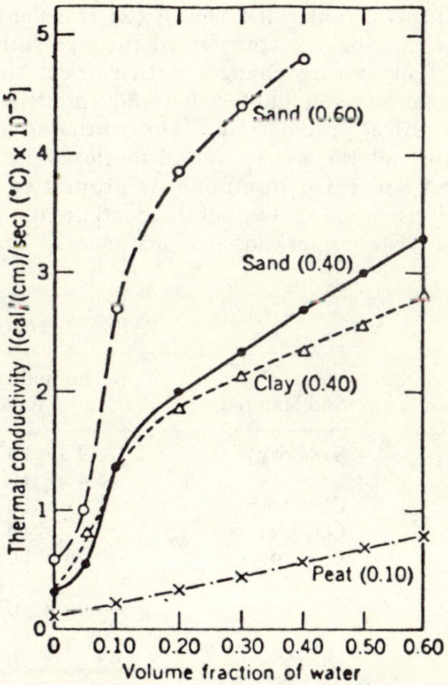

FIGURE 1. Thermal conductivity of soils as functions of the volume fraction of water (given on the abscissa) and of solids (given in parentheses). (From Baver et al., 1972)

the water content; at higher water content their relative importance reverses. This inverse relationship is due to the better contact between the particles that have a high thermal conductivity. Adding water increases the soil thermal conductivity by improving interparticle contact and by replacing the air in the soil pores.

Thermal Diffusivity. The soil's *thermal diffusivity* α is defined as $\alpha = \lambda/c$, the ratio between the thermal conductivity λ and the heat capacity (q.v.) c of the soil. This parameter denotes the temperature change occurring in a unit volume of soil due to heat transfer through it. The soil thermal diffusivity depends on the soil texture, mineralogical composition, and constituent volume fractions. Figure 2 and Table 2 give values of thermal diffusivity for various soils. Dimensional analysis of the units used for the thermal conductivity, heat capacity, and the thermal diffusivity definition yield the unit of thermal diffusivity which equals $cm^2\ s^{-1}$.

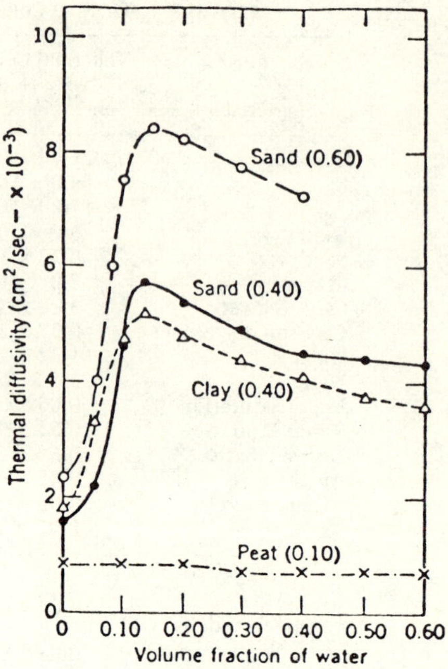

FIGURE 2. Thermal diffusivity as a function of volume fractions of water (given on the abscissa) and solids (given in parentheses). (From Baver et al. 1972)

Determination of Thermal Conductivity and Diffusivity

Though there are many methods for determining the thermal conductivity and diffusivity of the soil. All the methods can be divided into three categories: (1) steady-state heat flow (i.e., heat is transferred through soil while the temperature at each point within the soil space remains constant with time); (2) transient heat flow (i.e., heat is transferred through soil while the temperature changes with respect to position in the soil bulk space and time); and (3) theoretical predictions. Unfortunately, with almost all the experimental methods used in a moist soil for determining its properties, moisture as well as temperature gradients result. Steady-state methods are sufficiently accurate for measuring the thermal conductivity of dry soils, whereas those using the transient heat transfer are more useful because they require short time durations during which the moisture gradient development is minute due to minimization of water transfer in response to the temperature gradient.

Steady-State Methods. The basis of steady state methods is as follows: A soil sample is brought in contact with a heating unit of a given constant thermal output and geometry.

TABLE 2. Thermal Diffusivities of Various Soils.

Soil Material	Thermal Diffusivity (cm^2/s)	% Saturation (°C)	Source
Sand dry	1.9×10^{-3}	0	van Duin (1963)
	4.4×10^{-3}	100	van Duin (1963)
Clay dry	1.8×10^{-3}	0	van Duin (1963)
Clay wet	3.7×10^{-3}	0	van Duin (1963)
Loam dry	0.45×10^{-3}	0	Nakshabandi and Kohuke (1965)
	4.25×10^{-3}	75	Nakshabandi and Kohuke (1965)
Quartz	43×10^{-3} at 20°C	–	Van Wijk and de Vries (1963)
Soil minerals	15×10^{-3} at 20°C	–	Van Wijk and de Vries (1963)
Water	1.42 at 20°C	–	Van Wijk and de Vries (1963)

Temperatures are measured at different points in the soil sample until they reach a constant value. The temperature gradient is then established, and the thermal conductivity is computed from the relationship given in equation 1 where q is the thermal output per second of the heating body, A is its surface area, $\Delta T/\Delta X$ is the thermal gradient, and λ is the thermal conductivity.

$$\frac{q}{A} = \lambda \frac{\Delta T}{\Delta X} \qquad (1)$$

Once the thermal conductivity is known and the heat capacity of the sample at the point for which the temperature gradient was established is calculated, it is simple to compute the appropriate thermal diffusivity for the soil.

Transient Methods. Most transient methods are based on changing the temperature around a small probe of a heating body for a short period of time, from which the temperature distributions around the probes, the thermal conductivity, and/or diffusivity are computed.

A thin cylindrical probe consisting of a heating wire and thermocouple or thermistor to monitor temperature changes of the probe is inserted into the soil and heated. Equation 2 gives the temperature rise (Jackson and Taylor, 1965), where T_0 and T are initial temperature and temperature at time t, respectively, q is the thermal output per unit length of the probe, t is time, and c is a constant. A plot of $(T - T_0)$ against ln t yields a linear relationship the slope of which equals $q/4\pi\lambda$.

$$T - T_0 = \frac{q}{4\pi\lambda}(c + 1nt) \qquad (2)$$

Out of the value of the slope, thermal conductivity is computed. The thermal diffusivity can then be determined, provided the soil's heat capacity is known.

A planar probe having a known thermal output per unit area per unit time is inserted into the soil and allowed to warm the soil, while the temperature is measured at the surface and at a distance z from the plate. The relative temperature change at distance z and time t given by equation 3 is used to compute the thermal diffusivity (Jackson and Taylor, 1965), where T_0, T_s, and T are the initial temperature, soil probe surface temperatures, and temperature at time t, at distance z, respectively, and erf is the error function.

$$T - T_0 = (T_s - T_0)(1 - \text{erf}\ ^z/2\sqrt{\alpha t}) \qquad (3)$$

The results of the computation $(T - T_0)/(T_s - T_0)$ for a given t are then used to read the error-function value for the argument $(Z/2\sqrt{\alpha t})$. The thermal diffusivity is then calculated from the argument. The thermal conductivity can be determined if the heat capacity of the soil is known.

Theoretical Prediction of Thermal Conductivity

Theoretically it is possible to express the approximate thermal conductivity of the soil as a function of the conductivities and volume fractions of the soil constituents (de Vries, 1963). The theory requires that one of the constituents (e.g., air or soil water) will be taken as the medium in which the other soil constituents are embedded.

Equation 4 gives the thermal conductivity of the soil, where λ_i is the thermal conductivity, χ_i is the volume fraction of constituent i, and k_i is a parameter that depends on the shape of the soil granules and their relative conductivity with respect to that of the medium in which they are embedded.

$$\lambda = \frac{\sum_{i=0}^{n} k_i \chi_i \lambda_i}{\sum_{i=0}^{n} k_i \chi_i} \qquad (4)$$

The procedure to be followed has been described by de Vries (1963) in detail.

In accordance with the same data for the theoretical prediction of the thermal conductivity, the diffusivity can be derived provided the soil's heat capacity is known.

AMOS HADAS

References

Baver, L. D., et al., 1972. *Soil Physics,* 4th ed. New York: Wiley, 273.

Bowers, S. A., and Hanks, R. J., 1962. Specific heat capacities of soils and minerals as determined with a radiation calorimeter, *Soil Sci.,* 94, 392–396.

de Vries, D. A., 1963. Thermal properties of soils, *in* W. R. Van Wijk, ed., *Physics of Plant Environment.* Amsterdam: North Holland, 210–235.

Duin, R. H. A. van, 1963. The influence of soil management on the temperature wave near the surface, *Tech. Bull. Inst. Land Water Mgnt. Res., Wageningen,* 29, 1–26.

Jackson, R. D., and Taylor, S. A., 1965. Heat transfer, *in* C. A. Black, et al, eds., *Methods of Soil Analysis.* Madison: American Society of Agronomy, Special Monograph No. 9., 349–360.

Kersten, M. S., 1949. Thermal properties of soils, *Bull. Univ. Minn. Inst. Tech. Engr. Exp. Stn.,* 28, 1–227.

Laudolte-Börnstein, 1967. *Zahlenwerte und Funktionen, Band a.* Berlin: Springer-Verlag, 916–922.

Nakashabandi, G. A., and Kohuki, H., 1965. Thermal conductivity and diffusivity of soils as related to moisture tension and other physical properties, *Agric. Meteorol.,* 2, 271–279.
Rosenberg, N. J., 1974. *Microclimatical: The Biological Environment.* New York: Wiley, 315.
Van Wijk, W. R., and De Vries, D. A., 1963. Periodic temperature variations in a homogeneous soil, *in* W. R. Van Wijk, ed., *Physics of Plant Environment.* Amsterdam: North Holland, 102–143.

Cross-references: *Energy Balance; Heat Capacity; Thermal Regimes.*

CONSERVATION

Soil erosion is the accelerated movement of soil caused by cultural practices such as farming and urban development; it is distinguished from the much slower process of *geological erosion.* Soil conservation practices are those remedial actions that are taken to reduce human-induced erosion to within acceptable limits.

Soil Erosion

Sediment ranks above domestic sewage, industrial wastes, and chemicals as a major cause of water pollution. Sediment fills stream channels, reservoirs, and estuaries; necessitates expensive additional treatment of water supplies; and reduces the oxygen content of water needed to assimilate wastes. In the United States, for example, suspended solids reaching the nation's streams by way of surface runoff are estimated to be at least 700 times the volume of sewage discharges.

Sediment sources and amounts constantly change as patterns of living change. Sediment yields in areas undergoing urban growth may be 5 to 500 times greater than in rural areas. Exposure of bare soil—for example, during strip-mining operations, residential and commerical development, and highway construction—is a major local factor causing the high rate of sediment damage. Nevertheless, erosion on farms (Figures 1 and 2) is still a principal source of

FIGURE 1. A newly seeded alfalfa field damaged by excessive rains. (Photo U.S. Department of Agriculture, Soil Conservation Service)

sediment and requires continuing development of control practices.

The Urban Problem

More people with more money and more leisure time, living in an age of bigger bulldozers and faster transportation, have created a new kind of conservation problem. The demand for houses in suburbia and shopping centers to serve them with access by superhighways is rapidly despoiling the land resources upon which affluence depends.

In a broad sense, there is no difference between conservation problems in urban-suburban areas and rural areas. The objectives are the same: clean, controlled streams; abundant water; soil held in place; living space; and beauty of landscape, whether man-made or natural. The goals are sought by farmer and urbanite alike. In a narrower sense, however, there are differences that complicate the urban problem far beyond the classic concept of rural conservation. These differences are more than a matter of geography. They are a complex of physical, social, and economic conditions that overlap and conflict. Man himself is so dominant in the scene that he overshadows the soil, water, plant, and wildlife relationships typical of conservation in rural areas. He can and does alter the landscape to suit his needs or even his fancy. In his alterations he must concern himself with public health, clean air, and beauty of landscape as well as the more mundane problems of less populated areas.

FIGURE 2. Soil erosion on a field seeded to corn with the rows planted down slope. (Photo U.S. Department of Agriculture, Soil Conservation Service)

Although a complete list of complicating factors would be impossible to compile, some of the more obvious ones include:

1. The difference in social values that define a "good life."
2. The intensity of land use, with its associated high land value.
3. The existence of many forms of government, often without clearly defined authority and rarely in harmony with physical features affecting conservation.
4. The lack of understanding of each other's objectives by conservationists, urban planners, and government agencies concerned with urban and urban-fringe planning.
5. The wide variety of interests competing for the urbanites' attention, time, and money.
6. The historical development of conservation programs primarily for rural lands with little or no regard for urban areas.

As an increasing percentage of the population of a country becomes urbanized, it becomes more difficult for people to realize their dependence on the soil and the effect that soil erosion may have on their lives and on future generations. This also holds true to some extent for the people who live close to the land. It often happens that soil erosion occurs very slowly and the effects may take place over a very long period of time, sometimes generations. The down-stream effect, which may be just as disastrous as those on the land, takes the form of pollution and sediment in reservoirs, streams, and so forth.

Conservation Plans

Conservation is a continuing long-term partnership with the land. Every land steward or owner should have a conservation plan, from the owner of a city lot to the largest farmer or rancher. This plan should be put in writing and kept with other important papers and documents. It may be impractical to implement this plan all at once as it sometimes takes several years. Changes may also have to be made in the plan from time to time, particularly as new knowledge becomes available.

In some aspects of conservation planning, the entire watershed must be taken into consideration. A watershed can be as small as a few square feet to as large as the Mississippi River. It is defined as all the land that contributes water that flows by a given point. If a dam is built on a stream, then all the land above it that furnishes water to its lake constitutes the watershed of the lake. It is obvious that from the standpoint of the lake, any erosion in its watershed will contribute to sedimentation in the lake. More than half of all pollution consists of soil particles eroded from the watershed.

Wischmeier Universal Soil-Loss Equation

To evaluate the principles of soil conservation, a reliable method of predicting soil losses under various conditions is required. The *Wischmeier Universal Soil Loss Equation,* (U.S. Dept. of Agric. Staff, 1962) is used to estimate sheet and rill erosion:

$$A = RKLSCP$$

A is the predicted average annual soil loss expressed in tons per acre per year. R is the rainfall factor. It is the number of erosion-index units in a normal year's rain. The *erosion index* is a measure of the erosion force of specific rainfail. When other factors are constant, storm losses from rainfall are directly proportional to the product of the total kinetic energy of the storm times its maximum 30-minute intensity. K is the soil-erodibility factor. This factor is the erosion rate per unit of erosion index for a specific soil in cultivated continuous fallow on a 9% slope 72.6 ft long. Soil properties that influence erodibility by water are: (1) those that affect the infiltration rate, permeability, and total water capacity, and (2) those that resist the dispersion, splashing, abrasion, and transporting forces of the rainfall and runoff. Figure 3 shows a recent improvement in the solution of the soil-loss equation that enables it to be used on any unmapped soil and on construction sites and other disturbed areas.

L is the slope-length factor. It is the ratio of soil loss from the field slope length to that from a 72.6-ft length on the same soil type and gradient. Slope length is the distance from the point of origin of overland flow to (1) the point where the slope decreases to the extent that deposition begins or (2) the point where runoff enters a defined channel. Figure 4 shows a combination of the LS factors in the soil-loss equation. It can be seen easily from the slope length of the field that terraces across the field would have a great effect on reducing the soil losses. The use of terraces is the best mechanical device that can be applied to reduce soil losses.

S is the slope-gradient factor. This factor is the ratio of soil loss from the field gradient to that from a 9% slope. The relation of soil loss to gradient is influenced by density of vegetal cover and by soil-particle size.

C is the cropping-management factor on cropland and other land uses. It is the ratio of soil loss from a field with a specified cropping and management or plant cover to that from the fallow condition on which the factor K is evaluated. This factor measures the combined effect of all the interrelated cover, and management variables plus the growth stage and vegetal cover at the time of the rain.

P is the erosion-control-practice factor. This factor is the ratio of soil loss with contouring, stripcropping, or terracing to that with straight-row farming up and down the slope. T is the soil-loss tolerance expressed in tons per acre per year. Soil-loss tolerance is usually estimated when the Wischmeier equation is applied to a particular field.

To make A the predicted soil loss no greater than T, the tolerance that has been set for a specific field, the factors L, S, C, and P can be changed by applying more conservative practices.

The no-tillage cultured system for row crops, for example, reduces soil erosion to near the permanent sod level and usually results in higher yields than conventional tillage. The number of operations over the field are usually less than with conventional methods. A suitable sod must be killed with a local recommended herbicide, after which fertilizers and more herbicides must be applied.

Planting requirements are similar to those of conventional tillage. In this method a complete kill of the mulch cover must be obtained.

Terrace Systems

Terracing is the best mechanical method of preventing soil erosion. All types of terrace systems can be classified according to alignment, cross-section, grade, and outlet (American Society of Agricultural Engineers, 1978).

Classification by Alignment. Parallel terraces are designed with straight segments, and the adjacent parallel terrace ridges are spaced equidistantly throughout their length. On irregular topography it may be necessary to divide terraces into parallel sections with different spacings for each section. Cuts and fills along the terrace channel may be required to establish satisfactory grades.

Nonparallel terraces are designed with the terraces fitted to the contour of the land without regard to paralleling. In practice, adjustment is made to eliminate sharp turns and short point rows by installing additional outlets, using variable grade and vegetated adjustment or turn areas.

Classification by Cross-section. Broadbase terraces are constructed so that crops can be planted and machinery operated over the entire terrace. Ridge and channel slopes are designed to fit different machine widths.

Flat-channel terraces or Zingg conservation bench terraces are level benches constructed below a sloping runoff area. Flat-channel terraces are designed for maximum conservation of

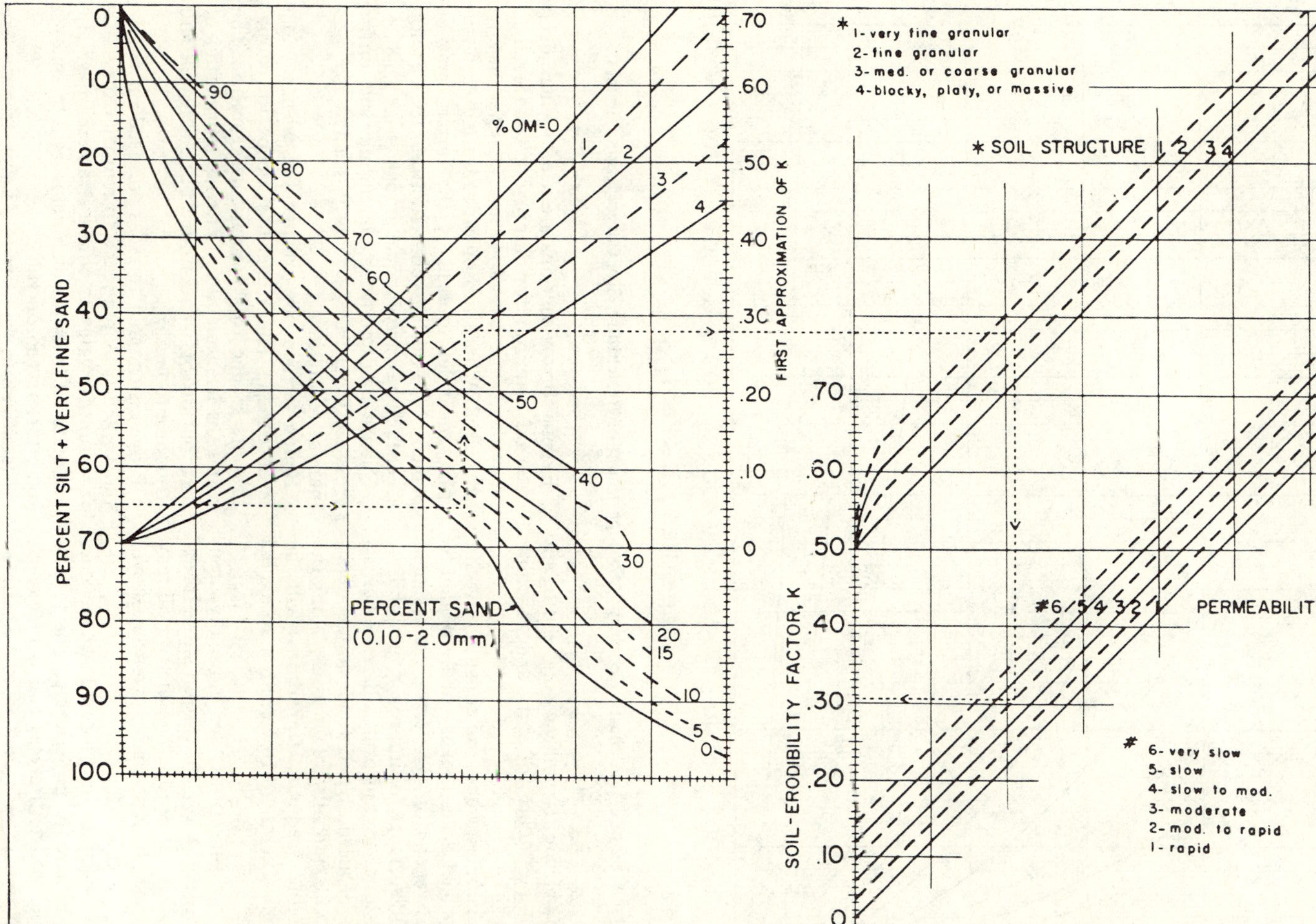

FIGURE 3. Procedure for finding *K*. With appropriate data, enter scale at left and proceed to points representing the soil's percent sand (0.10–2.0 mm), percent organic matter, structure, and permeability, *in that sequence.* Interpolate between plotted curves. The dotted line illustrates procedure for a soil having: si + vfs 65%; sand 5%, *OM* 2.8%, structure 2, permeability 4. Solution: $K = 0.31$. (After Wischmeier, Johnson, and Cross, 1971)

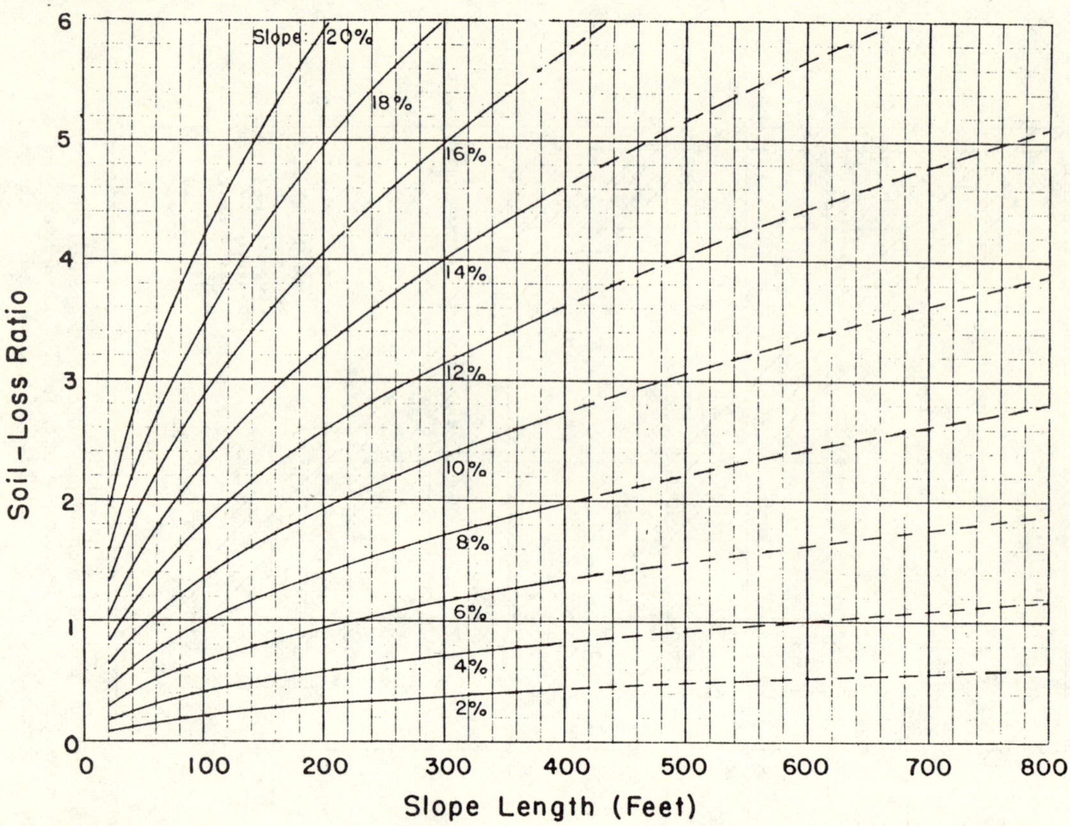

FIGURE 4. Slope effect chart. (Courtesy U.S. Department of Agriculture, Soil Conservation Service)

water. Channel widths will vary depending on the land slope, depth of cut, and machine width.

Steep-backslope terraces are constructed with a very steep backslope that is usually seeded to a permanent vegetation. The terrace is designed to shape land surfaces for improved farmability. When field slopes are uniform, a balanced cross-section can be used. When field slopes are irregular and cuts and fills are necessary along the terrace length, frequent transitions will be required from excess borrow section to excess fill sections.

Balanced cross-sections are normally constructed entirely from the lower side. If exposed subsoil is not productive, the topsoil may be moved downhill first then moved back over the excavated area.

Excess-fill sections are formed where cuts below the terrace are undesirable. The terrace ridge may be built entirely of soil from excess-borrow sections of the terrace or from a planned borrow area. A fill section is desirable when cuts below the terrace would result in increased row grades below the terrace or where terraces are built on shallow soils.

Excess-borrow sections are useful where deep cuts are required for improving terrace alignment; both the backslope and channel may be excavated. Where subsoil will respond to fertilizers, the ridge crossed by the terrace may be planed off during construction to develop acceptable terrace channel grades.

Narrow-base terraces, including the entire cross-section, are frequently seeded to permanent vegetation.

Bench terraces are constructed with a wide-level bench. The ridge has steep side slopes that are seeded to a permanent vegetation. The terrace is used mainly to provide more efficient distribution of water under both irrigated and dryland production.

Ridgeless-channel terraces are used on nearly flat to gently sloping land to remove excess surface runoff water at nonerosive velocities. The cut material from the channel is used to smooth the interterrace interval.

Classification by Grade. Level terraces are constructed where water conservation is the primary objective with erosion control as an accompanying benefit. They are usually used in low-rainfall regions to retain runoff for storage in the soil and may be used in high-rainfall regions on very permeable soils. The ends of the channel may be blocked or left open.

Graded terraces reduce field slope lengths and remove runoff with tolerable soil loss as the primary objective. Water conservation may be an accompanying benefit. The grade to the outlet may be uniform or variable.

Contouring and Stripcropping

Contouring is the practice of performing tillage operations on the contour or very nearly on a level line.

Stripcropping is the practice of laying out alternate strips on the contour of cultivated crops, close-growing crops, and hay crops. Three types of stripcropping are in general use. The most effective method is contour stripcropping (Figure 5). In this case the strip boundary is laid out on the contour or very nearly on the contour.

Buffer stripcropping is laying out strips on the exact contour. Any field irregularities are planted to permanent vegetation. Field stripcropping is the use of uniform-width strips laid out across the slope without regard to the contour.

Wind Erosion

Wind erosion is a problem in most areas; in semiarid and arid regions the problem is more severe than water erosion. A wind-erosion equation, similar in principle to the water-erosion equation, has been developed by the U.S. Soil Conservation Service (U.S. Department of Agriculture Staff, 1968). The amount of soil loss E, expressed in tons per acre per year can be expressed as:

$$E = f(I, K, C, L, V)$$

where I is the soil-erodibility index, K is the soil ridge-roughness factor, C is a climatic factor, L is field length along the prevailing wind-erosion direction, and V is equivalent quantity of vegetative cover.

Based on present techniques there is very little that can be done about the climatic factor C

FIGURE 5. Soil conservation practiced by cropping with alternate strips of close-growing crops, row crops, and grass waterways. (Photo U.S. Department of Agriculture, Soil Conservation Service)

and the soil-erodibility factor *I*. The field length *L* can be effectively changed by use of strip-cropping with strips placed at right angles to the prevailing wind if possible. The most effective method of wind-erosion control is a protective vegetative cover on the ground surface. The soil ridge-roughness factor *K* can be accomplished by leaving the surface rough and the soil as compact as possible.

BERNARD L. PARSONS

References

Anonymous, 1967. Sediment, its filling harbors, lakes and roadside ditches, *U.S. Dept. Agric. Bull. 325,* 15p.

American Society of Agricultural Engineers, Soil and Water Division Steering Committee. 1978. *A.S.A.E. Recommendation R 268.1, Design, Layout, Construction of Terrace Systems.* St. Joseph, Mich.: 4p.

Beasley, R. P., 1972. *Erosion and Sediment Pollution Control.* Ames: Iowa State University Press, 320p.

Bennett, H. H., 1955. *Elements of Soil Conservation.* New York: McGraw-Hill, 358p.

Lillard, J. H., and Smith, E. S., 1971. *No-tillage Machinery and Seedbed Requirements.* Blacksburg: Virginia Polytechnic Institute and State University, Publication 419, 8p.

Schwab, G. O., Frevert, R. K., Edminister, T. W., and Barnes, K. K., 1966. *Soil and Water Conservation Engineering.* New York: Wiley, 683p.

Stallings, J. H., 1964. *Soil Conservation.* Englewood Cliffs, N.J.: Prentice-Hall, 575p.

U.S. Department of Agriculture Staff, 1938. *Soils and Men; Yearbook of Agriculture 1938.* Washington, D.C.: Government Printing Office, 1232p.

U.S. Department of Agriculture Staff, 1955. *Water; Yearbook of Agriculture 1955.* Washington, D.C.: Government Printing Office, 751p.

U.S. Department of Agriculture Staff, 1957. *Soil; Yearbook of Agriculture 1957.* Washington, D.C.: Government Printing Office, 784p.

U.S. Department of Agriculture Staff, 1962. Rainfall-erosion losses from Cropland east of the Rocky Mountains, *U.S. Department of Agriculture Handbook* No. 282, 47p.

U.S. Department of Agriculture Staff, 1968. Wind erosion forces in the United States and their use in predicting soil loss, *U.S. Department of Agriculture Handbook No. 346,* 42p.

Willrich, T. L., and Smith, G. E., 1970. *Agricultural Practices and Water Quality.* Ames: Iowa State University Press, 415p.

Wischmeier, W. H.; Johnson, C. B.; and Cross, B. V., 1971. A soil erodibility nomogram for farmland and construction sites, *Jour. Soil and Water Conserv.,* 178-216, 189-193.

Cross-references: *Management of Soils; Soil Drainage; Soil Stabilization; Water Erosion; Wind Erosion; Vol. III: Eolian Transport; Erosion; Gully Erosion; Sheet Erosion.*

CROPS—See AGRONOMY

CULTIVATION—See TILLAGE

CYCLES—See CARBON CYCLE; NITROGEN CYCLE; PHOSPHORUS CYCLE; SULFUR TRANSFORMATIONS. Vol. IVA: CYCLES, GEOCHEMICAL; PHOSPHORUS CYCLE; SILICA CYCLE

D

DATING—See RADIOCARBON DATING. Vol. XII, Pt. 2: AGE DETERMINATION

DENITRIFICATION

Whether it occurs in soil or water or in a culture vessel, *denitrification* is defined as that type of biochemical reduction of soluble sodium or potassium nitrate that is carried out by bacteria and results in the release of elemental dinitrogen (N_2) into the atmosphere. Strictly chemical (abiological) denitrification contributes little if at all to the process in nature, and the oxygen of our atmosphere must be excluded from a system before the bacterial process will begin and continue. Denitrification is thus particularly likely to occur in the *anaerobic environment* of waterlogged soil or fresh-water or marine sediment—particularly if these environments contain a rich supply of organic material, which the heterotrophic bacteria can oxidize for energy. Reduced inorganic sulfur compounds may also be oxidized (serve as electron donors) during denitrification by autotrophic bacteria (i.e. *Thiobacillus* species, Table 1). *Heterotrophic organisms* are those that require an organic source of carbon and energy for growth. In contrast, *autotrophic organisms* are dependent either on light or on oxidation of inorganic or elemental substances for energy, and CO_2 as the sole source of carbon. There are thus four essential components for denitrification: (a) anaerobic conditions, (b) the appropriate autotrophic or heterotrophic bacteria, (c) suitable oxidizable inorganic or organic material (electron donors), and (d) a supply of nitrate or nitrite.

TABLE 1. Genera of Bacteria in Which Denitrifying Species Are Included.

Achromobacter
Alcaligenes[a]
Bacillus
Chromobacterium
Corynebacterium[b]
Flavobacterium
Hyphomicrobium
Moraxella
Paracoccus
Propionibacterium
Pseudomonas[b]
Spirillum
Thiobacillus

[a]One species does not reduce nitrate but will denitrify at the expense of nitrite.

[b]Certain species and strains yield nitrous oxide as a terminal product of denitrification.

When the phenomenon was first observed (Gayon and Dupetit, 1886) in France in 1886, European farmers were gravely concerned about its consequences. Knowing the advantages of using nitrate as a crop fertilizer, they found frightening the possibility that bacteria (Table 1) in the soil might starve their crops for nitrogen by liberating the element from a useful form. As indicated in Figure 1, unlike the desirable varieties of assimilatory and dissimilatory reduction of nitrate that take place continually in green plants and certain microorganisms (and result in the transformation of nitrate taken from the soil into ammonia that is utilized for production of proteins in these organisms), denitrification serves no useful nutritional function for plants—or indeed for man and animals that depend on plants for nutrition. The process is viewed, in fact, as a very significant agricultural defect. Concern for the immediacy of the threat to agriculture it represents has lessened as years have passed, but the true amount of nitrogen lost annually to denitrification is still not known.

For many years lack of a reliable, inexpensive, and easily applied method of assay made measurement of nitrogen loss from soils that were fertilized with nitrate so difficult that progress was very slow. But improvement of gas-chromatographic methods (Hollis, 1966; Barbaree and Payne, 1967) made development of useful analytical procedures possible, and both agronomic and biochemical experiments are now underway. Preliminary results from studies of denitrification in agricultural soils are shocking. Losses of nitrogen from nitrate fertilizer applied to various farm soils varies from a small amount for well-managed plots to remarkable quantities in certain uncontrolled situations. Improved schedules and techniques for fertilizer application to minimize loss are urgently needed. It is reassuring that

COMPREHENSIVE VIEW OF MICROBIAL REDUCTION OF NITROGENOUS OXIDES

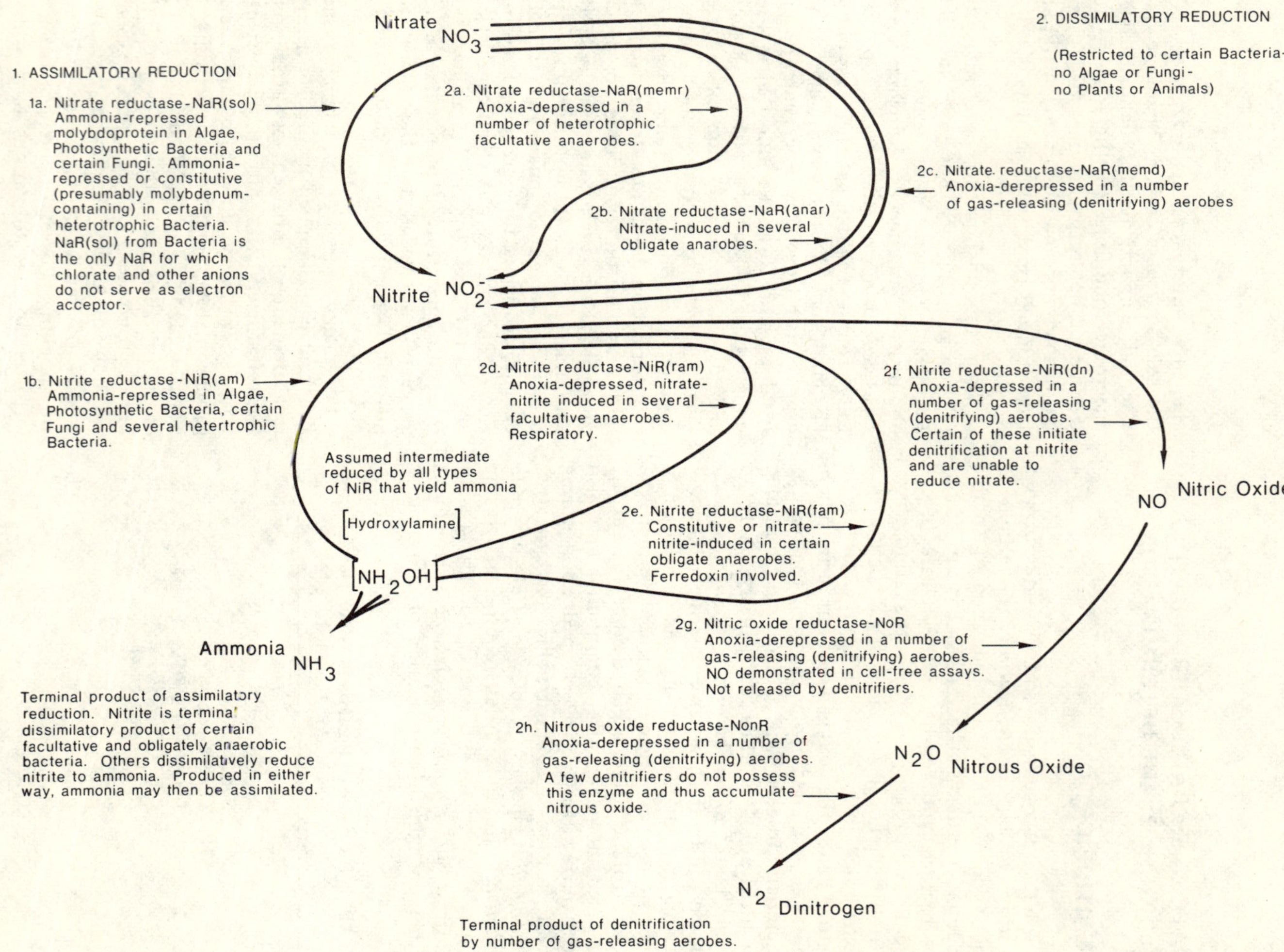

both biological nitrogen fixation and the application of nitrogen-rich manures continually renew and restore to the soil the stock of ammonia and nitrate needed for plant growth, despite the fact that fixation uses up much biological energy. Denitrification undermines attempts to maintain soil fertility and is an added burden that further consumes biological energy. The process should be better understood if it is ever to be controlled.

In biochemical terms, denitrification is simply one of several forms of anaerobic respiration that certain otherwise ordinary, widely distributed bacteria (but no other forms of life, large or small) carry out when their oxygen supply is limited. They much prefer oxygen-linked respiration and undertake denitrification only when forced to by lack of oxygen. Unlike fermentation, another anaerobic means of gaining energy that is used by many bacteria and yeasts, denitrification involves cytochrome pigments [see Lehninger (1975) for an explanation of essential differences between fermentation and respiration]. Respiratory

FIGURE 1. Summary of the events characterizing reduction of nitrogenous oxides by microorganisms. (1) *Assimilatory reduction:* can occur during either aerobic or anaerobic growth (arrows looping to the left). (a) NaR(sol) = the soluble enzyme, nitrate reductase; (b) NiR(am) = the enzyme, nitrate reductase, that yields ammonia as final product. (2) *Dissimilatory reduction:* occurs only during anaerobic growth (arrows looping to the right). *Nitrate reductases:* (a) NaR(memr) = the membrane-bound respiratory enzyme, nitrate reductase, formed by ordinarily aerobic bacteria that have the faculty for various types of anaerobic growth (fermentative and respiratory) when necessary; (b) NaR(anar) = the respiratory enzyme, nitrate reductase, produced by obligately anaerobic bacteria (those that cannot grow aerobically); (c) NaR(memd) = the membrane-bound respiratory enzyme, nitrate reductase, produced under oxygen-limiting conditions by denitrifying bacteria. *Nitrite reductases:* (d) NiR(ram) = presumably respiratory enzyme, nitrite reductase, produced by ordinarily aerobic bacteria that have the faculty for various types of anaerobic growth (fermentative and respiratory) when necessary; (e) NiR(fam) = the presumably respiratory, ferredoxin-linked enzyme produced by obligately anaerobic bacteria; (f) NiR(dn) = the enzyme, nitrite reductase, producted under oxygen-limiting conditions by denitrifying bacteria, yielding nitric oxide instead of ammonia as its product. *Nitric-oxide reductase:* (g) NoR = the enzyme, nitric-oxide reductase, produced under oxygen-limiting conditions by denitrifying bacteria, yielding nitrous oxide as its product. *Nitrous-oxide reductase:* (h) NonR = the membrane-bound enzyme, nitrous-oxide reductase, produced under oxygen-limiting conditions by denitrifying bacteria, yielding elemental dinitrogen as its product. Reductases 2c, 2f, 2g, and 2h comprise the denitrifying enzymes.

transport of electrons (Payne, 1973; Lehminger, 1975) that culminates in the following reductive conversions comprise denitrification (Figure 2): (2) nitrate to nitrite, (b) nitrite to nitric oxide, (c) nitric oxide to nitrous oxide, and (d) nitrous oxide to dinitrogen. Thus unlike aerobic respiration in which only one terminal oxidant, oxygen, is reduced, denitrification is characterized by the successive and orderly reduction, in sequence, of four oxidants. The bacterial cells must produce four sets of specialized enzymes (Payne, 1973) when they are needed for catalysis of each reduction (a–d), then dispose of these enzymes when oxygen again becomes available. The rapidity with which the bacteria can convert from one type of respiration to the other is one of the factors that makes denitrification particularly difficult to control.

Examination of the events represented on the left of the diagram in Figure 2 indicates that the same range of oxidizable substrates and electron-transport cofactors (Lehninger, 1975) are employed in the early events that characterize both ordinary aerobic and denitrifying respiration in bacteria. But beginning with nitrate reduction, each step on the right in the denitrifying scheme depicted in Figure 2 has unique properties; and release of dinitrogen is the final event. Each step is considered in turn in what follows.

Nitrate Reduction

In denitrifying bacteria, the enzyme nitrate reductase [NaR(memd)] is located in the cytoplasmic membrane (Forget and Dervartanian, 1972; Lehninger, 1975). It is one of a few molybdenum-containing enzymes found in nature and converts nitrate to nitrite at the expense of electrons transported directly from a *b*-type cytochrome. NaR(memd) of denitrifying bacteria is less complex than the assimilatory type of green-plant or fungal-nitrate reductase, which is not membrane bound (Forget and Dervartanian, 1972; Payne, 1973; Payne, Fitzgerald, and Dodgson, 1974). Strangely, chlorate and other anions are reduced by NaR(memd) and by most other nitrate reductases from bacteria, fungi, and plants. NaR(memd) is adversely affected by oxygen which inhibits both its production and its functioning once it is produced. Reduction of nitrate by NaR(memd) is linked to production of the energy currency of living cells, adenosine triphosphate (Lehninger, 1975).

Nitrite Reduction

Unlike the enzyme of green plants and fungi, which converts nitrite to ammonia, nitrite

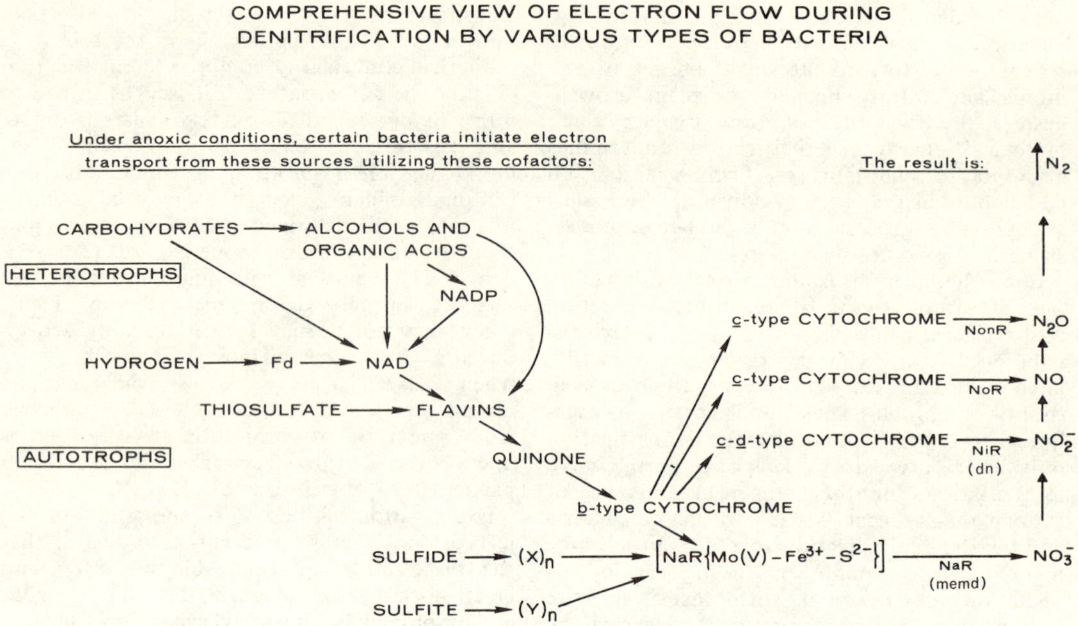

FIGURE 2. Composite accounting for the reactions by which various bacteria carry out denitrification. The sharp-pointed arrows leading left to right indicate flow of electrons; the broader-headed ones ascending on the right indicate the formation of progressively more reduced intermediates of nitrate conversion to and liberation of dinitrogen. Abbreviations: Fd is ferredoxin, a low electropotential electron-carrier protein associated as well with nitrogen fixation, photosynthesis, and anaerobic metabolism; NAD is nicotinamide adenine dinucleotide; and NADP is nicotinamide adenosine dinucleotide phosphate, two widely distributed electron-transport cofactors found in all living cells.

reductase [NiR(dn)] produced by denitrifying bacteria reduces nitrite to nitric oxide at the expense of electrons transported from *b*-type to unique *c*- and *c-d*-type cytochromes (Cox and Payne, 1973; Newton, 1969). The location of NiR(dn) in the cell is now known, but it does not remain firmly attached to membrane when cells are ruptured. One denitrifying *Alcaligenes* species (Table 1) lacks NaR(memd) and actually begins denitrification with NiR(dn)—but only if supplied with nitrite (Payne, 1973). Growing bacterial cells do not release nitric oxide into the atmosphere. Its production by NiR(dn) has been demonstrated, however, by experiments with extracts of bacteria that have been disrupted so that NiR(dn) may be solubilized and studied separately from the other enzymes of the cell. Oxygen inhibits production and functioning of NiR(dn). During the operation of the process catalyzed by NiR(dn), while nitric oxide formed by nitrite reduction is being released in cell-free enzyme mixtures, there is a time when the nitric oxide forms a complex with the heme structure of *c*-type cytochrome (Cox, Payne, and Dervartanian, 1971). At first glance, this does not appear an event of singular significance, for production of heme-nitric oxide complexes is well known to biophysical chemists. But in those formed experimentally with other cytochromes, the nitric oxide is irreversibly bound. The striking difference is that in the denitrifying bacteria, formation of the heme-nitric oxide complex is only transient, and nitric oxide is released from the binding (but not from the cell).

Nitric-oxide Reduction

Because whole bacterial cells other than those of the *Thiobacillus* species (Table 1) have not been shown to reduce nitric oxide that is supplied externally, proof of the existence of the enzyme nitric-oxide reductase (NoR) was actually obtained by analyses with extracts of ruptured denitrifying cells (Renner and Becker, 1970; Payne, Riley, and Cox, 1971). As is true for the other denitrifying enzymes, synthesis of NoR is repressed by oxygen. The enzyme does not remain tightly bound with the membrane when cells are ruptured, and oxygen inhibits its activity. Displaying an interesting additional dimension not shared with other denitrifying enzymes, NoR catalyzes coupling of two nitrogen atoms simultaneously with reduction of nitric oxide. In other words, NoR acts at once on two molecules of nitric oxide—

one of which it reduces and the other to which it couples the reduced molecule. A unique *c*-type cytochrome is associated with the enzyme; and while nitric oxide is undergoing reduction to nitrous oxide, a heme-nitric oxide complex forms transiently. It is different from that occurring during nitrite reduction (Cox, Payne, and Dervartanian, 1971). The presence of nitrate (and to a lesser degree, nitrite) suppresses the rate of activity of NoR (Payne and Riley, 1969). This ensures the orderly reduction of the ionic oxides, nitrate and nitrite, before the gaseous intermediates, nitric and nitrous oxide, are reduced. Then as a further means of regulation, nitrous oxide, the product of nitric oxide reduction, is not itself rapidly reduced as long as nitric oxide is present in abundance (Payne, Riley, and Cox, 1971). In some bacteria this slowdown in reduction results in temporary release of nitrous oxide from the cells into the environment. But most of it is soon reabsorbed and reduced.

Nitrous-oxide Reduction

This action completes the cycling of nitrogen begun when atmospheric dinitrogen is fixed by bacteria and blue green algae in soil or water. Production of the enzyme nitrous-oxide reductase (NonR) enables almost all denitrifying bacteria to reduce nitrous oxide and to release elemental dinitrogen into the atmosphere (Payne, Riley, and Cox, 1971; Matsubara, 1975). The presence of oxygen represses synthesis and inhibits the activity of NonR. One *Corynebacterium* species and selected strains of a *Pseudomonas* species (Table 1) lack NonR (Renner and Becker, 1970). For them, nitrous oxide is the terminal product of denitrification. In the much more common (i.e., the dinitrogen-releasing species), however, NonR is produced and functions anaerobically. It is a membrane-bound enzyme. A *c*-type cytochrome apparently functions in the transfer of electrons for the reduction of nitrous oxide by NonR but has not yet been separated from the bacterial extracts. A recent breakthrough in the study of NonR activity has given hope for rapid advancement in assessment of the impact of denitrification in soil. When denitrifying bacteria are incubated in the presence of acetylene [which, unlike the nitrogen-fixing bacteria (Hardy, Burns, and Holsten, 1973), denitrifiers do not reduce], NonR activity is inhibited (Federova, Milekhana, and Ilyukhina, 1973; Balderston, Sherr, and Payne, 1976). Since nitrous oxide is not known to be produced by any other biological phenomenon in nature, the ability to block its reduction and thus cause the nitrous oxide to accumulate in a test system seems likely to provide a useful means of studying denitrification in soils. Development of an assay based on this observation may overcome an analytical problem of long standing for soil scientists. That is, when release by denitrification of small quantities of dinitrogen is tested for, the possibility of contamination by atmospheric dinitrogen (which comprises 79% of our earth's atmosphere) always leaves the investigator somewhat uncertain about the validity of his results. Perhaps now the less equivocal measurement of the amount of nitrous oxide that accumulates behind an acetylene block applied to experimental soil plots will obviate this problem and remove doubts. If use of the acetylene-block method is successful, it may lead at long last to our securing reliable estimates of the yearly losses of dinitrogen resulting from ordinary denitrification throughout the world, as well as that following application of nitrate fertilizer to farmlands.

W. J. PAYNE

References

Balderston, W. L.; Sherr, B.; and Payne, W. J., 1976. Blockage by acetylene of nitrous oxide reduction by *Pseudomonas perfectomarinus, Appl. Environ. Microbiol.*, 31, 504–508.

Barbaree, J. M., and Payne, W. J., 1967. Products of denitrification by a marine bacterium as revealed by gas chromatography, *Marine Biol.*, 1, 136–139.

Cox, C. D., Jr., and Payne, W. J., 1973. Separation of soluble denitrifying enzymes and cytochromes from *Pseudomonas perfectomarinus, Can. Jour. Microbiol.*, 19, 861–872.

Cox, C. D., Jr.; Payne, W. J.; and Dervartanian, D. V., 1971. Electron paramagnetic resonance studies on the nature of hemoproteins in nitrite and nitric oxide reduction, *Biochim. Biophys. Acta*, 253, 290–294.

Federova, R. I.; Milekhana, E. I.; and Ilyukhina, N. I., 1973. Possibility of using the "gas exchange" method to detect ultra terrestrial life: Identification of nitrogen fixing organisms. *Akad. Nauk SSSR Izv. Ser. Biologicheskaia*, 6, 797–806.

Forget, P., and Dervartanian, D. V., 1972. The bacterial nitrate reductases: EPR studies on nitrate reductase A from *Micrococous denitrificans, Biochim. Biophys. Acta*, 256, 600–606.

Gayon, U., and Dupetit, G. 1886. Recherches sur la transformation des nitrates par les infiniments petits, *Mem. Soc. Sci. Phys. Nat. Bordeaux*, Ser. 3, 2, 201–307.

Hardy, R. W. F.; Burns, R. C.; and Holsten, R. D., 1973. Applications of the acetylene-etheylene assay for measurement of nitrogen fixation, *Soil Biol. Biochem*, 5, 47–81.

Hollis, O. L., 1966. Separation of gaseous mixtures using porous polyaromatic beads, *Anal. Chemistry*, 38, 309–316.

Lehninger, A. L., 1975. *Biochemistry*, New York: Worth, 417–508.

Matsubara, T., 1975. The participation of cytochromes in the reduction of N_2O to N_2 by a denitrifying bacterium, *Jour. Biochem.* (*Tokyo*), 77, 627–632.

Newton, N., 1969. The two-heme nitrite reductase of *Micrococcus denitrificans, Biochim. Biophys. Acta,* **185**, 316–331.

Payne, W. J., 1973. Reduction of nitrogenous oxides by microorganisms, *Bacteriol. Rev.,* **37**, 409–452.

Payne, W. J.; Fitzgerald, J. W.; and Dodgson, K. S., 1974. Methods for visualization of enzymes in polyacrylamide gels. *Appl. Microbiol.,* **27**, 154–158.

Payne, W. J., and Riley, P. S., 1969. Suppression by nitrate of enzymatic reduction of nitric oxide, *Proc. Soc. Expl. Biol. Med.,* **132**, 258–260.

Payne, W. J.; Riley P. S.; and Cox, Jr., O. D., 1971. Separate nitrite, nitric oxide and nitrous oxide reducing fractions from *Pseudomonas perfectomarinus, Jour. Bacteriol.,* **106**, 356–361.

Renner, E. D., and Becker, G. E., 1970. Production of nitric oxide and nitrous oxide during denitrification by *Corynebacterium nephridii, Jour. Bacteriol.,* **101**, 821–826.

Cross-references: *Aeration, Respiration, Atmosphere; Enzyme Activity; Fertilizers, Inorganic; Metabolic Activity; Nitrogen Cycle; Nitrogen Fixation; Nutrient Losses; Plant Nutrients; Soil Fertility; Vol. IVA: Nitrogen Cycle; Nitrogen: Element and Geochemistry.*

DENSITY—See BULK DENSITY; PARTICLE DENSITY

DENTAL CARIES AND SOILS

Dental decay (caries) is a disease associated with civilization; it is prevalent especially among children and adolescents. Because incidence differs among individuals in the same locality and also from one geographic area to another, investigations of various epidemiological factors, including soils, have been carried out. Possible relationships between the incidence of dental decay and soils are considered in this article.

Geographic Distribution of Caries

The dental status of inductees in the American armed forces during World War II suggested a geographic pattern, which in turn suggested that soils might be one related factor. Dental decay was highest among men from New England, followed in order by those from the eastern part of the central states, the western part of the central states, the southern states, and the southwestern states (reviewed by Dunning, 1953). The major kinds of soils differ among these regions. New England has Spodosols and Inceptisols (Podzols and related kinds); the eastern central states have Alfisols (Gray-Brown Podzolic soils); the western central states have Mollisols (Brunizems and Humic-Gley soils); the southern states have Ultisols (Red-Yellow Podzolic soils); and the southwestern states have Aridisols (Desert soils), (Soil Survey Staff, 1975). Possible relationships of dental caries to the regional character of soils have also been reported from New Zealand (Cadell, 1964; Losee, 1962).

Soils, Trace Elements, and Caries

Minor-element status of soils has been related in a few places to the incidence of dental decay. In the Hawke's Bay area of North Island, New Zealand, Ludwig, Healy, and Losee (1960) found that the incidence of caries was appreciably greater in Hastings than in Napier, cities about 12 mi apart with populations of 12,000 using the same sources for milk and water. A seaport, Napier lies next to the uplifted bottom (approximately 12 km^2) of a former salt-water lagoon, raised about 3 m above sealevel by an earthquake in 1931. Since 1945 part of the old lagoon has been used to produce vegetables, which are largely consumed in Napier. They contain more aluminum, molybdenum, and titanium and less barium, copper, manganese, and strontium than the vegetables grown near and consumed in Hastings (Healy, Ludwig, and Losee, 1961). Molybdenum levels are appreciably higher, and it was therefore postulated that greater intake of the element was responsible for the fewer dental caries in Napier. To test this hypothesis, diets with mineral ash obtained from vegetables high in molybdenum and from vegetables low in molybdenum were fed to rats. Animals receiving the larger amounts of molybdenum had fewer caries, lending support to the finding that the trace element was associated with less dental decay among children in Napier than in Hastings (Ludwig, Malthus, and Healy, 1962).

Earlier indications that molybdenum might affect the incidence of dental decay were shown by Adler and Straub (1953) from studies of dental caries in relation to water supplies in Hungary. Additional evidence has also been obtained in England where the name "teart pastures" is given to pastures on soils high in molybdenum. In such an area in Somerset, where soils are derived from sedimentary rocks and are alkaline in reaction, children 12 years of age had 29% less dental decay than comparable children in an adjacent control area (Anderson, 1966). In contrast to these findings, Curzon, Kubota, and Bibby (1971) did not find fewer caries prevalence in parts of California where soils were high in molybdenum. It may be significant that the California soils were derived from igneous rather than sedimentary rocks.

Elements other than molybdenum have been suggested as affecting caries. Thus in New Zealand Losee, Cadell, and Davies (1961) found that children in the Cheviot district had fewer cavities than normal, and they attributed this finding to the use of ground water supplies from soil and rock high in calcium and magnesium. An association of fewer carious cavities with calcareous soils has been observed elsewhere in New Zealand (Cadell, 1964). The composition of soils may in many instances be related to the underlying rock strata, and there is also the relationship of rocks, soils, groundwater, and trace-element intake.

In Finland lower caries prevalence was found in the population of areas with granite rather than diabase as the country rock (Präako, 1975). Differences in composition of soils overlying the two kinds of rock suggested that several elements, particularly fluorine and potassium, might affect the incidence of dental decay. High geologic strontium, occurring as strontionite and celestite, was ascribed as one cause of low caries prevalence in an area of northwest Ohio (Curzon et al, 1970). The exceptionally low levels of caries in this area could not be accounted for by drinking-water fluoride levels alone. Recently further studies have demonstrated low caries associated with high (5 mg/1) concentrations of strontium in drinking water in Wisconsin (Curzon and Iker, 1977).

Complex Effects

Available data suggest that several trace elements entering the food chain from soils may affect the incidence of dental caries. Direct effects of single elements may be less important than those of combinations of elements and any possible interaction between elements or other environmental factors. Such indications can be drawn from the amounts of some nine trace elements in the standard school lunches served in various parts of the United States (Murphy, Page, and Wath, 1971). Lunches were sampled in 300 schools to represent five regions of the continental United States. Plotting the mean amounts of the nine elements found in each region on a map suggested a relationship between their intake and the known caries prevalence. Larger amounts of four elements, aluminum, barium, boron, and strontium, were found in diets of the southwest and southeast of the United States compared with the northeast. The incidence of dental decay is by far the highest in New England. New England has a high proportion of Spodosols (Podzols), soils that have been subject to leaching by organic acids during their development.

Numerous inconsistencies can be found among the associations reported between the incidence of dental decay and the elemental composition of food, water, plants, and soils. Such inconsistencies could be due in part to unknown interactions that affect the absorption and utilization of elements in human metabolism. The role of soils is in turn but part of the story of the entry of the elements into the food chain, so the effects may be indirect and several steps removed from the soils themselves. Available evidence suggests that the soils, and the trace elements in them, do have a bearing on dental decay through food and water.

M. E. J. CURZON

References

Adler, P., and Straub, J., 1953. A water-bourne caries protective agent other than fluorine, *Acta Med. Acad. Sci. Hungarica,* 4, 221–227 (in Hungarian).

Anderson, R. J., 1966. Dental caries prevalence in relation to trace elements, *Brit. Dent. Jour.,* 120, 271–275.

Cadell, P. B., 1964. Geographic distribution of dental caries in relation to New Zealand soils, *Austral. Dent. Jour.,* 9, 32–38.

Curzon, M. E. J.; Adkins, B. L.; Bibby, B. G.; and Losee, F. L., 1970. Trace elements other than fluorine and dental caries, *Jour. Dent. Res.,* 49, 526–529.

Curzon, M. E. J.; Kubota, J.; and Bibby, B. G., 1971. Environmental effects of molybdenum on caries, *Jour. Dent. Res.,* 50, 74–77.

Curzon, M. E. J., and Iker, H. A., 1977. Strontium in drinking water and dental caries, Internat. Assoc. Dent. Res. Ann. Mtg., Copenhagen.

Dunning, J. M., 1953. The influence of latitude and distance from the sea coast on dental disease, *Jour. Dent. Res.,* 32, 811–842.

Healey, W. B.; Ludwig, T. G.; and Losee, F. L., 1961. Soils and dental caries in Hawke's Bay, New Zealand, *Soil Sci.,* 92, 259–266.

Losee, F. L., 1962. Soils and human health, with particular reference to dental health, *Trans. Jt. Mtg. Comm. IV and V, Internat. Soc. Soil Sci. (New Zealand), November 13-22,* 889–894.

Losee, F. L.; Cadell, P. B.; and Davies, G. N., 1961. Caries, enamel defects and soil: Owaka-Cheviot districts, New Zealand, *New Zealand Dent. Jour.,* 57, 135–143.

Ludwig, T. G.; Healy, W. B.; and Losee, F. L., 1960. An association between dental caries and certain soil conditions in New Zealand, *Nature (Lond.),* 186, 695–698.

Ludwig, T. G.; Malthus, R.S.; and Healy, W. B., 1962. An association between soil conditions and dental caries in rats, *Nature (Lond.),* 194, 456–458.

Ludwig, T. G., and Bibby, B. G., 1969. Geographic variations in prevalence of dental caries in the United States of America, *Caries Res.,* 3, 32–43.

Murphy, E. W.; Page, L.; and Watt, B. K., 1971. Trace minerals in Type A school lunches, *Jour. Am. Dietetic Ass.,* 58, 115–122.

Präako, A., 1975. Dental caries prevalence in the Rapakive granite and olivine diabase areas of Laitila, Finland, *Suomen Hammaslääkäriseuran Toimituksia,* 81, 1–55 (in Finnish).

Soil Survey Staff, 1975. Soil taxonomy: A basic system of soil classification for making and interpreting soil surveys, *U.S. Department of Agriculture Handbook No. 436,* 754p.

Cross-references: *Micronutrients; Plant Nutrients; Trace Elements Vol. IVA: Medical Geography; Medical Geology.*

DIFFUSION PROCESSES

The diffusion process arises from the random thermal motion of ions and molecules. In a single phase molecules and ions tend on balance to move from high to low concentrations. If movement in the x direction alone is considered, a diffusion coefficient D is defined as follows:

$$F = -D\frac{dC}{dx} \qquad (1)$$

where F, the flux, is the net amount crossing a section of unit area at right angles to the x direction, and C is the total concentration of the diffusate in the phase. Equation 1 is known as *Fick's first law of diffusion.*

The diffusion coefficient of molecules is related to their absolute mobility u in the same phase by the *Nernst-Einstein equation:*

$$D = ukT \qquad (2)$$

where k is the *Boltzmann constant,* and T the Kelvin scale temperature. This relation is obtained as follows: Since the force on a molecule is the gradient of its chemical potential μ its flux is given by

$$F = -Cu\left(\frac{d\mu}{dx}\right) \qquad (3)$$

For a single molecule

$$\mu \sim \mu_0 + kT\ln C \qquad (4)$$

Differentiating equation 4 with respect to x and substituting in equation 3 gives

$$F = -ukT\left(\frac{dC}{dx}\right) \qquad (5)$$

which yields the Nernst-Einstein relation by comparison with equation 1.

Diffusion of Ions

Because ions are charged, their diffusion is influenced by the movement of other ions and by external electric fields. The force on an ion is the gradient of its electrochemical potential. Thus the interpretation of the measured diffusion coefficient of an ion may be complex for salt solutions and for soils (Nye, 1966). In practice, an ion such as sodium is accompanied by an ion of opposite charge such as chloride, or it moves in the opposite direction to an ion of like charge such as potassium. The first process is called *salt diffusion,* and the second, *counterdiffusion.* In *self-diffusion* an ion counterdiffuses with its own isotope.

Range of Mobilities

Table 1 illustrates the very wide range of solute diffusion coefficients occurring in different parts of the soil. In the gaseous phase, molecules are about 10^4 times as mobile as they are in the aqueous phase. Ions adsorbed on purified clays vary greatly in mobility depending primarily on the thickness of the water layer between the aluminosilicate sheets.

Since the root mean square displacement of a diffusing particle in the x direction is $(2DT)^{1/2}$ (Jost, 1960), it may be deduced from Table 1 that a molecule of gas will diffuse about 1 m in 10^5 s (1.2 d). In solution it will move about 1 cm in 10^5 s. But the potassium ion in illite ($D = 10^{-23}$ cm^2 s^{-1}) will move only 1 nm in 5×10^8 s (16 year).

Diffusion in Soils

In soils the diffusing substance may be present in one, two, or all three of the gas, liquid, and solid phases.

Nonvolatile Solutes. Here only the liquid and solid phases are involved, conditions that

TABLE 1. Some Typical Diffusion Coefficients.

Phase	T(°C)	D(cm^2 s^{-1})
Gas phase		
O_2 into air	25	0.209
CO_2 into air	25	0.163
Chloropicrin into air	25	0.088
Liquid phase		
O_2 in water	25	2.26×10^{-5}
CO_2 in water	25	1.66×10^{-5}
NaCl in water	25	1.61×10^{-5}
Glucose in water	25	0.67×10^{-5}
Solid phase		
Na in montmorillonite gel	25	4×10^{-6}*
Na in vermiculite	20	6×10^{-9}*
K in illite	Lab	10^{-23}*
Soil		
Cl in sandy clay loam (40% H_2O by vol.)	Lab	9×10^{-6}*
Cl in sandy clay loam (20% H_2O by vol.)	Lab	1.4×10^{-6}*
Na in sandy clay loam (40% H_2O by vol.)	Lab	2.2×10^{-6}*
Na in sandy clay loam (20% H_2O by vol.)	Lab	0.5×10^{-6}*
PO_4 in sandy clay loam (40% H_2O by vol.)	Lab	3.3×10^{-9}*
PO_4 in sandy clay loam (20% H_2O by vol.)	Lab	0.3×10^{-9}*

*Indicates self-diffusion coefficient.

apply for most plant nutrient elements. If the soil has a well-defined geometry—for example, a cluster of close-packed solid spheres with the spaces between them filled with soil solution—then it would be possible to express the diffusive flux in terms of the concentration gradient, the individual diffusion coefficients of the solute in the liquid and solid and geometrical factors. This approach has been reviewed by Barrer (1969) and Crank (1975). The resulting expressions are complex because the flux is not simply the sum of independent fluxes through solid and liquid: Solute molecules are in rapid dynamic equilibrium between liquid in pores and adjacent solid surfaces and follow a combined pathway.

In a medium as heterogeneous as soil, it is simplest to treat it, for macroscale problems, as a homogeneous body in which it is legitimate to average over the microscale variations between particle and pore in the plane normal to the diffusion flux. An effective soil solute diffusion coefficient may then be defined by Fick's first law,

$$F = -D_e \frac{dC}{dx}$$

Here C is the concentration of diffusible solute in the whole soil, and includes both liquid and solid phases. Solute that does not exchange rapidly between solid and solution is not regarded as diffusible but rather as having a finite rate of reaction.

Effective Soil Diffusion Coefficient. In practice, the solute will move in the direction of diffusion through the pore solution, and it may also move in the adsorbed form by surface diffusion. Although its mobility in this form may be much lower than in solution, a much greater proportion of it may be adsorbed so that it cannot be neglected. The flux may be written as the sum of the flux by the liquid pathway, together with an excess term:

$$F = -D_1 \theta f_1 \frac{dC_1}{dx} + F_E \qquad (6)$$

where

D_1 = the diffusion coefficient of the solute in free solution
θ = the fraction of the soil volume occupied by solution
f_1 = an impedance factor (see below)
C_1 = the concentration of solute in the soil solution
F_E = the excess flux due to any surface mobility

The term F_E includes more than an exclusively solid pathway. It allows for any additional flux that arises because the solute is mobile when adsorbed on the solid. By definition,

$$F = -D_e \frac{dC}{dx}$$

thus

$$D_e = D_1 \theta f_1 \frac{dC_1}{dC} + D_E \qquad (7)$$

where D_E allows for any effect that mobility in the solid phase has on D_e.

Terms in Equation 7. The term θ, the volume fraction of liquids, is also the average cross section available for diffusion through the pore solution. The impedance factor f_1 takes account primarily of the tortuous pathway followed by the solute. This quantity has the effect of both increasing the path length to be traversed and of reducing the concentration gradient along the line of this path. The impedance factor may also include the "drag" caused by the increased viscosity of water within a few molecular layers of charged surfaces, an effect likely to be important only in dry soils. Anions will be excluded from very narrow pores by negative adsorption and large molecules by their size; hence connections between large pores may be cut off, and this effect is also included in f_1.

Figure 1 shows the relation between f_1 and θ for chloride ion in a sandy loan (Rowell, Martin, and Nye, 1967). In very dry soil f_1 is very low.

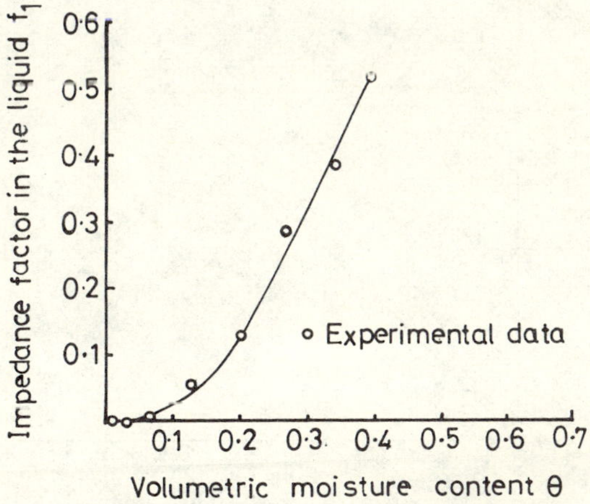

FIGURE 1. Relation between the impedance factor f_1 for chloride ion in a sandy soil, and the soil moisture content.

At a water potential of −100 bar, it was 2×10^{-4}, and at −15 bar, it was 10^{-2}. At water potentials between −1 bar and −0.1 bar, in most soils f_1 increases linearly with θ. It is important to note that over the field moisture range −0.1 to −10 bars, the product θf_1 and hence D_e decreases by at least two orders of magnitude, if dC/dC_1 is constant. This term dC/dC_1, called the *differential buffer power,* has a very wide range of values. For a nonadsorbed solute such as chloride or nitrate ions, $C = C_1\theta$, and hence $D_e = D_1 f_1$, so that in a moist soil $D_e \sim 2 \times 10^{-5} \times 0.5 \sim 10^{-5}$ cm² s⁻¹. On the other hand, for a strongly adsorbed solute such as phosphate on a ferruginous clay soil, dC/dC_1 may be 10^3, so that in a moist soil D_e, from the terms in equation 7, may be as low as $10^{-5} \times 0.3 \times 0.3 \times 10^{-3} \sim 10^{-9}$ $cm^2\ s^{-1}$. Thus in one day the root mean square displacement of nitrate in a moist soil may be as much as 1 cm, while that of phosphate may be as low as 10^{-2} cm. Since the relation between C and C_1 is usually nonlinear, the diffusion coefficient will vary with concentration. Figure 2 shows the slope of the sorption isotherm for potassium and the corresponding measured diffusion coefficent (Vaidyanathan, Drew, and Nye, 1969).

Self-diffusion coefficients are calculated from measurements of counterdiffusion between an isotopically labeled and unlabeled block of soil. In any one experiment the ratio C_1/C is the same for both isotopes and is constant; therefore, dC_1/dC is also constant.

Equation 7 has been found experimentally to express the diffusion coefficients of simple inorganic ions such as calcium and magnesium (Tinker, 1969), potassium (Vaidyanathan, Drew, and Nye, 1969; Nielsen, 1972), hydrogen (Farr, Vaidyanathan, and Nye, 1970), strontium (Mott and Nye, 1968), and phosphate (Rowell, Martin, and Nye, 1967; Olsen and Kemper, 1968); and nonvolatile organic solutes such as dimethoate (Graham-Bryce, 1969).

The significance of D_E has been found to be negligible for hydrogen, strontium, potassium, sodium, and phosphate in moist sandy loam; and for magnesium and calcium in a moist clay. The relative importance of D_E would be expected to increase when the proportion of ions in the soil solution decreased, and Rowell, Martin, and Nye (1967) have found significant values of D_E for sodium in a dry soil. The reason for the apparent low mobility of exchangeable ions in the few soils critically studied is unknown, though Nye (1979) has suggested that hydrous oxides (q.v.) adsorbed on clay surfaces may block diffusion pathways in soils, whereas such effects would be absent in purified clays.

Volatile Solutes. Since the mobility of molecules in the gas phase is about 10^4 times greater than in aqueous solution, it is to be expected that in a moist soil, gaseous diffusion will dominate if the distribution coefficient, $C_1/C_g << 10^4$. For example, at 0°C, C_1/C_g is for O_2, 0.048; CO_2, 1.71; SO_2, 80. When the distribution coefficient is about 10^4, in steady-state diffusion the gaseous phase may replace the solution without greatly affecting the overall diffusion coefficient, as illustrated by Graham-Bryce's (1969) measurements on disulfoton, a volatile organophosphorus insecticide, over a range of moisture levels (Figure 3).

The overall diffusive flux of a volatile solute may be expressed by an equation analogous to equation 6:

$$F = -D_g v_g f_g \frac{dC_g}{dx} - D_1 \theta f_1 \frac{dC_1}{dx} + F_E \qquad (8)$$

where v_g is the volume fraction of gas, and subscript g refers to the gas phase. In reality the

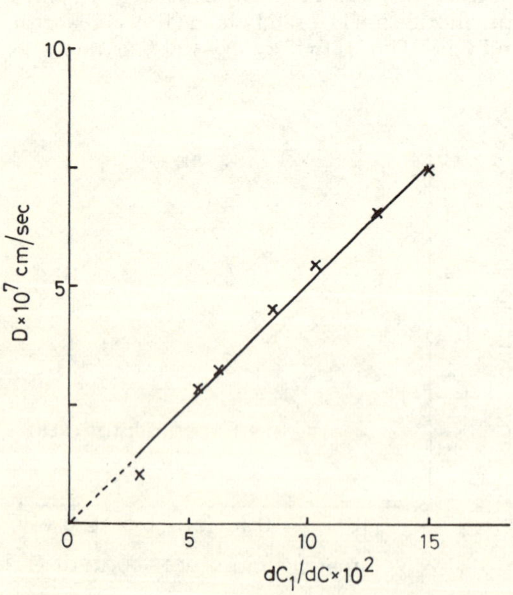

FIGURE 2. Diffusion coefficient for potassium in clay soil related to the slope of the adsorption isotherm. (After Vaidyanathan, Drew, and Nye, 1969)

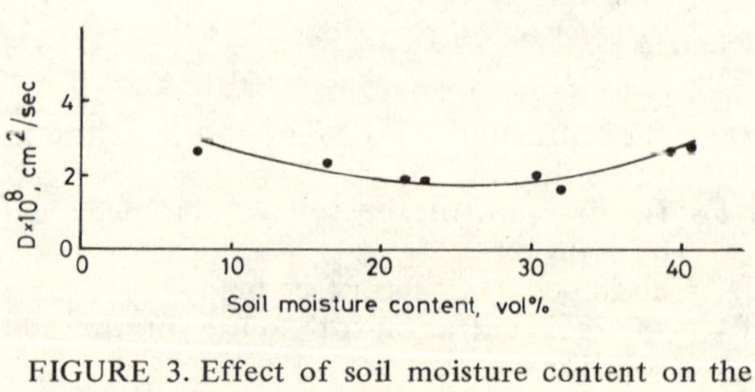

FIGURE 3. Effect of soil moisture content on the diffusion coefficient of disulfoton.

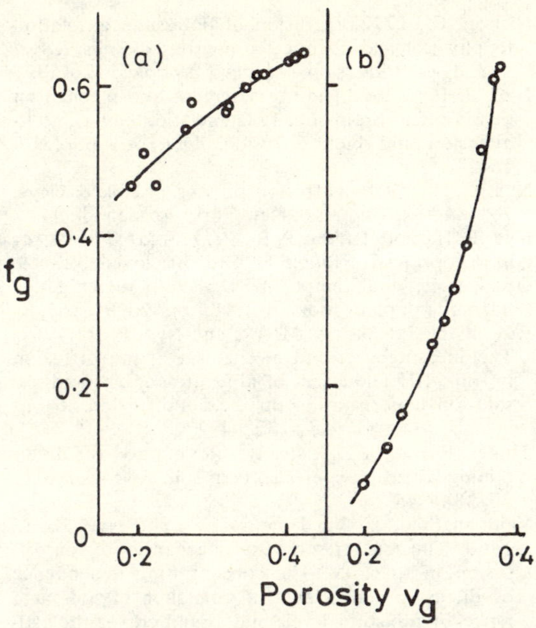

FIGURE 4. Effect on the impedance factor f_g of reducing air volume in sand by (a) compaction, and (b) adding water. (After Currie, 1970)

gaseous and solution pathways, while usually continuous, are not independent, and the true flux will be greater than these estimates because there will be an extra term for the interaction between them.

Impedance Factor f_g. In dry materials the *Bruggeman* (1935) *equation*, $f_g = v_g^n$, accounts for the relation between f_g and porosity. For spheres n is 0.5 and larger for more complicated shapes. Currie (1960) found $n = 1$ for dry soil crumbs. When water is added to dry material with a unimodal distribution of pore sizes, e.g., sand, it has a greater effect in lowering f_g than would consolidation of the dry material to the same v_g (Currie, 1970) (Figure 4). The reason is that the water is held at the narrowest constrictions of the channels between the pores, where it has the greatest blocking effect. For several types of material the variation of f_g on wetting is described by the equation (Currie, 1961)

$$f_g = v_g^n \left(\frac{v_g}{v_g + \theta} \right)^{2.5} \qquad (9)$$

In soils with a well-developed structure, the pore-size distribution is bimodal, corresponding to the micropores within the crumbs and the macropores between them. When water is added, the micropores are filled first and the value of f_g may increase as v_g is reduced because the gaseous pathway is less tortuous (Figure 5). On further wetting, f_g follows the same course as in sand. Considered in isolation, dry crumbs have low values of f_g. Currie (1970) gives $f_g = 0.33$ for a crumb from permanent pasture and 0.17 for a crumb from an overcultivated arable soil, compared with 0.67 for dry, unconsolidated material.

Effect of Mass Flow on Diffusion in the Soil Solution

In practice, solutes are moved both by diffusion and by mass flow of the soil solution caused by transpiration, evaporation, and drainage. Because water flows irregularly through the soil pores, it tends to disperse solutes. Frissel and Poelstra (1967) show that this process augments the molecular diffusion due to random thermal motion, and they replace the solution diffusion coefficient D_1 by a dispersion coefficient, $D_1{}^*$, given by

$$D_1{}^* = D_1 + \frac{v d \lambda}{\theta f_1} \qquad (10)$$

where d is the diameter of the particles, λ is a packing factor, and v is the water flux (cm^3 cm^{-2} s^{-1}). In soil, $d\lambda$ cannot be separated. The product had values of 0.7, 0.8, 6.0 cm in sandy, clay, and loess loam soils over a range of v from 0.6×10^{-6} to 230×10^{-6} cm s^{-1}.

In natural soils at flow rates that occur during infiltration (q.v.) and drainage (q.v.), it is clear that the dispersion coefficient will usually greatly exceed the molecular diffusion coefficient. For solute convected to root surfaces by

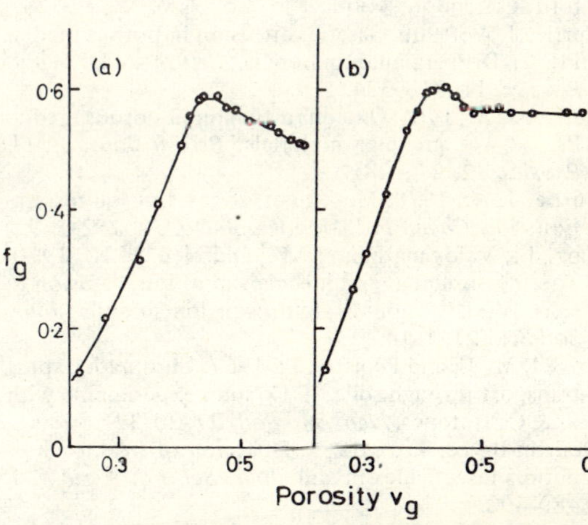

FIGURE 5. Effect of moisture level on the impedance factor f_g of soil crumbs (a) under continuous cultivation, and (b) under permanent pasture. (After Currie, 1970)

transpiration, it is unlikely that $D_1{}^*$ is appreciably greater than D_1. If v at the root surface is 10^{-6} cm s^{-1}, θf_1 is 0.1 and $d\lambda$ is 1.0 cm, then

$$\frac{v d\lambda}{\theta f_1} = 10^{-5} \text{ cm}^2 \text{ s}^{-1}$$

which is comparable to D_1. However, v decreases with radial distance from the root, and $d\lambda$ seems likely to be less over short distances around roots than in soil profiles in which long channels and cracks play an important part in creating dispersion.

Applications

The application of diffusion measurements and theory to soil aeration is described by Currie (1970); to mineral nutrition of plants by Nye and Tinker (1977); and to movement of salts by irrigation, drainage, and evaporation by Bresler (1969).

P. H. NYE

References

Barrer, R. M., 1969. Diffusion and permeability in heterogeneous media, *in* J. Crank and G. S. Park, eds., *Diffusion in Polymers.* New York: Academic, 452p.

Bresler, E., 1969. Solute movement in soils–Theory and applications, *in* B. Yaron, ed., *Irrigation in Arid Zones.* Israel: Volcani Institute, 306p.

Bruggeman, D. A. G., 1935. The calculation of various physical constants of heterogeneous substances. I. The dielectric constants and conductivities of mixtures composed of isotropic substances, *Annals Physik.,* **24,** 663–664.

Crank, J., 1975. *The Mathematics of Diffusion.* Oxford: Clarendon, 414p.

Currie, J. A., 1960. Gaseous diffusion in porous media. Pt. 2. Dry granular materials, *British Jour. Appl. Physics,* **11,** 318–324.

Currie, J. A., 1961. Gaseous diffusion in porous media. Pt. 3. Wet granular materials, *British Jour. Appl. Physics,* **12,** 275–281.

Currie, J. A., 1970. Movement of gases in soil respiration, *Soc. Chem. Ind. Monograph No. 37,* 152–169.

Farr, E.; Vaidyanathan, L. V.; and Nye, P. H., 1970. The measurement and mechanism of ion diffusion in soils. V. Diffusion of hydrogen ion in soils, *Jour. Soil Sci.,* **21,** 1–14.

Frissel, M. J., and Poelstra, P., 1967. Chromatographic transport through soils. II. Column experiments with Sr & Ca isotopes, *Plant and Soil,* **27,** 20–32.

Graham-Bryce, I. J., 1969. Diffusion of organo-phosphorus insecticides in soil, *Jour. Sci. Fd. Agric.,* **20,** 489–492.

Jost, W., 1960. *Diffusion in Solids, Liquids and Gases.* New York: Academic, 558p.

Mott, C. J. B., and Nye, P. H., 1968. Contribution of adsorbed strontium to its self-diffusion in a moisture-saturated soil, *Soil Sci.,* **105,** 18–23.

Nielsen, C., 1972. Diffusion of potassium in relation to physical and chemical properties of soils, *Royal Vet. Agric. Univ. Copenhagen Yearbook,* 142–159.

Nye, P. H., 1966. The measurement and mechanism of ion diffusion in soil. 1. The relation between self-diffusion and bulk diffusion, *Jour. Soil Sci.,* **17,** 16–23.

Nye, P. H., 1979. Diffusion of ions in soils & clays, *Adv. Agron.* (in press). New York: Academic.

Nye, P. H., and Tinker, P. B., 1977. *Solute Movement in the Soil-Root System.* Oxford: Blackwell, 323p.

Olsen, S. R., and Kemper, W. D., 1968. Movement of nutrients to plant roots. *Adv. Agron.,* **20,** 91–151.

Rowell, D. L.; Martin, M. W.; and Nye, P. H., 1967. The measurement and mechanism of ion diffusion in soils. III. The effect of moisture content and soil solution concentration on the self-diffusion of ions in soils, *Jour. Soil Sci.,* **18,** 204–222.

Tinker, P. B., 1969. A steady state method for determining diffusion coefficients in soil, *Jour. Soil Sci.,* **20,** 336–345.

Vaidyanathan, L. V.; Drew, M. C.; and Nye, P. H., 1969. The measurement and mechanism of ion diffusion in soils. IV. The concentration dependence of diffusion coefficients of potassium in soils at a range of moisture levels and a method for the estimation of the differential diffusion coefficient at any concentration. *Jour. Soil Sci.,* **19,** 94–107.

Cross-references: *Aeration, Respiration, Atmosphere; Electrochemistry; Evaporation; Exchange Phenomena; Flow Theory; Imbibition; Pore-size Distribution; Salt Leaching; Soil Solution; Soil Structure; Water Fluxes; Wetting Front.*

DISPERSION PHENOMENA

Dispersion is the opposite of aggregation (q.v.). It consists of the breaking down of soil aggregates, a process that ultimately stops with discrete particles of sand, silt, and clay. Dispersion phenomena have been studied mainly in relation to aggregation of soils but to some extent for making particle-size analyses (Greenland, 1965).

Dispersion Mechanisms

Destruction of Binding Agents. In various ways clay particles, or domains (clusters of clay particles), sand, and silts may be bound together by agents such as polysaccharide (q.v.), humic acids (see *Humus*), iron and aluminum oxides (see *Hydrous Oxides*), and calcium carbonate (Emerson, 1959). The binding action is attributed to the bonds between organic polymers such as polysaccharide or humic acid with clay surfaces through hydrogen ions, to polyvalent cation bridges, to sesquioxide humus complexes, and to van der Waals forces. Consequently, oxidation or microbial breakdown of the organic substances or the solution of iron and aluminum oxides or of carbonates can eliminate the binding agents and thus promote dispersion.

Deflocculation. The reverse process of *flocculation* (q.v.) is *deflocculation,* reducing the clumping of clay particles and thus permitting more ready dispersion of a soil mass as a whole. The surfaces of clay particles have dominantly negative charges, and the edges have dominantly positive charges. Clay particles are larger, and thus a clay particle has a net negative charge. Anions are therefore repelled so that concentrations near the particle surface are lower than in the bulk soil solution. In contrast, cations in the soil solution are attracted by the net negative charge, and some are adsorbed on the particle surfaces. Opposing this tendency is the normal diffusion of cations into the soil solution, and the resulting distribution equilibrium results in formation of a diffuse electrical double layer around a clay particle. The inner layer is the lattice surface with its negative charge. The outer layer consists of two sets of cations. One set comprises the cations adsorbed on the particle surface, known as the "Stern layer." The second and outer set comprises the cations in solution near the particle surface, known as the "Gouy diffuse layer" and frequently as the "counterion atmosphere" (Baver, Gardner, and Gardner, 1972; Russell, 1973).

The gradient in concentration of cations in the electrical double layer around a clay particle results in a difference in the electrical potential (electrokinetic or zeta potential) between the particle surface and the bulk solution. Thickness of the double layer and the zeta potential decrease together as the electrolyte concentration of the soil solution (q.v.) increases. The double layer and the zeta potential are inversely related to the valence of cations at equivalent concentration of electrolyte. Thick, diffuse double layers and high zeta potentials promote repulsion of one clay particle by another and thus encourage dispersion. Such dispersion is due to long-range coulombic forces.

In contrast to the repelling effects of the zeta potential and diffuse double layer, coulombic attraction may occur between the positively charged edges of one clay particle and the negatively charged surface of another. This promotes flocculation or tends to offset dispersion when the attraction is large enough. Similarly, short-range van der Waals forces also tend to offset dispersion and must be overcome. These forces operate over very small distances, however, and are of limited importance generally.

The types and concentrations of ions in solution thus govern the dominance of the attracting and repelling forces and the resulting flocculation or deflocculation of clays. For example, saline soils in arid regions are normally flocculated because of the large amounts of electrolyte present. In contrast, other soils in the same region low in salt and high in exchangeable sodium are deflocculated and dispersed.

The Dispersion Process

According to one model of a microaggregate, clay domains or clusters of particles (D) may be combined either electrostatically because of attractive forces between positive edges (E) of one domain and negative forces (F) of adjacent domains or because of derivatives of organic matter (OM) sandwiched between two adjacent domains (Emerson, 1959). Three types of bonding are possible in the latter case: F–OM–F, F–OM–E, and E–OM–E. Sand and silt (e.g., quartz) are bound together by organic matter and by domains or domain-organic matter complexes to form microaggregates. The number of possible arrangements of the basic components of the microaggregate are practically unlimited. Groups of microaggregates are combined to give soil structure (q.v.).

Considering one of these possible arrangements and following the pattern suggested by Edwards and Bremner (1967), dispersion can be considered the reversal of aggregation as follows:

$$[(Q - (OM) - P - D)_x - Q)_y]_z \rightarrow 2zyQ + zyxOM\downarrow + zyxD + zyxP\downarrow \rightarrow 2zyQ + zyxD$$

where

D = one domain or several domains bonded electrostatically (E–F)

P = polyvalent cations

x, y, z = finite whole numbers indicative of the various possible arrangements within the microstructure and the extent of aggregation.

The dispersion process may occur naturally or can be performed in the laboratory, as is done in particle-size analysis. In the laboratory it is generally achieved by mechanical agitation combined with use of oxidants, acids, and peptizing reagents. The organic matter is oxidized, and the polyvalent cations are replaced by monovalent cations.

Factors Inducing Dispersion

Disperson can be induced by various factors such as climatic regimes, inherent soil properties, composition of irrigation waters, and soil-management practices. These factors affect stability of soil aggregates in various ways and may act singly or in combination.

Climatic Factors. These can induce or permit dispersion in several ways. For example, the formation of binding agents such as silicate clays, humus, and polysaccharide is limited in arid and semiarid regions. The absence of such binding agents means that aggregation is limited in the soils and that dispersion is relatively easy. High levels of evaporation combined with limited rainfall can result in the formation of saline soils (see *Salt-affected Soils* in Vol. XII, Pt. 2), which tend to be flocculated. Limited leaching of saline soils can cause dispersion if much sodium is present. Surface crusts may also form. In contrast to arid and semiarid climates, those of humid regions promote the formation of binding agents. Thus, for example, aggregation is especially strong in the *A* 1-horizons of the Mollisols of the central prairies of the United States with their high contents of organic matter. At the same time substantial leaching is possible in humid climates with translocation of binding agents from upper to lower horizons to cause dispersion of the former, as occurs in the *A* 2-horizons of Podzols. Climate can also contribute to dispersion through the action of rainwater. Rapidly falling large drops can break down aggregates and thus effect dispersion. Small raindrops lack energy to break down aggregates on impact but may cause dry aggregates to slake as they become wet.

Soil Properties. Like climatic conditions, soil properties can operate in conflict to some extent in promoting and retarding dispersion. Soils with high sodium levels (ESP—exchangeable sodium percentage greater than 15%) become highly hydrated and disperse readily when wetted by rain or by irrigation water low in electrolytes. As the ESP increases, so does dispersion (Richards, 1954). Low levels of organic matter in soils usually result in poor aggregation, and those soils disperse more readily when wet than soils high in organic matter. Emerson (1959) attributes this tendency to inadequate amounts of organic matter of various forms to bind clay domains, silt, and sand into stable aggregates. Furthermore, low levels of organic matter normally mean that the microbial populations necessary to the conversion of fresh organic residues into forms that serve as binding agents are limited.

Soils low in one or both of iron or aluminum oxides disperse more readily than those that are high. The oxides, especially if formed from irreversibly dehydrated gels, are effective cementing agents (McIntyre, 1956). Partial removal of free iron oxides reduced aggregate stability of samples of a group of soils in Hawaii and accentuated the dispersing effect of the sodium ion (McNeal et al., 1968).

Irrigation Water. Normally addition of irrigation water dilutes the soil solution in saline soils and may thus promote dispersion of those that are highly alkaline. A common laboratory exercise in introductory courses in soils is to have students leach a sample of soil with a sodium carbonate solution and watch the gradual slowing and eventual stopping of percolation as the sample disperses. Long-time irrigation with bicarbonate waters (with high residual sodium carbonate) may increase dispersion. The presence of the bicarbonate may cause precipitation of calcium and magnesium as carbonates and thus allow a buildup of sodium (Eaton, 1950)

Soil-management Practices. Tillage is essential for seedbed preparation, as a rule, but it may cause dispersion. When performed at high moisture contents, tillage subjects the soil to shearing forces that mechanically destroy structure and enhance dispersion. Over long periods tillage may also promote dispersion through exposure of the soil surface to impact of raindrops, greater decomposition of organic matter because of improved aeration, and greater leaching of flocculating salts. Long-term cultivation has resulted in deterioration of the structure of many surface soils (Harris, Chesters, and Allen, 1966; Low, 1972).

Effects of Dispersion

Dispersion can have unfavorable effects on moisture regimes, aeration, seedling emergence, and root growth. Dispersion can reduce infiltration rates and hydraulic conductivity, thus increasing runoff (Hamid and Mustafa, 1975). Partial or complete plugging of pores through dispersion restricts aeration and can limit the respiration of seeds, microorganisms, and plant roots. Under some conditions dispersion results in the formation of surface crusts that are strong enough to hamper seedling emergence (Hanks, 1960; Parker and Taylor, 1965; Richards, 1953). Dispersion of soil will also permit greater compaction of the plow layer and the immediately underlying part of the profile, which can then limit extension and proliferation of root systems.

Control of Dispersion

A number of management practices can aid in the control of soil dispersion. These include methods of tillage, crop selection, mulches, and applications of various chemicals. Timeliness of tillage operations (see *Tillage*) in relation to soil moisture is important in many soils, especially those that tend to disperse readily. For some soils, zero tillage (q.v.) is required to prevent dispersion. The periodic growing of perennial grasses serves to maintain desirable structure and prevent dispersion of soils generally. The grasses have massive fibrous root systems, provide protection from raindrop impact, and furnish a continuous supply of secretions and residues within the soil. Grass-legume mixtures can have the same effect as perennial grasses (Harris, Chesters, and Allen, 1966). The use of a mulch (q.v.) protects the soil surface from the impact of raindrops and reduces erosion, all of which retard dispersion. Mulches also provide organic residues that can be converted by microorganisms into binding agents. Farmyard manure serves similar functions.

A number of chemicals have been used to promote aggregation and thereby prevent dispersion. Additions of lime improve aggregation and aeration of soil horizons. Sulfuric acid has improved aggregate stability in the surface layers of calcareous soils, whereas phosphate fertilizers have stabilized aggregates in surface layers of both acid and calcareous soils (Thien, 1976). Soil conditioners (q.v.) of various kinds have been useful in special situations (Allison, 1956). Conditioners are compounds with functional groups and characteristics similar to those of microbial polysaccharide. Examples are polyvinyl alcohol (PVA), hydrolysed polyacrylonitrite (HPAN), and vinyl-acetate maleic acid (VAMA). These are more effective in aggregating soil materials than polysaccharide, but they are also expensive and thus not given to wide-scale agricultural use (Baver, Gardner, and Gardner, 1972).

M. A. MUSTAFA

References

Allison, L. E., 1956. Soil and plant responses to VAMA and HPAN soil conditioners in the presence of high exchangeable sodium. *Soil Sci. Soc. America Proc.,* **20**, 147–151.

Baver, L. D.; Gardner, W. H.; and Gardner, W. R., 1972. *Soil Physics,* 4th ed. New York: Wiley, 498p.

Eaton, F. M., 1950. Significance of carbonates in irrigation waters, *Soil Sci.,* **69**, 123–133.

Edwards, A. P., and Bremner, J. M., 1967. Microaggregates in soils, *Jour. Soil Sci.,* **18**, 64–73.

Emerson, W. W., 1959. The structure of soil crumbs, *Jour. Soil Sci.,* **10**, 235–244.

Greenland, D. J., 1965. Interaction between clays and organic compounds in soils, Part I: Mechanisms of interaction between clays and defined organic compounds; Part II: Adsorption of soil organic compounds and the effects on soil properties, *Soils and Fert.,* **28**, 415–425, 521–531.

Hamid, K. S., and Mustafa, M. A., 1975. Dispersion as an index of relative hydraulic conductivity in salt-affected soil of the Sudan, *Geoderma,* **14**, 107–114.

Hanks, R. J., 1960. Soil crusting and seeding emergence, *7th Intnat. Congr. Soil Sci. Trans.* (Madison), **1**, 340–346.

Harris, R. F.; Chesters, G.; and Allen, O. N., 1966. Dynamics of soil aggregation, *Adv. Agron.,* **18**, 107–160.

Low, A. J., 1972. The effect of cultivation on the structure and other physical characteristics of grassland and arable soils (1945–1970), *Jour. Soil Sci.,* **23**, 363–380.

McIntyre, D. S., 1956. The effect of free ferric oxide on the structure of some terra rossa and rendzina soils, *Jour. Soil Sci.,* **7**, 302–306.

McNeal, B. L.; Layfield, D. A.; Norvell, W. A.; and Rhoades, J. D., 1968. Factors influencing hydraulic conductivity of soil in the presence of mixed salt solutions, *Soil Sci. Soc. America Proc.,* **32**, 187–190.

Parker, Jr., J. J., and Taylor, H. M., 1965. Soil strength and seedling emergence relations. I. Soil type, moisture tension, temperature, and planting depth effects, *Agron. Jour.,* **57**, 289–291.

Richards, L. A., 1953. Modulus of rupture as an index of crusting of soils. *Soil Sci. Soc. America Proc.,* **17**, 321–323.

Richards, L. A., ed., 1954. Diagnosis and improvement of saline and alkali soils, *U.S. Department of Agriculture Handbook No. 60,* 160p.

Russell, E. W., 1973. *Soil Conditions and Plant Growth,* 10th ed. London: Longmans, 849p.

Thien, S. J., 1976. Stabilizing soil aggregates with phosphoric acid, *Soil Sci. Soc. America Proc.,* **40**, 105–108.

Cross-references: *Aggregation, Aggregate Stability; Flocculation; Puddling; Soil Stabilization; Soil Structure.*

DRAINAGE—See SOIL DRAINAGE

E

ECOLOGY—See MICROBIAL ECOLOGY AND CLAY MINERALS; MICROHABITATS. Vol. XII, Pt. 2: ECOLOGY OF SOILS

EDAPHOLOGY

Edaphology (from the Greek word *edaphos,* which means "soil" or "ground") is the study of soil as a medium for plant growth. The large differences among soils in properties such as depth, texture, drainage (q.v.), organic content, mineralogy (q.v.), pH, and tilth (q.v.) make some soils naturally fertile and productive and some infertile, but all display some degree of responsiveness to technological and management input. The differences in the productivity of various soils is receiving more attention today than ever before as humanity seeks to produce more food and fiber. Indeed, modern agriculture is more concerned about the responsiveness of soils to technological and management input than about their inherent fertility.

Exchange Capacity

One of the most important soil characteristics that influences plant growth is the negative charge, or *exchange complex,* associated with clay and humus. These inorganic and organic colloids possess a negative charge that attracts positive-charged nutrient elements such as calcium (Ca^{2+}) and potassium (K^{+}) and others. These elements are held by the soil colloids in forms available to plant roots. The enormous surface areas provided by clay and humus are important storehouses for plant nutrients. Sandy soils, low in organic matter, have few exchange sites. They are considered "poorly buffered soils" and are generally infertile. Such soils can also become acid or basic very easily, another hazard in crop production.

Importance of Organic Matter

Although the organic matter content of mineral soils is usually less than 5%, it markedly influences soil properties and plant growth. Organic matter (q.v.) influences both physical and chemical properties. The loose, friable condition common to most productive soils generally results from high organic matter content. Thus through its beneficial effects on soil structure (q.v.), organic matter increases water infiltration (q.v.) and storage. In addition, organic matter is important in reducing soil erosion (q.v.) by wind and water. It supplies needed nutrients, energy for soil microbes, and as humus (q.v.), it provides a seat of electrical charges to hold plant nutrients against leaching.

Soil Life

That soil is a dynamic system becomes most evident when one considers the vast biological population it supports. Macroanimals such as insects and worms aerate the soil and pulverize plant residues. Microanimals such as nematodes and protozoa contribute to the breakdown of complex organic matter into simpler forms that will be still further mineralized by the vast population of microplants such as bacteria, fungi, and streptomycetes. A most important function of the soil fauna and flora involves biochemical processes such as nitrogen fixation, nitrification, and mineralization of organic matter with resultant release of plant nutrients such as nitrogen, phosphorus, and sulfur. Decomposition of organic residues prevents the accumulation of trash in surface soil that would make it unproductive. Soil organisms also take part in the synthesis of humus.

Soil Nitrogen

Nearly all soil nitrogen is associated with organic matter, which must be mineralized by microbes before the nitrogen becomes available to plants (see *Nitrogen Cycle*). At least a 20-fold difference exists in the amounts of nitrogen contained in agricultural soils as a group. Mean temperatures and rainfalls are important to inherent soil nitrogen status. Cool moist climates favor higher nitrogen levels than arid or tropical climates. Various management practices and amounts of nitrogen removed by crops result in different levels of soil nitrogen within a given soil type. Soils seldom supply enough nitrogen for good yields of cultivated crops after several years of farming. Nitrogen fertilizer is being used increasingly to make up the deficit, especially with the use of high yielding varieties. In Figure 1 the increased wheat growth following nitrogen fertilization is apparent. Turning under plant residues that are not too low in nitrogen aids decomposition, humus formation, and the ultimate release of nitrogen again for the next crop. Excess nitrate can be leached or in poorly aerated soils can be reduced by microbes and lost as nitrogen gas through a process called *denitrification* (q.v.). On the other hand, biological nitrogen fixation carried out by various soil organisms, both nonsymbiotically or, as in the case of legumes, symbiotically, adds much needed nitrogen to soils.

Phosphorus

After nitrogen, phosphorus is usually the most limiting available nutrient element necessary for optimum plant growth. In contrast to plant needs for phosphorus, average soil phosphorus levels are quite high. The problem, however, is that it is too often unavailable to plants. Under favorable conditions mineralization of organic matter not only releases nitrogen but also phosphorus in available forms. Phosphorus is often unavailable in soils because the inorganic form combines chemically with clay, aluminum, iron, and calcium. Soil pH and the degrees of leaching and weathering are all important in controlling the amounts of available phosphorus. Phosphate fertilization is commonly used for vegetables and many cereals. Many grasses and rice frequently do not respond to applied phosphate, and their response depends on the kind of soil. Soil tests designed to determine available inorganic forms of phosphate are valuable guides in predicting phosphate fertilizer requirements for a given crop.

FIGURE 1. Plots on the left show increased wheat growth following nitrogen fertilization in contrast to plots on the right show nonfertilized wheat.

Potassium

Although the total potassium contents of many soils are high, most potassium exists in complex mineral forms. Some of the potassium in soils is held by colloids in an exchangeable form and is available; but it can be exhausted under conditions of high crop yields where nitrogen and phosphate fertilizers have been used. This is particularly the case in sandy loams or peat soils. A soil test for potassium is a good measure of its availability to plants. When soils are fertilized with potassium, depending on the clay type, an appreciable amount of potassium can be "fixed" into unavailable forms. Thus soil potassium can be grouped into four categories: complex minerals, slowly available fixed "clay" potassium, exchangeable potassium, and solution potassium. The amount of each form and the conversion of one form to another due to cropping, fertilization, and weather conditions are important factors in the potassium nutrition of crops.

Secondary and Micronutrients

The availability of secondary nutrients such as sulphur, calcium, and magnesium and of micronutrients (q.v.) depends on soil properties and plant species. Although calcium deficiencies as such are uncommon, lime is often added to acid soils in the form of carbonates to reduce "sourness," or soil acidity. The beneficial effects of added calcium is not so much due to increased availability of that element but indirectly through reducing toxic amounts of aluminum, iron, or manganese so often associated with soil acidity. Magnesium deficiency may occur, especially in high rainfall areas, even though calcium and magnesium are the dominant cations held by soil colloids. In arid regions they are plentiful, assuring the soil high base status. Under high rainfall, however, calcium and magnesium can be leached and replaced by hydrogen and aluminum, reflected in their low pH.

Sulfur Reactions

Plants take up sulfur as sulfate. This form is subject to leaching in areas of high rainfall or excessive irrigation, and it may be leached below the root zone. Sulfur deficiencies are becoming more common as high-analysis, nitrogen-phosphate fertilizers are used and as environmental controls restrict the sulfur escaping into the atmosphere from industrial plants.

Micronutrient Uptake

Iron chlorosis in many crops is common on soils high in lime although plant varieties within a species may differ markedly in their sensitivity to lime. Deficiencies in other micronutrients such as copper, zinc, boron, and manganese are also more likely to occur in the presence of lime, though in some acid soils boron may be deficient. Plant species and even a given variety will determine the likelihood of a micronutrient deficiency. Moreover, the degree of soil weathering, mineralogy, and acidity affects their availability. Molybdenum is different from the other micronutrients because its availability is greater in basic or high-lime soils; it is more often deficient in strongly weathered and acid soils.

Saline Sodic Soils

In arid regions plant growth can be affected by the presence of salts or sodium in soils. Plant species display differing tolerances to these two conditions, but saline and/or high sodium levels are often the cause of crop failures.

Soil Moisture Regime

In spite of the need for all nutrients to be available in soils for optimum plant growth, water is often the most limiting. Soil properties that affect the moisture-supplying capacity depend on the amount, geometry, and distribution of the soil pores. Soil texture, structure, and tilth control the nature and volume of spaces between soil particles. Soil humus and organisms in the soil are also important in affecting soil porosity. The rate of movement of water into and through the soil to enter plant roots as well as total moisture storage affect plant growth. The kinds and amounts of pores affect the soil moisture regime. Moreover, soil aeration is important because plant roots require oxygen for satisfactory growth.

Conclusion

Soils differ markedly from one another as media for plant growth. At the same time soils are one of man's most valuable natural resources. They are a living part of nature that when properly managed, fertilized, and watered, provide the basis for production of abundant crops to supply man with food, fiber, and shelter.

MURRAY D. DAWSON

References

Brady, N. C., 1974. *The Nature and Properties of Soils.* New York: Macmillan, 639p.

Donahue, R. L.; Schickluna, J. C.; and Robertson, L. S., 1971. *Soils: An Introduction to Soils and Plant Growth.* Englewood Cliffs, N.J.: Prentice-Hall, 587p.

Kellogg, C. E., 1975. *Agricultural Development: Soil, Food, People, Work.* Madison: Soil Science Society of America, 233p.

Cross-references: *Agronomy; Buffers, Buffering; Denitrification; Enzyme Activity; Exchange Phenomena; Fauna; Fertilizers, Inorganic; Fertilizers, Organic; Humus; Leaching; Management of Soils; Nitrogen Cycle; Nitrogen Fixation; Phosphorus Cycle; Soil Fertility; Sulfur Transformations.*

ELECTROCHEMISTRY

Electrochemistry is the branch of physical chemistry (q.v.) that studies the relationship between chemical transformations and energy in reactions that involve electrical energy external to the system (Milazzo, 1963). One of the most important applications of electrochemistry deals with the electrolytic methods of chemical analysis. The areas of electrogravimetry, potentiometry, amperometry, polarography, voltammetry, and coulometry arise from the manipulation of potential, current, time, and concentration. Exhaustive descriptions of the analytical methods and theory of the methods listed have been published in many works (e.g. Milazzo, 1963; Denaro, 1965; Bockris and Reddy, 1970; Yeager and Salkind, 1973; and Bockris and Conway, 1975); a discussion of the methods and theories is beyond the scope of this review. Soil scientists have taken advantage of some of the electrolytic methods of chemical analysis and applied them to the measurement of soil properties.

Potentiometry and Potentiometric Titrations

Potentiometry allows the determination of the concentration of a solution constituent by measuring the potential difference or EMF (electromotive force) of a cell in that solution at zero current. The measurement involves the use of an indicator electrode reversible to the species to be determined, a reference electrode, and the *Nernst equation,* which relates the measured voltage to the solution activity of the unknown. Potentiometry provides a versatile analytical technique for the determination of many parameters such as activities, activity coefficients, solubility products, dissociation constants, equilibrium constants, transference numbers, oxidation states, etc.

Indicator electrodes reversible to various ions are used extensively in soils research, and their usefulness has not yet been fully exploited. The oldest and most common laboratory tool is the *hydrogen-ion selective glass electrode* (pH electrode). Actually, the modification of the glass compositions of the first glass electrodes led to the development of electrodes selective for a variety of cations. The glass electrodes have had considerable use in soils research for the determination of H^+ in water extracts and in soil or clay suspensions. Eisenman et al (1965) discusses in detail the theory and applications of H^+ selective glass electrodes. *Sodium-ion selective glass electrodes* have been used to determine sodium activity in soil and clay systems (Bower, 1961) and solutions (Bower, 1959; El-Swaify and Gazdar, 1969). Some deviations due to electrode sensitivity to K^+ was observed in the solution studies. Transference numbers, ionic mobilities, and diffusion coefficients were obtained for Na^+ and Li^+ in clay-water systems (Mokady and Low, 1966). Different combinations of sodium-ion and lithium-ion sensitive glass electrodes and Ag-AgCl electrodes were used for the determinations. A *potassium-ion selective glass electrode* was used to determine K^+ activity in 30 Michigan soil extracts (Mortland, 1961). Potential problems due to electrode reversibility to H^+, Na^+, Li^+, and $NH_4{}^+$ were discussed by Mortland (1961) and Eisenman et al (1965). The relative partial molal free energies of NaCl, KCl, and HCl were measured in Wyoming bentonite suspensions using a sodium-ion selective glass electrode, a potassium-ion selective glass electrode, and a hydrogen-ion selective glass electrode (Baker and Low, 1970). An *Ag-AgCl electrode* was used in conjunction with the ion selective glass electrodes so that the liquid junction potential-suspension effect controversy would be eliminated from the measurements. These electrodes have also been used to determine chloride activity in soils (Krupskiy, Aleksandrova, and Boriskova, 1973) and in water extracts of soils (Davey and Bembrick, 1969). The *nitrate-specific ion electrode* has been used for the determination of nitrate in soil extracts (Raveh, 1973). The distillation step in total nitrogen analysis of soil by the *Kjeldahl method* was eliminated by use of an ammonia electrode for determination of the amonium formed (Bremner and Tabatabai, 1972). The *ammonia electrode* was also found to perform satisfactorily in the determination of ammonium in soil extracts and water samples (Banwart, Tabatabai, and Bremner, 1972) and in the determination of cation exchange capacity (Miller, Riecken, and Walter, 1975). The *calcium-ion selective electrode* has been used in the determination of exchangeable Ca^{2+} in soil solutions (Doiron and Chamberland, 1973) but has limited overall usefulness because of its sensitivity to magnesium ions present in calcium-magnesium salt mixtures (El-Swaify and Gazdar, 1969). The calcium electrode was also used for measuring ionic activities in clay suspensions (Uehare and Mortland, 1960). Labile soil fluoride was determined by using a *flouride-ion specific electrode* to measure the amount of flouride extracted from soil in an anion-exchange resin (Larsen and Widdowson, 1971). These values were well correlated with isotopically exchangeable (labile) soil-fluoride values. The *sulfide-ion specific electrode* has been used to measure the soluble sulfide levels in submerged soils and for measuring sorption of sulfide by clay (Allam, Pitts, and Hollis, 1972). The potentiometric determination of boron in water, soil, and plant-tissue samples has been accomplished using a *tetrafluoroborate-ion specific electrode* (Carlson and Paul, 1969). Boron is first isolated on a column of the boron specific ion-exchange resin Amberlite XE-243. Addition of hydrofluoric acid to the column converts the boron to tetrafluoroborate ion, which is measured potentiometrically with the electrode. The $BF_3{}^-$ specific-ion electrode method is free of the interferences that are troublesome in other methods.

Potentiometric titrations involve the determination of changes of the electrode potential difference as changes in the concentration of an electroactive species occur. At the stoichiometric end point of the chemical reaction involved, the solution EMF will shift greatly. From the amount of titrant consumed and the stoichiometry of the reaction, the concentration of the desired substance may be calculated. For soil systems or any system in which the detected species is only partially ionized, the potentiometric methods using ion-sensitive electrodes are useful for measuring ion activities while potentiometric titrations are useful for measuring ion concentrations. Potentiometric titrations have been applied to soils research to some extent. The *glass electrode-calomel electrode combination* has been used in potentiometric titrations for the determination of the various properties of organic matter (Borggaard, 1974; Khana and Stevenson, 1962) aqueous salt solutions (Frink and Sawhney, 1967), ion-exchange resins (Rich, 1970), soil suspensions (Dewan and Rich, 1970), and clay suspensions (Kaddah and Coleman, 1967; Khan and Singhal, 1967; Rich, 1970; Shainberg and Dawson, 1967). Sharpe and Hoagland (1916) were probably the first to use this electrode combination for the acidity determinations of naturally occuring soils. Also a *lead-mercury amalgam indicator electrode* (Robbins, Carter, and James, 1973) and a *lead-ion* specific electrode (Goertzen and Oster, 1972) were used for the potentiometric titration of sulfate in water and soil extracts. The method of sulfate determination by Goertzen and Oster (1972) is less time consuming and less subject to interferences than previous methods.

Reduction Potential

The *reduction potential* of a chemical system is a measure of the tendency for reduction reactions to occur. It is electrochemically defined as

$$E_{red} = E^{\circ} + (RT/nF) \ln \frac{[ox]}{[red]}$$

The soils literature sometimes refers to the *oxidation potential* which is equal in magnitude, but opposite in sign of the reduction potential, i.e., the tendency for oxidation reactions to occur in that system.

Redox potentials are measured with a reference electrode in combination with a metallic electrode such as platinum, sensitive and reversible to oxidation-reduction conditions. This potential is then compared with a table of standard potentials, which provides a scale of reference. Very-well-oxidized soils may register a redox potential of 0.8 V while an extremely reduced soil may register –0.5 V. Principles of redox reactions (q.v.) are relatively simple when each reaction is considered singly with respect to the reactants and products in true solution. However, the soil system is very complex, and the effect of each phase's contribution to the total potential cannot be deduced. Soil iron, manganese, sulfur, organic matter, and numerous other ions at equilibrium in the soil each have some finite ratio of oxidation states and/or molecular species that combine to give the observed redox potential. The effect of each phase will also vary with factors such as water content, oxygen activity in the soil, and pH. Bohn (1971) presented an excellent discussion on the theory and measurement of redox potentials. Redox measurements have been widely used in saturated and flooded soil systems in the study of manganese (Mandal, 1962), iron (Mandal, 1962), phosphorus (Savant and Elus, 1964), nitrogen (Meek et al, 1970), arsenic (Deuel and Swoboda, 1972), pH (Ponnamperuma, Tianco, and Loy, 1966), oxygen status (Mandal, 1962), and soil aeration (Bonner and Ralston, 1968). Redox potentials have also been applied in soil-genesis studies such as the determination of the *B*-horizon formation of podzols (McKenzie, Whiteside, and Erickson, 1960) and the variance of redox potentials with depths corresponding to soil horizons (McKenzie and Erickson, 1954).

Conductance and Conductometric Titrations

The electrical conductance of any conductor is defined as the reciprocal of its resistance. Thus every measurement of conductance reduces to a measurement of resistance. In aqueous systems the electrical conductance depends on the concentration of electrolytes and the degree of dissociation of electrolytes as affected by temperature, viscosity, pressure, and electric field.

Conductometric titrations are of particular analytical importance for the detection of the end points of analytical reactions. End points are usually determined through sudden changes in conductance when a suitable reagent is added to the solution in question. Conductance measurements have been used by soil scientists for the determination of soil salinity (Kemper, 1959), soil moisture (Larsen and Widdowson, 1966), anion sorption (Mehlich, 1964), ion activity (Dutt and Anderson, 1964), ion-activity coefficients (Griffin and Jurinak, 1973), ionic strength (Griffin and Jurinak, 1973; Ponnamperuma, Martinez, and Loy, 1966), transference numbers (Shainberg and Kemper, 1972), cation-exchange capacity (Dakshinamurti and Chandool, 1966) isoconductivity values (Dakshinamurti and Chandool, 1966), and ion-activation energy (Brown and Miller, 1971). Conductometric titrations have been used in the titration of clays (Shainberg and Dawson, 1967), the determination of exchangeable aluminum (Rich, 1970), and for the estimation of soil acidity (Kissel and Thomas, 1969). Conductometric titrations data are often used in conjunction with potentiometric titration data (Khan and Singhal, 1967; Mitra and Kapoor, 1969; Rich, 1970; Shainberg and Dawson, 1967).

Polarography

Polarographic analysis involves the recording and interpretation of current versus voltage curves obtained by the electrolysis of solutions with a dropping mercury electrode and a reference electrode. The theory of operation is simple. A linearly varying direct current is applied to the dropping mercury electrode and current flow in the cell circuit is measured. This current is not dependent on the duration of the electrolysis. The intensity of the diffusion-limited current is proportional to the concentration of the electroactive species. Therefore, a series of substances can be analytically determined provided their values of reduction potential are different.

Polarography is used extensively for studying the analytical and physical chemistry of solutions. Such uses include: the qualitative and quantitative determination of many cations, anions, and organic substances; the determination of the structure of substances; studies of reaction kinetics; evaluation of equilibrium constants; and the stability constant of complexes. Polarography has been used by soil scientists for the determination of pesticides (Gajan, 1962), herbicides (Englehardt and McKinley, 1966), total cation-exchange capacity of soils (Holtzinger, McHeney, and Rhodes, 1954), nitrate concentrations in soil filtrates (Skyring, Carey, and Skerman, 1961), soil humic acids (Lindbeck and Young, 1966), and soil dissolved oxygen (Evans and Scott, 1955). The most recent polarographic methods have been applied to heavy metal determinations in soil extracts and digests (Marek, 1971; Rudin and Shipotovski, 1970, 1972; Stanton and Burger, 1967). A unique method of measuring oxygen concentrations was developed using a membrane-covered, stationary polarographic electrode (cathode) and a nonpolarizable reference electrode (anode) (Lemon and Erickson, 1955; Willey and Tanner, 1973). When operated in the "diffusion current" plateau region of the oxygen polarogram, virtually all the oxygen that diffuses to the polarographic electrode is reduced. The electrode current, being proportional to the concentration of oxygen, is then limited by the diffusion rate of oxygen to the electrode. The method has been used in oxygen-concentration studies and diffusivity determinations (Shearer, Millington, and Quirk, 1966).

Conclusion

Electroanalytical methods of analysis have been used to some extent for the study of the properties of soils. Within the last few years new, advanced techniques and instruments developed for stripping voltammetry, alternating-current polarography, pulse polarography (conventional direct-current pulse, derivative pulse, and differential pulse), and other areas of electrochemistry offer new tools to the soil scientist for studying interface properties, mass transport, intensity and capacity relations, etc. However, the soil scientist has as yet taken advantage of only a small amount of knowledge and technique that the discipline of electrochemistry has to offer.

JAMES DRAGUN
DALE E. BAKER

References

Allam, A. I.; Pitts, G.; and Hollis, J. P., 1972. Sulfide determination in submerged soils with an ion selective electrode, *Soil Sci.*, **114**, 456–467.

Baker, D. E., and Low, P. F., 1970. Effect of the sol-gel transformation in clay-water systems on biological activity. II. Sodium uptake by corn seedlings, *Soil Sci. Soc. America Proc.*, **34**, 49–56.

Banwart, W. L.; Tabatabai, M. A.; and Bremner, J. M., 1972. Determination of ammonium in soil extracts and water samples by an ammonium electrode, *Comm. Soil Sci. Plant Anal.*, **3** (6), 449–458.

Bockris, J. O'M., and Reddy, A. K. N., 1970. *Modern Electrochemistry,* Vols. 1 and 2. New York: Plenum, 1432p.

Bohn, H. L., 1971. Redox potentials, *Soil Sci.*, **112**, 39–45.

Bonner, F. T., and Ralston, C. W., 1968. Oxidation-reduction potential of saturated forest soils, *Soil Sci. Soc. America Proc.*, **32**, 111–112.

Bower, C. A., 1959. Determination of sodium in saline solutions with a glass electrode, *Soil Sci. Soc. America Proc.*, **23**, 29–31.

Bower, C. A., 1961. Studies on the suspension effect with a sodium electrode, *Soil Sci. Soc. America Proc.*, **25**, 18–21.

Bremner, J. M., and Tabatabai, M. A., 1972. Use of an ammonia electrode for determination of ammonium in Kjeldahl analysis of soils, *Comm. Soil Sci. Plant Anal.*, **3** (2), 159–165.

Brown, D. S., and Miller, R. J., 1971. Bentonite instability and its influence on activation energy measurements, *Soil Sci. Soc. America Proc.*, **35**, 705–710.

Carlson, R. M., and Paul, J. L., 1969. Potentiometric determination of boron in agricultural samples, *Soil Sci.*, **108**, 266–272.

Dahnke, W. C., 1971. Use of the nitrate specific ion electrode in soil testing, *Comm. Soil Sci. Plant Anal.*, **2**(2), 73–84.

Dakshinamurti, C., and Chandool, D. E., 1966. Isoconductivity values and cation exchange capacity of soils and clays, *Soil Sci.*, **102**, 123–130.

Davey, B. G., and Bembrick, M. J., 1969. The potentiometric estimation of chloride in water extracts of soils, *Soil Sci. Soc. America Proc.*, **33**, 385–387.

Denaro, A. R., 1965. *Elementary Electrochemistry.* Washington: Butterworth, 221p.

Deuel, L. E., and Swoboda, A. R., 1972. Arsenic solubility in a reduced environment, *Soil Sci. Soc. America Proc.*, **36**, 276–278.

Dewan, H. C., and Rich, C. I., 1970. Titration in acid soils, *Soil Sci. Soc. America Proc.*, **34**, 38–44.

Doiron, E. B., and Chamberland, E., 1973. Rapid determination of exchangeable calcium in soil with the calcium electrode, *Comm. Soil Sci. Plant Anal.*, **4**(3), 205–209.

Dutt, G. R., and Anderson, W. D., 1964. Effect of Ca-saturated soils on the conductance and activity of Cl^-, $SO_4^=$, and Ca^{++}, *Soil Sci.*, **98**, 377–382.

Eisenman, G.; Bates, R.; Mattock, G.; and Friedman, S. M., 1965. *The Glass Electrode.* New York: Wiley, 106p.

El-Swaify, S. A., and Gazdar, M. N., 1969. Evaluating the use of Na^+, Ca^{++}, and divalent cation electrodes in some soil extracting solutions, *Soil Sci. Soc. America Proc.*, **33**, 665–667.

Englehardt, J., and McKinley, W. P., 1966. Bipyridylium herbicides. Polarography of 1,1′-ethylene-2,2′-bipyridylium dibromide, *Jour. Agric. Food Chem.*, **14**, 377–380.

Evans, D. D., and Scott, A. D., 1955. A polarographic method of measuring dissolved oxygen in saturated soil, *Soil Sci. Soc. America Proc.*, **19**, 12–16.

Frink, C. R., and Sawhney, G. L., 1967. Neutralization of dilute aqueous aluminum salt solutions, *Soil Sci.*, **103**, 144–148.

Gajan, R. J., 1962. Application of oscillographic polarography to determination of organic phosphorus pesticide residues, *Jour. Assoc. Off. Agric. Chem.*, **45**, 401–406.

Goertzen, J. O., and Oster, J. D., 1972. Potentiometric titration of sulfate in water and soil extracts using a lead electrode, *Soil Sci. Soc. America Proc.*, **36**, 691–693.

Griffin, R. A., and Jurinak, J. J., 1973. Estimation of activity coefficients from the electrical conductivity of natural aquatic systems and soil extracts, *Soil Sci.*, **116**, 26–30.

Holtzinger, K. R.; McHeney, J. R.; and Rhodes, D. W., 1954. A polarographic method for determining the total cation exchange capacity of soils, *Soil Sci.*, **77**, 137–142.

Kaddah, M. T., and Coleman, N. T., 1967. Salt Displacement and titration of $AlCl_3$-treated trioctahedral vermiculites, *Soil Sci. Soc. America Proc.*, **31**, 328–332.

Kemper, W. D., 1959. Estimation of osmotic stress in soil water from the electrical resistance of finely porous ceramic units, *Soil Sci.*, **87**, 345–349.

Khana, S. S., and Stevenson, F. J., 1962. Metallo-organic complexes in soil. I. Potentiometric titration of some soil organic matter isolates in the presence of transition metals, *Soil Sci.*, **93**, 298–305.

Kissel, D. E., and Thomas, G. W., 1969. Conductimetric titrations with $Ca(OH)_2$ to estimate the neutral salt replaceable and total soil acidity, *Soil Sci.*, **108**, 177–179.

Krupskiy, N. K.; Aleksandrova, A. M.; and Boriskova, Zh. C., 1973. Electrometric methods of determining the activity of ions in soils: III. Method for determining chloride ion activity in soils, *Sov. Soil Sci.*, **5**, 749-752.

Larsen, S., and Widdowson, A. E., 1971. Soil fluorine, *Jour. Soil Sci.*, **22**, 210–221.

Larsen, S., and Widdowson, A. E., 1966. Determination of soil moisture by electrical conductivity, *Soil Sci.*, **101**, 420.

Lemon, E. R., and Erickson, A. E., 1955. Principle of the platinum microelectrode as a method of characterizing soil aeration, *Soil Sci.*, **79**, 383–392.

Lindbeck, M. R., and Young, J. L., 1966. Polarography and coulometry, in dimethylsulfoxide, of nitric acid oxidation products from soil humic acid, *Soil Sci.*, **101**, 366–372.

McKenzie, L. J., and Erickson, A. E., 1954. The use of redox potentials in studies of soil genesis, *Soil Sci. Soc. America Proc.*, **18**, 481–485.

McKenzie, L. J.; Whiteside, E. P.; and Erickson, A. E., 1960. Oxidation-reduction studies on the mechanism of B horizon formation in podzols, *Soil Sci. Soc. America Proc.*, **24**, 300–305.

Mandal, L. N., 1962. Levels of iron and manganese in soil solution and the growth of rice in water-logged soils in relation to the oxygen status of soil solution, *Soil Sci.*, **94**, 387–391.

Meek, B. D.; Grass, L. B.; Willardson, L. S.; and MacKenzie, A. J., 1970. Nitrate transformations in a column with a controlled water table, *Soil Sci. Soc. America Proc.*, **34**, 235–239.

Mehlich, A., 1964. Influence of sorbed hydroxyl and sulfate on liming efficiency, pH, and conductivity, *Soil Sci. Soc. America Proc.*, **28**, 496–499.

Milazzo, G., 1963. *Electrochemistry*. New York: Elsevier, 708p.

Miller, G. A.; Riecken, F. F.; and Walter, N. F., 1975. Use of an ammonia electrode for determination of cation exchange capacity in soil studies, *Soil Sci. Soc. America Proc.*, **39**, 372–373.

Mokady, R. S., and Low, P. F., 1966. Electrochemical determination of diffusion coefficients in clay-water systems, *Soil Sci. Soc. America Proc.*, **30**, 438–442.

Mortland, M. M., 1961. Using a glass electrode for estimating exchangeable potassium in soils, *Mich. Agric. Exp. Sta. Quart. Bull.*, **43**, 491–498.

Ponnamperuma, F. N.; Martinez, E.; and Loy, T., 1966. Influence of redox potential and partial pressure of carbon dioxide on pH values and the suspension effect of flooded soils, *Soil Sci.*, **101**, 421–431.

Ponnamperuma, F. N.; Tianco, E. M.; and Loy, T. A., 1966. Ionic strengths of the solutions of flooded soils and other natural aqueous solutions from specific conductance, *Soil Sci.*, **102**, 408–413.

Raveh, A., 1973. The adaptation of the nitrate-specific ion electrode for soil and plant analysis, *Soil Sci.*, **116**, 388–389.

Rich, C. I., 1970. Conductometric and potentiometric titration of exchangeable aluminum, *Soil Sci. Soc. America Proc.*, **34**, 31–38.

Robbins, C. W.; Carter, D. L.; and James, D. W., 1973. Potentiometric titration of sulfate in water and soil extracts using a lead-mercury amalgam indicator electrode, *Soil Sci. Soc. America Proc.*, **37**, 212–215.

Rudin, V. D., and Shipotovski, V. V., 1970. Determination of copper in soils by stripping analysis with accumulation, *Tr. Stavropol Sel'skokhoz Inst.*, **33**(2) 83–86.

Rudin, V. C., and Shipotovski, V. V., 1972. Determination of the levels of Zn and Co in soils by using inverse polarography, *Tr. Stavropol Sel'skokhoz Inst.*, **34**(2), 170–173.

Savant, N. K., and Elus, R., Jr., 1964. Changes in redox potential and phosphorus availability in submerged soil, *Soil Sci.*, **98**, 388–394.

Shainberg, I., and Dawson, J. E., 1967. Titration of H-clay suspensions with salt solutions, *Soil Sci. Soc. America Proc.*, **31**, 619–626.

Shainberg, I., and Kemper, W. D., 1972. Transport numbers and mobilities of ions in bentonite membranes, *Soil Sci. Soc. America Proc.*, **36**, 577–582.

Sharpe, L. R., and Hoagland, D. R., 1916. Acidity adsorption in soils as measured by hydrogen electrode, *Jour. Agric. Research*, **7**, 123–145.

Shearer, R. C.; Millington, R. J.; and Quirk, J. P., 1966. Oxygen diffusion through sands in relation to capillary hysteresis: I. Calibration of oxygen cathode for use in diffusion studies, *Soil Sci.*, **101**, 361–365.

Skyring, G. W.; Carey, B. J.; and Skerman, V. B. D., 1961. A polarographic method for the determination of nitrate in soils, *Soil Sci.*, **91**, 388–392.

Uehara, G., and Mortland, M. M., 1960. Activity measurements in water and clay-water systems with tertiary electrodes, *Soil Sci. Soc. America Proc.*, **24**, 261–264.

Van Raij, B., and Peech, M., 1972. Electrochemical properties of some oxisols and alfisols of the tropics, *Soil Sci. Soc. America Proc.*, 36, 587–593.

Willey, C. R., and Tanner, C. B., 1973. Membrane-covered electrode for measurement of oxygen concentration in soil, *Soil Sci. Soc. America Proc.*, 27, 511–515.

Yeager, E., and Salkind, A. J., 1973. *Techniques of Electrochemistry*. New York: Wiley, 1039p.

Cross-references: *Acidity; Activity Ratio,; Adsorption Phenomena; Conductivity, Electrical; Diffusion Processes; Exchange Phenomena; Physical Chemistry; Redox Reactions; Soil Chemistry; Soil Reaction.*

ELECTROKINETICS, ELECTROKINETIC PROPERTIES OF SOILS—See ELECTROCHEMISTRY

ELECTROLYTES—See FLOCCULATION; PHYSICAL CHEMISTRY. Vol. IVA: ELECTROLYTES, FLOCCULATION OF COLLOIDS

ELECTROOSMOSIS

Electroosmosis denotes the flow of a liquid, usually water or aqueous solution, in a capillary system, exemplified by moist soil, by living or dead animal tissues and membranes, or by a single water-filled glass capillary, under the action of an electric potential. This potential is commonly impressed through charged electrodes, but it may also result from the use of electrode metals of widely differing electron pressures or from compositional, pressure or temperature differences within the system. Electroosmosis was discovered by Reuss (1808) when he applied an electric potential to a wet clay mass by means of electrodes placed in water-filled glass tubes that had been pushed into the clay. The phenomenon was subsequently studied by a number of investigators, especially Wiedemann (1852, 1856) whose "laws" together with the results of Quincke's (1861) studies on electrophoresis formed the basis for Helmholtz' (1879) concept of the electric double layer, which still furnishes the starting point for all present theories on electroosmosis.

The physical model of the Helmholtz concept may be visualized as a capillary tube with a sheet-like system of immoveable negative charges within the solid internal surface of the tube and a parallel system of moveable positive charges within the adjoining liquid situated at a very small but not infinitesimal distance from the solid wall. Application of an electric potential in the longitudinal direction of the capillary moves the positive charges and the associated liquid at a rate that is zero next to the solid wall and a maximum at the locus of the positive charges. In soils the number of positive and negative charges per unit of internal surface is a function of the ion-exchange capacity of the solid soil constituents and of the number and types of exchange ions present. To account for the actual electroosmotic behavior of widely differing systems, the moveable portion of the Helmholtz double layer had to be loosened up, and for some systems the best agreement between theory and test results is obtained by assumption of uniform distribution of the counterions in the liquid phase.

In engineering practice, electroosmosis is used mainly for drainage and detouring of seepage water in fine-grained soils of low hydraulic conductivity (q.v.). For this use, pioneered by Casagrande (1939–1953), one determines experimentally the electroosmotic transmission constant which is the volume of water that passes in unit time through a unit cross-section under an electric potential of 1 v/cm normal to the cross-section. If one considers the number of electric charges present in 1 cm^3 of water-saturated, fine-grained soil and the forces exerted on them by an electric gradient of 1 v/cm, it is realized that electroosmosis is essentially a means to apply much greater forces for moving water in such soils than would be practical or even possible by mechanical means.

There exists a large literature on electroosmosis and its engineering applications. The early experimental and theoretical development is well covered and documented by Bikerman (1948) with extensive references to authors previous to that date; more recent theoretical and practical developments are documented by Kozan and Fenwick (1961) and Ingles and Metcalf (1973). Data and concepts concerning the relationship between electro- and thermoosmosis are presented by Winterkorn (1947).

HANS F. WINTERKORN

References

Bikerman, J. J., 1948. *Surface Chemistry for Industrial Research*. New York: Academic, 464p.

Ingles, O. G., and Metcalf, J. B., 1973. *Soil Stabilization*. New York: Wiley, 374p.

Kozan, G. R., and Fenwick, W. B., 1961. Electrical stabilization of fine-grained soils; summary reviews of soil stabilization processes, *U.S. Army Waterways Exp. Sta., Corps. Eng., Misc. Pap. No. 3–122, Report 7*, 29p.

Winterkorn, Hans F., 1947. Fundamental similarities between electroosmotic and thermoosmotic phenomena, *Proc. Highway Res. Board*, 27, 443–455.

Cross-references: *Capillary Pressure; Conductivity, Hydraulic; Electrochemistry; Envelope-pressure Potential; Soil Solution; Soil Stabilization.*

ENERGY BALANCE

An *energy balance* for the soil is an analysis of heat fluxes to and from a given soil volume. The energy balance at the soil surface is of particular interest because heat flux into and out of the soil is an important component of the energy balance of the earth's surface and therefore influences weather patterns, crop production, and microclimates of plants and animals. Heat flow in soil cannot be analyzed without knowing either soil surface temperature or heat flux across the soil surface as a function of time. Neither of these can be specified without solving the entire energy budget for the soil surface to determine the input of radiant energy and the loss of sensible and latent heat. The *soil heat flux* is the storage term for this system. An analysis of heat flow in soil requires a knowledge of the energy budget at the soil surface, the distribution of moisture and composition of solids in the soil profile, and a simultaneous solution of the moisture balance in the soil because moisture flow and distribution influence heat flow.

The differential equations governing heat transfer in soil are similar to heat-transfer equations for any solid medium. Horizontal temperature variations a few centimeters below the soil surface are generally small compared to variations normal to the surface, so it is usually necessary to consider only one dimensional heat flow. Heat flux density across any plane parallel to the soil surface is given by the *Fourier law:*

$$G = -\frac{\lambda dT}{dz} \qquad (1)$$

where G is the soil heat flux density, λ is the thermal conductivity of the soil, and T is temperature. The energy balance for any location in the soil is obtained by combining equation 1 with the continuity equation:

$$C\frac{\partial T}{\partial t} = \frac{-\partial G}{\partial z} \qquad (2)$$

where t is time and C is the volumetric soil heat capacity. Solutions to equation 2, subject to appropriate boundry and initial conditions, describe the soil energy balance and the temperature distribution in the soil.

The novel feature of soil heat-flow analysis is not the basic equations describing the process, but rather the behavior of the thermal conductivity (q.v.) and heat capacity (q.v.) functions. Soil is a mixture of solids, water, and air. The volume fraction of solids remains fairly constant with time, but the volume fractions of air and water are constantly changing. The heat capacity and thermal conductivity of the soil are therefore variable in space and time. The functional relationship between heat capacity and water content (q.v.) is

$$C = \sum_i x_{si} C_{si} + x_{\mathrm{w}} C_{\mathrm{w}} + x_{\mathrm{a}} C_{\mathrm{a}}$$

where x_{si} is the volume fraction of the ith solid component, x_{w} and x_{a} are the volume fractions of water and air, and the Cs are the corresponding heat capacities (Table 1).

The thermal conductivity function for the soil has two components. Sensible heat is conducted through the mineral grains of the soil and across water or air bridges between the grains. Latent heat is transferred by distillation of water from warm to cold regions of the soil. Since the thermal conductivity of water is 20 times that of air (Table 1), the thermal conductivity of the soil depends strongly on water content. Figure 1 shows relationships between thermal conductivity and water content for three soils. De Vries (1966) has developed a theory that can be used to determine the thermal conductivity function for any soil if the chemical makeup of the solid fraction is known.

The rate of distillation, and therefore the latent heat flux in soil, is a function of pore space, (see *Soil Pores*) soil temperature, and soil moisture. Since temperature gradients induce moisture flow in soils, the energy-balance problem cannot be solved without simultaneous solution of the moisture-balance problem. For moist soils the apparent thermal conductivity due to latent heat transport within a soil pore is

$$\lambda_{\mathrm{v}} = L\, D \frac{d\rho_{\mathrm{v}}}{dT}$$

TABLE 1. Thermal Properties of Soil Materials, Water, and Air at 10°C.

Substance	C(MJ m^{-3}K^{-1})	λ(W m^{-1}K^{-1})
Quartz	2.0	8.8
Other minerals	2.0	2.9
Organic matter	2.5	0.25
Water	4.2	0.57
Air	1.25×10^{-3}	0.025

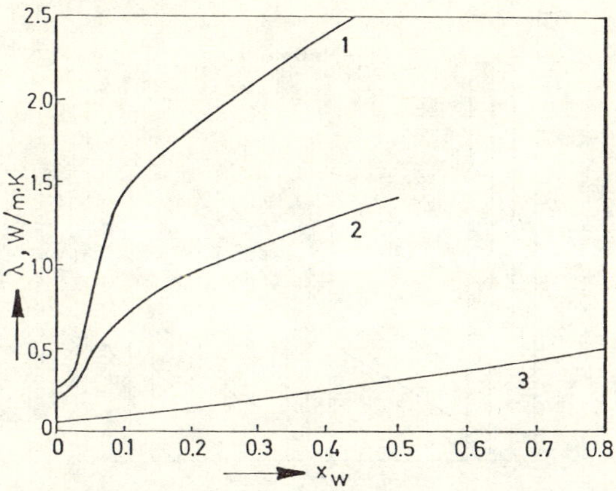

FIGURE 1. Thermal conductivity as a function of volumetric soil water content at 10 C. Curve 1: quartz sand with x_s = 0.55. Curve 2: loam with x_s = 0.5. Curve 3: peat soil with x_s = 0.2. (after de Vries, 1975).

where ρ_v is the saturation vapor density, L is the latent heat of vaporization for water, and D is the diffusion coefficient for water vapor in air. The effect of latent heat flux on soil heat flow is accounted for by replacing the thermal conductivity of air (λ_a) in the de Vries formula by the sum $\lambda_v + \lambda_a$. At temperatures near zero, λ_v is about the same size as λ_a, and latent heat transfer has little effect on soil thermal conductivity. However, at 60°C, λ_v is about equal to the thermal conductivity of water, so the thermal conductivity becomes independent of water content.

GAYLON S. CAMPBELL

References

de Vries, D. A., 1966. Thermal properties of soils, *in* W. R. van Wijk, ed., *Physics of Plant Environment.* Amsterdam: North Holland, 210–235.

deVries, D. A., 1975. Heat transfer in soils, *in* D. A. de Vries and N. H. Afgan, eds., *Heat and Mass Transfer in the Biosphere, I. Transfer Processes in Plant Environment.* New York: Wiley, 594p.

Cross-references: *Conductivity, Thermal; Heat Capacity; Thermal Regimes; Transport Processes; Water Content and Retention.*

ENERGY IN SOIL—See ENERGY BALANCE; ENZYME ACTIVITY; IONIC ACTIVITIES; METABOLIC ACTIVITY; RADIOISOTOPES SPECTRAL CHARACTERISTICS; THERMODYNAMICS, SOIL MOISTURE; WATER FLUXES

ENVELOPE-PRESSURE POTENTIAL

The *envelope-pressure potential* is a measure of the change in the energy status of the water in a porous medium due to the overall outside pressure on the system of which the water is a component. For one-dimensional systems—i.e., systems in which the variables change in only one direction—the overall pressure is equal to the overburden being the sum of the weight of the overlaying layer (e.g., soil) and the surface load. The envelope-pressure potential is then known as the *overburden potential.*

According to the definition by the International Soil Science Society, the envelope-pressure potential is the increment of the equilibrium liquid pressure (as measured, for example, by a tensiometer) on the application of an envelope pressure to a soil sample originally under zero envelope pressure while the air pressure in the system is kept constant and water is not allowed to enter or leave the system.

P. H. GROENEVELT

References

Anon., 1975. Soil physics terminology, *Internat. Soil Sci. Soc. Bull.*, 48, 16–22.

Cross-references: *Conductivity, Hydraulic; Flow Theory; Soil Water.*

ENZYME ACTIVITY

Enzymes are proteins, synthesized by living cells, that catalyze biological reactions. Like all catalysts, they (1) are effective in very small concentrations, (2) speed up a reaction until it reaches equilibrium, (3) are unchanged by the reaction, and (4) exhibit specificity for certain limited reactions. Thus many different enzymes are required to catalyze the numerous biological transformations in nature. During the course of reaction, the combination of the enzyme with its substrate lowers the energy of activation of the latter so that changes that otherwise might occur only at elevated temperatures can proceed under normal physiological conditions. The formation of the transient complex of enzyme E with substrate S occurs at a location in the enzyme molecule termed the *active center.* The configuration of the active center regulates the substrate specificity of the enzyme. The overall reaction resulting in the formation of products P can be summarized:

$$E + S \underset{k2}{\overset{k1}{\rightleftharpoons}} ES \overset{k3}{\rightarrow} E + P$$

where $k1$, $k2$, and $k3$ are the rate constants of

the individual reactions. The specificity imposed by the active center is but one restriction on enzyme-substrate interaction in nature. For instance, few enzyme reactions in soil take place in a manner similar to the homogeneous solutions of in vitro studies; rather they occur in a milieu that is heterogeneous and in which reactions at solid- or gel-liquid interfaces predominate.

Origin of Soil Enzyme Activity

Early studies indicate that enzymes are deposited in soil by microflora, microfauna, and plant debris. Later studies with sterile root systems also implicate root exudates and sloughed-off root cells. While microorganisms contribute largely to the accumulated activity, excretion by plant roots may be a major source in some instances and may also indirectly stimulate enzyme production. For instance, barley roots have high invertase and phosphatase activity whereas urease levels depend on organisms in the rhizosphere, i.e., the more biologically active soil adjacent to roots (Estermann and McLaren, 1961). The review by Skujins (1967) lists a wide range of enzymes deposited in the soil from both microbial and plant sources. Because of the deposition of enzymes by decaying roots and root debris and by lysed microbial cells, there is obvious difficulty in adequately distinguishing between truly extracellular activity and that derived from the release and accumulation of cytoplasm and cytoplasmic inclusions on death and lysis of cells.

The truly extracellular enzymes excreted by metabolically active cells are involved in the breakdown of high-molecular-weight compounds that cannot permeate the cell membranes. The cellulases, amylases, proteases, and phosphatases released by a wide range of bacteria and fungi are examples. Some of these accumulate at high background levels that are relatively constant compared to fluctuating seasonal changes in microbial numbers. Likewise, high levels of enzymes can occur where microbial proliferation is suppressed. This is evident in the high phosphatase activity in acid Histosols and its inverse relationship with microbial numbers (Figure 1).

The extracellular activity must be viewed, however, as only one component in biological transformations. Many enzymes more properly assigned to the study of soil microbiology are inducible in the presence of specific substrates. Some may be controlled by substrate or product concentration and thus regulated by inducer-repressor mechanisms. Their relevance to soil enzymology depends on the extent to which activity accumulates and persists in the soil per se.

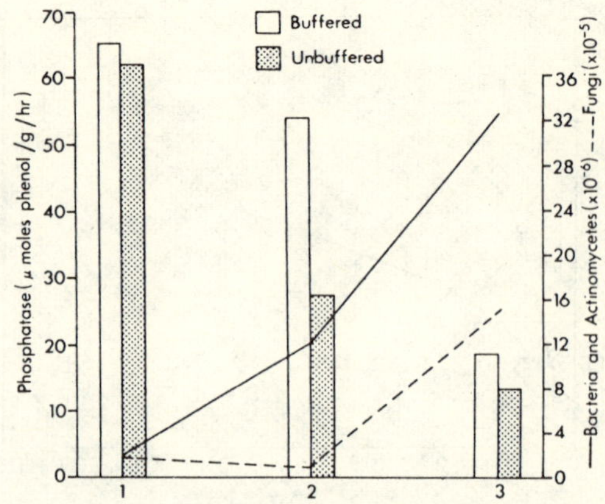

FIGURE 1. Phosphatase activity and microbial plate counts in Histosols: Sample pH 1, 3.8; 2, 5.3; 3, 5.9. Buffer pH (selected for maximum activity) 1, 4.5; 2, 6.5; 3, 6.5. (Based on Herlihy, 1972)

Methods of Detection

Measurements are necessarily indirect because of the stabilization of activity within the soil matrix. Extraction of the enzyme complex may yield only fractional amounts. The assay procedure involves incubating the soil with specific substrates and measuring their residual levels, extracting the reaction products or detecting volatile products such as evolution of $^{14}CO_2$ from ^{14}C-labeled substrate. Sorption of reaction products should be avoided or appropriate corrections applied.

The distinction between extracellular activity and activity associated with intact microbial cells requires that the technique should (1) preserve the stability of the cell membrane against liberation of intracellular enzyme or assimilation of substrate, (2) prevent further microbial synthesis of the enzyme, and (3) prevent inactivation of the activity being measured. Despite these problems, individual enzymes are more readily detected than other soil organic substances because small quantities effect measurable changes in added substrates. Reviews by Kiss, Dragan-Bularda, and Radulescu (1975); Kuprevich and Shcherbakova (1971); and Skujins (1967) include details and references to the numerous individual soil enzyme assays.

Measurement of accumulated extracellular activity requires selective sterilization of soils prior to assay. The chemical sterilizing agents include toluene, chloroform, phenol, and ether. Toluene has broadest application although it may act as a plasmolytic agent thereby releasing some enzymes and unmasking others such as β-galactosidase and alkaline phosphatase (see Skujins, 1967).

High energy radiation is an alternative to the

chemical agents that leave the physical and chemical properties of the soil almost unaltered. Irradiation of soils in the dry state minimizes autolysis or cell rupture. Hard X-rays, gamma rays (^{60}Co), or an electron beam operating in the range of 5–10 MeV may be used. Because of its exponential component, the decrease in activity is expressed as

$$\ln \frac{A}{Ao} = -kD$$

where A/Ao is the ratio of the residual to the initial enzyme activity, D is the radiation dose, and k is a constant for a particular enzyme. Much of the information regarding the radiation effects on soil enzymes is available from the research of McLaren and coworkers; the results of these and other investigators are summarized by Cawse (1975).

The use of radiation sterilization does not remove all anomalies. It is well established that urease activity may be stimulated because of improved accessibility of substrate to intracellular urease following damage to cell membranes. Advantage can be taken of this observation, however, to show that soils in which increasing radiation dosage does not increase activity have comparatively more extracellular urease than those in which an increase occurs. Thus the data in Figure 2 indicate that the Nipe and Lahaina soils exhibit intracellular and extracellular activity, respectively (Skujins and McLaren, 1969).

Types of Enzymes

Relevance and Function. Much of the original information regarding the relevance of soil enzymology to soil fertility and pedology is

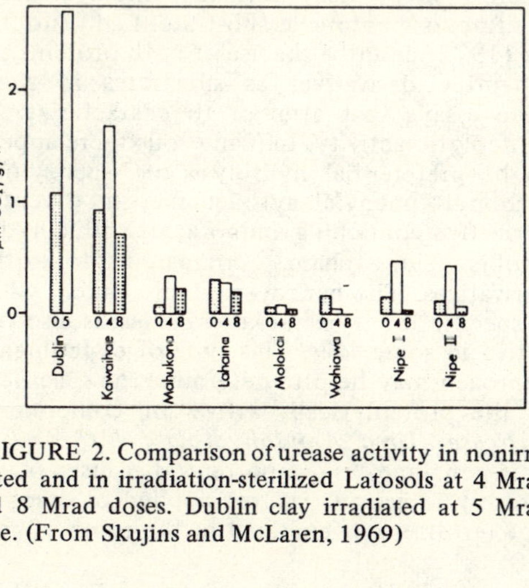

FIGURE 2. Comparison of urease activity in nonirradiated and in irradiation-sterilized Latosols at 4 Mrad and 8 Mrad doses. Dublin clay irradiated at 5 Mrad dose. (From Skujins and McLaren, 1969)

available from the work of Galstian and Kuprevich and coworkers in the USSR and Hofmann and Hoffmann in Germany (see Briggs and Spedding, 1963). Their emphasis on the variation of enzymes independently of microbial activity and between broad groups of soils testifies to the variation obtaining in very diverse ecosystems but does not necessarily form a basis for the contention that enzymatic tests should be applied to soil-fertility assays.

Macura (1974) considers the relationship between enzymes and fertility to be less relevant than recent work with lytic enzymes such as chitinase and laminarinase in the control of fungal pathogens. There is continued interest and relevance, however, in the transformations of organic nutrient sources and in enzymes associated with the respiratory activity of soil and its associated microflora. These latter systems include dehydrogenase and catalase, which occur widely in microorganisms. A quantitative method is available for the detection of life forms in soils and other biogeochemical environments based on the catalatic activity of ruptured microbial cells (Weetall, Weliky, and Vango, 1965).

The more specific transformations that are not necessarily associated with metabolically active cells include the polymer-cleaving enzymes such as amylase, cellulase, xylanase, dextranase, and the nonspecific soil carbohydrases active toward soil polysaccharides. This list alone is impressive in its inclusion of enzymes essential to the ecological balance between organic residues and atmospheric carbon dioxide. Two further examples emphasize the flexible nature of the soil as a multienzyme system. A carboxylesterase that partially degrades malathion can be shown to occur as a stable cell-free glycoenzyme in soils to which the insecticide is applied (Satyanarayana and Getzin, 1973). Likewise, continued deposition of the synthetic estrogen diethylstilboestrol, which is used to stimulate animal growth, results in the release of $^{14}CO_2$ from the ^{14}C-labeled hormone added to soil sterilized with gamma radiation (Gregers-Hansen, 1964).

The mineralization of nutrients from soil organic matter and their enzymatic release from some fertilizer sources is relevant to plant nutrition and soil fertility (q.v.). The more important examples include the hydrolysis of organic nitrogen to inorganic forms by proteolytic enzymes; of fertilizer urea to ammonia by urease; of organic phosphate and polyphosphate fertilizers by phosphatases and polyphosphatases; and of organic-sulfate esters by arylsufatase. The degradation of polyphosphates discharged into soils and waterways from both domestic and industrial sources indicates a possible role for soil enzymes in the control of environmental pollution.

While most available information relates to catabolism, the role of transferases in anabolism or synthesis of oligosaccharides is evident from the results of Drobnik, Hoffmann, and Kiss and coworkers (Skujins, 1967). Likewise, relatively little is known of the enzymatic contribution to the process of humus synthesis. Some information is available regarding the oxidation of o-diphenol to o-quinone by the appropriate diphenol oxidase. Condensation of quinones with amino acids and peptides may contribute to humic-acid (q.v.) formation.

Enzymes Found in Soil. Listings of the diverse enzymes detected are available and are constantly being updated (see Briggs and Spedding, 1963; Kiss, Dragan-Bularda, and Radulescu, 1975; Kuprevich and Shcherbakova, 1971; Skujins, 1967). Table 1 lists many important soil enzymes that are detected by various procedures and, consequently, may sometimes reflect activity localized in cells or in cytoplasmic inclusions released during assay as distinct from truly extracellular activity. Many metabolically important enzymes are excluded from the various lists possibly because priorities are assigned to the reactions of most relevance to soil metabolism. The continuation of this trend is evident from the recent identification of enzymes associated with nutrient cycling and ecological relationships. The carboxylesterase, which degrades malathion, was cited previously. Further examples include the hydrolysis by arylsulfatase of sulfate esters in soil organic matter (Tabatabai and Bremner, 1971) and the excretion of laminarinase by lytic microorganisms in soil zones enriched with fungal cell walls containing β-1-3,- glucan (Webley and Jones, 1968).

TABLE 1. Some Enzymes Found in Soils.

Oxidoreductases (E.C.1) dehydrogenases, catalase, peroxidases, urate oxidase, orth- and para-diphenol oxidase

Transferases (E.C.2) glycosyltransferases, aminotransferases

Hydrolases (E.C.3) carboxylesterase, lipases, acetylesterases, phosphatases, pyro- and metaphosphatase, phytase, nucleases, arylsulfatase, amylase (α– and β–), cellulase, laminarinase (lichenase), inulase, xylanase, dextranase, chitinase, glucosidase (α– and β–), galactosidase (α– and β–), β–fructofuranosidase, proteases, cathepsin and pepsin, trypsin, asparaginase, deaminase, urease, cyanase

Lyases (E.C.4) aspartate decarboxylase

(See Dixon and Webb, 1964, for systematic names and reactions catalyzed; and Skujins, 1967).

Important Enzyme Systems. The detailed study of specific enzymes is related a priori to their contribution to the more important transformations and also to ease of detection and use in the study of extracellular enzymes and model systems. The discussion of a few selected ones provides an insight into some aspects of soil enzymology.

Dehydrogenase Activity (Oxidoreductases E.C.1.). The quantitative dehydrogenase test developed by Lenhard and subsequently modified by others is sometimes used as a biochemically integrated measure of soil metabolism or respiratory activity of the soil microflora. No distinction is made between intra- and extracellular soil dehydrogenases because of their expected association with cellular components. The normal assay involves incubating the soil with a suitable hydrogen or electron acceptor such as triphenyltetrazolium chloride (TTC) and measuring the color change on mild reduction to the formazan (TTF). The reaction may be summarized

$$\text{TTC} \xrightarrow{2e^-H^+} \text{TTF}$$

Little is known of the nature of individual dehydrogenases in soil although a response to some specific substrates viz. glucose-6-phosphate is demonstrable (Galstian, 1974; Ross, 1971). Some workers report a direct relationship between dehydrogenase activity and respiration in the presence of decomposable substrates whereas others suggest its use as an indicator of endogenous microbial activity in unamended soils (Skujins, 1967). Many results are inconsistent with its use in either of these situations.

Proteases (Peptide Hydrolases E.C. 3.4). These enzymes, which hydrolyze proteins to peptides and amino acids, occur extracellularly. Their indirect assay normally employs casein, gelatine, or peptone as substrate. Ladd and Butler (1972) describe the use of both proteins and dipeptide derivatives as substrates in short-term assays that attempt to characterize soil proteolytic activity. In many soils there appears to be preferential hydrolysis of N-benzyloxycarbonyl phenylalanyl leucine, a dipeptide derivative containing amino acids with a hydrophobic side chain, compared to other derivatives. The benzoyl arginine amide, which is specific for trypsin-like enzymes, is also very active in some soils. This type of experimental approach may help to determine the specificity of the proteolytically active soil components.

Urease (Urea Amidohydrolase E.C.3.5.1.5). This enzyme is important because of its hydrolytic action on urea applied in commercial fertilizers or excreted in the urine of grazing animals. Further interest derives from the specificity of its reaction, even in soil, its ability to accumulate extracellularly, and its high radiation resistance. The reaction catalyzed is

$$NH_2CONH_2 + 2H_2O \rightarrow (NH_4)_2CO_3$$

The ammonium product or the unreacted urea substrate is measured by various colorimetric procedures. More sensitive methods detect the $^{14}CO_2$ released from ^{14}C-urea. Maximal activity usually occurs in the pH range of 6.5–7.0 and increases in the temperature range of 2°C–45°C. The enzyme may occur extracellularly or in association with ureolytic microorganisms. Activity is highest in the rhizosphere and is dependent on the plant species (Skujins, 1967).

Phosphatase (Orthophosphoric Monoester Phosphohydrolyase E.C.3.1.3.). Acid, neutral and alkaline phosphatases with different pH optima hydrolyze phosphate bonds in soil organic matter. Some of the early work on the deactivation of phosphatase by clay minerals measured the release of orthophosphate but did not account for the phosphate-fixing characteristics of the clay. However, the adsorption characteristics of both substrate and products need evaluation. The wide range of substrates used includes disodium phenylphosphate, glycerophosphate, phenolphthalein phosphate, sodium β-naphthylphosphate, nucleic acids, and phytin, although there is evidence that soil phosphatases are active only toward the lower esters of phytate. Kiss, Dragan-Bularda, and Radulescu (1975) conclude from results of Ladd and Paul and of Ramirez-Martinez and McLaren that substantial quantities of the enzyme occur extracellularly. The data in Figure 1 also suggest the accumulation of high extracellular levels of the acid phosphatase in Histosols under conditions which favor accumulation of partially decomposed organic matter.

State and Activity of Extracellular Enzymes in the Soil Matrix

Enzymes and Soil Colloidal Material. Activity is closely associated with the mineral and organic particulate material in the soil. Hoffmann's (1959) results from sedimentation tests indicate that saccharase, β-glucosidase, and amylase separate out with the humus-rich silt fraction in some soils. Also pure enzymes can be taken up from solution and become attached to clay and silt particles, partly by ion exchange and partly by nonionic bonding. The paper by McLaren, Peterson, and Barshad (1958) outlines experimental procedures for studying the mechanism of adsorption and reactions of enzymes and other proteins on kaolinite and montmorillonite. Sorption is low above the isoelectric point of the enzyme or protein. Proteins are amphoteric so that as pH rises, their net charge becomes less positive, resulting in decreased sorption by the negatively charged clay. The limitations in the use of pure colloid systems are recognized. Glucose oxidase is readily adsorbed by isolated clay minerals but only partly adsorbed when added to soils in quantities that simulate naturally occurring levels (Ross and McNeilly, 1972).

The soil organic phase is also involved in enzyme bonding. Laccases, which exhibit p-diphenol oxidizing activity toward p-phenylenediamine, isolated from soil and litter are strongly electronegative and likely to be weakly sorbed by humic colloids in a manner that may protect the active center (Mayaudon and Sarkar, 1975). Cationic groups in enzymes may also be involved with humic acid viz the α-amino groups of terminal amino acids. As a result proteases may be bound by a cation exchange mechanism involving amino-carboxyl salt linkages (Ladd and Butler, 1971).

The need to study enzyme-substrate-colloid complexes is obviously related to the extent to which activity can be partitioned between cellular metabolism and that localized in the soil matrix. There is indirect statistical evidence that under steady-state conditions, up to 90% of the total urease activity may be sorbed on the colloidal material (Paulson and Kurtz, 1969). Results from extraction procedures provide direct evidence that native soil urease is associated with the organo-mineral complex, either chemically attached to a humus polymer network in contact with clay minerals or complexed within the inner layers of organic matter (Burns, Pukite, and McLaren, 1972; Nannipieri et al., 1974).

Effects of Soil Colloids. Normally an enzyme added to soil is inactivated by biological attack or attenuated by its physicochemical reaction with soil constituents, although in pure systems the formation of a urease-bentonite complex imparts protection against inhibitors such as Cu^{2+}, and adsorption of uricase can stimulate the hydrolysis of uric acid (Durand, 1965). In contrast, amorphous allophane, halloysite, and montmorillonite are known to reduce the hydrolysis of starch by β-amylase to between 5% and 30% of the free enzyme activity (Aomine and Kobayashi, 1966). The magnitude of the decrease is influenced by the order of addition of enzyme and substrate (Figure 3). The higher activity when the substrate is first adsorbed may reflect increased duration of the free enzyme in solution, whereas the nonlinear reaction in this instance may be due to adsorption of enzyme molecules in place of the hydro-

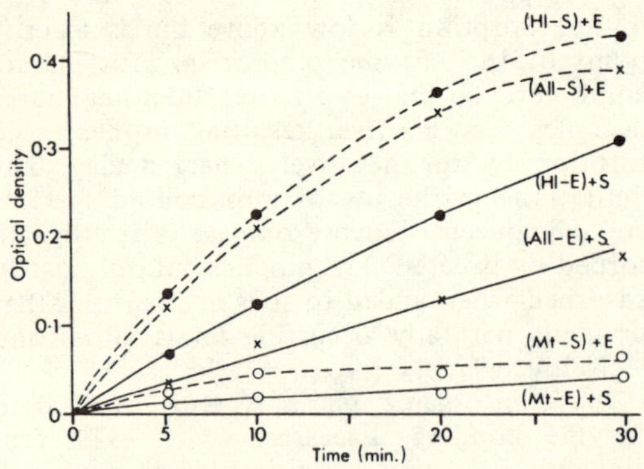

FIGURE 3. Influence of adsorption on enzymatic activity. All: allophanic clay; Mt: montmorillonitic clay; Hl: halloysitic clay; S: soluble starch; E: beta-amylase; ( ): complex. (From Aomine and Kobayashi, 1966)

lyzed starch molecules. These observations relate to saturated clays. In more dilute systems colloidal surfaces provide possible concentrating regions for the encounter of microbial exoenzymes with substrate (Zobell, 1946).

Protective Mechanisms and Persistence. The comparative resistance of soil enzymes against heat inactivation is but one example of their beneficial association with soil colloids (Galstian, 1974). The significance of the effect is also evident from the detection of urease activity in agricultural soils stored air-dry for 6–12 years and of urease and phosphatase in 9550-year-old, Alaska permafrost samples (Skujins and McLaren, 1969). It is likely that the incorporation into protective sites, which enhances persistence, is accompanied by a concurrently high loss of activity. For instance, the activity of urease added to soil diminishes both in the presence and absence of the proteolytic enzyme pronase, although neither the latter nor the proteolytic action of the soil itself affects the native soil urease (Burns, Pukite, and McLaren, 1972).

The Molecular Environment. The ecological conditions in soil can be evaluated both in terms of the microenvironment that reflects the point-to-point variation of solute concentrations in a matrix of sand, clay, and humus and in terms of the molecular environment wherein gradations in pH and reduction-oxidation potentials occur on particulate surfaces (McLaren, 1963). The influence of the molecular environment on both the pH profile and the redox dependence of the proteolytic enzyme papain is well illustrated by the proteolytic activity of soluble papain and of a papain-kaolinite complex (Figure 4).

The pH profile shifts to higher values in the enzyme-clay complex, which also decreases activity relative to the soluble enzyme in the presence of positively charged disulfides formed in thiol-disulfide exchange equilibria. These phenomena are the result of enzyme adsorption on the negatively charged clay surface. The negatively charged particle attracts a higher concentration of hydrogen ions than the bulk solution, which gives rise to a difference between the pH at the surface (pH_s) and that in the bulk solution (pH_b). These are related by the *Hartley-Roe equation,* which at 25°C has the form

$$pH_s = pH_b + \frac{\zeta}{60}$$

where ζ is the electrokinetic potential of the particle. The pH_s is lower than pH_b, which results in an upward shift in the pH optimum of an enzyme adsorbed on clay particles compared to the free enzyme (McLaren and Packer, 1970). In an analagous manner the preferential attrac-

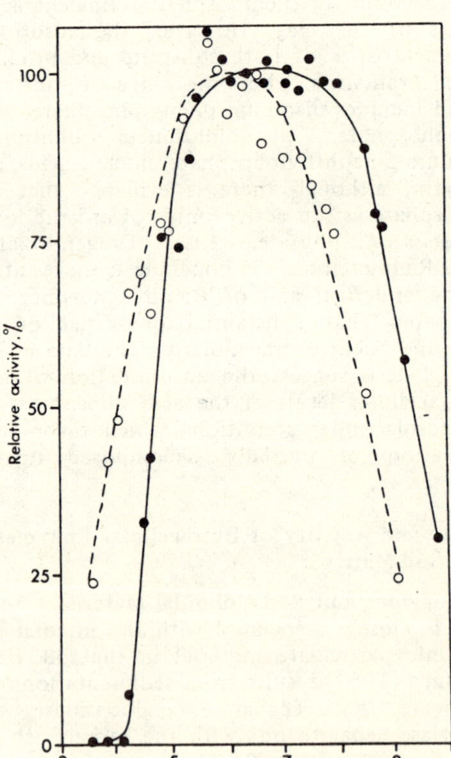

FIGURE 4(a). Papain activity (relative to maximum activity observed) with benzoyl arginine ethyl ester as a function of ph_b in the presence of dithioTHReitol. Ionic strength ca. 0.07. Open circles: soluble papain; solid circles: papain-kaolinite complex. (From Benesi and McLaren, 1975)

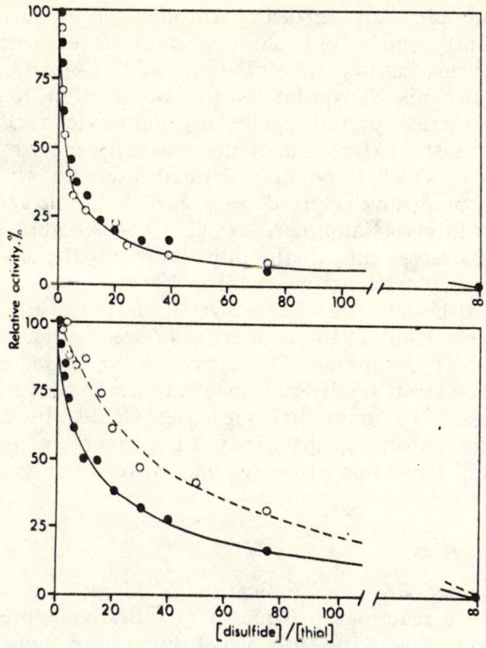

FIGURE 4(b). Activity of papain with benzoyl arginine ethyl ester (relative to maximum activity in assay) as a function of ratios of concentrations of oxidizing disulfide to reducing thiol. Open circles: soluble Papain (pH_b 6.1, ionic strength ca. 0.07); solid circles: papain-kaolinite complex (pH_b 6.9, ionic strength ca. 0.07). Upper diagram dithiodiglycol and beta-mercaptoethanol; lower diagram, dithiodiglycol and beta-amino ethylmercaptan. (From Benesi and McLaren, 1975)

tion of positively charged aminoethyl mercaptan increases the oxidizing potential within the molecular environment of the clay surface, whereas the relative concentration of the uncharged mercaptoethanol is unchanged by the presence of the clay. These pH and redox phenomena indicate that the reactions occur on the clay surface without elution of the enzyme into solution.

The influence of the molecular environment is also evident in the higher energy (Ea), enthalpy (ΔHa), and entropy (ΔSa) of activation of soil urease compared to microbial urease. This influence may reflect the steric effect of the soil matrix on energy required for formation of the enzyme-substrate complex. The free energy of activation (ΔFa) is similar for both sources of the enzyme, indicative perhaps of similar function but changed form of the soil enzyme (Table 2).

Extraction and Characterization

Purpose. The broad objective for extracting a soil enzyme relates to the need to demonstrate unequivocally its activity in the absence of microorganisms; to establish its function and form extracellulary in the heterogeneous soil system; to compare its properties and specificity with the pure enzyme; to study its stability and activity in the soil complex; and to delineate its location in the soil, i.e., the manner in which it is retained within the mineral and organic colloids.

Extracted Soil Enzymes. There are some descriptions of early success in this area such as Fermi's (1910) extraction of a proteolytically active fraction with phenol and the extraction and precipitation by Subrahmanyan (1927) of material which deaminates glycine (see Skujins, 1967). Many of the later successes involve the extraction of urease (Briggs and Segal, 1963; Burns, Pukite, and McLaren, 1972; Lloyd, 1975; Nannipieri et al., 1974). Generally, the procedures involve extraction with phosphate, pyrophosphate, or urea-salt solutions near neutral pH. The latter is a rather complex mixture that disrupts hydrogen bonds and because of the presence of urease in the extracted colloidal

TABLE 2. Calculated Values of Ea, ΔHa, ΔSa, and ΔFa of Bacterial Urease and Soil Urease.

Source	Ea (kcal/ mole)	ΔHa (kcal/ mole)	ΔSa (cal/deg/ mole)	ΔFa (kcal/ mole)
Bacterial urease (Larson and Kallio)	4.40	3.78	–27.56	11.33
Soil urease (Gould et al)	9.80	9.18	–6.52	11.12
Soil urease (Rackinskii and Pelttser)	22.65	22.05	+35.77	11.31

(From Dalal, 1975; copyright © 1975 by the Williams & Wilkins Co., Baltimore).

sol, is a self-destructing reagent (Burns, Pukite, and McLaren, 1972). Peroxidases may also be extracted with minimal autolysis or permeability changes in living cells during the extraction process (Bartha and Bordeleau, 1969). A protease can be extracted with tris-borate buffer, which has properties similar to the enzyme associated with plant fragments in soil (Ladd, 1972). Success is also evident in the extraction and partial purification of other enzyme systems including some involved in biocide degradation.

Relative Yield of Extractable Enzyme. The amount of extracted soil enzyme activity can vary widely. One report outlines the inability to extract urease from four mineral soils and the extraction of 8–11% of the activity from an alpine humus soil (Lloyd, 1975). In contrast, as much as 20% of the total urease activity or 30–40% of the extracellular activity is extractable from other mineral soils (Burns, Pukite, and McLaren, 1972; Nannipieri et al, 1974).

Form and Properties of Extracted Enzymes. The purified preparation of 2 mg urease/kg soil by Briggs and Segal (1963), while representing the first claim to extraction of a purified enzyme, remains unsubstantiated. Indeed, the existence of indigenous enzyme in sufficient quantity for crystallization after extraction is questioned (Lloyd, 1975). Also the prevailing view now is that activity is intimately associated with a colloidal structure that renders the enzyme more stable but less accessible to proteolytic attack, inhibition, and indeed, extracting agents. For instance, McLaren and coworkers consider that extracellular urease is embedded in, and perhaps chemically attached to, a humus polymer network in contact with clay particles. This model visualizes substrates and products diffusing through pores sufficiently small to exclude proteolytic enzymes (Burns, Pukite, and McLaren, 1972). Consistent with this viewpoint is the observation that extraction of nonspecific organic matter facilitates the extraction of the urease-organo complex, which is possibly formed internally with decomposing cell debris during humic-acid synthesis (Nannipieri et al, 1974). Figure 5 illustrates the distinction between the two organic matter components.

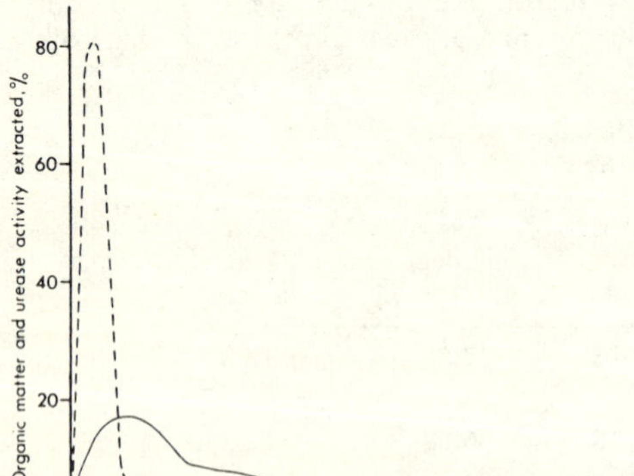

FIGURE 5. Effect of time on extraction of organic matter (– – –) and urease activity (——). (From Nannipieri et al, 1975)

Although the urease system is most intensively studied, the properties of some other enzymes are known. The activity of an extracted protease is relatively resistant to irradiation and attack by microbial proteases (Ladd, 1972). The substrate specificity of extractable and nonextractable proteases may differ because of their different origins.

Kinetics

General. The application of *kinetics* to enzyme reactions provides a quantitative expression of activity from which inferences may be made regarding the mechanism of the reaction that is catalyzed. When the enzyme concentration is constant and the initial substrate concentration increased, the *reaction velocity,* or rate of change of substrate to product, has the form of a rectangular hyperbola. At low substrate concentrations velocity is directly proportional to substrate concentration. At high substrate concentration a maximal velocity is reached that is independent of substrate concentration. Thus the reaction obeys first-order and zero-order kinetics at low and high substrate concentrations, respectively. The description of the initial reaction is provided by the *Michaelis-Menten equation*

$$v = \frac{V\,[S]}{Km + [S]}$$

where v is the observed reaction rate at a given substrate concentration [S], Km is the Michaelis constant, and V the maximum reaction rate at saturating concentration of the substrate. Substitution of $v = \frac{1}{2}V$ into the equation yields Km = [S], that is, the Michaelis constant is the substrate concentration required for half the maximum velocity of the enzyme-catalyzed reaction (see Dixon and Webb, 1964). It is evident from the preceding equation that changes in reaction velocity independent of [S] are due to changes in Km or V or both. A change in Km reflects only a change in the reaction of enzyme with substrate ($k1$ or $k2$; see introduction). A change in V reflects a change in the catalytic process ($k3$) where $k2 \gg k3$.

The Km approximates the dissociation constant of the enzyme-substrate complex where this reaction is rapid relative to product formation and is inversely related to the affinity of an enzyme for its substrate. In practice, variants of the basic equation are used that yield linear plots such as the double-reciprocal plot of Lineweaver and Burke:

$$\frac{1}{v} = \frac{\text{Km}}{V} \cdot \frac{1}{[\text{S}]} + \frac{1}{V}$$

The plot of $1/v$ versus $1/[\text{S}]$ gives an intercept on the ordinate equivalent to $1/V$ with a slope Km/V. This equation is suitable for use at high substrate concentrations.

While the in vitro determination of Km usually employs high saturation levels of substrate, the natural abundance of many substrates may be low relative to the quantity of accumulated enzyme. The need then arises to include a first-order rate term. Because of this need, an integrated form of the Michaelis-Menten equation, which includes both zero- and first-order terms, is perhaps the most generally useful and provides in particular for the comparatively prolonged soil assays when the initial velocity is not maintained due to decreasing substrate concentration. The general application of the integrated equation is discussed by Dixon and Webb (1964). An application of the standard first-order form to phosphatase reaction in soil is described in Brams and McLaren (1974):

$$(So) - (St) + 2.30\ \text{Km} \log\left(\frac{So}{St}\right) = V \cdot t$$

where $S0$ and St are the substrate concentrations, initially and after reaction time t, respectively, and the other symbols have the meanings already described.

A further difficulty posed by the heterogeneous soil system is the adsorption of substrate under non-steady-state conditions. Cervelli et al (1973) describe one approach to this problem: In the study of acid phosphatase the adsorption of the substrate, p-nitrophenyl phosphate, obeys the Freundlich adsorption isotherm

$$\frac{x}{m} = \text{K}C^{1/n}$$

where x is the weight of substrate adsorbed by a weight m of soil; C is the solution concentration at equilibrium; and K and n constants dependent on the system being studied. Substitution of these values into the *Lineweaver and Burke equation* yields:

$$\frac{1}{v} = \frac{\text{K}^n\text{Km}}{V} \cdot \frac{1/x^n}{m^n} + \frac{1}{V}$$

A plot of $1/v$ versus $1/(x/m)^n$ gives a straight line, which at $1/v = 0$ gives a corrected constant:

$$\text{K}^1\text{m} = \frac{(x/m)^n}{\text{K}^n}$$

McLaren and Packer (1970) describe the more complex treatment required where reactions occur at solid- or gel-liquid interfaces or where a soluble enzyme acts on insoluble substrate. Examples of the latter include the action of cellulase on cellulose and of soil enzymes on organic matter.

Experimental Values. Apparent Michaelis constants that are uncorrected for electrostatic and other effects are available for arylsulfatase, phosphatase, and urease. The Km values vary from 1.37×10^{-3} M to 5.69×10^{-3} M for arylsulfatase; 1.26×10^{-3} M to 4.58×10^{-3} M for phosphatase; and 1.30×10^{-3} M to 7.0×10^{-3} M for urease (Tabatabai, 1973; Tabatabai and Bremner, 1971). Shaking the soil-substrate mixtures reduces the values, which suggests that the reactions are diffusion controlled.

A corrected constant K^1m, which accounts for substrate adsorption during assay, gives lower values than the uncorrected Michaelis constant, Km (Cervelli et al, 1973). This result may reflect the diminished accessibility of substrate in a heterogeneous system. Table 3 shows both uncorrected and corrected values for acid phosphatase. The corrected values are close to those reported from plant and microbial sources of the enzyme.

Brams, Day, and McLaren (1975) also report a higher apparent value for the first-order reaction describing phosphatase activity in a soil column under steady-state conditions. The apparent value is about 70% higher than the true value because of electrostatic repulsion between negatively charged substrate and soil particles. It is evident, therefore, that Km

TABLE 3. Apparent (Km) and Corrected (K^1m) Rate Constants for Acid Phosphatase Activity of Some Soils.

Soils	Km (moles/l)	K^1m (moles/l)
S. Cataldo (silty clay)	1.0×10^{-2}	5.40×10^{-3}
Avane (Terra Rossa)	9.7×10^{-3}	1.21×10^{-3}
Orentano (peaty soil)	12.2×10^{-3}	0.35×10^{-3}
Campeda (Andosol)	16.4×10^{-3}	0.58×10^{-3}

(From Cervelli et al, 1973).

depends on both the nature of the soil matrix in which the enzyme activity is localized and on the reaction substrate.

MICHAEL HERLIHY

References

Aomine, S., and Kobayashi, Y., 1966. Effects of allophanic clays on the enzymatic activity of beta-amylase, *Soil Sci. Plant Nutr.* (*Jap.*) **12**, 7–12.

Bartha, R., and Bordeleau, L., 1969. Cell-free perodidase in soil, *Soil Biol. Biochem.*, **1**, 139–143.

Benesi, A., and McLaren, A. D., 1975. Redox dependence of papain enzyme activity on charged kaolinite surfaces and in solution, *Soil Biol. Biochem.*, 7, 379–381.

Brams, W. H.; Day, P. R.; and McLaren, A. D., 1975. Effect of hydrodynamic dispersion on phosphatase reaction in a soil column, *Soil Biol. Biochem.*, **7**, 223–225.

Brams, W. H., and McLaren, A. D., 1974. Phosphatase reactions in columns of soil, *Soil Biol. Biochem.*, **6**, 183–189.

Briggs, M. H., and Segal, L., 1963. Preparation and properties of a free soil enzyme, *Life Sci.*, **2**, 69–72.

Briggs, M. H., and Spedding, D. J., 1963. Soil enzymes, *Sci. Prog., Lond.*, **51**, 217–225.

Burns, R. G.; Pukite, A. H., and McLaren, A. D., 1972. Concerning the location and persistence of soil urease, *Soil Sci. Soc. America Proc.* **36**, 308–311.

Cawse, P. A., 1975. Microbiology and biochemistry of irradiated soils, *in* E. A. Paul, and A. D. McLaren, eds., *Soil Biochemistry*. New York: Marcel Dekker, 213–267.

Cervelli, S.; Nannipieri, P.; Ceccanti, B.; and Sequi, P., 1973. Michaelis constant of soil acid phosphatase, *Soil Biol. Biochem.*, **5**, 841–845.

Dalal, R. C., 1975. Effect of toluene on the energy barriers in urease activity of soils, *Soil Sci.*, **120**, 256–260.

Dixon, M., and Webb, E. C., 1964. *Enzymes*. New York: Academic, 950p.

Durand, G., 1965. Les enzymes dans le sol, *Rev. Ecol. Biol. Sol.*, **2**, 141–205.

Estermann, E. F., and McLaren, A. D., 1961, Contribution of rhizoplane organisms to the total capacity of plants to utilise organic nutrients, *Plant Soil*, **15**, 243–260.

Galstian, A. Sh., 1974. Enzymatic activities of soils, *Geoderma*, **12**, 43–48.

Gregers-Hansen, B., 1964. Decomposition of diethylstilboestrol in soil, *Plant and Soil*, **20**, 50–58.

Herlihy, M., 1972. Microbial and enzyme activity in peats, *Acta Hortic.*, **26**, 45–50.

Hoffmann, G., 1959. Distribution and origin of some enzymes in soil (in German), *Z. Pfl Ernahr Düng Bodenk.*, **85**, 94–104.

Kiss, S.; Dragan-Bularda, M.; and Radulescu, D., 1975. Biological significance of enzymes accumulated in soil, *Adv. Agron.*, **27**, 25–87.

Kuprevich, V. F., and Shcherbakova, T. A., 1971. Comparative enzymatic activity in diverse types of soil, *in* A. D. McLaren and J. J. Skujins, eds., *Soil Biochemistry*. New York: Marcel Dekker, 167–201.

Ladd, J. N., 1972. Properties of proteolytic enzymes extracted from soil, *Soil Biol. Biochem.*, **4**, 227–237.

Ladd, J. N., and Butler, J. H. A., 1971. Inhibition by soil humic acids of native and acetylated proteolytic enzymes, *Soil Biol. Biochem.*, **3**, 157–160.

Ladd, J. N., and Butler, J. H. A., 1972. Short-term assays of soil proteolytic enzyme activities using proteins and dipeptide derivatives as substrates, *Soil Biol. Biochem.*, **4**, 19–30.

Lloyd, A. B., 1975. Extraction of urease from soil, *Soil Biol. Biochem.*, **7**, 357–358.

Macura, J., 1974. Trends and advances in soil microbiology from 1924 to 1974, *Geoderma*, **12**, 311–329.

Mayaudon, J., and Sarkar, J. M., 1975. Laccases de *Polyporus versicolor* dans le sol et la litiere, *Soil Biol. Biochem.*, **7**, 31–34.

McLaren, A. D., 1963. Biochemistry and soil science, *Science*, **141**, 1141–1147.

McLaren, A. D., and Packer, L., 1970. Some aspects of enzyme reactions in heterogeneous systems, *in* F. F. Nord, ed., *Advances in Enzymology*. New York: Wiley-Interscience, 245–308.

McLaren, A. D.; Peterson, G. H.; and Barshad, I., 1958. The adsorption and reactions of enzymes and proteins on clay minerals: 4. Kaolinite and montmorillonite, *Soil Sci. Soc. America Proc.*, **22**, 239–244.

Nannipieri, P.; Ceccanti, B.; Cervelli, S., and Sequi, P., 1974. Use of 0.1 M pyrophosphate to extract urease from a podsol, *Soil Biol. Biochem.*, **6**, 359–362.

Paulson, K. N., and Kurtz, L. T., 1969. Locus of urease activity in soil, *Soil Sci. Soc. America Proc.*, **33**, 897–901.

Ross, D. J., 1971. Some factors influencing the estimation of dehydrogenase activities of some soils under pasture, *Soil Biol. Biochem.*, **3**, 97–110.

Ross, D. J., and McNeilly, B. A., 1972, Some influences of different soils and clay minerals on the activity of glucose oxidase, *Soil Biol. Biochem.*, **4**, 9–18.

Satyanarayana, T., and Getzin, L. W., 1973. Properties of a stable cell-free esterase from soil, *Biochemistry*, **12**, 1566–1572.

Skujins, J. J., 1967, Enzymes in soil, *in* A. D. McLaren, and G. H. Peterson, eds., *Soil Biochemistry*, New York: Marcel Dekker., 371–414.

Skujins, J. J., and McLaren, A. D., 1969. Assay of urease activity using ^{14}C-urea in stored, geologically preserved, and in irradiated soils, *Soil Biol. Biochem.*, **1**, 89–99.

Tabatabai, M. A., 1973. Michaelis constants of urease in soils and soil fractions, *Soil Sci. Soc. America Proc.*, **37**, 707–710.

Tabatabai, M. A., and Bremner, J. M., 1971. Michaelis constants of soil enzymes, *Soil Biol. Biochem.*, **3**, 317–323.

Webley, D. M., and Jones, D., 1968. Techniques for the study of localised microbial activity in soil, *Internat. Congr. Soil Sci., Trans.*, **3**, 657–664.

Weetall, H. H.; Weliky, N.; and Vango, S. P., 1965. Detection of microorganisms in soil by their catalatic activity, *Nature* (*Lond.*), **206**, 1019–1021.

Zobell, C. E., 1946. *Marine Microbiology*, Waltham, Mass.: Chronica Botanica, 240p.

Cross-references: *Adsoption Phenomena; Algae; Clay-organic Reactions; Exchange Phenomena; Humus;*

Microbial Ecology; Microhabitats; Nitrogen Cycle; Polysaccharide; Phosphorus Cycle; Rhizosphere; Soil Components, Organic; Soil Fertility; Sulfur Transformations.

EROSION–See WATER EROSION; WIND EROSION. Vol. IVA: SOIL EROSION

ESSENTIAL ELEMENTS–See MACRONUTRIENTS; MICRONUTRIENTS; PLANT NUTRIENTS; SOIL FERTILITY

EVAPORATION

Water returns directly into the atmosphere from the soil by the *evaporation process*. Associated with evaporation is the *transpiration process* by which water is taken up by the plant roots, moved through the plant system, and vaporized at the leaf surface. These two processes are related and where plants are growing, are difficult to differentiate in the field.

The amount of water returned to the atmosphere by evaporation depends on many factors such as the amount and frequency of precipitation, the soil type, and the presence of a mulch and plant cover. Under dryland agriculture where *summer fallow* (the practice of leaving the soil bare on alternate years) is practiced, evaporation may be as much as 60% of the precipatation. In contrast, in regions where complete plant cover occurs all year, evaporation may be as low as 5%.

The evaporation process that occurs after a soil is wetted involves two stages, a climate-controlled stage and a soil-controlled stage. After wetting, there is a period in which the amount of evaporation is controlled by climatic factors (first stage of evaporation). At this time evaporation from the soil is very nearly the same as a free-water surface. Evaporation involves a change in phase of the soil water (q.v.) from a liquid to a gas, which requires that about 585 cal/g of water be available for this process. This heat energy comes mostly from solar radiation, but it can also be extracted from the heat stored in the air or the soil. Thus the ratio of evaporation from the soil to that from a free-water surface is essentially as shown in Figure 1. The amount of energy reaching the soil surface, and thus the evaporation during stage 1, can be influenced by shading the soil with plants or with mulches (Figure 1). Shading modifies the climate at the soil surface.

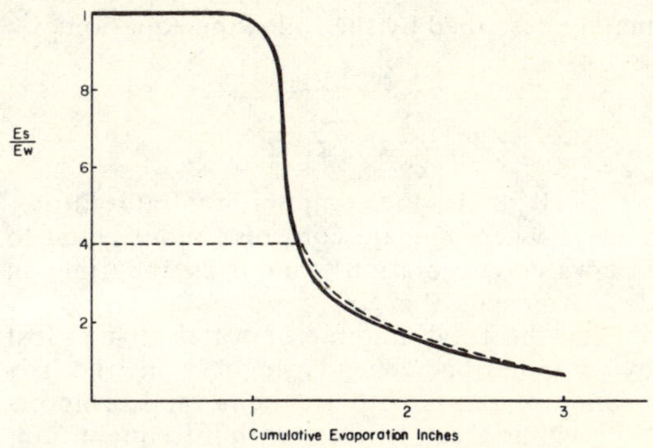

FIGURE 1. Ratio of evaporation from the soil Es, to evaporation from a free-water surface, Ew, as a function of cumulative evaporation from the soil. The solid line is for a bare soil; the dashed line is for a soil with a surface mulch.

During the first stage of drying, the surface-soil water content (q.v.) decreases as does the water content deeper in the profile. As this process continues, the surface-soil water content dries until it reaches a minimum, the air-dry content, which marks the end of stage 1 and the beginning of stage 2. Thereafter the soil cannot move water to the soil surface fast enough to equal the rate of heat input to the surface, and the ratio of soil evaporation to evaporation from a free-water surface falls. The ability of the soil to transmit water as the soil continues to dry decreases with time (Figure 1).

After wetting, the first stage of drying of a bare soil may last only a few hours during a summer day when solar radiation is high. If the drying environment is less severe or the soil is partially shaded or a mulch is present, this stage will last several days but will end for a given soil with a given amount of initial wetting when the same amount of water is lost by evaporation as was lost by the bare soil (Figure 1). Thereafter the evaporation rate is the same for the mulched as for the bare soil. Thus if the time between rains is long, little decrease in evaporation from the soil will result from the use of a mulch.

In a typical drying cycle under constant climatic conditions, about 2 cm of water will be lost in the first 2 days, another 2 cm will be lost in the next 3 days, and another 2 cm will be lost in the next 30 days. At the end of 35 days the rate of water lost by evaporation will be about 0.05 cm/day. This falling rate of evaporation (second stage) E can be approximately described by the following equation:

$$E = \frac{E_{wet}}{\sqrt{t+1}}$$

where E_{wet} is the evaporation rate during stage 1 where climate controls (about equal to free-water evaporation), and t is the time in days after stage 2 begins.

Thus the total amount of water that is lost by evaporation depends highly on the frequency of rains. With frequent rains, evaporation will be great, whereas with infrequent rains evaporation will be small. For example, in one experiment where 0.25 cm of water was added to a soil every day for 30 days, 96% evaporated, whereas only 44% evaporated when the same amount of water was added at one time.

The depth of wetting also influences the amount of water lost by evaporation. Experiments have shown that if the soil is wetted below about 30 cm, deeper wetting will cause no substantial increase in the total amount of water lost by evaporation.

In many areas of the world where water is insufficient for maximum plant growth, considerable efforts have been expended to decrease evaporation and thus increase the water available for plant transpiration. Most of these efforts have not proved practical. The most commonly used practice is to leave a mulch of plant straw on the soil surface. This practice has been shown to reduce evaporation to a limited extent, but the amount of mulch available in the water-deficient areas is much less than needed for maximum results. Another practice that has been tested experimentally is the use of plastic or gravel mulches between the rows of plants. These practices have been effective in decreasing evaporation but commonly cost much more than the increased crop yield derived from their use. Plastic mulches are used commercially in high-income crops to increase soil temperature and minimize weeds as well as minimize evaporation (see *Mulch*).

A problem associated with water evaporation where water tables are close to the surface is that of salt accumulation. In many areas of the world, especially irrigated areas, a water table may be sufficiently close to the surface that the second stage of drying will not occur (no continued falling off of the evaporation rate); rather, a constant rate of evaporation (which may be lower than the stage 1 rate) develops, which decreases as the depth to the water table decreases. A steady-state situation will then be set up where water is drawn up into the soil from the water table at the same rate that it is evaporated at the soil surface. If the soil water solution of the water table contains salts, this salt will move with the water to the soil surface and there be deposited because only the water and not the salt is evaporated. A salinization problem may then result.

R. J. HANKS

References

Hanks, R. J.; Gardner, H. R.; and Fairbourn, M. L., 1967. Evaporation of water from soils as influenced by drying with wind or radiation, *Soil Sci. Soc. America Proc.*, **31**, 593–598.

Hillel, D., 1971. *Soil and Water–Physical Principles and Processes.* New York: Academic, 288p.

Gardner, H. R., and Hanks, R. J., 1966. Effect of sample size and environmental conditions on evaporation of water from soil. *USDA-ARS. Cons. Res. Rept. No. 9.*, 14p.

Gardner, H. R., and Gardner, W. R., 1969. Relation of water application to evaporation and storage of soil water, *Soil Sci. Soc. America Proc.*, **33**, 192–196.

Staple, W. J., 1964. Dryland agriculture and water conservation. *Am. Soc. Agron. Spec. Pub. No. 4.*, 15–30.

Willis, W. O.; Haas, H. J.; and Robins, J. S., 1963. Moisture conservation by surface and subsurface barriers and soil configuration under semiarid conditions, *Soil Sci. Soc. America Proc.*, **27**, 577–580.

Cross-references: *Field Water Cycle; Imbibition; Irrigation; Permeability; Salt Leaching; Soil Water; Surface Soil Water Content; Thermal Regimes; Water Content and Retention; Water Fluxes; Wetting Front; Vol. XII, Pt. 2: Salinization.*

EXCHANGEABLE NUTRIENTS—See SOIL FERTILITY

EXCHANGE COMPLEX—See BASE SATURATION; EXCHANGE PHENOMENA

EXCHANGE PHENOMENA

Soil colloids have two basic properties that make them chemically active; a large surface area and a surface electric charge. Due to the requirement of electroneutrality, this surface electric charge must be balanced by an equal quantity of oppositively charged counterions. The general term, *exchange phenomena,* or more specifically *ion exchange* (q.v. in Vol. IVA) *in soils,* refers to the exchange between the counterions balancing the surface charge on the soil colloid and the ions in the soil solution.

Constant Charge Surfaces

The charge on soil colloids may result from either structural imperfections in the interior of the crystal structure or preferential adsorption of certain ions on particle surfaces (van Olphen, 1963). Structural imperfections due to ion substitutions or site vacancies frequently result in a permanent charge on soil colloidal particles. This type of colloid is considered to have a constant charge, variable potential surface. Such colloids are typified by, but may not be limited to, the so-called 2:1 type of clay minerals such as micas or smectites (often referred to as *montmorillonites*) where the charge results from isomorphous substitution of octahedrally or tetrahedrally coordinated cations (see *Clay Minerals*). The charge resulting from a given ion substitution may theoretically be either positive or negative (Ross and Hendricks, 1945). However, in the case of the 2:1 type clay minerals, which consist largely of Si^{4+} or Al^{3+} in tetrahedral coordination and Fe^{3+}, Fe^{2+}, Al^{3+}, and Mg^{2+} in octahedral coordination, ion size limitations generally result in a substitution of cations of lower valence for those of higher valence resulting in a net negative charge on the clay structure (Pauling, 1930; Marshall, 1949).

On the basis of known ion size limitations, coordination numbers, and a knowledge of the chemical composition for a given 2:1 type clay mineral, it is possible to calculate the chemical formula, assign atoms to the structural planes, and calculate unit cell dimensions, surface area, charge deficit per unit cell, surface charge density, and a theoretical cation exchange capacity (Marshall, 1949; Ross and Hendricks, 1945; van Olphen, 1963). However, because such calculations apply only to homogeneous samples of a given clay mineral, they cannot be generally applied to the complex mixture of clay minerals found in most soils. As a result, it is necessary to measure the extent of the surface charge experimentally by determining the quantity of counterions (cations) compensating the negative charge (Ensminger, 1944; Mehlich, 1948). The charge determined in this manner is commonly referred to as the *cation exchange capacity* (CEC) and if appropriate precautions are taken, is a measure of the negative charge balanced by cations available for exchange by the saturating cations.

Cation exchange capacity measurements generally involve washing the soil sample with a given salt solution until the negative charge is compensated by only the added cation species. The quantity of compensating cations, and hence the net negative charge or CEC, can then be determined by either extracting with an excess of a second salt solution and determining the quantity of counterions present (Jackson, 1969; Rich, 1961) or by labeling the saturating cation with an appropriate radioisotope and counting it directly without extraction from the soil (Beetem, Janzer, and Wahlberg, 1962; Francis and Grigal, 1971). In either case, it is common to wash out the excess salt before extraction or counting. Other than the possible retention of excess salt and loss of sample by dispersion (Jackson, 1969; Rich, 1962), hydrolysis of the saturating cation during the washing procedure constitutes the most likely discrepancy between the net negative charge and the measured CEC.

The charge or CEC measured in this manner is the charge neutralized by cations that are available for exchange with those in the saturating solution. Therefore, it is important to use a cation that is preferred or held more tightly by the surface than the cations being replaced. For this reason, Cs^+, Ca^2, Mg^{2+}, or Sr^{2+} are commonly used (Sawhney, Jackson, and Corey, 1959). Even when an excess of a highly preferred cation is used, part of the net negative charge on the surface may be blocked in such a manner that the compensating cations are not replaced by those in the saturating solution. This is frequently the case in micas and illites where the K^+ is "fixed" against normal exchange processes. In these cases the measured CEC is significantly less than the net charge deficit on the mineral structure.

pH Charge Dependent Surfaces

In contrast to ions in the flat surfaces of clay minerals that are fully coordinated, the surfaces of oxide and hydrous-oxide minerals contain ions that are not fully coordinated and hence are electrically charged (Parks, 1967). When placed in an aqueous environment such as exists in soils, the net charge on such surfaces is balanced by the adsorption of H^+ or OH^- ions resulting in an hydroxylated surface. Charge can develop on these hydroxylated surfaces either through amphoteric dissociation of the surface hydroxyl groups or by adsorption of H^+ or OH^- ions. Following Parks (1967), these reactions can be written as follows, where italicized symbols refer to species forming part of the surface.

$$M\,OH \rightleftharpoons M\,O^- + H^+ (aq) \quad (1)$$
$$M\,OH \rightleftharpoons M^+ + OH^- (aq) \quad (2)$$
$$M^+ H_2 0 \rightleftharpoons M\,OH^2 \quad (3)$$

Since the probability of bare M^+ existing at the surface is small, the basic dissociation or formation of a positively charged site probably occurs through a combination of the reactions shown in equations 2 and 3 (Mott, 1970).

$$M\,OH + H_2 0 \rightleftharpoons M\,OH_2{}^+ + (OH^-)_{aq} \quad (4)$$

In both the preceding mechanisms, the H^+ and OH^- ions that establish the surface charge are referred to as *potential determining ions* (PDI). Concentrations of the PDI and the net surface charge are obviously pH dependent, and there will be a pH value at which the net surface charge is zero; i.e., the density of positive and negative charge is equal. This pH is referred to as the *isolectric point* (IEP) or *zero point of charge* (ZPC) of the solid. Since the net charge on such surfaces depends on the H^+ or OH^- concentration in the equilibrium soil solution, it can be made positive or negative by raising or lowering the pH and is referred to as *pH dependent charge.*

Soils in the temperate regions tend to be dominated by the crystalline or constant charge-clay minerals while the high oxide content of many extensively weathered tropical soils results in a dominance of oxides that have pH dependent charge surfaces. All soils will, however, contain a mixture of both types of surfaces even though one type might tend to dominate the other. These mixtures may result from any one or all of the following:

1. Edge effects on crystalline clay minerals
2. Isomorphous substitution or site vacancies in Si-Al cogels
3. Oxide coatings or interlayers on crystalline clay minerals

Edges of crystalline clay minerals contain ions not fully coordinated, much the same as oxide surfaces. These edges also adsorb H^+ and OH^- to form hydroxylated surfaces, which in turn develop an electrical charge through amphoteric association or dissociation. Since all clay minerals have such edges, they all must have a mixture of constant and pH dependent charged surfaces. However, in the case of the 2:1 type clay minerals, the edges constitute about only 1% of the total available surface area (Dyal and Hendricks, 1950). Consequently, the pH dependent charge on these minerals has limited effect on their overall behavior.

The 1:1 or kaolinitic type clay minerals, however, tend to be aggregated into much larger particles through stacking in the *c* axis direction. Since the internal surfaces do not tend to separate except through rather specific intercalation reactions, they are not available for surface reactions. Consequently, these minerals have relatively low exchange capacities (Grim, 1969); and while much of the surface charge can be attributed to the amphoteric dissociation on the particle edges, it is difficult to rule out the possibility of a negative charge resulting from some isomorphous substitution (Schofield and Samson, 1953).

Anion Exchange

In contrast to the fixed-charge clay minerals, the charge on oxide surfaces and the edges of clay minerals may either be negative, resulting in a cation exchange capacity (CEC), or positive, resulting in an *anion exchange capacity* (AEC), depending on the pH. Anion exchange capacities exist at low pHs where there is adsorption of H^+ ions resulting in a positive charge on the mineral surface. This positive charge must be balanced by an equal amount of anions such as Cl^-, NO_3^-, and $SO_4^=$. As a result, soils in which these types of surfaces predominate will retain anions at low pHs and potentially alter their behavior in terms of leaching and availability for plant uptake. In contrast, soils in which negatively charged clay minerals predominate do not retain anions, and hence they are readily available for leaching and plant uptake. At high pHs, oxide surfaces will be negatively charged due to OH^- adsorption (or H^+ dissociation) and have a CEC that must be balanced by an equal amount of cations such as K^+, Na^+, Ca^{2+}, or Mg^{2+}.

Ion Exchange Process

Ion exchange refers to the exchange between the electrostatically held counterions balancing the surface charge on the soil particles and other ions in the soil solution and has the following general characteristics (Helfferich, 1962):

1. It is reversible.
2. It is diffusion controlled; i.e., the rate-limiting step is the diffusion of one counterion against another.
3. It is stoichiometric.
4. In most cases there is some selectivity or preference for one ion over the other by the surface.

All these characteristics can be readily demonstrated in the case of a pure montmorillonite clay suspension where the quantity of counterions equals the charge deficit on the crystal structure, one counterion species can be stoichiometrically exchanged for another, and the exchange of one ion for another occurs within the time limits of detection if the suspension is rapidly stirred to eliminate concentration gradients.

While these general processes also apply to soil systems, they are often more complex and less obvious as in the case of "fixed" K^+ on mica surfaces. This K^+ can be considered a counterion because it does balance the fixed charge on the mica structure resulting from isomorphous substitution, and it can be removed without altering the mineral structure itself. The exchange of another cation species for the interlayer K^+ is stoichiometric, and the exchange rate is controlled by the diffusion rate in the interlayer region of the clay particles. This exchange process is therefore reversible, stoichiometric, and diffusion controlled. Since the interparticle-diffusion rates are very slow (Jacobs, 1963) and since the K^+ is so highly preferred over other cations most commonly found in soils, ion exchange on mica-like surfaces is generally considered separately from ion exchange on the "more available" colloidal surfaces.

The total diffusion process, which involves movement of both solution and counterions against a concentration gradient, has been demonstrated to play a significant role in the movement of nutrient ions to plant root surfaces (Barber, Walker, and Vasey, 1963). The extent that "ion exchange" diffusion, or stoichiometric exchange of one counterion for another during movement along the colloid surface, contributes to the total movement or flux of ions through the soil, however, depends on the relative quantities of ions in the soil solution versus those balancing the surface charge. Solution diffusion is in general a much faster process than "ion exchange" diffusion.

Ion exchange selectivity, or the preference of one ion over another, has been the subject of a great deal of study. If these ions could be treated as point charges, then there would be no preference between ions of equal valence; ions do, however, have significantly different crystalline and hydrated sizes. Since electrostatic forces are involved in the retention of counterions, it can predicted from *Coulomb's law* (Pauley, 1953) that the ion having the smallest effective radius and highest valence will be preferred. It has been frequently demonstrated that there is a definite relationship between ion size and ion selectivity (Pauley, 1953; Cloos et al, 1965; Jenny, 1932; Gast, 1969). Since cations present in soils are hydrated, ions with the smaller hydrated size or the larger crystalline size are preferred over other ions having the same valence. This is reflected in the so-called *Lytropic series,* which shows the following order of monovalent cation preference by soil colloids: Cs > Rb > K > Na > Li. Close relationships have been demonstrated between selectivity and various ion size parameters including polarizability, Debye Huckle parameters of closest approach, and hydrated radii.

Ion exchange reactions in soils have been intensively studied for well over 100 years and very justifiably so. It is the mechanisms by which many nutrient ions are retained by soils against leaching losses and at the same time made generally available for meeting the nutrient requirements of plants.

ROBERT G. GAST

References

Barber, S. A.; Walker, J. M.; and Vasey, E. H., 1963. Mechanisms for the movement of plant nutrients from the soil and fertilizer to the plant root, *Agric. Food Chem.*, **11**, 204–207.

Beetem, W. A.; Janzer, V. J.; and Wahlberg, J. S., 1962. Use of cesium-137 in the determination of cation exchange capacity. *U.S. Geol. Surv. Bull.*, **1140-B**, 8p.

Cloos, P.; Fripiat, J. J.; Poncelet, G.; and Poncelet, A., 1965. Comparison entre les proprietes d'echange de la montmorillonite et d'une resine vis-á-vis des cations alcalins et alcaline-terraux. II. Phenomene de selectivité, *Bulletin de la Societé Chimeque de France,* **42**, 215–219.

Dyal, R. S., and Hendricks, S. B., 1950. Total surface of clays in polar liquids as a characteristic index, *Soil Sci.*, **69**, 421–432.

Ensminger, L. E., 1944. A modified method for determining base-exchange capacity of soils, *Soil Sci.*, **58**, 425–432.

Francis, C. W., and Grigal, D. F., 1971. A rapid and simple procedure using Sr-85 for determining cation exchange capacities of soils and clays, *Soil Sci.*, **112**, 17–21.

Gast, R. G., 1969. Standard free energies of exchange for alkali metal cations on Wyoming bentonite, *Soil Sci. Soc. America Proc.*, **33**, 37–41.

Grim, R. E., 1969. *Clay Mineralogy.* New York: McGraw-Hill, 396p.

Helfferich, F., 1962. *Ion Exchange.* New York: McGraw-Hill, 624p.

Jackson, M. L., 1969. *Soil Chemical Analysis–Advanced Course.* Madison: Department of Soils, University of Wisconsin, 895p.

Jacobs, D. G., 1963. The effect of collapse-inducing cations on the cesium sorption properties of hydrobiotite, *Internat. Clay Conf.*, (Stockholm), **1**, 239–248.

Jenny, Hans., 1932. Studies on the mechanism of ionic exchange in colloidal aluminum silicates, *Jour. Phys. Chemistry*, **36**, 2217–2258.

Marshall, C. E., 1949. *The Colloidal Chemistry of the Silicate Minerals.* New York: Academic, 195p.

Mehlich, A., 1948. Determination of cation and anion exchange properties of soils, *Soil Sci.*, **6**, 429–445.

Mott, C. J. B., 1970. Sorption of anions by soils, *in Sorption and Transport Processes in Soils,* Soc. Chem. Ind. Mon., **37**, 40–53.

Parks, G. A., 1967. Aqueous surface chemistry of oxides and complex oxide minerals. Isoelectric point and zero point of charge, *in* R. F. Gould, ed., *Adv. Chem. Series,* **67**, 121–160.

Pauley, J. L., 1953. Prediction of cation exchange equilibria, *Jour. Am. Chem. Soc.*, **76**, 1422–1425.

Pauling, L., 1930. The structure of micas and related minerals, *Proc. Natl. Acad. Sci. U.S.*, **16**, 123–129.

Rich, C. I., 1961. Calcium determination for cation

exchange capacity determinations, *Soil Sci.*, **92**, 226–231.

Rich, C. I., 1962. Removal of excess salt in cation exchange capacity determinations, *Soil Sci.*, **93**, 87–94.

Ross, C. E., and Hendricks, S. B., 1945. Minerals of the montmorillonite group. Their origin and relation to soils and clays. *U.S. Geol. Surv. Prof. Paper*, **205-B**, 23–79.

Sawhney, B. L.; Jackson, M.; and Corey, R. B., 1959. Cation exchange capacity determinations of soils as influenced by the cation species, *Soil Sci.*, **85**, 243–248.

Schofield, R. K., and Samson, H. R., 1953. The deflocation of kaolinite suspensions and the accompanying change-over from positive to negative chloride adsorption, *Clay Mineral Bull.*, **2**, 45–51.

Thomas, G. W., 1977. Historical developments in soil chemistry: Ion exchange, *Soil Sci. Soc. America Jour.* **41**(2), 230–238.

van Olphen, H., 1963. *Introduction to Clay Colloid Chemistry*. New York: Interscience, 301p.

Cross-references: *Adsorption Phenomena; Base Saturation; Buffers, Buffering; Clay Minerals, Silicates; Clay-organic Reactions; Clays, Physical Chemistry; Diffusion Processes; Electrochemistry; Humus; Hydrous Oxides; Ionic Activities; Soil Reaction; Soil Solution; Vol. IVA: Clay Minerals–Base Exchange.*

F

FALLOWING–See MANAGEMENT OF SOILS

FAUNA

In most parts of the world there is more life beneath the surface of the soil than above it. The following discussion introduces the animals, or *fauna,* of the soil and examines their relationships with the other soil organisms and the nonliving components of the soil.

From the zoologist's point of view, *soil* is the mineral substrate together with the dead organic matter; the decomposing organic remains that lie immediately above the soil make up the *litter layer.* It is impossible for the zoologist to think of the soil fauna without considering the litter layer because many soil animals are found both in it and below it, often moving from one to the other. While the protozoa are still regarded by many as constituting part of the animal kingdom, their biology is comparable more to microorganisms than to fauna, and thus they will not be considered here.

Growth of Soil Zoology

The scientific study of soil animals has developed into a discipline only since the 1950s, though some observations were made a long time before that. There have been of course fleeting written referrences to a few soil-inhabiting animals since pre-Biblical times, but, like other observations recorded in ancient texts, even Aristoteles' brief mention of earthworms is garbeled; thus soil zoology may perhaps be said to have begun in the middle of the sixteenth century, with the publication of a little book, *On Subterranean Animals,* published in Latin at Basel, Switzerland, by the Saxon teacher, physician, and scientist Georg Bauer, alias Georgius Agricola (1549). The author had an abiding interest in mining, and he referred to insects that lived underground such as certain bees and cockroaches, as well as crickets. In 1770, however, the English clergyman Gilbert White (1789) wrote much more informatively about earthworms and mole-crickets. Important early contributions were made by Darwin (1840, 1881) on earthworms, and by early Scandinavian authors, culminating in the works of Müller (1879, 1884), who considered the role of various invertebrates in humus formation. Kevan (1962) mentions some important subsequent contributors to the growth of soil zoology.

It was, however, not until Kühnelt (1950) in *Bodenbiologie* summarized the greater part of what was known about soil animals, and Franz (1950) in his *Bodenzoologie* reviewed the practical implications of the study of the soil fauna that soil zoology was recognized as a discrete discipline. Symposia dealing with soil animals have been held in many places since the first one in England (Kevan, 1955), and the proceedings of most of them have been published (e.g., Murphy, 1962; Doeksen and Drift, 1963; Rapoport, 1966; Graff and Satchell, 1967; d'Aguilar et al, 1971; Vaněk, 1975).

Other books largely or wholly devoted to soil zoology include those of Delamare Deboutteville (1951); Lawrence (1953); Eglitis (1954); Kühnelt (1954, 1961); Nosek (1954); Farb (1959); Kevan (1962); Kipenvarlitz (1961); Delamare Deboutteville and Rapoport (1962-1968); Schaller (1962, 1968); Bachelier (1963); Dunger (1964); Gilyarov (1964); Palissa (1964); Müller (1965); Loksa (1966); Burges and Raw (1967); Brauns (1968); Delamare Deboutteville (1970); Wallwork (1970, 1976); Pesson (1971); Phillipson (1971); Savory (1971); Andrews (1972); Gilyarov (1973); Tikhomirov (1973); and Richards (1974). Soil animals have been emphasized by some major general works on animal ecology (Tischler, 1955; MacFadyen, 1957; Balogh, 1958); plant ecology (Hartmann, 1952); pedology (Kübiena, 1953; Handley, 1954; Wilde, 1954; Russell, 1957; Russell and Russell, 1973) and even introductory general zoology (Moment, 1958).

Composition of the Soil Fauna

The soil fauna includes those animals that pass one or more *active* stages wholly or largely in soil or litter; they exclude those species that occur there only in a passive stage such as egg, cyst, or pupa or that use the earth only for hibernation, aestivation, or temporary shelter. Many, especially the small ones, are equally at home in soil and in fresh water or in soil and in certain habitats above ground. Illustrated keys to the main groups of animals are found in Kevan (1955) and Dunger (1964), and references to specialist works are listed in Kevan (1955), Kühnelt (1963), and Wallwork (1970). Nearly all the animal phyla that are not purely marine are represented in soil.

Nonarthropod Groups. Vertebrates are represented by a number of mammal species, particularly root-feeding rodents (e.g., pocket gophers such as *Geomys*, various mole-rats such as *Bathyergus* and *Stalacopus*, and rodent-moles such as *Spalax*) and carnivorous insectivores (e.g., true moles such as *Talpa* and *Scalopus*), whose burrow systems may be locally important in soil formation (Joffe, 1936). A few worm-like reptiles (amphisbaenid lizards and typholopid or uropeltid snakes) and amphibia (Gymnophiona) also occur in warmer parts of the world, but their influence on the soil is small.

Gastropod mollusks (slugs and snails) may abound in moist soils, but their importance in soil is also small.

Annelid worms include the most familiar members of the soil fauna, the earthworms, as well as their more numerous but less well-known relatives, the pot-worms, or Enchytraeidae. Almost all soil annelids belong to the subclass Oligochaeta (a largely aquatic group). Contrary to common opinion, not all earthworms belong to the family Lumbricidae. In the tropics and the Southern Hemisphere, members of the Megascolicidae and certain other families may be more frequently encountered in soil, although in some regions introduced Lumbricidae have tended to displace native species.

Each of the hundreds of earthworm species has its own particular requirements, feeding and burrowing habits. Most soils support several species, a fact not appreciated by Darwin (1881). Enchytraeidae, though largely aquatic, abound in certain soils, especially in raw humus. They resemble small pale earthworms, but they do not burrow.

Even more abundant in soil are the Nematoda. Most are free living, and about half of the known species are found in soil, though some occur also in fresh water. Nematodes, or "nemas," differ considerably in their food requirements. Some feed on diatoms and other algae or on protozoa; many (including the free-living stages of many animal and plant parasites) apparently feed on bacteria; some consume fungi, and others, like *Mononchus*, are voracious predators of small animals (including other nematodes and rotifers). Many species (e.g., *Rahbditis* and *Plectus*) that appear to be saprophagous, probably feed on the decomposed microorganisms and their liquified products rather than directly on the decaying organic matter.

Various other groups of "worms" (e.g., Turbellaria, or flatworms, and minute Gasterotricha) are also found in soil, but not generally in abundance.

Tardigrades are curious microscopic, short, eight-legged animals generally classified with the arthropods, although the relationship seems far from close. They are usually aquatic, but some are terrestrial, usually associated with mosses, although they may be found in moist soils. Their effect on the soil ecosystem is insignificant.

Arthropods. Over 80% of all known animal species are arthropods, and most of these are insects. They include the most diverse of the soil inhabitants.

The most abundant soil arthropods are mites (Arachnida; Acarina) and springtails (Collembola). Other major groups include millipedes (Diplopoda) and certain orders of insects, particularly termites (Isoptera), fly larvae (Diptera), beetles and their larvae (Coleoptera), and ants (Hymenoptera; Formicoidea). Other common groups, of lesser importance in the soil, include woodlice (terrestrial Isopoda; Crustacea); centipedes (Chilopoda); cutworms and other caterpillars (Lepidoptera); root-feeding bugs such as immature cicadas, aphids, and scale insects (Hemiptera-Homoptera); burrowing crickets (Grylloptera-Orthoptera-Ensifera of most authors); and spiders (Araneida).

Microarthropods, particularly the mites and Collembola, are usually less than 2 mm long. Many soils contain more than a million of these within a square meter. Like the nematodes, the mites constitute one of the most successful, ubiquitous, and yet (except for a few of direct economic or medical significance) neglected animal groups. Most are free living in soil and litter. In most soils they outnumber all other arthropods by more than three to one. The Cryptostigmata or "oribatids" are especially important in breaking up plant remains and therefore in humus formation. Most are hard bodied and slow moving, and they probably feed on fungi. Some live within small twigs, leaf petioles, and conifer needles (Jacot, 1939), feeding on the softer tissues or on the decomposer microorganisms found there. Their paler, soft-bodied relatives, the Astigmata, are better known as parasites of vertebrates or as "cheese" or "grain" mites, but some are particularly abundant in moist, decomposing vegetable matter and in certain damp soil situations.

Another group of mites, the Mesostigmata, tend to be slightly larger, rather flattened, and comparatively swift moving. Most prey on other microarthropods, nematodes, and enchytraeids. Some are parasitic or epizoic on other animals, or feed on dead-animal remains, including carrion, and on fecal matter (or on the decomposed mircroorganisms associated with these latter habitats).

The other major group of soil mites, the Prostigmata (including the Heterostigmata) is very large and diverse. While the best-known members of this group are the foliage-feeding Tetranychidae, or spider-mites, few of these are found in soil. The medically important Trombiculidae, or chigger-mites, however, spend nymphal and adult stages in the soil, as do their relatives the Trombidiidae. Trombiculidae are parasites of arthropods in the young stages. The commonest member of this group in the soil are the Tarsonemoidea (Heterostigmata). Many tarsonemoids are extremely minute and frequently predatory, although some species seem to feed on plant and animal remains.

The Collembola, or springtails, constitute numerically the bulk of the remaining soil arthropods. Most of the elongate species, the Arthropleona, live within the soil whereas most (but not all) of the globular ones, the Symphypleona, live above ground. Because they have six legs, they have usually been classified along with insects, although they are not closely related to them. The few species that are reasonably well known illustrate that even closely related species may have quite different requirements for food and space. Some have a fairly general diet of decaying leaves, but they may ingest decaying organic matter, humus, decomposed microorganisms, pollen, feces of other soil animals, the remains of dead animals, or, rarely, even live soil animals such as nematodes. Collembola are of considerable importance in the breakdown of organic matter and in the formation of humus.

Categories of Soil Animals

The soil fauna may be classified according to size into "macrofauna," "meso-" or "meiofauna," and "microfauna" (Haarløv, 1960; Kevan, 1962). Macrofauna can be seen readily with the naked eye, mesofauna can be seen but less clearly than macrofauna, and microfauna cannot be seen at all. The microfauna are less than 100 μm long and comprise mainly nematodes, rotifers, small turbellarians, and tardigrades; the mesofauna include many of the enchytraeid worms and the smaller mollusks and arthropods, and the macrofauna the earthworms, larger mollusks, arthropods, and vertebrates that are measured in in centimeters.

Based on their space requirements, the fauna may be alternatively divided into the "water fauna" (animals requiring free water, even if only a thin film, in which to swim), virtually the same as the microfauna; the "burrowers" (animals capable of making their own channels through the soil), including in large part, but not entirely, the macrofauna; and the remainder, or "air-space fauna," including most, but not quite all, of the meiofauna (exceptions being ants and termites, which can burrow) (Kühnelt, 1957; Kevan, 1962).

Members of the soil fauna are also classified according to their occurrence in the soil profile. *Epedaphon* inhabit the soil surface and may frequently enter the soil proper; *hemiedaphon* inhabit the litter and fermentation layers; and *euedaphon* inhabit the deeper, largely mineral, layers. The members of these three categories are increasingly more specialized morphologically or physiologically for life in the soil, and their study has provided interesting examples of the development of life forms.

Trophic relations form yet another possible basis for classification. "Primary" feeders include the truly phytophagous animals, which feed directly on living or at least undecomposed vegetable matter (e.g., rodents, certain mollusks, plant parasitic nematodes, and many insects including whitegrubs, wireworms, root maggots, aphids, and immature cicadas and certain mites); and the "low primaries," which feed on decaying plant material (e.g., mollusks, earthworms, enchytraeids, millipedes, woodlice, springtails, mites, some nematodes and many insects including termites and crickets) or on dead wood (e.g., termites, many beetle larvae, fly maggots and even some nematodes). It is often difficult to determine whether such animals are truly saprophagous or whether their actual nutritional requirements are provided by the fungi and other microorganisms ingested with the dead organic material. A few soil animals ingest much mineral matter with their food (e.g., earthworms, millipedes, many termites, and apparently some Collembola).

"Secondary" feeders prey on or parasitize primary feeders, i.e., they obtain their food from plants secondhand. They include some moles, carnivorous insects such as ants, many beetles and their larvae, some fly larvae, virtually all centipedes, all arachnida other than many of the mites, planarian worms (Turbellaria), many rotifers, some tardigrades, and many nematodes. They also include those animals that feed on fungi and bacteria, which obtain their nourishment from dead or living organic matter (e.g., most mites and Collembola, many small beetle and fly larvae, some termites and ants, a few earwigs and mollusks, and many nematodes).

Of special interest are certain Protura (small, elongate, soft-bodied, insect-like animals) that apparently feed on mycorrhiza (fungi living in symbiotic association with plant roots) (Sturm, 1959). "Low secondaries" include the necrophagous soil animals, which devour the remains of other animals (e.g., many beetles and their larvae, fly maggots, mites, some Collembola, and even a few mollusks and nematodes), and the coprophagous species, which consume the droppings of both vertebrates and invertebrates (e.g., certain lumbricid earthworms and enchytraeid potworms, termites, beetles and their larvae, fly maggots and some Collembola, mites, and nematodes). Many of these, however, are probably feeding on the microorganisms within the carrion of feces. The terms "tertiary" and "quaternary" feeders are used to describe those animals that devour the others in succession.

TABLE 1. Numbers/m^2 of Animals that May Occur in the Soil.

Animal	Number
Earthworms (Lumbrididae)	30–2000
Pot-worms (Enchytraeidae)	200–20,000
Slugs and snails (Mollukca)	100–8500
Millipedes and centipedes (Diplopoda and Chilopoda)	900–1700
Woodlice (Isopoda)	100–400
Spiders (Araneida)	180–840
Beetles and their larvae (Coleoptera)	500–1000
Fly maggots (Diptera)	ca. 1000
Ants (Hymenoptera)	200–500
Springtails (Collembola)	10,000–40,000
Mites (Acarina)	10,000–120,000
Nematodes (Nematoda)	1.8–120 million

After Stöckli, 1946; MacFadyen, 1957.

Numbers of Animals in the Soil

The techniques involved in sampling and extracting soil animal populations are almost as diverse as the fauna itself. They include hand sorting, sifting, repellence by heat or chemicals, funnels and various related devices, flotation, centrifugation, elutriation, and numerous other methods (Murphy, 1962). No one method is suitable for extracting even a fraction of the fauna and all methods are very time-consuming. The greatest stumbling block in the study of soil animals, however, is that so few of the species can be identified, for much taxonomic work remains to be done. Species that are difficult to distinguish morphologically may differ considerably in their biology.

To give some idea of the numbers of animals that may be present in soil, and by way of reassuring sceptics of their importance, the estimates from Stöckli (1946) as modified by MacFadyen (1957) for fauna beneath one square meter of European grassland are given in Table 1 (see also Figure 1).

The upper limits indicated in the table are not maximal, and the figures would differ considerably in different habitats. In woodland soils enchytraeids may number more than half a million per square meter. No less than 834,500 arthropods per square meter were recorded by Murphy (1953) from cultivated heathland. The estimated number of animals, excluding nematodes (and protozoans), may total well over 1.5 million per square meter, although figures of this kind should be treated with some reservation because they are necessarily based on very small samples.

Populations of different size groups follow the usual "pyramid of numbers," there being generally fewer larger animals than smaller ones. Numbers also fall off rapidly with depth, the bulk of the nonburrowing fauna living in the top 30 cm and the majority of these in the top 4 or 5 cm. Earthworms, symphylids (small centipede-like animals), and nematodes, however, have all been collected from below 2 m in depth; termites from 10 m or more down. It would appear from Haarløv's (1955, 1960) results that the available *Lebensraum* is far from completely occupied.

More important than mere numbers, however, are figures for biomass because they give a better idea of the actual amount of living animal protoplasm present in soil. In mull under deciduous woodland, it was estimated by Nef (1957) that there were approximately 96 g of animals per square meter, but only about 29 g in raw humus under spruce (a comparatively poor soil). In the mull about two-thirds of the biomass consisted of earthworms. In the raw humus soils, earthworms were scarce, but about one-third of the animal biomass consisted of enchytraeids and nearly half of arthropods; the rest was made up mainly of nematodes.

Modern workers often use respiratory activity as a measure of food consumption and therefore of energy release. Using the data referred to, Nef (1957) calculated that, in the mull soil the animals may consume nearly 50 mg of oxygen per hour per square meter at 13°C. A little over 20% of this oxygen was used by earthworms (relatively little as compared to their biomass), slightly less by nematodes, and rather more by mites alone; Collembola were the other large oxygen consumers, using about half that respired by mites. Where earthworms were absent, total respiration was only slightly less than for the nonearthworm fauna of mull soil. Mites consumed about the same amount of

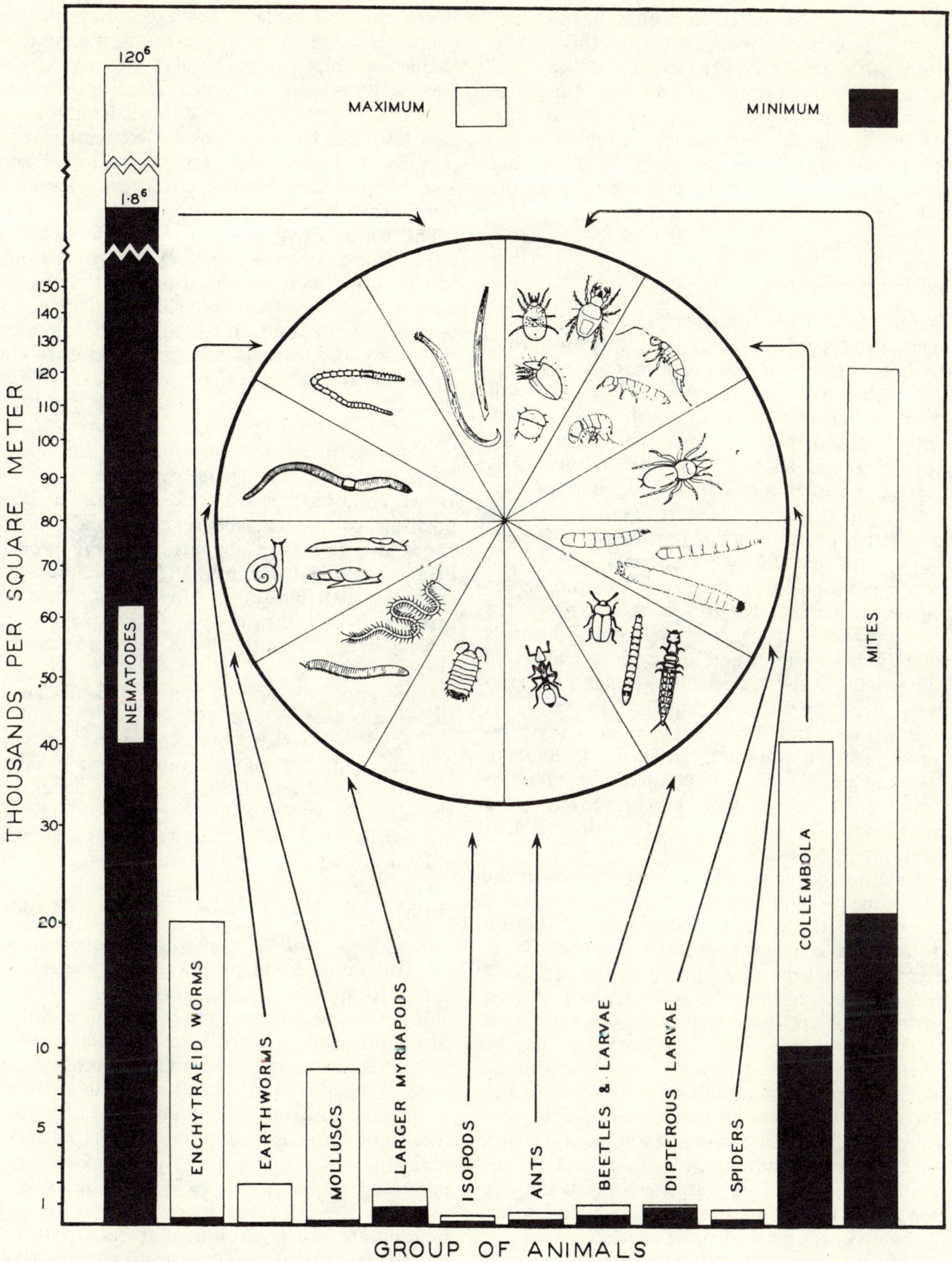

FIGURE 1. Numbers of animals per-square meter in soil of a European grassland, based on the estimates of Stöckli (1946) modified by Macfadyen (1957) and plotted on a logarithmic scale. Such figures are only indications, since they are mostly calculated from small samples. Of the groups shown, spiders and centipedes are wholly carnivorous and ants predominantly so; beetles, fly maggots, mites, and nematodes range widely in their diet according to species; the remaining groups feed largely on decaying organic matter, many Collembola being also closely associated with fungi and other microroganisms. (From Kevan, 1965; Copyright © 1965 by the Regents of the University of California)

oxygen in raw humus as in mull soil, but their relative importance was greater, as this constituted 40% of the oxygen used by the total fauna. Collembola used only about a quarter as much as mites, for they were much less numerous than in mull. Soil animals require a quantity of nutrients approximately equal to that provided by the litter produced annually by the surface vegetation.

Soil Environment

As in all other environments, the factors influencing the fauna may be divided into those that are physicochemical and those that are biotic.

The Physicochemical Environment. Soil structure and texture are naturally very important environmental factors, particularly for burrowing animals. Their ability to burrow is affected by structure (or arrangement of particles), and their rate of burrowing by texture (or relative proportions of particles of different sizes). With nonburrowing animals, one of the most critical factors is pore space, or the size of the microcaverns in the soil. If the microcaverns are very small, only the very smallest animals will be able to enter them; however, it is amazing how small are the crevices into which comparatively large animals can squeeze by virtue of their extensibility. Pore space is also very important for the aquatic microfauna. Their small size may permit them to enter the smallest of crevices, but they may be prevented from inhabiting them because of insufficient aeration.

Moisture content is of vital importance to the soil fauna. Absolutely dry conditions are rare except near the surface of deserts, and the soil atmosphere is usually saturated or nearly so. Free water is generally present in the form of a thin film surrounding the soil particles. If such a film is not continuous, however, the dispersal of members of the aquatic microfauna may be severely restricted. Many air-breathing animals require very moist conditions in which to survive (or at least remain active), even if they are not subaquatic. Earthworms and enchytraeids, for example, are among those that need at least small amounts of free water to prevent desiccation, and much of the remaining soil fauna probably needs a near-saturated atmosphere.

Temperature is probably of less importance to soil animals than it is to surface-dwelling species, not because they are insensitive to temperature changes, but because soil is an excellent insulator. Temperatures may vary throughout the year, especially in the upper layers; yet only near the surface, and then rarely, are fluctuations sudden or drastic. On or above the soil surface, however, they are frequently so. The slow rate of temperature change enables most species to migrate to greater depths or to enter a resistant phase, to avoid adverse effects. Many soil animals at high latitudes or elevations are extremely resistant to cold and may even reproduce at very low temperatures. Snow cover provides additional insulation.

The Biotic Environment. Biotic factors that affect the soil fauna may act directly or indirectly. For example, variations in the organic matter present affect physical structure and texture of the soil; and living and dead roots may provide channels by which nonburrowing animals may penetrate the soil as well as provide them with food. Chemical secretions produced by fungi or even the physical nature of their hyphae may render certain soils intolerable (e.g., for earthworms) or, by depressing plant growth, may affect the food supply. The composition of the surface vegetation also influences the soil fauna by affecting the type of litter available (Murphy, 1955) and may also influence distribution by providing more or less attractive oviposition sites for species with surface-living adults, e.g., elaterid beetles (Fox, 1959).

Food supply is one of the most significant of the biotic factors affecting life in the soil, as elsewhere. However, our knowledge of what particular species of soil animals eat is rather fragmentary. Predators and parasites depend on the abundance of all the foregoing animals and are in turn an important biotic factor regulating animal populations in the soil.

Human Activities

Man, as a kind of superbiotic factor, affects the soil fauna in many ways; for example, by cultivation, tree felling, drainage, irrigation, construction, destruction, industrial pollution, the application of fertilizers and plant protection chemicals, and pest control; and by "cultural" means such as "biological control." Biological control may be attempted by either the deliberate introduction of predatory or parasitic organisms from other regions or, as indicated by Solomon (1953), by the judicious use of pest-control materials or other means to manipulate the environment to favor naturally occurring control agents—a form of "integrated control." There are rather few examples of successful control of soil pests by biological agents.

The repercussions of human activities on the soil fauna and the consequent ultimate effects on the environment may be incalculable and, unfortunately, have hitherto received the scantest attention.

Interaction Between Soil Fauna and Soil Microflora

The role of soil animals in the breakdown of litter and in soil formation is not completely understood, but it is undoubtedly very significant. Its importance was indicated from a pedological point of view long ago by Müller (1879, 1884) and has been stressed more recently by Kubiëna (1948, 1953, 1955) and Hartmann (1951, 1952). Kubiëna (1955) considers animals essential to mixing and binding the particles of highly developed humus soils.

Clearly one of the most important roles of the soil fauna is that of breaking down humus particles and litter. This breaking-down results in an ever-increasing surface area on which fungi and other microorganisms can act. Witkamp and Drift (1961) note that the larger animals such as earthworms, millipedes, and some insects are more important in mechanical than in chemical breakdown. Lignous material, which cannot be digested by animals, requires the action of fungi to soften it (Jacot, 1939; Winston, 1956) before animals can fragment it and ultimate decomposition can take place (Birch and Clark, 1953). The animals that break down wood include the millipedes, dipterous and coleopterous larvae, and, of course, termites. The frass and feces produced by these (except in the special case of termites) and the fungi that attack wood are in turn consumed by microarthropods whose progressively smaller frass, or feces, and remains provide a milieu for other microorganisms. There is therefore an inseparable association between the soil fauna and the microflora.

Translocation of plant remains from the litter layer to lower horizons by earthworms and millipedes is probably also of very great significance in the breakdown of fragments since the latter environment appears to be more favorable for microbial activity (Macfadyen, 1957). The fauna may also be attracted or repelled by the metabolic by-products of the microflora.

In addition to providing food, some members of the microflora are parasitic on the fauna, and some "predaceous" fungi even catch nematodes and other members of the microfauna. On the other hand, members of the soil fauna may be important, even indispensable, in the dissemination of reproductive spores produced by the soil flora. This dissemination may be achieved either by means of the spores' adhering to the animals' bodies (especially in the case of animals with numerous setae) or by being passed unharmed through the animals' guts at a distance from the point of ingestion.

In connection with the interrelation between soil fauna and microflora, it cannot be overemphasized that simply because certain animals are known to feed on fungi or bacteria, this does not necessarily mean that they do so indiscriminately, nor is there any reason to suppose that they will select the most abundant of acceptable species (either beneficial or harmful from an anthropocentric point of view).

Conclusion

This brief summary provides but a glimpse into the hidden world of the soil. It is ironical that, while man has conquered the nearer heavens through space travel, he has, as yet, only the most rudimentary knowledge of the workings of the few centimeters of soil beneath his feet.

D. KEITH McE. KEVAN
S. B. HILL

References

Agricola [né: Bauer], G., 1549. De Animantibus Subterraneis, Basiliae (Basel). 79 + 16 pp. [reprinted 1614, Wittenberg; 80p + index].

Andrews, W. A., ed., 1972. *A Guide to the Study of Soil Ecology.* Toronto: Prentice-Hall, 198p.

Bachelier, G., 1963. *La Vie Animale dans les Sols.* Paris: O.R.S.T.O.M., 279p.

Balogh, J., 1958. *Lebensgemeinschaften der Landtiere, ihre Erforschung unter besonderer Berücksichtigung der zoologischen Arbeitsmethoden.* Verlag der Ungarischen Akademie der Wissenschaften, Budapest. Berlin: Akademie-Verlag, 560p.

Birch, L. C., and Clark, D. P. 1953. Forest soil as an ecological community with special reference to the fauna, *Quart. Rev. Biol.,* **1**, 13-26.

Brauns, A., 1968. *Praktische Bodenbiologie.* Stuttgart: Gustav Fischer, 470p.

Burges, A., and Raw, F., eds., 1967. *Soil Biology.* London: Academic, 532p.

D'Aguilar, J.; Athias Henriot, C.; Bessard, A.; Bouche, M.-B., and Pussard, M., eds., 1971. *Organismes du sol et production primaire. IV Colloquium Pedobiologiae.* Paris: Institut National de la Recherche Agronomique, 590p.

Darwin, C. R., 1840. On the formation of mould, *Trans. Geol. Soc. London,* **5**(2), 505-509.

Darwin, C. R., 1881. *The Formation of Vegetable Mould through the Action of Worms, with Observations on their Habits.* London: J. Murray, 326p. [Reprinted 1945, with a foreward by Sir A. Howard, *Darwin on Humus and the Earthworm.* London: Faber & Faber, 153p.]

Delamare Deboutteville, C., 1951. *Microfauna du sol des Pays Tempérés et Tropicaux.* Vie et Milieu, Suppl. 1, Paris: 360p.

Delamare, Deboutteville, C., ed., 1970. *Réactions des Microarthropodes aux Variations de L'Etat hydrique du Sol. & Techniques Relatives à L'Extraction des Arthropods du Sol.* Paris: Centre National de la Recherche Scientifique, 319p.

Delamare Deboutteville, C., and Rapoport, E., eds., 1962-1968. *Biologie de L'Amerique Australe. 4 vols. Etude sur Faune de Sol.* Paris: Editions du Centre National de la Recherche Scientifique, 657 + 399 + 725 + 473p.

Doeksen, J., and Drift, J. van der, eds., 1963. *Soil Organisms. Proceedings of the Colloquium on Soil Fauna, Soil Microflora and Their Relationships, Oosterbeek, the Netherlands, September 10-16, 1962.* Amsterdam: North Holland, 453p.

Drift, J. van der, 1951. Analysis of the animal community in a beech forest floor, *Meded. Inst. Toegep. Biol. Onders. Nat.,* **9,** 1-168.

Dunger, W., 1964. *Tiere im Boden. Neue Brehm-Bücherei No. 327 A.* Wittenberg Lutherstadt: Ziemsen Verlag, 265p.

Eglitis, V. K., 1954. *Fauna pochv Latviiskoi SSR Akademiya Nauk. Riga:* Latviiskoi, SSR, 261p.

Farb, P., 1959. *Living Earth.* New York: Harper and Brothers, 178p.

Fox, C. J. S., 1959. Influence of vegetation on distribution of wireworms, Agriotes spp. in grassland: a progress report, *Ann. Rev. Entomol. Soc. Ontario,* **89,** 47-49.

Franz, H., 1950. *Bodenzoologie als Grundlage der Bodenpflege.* Berlin: Akademie-Verlag, 316p.

Gilyarov, M. S., ed., 1964. *Opreditel' obitayuschikh v pochve lichinok nasekomy̆kh.* Moskva: Akademiya Nauk, SSSR, 919p.

Graff, O., and Stachell, J. E., 1967. *Progress in Soil Biology: Proceedings of the Colloquium on the Dynamics of Soil Communities.* Amsterdam: North Holland, 656p.

Haarløv, N., 1955. Vertical distribution of mites and Collembola in relation to soil structure. *In* D. K. McE. Kevan, ed., *Soil Zoology.* London: Butterworths, 167-179.

Haarløv, N., 1960. Microarthropods from Danish soils: Ecology, phenology, *Oikos,* **3,** 1-176.

Handley, W. R., 1954. Mull and mor formation in relation to forest soils, *Forest Commiss. Bull. (London) 23,* 115p.

Hartmann, F., 1951. *Der Waldboden, Humus-, Boden- und Wurzeltypen als Standortsanzeiger.* Wien: Österreiches Produktivatäts Zentrum, 151p.

Hartmann, F., 1952. *Forstökologie-Zustandserfassung und Standortsgemasse Gestaltung der Lebensgrundlagen des Waldes.* Wien: Georg Fromme, 461p.

Jacot, A. P., 1939. Reduction of spruce and fir litter by minute animals, *Jour. Forestry,* **37,** 858-860.

Joffe, J. S. 1936. *Pedology.* New Brunswick, N.J. Rutgers University Press, 575p.

Kevan, D. K. McE., ed., 1955. *Soil Zoology: Proceedings of the University of Nottingham Second Easter School in Agricultural Science.* London: Butterworths, 512p.

Kevan, D. K. McE., 1962. *Soil animals.* London: H. F. and G. Witherby, 237p.

Kevan, D. K. McE., 1965. The soil fauna–its nature and biology, in K. F. Baker and W. C. Snyder, eds., *Ecology of Soil-Borne Plant Pathogens: Prelude to Biological Control.* Berkeley, Los Angeles, Calif.: University of California Press, 33-51.

Kipenvarlitz, A. F., 1961. *Ismenenie pochvennoi fauny̆ nizinny̆kh bolot pod vliyaniem melioratzii sel'skokhozyaistvennogo osboeniya.* Minsk: SSR, Sel'khozgiz, Belorusskoi, 199p.

Kubiëna, W. L., 1948. *Enwicklungslehre des Bodens.* Wien: Springer Verlag, 215p.

Kubiëna, W. L., 1953. *The soils of Europe. Illustrated diagnosis and systematics.* Issued under the auspices of the Consejo Superior de Investigaciones Cientificas, Madrid. London: T. Mirby, 317p.

Kubiëna, W. L., 1955. Animal activity in soils as a decisive factor in establishment of humus forms, in D. K. McE. Kevan, ed., *Soil Zoology.* London: Butterworths, 73-82.

Kühnelt, W., 1957. *Biologia del suelo.* (Trans. E. Humbert) (2d ed. of Kühnelt, 1950.) Madrid: Consejo Superior de Investigaciones Cientificas, 267p.

Kühnelt, W., 1959. Über die Herkunft der Bodentierwelt des Festlandes, *Proc. XIV. Internat. Congr. Zool., London, 1958.* 353-360.

Kühnelt, W., 1961. *Soil Biology: With Special Reference to the Animal Kingdom. (Transl. N. Walker.)* (3rd ed. of Kühnelt, 1950). London: Faber and Faber, 397p.

Kühnelt, W., 1963. Soil-inhabiting Arthropoda, *Ann. Rev. Entomol.,* **8,** 115-136.

Lawrence, R. F., 1953. *The Biology of the Cryptic Fauna of Forests with Special Reference to the Indigenous Forests of South Africa.* Cape Town: A. A. Balkena, 408p.

Loksa, I., 1966. *Die Bodenzoozönologischen Verhältüisse der Flaumeichen-Buschwälder Südostmitteleuropas.* Budapest: Akademaia Kiadó, 437p.

Macfadyen, A., 1957. *Animal Ecology: Aims and Methods.* London: Sir Isaac Pitman and Sons Ltd., 264p. (2nd ed, 1963, 344p.)

Macfadyen, A., 1962. Soil arthropod sampling, *in* J. B. Cragg, ed., *Advances in Ecological Research,* **1,** 1-34.

Moment, G. B., 1958. *General Zoology.* Boston: Houghton Mifflin, 631p.

Müller, G., 1965. *Bodenbiologie.* Jena: Gustav Fischer, 889p.

Müller, P. E., 1879. Studier over Skovjord, som bidrag til skovdyrkingens theori. I. Om bogemuld og bøgemor paa sand og ler, *Tidsskr. Skovbr.,* **3,** 1-124.

Müller, P. E., 1884. Idem. II. Om muld og nor i egeskove og paa heder, *Tidsskr. Skovbr.,* **7,** 1-232.

Murphy, P. W., 1953. The biology of forest soils with special reference to the mesofauna or meiofauna, *Jour. Soil Sci.,* **4,** 155-193.

Murphy, P. W., 1955. Ecology of the fauna of forest soils, *in* D. K. McE. Kevan, ed., *Soil Zoology.* London: Butterworths, 99-124.

Murphy, P. W., ed., 1962. *Progress in Soil Zoology.* London: Butterworths, 398p.

Nef, L., 1957. Etat actuel des connaissances sur le rôle des animaux dan la decomposition de litières des forêts, *Agricultura,* **5**(2), 245-316.

Nosek, J., 1957. *Poznamky k ekologii pudny fauny hlediska biologie pudy.* Bastishna: Slovenskej Akadémie Vied v Bratislava; 154p.

Palissa A., 1964. *Bodenzoologie. Wisenschaftliche Taschenbücher, 17.* Berlin: Akademie-Verlag, 180p.

Pesson, P., ed., 1971. *La Vie dans les Sols.* Paris: Gauthier-Villars, 471p.

Phillipson, J., ed., 1971. *Methods of Study in Quantitative Soil Ecology; I.B.P. Handbook No. 18.* Oxford: Blackwell, 297p.

Rapoport, E., ed., 1966. *Progresos en Biología del*

Suelo: Actas del Primero Colloquio Latinamericano de Biología del Suelo, Bahia Blanca (Argentina) 13-17 Octubre 1965. Montevideo: Centr. Coop. Cient. UNESCO para América Latina, 715p.

Richards, B. N., 1974. *Introduction to the Soil Ecosystem.* Harlow: Longmans, 266p.

Romell, L. G., 1935. An example of myriapods as mull formers, *Ecology,* **16,** 67-71.

Russell, E. J., 1957. *The World of the Soil.* London: Collins, 237p.

Russell, E. J. and Russell, E. W., 1973. *Soil Conditions and Plant Growth,*10th ed. New York: Wiley, 849p. [See also earlier editions to 9th, 1950. 635p.]

Savory, T. H., 1971. *Biology of the Cryptozoa.* Watford, England: Merrow, 48p.

Schaller, F., 1962. *Die Unterwelt des Tierreiches. Verständl. Wiss. No. 78.* Berlin: Springer-Verlag, 126p.

Schaller, F., 1968. *Soil Animals.* Ann Arbor: University of Michigan Press, 145p.

Solomon, M. E., 1953. Insect population balance and chemical control of pests, *Chem. Ind. (London),* **1953,** 1143-1147.

Stöckli, A., 1946. Die biologische Komponente der Vererdung der Gäre und der Nährstoffpufferung, *Schweiz. Landwirtsch. Monatsh.,* **24,** 3-19.

Sturm, H., 1959. Die Nahrung der Pluturen, *Naturwissenschaften,* **46,** 90-97.

Tikhomirov, B. A., ed., 1973. *Biogeotsenozy̆ Taimy̆rskoi Tundry̆ i ikh Produktivnoct',* vol. 2. Leningrad: Nauka, 207p.

Tischler, W., 1955. *Synökologie der Landtiere.* Stuttgart: Fischer, 414p.

Vaněk, J., ed., 1975. *Progress in Soil Zoology. Proc. 5th Internat. Colloquium on Soil Zoology Held in Prague, September 17-23, 1973.* Prague: Junk, The Hague & Academia, 630p.

Wallwork, J. A., 1970. *Ecology of Soil Animals.* New York: McGraw-Hill, 283p.

Wallwork. J. A., 1976. *The Distribution and Diversity of Soil Fauna.* London: Academic, 355p.

White, G., 1789. *The Natural History and Antiquities of Selborne.* London: B. White & Son, (various editions), 468.

Wilde, S. A., 1954. Forest humus: Its genetic classification. Trans, *Wisconsin Acad. Sci.,* **43,** 137-167.

Winston, P. W., 1956. The acorn microsere with special reference to arthropods, *Ecology,* **37,** 120-132.

Witkamp, M., and Drift, J. van der, 1961. Breakdown of forest litter in relation to environmental factors, *Plant and Soil,* **15,** 295-311.

Cross-references: *Enzyme Activity; Humus; Metabolic Activity; Microhabitats; Soil Biochemistry; Soil Microbiology.*

FERTILIZERS, INORGANIC

Fertilizers are added to soils to improve their nutrient status. Although definitions of fertilizer materials differ from one region to another, the five legal classes for California (California Fertilizer Association, 1965) may be used as examples. *Agricultural minerals* are mineral substances or mixtures of mineral and organic materials that contain less than 5% available nitrogen (N), phosphorus (P), and potassium (K), singly or in combination. These materials do not include sand or soil. *Auxiliary soil chemicals* are synthetic polyelectrolytes, biocatalytic soil conditioners, lignin or humus preparations, wetting agents, bacterial inoculants, and other products designed to modify chemical and physical properties of soils. *Commercial fertilizers* contain 5% or more nitrogen (in any available form), available phosphorus (in terms of phosphorus pentoxide), and soluble potassium (in terms of potassium oxide), singly or in combination. Commercial fertilizers do not include manure, hay, straw, peat, or leaf mold. *Manure* consists of the excreta of any domestic animal or fowl; manure is not artificially mixed with any material except that which has been used as bedding for sanitary or feeding purposes or for the preservation of the manure (see *Fertilizers, Organic*) *Soil amendments* include hay, straw, peat, leaf mold, sand, diatomite, vermiculite, perlite, pumice, clay, and various potting mediums intended for use solely on the basis of their physical properties (see *Soil Conditioners*). California law stipulates tolerance limits, registration procedures, and penalties in case of violation.

World Use of Fertilizers

Worldwide use of the nutrient elements nitrogen (N), phosphorus (P), and potassium (K) in fertilizers is shown in Figure 1 (FAO, 1972). A rather steady increase in fertilizer consumption was interrupted by the Great Depression of the early 1930s. Subsequent to World War II, however, production and consumption have increased greatly, especially that of nitrogen. Prior to the war, more phosphorus and potassium than nitrogen were being used, but by 1970 the amount of nitrogen used was more than twice as large as those of the other two elements. Production of nitrogen, phosphorus, and potassium fertilizers exceeds consumption in developed countries, whereas the reverse is true in developing countries. Production and consumption are about equal except for potassium fertilizers in centrally planned countries (Figure 2). The total quantities of nitrogen fertilizer used worldwide have increased tenfold in the last 25 years and now greatly exceed the amounts of phosphorus and potassium fertilizers (Tables 1 and 2 on page 164).

Fertilizer application per unit area differs widely among countries, as shown in Table 1. Applications of nitrogen in the Netherlands are 440 kg/ha on arable land and 176 kg/ha on

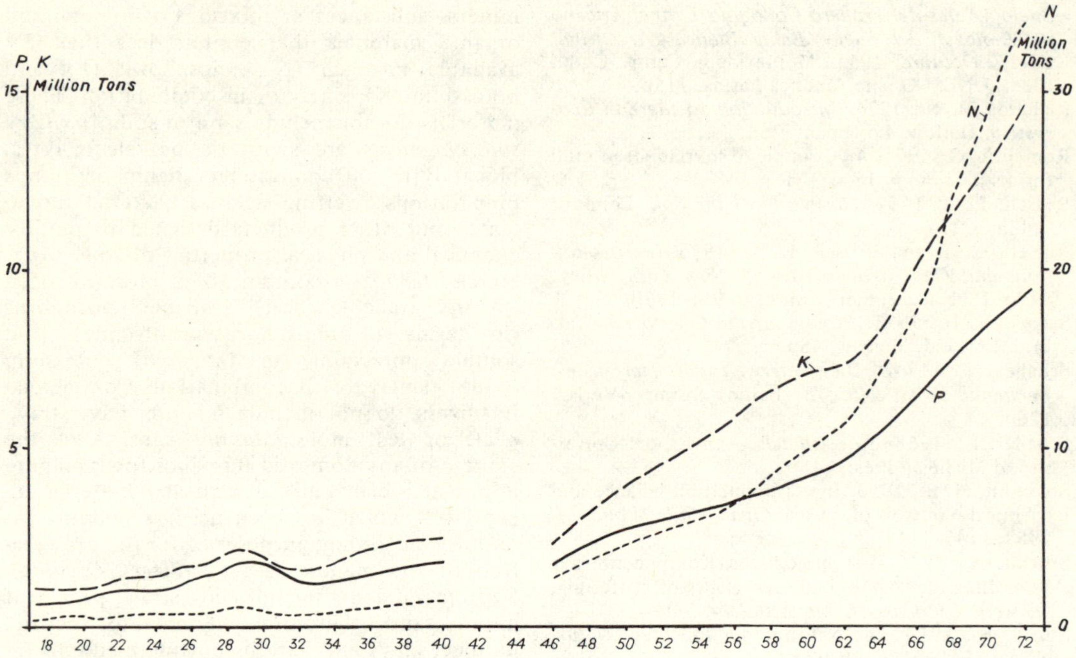

FIGURE 1. World consumption of nitrogen (N), phosphorus (P), and potassium (K) fertilizers. Drawn from recalculated data, plotted in elemental form. (FAO, 1972)

all agricultural land. The corresponding rates in Australia are 2.8 and 0.3 kg/ha. Generally, rates of application are higher in developed than in developing countries, although there are exceptions (see FAO 1972, and earlier editions). For the most part, applications of nitrogen fertilizers are now higher than those of phosphorus and potassium fertilizers (Tables 1 and 2.) Exceptions are related to local soils and agricultural practices. In New Zealand, for example, the use of nitrogen fertilizers is low as compared to phosphorus and potassium fertilizers. Consumption is higher for potassium than for nitrogen fertilizers in Ireland and Switzerland. Consumption is lower for phosphorus than for potassium fertilizers except in Oceania, New Zealand, and Australia.

Approximately one-fourth of the fertilizer used in the world each year is applied to soils in the United States (Table 2). In that country costs of fertilizer remained nearly constant during the 1950s and 1960s. Since 1973, however, the cost of nitrogen fertilizers has more than doubled because of increased costs of fuel.

Tests for Fertilizer Needs

Fertility maintenance and improvement must rest on information as to what fertilizers are needed, when, and at what rates. In practice, a farmer may rely entirely on past experience at the one extreme, or on sophisticated tests of various kinds at the other. Much effort has therefore gone into development of tests to measure the nutrient status of soils (see *Soil Tests*). Tests are chiefly chemical analyses and bioassays (Walsh and Beaton, 1973). Both kinds of tests are in use today, although neither is completely successful. Both are discussed in this section.

Chemical Analyses of Soil Samples. The nutrient status of soils depends on what can be called the *intensity* and *capacity factors.* Determination of these two factors is both difficult and uncertain. Many empirical tests have been devised, however, in efforts to measure the intensity and capacity factors. Such efforts have been directed especially at measuring phosphate status. Correlation of test results with field experience in the use of fertilizer on identified soils has permitted recommendations for fertilizer applications that have been quite useful for limited ranges of soils and crops.

The *intensity factor* is usually expressed as the nutrient concentration in the soil solution (q.v.). For practical purposes, it is determined in an equilibrium solution, but the water-to-soil ratio differs among investigators. Because nutrient concentrations in the soil solution cannot necessarily be inferred from extracts made at unnaturally high moisture contents, several

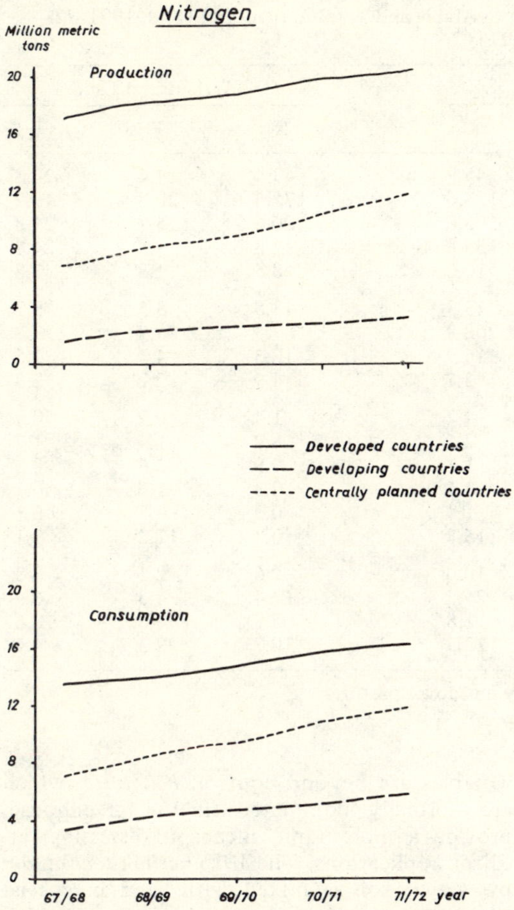

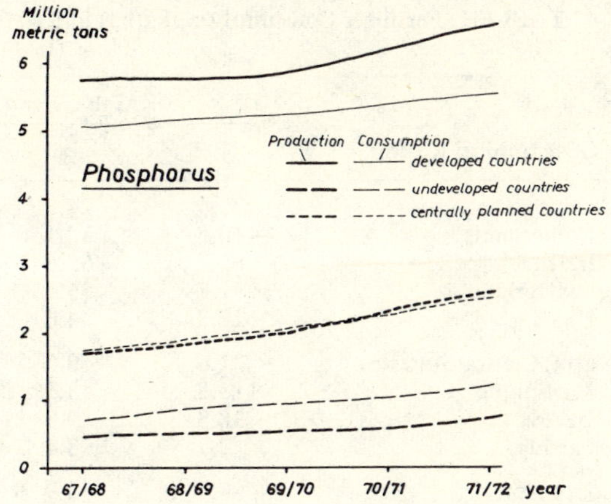

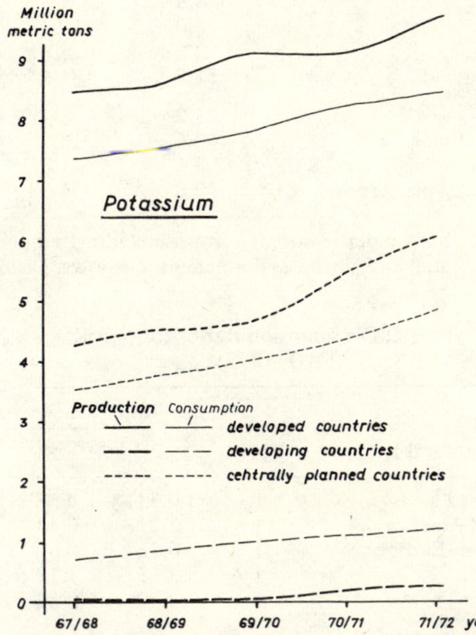

FIGURE 2. Production and consumption of fertilizers in developed, developing, and centrally planned countries: (a) nitrogen (N); (b) phosphorus (P); and (c) potassium (K). All values are given in elemental form.

methods have been developed to obtain samples of the soil solution under more nearly natural conditions through use of water-displacement and pressure-membrane apparatus.

The *capacity factor* represents reserves that can become available for plant growth and indicates how well a soil withstands depletion of nutrients. The capacity factor varies with the intensity factor in a way that can be determined only by experiment. Theoretically, information on the entire range of the capacity factor should be available so that it can be expressed like a buffer curve. In practice, however, a single-value expression for the capacity factor is commonly used. For example, soluble and readily exchangeable potassium represents the potassium reserve that can be made available for plant growth, whereas the potassium ion concentration (or its chemical potential) would be a measure of the intensity factor. The capacity factor is thus determined as the sum of soluble and exchangeable potassium ions. In other cases where many different compounds contribute to the supply, as is the case with phosphates, a single-value expression for the capacity factor is more difficult to obtain and less reliable. The number of extraction techniques for such nutrients is therefore constantly increasing (Tisdale and Nelson, 1966).

Biotests. Several biotests, with many modifications, have been developed to assess soil fertility. They are conducted in the field, in the greenhouse, or in the laboratory. With *field-test plots* fertilizers are added directly to soils in the field in different amounts and combinations. The plots are selected according

TABLE 1. Fertilizer Consumption (kg/ha) in Relation to Arable and Total Agricultural Land, 1971–72. (FAO, 1973)

Geographical Region	Arable Land N	Arable Land P	Arable Land K	Agricultural Land N	Agricultural Land P	Agricultural Land K
Europe	70.6	24.3	45.9	43.4	14.4	28.4
Netherlands	440.6	52.1	123.6	175.6	20.8	49.2
Ireland	85.7	65.6	103.6	20.6	15.7	24.8
Switzerland	97.7	55.4	139.8	17.3	9.8	24.8
USSR	22.3	4.6	10.0	8.5	1.8	3.8
North, Central America	31.6	8.0	13.3	13.5	3.5	5.7
Martinique	113.8	45.2	208.9	67.3	26.7	123.7
USA	38.3	9.9	16.9	16.3	4.3	7.5
Canada	7.7	3.4	3.7	4.9	2.2	2.3
South America	7.4	3.9	4.7	1.4	0.7	0.9
Asia	17.2	2.9	2.5	8.2	1.4	1.2
Africa	4.6	1.3	1.1	0.9	2.6	0.3
Oceania	3.0	10.4	3.4	0.3	1.0	0.3
Australia	2.8	7.6	1.4	0.3	0.7	0.1
New Zealand	11.9	181.6	115.8	0.7	11.2	7.1
World	23.2	6.4	10.0	7.6	2.1	3.2
Developed	40.6	13.9	21.3	12.4	4.3	6.5
Developing	8.6	1.8	1.8	2.8	0.6	0.6
Centrally planned	30.0	6.3	12.3	10.7	2.3	4.4

Arable land: under temporary crops, including temporary meadows, pastures.
Agricultural land: includes permanent meadows, pastures.

TABLE 2. Consumption Ratios for N:P:K (FAO, 1973)

Geographical Region	Ratio N	Ratio P	Ratio K
Developed areas	1	0.34	0.52
North America	1	0.27	0.44
Western Europe	1	0.36	0.60
Oceania	1	3.63	1.19
Others	1	0.38	0.52
Developing areas			
Africa	1	0.34	0.37
Latin America	1	0.31	0.39
Near East	1	0.18	0.03
Far East	1	0.14	0.17
Others	1	0.06	0.04
Centrally planned areas	1	0.21	0.41
Asia	1	0.14	0.04
Europe, USSR	1	0.24	0.56
World	1	0.28	0.43
World 1967–68	1	0.37	0.56
1968–69	1	0.30	0.46
1969–70	1	0.29	0.45
1970–71	1	0.28	0.43
1971–72	1	0.28	0.43

to some statistical pattern. The main advantage of field tests is that crops are grown under normal conditions in relatively large areas with few boundary effects. On the other hand, many variables are beyond control. *Pot tests,* which are normally conducted in the greenhouse, provide a more rapid means of assessing fertilizer applications. The *Mitscherlich technique,* by which soil is mixed with sand in special pots, and the *Neubauer technique,* by which nutrients are extracted by 100 rye seedlings from a soil-sand mixture in a definite ratio, are both used in certain countries (Stewart, 1932).

Microbiological Assays. The growth of *Aspergillus niger* is frequently used as a test for zinc, iron, molybdenum, copper, and potasssium and phosphorus. *Azotobacter* have been used in assays for calcium, phosphorus, and potassium. *Algae* have been used as test organisms for nitrogen, phosphorus, and potassium.

Diagnosis at the Growing Plant

Deficiency symptoms, which are often typical for a given element, occur when the supply is too low for satisfactory growth. By the time the symptoms appear, however, the time for corrective measures is usually past (Sprague, 1964).

Plant composition provides much information on the amounts of nutrient elements removed by crops (Table 3) but does not serve well as a basis for estimating fertilizer needs. The quantities taken up by plants range from those smaller than optimum for growth to luxury consump-

TABLE 3. Nutrient Removal in Elemental Form by Crops (kg/ha).

Crop		Nutrient N	P	K
Cereals				
Barley	Grain	125	20	35
	Straw	45	5	105
	Total	170	25	140
Corn	Grain	150	25	35
	Stores	75	10	150
	Total	225	35	185
Oat	Grain	75	10	20
	Straw	40	5	95
	Total	115	15	115
Rice	Grain	80	10	20
	Straw	35	5	45
	Total	115	15	65
Sorghum	Grain	85	15	30
	Stores	105	15	170
	Total	190	30	200
Sweet corn	Ears	30	5	15
	Stores	140	20	100
	Total	170	25	115
Wheat	Grain	85	15	15
	Straw	55	5	90
	Total	140	20	105
Tree fruits				
Apples	Fruit	55	5	65
	Leaves	75	10	60
	Total	130	15	125
Apricots	Fruit	50	5	65
	Leaves	90	5	75
	Total	140	10	140
Grapes	Fruit	50	15	70
	Leaves	65	5	45
	Total	115	20	115
Oranges	Fruit	95	15	130
	Leaves	40	5	35
	Total	135	20	165
Peaches	Fruit	40	5	75
	Leaves	70	15	60
	Total	110	20	135
Pears	Fruit	40	5	40
	Leaves	55	5	45
	Total	95	10	85
Prunes	Fruit	60	10	90
	Leaves	40	5	35
	Total	100	15	125
Forage plants				
Alfalfa		375	35	250
Bluegrass		90	15	75
Clover (timothy)		210	30	165
Bermudagrass		640	70	370
Corn silage		225	40	230
Fescue, tall		200	40	260
Grasses, average		95	15	85
Johnson grass		1000	95	585
Ladino clover		220	20	165
Lespedeza		145	15	65
Orchardgrass		200	30	185
Clover red		180	20	130
Vetch		435	50	300
Vegetables				
Asparagus		105	25	110
Beans	Pods	120	5	15
	Leaves, stems	55	5	45
	Total	175	10	60
Beans, snap	Beans	985	15	55
	Leaves, stems	215	5	250
	Total	1200	20	305
Broccoli		90	15	70
Cabbage		250	15	195
Cauliflower		125	15	70
Celery		260	90	625
Cucumber	Fruit	20	5	30
	Leaves, stems	50	5	50
	Total	70	10	80
Kale		65	10	50
Lettuce		100	15	170
Peas	Peas	35	5	15
	Leaves, stems	50	5	35
	Total	85	10	50
Pepper	Fruit	20	5	20
	Leaves, stems	30	10	30
	Total	50	15	50
Rutabagas		65	10	50
Spinach		135	20	55
Squash	Fruit	20	5	35
	Leaves, stems	75	5	80
	Total	95	10	115
Tomatoes	Fruit	190	25	335
	Vines	90	10	110
	Total	280	35	445
Rootcrops, Tubers, Bulbs				
Beets, table	Roots	110	5	110
	Tops	145	5	75
	Total	255	10	185
Carrots	Roots	100	15	95
	Tops	120	5	175
	Total	220	20	270
Onions	Bulbs	90	20	90
	Tops	15	5	30
	Total	105	25	120
Potatoes	Tubers	125	15	185
	Vines	100	10	100
	Total	225	25	285
Sweet potatoes	Roots	85	10	115
	Vines	45	5	45
	Total	130	15	160
Turnip	Roots	85	10	75
	Tops	135	5	50
	Total	220	15	125
Miscellaneous Crops				
Cotton	Lint	85	15	35
	Burrs, leaves, stalks	55	5	45
	Total	140	20	80
Flax	Grain	40	5	5
	Straw	45	5	10
	Total	85	10	15
Peanuts	Nuts	125	5	20
	Vines	125	15	95
	Total	250	20	115
Pumpkin	Fruit	60	5	65
	Leaves	40	5	35
	Total	100	10	100
Rape		130	30	130
Strawberries	Fruit	15	5	20
	Vines	95	15	55
	Total	110	20	75
Sugar cane	Stalks	225	50	320
	Tops	175	25	230
	Total	400	75	550
Tobacco	Leaves	65	5	85
	Stalks	35	5	95
	Total	100	10	180

(Compiled from Tisdale and Nelson, 1966).

tion when large amounts of a nutrient element are readily available.

Foliar analysis, which measures amounts of individual nutrient elements in leaf tissue, provides a better measure of the nutrient status and needs of the plant during growth than can the total amounts absorbed. The curves in Figure 3 are a schematic representation of yield as a function of nutrient content in the leaf tissue. Note that as the content increases in the deficiency range, the yield also increases. At relatively high foliar contents, yields remain constant over the range of luxury consumption. Yield is depressed again at very high nutrient levels when amounts are large enough to become toxic to some degree. The inflection in the curve where the amount of a nutrient element shifts from deficient to adequate is called the *critical level;* it is often taken as the nutrient concentration at which the yield is 90% of the maximum.

In foliar diagnosis a calibration curve, as shown in Figure 3, is established in greenhouse experiments for a particular nutrient and a certain crop. The same types of tissues are then sampled in the field, analyzed, and the nutrient concentrations compared to the previously established calibration curve. If foliar analyses fall on the ascending branch of the curve, a nutrient deficiency is indicated and remedies are necessary. Periodic analyses of tissues in the luxury consumption range can indicate the approach of deficient levels and how soon corrective measures must be made before production will be reduced (see *Law of the Minimum*).

Specific Fertilizers and Their Reactions in Soils (Table 4)

Nitrogen Fertilizers. *Anhydrous ammonia,* with 82.2% nitrogen, is the most concentrated nitrogen fertilizer. It can be added to irrigation water, applied directly in soils, or used in formulating mixed fertilizers. When applied to soils, preferably 10–15 days before planting, it must be buried immediately to a depth of at least 10–15 cm to avoid evaporation losses. Although anhydrous ammonia initially inhibits nitrification, it is eventually transformed to nitrate. *Aqua ammonia* is produced by dissolving anhydrous ammonia in water to give about 20% nitrogen. Due to the low vapor pressure, it is easy to handle and requires inexpensive equipment. *Ammonium sulfate,* which also has good handling properties, supplies sulfur as a secondary nutrient; the acidifying property of this fertilizer can, however, be a disadvantage. *Ammonium chloride* is used in rice fields where sulfate might be reduced to hydrogen sulfide, a compound that may be a cause of Akiochi (late summer) disease. In normal crops the chloride is more likely to cause salt injuries than the sulfate. *Ammonium nitrate,* a quick-acting

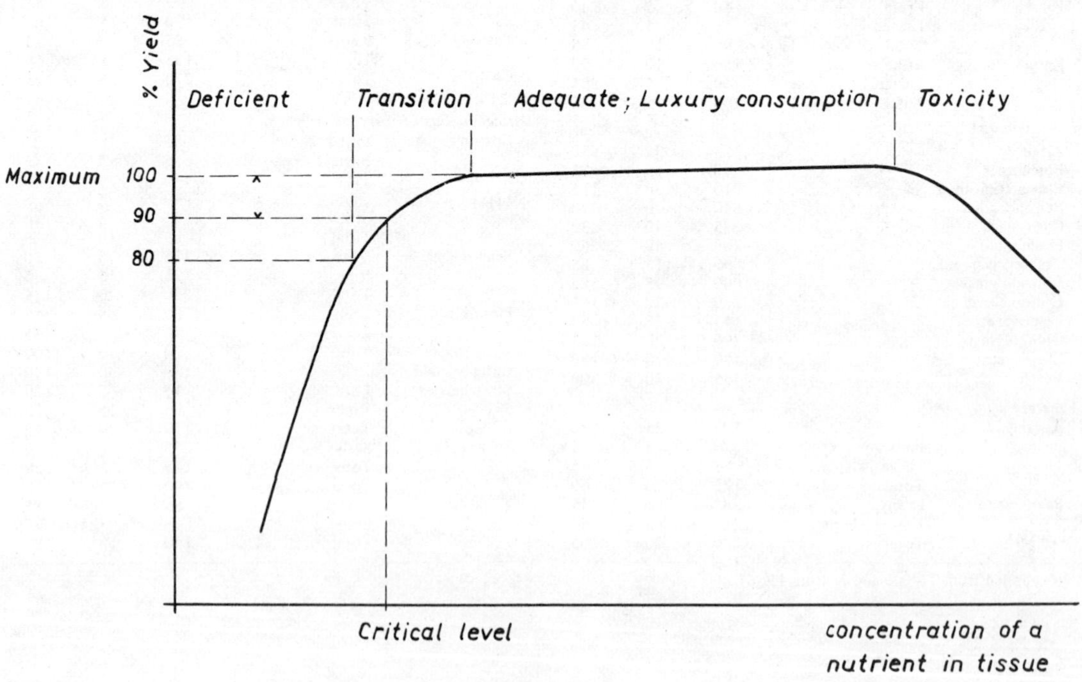

FIGURE 3. Schematic calibration curve for foliar diagnosis.

TABLE 4. Elemental Composition of Fertilizers.

Fertilizer Material	Formula	Elemental Composition N	P	K	Ca	Mg	S	Cl
Anhydrous ammonia	NH_3	82.2						
Aqua ammonia	NH_4OH	20						
Ammonium sulfate	$(NH_4)_2SO_4$	20.5					23.4	
Ammonium chloride	NH_4Cl	28.0						67
Ammonium nitrate	HN_4NO_3	32.5						
Ammonium nitrate + lime		20.0			7.1	4.1	0.6	
Ammoniated ordinary superphosphate		3–6	7		16.4	0.3	10.0	0.3
Monoammonium phosphate	$NH_4H_2PO_4$	11	20.9		1.4	0.3	2.6	
Diammonium phosphate	$(NH_4)_2HPO_4$	20	23					
Ammonium phosphate Ammonium sulfate	e.g. 40% $NH_4H_2PO_4$ 60% $(NH_4)_2SO_4$	16.5	8.7				15.4	
Nitric phosphates		14–20	6–8.7		5.7–7.1		0.4	
Calcium cyanamid	$CaCN_2$	22.0			38.6		0.6	
Calcium nitrate	$CA(NO_3)_2$	15.5			19.3	1.5		0.2
Sodim nitrate	$NaNO_3$	16.0						0.6
Potassium nitrate	KNO_3	13.4		37	0.7	0.3	0.2	1.2
Urea	$(NH_2)_2CO$	46.0						
Urea sulfur		40					10	
Urea formaldehyde		38						
Urea ammonium Urea nitrate solution	35.4% $(NH_2)_2CO$ 44.3% NH_4NO_3	32						
Ammoniumpolyphosphate		15	25					
Magnesium ammonium phosphates		8	17			14		
Superphosphate single	50% $CaSO_4$ 30% $CaH_4(PO_4)_2$		7.9–8.7		13–15		12	
Superphosphate double, triple	$CaH_4(PO_4)_2$		18–20		9–10		1	
Liquid phosphoric acid	H_3PO_4		23–24		0.2			
Superphosphoric acid			34					
Dicalcium phosphate	$CaHPO_4$		23		29			
Potassium phosphate			18–22		29–45			
Acidulated bonephosphate			6–6.5					
Rock phosphate			11–17		33–36			
Basic slag			3.5–8		32	3	0.2	
Defluorinated P Phosphate rock			9		20			
Magnesium silicate glass			10		20	8.4		
Rhenania phosphate			10		30	0.3		
Calcium metaphosphate	$Ca(PO_3)_2$		27		19			
Potassium metaphosphate	KPO_3		24–25	29–32				
Potassium carbonate	K_2CO_3			56				
Potassium bicarbonate	$KHCO_3$			39				
Potassium chloride	KCl			52				48
Potassium sulfate	K_2SO_4			44			18	
Sulfate of potash magnesia	$K_2SO_4 \cdot 2MgSO_4$		2	18–22	0.7	11	11–15	

(Compiled from Tisdale and Nelson, 1966).

source of nitrogen that is also long lasting, is hygroscopic and must be stored properly. If combined with combustible materials, it forms an explosive mixture. Lime is sometimes added to counteract the acidification that results from the oxidation of ammonium to nitrate in soils. *Sodium nitrate,* as Chilean nitrate, was at one time a major source of fertilizer nitrogen. Because it is derived from natural deposits, Chilean nitrate also contains some micronutrients, whereas synthetic sodium nitrate is largely devoid of trace elements. *Calcium nitrate,* in spite of its low nitrogen content, is occassionally preferred because of the beneficial effects of calcium. *Calcium cyanamid,* obtained from the reaction of N_2 with calcium carbide, releases nitrogen in the soil through a stepwise uptake of water:

$$CaCN_2 \rightarrow Ca(HCN_2)_2 \rightarrow H_2CN_2 \rightarrow CO(NH_2)_2$$

Because some intermediates in this reaction sequence are toxic, this fertilizer must be applied two to three weeks before a crop is planted. Calcium cyanamid can also be used as a weed killer. *Ammonium nitrate-sulfate,* which has a variable composition, is an excellent source of nitrogen and sulfur, supplying both simultaneously. It has good handling properties and stores well. *Ammonium phosphates* are good suppliers of nitrogen and phosphorus and have excellent physical properties. *Nitric phosphates* are adequate for most purposes but rather expensive. *Potassium nitrate* supplies potassium and nitrogen simultaneously, though the nitrogen content is only 13.4%. *Metal ammonium phosphates,* especially those containing the magnesium salt, have been used as low-level (8% nitrogen), slow-release fertilizers for nitrogen. *Urea* is a nitrogen-rich (46% nitrogen) organic fertilizer that is rapidly broken down in soils to ammonium ions, which are subsequently converted to nitrates. After rapid hydrolysis of surface-applied urea, some ammonia may be lost to the atmosphere. Biuret, an important condensation product of urea at high temperatures, is injurious to plants, but most commercial fertilizers do not contain toxic amounts. Urea has been used successfully as a foliar applicant in orchards. This fertilizer is hygroscopic, making it difficult to handle. The reaction of urea with formaldehyde produces *urea-formaldehyde,* a long-chain polymer that releases nitrogen slowly. Many other products have been used as a source of nitrogen for plant growth. Table 5 lists some of the more important organic forms (see *Fertilizers, Organic*). Still other products such as ammoniated sawdust have also been tried (Jacob, 1953).

Reactions of Fertilizer Nitrogen in Soils. The added nitrogen takes part in various reactions, depending partly on the form in which the element is added and partly on the nature of the soil. Thus, for example, added ammonia (NH_3) is usually converted to nitrate (NO_3) within a few weeks, although some ammonium ions are absorbed by the exchange complex, and others

TABLE 5. Average Composition of Some Common Natural Organic Materials.

	Percent Composition						
Organic Material	N	P	K	Ca	Mg	S	Cl
Activated sewage sludge	6.0	1.0		1.8	0.9	0.4	0.5
Blood, dried	13.0			0.4			0.6
Bone meal (raw)	3.5	19.8		22.5	0.6	0.2	0.2
Bone meal (steamed)	2.0	12.2		23.6	0.3	0.2	
Castor pomace	6.0	0.6	0.4	0.4	0.3		0.3
Cocoa meal	4.0	0.6	2.1	0.4	0.6		
Cocoa shell meal	2.5	0.4	2.5	1.1	0.3		
Cocoa tankage	2.5	0.6	1.0	12			
Cottonseed meal	6.6	1.1	1.2	0.4	0.9	0.2	
Fish scrap (acidulated)	5.7	1.3		6.1	0.3	1.8	0.5
Fish scrap (dried)	9.5	2.6		6.1	0.3	0.2	1.5
Garbage tankage	9.5	0.6	0.8	3.2	0.3	0.4	1.3
Peanut meal	7.2	0.6	1.0	0.4	0.3	0.6	0.1
Peanut hull meal	1.2	0.2	0.7				
Peat	2.7			0.7	0.3	1.0	1.1
Peruvian guano	13.0	5.5	2.1	7.9	0.6	1.4	1.9
Process tankage	8.2			0.4		0.4	
Soybean meal	7.0	0.5	1.3	0.4	0.3	0.2	
Tankage, animal	7.0	4.3		11.1	0.3	0.4	0.7
Tobacco stems	1.5	0.2	4.2	3.6	0.3	0.4	1.2
Whale guano	8.5	2.6		6.4	0.3		
Manure Source							
	N	P	K	Organic			
Dairy manure	0.7	0.1	0.5	30			
Goat manure	2.8	0.6	2.4	60			
Hog manure	1.0	0.3	0.7	30			
Horse manure	0.7	0.1	0.4	60			
Poultry	1.6	0.5	0.8	50			
Rabbit	2.0	0.6	1.0	50			
Sheep	2.0	0.4	2.1	60			
Steer	2.0	0.2	1.6	60			

Recalculated from Tisdale and Nelson (1966).

are fixed by certain clay minerals and so persist unchanged in soils. With the conversion of ammonia to nitrates, the acidity of a soil is increased. The greater acidity (lower pH) can be desirable in soils that are alkaline in reaction, but lime must be added in most soils of humid regions to offset the increased acidity that follows long continued use of nitrogen fertilizers. Conversion of ammonia to nitrate has a second effect in making the element more readily subject to loss from the soil by leaching.

Part of the added nitrogen, largely regardless of its original form, is absorbed sooner or later by microorganisms and incorporated into their tissues. These in turn become part of the soil organic matter, natural compounds in which nitrogen can be stored for long periods. Efforts to supplement the natural compounds with synthetic organic fertilizers have met with little success. Through the activities of microorganisms in soils, the natural compounds are decomposed and the nitrogen mobilized in a form that can be absorbed by plants or taken up again by microorganisms. The breakdown and resynthesis of soil organic matter occurs continuously and may result in net mobilization or immobilization, depending on the conditions in the soil (Bartholomew and Clark, 1965).

If aeration is poor, as in waterlogged soils, nitrates are commonly reduced, first to nitrites and finally to free nitrogen, which can then escape into the atmosphere as a gas (see *Nitrogen Cycle*).

Phosphorus Fertilizers. *Liquid phosphoric acid, green phosphoric acid,* produced by treating rock phosphate with sulfuric acid and separating the liquid phosphoric acid from the calcium sulfate, is occasionally used as a phosphorus fertilizer in irrigated agriculture when it can be added directly in the water. Phosphoric acid is also used in the manufacture of a number of solid fertilizers and liquid fertilizer mixtures. *Superphosphoric acid* is obtained by burning rock phosphate in an electric furnace to produce elemental white phosphorus, which reacts with water to give phosphoric acid. Enrichment with phosphorus pentoxide forms superphosphoric acid, which is actually a mixture of ortho-, pyro-, and polyphosphates. It is the basis for phosphorus-rich fertilizers, especially liquid mixtures. The high production costs are partly offset by lower storage and transportation expenses. *Superphosphates* are manufactured by treating rock phosphates with an acid so that different quantities of monocalcium phosphate are obtained:

$$Ca_3(PO_4)_2\ CaF_2 + 6H^+ \rightarrow Ca(H_2PO_4)_2 + 3Ca^{2+} + 2HF$$

Normal or single superphosphate is made by treating rock phosphate with sulfuric acid so that a mixture of monocalcium phosphate and gypsum in about equal amounts is formed. *Triple superphosphate,* also called "treble," "double," or "concentrated superphosphate," is manufactured by treating rock phosphate with phosphoric acid. *Nitric phosphates* are formed by treating rock phosphate with nitric acid. Ammonium or potassium can also be introduced as cations, if desired. Various N:P:K ratios are offered on the market. *Ammonium phosphates* are produced when ammonia is added to phosphoric acid, and various ratios of N:P are commercially available. *Ammonium polyphosphate* is a new fertilizer that is obtained by treating ammonia with superphosphoric acid. *Ammonium phosphate nitrates* are water-soluble mixtures of ammonium phosphates and ammonium nitrates. *Metal ammonium phosphates,* slow-release nitrogen-phosphorus fertilizers, are used for only special purposes. *Potassium phosphates* are completely water-soluble fertilizers that are sold mainly to home gardeners. *Dicalcium phosphate* is not widely used as a fertilizer but rather as a feed supplement. *Metaphosphates,* potassium and calcium salts of metaphosphoric acid, are high-phosphorus fertilizers that have low solubilities in water. Metaphosphates are slowly converted to orthophosphates in soils. *Rock phosphate* is frequently added directly to acid soils, but large quantities of finely ground rock must be applied. *Basic slag, Thomas slag,* is a by-product of steel manufacturing by the Bessemer and open-hearth processes. It is used widely in some countries, being about as efficient as superphosphate. *Bone products* are offered with various pretreatments. Because they are more expensive than other fertilizers, *bone meal* is used mainly in greenhouses and homes.

Reactions of Fertilizer Phosphorus in Soils. Fertilizer phosphates are gradually rendered unavailable in acid soils through the reaction with aluminum and iron compounds. In alkaline soils they are precipitated as calcium phosphates, which then undergo slow changes toward more insoluble forms (apatites). Highest solubilities of phosphates are observed at near neutral pH values. The fixation of phosphates is a serious problem in the management of many soils. Fixation effects can be reduced by local, concentrated placement of fertilizer and by adjustment of the application rate (Pierre, 1953).

Potassium Fertilizers. *Potassium chloride* (muriate of potash) is the most widely used potassium fertilizer in the United States. Potassium chloride is so soluble in water that sensitive crops can be injured by the fertilizer. *Potassium sulfate* (sulfate of potash) is preferred where chlorides may injure certain crops such as tobacco. This fertilizer has good handling properties and is frequently added to sulfur-deficient soils. *Potassium nitrate* supplies both potassium and nitrogen in highly soluble form as *potassium magnesium sulfate* does for potassium, magnesium, and sulfur. *Potassium phosphates* are excellent fertilizers, supplying both potassium and phosphorus. They are not widely used, however, because of high production costs. *Potassium hydroxide* is primarily applied as a special liquid fertilizer. *Manure salts* (kainit), potassium compounds with a high level of chlorine, are of uncertain composition and minor importance.

Reaction of Fertilizer Potassium in Soils. Potassium is adsorbed by the cation exchange complex, which greatly reduces leaching losses. In some soils the added potassium in fertilizers is, however, made less available through fixation between layers of certain clay minerals (Kilmer, Younts, and Brady, 1968).

Sulfur, Calcium, and Magnesium. The elements sulfur, calcium, and magnesium are considered secondary plant nutrients because they are needed less generally than nitrogen, phosphorus, and potassium. Nevertheless, these elements are essential to plant growth, and deficiencies do occur. Sulfur deficiency is alleviated by the addition of calcium sulfate (gypsum), elemental sulfur, sulfuric acid, or through mixed fertilizers that contain sulfur, e.g., ammonium sulfate. Adequate amounts of calcium are required in soils for satisfactory root growth and plant nutrition and also to help keep soils in good physical condition. Calcium deficiencies can be corrected by the addition of crushed limestone, dolomite, or some calcium-containing fertilizer such as calcium nitrate. Magnesium deficiencies can be corrected by adding dolomite (which contains both calcium and magnesium), magnesium oxide, magnesium sulfate (Epsom salts), or other fertilizers containing magnesium.

Micronutrients.

Deficiencies of micronutrients (q.v.) are of great local importance, and various means have therefore been found to correct or avoid them, including applications of the required elements to soils as salts or as components of fertilizers, in foliar sprays, as nails in tree trunks, and as chelates. Required quantities are so small in many instances that direct applications to soil are difficult and can be wasteful (Mortvedt, Giordano, and Lindsay, 1972). For example, needed molybdenum is in the range of 100 g/ha. Elements such as manganese, zinc, copper, and molybdenum have been added to soils in the form of salts. Those elements and copper have also been incorporated in small quantities into fertilizers.

Foliar sprays have been used successfully for applications of manganese, zinc, and boron. Zinc has also been added by driving galvanized nails into tree trunks and by drilling holes and filling them with salts. Special care is necessary in foliar applications of boron because of the narrow range between deficient and toxic levels. Applications of iron as salts in foliar sprays have not worked well.

In recent years correcting or avoiding iron deficiencies in plants has become much easier because of chelates. Chelates are compounds in which iron atoms are bound at two or more sites by organic molecules (ligands). Chelates can be applied to soils or used in foliar sprays. Iron-EDTA, ethylene-diamine-tetraacetic acid, has been in use since 1950. For calcareous soils, iron-EDDHA, ethylene-diamine-di (O-hydroxy-phenyl-acetic acid), is a much better source of iron. To be effective, an iron chelate must not decompose spontaneously and must also be resistant to microorganisms in soils. The iron-ligand must be more stable than competing ions over a considerable range in pH. At the same time the chelate must not be so stable that it cannot release iron to growing plants.

Small quantities of cobalt salts have been added to superphosphates now available in Australia and New Zealand. Direct applications of cobalt in foliar sprays are also made, a rate of 250 g/ha being enough generally to provide adequate levels in subterranean clover for grazing animals. Thus far cobalt has not been found essential for plant growth but it is essential to animals.

Fertilizer-Pesticide Combinations

Insects cause enormous crop damage by feeding on plants and acting as vectors of important diseases. Weeds cause losses in crop production amounting to several billion dollars per year in the United States alone. To reduce such losses, chemicals other than fertilizers are applied to both soils and plants. As a group, these are called *pesticides.* Depending on their functions, they are known as "insecticides," "herbicides," "fungicides," etc.

Combinations of fertilizers and pesticides are applied in some instances to effect labor savings. Mixtures may be no more than small amounts of fertilizer included in foliar pesticide sprays to add small quantities of nutrients that can be quickly absorbed. Combinations may not be suitable because both fertilizer and pesticide must be adjusted to the individual crop. Field mixing is often preferable to bulk mixing because amounts can then be adjusted to the specific crop needs. Chemicals of both kinds must be compatible so as to form homogeneous mixtures suitable for application. Other problems in mixing fertilizers and pesticides are health hazards from the latter. In the United States all fertilizer-pesticide combinations must be registered with the Environmental Protection Agency.

Controlled-release Fertilizers

The control of nutrient release from fertilizers assures plants of a continuous supply of nutrients, lowers leaching losses, reduces groundwater contamination, and increases efficiency of fertilizer utilization. It permits a single, labor-saving application of large quantities of nutrients without danger of salt injury.

Slow-release fertilizers are especially desirable for nitrogen and phosphorus, and to a less extent potassium. There are several methods of regulating the release of fertilizer nutrients, which include: coating fertilizer salts with a diffusion barrier between the salt and soil; using organic compounds that break down slowly to release nitrogen (urea formaldehydes, crotonylidene diurea, oxamid, dicyandiamid); applying inorganic compounds of low solubility (metal-ammonium phosphates, glassy frits for micronutrients); and adding processed waste products such as ammoniated sawdust or nitrogen enriched oxidized coal. Some control of the nutrient release is possible through use of inhibitors of microbial activity. Nitrification inhibitors, for example, reduce leaching losses by keeping nitrogen in the ammonium form. Compounds such as Guanyl thiourea and some substituted thiazoles and triazines are effective.

Controlled-release fertilizers must meet certain requirements in addition to those of regular fertilizers. The additional constituents, for example, must never produce toxic by-products and the release rate must be predictable, never exceeding the tolerance limits of a specified crop. A disadvantage of these fertilizers is that frequently the release of nutrients continues in the absence of crops, with a possibility of salinity problems (Olson et al, 1971).

Mixed Fertilizers

A *complete fertilizer* contains the three nutrient elements nitrogen, phosphorus, and potassium. In the United States laws require that fertilizer bags be clearly marked and that the contents of nitrogen, phosphorus, and potassium be given in sequence with phosphorus and potassium in terms of P_2O_5 and K_2O, respectively. Thus a 10:20:10 fertilizer contains 10% nitrogen, 20% P_2O_5, and 10% K_2O. Elemental compositions can be calculated from the following relations:

$$P = \frac{1}{2.3} P_2O_5$$

$$K = \frac{1}{1.2} K_2O$$

Mixed fertilizers allow more uniform and balanced fertility management with reduced application costs. During the past twenty years the use of bulk-blended dry fertilizers has greatly increased. Many fertilizers are compatible for blending, but some materials should not be mixed together such as urea and ammonium nitrate, which form a very hygroscopic mixture. Materials that are blended together must have similar particle-size distributions to avoid resegregation; particle shape and density are less critical factors.

Mixed liquid fertilizers have come into the picture only recently. The mixtures are produced by either dissolving various fertilizer materials in water or by neutralizing phosphoric acid with anhydrous ammonia followed by additions of more nitrogen and potassium carriers. Suspension fertilizers (slurries) are liquid fertilizers in which the nutrients are stored and shipped in rather concentrated form. Liquid fertilizers permit uniform application of a homogeneous mixture of soluble nutrients. Labor-saving combinations with pesticides are feasible, though they are less suitable for specific local conditions.

Timing and Rate of Fertilizer Applications

A sufficient supply of nutrients should always be available to permit optimum growth of crops (Barber, 1976). Thus the timing of fertilizer additions must take into account the type of fertilizer to be used, the crop, the soil, and the prevailing climatic conditions. Leaching losses from sandy soils are reduced with small but frequent applications of fertilizers. In many cases, however, only a single initial fertilization is possible because applications are not feasible after the crop is established. For some crops fertilization at exactly the right stage of growth is important. The timing of applications should be adjusted to crop needs insofar as feasible within farm operations. To facilitate their application, many dry fertilizers are now granulated or formed into pellets (Ignatieff and Page, 1958; Cooke, 1975).

To determine the rate and timing of fertilization, a farmer needs to appraise the release of nutrients from reserves in the soil; the removal by crops; and the expected losses from leaching (q.v.), immobilization (fixation), erosion, denitrification (q.v.), and uptake by weeds. All these factors are considered in terms of crop growth. Economic factors are also important and may be controlling.

Natural reserves of nutrient elements differ greatly among soils. Some are low in nitrogen, phosphorus, and potassium, whereas others are high, usually in one or two rather than all three elements. Decomposition of the native organic matter releases enough nitrogen for crops over a period of decades when soils such as the Mollisols of the Corn Belt of the United States are first plowed. Other soils such as the Ultisols along the Atlantic and Gulf Coasts contain little native organic matter. Additions of nitrogen are beneficial in the first years of cultivation. Additions are necessary for maximum crop production on many soils. Reserves of phosphorus and potassium, like those of nitrogen, are low in the Ultisols and relatively high in the Mollisols and also in the Aridisols of the dry southwestern parts of the country. Reserves of those two elements in soils depend greatly on the original levels in the parent rock. Thus many soils in the Bluegrass region of Kentucky are derived in part from phosphatic rock and are high in phosphorus.

Management practices can reduce the loss of plant nutrients through leaching, erosion, and uptake by weeds. The timing of applications in relation to growth of the crop, the placement of fertilizers in the soil, and the control of weeds bear directly on the efficiency of fertilizer use.

Placement of Fertilizers

The placement of fertilizers in the soil is important to maximizing utilization of applied nutrients. Through proper placement, losses can be kept to a minimum, and a larger share of the applied nutrients kept in forms available to plants.

A number of different methods of fertilizer application are widely used. *Broadcasting* (topdressing) consists of the uniform distribution of dry or liquid fertilizers on the soil surface. This method is used most widely in grain fields, pastures, range land, and orchards where extensive areas must be covered (Mayes, 1973). *Drilling* places fertilizer 3-5 cm below the seed at the time of sowing or at greater depths in sods. It is used in pastures and in establishing stands of alfalfa. *Banding* places the fertilizer in one band on one side or in two bands on both sides of the seed row at or slightly below seeding depth. Distances between the seeds and bands are adjusted to individual crops, but bands are commonly 3-7 cm to the side and about the same distance below the seeds. *Sidedressing* is banding of fertilizers after a crop has become established, at which time proper placement is especially important. The method is not well suited to phosphate fertilizers because of the limited mobility of phosphorus in soils. *Furrow placement* is really a type of banding in which the fertilizer is placed at the bottom of a furrow below the future sites of plants. *Starter fertilizers* are of two kinds: The first consists of coatings placed on seeds (popup fertilizers), and the second consists of a solution added to soil at the time of planting to provide both water and nutrients. *Foliar sprays* consist of dilute solutions of fertilizers applied to plants by ground rigs or from airplanes. They can be applied when fields cannot be entered, to add micronutrients, and for other specialized needs. *Application in irrigation water* consists of adding dry or liquid fertilizer to the water at some central point before it is distributed. This method can thus approximate broadcasting or furrow placement, depending on the method of irrigation (McVickar, 1970).

Optimum placement of fertilizers depends on a number of things, viz, (1) the mobility of nutrients, (2) the depth and spread of the rooting system of the crop, (3) climatic conditions, (4) the type of soil, (5) the kind of fertilizer being used, and (6) the amount to be applied per dressing.

J. J. OERTLI

References

Barber, S. A., 1976. Efficient fertilizer use, *in* F. L. Patterson, ed., *Agronomic Research for Food.* American Society of Agronomy Special Publication 26, 13-29.

Bartholomew, W. V., and Clark, F. E., eds., 1965. *Soil Nitrogen.* Madison: Soil Science Society of America, 615p.

California Fertilizer Association, 1965. *Western Fertilizer Handbook,* 4th ed. Sacramento, 200p.

Cooke, G. W., 1975. *Fertilizing for Maximum Yield.* London: Crosby Lockwood Staples, 298p.

Food and Agriculture Organization of the United Nations, 1973. *Annual Fertilizer Review,* 1972 (and earlier editions). Rome: FAO, 190p.

Ignatieff, V., and Page, H. J., eds., 1958. Efficient use of fertilizers, *FAO Agricultural Studies, No. 43,* 295p.

Jacob, K. D., ed., 1953. *Fertilizer Technology and Resources of the United States.* New York: Academic, 454p.

Kilmer, V. J.; Younts, S. E.; and Brady, N. C., eds., 1968. *The Role of Potassium in Agriculture.* Madison: American Society of Agronomy et al, 509p.

Mayes, D. A., ed., 1973. *Forage Fertilization.* Madison: American Society of Agronomy et al, 621p.

McVickar, M. H., 1970. *Using Commercial Fertilizers,* 3rd ed. Danville, Ill.: Interstate, 352p.

Mortvedt, J. J.; Giordano, P. M.; and Lindsay, W. L., eds., 1972. *Micronutrients in Agriculture.* Madison: Soil Science Society of America, 666p.
Olson, R. A.; Army, T. J.; Hanway, J. J.; and Kilmer, V. J., eds., 1971. *Fertilizer Technology and Use,* 2nd ed. Madison: Soil Science Society of America, 611p.
Pierre, W. H., and Norman, A. G., eds., 1953. *Soil and Fertilizer Phosphorus in Crop Nutrition.* New York: Academic, 492p.
Sprague, H. B., ed., 1964. *Hunger Signs in Crops; A Symposium,* 3rd ed. New York: McKay, 461p.
Stewart, R., 1932. The Mitscherlich, Wiessman, and Neubauer methods of determining the nutrient content of soils, *Commonwlth. Bur. Soil Sci., Tech. Comm. No. 25,* 46p.
Tisdale, S. L., and Nelson, W. L., 1966. *Soil Fertility and Fertilizers,* 2nd ed. New York: Macmillan, 694p.
Walsh, L. M., and Beaton, J. D., 1973. *Soil Testing and Plant Analysis.* Madison: American Society of Agronomy et al, 621p.

Cross-references: *Fertilizers, Organic; Law of the Minimum; Management of Soils; Nutrient Potential; Plant Nutrients; Soil Fertility; Soil Tests; Vol. XVI: Fertilizer Materials.*

FERTILIZERS, ORGANIC

Organic fertilizers include agricultural and urban wastes and industrial by-products that supply plant nutrients and organic matter in various concentrations. The term is synonymous with *manures,* although the latter has a more restricted meaning in some countries. Synthetic organic nutrient sources, e.g., urea formaldehyde, are conventionally classified with inorganic fertilizers (q.v.). Prior to the intensification of agriculture and the advent of commercial fertilizers, the organic materials had an important role in crop production. The decline in their use has been reversed, however, with the renewed interest in exploitation of wastes and by-products from agriculture and other sources due to (1) the increased energy costs in the 1970s, which have affected the manufacture and transport of commercial fertilizers; and (2) the emphasis on a balanced approach to environmental problems stimulated by the threat of pollution from both increasingly intensified animal production systems and urban wastes.

Composition

A broad subdivision of organic fertilizers in physical terms provides the categories (1) bulky solid; (2) nonbulky solid; and (3) slurry or material of less than 15% dry matter. The bulky solids are generally applied at levels of from 15–50 ton/ha. They include the traditional farmyard or barnyard manure of mixed arable and grassland farming systems, feedlot manure, compost, sewage sludge, and seaweed, which is used in some coastal areas. Farmyard manure is a combination of feces, urine, and litter such as straw, and it is usually partially decomposed by microbial action during storage prior to spreading on land. It is high in plant constituents and in microbial biomass derived from fecal matter. Feedlot manure, without added litter, has higher nutrient concentrations than farmyard manure. Sewage sludges are of two main types: *activated sludge* produced by injection of air during the primary treatment, and *digested sludge,* which results from an anaerobic process. The nitrogen content is generally higher in activated sludge. The phosphorus content varies widely while the potassium content is generally low. The nonbulky solids include the traditional organic fertilizers that are generally by-products of the agricultural industry. They comprise slaughter-house wastes such as dried blood, plant residues typified by the oil-seed cakes remaining after oil extraction, and guano, formed from droppings of marine birds, which was at one time an important source of nitrogen. These materials are normally applied at rates of up to 3 ton/ha and supply little organic matter (Hall, 1955).

The impact of intensive livestock-fattening units is evidenced in the greatly increased production of animal *slurry,* which is a mixture of feces, urine, and variable quantities of wash water. Thick-slurry is 10–15% dry matter and flows slowly; medium-slurry is 5–9% dry matter; thin-slurry of less than 5% dry matter is suitable for organic irrigation.

The data in Table 1 illustrate the major element composition of organic materials commonly used as nutrients for crops. In many instances composition varies widely and values may exceed those given. The concentration of nutrients in slurry, for instance, changes not only with type of animal and level of nutrition but also with dilution by wash water and with the conditions and duration of storage. Mean values serve mainly as a basis for comparing different sources. Where only one value is shown, it more likely reflects a lack of data than a relatively constant composition.

In addition to plant nutrients, some organic fertilizers contain toxic elements at levels that may be harmful to plants and food chains. Sludges and effluents are potential hazards because of their content of cadmium, chromium, cobalt, copper, lead, mercury, nickel, zinc, and other heavy metals. Chaney lists the critical levels of elements (μg/g) in sludge used for land application as 15 for mercury, 1000 for lead, and <2000 for zinc (Boswell, 1975). High levels of copper, up to 750 μg/g of dry matter, may

TABLE 1. Major Nutrient Contents for Some Organic Fertilizers.

Material	N	P	K	Moisture
Animal Wastes (% of fresh material)				
Farmyard manure	0.3–2.2[a]	0.04–0.92	0.4–1.2	8–86
	0.6[b]	0.13	0.5	76
Beef feedlot manure	0.6–3.5	0.3–0.9	0.8–2.4	5–40
Broiler litter	0.4–3.6	0.1–1.7	0.3–2.0	9–75
	2.3	0.9	1.0	32
Cattle slurry	0.2–1.9	0.04–1.1	0.08–1.8	86–89
Swine slurry	0.2–0.6	0.01–0.4	0.03–0.7	85–98
Urban Wastes (% of dry matter)				
Activated Sewage sludge	5.6	2.5	0.3	–
Digested sewage sludge	2.0	0.5	0.3	–
Miscellaneous (% of dry matter)				
Guano	16–20	5–6	2–3	–
Dried blood	10–14	–	–	–
Meat- and bone-meal	5–12	<8	–	–
Bone meal	0.4–4.5	6.5–14.5	–	–
Rape cake	5	1	1	–
Spent mushroom compost	0.6–1.1	0.2–0.4	0.3–1.2	–
Peat moss	0.8	0.1	0.1	–
Cereal straw	0.6	0.1	1.2	–
Seaweed	0.6–2.0	0.1	0.9–2.1	–

[a]Range
[b]Mean

occur in swine slurry because this element is added as a feed supplement.

The data in Table 2 outline the total contents of heavy metals reported by various authors for some wastes. It should be noted that available levels may be very much lower than total values. Loehr (1974) provides a comprehensive list of the organic matter, major and minor nutrients, and toxic materials in manures and wastes from agricultural and industrial sources.

Management and Use

The management of agricultural wastes involves such approaches as daily collection and spreading; storage of solid manure and spreading after partial decomposition by microorganisms; storage of slurry in an aerated lagoon or oxidation ditch; and anaerobic storage of slurry where microbial decomposition is maintained by oxygen derived from nitrates and sulphates. The type of waste management used depends primarily on the animal production system but may be modified by hygiene and environmental requirements. For instance, aerobic storage may be required for odor control where swine production units are adjacent to private dwellings.

The utilization of most bulky organic nutrient sources, including slurry, exists on two levels. The relatively nonintensive production units emphasize the return of nutrients to the soil as supplements to inorganic fertilizers. Most economical use of manure is then attained by its application at levels that satisfy crop demands for one or more major nutrient elements and where the requirement for other nutrients is satisfied by application of inorganic fertilizers. In contrast, manure disposal from intensive units is likely to be limited only by the threat to environmental pollution or by reduction in crop yields. In this context maximum application rates are determined not so much by the nutrient requirements of crops as by the

TABLE 2. Total Contents of Heavy Metals in Organic Wastes (μg/g fresh material).

Material	Cr	Cu	Pb	Mn	Ni	Zn
Digested sewage sludge	215	48–600	–	172	11–290	3200
Town refuse	–	90–570	102	–	–	250–2200
Pig slurry	–	4–365	–	–	–	2-323
Farmyard manure	–	60	–	150	–	200

potential groundwater pollution and the accumulation of toxic materials such as soluble salts and metals in soil. The nonbulky organic manures are normally confined to use in horticulture and are best justified for greenhouse conditions where nitrate leaching occurs or where high levels of soluble salts obtain (Cooke, 1967).

Approximate application rates (ton/ha) of fresh solid materials vary from 3 with the relatively nutrient-rich, nonbulky solids to 50 for farmyard manure, and 70 for beef feedlot manure. Emphasis on disposal of the latter using minimum land requirements may result in the use of much higher loadings. Frequent light dressings of slurry are recommended to prevent the sealing of soil surfaces and the anaerobic conditions created by heavy dressings. Application rates obviously vary widely because of the variable dry matter content. Annual applications (m^3/ha) of up to 50 probably utilize nutrients most efficiently, although loadings of between 200 and 400 are common where there is a need to dispose of large amounts. Much is known of the short-term availability of nutrients in slurry and solid manure. Insofar as variations in cropping systems and climate permit generalization, it is probable that not more than 30% of the nitrogen and 50% of the phosphorus is available from farmyard manure in the year of application. All the potassium is available because it exists in inorganic forms and only participates in a minor way in microbiological processes. Cattle and swine slurries have similar proportions of phosphorus and potassium and up to 65% of their nitrogen available. Materials of less than 2% nitrogen in the dry matter generally do not release nitrogen.

There is relatively little information available on residual or long-term nutrient effects. Here again nitrogen is most relevant, and it is particularly important to determine the time-course of its release from organic forms in areas where there is continued disposal of manures. The concept of a decay series proposed by Pratt and coworkers may be useful in such instances (see Azevedo and Stout, 1974). This approach provides for adjusting the application rates to maintain a known yearly nitrogen supply. For instance, dry corral manure of 2.5% nitrogen may have a decay series of 0.40, 0.25, 0.06, and 0.03. Thus 40% of the nitrogen applied is released in the first year, 25% of that remaining is released in the second year with 6% and 3% of the residual nutrient levels released in subsequent years. There is relatively more information available on the accumulation of organic matter, especially in the case of solid manures. Results from the decomposition of uniformly ^{14}C-labeled straw and cattle manure under field conditions clearly illustrate the relative resistance of some organic components to biological decomposition; five years after application, 21% of the organic carbon in straw and 31% of that in a mixture of straw and cattle excreta may remain in soil, much of it in the humic fraction (Oberländer and Roth, 1975).

MICHAEL HERLIHY

References

Azevedo, J., and Stout, P. R., 1974. Farm animal manures: An overview of their role in the agricultural environment, *Calif. Agric. Exp. Sta. Ext. Serv. Manual,* **44,** 1-110.

Boswell, F. C., 1975. Municipal sewage sludge and selected element applications to soil: Effects on soil and fescue, *Jour. Environ. Quality,* **4,** 267-272.

Cooke, G. W., 1967. *The Control of Soil Fertility.* London: Crosby Lockwood, 526p.

Hall, A. D., 1955. *Fertilizers and Manures,* 5th ed. London: Murray, 333p.

Loehr, R. C., 1974. *Agricultural Waste Management.* New York: Academic, 576p.

Oberländer, H. E., and Roth, K., 1975. Transformation of a ^{14}C-labelled manure of cattle slurry and straw in soil. A five years' small scale field trial (in German), *Bodenkultur,* **26,** 139-145.

Cross-references: *Fertilizers, Inorganic; Humus; Management of Soils; Metabolic Activity; Mulch; Organic Matter; Plant Nutrients.*

FIELD WATER CYCLE

The pattern of water into, through, and out of the soil profile has an important influence on physical, chemical, and biological processes that take place in the biosphere. For this reason, the pattern deserves special consideration. For example, the soil profile plays a significant role in the hydrologic cycle (q.v. in vol. IVA) in its partitioning of precipitation into infiltration (q.v.) and runoff and in its partial control of the earth's energy budget.

The Water Budget

The field water cycle is best understood by dividing it into several components, each of which can be considered separately. Figure 1 is a schematic diagram of the various processes as they are conventionally described. In mathematical form, these terms can be summed to describe the water budget of a soil profile:

$$P - R = I = E + T + D + \Delta W \qquad (1)$$

where P is the incident precipitation, including irrigation; R is the runoff. The difference I represents the water that infilters downward

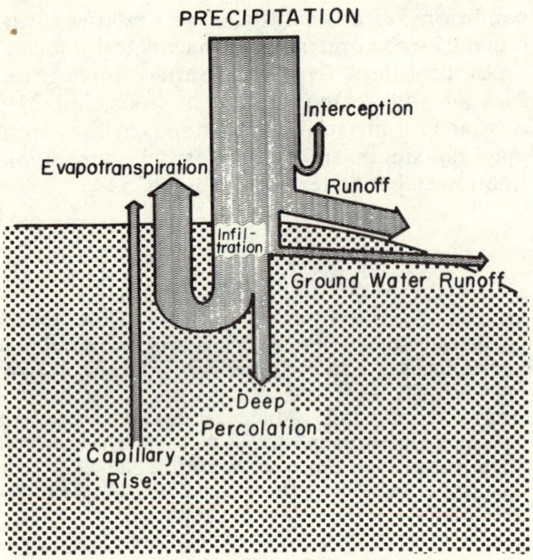

FIGURE 1. Schematic drawing of the portions of the hydrologic cycle involving the soil profile. (After Baver, Gardner, and Gardner, 1972)

through the soil surface. The fate of this water is described by the far right-hand side of equation 1. The symbol E stands for the loss of water from the soil surface by evaporation, and T stands for transpiration that water loses through evaporation from plant leaves. In practice, it is often difficult to separate E and T, and the symbol ET is frequently used to represent their sum. D represents drainage out of the soil profile. *Capillary rise* is the upward movement of water from below and can be considered as negative drainage. In any given soil profile, D may alternate between positive and negative values, depending on the rainfall distribution and ET. The final term in equation 1, ΔW, represents the change in the water stored in the soil profile. Over a yearly cycle, ΔW tends to be zero. On an annual basis the drainage and evapotranspiration from the profile tend to equal the infiltration.

Figure 2 shows the seasonal behavior of the various components of the water budget, except for the soil water storage, for a sandy soil. Unlike most field crops, onions do not provide substantial vegetative cover of the soil surface so that the evaporation (q.v.) is several times the transpiration.

Infiltration

The runoff rate R is determined by the difference between the precipitation rate P and the infiltration rate F. If water is ponded on the surface of a relatively dry soil profile, the infiltration rate may be initially extremely high, but then it will continuously decrease with time until it eventually levels off at a constant rate. The initially high infiltration rate is due to the *matric potential gradient* associated with the dry soil (see *Water Content and Retention*). With increasing water content in the soil, this matric potential gradient becomes so small as to be overshadowed by the gravitational potential gradients that remain constant. In a well-drained profile, the final steady rate equals the average permeability (q.v.) of the soil when the potentials are expressed in units of length, as is the case when calculated as potential energy per unit weight of water. A convenient empirical infiltration equation is

$$I = S(t^{1/2}) + Kt \qquad (2)$$

where S is a quantity known as the *sorptivity;* t is the time; I is the cumulative infiltration from $t = 0$ to time t; and K is the saturated permeability of the soil profile. It should be noted that equation 2 is not an exact solution of the water-flow equation but is often sufficiently close to give meaning to the parameter S which appears in it. The sorptivity S is proportional to the difference between the initial and final water contents and to the square root of the soil water diffusivity. In the field, air entrapment within the soil profile is not uncommon. Because the air tends to bubble up through and occupy the largest pores—that is, those pores that conduct the greatest quantities of water—this effect may reduce the infiltration rate substantially below that which would otherwise occur. Surface crusts tend to reduce the infiltration rate and usually cause the soil profile to be less than saturated during infiltration (Hadas et al, 1973).

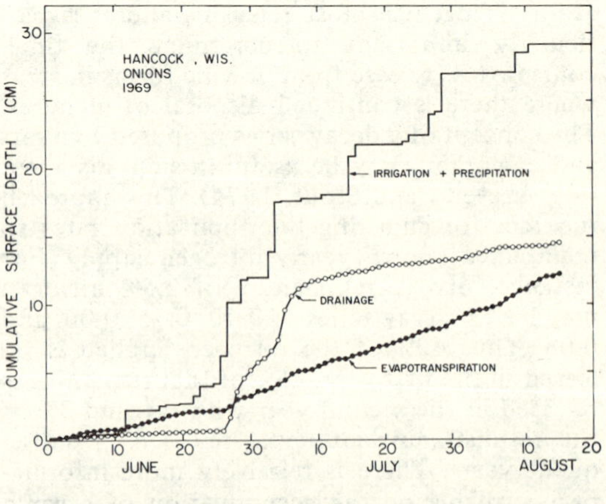

FIGURE 2. Cumulative precipitation, evapotranspiration, and drainage from the root zone of a sandy soil over the summer season. The crop is seed onions in rows.

Redistribution

After infiltration has ceased, the soil water continues to move downward within the soil profile. This downward movement, or *redistribution,* becomes increasingly slower with time and in many field situations may appear to cease altogether. Because of the gravitational gradient, which is always downward and always the same everywhere, the soil water content in a deeply wetted profile is found to be roughly the same at each depth, all the while draining ever more slowly. When the redistribution or drainage rate becomes negligibly small, the soil is said to be at *field capacity*. This term is not precise because that which is negligible in one context may be critical in another. In general, in a soil profile that is not underlain by a high water table, the redistribution rate will become relatively slow after two or three days. However, careful measurements will often reveal significant drainage over periods of weeks or even months.

Evaporation

When the soil water content is high, the evapotranspiration tends to be limited by the solar energy available at the soil surface necessary to transform the liquid water to vapor (Hagan, Haise and Edminster, 1967). As each unit of water evaporates, another from below moves up toward the soil surface in response to the matric potential gradient set up by the drying of the soil surface. The evapotranspiration in this circumstance is therefore limited by atmospheric factors and is proportional to the potential evaporation, which characterized the meteorological conditions. In the absence of a water table within a reasonable distance of the soil surface, the soil profile cannot supply water at the potential rate indefinitely (Hagan, Haise, and Edminster, 1967). In fact, within only a day or two after a rainfall, the constant rate phase of evaporation for most soils can no longer be maintained. The second, or falling, rate stage of evaporation then begins (Gardner and Gardner, 1969). During this stage evaporation is limited by water transport upward to the soil surface and becomes governed by soil properties rather than atmospheric conditions. In fact, under moderate to high potential evaporation, the falling rate stage begins within a day or so after cessation of infiltration and decreases inversely roughly as the square root of time. Consequently, the cumulative evaporation then increases in proportion to the square root of time. If the depth of wetting of the soil profile is only a centimeter or two, the rate of evaporation falls off more rapidly than the square root of time (Gardner and Gardner, 1969).

Capillary Rise

If there is a water table within a meter or two of the soil surface, it may be possible for a given evaporation rate to be maintained indefinitely by capillary rise. This maximum possible soil rate depends on the soil water properties and the depth to the water table. As a reasonable approximation, the maximum rate that can be maintained by a given soil is given by

$$E_{\max} = A \frac{K}{(d)^n} \tag{3}$$

where A and n are soil characteristic constants, and K is the saturated permeability of the soil (Baver, Gardner, and Gardner, 1972). The factor n is about 2 or 3 for clay soils and may be as high as 10 or more for sands. Figure 3 shows the effect of water-table depth on the steady-state evaporation of water from a fine sandy loam soil. When the water table is within 2m of the soil surface, the relatively low potential rate of evaporation of just less than 2 mm/day is maintained, as indicated by the horizontal line segment. The circles represent actual measured evaporation rates. As the water table is lowered below 2 m, the soil profile immediately becomes limiting, and the actual evaporation rate drops approximately as the square of the depth to the water table. Water movement up into a plant root zone may be treated in the same manner.

A surface crust or mulch influences the evaporation rate, though not to the same extent as in the case of infiltration. Such a crust tends to

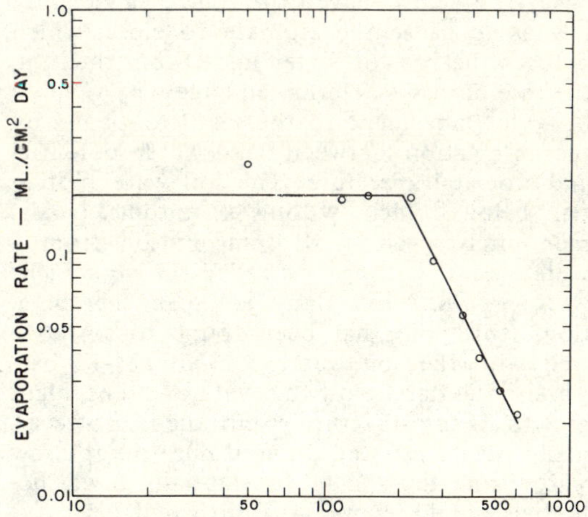

FIGURE 3. The steady-state evaporation rate from a fine sandy loam soil as a function of the depth to the water table. The potential evaporation rate is about 1.8 mm/day.

restrict the rate of vapor movement from the soil to the atmosphere and thus reduces and prolongs the constant rate stage of evaporation. Over a long period of time, i.e., when the falling rate stage is at least as long as the constant rate stage, the cumulative evaporation is relatively insensitive to the presence of a mulch or crust. However, under a mulch (q.v.) the water content near the soil surface is much higher than it would be in the absence of the mulch. This may have important consequences with respect to biological processes and soil microflora and fauna (q.v.).

In the higher latitudes an additional factor must be considered in the field water cycle. If the soil surface freezes, the profile responds in a manner similar to that during evaporation. The frozen surface layer acts in a manner similar to a dry zone, and water begins to move upward. This new water, if it moves sufficiently rapidly, joins the frozen zone in soil pores that must expand to accommodate the freezing water. The expanding pores result in "frost heaving," which can exert extremely high pressures at the surface, which may result in rupture of highway pavements or damage to perennial crops. In the spring, as soon as the last of the frozen water melts, this formerly frozen water drains rapidly.

The presence of vegetative cover enhances the amount of water that leaves the soil profile in response to evaporative forces. The plant root system conducts water much more readily than moderate to dry soil. For this reason, a given evaporation rate can be maintained by vegetation for a longer time than for the same soil with a bare surface. Eventually, however, the water potential in the plant leaves, which must always be below that in the soil, is sufficiently low as to cause the stomata to close, which reduces the rate of water loss. From then on, the rate of loss is determined largely by the rate of water movement to the plant roots and by the interaction between leaf water potential and stomatal aperature. The soil water potential below which wilting symptoms persist indefinitely even if all transpiration from a wilted plant is stopped is called the *permanent wilting point.* For many soil-plant combinations, this point has been found to correlate well with the soil water content at −15 bars water potential. Below this water content, plant growth is most severely curtailed. However, many plants can and do continue to withdraw water from the soil to substantially lower potentials and do recover when irrigated.

W. R. GARDNER

References

Baver, L. D.; Gardner, W. H.; and Gardner, W. R., 1972. *Soil Physics.* New York: Wiley, 498p.

Bear, J.; Zaslavsky, D.; and Irmay, S., 1968. *Physical Principles of Water Percolation and Seepage.* Paris: UNESCO, 465p.

Bruce, R. R., ed., 1973. *Field Soil Water Regime.* Madison: Soil Science Society of America, Special Publication Number 5, 212p.

Gardner, H. R., and Gardner, W. R., 1969. Relation of water application to evaporation and storage of water. *Soil Sci. Soc. America Proc.* 33, 192–196.

Hadas, A.; Swartzendruber, D.; Rijtema, P. E.; Fuchs, M.; and Yaron, B., eds., 1973. *Physical Aspects of Soil Water and Salts in Ecosystems.* New York: Springer-Verlag, 460p.

Hagan, R. M.; Haise, H. R.; and Edminster, T. W., eds., 1967. *Irrigation of Agricultural Lands.* Madison: American Society of Agronomy, 1180p.

Hillel, D., 1971. *Soil and Water.* New York: Academic, 288p.

Kirkham, D., and Powers, W. L., 1972. *Advanced Soil Physics.* New York: Wiley, 534p.

Swartzendruber, D., 1969. The flow of water in unsaturated soils, *in* R. J. M. de Wiest, ed., *Flow Through Porous Media.* New York: Academic, 530p.

Van Schilfgaarde, J. ed., 1974. *Drainage for Agriculture.* Madison: America Society of Agronomy, 700p.

Cross-references: *Capillary Pressure; Conductivity, Hydraulic; Sail Drainage; Energy Balance; Evaporation; Flow Theory; Imbibition; Infiltration; Moisture Management; Percolation; Pore Size, Drainable; Water Content and Retention; Water Movement; Wetting Front. Vol. IVA: Energy Budget of the Earth's Surface; Hydrologic Cycle; Water Balance.*

FLOCCULATION

This is the aggregation of a discontinuous (disperse) phase in a continuous one, commonly of solid particles in a liquid. The resulting aggregates are *flocs,* from the Latin "floccus" (a lock of wool). The late Latin diminutive "flocculus" gave "flocculate" and "floccule." "Floccule" usually implies a small floc or some kind of smaller unit in a floc. The term *flocculation* has tended to displace "coagulation," which is sometimes preferred when the process is at least partially reversible.

In general, this discussion concerns the flocculation of crystalline particles (the discontinuous phase) in aqueous electrolyte solutions (the continuous phase) and thus with the flocculation of lyophobic (here "hydrophobic") colloidal systems. The terms "lyophobic" and "hydrophobic" are misleading because the discontinuous phase does not "fear" the ambient liquid and is solvated, although less markedly than in lyophilic systems where interaction with the continuous phase helps to maintain the dispersed state, the reverse of the flocculated condition. Even a lyophobic discontinuous phase is partially stabilized by adsorption of the continuous phase because its aggregation involves a positive change in desolvation free energy (Shaw, 1970). Its stability, however, depends mainly on the ability of interparticle electrical repulsion to overcome the interparticle attraction at the closest mutual approach attainable by the particles during their thermal kinetic (Brownian) movement.

The relative effectiveness of the repulsive and attractive forces at particular interparticle distances changes with the nature and/or concentration of the electrolytes in the continuous phase, and thus the tendency to dispersion or flocculation varies concomitantly. Hydrophobic colloial particles, with which clay minerals have affinities, can be flocculated by small concentrations of electrolytes. For instance, if the particles have negative surface charges, the cations, rather than the anions, of the ambient electrolyte solutions are more significant in causing flocculation. Flocculating power increases rapidly with their charge number, but their chemical nature is less significant (van Olphen, 1963, 1977).

The stability of hydrophilic colloids is little affected by electrolytes, although precipitation ("salting out") can occur at high electrolyte concentrations, presumably because the ions introduced compete with the disperse phase by virtue of their own hydration for the solvating liquid that promotes its stability.

Flocculation and Its Consequences

Because the tendency to dispersion or flocculation is controlled by interfacial forces, it becomes more apparent as the interfacial area between the discontinous and continuous phases increases in a unit volume of the system. The tendency is usually obvious when the particles in a liquid are ca 1.0 μm–1.0 nm (10^{-6} – 10^{-9} m) equivalent spherical diameter (esd). These limits are arbitrary, and 1.0 nm, for instance, supposedly approximates to the diameters of large molecules. However, macromolecules and association colloids (micelles) have effective diameters well above 1.0 μm, yet both disperse spontaneously in appropriate continuous phases (*solvents*) and can be classed as *lyophilic colloids*.

Particles ca 1.0 μm esd are usually dense enough to settle slowly in water. Those with diameters near 1.0 nm esd stay in suspension indefinitely as long as they remain dispersed rather than flocculated because they are small enough to share markedly in the thermal kinetic energy of the system.

The effects of repulsive or attractive interfacial forces can be nullified, wholly or in part, by those of inertial forces. This tendency develops at smaller particle sizes as the density of the particle becomes increasingly greater than that of the ambient liquid. Nevertheless, clean, unwetted polystyrene spheres, slightly denser than water and 1.0 μm diameter, flocculate spontaneously on the surface of the liquid because inertial forces remain ineffective until a wetting agent is added. The spheres then disperse (deflocculate or become peptized) and subsequently sink (Sennett and Olivier, 1965).

Flocculation increases the size of the largest kinetic units in a system. The enlarged units may remain suspended (dispersed) if they still share sufficiently in the thermal kinetic energy of the system, but commonly they soon become big enough to sink and form a sediment. This sediment is markedly porous because interparticle attraction persists and causes both bridging by particles and the adhesion of floc to floc. Thus many particles cannot assume the positions of lower potential energy, which would be appropriate to a system predominantly influenced by gravitational forces. If the particle concentration is sufficient, the flocs may become linked throughout the entire volume of the system, which thus changes to a gelatinous mass (*gel*). This is exemplified by some systems of montmorillonite and aqueous electrolytes. Some gelled materials (q.v. in Vol. IVA: *Colloids; Mineraloids*) often develop lower apparent viscosities when disturbed because redispersion, usually incomplete, is producible mechanically. The resulting dispersion (*sol*) reverts to a gel when mechanical disturbance ceases. This cyclic behavior characterizes thixotropy (see *Thixotropy, Thixotropism*).

Sediments from flocculated systems often show plastic, pseudoplastic, or thixotropic behavior if the supernatant liquid is first removed. Those from dispersed systems, when freed from the supernatant liquid, are commonly "dilatant" in the sense defined by Osborne Reynolds. Modern usage of this term does not always accord with the original definition (Roscoe, 1953).

Particles settling from dispersed systems, because they are mutually repulsive and unlikely to adhere, are sufficiently affected by gravitational forces to take up positions of relatively low potential energy and thus acquire some measure of close packing. Shearing of the resulting sediment produces regions in which the particles "come out of mesh" and the system expands (dilates) in consequence. However, dilatancy can be shown by aggregates of relatively coarse flocculated particles because here inertial forces suffice to produce relatively close packing. Cleavage fragments of feldspar exemplify the phenomenon (Williamson, 1946).

Flocculated, clay-rich, subaqueous sediments, especially if they contain montmorillonite, can crack beneath their supernatant liquid if, for instance, its salinity is changed (Burst, 1965). Thus ancient mudcracks in sedimentary rocks need not imply an earlier exposure to atmospheric drying.

Fundamental Approaches to the Flocculation of Lyophobic Colloids

These approaches are discussed by Sennett and Olivier (1965), Shaw (1970), and van Olphen (1963), but further details are provided by Derjaguin and Landau and by Verwey and Overbeek whom these authors cite. If a particle in an electrolyte solution has a negative surface charge—and a negative charge is common and arises by several mechanisms—this charge is partially screened by positive counterions (cations), which are components of a diffuse electrical double layer. The cations necessary to screen the remaining charge are close to the particle surface as components of the *Stern layer.* The anions, which necessarily accompany the cations in an electrolyte solution, constitute the coions or simili-ions and concentrate away from, rather than towards, the negatively charged particle surface.

Flocculation, thermodynamically considered, implies a state energetically lower than that of the corresponding dispersed system, yet it is not always rapid, and some systems remain dispersed for years. Although the thermal kinetic energy of the particles promotes mutual collisions, the collision frequency is manifestly not the sole factor controlling flocculation rate.

Flocculation results from the predominance of attractive forces at attainable interparticle distances. Such forces include those categorized as van der Waals, but especially their London components, which become increasingly effective as mass increases; the masses of colloidal particles suffice to make these forces significant. However, coulombic attractions are involved, for instance, in the flocculation of clay minerals.

If only the most thermokinetically energetic particles achieve the interparticle distances at which attractive forces predominate, flocculation is slow. If none achieves them, the system remains dispersed, or if very many, it flocculates rapidly. Opposition to flocculation results from interaction between the electrical double layers that surround each particle, and mutual approach of the particles causes an increasing repulsive force that may or may not be overcome by the attractive forces that promote flocculation. In essence, the situation can be represented by energy barriers of various heights that must be surmounted by virtue of an activation energy supplied thermokinetically by the mutually approaching particles. Appropriate calculations yield a family of curves (Figure 1).

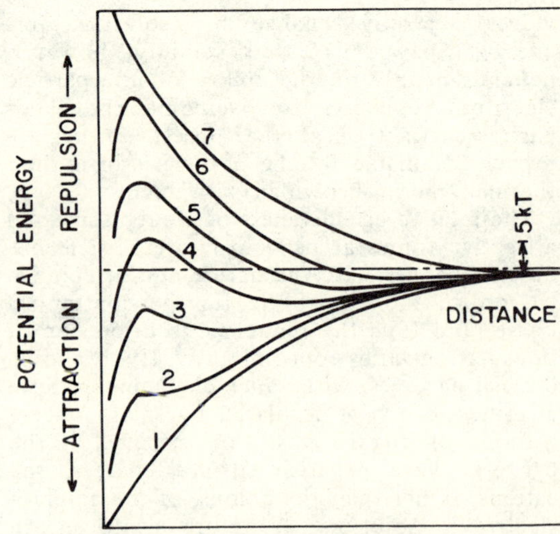

FIGURE 1. Net potential energy of interaction between particles in a colloidal suspension. Curve 1: attraction in the absence of effective repulsion, implying high electrolyte concentration and rapid flocculation. Curves 2–7: effects of decreasing electrolyte concentration and hence increasing double-layer thickness and longer ranging repulsive forces. Curve 5: the ca 15 kT barrier imparts considerable stability to the dispersion. Curves 3–4: rapid and slow flocculation, respectively. Some curves have a secondary minimum at ca 200 Å interparticle separation that may be deep enough (curve 4) to cause reversible flocculation (coagulation) of a considerable number of particles, especially if these are rod- or plate-like.

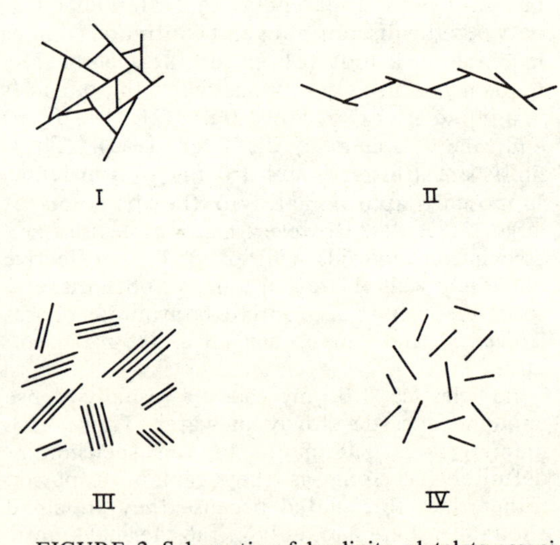

FIGURE 2. Schematic of kaolinite platelets associated in: I card-house floc formed under condition *a* of Figure 3; II adlineated floccule, presumably related to a card-house floc but with fewer edge-to-face contacts; III card-pack flocs. In IV the platelets are dispersed as suggested by condition *c* and perhaps *b* of Figure 3.

Extensions of the mathematical arguments that lead to Figure 1, and to cognate representations, explain the *Schulze-Hardy rule* (Sennett and Olivier, 1965), and in addition, the compression of the electrical double layers that may result in an interparticle attraction effective enough to cause flocculation. Such compression can develop, for instance, in the presence of a sufficiently concentrated electrolyte, although smaller concentrations might favor dispersion rather than flocculation.

Chemical reaction between particles and their surrounding electrolyte solution complicates flocculation behavior. Clay minerals, in water or in aqueous electrolyte solutions, commonly behave in this way, and elements that were originally constituents of their crystal structures become ions, usually complex, in the ambient solution. Thus a clay mineral particle can contribute, by continued chemical reaction, to its own counterions and coions, with corresponding alterations in the flocculation state of the system, sometimes made apparent by simultaneous changes in the apparent viscosity. The apparent viscosity commonly increases as flocculation progresses.

Flocculation of Kaolinite

Kaolinite platelets can associate in card-house, card-pack (Figure 2), or other types of floc. The type of floc varies with the nature and concentration of the electrolytes present and with the corresponding pH values. Okuda and Williamson (cited in Williamson, 1974) photographed both card-house and card-pack flocs existing undisturbed in suspensions. The reality of such flocs cannot be doubted, but it is now realized that several types of open- or close-textured floc can be postulated, and electron microscopical techniques have confirmed the existence of some of them (Smart, 1971; Williamson, 1974), although the mechanisms for their production are not altogether clear. The viscometric, sedimentation volume, or optical data thought to imply the presence of the card-house flocs postulated by Schofield and Samson (cited in Williamson, 1974) could be interpreted equally well in terms of other types of open-textured floc. Okuda and Williamson showed that card-house flocs became attached to negatively charged air bubbles by the edge faces of their constituent crystals; demonstrably, these faces were positively charged. A possible mechanism for the origin of the positive charge, implied by Schofield and Samson but not universally accepted, is indicated in Figure 3.

As the positive edge faces of kaolinite visibly adhered also to the basal surfaces of adjacent crystals (Okuda, cited in Williamson, 1974), the latter surfaces are negatively charged. It had long been postulated that such a negative charge existed and was independent of pH. It could arise because of very minor substitutions in the kaolinite structure, such as Al^{3+} for Si^{4+} or Mg^{2+} for Al^{3+}. This concept is not easily correlated with recent analytical data (Jepson and Rowse, 1975; see Ferris and Jepson, 1975; and Rand and Melton, 1975). Card-pack flocs tend to develop in the presence of multivalent cations or when monovalent cations are present in excess even sodium ions, which at low concentrations usually favor dispersion rather than

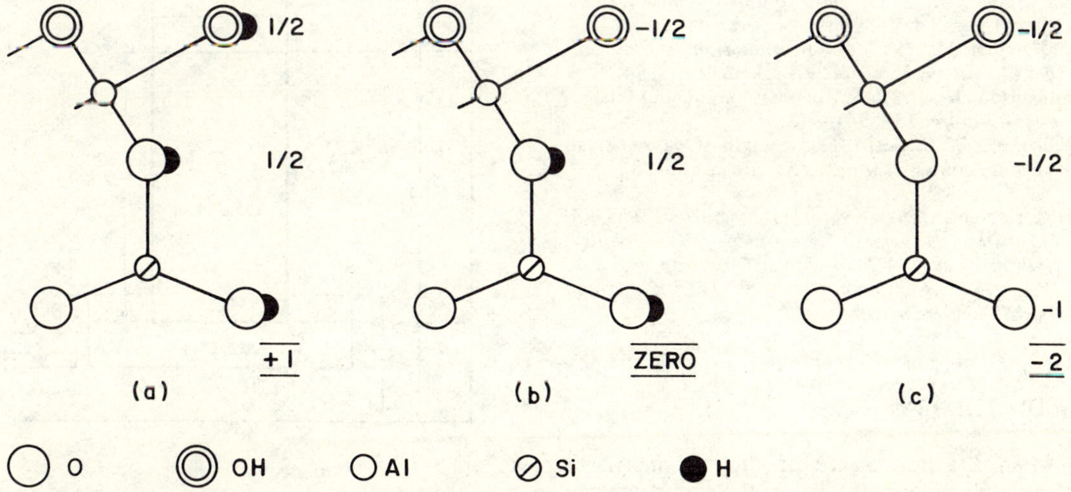

FIGURE 3. Hydrions attached, if at all, to oxygens at the edge faces of kaolinite when the ambient solution is (a) acid, (b) slightly alkaline, and (c) strongly alkaline.

flocculation. Okuda and and Williamson photographed such flocs, which had formed in approximately neutral systems containing calcium or lead cations.

Clay pastes, because of their high concentration of solids, are unlikely to contain open-textured flocs. Moreover, shearing of the paste could collapse such flocs if they existed. For instance, extruded plastic clay, whether dispersed or flocculated, tends to contain clay mineral platelets arranged as if in card-pack flocs. Even so, the structural arrangement in the dried flocculated clay may be somewhat more porous and mechanically weaker.

W. O. WILLIAMSON

References

Burst, J. F., 1965. Subaqueously formed shrinkage cracks in clay, *Journ. Sed. Petrology,* **35,** 348–353.

Jepson, W. B., and Rowse, J. B. 1975. The composition of kaolinite–an electron microscope microprobe study, *Clays Clay Min.,* **23,** 310–317.

Ferris, A. P., and Jepson, W. B., 1975. The exchange capacities of kaolinite and the preparation of homoionic clays, *Jour. Coll. Interfce. Sci.,* **51,** 245–259.

Rand, B., and Melton, I. E., 1975. Isoelectric point of the edge surface of kaolinite, *Nature,* **257,** 214–216.

Roscoe, R., 1953. Suspensions, *in* J. J., Hermans, ed *Flow Properties of Disperse Systems.* Amsterdam: North Holland, 10–12.

Sennett, P., and Olivier, J. P., 1965. Colloidal dispersions, electrokinetic effects and the concept of zeta potential, *Ind. Eng. Chem.,* **57,** 32–50.

Shaw, D. J., 1970. *Introduction to Colloid and Surface Chemistry.* London: Butterworths, 236p.

Smart, P., 1971. Soil structures in the electron microscope, structure, solid mechanics and engineering design, Part I, *Proc. Southampton 1969 Civil Eng. Materials Conf.,* 249–255.

van Olphen, H., 1963. *Clay Colloid Chemistry.* New York: Interscience, 301p.

van Olphen, H., 1977. *An Introduction to Clay Colloid Chemistry,* 2d ed. New York: Wiley, 318p.

Williamson, W. O., 1946. Rheology of aqueous felspar-pastes, *Nature,* **157,** 341–342.

Williamson, W. O., 1974. The strength of clays at small moisture contents, *Minerals Sci. Eng.,* **6,** 3–18.

Cross-references: *Aggregate Stability; Clays, Physical Chemistry; Dispersion Phenomena; Hydrophility, Hydrophobicity; Thixotropy, Thixotropism. Vol. IVA: Colloids; Mineraloids. Vol. IVB: Clays and Clay Minerals; Thixotropy.*

FLOW THEORY

When the pore space of the soil matrix is completely filled with water, the soil is referred to as *water saturated.* Thus *saturated flow* may be defined as the movement of water through water-saturated soils. The theory governing saturated water flow in soils was first formulated by Darcy (1856). He discovered that water flow in vertical columns of water-saturated, sand filter beds is proportional to the head loss (or potential difference) between the ends of the column and to a factor known as the *hydraulic conductivity* (q.v.), K. Darcy's apparatus is illustrated in Figure 1, where L is the length of the sand column, A is the cross-sectional area, H_1 and H_2 refer to the height of standing water (or hydraulic heads) for the upper and lower ends of the column, respectively. Darcy found that the volume of water flow through the column per unit time can be expressed as

$$Q = K A \frac{H_1 - H_2}{L} = KAi \qquad (1)$$

where i (dimensionless) is called the *hydraulic gradient.* Rearranging equation 1, Darcy's law may be written in the form,

$$v = \frac{Q}{A} = Ki \qquad (2)$$

where v is the water flow velocity that represents the volume of water flow per unit area of soil per unit time. It is clear from equation 2 that the hydraulic conductivity K (the proportionality factor) has the dimensions of water

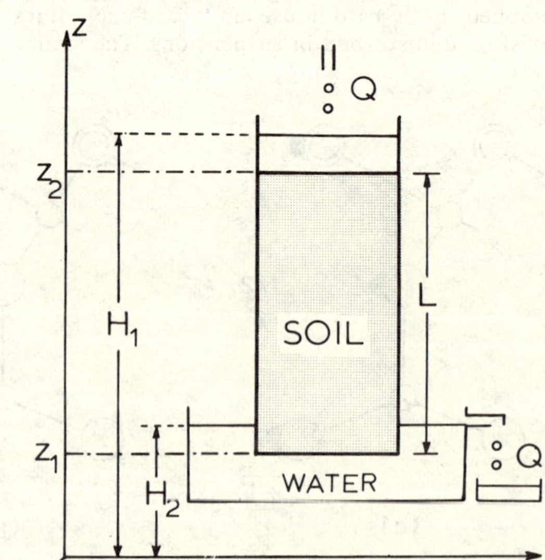

FIGURE 1. Diagram of soil column illustrating water flow in the vertical direction.

velocity v. Furthermore, K is characteristic of the soil or the porous media, which indicates the ability of the soil to conduct water. In fact, from equation 2 the hydraulic conductivity K is the water flow velocity in the soil (or porous media) when the hydraulic gradient i is unity.

It is convenient to express Darcy's law (equations 1 or 2) in differential forms, which is achieved by rewriting (1) as

$$v = K\frac{H_1 - H_2}{z_1 - z_1} = -K\frac{H_2 - H_1}{z_2 - z_1} = K\frac{\partial H}{\partial z} \quad (3)$$

where the partial derivative represents the change of H with respect to soil depth. Since equation 3 represents water flow velocity in the z direction, i.e., vertically downward in the soil column (Figure 1), we can write (3) as

$$v_z = -K_z\frac{\partial H}{\partial z} \quad (4)$$

where v_z and K_z are the water flow velocity and hydraulic conductivity of the soil in the z direction. Note that the negative signs in equations 3 and 4 indicate that the water flow is in the negative z direction. Similarly, for water flow in the x direction and y direction, Darcy's equation can be written, respectively, as

$$v_x = -K_x\frac{\partial H}{\partial x} \quad (5)$$

and

$$v_y = -K_y\frac{\partial H}{\partial y} \quad (6)$$

When the hydraulic conductivity of the soil is uniform, that is,

$$K_x = K_y = K_z \quad (7)$$

the soil is referred to as *isotropic;* otherwise, the soil is *anisotropic.*

The Continuity Equation

Darcy's law (equations 4, 5, and 6) is useful in describing flow patterns in the soil when the water movement (q.v.) is one dimensional, i.e., when water flow is in one direction only. Although Darcy's law is still valid, it is often difficult to apply directly to general soil-drainage problems in which water flow is predominantly in two or three dimensions.

The continuity equation is a general representation of water flow in the soil system and accounts for hydraulic head and soil water content changes with time at any location in the soil. To derive the continuity equation, let us examine the flow of water through a small volume element of a saturated soil. For simplicity, we consider the volume element to be a small rectangular parallepiped with dimensions Δx, Δy, and Δz as shown in Figure 2. Assume that v_x is the water flow in the x direction (i.e., volume of water flow per unit area and time) entering the face $ABCD$ of the volume element. Therefore the *inflow rate,* or the total volume of water flowing into $ABCD$ per unit time, is

$$\text{Inflow rate} = v_x\,\Delta y\,\Delta z \quad (8)$$

Similarly, if $v_x + \Delta x$ is the water flux in the x direction for water flow leaving the face $EFGH$, the total volume of water leaving $EFGH$ per unit time, i.e., the *outflow rate,* is

$$\text{Outflow rate} = v_{x+\Delta x}\,\Delta y\,\Delta z \quad (9)$$

From elementary calculus $v_{x+\Delta x}$ can be evaluated (approximately) from

$$v_{x+\Delta x} = v_x + \frac{\partial v_x}{\partial x}\Delta x \quad (10)$$

where $\partial v_x/\partial x$ is the rate of change of v_x in the x direction. From equations 8, 9, and 10, the net volume of water flow (inflow-outflow) per unit time in the volume element from flow in the x direction is

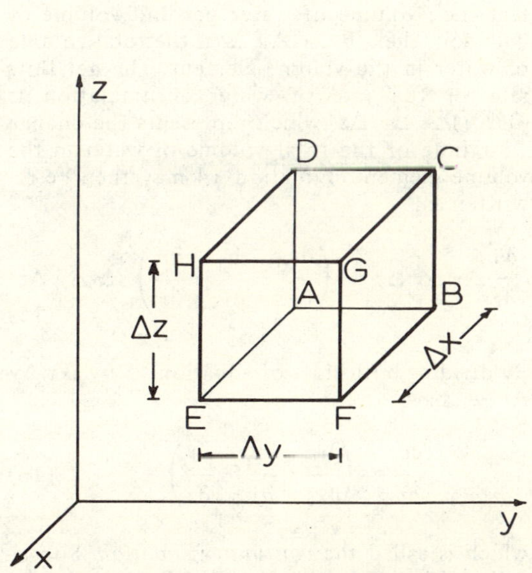

FIGURE 2. Rectangular volume element in water-saturated soil.

$$\text{Inflow rate-outflow rate} = (v_x - v_{x+\Delta x})\,\Delta y\,\Delta z$$

$$= (v_x - v_x - \frac{\partial v_x}{\partial x}\Delta x)\,\Delta y\,\Delta z$$

$$= -\frac{\partial v_x}{\partial x}\Delta x\,\Delta y\,\Delta z \qquad (11)$$

Similarly, the net volume of flow per unit time from flow in the y direction is

$$-\frac{\partial v_x}{\partial y}\Delta x\,\Delta y\,\Delta z \qquad (12)$$

and from flow in the z direction is

$$-\frac{\partial v_z}{\partial z}\Delta x\,\Delta y\,\Delta z \qquad (13)$$

Adding (11), (12), and (13) yields the net volume of water (inflow-outflow) per unit time for the entire volume element from water flow in the x, y, and z directions.

$$\text{Net rate of flow} = -\left(\frac{\partial v_x}{\partial x} + \frac{\partial v_y}{\partial y} + \frac{\partial v_z}{\partial z}\right)\Delta x\,\Delta y\,\Delta z \qquad (14)$$

This net rate of water flow is the volume of water gained or lost within the volume element per unit time. It is often called the *rate of water accumulation.* If θ is the volumetric water content, i.e., volume of water per unit volume of bulk soil, then, $\theta\,\Delta x\,\Delta y\,\Delta z$ is the total volume of water in the volume element. The net flow rate or the rate of water accumulation is $(\partial\theta/\partial t)\Delta x\,\Delta y\,\Delta z$, which represents the change with time of the total volume of water in the volume element. Equation 14 may then be rewritten as

$$\frac{\partial\theta}{\partial t}\Delta x\,\Delta y\,\Delta z = -\left(\frac{\partial v_x}{\partial x} + \frac{\partial v_y}{\partial y} + \frac{\partial v_z}{\partial z}\right)\Delta x\,\Delta y\,\Delta z \qquad (15)$$

By dividing both sides of equation 15 by $\Delta x\,\Delta y\,\Delta z$, we have

$$\frac{\partial\theta}{\partial t} = -\left(\frac{\partial v_x}{\partial x} + \frac{\partial v_y}{\partial y} + \frac{\partial v_z}{\partial z}\right) \qquad (16)$$

which is called the *continuity equation.* Substituting the differential forms of Darcy's law for v_x, v_y, and v_z (equations 4, 5, and 6) into the continuity equation 16, we have

$$\frac{\partial\theta}{\partial t} = -\left(\frac{\partial}{\partial x}K_x\frac{\partial H}{\partial x} + \frac{\partial}{\partial y}K_y\frac{\partial H}{\partial y} + \frac{\partial}{\partial z} + K\frac{\partial H}{\partial z}\right)$$

If the soil is isotropic, that is, $K_x = K_y = K_z = K$, and K is constant, (17) becomes

$$\frac{\partial\theta}{\partial t} = -K\left(\frac{\partial^2 H}{\partial x^2} + \frac{\partial^2 H}{\partial y^2} + \frac{\partial^2 H}{\partial z^2}\right) \qquad (18)$$

Furthermore, if the soil is water saturated, that is, if θ does not change with time, then

$$\frac{\partial^2 H}{\partial x^2} + \frac{\partial^2 H}{\partial y^2} + \frac{\partial^2 H}{\partial z^2} = 0 \qquad (19)$$

which is known as *Laplace's equation* (see Childs, 1969).

Soil Drainage

Soil drainage (q.v.) may be defined as the removal of excess water from the soil root zone (cf. *Rhizosphere*). The drainage problems encountered under field conditions are unlimited. This discussion concerns describing the water flow pattern (*hydraulic head distribution)* and water-discharge rate for a given soil-drainage condition. These conditions are referred to as the "boundary conditions," which include, for example,: rainfall or irrigation intensity, height of the water table, depth and spacings of drains, depth to an impervious layer. For a given soil-drainage problem, the boundary conditions are first outlined. A solution of the problem is then attempted by solving the water flow equation (16 or 19) for the outlined conditions. For simplification purposes, several assumptions are frequently made to arrive at a solution for a drainage problem. Therefore it is essential to use judgment in applying the drainage theory.

To illustrate common drainage problems, Figure 3 shows a schematic of a uniform soil of a hydraulic conductivity K, having open or surface drains at spacings S. The soil is underlain by an impervious layer at depth D from the soil surface. Furthermore, the soil is receiving steady rainfall or irrigation with intensity v, where v is the volume of water per unit soil surface area per unit time. This water intensity maintains a water table with maximum height B midway between the drains and a water level A in the open drain. The corresponding drainage problem for soils having subsurface drains is illustrated in Figure 4. As shown in Figures 3 and 4, the water table may be regarded as the boundary separating the saturated soil region (below) and the unsaturated soil region (above). Therefore, below the water table, water flow is

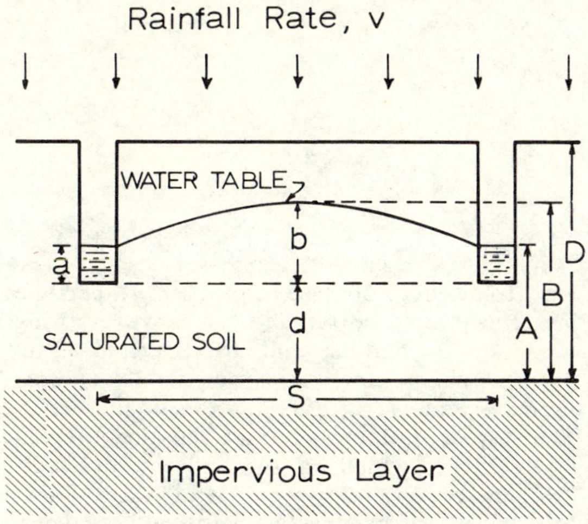

FIGURE 3. Schematic of surface drainage in a uniform soil underlain by an impervious layer. A water table between the open drain is maintained by a steady rainfall rate v.

under saturated conditions, whereas water flow is under unsaturated conditions in the soil region bounded by the soil surface and the water table. Note also that in the saturated soil region, the flow path or water flow lines, that which indicate the direction of water flow to the open or surface drains, are curvilinear in nature.

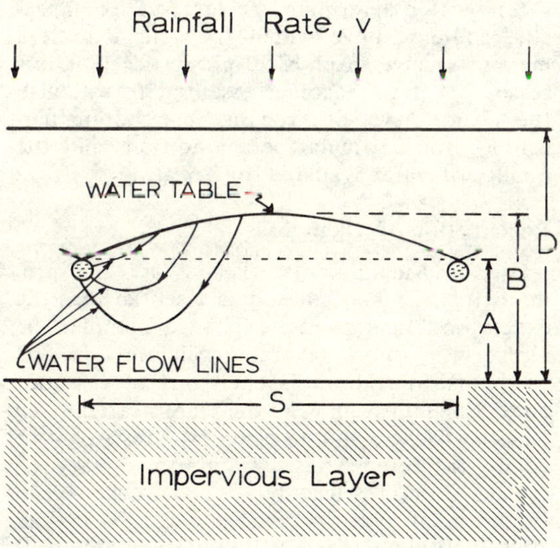

FIGURE 4. Schematic of subsurface draining in a uniform soil. A water table between the subsurface drains is maintained by a steady rainfall rate v.

A simple formula, often used to determine the drain spacing S for soils having surface drains, Figure 3, is

$$S^2 = \frac{4K}{v}(b^2 - a^2 + 2bd - 2ad) \tag{20}$$

This formula clearly shows the dependence of S on the hydraulic conductivity of the soil K and the irrigation or rainfall intensity v. A similar formula for soils having subsurface drains (Figure 4) is

$$S^2 = \frac{4K}{v}(B^2 - A^2) \tag{21}$$

The above drain-spacing formulas (equations 20 and 21) are obtained on the basis of a certain simplification generally known as the *Dupuit-Forchheimer assumptions* (see Kirkham and Powers, 1972). Although these formulas do not provide an accurate description of the flow patterns, they are commonly used and generally yield adequate results (especially where B/S is small). More accurate drain-spacing formulas for the drainage problems shown in Figures 3 and 4 and many others can be found in Harr (1962), Luthin (1957), and Kirkham and Powers (1972).

H. M. SELIM

References

Childs, E. C., 1969, *An Introduction to the Physical Basis of Soil Water Phenomena.* New York: Wiley-Interscience, 493p.

Darcy, H., 1856. *Les fontaines publiques de la ville de Dijon.* Paris: Dalmont.

Harr, M. E., 1962. *Groundwater and Seepage.* New York: McGraw-Hill 315p.

Kirkham, D., and Powers, W. L., 1972. *Advanced Soil Physics.* New York: Wiley-Interscience, 534p.

Luthin, J. N., ed. 1957. *Drainage of Agricultural Lands,* Monograph 7. Madison: American Society of Agronomy, 620p.

Cross-references: *Capillary Pressure; Conductivity, Hydraulic; Field Water Cycle; Imbibition; Infiltratation; Percolation; Permeability: Soil Drainage; Water Fluxes; Water Movement.*

FORENSIC SOIL ANALYSIS—See VOL. XIII: FORENSIC GEOLOGY

FUNGI—See SOIL MICROBIOLOGY

G

GRANULATION—See AGGREGATION

GYPSUM REQUIREMENT

Soils that have accumulated sufficient sodium on the exchange complex to interfere with the normal growth of most crop plants are referred to as *alkali soils* (Figure 1). The term *sodic soil* is often used interchangeably with *alkali soil.* Exchangeable sodium is that fraction of the total sodium in the soil that is adsorbed on the exchange complex and is expressed as a percentage of the total cation exchange capacity of the soil (exchangeable sodium percentage, ESP). Alkali, or sodic, soils usually do not contain any appreciable quantity of neutral soluble salts, and the electrical conductivity (q.v.) of the saturation extract is usually less than 4 mmho/cm at 25°C. In the absence of excess neutral soluble salts, excess exchangeable sodium imparts poor physical properties and high pH to these soils. The high pH results from the hydrolysis of exchangeable sodium or sodium carbonate, which is often present in appreciable quantities in highly deteriorated soils. The sodium hydroxide produced by hydrolysis dissolves the organic matter, which is responsible for the black surface of many alkali soils.

In the absence of neutral salts, alkali soils tend to deflocculate (see *Flocculation*), become impermeable to water and air, have a strong tendency to puddle, and not reaggregate on drying. When wet, the individual colloidal particles repel each other and remain dispersed. The dispersed finer soil fractions clog the soil pores (q.v.), which in turn adversely affects the penetration and movement of water. When dry, these soils are massive, hard, and compact, and they have a tendency to form cracks (Figure 2). Poor soil physical conditions and high pH resulting in nutrient imbalances are the principal causes of poor productivity of soils; in extreme cases, no crop growth is possible.

FIGURE 1. Barren alkali soil with surface accumulation of sodium salts. (Photo K. B. Singh, Karnal, India)

The critical value of exchangeable sodium percentage, above which the soil undergoes sharp deterioration in physical properties is a matter of practical concern. Workers at the U.S. Salinity Laboratory have tentatively suggested an ESP of 15 as the limiting value for characterizing alkali soils. Although this limit is a useful general guide, several factors including soil texture, organic matter content, nature of the clay minerals, and quality of irrigation water determine the level of exchangeable sodium above which the soils exhibit a marked change in physical properties. For this reason it often becomes essential to establish the critical level of exchangeable sodium percentage for particular soil conditions and the quality of water available for irrigation.

Reclamation of Alkali Soils

Use of Amendments. To restore the productivity of alkali soils, it is essential that the excess sodium on the exchange complex be replaced by more favorable calcium ions and the replaced sodium leached out of the root zone. Calcium is an essential plant nutrient (see *Macronutrients*) and an essential constituent of good structured soils. Calcium necessary to replace sodium is usually applied to the soil in the form of gypsum ($CaSO_4 \cdot 2H_2O$) or calcium chloride ($CaCl_2 \cdot 2H_2O$). Other chemicals used to amend alkali soils include acids, e.g., sulfuric acid (H_2SO_4), or chemicals that ultimately produce acid when applied to the soil, e.g., sulfur, lime sulfur, iron sulfate ($FeSO_4 \cdot 7H_2O$), and aluminum sulphate $Al_2(SO_4)_3 \cdot 18H_2O$. The acid reacts with lime ($CaCO_3$)

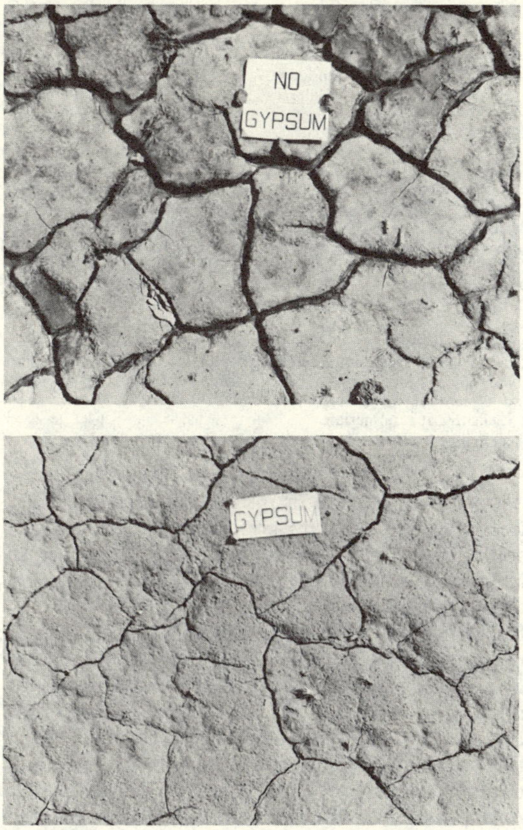

FIGURE 2. Surface physical properties of alkali soils. The untreated soil in (a) is characterized by a shiny surface with large cracks, both due to the setting of dispersed particles. The reclaimed soil in (b) exhibits better aggregation of flocculated soil materials after being treated with gypsum. (Photo K. B. Singh, Karnal, India)

to yield soluble calcium, which in turn replaces exchangeable sodium. The following equations depict the effect of an amendment.

$$2S + 3O_2 + 2H_2O \rightleftharpoons 2H_2SO_4$$ (Reaction brought about by sulfur bacteria in the soil)

$$CaS_5 + 8O_2 + 4H_2O \rightleftharpoons CaSO_4 + 4H_2SO_4$$

$$FeSO_4 + H_2O \rightleftharpoons H_2SO_4 + FeO$$

$$H_2SO_4 + CaCO_3 \rightleftharpoons CaSO_4 + H_2O + CO_2$$

$$\begin{matrix}Na\\Na\end{matrix}\text{ Clay } + CaSO_4 \rightleftharpoons Ca\text{ Clay } + \underset{\text{Leached}}{\underset{\downarrow}{Na_2SO_4}}$$

Table 1 gives the quantity of different amendments required for reclamation relative to 1 ton

TABLE 1. Equivalence of Different Amendments to Pure Gypsum.

Amendment	Tons
Gypsum ($CaSO_4 \cdot 2H_2O$)	1.00
Calcium chloride ($CaCl_2 \cdot 2H_2O$)	0.85
Sulfur	0.19
Sulfuric acid	0.57
Iron sulfate ($FeSO_4 \cdot 7H_2O$)	1.62
Aluminum sulfate [$Al_2(SO_4)_3 \cdot 18H_2O$]	1.29
Calcium polysulfide (CaS_5) (24% sulfur)	0.77

pure gypsum. In practice, gypsum is the most commonly used amendment throughout the entire world because it is economical and abundant.

The quantity of gypsum necessary to reclaim an alkali soil (Figure 3) depends on such factors as the soil texture and its mineralogical make-up, extent of soil deterioration (ESP), and the intended crop growth. In many laboratories the quantity of gypsum required for reclaiming an alkali soil is determined by the *gypsum requirement* (*GR*) *test* (Richards, 1954). The test is performed by mixing a small soil sample (5 g) with a relatively large volume of saturated gypsum solution and measuring the calcium lost from the solution after reaction with the soil. Sodium salts in an alkali soil are so diluted by this treatment that nearly complete displacement of exchangeable sodium by calcium from the gypsum solution occurs. The decrease in calcium from the solution when expressed on the basis of tons of $CaSO_4 \cdot 2H_2O$ per foot of soil is the gypsum requirement of the soil. Many alkali soils contain, in addition to excessive quantities of exchangeable sodium, appreciable amounts of soluble sodium carbonate. In such cases the gypsum requirement test evaluates the amount of calcium required to replace the exchangeable sodium plus that required to neutralize all the soluble sodium carbonate in the soil. Recent studies have shown (Abrol and Dahiya, 1974) that under field conditions a major fraction of the soluble carbonates will leach without reacting with the applied gypsum. For this reason, a modification in the method of determining the gypsum requirement of soils has been proposed (Abrol, Dahiya, and Bhumbla, 1975). In the modified procedure, the soil is washed free of soluble carbonates before proceeding with the GR test. The modified procedure gives a more realistic estimate of the gypsum needs of sodic soils containing varying amounts of soluble carbonates.

Choice of Crops for Alkali Soils. Proper selection of crops and crop rotations during

FIGURE 3. Alkali soils can be unproductive unless reclaimed. As shown in (a), at very high soil ESP, even the rice crop fails unless gypsum is applied. The plot with a good crop, as indicated in (b), received gypsum at the rate of 15 ton/ha. (Photo K. B. Singh, Karnal, India)

reclamation is important in reclamation projects. Crops vary widely with regard to their tolerance to adverse physical conditions and nutritional disorders resulting from excess exchangeable sodium in soils. In the initial years of reclamation, therefore, it is advisable to grow such crops in rotation as would yield economic return under adverse soil conditions. Table 2 lists crops according to their tolerance to exchangeable sodium. Since the ESP of subsurface soil layers modifies the water relations of soils and the depth and proliferation of the root systems, it must be taken into account when making crop selections.

Crop-management Practices. Apart from the use of amendments and crop selection, the

TABLE 2. Relative Tolerance of Crops to Exchangeable Sodium.

Sensitive[a]	Moderately Tolerant	Tolerant
Avocado	Alfalfa	Beets
Beans	Barley	Bermuda grass
Chickpeas	Clover	Crested and fairaway wheat grass
Citrus	Cotton	Para grass
Cotton[b]	Millets	Rice
Cowpeas	Oats	Rhodes grass
Corn	Safflower	Spinach
Deciduous trees	Sugarcane	Tall wheat grass
Maize	Wheat	
Nuts		

[a]Crop yields are seriously affected if the soil ESP is more than about 10, 35, and 55 in the case of sensitive, moderately tolerant, and tolerant crops, respectively.
[b]Sensitive at germination.

adoption of certain crop-management practices is necessary to obtain good crop yields in alkali soils. These practices often differ from those recommended for soils unaffected by exchangeable sodium and have to be evaluated for different agroclimatic conditions. Practices that require special consideration include plant population, sowing method, nature and amount of fertilizers including micronutrients, (q.v.) and the method and frequency of irrigation. With the adoption of modified agronomic practices coupled with the application of amendments, it is possible to obtain good yields of selected crops in the initial years of reclamation, even in high deteriorated soils (Abrol, Dargan, and Bhumbla, 1973.

I. P. ABROL

References

Abrol, I. P.; Dargan, K. S.; and Bhumbla, D. R., 1973. Reclaiming alkali soils, *Centr. Soil Salinity Res. Inst. (Karnal, India), Bull. 2,* 58p.

Abrol, I. P., and Dahiya, I. S., 1974. Flow associated precipitation reactions in saline sodic soils and their significance, *Geoderma,* **11**, 304–312.

Abrol, I. P.; Dahiya, I. S.; and Bhumbla, D. R., 1975. On the method of determining gypsum requirement of soils, *Soil Sci.,* **120**, 30–36.

Pearson, G. A., 1960. Tolerance of crops to exchangeable sodium. *U.S. Dept. Agric. Inf. Bull. 216,* 4p.

Richards, L. A., ed., 1954. Diagnosis and improvement of saline and alkali soils, *U.S. Department of Agriculture Handbook,* 60, 160p.

Cross-references: *Aggregation; Conductivity, Electrical; Exchange Phenomena; Flocculation; Irrigation Water; Leaching; Management of Soils; Salt Leaching; Soil Pores. VOL. XII: Salt-affected Soils.*

H

HEAT CAPACITY

The specific heat capacity of any substance is defined as the quantity of heat in calories required to increase the temperature of 1 g of the substance by 1°C. The unit is the cal g^{-1} °C^{-1}. The heat capacity of a substance can be defined as the product of its specific heat capacity and its mass; the unit is the Cal cm^{-3} °C^{-1}

Soils are mixtures of various constituents, each having its specific heat capacity. Thus the specific heat capacity of the soil is the summation of the specific heat of each of its constituents times their corresponding masses according to equation 1.

$$C_s = \sum_{i=1}^{n} C_{si} M_i \, [\text{cal g}^{-1}\,{}^\circ\text{C}^{-1}] \qquad (1)$$

where C_s and C_{si} are the specific heat values of the soil and the soil constituent i, having a mass M_i.

The heat capacity of a soil constituent equals its specific heat times its density, and the soil's heat capacity will equal the sum of the heat capacities of the different constituents according to equation 2 where

$$C = \sum_{i=1}^{n} C_i \chi_i \, [\text{cal cm}^{-3}\,{}^\circ\text{C}^{-1}] \qquad (2)$$

C and C_i are the heat capacities of the soil and of constituent i occupying a volume fraction χ_i.

Table 1 gives values of specific heat and heat capacities for the soil constituents are given in Table 1. Equation 2 can be simplified by taking average C_i values for respective constituents, namely, 0.48 Cal cm^{-3} °C^{-1} for soil solids, 1.0 Cal cm^{-3} °C^{-1} for water, and 0.60 Cal cm^{-3} °C^{-1} for the soil organic matter (q.v.), respectively. The air content and its addition to the soil's heat capacity are ignored due to air's very low heat capacity. Therefore, equation 2 can be simplified and brought to the form of equation 3 provided that the soil is not made of crushed granite, iron or aluminum oxides. Equations 2 or 3 are used to calculate the soil's heat capacity from its bulk density, texture, and mineralogical constituents

$$C_{soil} = 0.48\,\chi_{solids} + \chi_{water} + 0.6\,\bar{\chi}_{o.matter} \qquad (3)$$

It should be noted that for a rigid soil, only the water content changes with time; thus, the heat capacity of a soil can be expressed as a linear function of the water content or of any of the soil constituents.

AMOS HADAS

TABLE 1. Heat Capacities of Selected Soil Constituents.

Material	C_s (cal g^{-1} °C^{-1})	P (g/cm^{-3})	C	Source
Quartz	0.91	2.65	0.506	1
	0.190		–	3
Granite	0.192	2.6	0.499	2
Dolomite	0.222	2.9	0.643	1
Basalt	0.213	3.0	0.639	2
$CaCO_3$	0.208	2.71	0.563	1
$CaSO_4$	0.195	2.45	0.477	1
Fe_2O_3	0.165	5.24	0.865	1
Al_2O_3	0.217	3.7	0.803	5
Kaolin	0.224	2.6	0.582	1
	0.223			5
Clay	0.27	–	–	4
Silty clay	0.26	–	–	4
Clay minerals	0.181	2.65	0.48	6
Humus	0.477	1.26	0.601	6
Air	0.23	0.00125	0.00029	6
Water	1.00	1.00	1.00	

1. Ulrich (1894), taken from de Vries (1963).
2. Landholt and Börnstein (1967), from de Vries (1963).
3. Kersten (1949).
4. Bower and Hanks (1962).
5. Land (1878), taken from de Vries (1963).
6. de Vries (1963).

References

Bowers, S. A., and Hanks, R. J., 1962. Specific heat capacities of soils and minerals as determined with a radiation calorimeter, *Soil Sci.*, 94, 392–396.

Kersten, M. S., 1949. Thermal properties of soils, *Bull. Univ. Minn. Inst. Tech. Eng. Exp. Stn.*, 28, 1–228.

De Vries, D. A., 1963. Thermal properties of soils, *in* W. R. Van Wijk, ed., *Physics and Plant Environment.* Amsterdam: North Holland, 210–235.

Cross-references: *Bulk Density; Thermal Regimes; Thermodynamics; Water Content and Retention. Vol. IVA: Enthalpy.*

HERBICIDE DIFFUSION

Herbicides are chemicals that kill plants or inhibit their normal growth. Not all herbicides are fully understood by scientists; however, many of these chemicals express their bioactivity through the soil. Regardless of the method of application, portions of all herbicidal sprays reach the soil, which has led to considerable research on the persistence of herbicides in soils. One aspect of persistence that directly influences the bioactivity of a herbicide in soil is that of transport and redistribution. *Transport* is important in other aspects of persistence such as volatilization, microbial decomposition, and plant uptake.

Mathematical Development of Diffusion Equations

Diffusion is an important transport mechanism by which herbicides redistribute in soil. Molecular herbicide diffusion is the process of dissipating any inequality of concentration or activity by the random thermal movement of molecules or ions. Diffusion is a slow process in comparison to transport by soil water (mass flow), but over short distances, it may become very important. The mathematics describing diffusion processes is well advanced (Crank, 1956; Carslaw and Jaeger, 1959; Jost, 1960), and most of the herbicide-diffusion problems in soils have been solved using classical diffusion theory. The modern theoretical treatment of diffusion processes is credited to Fick (1855), who recognized an analogy with heat conduction in solids. There is agreement on the mathematical theory of diffusion, but there seems to be little agreement on the symbols used for several of the parameters; as a result, this article relies on the terminology and symbols used by Olsen and Kemper (1968) and Scott, Phillips, and Paetzold (1974).

Quantitative measurements of the rate at which diffusion of herbicides occurs are expressed in terms of a *diffusion coefficient*. The diffusion coefficient is a quantitative expression of the difficulty that diffusing molecules have in moving through different media. The depth of penetration X of a herbicide diffusing in soil can be estimated by the equation $X = 2\sqrt{Dt}$. Under steady-state conditions, the diffusion coefficient in solution can be defined as the quantity of herbicide transferred across a unit cross-sectional area per unit time per unit gradient of concentration at the section. Thus if J is the flux of herbicide transferred across unit area perpendicular to the direction of flow per unit time, C the concentration of diffusible herbicide molecules in solution, and X the space coordinate, then

$$J = -D_o \frac{dC}{dX} \quad (1)$$

is the definition of the aqueous diffusion coefficient D_o. This definition can be ambiguous unless the frame of reference is specified. Generally the concentration C has the same quantity reference as J (moles or grams), and the volume has the same length reference as A and X (cm). As a result, the diffusion coefficient is usually presented with dimensions of square centimeters per unit length of time. The negative sign in equation 1 indicates that diffusion occurs in the direction of decreasing concentrations.

The magnitude of the diffusion coefficient of a herbicide in soil is a measure of the various physical and chemical factors that affect the transport rate. Olsen and Kemper (1968) showed that for ions or molecules restricted to the solution phase, the porous diffusion coefficient D_p can be related to the diffusion coefficient in aqueous solution, D_o, by

$$D_p{}^m = D_o \left(\frac{L}{Le}\right)^2 \theta\alpha\gamma \quad (2)$$

where $(L/Le)^2$ is the ratio of diffusion pathlength in soil to that in aqueous solution; α is the ratio of the soil water mobility where the herbicide is diffusing to that of pure water; θ is the volumetric soil water content; and γ is the electrical interaction of the ion or molecule. The quantities $(L/Le)^2$, θ, α, and γ can take on values between 0 and 1. Collectively they are known as the *transmission factor*. The superscript m indicates that the measurements are based on the volume of the entire soil mass. In many systems it is more convenient to measure the herbicide concentration with respect to the solution phase. Therefore, if θ is the volumetric soil water content, then

$$\theta C = C^m \quad (3)$$

and *Fick's first law* for herbicide diffusion through soil may be written as

$$J = -D_p{}^m \theta \frac{dC}{dX} \quad (4)$$

In porous materials such as soils, herbicides interact with soil surfaces. As a result, herbicide diffusion can occur in the adsorbed, liquid, and vapor phases. Thus the total flux through the soil would be the sum of the fluxes in each phase. The concentration and mobility of the herbicide in each of these soil phases will vary depending on the relative partitioning between these phases and the interaction between the physical and chemical properties of the pesticide and the soil. As shown by Scott, Phillips, and Paetzold (1974), the apparent diffusion coefficient D_e of a herbicide in soil will depend on the sum of the diffusion rates in each soil phase. Assuming that diffusion occurs only in the solution and adsorbed phases and the adsorption isotherm is linear, D_e can be defined as

$$D_e{}^m = \frac{D_p{}^m}{(\theta + Kd\rho_b)} = \frac{D_p{}^m}{B\theta} \quad (5)$$

where Kd is the slope of the adsorption isotherm, ρ_b is the soil bulk density, and B is the capacity factor or retardation factor. B can be further defined as

$$B = 1 + R \quad (6)$$

where R equals $Kd\rho_b$ divided by the water content θ. The value of D_p is a measure of the ease with which a herbicide diffuses through the pores of the medium.

For transient state (nonsteady) conditions, the conservation-of-mass equation has to be satisfied. This equation implies that the rate of change of herbicide concentration equals the net rate of flow over the boundaries of the volume element. Combination of the equation of conservation of mass with equation 4 yields

$$\frac{\partial C}{\partial t} = \frac{\partial}{\partial X}\left(D_e{}^m \frac{\partial C}{\partial X}\right) \quad (7)$$

which is known as *Fick's second law*. If the herbicide-diffusion coefficient is shown to be independent of concentration, equation 7 simplifies to

$$\frac{\partial C}{\partial t} = D_e{}^m \frac{\partial^2 C}{\partial X^2} \quad (8)$$

which can be solved analytically for several initial and boundary conditions.

Magnitude of Herbicide Diffusivities

The apparent diffusion coefficient of a herbicide in soil is a complex quantity and has been shown to be influenced by several physical and chemical properties of soil and herbicide. Table 1 lists the chemical names of several herbicides.

First to be considered is the diffusion of the herbicide in a bulk quantity of the phase considered. In porous materials such as soils, these would include the vapor, solution, and adsorbed phases. According to Hartley (1964), the diffusion coefficients of herbicides in air are 10,000 times greater than in aqueous solution. He postulated that with the less polar, nonionized herbicides, air diffusion may be much more important than diffusion in aqueous solution. Saxena et al (1974) used the agar and cell methods to measure the self-diffusion coefficients of 2, 4-D in aqueous solution. The observed diffusivity was approximately 0.80 × 10^{-5} cm^2/s at 25°C. Scott and Phillips (1973) used the capillary tube technique to determine self-diffusion coefficients of several herbicides in aqueous solution. The magnitude of the values were approximately 0.6 × 10^{-5} cm^2/s at 23°C, which agreed fairly well with published data on molecules of similar molecular weight. No relation was observed between the configuration of the molecule and diffusivity. They also used the agar technique to determine the self-diffusion coefficient of chlorpropham. Extrapolation of the diffusion coefficient to 0% agar gave a value approximately equal to 0.70 × 10^{-5} cm^2/s at 25°C. Brown (1974) also used the capillary tube technique to determine the diffusivity of metribuzin in aqueous solutions. The magnitude of the diffusivity was 0.94 × 10^{-5} cm^2/s at 25°C. Table 2 summarizes the results of these three papers. Scott, Phillips, and Paetzold (1974) derived an equation from Fick's second law from which the diffusion coefficient in the adsorbed

TABLE 1. Chemical Designations of Herbicides Mentioned in Text.

Common Name	Chemical Name
Atrazine	2-chloro-4-ethylamino-6-isopropylamino-*s*-triazine
2,4-D	2,4-dichlorophenoxyacetic acid
Fluometuron	3-(*m*-trifluoromethylphenyl)-1, 1-dimethylurea
Prometryne	2,4-bis(isopropylamino)-6-methylthio-*s*-triazine
Propazine	2-chloro-4,6-bis(isopropylamino)-*s*-triazine
Metribuzin	4-amino-6-*tert*-butyl-3-(methylthio)-*as*-triazin-5(4H)one
Prometone	2,4-bis(isopropylamino)-6-methoky-*s*-triazine
Diphenamid	N,N-dimethyl-2,2-diphenylacetamide
Chlorpropham	isopropyl *m*-chlorocarbanilate
Trifluralin	α,α,α-trifluoro-2,6-dinitro-*N*,*N*-dipropyl-*p*-toluidine
Benefin	N-butyl-N-ethyl-α,α,α-trifluoro-2,6-dinitro-*p*-toluidine
Propachlor	2-chloro-N-isopropyl acetanilide

TABLE 2. Self-diffusion Coefficients and Molecular Weights of Several Herbicides in Aqueous Solution. The Methods Used Include Capillary Tubes (CT), Diffusion Cells (DC), and Agar (A).

Herbicide	Method	Molecular Weight	Temperature (°C)	Self-diffusion Coefficient ($D \times 10^5$ cm²/s)
Fluometuron	CT	232.1	23	0.60 ± 0.04 (13)
Diphenamid	CT	239.1	23	0.57 ± 0.01 (13)
Prometone	CT	225.1	23	0.66 + 0.07 (13)
Prometryne	CT	241.4	23	0.68 ± 0.06 (13)
Simazine	CT	201.7	23	0.58 ± 0.19 (13)
Atrazine	CT	215.7	23	0.66 ± 0.10 (13)
2,4-D	CT	221.0	23	0.58 ± 0.03 (13)
	DC	221.0	25	0.79 (11)
	A	221.0	25	0.80 (11)
Chlorpropham	CT	213.7	23	0.64 ± 0.13 (13)
	A	213.7	25	0.70 (13)
Metribuzin	CT	214.3	25	0.94 ± 0.01 (2)

phase D_a^m may be calculated. The assumptions used to derive their equation, equation 9, were (1) no precipitation and volatilization, (2) a linear adsorption isotherm, and (3) no net change in herbicide concentration in the adsorbed phase. The equation is

$$D_a^m = \frac{1}{R} \left[(1 + R) D_e^m - D_s^m \right] \tag{9}$$

where D_s^m is the diffusion coefficient in the solution phase, and the other quantities have been defined previously. Based on the preceding assumptions, their results showed that the magnitude of D_a^m was less than D_s^m and decreased with a corresponding decrease in water content. Negative values of D_a^m were calculated at the two lower water contents. Since this result was not physically possible, they attributed it to an increase in the amount of metribuzin in the adsorbed phase as the soil water content decreased.

The second factor influencing the magnitude of the apparent diffusion coefficient is the volume fraction of a phase and related geometric characteristics. This factor includes the effects of soil moisture content and bulk density. As a result, herbicides diffuse more slowly in soil than in air or free solution because the pathway through the pores is restricted and tortuous. The exact geometry of the pathway depends on the nature of the pore space (see *Soil Pores*) formed by the particles and on the water content. The exact way these factors influence diffusion of a herbicide depends on its solubility and volatility that govern the relative amounts moving in the soil solution (q.v.), vapor and adsorbed phases. Lindstrom, Boersma, and Gardiner (1968) measured the diffusion coefficient of 2,4–D in nine saturated Oregon soils. Soil columns 10 cm long were packed with soil to a bulk density of approximately 1.1 g/cm³, wetted and equilibrated for a day. They applied ^{14}C-labeled 2,4–D, and then after three days, they used the distribution of chemical to calculate the diffusion coefficient. The diffusivity in these soils ranged from a low of 0.06×10^{-6} cm²/s in a Mulkey loam to a high of 4.3×10^{-6} cm²/s in a quartz sand.

Diffusion coefficients of several herbicides through Lanton silty clay loam at different soil water contents were measured by Scott and Phillips (1972). Measurements were usually made at three soil water contents. The diffusivities for prometone, atrazine, simazine, prometryne, diphenamid, and chlorpropham all increased as the soil water content increased. Trifluralin was not greatly affected by soil water content, which was attributed to a considerable contribution from vapor phase diffusion. Lavy (1970) observed that the diffusion rates of atrazine, simazine, and propazine increased with increasing soil moisture content. Bode et al (1973) studied the diffusion of trifluralin in Mexico silt loam as a function of soil moisture, bulk density, and temperature. Trifluralin was found to move primarily in the vapor phase. Values of D_e increased from air dry soil (3.1% water content by weight) to 8.1% and then decreased as moisture content increased. A 15-term equation was developed to predict the logarithm of D_e as a function of these soil physical properties. Values of D_e ranged from 3.8×10^{-11} cm²/s to 2.8×10^{-11} cm²/s. Ritter, Johnson, and Lovely (1973) reported results of diffusion studies of atrizine and propachlor in Ida silt loam with 2% organic matter. Increasing soil moisture from 8 to 23% (by weight) and bulk density from 0.9 to 1.3 g/cm³ decreased the movement of both herbicides.

The third main parameter is partition of the herbicide between the soil phases. This aspect influences the concentration of herbicide in the phases and includes the effects of adsorption by soil surfaces, volatilization, and solubility in soil water. Thus the partition of the herbicide between the soil phases depends on both soil and chemical properties. Walker and Crawford (1970) studied the diffusion of two triazines, propazine and prometryne, in six soils of varying soil organic matter contents. The magnitude of the diffusion coefficients varied from 1.5×10^{-7} cm²/s for propazine in a sandy soil to 3.1×10^{-9} cm²/s for prometryne in an organic soil. The diffusion coefficients decreased as the value of the distribution coefficient, *Kd*, increased being closely approximated by a slope of –1. Their calculations using measured diffusion coefficients indicated that the distances moved by diffusion can be only a few millimeters per month. Lavy (1970), using the autoradiograph technique, observed the diffusivities of three chloro *s*- triazines in eight soils. The average diffusion coefficients were 1.52, 0.45, and 0.24 × 10^{-7} cm^2/s for atrazine, simazine, and propazine, respectively. Diffusion rates were highly correlated with total surface area of the soil. Lindstrom, Boersma, and Gardiner (1968) found that a plot of retentive ability of 2,4-D versus percent organic matter for nine soils demonstrated a strong relationship between these two factors despite a wide range in pH, clay content, and cation exchange capacity. Values of the porous diffusion coefficient ranged from a low of 1.24 × 10^{-6} cm^2/s to a high of 7.05 × 10^{-6} cm^2/s. Scott and Phillips (1972) studied the relationship between the distribution coefficient of several herbicides having differing physical and chemical properties in a Lanton silty clay loam and the one-day self-diffusion coefficient. When the self-diffusion coefficient of all herbicides was compared with the amount adsorbed, the correlation coefficient r was insignificant. However, when the volatile herbicides were deleted, a highly significant r of −0.85 was obtained. This indicated the adsorption of these herbicides by the Lanton soil significantly affected the amount in the solution phase—and thereby the rate of diffusion through the soil—and also that the diffusivity decreased in proportion to the extent of adsorption.

The last main aspect is the influence of processes that are time dependent. These include primarily those processes associated with microbial degradation. Scott and Phillips (1972) conducted self-diffusion experiments as a function of time with several ^{14}C-herbicides in a Lanton silty clay loam. The soil moisture content and temperature was 25% (by weight) and 25°C, respectively. Their results showed that diffusion time significantly affected the self-diffusion coefficients of 2,4-D, chlorpropham, diphenamid, and prometone at the 5% probability level. Generally, these herbicides are among those that are rapidly degraded in warm, moist soils that contain moderate amounts of organic matter. The self-diffusion coefficients of fluometuron, atrazine, simazine, prometryne, and benefin were found not to be significantly affected by diffusion time. This was attributed to a slow rate of degradation of the parent molecule or to a similar mobility of the radioactive degradation products. These results are repeated in Table 3. Tucker (1969) showed that increasing the diffusion time from 24 to 192 hours did not have a significant effect on the self-diffusion coefficient of fluometuron in agar but did affect the rate of diffusion in a Crowley silt loam.

H. D. SCOTT

TABLE 3. Influence of Diffusion Time on the Magnitude of the Self-diffusion Coefficients of Several Herbicides in a Lanton Silty Clay Loam. The Moisture Content and Temperature was 25% and 25°C, Respectively.

	Diffusion Time 3 (Days)					
Herbicide	1	2	3	5	7	8
2,4-D	7.3*	5.0	3.5			2.2
Fluometuron	2.5	1.8	1.7	1.9		
Diphenamid	1.0	2.0	1.6		0.8	
Chlorpropham	1.0	0.8	0.5		0.4	
Atrazine	1.5	1.1	1.3			1.2
Simazine	1.3	1.2	1.1			1.3
Prometone	2.9	3.6	1.6			0.7
Prometryne	0.6	0.6	0.8		0.4	
Trifluralin	0.3	0.1	0.3		0.1	
Benefin	0.1	0.1	0.1		0.1	

*Diffusion coefficient ($D \times 10^7$ cm^2/sec).

References

Bode, L. E.; Day, C. L.; Gebhardt, M. R., and Goering, C. E., 1973. Prediction of trifluralin diffusion coefficients, *Weed Sci.*, **21**, 485–489.

Brown, D. A., 1974. A capillary tube diffusion cell for measuring ion diffusion in aqueous solutions, *Soil Sci. Soc. America Proc.*, 38, 533–535.

Carslaw, H. S., and Jaeger, J. C., 1959. *Conduction of Heat in Solids.* Oxford: Oxford University Press, 510p.

Crank, J., 1956. *The Mathematics of Diffusion.* Oxford: Oxford University Press, 347p.

Hartley, G. S., 1964. Herbicide behavior in soil, *in* L. J. Audus, ed., *The Physiology and Biochemistry of Herbicides.* New York: Academic, 111–161.

Jost, W., 1960. *Diffusion in Solids, Liquids, Gases.* New York: Academic, 558p.

Lavy, T. L., 1970. Diffusion of three chloro s-triazines in soil, *Weed Sci.*, **18**, 53–56.

Lindstrom, F. T.; Boersma L., and Gardiner, H., 1968. 2,4-D diffusion in saturated soil: A mathematical theory, *Soil Sci.*, **106**, 107–113.

Olsen, S. R., and Kemper, W. D., 1968. Movement of nutrients to roots, *Adv. Agron.*, **20**, 91–151.

Ritter, W. F.; Johnson, H. P.; and Lovely, W. G., 1973. Diffusion of atrazine, propachlor, and diazinon in a silt loam soil, *Weed Sci.*, **21**, 381–384.

Saxena, S. K.; Boersma, L.; Lindstrom, F. T.; and Young, J. L., 1974. The self-diffusion coefficients of ^{45}Ca and 2,4-dichlorophenoxyacetic acid, *Soil Sci.*, **117**, 14–20.

Scott, H. D., and Phillips, R. E., 1972. Diffusion of selected herbicides in soil, *Soil Sci. Soc. America Proc.*, 36, 714–719.

Scott, H. D., and Phillips, R. E., 1973. Self-diffusion coefficients of selected herbicides in water and estimates of their transmission factors in soil, *Soil Sci. Soc. America Proc.*, 37, 965–967.

Scott, H. D.; Phillips R. E.; and Paetzold, R. F., 1974. Diffusion of herbicides in the adsorbed phase, *Soil Sci. Soc. America Proc.*, 38, 558–562.

Tucker, M. R., 1969. *The Effect of Nitrogen upon the Self-diffusion of Fluometuron in Soils.* M.S. Thesis (unpubl.). Fayetteville: University of Arkansas.

Walker, A., and Crawford, D. V., 1970. Diffusion coefficients of two triazine herbicides in six soils, *Weed Res.*, **10**, 126–132.

Cross-references: *Clay-organic Reactions; Diffusion Processes; Pollution.*

HUMIC ACID-FULVIC ACID RATIO

The need to subdivide the types of humus found in different soils for soil classification and characterization has caused the soil chemist many headaches. Soil humus is difficult to analyze and has been insufficiently studied in the past. Most earlier chemists felt that the ratio of carbon to nitrogen was the best criterion (see *Carbon-Nitrogen Ratio*), partly because nitrogen was vital to plant growth and partly because the carbon content should be a fair indication of the total humus content of the soil. Humic acid and fulvic acid had been described, but there was considerable doubt as to whether fulvic acid was a distinct group.

In 1949 Tyurin showed that the C:N ratio was highest in the Chernozems and decreased somewhat both north and south. The range of C:N ratios was, however, not very different in most of the soils, and artificial fertilizers could cause marked discrepancies. Actual ranges within a small area proved to be considerable. Although the C:N ratio was used in earlier versions of the U.S. System of Soil Classification, it has now been abandoned.

Tyurin and his coworkers have also shown that the humic acid–fulvic acid ratio is a much more diagnostic feature. It is also probably less affected by the use of artificial fertilizers. It appears that humic-acid content changes regularly with soil type (Table 1), and Tyurin regarded the conditions favoring humic-acid production as being the same as those favoring the total humus content. The fulvic-acid content was taken to be the sum of the organic substances in the acid solution after precipitating humic acid from the alkali extract. Changes in the fulvic acid are more irregular, but there is a strong inverse relationship between the fulvic-acid and humic-acid content. It is this inverse relationship that makes the ratio so diagnostic.

Table 1 provides data on C:N ratio, average humus content and humic acid-fulvic acid ratios for the main soil types of the USSR, (after Kononova, 1961; Muresanu, Czeizler, and Valceanu, 1963). Currently work is being conducted to determine whether the ratios can be applied to the full range of Canadian soils (J. Dormaar, personal communication, 1976) and will be reported at the next conference of the International Society of Soil Science in 1978 (Edmonton, Alberta, Canada). Recent studies (Akramov and Aliyev, 1974; Schnitzer and Skinner, 1974; Neyroud and Schnitzer, 1974) have shown that the humic and fulvic acids are chemically distinct. The fulvic acids are more phenolic in character, and it appears that oxidation of humic substances produces phenolic and benzenecarboxylic acids as major products in addition to smaller amounts of aliphatic dicarboxylic acids. Their proportions vary with climate. Whether these products arise from more complex structures of lignin or soil microorganisms, or whether they occur

TABLE 1. Composition of the Humus Fraction for Some Soils in the USSR and Romania.

Soil Type	Average Humus (%)	% of Total Humus: Humic Acids	Fulvic Acids	Humic Acid / Fulvic Acid Ratio	C:N Ratio
Podzol	–	–	–	0.1	–
Tundra soils	1.0	10	30	0.3	–
Podzolic soils	2.5–3.0	12–15	25–28	0.6	9.0–10.8
Brown forest podzolized soils	4.0–8.0	25–30	30–35	0.7–0.9	–
Gray forest soils	4.0–6.0	25–30	25–27	1.0	10.7
Krasnozems	4.0–6.0	15–20	30	0.6	18.9
Chernozems	5.5–10.0	28–40	20	1.4–2.5	8.6–11.5
Chestnut soils	2.0–4.0	29–35	20	1.4–1.7	8.2–11.3
Brown soils	1.5–2.0	16	23	0.7	–
Serozems	0.8–2.0	17–20	24–25	0.7–0.8	7.8–8.4
Alluvium	–	–	–	0.69	–
Glei soils	–	–	–	1.31	–

(After Kononova, 1961; Muresanu, Czeizler, and Valceanu, 1964).

in the forms in which they were isolated but held together by hydrogenbonding or Van der Waals forces, remains to be determined. Certainly the humic acid–fulvic acid ratio criterion shows greater promise than any other yet suggested for organic matter characterization.

STUART A. HARRIS

References

Akramov, Yu, and Aliyev, S. A., 1974. Elementnyy sostav guminovykh kislot pochv [Elemental composition of humic acids in soils], *Akademiya Nauk Tadzhikskoy SSR, Doklady, Dushanbe.*, **17**(6), 46–48.

Kononova, M. N., 1961. *Soil Organic Matter.* New York: Pergamon, 450p.

Muresanu, P. L.; Czeizler, G.; and Valceanu, N., 1964. Investigations on physical and chemical properties of humus of some soil types, *8th Internat. Congr. Soil Sci. Trans.* (*Bucarest*), **II**, 23.

Neyroud, J. A., and Schnitzer, M., 1974. The chemistry of high molecular weight fulvic acid fractions, *Can. Jour. Chem.*, **24**, 4123–4132.

Schnitzer, M., and Skinner, S. I. M., 1974. The low temperature oxidation of humic substances, *Can. Jour. Chem.*, **24**, 1072–1080.

Cross-references: *Humus; Carbon-Nitrogen Ratio.*

HUMUS

The *term* humus, as normally used, refers to the organic compounds in soils and sediments exclusive of undecayed plant and animal residues. Two groups of compounds are considered to be present: (1) *nonhumic substances,* consisting of the metabolic products of mircoorganisms, and (2) *humic substances,* a series of high-molecular-weight, brown to black substances formed by secondary synthesis reactions (see *Soil Components, Organic*). Included in the first group are constituents of the well-known classes of organic chemistry such as amino acids, carbohydrates, fats, waxes, resins, and organic acids. Compounds in the second group are distinctive to the soil or sediment environment because they have characteristics unlike polymers from other natural sources. The two groups are not easily separated because some nonhumic substances such as carbohydrates may be covalently bound to the humic matter (Dubach and Mehta, 1963).

Humus probably contains most, if not all, of the biochemical compounds synthesized by living organisms. However, many of these compounds are present in exceedingly small amounts. In general, from 50 to 85% of the organic matter in sediments and terrestrial soils can be recovered as brown to black, colloidal pigments by alkali extraction.

Fractionation of Humus

The methods used to fractionate humus include (1) those based on the solubility characteristics of the major compounds occurring in plant tissues, and (2) those based on the division of the humus proper into subclasses possessing similar solubility characteristics.

In the first method a single sample of soil, sediment, or compost is given a series of treatments designed to dissolve specific classes of organic compounds, namely, ether extraction for fats, waxes, and oils (0.5 to 5%); alcohol extraction for resins (0.6 to 3%); hydrolysis with 2% HCl for "hemicellulose" (6 to 12%); digestion with 80% H_2SO_4 for cellulose (3 to 5%); and analysis of the final soils residue for nitrogen and carbon to give "protein" (30 to 40%) plus "lignin-humus" (35 to 50%). A major limitation of this approach is that most of the organic matter is recovered in the vague fractions designated as "protein" (N × 6.25) and "lignin-humus." Use of the two terms to describe major humus fractions is probably unfortunate because not all the nitrogen occurs as "protein," and the "lignin humus" may not be derived from lignin. Also the fractions labeled "hemicellulose" and "cellulose" may contain other components.

Improved isolation techniques and hydrolysis procedures have shown that from 10 to 25% of the soil humus occurs as proteinaceous constituents, part of which may be in the form of amino acids or peptides covalently bound to the humic substances (Parsons and Tinsley, 1975). Proteins per se have yet to be isolated from soils. An equal quantity of the humus exists as carbohydrates, with about one-third being present as intact polysaccharides (Greenland and Oades, 1975). The lipid fraction of humus is exceedingly complex and contains a variety of compounds ranging from simple organic acids to intermediate chlorophyll-degradation products to complex waxes and resins (Braids and Miller, 1975; Morrison, 1969; Stevenson, 1966).

The second fractionation method is based on the extraction of organic matter (q.v.) from soil with base and further subdivision of the extracted material by partial precipitation with mineral acids and by partial solubility in alcohol. By this scheme the following groups are obtained: the *humin fraction,* insoluble in alkali; the *humic-acid fraction,* soluble in alkali, insoluble in acid; and the *fulvic-acid fraction,* soluble in alkali, soluble in acid. Other groups are: the *hymatomelanic-acid fraction,* alcohol-soluble part of humic acid; and the *β-humus fraction,* fulvic-acid material, insoluble at pH 4.8. The method of fractionation is given in Figure 1.

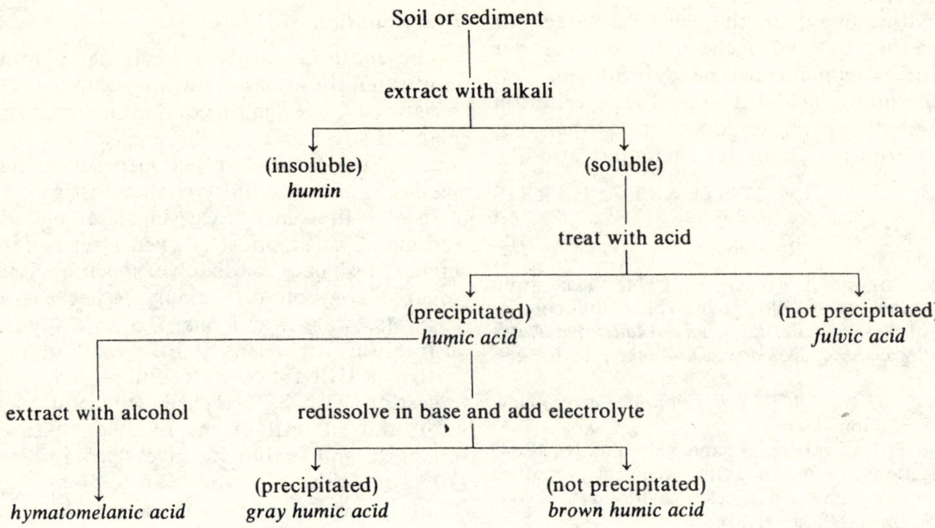

FIGURE 1. Fractionation of humus based on solubility characteristics.

German scientists further subdivide humic acid into two groups by partial precipitation with electrolytes. The first group, the brown humic acids (braunhuminsäuren), are not easily coagulated by electrolytes and are characteristic of the humic acids in brown coals, peats, and podzolic soils. The second group, the gray humic acids (grauhuminsäuren), are easily coagulated and are characteristic of humic acids in Chernozems and Rendzinas (Kononova, 1961).

The relative part of the humic matter that occurs as humic acid varies widely. The humic acid–fulvic acid ratio (q.v.) is higher in grassland soils than in forest soils, and it is higher in the surface layer than in the subsoil. The so-called humin fraction may not represent a separate group but may be present as humic and fulvic acids that occur in stable combination with mineral matter.

Properties of Humic Substances

Nearly all humic substances are weakly dissociated, colloidal acids. They exhibit buffering in acid, neutral, and slightly alkaline pH ranges. A wide variety of functional groups have been reported to be present (carboxyl; phenolic-, alcoholic-, and enolic OH; quinones and hydroxyquinones; lactones; esters). Most humic preparations contain trace amounts of methoxyl groups (OCH_3). The elemental composition and distribution of functional groups in the humic substances isolated from a Podzol are given in Table 1. Typical molecular formulas reported for humic substances are given in Table 2.

TABLE 1. Elemental Composition and Functional Groups in Humic Preparations from a Podzol Soil. (From Wright and Schnitzer, 1960)

	Elemental Composition (%)[a]				Functional groups (me/100 g)[a]					
Preparation	C	H	N	O + S	COOH	Phenolic OH	Alcoholic OH	C = O	OCH_3	Total Acidity
Fulvic acid, *Ao* horizon	49.93	4.67	1.27	44.13	610	280	460	310	30	890
Fulvic acid, *Bh* horizon	46.94	3.48	2.31	47.27	720	360	280	–	20	1,080
Humic acid, *Ao* horizon[b]	58.13	5.17	2.14	34.56	280	290	300	300	50	570
Humic acid, *Bh* horizon[c]	56.69	4.94	2.47	35.90	370	290	–	–	40	660

[a]Dry ash-free basis.
[b]Originally listed as *Ao* humic-II.
[c]Originally listed as *Bh* humic.

TABLE 2. Molecular Formulas for Humic Substances.

Source	Formula	References
Humic acid (Japan)	C265H258076 $(OCH_3)(COCH_3)$ $(OH)_{12}(COOH)_{12}$	Hosoda and Takota (1957)
Electrodialyzed humic material from podzol *Ao* horizon	$C_{75}H_{33}O_{17}N_3$ $(COOH)_3$ $(OH)_{12}(CO)_2$	Schnitzer and Desjardins (1962)
Electrodialyzed humic material from podzol *Bh* horizon	$C_{21}H_{12}(COOH)_6$ $(OH)_5(CO)_2$	Schnitzer and Desjardins (1962)

Many investigators now believe that all dark-colored humic substances are part of a system of closely related, but not completely identical, high-molecular-weight polymers (Stevenson and Butler, 1969). According to this concept, differences between the various representatives (humic acids, fulvic acids, etc.) can be explained by variations in molecular weight, numbers of functional groups (carboxyl, phenolic OH, etc.), and extent to which polymerization reactions have occurred. The postulated relationships are depicted in Figure 2, in which it can be seen that carbon and oxygen contents, acidity, and degree of polymerization all change systematically with increasing molecular weight. The low-molecule-weight fulvic acids have higher oxygen but lower carbon contents than the high-molecular-weight humic acids. Also the total acidities of fulvic acids (usual range 900 to 1400 meq/100 g) are considerably higher than for humic acids (usual range 400 to 870 meq/100 g). In as much as fulvic acids are less susceptible to precipitation with acids and polyvalent cations than humic acids, they are the constituents usually responsible for the brownish color of many natural waters (see Gjessing, 1976; Schnitzer and Khan, 1972).

A diagram illustrating the distribution of oxygen-containing functional groups in lignites

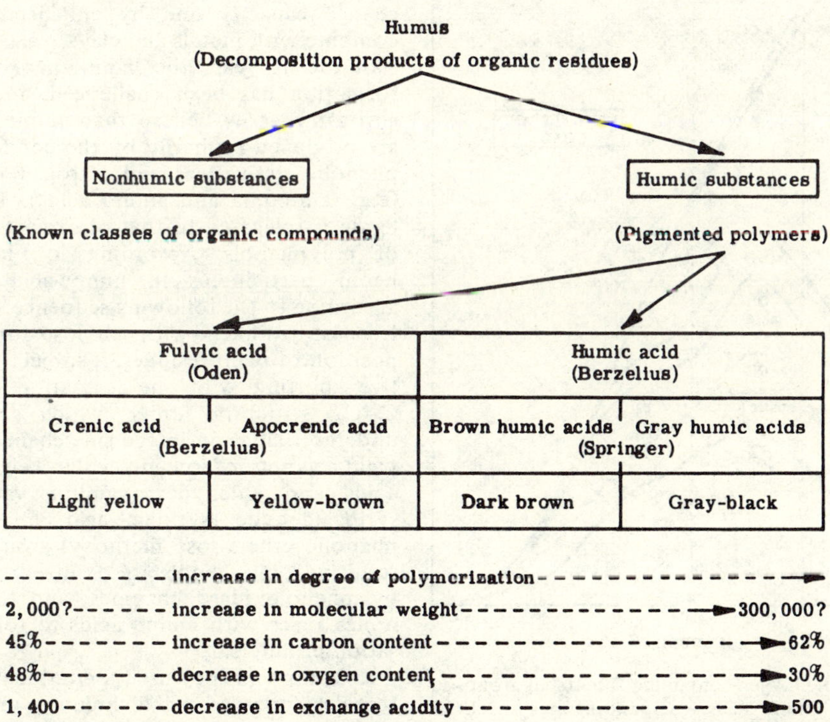

FIGURE 2. Classification and chemical properties of humic substances. (Scheffer and Ulrich, 1960)

and coals as related to their carbon and oxygen contents has been prepared by Blom, Edelhausen, and van Krevelen (1957). From similar data published in the literature for humic and fulvic acids, it has been possible to extend this diagram to pigments having higher oxygen, but lower carbon, contents. This modification is shown in Figure 3. On the logical assumption that humic and fulvic acids are involved in the coalification process, the COOH groups disappear first, followed in order by OCH_3 and OH groups.

From the foregoing, it is apparent that humic substances consist of a heterogeneous mixture of compounds. Each fraction (humic acid, fulvic acid, etc.) must be regarded as being made up of a series of different sizes, few having precisely the same structural configuration or array of reactive groups. For humic acids, contemporary investigators favor a "type" molecule consisting of micelles of polymeric nature, the basic structure of which is an aromatic ring of the di- or trihydroxyphenol type bridged by -O-, $-CH_2-$, -NH-, -N=, -S-, and other groups and containing both free OH groups and the double linkage of quinones. The hypothetical structure for humic acid shown in Figure 4 contains many of the requirements for a "typical" humic acid, including free and bound phenolic OH groups, quinone structures, nitrogen and oxygen as bridge units, and COOH groups variously placed on aromatic rings. Despite early reports to the contrary, most x-ray diffraction studies indicate that humic substances are amorphous.

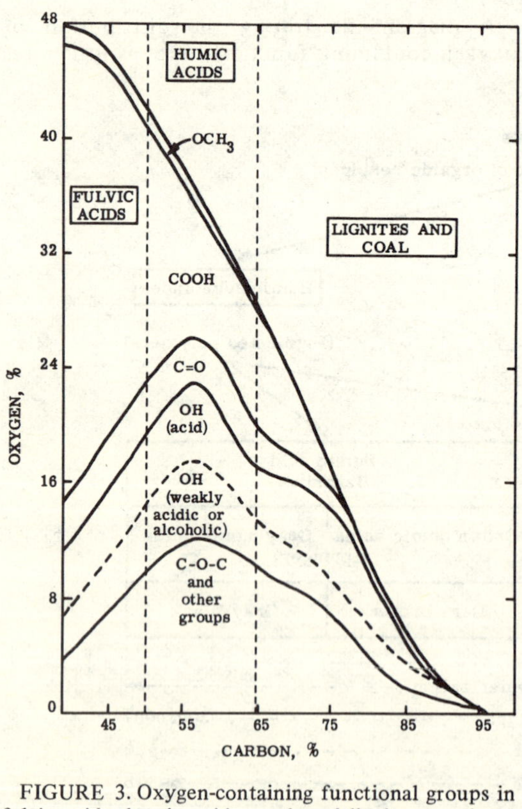

FIGURE 3. Oxygen-containing functional groups in fulvic acids, humic acids, coal, and lignites as related to carbon and oxygen contents. (After Blom, Edelhausen, and van Krevelen, 1957)

Origin of Humic Substances

The classical theory of the origin of humic substances is that they are derived from plant lignins. Waksman (1936), for example, popularized the view that humic acids are formed by condensation of lignins and proteins (the ligno-protein theory). An extension of this theory is that humic substances are formed from lignins by autooxidation and ammonia fixation. Support for the lignin theory of humus formation has been provided by the finding that lignins are highly resistant to attack by microorganisms and that oxidized lignins have many of the properties of humic acids. Assuming a lignin origin, the major changes in lignin involve the oxidation of side chains (introduction of carboxyl groups) and loss of methoxyl groups (increase in phenolic OH groups). These changes would be reflected by an increase in base exchange capacity and by enhanced ability to combine with metals and clays.

In recent years the lignin theory of humus formation has been challenged, and many investigators now believe that humic substances are produced primarily by the condensation of phenolic derivatives and nitrogen compounds (e.g., ammonia and amino acids). The role of lignin is relegated to that of serving as a source of polyphenols. According to Flaig (1966), lignin participates in humic-acid formation according to the following sequence: (1) Lignin, freed of its linkage with cellulose during decomposition of plant residues, is subjected to oxidative splitting with the formation of primary C_6-C_3 structural units, which subsequently undergo oxidation in the side-chain position to yield a variety of low-molecular-weight phenolic acids and aldehydes (vanillin, vanillic acid, syringaldehyde, syringic acid, etc.); (2) the phenolic ethers lose methoxyl groups, and the resulting polyphenols are oxidized to quinones by phenoloxidase enzymes; and (3) the quinones react with amino acids to form condensation products.

Kononova (1966) used a combination of histological microscopic techniques and chemical analyses to study the decomposition of plant remains and concluded that humic substances

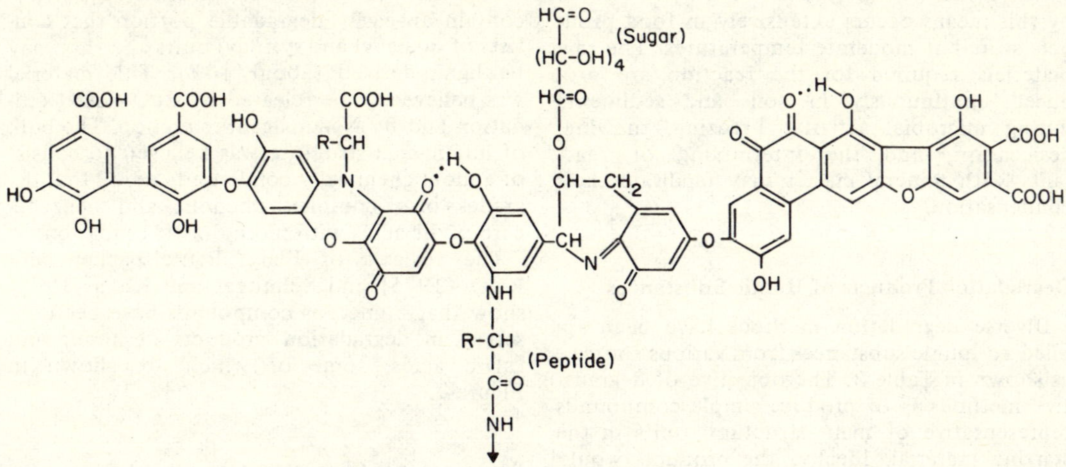

FIGURE 4. Hypothetical structure for humic acid.

were being formed by cellulose-decomposing myxobacteria *before* the lignin had begun to decompose. She postulated that polyphenols, synthesized by the myxobacteria, and phenol derivatives, released during the subsequent decay of lignin, served as source materials for humus formation.

Many microorganisms, including bacteria, actinomycetes, and fungi, produce humic-like substances when grown on laboratory culture media. In some instances phenoloxidase enzymes and free polyphenols have been detected in the metabolic products. According to Martin and Haider (1971), microscopic fungi of the Imperfecti group play a significant role in the synthesis of humic substances in soil. Their studies have shown that such fungi as *Hendersonula toruloidea, Epicoccum nigrum, Stachybatrys atra, S. chartarum,* and *Aspergillus sydowi* degrade lignin as well as cellulose and other plant constituents, and in the process they synthesize appreciable amounts of humic-acid–like polymers in which part of the phenolic units originate from lignin and part through synthesis by the fungi. Leaf polyphenols are believed to be important in humus formation in forest soils. Possible sources of polyphenols for the synthesis of humic acids are summarized in Figure 5.

The idea that carbohydrates also participate in the formation of humic substances has persisted since the middle of the last century. The reactions involved (Figure 6) are similar to those that take place during the formation of brown nitrogenous polymers by condensation of carbonyl and amino derivatives (the Maillard reaction). The formation of colored polymers

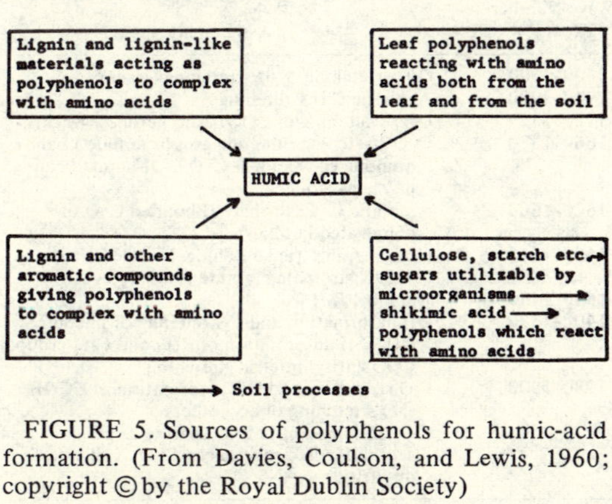

FIGURE 5. Sources of polyphenols for humic-acid formation. (From Davies, Coulson, and Lewis, 1960; copyright © by the Royal Dublin Society)

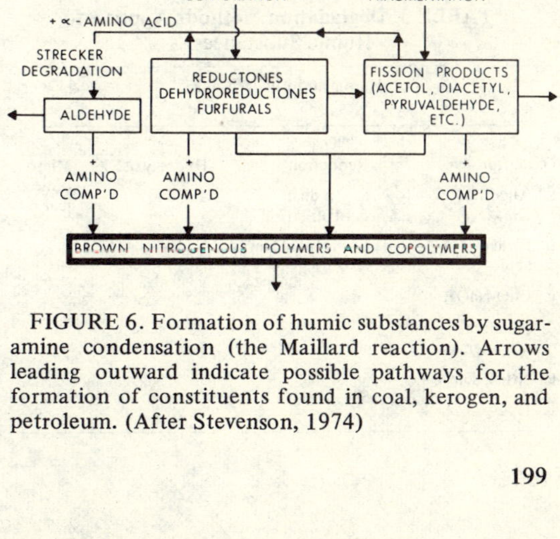

FIGURE 6. Formation of humic substances by sugar-amine condensation (the Maillard reaction). Arrows leading outward indicate possible pathways for the formation of constituents found in coal, kerogen, and petroleum. (After Stevenson, 1974)

by this means occurs extensively in food products stored at moderate temperatures. The raw materials required for the reaction are produced continuously in soils and sediments during microbial activity. Freezing, thawing, desiccation, and the intermixing of reactants with mineral matter may facilitate their condensation.

Degradation Products of Humic Substances

Diverse degradation methods have been applied to humic substances from various sources, as shown in Table 3. The objective of degradative methods is to produce simple compounds representative of main structural units in the starting material. Ideally, the products would provide information from which "type" structures and formulas could be devised. Since both plant and microbial products serve as precursors for humus synthesis, the array of compounds produced should reflect the conditions of climate and vegetation under which the humus was formed.

Serious problems are encountered in the application of degradation methods for characterizing humic and fulvic acids. With mild procedures, yields of identifiable products are extremely low, usually on the order of a few percent or less of the starting material. On the other hand, with drastic degradation methods, the material is broken down into such small fragments (e.g., CO_2, H_2O, acetic acid, and oxalic acid) that any resemblance to the original structure is lost. The ideal method would produce high yields of degradative products of moderate molecular complexities while minimizing the formation of artifacts. A combination of degradative procedures with increasing severity will undoubtedly be required, as recently done by Schnitzer and de Serra (1973). According to these investigators, humic acids contain an easily degradable portion that consists of quaiacyl and syringyl units and that may be lignin-derived (about 10%). This material was believed to be released by CuO-NaOH oxidation and by Na-amalgam reduction. The bulk of humic-acid structure was believed to consist of a more chemically condensed "core" that degrades into complex phenolic and benzene-carboxylic acids with more drastic oxidation.

The reviews of Flaig, Beutelspacher, and Rietz (1975) and Schnitzer and Khan (1972) show that numerous compounds have been observed in degradation products of humic and fulvic acids, some of which are shown in Figure 7.

TABLE 3. Degradation Methods Applied to Humic Substances.

Humic and Fulvic Acids			
Oxidation	Reduction	Hydrolysis	Microbiological
a. Alkaline-KM_nO_4	a. Zn dust distillation	a. Acid	
b. Peracetic acid	b. Na-amalgam	b. Base	
c. CuO-NaOH		c. Water	
d. Alkaline nitrobenzene			
e. Nitric acid			
f. H_2O_2			

Spectroscopic Characteristics

Attempts to characterize humic substances by ultraviolet and infrared absorption spectroscopy have been only partially successful. In the ultraviolet and visible wavelength regions, a small shoulder is usually observed between 250 and 280 mμ, and absorption increases toward the shorter wavelengths. Though extinction coefficients vary considerably, preparations having high values are believed to contain more condensed aromatic rings than those having low values. The extinction coefficients of humic acids have been found to follow the order: Chernozem > Gray Forest soil > Chestnut soil > Sierozem > Sod-podzolic soil > Kraznozem > strongly podzolized soil (Kononova, 1966).

The infrared spectra of humic preparations indicate that they have highly unsaturated structures and that carboxyl and hydroxyl groups are present. Absorption bands suggestive of amides, esters, quinones, and ethylenic

TABLE 4. Main Infrared Absorption Bands of Humic Substances.

Wave Number (cm^{-1})	Assignment
3400–3300	O-H stretching, N-H stretching (trace)
2940–2900	Aliphatic C-H stretching
1725–1720	C=O stretching of COOH and ketones (trace)
1660–1630	C=O stretching of amide groups (amide I band), quinone C=O and/or C=O of H-bonded conjugated ketones
1620–1600	Aromatic C=C, strongly H-bonded C=O of conjugated ketones?
1590–1517	COO^- symmetric stretching, N-H deformation + C=N stretching (amide II band)
1460–1450	Aliphatic C-H
1400–1390	OH deformation and C-stretching of phenolic OH, C-H deformation of CH_2 and CH_3 groups, COO antisymmetric stretching
1280–1200	C-O stretching and OH deformation of COOH, C-O stretching of aryl ethers
1170–950	C-O stretching of polysaccharide or polysaccharide-like substances, Si-O of silicate impurities

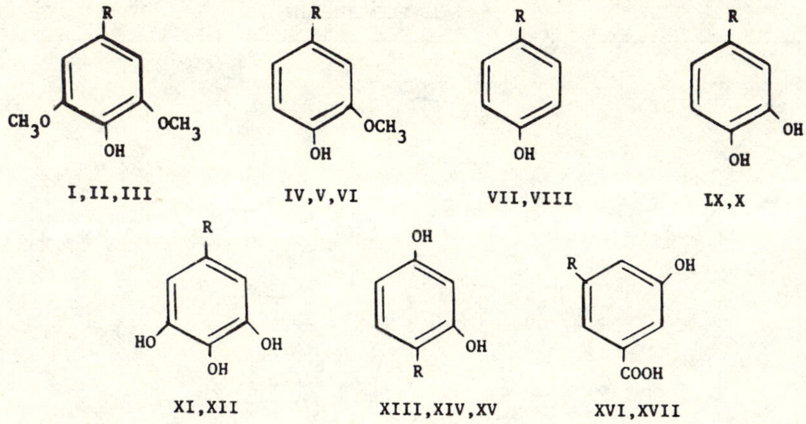

I.	R = CHO	Syringaldehyde
II.	R = COOH	Syringic acid
III.	R = CH_2-CH_2-COOH	Syringylpropionic acid
IV.	R = CHO	Vanillin
V.	R = COOH	Vanillic acid
VI.	R = CH_2-CH_2- COOH	Guaiacylpropionic acid
VII.	R = COOH	p-Hydroxybenzoic acid
VIII.	R = CHO	p-Hydroxybenzaldehyde
IX.	R = H	Catechol
X.	R = COOH	Protocatechuic acid
XI.	R = H	Pyrogallol
XII.	R = COOH	2,3,4-Trihydroxybenzoic acid
XIII.	R = H	Resorcinol
XIV.	R = COOH	2,4-Dihydroxybenzoic acid
XV.	R = OH	Phloroglucinol
XVI.	R = H	m-Hydroxybenzoic acid
XVII.	R = OH	3,5-Dihydroxybenzoic acid

FIGURE 7. Products obtained by chemical degradation of humic acids. Numbers I through XII are compounds of possible lignin origin while those from XIII to XVII represent flavonoids or products of microbial metabolism.

moieties are frequently intimated, but independent confirmation for these groups has not been forthcoming. The spectra of humic materials invariably lack the sharp bands characteristic of pure compounds.

Infrared spectra of humic and fulvic acids are shown in Figures 8 and 9, respectively. Assignments to the various absorption bands are tabulated in Table 4. Main absorption bands are in the regions of 3300 cm^{-1} (H-bonded OH groups), 2900 cm^{-1} (aliphatic C-H stretching); 1720 cm^{-1} (C=O stretching of COOH and ketonic C=O); 1610 cm^{-1} (aromatic C=C and H-bonded C=O); and 1250 cm^{-1} (C-O stretching and OH deformation of COOH groups). In addition, small bands are often evident at about 1500 cm^{-1} (aromatic C=C); 1460 cm^{-1} (C-H deformation of CH_2 or CH_3 groups); and 1390 cm^{-1} (O-H deformation, CH_3 bending, or C-O stretching).

Humic and fulvic acids have somewhat similar spectra, the main difference being that the intensity of the 1720 cm^{-1} band is considerably stronger in the fulvic acids because of the occurrence of more COOH groups. Also the 1600 cm^{-1} band is centered at a higher wave number (near 1640 cm^{-1}).

Stevenson and Goh (1971) classified the infrared spectra of humic substances into three general types, as follows:

Type I. Spectra of this type are typical of those generally shown for humic acids. Strong bands are evident near 3400, 2900, 1720, 1600,

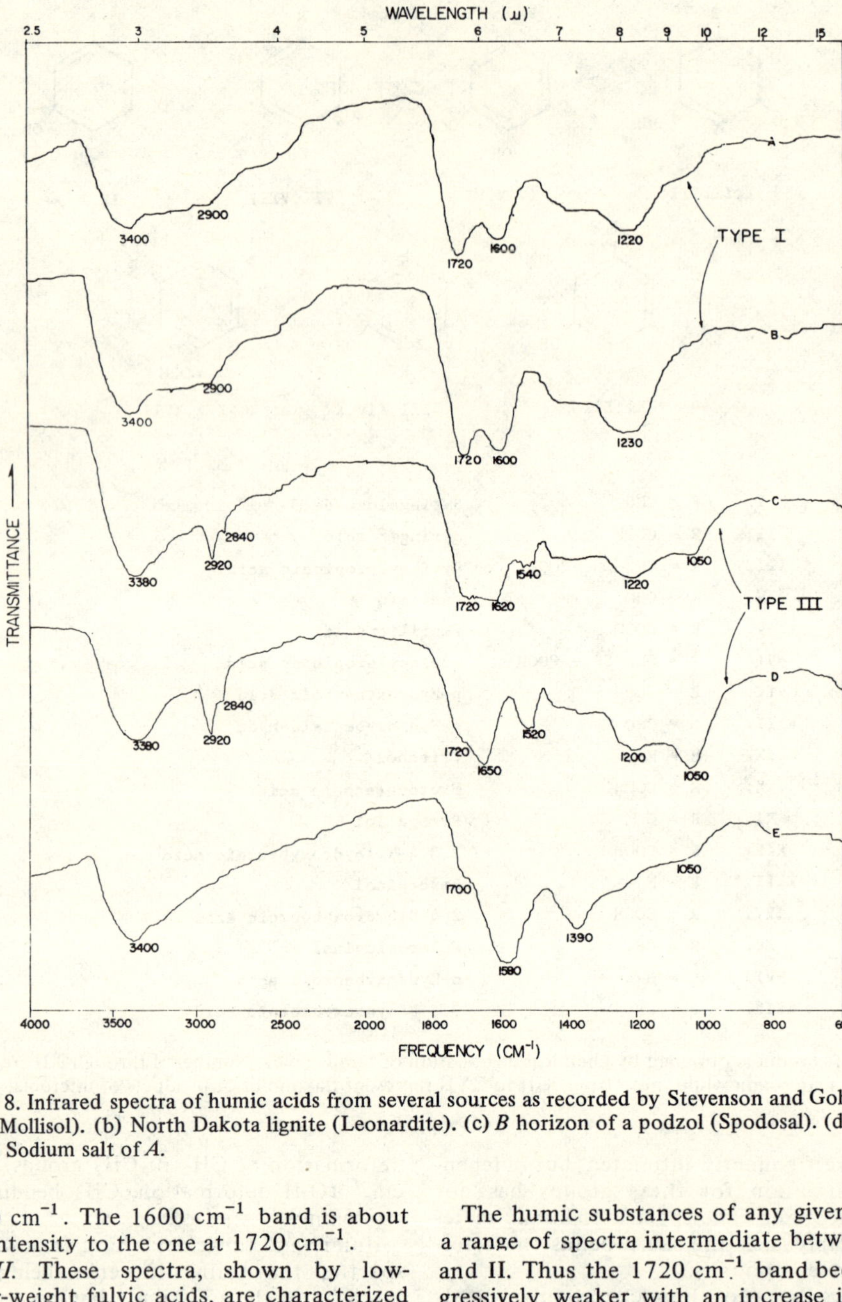

FIGURE 8. Infrared spectra of humic acids from several sources as recorded by Stevenson and Goh (1971). (a) Brunizem (Mollisol). (b) North Dakota lignite (Leonardite). (c) *B* horizon of a podzol (Spodosal). (d) Mud Lake, Florida. (e) Sodium salt of *A*.

and 1200 cm^{-1}. The 1600 cm^{-1} band is about equal in intensity to the one at 1720 cm^{-1}.

Type II. These spectra, shown by low-molecular-weight fulvic acids, are characterized by very strong absorption near 1720 cm^{-1}. A second unique feature is that absorption in the 1600 cm^{-1} region is weak and centered near 1640 cm^{-1}

Type III. In addition to the major absorption bands shown by types I and II, relatively strong bands are evident near 1540 cm^{-1}. Absorption near 2900 cm^{-1} (aliphatic C-H absorption) is also more pronounced. A unique feature of these spectra is the presence of bands indicative of proteins and carbohydrates.

The humic substances of any given soil show a range of spectra intermediate between types I and II. Thus the 1720 cm^{-1} band becomes progressively weaker with an increase in color intensity, indicating a progressive decrease in COOH content with increasing molecular weight. Spectra of type III are typical of lake humic acids (Ishiwatari, 1970).

Electron paramagnetic studies have shown that humic and fulvic acids contain stable free radicals, presumably due to a mixture of quinones, semiquinones, and quinhydrones (Steelink and Tollin, 1967). The chemical species that may be involved are illustrated in Figure 10. Other possible souces of the free

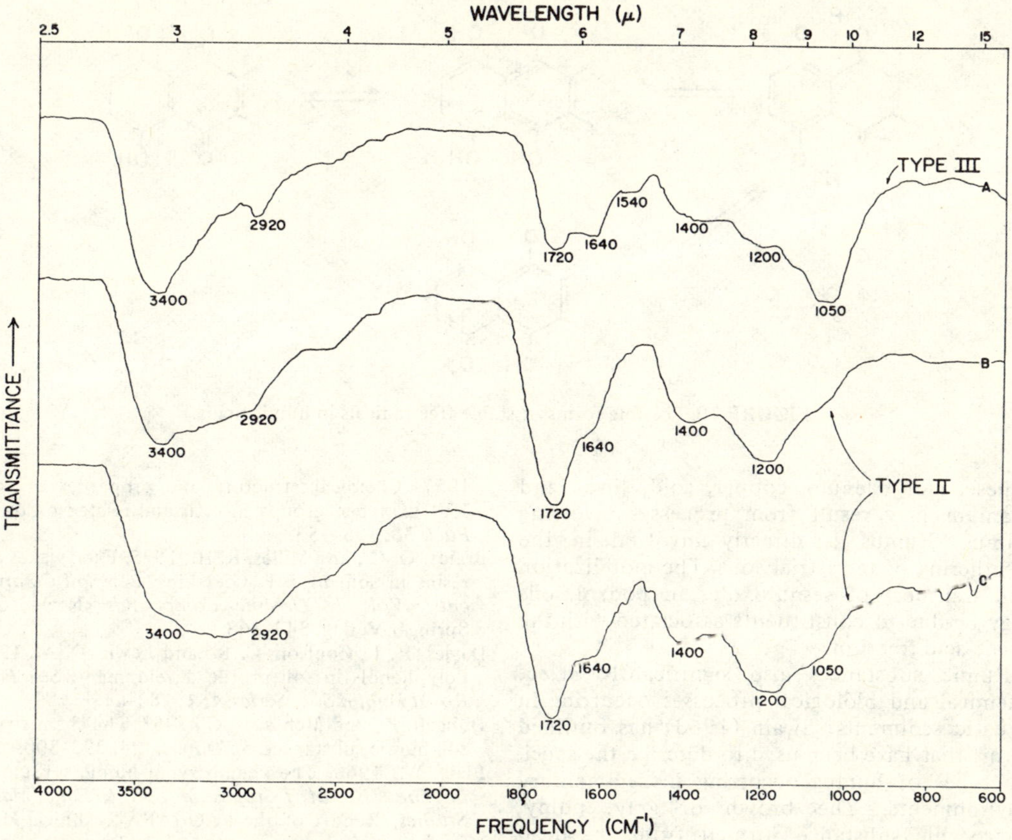

FIGURE 9. Infrared spectra of fulvic acids from a prairie grassland soil as recorded by Stevenson and Goh (1971). (a) Colloid (nondialyzable) preparation. (b) Dialyzable component I. (c) Dialyzable component II.

radicals include adsorbent complexes, trapped radicals, and polynuclear hydrocarbons.

The results of a rather extensive study by Atherton et al (1967) on the ESR spectra of humic acids can be summarized as follows:

1. All humic acids gave spectra of breadth 1.75 to 1.9 gauss, which were susceptible to oxygen effects.
2. The signals were eliminated by reduction with sodium dithionite and recovered by exposure to air, thereby implicating semiquinone ion radicals.
3. The spectra could be divided into two types depending on the pH of the soil or sediment from which the humic acid was derived. Humic acids from acidic environments (pH 2.8–4.3) such as mor humus layers, bog peats, and humus-containing horizons of podzols, showed four-lined spectra. On the other hand, humic acids from more basic environments (pH 4.3–7.2) such as mull humus layers, fen peats, and most soils, gave ill-defined, structureless spectra without distinct peaks.
4. The spectra depended on the overlying vegetation only in so far as this corresponded to soil pH (see item 3). The nature of the signal was also independent of depth or age of the soil.

Humus as a Geochemical Agent

Geologically speaking, humus is important as the mother substance for coal and oil shales and as a transporting agent and precipitant for a number of elements (Manskaya and Drozdova, 1968). Kerogen, the most common form of organic carbon in the earth, is probably derived from humic substances. The hydrocarbons in petroleum may have originated in part from humus, but this possibility has not been adequately evaluated.

Humus substances are believed to be directly involved in the formation of underclays and bauxites because they assist in the removal of silicon, iron, and bases from the weathering aluminum-bearing rocks (Swain, 1963). Iron is probably the most important of the elements that are transported and concentrated through the influence of humus. The iron found in many iron formations and bog ores may have moved to the area of deposition in combination with organic matter. Concentrations of many other elements, including phosphorus, man-

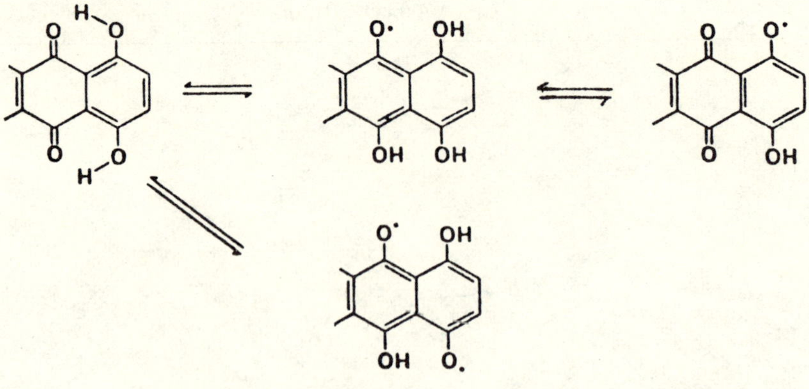

FIGURE 10. Possible forms of stable free radicals in humic acids.

ganese, molybdenum, copper, gold, silver, and uranium may result from processes involving humus. Humus is directly involved in the weathering of terrestrial soils. The mobilization and transport of sesquioxides in podzol soils may be due to constituents associated with the fulvic-acid fraction.

Humic substances also significantly affect chemical and biological processes occurring in aquatic sediments. Swain (1963) has outlined terms that have been used to describe the specific types of humus occurring in sedimentary environments. The brown or gray, pulpy, coprogenic substance formed from the dead remains of microscopic plants and animals in the top muds of eutrophic lakes and marshes is called *copropel.* A black mass of humus occurring in deeper hypolimnetic areas of lakes and bays is termed *sapropel.* A pondweed type of sapropel believed to originate from cellulose-rich plants is called *förna.* A deposit in dystrophic lakes consisting of an allochthonous precipitate of humic acids and detritus is known as *dy.* Marine slime resulting from settled planktonic detritus has been referred to as *pelogoea,* and an amorphous, gummy concentration of humic substances beneath or within certain peat bogs is *dopplerite.*

The matted layer of partially decomposed leaf litter on the forest floor is called *mor*; when the leaf litter is mixed with mineral matter so that decomposition occurs in the superficial layers of the soil instead of on the surface, the material is referred to as *mull.*

F. J. STEVENSON

References

Atherton, N. M.; Cranwell, P. A.; Floyd, A. J.; and Haworth, R. D., 1967. Humic acid-1. ESR spectra of humic acids, *Tetrahedron,* **23,** 1653–1667.

Blom, L.; Edelhausen, L.; and van Krevelen, D. W. 1957. Chemical structure and properties of coal. XVIII-Oxygen groups in coal and related products, *Fuel,* **36,** 135–153.

Braids, O. C., and Miller, R. H., 1975. Fats, waxes and resins in soil, *in* J. E. Gieseking, ed., *Soil Components: Vol. 1. Organic Compounds.* New York: Springer-Verlag, 343–368.

Davies, R. I.; Coulson, C. B.; and Lewis, D. A., 1960. Polyphenols in soil profile development, *Sci. Proc. Royal Dublin Soc. Series A,* **1,** 183–89.

Dubach, P., and Mehta, N. C., 1963. The chemistry of soil humic substances, *Soils Fert.,* **24,** 293–300.

Flaig, W., 1966. The chemistry of humic substances, *in The Use of Isotopes in Soil Organic Matter Studies,* Report of the FAO/IAEA Technical Meetings, New York: Pergamon, 103–127.

Flaig, W.; Beutelspacher, H.; and Rietz, E., 1975. Chemical composition and physical properties of humic substances, *in* J. E. Gieseking, ed., *Soil Components: Vol. 1. Organic Components.* New York: Springer-Verlag, 1–211.

Gjessing, E. T., 1976. *Physical and Chemical Characteristics of Aquatic Humus.* Ann Arbor, Michigan: Ann Arbor Science Press, 120p.

Greenland, D. J., and Oades, J. M., 1975. Saccharides, *in* J. E. Gieseking, ed., *Soil Components: Vol. 1. Organic Components.* New York: Springer-Verlag, 213–261.

Hosoda, K., and Takota, H. J., 1957. Studies on humus of black soil. 7. The chemical structure of humic acid, *Jour. Sci. Soil Tokyo,* **28,** 64–68.

Ishiwatari, R., 1970. Structural characteristics of humic substances in Recent lake sediments, *in* G. D. Hobson and G. C. Speers, eds., *Advances in Organic Geochemistry, 1966.* London: Pergamon, 283–311.

Kononova, M. M., 1966. *Soil Organic Matter.* London: Pergamon, 450p.

Manskaya, S. M., and Drozdova, T. V., 1968. *Geochemistry of Organic Substances.* London: Pergamon, 330p.

Martin, J. P., and Haider, K., 1971. Microbial activity in relation to humus formation, *Soil Sci.,* **111,** 54–63.

Morrison, R. I., 1969. Soil lipids, *in* G. Englinton and M. T. J. Murphy, eds., *Organic Geochemistry.* New York: Springer-Verlag, 558–575

Parsons, J. W., and Tinsley, J., 1975. Nitrogenous substances, *in* J. E. Gieseking, ed., *Soil Components: Vol. 1. Organic Components.* New York: Springer-Verlag, 263–304.

Scheffer, F., and Ulrich, B., 1960. *Humus und Humusdüngung, Bd. 1.* Stuttgart, Germany: Ferdinand Enke, 266p.

Schnitzer, M., and Desjardins, J. G., 1962. Molecular and equivalent weights of the organic matter of a podzol, *Soil Sci. Soc. America Proc.*, **26**, 362–365.

Schnitzer, M., and de Serra, M. I., 1973. The chemical degradation of a humic acid. *Can. Jour. Chem.*, **51**, 1554–1566.

Schnitzer, M., and Khan, S. U., 1972. *Humic Substances in the Environment.* New York: Marcel Dekker, 327p.

Steelink, C., and Tollin, G., 1967. Free radicals in soil, *in* A. D. McLaren and G. H. Peterson, eds., *Soil Biochemistry.* New York: Marcel Dekker, 147–169.

Stevenson, F. J., 1966. Lipids in soil, *Jour. Amer. Oil Chemists' Soc.*, **43**, 203–210.

Stevenson, F. J., 1974. Nonbiological transformations of amino acids in soils and sediments, *in* B. Tissot and F. Brenner, eds., *Advances in Organic Geochemistry, 1973.* Paris: Editions Technip, 701–714.

Stevenson, F. J., and Butler, J. H. A., 1969. Chemistry of humic acids and related pigments, *in* G. Englinton and M. T. J. Murphy, eds., *Organic Geochemistry.* New York: Springer-Verlag, 534–557.

Stevenson, F. J., and Goh, K. M., 1971. Infrared spectra of humic acids and related substances. *Geochim. et Cosmochim. Acta,* **35**, 471–483.

Swain, F. W., 1963. Geochemistry of humus, *in* I. A. Greger, ed., *Organic Geochemistry.* New York: MacMillan, 87–147.

Waksman, S. A., 1936. *Humus.* Baltimore: Williams and Wilkins, 496p.

Wright, J. R., and Schnitzer, M., 1960. Oxygen-containing functional groups in the organic matter of the Ao and Bh horizons of a podzol. *7th Internat. Congr. Soil Sci. Trans.*, **2**, 120–127.

Cross-references: *Exchange Phenomena; Humic Acid-Fulvic Acid Ratio; Metal-complexing Macromolecules; Mulch; Soil Components; Organic; Soil Microbiology.*

HYDROPHILITY, HYDROPHOBICITY

The terms *lyophilic* ("liquid loving") and *lyophobic* ("liquid hating") describe the tendency of a surface, particle, or functional group to become wetted or solvated. If the liquid medium is aqueous, the terms *hydrophilic* and *hydrophobic* are used (Shaw, 1966). The term *sol* is used to distinguish colloidal suspensions from macroscopic suspensions in general, but when the dispersion medium is aqueous, the term *hydrosol* is frequently used. The two major classes of hydrous colloidal systems—hydrophilic and hydrophobic colloids—were formerly called "emulsoids" and "suspensiods," respectively.

A material that displays an affinity for water (or solvent) is described as *hydrophilic* (lyophilic). A material that tends to repel water is described as hydrophobic (lyophobic). In colloid chemistry these terms characterize the stability of disperse systems such as sols, suspensions, and emulsions and are particularly useful when considering the phenomenon of surface activity. Hydrophilic sols or colloids, which are prepared from organic macromolecular substances such as natural or synthetic gums, are also referred to as *macromolecular colloids* or *polyelectrolyte solutions* (van Olphen, 1977). The disperse systems of soil minerals (clay minerals, oxides, and derivatives of humic acid), however, due to their stabilities, belong to hydrophobic systems.

Table 1 compares the principal properties of hydrophilic and hydrophobic disperse systems and surfaces. The surface particles in hydrophobic disperse systems may be either hydrophilic (e.g., a suspension of clay minerals) or hydrophobic (e.g., carbon black). Because true or absolute hydrophobic surfaces do not exist, hydrophobicity may be considered a very low degree of hydrophility. Oil, for exam-

TABLE 1. Surfaces and Hydrophilic Disperse Systems.

Surface			Disperse Systems (Sols, Suspensions)		
	Surface			Systems	
Properties	Hydrophilic	Hydrophobic	Properties	Hydrophilic	Hydrophobic
Heats of immersion $Q_1/Q_2 = \beta$	> 1	< 1	Stability	High	Low
Contact angles $\theta_1/\theta_2 - \alpha$	$\rightarrow 1$	$>> 1$	Viscosity	High	Low
Wetting with water	Wettable	Nonwettable	Reversibility	Reversible	Irreversible

Q: heat of immersion.
θ: contact angle.
Subscripts 1 and 2 correspond to water and nonpolar liquid, respectively.

ple, contains soluble water, and it therefore is also hydrophilic, though to a very small degree. Highly hydrophilic surfaces carry no hydrogen bonds or groups with lone electron pairs or ion or free radicals.

The equilibrium of the water drop on a solid surface is defined by the following relationship:

$$\gamma_{sv} = \gamma_{sl} + \gamma_{lv} \cdot \cos\theta$$

where γ represents tensions at interfaces of solid-vapor (sv), solid-liquid (sl), and liquid-vapor (lv), with θ as the contact angle (Figure 1). At wetting, inequality exists:

$$\gamma_{sv} > \gamma_{sl} + \gamma_{lv}$$

as it does at nonwetting (repellency) where $\gamma_{sv} < \gamma_{sl} - \gamma_{lv}$. Water does not spread over the hydrophobic surface because the intermolecular forces in water are composed of dispersion and polar forces (Zettlemoyer, 1968) where

$$\gamma \text{ total } H_2O = \gamma \text{ disp. } H_2O + \gamma \text{ polar } H_2O$$

Because

$$\gamma \text{ total } H_2O = 72 \text{ erg cm}^{-2} \quad \text{and} \quad \gamma \text{ disp } H_2O = 22 \text{ erg cm}^{-2}$$

and the interaction of water with hydrophobic surfaces occurs through dispersion forces, the main part of intermolecular forces (72 - 22 = 50 erg cm^{-2}) remains free, and the water drop on the hydrophobic surface forms a ball. Hydrocarbons, on the other hand, spread easily on a hydrophobic surface because all their intermolecular forces are componsed of dispersion forces:

$$\gamma \text{ total CH} = \gamma \text{ disp CH}$$

Thus surface hydrophility depends on the affinity of the surface molecules, groups, atoms, and ions to water. The functional groups – OH, – COOH, and – NH_2 increase surface hydrophility whereas – CH_3, = CH_2, and – CF_3 diminish it. All substances that possess high intensity of interatomic or interionic forces are hydrophilic (e.g., crystals of ionic structure, oxyhydrates, most silicates).

Hydrophility and hydrophobicity can be evaluated by (a) measurements of contact angles in air and in liquids, (b) measurements of the heats of adsorption and immersion, (c) adsorption of gases and vapors, (d) measurements of time of wetting, (e) measurements of infiltration rate, and (f) adsorption of high-molecular aliphatic alcohols from water. The evaluation of hydrophility and hydrophobicity presents some difficulties for dispersed (powdered) systems, especially those that are heterogeneous and rough.

Inorganic Soil Minerals

In the study of contact angles, the plane surface is determined from dispersed material by obtaining (a) bricks through the use of pressure and (b) by evaporation of a suspension over a glass plate. The contact angle is measured by projecting the image of a water drop on a screen. Contact angles determined in this way are normally less than 30° when a water drop is placed on most soil minerals and greater than 50° for the humic-acid fraction (Tschapek, 1949). On the other hand, studies of the hydric properties of soil samples indicate that some soil particles are probably coated with hydrophobic materials such as magnesium soap, fungi, or residues from wild fires (Roberts and Carbon, 1972; Wander, 1949; and Debano and Letey, 1969). The determination of the density of hydrophobic sites on soil minerals has recently been attempted by Tschapek and Wasowski (1976). They found that because the molecule 1-heptanol has an amphophilic structure i.e., possess hydrophilic (OH) as well as hydrophobic

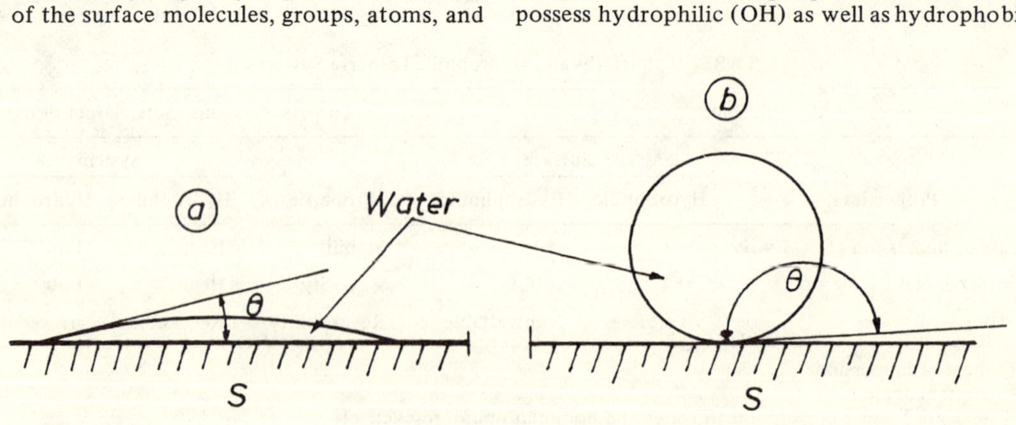

FIGURE 1. Profiles of a water drop on (a) hydrophilic ($\theta < 90°$) and (b) hydrophobic ($\theta > 90°$) surfaces.

TABLE 2. Contact Angles, Densities of Hydrophobic Sites, and Behavior of Water Droplet on Disperse Powers.

Mineral	Density of Hydrophobic Sites per 100 Å^2	Contact Angle θ	Behavior of Water Droplet
Carbon black	2.4	180°	Forms ball
Humic acid	2.0		
Silica hydrophobized	1.1		
Bentone 34	0.72		
Bentonite	0	0°	Penetrates in < 5 s
Kaolin	0		
Silica	0		
Alumina	0		

(CH_3) radicals, it is useful in evaluating the probable degree of coating of soil minerals. The hydrophilic-hydrophobic balance (HHB) of the 1-heptanol molecule, according to the Davis ratio (Davis, 1957) of 5.6, indicates the impossibility of its adsorption from water on hydrophilic surfaces. Assuming that every adsorbed 1-heptanol molecule corresponds to one hydrophobic site, it is possible to determine the density of surface hydrophobic sites (Tschapek and Wasowski, 1976). Table 2 contains data resulting from studies of contact angles and 1-heptanol adsorption. These experiments indicate that clay minerals such as kaolinite or montmorillonite (smectite) are hydrophilic. The dehydroxylation of oxides increases the density of hydrophobic sites, especially when rehydroxylation is difficult but does not change when rehydroxylation is easy. Silanol ≡ SiOH is hydrophilic whereas siloxane ≡ Si-0-Si ≡ is hydrophobic. The splitting of siloxane with its conversion into silanol, i.e., hydroxylation of dehydroxylated silica, takes place only at a high pH. The situation is different with dehydroxylated alumina because in the presence of water it hydroxylates.

Humic Acid

Humic acid is soluble in water only after its ionization, i.e., after its transformation into alkaline humate. The fact that humates are surface active indicates an amphophilic structure. Humic acid comprises both polar (hydrophilic), e.g., carboxyl (COOH) and hydroxyl (OH), and nonpolar (hydrophobic), e.g., aliphatic (CH_2) and carbonyl (CO), functional groups (see *Humus*). The high value of β (ratio of the heat of immersion in water to that in a nonpolar liquid), however, indicates that the hydrophilic part of the humic-acid molecule is dominant, although insufficient for its solution in water. From 1-heptanol and 1-butanol adsorption experiments (Tschapek, Ardizzi, and Bussetti, 1973; Tschapek and Wasowski, 1976) it was concluded that humic acid possesses nearly 2 hydrophobic sites per 100 Å^2. For comparison, carbon black possesses 2.5 hydrophobic sites per 100 Å^2 whereas montmorillonite (smectite) and kaolin, for example, do not possess hydrophobic sites. Alkaline humates, which are surface-active substances, stabilize hydrophobic suspensions in water by hydrophilizing their surfaces. Thus water suspensions containing highly hydrophobic substances such as carbon black are extremely unstable but with the addition of sodium humate, for example, the suspension is stabilized.

Management of Water-repellent Soils

Management of water-repellent soils involves methods that control soil hydrophility or hydrophobicity. This is normally accomplished by applying waterproof substances or surfactants (see *Soil Conditioners*). Many questions remain on the actions of these materials on soils, and additional research is required for a better understanding of the problems involved. To increase the wettability of soil, i.e., decrease the water repellence, cationic and nonionic wetting agents can be used. According to Davis (1957), the HHB of wetting agents must be between 7 and 9. The use of cationic wetting agents is, however, generally impractical (Debano and Letey, 1969). Wetting agents decrease the tension at the water-air interface, which increases the spreading coefficient S

$$S = \gamma_{sv} - (\gamma_{sl} + \gamma_{lv})$$

where γ is the tension at the solid vapor (sv), solid liquid (sl), and liquid vapor (lv) interface. To decrease soil wettability, the soil particles should be coated with a substance capable of increasing their hydrophobicity. Hydrophobization of a mineral surface may be achieved by substitution of organic cations for metallic ions, by complex formation on the surface, and by chemical reation on the surface with derivatives of silanol.

M. TSCHAPEK

References

Anon., 1973. *Hydrophobization.* Kiev: Institute of Colloid and Water Chemistry, 240p.

Davis, J. T., 1957. A quantitative kinetic theory of emulsion type. I. Physical chemistry of emulsion agents, *Proc. 2nd Internat. Congr. Surface Activity,* 1, 426–438.

Debano, L. E., and Letey, J., eds., 1969. *Water Repellent Soils.* Riverside, Calif.: University of California Press, 354p.

Johnson, H. E., and Dettre, R., 1969. Wettability and contact angles, *Surface and Colloid Sci.,* 2, 85–153.

van Olphen, H., 1977. *An Introduction to Clay Colloid Chemistry*. New York: Wiley, 318p.

Roberts, F. J., and Carbon, B. A., 1972. Water repellence in sandy soils of southwestern Australia. II. Some characteristics of the hydrophobic skins, *Aust. Jour. Soil Res.*, **10**(1), 35–42.

Shaw, D. J., 1966. *An Introduction to Colloid and Surface Chemistry*. London: Butterworths, 186p.

Tschapek, M., 1949. *Quimica Coloidal del Suelo*. Buenos Aires: Museo Argentino de Ciencias Naturales "Bernardino Rivadavia," 344p.

Tschapek, M.; Ardizzi, G. P.; and Bussetti, S., 1973. Wettability of humic acid and its salts, *Z. Pfl. Bod.*, **135**, 16–31.

Tschapek, M., and Wasowski, C., 1975. Adsorption of aliphatic alcohols by soil minerals as a method of evaluating their hydrophibic sites, *Jour. Soil Sci.*, **27**, 175–183.

Tschapek, M., and Wasowski, C., 1976. The surface activity of humic acid, *Geochim. et Cosmochim. Acta*, **40**, 1343–1345.

Wander, I. W., 1949. An interpretation of the causes of water-repellent sandy soils found in citrus groves in central Florida, *Science*, **110**, 299.

Zettlemoyer, A. C., 1968. Hydrophobic surfaces, *Jour. Colloid and Interface Sci.*, **28**, 343–368.

Cross-references: *Clay Minerals, Silicates; Flocculation; Humus; Imbibition.*

HYDROUS OXIDES

Only a limited number of hydrous-oxide minerals are commonly found in soils. Examples of such minerals are oxides of aluminum, iron, and manganese (Table 1). They are important components of soils because they may influence soil structure (q.v.) through their bonding action and may affect soil fertility through their reactions with plant nutrients (q.v.) such as phosphorous and micronutrients (q.v.). Such oxides are commonly of pedogenic origin (see vol. XII: *Pedosesquioxides*) and may be major components of some soil types (e.g., gibbsite in laterites). Average crystal size is usually small

TABLE 1. Crystalline Hydrous-Oxide Minerals Commonly Occurring in Soils.

Mineral Name	Chemical Formula	Properties and Occurrence
Gibbsite	$Al(OH)_3$	Soft white or pale tinted green, pink, brown mineral, slowly soluble in concentrated acids. A major constituent of some bauxites and laterites.
Boehmite	AlO.OH	White or pale tinted mineral, insoluble in acids. A major constituent of some bauxite deposits.
Diaspore	AlO.OH	Hard white or pale tinted grey, brown, green mineral, slowly soluble in hydrofluoric acid. A major constituent of some bauxite deposits.
Lepidocrocite	FeO.OH	Reddish brown mineral, shows orange streak and is soluble in hydrochloric acid. Commonly occurs in soils showing impeded drainage.
Goethite	FeO.OH	Yellowish brown to red mineral with yellow streak, soluble in hydrochloric acid. A common weathering product in many soil types.
Birnessite Lithiophorite Todorokite	$Mn_7O_{13}.5H_2O$ $(Al, Li)MnO_2(OH)_2$ $(Mn, Mg, Ca, Ba, Na, K)_2 Mn_5O_{12}.3H_2O$	Black or dark brown microcrystalline minerals occurring in many soil types as disseminated pigment, coatings, or nodules. Nodules are best developed in soils subject to seasonal waterlogging. Very variable in composition, often containing abundant Fe, Co, Ni, Mo, Cr, V, Zn, and Cu.

(After Mackenzie, 1975; McKenzie, 1972)

(about 1 μm or less), and structural disorder and isomorphous substitution commonly occur. Hydrous-oxide minerals may be present in soils in the form of coatings or as concretions composed of one or more mineral species. Often attached to the surfaces of other soil minerals, they may also occur as individual crystals disseminated throughout the soil. The identification of hydrous oxides in soils is best undertaken by X-ray powder diffraction procedures (Rooksby, 1972). However, the presence of amorphous hydrous oxides of aluminum, iron, manganese, and silicon, which are present in some soils, is not revealed by this technique.

R. J. GILKES

References

Mackenzie, R. C., 1975. The classification of soil silicates and oxides, *in* J. E. Gieseking, ed., *Soil Components, Vol. 2, Inorganic Components.* New York: Springer-Verlag, 1-25.

McKenzie, R. M., 1972. The manganese oxides in soils–A review, *Zeit. Pfl. Boden.*, **131**, 221–242.

Rooskby, H. P., 1972. Oxides and hydroxides of aluminum and iron, *in* G. Brown, ed., *The X-Ray Identification and Crystal Structures of Clay Minerals.* London: Mineralogical Society, 544p.

Cross-references: *Mineral Analysis; Soil Mineralogy. Vol. IVA: Oxides and Hydroxides. Vol. IVB: Soil Minerals. Vol. XII, Pt. 2: Pedosesquioxides.*

HYGROSCOPICITY, HYGROSCOPIC CONSTANT

Hygroscopicity is the property of small-particle systems to take up moisture from the atmosphere through strong sorption forces on the particle surfaces and through capillary condensation due to the lowering of the water-vapor pressure above concave capillary menisci. Hygroscopicity differs from *deliquescence* in that it does not result in liquefaction even at relatively high atmospheric humidity.

Hygroscopic moisture, as defined by the soils engineer, is the water content of an air-dry soil as determined by oven-drying at 105–110°C and expressed as a percentage of the oven-dry weight. This parameter is practically important for field and laboratory work; however, because of its dependence on environmental conditions, especially the relative humidity of the atmosphere, it cannot serve as a fundamental soil characteristic. For the latter role, the environmental influence must be either eliminated or standardized.

The *hygroscopic constant* (or coefficient) W_{hc} is defined as that moisture content of a soil material at which further addition of water will not release additional heat of wetting. This moisture content corresponds closely to that reached by an oven-dry soil sample at room temperature in the well-stirred atmosphere within a desiccator above a 10% aqueous solution of sulfuric acid. This moisture content corresponds to a relative humidity (RH) of about 95%. Because at this and lower relative humidities, the hygroscopic moisture is directly proportional to the RH value, tests may be performed at higher sulfuric-acid concentrations corresponding to lower relative humidities. This test is especially advantageous because at lower humidities the contribution by capillary condensation is minimized if not eliminated, and the hygroscopicity values obtained are significant soil parameters expressive of the specific surface and the physicochemical characteristics of the soil constituents. This significance is lessened or even destroyed by the presence of soluble salts. The close relationship between W_{hc} and heat of wetting is exemplified by the fact that the heat of wetting in calories per gram of oven-dry soil is about $0.5 \times W_{hc}$ (10% H_2SO_4) and $0.75 \times W_{hc}$ (30% H_2SO_4).

Capillary condensation, which can contribute significantly to water-vapor sorption at high relative humidities, may be understood from, but not calculated by, the expression $dp/p = 2\,\sigma M/(r\delta RT)$ where dp is the change in vapor pressure above a meniscus with radius of curvature r, p is the vapor pressure over a free-water surface, R is the gas constant, and T the absolute temperature in °K; σ, M. and δ are the surface tension, molecular weight, and density of the water. An approximate measure of the contribution of capillary condensation to total water sorption is the difference between hygroscopicity measured at or close to 100% RH and the value obtained for this RH by linear extrapolation from hygroscopicity data determined at less than 95% RH.

HANS F. WINTERKORN

References

Anon., 1952. Landolt-Börnstein: Zahlenwerte und Funktionen, *Bodenkunde,* 3 (3237), 358–368.

Baver, L. D., and Winterkorn, H. F., 1935. Sorption of liquids by soil colloids, *Soil Sci.,* **40**, 403–419.

Winterkorn, H. F., and Fang, H. Y., 1975. Soil technology and engineering properties of soils, *in* H. F. Winterkorn and H. Y. Fang, eds., *Foundation Engineering Handbook.* New York: Van Nostrand Reinhold, 67–120.

Cross-references: *Capillary Pressure; Heat Capacity; Water Content and Retention.*

HYSTERESIS—See CAPILLARY PRESSURE; PERMEABILITY

I

IMBIBITION

The term *imbibition* has several meanings, sometimes being used as a synonym for *wetting* or *infiltration* (q.v.). Usage, however, seems to have attached to the word *imbibition* a more restricted meaning. Imbibition is, precisely, the phenomenon of water penetration into a relatively dry porous medium, which is therefore *wetted,* from a saturated source of water (e.g., a water container or a saturated porous medium such as a porous plate) in contact with the porous medium (soil) under the *sole* motor action of the *capillary* forces (see *Capillary Pressure*). The thorough understanding of the reason why water at atmospheric pressure will penetrate a porous medium and displace another fluid such as air, also at atmospheric pressure, is of capital importance to gain a sound physical understanding of water movement (q.v.) in soils. To the extent that a porous medium, as a first rough approximation, can be visualized as a bundle of unconnected cylindrical tubes, some small, some large, there is merit to the study of imbibition in a tube. This oversimplified model of the porous medium is an obviously inaccurate picture of an extremely complex reality. Nevertheless, it provides a convenient and simple introduction to ulterior, more complicated developments.

If the water pressure in the source of water supply is equal to atmospheric pressure, the imbibition is said to be *free.* If that water pressure is greater than atmospheric, the imbibition is *forced;* if it is less than atmospheric, the imbibition is said to be *restrained.*

Free (Capillary) Imbibition in a Tube

Consider a preferentially water-wet, small-diameter tube, initially filled with air. The tube is brought in contact with an ample supply of water at atmospheric pressure (Figure 1). The initial water-air interface may be considered to be the plane *ab.* For an infinitesimal amount of time, the force diagram will be as shown in Figure 1(b). The interface will be in equilibrium except at the contact with the solid. There, a net impulse of magnitude proportional to $\sigma_{as} - \sigma_{ws}$ causes the point *a* of the interface to move into the tube. The term σ_{as} is the air-solid surface tension (see *Capillary Pressure* for a definition of surface tension), and σ_{ws} is the water-solid surface tension. Because the other points of the interface are in equilibrium, the interface curves in, and it very quickly assumes the shape shown in Figure 1(c), where α is the equilibrium wetting angle. The interface is now in equilibrium because the surface tension force and the capillary-pressure force balance one another. Because the pressure drop in the moving-air phase is essentially negligible compared to that in water, the pressure in the air just ahead of the interface is essentially atmospheric. Just behind the interface, the pressure in the water is less than atmospheric by an amount equal to the capillary pressure. At the inlet the pressure is atmospheric. As a result, there exists an unbalanced pressure gradient in the water phase, which causes the water to flow and which sustains the motion initially created by unbalanced surface-tension forces. Viscous shear (particle *A,* Figure 1) will resist the driving force tending to enforce a uniform flow. Actu-

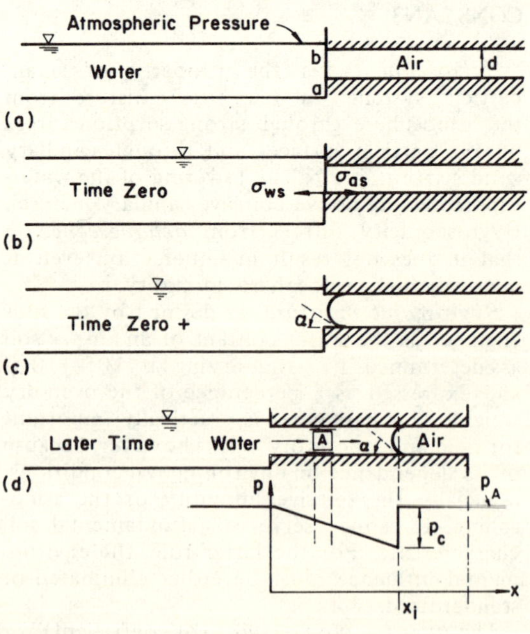

FIGURE 1. Initiation and evolution of imbibition in a capillary.

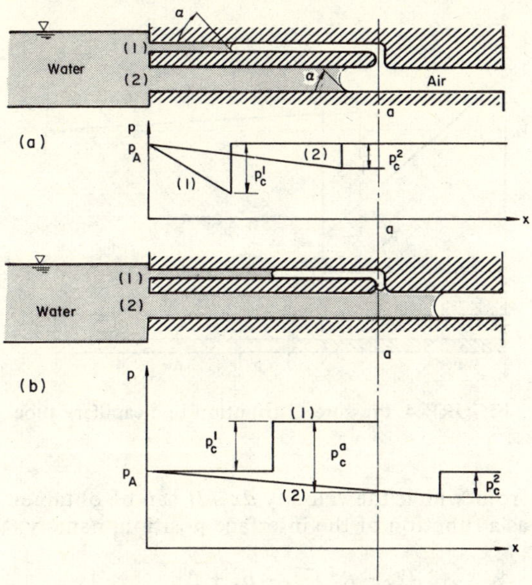

FIGURE 2. Mechanism of air entrapment in a wetting medium.

ally, flow is constantly retarded because the pressure gradient in the water phase, p_c/x_i, itself decreases with time.

In summary, motion is originated by preferential wettability and later sustained by the induced pressure gradient in the wetting phase.

Entrapment of Air

As water moves into the porous medium, it displaces most of the air ahead of its path, but it also manages to trap a fraction of it. How such entrapment takes place can be qualitatively understood by considering a system of interconnected tubes. Figure 2 illustrates the geometry of the simplest of such systems.

First consider an experiment of the same nature as that just described. Assume that atmospheric pressure prevails at both ends of the tubes (Figure 2). Assuming that the interface has the shape of a spherical cap, the capillary pressure at the interface in tube *i* is

$$p_c{}^i = \left(\frac{2\sigma}{R_i}\right) \cos\alpha \qquad (1)$$

Assuming again a negligible pressure drop in the gas phase, the pressure gradient in the water phase is

$$\left(\frac{2\sigma}{R_i}\right)\left(\frac{\cos\alpha}{x_i}\right)$$

where x_i is the location of the water-air interface in tube *i*. Application of *Poiseuille's equation* for laminar flow in a cylindrical tube (Schlichting, 1960, p. 66) yields the velocity of the interface:

$$\frac{dx_i}{dt} = \frac{R_i \sigma \cos\alpha}{4\mu_w x_i} \qquad (2)$$

and after integration:

$$x_i{}^2 = \frac{R_i \sigma t \cos\alpha}{2\mu_w} \qquad (3)$$

where μ_w is the (dynamic) viscosity of water. Under free imbibition, it follows from equation 2 that water flows more rapidly in the larger tube. Thus water will reach point *a* through tube 2 first. Immediately after the water has passed section *a* in tube 2, water imbibes tube 1 due to the rock's preferential wettability to water. An interface starts to propagate into tube 1. It will come to a stop at an equilibrium position somewhere within the constriction between tubes 1 and 2 where the radius is R_a such as to satisfy:

$$p_c{}^a = \frac{2\sigma\cos\alpha}{R_a} \qquad (4)$$

Air is then entrapped. Note that the existence of a constriction between tube 1 and tube 2 is a necessary condition for entrapment. Under free imbibition the maximum value for $p_c{}^a$ is $p_c{}^1 + p_c{}^2$. Thus at the point of maximum constriction, the radius must be somewhat less than $(R_1R_2)/(R_1 + R_2)$. Under such conditions, entrapment is stable. In the event of an external pressure gradient such as obtained by a column of water (Figure 3), the constriction must be more severe to constrain air within the smaller tube. The limiting value for R_a is the one that gives

$$p_c{}^a = p_c{}^1 + p_c{}^2 + \Delta p \qquad (5)$$

Thus

$$R_a \leqslant \left(\frac{1}{R_1} + \frac{1}{R_2} + \frac{\Delta p}{2\sigma\cos\alpha}\right)^{-1} \qquad (6)$$

The larger the imposed pressure gradient, the less chance there is to entrap air. In the limit for an infinite pressure gradient, there would be no entrapment at all. Under restrained imbibition ($\Delta p < 0$) the opposite is true. It follows that the degree of air entrapment (or the re-

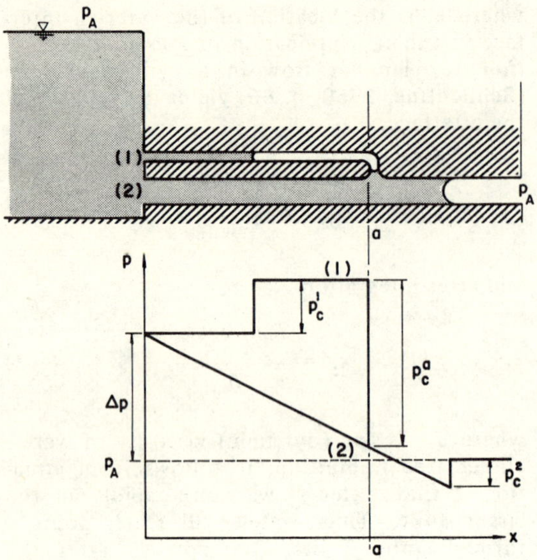

FIGURE 3. Mechanism of air entrapment under forced imbibition.

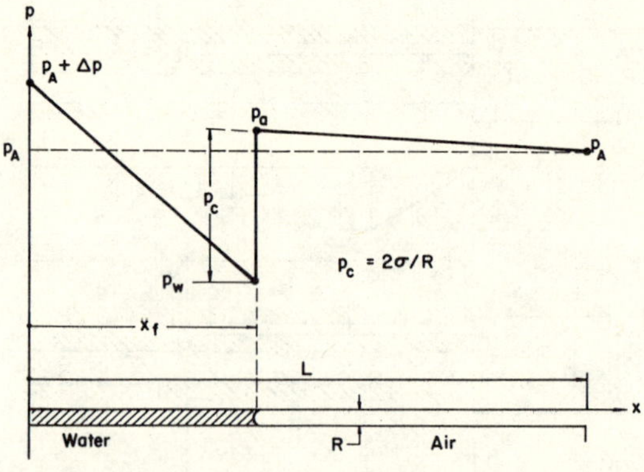

FIGURE 4. Pressure distribution in a capillary tube.

sidual air saturation, or the displacement efficiency) depends not only on the geometric characteristics of the medium but also on the rate at which the displacement proceeds.

Imbibition Rate in a Tube

Let the position of the water-air interface be located at x_f at time t. The pressure distribution in the tube is indicated in Figure 4. Application of Poiseuille's equation in the water and in the air phases yields for the water and air velocities in the tube, respectively,

$$u_w = \frac{dx_f}{dt} = \frac{R^2}{8\mu_w}\frac{(p_A + \Delta p - p_w)}{x_f} \tag{7}$$

and

$$u_a = \frac{dx_f}{dt} = \frac{R^2}{8\mu_a}\frac{(p_a - p_A)}{(L - x_f)} \tag{8}$$

where L is the length of the tube. As the two velocities must be equal, it follows by equating u_w and u_a that the unknowns p_w and p_a must be related by the equalities:

$$\frac{p_a - p_A}{\mu_a (L - x_f)} = \frac{p_A + \Delta p - p_w}{\mu_w x_f}$$

$$= \frac{p_c + \Delta p}{\mu_a (L - x_f) + \mu_w x_f} \tag{9}$$

from which the velocity dx_f/dt can be obtained as a function of the interface position, namely,

$$\frac{dx_f}{dt} = \frac{R^2}{8}\left[\frac{p_c + \Delta p}{x_f(\mu_w - \mu_a) + \mu_a L}\right] \tag{10}$$

After integration, one obtains

$$x_f = \frac{\sqrt{(\mu_a L)^2 + (\mu_w - \mu_a)(R^2/4)(p_c + \Delta p)t} - \mu_a L}{\mu_w - \mu_a} \tag{11}$$

from which the imbibition rate $I = dx_f/dt$ can be deduced as a function of time:

$$I = \frac{1}{2}\frac{R^2/4(p_c + \Delta p)}{\sqrt{(\mu_a L)^2 + (\mu_w - \mu_a)(R^2/4)(p_c + \Delta p)t}} \tag{12}$$

If the viscosity of air is desregarded compared to that of water ($\mu_w/\mu_a \cong 50$), equation 12 reduces to the simpler form:

$$I = \frac{1}{2}\frac{S_P}{\sqrt{t}} \tag{13}$$

where

$$S_P = \left[\frac{R^2}{4\mu_w}(p_c + \Delta p)\right]^{1/2} \tag{14}$$

can be called the "Poiseuille *sorptivity.*" The sorptivity characterizes the ability of the *tube* to imbibe *water.* Under forced, free, or restrained imbibition, Δp is positive, zero, or negative, respectively. Equation 13 yields an infinite imbibition rate at time zero, which is not physically realistic. However, if the terms in μ_a are not disregarded, as in equation 12, then the imbibition rate is not infinite (singular) at time zero. Equation 13 is a good approximation for equation 12 for *large* values of *time,* and it shows that the (asymptotic) imbibition rate decreases like the *inverse of the square root of time.* On the other hand, for small times the Taylor expansion of equation 12, truncated after the first-order term, yields

$$I = \frac{R^2(\Delta p + p_c)}{8\mu_a L} - \frac{R^4(\Delta p + p_c)^2(\mu_w - \mu_a)t}{64(\mu_a L)^3} \quad (15)$$

which shows that the imbibition rate *decreases linearly with time for small times* and is finite at time zero.

Free Imbibition Rate in a Bundle of Tubes

Let $n(r)dr$ represent the number of tubes of radius between r and dr. Defining

$$\delta(r) = \frac{n(r)}{N}, \delta(r)$$

is the density function of the number of tubes of size r. From its definition the integral of $\delta(r)$ between the limits of zero and infinity is one. For a bundle of tubes, the *average* imbibition rate (disregarding air viscosity) per unit bulk cross-section area is deduced from equations 13 and 14 to be:

$$I = \frac{1}{2\sqrt{t}}\left\{\phi\sqrt{\frac{\sigma}{2\mu_w}}\,\frac{\int_0^\infty r^2\sqrt{r}\,\delta(r)dr}{\int_0^\infty r^2\delta(r)dr}\right\} \quad (16)$$

where ϕ is the porosity of the tube system.

From the analogy of Poiseuille's and Darcy's law, the *conductivity* of a tube filled with water can be defined by the relation:

$$K(R) = \frac{R^2\rho_w g}{8\mu} \quad (17)$$

Introducing this notation and that of the capillary-pressure head, equation 16 takes the form:

$$I = \frac{1}{2\sqrt{t}}\left\{\frac{\phi\int_0^\infty \sqrt{2K(r)h_c(r)}\,K(r)\delta(r)dr}{\int_0^\infty K(r)\delta(r)dr}\right\} \quad (18)$$

The application of the mean theorem to the integral in the numerator of equation 18 yields:

$$I = \frac{1}{2}\phi[\overline{2K(r)h_c(r)}]^{1/2}\frac{1}{\sqrt{t}} \quad (19)$$

where the bar over the term $2K(r)h_c(r)$ means the appropriate mean value of this quantity. Equation 19 suggests that there is a bundle of tubes of *uniform* size R, which is equivalent to the bundle of tubes of different sizes, such that,

$$[K(R)h_c(R)]^{1/2} = [\overline{K(r)h_c(r)}]^{1/2} \quad (20)$$

Imbibition Rate in a Porous Medium

The results of the previous section are not of great practical interest because a porous medium is a far more complex system. Besides, the $\delta(r)$ distribution is unknown. Nevertheless, the equations have provided insight to the solution of the problem of predicting the imbibition rate in a *real* porous medium.

The Water-content Equation (Without Air Effect). Away from the water source the soil is dry; its *water content* (q.v.) is low. Close to the source it is high. At a fixed location x_0 (Figure 5), the soil gets wetter as time passes. In general, the water content θ is a function of x and t. The variation of θ in time at a fixed location x is related to the net inflow of water at that time through the principle of conservation of mass, which expressed in mathematical symbols, reads:

$$\frac{\partial\theta}{\partial t} + \frac{\partial}{\partial x}(v_w) = 0 \quad (21)$$

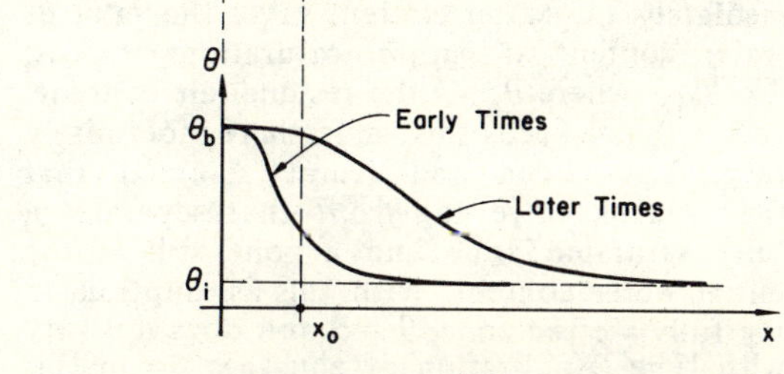

FIGURE 5. Water-content evolution in a horizontal soil slab.

where ν_w is the water velocity in a Darcy sense (i.e., a volume flow rate per unit bulk cross-section area in a direction perpendicular to flow). The water velocity is given by Darcy's law, namely,

$$\nu_w = -K(\theta)\frac{\partial h_w}{\partial x} \tag{22}$$

or

$$\nu_w = K(\theta)\frac{\partial h_c}{\partial x} = \left[K(\theta)\frac{dh_c}{d\theta}\frac{\partial \theta}{\partial x}\right] \tag{23}$$

if one assumes that air pressure is atmospheric everywhere in the medium. The term $K(\theta)$ is the *unsaturated hydraulic conductivity.* It measures the conductivity of the medium to water at various water contents. Substitution of equation 23 in equation 21 yields the *water-content equation,* also called *Richard's equation:*

$$\frac{\partial \theta}{\partial t} + \frac{\partial}{\partial x}\left[K(\theta)\frac{\partial h_c}{\partial x}\right] = 0 \tag{24}$$

Defining a new positive function of water content, the *diffusivity,* namely,

$$D(\theta) = -K(\theta)\frac{dh_c}{d\theta} \tag{25}$$

the water-content equation takes the form referred to as the *diffusivity equation,* namely,

$$\frac{\partial \theta}{\partial t} - \frac{\partial}{\partial x}\left[D(\theta)\frac{\partial \theta}{\partial x}\right] = 0 \tag{26}$$

Approximate Solutions of the Water-content Equation. *The Green and Ampt Solution:* It is assumed that the initial water content of the soil is uniform at value θ_i. At time zero, the soil is placed in contact with the water supply. Immediately the water content takes the value $\tilde{\theta}$, water content at *natural* saturation (that is, $\Phi - \theta_{ar}$, where θ_{ar} is the residual air content) at $x = 0$ and keeps this value there indefinitely thereafter. Green and Ampt assumed that there exists a wetting *front* that separates a fully saturated zone and a zone still at the initial water content. With this assumption, in the fully wetted zone, $\theta = \tilde{\theta}$ and does not vary with time. Application of equation 24 in this zone yields $\partial h_c/\partial x$ = constant. In other words, the capillary pressure increases linearly with the distance x, which means that the water pressure

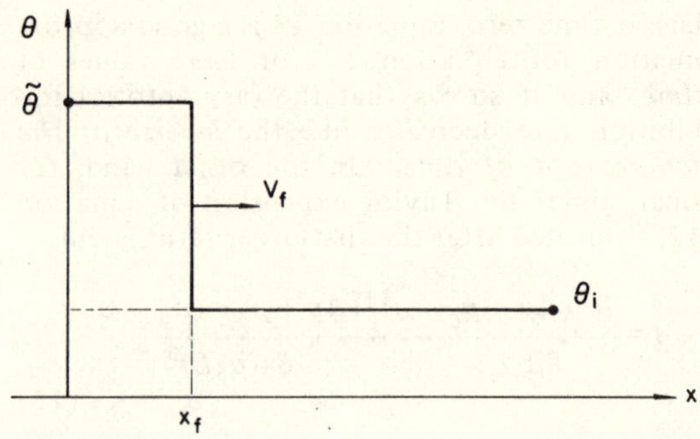

FIGURE 6. The Green and Ampt (piston) displacement model.

decreases linearly with x as in the case of a capillary tube. The same reasoning therefore yields for the velocity of propagation of the front, V_f (see Figure 6):

$$V_f = \frac{dx_f}{dt} = \frac{\tilde{K}H_f}{x_f(\tilde{\theta} - \theta_i)} \tag{27}$$

where H_f is the capillary-pressure head at the front. After integration for x_f, one obtains:

$$x_f = \left(\frac{2\tilde{K}\,H_f\,t}{\tilde{\theta} - \theta_i}\right)^{1/2} \tag{28}$$

and the imbibition rate is finally:

$$I = \frac{1}{2}\,[2(\tilde{\theta} - \theta_i)\tilde{K}\,H_f]^{1/2}\frac{1}{\sqrt{t}} \tag{29}$$

From equation 29 one deduces the *Green and Ampt sorptivity:*

$$S_f = [2(\tilde{\theta} - \theta_i)\tilde{K}H_f]^{1/2} \tag{30}$$

In a capillary tube the capillary pressure on the water side of the meniscus is well defined: $2\sigma/r$. Unfortunately, H_f, the capillary-pressure head at the wetting front, is not.

From the Green and Ampt model it is not possible to relate H_f to the soil characteristic curves. The term H_f is in a sense some sort of *average* capillary pressure. It can be determined for a given soil from experiments.

It has been suggested that a *weighted* average value of capillary-pressure head should be used with the relative permeability to water k_{rw}, which is a function of h_c, as the weight. Then one obtains an effective capillary drive, H_b, defined by the equation:

$$H_b = \int_0^{h_{ci}} k_{rw}\,dh_c \tag{31}$$

where $h_{ci} = h_c(\theta_i)$, the initial capillary-pressure head, leading to the *Bouwer sorptivity:*

$$S_b = [2(\tilde{\theta} - \theta_i)\tilde{K}H_b]^{1/2} \quad (32)$$

Experience has shown that this sorptivity led to quantitatively good predictions of the imbibition rate.

Gardner's Solution: For a constant diffusivity $\bar{D}$, independent of water content, the diffusivity equation, equation 26, is linear with constant coefficients. Its solution for a uniform initial water content θ_i and a fixed boundary value θ_b [Figure 5(b)] is:

$$\frac{\theta - \theta_i}{\theta_b - \theta_i} = \operatorname{erfc}\left(\frac{x}{\sqrt{4\bar{D}t}}\right) \quad (33)$$

where erfc( ) is the complementary error function defined as:

$$\operatorname{erfc}(u) = 1 - \frac{2}{\sqrt{\pi}}\int_0^u e^{-\xi^2}\, d\xi \quad (34)$$

From equation 33 one can deduce the imbibition rate:

$$I = \frac{1}{2}(\theta_b - \theta_i)\sqrt{\frac{4\bar{D}}{\pi}}\,\frac{1}{\sqrt{t}} \quad (35)$$

from which one deduces the *Gardner sorptivity:*

$$S_G = (\theta_b - \theta_i)\sqrt{\frac{4\bar{D}}{\pi}} \quad (36)$$

The solution is exact in the sense that equation 33 is an exact solution of equation 26 when D is constant, but in reality, it is a very approximate one because the diffusivity varies tremendously with water content. Defining the average diffusivity $\bar{D}$ as

$$\bar{D} = \frac{\int_{\theta_i}^{\theta_b} w(\theta)D(\theta)d\theta}{\theta_b - \theta_i} = \frac{\tilde{K}\int_{\theta_i}^{\theta_b} wk_{rw}dh_c}{\theta_b - \theta_i} \quad (37)$$

where $w(\ )$ is an appropriate weight function, equation 36 takes the form:

$$S_G = \left[\frac{4}{\pi}(\theta_b - \theta_i)\tilde{K}\int_{h_{cb}}^{h_{ci}} wk_{rw}dh_c\right]^{1/2} \quad (38)$$

As with the Green and Ampt approach, the problem is the choice of the weight w.

Parlange's Solution. Figure 7 indicates that as imbibition proceeds, there is a region near the water-entry (boundary) face where the water content no longer changes value significantly with time. In that region, $\partial\theta/\partial t$ is essentially zero. From equation 21 one concludes that in this region $\partial v_w/\partial x$ is essentially zero or in other words, that the water flux is constant with distance in this region. Symbolically,

$$v_w = -D\frac{\partial\theta}{\partial x} = v_w\Big|_{x=0} = I \quad (39)$$

Naturally the imbibition rate I still varies with time. Equation 39 for a fixed t is an ordinary differential equation in θ and x that can be integrated with the result:

$$x = \frac{1}{I(t)}\int_{\theta}^{\theta_b} D(\alpha)d\alpha = \frac{1}{I(t)}\int_{h_{cb}}^{h_c} k_{rw}\, dh_c \quad (40)$$

Equation 40 provides the position x of a given water content θ at a given time t. It is a good approximation to the exact profile near the water-entry face, away from the wetting front region.

The cumulative infiltration from time 0 to time t, W, is the cross-hatched area shown on Figure 7, or,

$$W = \int_0^{\infty} (\theta - \theta_i)\, dx \quad (41)$$

Integration by parts of equation 41 yields

$$W = x(\theta - \theta_i)\Big|_0^{\infty} - \int_{\theta_b}^{\theta_i} x d\theta = \int_{\theta_i}^{\theta_b} x d\theta \quad (42)$$

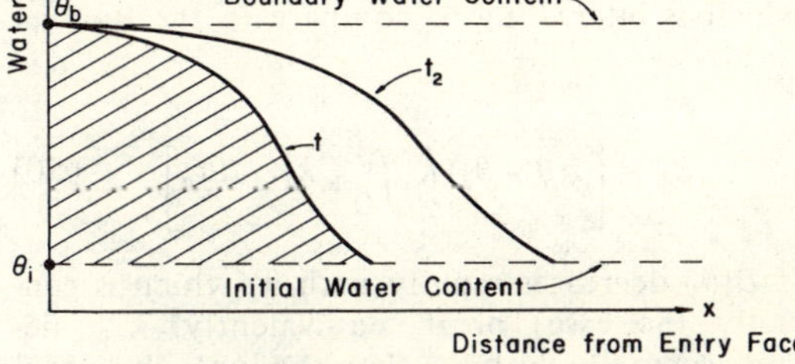

FIGURE 7. Time evolution of water-content profile in a horizontal soil.

Substitution of equation 40 for x in equation 42 gives

$$W = \frac{1}{I}\int_{\theta_i}^{\theta_b}\left(\int_{\theta}^{\theta_b} D(\alpha)d\alpha\right)d\theta \quad (43)$$

Integration by parts of the definite integral in equation 43 yields

$$IW = \int_{\theta_i}^{\theta_b}(\theta - \theta_i)\, D(\theta)\, d\theta \quad (44)$$

and since $I = dW/dt$, equation 44 can be integrated for $W(t)$ with the result

$$W = \left[2\int_{\theta_i}^{\theta_b}(\theta - \theta_i)\, D(\theta)\, d\theta\right]^{1/2} t^{1/2} \quad (45)$$

from which the infiltration rate is deduced:

$$I = \frac{1}{2}\,\frac{S_p}{\sqrt{t}} \quad (46)$$

where the *Parlange sorptivity* is

$$S_p = \left[2\int_{\theta_i}^{\theta_b}(\theta - \theta_i)\, D(\theta)\, d\theta\right]^{1/2} \quad (47)$$

From the definition of the diffusivity, equation 25, and of the relative permeability equation 47 can be rewritten in the form,

$$S_p = \left[2(\theta_b - \theta_i)\tilde{K}\int_{h_{cb}}^{h_{ci}}\left(\frac{\theta - \theta_i}{\theta_b - \theta_i}\right)k_{rw}\, dh_c\right]^{1/2} \quad (48)$$

In particular, for $\theta_b = \tilde{\theta}$ (case of free imbibition), the Parlange sorptivity takes the form

$$S_p = \left[2(\tilde{\theta} - \theta_i)\tilde{K}\int_{0}^{h_{ci}}\left(\frac{\theta - \theta_i}{\tilde{\theta} - \theta_i}\right)k_{rw}\, dh_c\right]^{1/2} \quad (49)$$

which is interesting to compare to the Bouwer sorptivity:

$$S_b = \left[2(\tilde{\theta} - \theta_i)\tilde{K}\int_{0}^{h_{ci}}k_{rw}\, dh_c\right]^{1/2} \quad (50)$$

If $D(\theta)$ decreases rapidly with θ (which is generally the case) or if, equivalently, k_{rw} decreases rapidly with h_c, then the low values of θ (or high values of h_c) contribute little to the value of the integral

$$\int_{0}^{h_{ci}}(\theta - \theta_i)\, k_{rw}\, dh_c,$$

which can be well approximated by the expression

$$(\tilde{\theta} - \theta_i)\int_{0}^{h_{ci}}k_{rw}\, dh_c.$$

Thus generally the Bouwer and the Parlange sorptivities will differ little. Both will lead to fair prediction of infiltration rates. It should be pointed out that Parlange sorptivity is more general than Bouwer's since it applies for arbitrary values of water content at the entry face θ_b (restrained imbibition), whereas Bouwer's sorptivity applies only for $\theta_b = \tilde{\theta}$, that is, for the case when the entry face is saturated. Similarly, equation 48 is more general than equation 47 as it applies also for h_{cb} in the capillary fringe, whereas equation 47 applies only for a boundary capillary pressure in excess or equal to the imbibition entry pressure (see *Capillary Pressure* for definitions).

Numerical Solution of the Water-content Equation

It can be readily verified that the assumed special relation between x, θ, and t, given by the equation,

$$x(\theta, t) = u(\theta)\sqrt{t} \quad (51)$$

will transform the partial differential equation, equation 26, into the ordinary differential equation for the unknown function of θ, $u(\theta)$:

$$-\frac{u}{2} = \frac{d}{d\theta}\left(D\,\frac{d\theta}{du}\right) \quad (52)$$

with boundary conditions $u = 0$ for $\theta = \theta_b$ $(=\theta_N)$ and $u = \infty$ for $\theta = \theta_i$ $(=\theta_0)$. Equation 52 can be rewritten by integrating with respect to θ in the form,

$$\int_{\theta_0}^{\theta} u\, d\theta = -2D\,\frac{d\theta}{du} \quad (53)$$

This equation can be solved by an efficient finite-difference technique introduced by Philip (1955). The nomenclature associated with this scheme is displayed in Figure 8. The cross-hatched area represents the discretized form of the integral

$$\int_{\theta_0}^{\theta_{n+1/2}} u\, d\theta$$

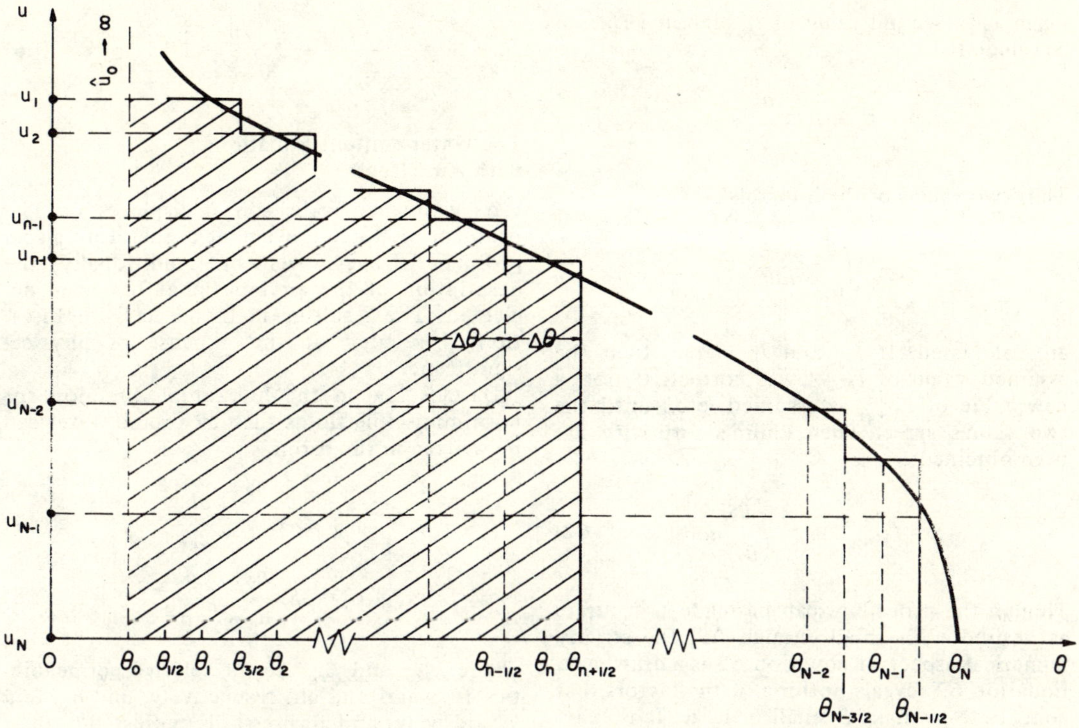

FIGURE 8. Finite difference nomenclature in numerical solution of the diffusivity equation.

and for convenience, one defines:

$$I_{n+\frac{1}{2}} = \frac{\int_{\theta_0}^{\theta_{n+\frac{1}{2}}} u d\theta}{\Delta\theta} \qquad (54)$$

Defining an average diffusivity in an interval as

$$\bar{D}_{n+\frac{1}{2}} = \frac{\int_{\theta_n}^{\theta_{n+1}} D(\theta)\theta}{\Delta\theta} \qquad (55)$$

then the finite-difference form of equation 53 leads to the equations:

$$u_n = u_{n+1} + \frac{2\bar{D}_{n+\frac{1}{2}}}{I_{n+\frac{1}{2}}} \qquad n = N-1, N-2 \ldots 2, 1 \qquad (56)$$

and from the very definition of the integral,

$$I_{n-\frac{1}{2}} = I_{n+\frac{1}{2}} - u_n \qquad n = N-1, N-2, \ldots 3, 2, 1 \qquad (57)$$

The iterative procedure is to *assume* a value for $I_{N-\frac{1}{2}}$, say $I^u{}_{N-\frac{1}{2}}$, calculate u_{N-1} from equation 56, then $I_{N-\frac{3}{2}}$ from equation 57, then u_{N-2} again from equation 56, etc., until u_1 and $I_{\frac{1}{2}}$ are obtained. In the range θ_0, θ_1 because $u_0 = \infty$, one cannot proceed with the finite-difference method, but rather one calculates a value of I_1 from an approximate analytical solution, assuming D constant in that interval (thus similar to Gardner's solution). This solution is

$$\frac{\theta - \theta_0}{\Delta\theta} = \frac{\operatorname{erfc}\left(\frac{u}{\sqrt{4\hat{D}}}\right)}{\operatorname{erfc}\left(\frac{u_1}{\sqrt{4\hat{D}}}\right)} = \frac{\operatorname{erfc}\left(\frac{u}{\sqrt{4\hat{D}}}\right)}{E_1} \qquad (58)$$

where $$\hat{D} = \frac{\int_{\theta_0}^{\theta_1} D(\theta) d\theta}{\Delta\theta} \qquad (59)$$

From equation 58 one can calculate I_1 by integration with the result:

$$I_1 = u_1 + \frac{2\hat{D}}{u_1} A\left(\frac{u_1}{\sqrt{4\hat{D}}}\right) \qquad (60)$$

where $A(\nu)$ is the special function 2ν [$i$erfc (ν)/ erfc(ν)], which has been tabulated by Philip.

From I_1 a second value of $I_{½}$ denoted $\hat{I}_{½}$ can be calculated:

$$\hat{I}_{½} = I_1 - \frac{u_1}{2} \tag{61}$$

Thus two values of the integral

$$\int_{\theta_0}^{\theta_1} u d\theta$$

are calculated. If $I_{½}$ and $I_{½}$ agree, then the assumed value of $I_{N-½}$ was correct. If not, a new value of $I_{N-½}$ is assumed, etc., until the two values agree. Then Philip's sorptivity has been obtained:

$$S = I_{N-½}\,\Delta\theta = \int_{\theta_0}^{\theta_b} u d\theta \tag{62}$$

Though the procedure can be made as accurate as desired with small enough $\Delta\theta$, the strictly numerical aspect of equation 62 is a drawback. Equation 62 reveals nothing of the factors that contribute to make S small or large. It is exact but *silent.* By contrast, the Bouwer and Parlange sorptivities display explicitly, tellingly, the physical factors that constitute the sorptivity and naturally are easier to compute. This latter advantage is not significant because Philip's procedure can be programmed easily for even the smallest computers and run at the cost of a few cents.

For the first estimate of $I_{N-½}$, Philip suggested the use of Gardner's solution with an average D value, D_a defined as:

$$D_a = \frac{2\int_{\theta_0}^{\theta_N} (\theta-\theta_0)D(\theta)d\theta}{(\theta_N - \theta_0)^2} \tag{63}$$

From Gardner's solution one obtains Philip's *initial sorptivity:*

$$S_i = \left[\frac{8}{\pi}\int_{\theta_i}^{\theta_b} (\theta - \theta_i)D(\theta)d\theta\right]^{½} \tag{64}$$

$$\text{or } S_i = \left[\frac{8}{\pi}(\theta_b - \theta_i)\tilde{K}\int_{h_{cb}}^{h_{ci}} \frac{\theta - \theta_i}{\theta_b - \theta_i} k_{rw} dh_c\right]^{½} \tag{65}$$

Comparison of Philip's initial sorptivity to Parlange's sorptivity shows that the two are in a constant ratio, namely,

$$\frac{S_i}{S_p} = \sqrt{\frac{4}{\pi}} = 1.13 \tag{66}$$

The Water-content Equations with Air Effect

As water penetrates soil, it displaces air. The problem of water movement is therefore also a problem of air movement. Traditionally, the description of the air movement has been neglected. The resulting mathematical simplification is secured at the expense of physical significance.

Darcy's law in the horizontal direction for two immiscible fluids such as air and water can be written in the form:

$$v_w = -\tilde{K}\, k_{rw} \frac{\partial h_w}{\partial x} \tag{67}$$

$$v_a = -\tilde{K} \frac{\mu_w}{\mu_a} k_{ra} \frac{\partial h_a}{\partial x} \tag{68}$$

where k_{rw} and k_{ra} are the relative permeabilities to water and air, respectively, and h_w and h_a are water and air pressures expressed both as equivalent water heights. Typical relative permeability curves are shown in Figure 9.

It is convenient to define the *total* velocity v as the *algebraic* sum of the water and air velocities. By adding equations 67 and 68 and using the fact that $h_a - h_w = h_c$ one obtains the relation,

$$v = -\tilde{K}\left(k_{rw} + \frac{\mu_w}{\mu_a} k_{ra}\right)\frac{\partial h_a}{\partial x} + \tilde{K}\, k_{rw} \frac{\partial h_c}{\partial x} \tag{69}$$

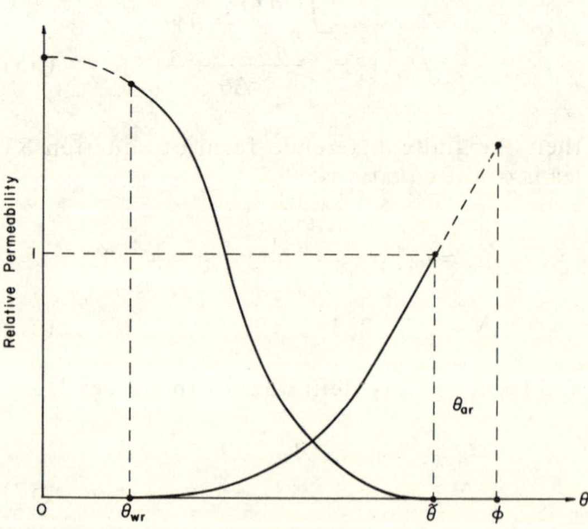

FIGURE 9. Typical relative permeability curves.

or multiplying through by

$$\mu_{rT} = \left[k_{rw} + \frac{\mu_w}{\mu_a} k_{ra} \right]^{-1}$$

(the *total* relative viscosity),

$$\mu_{rT}\, \nu = -\tilde{K} \frac{\partial h_a}{\partial x} + \tilde{K} f_w \frac{\partial h_c}{\partial x} \qquad (70)$$

where the *little* f_w function of water content is defined by the relation

$$f_w = \frac{k_{rw}}{k_{rw} + (\mu_w/\mu_a)\, k_{ra}} = \frac{1}{1 + (\mu_w/\mu_a)(k_{rw}/k_{ra})} \qquad (71)$$

A typical f_w curve is shown on Figure 10, and a typical $\mu_{rT} = [(f_w/k_{rw})]$ curve is shown on Figure 11. Disregarding air compressibility, assuming that air can escape readily ahead of the wetting front, the expression of mass conservation for water and air yields the equations:

$$\frac{\partial \theta}{\partial t} + \frac{\partial \nu_w}{\partial x} = 0 \qquad (72)$$

$$\frac{\partial \theta_a}{\partial t} + \frac{\partial \nu_a}{\partial x} = 0 \qquad (73)$$

where θ_a is the air content. Adding these two last equations yields the result,

$$\frac{\partial(\theta + \theta_a)}{\partial t} + \frac{\partial \nu}{\partial x} = 0 \qquad (74)$$

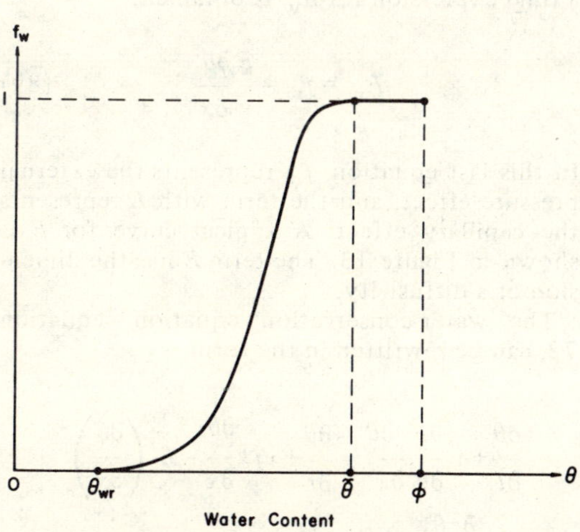

FIGURE 10. Typical *little*-f_w curve.

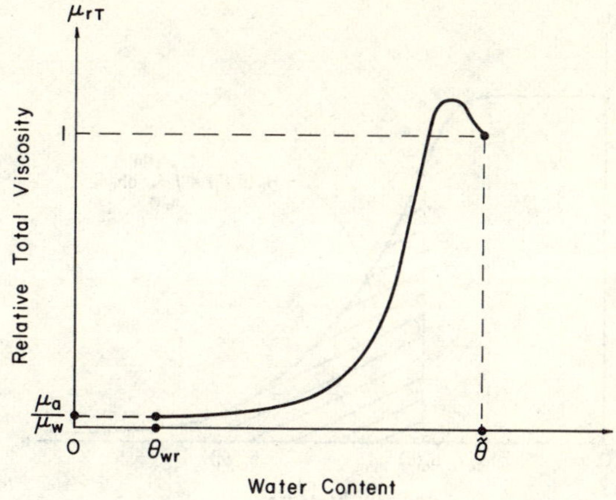

FIGURE 11. Typical relative total viscosity

For a nondeforming soil (ϕ = constant), equation 74 reduces to the result,

$$\frac{\partial \nu}{\partial x} = 0 \qquad (75)$$

In other words, the total velocity is space invariant. This result is important because it means that ν is only a function of time, whereas θ and ν_w are functions of both time and x coordinate. Thus it is *easier to work with* ν *than either* θ *or* ν_w.

Equation 70 is exact. No assumptions of air incompressibility or porous medium deformability were made in its derivation. Integration of the two sides of equation 70 with respect to x between the water-entry face and just ahead of the wetting front at position x_2 yields:

$$\nu = \frac{\tilde{K} \int_{h_{cb}}^{h_{ci}} f_w dh_c}{\int_0^{x_2} \mu_{rT} dx} \qquad (76)$$

or defining the *effective capillary drive* as

$$H_c(\theta, \theta_i) = \int_{h_c(\theta)}^{h_c(\theta_i)} f_w\, dh_c \qquad (77)$$

ν takes the form,

$$\nu = \frac{\tilde{K} H_c(\theta_b, \theta_i)}{\int_0^{x_2} \mu_{rT} dx} = I \qquad (78)$$

A typical curve of f_w versus h_c is shown in Fig-

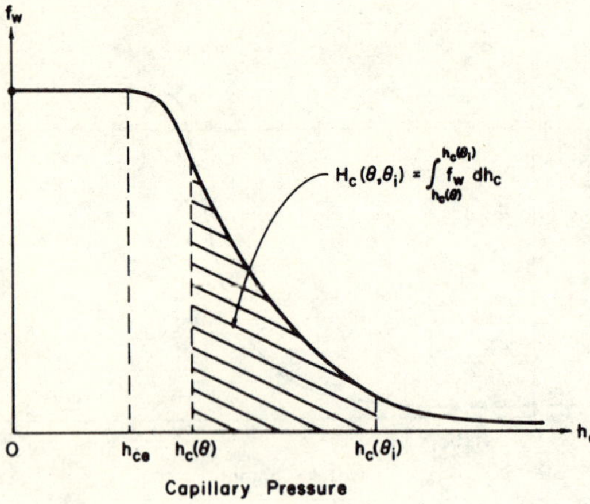

FIGURE 12. Typical curve of *little-*f_w versus capillary pressure.

ure 12. Equation 78 provides the imbibition rate into the soil. It must be noted that the numerator can be calculated exactly without the prior determination of a water-content profile. In fact, if one assumes a piston displacement (i.e., a Green and Ampt model for the wetting front), equation 78 immediately yields,

$$I = \frac{\tilde{K} H_c}{x_f} \qquad (79)$$

where

$$H_c = \int_0^{h_{ci}} f_w \, dh_c \qquad (80)$$

Noting the complete similarity between equation 79 and the Green and Ampt one,

$$I = \frac{\tilde{K} H_f}{x_f} \qquad (81)$$

deduced from equation 27, one can write immediately the *correct wetting front suction* sorptivity:

$$S_c = \left[2(\tilde{\theta} - \theta_i)\tilde{K} \int_0^{h_{ci}} f_w dh_c \right]^{1/2} \qquad (82)$$

We are quick to point out that this sorptivity is *not* exact. It is only a Green and Ampt sorptivity using a correct wetting front suction rather than the empirical H_f or the Bouwer estimate H_b. It is still not an exact sorptivity because the denominator of equation 78, the relative viscous resistance, was calculated using an assumed crude profile, a piston profile, rather than the correct profile.

To determine the correct profile, the derivation of the water-content equations must be completed. It is convenient to define the ratio of water velocity to total velocity as the fractional-flow function:

$$F_w = \frac{v_w}{v} \qquad (83)$$

An explicit form of F is obtained from equations 67 and 68, rewritten in the form

$$\frac{v_w}{\tilde{K} k_{rw}} = -\frac{\partial h_w}{\partial x} \qquad (84)$$

$$\frac{v_a}{\tilde{K}(\mu_w/\mu_a) k_{ra}} = -\frac{\partial h_a}{\partial x} \qquad (85)$$

Subtracting yields

$$\frac{v_w}{\tilde{K} k_{rw}} - \frac{(v - v_w)}{\tilde{K}(\mu_w/\mu_a) k_{ra}} = \frac{\partial h_c}{\partial x} \qquad (86)$$

Collecting terms in v_w and dividing by v yields

$$F_w = \frac{v_w}{v} = f_w + \frac{\tilde{K}}{v}(1 - f_w)\, k_{rw} \frac{\partial h_c}{\partial x} \qquad (87)$$

Defining the nonnegative function of water content

$$E = -\tilde{K} k_{rw} (1 - f_w) \frac{dh_c}{d\theta} \qquad (88)$$

a final expression for F_w is obtained:

$$F_w = f_w - \frac{E}{v} \frac{\partial \theta}{\partial x} \qquad (89)$$

In this last equation, f_w represents the external pressure effect, and the term with E represents the capillary effect. A typical curve for E is shown in Figure 13. The term E has the dimension of a diffusivity.

The water-conservation equation, equation 72, can be rewritten in the form,

$$\frac{\partial \theta}{\partial t} + v \frac{\partial F}{\partial \theta} \frac{\partial \theta}{\partial x} = \frac{\partial \theta}{\partial t} + v f' \frac{\partial \theta}{\partial x} - E' \left(\frac{\partial \theta}{\partial x} \right)^2 - E \frac{\partial^2 \theta}{\partial x^2} = 0 \qquad (90)$$

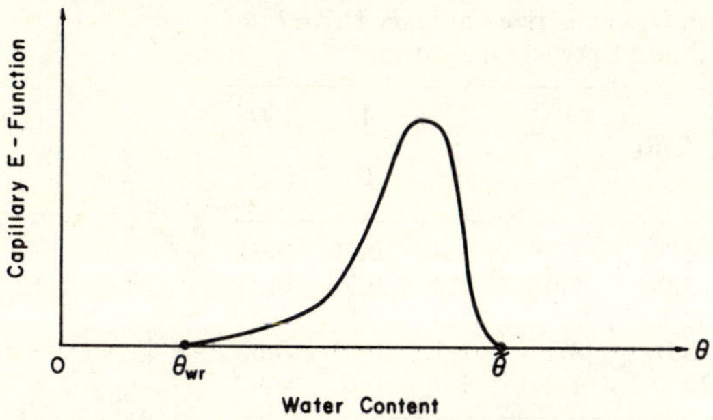

FIGURE 13. Typical curve of the capillary E function of water content.

This nonlinear equation contains two unknowns, θ and v, and a second equation is required for solution, namely, equation 76. To summarize, with inclusion of air effects two equations must be solved simultaneously:

$$\frac{\partial\theta}{\partial t} + vF'\frac{\partial\theta}{\partial x} - \frac{\partial\theta}{\partial t} - \frac{\partial}{\partial x}\left(E\frac{\partial\theta}{\partial x} - vf\right) = 0 \quad (91)$$

and

$$v = \frac{\tilde{K}H_c}{\int_0^L \mu_{rT}dx} \quad (92)$$

one partial differential equation, and one ordinary differential equation. (The prime indicates differentiation with respect to θ at a fixed time.) It can be verified readily that the assumed special relation

$$x(\theta, t) = u(\theta)\sqrt{t} \quad (93)$$

will transform the partial differential equation 91 into the ordinary differential equation for the unknown function of θ, $u(\theta)$:

$$-\frac{u}{2} = \frac{d}{d\theta}\left(E\frac{d\theta}{du} - \frac{S}{2}f\right) \quad (94)$$

with boundary conditions $u = 0$ for $\theta = \theta_b$, and $u = \infty$ for $\theta = \theta_i$, and the ordinary differential equation 92 into the algebraic relation,

$$S = \frac{2\tilde{K}H_c}{\int_0^\infty \mu_{rT}du} \quad (95)$$

A procedure similar to that of Philip for the Richards equation can be followed leading to the system of equations:

$$\int_{\theta_0}^{\theta} ud\theta = -2E(\theta)\frac{d\theta}{du} + Sf(\theta) \quad (96)$$

$$\int_{\theta_0}^{\theta_b} -\mu_{rT}\left(\frac{du}{d\theta}\right)d\theta = 2\tilde{K}H_cS \quad (97)$$

The finite-difference forms of equation 96 analogous to equations 56 and 57 are

$$u_n = u_{n+1} + \frac{2\bar{E}_{n+½}}{I_{n+½} - (S/\Delta\theta)\bar{f}_{n+½}}$$
$$n = N-1, N-2, \ldots 2, 1 \quad (98)$$

$$I_{n-½} = I_{n+½} - u_n$$
$$n = N-1, N-2, \ldots 2, 1 \quad (99)$$

Again this procedure is strictly numerical. An approximate but more practical solution can be obtained.

From equation 91 it follows that the velocity of propagation of a given water content is

$$\left(\frac{dx}{dt}\right)_\theta = vF'_{\mathrm{w}} \quad (100)$$

In other words, the velocity is proportional to the slope of the curve F. Since the profile deforms strictly by stretching (see equation 93), the curve F does not change with time, and the solution for x is

$$x(\theta, t) = F'(\theta)\,W(t) \quad (101)$$

where $W(t)$ is the cumulative infiltration. Since the low water contents must travel faster than the high ones, the F curve must be concave towards the θ axis. By definition of F (equation 89), F is always greater than f since the term $-(E/v)(\partial\theta/\partial x)$ is positive. Thus a lower limit for the F curve is the curve f_{w} for $\theta > \theta_f^-$, and the tangent line shown in Figure 14 in the range

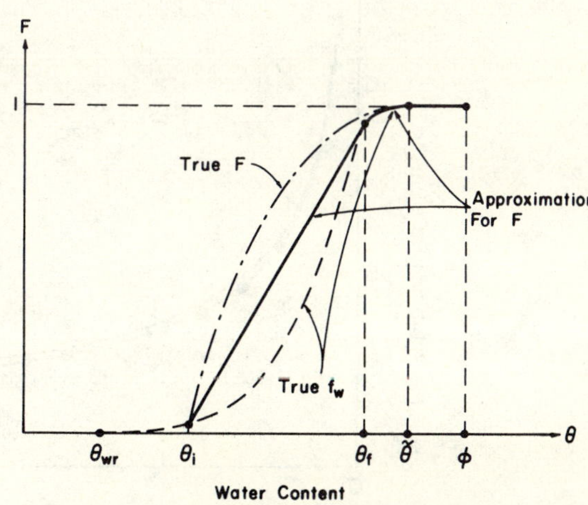

FIGURE 14. Fractional flow function of water content and its time evolution during imbibition.

TABLE 1. Values of Critical Pressure Head H_b, Effective Capillary Drive H_c, Viscous Resistance Correction β, and Derived Quantities.

Soil Type	H_b (cm)	H_c (cm)	$\frac{H_c}{H_b}$	β	$\frac{1}{\beta}$	$\frac{1}{\beta}\frac{H_c}{H_b}$
Plainfield Sand	11.7	14.7	1.26	1.4	0.7	0.9
Columbia Sandy Loam	23.8	25.0	1.05	1.4	0.7	0.75
Guelph Loam	31.4	32.1	1.02	1.3	0.8	0.8
Ida Silt Loam	7.4	5.0	0.7	1.1	0.9	0.65
Yolo Light Clay	22.4	23.1	1.03	1.7	0.6	0.6

$\theta_i \leqslant \theta \leqslant \theta_f^-$. Using this lower limit for F as an approximation, one obtains the continuous solution for x:

$$x = Wf' \qquad \theta_f^- \leqslant \theta \leqslant \tilde{\theta} \tag{102}$$

and a front of size $\theta_f^- - \theta_i$ located at $xf = Wf'(\theta_f^-)$. In other words, we use the following approximations for F_w:

$$F_w = f_w \qquad \theta_f^- \leqslant \theta \leqslant \tilde{\theta} \tag{103}$$

$$F_w = f_i + \left[\frac{f_w(\theta_f^-) - f_i}{\theta_f^- - \theta_i}\right](\theta - \theta_i) \tag{104}$$

Because the effective capillary drive is already known exactly, this approximate profile is necessary only to calculate the viscous resistance term. This term, the denominator of equation 92, takes the form

$$W\int_0^{xf^-} \mu_{rT} f''\, d\theta = W\int_{\theta_f^-}^{\tilde{\theta}} -f''\mu_{rT} d\theta \tag{105}$$

Let

$$\beta = (\tilde{\theta} - \theta_i)\int_{\theta_f^-}^{\tilde{\theta}} -\frac{ff''}{k_{rw}}\, d\theta \tag{106}$$

then equation 92 takes the form

$$W\frac{dw}{dt} = \frac{(\tilde{\theta} - \theta_i)\tilde{K}H_c}{\beta} \tag{107}$$

or

$$W = \left[\frac{2(\tilde{\theta} - \theta_i)\tilde{K}H_c}{\beta}\right]^{1/2} t^{1/2} \tag{108}$$

from which we deduce a new sorptivity:

$$S_\beta = \left[\frac{2(\tilde{\theta} - \theta_i)\tilde{K}H_c}{\beta}\right]^{1/2} \tag{109}$$

where β is a viscous correction factor. Table 1 lists values of β for various soils. Figure 15 shows a comparison of predictions of imbibition rate using the sorptivity S_β and of experimental observations.

H. J. MOREL-SEYTOUX

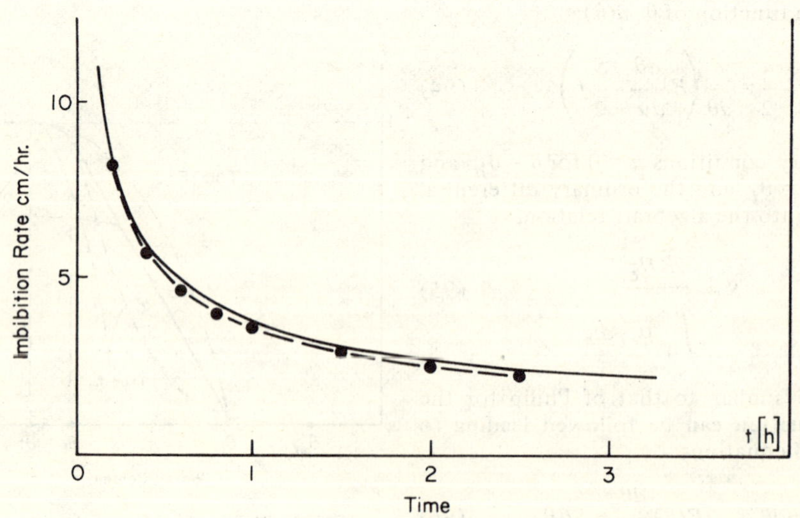

FIGURE 15. Comparison of observed (– • –) and predicted (solid line) imbibition rates as a function of time.

References

Bouwer, H., 1964. Unsaturated flow in ground water hydraulics, *Jour. Hydraulics Div. Proc. ASCE*, **90**, 121–144.

Gardner, H. R., 1959. Solution of the flow equation for the drying of soils and other porous media, *Soil Sci. Soc. America Proc.*, **23**, 183–187.

Green, W. H., and Ampt, C. A., 1911. Studies on soil physics: 1. Flow of air and water through soils, *Jour. Agric. Sci.*, **4**, 1–24.

Morel-Seytoux, H. J., and Khanji, J., 1974. Derivation of an equation of infiltration, *Water Resour. Res. Jour.*, **10**(4), 795–800.

Morel-Seytoux, H. J., and Khanji, J., 1975. Prediction of imbibition in a horizontal column, *Soil Sci. Soc. America Proc.*, **39**(4), 613–617.

Parlange, J. Y., 1971. Theory of water movement in soils: I. One-dimension 1 absorption, *Soil Sci. Jour.*, **111**, 134–137.

Philip, J. R., 1955. Numerical solution of equations of the diffusion type with diffusivity concentration dependent, *Faraday Soc. Trans.*, **51**, 885–892.

Schlichting, H., 1960. *Boundary Layer Theory*, New York: McGraw-Hill, 647p.

Cross-references: *Capillary Pressure; Conductivity, Hydraulic; Field Water Cycle; Infiltration; Percolation; Permeability; Pore Space, Drainable; Soil Solution; Soil Water; Thermodynamics; Water Content and Retention; Water Fluxes; Water Movement; Wetting Front.*

INFILTRATION

The term *infiltration* has several meanings, sometimes being used as a synonym for *wetting, imbibition* (q.v.), or *percolation* (q.v.). However, the concept of infiltration specifically relates to the phenomenon of water penetration from the surface of the ground into the sub jacent soil. As water penetrates the soil, its distribution in space and time varies. The description of the evolution of the water (see *Water Content and Retention*) in the soil resulting from the occurrence of a rain or of a pond of water at the surface is also sometimes considered part of the infiltration phenomenon. Mostly, however, infiltration is thought of as the phenomenon of water crossing from the air side to the soil side of the air-soil interface. Certainly, it is this aspect of infiltration that is of greatest concern in hydrology, and thus it is the aspect discussed here.

Significance in the Hydrologic Cycle

If water has a choice, so to speak, it will infiltrate rather than run off on the land surface. Rain will infiltrate until it exceeds the so-called *infiltration capacity* of the soil. Infiltration capacity is the maximum rainfall rate the soil can fully absorb at a given time, given the previous history of soil moisture and rainfall pattern. A number of factors affect infiltration capacity, which varies greatly among soils depending on their permeability and capillary characteristics. For a given soil, the infiltration capacity will vary according to whether the soil was initially dry when the rainfall started or already fairly wet from a relatively recent rain. The ability to predict the infiltration capacity of the soil and the actual infiltration rates is a prerequisite to the prediction of surface runoff and therefore streamflow. Depending on the infiltration capacity of the soil at the time of rainfall, a storm may lead to an inoffensive streamflow change or on the contrary, to a devastating flood.

Infiltration in a Capillary Tube

Consider the highly simplified situation shown in Figure 1. Due to intense rain, water is continuously available at the surface. Water covers the surface with a film of thickness H. At time zero the tube is dry, and immediately as water becomes available at the soil surface, a meniscus forms in the tube and propagates in the tube under the action of the capillary-pressure forces (for a detailed discussion of this phenomenon, see *Imbibition*). Later the distribution of pressure (*expressed as a water height*) in the two phases, h_w and h_a, is as shown in Figure 1. The term h_w^- is the water pressure on the water side of the interface, h_a^+ is the air pressure on the air side of the interface, and h_A is the atmospheric pressure. The velocity of water (or air) in a vertical tube of length L is given by *Poiseuille's law*:

$$u_w = \frac{R^2}{8\mu_w} \frac{\Delta\Phi}{L} \tag{1}$$

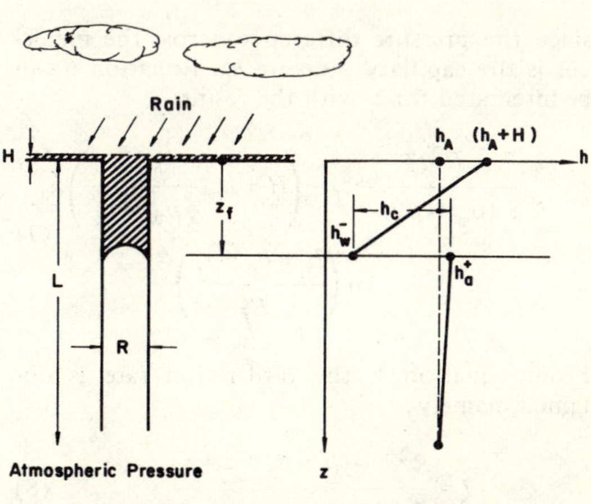

FIGURE 1. Infiltration in a capillary tube.

where $\Delta\Phi$ is the drop in total head Φ, and μ_w is the viscosity of water. The total head is defined as

$$\Phi = h_w - z \tag{2}$$

where z is the vertical coordinate oriented positive downward. Application of Poiseuille's law for the water phase yields

$$\begin{aligned} u_w &= \frac{R^2}{8\mu_w} \frac{[h_A + H - (h_w^- - z_f)]}{z_f} \\ &= \frac{R^2}{8\mu_w} \frac{[h_A + H + z_f - h_w^-]}{z_f} \end{aligned} \tag{3}$$

since $z = 0$ at the entrance to the tube and there the water pressure is hydrostatic, i.e., equal to H. Similarly, one obtains for u_a (neglecting the small gravity effect on air),

$$u_a = \frac{R^2}{8\mu_a} \frac{[h_a^+ - h_A]}{(L - z_f)} \tag{4}$$

Since the velocities of water and air are both equal to dz_f/dt, equating equations 3 and 4 yields

$$\begin{aligned} \frac{dz_f}{dt} &= \frac{R^2}{8} \frac{(h_A + H + z_f - h_w^-)}{\mu_w z_f} \\ &= \frac{R^2}{8} \frac{(h_a^+ - h_A)}{\mu_a(L - z_f)} \end{aligned} \tag{5}$$

or

$$\begin{aligned} \frac{dz_f}{dt} &= \frac{R^2}{8} \frac{[H + (h_a^+ - h_w^-) + z_f]}{\mu_w z_f + \mu_a(L - z_f)} \\ &= \frac{R^2}{8} \frac{(H + h_c + z_f)}{z_f(\mu_w - \mu_a) + \mu_a L} \end{aligned} \tag{6}$$

since the pressure difference across the meniscus is the capillary pressure h_c. Equation 6 can be integrated for z_f with the result:

$$\frac{R^2 t}{8(\mu_w - \mu_a)} = z_f - \left(H + h_c - \frac{\mu_a L}{\mu_w - \mu_a}\right) \ln\left(\frac{H + h_c + z_f}{H_f}\right) \tag{7}$$

From equation 6 the infiltration rate is obtained, namely;

$$I = \frac{R^2}{8} \frac{(h + h_c + z_f)}{z_f(\mu_w - \mu_a) + \mu_a L} \tag{8}$$

In equation 8 the infiltration rate is defined in terms of the position of the meniscus (the *wetting front*). The infiltration rate at a given time is obtained by calculating z_f from equation 7 for a given value of t and then evaluating I from equation 8 for this value of z_f.

Infiltration in a Soil

The Water-content Equation (Without Air Effect). Away from the water source the soil is dry; its *water content* (q.v.) is low. Close to the source it is high. At a fized location z_0 (Figure 2), the soil gets wetter as time passes. In general, the water content θ is a function of z and t. The variation of θ in time at a fixed location z is related to the net inflow of water at that time through the principle of conservation of mass, which expressed in mathematical symbols, reads

$$\frac{\partial\theta}{\partial t} + \frac{\partial}{\partial z}(v_w) = 0 \tag{9}$$

where v_w is the water velocity in a Darcy sense (i.e., a volume flow rate per unit bulk cross-section area in a direction perpendicular to flow). The water velocity is given by *Darcy's law*, namely,

$$v_w = -K(\theta)\frac{\partial h_w}{\partial z} + K(\theta) \tag{10}$$

$$\text{or} \quad v_w = K(\theta)\frac{\partial h_c}{\partial z} + K(\theta) = \left[K(\theta)\frac{dh_c}{d\theta}\right]\frac{\partial\theta}{\partial z} \tag{11}$$

if one *assumes* that air pressure is atmospheric everywhere and for all times in the medium.

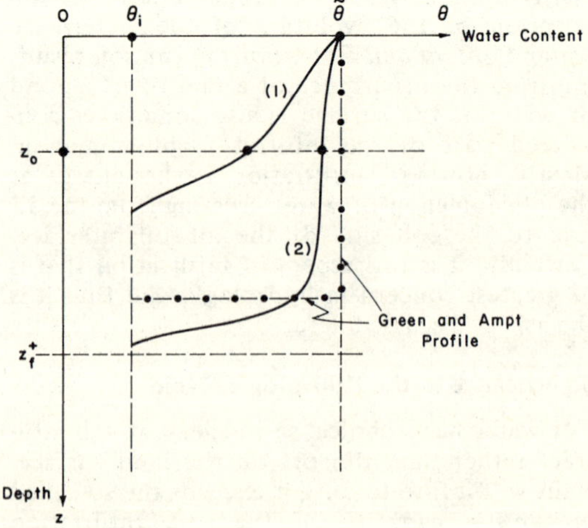

FIGURE 2. Evolution of water contents in a soil during infiltration.

The term $K(\theta)$ is the *unsaturated hydraulic conductivity.* It measures the conductivity of the medium to water at various water contents. Substitution of equation 11 in equation 9 yields the *water-content equation,* also called *Richard's equation:*

$$\frac{\partial \theta}{\partial t} + \frac{\partial}{\partial z}\left(K(\theta)\frac{\partial h_c}{\partial z}\right) + \frac{\partial K(\theta)}{\partial z} = 0 \qquad (12)$$

Defining a new positive function of water content, the *diffusivity,* namely,

$$D(\theta) = -K(\theta)\frac{dh_c}{d\theta} \qquad (13)$$

the water-content equation takes the form

$$\frac{\partial \theta}{\partial t} - \frac{\partial}{\partial z}\left[D(\theta)\frac{\partial \theta}{\partial z}\right] + \frac{\partial K(\theta)}{\partial z} = 0 \qquad (14)$$

Approximate Solutions of the Water-content Equation

The Green and Ampt Solution. It is assumed that the initial water content of the soil is uniform at value θ_i. At time zero the soil is placed in contact with an ample water supply. Immediately the water content takes the value $\tilde{\theta}$, water content at *natural* saturation (that is, $\theta - \theta_{ar}$, where θ_{ar} is the residual air content) at $z = 0$ and keeps this value there indefinitely thereafter. Green and Ampt (1911) assumed that there exists a wetting *front* (q.v.) that separates a fully saturated zone and a zone still at the initial water content. With this assumption, in the fully wetted zone, $\theta = \tilde{\theta}$ and does not vary with time.

The same reasoning discussed in the section *Imbition* leads to the velocity of propagation of the front V_f:

$$V_f = \frac{dz_f}{dt} = \tilde{K}\,\frac{(H + H_f + z_f)}{z_f(\tilde{\theta} - \theta_i)} \qquad (15)$$

The infiltration rate is

$$I = \tilde{K}\,\frac{(H + H_f + z_f)}{z_f} \qquad (16)$$

where H_f is the capillary-pressure head at the front. After integration for z_f, one obtains

$$\frac{\tilde{K}t}{\tilde{\theta} - \theta_i} = z_f - (H + H_f)\,\ell n\left[1 + \frac{z_f}{H + H_f}\right] \qquad (17)$$

which is the well-known Green and Ampt equation. Equation 17 can be rewritten in terms of the cumulative infiltration up to time t, W, since $W = z_f(\tilde{\theta} - \theta_i)$, namely,

$$\tilde{K}t = W - (H + H_f)(\tilde{\theta} - \theta_i)\;\ell n\left[1 + \frac{W}{(H + H_f)(\tilde{\theta} - \theta_i)}\right] \qquad (18)$$

In a capillary tube the capillary pressure on the water side of the meniscus is well defined: $2\sigma/r$. Unfortunately, H_f, the capillary-pressure head at the wetting front, is not (see *Wetting Front*).

From the Green and Ampt model, it is not possible to relate H_f to the soil characteristic curves. In a sense, H_f is some sort of *average* capillary pressure. If can be determined for a given soil from experiments.

It has been suggested that a *weighted* average value of capillary-pressure head should be used with the relative permeability to water, k_{rw}, which is a function of h_c, as the weight. Then one obtains an effective capillary drive, H_b, defined by the equation;

$$H_b = \int_0^{h_{ci}} k_{rw}\,dh_c \qquad (19)$$

where $h_{ci} = h_c(\theta_i)$, the initial capillary-pressure head. This estimated value of the effective capillary drive leads to relatively good predictions of the infiltration rates.

Philip's Solution. Philip (1957) recognized that in the case of imbition (i.e., horizontal water penetration), the water-content equation, which reduces then to the diffusivity equation,

$$\frac{\partial \theta}{\partial t} - \frac{\partial}{\partial z}[D(\theta)\frac{\partial \theta}{\partial z}] = 0 \qquad (20)$$

could be reduced to an ordinary differential equation,

$$\frac{u_0}{2} + \frac{d}{d\theta}\left(D\,\frac{d\theta}{du_0}\right) = 0 \qquad (21)$$

by the similarity transformation,

$$z = u_0\,(\theta)\sqrt{t} \qquad (22)$$

This approach and the numerical method of solution of equation 21 are discussed in detail in the section *Imbibition.*

The next step is to view equation (14) as a combination of equation 20 and a pertubation term: $\partial K(\theta)/\partial z$. One then expects that the solution of equation 14 can be fruitfully put in the form $z = z_0 + \Delta z$. Naturally, no gain is achieved if Δz is not a separable function of θ and t. Thus the procedure is to look for a solution of equation 14 in the form $z = u_0(\theta)\sqrt{t} + u_1(\theta)g(t) + \epsilon(\theta, t)$. After substitution in equation 14, one obtains an equation for u_1, one disregards the terms in ϵ, and one selects the function $g(t)$, which reduces the approximate remaining equation to an ordinary differential equation. Repeating the operation several times, one obtains a solution in the form,

$$z(\theta, t) = u_0(\theta)t^{1/2} + u_1(\theta)t + u^2(\theta)t^{3/2} + \cdots \tag{23}$$

From this solution one can deduce an equation for the rate of infiltration I as a function of time:

$$I = \frac{C}{\sqrt{t}} + A + D\sqrt{t} + Et + Ft^{3/2} + \cdots \tag{24}$$

where the constants $C, K, \cdots$, depend on the soil characteristics and the boundary conditions θ_i and $\bar{\theta}$.

Equation 24 predicts that at small times the infiltration rate is high, which is appropriate. For large times the infiltration rate increases indefinitely. As this result is totally contradictory to experimental evidence, this indicates that the method of solution does not converge for large times, which is clear from equation 23. It is accepted practice now to truncate equation 24 after the constant term A and to interpret A to be the saturated hydraulic conductivity at natural saturation.

The inconsistency of the results arises from the implicit assumption of the perturbation procedure that gravity is small compared to capillarity. At large times (of the order of only one hour in some instances), the gravity force will actually become the dominant one. Nevertheless, Philip's method has given good predictions of infiltration using three terms in the series of equation 22 particularly, as one would expect, for tight soils with low conductivity and high capillary drive.

The Water-content Equations (With Air Effect). As water penetrates the soil, it displaces air. The problem of water movement is also a problem of air movement. The treatment of the air movement is parallel to that of water. Indeed, Darcy's law for water and air can be written very *symmetrically* in the form,

$$v_w = -\frac{kk_{rw}}{\mu_w}\left(\frac{\partial p_w}{\partial z} - \rho_w g\right) \tag{25}$$

$$v_a = -\frac{kk_{ra}}{\mu_a}\left(\frac{\partial p_a}{\partial z} - \rho_a g\right) \tag{26}$$

where v_w and v_a are the fluid velocities (in the Darcy sense); k is the intrinsic permeability; k_{rw} and k_{ra} are the fluid relative permeabilities; p_w and p_a are the fluid pressures; ρ_w and ρ_a are their specific masses; and g is the acceleration of gravity. If *both* water and air pressures are expressed as equivalent *water* heights, the expressions for the fluid velocities become,

$$v_w = -\bar{K}k_{rw}\frac{\partial h_w}{\partial z} + \bar{K}k_{rw} \tag{27}$$

$$v_a = -\bar{K}\frac{\mu_w}{\mu_a}k_{ra}\frac{\partial h_a}{\partial z} + \bar{K}\frac{\mu_w}{\mu_a}\frac{\rho_a}{\rho_w}k_{ra} \tag{28}$$

where the relative permeabilities are relative to $\bar{K}$ or, in other words,

$$k_{rw} = \frac{K(\theta)}{\bar{K}} \tag{29}$$

It is convenient to define the *total* velocity v as the *algebraic* sum of the water and air velocities. By adding equations 27 and 28 and using the fact that $h_c = h_a - h_w$, one obtains the *exact* relation,

$$v = \frac{\bar{K}\left[-\dfrac{\partial h_a}{\partial z} + f_w\dfrac{\partial h_c}{\partial z} + f_w\right]}{\mu_{rT}} \tag{30}$$

where f_w is the *little-f* function of water content, defined as

$$f_w = \frac{1}{1 + (k_{ra}/k_{rw})(\mu_w/\mu_a)} \tag{31}$$

and μ_{rT} is the *total relative* viscosity (a function of water content) defined as

$$\mu_{rT} = \frac{f_w}{k_{rw}} \tag{32}$$

Typical curves of f_w and μ_{rT} are given in the section on *Imbibition*.

The equations of conservation for water and air are, very symmetrically,

$$\frac{\partial\theta}{\partial t} + \frac{\partial v_w}{\partial z} = 0 \tag{33}$$

$$\frac{\partial \theta_a}{\partial t} + \frac{\partial v_a}{\partial z} = 0 \qquad (34)$$

where θ_a is the air content in the soil. Adding equations 33 and 34 yields the result

$$\frac{\partial(\theta + \theta_a)}{\partial t} + \frac{\partial v}{\partial z} = 0 \qquad (35)$$

For a nondeforming soil (ϕ, porosity = constant), equation 35 reduces to the result

$$\frac{\partial v}{\partial z} = 0 \qquad (36)$$

In other words, the total velocity is *invariant* in *space*. This result is important because it means that v is a function only of time, whereas θ and v_w are functions of both time and z coordinate. Thus it is *easier* to work with v than with either θ or v_w. Taking advantage of the fact that v is not a function of z, the integration of equation 30 between any two levels 1 and 2 yields the *v-integral* equation

$$v = \frac{\tilde{K}\left[h_{a1} - h_{a2} + \int_1^2 f_w dh_c + \int_{z_1}^{z_2} f_w dz\right]}{\int_{z_1}^{z_2} \mu_{rT} dz} \qquad (37)$$

With air effects, the water-content equations are *two:* the partial differential equation (33) and the equation (37). In the case of infiltration into a dry soil with a constant head of water on the soil surface (*ponded* infiltration), application of equation 37 between the soil surface and a position z_2 just downstream from the wetting front yields the infiltration rate I, as

$$I = \frac{\tilde{K}\left[H + H_c + \int_0^{z_2} f_w dz\right]}{\int_0^{z_2} \mu_{rT} dz} \qquad (38)$$

where $$H_c = \int_0^{h_{ci}} f_w dh_c \qquad (39)$$

is the *effective capillary drive* (see *Wetting Front*). Naturally, the exploitation of equation 38 requires a knowledge of the water-content profile. With the Green and Ampt assumption (Figure 2), the profile shape is assumed to be known a priori. For this assumed profile, equation 38 reduces to a formula derived previously, namely, equation 16, which can be rewritten as

$$I = \frac{\tilde{K}\left[H + H_c + (W)/(\tilde{\theta} - \theta_i)\right]}{(W)/(\tilde{\theta} - \theta_i)} \qquad (40)$$

because in a saturated zone, $f_w = 1$, and $\mu_{rT} = 1$. Equation 40 can be integrated to yield an equation identical to equation 18 except that H_f, then unknown, is now replaced by the known quantity H_c. Naturally, the Green and Ampt assumed profile does not account for any *simultaneous* flow of water and air. One may legitimately ask whether that profile is not too crude to accurately represent the phenomenon. The problem is to find an approximate water-content profile that accounts correctly for the viscous interaction of the two flowing phases. To answer this question, it is convenient to discuss first the *fractional-flow* concept.

The Fractional-flow Function. The *total* velocity is defined as the *algebraic* sum of the water and air velocities:

$$v = v_w + v_a \qquad (41)$$

In the total volumetric stream of fluids, only a fraction of the stream is water, and *by definition,* the *fractional-flow* function is

$$F_w = \frac{v_w}{v} \qquad (42)$$

In this form equation 42 is not too useful, but using equations 27, 28, and 41, one can eliminate the water and air pressures and obtain the more explicit result:

$$F_w = f_w + \frac{G_w}{v} - \frac{E_w}{v}\frac{\partial \theta}{\partial z} \qquad (43)$$

where the two nonnegative functions of water content

$$G_w = \tilde{K} k_{rw} (1 - f_w) \qquad (44)$$

and $$E_w = -G_w \frac{dh_c}{d\theta} \qquad (45)$$

are defined. In equation 43 the term f_w represents an external pressure effect, the term with G_w represents the gravity effect, and the term with E_w represents the capillary effect. A typical curve for E_w is given in the section *Imbibition,* and the curve of G_w is very similar. The term G_w has the dimension of a *velocity,* whereas E_w has the dimension of a *diffusivity.*

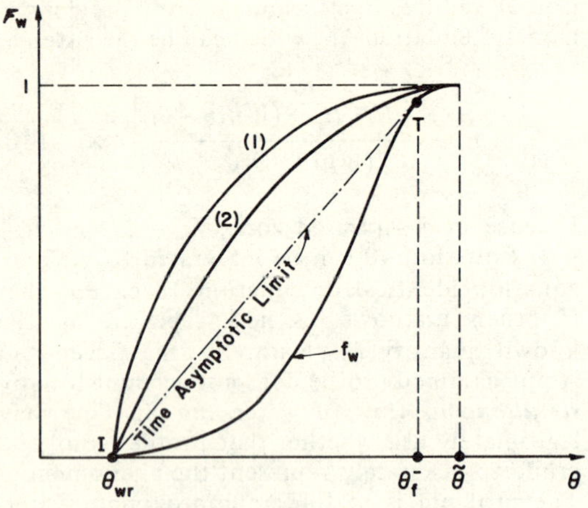

FIGURE 3. Fractional-flow functions corresponding to water-content profiles at various times.

The F_w curve depends on the water-content profile since it depends on $\partial\theta/\partial z$. For the two typical water-content profiles of Figure 2, the corresponding F_w curves are shown in Figure 3.

Movement of Water Contents. The water-content equation can be rewritten in the form

$$\frac{\partial \theta}{\partial t} + \frac{\partial}{\partial z}(\nu F_w) = 0 \tag{46}$$

or

$$\frac{\partial \theta}{\partial t} + \nu F_w' \frac{\partial \theta}{\partial z} = 0 \tag{47}$$

where F_w' means the derivative of F_w with respect to θ at a *fixed time* and graphically is the *slope* of the curve of F_w versus θ at a given time. It follws from equation 47 that the *velocity of propagation* of a given water content is

$$\left(\frac{dz}{dt}\right)_\theta = \nu \left.\frac{\partial F_w}{\partial \theta}\right|_t = \nu F_w' \tag{48}$$

It follows that if the water-content profile evolves strictly by translation, all water contents move at the same speed, and F_w' must be constant, that is, the curve F_w must be straight. As shown in Figure 3, as time proceeds, the curve F_w tends to straighten up. In fact, as F_w is always greater than f_w during infiltration—since all terms in equation 43 are positive and must be convex because otherwise the water-content profiles would exhibit multiple values—one concludes that for large times, F_w tends to a limit shown in Figure 3. This limit is the tangent from point I to the curve f_w, namely, line IT, and beyond T it is the curve f_w itself. With this approximate choice for F_w, namely,

$$F_w = \left[\frac{f_w(\theta_f^-) - f_w(\theta_i)}{\theta_f^- - \theta_i}\right](\theta - \theta_i) + f_w(\theta_i)$$

$$\theta \leqslant \theta \leqslant \theta_f^- \tag{49}$$

and

$$F_w = f_w \qquad \theta_f^- \leqslant \theta \leqslant \tilde{\theta} \tag{50}$$

Equation 48 can be integrated to yield the profile

$$z(\theta, t) = f_w'(\theta) \int_0^t \nu(\tau)d\tau = f_w'(\theta)\, W(t) \tag{51}$$

where $W(t)$ is the cumulative infiltration up to time t. Substitution of equation 41 in the denominator of equation 38 plus the assumption that $f_w = 1$ behind the wetting front leads to the equation

$$I = \frac{\tilde{K}\left[H + H_c + (W)/(\tilde{\theta} - \theta_i)\right]}{\beta\left[(W)/(\tilde{\theta} - \theta_i)\right]} \tag{52}$$

where β (dimensionless), the *viscous correction* factor, is (like H_c) solely a function of the soil and fluid characteristics and defined by the relation

$$\beta = (\tilde{\theta} - \theta_i) \int_{\theta_f^-}^{\tilde{\theta}} - f_w''(\theta)\mu_{rT}(\theta)\, d\theta \tag{53}$$

Equation 52 can be integrated to yield

$$\frac{\tilde{K}t}{\beta} = W - (H + H_c)(\tilde{\theta} - \theta_i)\, \ell n\left[1 + \frac{W}{(H + H_c)(\tilde{\theta} - \theta_i)}\right] \tag{54}$$

which is the same as equation 18 except for the substitution of H_c for H_f and the presence of the viscous correction factor β. (Values of β for various soils are given in the section on *Imbibition*). Though β is not much larger than 1, it averages around 1.3 and is not negligible.

Effect of Air Compression. The previous results were derived for the situation when air can escape freely ahead of the wetting front. If, on the other hand, the soil column is of limited depth D to an impervious stratum or to a water table, air compression will result. A similar analysis to that discussed in the previous section (Morel-Seytoux and Khanji, 1975) leads to the result for cumulative infiltration:

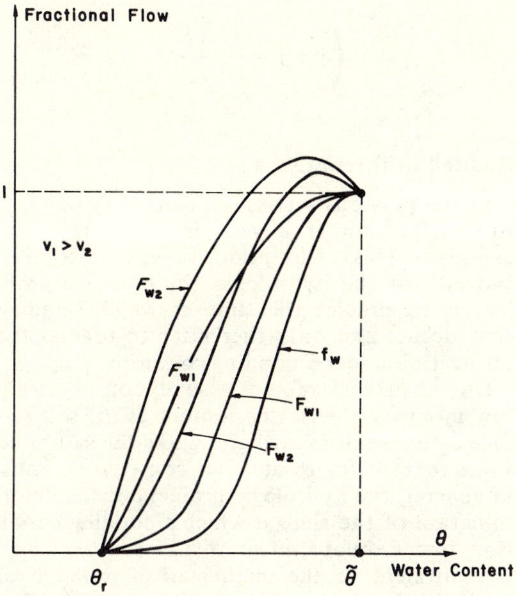

FIGURE 4. Fractional-flow functions for various values of total velocity.

$$\frac{\tilde{K}t}{\beta} = \frac{W}{\left[1-(H_A)/(D)\right]} - \frac{(\tilde{\theta}-\theta_i)(H+H_c)}{\left[1-(H_A)/(D)\right]^2} \ln\left\{1+\frac{\left[1-(H_A)/(D)\right]W}{(\tilde{\theta}-\theta_i)(H+H_c)}\right\} \qquad (55)$$

where H_A is atmospheric pressure expressed as a water depth. Equation 55 has the same structure as the Green and Ampt equation with an additional two correction terms: β for the total viscous-resistance effect and $\left[1-(H_A)/(D)\right]$ for the compression effect. Equation 55 can be used only as long as W does not exceed the value

$$\frac{D(H+H_c)(\tilde{\theta}-\theta_i)}{H_A + D(\beta-1)}$$

due to the appearance of the *counterflow* phenomenon. At high infiltration rates both water and air flow downward, but as the infiltration rates starts to decrease, air starts to flow upward, that is, in a direction opposite to the water flow.

Effect of Air Counterflow. In an infiltration process, $\partial\theta/\partial z$ is negative everywhere in the medium (see Figure 2). From the expression of the *fractional-flow* function, given by equation 43, it is clear that $F_w \geqslant \mathrm{F}_w = f_w + (G_w)/(v) \geqslant f_w$. In particular, at a fixed value of θ, F_w increases as v decreases. Figure 4 displays several F_w curves for two values of v: one large, v_1, and one small, v_2. For large v (case 1), the F_w and F_w curves do not cross the horizontal line of ordinate 1. However, for small values of v (case 2), the quantity G_w/v can become quite large, and F_w will exceed 1 in the range of high water contents. Because, by definition, $v_w = F_w v$ and $v_a = v - v_w$, it follows that $v_a = (1 - F_w)v$ is negative in the range of the high water contents. This means that in the vicinity of the soil surface, air is moving *upward*.

It is of interest to find out for which value of v counterflow will start. In the water-invaded zone near the surface (see Figure 2), when infiltration has proceeded for a relatively long time (curve 2), f_w is essentially 1, and the capillary-pressure gradient is negligible. Thus equation 30 reduces to

$$v = \frac{\tilde{K}}{\mu_{rT}}\left[1-\frac{\partial h_a}{\partial z}\right] = \frac{\tilde{K}}{\mu_{rT}}\left[1-\frac{\partial h_w}{\partial z}\right] = K_T\left[1-\frac{\partial h_w}{\partial z}\right] \qquad (56)$$

Equation 56 can be interpreted as Darcy's equation for the instantaneous velocity of a single fictitious fluid (the *total* fluid) in a heterogeneous column of conductivity $K_T(z)$, which varies with z since μ_{rT}, the total relative velocity, varies with θ (see *Imbibition*). Because at the surface $\theta = \tilde{\theta}$, then $K_T(0) = \tilde{K}$. The curve $K_T(z)$ has a minimum since μ_{rT} has a maximum for a value of $\theta < \tilde{\theta}$. Let us suppose now that $v(=I)$ drops below the value $\tilde{K}$, which will happen even in the case of an infinite column because $\beta > 1$. As soon as I becomes less than $\tilde{K}$, it follows that at the surface $\partial h_w/\partial z > 0$. The water-pressure gradient will be zero at a depth z_c below the surface where $K_T(z_c) = I$. In this zone $(0 \leqslant z \leqslant z_c)$, water pressure increases with depth. Water still flows downward in spite of it due to its weight. On the other hand, in this same zone the increase of air pressure with depth (approximately the same as for water since the capillary pressure gradient is negligible in this zone) cannot be counteracted by the weight of air. Consequently, in this zone air flows upward. The air counterflow starts as soon as the infiltration rate drops below the value $\tilde{K}$. This does not mean that the buoyancy force is not felt until I dips below $\tilde{K}$. In fact, its effect on the infiltration rate is felt significantly before this moment. When the counterflow effect is included in the analysis (Morel-Seytoux and Khanji, 1975), the equation for cumulative infiltration versus time is

$$\ln\left\{1-\frac{\gamma W}{(\tilde{\theta}-\theta_i)(H+H_c)}+\frac{\gamma(\gamma+1)\tilde{K}t}{(\tilde{\theta}-\theta_i)(H+H_c)\beta}\right\}$$

$$=-\gamma\ln\left[1+\frac{W}{(\tilde{\theta}-\theta_i)(H_c+H)}\right] \tag{57}$$

or

$$\frac{\tilde{K}t}{\beta}=\frac{W}{\gamma+1}+\frac{(\tilde{\theta}-\theta_i)(H+H_c)}{\gamma(\gamma+1)}$$

$$\left\{\left[1+\frac{W}{(\tilde{\theta}-\theta_i)(H+H_c)}\right]^{-\gamma}-1\right\} \tag{58}$$

where γ (a dimensionless parameter) is given by the expression

$$\gamma=(\tilde{\theta}-\theta_i)\int_{\theta_f^-}^{\tilde{\theta}}\frac{G''}{\tilde{K}}\mu_{rT}d\theta \tag{59}$$

From equation 58 one can deduce that the limiting value of the infiltration rate for large times is $[(\gamma+1)/(\beta)]\,(\tilde{K})$. As one would expect, the counterflow effect tends to compensate for the total viscous effect.

Joint Effect of Air Compression and Counterflow. The equation for cumulative infiltration versus time is

$$\frac{\tilde{K}t}{\beta}=\frac{W}{\gamma+1-(H_A)/(D)}+\frac{(\tilde{\theta}-\theta_i)(H+H_c)}{\gamma[\gamma+1-(H_A)/(D)]}$$

$$\left\{\left[1-\frac{W[(H_A)/(D)-1]}{(\tilde{\theta}-\theta_i)(H+H_c)}\right]^{\frac{\gamma}{[(H_A)/(D)-1]}}-1\right\} \tag{60}$$

The infiltration will come to a halt when the cumulative infiltration reaches the critical value

$$W_c=\frac{(\tilde{\theta}-\theta_i)(H+H_c)}{(H_A)/(D)-1} \tag{61}$$

The corresponding critical time is

$$t_c=\frac{\beta(\tilde{\theta}-\theta_i)(H+H_c)}{\gamma\tilde{K}[(H_A)/(D)-1]} \tag{62}$$

This is possible only if $D < H_A$. Otherwise, the infiltration will proceed. Its limiting value is then

$$\left(\gamma+1-\frac{H_A}{D}\right)\frac{\tilde{K}}{\beta}$$

Rainfall Infiltration in a Soil

In the previous section immediate ponding at the surface under a very intensive rain was assumed. This assumption is very restrictive. Indeed, for the hydrologist the practical problem is to predict the time at which ponding first occurs and only thereafter to predict the infiltration under a ponding condition.

Under natural rainfall, usually of relatively low intensity, the water content at the soil surface θ_u does not immediately take the saturated value (except for residual air content) $\tilde{\theta}$. Thus, in general, the hydrologic problem of the determination of the time at which a ponding condition first develops, in short, the *ponding time,* is equivalent to the mathematical problem of finding the unknown function $\theta_u(t)$. The "ponding" time is the solution of the equation

$$\theta_u(t_p)=\tilde{\theta} \tag{63}$$

The problem is difficult and an exact solution not possible. Mein and Larson (1973), assuming that the water displacement of air is piston-like from the beginning of rainfall, derived an equation for t_p, namely,

$$t_p=\frac{(\tilde{\theta}-\theta_i)H_b}{r[r/\tilde{K}-1]} \tag{64}$$

where r is the (constant) rainfall rate, $\tilde{K}$ the hydraulic conductivity at natural saturation, and H_b is Bouwer's estimate of the effective capillary drive.

Smith (1972), using a computer model as a theoretical infiltrometer, ran many experiments. He then attempted to correlate the calculated (observed) ponding times with the rainfall intensity. Based on the data generated, he concluded that for a given soil and a given initial water content, t_p could be "well described by a power decay curve such that"

$$t_p=a_1r^{-a_2} \tag{65}$$

where a_1 and a_2 are constants, characteristic of the soil and dependent on θ_i. If equation 64 is valid, then without going into details, it is clear that a_2 should have a value on the order of 2. Indeed, the least-square values Smith obtained do not deviate too much from 2.

The integration of the total velocity between the surface ($z = 0$) and a depth z_2 just below the wetting zone yields

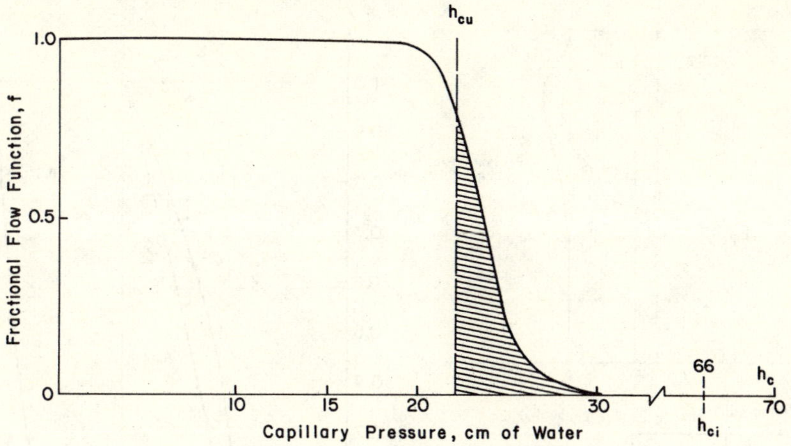

FIGURE 5. Curve of little-f function versus capillary pressure.

$$v = \frac{\tilde{K}\left[H_c(\theta_u, \theta_i) + \int_0^{z_2} f_w dz\right]}{\int_0^{z_2} \mu_{rT} dz} \qquad (66)$$

where H_c (θ_u, θ_i) is the generalized *effective capillary drive* defined as

$$H_c(\theta_u, \theta_i) = \int_{h_{cu}}^{h_{ci}} f_w dh_c = \int_{h_{cu}}^{\infty} f_w dh_c - \int_{h_{ci}}^{\infty} f_w dh_c \qquad (67)$$

Since the curve of f_w versus h_c (see Figure 5) is known for a given soil, the term H_c (θ_u, θ_i) depends only on the unknown θ_u and *not on the shape of the moisture profile between* θ_i *and* θ_u. Thus the capillary term is known exactly if θ_u is known. On the other hand, the gravity term, that is, $\int_0^{z_2} f_w dz$ depends on the shape of the profile. This term can be approximated using a piston profile and takes therefore the form $[W(1 - f_i)]/[\tilde{\theta} - \theta_i]$, where f_i is abbreviated notation for $f_w(\theta_i)$ and W is cumulative infiltration (expressed as a water height.) Equation 66 takes the more explicit form,

$$v = \frac{\tilde{K}\left[H_c(\theta_u, \theta_i) + [W(1 - f_i)]/(\tilde{\theta} - \theta_i)\right]}{\int_0^{z_2} \mu_{rT} dz} \qquad (68)$$

Figure 6 displays the shape of the total relative viscosity as a function of water content for an actual soil. The determination of the denominator, the viscous-resistance term, is the big problem. To evaluate it, one must know the water-content profile or at least a good approximation for it.

Water-content Profile Approximation. The solution of equation 48 is difficult in the general case. However, in the case of a given rainfall rate, the value of the total velocity is approximately known. Figure 7 shows a water-content profile and an air-pressure profile for a given time. For a column open at its bottom, the air pressure is atmospheric at both ends, and the air-pressure distribution must be as shown in Figure 7. The implication is that some air flows downward and some upward. From the definition of the total velocity, it follows that

$$v = r + v_a|_{z=0} \qquad (69)$$

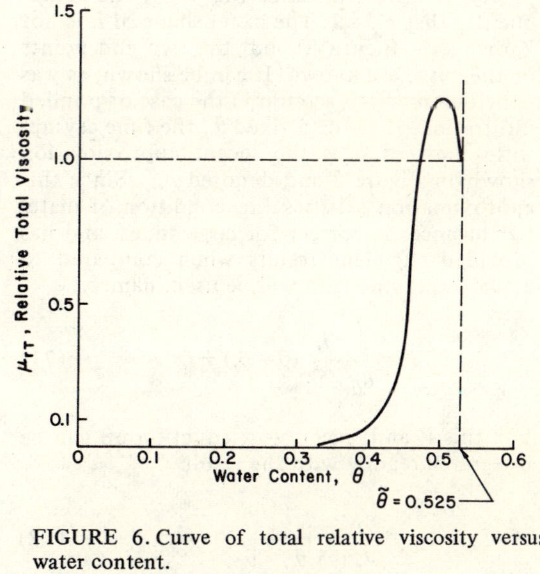

FIGURE 6. Curve of total relative viscosity versus water content.

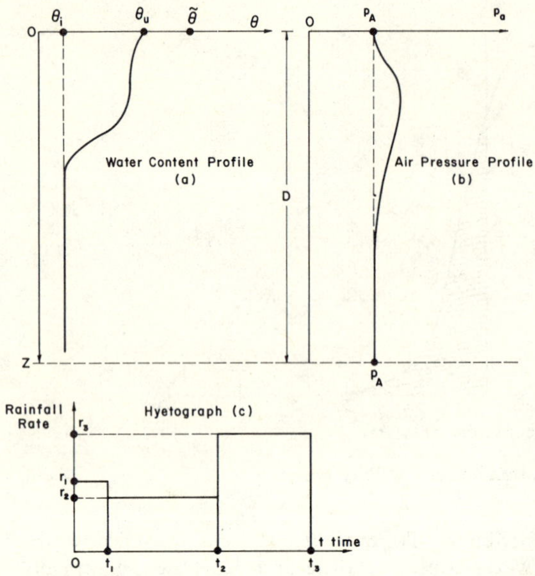

FIGURE 7. Water-content and air-pressure profiles during rain infiltration.

Since $v_a|_{z=0}$ is negative, it follows that v is less than r and equal to r only if the air velocity at the surface is zero. By definition, $v_w = vF$, and in particular, at the soil surface,

$$r = vF(\theta_u) = vF_u \tag{70}$$

If one assumes that $v = r$, then $F_u = 1$. If v is $<r$, the $F_u = (r)/(v) > 1$. The implication is that an error in v in one direction is compensated by an error in F_u in the other direction and the product vF_u remains "exact." For simplicity, it will be assumed that $v = r$, and consequently that $F_u = 1$. The exact shape of F is not known (see Figure 8), but the two end points of the curve are known. It can be shown, as was done in a previous section in the case of ponded infiltration, that for a fixed θ_u the time asymptotic limit of F is the secant approximation shown in Figure 8 and denoted F_∞. Since this approximation satisfies the condition of material balance, is correct for large times, and has provided excellent results when compared to actual experiments, it will be used, namely,

$$F = \frac{1 - f_i}{\theta_u - \theta_i}(\theta - \theta_i) + f_i \tag{71}$$

For this F and since $v = r$, equation 68 can be integrated readily with the result

$$z_\theta = \int_{T(\theta)}^{t} \frac{1 - f_i}{\theta_u - \theta_i} r(\tau) d\tau \tag{72}$$

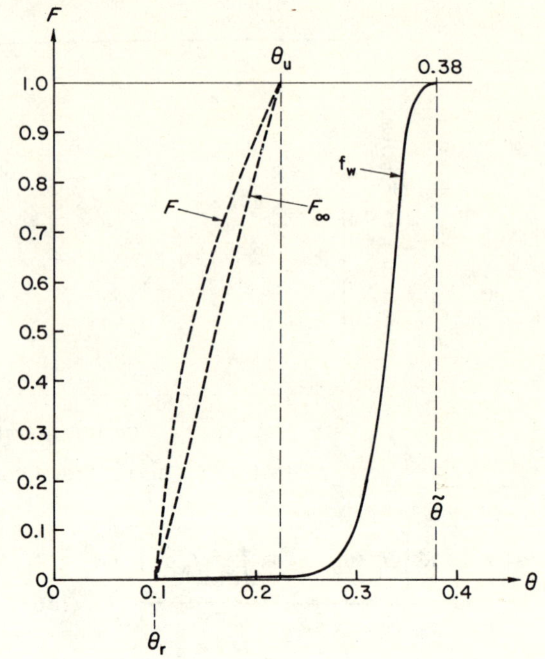

FIGURE 8. Fractional-flow function during rain infiltration.

where $T(\theta)$ is the time of the first appearance of the water content θ at the soil surface. Since θ_u is water content at time t at the surface, it follows by definition that

$$T(\theta_u) = t \tag{73}$$

Using equation 72 to calculate $(\partial z)/(\partial\theta)|_t$, one obtains for the viscous-resistance term

$$\int_0^{z_2} \mu_{rT}(\theta) dz = \int_{\theta_u}^{\theta_i} \mu_{rT}(\theta) \left.\frac{\partial z}{\partial \theta}\right|_t d\theta = \frac{r(1 - f_i)}{(\theta_u - \theta_i)} \int_{\theta_i}^{\theta_u} \mu_{rT} \frac{dT}{d\theta} d\theta \tag{74}$$

Substitution of this result into equation 68 leads to the fundamental equation for the unknown θ_u at a given time:

$$\frac{(\theta_u - \theta_i) H_c(\theta_u, \theta_i)}{1 - f_i} + \frac{(\theta_u - \theta_i)W}{(\tilde{\theta} - \theta_i)} = \frac{r^2}{\tilde{K}} \int_{\theta_i}^{\theta_u} \mu_{rT} \frac{dT}{d\theta} d\theta \tag{75}$$

Differentiating equation 75 with respect to time yields

$$\frac{d\theta_u}{dt}\left\{\frac{H_c(\theta_u, \theta_i)}{1 - f_i} + \frac{(\theta_u - \theta_i)}{1 - f_i}\frac{d}{d\theta_u}\left[H_c(\theta_u, \theta_i)\right] + \frac{W}{\tilde{\theta} - \theta_i}\right\} = r\left\{\frac{r}{\tilde{K}}\mu_{rT}(\theta_u) - \frac{\theta_u - \theta_i}{\tilde{\theta} - \theta_i} + \frac{2}{\tilde{K}}\frac{dr}{dt}\int_{\theta_i}^{\theta_u}\mu_{rT}dT\right\} \tag{76}$$

which is a nonlinear ordinary differential equation for the unknown $\theta_u(t)$. Equation 76 looks rather formidable, and for brevity H_c will be written for $H_c(\theta_u, \theta_i)$. Note, however, that the coefficient of $(d\theta_u)/(dt)$ is a linear function of W, say, $a(\theta_u) + b(\theta_u)W$ where $a(\theta)$ and $b(\theta)$ are deduced from the basic soil characteristics. Similarly, the right-hand side is an explicit function of r and θ_u except for the last integral.

Solution for Constant Rainfall Rate. In this solution method the bothersome integral on the right-hand side of equation 76 disappears. The remaining equation could then be solved numerically without too much difficulty. More insight, however, is obtained by integrating it approximately but analytically.

As long as rain infiltrates totally, that is, until ponding time, $v = I = r =$ constant. The denominator of equation 68 will increase linearly with time. To maintain a constant ratio between the numerator and the denominator, it is necessary for $H_c(\theta_u, \theta_i)$ to also increase with time. From Figure 5 and a general knowledge of the shape of the capillary-pressure curve (see *Capillary Pressure*), one can infer that in the range of intermediate water contents, $H_c(\theta_u, \theta_i)$ does not vary linearly with θ_u but is a much slower function of θ_u. Consequently, θ_u must vary with time more rapidly than linearly. Thus θ_u will of necessity rise quickly to values near $\tilde{\theta}$. Thus except for very early and very late times, it is legitimate to evaluate the *coefficients* in equation 76 in the neighborhood of $\theta_u = \tilde{\theta}$, and for simplicity at $\theta_u = \tilde{\theta}$. The resulting simplified equation takes the form

$$\left\{\frac{H_c}{1 - f_i} + \frac{W}{\tilde{\theta} - \theta_i}\right\}\frac{d\theta_u}{dt} = r\left[\frac{r\bar{\mu}_{rT}}{\tilde{K}} - 1\right] \tag{77}$$

where $\bar{\mu}_{rT}$ is an average value of μ_{rT} in the high range of water contents. Equation 77 is linear and can be integrated readily (Morel-Seytoux, 1975). Ultimately one obtains the ponding time

$$t_p = \frac{(\tilde{\theta} - \theta_i)H_c}{(1 - f_i)r}\left[e^{\frac{\tilde{K}}{\bar{\beta}r - \tilde{K}}} - 1\right] \tag{78}$$

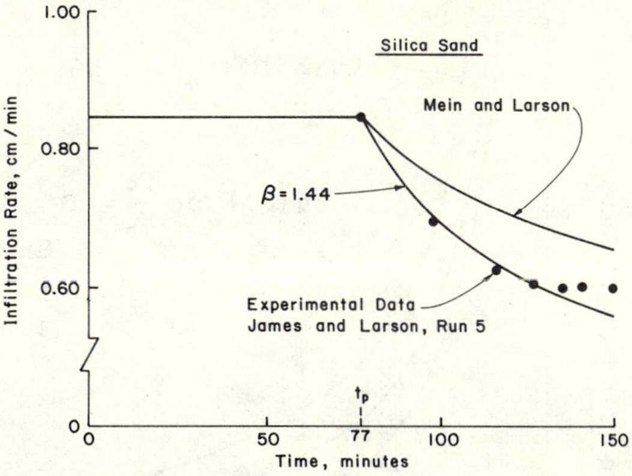

FIGURE 9. Comparison of observations and predictions of infiltration rates after ponding.

where $\bar{\beta}$ is an average between 1 and β defined previously (equation 53). The simple arithmetic average $\bar{\beta} = (1 + \beta)/(2)$ has led to good agreement with experimental results (Figure 9). Predictions of ponding times (or equivalently of cumulative infiltration at ponding time) by equation 78 compare favorably with experimental results (Figure 10).

After ponding, the infiltration rate is given by the relation

$$I = \frac{dW}{dt} = \frac{\frac{\tilde{K}}{\bar{\beta}}\left[H_c(\tilde{\theta}, \theta_i) + \frac{(1 - f_i)W}{(\tilde{\theta} - \theta_i)}\right]}{\frac{W(1 - f_i)}{\tilde{\theta} - \theta_i} - \frac{W_p(1 - f_i)}{(\tilde{\theta} - \theta_i)}(1 - \frac{1}{\bar{\beta}})} \tag{79}$$

or in integrated form,

$$\frac{\tilde{K}}{\bar{\beta}}(t - t_p) = W - W_p - \left[\frac{(\tilde{\theta} - \theta_i)}{1 - f_i}H_c(\tilde{\theta}, \theta_i) + W_p(1 - \frac{1}{\bar{\beta}})\right]\ell n\left[\frac{1 + \frac{(1 - f_i)W}{(\tilde{\theta} - \theta_i)H_c(\tilde{\theta}, \theta_i)}}{1 + \frac{W_p(1 - f_i)}{(\tilde{\theta} - \theta_i)H_c(\tilde{\theta}, \theta_i)}}\right]$$

where W_p is the cumulative infiltration at ponding. Naturally, the preceding formula is valid only for $t > t_p$.

Solution for Variable Rainfall Rate. For a succession of rainfall intensities such as shown in Figure 7, a ponding-time formula can be derived, namely,

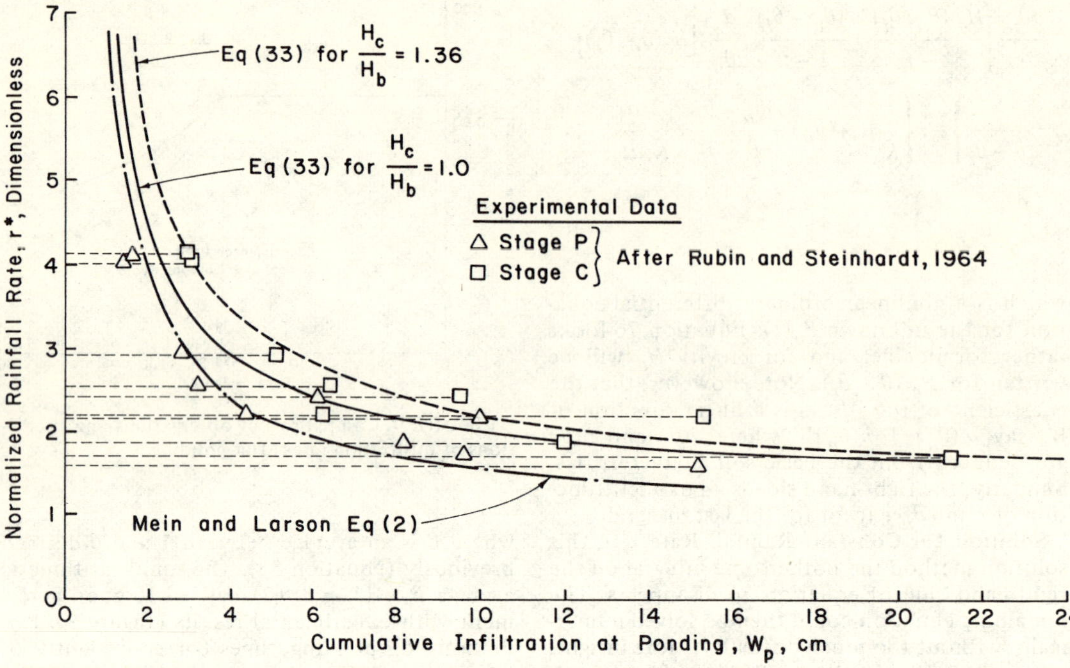

FIGURE 10. Comparison of observations and predictions of ponding times.

$$t_p = t_{j-1} + \frac{1}{r_j}\left\{\frac{(\tilde{\theta} - \theta_i)H_c}{(1 - f_i)} + W_{j-1}\right\} \left\{\frac{\tilde{K}}{e^{\bar{\beta} r_j} - \tilde{K}}\left[\prod_{\nu=1}^{j-1} \left(\frac{1 + \dfrac{W(1 - f_i)}{(\tilde{\theta} - \theta_i)H_c}}{1 + \dfrac{(1 - f_i)W_{\nu-1}}{(\theta - \theta_i)H_c}}\right)^{-\left(\frac{\bar{\beta} r_\nu - \tilde{K}}{\bar{\beta} r_j - \tilde{K}}\right)} - 1\right]\right\} \quad (81)$$

where t_{j-1} is the beginning time of the rainfall intensity r_j, and W_{j-1} is the cumulative infiltration up to time t_{j-1}. Equation 81 applies if the ponding time occurs in the time interval (t_{j-1}, t_j).

For unsteady rainfall with various rates r_j, in order to determine the ponding time, one uses equation 81 for the case $j = 1$, namely,

$$t_p{}^1 = \frac{(\tilde{\theta} - \theta_i)H_c}{r_1(1 - f_i)}\left\{e^{\frac{1}{\bar{\beta} r_1{}^* - 1}} - 1\right\} \quad (82)$$

If $t_p{}^1$ is $< t_1$, then the ponding time is given by equation 82. If $t_p{}^1$ is $> t_1$, then ponding does not occur during the first constant rainfall rate time interval. One then recalculates t_p from equation 81 for $j = 2$, namely,

$$t_p{}^2 = t_1 + \frac{1}{r_2}\left\{\frac{(\tilde{\theta} - \theta_i)H_c}{1 - f_i} + W_1\right\} \left\{\frac{1}{e^{\bar{\beta} r_2{}^*} - 1}\left[1 + \frac{(1 - f_i)W_1}{(\tilde{\theta} - \theta_i)H_c}\right]^{-\left(\frac{\bar{\beta} r_1{}^* - 1}{\bar{\beta} r_2{}^* - 1}\right)} - 1\right\} \quad (83)$$

If $t_p{}^2$ is $< t_2$ then the ponding time is given by equation 83. If $t_p{}^2$ is $> t_2$, then ponding does not occur during the second constant rainfall rate time interval. One then recalculates t_p from equation 81 for $j = 3$, namely,

$$t_p{}^3 = t_2 + \frac{1}{r_3}\left\{\frac{(\tilde{\theta} - \theta_i)H_c}{1 - f_i} + W_2\right\}$$

$$\left\{\frac{1}{e^{\bar{\beta} r_3{}^*} - 1}\left[1 + \frac{(1 - f_i)W_1}{(\tilde{\theta} - \theta_i)H_c}\right]^{\frac{\bar{\beta}(r_2{}^* - r_1{}^*)}{\bar{\beta} r_3{}^* - 1}}\right.$$

$$\left.\left[1 + \frac{(1 - f_i)W_2}{(\tilde{\theta} - \theta_i)H_c}\right]^{-\frac{\bar{\beta} r_2{}^* - 1}{\bar{\beta} r_3{}^* - 1}} - 1\right\} \tag{84}$$

In this form it is clear that if $r_2 = r_1$ and thus $t_2 = t_1$, equation 84 reduces to equation 83 as it should. This also shows that successive rainfall rates need not be different. The process is repeated until the ponding moment is located in the proper interval. Once $t_{j-1} < t_p \leqslant t_j$ is known, then the cumulative infiltration up to ponding is

$$W_p = \sum_{i=1}^{j-1} r_i(t_i - t_{i-1}) + r_j(t_p - t_{j-1}) \tag{85}$$

and W can be obtained from equation 80 for $W \geqslant W_p$ as a function of time. At a given time, the infiltration rate is given by equation 79.

From a theoretical standpoint, equation 81 should be particularly good if the rainfall intensities are high due to the assumptions made in its derivation that the coefficients in the differential equation for $\theta_u(t)$, water content at the soil surface, are evaluated at $\theta = \tilde{\theta}$. The variable θ_u will rise quickly to values close to $\tilde{\theta}$ for high rainfall rates, but its rise may be slow for low rainfall rates. In the latter case, equation 81 may be in serious error. For r^* slightly $> 1/\beta$, the rise of θ_u with time is probably underestimated. This fact is confirmed when using equation 82 for a single rain of intensity r and equation 83 for two intensities: one of value $\tilde{K}/\bar{\beta}$ and duration t_1 and a later one of value r. From equation 82 one obtains

$$t_p{}^1 = \frac{a}{r}\left[\frac{1}{e^{\bar{\beta} r^*} - 1} - 1\right] \tag{86}$$

where $$a = \frac{(\tilde{\theta} - \theta_i)H_c}{(1 - f_i)}$$

and from equation 83 one obtains

$$t_p{}^2 - t_1 = \frac{1}{r}(a + W_1)\left[\frac{1}{e^{\bar{\beta} r^*} - 1} - 1\right] \tag{87}$$

or $$t_p{}^2 - t_1 = \frac{W_1}{r}\left[\frac{1}{e^{\bar{\beta} r^*} - 1} - 1\right] + t_p{}^1 \tag{88}$$

From a physical standpoint, one expects $t_p{}^2 - t_1$ to be $< t_p{}^1$, but clearly from equation 88 it is predicted otherwise. Thus for low antecedent rainfall intensities, that is, for r^* greater than $1/\bar{\beta}$ and close to $1/\bar{\beta}$, equation 81 is in error, and possibly in serious error if r^* is itself small and W_1 is large. However, once r^* significantly exceeds $\tilde{K}\,\bar{\beta}$, the compensating factor

$$[1 + (W_1)/(a)]^{-\frac{\bar{\beta} r_1{}^* - 1}{\bar{\beta} r_2{}^* - 1}}$$

comes into play, and results are in the correct direction, though they may still not be very accurate.

The basis for the improvement of equation 81 naturally lies in a reexamination of the assumptions leading to it, namely, the steps leading from equation 76 to 77. For constant rainfall rate the use of a time-average for $\mu_{rT}(\theta_u)$, namely, $\bar{\beta}$, is justified. However, for a succession of rainfall rates, in the interval of time when ponding occurs, the value of $\mu_{rT}(\theta_u)$ is on the average closer to β than to $\bar{\beta}$. Thus formula 81 tends to overestimate t_p. An obvious correction to equation 81 in the right direction is

$$t_p = t_{j-1} + \frac{1}{r_j}\left\{\frac{(\tilde{\theta} - \theta_i)H_c}{1 - f_i} + W_{j-1}\right\}$$

$$\left\{\frac{\tilde{K}}{e^{\beta r_j} - \tilde{K}}\left[\prod_{\nu=1}^{j-1}\left(\frac{1 + \dfrac{W_\nu(1 - f_i)}{(\tilde{\theta} - \theta_i)H_c}}{1 + \dfrac{(1 - f_i)W_{\nu-1}}{(\tilde{\theta} - \theta_i)H_c}}\right)^{-\left(\frac{\beta r_\nu - \tilde{K}}{\beta r_j - \tilde{K}}\right)}\right] - 1\right\} \tag{89}$$

Infiltration Formulas in Hydrologic Practice

Naturally, practical hydrology could not wait for the fully mature development of scientific hydrology to solve hydrologic problems. This is particularly true since the birth of scientific hydrology as a discipline was only recent as compared to other disciplines. The International Association of Hydrological Sciences is less than a quarter century old.

There are two types of practical infiltration formulas: (1) those whose functional form was derived from solutions (albeit approximate) of the basic equations and (2) those whose functional form was selected a priori to perform in qualitative agreement with the known observational facts. In the first category are the formulas of Green and Ampt, Morel-Seytoux, and Philip. The Green and Ampt formula for cumulative infiltration is

$$W - (H + H_f)(\tilde{\theta} - \theta_i) \ln \left(1 + \frac{W}{(\tilde{\theta} - \theta_i)(H + H_f)}\right) = \tilde{K}t \quad (90)$$

The parameters H_f, $\tilde{K}$, and $\tilde{\theta}$ are characteristics of the soil, whereas H is a ponding depth variable and θ_i describes the initial moisture state of the system. Under natural conditions H can be taken as zero because H is usually negligible compared to H_f. Morel-Seytoux's Formula

$$W - (H + H_c)(\tilde{\theta} - \theta_i) \ln \left[1 + \frac{W}{(\tilde{\theta} - \theta_i)(H + H_c)}\right] = \frac{\tilde{K}}{\beta} t \quad (91)$$

differs from the Green and Ampt formula only in the introduction of the viscous correction factor β. Philip's formula is:

$$W = \tilde{K}t + St^{1/2} \quad (92)$$

The parameters $\tilde{K}$ and S are characteristics of the soil.

In the second category belong the formulas of Kostiakov, Horton, and Holtan, namely, and in that order,

$$W = At^n \qquad 0 < n < 1 \quad (93)$$

where A and n are empirical parameters,

$$W = \tilde{K}t + \frac{1}{k}(I_0 - \tilde{K})(1 - e^{-kt}) \quad (94)$$

where I_0 and k are parameters, and *Holtan's formula* for the infiltration rate:

$$I = I_c + a(W_M - W)^m \quad (95)$$

where W_M, I_c, and a are parameters. In Holtan's formula, W_M is interpreted as a maximum available storage quantity.

Most of the formulas listed behave similarly for small and large times. For large times equations 90, 91, 92, 94 and (and even 95 if we interpret to mean by large times $W = W_M$) all lead to a constant infiltration rate: $\tilde{K}$, $\tilde{K}/\beta$, or I_c. In the case of Holtan's equation, I_c is interpreted to mean the hydraulic conductivity at natural saturation of a deeper, less pervious stratum. Thus I_c is only a fraction of $\tilde{K}$. We note from equation 55 that the limiting infiltration rate is $(\tilde{K})/(\beta)$ $[1 - (H_A)/(D)]$, which can be any fraction of $\tilde{K}/\beta$ with a proper choice of D. Thus the low value of I_c compared to $\tilde{K}$ in Holtan's formula could be interpreted just as well to be due to air compression rather than due to the existence of a less pervious, deep stratum. From the remark that follows equation 56, it is clear that the extension of the simplistic Green and Ampt approach has been validated by the more rigorous two-phase approach so that simple formulas to account for air effect, air compression, and the presence of a deeper, less pervious stratum can be derived.

For small times again equations 90, 91, and 92 indicate that the infiltration decreases like $t^{-1/2}$. Essentially all the information in equation 92 is contained in equations 90 and 91, but equations 90 and 91 provide information regarding the *intermediate* times. Equation 92 assumes without justification that the addition of the small-time and large-time behaviors provide the intermediate time behavior. Equation 91 makes no such assumption and has a sounder physical basis. It is also an experimental fact that the Green and Ampt formula gives better fit to experimental observations than Philip's formula.

Strictly speaking, the preceding equations apply only for ponded infiltration. Except equations 92 and 93, all equations generalize to the case of rainfall infiltration. In equation 94, t then is time since ponding, W is cumulative infiltration since ponding, and $I_0 = r$. The generalization of equation 91 after ponding is given by equation 80 and that of equation 90 by the same equation with $\beta = 1$. Theoretically, equations 94 and 95 could be used to calculate a ponding time. The *Hortan ponding time* would be the solution of the equation

$$\left(\frac{r - \tilde{K}}{I_0 - \tilde{K}}\right) t_p = \frac{1}{k}(1 - e^{-kt_p}) \quad (96)$$

This ponding time shows the influence of rainfall intensity in agreement with experimental evidence. From equation 95 one deduces a ponding time:

$$t_p = \frac{W_M}{r} - \left(\frac{r - I_c}{a}\right)^{1/m} \quad (97)$$

This ponding time shows the proper decrease of t_p with r, but it does not indicate that ponding will not occur if r is less than $\tilde{K}$.

Two simple formulas for ponding time have been proposed for constant rainfall: the *Mein and Larson (1973) formula:*

$$t_p = \frac{(\tilde{\theta} - \theta_i)H_b}{r\left(\frac{r}{\tilde{K}} - 1\right)} \quad (98)$$

and Morel-Seytoux' formula:

$$t_p = \frac{(\tilde{\theta} - \theta_i)H_c}{r}\left[\frac{\tilde{K}}{e^{\bar{\beta}r} - \tilde{K}} - 1\right] \quad (99)$$

Both formulas show the proper dependence of t_p with r and give similar results. The Green and Ampt assumptions used to derive equation 98 do not permit generalization of variable rainfall rates. With the two-phase approach (Morel-Seytoux, 1975), equation 99 can be generalized with the result given in equation 89. Though equation 89 looks rather formidable, it requires no more parameters than equation 99 since $\bar{\beta} = (1 + \beta)/2$. The result is limited to the situation when the rainfall intensities vary but nevertheless exceed the limiting value $\tilde{K}/\beta$. Extensions which remove this limitation have appeared in the literature (Morel-Seytoux, 1978) since this article was written.

H. J. MOREL-SEYTOUX

References

Bouwer, H., 1969. Theory of seepage from open channels, *in* V. T. Chow, ed., *Advances in Hydroscience,* **5**, 121–172.

Green, W. H., and Ampt, C. A. 1911. Studies on soil physics. I. Flow of air and water through soils, *Jour. Agric. Sci.,* **4**, 1–24.

James, L. G., and Larson, C. L., 1974. Modeling infiltration and redistribution of soil water during intermittent applications. Paper presented at the 1974 Winter Meeting of the Am. Soc. Agric. Eng., Chicago, Dec. 10–13, 1974, 26p.

Mein, R. G., and Larson, C. L., 1973. Modeling infiltration during a steady rain, *Water Res. Res. J.,* **9**, (2), 384–394.

Morel-Seytoux, H. J., 1973. Two-phase flows in porous media, *in* V. T. Chow, ed., *Adv. Hydrosci.,* **9**, 119–202.

Morel-Seytoux, H. J., 1976. Derivation of equations for rainfall infiltration, *J. Hydrol.,* **31**(3/4), 203–219.

Morel-Seytoux, H. J., 1978. Derivation of equations for variable rainfall infiltration, *Water Res. Res. J.,* **14**(4), 561–568.

Morel-Seytoux, H. J., and Khanji, J., 1975. Equation of infiltration with compression and counterflow effects, *Bull. Internat. Assoc. Hydrol. Sci.,* **20**(4), 505–517.

Philip, J. R., 1957. The theory of infiltration. I. The infiltration equation and its solution, *J. Soil Sci.,* **83**, 345–357.

Rubin, J., and Steinhardt, R., 1964. Soil water relations during rain infiltration. III. Water uptake at incipient ponding, *Soil Sci. Soc. America Proc.,* **28**(5), 614–619.

Smith, R. E., 1972. The infiltration envelope: Results from a theoretical infiltrometer, *Jour. Hydrol.,* **17**, 1–21.

Whisler, F. D., and Bouwer, H., 1970. Comparison of methods for calculating vertical drainage and infiltration for soils, *Jour. Hydrol.,* **10**(1), 1–19.

Cross-references: *Capillary Pressure; Conductivity, Hydraulic; Imbibition; Flow Theory; Percolation; Pore Space, Drainable; Puddling; Soil Water; Surface-soil Water Content; Water Content and Retention; Water Fluxes; Water Movement; Wetting Front. Vol. IVA: Hydrology.*

INFORMATION SYSTEMS

Soil information systems belong to the family of geographical information systems. These systems differ from others in that data are referenced geographically, thereby allowing for analysis, retrieval, and display of data according to location. A location identifier may be a geographical coordinate but may also be an arbitrary code that relates the data to some spatial designation. Tomlinson (1971) lists the four main types of location identifiers as *external index, coordinate reference, arbitrary grid,* and *explicit boundary.*

Definition

Information is knowledge derived from reading, observation, or instruction; information is the meaning assigned to data by the known conventions used in the data representation (Anon., 1964); information is data retrieved and processed for a particular use (Tomlinson, 1972). A *system* is a group of entities and activities connected and satisfactorily bounded to interact for a common purpose or purposes (Thomas and Shafer in Tomlinson, 1972).

A soil information system is a collection of activities consisting of systematic, geographically referenced soil data collection, storage, retrieval, and analyses using procedures of data processing. Information for specific purposes is derived from the data stored, using standard soil science conventions.

Soil information systems are often called *soil data banks* or confused with soil data processing. The term "data bank" refers primarily to a depository of data for future use, whereas "soil data processing" refers chiefly to the manipulation of small, special-purpose sets of data by individual researchers. Neither of these activities needs to depend on the other. An information system involves both of these functions as well as systematic data collection.

Development

Soil information systems developed through research into the use of computers in soil science. Bidwell and Hole (1963), Russell and Moore (1967), Norris (1972), Protz, Arnold, and Presant (1968) and Webster and Burrough (1974) used various statistical and other numerical techniques in studies of soil characterization and variability. Such research properly falls in the realm of soil data processing, as compared to those reported by John et al (1969) and Swanson (1970), which involves soil data banks and processing of soil data. All of these were local attempts to organize specific, small sets of data for certain, narrowly defined purposes.

Within the last decade, interest has broadened from specific-purpose files to general-purpose files. Also the scope changed from one of file manipulation to consideration of a total man-machine system of components beginning with data collection in the field through specific retrieval and analysis capabilities. Systems began to take on national and international dimensions. Work on the Canada soil information system began in 1972 (Dumanski, Kloosterman, and Brandon, 1975). At the same time a group in France was actively standardizing soil data terminology (ORSTOM, 1969) and had begun work on development of an information system. Similar activities were going on in the United States (Swanson, 1973), the Netherlands (Schelling, 1972) and New Zealand (Lee, Newman, and Gibson, 1976). At the international level there have been recommendations for development of an international soil information system (FAO, 1972). A working group on soil information systems has been established by the International Society of Soil Science (Bie, 1975).

Requirements

Dumanski, Kloosterman, and Brandon (1975) have reviewed the requirements of a soil information system. These are orderliness in data collection and storage, accessibility, editing and updating, simplicity, flexibility, and feedback. Of equal importance is the need to predefine the users of the system and their needs and expectations, as well as to develop a permanent relationship between users and specialists including those in data gathering and data processing.

A soil information system requires a high degree of standardization on the part of those contributing data as well as those expecting to use the data. Definitions must be agreed on for the geographical area within which the system operates, and such definitions should be consistent with international terminology. Lacking this, data flow will be restricted by national boundaries.

The diverse nature of the computer hardware industry and the specificity of software packages pose a problem for soil information systems. This problem can restrict information transfer between systems and can result in considerable duplication of effort.

Components

Although soil information systems have developed independently in a number of countries to result in a variety of designs, all systems have the components of input, manipulation, and retrieval in common (Bie and Schelling, 1975). Coincident to these components is the fundamental role played by project management.

From the standpoint of information systems, there are two types of soil data. These are commonly called "hard," or point, data and map data. Each type requires system components that reflect the nature of the data.

Data Input. Data input for "hard" data includes the procedures for data collection, checking and editing; conversion to a computer-compatible form; and submission to the system. Method of input differs, ranging from a free language approach used in the Netherlands to a highly structured and controlled approach in France, Canada, and the United States. The data-collection documents may be computer punch cards, notebooks, specially prepared forms for keypunching or forms for optical character reading (Kloosterman, 1975). In all systems, data quality, ease of use, and accessibility are important considerations.

Soil maps can be input in grid format (Nichols and Bartelli, 1974), or as polygons in which case they must be either digitized or electronically scanned (Tomlinson, 1967). Input as grid cells consists of superimposing a grid of predetermined size onto a soil map and coding the map unit notation(s) pertinent to each grid cell. The approach is relatively simple to program, and it is ideally suited to map output on a line printer. On the other hand, it is tedious and error prone on input. The size of grid used strongly affects the degree of data generalization.

There is no data generalization if the map is input as a collection of polygons, but this can be done only with special equipment. Digitizing is accomplished by following each line or line segment with a special sensor; each line is transformed into a series of x:y coordinates relative to the reference axis. Scanning consists of rapidly passing a concentrated energy source across the map in a series of minute, parallel passes whereby the lines are represented as a collection of on-off bits. In further processing, these bits are transformed into a collection of related x:y coordinates. The procedure chosen for inputting the soil map depends on the objectives of the program and the manpower and budgets available.

Manipulation. Data manipulation consists of data-base management and information output. It is closely related to data retrieval in that it processes the data into a form permitting future access by the system and provides for data validation and file updating.

A fundamental difference exists between file management and data-base management. The former relates to management of simply related data with a small set of output requirements, whereas the latter implies management of data of complex structure with a wide variety of unrelated uses. In either case the capabilities of define-load, update, and retrieval are necessary features of a data-management system (Gordon, 1975).

A number of commercial data-base management systems are available, but many of these were developed for business applications and are not well suited to the more complex structure of soil data. As a result, some groups have begun to develop systems tailored to the requirements of soil science. Similar efforts have been going on in related disciplines such as geology (Gordon, 1975).

Data-base management systems available for digitized soil maps have generally been developed within geographic and cartographic fields. Many of these are specific to the machines on which they operate and to the kinds of maps (thematic, topographic, etc.) for which they were designed. Soil maps have been computerized only recently, and techniques employed have borrowed heavily from these related fields. Although first-generation systems have been somewhat uneconomic, cost-benefit ratios are improving as better systems are being designed using third-generation computer technology. A few countries, including the United States, the Netherlands, and Canada, have developed considerable expertise in this field.

Retrieval. The function of data retrieval is to retrieve and summarize data in table, paragraph, or map form. The abilities to sort, select, summarize, combine, manipulate, and compare data sets are important features of a retrieval system. These apply equally to "hard" data and to map data. Some systems provide this capability interactively, but most provide these features only in batch mode.

The Management Team. A soil information system requires a combination of functions performed by soil scientists and computer technologists. In the first two decades of computer use, some first-generation systems were built by soil scientists working with small, single-purpose files with little assistance from computer specialists. As systems evolved from single-purpose, single-file systems to complex multifile, multipurpose systems, management teams consisting of soil scientists, system designers, computer programmers, and clerical staff were put together (Dumanski, Kloosterman, and Brandon, 1975). These teams analyze needs and requirements collectively, and although they can be difficult to maintain as a cohesive unit, they can also devise better systems.

Role of Soil Information Systems

The primary role of a soil information system is data-management service (Schelling, 1975). The system is a working tool for the science, a vehicle for dissemination of information to other disciplines, and a channel for interfacing with other systems.

Traditional, manual data-handling methods are slow and thus limit the volume that can be used. In many cases the volume of data collected in a project could have been large enough to require preclassification and grouping at early stages, often resulting in the loss of primary data. Computer-oriented data management overcomes some of these deficiencies. It provides a storehouse of basic data. Further, it permits a scientist to maintain contact with his basic data up to the point of publication, thereby precluding the necessity of early data classification. As new data are generated, the original store is expanded. This capacity facilitates the resorting and resummarizing of data in a variety of ways at different times as maps, summaries, or tables in response to requests generated by environmental and land-management agencies. By developing lines of communication and interfaces with other resource-oriented data systems, soil information systems contribute to the development of the comprehensive data bases necessary for effective land evaluation, planning, and management.

Currently soil information systems are not exploiting the full potential of computers in data handling. This is due partly to their youthful stage of development but also in part to persistence of approaches developed when all data were handled manually. In the literature the terms "typical," "representative," and "benchmark" represent mental attempts to summarize large numbers of observations. In cartography, automation consists of mechanizing only those functions that were formerly done by hand. Some changes in approach in soil science will be necessary before the full potential of soil information systems can be realized.

J. DUMANSKI
B. KLOOSTERMAN

References

Anon., 1964. *Definitions of Computing Terms.* Edmonton: Department of Computing Science, University of Alberta, 77p.

Bidwell, O. W., and Hole, F. D., 1963. Numerical taxonomy and soil classification, *Soil Sci.,* 97, 58–62.

Bie, S. W., ed., 1975. *Soil Information Systems.* Wageningen, The Netherlands: Center for Agricultural Publishing and Documentation, 87p.

Bie, S. W., and Schelling, J., 1975. Soil information systems in 1975, *in* S. W. Bie, ed., *Soil Information Systems.* Wageningen, The Netherlands: Center for Agricultural Publishing and Documentation, 3–12.

Dumanski, J.; Kloosterman, B.; and Brandon, S. E., 1975. Concepts, objectives and structure of the Canada Soil Information System, *Can. Jour. Soil Sci.,* 55, 181–187.

FAO, 1972. *Report of FAO/Unesco Ad Hoc Consultation on Computerized Soil Data Interpretations for Development Purposes.* Rome: FAO, 55p.

Gordon, T., 1975. Data management: General systems and particular needs, *in* S. W. Bie, ed., *Soil Information Systems.* Wageningen, The Netherlands: Center for Agricultural Publishing and Documentation, 43–47.

John, M. K.; Lavkulich, L. M.; Jungen, J. R.; and Leigh, J. L., 1969. Interpretation of soil data by computer, *Can. Jour. Soil Sci.,* 49, 177–179.

Kloosterman, B., 1975. Input of data to soil information systems, *in* S. W. Bie, ed., *Soil Information Systems.* Wageningen, The Netherlands: Center for Agricultural Publishing and Documentation, 22–30.

Lee, R.; Newman, M. J.; and Gibson, A. R., 1976. Computer processing of soil profile data from surveys in New Zealand, *Geoderma,* 16, 201–211.

Nichols, J. D., and Bartelli, L. J., 1974. Computer-generated interpretive soil maps, *Jour. Soil & Water Conserv.,* Sept.–Oct., 232–235.

Norris, J. N., 1972. The application of multivariate analysis to soil studies: III. Soil variation, *Jour. Soil Sci.,* 23, 62–75.

ORSTOM, 1969. *Glossaire de Pedologie: Description des horizons en vue du traitement informatique.* Paris: Office de la Réchérche Scientifique et Technique Outre-Mer, 82p.

Protz, R.; Arnold, R. W.; and Presant, E. W., 1968. The approximation of the true modal profile with the use of the high speed computer and landscape control, *9th Internat. Cong. Soil Sci. Trans., (Adelaide),* 4, 193–204.

Russell, J. S., and Moore, A. W., 1967. Use of a numerical method in determining affinities between some deep sandy soils, *Geoderma,* 1, 47–69.

Schelling, J., 1972. Earth science information system for the Netherlands, *in Report of FAO/Unesco Ad Hoc Consultation on Computerized Soil Data Interpretations for Development Purposes.* Rome: FAO, 29–30.

Schelling, J., 1975. The role of soil information systems, *in* S. W. Bie, ed., *Soil Information Systems.* Wageningen, The Netherlands: Center for Agricultural Publishing and Documentation, 13–16.

Swanson, D. W., 1970. Use of automatic data processing in the soil survey. USDA, SCS, *Soil Survey Technical Notes,* 2–5.

Swanson, D. W., 1973. Proposed coding system for the pedon data record of the National Cooperative Soil Survey. Washington, D.C.: USDA, Soil Conservation Service, 163p.

Tomlinson, R. F., 1967. An introduction to the geo-information system of the Canada Land Inventory. Ottawa: Department of Forestry and Rural Development, 23p.

Tomlinson, R. F., 1971. Geo-information systems and the use of computers in handling land use information, *Proc. U.S. Geol. Survey Conference on Land Use Classification Systems.* Washington, D.C.: U.S. Geol. Survey, 38–73.

Tomlinson, R. F., 1972. Introduction: Environment information systems, *Proc. 2nd Symposium Geographic Data Handling, Internat. Geogr. Union Comm. Geographic Data Sensing and Processing.* Paris: UNESCO, 1–11.

Webster, R., and Burrough, P. A., 1974. Multiple discriminant analysis in soil survey, *Jour. Soil Sci.,* 25, 120–134.

INTERNATIONAL SOCIETY OF SOIL SCIENCE

The International Society of Soil Science (ISSS) is a nongovernmental organization comprising groups of individuals and institutions engaged in the study of soil science. The objectives of the society are to ensure international links within the field, to foster the advancement of its different branches, and to promote applications of its findings.

The ISSS was founded in Rome in 1924 following a series of international activities that started in 1909 with the International Agro-Geological Conference which convened in Budapest. The ISSS comprises seven commissions: (I) soil physics; (II) soil chemistry; (III) soil biology; (IV) soil fertility and plant nutrition; (V) soil genesis, classification, and cartography; (VI) soil technology; and (VII) soil mineralogy. A subcommission has also been established to deal with salt-affected soils. Specialized working groups deal with micromorphology, soil conditioning, nomenclature, information systems, remote sensing, land evaluation, desertification, cryogenic soils, soil zoology, and forest soils.

The ISSS conducts its activities through a general congress held every four years; through seminars and conferences organized by its committees, and through sessions of its specialized working groups. The first congress was held in Washington, D.C., in 1927. Other international congresses were held in Leningrad (1930), Oxford (1935), Amsterdam (1950), Leopoldville (1954), Paris (1956), Madison (1960), Bucarest (1964), Adelaide (1968), Moscow (1974), and Edmonton (1978). The 12th congress will be held in New Delhi, India, in 1982.

The ISSS publishes the proceedings of the congresses and conferences, and twice a year it publishes a news bulletin. Membership in the ISSS is granted on an individual basis or through a national society of soil science that is affiliated with the ISSS. As of 31 December 1978, the ISSS had 5600 members in fifty-one national societies from 108 countries. The society is governed by the general assembly, which meets during the congresses, by the council that is composed of the members of the bureau, by the chairmen of the commissions, and a representative of each of the affiliated national societies. The bureau is composed of the president and vice-president, who are nationals of the host country for the next congress; the past president; the secretary-general; the deputy secretary-general, and the treasurer.

The ISSS has consultative status with FAO, UNESCO, and WMO. It is a scientific associate of ICSU and maintains working relationships with a number of other international organizations.

The address of the International Society of Soil Science is c/o International Soils Museum, 9, Duivendaal, Wageningen, 6140, The Netherlands.

R. DUDAL

ION EXCHANGE—See EXCHANGE PHENOMENA. Vol. IVA: ION EXCHANGE

IONIC ACTIVITIES

The uptake of a chemical element depends on its availability or the readiness with which a small increment of the quantity present moves into growing plants. The movement of an ionic species from soil to plant roots is governed by intensity, relative intensity, and capacity factors that differ for each essential element and that vary among soils. Further, the interactions among soils, plant species, and even varieties within species are different depending on limiting factors for each chemical element. In addition to species, variety or hybrid, and soil factors, other major parameters of crop production include phenology; soil and water management; fertilizer, manure, and lime placement and time of application; and many cultural and management practices that affect the production capability of a soil. Soil properties associated with soil type and production capability include depth, drainage, structure, and colloidal composition as they affect air-water relations and ionic retention, release, and movement.

From information supplied by the grower or predicted from climatic records, soil maps, and yield response data, a yield potential is predicted for a particular crop on each soil sample subjected to a soil test. Since it is the fertilizer recommendation that the grower uses in management, soil testing is generally looked on as a "test" to determine fertilizer and lime needs for the crop to be grown. Changes in soil test levels are important as they relate to treatment requirements for efficient production in the following years, environmental quality, quality of feed and food, and energy efficiency.

The amount of an element to be added to a soil for the growth of a crop is a function of crop removal in relation to adsorption and dissolution of soil minerals and/or man-induced changes from applications of fertilizers and other materials (manure, sewage sludge, industrial wastes). The concept of cation exchange has been helpful in studying the available supply of K^+, Ca^{2+}, Mg^{2+}, and other cations in soils (see *Exchange Phenomena*). The exchangeable amounts of an ion M represents a "pool" that can maintain the availability of the ion over the growing season:

$$\begin{array}{ccccc} & & \text{Solution-M} & & \\ & \rightleftarrows & & \rightleftarrows & \\ \text{Solid phase-M} & & \upharpoonleft\downharpoonright & & \text{Plant-M} \\ & \rightleftarrows & & \rightleftarrows & \\ & & \text{Exchangeable-M} & & \end{array} \quad (1)$$

where solid phase concentrations represent the total amounts within the silicate and nonsilicate minerals plus organic matter; solution-M represents the concentration or activity of the ion relative to those in aqueous solutions; exchangeable-M refers to the quantity held by electrostatic forces; and plant-M represents the supply of the ion at the root surface, within the cell sap, or the total composition of the aerial part of the plant or other plant-analysis parameters. With respect to any solute MA and any system at equilibrium, the activity and relative partial molal free energy must be equal at all distances from the solid phase:

$$(\bar{G} - \bar{G}^{\circ})_{MA} = RT\ln\frac{\bar{A}_{MA}}{\bar{A}^{\circ}_{MA}} \qquad (2)$$

Theoretical considerations relating important parameters of ionic availability have been reported by Baker (1971, passim); Sulaiman and Kay (1972); Baker and Low (1970); Oliver and Barber (1966); and Olsen and Kemper (1968). The concept of ion activity and soil parameters relating to intensity, relative intensity, and capacity factors has been discussed by Khasawneh (1971).

The relative intensity factor is important for ions that interact with respect to their uptake by plants. Some interactions appear to operate predominantly at the plant absorption site while others result from electrostatic effects of soil colloids that maintain constant ionic ratios at the sites of absorption by plants that can be evaluated by Donnan equilibria measurements involving ionic activity ratios including:

$$\frac{K}{\sqrt{Mg}}, \frac{Mg}{Ca}, \frac{Cd}{Zn}, \frac{H}{\sqrt{Ca + Mg}}, \frac{K}{\sqrt{Ca}}, \frac{H}{\sqrt[3]{Al}}, \text{etc.} \qquad (3)$$

Donnan equilibria, ionic ratios, and ionic interactions have been reviewed by Baker (1972); Hiatt and Leggett (1974); Wiklander (1965); and Woodruff (1955). Biological aspects of the root and its environment and soil physical and chemical aspects of the root environment (see *Rhizosphere*) are discussed by several scientists in "*The Plant Root and Its Environment*" edited by E. W. Carson (1975).

From discussions in the literature, it is evident that plants differentially affect elemental availability within soils. For example, buckwheat and millet differ in their capacities to remove phosphorus from soils (Baker, 1964), but the soil parameters limiting phosphorus availability to both crops are the same. Intensity, capacity, and relative intensity factors function in soils to regulate the availability and quantity of an ion removed by a crop over the growing season.

The objectives of the remaining presentation is to discuss the concept of ion activities and ratios as their measurements relate to factors affecting the amounts of an element to be applied to soils.

Determination of Soil Lime Requirements

The intensity, capacity, and relative intensity concept is not new with respect to soil acidity (q.v.). By definition pH relates to intensity, or relative partial molal free energy as shown by the following equation:

$$(\bar{G} - \bar{G}^{\circ})_{H^+} = RT\ln\frac{\bar{A}_{H^+}}{\bar{A}^{\circ}_{H^+}} = -1364 \text{ pH} \qquad (4)$$

where the units are calories per mole at 25°C, and the standard state is taken as unit activity of H^+. The capacity factors involve exchangeable, reserve, potential, or titratable acidity expressed in milliequivalents per 100 g of soil. Relative intensity refers to acid-base balance and is expressed as a percent base saturation of the cation exchange capacity (CEC) or as a lime potential (pH − ½ pCa + Mg). These relationships have been the subject of several discussions and research reports (McLean, Dumford, and Cononel, 1966; McLean, 1973, 1975; Peech, 1965; Schofield and Taylor, 1955; Shoemaker, McLean, and Pratt, 1962; and Taylor, 1960).

As pointed out recently (Baker, 1976), measurement of lime requirement (q.v.) or exchangeable acidity using buffers is an example of how ionic activity data may be used to determine a corrective treatment for a soil. Theoretically, a buffer solution used to determine soil lime requirements should meet the following conditions:

1. The initial pH of the buffer should equal the pH desired at the surface of plant roots under field conditions. Since the desired pH varies for soils and crops, there has been some controversy with respect to the desired pH.
2. The buffer solution must be capable of reaching a reversible equilibrium with all forms of soil acidity. In other words, only exchange reactions among cations should be involved with little or no soil adsorption of the buffer acids or their salts.
3. The change in total acidity with a change in pH of the buffer must be known and be approximately equal to that of the soils tested. As pointed out for Woodruff's (1948) buffer by Baker (1976), the linear change in pH with amounts of H^+ adsorbed from soils is extremely advantageous.

If the buffer capacity or CEC of the soils tested is approximately equal to that of the buffer, then the soil lime requirement will be approximately equal to double the amounts of acidity adsorbed from the soil by the buffer. When used in the conventional manner on soils with a CEC of approximately 10 meq/100 g, the Woodruff buffer will indicate a lime requirement of 1 ton calcium carbonate per acre furrow slice for each tenth unit decrease in pH. If a pH lower than 7.0 is desired, the lime requirement would be 1000 lb of calcium carbonate per acre for each tenth unit decrease in buffer pH plus another 1000 lb per acre for each tenth unit decrease below the desired pH. For example, if the Woodruff buffer pH at equilibrium is 6.2 and the desired soil pH is 6.8, for a soil with a CEC of approximately 10 meq/100 g, the lime requirement would be 7 tons of calcium carbonate per acre.

From the relationships among pH, buffer pH, and lime requirements of soils, it has been concluded that a corrective treatment may be determined more readily than the desired soil pH. A similar dilemma exists for all chemical elements important in soil-plant relationships.

Requirements for Macrocations

Historically, soil testing with respect to cations including hydrogen, potassium, calcium, and magnesium has evolved from two approaches. The approach most commonly used today developed from efforts to statistically correlate test levels obtained by extracting soils with salts, acids, and combinations with crop responses to various extractable cations (Doll and Lucas, 1973). This approach assumes that plant uptake of an ion M is a function of the amount released to the extractant without regard to the quantities existing in the different forms as illustrated in equation 1. The other approach has involved studies to elucidate the parameters affecting the equilibria of equation 1 as they affect ionic activities, ratios, and capacity factors, which relate to ionic uptake over time. Several of these parameters become evident from studies of diffusion (Baker and Low, 1973; Olsen and Kemper, 1968) in porous media like soils. The equation of Makady and Low (1966) relates the diffusive flux J_i of the ion i to the cross-sectional area A of the diffusion path; a geometry or tortuosity factor B; an ionic mobility factor $\bar{\lambda}_i$ to the labile concentration $\bar{C}_i$; and a free-energy gradient, $d\bar{G}_i/dx$:

$$J_i = -ABC_i \bar{\lambda}_i \frac{d\bar{G}_i}{dx} \quad (5)$$

where x is the distance in the direction of diffusion. The application of this equation has been studied by Baker and Low (1970). An excellent example of studies involving the interrelationships among soil parameters important in ion diffusion is that of Sulaiman and Kay (1972). Ionic diffusion increases as the cross-sectional area increases, increases as the labile or exchangeable amount of an ion increases, decreases as the apparent mobility of the ion decreases, and decreases as the relative partial molal free energy increases and free-energy gradient decreases between the bulk soil solution and the root surface. Soil parameters affecting ionic mobility and relative partial molal free energy are not easily distinguished or measured. As the water content (see *Water Content and Retention*) of a soil decreases, the area of diffusion decreases as well as the fluidity. The reduced fluidity of water can be related to a greater viscosity of the water or to a reduced mobility of the ion. The results by Low and his associates (1926*a;b*) favor the use of increased viscosity to account for the effects of reduced water fluidity on ionic diffusion. The electrostatic restriction is another important parameter that adds to the nonideal behavior of ions in suspensions compared with those in true solutions. When both molecular interaction and electrostatic restriction parameters are combined into one term, an *activity coefficient* γ_i is obtained that relates solute activities in soil suspensions to those of true solutions:

$$\bar{A} = \gamma \bar{C} \quad (6)$$

For a new approach to soil testing (Baker 1971, 1973), an effort is being made to determine corrective soil treatments for all ions using an approach analogous to that involving pH and buffer pH in the determination of lime requirements (see *Lime, Liming*). The testing solution contains cations at activities and activity ratios that provide minimum uptake for optimum yields. Compared to the properties of buffer solutions used to determine soil lime requirements, the testing solution for all cations should meet the following conditions:

1. The initial pK, pCa, pMg, pH, and activity ratios should equal the ideal values under field conditions associated with efficient crop production.
2. The cations must reach a reversible equilibrium with the exchangeable cations and other soluble fractions as indicated in equation 1.
3. The change in total amount of each ion must be known for the soil-solution ratio in use. For the conditions under study (Baker, 1976), the equivalents of cations in solution are approximately equal to the CEC of the soils tested.

Using the solution concentrations of 50×10^{-4} M $CaCl_2$, 10×10^{-4} M $MgCl_2$, and 2.5×10^{-4} M KCl at pH 7.3, or other combinations for crops other than field corn, and forage legumes that require rather high soil fertility levels for optimum production, the following assumptions are made relative to soil parameters as they affect solution chemistry:

The anion Cl^- adsorption and distribution remain relatively constant over soils when the soil-solution ratio remains constant. For soils tested, ionic activities are affected by similar electrostatic forces. Since availability is evaluated from solution concentration of equation 1, it is not possible to differentiate electrostatic effects from reduced mobility or water fluidity effects under field conditions. However, the surface properties or miscrostructure of soil colloids, which affect one of these parameters, will have an analogous effect on the other. Thus it is assumed that by adjustment of the soil-solution ratios, $\bar{A}_i$ is measured equally well for different soils by the concentration C in the testing solution:

$$\bar{A}_i = C_i + N\Delta C_i \qquad (7)$$

where $\bar{A}_i$ is the activity of ion i under field conditions; C_i is the initial concentration in the testing solution; and N is a function of the buffer capacity of the soil in relation to the soil to testing solution ratio.

For soils differing substantially with respect to CEC and/or mean free bonding energies for an ion i (Marshall, 1964), one should not expect a high correlation between soil test results and plant responses. With respect to most elements essential for plants, soil testing can provide for efficient use of fertilizers with no obvious crop response to an additional increment of a particular element. The concept of cation adsorption from solution as a measure of soil requirements has been discussed in some detail at a recent symposium, "Soil Testing: Correlating and Interpeting Analytical Results," at the 1975 meetings of the American Society of Agronomy. For potassium and magnesium using a soil-solution ratio of 1:10 and a solution composition of Ca + Mg + K at pH 7.3, which about equals the CEC of the soil being tested, the corrective treatments may be calculated from the equations:

$$\text{K (requirement in ppm)} = 78 \times 10^4 \Delta C_K$$

$$\text{Mg (requirement in ppm)} = 48 \times 10^4 \Delta C_{Mg}$$

where ΔC_K and ΔC_{Mg} refers to the change in testing solution concentration of the respective elements.

Requirements for Microcations and Phosphorus

For the microelements–iron, manganese, copper, zinc, and other heavy metals–uptake by plants is generally limited more by their activity in the soil solution (q.v.) than by the adsorbed or labile quantities present. In fact, iron is not a trace element with respect to abundance in the earth's crust. For these elements, an effort is being made to calculate their activities under field conditions from their equilibrium concentrations or test levels using chelators such as DTPA (diethylenetriaminepentaacetic acid) with the solution used to determine corrective treatments for the macrocations. For this work it is assumed that the chemistry of the chelator with respect to formation constants and pH will reflect the activities of the ions in the soil under field conditions. Correlations relating copper uptake by corn hybrids to pCu in the testing solution indicates that plants tolerate solution copper where pCu ranged from 6–13 (Dragun, 1975). The microcations are somewhat analogous to phosphate with respect to their movement and uptake from soils. Their activities can be extremely low, and uptake will be adequate if the activities at the absorption sites remain at near the initial levels. Under these conditions the partial molal free-energy gradient is high, but the distance the ions diffuse is very small (Baker, 1976). Uptake will be limited by $\bar{A}_i$ if the pH becomes high and/or if $\bar{C}_i$ becomes very low. It is these two parameters of equation 5 that are evaluated for microcations by the approach under development (Baker, 1971, 1973, 1976).

Electrostatic restriction in soils for microelements and the phosphate radical is not analogous to that for the macrocations. Solubility phenomena in relation to adsorption and/or coprecipitation are important interrelations. The concept and importance of including the labile pool, which is somewhat analogous to $\bar{C}_i$ of equation 6, and the exchangeable amounts of the macrocations will be increasingly important for elements where the quantity removed with time is relatively great compared to the equilibrium activity. For these ions the relative partial molal free energy can be relatively high (low activity) so long as the free-energy gradient of equation 5 remains small. Theoretically, it is possible to explain why the Bray No. 1 P test (Bray and Kurtz, 1945) is superior to measurements of the P solubility (Baker, 1973).

Requirements for Nitrogen and Sulfur

The electrostatic restriction parameter for the movement of NH_4^+ in soil is analogous to that for K^+. However, through nitrification most nitrogen is absorbed by crops as nitrate, NO_3^- (see *Nitrogen cycle*). The electrostatic restriction on the movement of NO_3^- and $SO_4^=$ in most soils is negligible, so the solution activity of these ions fluctuates greatly over the growing season as the amounts released by microorganisms or added from fertilizers are made to equal the amounts removed by the crop. The concept of ionic activities is not useful in predicting crop requirements for these elements. However, for water quality NO_3-N concentrations greater than 10 ppm are to be prevented by management systems that recycle NO_3-N within vegetation during the growing season. Monitoring of soils for environmental quality and animal and human health has been reviewed (Baker and Chesnin, 1975).

Summary and Conclusions

Ionic activities and ratios in soils, suspensions, and solutions are comparable and are of major biological significance. Corrective treatments may be determined directly that adjust activities and activity ratios to levels that require the least amounts of each chemical element for crop production. Crop fertilization should be managed separately from corrective treatments of soils, especially for nitrogen and potassium where crop removal accounts for a substantial amount of the available supply of the element within soils.

In conclusion, soil-testing methods analogous in theory and application to existing tests for pH and lime requirement using buffer solutions are possible for many adsorbed elements in soils that are important in plant nutrition and animal and human health.

DALE E. BAKER
RICHARD M. ESHELMAN

References

Baker, D. E. 1964. A study of isotopic dilution as a method for relating phosphorus retention to availability of phosphorus in widely different soils, *Soil Sci., Soc. America Proc.*, **28**, 511-517.

Baker, D. E., 1971. A new approach to soil testing, *Soil Sci.*, **112**, 381-391.

Baker, D. E., 1972. Soil chemistry of magnesium, *in* J. B. Jones, Jr.; M. C. Blount; and S. R. Wilkinson, eds., *Magnesium in the Environment.* Reynolds: Taylor County Printing Co., 380p.

Baker, D. E., 1973. A new approach to soil testing: II. Ionic equilibria involving H, K, Ca, Mg, Mn, Fe, Cu, Zn, Na, P and S, *Soil Sci. Soc. America Proc.*, **37**, 537-541.

Baker, D. E., 1976. Ion activities and ratios in relation to corrective treatments of soils. T. Peck, d. *Proceeding of Symp. on Soil Testing: Correlations and Interpreting Analytical Results.* Madison: American Society Agronomy, 117p.

Baker, D. E., and Chesnin, L., 1975. Chemical monitoring of soils for environmental quality and animal and human health, *in* N. Brady, ed., *Adv. Agron.*, **27**, 307-374.

Bray, R. A., and Kurtz, L. T., 1945. Determination of total, organic and available forms of phosphorus in soils, *Soil Sci.*, **59**, 39-45.

Carson, E. N., ed., 1974. *The Plant Root and Its Environment.* Charlottesville: University Press of Virginia, 691.

Doll, E. C., and Lucas, R. E., 1973. Testing soil for potassium, calcium and magnesium, *in* L. M. Walsh and J. D. Benton, eds., *Soil Testing and Plant Analysis.* Madison: Soil Science Society of America, 428p.

Dragun, J., 1975. *Substrate Copper Activities and Uptake by Corn Hybrids.* M.S. Thesis. The Pennsylvania State University, Department of Agronomy, 86p.

Hiatt, A. J., and Leggett, J. E., 1974. Ionic interactions and antagonisms in plants, *in* E. W. Carson, ed., *The Plant Root and Its Environment.* Charlottesville: University Press of Virginia, 691p.

Khasawneh, F. E., 1971. Solution ion activity and plant growth, *Soil Sci. Soc. America Proc.*, **35**, 426-436.

Low, P. F., 1962*a*. Effect of quasi crystalline water on rate processes involved in plant nutrition, *Soil Sci.*, **93**, 6-15.

Low, R. F., 1962*b*. Influence of absorbed water on exchangeable ion movement, *Clays Clay Minerals*, **9**, 219-228.

McLean, E. O., 1973. Testing soils for pH and lime requirements, *in*, L. M. Walsh and J. D. Beaton, eds., *Soil Testing and Plant Analysis.* Madison, Wisc.: Soil Sci. Soc. America, 78-95.

McLean, E. O., 1975. Recommended pH and lime requirement tests, *NCR Pub. 221, N. Dak. Agr. Exp. Sta. Bull. 499*, 23p.

McLean, E. O.; Dumford, S. W.; and Cononel, F., 1966. A comparison of several methods for determining lime requirements of soils, *Soil Sci. Soc. America Proc.*, **30**, 26-30.

Marshall, C. E., 1964. *The Physical Chemistry and Mineralogy of Soils.* New York: Wiley, 244p.

Mokady, R. S., and P. F. Low, 1966. Electrochemical determination of diffusion coefficients in clay-water systems, *Soil Sci. Soc. America Proc.*, **30**, 438-442.

Oliver, S., and Barber, S. A., 1966. An evaluation of the mechanisms governing the supply of Ca, Mg, K, and Na to soybean roots (*Glycine max.* L.), *Soil Sci. Soc. America Proc.*, **30**, 82-86.

Olsen, S. R., and Kemper, W. D., 1968. Movement of nutrients to plant roots, *Adv. Agron.*, **20**, 91-151.

Peech, M., 1965. Hydrogen ion activity, *in* C. A. Black, ed., *Methods of Soil Analysis: II. Chemical and Microbiological Properties.* Madison: American Society of Agronomy, 914-926.

Schofield, R. K., and Taylor, A. W., 1955. The measurement of soil pH, *Soil Sci. Soc. America Proc.*, **19**, 164-167.

Shoemaker, H. E.; McLean, E. O.; and Pratt, P. F., 1962. Buffer methods for determination of lime requirement of soils with appreciable amounts of aluminum, *Soil Sci. Soc. America Proc.*, **25**, 274-277.

Sulaiman, W., and Kay, B. D., 1972. Measurement of the diffusion coefficient of boron in soil using a single cell technique, *Soil Sci. Soc. America Proc.*, **36**, 746-752.

Taylor, A. W., 1960. The significance of ratios of ion activities, *7th Internat. Congr. Soil Sci. Trans.*, **2**, 342-346.

Wicklander, L., 1965. Cation and anion exchange phenomena, *in* F. E. Bear, ed., *Chemistry of the Soil.* London: Reinhold, 163-205.

Woodruff, C. M., 1948. Testing soils for lime requirement by means of a buffer solution and glass electrode, *Soil Sci.*, **66**, 53-63.

Woodruff, C. M., 1955. The energies of replacement of calcium by potassium in soils, *Soil Sci. Soc. America Proc.*, **19**, 167-170.

Cross-references: *Adsorption Phenomena; Buffers, Buffering; Diffusion Processes; Electrochemistry; Exchange Phenomena; Lime, Liming; Nutrient Potential; Soil Solution; Water Content and Retention.*

IRRIGATION

Ancient civilizations in Mesopotamia, Egypt, India, China, and the Americas began to irrigate early—about 7000 years ago in Mesopotamia—as farming spread into low-rainfall areas. Extensive systems of canals or tunnels were dug to send water to the fields. Ancient records from Egypt and China have described the highly organized efforts required for the canal system construction and maintenance. The decline of some early civilizations that depended on irrigation appears to have been hastened in some cases by the silting of canals and the salination of fields (Marr, 1967).

Ancient civilizations relied on various forms of surface-irrigation systems in which water from a canal or ditch was distributed over the fields. Sprinkler methods were developed early in this century (Israelsen and Hansen, 1962) whereas drip (trickle) irrigation in the field was developed only during the last decade (Bernstein and Francois, 1973*a*). Subirrigation, by which water is supplied to fields by maintaining ditch levels to control water tables, was practiced by Indians in Mesoamerica; present-day systems, however, were developed in conjunction with drainage of waterlogged soils in low-lying areas (Israelsen and Hansen, 1962; Criddle and Kalisvaart, 1967).

Methods

The *wild-flooding* irrigation method involves running water over a field with little or no prior land preparation or control. It is used mainly on pasture lands, often with water diverted from a nearby stream. This method is rather inefficient.

Border-strip irrigation fields (Figure 1) are divided by low levees into strips 9–18 m wide and 100–400 m long. The land surface is leveled

FIGURE 1. Water is supplied to this field by the border strip irrigation method. Aluminum syphon tubes thrown over the bank of the head ditch supply an even sheet of water across the strip with minimal risk of soil erosion or the need to cut the bank of the head ditch. (Photo: U.S. Department of Agriculture, Soil Conservation Service)

FIGURE 2. Furrow irrigation in Nebraska. The 25 cm gated portable pipe is laid transverse to the furrows and water is discharged to each corn row via a special emitter. (Photo: U.S. Department of Agriculture, Soil Conservation Service)

across a strip with slopes along the length generally ranging from 0.2–0.4%, but up to 6% for pastures. Water should reach the end of the strip in one-fourth or less the time required for the desired water penetration at a given point once the soil is wetted. Even so, contact time is variable along the length of the strip causing unequal water penetration and decreased *field-irrigation efficiency,* viz. the percentage of total water applied that is used by the crop and evaporated from the soil. Deep percolation (q.v.) below the root zone (see *Rhizosphere*) and water loss from the lower end of the field (tailwater) reduce field-irrigation efficiency. The border-strip method is used for pasture and hay crops, small grains, and other field crops.

In the system of *border checks* (or *level basins*), the border strip may be divided by cross-levees to form level or nearly level basins. Additional ditches are needed to bring the water to the individual basins or checks. Irrigation efficiencies may be much improved as compared to those of border strips. This method is used to irrigate field crops and orchards.

In *furrow irrigation* (Figure 2), water is channeled in small ditches or furrows having average downfield slopes of 0.5–3%. The furrows are usually 10–20 cm deep and 100–400 m long with centers spaced 75–150 cm apart. Generally, crops are grown either in a single row in the center of the raised bed between two furrows or in double rows along each side of wider beds. The method is used for row crops (cotton, maize, vegetables, etc.) and orchards. Efficiencies are comparable to or somewhat better than those for border-strip irrigation.

Though *sprinkling methods* can be adapted to most crops and soils, they are particularly well suited to shallow soils, soils with extremely low or high water-intake rates, and lands so uneven that leveling is impractical. Unlike surface irrigation, which often permits whole farms to be irrigated during a relatively short period of time, many sprinkler systems require almost continuous irrigation in some part of the farm during periods of peak water use by the crop.

Sprinkler systems differ greatly with regard to piping patterns and size and distribution of sprinkler heads. *Stationary,* or *solid-set,* systems are left in place during the period of crop growth whereas *portable* systems require the movement of main pipes as well as laterals, with attached sprinklers, to different fields or parts of a field during the course of irrigation (Figure 3). *Semiportable* systems combine stationary main lines and portable laterals. A recent development is the self-propelled *center pivot* system that can irrigate quarter-section (250ha) plots. The circularly irrigated fields are a prominent feature in the landscape (Figure 4).

Irrigation efficiencies for sprinkler systems, which are generally higher than for surface methods, approach and may even exceed 85% under optimum conditions. Maximum uniformity of water penetration and efficiency are obtained when the distance between sprinklers is about 55% of the diameter of the area wetted by a single sprinkler. Furthermore, application rates should not exceed the minimum infiltration rates of soils and should not allow water to accumulate on the ground surface.

Foliar absorption of salt with consequent damage to plants is a hazard of sprinkler irrigation. Susceptible plants such as citrus and stone fruits can be injured by low levels of sodium or chloride in irrigation water (Bernstein, 1975).

The drip-, or trickle-, irrigation method uses flexible plastic tubing or rigid plastic pipe to carry water along plant rows. Holes in the flexible (usually double-walled) tubing or special emitters in the pipe permit a steady drip or slow trickle of water. It is usually applied in small amounts each day so as to replenish moisture in the root zone without causing waterlogging. Advantages of the method include the increased irrigation efficiency due to the elimination of deep percolation, except as needed for salinity control, and the formation of a small wetting zone around emitters, which results in limited weed growth and efficient fer-

FIGURE 3. This portable irrigation gun is attached to a flexible pipe and automatically moves ahead at a set pace which controls the application rate. This design can supply up to 2000 liters per minute over a 100 meter semi-circle. (Photo: U.S. Department of Agriculture, Soil Conservation Service)

tilizer application via the irrigation water. Emitter clogging, increased susceptibility to crop damage during a system breakdown, erratic drip rates due to small elevational differences in the field, and the need to leach salts following harvest are disadvantages of drip, or trickle, irrigation (Bernstein and Francois, 1973*a*).

Subirrigation may be effected via buried perforated pipe, but it is generally managed by careful control of a relatively shallow water table by maintaining the appropriate level of water in open ditches (Criddle and Kalisvaart, 1967). The method is best adapted to humid areas (e.g., the Netherlands) where rainfall prevents salt accumulation in the root zone. In drier regions supplemental surface irrigation, usually by sprinkling, is frequently required to reduce salt accumulation in the soil. Subirrigation is often used in low-lying areas that require artificial drainage by ditches during most of the year. Water is added to the ditches during dry periods when water use exceeds rainfall.

Choice of Methods

Water Supply. Some water supplies such as farm wells are continuously available; others, however, may be available only intermittently, when several farms share a common source. Most sprinkler- and drip-irrigation systems require continuous supplies; surface irrigation methods may, in contrast, use intermittent supplies. Water quality is also an important consideration because water that is satisfactory for surface methods may cause salt injury when sprinkled on susceptible crops (see *Irrigation Water*) (Bernstein, 1975). Because the cost of water is becoming an increasingly important factor in irrigation schemes, water should be conserved, and the most efficient methods such as border check, sprinkling, and drip irrigation should be favored.

Soil Properties. Few soils have properties ideal for irrigation. Even for those soils that can be irrigated, methods must be adjusted to the

FIGURE 4. This self-propelled center pivot system is being used to irrigate two quarter sections of corn in Minnesota. One well serves both fields with water being pumped underground through steel pipes to each pivot point. Irrigation equipment is moved from one center to the other along a trail that bisects both fields. (Photo: U.S. Department of Agriculture, Soil Conservation Service)

nature of the soil as well as to the crop and available equipment. For example, shallow or coarse-textured (sandy) soils retain only small amounts of water, making frequent light water applications necessary, which is possible with sprinkler and drip, or trickle, methods. Steep soils cannot be irrigated with surface methods, but many can with sprinklers at application rates as low as 2 mm/hr, which minimizes the erosion risk.

Soils with uniform profiles, medium textures, and low salinity are most desirable for surface irrigation because of the careful grading required to permit uniform spreading of water. Soils with important components of sand or gravel in their profiles and those with unfavorable chemical properties are poorly suited for surface irrigation.

Crops. Shallow-rooted crops or those sensitive to prolonged soil water surpluses or deficits may require frequent light irrigations, often achieved more easily by sprinkling or drip methods than by surface methods. On the other hand, citrus and stone fruits absorb salt through their leaves so rapidly that *leaf burn* (salt injury) occurs when leaves are wetted by sprinklers. Such injuries do not occur when surface methods are used (Bernstein, 1975).

Furrow irrigation of row crops may inhibit seed germination due to the concentration of salts in raised planting beds. Planting on sloping beds, however, alleviates salting problems because the advancing water front sweeps the salt past the seed into the peak of the bed (Bernstein, MacKenzie, and Krantz, 1955). Sprinkling row crops during germination not only reduces salt accumulation but also maintains a lower temperature more favorable for germination of some crops while requiring only a fraction of the water needed for repeated or prolonged furrow irrigations. Because fertilizers can be applied by sprinklers (or drip irrigation) and because sprinklers may be used to prevent frost damage (Christiansen and Davis, 1967), sprinkling is sometimes favored over other irrigation methods.

Economic Factors. The choice of an irrigation system is influenced by such factors as availability and cost of labor especially when compared to capital investments required for alternative systems. When labor is scarce or costly, net returns may be greater when labor-saving equipment or systems are installed. When water is cheap, inefficient irrigation methods may yield higher returns than more efficient but more costly methods.

Comparative Irrigation Efficiencies, Yield Potentials, and Salinity Hazards

Irrigation Efficiencies. Because *contact time,* the time during which water is entering the soil, usually varies for different parts of a surface-irrigated field, water infiltration will differ and efficiencies will be decreased, especially for methods that irrigate large areas from a single point or line of supply (border-strip and furrow methods). Irrigation efficiencies for surface methods range from about 30–75%, depending on soil properties, adequacy of leveling, and adjustment of irrigation stream size to minimize differences in contact time. Sprinkler systems give essentially uniform contact time over the irrigated area and may minimize the effects of variable soil infiltration rates.

Ideally, drip irrigation can achieve nearly 100% irrigation efficiency, but variable emitter performance may limit practical efficiencies to about the same level as that of sprinklers (85%). Subirrigation may also achieve high efficiencies, but the suitability of subirrigation is limited by the need for salinity control.

With highly efficient irrigation systems, care must be taken to avoid underirrigation, which can lead to salt accumulation in the root zone. In addition to meeting the consumptive water needs of the crop, the extra water needed for leaching salt (the *leaching requirement;* see *Salt-Leaching*) must be applied either regularly or at intervals (U.S. Salinity Laboratory Staff, 1954; Bernstein and Francois, 1973*b*).

Yield Potentials. Bernstein and Francois (1973*a*) studied efficiencies of drip, furrow, and sprinkler irrigation in small field plots on sandy loam soil in which uniformity of application was close to 100% for all three methods. Even then, furrow irrigation required 50% more water than drip irrigation because of unavoidable deep percolation losses that occurred during the early stages of crop growth. Sprinkling required an intermediate amount of water. Once the root systems developed and permeated the soil, irrigation efficiencies became nearly equal for all three irrigation methods. In the field, however, efficiency of furrow irrigation would be less than in the small plots because of nonuniform water penetration.

When water of good quality (see *Irrigation Water*) was applied in the amounts needed for each irrigation method, yields were essentially the same for all three irrigation methods, indicating that the greater frequency of irrigation by the drip method did not significantly increase yields compared to optimally operated sprinkler or furrow systems. With saline water, drip irrigation produced somewhat better yields than furrow irrigation because soil water salinity was kept at a minimum by the daily applications. Less frequent furrow irrigation allows soil water salinity to increase between irrigations as soil water content decreases. Sprinkling with saline water caused almost complete loss of yield because of direct salt damage to the crop.

Salinity Hazards. No irrigation method is free of salinity hazards. Moreover, increased salinity may follow from either overirrigation or underirrigation. The latter is more often responsible, inasmuch as it permits progressive accumulations of salts in soils (see Vol. XII, Pt. 2: *Salt-Affected Soils*). Under-irrigation may follow under any of the current methods. Amounts of water supplied by sprinklers, for example, may be too small to leach out salts (see *Leaching*). Poorly designed border-strip, furrow, or basin systems may leave parts of a field underirrigated and thus allow build-ups of salinity. Within a field, salts may also accumulate in bodies of soil that have low infiltration rates, even though water has been spread uniformly. Precaution is therefore continuously necessary with any method to prevent salinity problems in irrigated areas.

LEON BERNSTEIN

References

Bernstein, L., 1975. Effects of salinity and sodicity on plant growth, *Ann. Rev. Phytopath.,* **13,** 295–312.

Bernstein, L., and Francois, L. E. 1973*a*. Comparisons of drip, furrow and sprinkler irrigation, *Soil Sci.,* **115,** 73–86.

Bernstein, L., and Francois, L. E., 1973*b*. Leaching requirement studies: Sensitivity of alfalfa to salinity of irrigation and drainage waters, *Soil Sci. Soc. America Proc.,* **37,** 931–943.

Bernstein, L.; MacKenzie, A. J.; and Krantz, B. A., 1955. The interaction of salinity and planting practice on the germination of irrigated row crops, *Soil Sci. Soc. America Proc.,* **19,** 240–243.

Christiansen, J. E., and Davis, J. R. 1967. Sprinkler irrigation systems, *in* R. M. Hagan; H. R. Haise; and T. W. Edminster, eds., *Irrigation of Agricultural Lands.* Madison: American Society of Agronomy, Monograph, **11,** 885–904.

Criddle, W. D., and Kalisvaart, C., 1967. Subirrigation systems, *in* R. M. Hagan; H. R. Haise; and T. W. Edminster, eds., *Irrigation of Agricultural Lands.*

Madison: American Society of Agronomy, Monograph, **11,** 905-921.

Hagan, R. M.; Haise, H. R.; and Edminster, T. W., eds., 1967. *Irrigation of Agricultural Lands.* Madison: American Society of Agronomy, Monograph, **11.**

Marr, P. D., 1967. The social context of irrigation, *in* R. M. Hagan; H. R. Haise; and T. W. Edminster, eds., *Irrigation of Agricultural Lands.* Madison: American Society of Agronomy, Monograph, **11,** 12-22.

Israelsen, O. W., and Hansen, V. E., 1962. *Irrigation Principles and Practices.* New York: Wiley, 447p.

U.S. Salinity Laboratory Staff, 1954. Diagnosis and improvement of saline and alkali soils, *U.S. Department of Agriculture Handbook,* **60,** 160p.

Cross-references: *Infiltration; Irrigation Water; Management of Soils; Moisture Management; Percolation; Permeability; Salt-leaching; Soil Drainage.*

IRRIGATION WATER

The *quality* of irrigation water, i.e., its concentration of ionic constituents and their composition, has been shown to profoundly affect not only crop growth but also soil properties. When a soil has been irrigated with a given water for a sufficiently long time, a dynamic chemical equilibrium between the water and soil will take place, resulting in definite fluctuating patterns for the concentration of soil solution (q.v.) and relative abundance of exchangeable cations. Whether soil physical properties will be affected by the nature of such chemical equilibrium depends on both the water and soil under consideration. Water parameters of importance in this regard and their definitions are as follows.

1. *Electrical conductivity* (in millimhos/cm at 25°C). This is commonly used as an index of total salt concentration in parts per million (ppm) or milliequivalents per liter (meq/l). For a more detailed discussion, see the section on *Conductivity, Electrical.*
2. *Cationic constituents* (in meq/l). These generally include Ca^{2+}, Mg^{2+}, and K^+. Subsequent parameters have been devised based on cationic contents and some have special significance in predicting irrigation water effects on soils. Most important among these is the *sodium adsorption ratio (SAR)* which, using cation concentrations as meq/l, is defined as:

$$SAR = \frac{Na^+}{\sqrt{\dfrac{Ca^{2+} + Mg^{2+}}{2}}}$$

For most soils SAR has been shown to be linearly related to the *exchangeable sodium percentage (ESP),* which represents the percent exchangeable sodium of the soil's total cation exchange capacity (CEC), and whose magnitude is related to physical properties of many soils (see *Leaching; Salt Leaching*).

3. *Anionic constituents* (in meq/l). These generally include Cl^-, $SO_4^=$, HCO_3^-, and $CO_3^=$. In certain waters such as groundwaters resulting from leaching accompanying weathering under tropical climates, some ionic forms of silicon may be present in significant amounts. The relative abundance of various anions in irrigation water affects the eventual soluble salt level in soil solution and subsequent cation balance in the solution. Expressions proposed to account for these effects include the "effective salinity," which represents the total salt concentration of irrigation water minus the fraction expected to precipitate as calcium- or magnesium-carbonates or as calcium-sulphates, and the "residual sodium carbonate," which expresses the sodium hazard of the water after such precipitations have occurred. The latter is represented by the equation $[(CO_3^= + HCO_3^-) - (Ca^{2+} + Mg^{2+})]$, in which all constituents are expressed as meq/l. In addition to these indirect effects, certain anions have been shown to directly affect the structural properties of soils rich in certain constituents.

The effects of ionic constituents of irrigation water on soil physical properties may be related to one or more of the preceding factors. The relative importance of electrolyte concentration, cationic constituents, or anionic constituents depends on the nature and relative proportions of different soil constituents. Colloidal-chemical theory allows for predicting these interactions, and related techniques offer simple experimental methods for evaluation of soil physical responses to changes in solution concentration and composition. These include assessment of colloidal stability and flocculation (q.v.) behavior, of swelling characteristics, and of electrokinetic behavior. Much information also exists, however, on soil physical responses to changes in irrigation water quality using direct indices. Most common among these indices are the stability of soil aggregates (see *Aggregate Stability*), water-transmission and retention characteristics (see *Water Movement; Water Content and Retention*), and rheological properties. The ensuing review of water-quality effects on soils will center around these indices.

Effect of Total Electrolyte Concentration

Contrary to known effects on plants, higher salt levels in irrigation water generally induce more favorable soil physical properties than do lower levels. This is more so when the sodium adsorption ratio of water is high. Colloidal-chemical principles predict, and experimental evidence confirms, that waters with high electrolyte concentrations induce the formation of flocculated soil clays and subsequent stable aggregation, high permeability, and open free drainage. To exploit this phenomenon, direct use has occasionally been made of saline water from seas or lakes as an initial step for enhancing the reclamation of sodic (alkali) soils. Certain precautions are necessary to preserve the favorable properties achieved with such applications when dilution of soil solution takes place, e.g., as a result of rainfall or subsequent irrigation(q.v.) with low salt water. Such precautions are more necessary if the initial salt-water application had resulted in allowing monovalent cations to prevail on the soil's exchange sites, due to high SAR values, and when the soil mineralogical composition does not provide for aggregate stability against dispersion. According to the concept of "threshold concentration" (Quirk and Schofield, 1955), there is an electrolyte concentration level for a water with a given proportion of sodium ions, below which significant detrimental changes may occur in the physical properties of certain soils. This concentration is expected and has been found to be higher for waters containing high sodium proportions (high SAR) than for those containing high divalent cation proportions (low SAR) and for soils containing an abundance of swelling minerals (e.g., montmorillonite) than for soils containing an abundance of nonswelling minerals (e.g., kaolinite). Addition of gypsum is perhaps the most practical method of increasing the salt concentration in irrigation water while also decreasing the relative abundance of sodium ions.

Effects of Cations

Voluminous data are now available on the effects of adsorbed, or exchangeable, cations on the physical properties of soils. Several recent reviews of this subject are also available (e.g., Kelley, 1964; Wilcox and Durum, 1967; El-Swaify, Ahmed, and Swindale, 1970; Kovda, 1973; and McNeal, 1974). Among the common cations in irrigation waters, the favorable effects of Ca^{2+} and detrimental effects of Na^+ are now well established for soils from arid and semiarid regions. However, the effects of Mg^{2+} and K^+ are subject to diverse opinions. Most investigators have shown similarities between the effects of Mg^{2+} and K^+ and those of Ca^{2+}(Quirk and Schofield, 1955; and El-Swaify, Ahmed, and Swindale, 1970; see also Figure 1). Many references have been made, however, to the possible roles of both, particularly of Mg^{2+}, in damaging soil structure (e.g., Bakker and

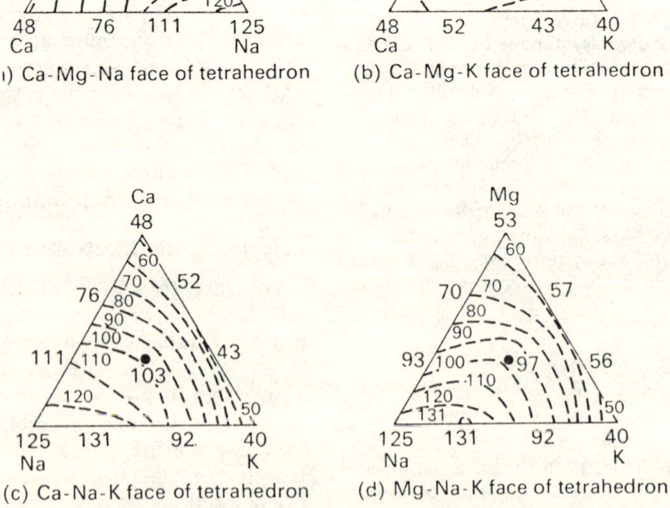

ı) Ca-Mg-Na face of tetrahedron

(b) Ca-Mg-K face of tetrahedron

(c) Ca-Na-K face of tetrahedron

(d) Mg-Na-K face of tetrahedron

FIGURE 1. Response surfaces for effects of simultaneously present exchangeable cations on water-retention characteristics of a tropical Vertisol. Figures on triangle corners represent water retention by monoionic systems. Figures on triangle sides represent water retention by bi-ionic systems. Figures inside triangles represent water retention by tri-ionic systems. (After El-Swaify, Ahmed, and Swindale, 1970)

Emerson, 1973). Such diverse findings are probably due to differences among soils used by different workers (particularly in the types and proportions of various organic and inorganic constituents), to unequal controls on electrolyte concentration and pH levels during the course of experiments, and to differences in the history of soils subjected to these investigations.

A clear illustration of differences among soils may be made from comparing the findings of Quirk and Schofield (1955) and McNeal and Coleman (1966) on the one hand and those of El-Swaify and Swindale (1968) on the other. Hydraulic conductivities (q.v.) of sodium-saturated illitic, micaceous and montmorillonitic soils used by the first authors declined to nearly zero when electrolyte concentration was lowered to between 0.01 and 0.04 M. On the other hand, appreciable hydraulic conductivities were maintained by sodium-saturated kaolin- and oxide-enriched soils used by the latter authors even when electrolyte concentrations were lowered to nearly 0.01 M. Furthermore, "chemical-hammer" tests for the latter soils showed only small reductions in hydraulic conductivity when the ESPs of the soils were elevated to 57 (using equilibration with seawater) and then all soluble salts from the soil removed by distilled water washing (Table 1). These findings indicate that no general criteria for soil tolerance to detrimental ions in irrigation water are expected to be successful in all soils.

Less research has been conducted on soils containing several exchangeable cations simultaneously than on monoionic soils. Most research had shown that the mutual presence of several cations on the exchange complex affects soil physical properties in a manner that is intermediate to the effects of the individual cations (El-Swaify, Ahmed, and Swindale, 1970). Figure 1 illustrates this behavior for water-retention characteristics in a tropical Vertisol. Nevertheless, there are certain findings that indicate the possible existence of "threshold" proportions of exchangeable cations at which sudden changes in soil properties occur. The most common application of this concept is the choice of ESP = 15 by the U.S. Salinity Laboratory Staff (1954) as the limit above which sodium is considered to dominate the behavior of soils under irrigation. However, based on recent, well-defined permeability experiments, it is now known that different soils have different tolerances to exchangeable sodium, some of which are in the vicinity of 15%, some are more than 15% (El-Swaify and Swindale, 1968; McNeal and Coleman, 1966), and some are less (Bakker and Emerson, 1973). Absence of swelling clays or presence of certain constituents that suppress swelling (such as amorphous or sesquioxidic materials that occur in abundance in highly weathered tropical soils) favors increased soil tolerance to sodium (El-Swaify and Emerson, 1975).

Literature available on other physical properties generally supports the findings reported above for effects of exchangeable cations on soil permeability (q.v.). Thus Kemper and Koch (1966) reported that aggregate stabilities for soils from western United States and Canada were virtually nonexistent at ESP values from 15 to 20. On the other hand, aggregates from Australian Red Brown Earths dispersed at ESP = 3 to 5 (Bakker and Emerson, 1973). In contrast, those from highly weathered tropical soils remained stable in the absence of electrolyte even at ESP = 100 (El-Swaify, 1970; see also Figure 2). El-Swaify, Ahmed, and Swindale (1970) and Kovda (1973) reviewed more available information on cation effects on other physical properties.

Effects of Anions

Anion effects on soil physical properties have received only scattered attention in the literature (e.g., Longenecker, 1960; and El-Swaify,

TABLE 1. Effects of Alternate Applications of Seawater and Distilled Water on Hydraulic Conductivities of a Tropical Oxisol and a Tropical Aridisol.

Soil and Horizon	Hydraulic Conductivity (cm/hr)				
	Seawater Application	Dist. Water Application	Seawater Application	Dist. Water Application	Seawater Application
Oxisol, surface	3.46	1.45	2.72	1.17	2.03
Oxisol, subsurface	5.39	1.70	3.05	1.34	3.51
Aridisol, surface	1.24	0.52	0.81	0.49	0.76
Aridisol, subsurface	1.93	0.79	1.56	0.65	1.49

After El-Swaify and Swindale, 1968

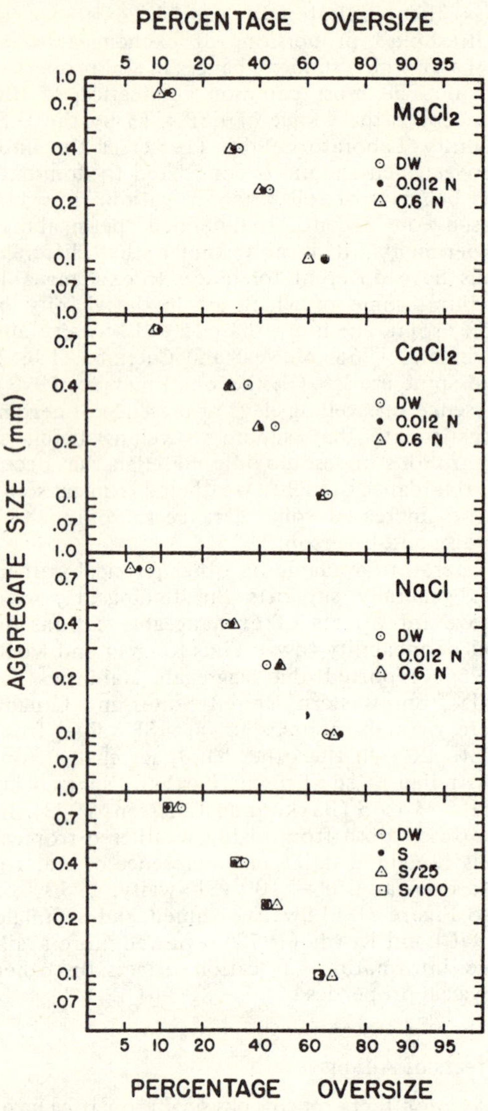

FIGURE 2. Effects of various ionic treatments on the distribution of aggregates of various diameters in the plow layer of a tropical Oxisol. Legends represent the solution in which wet-sieving analyses were made (DW, distilled water; S, seawater with denominators referring to dilution ratio). (After El-Swaify, 1970)

1973). This lack of attention is due to the fact that conventional application of colloid-chemical principles to soil clays, which normally carry negative electric charges, resulted in the belief that structural properties are primarily determined by ions that carry opposite charges, i.e., cations. Thus as discussed in the preceding section, cation effects on soils have received considerably more attention from researchers than have anion effects. However, anions are expected to play important roles in the flocculation or dispersion of soil colloids, and subsequent responses in soil physical properties, because of at least three reasons. The first reason is that certain anions have neutralizing or charging effects at the surface- or edge-sites of certain soil constituents such as kaolins, metal oxides and hydroxides, and other amorphous organic and inorganic materials. Second is the ability of certain anions to cause pH changes with subsequent modification of charge characteristics of these same constituents. And third is the anions' abilities to modify flocculation or dispersion effects of cations by controlling their effective concentration, i.e., activities.

In a recent study (El-Swaify, 1973), it was found that the effects of anions in irrigation water on soil structure were determined by one or more of the preceding factors depending on the soil under consideration. The study explored the effects of chlorides, sulphates, bicarbonates, and silicates on the hydraulic conductivities of Aridisols, Inceptisols, Oxisols, and Vertisols. Anion effects were increasingly important as the proportions of amorphous materials and/or metal oxides or hydroxides in a soil increased. In such a soil, typified by the volcanic ash Inceptisol (Figure 3), effects were primarily due to the first and second factors cited above. When the proportion of 2:1 clays increased, as was the case for the Vertisol, cation activity changes, reflected by changes in electrolyte concentration, produced more significant structural changes than did changes in anion types.

Additional Considerations for Irrigation Water Quality

The management of irrigation water–i.e., determination of quantity, frequency, and method of water application–represents an additional dimension that must be used in overall judgment of quality. If the physical condition of a soil allows free downward movement of applied water, then an application of the "salt-balance principle" shows that there is little danger of accumulating soluble salt in the root zone when poor-quality water is used in sufficient quantity and with an appropriate application method. Such a method would be aimed at avoiding excessive ponding during irrigation and allowing a sufficient "leaching fraction" (U.S. Salinity Laboratory Staff, 1954). Another prerequisite for using waters with marginal quality for irrigation is that the soil possesses high resistance to the detrimental effects of sodium ions. This is an important consideration because structural deterioration may result from lowering the electrolyte con-

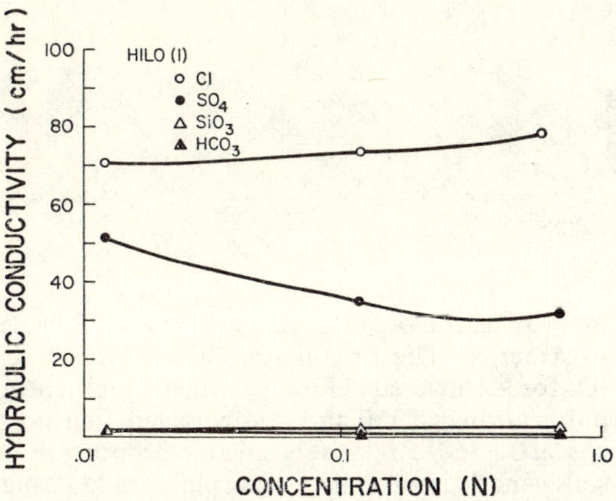

FIGURE 3. Changes in hydraulic conductivity of a sodium-saturated surface tropical Inceptisol in relation to electrolyte concentration and anions in solution. (After El-Swaify, 1973)

centration by unexpected rainfall of the solution surrounding a highly sodium-saturated soil (El-Swaify and Swindale, 1968).

These considerations, as well as differences among soils in responding to the effects of ionic irrigation water constituents as discussed in the preceding sections, explain why any attempted generalizations to establish irrigation water-quality standards have proved inadequate. Rather, standards are needed for each region that best suit soils of that region and prevailing irrigation management practices therein.

S. A. EL-SWAIFY

References

Bakker, A. C., and Emerson, W. W., 1973. The comparative effects of exchangeable calcium, magnesium, and sodium on some physical properties of Red-Brown Earth's subsoils: III. The permeability of Shepparton soil and comparison of methods, *Aust. Jour. Soil Res.,* **11,** 159-165.

El-Swaify, S. A., 1970. The stability of saturated soil aggregates in certain tropical soils as affected by solution composition, *Soil Sci.,* **109,** 197-202.

El-Swaify, S. A., 1973. Structural changes in tropical soils due to anion in irrigation water, *Soil Sci.,* **115,** 64-72.

El-Swaify, S. A., and Swindale, L. D., 1968. Hydraulic conductivity of some tropical soils as a guide to irrigation water quality, *9th Internat. Cong. Soil Sci. Trans.,* **I,** 381-89.

El-Swaify, S. A.; Ahmed, S.; and Swindale, L. D., 1970. Effects of adsorbed cations on physical properties of tropical Red and tropical Black Earths: II. Liquid limit, degree of dispersion, and moisture retention, *Jour. Soil Sci.,* **21,** 188-198.

El-Swaify, S. A., and Emerson, W. W., 1975. Effects of precipitated aluminum and iron hydroxides on the physical properties of soil clays. I. Swelling and aggregate stability after drying, *Soil Sci. Soc. America Proc.,* **39,** 1056-1063.

Kelley, W. P., 1964. Soil properties in relation to exchangeable cations and kinds of exchangeable material, *Soil Sci.,* **98,** 408-412.

Kemper, W. D., and Koch, E. J., 1966. Aggregate stability of soils from western United States and Canada, *A.R.S., U.S. Dept. Agric., Tech Bull. No. 1355,* 52p.

Kovda, V. A., ed., 1973. Quality of irrigation water, *in Irrigation, Drainage, and Salinity–An International Sourcebook.* Paris UNESCO: Hutchinson, 377p.

Longenecker, D. E., 1960. Influence of soluble anions on some physical and physico-chemical properties of soils, *Soil Sci.,* **90,** 185-191.

McNeal, B. L., 1974. Soil salts and their effects on water movement, *in* J. Van Schilfgaarde, ed., *Drainage for Agriculture.* Madison: American Society of Agronomy, 409-431.

McNeal, B. L., and Coleman, N. T., 1966. Effect of solution composition on soil hydraulic conductivity, *Soil Sci. Soc. America Proc.,* **30,** 308-313.

Quirk, J. P., and Schofield, R. K., 1955. The effect of electrolyte concentration on soil permeability, *Jour. Soil Sci.,* **6,** 163-178.

U.S. Salinity Laboratory Staff, 1954. Diagnosis and improvement of saline and alkali soils, *U.S. Department of Agriculture Handbook No. 60,* 160p.

Wilcox, L. V., and Durum, W. H., 1967. Quality of irrigation water, *in* R. M. Hagan, ed., *Irrigation of agricultural lands, Agron.,* **11,** 104-122.

Cross-references: *Activity Ratio, Conductivity, Electrical; Conductivity, Hydraulic; Dispersion Phenomena; Electrochemistry; Exchange Phenomena; Flocculation; Irrigation; Leaching; Permeability; Soil Drainage; Soil Solution; Water Content and Retention; Water Movement.*

J

JOURNALS, SOIL SCIENCE

As with other scientific disciplines, the primary results of laboratory studies and field investigations in the subject of soil science are published in diverse periodicals and technical journals, research reports and theses, conference proceedings, and government publications. Soil science literature has been expanding steadily and now comprises some 4000 items per year, increasing by 50 to 80 items per year.

Only a part of the literature is concentrated in those journals specific to soil science (Table 1); a large portion is dispersed in the general geological, chemical, engineering, and especially agronomical literature. To keep in touch with the developments is possible mainly through abstract journals, the most comprehensive of which is *Soils and Fertilizers* (Commonwealth Bureau of Soils). Smaller sections on soils appear in all major abstract journals.

Primary Literature

Historically the first soil science journal was the Russian language *Pochvovedenie,* which commenced publication in 1899. Now published monthly by the Soviet Academy of Sciences (volumes not numbered), it consists of some 20 articles and 150 pages per issue. A cover-to-cover English translation, sponsored by the Soil Science Society of America and subsidized by the American National Academy of Science, was started in 1958. Now produced commercially by Scripta Publications Corporation under the name *Soviet Soil Science,* it contains articles selected from *Pochvovedenie* and other journals.

The American monthly *Soil Science* was founded in 1916 by Jacob G. Lipman at Rutgers University. It has an international character and has published many of the most significant soil science papers. Published in two volumes each year by The Williams & Wilkins Company of Baltimore, it has reached volume 126 (1978) and continues to be edited and partially subsidized by Rutgers University. *Soil Science Society of America Journal,* published bimonthly, is the major and most comprehensive American soil science journal. With an agronomic bias, it has especially strong sections in soil physics and soil chemistry. It was started in 1936 as the organ of the Soil Science Society of America. The British *Journal of Soil Science* (Oxford University Press), which commenced publication in 1950 and is now issued quarterly, has attracted a relatively greater proportion of soil genetic and soil geomorphic studies than the other soil journals. A recent significant addition is *Geoderma* (Elsevier, since 1967). Published in two volumes each year, *Geoderma* has an international editorial board. It covers the different field of pedology and has a current literature section.

Australia, Canada, Belgium, France, Poland, and Rumania publish soil science journals sponsored by their national research organizations or societies. In some other countries mainly the applied or practical aspects of soil use receive attention in joint journals with agronomy and plant nutrition. In all, some 800 periodicals contain occasional soil science articles.

The small number of more specialized and international soil journals includes *Géotechnique* (Institution of Civil Engineers, London, since 1951), which is an international journal of soil mechanics; *Soil Biology and Biochemistry* (Pergamon, since 1969), which deals with the biochemical interactions of soil organisms with the soil environment; and the recently started *Catena* (1973), an interdisciplinary journal of pedology-hydrology-geomorphology (Catena Verlag, W. Germany) that caters especially to the longer, more comprehensive interdisciplinary articles.

The International Society of Soil Science struggled for some time to publish its own journal (*Soil Research,* 1928-1938), but has now settled to publish semiannually an information *Bulletin of the ISSS* for its membership, which is greater than 4000. Transactions of the *International Congresses of Soil Science* (1st Congress 1927, Washington, D.C.; 11th Congress, 1978, Edmonton) are usually publishd in several volumes, the recent ones containing some 400 papers.

Of major importance and impact are the monographs published as proceedings of specialized symposia or conferences, organized by various national institutions or by commissions of the International Society of Soil Science. They include the *Transactions of the International Soil Conference in New Zealand* (Joint

TABLE 1. Major Soil Science Journals (1974–1975).

	Country	First Issue	Nos/Vols per Year	Articles per Year	Pages per Year	Copies Issued
Soil Science Society of America Journal	U.S.	1936	6/1	200	1000	6000
Soil Science	U.S.	1916	12/2	130	900	3500
Geoderma	Internat.	1967	8/2	60	700	900
Journal of Soil Science	U.K.	1950	4/1	50	500	2000
Pédologie	Belgium	1951	3/1	20	300	
Soviet Soil Science (translations)	U.S.	1958	6/1	90	750	
Canadian Journal of Soil Science	Canada	1957	4/1	60	500	1300
Australian Journal of Soil Research	Australia	1963	2/1	20	200	1200
Soils and Fertilizers	U.K.	1938	12/1	5700 (abstracts)	500	2000

Meeting Comm. IV and V, 1962); *Soil Micromorphology* (ed. A. Jongerius, Elsevier, 1964); *Experimental Pedology* (ed. E. G. Hallsworth or D. V. Crawford, Butterworths, 1965); *Soil Clay Mineralogy* (ed. C. E. Rich and G. W. Kunze, University of North Carolina Press, 1964); *Land Evaluation* (ed. G. A. Stewart, Macmillan, 1968); *Paleopedology* (ed. D. H. Yaalon, Israel Universities Press, 1971); *Gley and Pseudogley* (ed. E. Schlichting and U. Schwertmann, Verlag Chemie, 1973); and *Soil Microscopy* (ed. G. K. Rutherford, Limestone Press, Kingston, 1974). Occasionally special journal numbers are devoted to a specific topic or a symposium.

Review Journals and Articles

Of considerable value and help in keeping abreast of the developments in soil science are review articles that assess critically the primary literature. There are no journals devoted solely to soil science reviews, but *Advances in Agronomy* (Academic Press) and *Earth Science Reviews* (Elsevier) carry reviews on specific soil subjects. Sporadically some of the soil journals also carry review articles.

Abstracting Services

Several abstracting services are available to start a search for information on some special topic. The most comprehensive service for abstracting soil science publications is that of *Soils and Fertilizers* (Commonwealth Bureau of Soils, Rothamsted, Harpenden), which now publishes some 5700 abstracts per year, about 60% of which are related to soils proper; the rest deal with plant nutrition and fertilizers. The bureau also compiles from its files *Annotated Bibliographies* (available separately) on selected topics covering a period of years. *Chemical Abstracts, Biological Abstracts, Bibliography and Index of Geology*, and *Geo-Abstracts A—Landforms and the Quaternary*, all carry special sections on soils but publish less than 1000 soil abstracts per year.

D. H. YAALON

References

Jacks, G. V., 1966. The literature of soil science, *Soils Fert.*, **29**, 227–230.

Jenkins, D. A., and Tully, R. I. J., 1973. Soil Science, *in* D. N. Wood, ed., *Use of Earth Sciences Literature*. London: Butterworths, 384–399.

Yaalon, D. H., 1964. Has soil research national characteristics?, *Soils Fert.*, **27**, 89–93.

Zonn, S. V., 1974. 75th anniversary of the journal *Pochvovedenie, Pochvovedenie,* **12**, 6–19 (in Russian).

L

LAW OF THE MINIMUM

In the mid-1800s the pioneer agricultural chemist Justus von Liebig developed a general model to describe the response of plants to applied nutrients. Liebig's concepts, which have come to be known as the *Law of the Minimum,* were summarized by him in 1843 as follows: "The crops on a field diminish or increase in exact proportion to the diminution or increase of the mineral substances conveyed to it in manure . . . By the deficiency or absence of one necessary constituent, all the others being present, the soil is rendered barren for all those crops to the life of which that one constituent is indispensable" (quoted in Russell, 1961).

In 1905 Blackman, a British plant physiologist, paraphrased Liebig's statements into the *Principle of Limiting Factors,* expressed as follows: "When a process is conditioned as to its rapidity by a number of separate factors, the rate of the process is limited by the pace of the 'slowest' factor." Swanson (1963) provided the following concise statement of the modern interpretation of Liebig's hypothesis: "Production is assumed to increase at a constant rate with respect to each factor until [unless] some other factor is limiting."

Meyer and Anderson (1952) suggested that essentially all models of plant growth might be construed as variations of the law of the minimum. This is implicit, for example, in their discussion of Mitscherlich's formulation, "The increase in any crop produced by a unit increment of a deficient factor is proportional to the decrement of that factor from the maximum" (Mitscherlich, 1909). The soil-fertility literature has, however, generally considered that Mitscherlich's hypothesis is an example of the so-called *Law of Diminishing Returns,* in contrast to the constant returns implied by Liebig. The literature on photosynthesis tends to present a similar dichotomy between Blackman and his principal critic, Harder (cited in Thomas, 1965). Therefore, the remainder of this discussion will consider the Law of the Minimum in the restricted sense of Swanson's definition.

Graphically, the Law of the Minimum is usually expressed as shown in Figure 1. More generally, the concept can be shown as in Figure 2, where each factor either limits or has its response limited by another factor. If, for example, factor *B* is present in sufficient quantity to produce a yield of only 2, there can be no response to factors *A* and *C.* If factor *B* is provided in amounts greater than necessary to reach yield 2, it will produce no further response, unless factor *A* is provided, etc.

An analogy is that of the incomplete barrel, as shown in Figure 3. Here the holding capacity is limited by the length of the shortest stave. If the length of stave *A* is increased to that of *C,* first *B* and then *D* will limit capacity; but if the length of these is increased to that of *F,* then the length of *A* will again become limiting, as will that of *C,* and later *E.*

Mathematically, the law of the minimum can be stated as follows:

$$\text{Yield} = \text{minimum value of: } a_i + b_i \text{ factor } i \text{ or} \ldots a_n + b_n \text{ factor } n$$

or

$$\frac{\Sigma(a_n + b_n \text{ factor } n^*)}{N}$$

where * is the maximum required of factor *n.*

An example is that of a typical response of potatoes to *N,* P_2O_5, and K_2O in Colombia. The individual response equations are given below where the number in parentheses represents the yield "plateau" determined by uncontrolled factors such as climate, genetic potential, and management. This "maximum" is also shown in Equation 4. (The amounts of each nutrient estimated as necessary to achieve this maximum yield are obtained by subtracting the intercept from the maximum and dividing the result by the linear coefficient of the nutrient, i.e., by its "slope.")

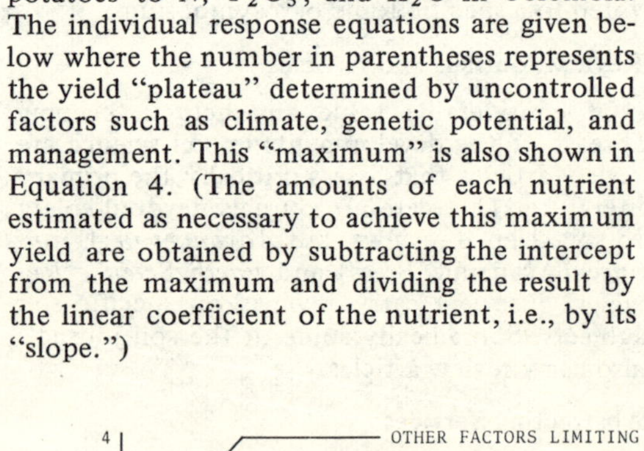

FIGURE 1. Condensed Liebig graph.

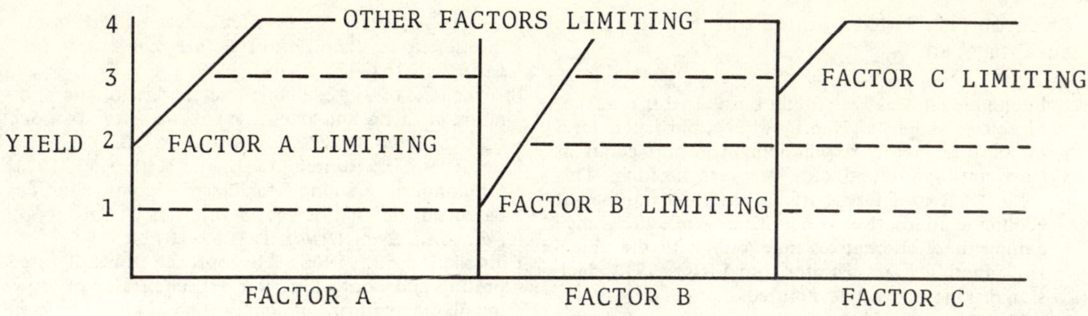

FIGURE 2. Expanded Liebig graph.

$$\text{Yield} = 19{,}200 + 70.0N \; (26{,}200) \quad (1)$$
$$\text{Yield} = 11{,}500 + 56.3P_2O_5 \; (26{,}200) \quad (2)$$
$$\text{Yield} = 24{,}100 + 41.3K_2O \; (26{,}200) \quad (3)$$
$$\text{Yield} = 26{,}200 \text{ (max)(with } 100N, \; 261P_2O_5, \text{ and } 51K_2O) \quad (4)$$

Solving each individual equation for yield, based on the nutrients indicated in the treatments columns of Table 1, one obtains the minimum (i.e., predicted) value for each treatment, which are underlined in the table.

An equivalent method has been presented by Swanson (1963) for linear programming terminology. In his system the predicted value for the "no-treatment" yield level is the minimum of:

$$Y = \frac{b_1}{a_1}$$
$$Y = \frac{b_2}{a_2}$$
$$Y = \frac{b_3}{a_3}$$
$$Y = \max$$

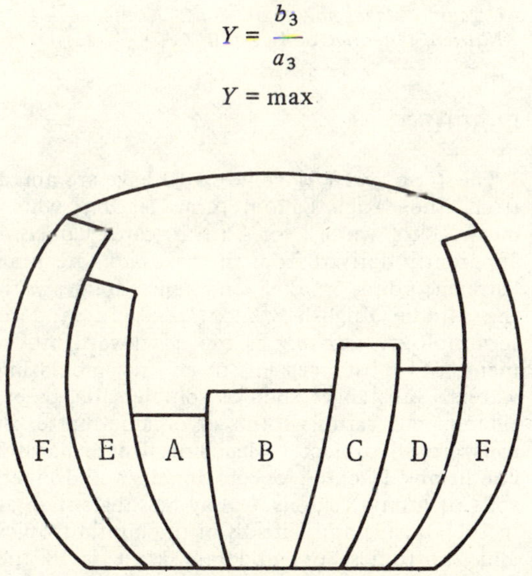

FIGURE 3. Barrel analogy of the law of the minimum.

where b_i is the amount of factor i available in the soil, and a_i is the requirement of the crop for factor i per unit of yield.

An approximate manual, i.e., graphical, fitting of the Liebig model is relatively simple (Boyd, 1970), but a rigorous least-squares fit is rather complicated. This may account for the unpopularity of the Liebig model in today's computer culture. However, a computer subprogram for this purpose will soon be incorporated in SAS (Barr and Goodnight, 1972). The Liebig model equals, and often surpasses, more common models in terms of predictability (R^2). More important, perhaps, are the indications that the Liebig model:

1. Results in recommendations substantially less than those indicated by curvilinear models.
2. Results in recommendations closer to observed farmer practice.
3. Results in recommendations that are more consistent, i.e., exhibit less variance.

These preliminary conclusions, it should be stressed, are based on graphical interpretations of individual experiments in homogeneous areas. The averaging of experiments from dissimilar sites tends, however, to result in curvilinear patterns.

The preceding recommendations are optimum in the physical sense, i.e., to achieve maxima. In addition, the law of the minimum has distinct advantages from the point of view of (a) economics, (b) statistics, and (c) extension, as

TABLE 1. Example of Liebig Calculations.

Treatments						
N	P_2O_5	K_2O	Eq. 1	Eq. 2	Eq. 3	Eq. 4
0	150	25	<u>19,200</u>	19,945	25,132	26,200
60	0	40	23,400	<u>11,500</u>	25,752	26,200
90	200	0	25,500	25,575	<u>24,100</u>	26,200
120	300	75	27,600	28,390	27,198	<u>26,200</u>

Swanson (1963) has pointed out. Briefly, these advantages are:

a. Economic choices are strictly limited–if it is advantageous to apply a given input at a particular level of another factor, it will remain advantageous at least until another factor becomes limiting. This means that in a three-variable case, the number of economic alternatives is only four. While the choice among these alternatives may vary with the objective function (Cate, Hunter, and Fitts, 1971), decision making is greatly simplified.
b. Statistical experimental design can emphasize precision with respect to single factors if other factors are maintained at a sufficient level (determined by other experimental results). Since interactions are considered either absolute or negligible, the complexities of factorial experiments can be minimized.
c. The teaching task is greatly simplified since farmers can be expected to readily absorb and integrate the simplified model and its interpretations.

There is a fundamental body of research data in the plant physiology literature that may provide a biological justification for the Liebig model. The original work was done by Specht (1960), who found that the coefficient of variation of plant-tissue analyses sharply decreases just before and just after, respectively, adequate or excessive amounts of a given nutrient are applied. Specht postulated that plants have a self-governing mechanism, which, in an almost teleological fashion, concentrates the organism's efforts on absorbing or eliminating the nutrient most limiting to growth. This would explain the Liebig model discontinuities. Friis-Nielson (1966) has corroborated Specht's findings. A corollary of this work is that the Liebig model should, perhaps, take into account the possibility of negative responses, as well as the chance of Steenbjerg's (1952) sigmoid curves.

In a lighter vein, the Law of the Minimum might be summarized in terms of maximization of consumer satisfaction. For example, if one has two beers and no cigarettes, one's joy is limited by the lack of nicotine. Conversely, if one has two cigarettes but no beer, one needs beer. At some point, however, satisfaction (while still dependent on more beer and cigarettes in the proper proportion) can only be further increased by the addition of another variable, e.g., pizza.

ROBERT B. CATE

References

Barr, A. J., and Goodnight, J. H., 1972. *A User's Guide to the Statistical Analysis System.* Raleigh: North Carolina State University, 260p.

Blackman, F. F., 1905. Optima and limiting factors, *Ann. Botany (London),* **19,** 281–295.

Boyd, D. A., 1970. Some recent ideas on fertilizer response curves, *Proc. 9th Cong. Internat. Potash Inst. (Antibes),* 461–473.

Browne, C. A., 1942. Liebig and the law of the minimum, *in* Liebig and after Liebig. *Am. Assn. Adv. Sci. (AAAS) Publ. No. 16,* F. R. Moulton, ed.

Cate, R. B., Jr.; Hunter, A. H.; and Fitts, J. W., 1971. Economically sound fertilizer recommendations based on soil analyses, *Proc. Internat. Symp. on Soil Fert. Eval. (New Delhi),* **1,** 1065–1071.

Friis-Nielson, B., 1966. An approach towards interpreting and controlling the nutrient status of growing plants by means of chemical plant analyses, *Plant Soil,* **24,** 63–80.

Meyer, B. S., and Anderson, D. B., 1952. *Plant Physiology.* New York: Van Nostrand, 612–613.

Mitscherlich, E. A., 1909. Des gesetz des minimums und das gesetz des abnehmenden bodenertrags, *Landw. Jahrb.,* **38,** 537–552.

Russell, E. J., 1961. *Soil Conditions and Plant Growth.* London: Longmans, 688p.

Specht, A. W., 1960. Evidence for, and some implications of a variation-control mechanism in plant composition, *Soil Sci.,* **89,** 83–91.

Steenbjerg, F., 1952. Verification of chemical soil and plant analyses: General considerations, *Trans. Internat. Soil Sci. (Dublin),* **1,** 309–322.

Swanson, E. R., 1963. The static theory of the firm and three laws of plant growth, *Soil Sci.,* **95,** 338–343.

Thomas, M. D., 1965. Photosynthesis (carbon assimilation): Environmental and metabolic relationships, *in* F. C. Steward, ed., 9–202. Vol. IVA *in Plant Physiology, A Treatise.*

Waugh, D.; Cate, R. B., Jr.; and Nelson, L. A., 1973. Discontinuous models for rapid correlation, interpretation, and utilization of soil analysis and fertilizer response data. North Carolina State University. Technical Bulletin No. 7, 77p.

Cross-references: *Fertilizers, Inorganic; Fertilizers, Organic; Management of Soils; Nutrient Losses; Nutrient Potential; Soil Fertility.*

LEACHING

The term *leach* is believed to have stemmed from the Anglo-Saxon term *leccan,* which means "to water" or "to irrigate." *Leccan* apparently derived from the base *lacu,* meaning "stream, pond, or lake," and hence merges with the Middle English *lak* or *lakke.* In modern terminology, *leaching* is the downward movement of substances dissolved in percolating waters. Substances such as soluble salts, bases, silicon, and various forms of organic matter in soils are all subject to leaching. Under natural conditions leaching occurs in all well-drained soils of humid regions. It may be substantial, as in the Oxisols and Ultisols of the humid tropics and subtropics, or moderate as it is in the Spodosols and Alfisols of cool-temperature and cool regions (Simonson, 1970). The Oxisols and Ultisols, subject to thorough leaching for long periods of time, are naturally strongly acid and mineral deficient, i.e., low in bases and other nutrient elements. General world distribution of Oxisols and Ultisols, which indicates regions of maximum soil leaching, is shown on the endpapers of this volume. In contrast to the Oxisols, Ultisols, and other soils of humid regions, those of arid and semiarid regions (Aridisols and others) are subject to slight leaching or in extreme cases, to none at all. Some soils in arid regions are therefore high enough in soluble salts so that artificial leaching by irrigation waters becomes a necessary first step in agricultural production.

Measurement and Evaluation

The extent to which a soil profile is leached depends on the climatic regime, vegetative cover, soil texture (see *Particle-size Distribution*) and ionic mobilities of the elements. Leaching losses indirectly influence soil properties such as the degree of base saturation (q.v.), soil reaction (q.v.), biological activity (see *Metabolic Activity*), and nutrient availability (see *Nutrient Potential; Soil Fertility*). These soil properties themselves may in turn be used to infer the degree of leaching. Low base saturation, an acid soil reaction (pH less than 5.0), and low levels of biological activity and nutrient availability are, for example, characteristic of many unproductive, highly leached soils.

Estimates of leaching losses have been made from the composition of drainage waters (Johnson et al, 1968) but are more directly measured by controlled lysimeter studies (Joffe, 1940; Stauffer and Rust, 1954; Harrold and Dreibelbis, 1958). The quantity and composition of leachate waters from lysimeters (q.v.) have been used in studies of nutrient cycles, losses of soluble fertilizers, and concentrations of nutrients and pollutants in water as it leaves the soil.

Knowledge of leaching processes has been important in a number of disciplines and is now becoming important generally. Because of the effects of leaching on the nutrient status of soils (see *Nutrient Losses*), it bears directly on crop production (see *Agronomy*) and is thus of direct importance to agronomists. Pedologists study leaching because of its importance to differentiation of horizons during soil genesis (Thorp, Strong, and Gamble, 1957). The time required for the development of soil horizons has been estimated from removal of carbonates by leaching. Iron and silicon can be leached out of soils and migrate laterally for great distances before they precipitate, sometimes as massive crusts (Stephens, 1971). Exploration geochemists are interested in the mobility of elements and the nature of the leaching processes because of the applications of such information to locating mineral deposits (see vol. XIII: *Exploration Geochemistry*). Geochemists also need to know whether assay values reflect enhanced or depleted concentrations. Anomalously high concentrations of elements are of special importance because they often indicate mineral deposits (Levinson, 1974). Elemental dispersion, related to both leaching and lateral migration, is thus of immediate and practical interest to geochemists. More and more people are becoming concerned about land disposal of wastes and the possibility of their leaching from soils into streams and domestic water supplies. The mobility of industrial and agricultural pollutants in soils is another matter of growing concern, closely related to leaching processes.

Leaching Mechanisms

The full set of processes in leaching of substances within and from soils is complex. Some substances dissolve readily in water, e.g., salts such as sodium sulfate. Solution can then be followed by leaching in percolating water. Other substances such as the alkali and alkaline earth elements in minerals and in exchangeable form are mobilized by hydrolysis due to dilute acids such as the carbonic acid formed when carbon dioxide is dissolved in water. Bicarbonates resulting from such reactions are relatively soluble. During such hydrolysis, silicon can also be released and leached, largely in the form of silicic acid (McKeague and Cline, 1963).

More complex reactions occur in the leaching of elements such as aluminum and iron. As organic matter decays in soils, partial decomposition products have acid radicals or functional groups capable of forming chelates and complexes with aluminum, iron, and other elements (Atkinson and Wright, 1957). As a rule, iron is also reduced to the ferrous form before it is chelated or complexed. Once in the form of a chelate or complex, aluminum and iron are both subject to leaching although not to as large a degree as the alkali and alkaline earth elements or silicon.

Leaching involves several complex sequential processes. Through *soluviation*, ordinary dissolution in water of elements from minerals, and *cheluviation*, chelation or complexing by organic molecules, elements are prepared for transport (Atkinson and Wright, 1957). After elements have been dissolved, they can be leached completely out of the soil or translocated downward within the profile. Transfer of substances down and out of upper horizons is known as *eluviation;* it covers substances moving in both solution and suspension. The accumulation of substances by downward transfer from above in a soil profile is called *illuviation* (Birkeland, 1974).

Leaching mechanisms involve many variables. Their complexity can be indicated by a partial list of important factors: (1) nature of the soil medium, (2) presence and effects of microorganisms, (3) formation of complex ions, (4) membrane effects and exchange phenomena associated with silicate clay minerals, (5) carbon dioxide pressure, and (6) porosity and hydraulic conductivity of the soil. These variables and others govern the relative mobilities of elements in soils, i.e., whether they will move, how far, and in what direction (downward or laterally). Leaching is not the same under acid, neutral, and alkaline conditions, nor is it under oxidizing versus reducing conditions (Andrews-Jones, 1968). The stabilities of minerals (Garrels and Christ, 1965) in the different environments provided by different soils also affect leaching mechanisms.

Acid-leaching Medium. Leaching under acid conditions occurs in oxidized soils when pH ranges from about 4.0 to 6.5, and when Eh ranges from +0.4 to +0.7. Under those conditions, the soil solution (q.v.) tends to be enriched in carbon dioxide and also to contain other products of the decay of organic matter. Sodium and calcium are the first elements to be leached, followed closely by magnesium and then by potassium. Humus and silicon are also leached to some extent, whether large or small, whereas aluminum and iron are generally removed to only small degrees.

Acid leaching is most effective in well-drained soils of the humid tropics where decay of the large quantitites of organic residues added by the vegetation produces large amounts of dioxide plus various organic acid radicals. The soils are largely leached of monovalent and divalent cations. Silicon released by weathering of silicates is also lost, whereas oxides of aluminum and iron tend to accumulate as residues. Some share of those elements is also lost by leaching but far less than of more soluble ones. Humid, cool, and temperate regions–in contrast to the tropics–are marked by slower and less complete leaching of all elements.

Neutral-leaching Medium. Soils of semiarid and arid regions are mostly neutral rather than acid or alkaline in reaction (range in pH between about 6.5 and 8.0). The exchange complex then carries divalent cations (calcium and magnesium), for the most part, with little hydrogen or sodium. Leaching removes some of the alkaline earth (calcium and magnesium) and alkali (sodium and potassium) elements from the upper horizons of soils. Those are largely redeposited in deeper horizons in the forms of carbonates and sulfates, the latter usually below the former. A small share of the leached alkaline earth and alkali elements is removed completely from the soil. Leaching of aluminum, iron, humus, and silicon is negligible, on the whole, although both humus and silicon can be transferred downward under favorable circumstances. The primary example of soils with leaching in a neutral medium are the Mollisols of subhumid, cool-temperate grasslands.

Alkaline-leaching Medium. Soils with alkaline reactions (pH range of about 8.0 to 9.5) are largely, but not entirely, restricted to arid regions where leaching itself is also restricted. A few soils in semiarid, subhumid, and even humid, regions have alkaline *B* horizons because of high proportions of exchangeable sodium. Such soils are minor components of the patterns in the regions of their occurrence. Under alkaline conditions, humus and silicon dissolve readily and can be leached if enough water is available. Aluminum may also form the rather soluble sodium aluminate under strongly alkaline conditions. On the whole, however, leaching is limited in soils with alkaline reactions. Soluble salts such as sodium sulfate and less soluble ones such as calcium sulfate persist in a number of the soils, either throughout the profiles or in deeper horizons. The least leached soils contain minerals such as halite and glauberite (Perel'man, 1967). The soils and soil horizons with alkaline reactions due to exchangeable sodium and without soluble salts are commonly marked by corrosion of quartz and feldspar grains plus some leaching of humus, silicon, and aluminum. All in all, however, the sum total of leaching under alkaline conditions is small compared to those under acid and neutral conditions, especially the former.

Reduced-leaching Medium. Reducing conditions prevail in soils when the Eh value is less than +0.2 V and the pH is near 7.0. Such conditions can occur in waterlogged soils (see Vol. XII, Pt. 2: *Gleys, Gleization*) that lack dissolved oxygen and contain reducing agents. Anaerobic bacteria thrive in such soils, obtaining their oxygen by breaking down various substances, chiefly forms of organic matter. Carbon dioxide, methane, and organic acids are released as waste products, normally dissolving in soil water. Both iron and manganese are reduced to the divalent state and then become highly mobile while reducing conditions prevail. Both elements may be concentrated in mottles or concretions or lost from the soil completely. For the most part, one or two horizons rather than the entire profile are affected by reducing conditions, and those are called "gleyed horizons." In addition to iron and manganese, gleyed horizons may lose nitrogen, phosphorus, calcium, and carbonates. Some silicon, sodium, and potassium are also lost. Perel'man (1967) lists the order of loss of elements from gleyed horizons as follows:

$$Ca < Fe < P < Cl < K < Si < Na$$

Depletion of bases and carbonates from the solum has progressed far enough in some wet soils so that half the exchange capacity is occupied by aluminum (McKeague, Day, and Clayton, 1973).

Leaching Losses

The great majority of soils lose substances through leaching, and quantities lost range from very large to very small. The magnitude of leaching losses depends on a combination of time and intensity. Other things being roughly equal, leaching losses during the same time interval are greater for soils of humid-tropical than humid temperate regions. Intensity of leaching is greater in the humid tropics than elsewhere in the world. Greater intensity can be offset by a longer period of leaching, however, and some soils of temperate regions are therefore strongly leached. Moreover, some soils in the humid tropics did not form long enough ago to permit thorough leaching. In contrast to soils of humid regions, those of semiarid and arid regions are not strongly leached and many are only slightly leached. Losses of chlorides and sulfates, bicarbonates, nitrogen, and organic matter (see *Nutrient Losses*) are of primary concern because their presence or absence is indicative of nutrient status or can have major effects on the management required and on soil productivity.

Losses of Chlorides and Sulfates. In humid climates, salts such as sodium chloride, sodium sulfate, and calcium sulfate are leached from the soil profile into streams or groundwaters (Figure 1). In arid regions, on the other hand, the highly soluble sodium chloride and sulfate remain in the profile, moving down only with percolating water as far as it goes. Deeper profiles thus commonly accumulate several parts per thousand of soluble salts. Calcium sulfate, with a solubility of only 2 parts per thousand, is less mobile than sodium chloride or sulfate and thus may occur as crystalline gypsum closer to the soil surface. These salts are neutral, and their removal does not affect pH.

Loss of Bicarbonates. The bicarbonate ion is usually associated with the cations, calcium, magnesium, potassium, and sodium released by the weathering of organic and inorganic compounds. In more humid regions, calcium and sodium are lost rapidly and completely, whereas magnesium and potassium are retained in clay minerals. The progressive loss of these bicarbonates, which act as bases, leads to acidification of soils (see *Acidity; Soil Reaction*). Under arid conditions, however, calcium bicarbonate is not completely leached from the soil profile but is moved only to the depth of penetration of percolating water (Figure 1). Because of the limited partial pressure of carbon dioxide and the low solubility of calcium carbonate in an alkaline medium, calcium carbonate forms from the bicarbonate and precipitates in the deeper profile. Precipitated carbonates take a variety of forms such as finely divided powder, pseudomycelia, soft or hard nodules of many sizes, and continuous cemented layers known widely as *calcrete, caliche* (q.v. in vol. VIA) and less widely as *petrocalcic horizons.*

Loss of Nitrogen. Although nitrates constitute but a small part of the nitrogen in a soil, they form a major share of that lost through leaching (Allison, 1955). Other forms such as ammonia (NH_3) are not easily leached because they are either absorbed by the exchange complex, held within minerals, or form integral constituents of the organic matter, which contains by far the greatest share of the nitrogen in soils (Stout and Burau, 1967).

The magnitude of losses of nitrogen through leaching is governed by factors such as the amount and distribution of rainfall, soil temperature, particle-size distribution, and consequent hydraulic conductivity of individual horizons in profiles, contents of organic matter, cropping systems, and tillage practices. Even without cropping, some loss occurs. Thus, for example, records kept at the Rothamsted Experimental Station in England from 1870 to 1915 for a group of unfertilized and uncropped plots indicate that leaching carried out 30 kg/ha/yr, slightly less than the additions (Allison, 1955). More recently Russell (1973) reported that as much as 85% of the nitrate added in combination with glucose to a soil at Rothamsted could be lost by leaching in 5 days. General indications are that rates of loss are lower under acid (pH 4.8) than under slightly alkaline (pH 8.0–8.6) conditions. Rates of loss are also lower from cold than warm soils. Leaching out of nitrates also means losses of bases such as calcium and magnesium with resulting more acid soil conditions as time passes (National Research Council, 1972).

Loss of Organic Matter. Some of the intermediate products in the complex processes of conversion of organic residues in soils to carbon dioxide, water, and mineral salts or to humus are soluble enough to be subject to leaching. Short-distance leaching of substances such as

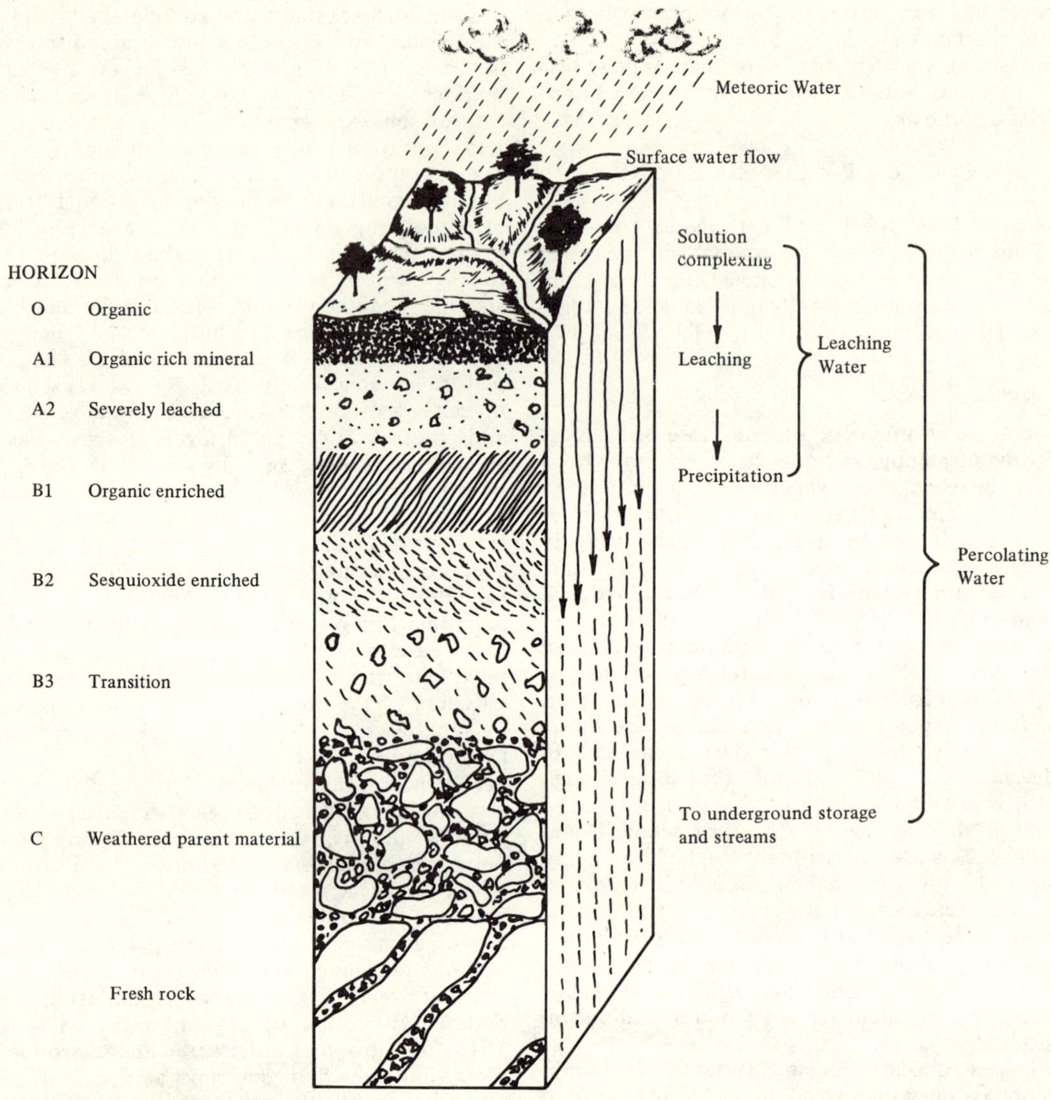

FIGURE 1. A generalized landscape-Spodosol profile. The notes on the left side of the diagram suggest a dominant characteristic for each horizon. The right or "wet" side of the diagram indicates water movement through the profile and the horizons subject to leaching and precipitation. Some leaching affects the entire profile.

polyphenols and fulvic acids is of prime importance in the formation of *A* and *B* horizons in Spodosols or Podzols (Atkinson and Wright, 1957). The organic substances are leached from the *A* into the *B* horizons. Evidence that such substances can be carried out of sandy Spodosols has been obtained in eastern North Carolina (Holshey, Daniels, and Gamble, 1975), where "black-water" streams and swamps are eventual sinks for leached organic matter. Leaching of organic substances occurs in soils of humid regions generally, indicated by coatings on natural partings to depths of two meters or more. Leaching of organic matter is far more limited in soils of semiarid regions and seems to be negligible in soils of arid regions. Primary losses of organic matter from soils, however, are through decay.

Leaching-Reclamation

Artificial leaching is part of the reclamation procedure for most saline, saline-sodic, and sodic soils. Some saline soils can be reclaimed by simply leaching out the soluble salts. Such salts must be neutral and in part be compounds of calcium and magnesium. Irrigation water of suitable quality can then be added to flush out the salts, after which the soils will be suitable for cultivation. Flushing of salts also requires that adequate drainage be provided. The amount of water required for artificial leaching depends on the quantity of salts in the soil, their composition, and the ease with which water can move through the soil.

One effort to summarize in a single statement the effects of the several physical properties of soil on artificial leaching is called the *leaching efficiency*. This is defined as the ratio between (1) the amount of irrigation water necessary for reduction of the ionic concentration of the soil solution to a given level and (2) the amount of water the soil would hold at field capacity (Buras, 1974).

Leaching out of salts will not of itself reclaim saline-sodic and sodic soils. While the soluble salts are present in those soils, their physical properties are favorable. The soils remain flocculated and crumbly, except for sodic soils with small amounts of soluble salts. If the salts are removed, the amounts of exchangeable sodium present in saline-sodic and sodic soils are commonly high enough to cause dispersion, severe stickiness, and other highly undesirable characteristics. Replacement of the exchangeable sodium is necessary in addition to removal of the soluble salts for reclamation. Sodium can be replaced by calcium or aluminum if concentrations in irrigation waters are high enough and if pH of the soil is below 9.0. Formulas have been developed by the U.S. Salinity Laboratory to predict the amounts of exchangeable sodium that can be removed by leaching with water of a given quality (Richards, 1954).

Leaching can thus serve a useful purpose in the reclamation of saline and saline-sodic soils. The success of the practice of artificial leaching for reclamation depends on the amounts and composition of the salts present, the general nature of the soil and its constituents, the quality of available irrigation water, and the possibilities of providing drainage to permit flushing out of salts.

CHARLES W. FINKL, JNR.

References

Allison, F. E., 1955. The enigma of soil nitrogen balance sheets. *Adv. Agron.*, 7, 213–250.

Andrews-Jones, D. A., 1968. The application of geochemical techniques to mineral exploration, *Colo. Sch. Mines Mineral Indust. Bull.*, **11**, 31p.

Atkinson, H. J., and Wright, J. R., 1957. Chelation and the vertical movement of soil constituents, *Soil. Sci.*, **84**, 1–11.

Birkeland, P. W., 1974. *Pedology, Weathering, and Geomorphological Research.* New York: Oxford, 285p.

Buras, N., 1974. Water management systems, *in* J. Van Schilfgaarde, ed., *Drainage for Agriculture.* Madison: American Society of Agronomy, 675–694.

Garrels, R. M., and Christ, C. L., 1965. *Solutions, Minerals, and Equilibria.* New York: Harper and Row, 450p.

Harrold, L. L., and Dreibelbis, F. R., 1958. Evaluation of agricultural hydrology by monolith lysimeters (1944–1955), *U.S. Dept. Agric. Tech. Bull. 1179*, 166p.

Holshey, C. S.; Daniels, R. B.; and Gamble, E. E., 1975. Thick Bh horizons in the North Carolina Coastal Plain: II. Physical and chemical properties and rates of organic additions from surface sources, *Soil Sci. Soc. America Proc.*, **39**, 1182–1187.

Joffe, J. S., 1940. Lysimeter studies: IV. Movement of anions through the profile of a gray-brown podzolic soil, *Soil Sci.*, **50**, 57–63.

Johnson, N. M.; Likens, G. E.; Bormann, F. H.; and Pierce, R. S., 1968. Rate of chemical weathering of silicate minerals in New Hampshire, *Geochim. et Cosmochim. Acta*, **32**, 531–541.

Levinson, A. A., 1974. *Introduction to Exploration Geochemistry.* Calgary: Applied, 612p.

McKeague, J. A., and Cline, M. G., 1963. Silica in soils, *Adv. Agron.*, **15**, 339–396.

McKeague, J. A.; Day, J. H.; and Clayton, J. S., 1973. Properties and development of hydromorphic mineral soils in various regions of Canada, *in* E. Schlichting and U. Schwertmann, eds., *Pseudogley and Clay: Genesis and Use of Hydromorphic Soils.* Weinheim: Verlag Chemie, 207–218.

National Research Council, 1972. *Accumulation of Nitrate.* Washington, D.C.: National Academy of Science, 106p.

Perel'man, A. I., 1967. *Geochemistry of Epigenesis.* New York: Plenum, 264p.

Richards, L. A., eds., 1954. Diagnosis and improvement of saline and alkali soils, *U.S. Department of Agriculture Handbook 60*, 160p.

Russell, E. W., 1973. *Soil Conditions and Plant Growth.* London: Longmans, 849p.

Simonson, Roy W., 1970. Loss of nutrient elements during soil formation, *in* O. P. Engelstad, ed., *Nutrient Mobility in Soils: Accumulation and Losses.* Madison: Soil Science Society of America, 21–45.

Stauffer, R. S., and Rust, R. H., 1954. Leaching losses, runoff, and percolate from eight Illinois soils, *Agron. Jour.*, **46**, 207–211.

Stephens, C. G., 1971. Laterite and silcrete in Australia: A study of the genetic relationships of laterite and silcrete and their component materials, and their collective significance in the weathered mantle, soils, relief and drainage of the Australian continent, *Geoderma*, **5**, 5–52.

Stout, P. R., and Burau, R. G., 1967. The extent and significance of fertilizer buildup in soils as revealed by vertical distributions of nitrogenous matter between soils and underlying water reservoirs, *in* N. C. Brady, ed., *Agriculture and the Quality of Our Environment,* American Association for the Advancement of Science Publ. No. 85, 283–310.

Thorp, J.; Strong, L. E.; Gamble, E., 1957. Experiments in soil genesis—The role of leaching, *Soil Sci. Soc. America Proc.*, **21**, 99–102.

Cross-references: *Infiltration; Irrigation; Lysimeters; Management of Soils; Moisture Management; Nutrient Losses; Percolation; Salt Leaching; Soil Drainage; Soil Solution; Water Movement.*

LIME, LIMING

The following discussion of lime and liming is premised on a three-part interrelationship of terms, as follows: (1) *lime,* the basic material used for neutralizing; (2 *lime requirement,* measured amount of soil acidity; and (3) *liming,* preparation and application of such materials. In humid climates permeability of water into the soil from precipitation exceeds the soil's capacity to retain the water. Hence percolation (q.v.) through the soil occurs, leaching out basic cations and leaving behind acid cations. In the course of time, such soils require the reversal of this process for optimum growth of crops. The addition of basic compounds in the form of lime is a normal practice on these soils, and millions of acres are thus treated each year.

Lime

Liming Materials. This group of basic materials is used for neutralizing soil acidity. There is quite a number of liming materials in use, but the one making up by far the largest portion of the total is ground limestone rock. This material, often called *agricultural ground lime,* consists of calcite mineral ($CaCO_3$) plus relatively small amounts of impurities such as clay (shale) and silica. Most liming materials produced from ground limestone also contain amounts of $MgCO_3$ ranging from only traces to as much as the equivalent of the $CaCO_3$ itself. The latter ($MgCO_3 \cdot CaCO_3$) is called *dolomite,* and limestones with significant quantities of $MgCO_3$ are called *dolimitic limestones.* Dolimitic limestones have special significance for application to soils low in exchangeable magnesium.

Ground limestone ($CaCO_3$) is only slightly basic (pH 8.5 in pure water). It is a salt of a weak acid and a relatively weak base, and it is relatively insoluble in water. Hence it dissolves in soils only as it comes in contact with acidic components. It can, however, be heated in a kiln to drive off the CO_2 content to produce so-called *quick lime* or *caustic lime* (CaO):

$$CaCO_3 \xrightarrow{\text{heat}} CaO + CO_2 \uparrow \qquad (1)$$

Quick lime can be hydrated or slaked by addition of water or by exposure to moist air to form *hydrated lime* [$Ca(OH)_2$]:

$$CaO + H_2O \rightarrow Ca(OH)_2 \qquad (2)$$

Both quick lime and hydrated lime are relatively soluble in water. They are usually bagged and sheltered from the atmosphere. Consequently, they are considerably more expensive than ground limestone. Yet they are still used as liming materials to a limited extent because of their relatively high solubility and hence rapid rate of reaction with soil acids.

Other liming materials used in areas where they are available locally are: ground shells (mollusks); marl (loosely consolidated earthy material composed largely of shell fragments); waste or by-product lime from various industries; and agricultural slag (ground or naturally fine fused silicate with residual hydrated lime produced from smelting of iron ore).

Lime Requirement

Definitions. This quantity is the amount of lime required to neutralize an acid soil to a selected pH or to a less toxic condition for plants. Since the amounts and kinds of soil acids formed in a given period of time vary with such factors as parent material, slope, moisture, temperature, and vegetation, lime requirements of acid soils vary widely. Lime requirement is normally expressed as $CaCO_3$ equivalent in kilograms per hectare (or tons per acre) or in milliequivalents per hundred grams of soil.

Normally the basis of selection of the point to which the soil is to be neutralized is the pH for most favorable plant growth. It may, however, be based on other criteria such as inactivation (precipitation) of exchangeable aluminum, and it is sometimes adjusted upward or downward on the basis of economic considerations. There seems to be little economic advantage in most places for liming soils above pH 6 for nonacid sensitive crops such as corn and soybeans, while acid-sensitive legumes such as alfalfa generally respond well to liming to pH 6.5 or greater. In the subtropics and tropics, even lower pH values may be most desirable. Also, peat and muck soils do not generally produce higher yields when limed above pH 5.2. This subject, with appropriate literature citations, was recently considered for mineral soils by McLean (1972).

Lime Requirement as a Capacity Factor. By far the largest fraction of the total acidity or lime requirement of a soil consists of undissociated hydrogen and aluminum ions of various types adsorbed to the negative charges (cation exchange sites). Hence the lime requirement is the capacity factor of soil acidity. In contrast, the very small fraction, i.e., 1/10,000 to as little as 1/100,000, of the total acidity is dissociated hydrogen ions measured and expressed as pH. This is the intensity factor of soil acidity (see *pH, Field Determination*).

The Significance of the Lime Requirement. Maintenance of the appropriate pH or base saturation in a soil by addition of lime at the rate indicated by the lime requirement produces an idealized plant root environment, which generally effects maximum use of applied fertilizers by plants.

An acid soil frequently has concentrations of aluminum, iron, or manganese, which are sufficiently high to be toxic to plants. Any amount of lime added will generally decrease the solubilities and hence the toxicities of these elements. Too much lime added may decrease the availability of iron and manganese sufficiently to cause deficiencies of these plant-essential elements. Hence a method of measuring the lime required for adjusting the soil to some specified pH or base-saturation (q.v.) level (or to a degree of inactivation of toxic substances) is a worthwhile soil-management tool. Since the ionized hydrogen is a relatively small fraction of the total acid present and since this fraction varies greatly from one soil to another, it is evident that pH is generally not a good indication of the lime requirement of a soil.

Factors Affecting Lime Requirement

Total acidity of a soil varies with degree of weathering to which it has been subjected, the number of cation exchange sites (organic matter and clay contents) on which the acid cations formed can be adsorbed, forms of acidity present, and past liming and fertilizer practices. Lime requirement is not appreciably affected by the various factors that alter pH through displacement or dissocation of hydrogen ions.

Degree of Weathering. To the extent that cropping, leaching, and erosion remove basic cations more rapidly than dissolution of weatherable minerals replaces them, a soil tends to become acid. As basic cations are removed and only partially replaced by breakdown of basic minerals, acidic cations accumulate in their places. Primary sources of H^+ are various weathering reactions such as dissociation of carbonic acid formed when carbon dioxide is dissolved in water or oxidation of pyrites to sulfuric acid. Increased time of exposure in humid climates generally causes greater accumulation of soil acidity and hence greater lime requirement.

Type of Parent Material. Rocks or minerals that weather into soil components may be either basic or acidic in nature. Those that are basic are either oxides or carbonates of basic cations, or basic cation-containing aluminosilicates. During the weathering process, basic rocks and minerals release basic cations that are gradually replaced by the acidic cations, H^+ and Al^{3+}, as the leaching processes continue. In contrast, acidic rocks and minerals contain a predominance of quartz or silica and relatively small amounts of basic cations. Hence weathering of these materials releases silicic acids of various kinds and accumulates soil material of very low base status even without, or prior to, leaching of bases from the soil.

Clay Content. The greater the concentration of clay in the soil, the greater the amounts of acidic cations that can be adsorbed to cation exchange sites. Also the more cation exchange sites are present, the more polymerized hydroxy-aluminum ions accumulate between the layers of the clay crystal (Jackson, 1960). Yet less water may percolate through a fine-textured than a coarse-textured soil. Hence a fine-textured soil usually has a higher base saturation than a similarly exposed coarser-textured soil. However, a very old, fine-textured soil may be so highly weathered as to be almost devoid of basic cations and thus almost completely saturated with acidic cations.

Organic Matter Content. The greater the concentration of organic matter in a soil, the greater the tendency for acidic cations to accumulate on exchange sites. It is thus not surprising that content of organic mater and lime requirement are highly correlated in soils (Keeney and Corey, 1963). Organic matter not only adsorbs acidic cations in exchangeable form but also complexes or chelates acidic cations in forms not readily displaced by ordinary exchange reactions (Clark and Nichol, 1966). The Fe^{3+}, Al^{3+}, and hydroxy-aluminum and -iron ion compexes involving the carboxyl groups of the organic matter cause the latter to act as a weaker acid than it otherwise would be.

Forms of Acidity Present. Several forms of acidity account for the total acidity of the soil, not just undissociated H^+. In fact, even though hydrogen ions replace basic cations during acid weathering, they do not accumulate without limit (Jackson, 1960). When a certain concentration builds up in the soil, hydrogen ions react with the clay crystal becoming adsorbed internally and simultaneously releasing Al^{3+}. Removal of the hydrogen ions from solution raises the pH causing partial neutralization of the Al^{3+} ions. Consequently, hydroxy-aluminum ions $Al(OH)^{2+}$ and $Al(OH)_2{}^+$ are formed. The latter ions are not only subject to complexation by organic matter as noted but they also polymerize on clay surfaces as continuous layers or as islands of nonexchangeable acidity (Schwertmann and Jackson, 1964; de Villiers and Jackson, 1967).

The cationic nature (positive charges) of the hydroxy-aluminum neutralizes the cation exchange sites of the clay and organic matter, and their large size precludes their exchange with other cations. Hence a corresponding decrease in the effective cation exchange capacity of an acid soil is noted (de Villiers and Jackson, 1967). Hydroxy-iron evidently plays the same role as hydroxy-aluminum in acid soils, except at a lower pH (Coleman and Thomas, 1964). The various hydroxy-aluminum and -iron polymers or complexes, being only partially neutralized and thus acidic cations, markedly increase the total lime requirement of a soil. Yet their relative inactivity and/or inaccessibility to the buffers used in lime-requirement tests may cause measured lime requirement to fall short of actual requirement. This, however, can be avoided by initial modification of the buffer used in the test so that indicated and actual values are approximately the same (Shoemaker, McLean, and Pratt, 1962).

Many types of OH groups exist in organic matter and clay minerals, which ionize hydrogen from coordinate covalent bonds as the pH is increased by addition of a base such as lime. Charges from which hydrogen ions dissociate are called *pH dependent* because they are effective for adsorbing other cations only at higher pH values. The pH-dependent charges adsorb calcium and magnesium from lime as H^+ is simultaneously neutralized:

(1) $\underset{\text{(Insol.)}}{CaCO_3} + H_2O \rightarrow \underset{\text{(Sol.)}}{CaHCO_3} + OH^-$ (3)

(2) $\boxed{\begin{matrix}\text{Soil} & - \\ \text{particle} & -\end{matrix}}\begin{matrix}H \\ H\end{matrix} + \begin{matrix}CaHCO_3 \\ OH^-\end{matrix} \rightarrow$

$\rightarrow \boxed{\begin{matrix}\text{Soil} & - \\ \text{particle} & -\end{matrix}}\,Ca$

$+ \begin{matrix}H_2CO_3 \\ H_2O\end{matrix} \nearrow \begin{matrix}CO_2 \\ H_2O\end{matrix}$ (4)

Small decrease in pH reverses the process so that hydrogen ions are readsorbed to the hydrogen preferring, pH-dependent charges, and basic cations are simultaneously released into solution where they may be taken up by plants or leached into lower horizons of the profile (McLean, 1970).

Neutralization Sequence. The predominant form of acidic cation in a soil differs with pH. A very acid soil of, say, pH 4.0, might have a small amount of ionic hydrogen (H^+ or H_3O^+) in solution and adsorbed to the exchange sites of the soil. But such a soil would likely have a large amount of Al^{3+} and a relatively small amount of hydroxy-aluminum ions adsorbed to exchange sites. In contrast, another soil of slightly higher pH may have a predominance of hydroxy-aluminum. If lime is added and mixed uniformly with the soil of pH 4.0 in sufficient quantity to increase the pH to 7.0, the various forms of acidic cations would be neutralized sequentially in the following order as the pH increases: H^+, Al^{3+}, $Al(OH)^{2+}$, $Al(OH)_2{}^+$, and other pH-dependnet hydrogen (Schwertmann and Jackson, 1964; Coleman and Thomas, 1967). If lime is not uniformly mixed with soil, microvolumes of soil might go through all the preceding steps in the neutralization process while other soil volumes remain at the initial pH.

Methods of Measuring Lime Requirements

Field Studies of Lime-application Rates. As suggested previously, the total acidity and hence the lime requirement of soils under natural conditions differ depending on the degree of weathering they have undergone and the capacities they have for holding the acidic cations. Before the development of soil tests for determining the lime requirement of individual soils, the amount of lime for a given soil condition and crop, or cropping sequence, was based on field experiments under comparable conditions. Such studies are still important for correlating soil test data with field response. In addition, they are important for separating direct effects on crops of soil base status (pH) from indirect effects of the condition on one crop that is reflected in a subsequent crop. For example, considerable increase in base saturation may have little direct effect on crop yields when adequate fertility including nitrogen is provided. But under conditions where nitrogen is more limiting, the same increase in bases may have a pronounced effect on the amount of nitrogen fixation from the air by legumes such as alfalfa. If so, the fixed nitrogen may markedly affect yields of a subsequent corn crop. Hence the added lime has little direct effect on corn yields, but by subsequently causing more nitrogen fixation by the alfalfa, corn yields are appreciably increased.

Soil-lime Incubations. The need for more rapid methods of determining lime requirements of soils for wider ranges in soil conditions than was provided by field studies lead to the development of the incubation method. This method consists of mixing increments of lime with aliquots of each soil, equilibrating with moistened soil, and measuring soil pH so that the amount of lime required to bring it to a selected pH is determined. Since microbial activity (q.v.) is intense under incubation conditions, salts—particularly nitrates of calcium, magnesium, potassium, etc.—accumulate in the soil. Hence they should be leached out or otherwise taken into account in estimating lime requirement because they displace hydrogen ions into solution lowering the pH more than would be true in field soil where plant uptake and leaching remove most of the generated salts.

Soil-base Titrations. Acidic cations in a soil can be titrated with a base just as any acid is titratable with a base. However, most of the acidity of a soil is not accessible for instantaneous reactions with base. Hence direct titration is slow and somewhat frustrating. Such titrations can be speeded up greatly by containing the soil in a relatively concentrated solution such as 1N KCl. Another type of tritration, which is slower but generally more satisfactory, is: (1) equilibrate aliquots of a given soil with progressively larger increments of a standard base such as $Ca(OH)_2$ and with correspondingly smaller amounts of water added to give a constant volume, and (2) measure soil pH after carbonation, aeration, and standing several days (Bradfield and Allison, 1933).

Soil-buffer Equilibrations. The dilemma of direct titration of soil acidity is how to complete the titration in a reasonable time and yet add the base at a rate sufficiently slow to neutralize it without subjecting the soil to a much higher pH than is encountered in normal liming practice. Thus various buffer-equilibration methods were developed that permit neutralization of the acidity at a relatively low and nearly constant pH. Some of the methods provide for removal of the buffer from the soil, and for its titration with base back to the pH of the buffer prior to its reaction with the soil (Schofield, 1933; Mehlich, 1939). Other methods merely measure the change in pH of a buffer caused by the acids in the soil, this change in buffer pH being a measure of the lime requirement of the soil (Adams and Evans, 1962; Brown, 1943; Shoemaker, McLean, and Pratt, 1962; Woodruff, 1948). The Woodruff, Shoemaker, McLean, and Pratt (SMP) and the Adams and Evans methods are used rather widely in soil-testing laboratories as the bases for recommending lime applications for acid soils.

The Woodruff method as originally constituted (Woodruff, 1948) generally underestimates the lime requirements of soils having lime requirements of 2242 kg/ha (1 T/acre) or more and may indicate less than half the actual lime requirement in soils of very high lime requirement (McLean, Dumford, and Coronel, 1966). A new buffer now used in the Woodruff method indicates considerably more lime than the original buffer, but unpublished data show that it has a tendency to indicate slightly too much in soils of high lime requirement and considerably too much in soils in low lime requirement. The SMP method has been shown to be especially well adapted for soils with a greater than 4482 kg/ha (2 T/acre) lime requirement, having a pH below 5.8, containing <10% organic matter, and having appreciable quantities of soluble aluminum (McLean, Dumford, and Coronel, 1966). The Adams and Evans method is designed to measure comparatively small magnitudes of lime requirment in coastal plain soils of low cation exchange capacity.

Liming

Defintion. Liming is the practice of adding liming materials to acid soils for the purpose of maintaining a favorable environment for plant root growth. More favorable root environment may be a consequence of more desirable soil pH or of inactivation of otherwise toxic elements. Addition of lime to acid soils affects them in two ways simultaneously, i.e., it augments the calcium supply of the soil and neutralizes soil acids. Both aspects of liming may improve the soil as an environment for plant roots by enhancing the availability of other plant nutrients and by depressing the toxicity of elements such as aluminum and manganese.

Use of Calcitic Versus Dolomitic Lime. Most temperate region soils where liming is practiced have adequate amounts of exchangeable calcium for plant needs without addition of lime. Hence lime is generally added as an ameliorant to improve the root environment rather than to supply calcium. However, many of these same acid soils may not have adequate amounts of exchangeable magnesium for plant needs. In such cases, dolomitic lime is recommended so that magnesium is provided for the plant as the soil is concomitantly neutralized. If exchangeable magnesium is found to be low and little soil acidity exists to dissolve the relatively insoluble dolomitic lime, it may be necessary to apply a much more expensive soluble source of magnesium such as epsom salts ($MgSO_4 \cdot 7H_2O$) to provide adequate magnesium to meet the needs of the crop.

Processing of Lime. Agricultural ground limestone is prepared by quarrying the limestone rock, breaking it up, grinding it, screening it, and piling it according to fineness and purity. Since limestone depositis vary in amounts of impurities they contain, $CaCO_3$ contents and hence the acid-neutralizing power of the lime vary accordingly. In general, the best grade of lime is the one with the least amounts of impurities and with the finest grinding. Fineness of grinding is not without limit, however, since additional benefits of still finer ground lime may not equal extra costs of grinding, and a more pure limestone deposit simply may not be available. Compensations for shortcomings in purity and fineness of lime can be made by increased rates of application.

Quarries are located on limestone rock deposits exposed to the ground surface by natural outcrops or by excavation of overburden. Usually the crushing, grinding, screening, and storage operations are performed near the quarry. Since ground limestone is relatively insoluble and the product is a relatively low-cost material, agricultural ground lime is generally stored (or piled) outside, from which point it is loaded, hauled, and spread on the soil with specially equipped trucks designed for bulk spreading. Only premium-grade lime materials such as finely ground, relatively pure calcite lime, hydrated lime, or quick lime are normally bagged, shipped, and spread on soils with more conventional low-volume spreaders.

Due to the relatively large proportion of the total cost of lime being for transportation, and to the generally wide distribution of limestone deposits over the earth's surface, agricultural lime is generally produced and distributed by many small, widely scattered operators rather than by a few large, centrally located ones. In some areas, agricultural lime is produced by large companies that also process lime for smelting of iron ore, for preparation of cement, and for construction of roads. These larger companies may also have kilns for preparation of quick or caustic lime by heating to drive off the CO_2 from $CaCO_3$. They may then prepare hydrated lime from the caustic lime by reaction with water.

Soil-lime Reactions. As mentioned under the section "Factors Affecting Lime Requirement," the specific reactions that occur in acid soils when lime is added depend on the forms of acids present, which in turn depend on pH and other factors. Also the specific reactions that occur depend on the type of lime added. In general, acid soils have varying amounts hydrogen ions or hydronium (H_3O^+) ions and aluminum ions in association with varying amounts of OH ions either as monomers or polymers. The Al^{3+}, $Al(OH)^{2+}$, $Al(OH)_2^+$ ions may be: (i) adsorbed as monomers or polymers to negative charges of the cation exchange sites of the soil, or (ii) complexed or chelated by various complex- or ligand-forming substances such as organic matter. The following reactions are typical of those occurring when the various types of lime are added.

1 [Soil particle –H –H] $+ CaCO_3 \rightarrow$ [Soil particle –Ca –] $+ 2H_2CO_3 \nearrow 2CO_2$ $2H_2O$

2 [Soil particle –H –H] $+ Ca(OH)_2 \rightarrow$ [Soil particle –Ca –] $+ 2H_2O$

3 [Soil particle –H –H] $+ CaO \rightarrow$ [Soil particle –Ca –] $+ H_2O$

4 [Soil particle – –] $Al(OH) + CaCO_3 \xrightarrow{H_2O}$ [Soil particle – –] $Ca + Al(OH)_3$ $\uparrow CO_2$

5 [Soil particle – –] $Al(OH) + Ca(OH)_2 \rightarrow$ [Soil particle – –] $Ca + Al(OH)_3$

6 [Soil particle – –] $Al(OH) + Ca(OH)_2 \rightarrow$ [Soil particle – –] $Ca + Al(OH)_3$

7 [Soil particle – –] $Al(OH) + CaO \xrightarrow{H_2O}$ [Soil particle – –] $Ca + Al(OH)_3$

The result of the reaction in each case is: (i) increased calcium saturation of the soil particle and (ii) neutralization of either the hydrogen or aluminum that cause the soil to be acid. It is evident then that either the decreased amounts of acidic ions or increased amounts of basic cations, or both, may be the cause of increased crop yields when lime is applied. Also because the leaching action of water percolating through the soil removes the basic cations and causes acidic cations to accumulate, the practice of liming is a continuing need for soils of the humid areas. The amounts of lime required to replace the bases lost each year may be as little as a few hundred kilograms per hectare to as much as two metric tons per hectare. The actual amount of lime required to replenish the bases lost by a given soil depends on such factors as: type of soil minerals and their rate of weathering, texture and structure of the soil, percolation-restricting horizons (layers) of the soil, amount of excess precipitation over that required to saturate the soil, amounts of water evaporated or transpired by crops, and amounts of acid-forming fertilizers applied.

Extent of Lime Usage for Crops. Statistics are not available for worldwide use of lime for crop-production improvement on acid soils. The amounts used in the United States are however, available and are perhaps indicative of the extent of lime usage in other developed countries where acid soils are cropped intensively. The amount of agricultural lime used annually in the United States during 1974, the most recent year for which figures are available, was 27.8 million tons. Of this, six north-central states (Illinois, Iowa, Missouri, Ohio, Wisconsin, and Indiana) accounted for nearly half (45.4%). Also if five southeastern states (Kentucky, Georgia, Alabama, Tennessee, and Virginia) and the border state Pennsylvania are also included, more than three fourths of the total for the United States is accounted for (75.4%).

Of the total limestone crushed or broken each year, by far the largest amount goes into aggregates for roads, buildings, etc., with smaller amounts used for agricultural ground limestone, fluxing stone for smelting minerals, riprap for various uses, and railroad ballast. Agricultural limestone and fluxing stone each account for a little less than 1/20 the total.

The amount of lime applied in the United States in 1974 was only 29.4% of the estimated annual needs, indicating that more lime should have been applied. Indeed, more basic cations leach from cropped soils of the humid regions each year than are applied in lime. For example, in a north-central state such as Ohio, which uses a relatively large amount of lime, that spread amounted to only 300 lb of lime per crop acre. Present losses in excess of applications coupled with great deficients from extensive losses over the many years of the past make increased rates of liming a matter of urgency.

E. O. McLEAN

References

Adams, F., and Evans, C. E., 1962. A rapid method for measuring lime requirement of Red-Yellow Podzolic soils, *Soil Sci. Soc. America Proc.*, **26**, 355–357.

Bradfield, R., and Allison, W. B., 1933. Criteria of base saturation in soils, *2nd Internat. Congr. Soil Sci. Trans.* **A**, 63–79.

Brown, I. C., 1943. A rapid method of determining exchangeable hydrogen and total exchangeable bases of soils, *Soil Sci.*, **56**, 353–357.

Clark, J. S., and Nichol, W. E., 1966. The lime potential, percent base saturation relations of acid surface horizons of mineral and organic soils, *Canadian Jour. Soil Sci.*, **46**, 281–315.

Coleman, N. T., and Thomas, G. W., 1964. Buffer curves of acid clays as affected by the presence of ferric iron and aluminum, *Soil Sci. Soc. America Proc.*, **28**, 187–190.

Coleman, N. T., and Thomas, G. W., 1967. The basic chemistry of soil acidity, *in* R. W. Pearson and F. Adams, eds., *Soil Acidity and Liming*, vol. 12. Madison: American Society of Agronomy, 1–41.

Jackson, M. L., 1960. Structural role of hydronium in layer silicates during soil genesis, *7th Internat. Congr. Soil Sci. Trans*, (Madison), **2**, 445–455.

Keeney, D. R., and Corey, R. B., 1963. Factors affecting the lime requirements of Wisconsin soils, *Soil Sci. Soc. America Proc.*, **27**, 277–280.

McLean, E. O., 1970. Lime requirements of soils–Inactive toxic substances or favorable pH range?, *Soil Sci. Soc. America Proc.*, **34**, 363–364.

McLean, E. O., 1971. Potentially beneficial effects from liming: Chemical and physical, *Soil Crop Sci. Soc. Florida*, **31**, 189–196.

McLean, E. O., Dumford, S. W., and Coronel, F., 1966. A comparison of several methods of determining lime requirements of soils, *Soil Sci. Soc. America Proc.*, **30**, 26–30.

Mehlich, A., 1939. Use of TEA-OAc-$Ba(OH)_2$ buffer for the determination of some base exchange properties and lime requirement of soil, *Soil Sci. Soc. America Proc.*, **3**, 162–166.

Schofield, R. K., 1933. Rapid method of examining soils, *Jour. Agric. Sci.* **23**, 252–254.

Schwertmann, U., and Jackson, M. L., 1964. Influence of hydroxy-aluminum ions on pH-titration curves of hydronium-aluminum clays, *Soil Sci. Soc. America Proc.*, **28**, 179–182.

Shoemaker, H. E., McLean, E. O., and Pratt, P. F., 1962. Buffer methods for determination of lime requirement of soils with appreciable amount of exchangeable aluminum, *Soil Sci. Soc. America Proc.*, **25**, 274–277.

Villiers, J. M. de, and Jackson, M. L., 1967. Cation-exchange capacity variations with pH in soil clays, *Soil Sci. Soc. America Proc.*, **31**, 473–476.

Woodruff, C. M., 1948. Testing soils for lime requirement by means of a buffer solution and the glass electrode, *Soil Sci.*, **66**, 53–63.

Cross-references: *Acidity; Base Saturation; Buffers, Buffering; Exchange Phenomena; Ionic Activities; Management of Soil; Soil Fertility; Soil Reaction; Soil Stabilization.*

LYSIMETERS

Researchers throughout the world have used *lysimeters* to investigate the concentration of nutrients and pollutants in the leachate water from the soil profile and to investigate the water balance at the earth's surface. Lysimeters consist of an isolated mass of soil equipped with drainage devices and in some cases weighing mechanisms. A variety of materials including steel, concrete, and fiberglass have been used to enclose the soil. Lysimeters range in size from as small as a 120l trash can to as large as 8m in diameter enclosing a 0.5m thick layer

FIGURE 1. An undisturbed monolith lysimeter being prepared for field installation. In this clay soil, angle iron was sufficient to retain the profile in the casing. (Courtesy of the Texas Water Resources Institute)

of soil. Lysimeters deep enough to include the root zone of perennial vegetation have also been constructed.

Many of the lysimeters used for research have been filled with disturbed soil. Precautions are commonly taken to replace the soil layers in the same order as they were found in nature. To achieve this, the soil must often be dried. It may need to be ground and sieved and packed back into place to achieve densities similar to those found in the field. While the repacked soil may come to sufficient equilibrium to simulate the water-balance behavior of field soils, recent evidence indicates that the mobility of ions in a disturbed soil may be considerably different from that found in nature. Therefore, some lysimeters have been constructed to enclose undisturbed soil monoliths (Figure 1). This is done by forcing a casing down around the soil similar to the procedure used in piling operations. Once the casing is forced to the desired depth, a bottom plate is pushed under the casing usually by hydraulic jacks. The entire assembly may then be removed for installation of drainage devices or for possible placement in another location.

Drainage devices are an essential component of all lysimeters. Porous ceramic filters are most often used. They may be placed in a layer of soil or other porous material at the bottom of the profile, or they may be imbedded directly in the soil. The latter procedure is preferable because the former may result in the development of an artificial, perched water table. Some lysimeters have been constructed with sand or gravel layers equipped with gravity drains at the bottom. These result in unnatural soil moisture profiles.

For leachate studies, lysimeters have most often been moved to central locations where they have been placed adjacent to trenches in which the collection bottles are located or near above ground enclosures where the leachate is collected by vacuum. The data from such lysimeters have proven valuable in understanding the nutrient cycle and the loss of soluble fertilizers from agricultural soils. Valuable information on the ability of soils to absorb and inactivate land-disposed waste has also been gained by this technique.

A varicty of weighing devices including liquid displacement, load cells, and field hoists have been used to determine water loss by evapotranspiration. Lysimeters used for this purpose are most often installed in a field surrounded by a vegetation similar to that under investigation to assure that the water loss will be representative of that in an extensive field and not just an isolated plot. Weighing lysimeters (Figure 2) have been developed with a resolution equivalent to a water depth of 0.1 mm, During periods of high evaporative demand, water loss may be recorded as often as every quarter hour. Such data has been of great value

FIGURE 2. A weighing lysimeter before the area backfill has been put in place. The soil profile is enclosed in a casing and positioned on a balance. An outercasing is used to isolate the lysimeter from the backfill and to protect the scale mechanism and the counterweights located in the narrow shaft. This lysimeter was designed by Dr. C. H. M. van Bavel.

in advancing our understanding of the water and energy balance at the earth's surface and has served as a means of chekcing predictions of evapotranspiration calculated from meteorological data.

K. W. BROWN

References

van Bavel, C. H. M., and Myers, L. E., 1962. An automatic weighing lysimeter, *Agric. Engin.*, **43**, 580–583, 587–588.

Brown, K. W.; Gerard, C. J.; Hipp, B. W., and Ritchie, J. T., 1974. A procedure for placing large undisturbed monoliths in lysimeters, *Soil Sci. Soc. America Proc.*, **38**, 981–983.

Harrold, L. L., and Dreibelbis, F. R., 1953. Water use by crops as determined by weighing monolith lysimeters, *Soil Sci. Soc. America Proc.*, **17**, 70–74.

Harrold, L. L., and Dreibelbis, F. R., 1958. Evaluation of agricultural hydrology by monolith lysimeters (1944–1955), *US Dept. Agric. Tech. Bull. 1179*, 166p.

Rosenberg, N. J.; Hart, E. H.; and Brown, K. W., 1968. Evapotranspiration–A review of research, *Nebr. Agric. Expt. Sta., Miscellaneous Publication No. 20*, 78p.

Cross-references: *Energy Balance; Evaporation; Leaching; Nutrient Losses; Plant Nutrients; Soil Water; Water Content and Retention. Vol. IVA: Evapotranspiration.*

M

MACRONUTRIENTS

Plants contain small amounts of 90 or more elements, but only 16 elements are known to be essential to higher plants (Epstein, 1971). An element is not considered essential unless a deficiency of it makes it impossible for the plant to complete its life cycle. Essential elements are directly involved in the nutrition of the plant and are not related to possible effects in correcting some unfavorable microbial or chemical condition of the soil (Gauch, 1972). It is customary to divide the essential nutrient elements into two groups on the basis of the quantity occurring in plants. Those present in trace amounts are classed as *micronutrients* (q.v.). The essential elements that occur in greater concentrations are called *macronutrients.* Carbon (C), hydrogen (H), oxygen (O), nitrogen (N), phosphorus (P), potassium (K), calcium (Ca), magnesium (Mg), and sulfur (S) are generally regarded as macronutrients. Even though some of these elements have been known since ancient times, their essentiality has been established only within the last century (Table 1).

Because nitrogen, phosphorus, and potassium are commonly supplied to the soil as farm manure and as commercial fertilizers, they are also referred to as *primary,* or *fertilizer, elements.* Similarly, calcium, magnesium, and sulfur are sometimes referred to as *secondary elements.* Calcium and magnesium, which are added to soils in limestone, are also called *lime elements.*

Source and Availability

The carbon, hydrogen, and oxygen contained in plants, which ordinarily range from 94–99.5% of fresh plant material, are derived from atmospheric carbon dioxide and soil water. Because higher plants obtain most of their carbon and hydrogen from air by photosynthesis and some oxygen from water, these three elements are not usually considered as mineral-nutrient elements (Tisdale and Nelson, 1966). The remaining macronutrients (nitrogen, potassium, phosphorus, calcium, magnesium, and sulfur) are obtained in the form of ions from the soil (Table 2).

Mineral or plant nutrient elements are released to the soil from rocks or other parent materials by weathering and soil-forming processes. The elements gradually become available for plant use through various processes of simplification, i.e., the more complex and less active forms are broken down into simpler and more available forms (see vol. IVB: *Soil Minerals*). The main forms in which macronutrients occur in soils are listed in Table 2.

Nitrogen. The nitrogen in soils is very largely present in complex forms in organic matter (q.v.) or humus (q.v.). Most plants, however, assimilate nitrogen in the forms of either or both of nitrate (NO_3^-) and ammonium (NH_4^-) ions. Exceptions are plants with nitrogen-fixing organisms in root nodules such as members of the large family of Leguminosae. Vigorous vegetative growth and deep green colors are associated with adequate supplies of nitrogen. A stunted yellow appearance often indicates a nitrogen-deficient plant. Nitrogen may be lost from the soil by leaching (see *Nutrient Losses*) or volatilization of ammonia, and it may become unavailable to plants by denitrification (see *Nitrogen Cycle*).

Phosphorus. Native phosphorus in soils, which probably originated from decomposition of the mineral apatite, may occur as finely divided fluorapatite, hydroxyapatite, or chlorapatite; as iron or aluminum phosphates; in combination with the clay fraction; or as organic complexes. Phosphorus is absorbed by plants mainly in the ionic ($HPO_4^=$, $H_2PO_4^-$) form, but to a limited extent in organic forms. An adequate supply of phosphorus is essential for seed formation and stimulation of root growth, and it is associated with early maturity in crops. The immobilization or fixation of phosphorus, which may result in deficiencies, is related to several factors such as the presence of 1:1-type clay minerals (see *Clay Minerals, Silicates*), presence of hydrous oxides of iron and aluminum and alkaline ($pH > 7.0$) or strongly acidic ($pH < 5.5$) soil conditions (Kardos, 1964; Russell, 1973). A phosphorus deficiency, which may, for example, be related to immobilization by reactions with calcium, aluminum, and iron-compounds (see *Phosphorus Cycle*), is normally accompanied by stunted growth and deep green color.

TABLE 1. Macronutrients Essential for Plant Growth.

Essential Element for Plants	Discovered by	Year Discovered	Proved Essential to Plants by	Year Proved Essential
Carbon	*	*	Sachs	1882
Hydrogen	Cavendish	1766	Sachs	1882
Oxygen	Joseph Priestly	1774	De Saussure	1804
Nitrogen	Lavoisier	1792	Rutherford	1872
Phosphorus	Brand	1669	Posternak	1903
Potassium	Davy	1807	Schimper	1890
Calcium	Davy	1807	Sal-Horstmar	1856
Magnesium	Davy	1808	Willstatter	1906
Sulfur	*	*	Peterson	1911

*Element known since ancient times.
(After Tamhane et al, 1966)

Potassium. The native potassium contained in soils is derived from the weathering of minerals such as biotite, muscovite, and potash feldspars. Potassium is also present in soils in the form of secondary minerals such as smectites, illites, vermiculites, and chlorites. The total potassium content of many soils is quite high, but the quantity present in a form easily absorbed by plants may be quite small. Potassium is absorbed in ionic form (K^+) and may be supplied in a number of inorganic salts (Table 2). Potassium is considered essential for the production and translocation of carbohydrates and seems to be intimately related to the nitrogen metabolism of plants. A potassium deficiency, due to fixation or leaching, may block the conversion of amino acids to proteins and is usually associated with a temporary increase in carbohydrates (Tisdale and Nelson, 1966).

Calcium. The calcium in soils is derived from a number of minerals such as calcite, dolomite, apatite, calcium feldspars, and amphiboles. Like potassium, calcium is required by all higher plants and is absorbed in ionic form (Ca^{2+}). Calcium is required for the growth of apical meristems and for the formation of flowers. Unlike nitrogen, phosphorus, potas-

TABLE 2. Forms of Macronutrients in Soils.

Macronutrient	Principal Complex Forms	Available Forms	
Nitrogen	Proteins, amino acids, and similar forms	Ammonium salts	NH_4^+
		Nitrite salts	NO_2^-
		Nitrate salts	NO_3^-
Phosphorus	Apatite, and secondary Ca, Fe, Al phosphates	Phosphates of Ca, K, Mg, etc.	HPO_4^-
	Phytin, nucleic acid, and other organic combinations	Soluble organic forms	$H_2PO_4^-$
Potassium	Feldspars and mica aluminum silicates such as clays, especially illite	Potassium ions adsorbed by colloidal complex; Potassium salts such as sulfates, carbonates, etc.	K^+
Calcium	Feldspars, hornblende, calcite, and dolomite	Calcium ions adsorbed by colloidal complex; Simple calcium salts	Ca^+
Magnesium	Mica, hornblende, dolomite, and serpentine; Montmorillonite, chlorite, and vermiculite	Magnesium ions adsorbed by colloidal complex; Numerous simple salts	Mg^+
Sulfur	Pyrite and gypsum	Various sulfites and sulfates of Ca, Na, K, Mg, etc.	SO_3^-
	Organic forms		SO_4^-

(After Brady, 1974)

sium, magnesium, and sulfur, calcium is rather immobile in the plant and is not translocated from older to younger regions when a deficiency occurs. Shortages of calcium, which are rather common in coarse-textured soils of humid regions, especially those formed in parent materials low in calcium, are therefore first evident in younger plants (Sprague, 1964). Low calcium levels in the *B* horizons of Ultisols (red-yellow Podzolic soils) of warm-temperate, humid regions hinder penetration by plant roots.

Magnesium. The magnesium in soils is derived from the decomposition of minerals such as biotite, chlorite, hornblende, dolomite, olivene, and serpentine. Magnesium is available to plants in ionic form (Mg^{2+}) as ions absorbed by the colloidal complex or as simple salts (Table 2). The only mineral constituent of the chlorophyll molecule, magnesium is required by all autotrophic green plants for photosynthesis. Magnesium is also related to phosphorus metabolism and is considered to be specific in the activation of a number of plant enzyme systems. Deficiency symptoms are most characteristic of coarse-textured soils of humid regions. Magnesium deficiency is seldom a problem in soils of arid regions.

Sulfur. The native sulfur in soils is partly in the form of soluble sulfates and sulfites and partly in complex organic forms. In some poorly drained soils, sulfur may be present in sulfides. The variety of organic compounds containing sulfur in soils is large, but changes proceed rapidly so that the identification of the forms is difficult. Sulfur is generally absorbed by plants as the $SO_4^{=}$ ion, though small amounts may be absorbed through the leaves as SO_2. In addition to maintenance of an adequate growth rate, sulfur is also essential for nodule formation on legume roots and is required for fruiting of some species. A sulfur deficiency, which may be caused by immobilization or the production of other unavailable forms (see *Sulfur Transformations*), is expressed in stunted growth and chlorosis.

Nutrient Levels and Their Management

The need for all macronutrients has been well established, but the role of individual elements in plant nutrition (see *Plant Nutrients*) and soil fertility (q.v.) are still not fully understood. It is known that all forms of any element are not necessarily available to plants. Collectively, the elements are absorbed in combined inorganic and organic forms, as gas, as ions in the soil solution (q.v.), and from the exchange complex. The total amount of a macronutrient in the soil does not indicate the quantity available to plants. If a given macronutrient occurs in a form or forms largely unavailable to plants, the element may reduce uptake of other elements and may limit growth (see *Law of the Minimum*). Levels and forms of nitrogen, phosphorus, and sulfur in particular are related to nutrient cycles (see *Nitrogen Cycle; Phosphorus Cycle; Sulfur Transformations*), which in turn are associated with levels of biological activity in soils (see *Metabolic Activity*).

To preclude deficiencies or to correct them, plant nutrients are added to soils as fertilizers (q.v.) or amendments such as lime (see *Lime, Liming*). Soil-fertility investigations are thus concerned not only with amounts but also with the forms of macronutrients in soils, the availabilities of the different forms to plants (see *Nutrient Potential*), and their interactions with other elements.

CHARLES W. FINKL, JNR.

References

Brady, N. C., 1974. *The Nature and Properties of Soils.* New York: Macmillan, 639p.

Epstein, E., 1971. *Mineral Nutrition of Plants: Principles and Practices.* New York: Wiley, 412p.

Gauch, H. G., 1972. *Inorganic Plant Nutrition.* Stroudsburg, Penn.: Dowden, Hutchinson & Ross, 488p.

Kardos, L. T., 1964. Soil fixation of plant nutrients, *in* F. E. Bear, ed., *Chemistry of the Soil.* New York: Reinhold, 369–394.

Russell, E. W., 1973. *Soil Conditions and Plant Growth.* London: Longmans, 849p.

Sprague, H. B., ed., 1964. *Hunger Signs in Crops.* New York: McKay, 461p.

Tamhane, R. V.; Motiramani, D. P.; Bali, Y. P.; and Donahue, R. L., 1966. *Soils: Their Chemistry and Fertility in Tropical Asia.* New Delhi: Prentice-Hall of India, 475p.

Tisdale, S. L., and Nelson, W. L., 1966. *Soil Fertility and Fertilizers.* New York: Macmillan, 694p.

Wallace, T., 1963. *The Diagnosis of Mineral Deficiencies in Plants.* New York: Chemical Publishing Company, 288p.

Cross-references: *Law of the Minimum; Micronutrients; Nutrient Potential; Plant Nutrients; Soil Fertility.*

MANAGEMENT OF SOILS

The study of soil management can be a fascinating experience. It can also be a nightmare of disjointed, hard-to-synthesize fragments of information. No modern college-level textbooks are available today. This state of affairs brings a need for reliance on numerous articles published in an array of scientific journals. It is therefore difficult for a person not accustomed to interpreting scientific data to synthesize the available information and to study the area of soil management.

In spite of this gloomy setting, it is essential that people be stimulated to study and to research the area of soil management. Kellog (1957) gives the basic impetus for the necessity of advances in soil management when he says, "Only a few of our soils in the United States have within themselves the dynamic to remain productive." The responsibility therefore for an adequate and continuous food supply is highly dependent on being able to "manage the soil." In this brief treatment of the subject, an attempt will be made to inform the reader of the many facets of soil management.

A logical approach to the study of soil management demands a definition of the study area. *Soil management* can be defined simply as "the manner in which man uses the soil to produce food, fiber, and forage." It can therefore be judged as good or bad only when lated to a standard. What constitutes an adequate standard for judging the merits of a management system?

To answer that question, one must identify the goals of soil management. Two major goals exist for a soil-management system. The first is to prevent soil deterioration. Examples of soil deterioration are excessive erosion, fertility depletion, and salt accumulations within the rooting zone. The second goal is to attempt improvements in the soil system that result in increased crop production. Examples of improvements are conservation of moisture, irrigation development, and increased water-use efficiency by proper fertilization. These two goals are very broad, and therefore many aspects of science must be integrated to attain them.

Articles by Anderson and Peterson (1973), Peterson and Vetter (1971), Herron and Erhart (1965), and Edwards et al (1973) provide examples of a fundamental aspect of soil management that must always be foremost—namely, that management practices have a cumulative effect on the soil's ability to continue to be

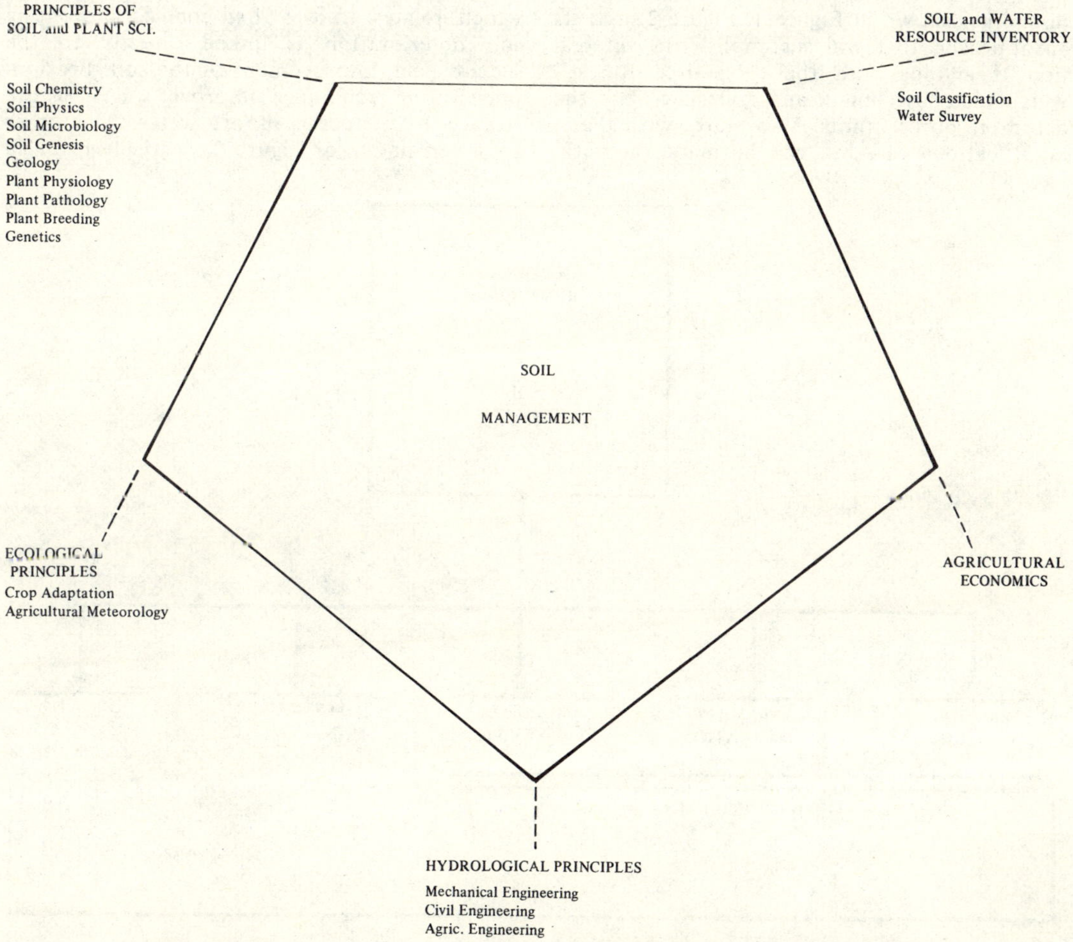

FIGURE 1. The scientific and technological inputs of a soil-management system.

productive over long periods of time. Long-range effects are of great importance to sound soil-management practices. The articles cited discuss the effects of monoculture, of cultivation on soil nitrogen, of animal manures on crop yield, of animal manures on soil chemical and physical properties, and of nitrogen fertilizer with and without animal manures.

Pierre (1958) set the stage for soil management by relating soil classification to the basic biological and physical sciences. He observed that the strength of this particular discipline of soil science rested in its ability to continually incorporate concepts and findings from the basic areas of chemistry, physics, microbiology, and mineralogy. The model in Figure 1 extends Pierre's concepts to the area of soil management. Note that the model shows many important relationships outside the areas of soil and crop sciences. Sciences such as engineering, meteorology, and economics are all vital to a management system.

The approach to the study of soil management involves integrating the material from the disciplines shown in Figure 1. Figure 2 suggests an approach that will accomplish the integration, beginning with the "Resource Block," which includes climate and soil. These are the natural-resource inputs. Man starts with them. Modifications can and will be made on both; but for the most part, they are the constants of the system. Agricultural meteorology and soil classification are the sciences involved in the description of these resources. Man now enters the scheme to impose a "management system," which has as its goal the production of food, fibers, or forage. The sciences of ecology, genetics, plant breeding, and various engineering technologies are all involved in decision making at this level. Even marketing potential, etc., for the eventual production must be considered.

The diagram shows that the cropping system imposed is directly related to the soil and climate resources. The second row in Figure 2 enumerates the major divisions of soil-management emphasis. The first three–erosion control, organic matter conservation, and moisture conservation–are the classical topics of soil management. The study of one or more of these three areas is directly related to the others. For example, control of erosion leads to direct savings in soil organic matter and soil moisture. Furthermore, the control of erosion, which relates to the first goal of preventing soil deterioration is linked directly to the second goal, improvement of the soil. Erosion-prevention techniques improve soil moisture storage by reducing runoff water. The other topics in line 2 of Figure 2, pesticide manage-

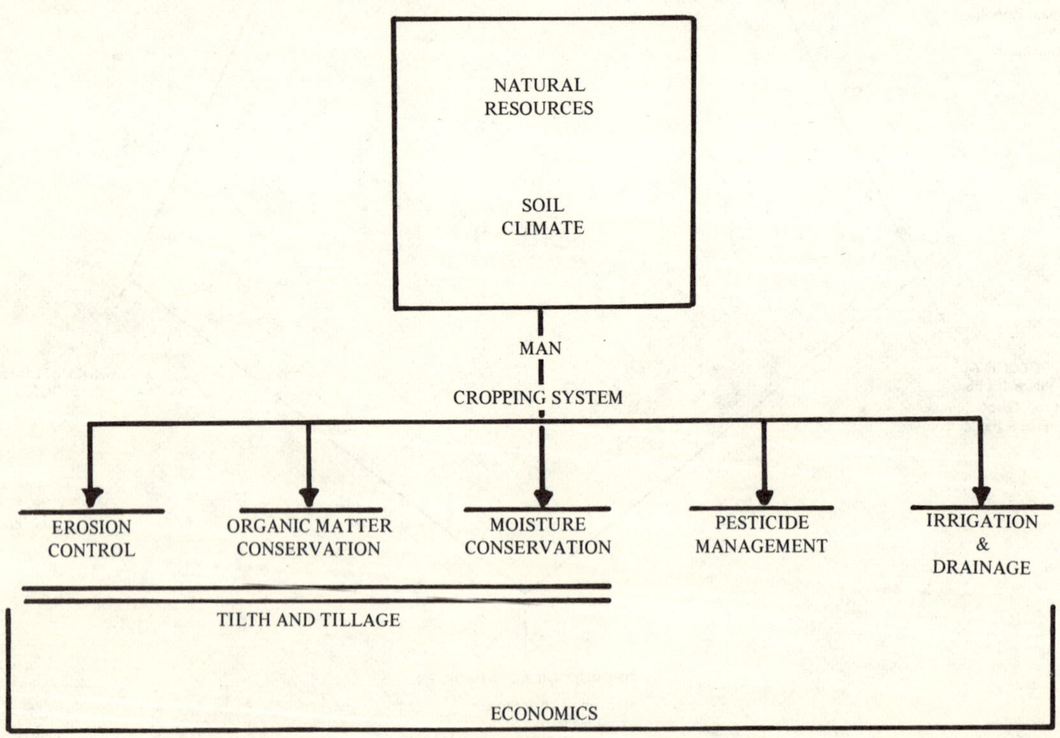

FIGURE 2. The major components of a soil-management system.

ment and irrigation and drainage, are equally important but have not been studied to as great an extent as the first three. In fact, pesticide management is a relatively new concept in soil management. Soil-manipulation techniques or tillage methods are very intimately involved in management. As indicated in the diagram, tillage techniques need to be considered as they affect all three of the classical management areas.

Finally, basic economic principles must underlie any sound management system. This is true in even the most primitive civilizations. If the people cannot live on the food or income produced, the system cannot survive. In more highly developed areas of the world, man's ability to manage has even influenced the potential value of the land. Priest, Whiteside, and Heneberry (1963) developed a land-evaluation method that hinges on soil characteristics reported in soil surveys and management practices as determined by average yields attained by farmers using that land. Thus man's ability to cope with the problems of a particular soil eventually determines its economic value to society.

Man has been given the responsibility of using the natural resources of climate, water, and soil to provide the food and fiber needed to sustain life on earth. Soil management is the integrated science that has the responsibility for success or failure in this monumental task. Tomorrow's food supply depends on today's soil-management decisions.

G. A. PETERSON

References

Anderson, F. N., and Peterson, G. A., 1973. Effects of continuous corn, manuring, and N fertilization on yield and protein content of the grain and on the soil N content, *Agron. Jour.*, 65, 697–700.

Edwards, W. M.; McGuinness, J. L.; Van Doren, D. M., Jr.; Hall, G. F.; and Kelley, G. E., 1973. Effect of long term management on physical and chemical properties of the Coshocton watershed soils, *Soil Sci. Soc. America Proc.*, 37, 927–930.

Herron, G. M., and Erhart, A. B., 1965. Value of manure on an irrigated calcareous soil, *Soil Sci. Soc. America Proc.*, 29, 278–281.

Kellogg, C. E., 1957. We seek; We learn, *in Soil, U.S. Department of Agriculture 1957 Yearbook of Agriculture*, 1-11.

Peterson, G. A., and Vetter, D. J., 1971. Soil N budget of the wheat-fallow area, *Nebr. Quart.*, 17, 23–25.

Pierre, W. H., 1958. Relationship of soil classification to other branches of soil science, *Soil Sci. Soc. America Proc.*, 23, 167–170.

Priest, T. W.; Whiteside, E. P.; and Heneberry, W. H., 1963. Use of soil management groups and related information in evaluation of farmlands and their utilization, *Soil Sci. Soc. America Proc.*, 27, 335–339.

Cross-references: *Agronomy; Aeragronomy; Conservation; Fertilizers, Inorganic and Organic; Gypsum Requirement; Irrigation; Lime, Liming; Moisture Management; Mulch; Nutrient Losses; Profiles, Physical Modification; Salt Leaching; Sludge Disposal; Soil Conditioners; Soil Drainage; Soil Fertility; Soil Stabilization; Soil Tests; Soils, Nonagricultural Uses; Tillage; Water Erosion; Wind Erosion; Zero-tillage.*

MASS ACTION AND EQUILIBRIUM CONSTANT—See THERMODYNAMICS, SOIL MOISTURE. VOL. IVA: MASS ACTION AND EQUILIBRIUM CONSTANT

MECHANICAL ANALYSIS—See PARTICLE-SIZE DISTRIBUTION. VOL. XIII: SOIL MECHANICS

MECHANICAL PROPERTIES OF SOILS —See SOIL MECHANICS; TILLAGE

METABOLIC ACTIVITY

The *metabolic acitvity* occurring in soil is the total of manifestations of biological transformations of material and energy. It is the result of various life processes of organisms that form an inseparable, constant, and essential component of soil. Soil organisms (*edaphon*) comprise the microorganisms (*microedaphon*) such as bacteria, actinomycetes, fungi, blue green algae, algae and protozoans, and the so-called *macroedaphon*, comprising worms, mollusks, arthropods, vertebrates, etc. The living component of soil further contains the living organs of higher plants (roots). A role in the matabolic activity of soil may be played by enzymes that are present free in the soil, either actively secreted by microorganisms or as passive constituents of destroyed or perished cells and tissues. It is estimated that the mass of the micro- and macroedaphon in soil amounts to as much as 8% of the total organic matter (on a dry weight basis). The microedaphon predominates in the metabolic total both quantitatively and qualitatively.

According to the type of nutrition, soil contains *autotrophic* as well as *heterotrophic* organisms. Autotrophic organisms will thrive on mineral substances plus a source of energy that may be either light (*photoautotrophy*) or chemical reactions (*chemoautotrophy*). Heterotrophic organisms derive their nutrition and energy from preformed organic compounds—they include saprophytes, phagotrophs, and parasites. Some organisms may derive their nutrition either autotrophically or heterotrophically, depending on the amount of available organic substances in the environment. Combination of autotrophic and heterotrophic nutrition in a single organism is called *mixotrophy*. Soil organisms may also be divided into *producers* and *degraders,* and one may define groups displaying special pathways and metabolic products. According to particular metabolic activity, one may define groups of organisms capable, for example, of decomposing cellulose and designate them as *cellulose degraders;* those producing methane as *methane formers;* those converting ammonia to nitrites as *ammonia oxidizers.* There may be degraders of lignin, pectin, chitin; oxidizers of nitrite, reducers of nitrate, oxidizers or reducers of sulfur or iron compounds, etc. According to Quastel (1946), soil may be considered a complex, comparable in its metabolic activity to tissues or organs of higher organisms; it may receive organic and mineral nutrients, water, and air; it may be the site of enzyme reactions. Soil has also been characterized as a bio-organo-mineral complex in which the active constituents are formed by soil microorganisms along with colloid organic as well as inorganic materials, clay minerals, and the soil solution.

Since microorganisms play the dominant role in the metabolic activity of soil, the history of investigation in the field is inseparably connected to the history of soil microbiology and to the names and works of its pioneers of the end of the nineteenth and the beginning of the twentieth centuries: e.g. P. A. Kostychev, D. Ivanovsky, S. N. Winogradsky, V. L. Omeliansky, M. W. Beijerinck, F. Löhnis, J. Stoklasa, and S. A. Waksman. During the last 50 years, from the time Winogradsky (1949) defined the position of soil microbiology as a branch of science and pointed out the main directions of its development, the biochemical and ecological approaches to the problems have gained importance. Studies of soil enzymes date back to 1950 when Hofmann and Seegerer published their data. Since that time much attention has been devoted to the problem because the aforementioned authors reached the conclusion that enzyme activity (q.v.) is associated with the properties of soil and may represent a criterion of soil fertility. Even if the validity of these postulates is not universal, much work has been accomplished on their initiative concerning soil enzyme systems, their assay, origin, and state, and function of enzyme activity in soil, and the relationships between enzyme activity and the properties of soil. Soil microbiology, enzymology, and biochemistry thus provided the spadework information on the properties and activity of soil microorganisms that was the basis for further research of the possibilities of increasing soil fertility and other useful properties of soil as a renewable natural resource.

Principal Metabolic Processes in Soils

The metabolic pathways functioning in soils are rather varied and reflect the heterogeneity of the participating organisms and of their metabolic properties, as well as the wide spectrum of organic and inorganic substrates that can be metabolically transformed. Metabolic reactions are invariably catalyzed by enzymes. According to existing knowledge, organisms can attack and assimilate more than 70 elements (either as such or combined in compounds) present in the earth crust. The activity of soil organisms can be considered a part of the flow of matter and energy in the biosphere. These processes are occasionally designated "biogeochemical" and may be divided into several categories. A certain type of transformation may be present in more than one category. The categories include (a) *mineralization,* i.e., conversion of the organic form of a certain element to the inorganic one; (b) *immobilization,* i.e., conversion of the inorganic form of an element into an organic complex as the result of assimilation of the element by cells and its incorporation into the protoplasmic combination; (c) *oxidation,* usually associated with the energy metabolism of cells, much like the reverse process (d) *reduction;* (e) *volatilization,* i.e., transformation of the nongaseous form of a compound into a gaseous one; (f) *fixation,* i.e., assimilation of a gaseous compound; (g) *precipitation* as well as (h) *solubilization,* connected with the formation of compounds affecting the solubility of substances in the medium (e.g., chelating, complexing agents, and acids), which may result in the formation of geological deposits (sulfur, sulfides, oil, etc.); (i) formation of structures affecting the adsorption processes in the environment; and (j) *isotopic fractionation,* i.e., a selective and more rapid use of a natural isotope occurring in mixture with others.

Irrespective of the type of nutrition, all organisms exhibit a set of processes summarily designated *respiration.* In these dissimilations, energy is liberated and used for growth, division, movement, water and substrate transport, and other life activities. Respiration may occur in the presence of oxygen (aerobically) or in its absence (anaerobically), and it is effected by oxidation-reduction chains. Under anaerobic conditions the energy yield is usually lower, organic compounds are oxidized usually more slowly and incompletely, and the outside medium may contain residues poor in oxygen and is mostly noxious.

Through their activity, soil organisms affect the cycles of carbon (q.v.), nitrogen (q.v.), phosphorus (q.v.) and other elements in nature. The most important metabolic processes in soil are transformations of organic materials. The entire organic fraction of soil originates either directly or indirectly from the residues of higher plants and to a lower extent, from the mass and metabolic products of autotrophic microorganisms. Other organisms use the organic material as the source of carbon and nitrogen compounds required for the building of their bodies and as the source of energy. These organisms too enrich the soil organic components by their excretions and metabolic products and after they perish, by their own bodies. One must take into consideration the enrichment of soil with organic materials introduced purposely by man. The organic material is present in the soil in different stages of transformation, together with all the products of organisms living in the soil. Qualitatively speaking, the spectrum of organic compounds in the soil is very broad, but their participation and significance may differ from one group to another, and their role may vary depending on various factors. The most important are those that arrive in the soil in the largest amounts and thus, in the function of substrate, determine the direction of metabolic processes. This holds especially for the sugars (e.g., cellulose, hemicelluloses, starch, chitin); lignin and its derivatives, which play an important role in the formation of humus components; and finally organic nitrogen-containing compounds such as proteins. Degradation of organic material liberates biogenic elements that serve as nutrition for plants. The principal product formed is carbon dioxide (in a medium fertile soil 5–10 ton/ha in 3 months), some one third of which dissolves in water to carbonic acid, which facilitates the solution of nutrients, the rest escaping into the atmosphere and permitting an increased plant assimilation. The amount of carbon dioxide produced by medium-fertile soil covers its consumption by green plants.

Another important metabolic process is the conversion of some of the organic matter to humus (*humification*) through resynthesis from simpler degradation products. Humification involves a complicated set of reactions in which the metabolism of aromatic structures plays a major role. This process may be effected sequentially by different groups of microorganisms (fungi, nonsporulating and sporulating bacteria, cellulolytic myxobacteria, actinomycetes) as well as by tiny animals. These microorganisms disintegrate the plant residues, thereby increasing their active surface, which facilitates the degradative action of microorganisms, and enriching the organic complement by the excreted nitrogen. Some animals—in particular, earthworms—form true humus substances in their digestive tract, thus contributing to the formation of the crumbly structure of soil.

Of the various transformations of nitrogen, the most important ones are mineralization of organic nitrogenous substances such as *ammonification* (i.e., gradual breakdown, say, of proteins accompanied by liberation of ammonia into the surrounding soil); immobilization of nitrogen; *nitrification* (formation of nitrites and nitrates from ammonia); *denitrification;* and symbiotic and nonsymbiotic fixation of atmospheric nitrogen. Of the metabolic conversions of other elements, mention should be made of the mineralization and immobilization of phosphorus and sulfur, transformations of iron, potassium, and manganese. The character of the parent rock material and of mineral components of soil can play a pronounced role here. Gradually, more and more importance is acquired by transformations of substances introduced into soil artificially, e.g., pesticides or pollution effluents from various sources. The great majority of organic and inorganic compounds, whether natural or artificial, can be attacked and metabolized by soil organisms under suitable conditions.

Environmental Conditions and Metabolic Activity of Soils

The soil metabolic activity is closely related to the physical, chemical, and biological conditions of the soil milieu. To the most important environmental factors belong the amount of water and air in soil, the content of organic and mineral nutrients, presence of clay minerals and the associated oxidation-reduction, osmotic and adsorption conditions, and acidity or basicity of soil, temperature, light, amount, composition, activity and mutual relationships between the organisms present. The environmental conditions can act at the subcellular (molecular) level, at the level of the cell, or at the level of the entire population of soil organisms. In the case of the subcellular level, modifications of environmental conditions can result in a change of the course of a single metabolic reaction, the synthesis or activity of a single enzyme. An example of this situation is the repression of synthesis of nitrate reductase in the presence of oxygen. In the case of the single cell, the activity of a given single cell is affected. An example of this situation is in the changes of respiratory activity of cells due to their adsorption on negatively charged particles in the soil. Finally, the activity and dynamics of individual species or physiological groups of soil organisms or of their entire population may be affected. There are several examples of this: The presence of some phenolic compounds may curtail the growth of the micropopulation of sporulating bacteria; in the close vicinity of plant roots, in the so-called rhizosphere (q.v.) continuously enriched by organic compounds in the form of root exudates, the amount, composition, activity, and properties of microorganisms differ substantially from those in nonrhizosphere soil; and alternate flooding of rice fields results in complex changes in the composition and activity of all soil organisms.

The metabolic activity of soil is influenced by the results of very complicated direct and indirect relationships between the various environmental factors. The effect of one factor on the metabolic activity of soil may be influenced by a change of another environmental factor either additively or antagonistically. Since metabolic activity usually results in environmental changes, there may exist an interaction between metabolic processes occurring in parallel. Continuous changes in the set of biological and nonbiological conditions of the soil environment and their response may be characterized as a dynamic steady state.

The nature of metabolic reactions in soil and their dynamic aspects are both temporal and vectorial. Individual metabolic changes proceed in a time sequence. One reaction may be essential for another, the activity of one group of organisms sets the stage for the operation of other organisms, which may be reflected in the appearance of whole metabolic pathways, of successions of microbial populations, and of trophic and energetic chains at all levels of the ecosystem. The vectorial aspect follows from the fact that the gaseous and the liquid components of soil usually undergo diffusion or another type of movement. Along with the movement of the organisms themselves, this is an essential condition for the transport of ions, of dissolved substances, or of living or nonliving particles. An example is the gradual oxidation of the ammonium ion to nitrites and nitrates in the nitrification process in which the first reaction is carried out by bacteria of the genus Nitrosomonas, the second by bacteria of the genus Nitrobacter. When these compounds are carried by the flow of water in soil in a certain direction, the reaction sequence $NH_4^+ \rightarrow NO_2^- \rightarrow NO_3^-$ may be expressed as a function of distance and of time. Another example is derived from the above-mentioned succession of the various microbial groups with different metabolic activities in the humification process, or generally the gradual changes in the quality and quantity of metabolic activity in the soil-forming process. The metabolic and the population succession may be seen clearly in the soils of rice fields after their inundation when the green algae are the first to grow actively, to be succeeded by sulfate-reducing bacteria that form black ferrous sulfide, further by purple bacteria (Thiorhodaceae), and finally by the nonsulfur purple bacteria (Athiorhodaceae). The availability of organic substrates in the rhizosphere may also be characterized as the function of distance from their source, i.e., from the plant roots. This causes changes not only in quantity (in the direction from the root, the amount and activity of microorganisms decreases) but also in quality (in the root vicinity there is an accumulation of microflora with more complex nutritional requirements, with different composition and metabolic capacities than are found outside the rhizosphere). Moreover, the temporal changes in the system are influenced by the quantity and the quality of root exudates in the course of plant growth and vegetation stand development.

When evaluating the signs of metabolic activity in soil, one should consider the mutual relationships with the surroundings in their full complexity and dynamic nature. Individual environmental factors do not affect metabolic activity separately but in combination. One factor, however, may predominate in the complex. Thus the effects of temperature and of water content in soil result in periodic maxima and minima of metabolic activity. In arable soils of the temperate zone, there is a regular spring and lower fall maximum, along with a summer and winter minimum. In forest soils where shading results in a slower heating and drying, the spring maximum is shifted into the summer month. In addition, a "major" periodicity may be observed, manifested by the variation of average values of activity and development of microorganisms in the course of years, as well as a "minor" periodicity within a single day. By analogy, changes in weather are reflected directly in the metabolic activity of soil. The complex effect of several factors such as the soil type, vegetation cover, content and quality of humus, and aeration is displayed clearly in the vertical distribution of metabolic activity in the soil profile. The intensity of metabolic activity generally decreases in the downward direction due to the decreasing amount of organic material and availability of oxygen. The diminishing quantity of microorganisms and of their activity is apparently associated with the soil type (e.g., Chernozem, Brown Soil, Podzol) because the thickness of the biologically active soil layer decreases in this order. The biological profile is attenuated in Podzols with a structurally and physically unfavorable elluvial A_2 horizon and the illuvial *B* horizon. The lowest, so-called *C* horizon, contains substances practically unaffected by microbial activity. Even greater differences in the profile distribution may be found in the individual types of metabolic activity. Thus, e.g., the processes of nitrification, fixation of atmospheric nitrogen, and decomposition of cellulose may be detected in Podzols only in a thin surface horizon because at greater depth the processes are inhibited by the lack of air, organic compounds, and an acid reaction.

A given soil type is characterized, among other things, by the amount and quality of structural aggregates. It was shown that the intensity of metabolic processes is closely related to the size and hence to the specific surface area of the particles present. Processes occurring within the individual structural aggregates may differ qualitatively from those occurring on the surface. The rate of oxygen consumption on the aggregate surface is greater than inside the aggregate because of easier diffusion and hence greater availability. Under suitable conditions, aerobic processes predominate on the aggregate surface while anaerobiosis prevails inside. A certain type of enzyme activity may be associated with particles of a given size. Thus invertase activity is found mainly in the fine silt fraction while dehydrogenases occur mainly in the clay fraction.

Metabolic activity of cultivated soils may differ both as a whole and in some particular aspect from the activity in noncultivated soil due to conditions altered by human interference (different contents of organic and mineral substances after fertilizing; modification of the favorable water-air ratio by plowing and other mechanical treatment of soil, or by irrigation; different soil structure; different vegetation cover, etc.). Cultivated soils are generally richer in microorganisms and more active biologically than virgin soils. In forest soils the set of conditions affecting metabolic activity differs in many respects from arable soils. The forest soil contains less water in the horizon with the highest number of roots, the highest moisture content in the superficial layer, lower variation of temperature, lower light intensity, lower air capacity, higher acidity. It is enriched annually with greater amounts of organic matter (leaf litter), containing poorly degradable compounds (lignin, fats, waxes, resins, etc.), and undergoing breakdown in completely organic surroundings and frequently with difficulty. This results in striking differences in the intensity and course of many metabolic processes occurring in field and forest soils.

The complex influence of the climate and of the soil type causes horizontal and vertical geographic distribution of soil organisms and hence of metabolic activity of soils. The number of microorganisms and metabolic activity of soil increases from polar toward equatorial regions, and the proportion of bacteria decreases in favor of actinomycetes. Bacilli requiring organic oxygen dominate in the northern soils. Similar changes in metabolic activity can be observed with respect to altitude.

Detection of Metabolic Activity in Soils

In view of the multifarious expressions of metabolic activity, one cannot determine it as a whole complex but simply in some characteristic processes, reactions, and their time phases (mineralization of organic compounds, increase of biomass, ammonization, nitrification, respiration, etc.). The course and intensity of a metabolic reaction is usually assessed from the decrement of the starting substance (substrate) or from the formation of intermediates or of the final product. A number of methods have been developed for demonstrating various aspects of soil metabolic activity. Many are derived from methods common in other fields of biology or biochemistry (respirometric and enzymological techniques, the use of isotopes, etc.); others are specific to work with soil metabolism (e.g., the percolation methods and the continuous-flow technique). The most important and the best developed are those techniques that have to do with general metabolic processes such as respiration or those that make it possible to study the intensity of microbial activity associated with the supply of mineral nutrients for plants and with other desirable properties underlying soil fertility (e.g., mineralization and humification processes). An example in this context is the estimation of respiration as one of the more general functions of living organisms. It is expressed as respiratory capacity according to the amount of oxygen (O_2) used for respiration or according to the amount of respiratory carbon dioxide (CO_2) evolved. The respiratory capacity depends on the activity of the microflora, on the amount, quality and availability of organic substances and mineral nutrients, moisture, pH, the oxidation-reduction potential, temperature, and other environmental factors. Thus it can be used for determining the deficient factor when the remaining factors are at their optimum. The method of estimating respiratory capacity is useful for defining the physiologically available plant nutrients, mainly nitrogen and phosphorus. It is an important factor for assessing the dynamics of biological processes in soil, degradation of organic compounds, and course of the soil-forming process. The higher the respiration intensity, the more nitrogen and other nutrients are released in the mineral form into soil solutions. Excessive intensity of mineralization processes results in a powerful mobilization of mineral nutrients of plants as well as in a disturbance of the equilibrium of the organic component and especially humus in soil and in a deterioration of soil fertility. The *respiratory capacity* of a soil sample without any additions is designated the *actual capacity;* respiratory capacity of a sample enriched with readily available and degradable substrate is designated the *potential capacity*. The potential respiratory capacity expresses the mineralization activity of the microflora of a soil sample; the actual respiratory capacity then expresses to what extent this capacity is used under the given conditions. The respiratory capacity in the collected samples is determined in the laboratory (or with greater difficulty, directly in the field) by measuring the consumption of O_2 or the production of CO_2. The uptake of oxygen is estimated by manometric or volumetric methods, by potentiometry or polarography. Formation of CO_2 is determined manometrically, volumetrically, gravimetrically, interferometrically, and conductometrically, and by titration. For aerobic respiration, one can use both the consumption of O_2 and production of CO_2; for anaerobic respiration only the formation of CO_2 is available.

Enzymological methods are widely applied to estimating the metabolic activity in soil. Activity estimations of extracellular enzymes as a component of soil metabolism is advantageous due mainly to their simplicity and to their exclusion of living and multiplying microorganisms. However, because it has not been possible to reliably determine the origin, localization, state, and role of enzymes in soil, one cannot draw conclusions about the relationship between enzyme activity and the total biological activity of soil, and between soil fertility and the potential significance for practical agriculture. Existing data indicate that a positive correlation between the activity of free and extracellular enzymes, on the one hand, and microbial counts in soil or respiratory capacity, on the other, is rather exceptional. Even if such a correlation exists (and it probably does), new and more reliable methods are required to demonstrate it unequivocally.

Estimation of the metabolic transformations in soil, involves the use of compounds containing either radioactive or heavy isotopes of some elements. The most common isotopes used are ^{14}C, ^{32}P, ^{35}S, and ^{13}C and ^{15}N. These elements are among the most important in biochemical processes, and their fate can offer a clear picture of the behavior of the given compound in soil and define the origin of metabolic products found in soil. In addition to the above-mentioned isotopes, a number of others are used in other fields of biology and biochemistry. The isotope content is determined by measuring its radioactivity; it can also be determined in a mass spectrometer. The development of gas chromatography made it possible to study in greater detail changes of gaseous metabolites in soil.

Metabolic transformations are mostly examined in *in situ,* batch, cultures. Using stationary cultures, substrate is offered to a soil sample in a single dose, and its fate is then monitored in various ways. A special form of these experiments, which is specific for soil microbiology, is the percolation technique. A column of soil is continuously perfused with the same solution and simultaneously aerated. The composition of the percolation (perfusion) solution changes by the activity of soil microorganisms and may be analyzed at suitable time intervals without disturbing the soil sample. Compounds, the transformation of which is followed, are added directly to the soil or to the percolation solution. The technique was introduced into soil microbiology by Lees and Quastel (1946) and Audus (1946). It meant distinct progress on the soil microbiology technique, and it permitted a deeper understanding of the mechanisms of soil metabolism. Still, like all other methods used so far, it studies soil as a closed system. Under natural conditions, soil is an open system with various substances arriving into it and others removed from it. For this reason, much attention was paid to the continuous-flow method, as introduced into soil microbiology by Macura and Málek (1958). A solution of constant substrate concentration is passed through the soil sample column at a constant and controllable rate. The solution can be collected and analyzed after passing through the sample. The soil is continuously aerated, and the composition of gases escaping from it is analyzed. After a certain time, a steady state is reached when substrate is metabolized at a constant rate. Thus the factor of supplying a changing substrate concentration can be eliminated. Soil can be considered an open system in this case. The method may be considered a special form of selective cultivation, permitting examination of the metabolic properties of accumulated microbial community, and mathematical expression of them from both the temporal and the vectorial point of view.

Significance of Metabolic Activity in Soils

The metabolic activity of soil determines the most important soil properties such as biological activity and fertility. Some aspects of metabolic activity are reflected in these properties. During metabolic processes in soil, nutrients are liberated and made available for plants and may accumulate. Metabolic activity cannot be neglected in the formation of a suitable soil structure and in soil-forming processes in general. Recently the questions of decontamination of polluted soil have come into prominence as they cannot be satisfactorily solved without regard to the metabolic properties of soil organisms.

Understanding of the basis of metabolic transformations and of their course under the complex conditions of the soil offers a basis for looking for new possibilities of interfering with these processes so as to make them useful for human activity. In practice, two approaches are used for affecting the course and intensity of metabolic processes in soil. The first approach is to enrich soil with a certain organism that possesses the required metabolic properties and can encourage certain soil activity. An example is inoculation of soil with bacteria capable of fixing atmospheric nitrogen, enriching the soil with it, and increasing the store of this essential nutrient for plants. To this end, inoculation was carried out with preparations of *Azotobacter* and silicate bacteria or various mixtures of microorganisms. The other approach is to alter some environmental factors that can affect metabolic activity, e.g., by suitable soil processing; by agrotechnical, amelioration and irrigation measures; organic and mineral-fertilization systems; and adapting crop rotation, and adding calcium.

F. KUNC

References

Alexander, M., 1977. *Introduction to Soil Microbiology*, 2nd ed. New York: Wiley, 467p.

Alexander, M., 1965. Biodegradation: Problems of molecular recalcitrance and microbial fallibility, *Adv. Appl. Microbiol.*, **7**, 35–80.

Alexander, M., 1971. *Microbial Ecology*. New York: Wiley, 511 p.

Audus, L. J., 1946. A new soil perfusion apparatus, *Nature*, **158**, 419.

Bartholomew, W. V., and Clark, F. E., eds., 1965. *Soil Nitrogen*. Madison: American Society of Agronomy, 287–502.

Dickinson, C. H., and Pugh, G. J. F., eds., 1974. *Biology of Plant Litter Decomposition*, 2 vols. New York: Academic Press, 288 + 610p.

Dommergues, Y., and Mangenot, F., 1970. *Écologie Microbienne du Sol*. Paris: Masson et Cie., 796p.

Domsch, K. H., 1962. Bodenatmung. Sammelbericht über Methoden and Ergebnisse, *Zentralblatt f. Bakteriologie, etc., II.*, **116**, 33–78.

Durand, G., 1965. Les enzymes dans le sol, *Rev. Ecol. Biol. Sol*, **2**, 141–205.

Galstyan, A. S.; Tatevosian, G. S.; and Havoundjian, Z. S., 1968. Fixation of enzymes by soil fractions of different particle sizes, *9th Internat. Congr. Soil Sci. Trans. (Bucharest)*, **3**, 281–288.

Gray, T. R. G., and Parkinson, D., eds., 1968. *The Ecology of Soil Bacteria*. Liverpool: Liverpool University Press, 681p.

Hattori, T., 1973. *Microbial Life in the Soil. An Introduction*. New York: Marcel Dekker, 427p.

Hofmann, E., and Seegerer, A., 1950. Der Fermentgehalt des Bodens als Massstab seiner biologischen Aktivität, *Biochem. Z.*, **321**, 97.

Kearney, P. C., and Kaufman, D. D., eds., 1969. *Degradation of Herbicides*. New York: Marcel Dekker, 394p.

Kononova, M. M., 1961. *Soil Organic Matter. Its Nature, its Role in Soil Formation and in Soil Fertility*. Oxford: Pergamon, 450p.

Lees, H., and Quastel, J. H., 1946. Biochemistry of nitrification in soil, 1., Addendum by H. Lees: A soil perfusion apparatus, *Biochem. Jour.*, **40**, 803–815.

Macura, J., and Málek, I., 1958. Continuous flow method for the study of microbiological processes in soil samples, *Nature*, **182**, 1796–1797.

McLaren, A. D., and Peterson, G. H., eds., 1967. *Soil Biochemistry*, vol. 1. New York: Marcel Dekker, 528p.

McLaren, A. D., and Skujins, J., eds., 1971. *Soil Biochemistry*, vol. 2. New York: Marcel Dekker, 576p.

Mishustin, E. N., 1956. *Microorganisms and Soil Fertility* (In Russian). Moscow: Izd. Akad. Nauk USSR, 246p.

Novogrudsky, D. M., 1956. *Soil Microbiology* (In Russian). Alma Ata: Izd. Akad. Nauk Kaz. SSR, 397p.

Pochon, J., and de Barjac, H., 1958. *Traité de Microbiologie des Sols. Applications Agronomiques*. Paris: Dunod, 685p.

Postgate, J., 1972. *Biological Nitrogen Fixation*. Watford: Marrow, 61p.

Quastel, J. H., 1946. *Soil Metabolism*. A lecture before the Royal Institute of Chemistry of Great Britain and Ireland, London.

Rosswall, T., ed., 1973. Modern methods in the study of microbial ecology, *Bull. Ecol. Res. Comm. (Stockholm)*, **17**, 13–28, 179–479.

Stoklasa, J., and Doerell, E. G., 1926. Handbuch der biophysikalischen und biochemischen Durchforschung des Bodens. Berlin: Parey.

Stotzky, G., 1972. Activity, ecology, and population dynamics of microbes in soil, *Crit. Rev. Microbiol.*, **2**, 59–137.

Waksman, S. A., 1938, *Humus. Origin, Chemical Composition and Importance in Nature*. London: Bailière, Tindall and Cox, 526p.

Waksman, S. A., 1952. *Soil Microbiology*. New York: Wiley, 356p.

Winogradsky, S. N., 1949. *Microbiologie du Sol. Problèmes et Méthodes.* Paris: Masson et Cie., 861p.

Cross-references: *Algae; Aeration, Respiration, Atmosphere; Carbon Cycle; Diffusion Processes; Enzyme Activity; Humus; Microbial Ecology and Clay Minerals; Microhabitats; Nitrogen Cycle; Phosphorus Cycle; Polysaccharide; Rhizosphere; Sludge Disposal; Soil Components, Organic; Soil Fertility; Soil Microbiology; Sulfur Transformations.*

METAL-COMPLEXING MACROMOLECULES

Organic macromolecules in soils are found mainly in the form of phenol polymers with linked amino acids, peptides, and other organic residues. They contain many functional groups, the principal ones being carboxilic, phenolic, and alcoholic OH, primary and secondary amine, etc. Oxygen, nitrogen, and less frequently sulphur and phosphorus can be found in the groups; these elements can behave as electron-donor atoms (bases) when interactions with electron-acceptor atoms (acids), i.e., metal ions, occur. Soil humus (q.v.) represents a mixture of numerous organic compounds, which lasts long in soil in close association with the inorganic fraction. Two types of polymers, humic acids and polysaccharides (q.v.), may, however, constitute up to 90% of the total soil humus (Felbeck, 1971). The term *humic acid* denotes the humus fraction that is soluble in alkali, precipitated by acid, and brown to almost black in color (see *Humus*). Humic acid-type molecules and polysaccharides with a relatively small molecular weight constitute a fraction that is soluble in alkaline and acidic solutions and is called *fulvic acid.* The molecular weight of humic acids vary from a few thousands to 200 thousands and are amorphous disks or flat ellipsoidal particles having a spongy structure (Orlov and Glebova, 1972). Humic acids in alkaline media form a solution of negatively charged hydrophillic colloidal particles exhibiting properties characteristic of these types of colloids (Schnitzer and Khan, 1972).

The *resistant soil organic matter* (SOM), or humus, is believed to be of microbial origin either through decomposition of animal and vegetal debris or through direct biosynthesis. Autoxidative and enzymatic polymerization of various precursors such as simple phenols, organic acids, and amino acids are also considered important mechanisms influencing soil humus formation (Haider and Martin, 1975; Haider, Martin, and Filip, 1974; Martin and Haider, 1971).

Metal ions in a soil system originate in the pure inorganic parent material from which the soil is derived, or they are imported from other environments by runoff or drainage waters. In the beginning of a soil-formation process, the action of simple metal-complexing substances, i.e., organic acids and simple phenols, excreted by some pioneering microorganisms (Zajic, 1969), together with other weathering processes of physical and chemical nature, disrupt the crystal lattices of the parent materials freeing metal ions to the incipient soil system.

Organic macromolecules in soil exhibit a strong metal ion-binding ability that is performed mainly by coordinated linkages between the functional groups or active sites of the macromolecule and the metal ions (Kononova, 1961). It seems that macromolecules bind the metal ions according to the nature of the electron-donor atoms of the functional groups available for reaction on the active surface of the colloidal particles (Zunino, 1975). Thus O-containing sites would preferentially bind metals that can be classified as class-A metals according to the concepts given by Ahrland, Chatt, and Davies (1958), i.e., alkali and alkaline-earth elements, Nitrogen, sulphur, and phosphorus-containing groups would preferentially bind class-B and transition metals, i.e., iron, manganese, copper, and zinc.

Several metallic elements are essential for plants and microorganisms to develop properly. These include the so-called macronutrients (q.v.) calcium, magnesium, and potassium and micronutrients (q.v.) such as iron, copper, manganese, and zinc. The essential metal ions are absorbed by the plant roots from the soil solution (q.v.) whereas these metals are present as free metal ions, aquo complexes, and/or simple organic complexes. Highly polymerized organic macromolecules in soil would act as the main reservoir or source of some essential metals for the soil solution thus governing the availability to plants of these metals.

The sequence shown in Figure 1 was proposed by Zunino (1975) to describe the behavior of metal-complexing macromolecules in soil in connection to the translocation of metal ions from rocks to biological systems. The gradual translocation of metals would resemble a cascade process wherein competition between soluble complexants and insoluble metal-complexing macromolecules primarily govern the availability of metal ions to living organisms. The macromolecule-metal associations highly saturated in metal ions would ultimately constitute the actual pool or reservoir of essential metals to the living world. Also metal-binding macromolecules avoid losses by leaching

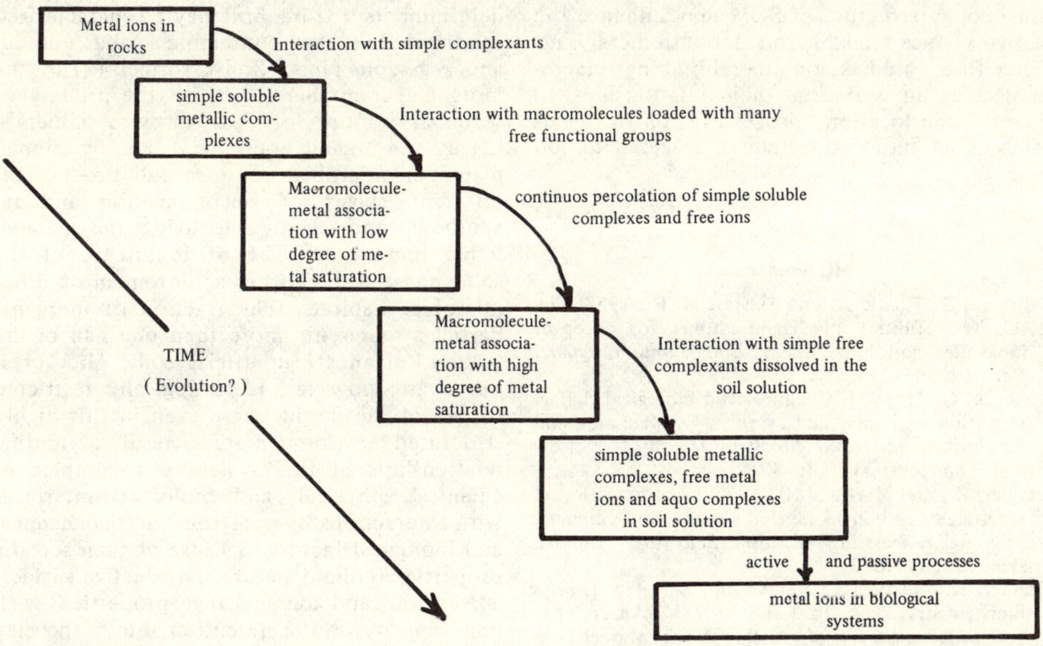

FIGURE 1. Behavior of metal-complexing macromolecules in soil in relation to the translocation of metal ions from rocks to biological systems. (After Zunino, 1975)

(q.v.) of essential metals from the soil profile to groundwaters.

Humic acid-like polymers synthesized by soil microbes constitute a good model that can be used to study the interactions between metal ions and macromolecules (Zunino et al, 1975). It has been demonstrated that the adsorption of cations on these macromolecules follow the Langmuir adsorption isotherm (Figure 2).

From this type of curve, a *maximum metal-binding ability* (MBA) can be calculated for the polymers (Table 1). The MBA for the several

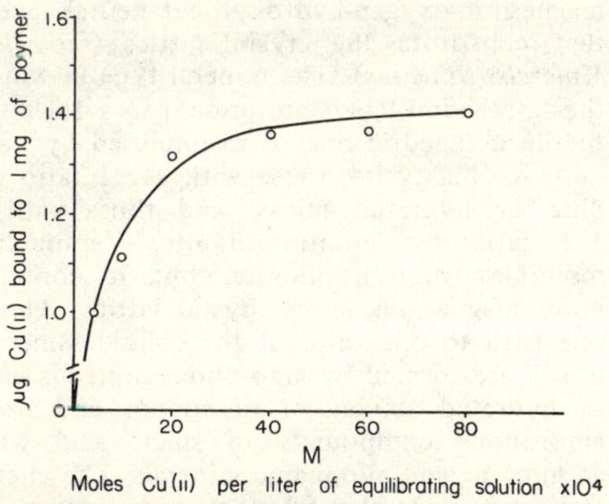

FIGURE 2. Adsorption of copper (II) onto a humic acid-like polymer from *Eurotium echinulatum*.

metals studied ranged from 0.08 to 2.26 atom-micrograms (μg) of metal per milligram (mg) of macromolecule. For the microbial polymers the sequence of MBA magnitudes was: Cu > Ca > Mg > Zn. This sequence is explained on the basis of the electron-acceptor properties of these elements. The data of Table 1 also show that different microbial species present different MBA. The microbial ecology in soil may thus influence the translocation of metals.

The spatial geometry and the nature of the electron-donor atoms of the active sites in the native SOM macromolecules may confer to the soil humus a high degree of specificity to bind certain class-B and transition metal ions. Also

TABLE 1. Maximum Metal-Binding Ability of Some Microbial Polymers.

MBA (Atom-micrograms of Metal, mg^{-1} of Polymer)				
Polymer Source	Cu(II)	Ca(II)	Mg(II)	Zn(II)
Stachybotris atra, humic acid	0.29	0.27	0.10	0.10
Hendersonula toruloidea, humic acid	0.59	0.44	0.22	0.15
Eurotium echinulatum, humic acid	1.45	0.68	0.41	0.30
Hansenula holstii, polysaccharide	0.19	0.05	0.18	0.08

From Zunino (1975)

the spongy structure of SOM may enhance the active surface available to react with metal ions. Therefore, studies on metal-binding macromolecules in soil and their relationships to metal translocation processes constitute at present an important field of research in soil science.

HUGO ZUNINO

References

Ahrland, S.; Chatt, J.; and Davies, M. R., 1958. The relative affinities of ligand atoms for acceptor molecules and ions, *Chem. Soc. London, Quart. Rev.*, **12**, 265–276.

Felbeck, G. T., 1971. Chemical and biological characterization of humic matter, *in* A. D. MacLaren and S. Skujins, eds., *Soil Biochemistry, Vol. 2.* New York: Marcel Dekker, 36–57.

Haider, K., and Martin, J. P., 1975. Decomposition of specifically carbon-14 labeled benzoic and cinnamic acid derivatives in soil, *Soil Sci. Soc. America Proc.*, 39, 657–662.

Haider, K.; Martin, J. P.; and Filip, Z., 1974. Humus biochemistry, *in* E. Paul and A. D. MacLaren, eds., *Soil Biochemistry, Vol. 4.* New York: Marcel Dekker, 195–244.

Kononova, M. M., 1961. *Soil Organic Matter.* New York: Pergamon Press, 450p.

Martin, J. P., and Haider, K., 1971. Microbiological activity in relation to soil humus formation, *Soil Sci.* **111**, 54–63.

Orlov, D. S., and Glebova, G. I., 1972. Electron Microscopy of humus. *Soviet Soil Sci.*, **4**, 445–451.

Schnitzer, M., and Khan, S. U., 1972. *Humic Substances in the Environment.* New York: Marcel Dekker, 327p.

Zajic, J. E., 1969. *Microbial Biogeochemistry.* New York: Academic Press, 345p.

Zunino, H., 1975. Metal-ion binding ability of some polymers synthesized by soil microorganisms and a discussion of their role in the translocation of essential elements from rocks to biological systems. Ph.D. Dissertation, University of California, Riverside.

Zunino, H.; Peirano, P.; Aguilera, M.; and Schalscha, E. B., 1975. Measurement of metal-complexing ability of polyfunctional macromolecules: A discussion of the relationship between the metal-complexing properties of extracted soil organic matter and soil genesis and plant nutrition, *Soil Sci.*, **119**, 210–216.

Cross-references: *Adsorption Phenomena; Exchange Phenomena; Humus; Soil Components, Organic; Transport Processes.*

MICROBIAL ECOLOGY AND CLAY MINERALS

Microbial ecology examines the relationships between microorganisms and their environment. The clay minerals represent an important component of the solid phase of soil, and they determine its texture. Soil may be characterized as a tripartite system containing a solid, a liquid, and a gaseous phase. Soil is formed (1) by the inorganic component, i.e., by the primary or secondary more or less disperse minerals, (2) by the organic component, i.e., by animal, plant, or microbial residues and products at different stages of decomposition and resynthesis, and (3) by the living and metabolizing microbiota. The predominance of the solid phase distinguishes soil from most other microbial habitats. The fraction of inorganic particles represents more than one half of the volume of most superficial soils. Microorganisms are, however, physiologically restricted to life in the liquid phase even in this highly structured environment. The result of mutual relationships in this system is a complex of chemical, physical, and biological interfaces with interactions of numerous physicochemical and biological factors. Because of their specific properties (colloid nature, large active surface, adsorption, and ion-exchange properties, swelling capacity, water retention, etc.), the clay minerals can strikingly influence the ecology, activity, and population dynamics of soil microorganisms, either through a direct influence on soil microbiota and/or by indirectly affecting other environmental conditions.

Ecologically Important Properties of Clay Minerals

Soil is a matrix of organic and inorganic colloid systems, a greater part of the latter being formed by clay minerals. The clay fraction is usually represented by crystalline or amorphous aluminosilicate minerals with different contents of magnesium, calcium, iron, potassium, sodium, etc. The individual types of crystalline minerals differ in the ratio of silicon-oxygen tetrahedral layers to the aluminum or magnesium-oxygen-hydroxyl octahedral layers that constitute the crystal lattice (see *Clay Minerals, Silicates*). The mineral type in which the tetrahedral layers are present in a 1:1 ratio to the octahedric ones is exemplified by kaolinite or halloysite, those with a 2:1 ratio by illite or hydrated micas, and those with a 1:2 ratio by montmorillonite. Vermiculite resembles montmorillonite, but it contains more magnesium in its crystal lattice. About one fifth to one third of the colloid minerals in soil are formed by amorphous minerals such as hydrated oxides of aluminum and iron, amorphous compounds of silicic acid with aluminum, and allophanic minerals. The chemical and mineralogical composition suggests differences in the physical and chemical properties of the individual types of clay minerals such as the magnitude of surface charge, the cation exchange capacity, specific surface, and water retention, swelling capacity.

The clay fraction includes mineral particles of colloid nature with diameters less than 2 μm. Due to this small size and to the dispersion of particles, the clay component of soil is characterized by a substantial specific surface. The clay particles carry an electric charge on their surface. The nature of the charge and the resulting ratio of attractive and repulsive forces will determine whether the particles will aggregate, coagulate, and form aggregates with other clay particles with organic colloids or with microbial cells. The mechanisms involved in these interactions can be the van der Waals forces and electrical double layers. The formation of a surface charge is assisted by ionogenic groups of interfacial macromolecules; the magnitude of the surface charge depends on the number of these groups and on the amount, type, and valency of the cations associated with it. Clay minerals can reversibly adsorb cations and anions; the adsorbed ions can, under certain conditions, be replaced by others present in solution. The adsorptive capacity of the clay minerals is expressed by the *cation exchange capacity* (CEC), i.e., the maximum amount of exchangeable cations the clay mineral can bind. The CEC differs strikingly for individual types of minerals; for kaolinite it ranges 1–15 mequiv/100 g clay (milliequivalents per 100 g of clay); for illite, 15–35 mequiv/100 g; for montmorillonite, 70–150 mequiv/100 g clay. The CEC value is proportional to the effective surface of clay; i.e., it rises with decreasing particle size. Adsorption and cation exchange can occur not only at the surface but also at the edges and at the accessible internal surface between the crystal lattice layers. In this manner one can explain the higher CEC values in clays of the 1:2 types such as montmorillonite. The force with which the ions are bound depends on their type and valency. Univalent cations (Na^+, K^+, NH_4^+) are more readily exchangeable than the divalent and tervalent ones (Ca^{2+}, Mg^{2+}, Fe^{3+}, Al^{3+}). The affinity of a given ion for the clay depends also on its concentration in solution. The degree of saturation of clay with a certain cation determines in a decisive way the relation of the soil clay fraction to the other components and properties of the environment (see *Exchange Phenomena*).

Clay minerals can absorb and retain water and change plastically their volume, which is reflected by a rather complicated physical phenomenon designated as *swelling*. The swelling capacity is different for different types of clays, being associated with their crystal structure. High values are found in minerals of the 1:2 type such as montmorillonite or vermiculite, where not only a surface adsorption of water takes place (increase of the diffuse enveloping layer) but where water can invade between the individual elementary layers that are pushed apart as it enters. The swelling intensity depends on the type of cations adsorbed on the minerals. Thus montmorillonite with a certain content of Ca^{2+} can absorb 200–300% water (w̃/w) while the Na^+-montmorillonite can absorb as much as 600% water. On the other hand, soil minerals such as kaolinite and illite contribute to soil swelling only negligibly. The clay minerals thus play an important role in the formation and regulation of the water regime in soils, e.g., by affecting the reserves, retention, permeability (q.v.), and hence its availability for soil organisms. The ability of clay minerals to retain water is probably one of the main causes why most of the microbial activity is associated with the clay mineral fraction. Water-coated particles can exist isolated but usually form aggregates with the aqueous film surrounding the aggregate and forming a link to the other aggregates. This is one of the basic mechanisms of formation of the so-called microhabitats (q.v.) in soil. The volume changes of clay particles caused by swelling and shrinking further influence the physical properties of soil (e.g., its porosity).

The role of ecologically important properties of the soil clay fraction depends on the amount of clay particles, on the type of mineral, on the participation of the various types and on their interactions with the other soil environmental factors. The more pronounced effects are generally observed in clays with a higher cation exchange and swelling capacity.

Direct Interactions Between Microorganisms and Clays

The clay particles interact with microbial cells whenever they come into direct contact. Depending on the relative size of the two components, the cells may adhere to the clay mineral (if the cells are smaller) or the clay particles to the cell surface. The site of contact between the adsorbent particle and the microbial cell may be the cell wall, a mucous capsule, flagella, fimbriae, etc. The adsorbed cells can act as connecting bridges or layers between clay particles and thus contribute to the formation of *organomineral structural aggregates,* which are a typical feature of the soil microenvironment. On the surface of larger particles, microorganisms can form microcolonies. The mechanisms permitting contact between microbial cells and clay particles are rather complicated. There are phenomena of physical and chemical adsorption, i.e., depending on the surface charge of the participating components, on formation of ionic bonds, binding of water by living and mineral particles, ability of microorganisms to produce various mucous substances, etc. Factors bearing on the adhesion of microorganisms and clay mineral particles are numerous and depend not only on the number and nature of microorganisms (species, age, cell size, motility, surface charge, etc.) but also on the amount and properties of the clay component (clay type, particle size, surface charge, cation exchange capacity, saturation by certain ionic species, amount of retained water, etc.). There may be an indirect role of various environmental conditions (e.g., soil acidity since the surface charge density of bacterial cells as well as of clay particles is a function of pH). The interplay between the forces of attraction and repulsion between particles determines whether the particles will adhere. The interaction between bacterial cells and clay particles can develop either by a slow, long-range process, or there may be rapidly formed bacterium-clay complexes or aggregates if the repulsion factors are negligibly small. An example of this type is a rapid formation of complexes caused by interaction between ionogenic groups, e.g., amino or carboxyl, at the bacterial surface and the oppositely charged clay face or its ionic shell. The contact is usually reversible but desorption is usually not complete. A change of a certain environmental condition resulting in desorption will not affect all the mechanisms contributing to the adsorption.

The mutual adhesion between soil microorganisms and clay minerals results in pronounced alterations in the physiological character of the microorganisms. Some biochemical processes may be completely suppressed in the adsorbed cells while others may be stimulated. In some cases there may be no effect whatever. Adsorbed microorganisms were found to grow more rapidly and to respire more vigorously than nonadsorbed ones. The activity of adsorbed enzymes can be decreased or increased as compared with the activity of nonadsorbed enzymes since the conditions of formation of the enzyme-substrate complex are different. Adsorption can influence the availability of substrate and the activity of enzyme. The surface ionic layer of adsorbent can directly affect the pH of the medium (deflect it from the biochemical optimum) or alter the diffusion of substrate or nutrients and thus support or antagonize a certain biochemical reaction. The adsorbed particles can mask some of the surface area across which metabolic products and exoenzymes are released and nutrients taken up, which results in an impaired metabolic activity (q.v.) of the cell. The ability of clay minerals to bind water on their surface or within their crystal structure can have a direct bearing on the metabolism of adsorbed cells. This affects the concentration and availability of substrates and nutrients, as well as of biologically active agents produced by the cells (e.g., antibiotics, toxins, gibberellins), the osmotic conditions, etc. Microbial cells adsorbed to clay particles better resist the unfavorable effects of drying; this protective effect of clays is apparently associated with their ability to bind water.

Indirect Effect of Clay Minerals on Soil Microflora

The clay minerals affect the ecology of soil microorganisms indirectly by altering other environmental conditions. The presence of clays may be reflected in changes of physical, chemical, and biological factors influencing the microbial populations and in changes of the mutual relationships between these factors. The change of one environmental factor acts simultaneously and/or subsequently on other factors and may reach back to the properties of the clay component of soil itself. Due to this effect, the practically infinite number of permutations of factors have a greater summary effect on soil microorganisms than a single factor would display in a less complex medium. The relative significance of a given factor follows from the sum of changes in the whole heterogenous, discontinuous, and structured soil complex. Even the clay minerals themselves may be considered as a complex ecological factor. All these complicated relationships must be taken into account when considering the possible effects of clay minerals on the individual environmental factors. Every summation is necessarily simplifying and may serve solely as an illustration of the complexity of these phenomena.

One of the most important ecological factors that can be substantially affected in soil by the presence of clay minerals is moisture. The water-holding capacity of soil is increased by the retentive capacity of clays, and drying is slowed down. The water content is directly linked to the concentration of substances in the soil solution (organic and inorganic nutrients, extracellular enzymes, growth factors and biologically active compounds such as antibiotics and various inhibitors) and to their availability and effectivity for soil microorganisms, to osmotic phenomena, the gaseous phase volume, and to oxidation-reduction conditions in soil. These soil properties are appreciably affected by clay minerals owing to their ion exchange and adsorption properties. These play an important role for the extent of the cation exchange and buffering capacity of soil, for soil acidity (q.v.), and ionic composition of the soil solution (q.v.) in general. They can represent the cause of decreased accessibility of organic substrates to soil microorganisms and of stabilization of the soil organic fraction. Clays will determine in a decisive manner the soil structure, formation of soil aggregates, their size and shape, and hence the soil porosity. The number and diameter of pores is then critical for the extent and the rate of movement of the liquid and gaseous soil phases. Changes in the volume of the clay particles due to swelling and shrinking are instrumental in the formation of the pores. It depends on the presence of clay minerals whether a soil can be classified as "heavy" or "light" and what type of spatial and temperature conditions will prevail in it as would affect the physiological processes of a microbial population. Last but not least, clay minerals are important in the formation, properties, and dynamics of soil aggregate microenvironments as the most characteristic sites of existence of soil microorganisms.

Changes in environmental conditions caused by clay minerals will be reflected in the living component of the soil ecosystem. It may be suppressed, enhanced, or unaltered in its metabolic activity, be it in the carbon cycle, nitrogen cycle, or cycles of other elements or in the processes of growth and reproduction of microorganisms responsible for the given process. The clay minerals affect positively or negatively such important processes as the mineralization of organic substances in soil, humification, turnover of proteins, amino acids, urea, ammonia, formation of nitrates, fixation of atmospheric nitrogen, biological transformations of sulfur and phosphorus, and degradation or inactivation of pesticides. Clays contribute to the protection of soil microorganisms from various environmental stresses such as desiccation and increased temperature, hypertonic osmotic pressure, various kinds of radiation. Clay particles may alter the antagonistic or competitive relationships among microorganisms by adsorbing or inactivating antibiotics, toxins, or other biologically active compounds, by changing the availability of nutrients, substrates, or metabolic products. Soil-borne viruses can be protected from decomposition; adsorption of bacteriophages and actinophages can prevent the lysis of their potential hosts. Such effects of clay minerals are usually accompanied by changes in the character and composition of the soil microbial population since the altered environmental conditions can be favorable for development and activity of a certain group of microorganisms and unfavorable for another one. The action of clay minerals on the individual components of soil microflora is selective and differentiated as follows from the heterogeneity of clay types, of the mechanisms of their effects and of environmental conditions that are affected by them. An example of such effects may be the significantly lower incidence of pathogenic fungi (*Fusarium oxysporum f. cubense, Histoplasma capsulatum*) of bacteria causing enzootic leptospirosis and of various human pathogens in soils containing montmorillonite than in localities without it.

Experimental Approaches

The complex character of the system formed by clay minerals, microorganisms, and the soil environment requires that the experimental approach to their investigation be integrated. It is necessary to assimilate and apply the concepts and techniques from different scientific fields (physics, chemistry, biology, geology, mineralogy, pedology, agriculture). All the components of the ecosystem (i.e., microorganisms, clays, and the soil environment) must be studied individually, in different combination, and in the whole complex at various levels of experimental and system complexity from the natural conditions of field experiments to the artificial model systems. In connection with surface interaction studies, one can investigate the adsorption of clays, cells, and metabolites, the physical and chemical characteristics of clay minerals, of clay-organic complexes, electrophoretic mobility; one can also use electron microscopy. Under pure-culture conditions, one follows microbial growth and metabolism. Soil plate studies permit the examination of growth rates, interactions between microorganisms, genetic transfer, both in sterilized and reinoculated and in nonsterile soil samples that can be used for studying autotrophic and heterotrophic metabolic processes. Greenhouse and field studies, and observations when research is done under the most complex natural conditions, complete this experimental hierarchy.

Integrated and multipronged research make it possible to evaluate more completely the observed phenomena and their mechanisms, to check the extent to which the various phenomena observed in simpler, laboratory conditions in vitro can occur in the more complete natural system in soil in situ, and to explain what mechanisms will participate in processes observed under natural conditions. A detailed understanding of the mechanisms of a certain phenomenon and of the conditions under which it can take place allows a rational control by man. Solution of these problems then facilitates the control of desirable or undesirable microorganisms and of their activity under natural conditions by altering the composition of the soil clay fraction. Their potential significance for agriculture, public health, and environmental control is obvious.

F. KUNC

References

Alexander, M., 1977. *Introduction to Soil Microbiology,* 2nd ed. New York: Wiley, 467p.

Alexander, M., 1964. Biochemical ecology of soil microorganisms, *Ann. Rev. Microbiol.,* **18,** 217–252.

Alexander, M., 1971*a*. Biochemical ecology of soil microorganisms, *Ann. Rev. Microbiol.,* **25,** 361–392.

Alexander, M., 1971*b*. *Microbial Ecology.* New York: Wiley, 511p.

Brock, T. D., 1966. *Principles of Microbial Ecology.* Englewood Cliffs, N.J.: Prentice-Hall, 306p.

Dommergues, Y., and Mangenot, F., 1970. *Écologie Microbienne du Sol.* Paris: Masson et Cie., 796p.

Grim, R. E., 1953. *Clay Mineralogy.* New York: McGraw-Hill, 596p.

Hattori, T., 1973. *Microbial Life in the Soil. An Introduction.* New York: Marcel Dekker, 427p.

Marshall, K. C., 1971. Sorptive interactions between soil particles and microorganisms, *in* A. D. McLaren and J. Skujins, eds., *Soil Biochemistry,* vol. 2. New York: Marcel Dekker, 409–445.

Marshall, K. C., 1975. Clay mineralogy in relation to survival of soil bacteria, *Ann. Rev. Phytopathology,* **13,** 357–373.

McLaren, A. D., and Skujins, J., 1968. The physical environment of micro-organisms in soil, *in* T. R. G. Gray and D. Parkinson, eds., *The Ecology of Soil Bacteria.* Liverpool: Liverpool University Press, 3–24.

Pochon, J., and deBarjac, H., 1958. *Traité de Microbiologie des Sols. Applications Agronomiques.* Paris: Dunod, 685p.

Stotzky, G., 1967. Clay minerals and microbial ecology, *Trans. N.Y. Acad. Sci., II.,* **30,** 11–21.

Stotzky, G., 1972. Activity, ecology, and population dynamics of microbes in soil, *Crit. Rev. Microbiol.,* **2,** 59–137.

Stotzky, G., 1973. Techniques to study interactions between microorganisms and clay minerals in vivo and in vitro, *in* T. Rosswall, ed., Modern methods in the study of microbial ecology, *Bull. Ecol. Res. Comm. (Stockholm),* **17,** 17–28.

Zvyagintsev, D. G., 1973. *Interactions of Microorganisms and Solid Surfaces* (in Russian). Moscow: Publ. House of Moscow University, 176p.

Cross-references: *Adsorption Phenomena; Clay Minerals, Silicates; Clay-Organic Reactions; Enzyme Activity; Exchange Phenomena; Humus; Metabolic Activity; Microhabitats.*

MICROFAUNA—See FAUNA; SOIL MICROBIOLOGY

MICROFLORA—See ALGAE; SOIL MICROBIOLOGY

MICROHABITATS

Understanding the microorganisms in soil depends in turn on understanding the size, extent, and nature of the microhabitats and the range of influence of individual species (Burges, 1958). Microhabitats are the voids of various kinds within a soil horizon in which bacteria, fungi, and the like can live.

Living Spaces for Soil Microorganisms

Sizes of Pores. Soil microorganisms live in the pore spaces; thus the sizes, shapes, and distribution of the pores are of great importance to those organisms. Pore-size distribution (q.v.) in turn is closely related to soil structure (q.v.), the arrangement of sand, silt, clay, and organic matter into small aggregates or clumps, and the combination of those into larger aggregates. According to Vershenin (1966), soil structures can be classified on the basis of unit sizes of aggregates as follows: (1) fine microstructure (less than 0.01 mm); (2) coarse microstructure (0.25–.01 mm); (3) macrostructure (10–0.25 mm); and (4) megastructure (greater than 10 mm). Names such as *crumbs* and *granules* are given to units of macrostructure and lumps or clods to units of megastructure. The two smaller sizes are collectively called *microstructure,* and individual structural units may be called *aggregates* (without a modifier).

Like the structure of a soil horizon, the fauna can be classified according to size, as illustrated in Figure 1. The *megafauna, macrofauna,* and *mesofauna* can live only outside units of the megastructure. *Microfauna* live outside units of the microstructure. Algae and fungi live mainly outside units of the fine microstructure, and only bacteria can live within those units.

Moisture Retention. Requirements of individual species of microorganisms for water differ. They can therefore be divided into three groups: (1) *hygrophiles* requiring high humidity, (2) *mesophiles* requiring moderate humidity, and (3) *xerophiles,* requiring little humidity. Bacteria and algae generally require high humidity. Comparative growth rates thus depend partly on water supply, and moisture retention within a soil horizon is related to pore-size distribution (see *Water Content and Retention*).

Water is held strongly in smaller pores such as those in the smallest aggregates. Bacterial cells

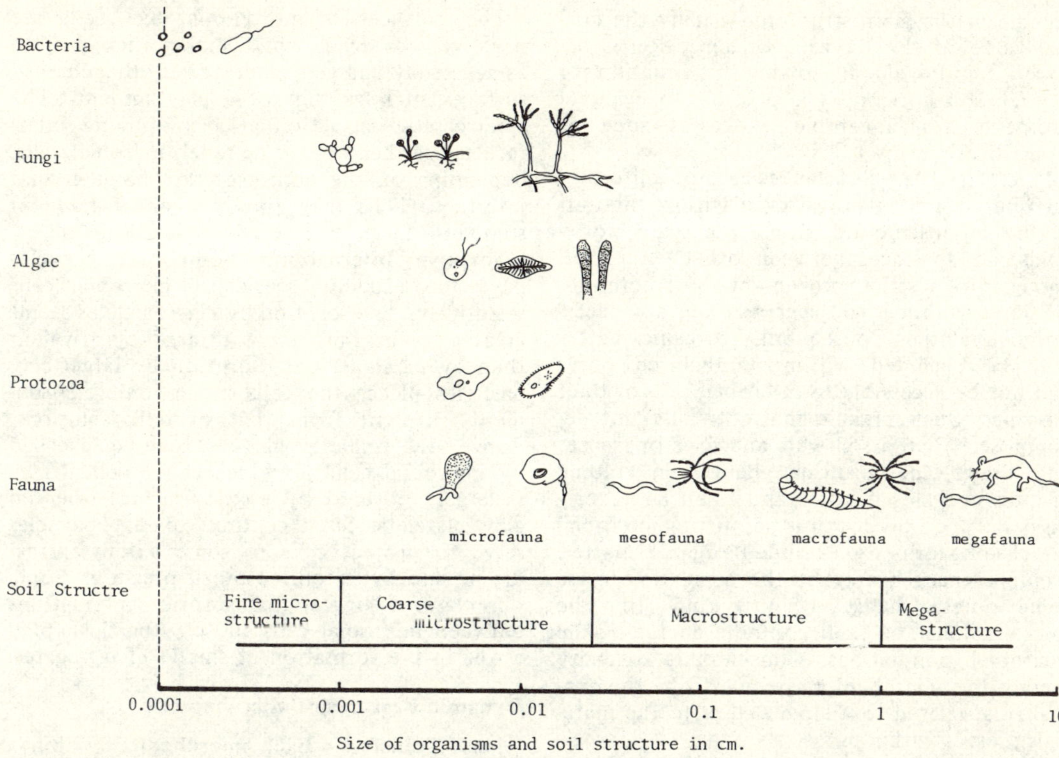

FIGURE 1. Diagram of the sizes of soil organisms and of the structural units or aggregates in a soil horizon. Most organisms live exclusively in pores, which are individually a fractional part of the smallest aggregate.

are therefore more abundant inside than outside the microstructure under moderate or low moisture levels. Under such conditions, fungi may grow luxuriantly outside the microstructure without being susceptible to vigorous attack by bacteria (Figure 2). When the moisture content of a soil horizon is high and temperatures are favorable, bacterial cells grow rapidly both inside and outside the microstructure. Fungal mycelia may then be attacked frequently by bacteria.

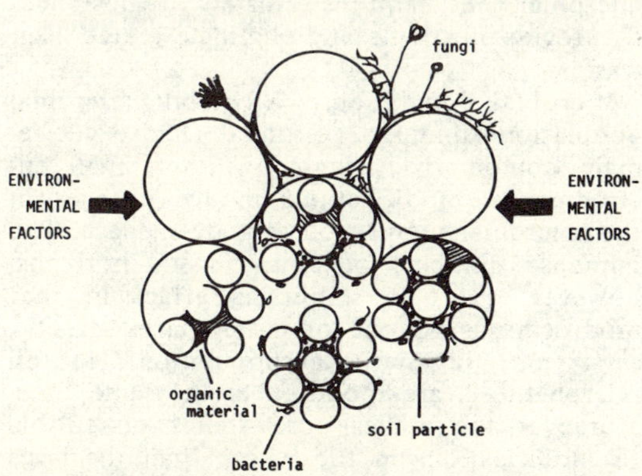

FIGURE 2. Diagram indicates microbial habitats in a soil aggregate. Larger circles represent silt grains.

Oxygen Diffusion. The availability of oxygen, like that of water, greatly affects the lives of the organisms within the microstructure. Thus the nature of that structure can determine whether aerobic or anaerobic metabolisms will prevail. At high levels of oxygen, aerobic metabolisms function (see *Metabolic Activity*). If the oxygen concentration in the pore spaces falls to very low levels, anaerobic metabolisms function. Amounts of oxygen available to microorganisms within the microstructure depend on diffusion of the gas among and into the aggregates. Greenwood and Berry (1962) developed an equation to define the volume (V_0) of the spherical space within which anaerobic processes could operate, viz.:

$$6DC_1 = M \cdot R_1^{\,2} \left[1 + 2\frac{V_0}{V_1} - 3\left(\frac{V_0}{V_1}\right)^{2/3} \right]$$

where R_1 is the radius of a unit in the microstructure, V_1 the volume of a unit in the microstructure, D is the diffusion coefficient, C_1 is the oxygen content outside the microstructure, and M is the rate of oxygen uptake per unit of volume of the microstructure. Usually the critical value of the oxygen concentration is between 2 and 6 μM. By solving this equation for a given soil horizon, it is possible to evaluate prospects for anaerobic processes such as denitrification (q.v.).

Pore Materials. Substances are not uniformly distributed among pores, even within one unit of the microstructure. Some pores are closed voids, and passages between others are small enough to restrict movement. Furthermore, all pores within a soil aggregate are not necessarily inhabited. Consequently, organic matter may be occluded within uninhabited pores and not be accessible to microbial cells or their enzymes. Such organic matter would not decompose. When a soil wets and drys or freezes and thaws, the resulting changes in volume can disrupt closed pores and open larger passages between others, thus permitting microbial attack on formerly occluded organic matter. Volume changes may at the same time close some pores. Changes of this kind alter the microhabitats and their contents, including the microbial populations. Considerable diversity among the microhabitats seems to be the general rule, related to a large extent to the materials present within pores.

Effects of Colloidal Particles

Ionic Interactions. The hydrogen ions produced by the metabolisms of microorganisms are exchanged for bases held by clay minerals and organic matter in the pore walls of microhabitats (see *Exchange Phenomena*). Such exchanges tend to stabilize pH within pores and thus permit continued existence of the microbial cells. Organic compounds such as amino acids, proteins, and humic substances are adsorbed on clay mineral surfaces. Subsequently, those substances are less generally susceptible to microbial decomposition. Reactions between clay particles and such substances tend to protect them from attack by microbial cells.

Clay minerals in pore walls affect the concentrations of ions in solution near their surfaces. Concentrations of cations such as hydrogen tend to be greater near the surface of a clay micelle than in the bulk solution a short distance away. In contrast, the concentrations of anions such as the negative radicals of organic acids are lower near clay mineral surfaces. These differences in ionic concentrations, related to clay mineral surfaces, directly affect the activities of microbial cells and their enzymes. Small changes in ionic composition of the solution surrounding microbial cells may induce changes in their metabolism and morphology.

The surfaces of most microbial cells are negatively charged, which makes them behave as negatively charged spheres near other charged surfaces such as that of a clay mineral. The electrostatic attraction at a solid-liquid interface can either attract or repel microbial cells, depending on the nature of the charge at that interface. Cells may thus be concentrated at some interfaces.

Sorptive Interactions. Some soil particles have the capacity to adsorb microbial cells selectively. The sorption by clay particles seems greater when they are saturated by trivalent than by divalent ions. Sorption by larger clay particles places the cells in an ionic environment different from that of bulk solutions. Conversely, fine clay particles may be adsorbed by microbial cells, in which case a cell surface is largely covered by edge-to-surface–oriented clay platelets. Such coatings of clay particles serve to protect cells to some extent against drying and hypertonic osmotic pressures. Some observations suggest that sorption interactions between microbial cells and clay particles play a role in the formation of small soil aggregates.

Dynamic Features of Microhabitats

Mass Transfers. Each microhabitat within a soil horizon is not a closed system, even though many differences can be found among the multitude present. Solutions of various kinds can move through the larger pores and the various channels. Substances can then diffuse into smaller pores. Some microbial cells may also be transferred from one set of pores to another by percolating water. Substances produced by microorganisms in their life processes such as nitrate from ammonia by nitrifiers move out of pores, and some part of the substance may be lost from the soil as a whole. The process of nitrification (see *Nitrogen Cycle*), as with any other life processes of microbial cells, requires transfers of substances to provide nutrients and eliminate metabolism wastes.

Microbial Populations. All in all, microbial populations undergo continual change yet remain comparatively stable. Rather small environmental modifications induce changes in the microbial population in a given space. The immense numbers of pores in soil horizons, however, exert a stabilizing effect in that modifications of one or more spaces affect a small part of the total population. Not all microhabitats are modified at the same time. Moreover, the smallest pores are less susceptible to modification on the whole than the large ones. The microbial population living in the larger spaces, e.g., those among units of the macrostructure and megastructure, is subject to extreme fluctuations in numbers. In many instances recovery seems to be as rapid as the disappearance.

Natural Soil Aggregates. Changes in the structural units or natural aggregates proceed more slowly than those in the microbial population. Over time, however, the structural units are also changed in various ways, which then alter microhabitats. Those can be destroyed and new ones constructed by modifications of soil aggregates of various sizes. Thus the microhabitats in a soil horizon are, like the microbial population, in a state of dynamic equilibrium.

TSUTOMU HATTORI

References

Burges, A., 1958. *Microorganisms in the Soil.* London: Hutchinson University Library, 173p.

Greenwood, D. J., and Berry, G., 1962. Aerobic respiration in soil crumbs, *Nature (London),* **195,** 161-163.

Hattori, T., 1973. *Microbial Life in the Soil: An Introduction.* New York: Marcel Dekker, 209-312.

McLaren, A. D., 1970. Temporal and vectorial reactions of nitrogen in soil: A review, *Canadian Jour. Soil Sci.,* **50,** 97-109.

Vershenin, P. V., 1966. Solid phase of soil as the basis of its physical regime, *in* A. F. Joffe and I. B. Revut, eds., *Fundamentals of Agrophysics.* Jerusalem: Israel Program for Scientific Translations, p. 187.

Cross-references: *Aeration, Respiration, Atmosphere; Aggregation; Algae; Exchange Phenomena; Humus; Metabolic Activity; Microbial Ecology and Clay Minerals; Microorganism Transport; Pore-size Distribution; Soil Microbiology; Soil Pores; Soil Solution; Soil Structure.*

MICRONUTRIENTS

Of the total number of elements in soils, only sixteen are recognized as essential for the healthy development of plants, animals, or microorganisms. Those elements that are essential for plant growth are called *plant nutrients* (q.v.). Individual nutrient concentrations that exceed 1000 ppm are characteristic of the following *macronutrients* (q.v.): hydrogen (H), carbon (C), oxygen (0), nitrogen (N), potassium (K), calcium (Ca), magnesium (Mg), phosphorus (P), and sulfur (S). Present in smaller amounts, the *micronutrients* are chlorine (C1), boron (B), iron (Fe), manganese (Mn), zinc (Zn), copper (Cu), and molybdenum (Mo). They are also sometimes called *minor nutrients* (Berger et al, 1961) and differ from trace elements (q.v.) in that the latter are not necessarily essential to plant growth. In view of the fact that only a limited number of higher plants have been tested, it is possible that other elements occurring in the soil in small amounts will eventually be added to the list. There is some evidence, for example, that trace amounts of cobalt (Co), Sodium (Na), silica (Si), and selenium (Se) may be required by some higher plants (Mitchell, 1964), and iodine (I) and vanadium (V) by certain species of soil algae (Epstein, 1971).

Source of Micronutrients

Micronutrients in soils are derived mainly from parent rocks in situ or from transported parent materials. Because some micronutrients are fairly mobile in certain forms, i.e., they can be transported relatively long distances in groundwaters, their concentration in any given profile may not be indicative of underlying primary sources. Organic matter is an important secondary source of micronutrients.

Inorganic Forms. Micronutrients, like other elements, are released to the soil system through various weathering processes. Because the sources of micronutrients differ from area to area and because knowledge of specific compounds in which they occur is incomplete, only general statements can be made about micronutrient sources. The main primary sources are given in Table 1. Because some of the micronutrients can be used to track down the location of near-surface mineralized zones, exploration geochemists have recently taken a keen interest in the derivation of trace elements in soils (Levinson, 1974). Much of their future work will consequently be of interest to soil scientists.

Organic Forms. Micronutrients are frequently held as complex combinations by organic colloids or within the structure of organic compounds. Many profiles have the greatest concentrations of micronutrients in the surface soils because surficial enrichment is associated with litter fall. The organic compounds in soils that form stable complexes with metal ions include polyphenols, amino acids, peptides, proteins, polysaccharides (q.v.), and the humic and fulvic acids (Stevenson and Ardankani, 1972). These and other chelating

TABLE 1. Major Natural Sources of Micronutrients.

Element	Major Forms in Nature
Iron	Oxides, sulfides, and silicates
Manganese	Oxides, silicates, and carbonates
Zinc	Sulfides, oxides, silicates
Copper	Sulfides, hydroxy carbonates
Boron	Borosilicates, borates
Molybdenum	Sulfides, molybdates
Chlorine	Chlorides
Cobalt	Silicates

After Brady (1974)

agents, which are thus important secondary sources of micronutrients, also affect the release and availability of micronutrients from metal-chelate complexes (see *Metal-complexing Macromolecules*).

Distribution and Occurrence in Profiles

The distribution of micronutrients in profiles depends on the elemental abundance in the parent materials and on the nature of the weathering and soil-forming processes. Differences in micronutrient concentrations should thus be expected on the basis of these two fundamental processes. Even though it would be desirable to make generalizations about the distributions of micronutrients in soil profiles, these would have limited validity because of the great number of variables involved. Factors such as the leaching regime (see Leaching), pH and Eh conditions, mechanical translocation, biological activity, and partitioning of elements contribute to the various concentrations of micronutrients in soils. General estimates of total micronutrient contents of mineral soils (Vinogradov, 1959; Mitchell, 1964, 1972; Andrews-Jones, 1968) do, however, give some idea of concentration levels. The values listed in Table 2 indicate the concentration ranges frequently encountered in mineral soils. These values are rough approximations of total concentration levels.

Investigation of the availability of micronutrients to plants requires some knowledge of the phase or phases in which elements can be expected to occur. Micronutrients can, for example, be associated with inorganic exchange materials such as clay minerals or with precipitated iron, aluminum or manganese oxides and organic complexes. The effect of adsorption and binding reactions by soil colloidal particles on the availability of micronutrients has been described through use of the Freundlich and Langmuir equations (see *Adsorption Phenomena*). Ellis and Knezek (1972) emphasize the need to relate adsorption and desorption of micronutrients to levels in soil solution (q.v.) in order to control toxicities and deficiencies. Precipitated oxides, which coat particles of all size fractions, frequently contain the highest concentration of micronutrients due to scavenging effects (Horsnail and Elliott, 1971). In fact, micronutrients are frequently found to be more abundant in the oxide and sesquioxide phases than in the original soil or any of the cleaned soil fractions (Mitchell, 1964). Because soil organic matter forms stable combinations with metal ions, micronutrients are also frequently associated with the organic phase (Hodgson, 1963). Organic substances are important in the transportation, concentration, and release of micronutrients. Chelating agents (see *Metal-complexing Macromolecules*) are particularly important because they function as vehicles for the movement of micronutrients to roots (Stevenson and Ardakani, 1972).

In addition to occurring in different phases and size fractions, micronutrients may also occur in greater concentrations in different parts of the profile. They are, for example, sometimes mainly associated with organic-rich surface horizons in Chernozems (Mollisols) (Mitchell, 1964), with *B* horizons in lateritic soils (Zeissink, 1971) and with *C* horizons in some soils. Because distribution patterns for the same element differ among profiles of the same great soil group, general characterization of profile distribution is not feasible.

TABLE 2. Common Concentration Levels of Micronutrients in Mineral Soils.

Micronutrient	Normal Range (ppm)	Extreme Values (ppm)
Iron	15,000–40,000	10–80,000
Manganese	300–7,000	12–10,000
Chlorine	30–450	5–800
Vanadium	12–700	1–900
Zinc	10–500	4–10,000
Copper	3–100	0.1–1300
Boron	3–100	0.1–1300
Cobalt	1–60	0.1–600
Molybdenum	0.4–7	0.1–80
Selenium	0.1–3	0.1–70

After Vinogradov (1959); Mitchell (1964, 1972); Andrews-Jones (1968)

Micronutrient Balance in Soils

Even though micronutrients occur in soils in very small amounts, often only trace amounts, they are still essential for plant growth. Minute changes in concentrations at these low levels can result in deficient or toxic conditions. Because the micronutrient concentration is delicately balanced, any imbalance can drastically affect plant growth and animal nutrition. Removal by crops or livestock or additions may interfere with the micronutrient status of soils. Other management practices such as irrigation, drainage, tillage, fertilization with macronutrients only, deforestation, and vegetation burning can also affect the micronutrient balance. In addition to human interference, natural micronutrient imbalances in strongly leached acid sandy soils, muck soils, and strongly alkaline soils are also apt to limit crop growth. Still other factors such as inherently low levels of trace elements in rocks and soil parent materials, kinds of profiles (pedogenic regimes), plant species and varieties, and climatic and weather conditions interact to give rise to regional and local patterns of micronutrient problems (Kubota and Allaway, 1972). Antagonistic effects of some elements on the adsorption of micronutrients are another cause of imbalance. For example, excess copper or sulfate may adversely affect the uptake of molybdenum; iron deficiency is encouraged by an excess of zinc, manganese and copper; heavy nitrogen fertilization intensifies copper efficiency; and excess lime may reduce boron uptake.

Thus even though the amounts of micronutrients required by plants are small, the needs tend to be rather specific. When the micronutrient concentration drops below a certain limit, referred to as the *critical level* (See *Plant Nutrients*), deficiency symptoms occur in plants. Though the critical level for each micronutrient differs from plant to plant, specific low levels are commonly associated with general deficiency symptoms (Epstein, 1971; Sprague, 1964; Wallace, 1961). Micronutrient deficiencies in soils are normally counteracted by incorporating the necessary element or elements into fertilizers in amounts sufficient to correct the symptoms (Table 3).

Most micronutrient imbalances result in deficiencies of some element or elements (Beeson, 1945; Berger, 1962). Most remedial measures are thus designed to correct abnormally low levels. Under more unusual conditions some micronutrients may be present in the soil in such large amounts that they become toxic. Molybdenosis in ruminants and selenium toxicity are examples of problems associated with high levels of two micronutrients. Natural causes leading to micronutrient toxicity problems are usually related to high concentrations of trace elements in groundwaters used for irrigation, in rocks, and in soil parent materials. Micronutrient excesses may also arise from accidental environmental contamination, heavy fertilizer applications, or repeated application of insecticides or fungicides (see *Pollution*).

Toxicity problems, which have been reported for most micronutrients, differ in their severity and geographic distribution. Salient aspects of the problem as it exists in the United States have been summarized by Cannon (1969). Because micronutrient toxicities are commonly associated with high levels of molybdenum and selenium, the problems are briefly indicated here.

Molybdenum toxicity may be a serious endemic nutritional problem of ruminant animals in areas of poorly drained neutral to alkaline soils formed in granitic alluvium of wet, narrow floodplains and alluvial fans of small streams. The molybdenum-toxicity problem appears to be related to several factors, viz. (1) the granitic alluvium from which the soils are derived is high in molybdenum, (2) soil wetness accentuates the uptake of soil molybdenum by plants, and (3) these areas are used as pastures for ruminants (Kubota et al. 1961, 1967). Within the United States, selenium toxicity occurs mainly where soils were derived from sedimentary rocks of Permian or Cretaceous age (Rosenfeld and Beath, 1964). The native selenium in such soils is not available to all plants but can be absorbed by certain species, especially those of the *Astragalus* genus. After those plants die and return the selenium to the soil, the element can be absorbed by other species. Animals grazing on such areas can then suffer acute cases of selenium toxicity (Olson, 1969).

TABLE 3. Micronutrient Deficiency and Recommended Rates of Application.

Micronutrient	Area Deficient in U.S. (millions of hectares)	Critical Level (ppm)	Recommended Application Rates (kg of element/ha)
Iron	1.54	50–80	0.5 10
Manganese	5.26	15–35	5–30
Zinc	2.63	15–30	0.5–20
Copper	0.24	3–8	1–20
Boron	4.85	20–30	0.5–5
Molybdenum	0.61	0.01–0.3	0.05–1

After Brady (1974)

Conclusion

Maintenance of a proper balance among micronutrients in soils is especially important. Fortunately, this is not a problem in a majority of the soils producing food for the world as a whole. Deficiencies are of consequence, however, in a number of soils, and those are widely distributed over the world. Toxic levels of certain micronutrients are less common than deficiencies, but they also occur in a number of places. With increasing use of high-analysis fertilizers, development of improved crop varieties, and growing knowledge of plant nutrition, progressively more attention is being given to the micronutrient status and balance in soils.

CHARLES W. FINKL, JNR.

References

Andrews-Jones, D. A., 1968. The application of geochemical techniques to mineral exploration, *Colo. Sch. Mines, Min. Indust. Bull.*, **11**(6), 31p.

Aubert, H., and Pinta, M., 1977. *Trace Elements in Soils.* New York: Elsevier, 394p.
Beeson, K. C., 1945. The occurrence of mineral nutritional diseases of plants and animals in the United States, *Soil Sci.*, **60**, 9-13.
Berger, K. C., 1962. Micronutrient deficiencies in the United States, *Jour. Agric. Food Chem.*, **10**, 178-181.
Berger, K. C.; Gammon, N.; Rhodes, H. F.; Chesnin, L.; Bingham, F. T.; Reisenauer, H. M.; and Allaway, W. A., 1961. Are minor elements important? *Crops Soils*, **13**, 7-10.
Brady, N. C., 1974. *The Nature and Properties of Soils.* New York: Macmillan, 639p.
Cannon, H. L., 1969. Trace element excesses and deficiencies in some geochemical provinces of the United States, *in* D. D. Hemphill, ed., *Trace Substances in Environmental Health III.* Columbia: University of Missouri Press, 21-43.
Ellis, B. G., and Knezek, D., 1972. Adsorption reactions of micronutrients in soils, *in* J. J. Mortvedt; P. M. Giordano; and W. L. Lindsay, eds., *Micronutrients in Agriculture.* Madison: Soil Science Society of America, 59-78.
Epstein, E., 1971. *Mineral Nutrition of Plants: Principles and Perspectives.* New York: Wiley, 412p.
Hodgson, J. F., 1963. Chemistry of the micronutrient elements in soils, *Adv. Agron.*, **15**, 119-159.
Horsnail, R. F., and Elliott, I. L., 1971. Some environmental influences on the secondary dispersion of molybdenum and copper in Western Canada, *Geochem. Explor., CIM Spec. Vol. 11*, 166-175.
Kubota, J.; Lazar, V. A.; Langan, C. N.; Beeson, K. C., 1961. The relationship of soils to molybdenum toxicity in cattle in Nevada, *Soil Sci. Soc. America Proc.*, **25**, 227-232.
Kubota, J.; Lazar, V. A.; Simonson, G. H.; and Hill, W. W., 1967. The relationship of soils to molybdenum toxicity in grazing animals in Oregon, *Soil Sci. Soc. America Proc.*, **31**, 667-671.
Kubota, J., and Allaway, W. H., 1972. Geographic distribution of trace element problems, *in* J. J. Mortvedt; P. M. Giardano; and W. L. Lindsay, eds., *Micronutrients in Agriculture.* Madison: Soil Science Society of America, 525-554.
Levinson, A. A., 1974. *Introduction to Exploration Geochemistry*. Calgary: Applied, 612p.
Mitchell, R. L., 1964. Trace elements in soils, *in* F. E. Bear, ed., *Chemistry of the Soil.* New York: Reinhold, 320-368.
Mitchell, R. L., 1972. Trace elements in soils and factors that affect their availability, *Geol. Soc. Amer. Bull.*, **83**, 1069-1076.
Olson, O. E., 1969. Selenium as a toxic factor in animal nutrition, *Proc. Georgia Nutr. Conf., Univ. Georgia, Feb., 1969*, 68-78.
Rosenfeld, I., and Beath, O. A., 1964. *Selenium.* New York: Academic, 411p.
Scheffer, F., and Schachtschabel, P., 1970. *Lehrbuch de Bodenkunde.* Stuttgart: Enke, 349p.
Sprague, H. B., ed., 1964. *Hunger Signs in Crops.* New York: McKay, 461p.
Stevenson, F. J., and Ardankani, M. S., 1972. Organic matter reactions involving micronutrients in soils, *in* J. J. Mortvedt; P. M. Giordano; and W. L. Lindsay, eds., *Micronutrients in Agriculture.* Madison: Soil Science Society of America, 79-114.
Stiles, W., 1961. *Trace Elements in Plants.* Cambridge: Cambridge University Press, 249p.
Vinogradov, A. P., 1959. *The Geochemistry of Rare and Dispersed Chemical Elements in Soils.* New York: Consultants Bureau, 209p.
Wallace, T., 1961. *The Diagnosis of Mineral Deficiencies in Plants.* New York: Chemical Publishing Company, 288p.
Zeissink, H. E., 1971. Trace element behavior in two nickeliferous laterite profiles, *Chem. Geol.*, **7**, 25-36.

Cross-references: *Fertilizers, Inorganic; Macronutrients; Metal-complexing Macromolecules; Plant Nutrients; Soil Fertility; Trace Elements.*

MICROORGANISMS—See DENITRIFICATION; ENZYME ACTIVITY; METABOLIC ACTIVITY; MICROORGANISM TRANSPORT; NITROGEN CYCLE; NITROGEN FIXATION; SULFUR TRANSFORMATIONS

MICROORGANISM TRANSPORT

The increasing reliance of soil as a medium for receiving wastes has resulted in a growing concern for the potential movement of microbes through the soil mantle. Although our understanding of microbial movement is not advanced, some modes of transport have been suggested. For convenience, this topic can be divided into two major categories involving transport via microbial growth and transport requiring the intervention of external forces.

Growth-related Transport

For organisms with *hyphal morphology*, movement within the soil is easy to visualize. As growth occurs, the organism literally extends itself from place to place. Movement can be visualized as occurring from one moisture film to another across a void. The fungi and the actinomycetes probably move by this mechanism. Still another form of transport is mass colony development. Extensive colony development is not common, even though organisms have a point distribution in soil. It is likely that colony development contributes only slightly to the total microorganism movement in soils.

While not specifically growth related, the active movement of motile organisms should be considered in this section. Numerous bacteria exhibit motility when grown in vitro. Presumptive evidence indicates that these same organisms possess the ability to move in situ, although Gray and Williams (1971) suggest that motility might be restricted to root surfaces (see *Rhizosphere*). However, not all microorganisms are motile or even pass through a motile stage.

Dispersion by External Forces

Microorganisms can be the passive passengers of other soil macroorganisms (see *Fauna*). Due to the surface configuration of the host or the microbe, attachment can occur, particularly if the surface is sticky or spiny. As the macroorganism moved through the soil, the microbe could be transported from place to place, perhaps knocked off as the host comes in contact with a portion of the solid matrix. It is also well known that numerous organisms are ingested by certain organisms. After transport the organism can be expelled into a new environment. Often with the ingestion of organic matter and egestion of feces, the fecal matter could even serve as an energy and nutrient source for the organism in its new environment.

Perhaps of greatest significance involves the dispersal of microbial fragments through soil by the intervention of external forces. In this instance, microorganisms can be transported by percolating water stemming from precipitation, irrigation, or application of waste-laden water to soil. From an environmental quality viewpoint, this transport mechanism may be the most important because it is independent of motility or microbe growth.

Recent studies indicate that this kind of dispersal is similar to miscible displacement described for anions (Wollum and Cassel, 1978). We have found that the relative concentration of organisms in the effluent of a soil column depends on both column length and the pore water velocity ($\bar{v}$). Maximum organism concentration was found for the column having the greatest $\bar{v}$, and concentration decreased in the effluent with decreasing $\bar{v}$. As $\bar{v}$ decreased, there was a broadening in the displacement peak and a greater volume of effluent collected before the maximum concentration was attained. In another study Dazzo, Smith, and Hubbell (1972) reported a complete removal of fecal coliforms in columns longer that 48 cm. Our data, however, indicates that organisms were detected in the effluent from columns 152 cm long. In all the displacement runs, there was extensive tailing of the displaced organism. Wollum and Cassel (1978) have also compared the displacement of two microorganisms with different surface properties and sizes. One organism consisted of smooth, nonsticky-surfaced conidia of a streptomycete and the other a bacterium that produces copious quantities of an extracellular polysaccharide (q.v.). The bacterium was larger than the conidia of the streptomycete. The conidia were more readily moved and occurred in greater concentrations than the bacterium in the effluent. This is in agreement with the work reported by Bitton, Tahav, and Henis (1974). They found that the movement of a capsulated *Klebsiella aerogenes* was restricted when compared with the movement of a noncapsulated form of *K. aerogenes*.

Not only do microbial properties affect transport but soil properties themselves exert a significant regulation in transport. Bitton, Tahav, and Henis (1974) showed that the retention of microbes was greater in clayey soils as compared to sandy soils. This is to be expected since clayey soils have a greater surface area; consequently, the potential for a surface mediated interaction between the microbe and solid phase would be greater for clayey soils. Also, in a clayey soil the size of individual pores would be much less, causing a significant restriction to movement within the soil matrix. Other properties such as clay type and the size of aggregates, undoubtedly play a significant role in governing the ease of microbial transport within the soil system.

Although microorganisms move within the soil matrix, more than 95% of applied organisms are trapped within the soil (Wollum and Cassel, 1978). These organisms are generally distributed throughout the soil with the greatest proportion in the upper part of the soil. However, microorganisms can be detected in the effluent even after prolonged leaching of the soil. This observation suggests that microorganisms unlike anions, might be trapped in small pores and a fraction of them released over a period of time into the water flowing through the porous medium.

A. G. WOLLUM, II

References

Bitton, G.; Tahav, N.; and Henis, Y., 1974. Movement and retention of *Klebsiella aerogenes* in soil columns, *Plant Soil*, **40**, 373–380.

Dazzo, F.; Smith, P.; and Hubbell, D., 1972. Vertical dispersal of fecal coliforms in Scranton fine sand, *Soil Crops Sci. Soc. Florida Proc.*, **32**, 99–102.

Gray, T. R. G., and Williams, S. T., 1971. *Soil Microorganisms.* Edinburgh: Oliver and Boyd, 240p.

Wollum, A. G., II, and Cassel, D. K., 1978. Transport of microorganisms in sand columns, *Soil Sci. Soc. America Jour.* **42**, 72–76.

Cross-references: *Microbial Ecology and Clay Minerals; Microhabitats; Pollution; Rhizosphere; Soil Microbiology.*

MICROSTRUCTURE, MANIPULATION

The structure of fine soils is mainly studied using the techniques of electron microscopy. To date, greater progress has been made with engineering aspects of "geological" soils, partly for historical reasons and partly because "pedological" soils often require high resolution scanning electron microscopy and X-ray microanalysis. Before discussing this progress, the nature of soil particles and the characteristics of pores will be considered. The terminology used here follows Smart (1969), although new schemes are also being considered by engineers (Collins and McGown 1974). Various methods have been reviewed by Smart (1973).

Particles

Beutelspacher and van der Marel (1968) and Gard (1971) have reviewed the electron microscopy of clay particles. Even though some minerals have characteristic shapes, there is no one-to-one relationship between morphology and mineralogy. For engineering purposes, the classification given in Table 1 is most useful. The terms *k-clay, L-clay,* and *M-Clay* correspond to the engineers' concepts of kaolinite, illite, and montmorillonite, respectively.

Because particles are often "dispersed" and "cleaned" for microscopy, there are doubts concerning their true size and shape. Weir (1960), for example, demonstrated the breakdown of smectite particles in water, and Tovey and Wong (1973) the breakdown of kaolinite particles during oven drying. Crystal structure may be studied by electron diffraction (Gard, 1971), with which Mering et al (in Gard, 1971) compared Camp-Berteaux and Wyoming bentonites. In water, one forms particles 100 Å in diameter, the other thin extensive flakes. These findings suggest that reclamation, drainage, wetting cycles, and freezing may induce physical changes in some clay particles.

TABLE 1. Tentative Classification of Small Soil Particles.

Name	Definition
Granules	All diameters approximately equal (a restricted usage)
k-clay	Rigid platy particles with small colloidal effect
L-clay	Rigid platy particles with large colloidal effect
M-clay	Flexible platy particles with large colloidal effect
Laths	Laths
Tubes	Tubes

Pictures of dispersed and cleaned particles and early hypothetical sketches of soil structure inferred that clay particles behaved like slippery rigid pennies. Whilst there is some support for this view for dispersed kaolin (and illite perhaps), it is not general, as evidenced, for example, by Weir's (1960) pictures of folded smectite plates. Further, the springiness of large mica flakes has long been recognized.

Contrast effects, e.g., Moire fringes, may indicate crystal defects, etc. (Gard, 1971). High-voltage electron microscopes are useful for studying these effects ((Smalley, Cabrera, and Hammond, 1973). Wet or vacuum decoration may locate electrostatic charges within crystals. Care is needed because (1) microtomy or rough treatment may cause artefacts, (2) the electron beam may break particles down causing artefacts, (3) regular features may be growth steps, and (4) irregular decoration features occur if the positions at which the decorating material lodges have no particular significance. These methods have shown that under suitable circumstances, clay plates have positive edges. They may also show the extent of local concentrations of negative charges; but such concentrations may be insignificant in chemical and mechanical problems, because they seem to be small and free to move within the crystal.

Another problem in determining the true nature of clay particles is that of differentiating them from stacks, packets, and quasicrystals. *Stacks* are groups of plates cemented together tightly, often by a true mineral-to-mineral bond. They are found as growth features in kaolin. *Packets* are small groups of plates arranged face to face with water between them. In micrographs, packets are indistinguishable from stacks except that the arrangement is irregular in packets. *Quasicrystals* are expanded smectite particles; the individual layers of smectite 10 or 20 or 30 Å thick are separated by water layers. It is debatable whether a quasicrystal is a particle or whether each of the thin smectite layers should itself be treated as a particle. The identification is complicated when a limited amount of water is available for expansion so that only part of the quasicrystal is expanded. It thus seems preferable to avoid a rigid definition of "a particle" but allow it to include aggregates and crumbs. Although a stack is clearly a particle, severe mechanical disturbances can reduce it to its component plates, in which case one particle is turned into several.

Pores

Pores are important elements of structure. Figure 1 summarizes some characteristics.

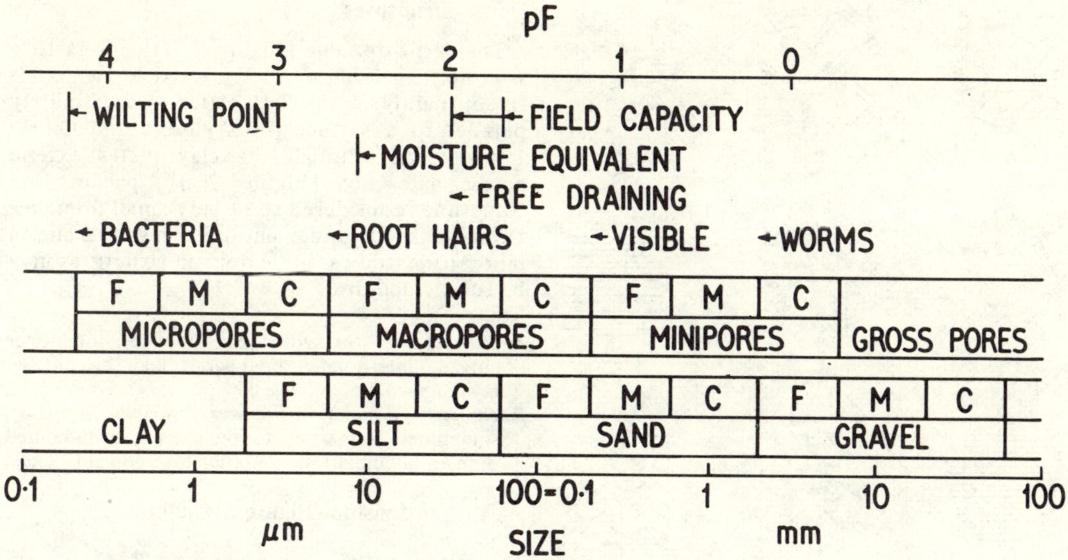

FIGURE 1. Pore-size classification.

It was drawn using:

$$hd = 0.3$$

where pF = log h and the equivalent cylindrical diameter is d cm. This corresponds to a surface tension of 73.5 dyn/cm at 15°C and is accurate within ±3% for ±15°C. Two values are shown for field capacity because there is no precise meaning of the term "free-draining". A tentative pore-size classification is shown together with the 2–6-system sometimes used for particle size (see *Soil Pores*). The major divisions at 200 μm, 6 μm, and 0.2 μm correspond approximately to the limits of resolution of the human eye, macrophotography, and optical microscopy, respectively. Pores larger than 200 mm could be called *caverns*; those smaller than 0.2 μm ultramicropores. Coarse, medium, and fine subdivisions would be required. Thus coarse macropores, 60–200 μm, would be freedraining, and medium macropores, 20–60 μm might be moderately freedraining depending on the circumstances. The 60 μm boundary would correspond to the definition of field capacity at pF 1.7 rather than pF 2.0(0.01 bars). As a rule of thumb, micropores would be available to bacteria, and macropores to root hairs. Silt (fine-sand) particles, less than 60 μm diameter, might be able to move through minipores; if so, these pores would be involved in internal erosion. Minipores of biological origin would be biopores. Usually in cultivated soils, gross pores would result only from cultivation. It is thus suggested that the sizes of pores should be measured at their minimum diameters or by desorption.

Open Structures

An early endeavor was to investigate the so-called *single-plate card-house structure* illustrated in Figure 2(a). Despite its popularity, surprisingly little work has been done to confirm the existence of this type. Two series of experiments should, however, be mentioned. Pusch (1970) observed the behavior of clay particles in a drop of water while the water was evaporated inside a high-voltage electron microscope; and O'Brien (1972) conducted a series of experiments on fresh sediments prepared in the laboratory. Smart (1969) summarized the types of open structures that had been observed in electron micrographs of undisturbed soils. The *book-house structure* [Figure 2(b)], a multiple-plate, card-house structure, has been frequently observed. *Flocculent structures* [Figure 2(c)], in which chains or tabular aggregates of particles form an open network, are also fairly common. Book-house structure, salt-flocculated structure, and lamellar structure seem to be synomyns; and the elongated flocs of flocculent structure are identical with or similar to stepped clusters. In some cases random clusters of particles have been seen mixed in among the chains, *Pusch's structure* [Figure 2(d)], or the clusters have occurred alone. The single-plate, card-house structure appears to be rare; one of the most "convincing" micrographs is of chlorite formed

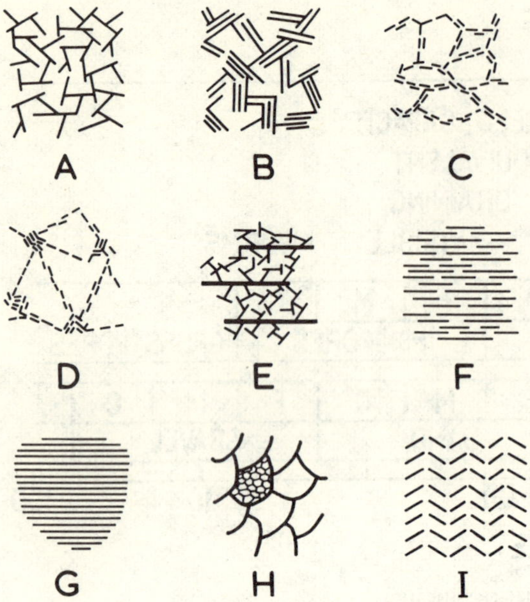

FIGURE 2. Types of structures. (a) Card-house structure, two-dimensional analog, each stroke represents a platy particle. (b) Book-house structure, analog. (c) Flocculent structure, cross-section. (d) Pusch's structure, cross-section. (e) Brownian structure, analog. (f) Domain, cross-section. (g) Packet, cross-section. (h) Kidney structure, sketch. (i) Herringbone structure, cross-section.

in situ (Hayes, 1970). Smart (1975) shows a single-plate, card-house structure that is tentatively attributed to microbiological activity. This suggests that in cultivated soils with high bulk densities, crumbs are formed as follows: (1) relatively large pores are first formed by cultivation; (2) bacteria then rearrange the particles near the surfaces of these pores one by one to form open structures stabilized by soft organic matter; and (3) this organic matter finally condenses, drawing the particles together and forming stronger bonds. *Brownian structure* [Figure 2(e)], which is fairly rare, would result if larger particles were sedimented quickly with horizontal attitude while smaller particles sink slowly with random orientation, in which case the larger may collect flocs of smaller particles before reaching the bottom. Open structures may also occur in rice paddies, although no direct observations are known to the writer. Otherwise, these open structures occur either (1) in an uncemented form, in which case they are unconsolidated muds and the structure would change on reclamation, or (2) the structures are cemented, in which case one obtains a quick clay (q.v. in Vol. XIII).

Dense Structures

The term *domain* [Figure 2 (f)] refers to a volume larger than a clay particle, which is filled mainly with clay plates approximately parallel to a surface (cf. Aylmore and Quirk, 1960). Even though the clay plates extend across a packet [Figure 2(g)], packets are sometimes considered to be very small domains. Domains occur under natural conditions and in laboratory studies, these domain structures may be subdivided into:

1. *Turbostratic structures:* large domains, sometimes curved, interdomain boundaries may be diffuse, few large voids.
2. *Discrete domain structures:* smaller domains, sometimes too small to be curved, well-defined domain boundaries, definite interdomain voids.

A further division (Figure 3) includes:

A. *Overlap domains:* the clay plates are separate particles.
B. *Intergrown domains:* the crystal lattices are continuous from plate to plate, perhaps imperfectly.

Overlap domains occur in failure zones and are likely in some cutans. Intergrown domains may occur in other cutans or inside peds. The concept was suggested by micrographs of shales. First-class replicas, or perhaps vacuum decoration, are needed. Figure 2(h) illustrates a *kidney structure,* which should perhaps be regarded as a subdivision of an intergrown domain structure, and which is thought to result from hydrothermal action.

Closely packed random structures have also been observed. These are denser than card- and book-house structures. If card-house structures resulted from stable edge-face contacts, the angles of contact would be distributed about a certain value, and the structure would be

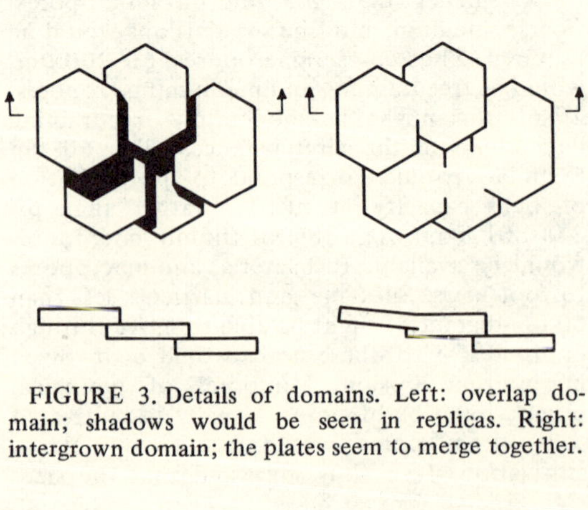

FIGURE 3. Details of domains. Left: overlap domain; shadows would be seen in replicas. Right: intergrown domain; the plates seem to merge together.

systematic, not random. Close-packed random structures, card-house structures, and (probably) book-house structures, will extinguish between crossed nicols.

Stress Effects

Consolidation, deformation, failure, and compaction are discussed in this section. Within limits, consolidation is relevant to wet subsoils, deformation and failure to cultivation, and compaction to traffic.

Consolidation. Consolidation follows from a decrease in volume of saturated clay as water is expelled by pressure. In laboratories the sample is often confined laterally, and the pressure doubled daily. There are two models:

1. Mechanical: the particles slip, bend, or break (very thin layers of absorbed water may separate the particles).
2. Physicochemical: double-layer effects dominate.

Agreement between some observations and Gouy-Chapman theory support the latter (Warkentin, Bolt, and Miller, 1957), but this is not general. Olson and Mesri (1970) suggested the following discriminatory tests:

1. If the pore fluid is water, changes in either the type or concentration of the electrolyte will affect physicochemical but not mechanical properties.
2. Replacement of the pore water by a fluid in which double layers cannot develop would result in substantial reductions in void ratio if physicochemical properties govern, whereas such a substitution will usually increase surface friction and reduce compressibility if mechanical characteristics control.

Olson and Mesri (1970) concluded that mechanical effects dominate in sand, muscovite, and kaolinite, and at high pressures in illite; whereas physicochemical effects dominate in smectite, and at low pressures and during rebound in illite. The following complexities must be superimposed on this diagnosis:

Reorientation of clay plates slipping under pressure has been measured using optical microphotometry, X-ray texture diffraction, and electron microscopy (Tchalenko, Burnett, and Hung, 1971; McConnachie, 1974; and others). In flocculated kaolin at high water contents, considerable reorientation occurred below 0.15 kg/cm^2 approximately. Thereafter, there was less, if any, reorientation up to 1000 kg/cm^2, at which pressure the preferred orientation was still imperfect. At low pressures, any natural bonding prevents both reorientation and consolidation (Jarrett, 1972). Two cautions apply to these interpretations. Jarrett found that the measurements depend upon the impregnant used in preparing samples, and Brindly (1953) reported that he was unable to find preferred orientation in compressed halloysite.

Figure 4 gives an impression of the size of domains in a consolidated kaolin. There is triple anisotropy: domain lengths are larger than breadths; horizontal domains are larger than vertical; and there is also an overall preferred horizontal orientation of the particles. The domains themselves were irregular in shape, and there were no interdomain voids. In another kaolin, McConnachie (1974) found that domain size decreased little with pressure. At 400 kN/m^2, his domains were approximately 1 μm × 0.4 μm on average. McConnachie's domains were discrete, smaller, and tighter, and there were interdomain voids between them, as compared to Figure 4. The differences between these kaolins are probably due to physicochemical factors. In contrast to this work on kaolin, Blackmore and Miller (1961), in a study of calcium-montmorillonite, using X-ray analysis, inferred from line broadening that the number of single clay sheets per tactoid (quasicrystal, domain) increased linearly with log pressure. It is not clear whether sheets were added one by one or whether quasicrystals coalesced.

McConnachi (1974) also mapped the interdomain voids at higher pressures. These were irregular in outline, anisotropic in shape and anisotropic in orientation, and decreased in size with pressure. Voids of this nature would control the saturated hydraulic conductivity and the air-entry characteristics, both of which usually decrease with pressure. From preliminary measurements on replicate samples consolidated to 1 kg/cm^2, A. J. Lovatt (unpub) found that the horizontal permeability was 2.46 times greater than the vertical, presumably

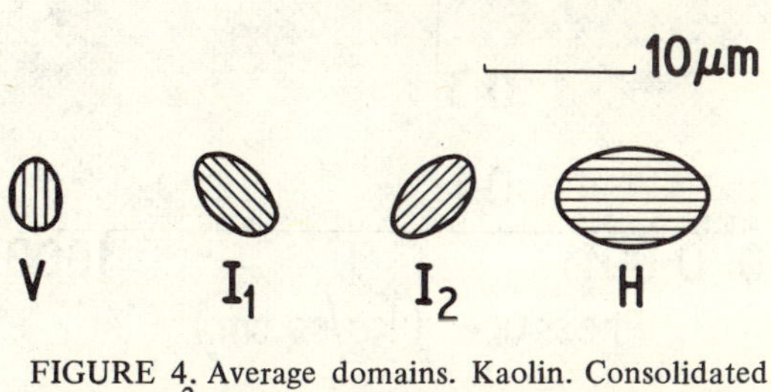

FIGURE 4. Average domains. Kaolin. Consolidated to 4 kg/cm^2. Major and minor diameters of these ellipses equal the average lengths and breadths of nearly vertical, inclined, and horizontal domains, *V*, *I*, and *H*, respectively. [Based on Smart (1966) Figure 3]

because of the anisotropy of the interdomain voids. The geometric mean breadth of McConnachie's voids, as seen in two dimensions in ultrathin sections, fell to 0.2 μm at 20 kg/cm^2 (300 lb/in^2). These observations suggest that micropores accessible to bacteria may survive the stresses of cultivation but that these might be anaerobic for much of the time. Figure 5 shows the intradomain void ratio of McConnachie's kaolin. It suggests that at high pressures the domains are themselves compressed as the mechanical interparticle forces due to the applied load overwhelm the physicochemical interparticle forces.

Secondary Consolidation. During consolidation, the expulsion of water is usually accompanied by an increase of pore water pressure. However, in some soils the volume continues to decrease after the excess pore water pressure has disappeared. This continued consolidation is called *secondary consolidation;* it may progress at a diminishing, steady, or increasing rate, and it is similar to or identical with creep or viscous flow. Three explanations have been offered:

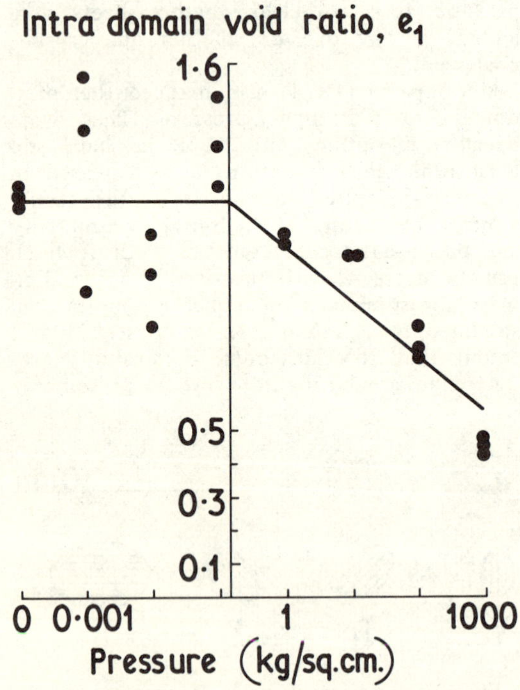

FIGURE 5. Intradomain void ratio during consolidation of kaolin. The term e_i is the intradomain void ratio, i.e., the volume of voids within domains divided by the volume of solids, calculated from void ratio and packing density using data of McConnachie (1974).

1. When the particles slip into fresh positions, they are at first held apart by their absorbed water layers, which are slowly squeezed out.
2. Secondary consolidation is caused by pressure recrystallization of the particles themselves.
3. The initial structure consists of groups of particles with large voids between. During primary consolidation, the groups slip, setting up pore water pressures, but during secondary consolidation, there is yielding only within the groups, any intragroup pore water pressure increase being inaccessible to measuring instruments.

Creep, whether involving change of size or of shape or of both, may lead to a deterioration of structure in wet, heavy subsoils. Natural structures are unlikely to be affected unless the conditions under which they were formed are altered; but mole drains and the structures formed by subsoilers would be liable to this form of deterioration.

Deformation and Failure. *Deformation,* i.e., change of shape less drastic than failure, has received less attention than either consolidation or failure. During deformation, dense sands are known to dilate and loose sands to contract; clays also dilate or contract according to circumstances. Various apparatus and procedures are used for shear testing. Almost all apply one stress to deform the soil while using other stresses to confine it. When the confining stress is itself much too low to break the initial structure of a soil sample, saturated clays dilate when deformed under drained conditions whereas they contract under much higher confining stresses. In unsaturated soils, the moisture tension will tend to act as a confinirg stress, and water scarcity may result in the absorbed layers becoming more ordered and thus increase the rigidity and the tendency to dilate; but shrinkage cracks would weaken the soil. Presumably, dispersing conditions, whether due to changes in pH or ionic composition or to dispersing agents, will tend to permit clay particles to slip individually, thereby facilitating contraction; whereas flocculating conditions would tend to cause the particles to move in groups and produce dilation. Preliminary results from work in hand tend to support this view; but these results also suggest that there may be an intermediate state in which the groups retain their identity but fail to bond to each other so that the soil behaves as if dispersed. Figure 2 (i) illustrates a *Herring-bone structure,* or microkinking, which may result from deformation under certain conditions.

Complete preferred orientation of clay particles occurs within *slickensided failure zones* (Smart, 1966; Foster and De, 1971; Morgenstern and Tchalenko, 1967). Depending on the stress and strain conditions, these zones occur singly, singly with complementary zones at approximately $45^\circ + \phi/2$, in oblique sets at this same inclination, or possibly as omnisepic fabric (ϕ = angle of internal friction). There may be water migration either into or out of the failure zone. Within failure zones the particles seem to have been separated from each other and to be moving singly, i.e., there has been mechanical dispersion. Comminution of the particles has been postulated, but it is difficult to demonstrate. Recrystallization has been observed in wear surfaces of alumina and snow-rolls have been found under avalanches, but similar features in soil are almost nonexistent.

Shear failures are more damaging to the structure of the soil than are tensile failures, so that implements should be designed to produce tensile failures either directly or induced by dilation, or to minimize shear strains. For example, digger ploughs probably induce tensile failures in dry soils, but when very wet soils must be ploughed, there is no hope of inducing tensile failure. Helical mouldboards were used to invert the furrow slice without inducing shear strain except near the share. Similarly, although Suffolk coulters and triple-disk coulters work well for seed drills under suitable conditions, single-disk coulters and arrow coulters could be used under wetter conditions with less risk of smearing the bottom of the seed furrow.

Compaction. This section discusses *compaction* of unsaturated soils by various combinations of compressive and shearing stresses applied for brief periods, e.g., by wheels or in laboratory tests. The response of a soil to compaction depends on both the compaction method and the moisture content. All other things being equal, the amount of compaction is greatest at an "optimum" moisture content; the terminology comes from engineers. In general, soils compacted "wet-of-optimum" are more liable to either swell or to collapse when wetted, have better permeability, a lower air-entry value, greater (ultimate) strength at the moulding moisture content, and greater rigidity both at the moulding moisture content and when swollen. Moulding moisture content, however, seems to have little effect on strength when swollen; judging from tests on foundry sands, dry strength increases with the moulding moisture content. There are reports of failure by piping erosion of earth dams that had been compacted dry-of optimum. With wet-of-optimum compaction, there is generally an increase in local preferred orientation of clay plates, sometimes an increase in overall preferred orientation, sometimes the formation of short failure planes, and generally a uniform large-scale structure. With dry-of-optimum compaction, crumbs tend to persist and relatively large intercrumb mini or macropores are formed. Because there are no large intercrumb pores wet-of-optimum, it follows that the intracrumb pores, formed by compaction, dry-of-optimum, are smaller than the pores in a similar sample compacted to the same bulk density wet-of-optimum. Some modifications to these generalizations are to be expected, for example, in strongly flocculated soils or highly plastic soils that resist compaction.

In general, however, it seems that the effect of the compaction method increases in the following order: static, vibratory, impact, kneading compaction, the differences being more pronounced wet-of-optimum. Thus dynamic and shearing effects are generally more severe. Machines traveling faster usually produce less compaction, presumably because the stresses they produce act for shorter times.

Several phenomena may be involved in resisting compaction. It seems that wet-of-optimum, the compacting stresses are applied for too brief a time to expel appreciable quantities of water so that the volume may be reduced almost to the sum of the volumes of the solids and of the water, a little air being trapped as isolated bubbles. Dry-of-optimum, a high pore water tension will tend to strengthen the crumbs. Further, the small amount of water will have a higher electrolyte concentration because that is independent of the water content, so that the double layers will contract and may become ordered, and tendencies to flocculate will increase. Further, interparticle contacts may also form and strengthen the crumbs. If dry-of-optimum during compaction, the crumbs probably break and slip to fill the largest voids. The relatively large voids left unfilled explain the good aeration and permeability (q.v.) and allow root penetration. They also explain the piping failure of dams; with the velocity of water flow through a large void high, the probability of eroding particles from the sides of the void will also be high. Similar internal erosion may also occur in cultivated soils and in the backfills over underdrains. Further, when crumbs are formed by fracturing, either during compaction dry-of-optimum or by cultivation, many could be left partly fractured so that particles are more likely to be eroded from them or the crumbs are more likely to collapse entirely when wetted. Soils tend to become mechanically dispersed during compaction, wet-of-optimum; the clay particles become separated from each other and slip into domains or shear zones. This structure explains the small rigidity when saturated. When dried, the particles would move closer together face to face, and a dense, rigid state would result.

Further Topics

This brief review omits reference to very early electron microscopy of inorganic clay-chemical interactions, recent work on organic clay-chemcial interactions, natural inorganic dispersants (Söderblom, 1966), soil formation, and bacteria (Penner, Gillot, and Eden, 1970). Smart (1969) lists many of the earlier papers. The Proceedings of the International Symposium of Soil Structure, Gothenburg, 1973 (obtainable from Swedish Geotechnical Institute, Banérgatan 16, 115 26 Stockholm) and Soil Microscopy, Proceedings of the 4th International working meeting on Soil Micromorphology, 1973 (G. K. Rutherford, ed. Limestone Press, Kingston, Ontario) together provide a comprehensive list of later work.

PETER SMART

References

Aylmore, L. A. G., and Quirk, J. P., 1960. Domain or turbostratic structure of clays, *Nature,* **187,** 1046-1048.

Beutelspacher, H., and van der Marel, H. W., 1968. *Atlas of Electron Microscopy of Clay Minerals and Their Admixtures–A Picture Atlas.* Amsterdam: Elsevier, 333p.

Blackmore, A. V., and Miller, R. D., 1961. Tactoid size and osmotic swelling in calcium montmorillonite, *Soil Sci. Soc. America Proc.,* **25,** 169-173.

Brindley, G. W., 1953. An X-ray method for studying orientation of micaceous minerals in shales, clays and similar materials, *Min. Mag.,* **30,** 71-78.

Collins, K., and McGown, A., 1974. The form and function of microfabric features in a variety of natural soils, *Geotechnique,* **24,** 223-254.

Foster, R. H., and De, P. K., 1971. Optical and electron microscopic investigation of shear induced structures in lightly consolidated (soft) and heavily consolidated (hard) kaolinite, *Clays Clay Min.,* **19,** 31-47.

Gard, J. A., ed., 1971. *The Electron-optical Investigation of Clays.* London: Mineralogical Society, 383p.

Hayes, J. B., 1970. Polytypism of chlorite in sedimentary rocks, *Clays Clay Min.,* 18, 285-306.

Jarrett, P. M., 1972. The effects of soil structure on the engineering behaviour of a sensitive clay, *Quart. Jour. Eng. Geology,* 5, 103-110.

McConnachie, I., 1974. Fabric changes in consolidated kaolin, *Geotechnique,* 24, 207-222.

Morgenstern, N. R., and Tchalenko, J. S., 1967. Microscopic structures in kaolin subjected to direct shear, *Geotechnique,* 17, 309-328.

O'Brien, N. R., 1972. Microstructure of a laboratory sedimented flocculated illitic sediment, *Can. Geotech. Jour.* 9, 120-122.

Olson, R. E., and Mesri, G., 1970. Mechanisms controlling compressibility of clays, *Am. Soc. Civil Eng. Proc.,* **96 SM6,** 1863-1878.

Penner, E.; Gillot, J. E.; and Eden, W. J., 1970. Investigation of heave in Billings shale by mineralogical and biogeochemical methods, *Can. Geotech. Jour.,* 7, 333-338.

Pusch, R., 1970. Clay microstructure. Stockholm: Nat. Swedish Inst. Building Research Document D8. Svensk Byggtjänst, 76p.

Smalley, I. J.; Cabrera, J. G.; and Hammond, C., 1973. Particle nature in sensitive soils and its relation to soil structure and geotechnical properties, *Proc. Internat. Symp. Soil Structure, Gothenburg* (Swedish Geotechnical Institute, Stockholm), 184-193.

Smart, P., 1966. Particle arrangements in kaolin, *Clays Clay Min.,* **15,** 241-254.

Smart, P., 1969. Soil structure in the electron microscope, *in* M. Te'eni, ed., *Structure, Solid Mechanics and Engineering Design, Vol. 1.* London: Wiley-Interscience, 249-255.

Smart, P., 1973. Electron microscope methods in soil micromorphology, *Proc. 4th Int. Work. Meet. Soil Micromorphology, Kingston,* Ontario, 190-206.

Smart, P., 1975. Soil microstructure, *Soil Sci.,* **119,** 385-393.

Söderblom, R., 1966. Chemical aspects of quick-clay formation, *Eng. Geol.,* **1,** 415-431.

Tchalenko, J. S.; Burnett, A. D.; and Hung, J. J., 1971. The correspondence between opitcal and X-ray measurements of particle orientation in clays, *Clay Min.,* **9,** 47-70.

Tovey, N. K., and Wong, K. Y., 1973. The preparation of soils and other geological materials for the S.E.M., *Proc. Internat. Symp. Soil Structure, Gothenburg* (Swedish Geotechnical Institute, Stockholm), 59-67.

Warkentin, B. P., Bolt, G. H., and Miller, R. D., 1957. Swelling pressure of montmorillonite, *Soil Sci. Soc. America Proc.,* **21,** 495-497.

Weir, A. H., 1960. A study in the relationships between certain physical properties and structure and composition of montmorillonitic group minerals. Ph.D. thesis, Rothamsted, London.

Cross-references: *Aggregation; Pore-size Distribution; Management of Soils; Microaggregation; Profiles, Physical Modification; Soil Mechanics; Soil Pores; Soil Structure.*

MINERAL ANALYSIS

Resistant primary minerals occurring in soils are generally present in the silt and sand size fractions. Soil mineralogists are, however, generally more interested in the secondary minerals present in the fine fractions of soils since these are the most effective in determining soil chemical and physical properties. In particular, adsorption and fixation of ions from solution often take place at the surfaces of these very small crystals. Thus large surface areas of clay-size materials usually dominate exchange phenomenon (q.v.) in soils. The mineralogy of the clay fraction may also indicate the type and degree of weathering (q.v. in vol. xii) that has occurred during soil formation.

Only a limited group of minerals is normally present in significant amounts in the fine fractions of soils. These are the clay minerals and the oxides of iron, aluminum, titanium, and manganese. The structures of these minerals are described elsewhere in this volume (see *Clay Minerals Silicates; Hydrous Oxides*). Clay minerals frequently have disordered, substituted structures, which, combined with the effect of small crystal size, may make them difficult to identify by the commonly used physical techniques. To increase analytical sensitivity, the clay fraction (less than 2 μm equivalent spherical diameter) is usually separated from the soil by dispersion and sedimentation.

Separation of the Clay Fraction

Dispersion of soils in water is best achieved by violent stirring in a milkshake mixer or similar device after addition of a dispersing agent. Ten grams of soil in 200 ml of distilled water containing 5 ml of 10% sodium tripolyphosphate and 0.5 ml of normal caustic soda solution give rapid dispersion of most soil types and generally yield adequate clay for mineralogical analysis. Organic soils may require pretreatment with hydrogen peroxide, and calcareous soils pretreatment with dilute hydrochloric or acetic acids to remove cementing materials and permit complete dispersion of the clay fraction. The dispersed soil is next allowed to stand until the required settling time has elapsed when samples of the desired size fraction may be taken with a pipette from a standard depth. Some settling times for different size particles are given in Table 1. Corrections should be made to these values given for use at different temperatures and for minerals with specific gravities significantly different from 2.65 Particles smaller than 2 μm may be separated with the aid of a centrifuge.

TABLE 1. Settling Times Required for 2.65 SG Particles of Different Equivalent Spherical Diameters to Settle 10 cm in Water at 25°C.

Particle Diameter	Settling Time
50 μm	0.66 min
40	1.05
30	1.86
20	4.2
10	16.8
5	67
2	420

Prior to analysis, the clay sample should be washed free of salts by repeated centrifugation or dialysis in distilled water. The clay fraction may then be dried at 105°C or at room temperature if the presence of temperature-sensitive minerals such as hydrated halloysite is suspected.

X-ray Diffraction Analysis

Most laboratories concerned with the study of clays make use of X-ray diffraction as a basic analytical tool. A cobalt, iron, or chromium X-ray tube energized by a 3-kW generator will provide a high-intensity source of X-rays at energies unlikely to cause specimen fluorescence and will give optimum conditions for recording patterns from poorly diffracting soil minerals. An initial diffraction pattern of a randomly oriented powder preparation of the clay fraction is usually obtained using either a Debye Scherrer powder camera or diffractometer to identify oxide and other accessory minerals. The JCPDS powder diffraction file of standard patterns and its several indexing systems can be used to identify unknown minerals (JCPDS, 1974). The three strongest lines of some common accessory minerals found in soils are given in Table 2 as an aid to rapid identification. In some cases the random

TABLE 2. Spacings and Intensities of the Three Strongest Diffraction Lines of Some Common Accessory Minerals Found in Soils. Minerals Marked with an Asterisk Vary in Chemical Composition Resulting in a Range of Spacings and Relative Intensities. Note That the Values Given Are Only Representative.

Mineral	Spacings (Å)			Relative Intensities		
Gypsum	7.56	3.06	4.27	10	6	5
Lepidocrocite	6.27	3.29	1.93	10	6	3
Boehmite	6.11	3.16	2.34	10	5	4
Gibbsite	4.85	4.37	2.38	10	4	3
Goethite*	4.18	2.69	2.45	10	3	3
Diaspore	3.98	2.07	2.31	10	5	5
Anatase	3.51	1.98	2.38	10	3	2
Anhydrite	3.50	2.85	2.33	10	3	2
Aragonite	3.40	1.98	3.27	10	7	5
Quartz	3.34	4.26	1.82	10	4	2
Orthoclase*	3.31	3.77	4.22	10	8	7
Zircon	3.30	4.43	2.53	10	5	5
Rutile	3.24	1.68	2.49	10	5	4
Albite*	3.21	3.18	4.04	10	5	5
Hypersthene*	3.18	2.88	2.56	10	5	4
Microcline*	3.24	4.21	3.83	10	6	5
Jarosite*	3.08	5.09	3.11	10	7	4
Calcite	3.03	2.28	2.09	10	2	2
Dolomite	2.89	2.19	1.78	10	3	3
Apatite*	2.81	2.71	1.84	10	6	6
Olivine*	2.79	2.53	2.48	10	6	6
Hornblende*	2.70	3.09	3.38	10	9	9
Hematite	2.69	2.51	1.69	10	8	5
Magnetite	2.53	1.49	2.97	10	4	3
Maghemite	2.51	1.47	2.95	10	4	3
Alunite	1.90	3.01	1.75	10	9	9

powder pattern may also be used to determine degree of order in kaolinites, polymorphs of micas, di- or trioctahedral nature of clay minerals and other structural characteristics. Figures 1(c) and 1(b) show a powder photograph and diffractometer trace of a randomly oriented soil clay fraction. The superiority of the diffractometer for the measurement of long spacings, resolving adjacent reflections, and recording the shape of reflections is clearly shown. Powder cameras are particularly useful for the rapid speed at which they can record the whole diffraction pattern and their ability to provide patterns from a few milligrams of material. In complex mixtures the clay minerals are best identified through observation of their characteristic basal (001) reflections. For this reason, diffractometer traces of basally oriented clay preparations giving greatly enhanced intensities for basal reflections are widely used [Figure 1(c)]. Oriented clays may be prepared by drying a suspension onto a glass slide or by suction onto a suitable filter material. Thin, 2-in square porous ceramic plates have proved to be most satisfactory for this technique. They are easily mounted in the diffractometer, and the clay may be readily leached with solutions if the effect of ion exchange or organic molecule absorption on the diffraction pattern is to be determined as an aid to identification. Basal spacings of the common clay minerals after magnesium saturation, glycerol solvation, and heating of potassium exchanged clay to 550°C are given in Table 3. The confusion that may occur between clay minerals with similar basal spacings is resolved through observation of their different behaviors in response to these treatments.

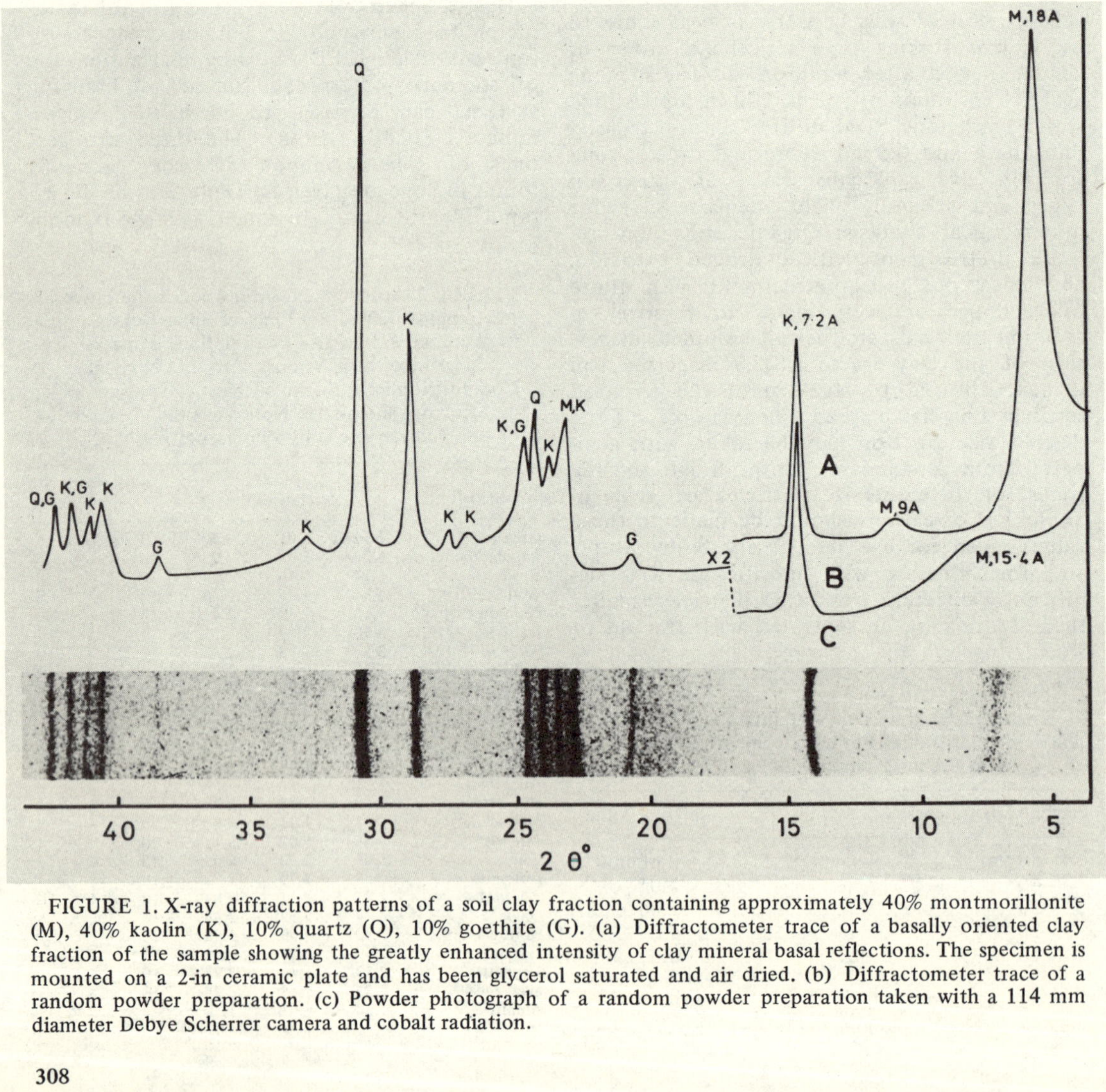

FIGURE 1. X-ray diffraction patterns of a soil clay fraction containing approximately 40% montmorillonite (M), 40% kaolin (K), 10% quartz (Q), 10% goethite (G). (a) Diffractometer trace of a basally oriented clay fraction of the sample showing the greatly enhanced intensity of clay mineral basal reflections. The specimen is mounted on a 2-in ceramic plate and has been glycerol saturated and air dried. (b) Diffractometer trace of a random powder preparation. (c) Powder photograph of a random powder preparation taken with a 114 mm diameter Debye Scherrer camera and cobalt radiation.

TABLE 3. Cation Exchange Capacities and Basal Spacings of the Common Clay Minerals After Magnesium Saturation, Glycerol Solvation, and Potassium Saturation Followed by Heating at 550°C. The Values in This Table Are Only Representative With Some Specimens Having Spacings and CEC Values Significantly Different From Those Shown.

	Basal Spacing (Å)			Cation Exchange Capacity
	Magnesium	Glycerol	Potassium/550°C	m equiv/100 gm
Kaolinite	7.2	7.2	Amorphous	2–15
Halloysite	10.1	10.7	Amorphous	20–25
Montmorillonite	15.4	17.8	9.9	80–130
Illite	10.0	10.0	10.0	20–30
Vermiculite	14.3	14.5	10.2	100–150
Chlorite	14.2	14.2	14.2	10–40

The identification of interstratified or mixed-layer clay minerals, which are frequently found in soils, is more difficult. Several techniques are, however, available for the determination of the characteristics and relative proportions of the different layer types present within these minerals. These techniques are described in the references given below. Quantitative analysis of minerals in mixtures may be attempted by X-ray diffraction, providing suitable standard minerals are available. Corrections must be made for differences in X-ray mass absorption coefficient between standard and unknown samples using either an internal standard mineral or measured mass absorption coefficients. Comprehensive reviews dealing with the identification of clay minerals by XRD may be found in Brown (1961) and Thorez (1975).

Electron Optical Techniques

The development of the transmission electron microscope (TEM) has permitted the direct observation of the size and shape of clay-fraction minerals. The construction of a modern TEM is analogous to that of an optical projection microscope. Electrons from a heated tungsten filament source are accelerated by voltages of 25–100 kV; they pass through the specimen and are focused by a series of electromagnetic lenses onto a fluorescent screen. Images of crystals are formed at variable magnifications up to a maximum of about 200,000. They are supported on thin carbon films. Images can be recorded on a photographic plate. Further photographic enlargement is possible so that crystal lattice spacings of less than 3 Å can be resolved under favorable circumstances. Because the column of the TEM is under high vacuum and considerable energy can be absorbed by crystals from the electron beam, some minerals may dehydrate, losing water or structural hydroxyls during observation. Use of a cooled specimen stage and a low-beam current will minimize this effect. Suitable specimens for TEM observation may be obtained by allowing a drop of very dilute, salt-free suspension to rest on a thin carbon film covering a metal specimen grid for a few seconds before removing the drop with the edge of a filter paper. If the drop is allowed to completely evaporate on the grid, clay crystals tend to form aggregates that prevent observation of the nature of individual crystals. The specimen grid is often shadowed with a thin (5–20 Å) layer of evaporated metal or carbon at a low angle of incidence to permit measurement of the thickness of individual crystals. Examples of the morphology of several clay minerals are shown in Figure 2. It is evident that small and otherwise unsuspected amounts of halloysite, rectorite, or sepiolite could be detected in a mixture by virtue of their distinctive crystal morphologies. Most clay minerals, however, differ greatly in their crystal morphology so that electron microscopy cannot be considered for the routine identification of soil minerals in mixtures. Studies of crystal morphology with the TEM are, however, of great value in understanding mechanisms of mineral alteration and chemical processes occurring on mineral surfaces within soils.

Electron-diffraction patterns from very small single crystals may be obtained with most TEMs by use of selector apertures and altered-focusing conditions [Figure 2(f)]. Electron-diffraction patterns from montmorillonite crystals of a single-unit cell thickness can be recorded because of the high efficiency of electron scattering which is some 10,000 times as effective as X-ray scattering. Single-crystal electron-diffraction patterns can be used for the identification of unknown minerals, mea-

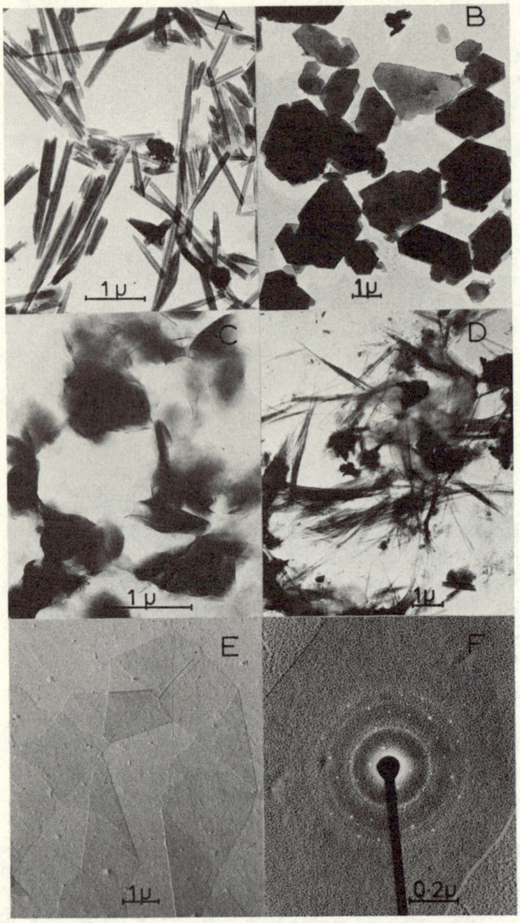

FIGURE 2. Electron micrographs of some clay minerals. (a) Tubular halloysite from a lateritic saprolite derived from dolerite, Western Australia. (b) Euhedral kaolinite from the pallid zone of a laterite over pegmatite, Western Australia (shadowed). (c) Fluffy aggregates of montmorillonite from a calcareous black earth, Western Australia. (d) Fibrous sepiolite and montmorillonite from a red, calcareous desert soil, Western Australia. (e) Folded ribbon-like rectorite, an unusual regular mixed-layer clay mineral, Morning Star Mine, Australia (shadowed). (f) Selected area electron-diffraction pattern of rectorite superimposed over a micrograph of the portion of the shadowed crystal from which the pattern was obtained.

surement of unit-cell parameters, and determination of the directions of elongation, crystal faces, twin planes, etc., of small crystals. Because of their platey morphology and similarity in structure in projection onto the basal plane, most clay minerals exhibit similar electron-diffraction patterns. Consequently, it may prove possible to conclude only that a mineral is di or trioctahedral through measurement of its *b*-axis dimension. Further X-ray diffraction studies of basal reflections would be required to exactly determine the clay mineral.

The shape and arrangement of small crystals and pores in soils may be studied with a scanning electron microscope (SEM) that provides images with a high depth of focus and requires minimal sample preparation [Figure 3(a)]. Similar information may be obtained with a TEM by use of a surface replicating technique [Figure 3(b)]. This procedure is best suited to the study of the finer details of soil structure and mineral morphology.

The characteristic X-rays emitted from minerals subjected to electron bombardment in the electron microscope (TEM and SEM) may be analyzed with either energy or wavelength dispersive X-ray detectors. This technique provides a semiquantitative chemical analysis of individual crystals. An excellent recent summary of the application of electron optical techniques to the study of clays may be found in the monograph edited by Gard (1971).

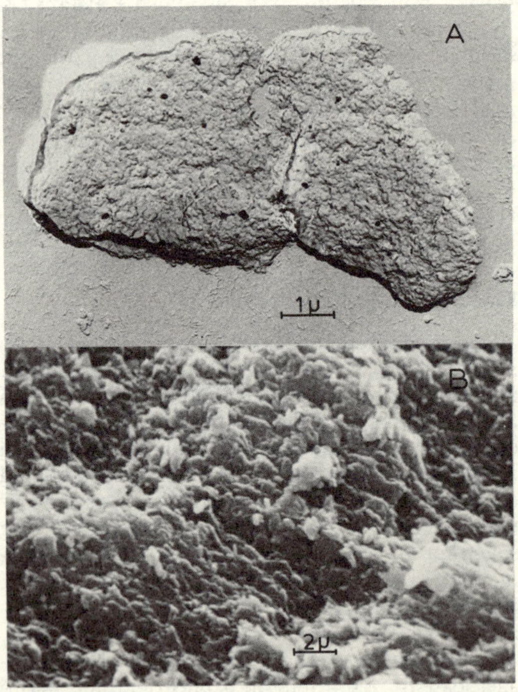

FIGURE 3. Electron micrographs. (a) Transmission electron micrograph of a shadowed single-stage carbon-platinum replica of an illite aggregate. (b) Scanning electron micrograph of a fracture surface of a kaolinitic soil crumb showing individual kaolinite crystals.

Infrared Absorption Spectrometry

Minerals absorb infrared radiation (IR) at characteristic frequencies corresponding to the vibrational frequencies of different groups of atoms within their structures. Clay samples for IR analysis are usually dispersed in thin potassium-bromide disks at concentrations determined by the strength of the absorption to be measured. Concentrations are usually about 1% clay. Absorption by potassium bromide does not interfere with absorption by the mineral in the range of frequencies normally investigated (400–4000 cm^{-1}). IR analysis has been widely used to study isomorphous substitution, hydroxyl configuration, and clay-organic complexes. It is not commonly used as a routine analytical technique for identification of minerals in clay fractions but may provide greater sensitivity for the detection of gibbsite, goethite, quartz, kaolinite, and halloysite in some mixtures. Infrared absorption spectrometry is particularly valuable in the study of poorly crystalline materials that give diffuse X-ray diffraction patterns. Highly disordered structures may still show good IR patterns that do not rely on the existence of regular atomic order. A recent review of the application of IR spectrometry to mineral analysis may be found in the monograph edited by Farmer (1974).

Thermal Analysis

A variety of thermal techniques are available for the study of soil minerals, but as in the case of IR analysis, they are generally most useful in detailed investigations of mineral properties rather than in the routine identification of the components of complex mixtures. Differential thermal analysis (DTA) involves the measurement of small changes in temperature that occur in a sample relative to an inert reference material as their temperature is steadily increased in a small furnace. Exothermic changes in the sample such as recrystallization and oxidation will cause a brief relative increase in temperature. Dehydration, dehydroxylation, and crystallographic inversions are endothermic and result in a brief relative cooling. The temperatures at which these changes occur and the sharpness of the transformations characterize a particular mineral. The curves obtained may be influenced by the response of the particular apparatus used, so comparisons of unknowns with standard reference minerals should be made. Thermogravimetric analysis (TGA) and differential thermogravimetric analysis (DTGA) involve the determination of weight losses and gains by specimens at different temperatures. Combined DTA and TGA apparatus is available from several manufacturers. Both theoretical and experimental aspects of thermal analysis have been thoroughly reviewed in two monographs edited by Mackenzie (1969, 1972).

Chemical Analysis

Total chemical analysis and measurement of cation exchange capacity (CEC) play major roles in the complete identification of clay minerals. The X-ray diffraction pattern may frequently indicate no more than the general structural group to which a mineral belongs, e.g., dioctahedral montmorillonite. A total chemical analysis of a pure sample of the mineral will permit the calculation of a structural formula indicating which ions occupy the various structural sites. This analysis permits an assignment of the mineral to the correct species within the group. Measurement of the CEC of a pure mineral acts as a check that the calculated distribution of ions within the mineral structure is correct since only ions in particular sites should be exchangeable. Measurement of CEC is also a valuable guide to the types of clay minerals that are present in soil clay fractions. XRD results should always be consistent with the measured CEC. Typical values of CEC for the major clay mineral groups are included in Table 3.

Surveys of the range of clay mineral compositions and other properties have been published by Weaver and Pollard (1973), Grim (1968), and Caillère and Hénin (1963).

R. J. GILKES

References

Brown, G., ed., 1961. *The X-Ray Identification and Crystal Structures of Clay Minerals.* London: Mineralogical Society, 544p.

Caillère, S., and Hénin, S., 1963. *Mineralogie des Argiles.* Paris: Masson et cie, 355p.

Farmer, V. C., ed., 1974. *The Infrared Spectra of Minerals.* London: Mineralogical Society, 539p.

Gard, J. A., ed., 1971. *The Electron-optical Investigation of Clays.* London: Mineralogical Society, 383p.

Grim, R. E., 1968. *Clay Mineralogy.* New York: McGraw-Hill, 596p.

Joint Committee on Powder Diffraction Standards, 1974. *Selected Powder Diffraction Data for Minerals, Publication DBM-1-23.* JCPDS, Swarthmore, Penn. 833p.

Mackenzie, R. C., ed., 1969. *Differential Thermal Analysis, Volume 1: Fundamental Aspects.* London: Academic Press, 775p.

Mackenzie, R. C., ed., 1972. *Differential Thermal Analysis, Volume 2: Applications.* London: Academic Press, 607p.

Thorez, J., 1975. *Phyllosilicates and Clay Minerals: A Laboratory Handbook for their X-ray Diffraction Analysis.* Dison, Belgium: G. Lelotte, 580p.
Weaver, C. E., and Pollard, L. D., 1973. *The Chemistry of Clay Minerals.* Amsterdam: Elsevier, 213p.

Cross-references: *Clay Minerals, Silicates; Hydrous Oxides: Particle-size Distribution; Soil Mineralogy. Vol IVA; Differential Thermal Analysis; Electron Microscopy; Infrared Analysis; Mineralogy; Neutron Activiation Analysis; X-ray Diffraction Analysis. Vol. IVB; Soil Minerals.*

MINERALIZATION–See CARBON CYCLE; NITROGEN CYCLE; NITROGEN FIXATION; SULFUR TRANSFORMATIONS

MINERALS–See CLAY MINERALS, SILICATES; HYDROUS OXIDES; SOIL COMPONENTS, INORGANIC. VOL. IVB: SOIL MINERALOGY; also individual minerals. VOL. XII, Pt. 2: PEDOSESQUIOXIDES

MOISTURE MANAGEMENT

Soil moisture is water contained in soil pores (q.v.) and absorbed on soil particles. Even though some water is combined chemically with soil materials, this aspect will not be considered here. Moisture content is expressed as a percent of the dry soil weight (oven-dried at 105 to 110°C) or of the bulk soil volume.

Importance of Soil Moisture

Soil moisture promotes innumerable chemical, physical, and biological activities and reactions. It also dissolves and carries plant nutrients (q.v.) and is a nutrient itself (Brady, 1974).

Water is the single most important factor, besides possibly sunlight and air, that influences the productivity of soil-grown plants. Soil moisture excesses may decrease or prevent plant growth through reduced soil aeration, while moisture deficiencies cause plant wilting and death. For efficient production, a constant moisture supply is needed throughout the growing season to meet the plant's water needs and to aid in nutrient dissolution, transport, and absorption. Also soil moisture influences tillage (q.v.) and harvest operations.

Moisture influences soil compactability and load-bearing capacity and thus is of major concern for road, earthen-dam, and building construction. For optimum compaction, the moisture content must be within a rather narrow limit, depending on the relative proportions of sand, silt, and clay particles in a soil (soil texture). Load-bearing capacity increases as soils become dry. Some soils shrink or swell considerably when drying or wetting, respectively. These soils are undesirable foundation materials for roads and building. When present at construction sites, they may be removed, covered with stable materials, or stabilized with lime (calcium).

Soil Moisture Types and Properties

Soil moisture is often classified as *gravitational, plant (readily) available,* and *plant unavailable water.* At saturation, all soil pores and voids are filled with water, much of which, however, drains from soil due to gravity after water application when the soil drains freely. Rapid drainage stops in one to three days in coarse-textured soils; fine-textured soils drain slower. All soils may continue to drain slowly for a much longer time. Water freely draining from soil is called *gravitational water.* Plants may absorb some of this water before it drains from the soil.

The soil moisture content remaining when free drainage stops is called *field capacity.* It is considered the upper limit of a soil's capacity to store water for plant use and is approximated under laboratory conditions by allowing a saturated sample to reach equilibrium with ⅓ bar or ⅓ atm of pressure applied in a ceramic plate apparatus. For medium- to coarse-textured soils, *field capacity* is approximated by the *moisture equivalent,* which is obtained by subjecting a 1-cm-thick saturated soil sample to a centrifugal force of 1000 times gravity for 30 min. The lower limit of available moisture is at the *wilting point* when plants cannot absorb water fast enough to maintain turgidity. It is approximated in the laboratory by allowing a saturated sample to reach equilibrium with an applied pressure of 15 bars (or atm) in a pressure membrane or ceramic plate apparatus.

The moisture content at which plants do not regain turgidity unless water is added to soil is the *incipient wilting point,* while the point at which plants do not regain turgidity even when water is added is the *permanent wilting point* (Peters, 1965). The difference between the points is called the *wilting range.* In this range, plant roots continue to absorb moisture, but too slowly to permit normal plant growth.

Unavailable moisture is held in soil by forces strong enough to prevent significant absorption by plant roots. This moisture is held in fine pores and as extremely thin layers on soil particles.

Factors Influencing Soil Moisture Contents

Soil moisture contents are influenced, among other factors, by soil texture, structure, and organic matter content. Coarse-textured (sandy) soils generally contain low amounts of moisture, often less than 10% by weight at field capacity. In contrast, some clay soils contain over 40% moisture by weight. At the permanent wilting point, moisture contents range from about 3% for sandy soils to about 25% for some clay soils. Besides the amount of clay, the types of clay minerals also influence the moisture content of soils.

Soil structure, (q.v.) namely, the combination or arrangement of primary soil particles into secondary particles, influences moisture contents through pore formation. Improved structural development promotes pore formation. Hence the degree of structural development influences soil pore space (see *Pore Space Drainable*), which in turn influences the amount of moisture a soil may retain. Organic matter (q.v.) has a high affinity for moisture. Therefore, soils high in organic matter retain more moisture than similar soils with less organic matter.

After precipitation or irrigation (q.v.), gravitational water percolates through the soil profile. A coarse-textured layer in the profile or an impervious or slowly permeable layer or bedrock sufficiently near the surface may restrict percolation (q.v.), thus increasing moisture contents in soil above the restricting layer.

The foregoing major factors, which influence the potential moisture-holding capacity of a soil, cannot be readily altered by management. Factors such as irrigation amount and frequency, infiltration (q.v.) and runoff, evaporation (q.v.), weed control, and tillage (q.v.), which determine whether a soil will be filled to capacity, can be controlled to some extent.

Techniques for Controlling Soil Moisture

Soil Moisture Excesses. Excess soil moisture may be detrimental to timely performance of field operations for crop production, to plant growth and development, to soil compaction for construction purposes, and to traffic of off-road vehicles. To decrease the moisture content of soils frequently plagued by excesses, open drainage ditches or tile drains may be installed. Drainage of agricultural lands has been thoroughly discussed in a book edited by Van Schilgaarde (1974) (see also *Soil Drainage*).

Excess soil moisture at critical times may be reduced by (a) allowing plants to grow, thus removing the moisture by transpiration; (b) removing surface residues to hasten evaporation; and (c) plowing, stirring, or mixing the soil to hasten evaporation by exposing moist soil to the atmosphere. Excess moisture contents can be avoided by reducing or preventing water entry into soils by (a) using surface covers or soil water-proofing materials; (b) smoothing the surface to enhance runoff and reduce infiltration; and (c) constructing dikes or diversions to hold or route surface waters away from affected areas.

Soil Moisture Deficiencies. Moisture deficiencies at construction sites can be overcome by applying water to obtain the desired moisture content. Since the area or soil volume is generally limited and moisture content is critical for relatively short periods only, no major difficulties generally are encountered.

Of greater importance and concern is the moisture content of vast areas of croplands (field, range, orchard, forest, etc.). Where water is readily available and plentiful, moisture deficiencies can be overcome by irrigation if the land is suitable. Irrigation principles and practices have been widely reviewed and discussed (Hagan, Haise, and Edminster, 1967).

Although large land areas are irrigated, plants on the vast majority of land depend on precipitation for their water supply. Hence an important phase of crop production is to manage soils and crops to obtain maximum benefits from precipitation. While in some regions precipitation is adequate in amount and distribution to sustain favorable crop production, it is limited and erratic in others resulting in droughts of varying duration. Since plants depend on soil moisture in the interval between precipitation events, soils must retain sufficient moisture to sustain the plants. Major factors responsible for inadequate soil moisture storage for sustained plant growth and possible techniques for overcoming or avoiding the undesirable conditions are discussed in the following sections.

Inadequate Soil Moisture Storage Capacity. Moisture storage capacities are inherently low in sandy and shallow soils. Unless precipitation is frequent, production is generally low. Hence these soils are best adapted for low-moisture-requirement crops or crops capable of withstanding droughts of varying duration without being severely affected. Mulches and reduced tillage reduce evaporation, thus permitting greater moisture use by plants (Unger and Phillips, 1973). Where a dense subsurface horizon restricts moisture storage and plant rooting, soil profile mixing increases storage, plant rooting, and crop yields.

Precipitation Rates Greater than Infiltration Rates. High intensity precipitation may cause runoff even though the soil is not wet to field capacity. This results from surface sealing due to the formation of a dense soil layer at the surface or from lower potential infiltration rates than precipitation rates. Practices for increasing infiltration (q.v.) include tillage to disrupt dense surface layers, create depressions at the surface for temporary water storage, and increase depth of loosened soil; contour plowing to hold water on the surface; leveling and diking to prevent runoff; and maintaining residues on the surface to dissipate raindrop energy, thus reducing soil dispersion and surface sealing. Surface residues also retard runoff, thereby permitting more time for infiltration. (see *Conservation*). No-tillage cropping practices (see *Zero-tillage*), where crops are grown with no soil distrubance other than that necessary to place seeds in soil, reduce runoff and evaporation and increase infiltration and thereby increase moisture storage (Unger and Phillips, 1973).

Inadequate Precipitation. In arid and semiarid regions, precipitation erratic in amount and distribution limits production when crops depend mainly on growing season precipitation. When soil moisture storage is adequate at the start of a crop's growing season, production is often good with low seasonal precipitation because the plants absorb soil moisture. To maximize moisture storage during fallow (interval between crops), runoff and evaporation should be low. In addition, weeds should be controlled to prevent moisture use. Some regions have distinct wet and dry seasons. However, for other climatic and economical reasons, growing crops at times other than during the major wet season may be more desirable. Consequently, moisture must be stored in soil for subsequent crop use.

PAUL W. UNGER

References

Brady, N. C., 1974. *The Nature and Properties of Soils.* New York: MacMillan, 639.

Hagan, R. M.; Haise, H. R.; and Edminster, T. W., eds., 1967. *Irrigation of Agricultural Lnads.* Madison: American Society of Agronomy, 1180p.

Peters, D. B., 1965. Water availability, *in* C. A. Black, ed., *Methods of Soil Analysis.* Madison: American Society of Agronomy, 279-285.

Unger, P. W., and Phillips, R. E., 1973. Soil water evaporation and storage, *in Conservation Tillage.* Ankeny, Iowa: Soil Conservation Society of America, 42-54.

Van Schilfgaarde, J., ed., 1974. *Drainage for Agriculture.* Madison: American Society of Agronomy, 700p.

Cross-references: *Conservation; Evaporation; Field Water Cycle; Infiltration; Irrigation; Management of Soils; Microstructure, Manipulation; Mulch; Profiles, Physical Modification; Soil Drainage; Soil Water; Tillage; Water Content and Retention; Water Movement.*

MULCH

The origin of the English word *mulch* has been traced to the German vernacular term *molsch,* meaning "soft and beginning to decay." Materials of many kinds are spread over soil surfaces as mulches including straw, hay, crop residues, paper, petroleum products, and even gravel and crushed rock. Cover crops are also grown to provide mulches. Different kinds of mulches and their effects on physical, chemical, and biological properties of soils as well as some crop responses are discussed in this article. The granular surface layers formed in some soils as they dry are excluded, although they are called "mulches," and such soils are said "to mulch themselves."

Mulching Materials

Among the large variety of materials used as mulches, examples chosen are straw or hay, cover crops (live mulches), sawdust, petroleum products, plastic sheets, and gravel or crushed rock. Each of these is described briefly. Straw and cover crops are by far most widely used.

Straw or Hay. Use of these materials seems to be as old as agriculture itself (Jacks, Brind, and Smith, 1955). The straw or hay is spread over the soil surface, thickness of the layer depending on soil, slope, and other factors. The history and uses of straw mulches have been reviewed by Rowe-Dutton (1957) and by McCalla and Army (1961). Such mulches are used primarily for the control of erosion and for moisture conservation (Adams, 1966; Lal, 1975).

One variant of straw mulches is the practice of leaving the stubble standing after harvest of a crop such as wheat. This is followed with surface tillage rather than with a turning plow, and the next crop is then planted in the stubble. Called "stubble-mulch farming" (Duley and Russel, 1948), this practice is followed to some extent in the Great Plains of North America.

Cover Crops. Many kinds of plants are grown to provide cover for soils, especially in the tropics and subtropics with their long growing seasons. The cover crops protect soils from raindrop impact and also tie up nutrient elements to prevent leaching losses. Within a cover crop stand, an arable crop is planted by opening a small strip for placement of seed and fertilizer. Herbicides are used if necessary to kill cover crop plants in the seedling strip.

In fields or groves of plantation crops such as rubber, cacao, and oil palm in tropical regions, the cover crops are planted between the rows. Examples of cover crops are *Desmodium, S. Humilis, Paspalum notatum,* and *Centrosema.* Some of these plants are legumes and fix atmospheric nitrogen, but they all compete with the commercial crops for available soil moisture and can thus reduce yields.

Sawdust. This must be placed on the surface and not mixed with the soil if it is to be effective. Use of sawdust as a mulch is largely restricted to gardens and other small areas; applications to farm fields are limited. Sawdust also acts as a soil conditioner (q.v.) and provides organic matter. It is high in carbon and low in nitrogen, however, so that nitrogen deficiencies can easily follow when the sawdust decomposes and microorganisms compete with plants for available nitrogen (Allison and Anderson, 1951).

Gravel. Thin layers of gravel or crushed rock are placed on the soil surface as mulches in some instances. They are used, for example, in ornamental plantings in city parks and around homes (Richards, 1965) and on highway backslopes and shoulders. Gravel mulches reduce soil loss (Adams, 1966; Fairborn, 1973). At the same time they interfere with tillage operations.

Petroleum Products. Petroleum products are used primarily in the form of emulsions sprayed on the soil surface. A 50% aqueous emulsion of petroleum resins added to the soil surface dries into a black film ranging from 0.3 to 0.5 mm in thickness. Afterward, daytime temperatures of the soil are higher, which may be either beneficial or harmful. If daytime temperatures are usually low, seed germination and seedling emergence and survival can be improved. Conversely, if daytime temperatures are high, seed germination and seedling emergence and survival suffer. When applied as bands between crop rows, a petroleum resin mulch can reduce evaporation from the soil by as much as 10% and thus improve the moisture regime. The critical band width seems to be near 15 cm (Lippert, Takatori, and Whiting, 1964). Hillel, Schwartz, and Steinhardt (1969) studied the effects of light fuel oil and of an emulsion of latex, asphalt, and wax on runoff and erosion. Fuel oil encreased runoff more than did the emulsion. These mulches also reduced infiltration rates on soils of sand and loamy sand texture but not on loamy soils.

Water. Placed in small plastic bags, water is used in special situations. The filled bags, about 10 cm in height, are then placed between crop rows (Bowers, 1968). Because of the high specific heat of water, the mulch acts as a sink for heat during the day and as a source of heat at night (Johnson, Hedden, and Wilson, 1966).

Plastic Sheets. In use since the late fifties, these differ in reflectivity and can thus alter soil temperature and radiant energy levels beneath the crop canopy. Sheets of various kinds have been used in temperate regions to increase soil temperatures in early spring. In a greenhouse test, ployethylene sheets increased soil temperature 20°C (Vick, 1961). During daytime hours, temperatures were similar in bare soil and under reflective sheets but much higher under clear and black plastic sheets. A mulch of transparent plastic sheets increased soil temperatures under maize in the Moscow region and under cotton in the Tashkent region, USSR (Zakharov and Semikina, 1964). In general, transparent plastic sheets increase soil temperatures by 6 to 8°C on clear days before the crop canopy becomes complete and from 2 to 6°C afterward. In contrast to possible benefits of high temperatures in the spring in cool regions, plastic sheets can decrease yields of maize in the tropics because soil temperatures become too high for the crop. As an alternative to plastic, Harrison-Murray and Lal (1975) placed aluminum foil on the soil surface and reported a 50% increase in the yield of maize, attributing it to the reflective radiation. Plastic sheets act as a vapor barrier and thus influence soil moisture regimes. Moisture contents of the surface soil have been greater by 5 to 7% under plastic sheets than in the open.

Crop Responses

The effectiveness of mulches depends on the combination of climate, crop, and soil. Crop responses thus have a wide range for the world as a whole and may have a considerable range within limited areas. Because responses differ from region to region, examples from the temperate and tropical zones are discussed separately. Also discussed separately is the behavior of roots in mulched soils so as to give that topic special emphasis.

Temperate Zones. Stubble mulches that increase soil temperatures in the spring improved the yield of maize in the north-central United States (Van Doren and Triplett, 1973). Earlier emergence and earlier maturity of sweet corn were obtained with plastic mulches as compared to no mulch (Miller and Bunger, 1963). Sorghum produced higher yields when straw or plastic mulches were used than on unmulched plots, and the higher yields were attributed to more favorable soil temperatures (Adams, 1962).

Vegetables and fruits as well as field crops

can also yield better when mulches of various kinds are used. Yields of tomatoes, lettuce, and kohlrabi were greater with plastic mulches in some trials in Canada (Chipman, 1961). Higher yields of potatoes were obtained with plastic mulches in Florida (Hensel, 1962). Higher yields of radishes, onions, lettuce, and beans were obtained with petroleum mulches than without them in experiments in Great Britain (Sale, 1966). Increased growth and earlier fruiting were obtained by the use of black plastic mulches for grapes (Moore, 1963). Earlier maturity of strawberries as well as larger numbers and sizes of fruit were obtained in Belgium with black plastic mulches (Linden, 1963). Straw mulches improved the growth of apple trees in England (Dancer, 1964), whereas mulching with plastic sheets improved the growth and survival of citrus seedlings in Florida (Cahoon, Morton, and Lee, 1963).

Tropical Zones. Mulches consisting of organic matter promise to be of special importance in the tropics. In addition to providing soil cover, they also serve to add organic matter.

As in temperate zones, mulching has increased the yield of maize in Nigeria (Okigbo, 1972; Harrison-Murray and Lal, 1975). The effects of mulching were better than those of fertilizers on cotton yields in a ten-year study in Zaire (Jurion and Henry, 1969). Cotton also responded favorably to petroleum mulching in Madras (Kolandaiswamy, Ranganathan, and Sirinivashran, 1967). Soil temperature was higher in the soil around the seedlings, and crusting was also lower in the mulched plots. In Nigeria the yield of casava (Ayanaba, Okigbo, and Okigbo, 1974) and tomatoes (Quinn, 1974) were higher in mulched than in unmulched plots. Jones, Robinson, and Wallis (1961) obtained increased yields of coffee in Kenya by applying mulches. Improved growth of young tea and reduced runoff and erosion have been reported from India (Manipura, 1972). The increased yields of various crops in tropical zones seem due in part to improvement of moisture regimes and in part to the addition of nutrient elements.

Root Growth. The influence of mulches on root growth is indirect through effects on moisture and temperature regimes, aeration, bulk density (q.v.), and pore-size distribution (q.v.). Examples of these indirect effects are given in a later section about effects on physical properties. The indirect effects are illustrated by trials with plastic mulches in Sudan (Baumer, 1964). As compared to plots with no mulch, those with plastic sheets had deeper root penetration, greater numbers of secondary roots, better distribution of secondary roots along the main ones and about 20% more roots by weight. Differences were attributed largely to more moisture under the plastic mulch.

Physical Effects

Mulches can affect runoff and erosion, moisture regimes, and soil structure and aeration. Major effects are usually on runoff, erosion, and moisture regimes.

Runoff and Erosion. Mulches minimize raindrop impact and the sealing of soil surfaces, which in turn reduce runoff and the amount of sediment it can carry. The reduction in runoff follows from greater infiltration (q.v.) of water because the surface does not seal and because mulches slow down water movement (Juo and Lal, in press). The reduced erosion follows from the reduced volume and velocity of runoff water and its consequent lower capacity to carry sediments when moving through and over a mulch.

Moisture Regimes. The effects on soil moisture regimes are due to increased infiltration and reduced evaporation. As already indicated, greater infiltration follows from the more open soil surface and the reduced velocity of runoff water. Amounts of moisture under mulches can be 5–10% greater than in unmulched areas.

After water has entered the soil, losses by evaporation can be reduced to some extent. The evaporation (q.v.) proceeds in three stages (Lemon, 1956). The first stage is a rapid loss when water is present in the surface layer in amounts large enough to meet the evaporative demand of the atmosphere. Plastic sheets and petroleum mulches can restrict evaporation sharply at this stage, but they also reduce infiltration. The second stage is marked by a rapid decline in evaporation after the surface layer has dried out so that water must be supplied to it by capillary rise from below. Soil conditions then control loss through evaporation. The third stage is an extremely slow but constant rate of evaporation after moisture levels in soils are quite low. Mulches are effective only in the first and second stages of evaporation (Willis, 1962). A minor effect of mulches on moisture regimes is the condensation of little water during temperature reversals at night (Charles, 1964).

Soil Structure and Aeration. Mulches affect structure and aeration chiefly through reducing raindrop impact and preventing crusting. Structure of the surface layer is thus more stable, especially under mulches of organic materials. Such mulches also provide food for organisms such as earthworms, and the population may therefore increase. The activities of earthworns and other fauna provide channels within the soil through which air and water can move.

Aeration may be imporved or reduced by mulches. The soil plots with organic matter mulches tend to be higher in oxygen and lower in carbon dioxide than unmulched plots (Lal, 1974). On the other hand, a bitumen mulch tends to increase the concentration of carbon dioxide in the soil atmosphere (Kruger and Benkenstein, 1972).

Chemical Properties

Effects of different types of mulches on the chemistry of soils differ. Known effects of mulches of straw and plant residues are to increase amount of organic matter and nitrogen in the surface layers of soils. In Nigeria, for example, the organic matter level of the surface soil was increased more by leaving crop residues on the surface than by plowing them under (Juo and Lal, in press). Increases in organic carbon and nitrogen also occurred in some trials with wheat straw applications in Washington state, (Black, 1973*a*, 1973*b*). Larger amounts of organic carbon increase cation exchange capacity (see *Exchange Phenomena*) and the water content and retention (q.v.) of soils.

Effects of plastic and straw mulches on nitrate levels of soils differ appreciably. Leaching (q.v.) losses were lower and nitrates higher in plots under plastic sheets than in the controls (Clarkson, 1960). Increased levels of nitrate were attributed to higher temperature under plastic sheets by Baumer (1964). Loss of nitrogen through volatilization was also reduced by plastic mulch (Nagai, 1967). Straw mulches have an appreciably greater effect on microorganisms than do plastic mulches. The straw provides an energy source for the organisms and their extra growth can tie up nitrogen that would otherwise be available for a crop (Burrows, 1960).

Temperature

Modification of soil temperature is one of the more important effects of mulches. The nature and magnitude of modifications depend on the kind of mulch, the kind of soil, and the weather. Most mulches act as insulation and thus damp down the effects of sudden fluctuations in air temperature. Mulches can also reduce heat losses during cold seasons in temperate regions and during nights in the tropics. Conversely, mulches can keep the soil temperatures lower in summers in temperate zones and in daytimes in the tropics. The mulches thus tend to reduce the annual and diurnal fluctuations in soil temperatures. They have been used to advantage in winters with suboptimal temperature regimes and in summers with supraoptimal regimes. The effects of mulches on soil temperatures are not necessarily beneficial. In trials in Nigeria, for example, soil temperatures were highest in the plots under transparent plastic and the yields of maize were lower on those plots.

Plastic and crushed-stone mulches affect air temperatures near the soil as well as the temperature within the soil (Shadbolt, McCoy, and Whiting 1962). Materials such as colored polyethylene films and aluminum foil reflect much of the incoming radiation and can thus directly affect plant growth. Plants receive some reflected as well as direct radiation where such mulches are in use.

Biological Effects

Organic mulches such as straw, hay, and cover crop residues have both direct and indirect effects on microbial life and fauna in the soil. Direct effects lie in providing additional energy sources for organisms. Indirect effects are the modifications of the moisture and temperature regimes. Plastic sheets, petroleum products, gravel, and crushed stone have indirect effects only.

Microorganisms such as bacteria, fungi and actinomycetes, and soil fauna such as termites, earthworms and millipedes all take part in the decomposition of the organic matter added as mulches. Those provide energy for microbial life and fauna to stimulate their growth and increase their numbers. Consequently, organic residues are decomposed rapidly in the humid tropics where moisture is generally available and temperatures are high. Maintaining an adequate mulch is difficult under those conditions. In contrast, decomposition of mulches is much slower in the temperate zone, as a general rule.

The nature of products formed by decomposition of organic matter added in the form of mulches is affected by aeration of the soil (Burges and Raw, 1967). For the most part, decomposition of mulching materials is aerobic, but in some instances it can be anaerobic, at least in part. Plant growth inhibitors may then be formed, as, for example, oxalic acid (McCalla, 1958). Even under aerobic conditions, some organic compounds have acid radicals formed as intermediate products in the long train that begins with plant residues and ends with carbon dioxide, water, and mineral salts. The large variety of compounds in the many stages of decay of organic mulches parallels the fate of residues added by native vegetation. Many of the intermediate products of decay are similar to those formed under native vegetation. They also influence the physical properties of the soil, provide plant nutrients, and in the long run affect the pathways of soil development.

R. LAL

References

Adams, J. E., 1962. Effect of soil temperature on grain sorghum growth and yield, *Agron. Jour.* **54**, 257-261.

——, 1966. Influence of mulches on runoff, erosion and soil moisture depletion, *Soil Sci. Soc. Amer. Proc.*, **30**, 110-114.

Allison, F. E., and Anderson, M. S., 1951. The use of sawdust for mulches and soil improvement, *U.S. Depart. Agric. Circular*, **891**, 19p.

Ayanaba, A., and Okigbo, B. N., 1975. Mulching for improved soil fertility and crop production, *FAO Soils Bull.*, **27**, 97-115.

Baumer, M., 1964. Plastic mulch, nitrogen fixation and root development, *Sudan Jour. Vet. Sci. Anim. Husb.*, 5(2), 38-39.

Black, A. L., 1973*a*. Crop residue, soil water, and soil fertility related to spring wheat production and quality after fallow, *Soil Sci. Soc. America Proc.*, 37(5), 754-758.

Black, A. L. , 1973*b*. Soil property changes associated with crop residue management in a wheat fallow rotation, *Soil Sci. Soc. America Proc.*, 37(6), 943-946.

Bowers, S. A., 1968. Influence of water mulches on soil temperature and sweet corn and green bean production, *Soil Sci.*, **105**, 335-345.

Burges, A., and Raw, E., (ed.), 1967. *Soil Biology*. London: Academic Press 532p.

Burrows, W. C., 1960. Mulch influence on soil temperature and corn growth, *Diss. Absts.*, **20**, 3451-3452.

Cahoon, G. A.; Morton, E. S.; and Lee, B. W., 1963. The use of plastic ground cover in citrus nurseries, *Amer. Soc. Hort. Sci. Proc.*, **83**, 309-319.

Charles, G., 1964. Effect of petroleum mulch on evaporation and secondary supplies of humidity in soils under Mediterranean conditions, *Bull. Assoc. France Et. Soil No. 6*, 250-254.

Chipman, E. W., 1961. Studies on tomato response to mulching on ridged and flat rows, *Canada Jour. Plant Sci.*, **41**, 10-15.

Clarkson, V. A., 1960. Effect of black polyethylene mulch on soil and microclimate, temperature and nitrate level, *Agron. Jour.*, **52**, 307-309.

Dancer, J., 1964. The influence of soil moisture and temperature on the growth of apple trees. II. Some observations on moisture and temperature conditions under grass arable and mulch, *Hort. Res.*, **4**, 3-13.

Duley, F. L., and Russel, J. C., 1948. Stubble-mulch farming to hold soil and water, *U.S. Depart. Agric. Farmer's Bull. 1997*, 32p.

Fairborn, M. L., 1973. Effect of gravel mulch on crop yields, *Agron. Jour.*, **65**(6), 925-928.

Harrison-Murray, R. S., and Lal, R., 1975. Soil temperature response of various crops, *Proc. Crop Physiology Program Formulation Conf. IITA, April 1975*, 77-84.

Hensel, D. R., 1962. Plastic mulching of potatoes, *Soil Crop Sci. Soc. Fla. Proc.*, **22**, 232-238.

Hillel, D.; Schwartz, A.; and Steinhardt, R., 1969. Laboratory tests of sprayable material for runoff inducement on a loessal soil, *Israel Jour. Agric. Res.*, **19**, 3-9.

Jacks, C. V.; Brind, W. D.; and Smith, R., 1955. Mulching, *Commonwealth Bureau of Soil Sci. Tech. Comm. No. 49*, 87p.

Johnson, W. H.; Heddon, D. K.; and Wilson, J. D., 1966. How liquid mulches affect moisture retention, temperature and seedling growth, *Agric. Engin.*, **47**, 196-199.

Jones, P. A.; Robinson, J. B. D.; and Wallis, J. A. N., 1961. Fertilizers, manure and mulch in Kenya coffee growing, *Kenya Coffee Bull.*, **Dec. 1961**, 441-459.

Juo, A. S. R., and Lal, R., 1977. Effect of fallow and continuous cultivation on the chemical and physical properties of an Alfisol in the tropics, *Plant and Soil*, **45**, 18p.

Jurion, F., and Henry. J., 1969. Can primitive farming be modernized? *INEAC Hons. Series 1964*, 457p.

Kolandaiswamy, S.; Ranganathan, A. K.; and Sirinivashran, V., 1967. Preliminary trial on agricultural mulch on cholam and cotton, *Madras Agric. Jour.*, **54**, 195-197.

Kruger, W., and Benkenstein, H., 1972. Effect of a layer of bitumen mulch on water infiltration, air composition, and condensation processes in soil, *Archiv für Acker-und Pflauzenbau und Bodenkunde*, **16**(4/5), 391-398.

Lal, R., 1974. Soil temperature, soil moisture, and maize yield from mulched and unmulched tropical soils, *Plant and Soil*, **40**, 129-143.

——, 1975. Role of mulching techniques in tropical soil and water management, *IITA Tech. Bull.* **1**, 38p.

Lemon, E. R., 1956. The potentialities for decreasing soil moisture evaporation loss, *Soil Sci. Soc. America Proc.*, **20**, 120-125.

Linden, R., 1963. Soil protection with plastic mulches, *Ann. Gembl.*, **69**, 601-608.

Lindermann, R. G., 1970. Plant residue decomposition products and their effects on host roots and fungi pathogenic to roots, *Phytopathology*, **60**, 19-22.

Lippert, L. F.; Takatori, F. H.; and Whiting, F. L., 1964. Soil moisture under bands of petroleum and polythene mulches, *Am. Soc. Hort. Sci. Proc.*, **85**, 541-546.

Manipura, W. B., 1972. Influence of mulch and cover crops on surface runoff and soil erosion on tea lands during the early growth of replanted tea, *Tea Quality*, **43**(3), 95-102.

McCalla, T. M., 1958. Microbial and related studies of stubble mulching, *Jour. Soil and Water Conserv.*, **13**(6), 225-258.

McCalla, T. M., and Army, T. J., 1961 Stubble mulch farming, *Adv. in Agron.*, **13**, 126-194.

Miller, D. E., and Bunger, W. C., 1963. Use of plastic soil covers in sweet corn production, *Agron. Jour.*, **55**, 417-419.

Moore, R. C., 1963. Plastic mulch aids growth of young grape vines and cuttings, *Biokemia*, **1**(1), 21-33.

Nagai, K., 1967. The volatilization of nitrogen during nitrification in soils under vinyl-cover culture, *Jour. Soil Manure, Tokyo*, **38**, 339-341.

Okigbo, B. N., 1972. Essais sur le mais dans pla planie de Nsukka (Nigeria) III. Les effects de differentes sorts de mulch d sur le rendement du mais en climat tropicale humide, *Agron. Trop.*, **10**, 1037-1047.

Quinn, J. G., 1974. An evaluation of methods of mulching and stalking tomatoes grown during the rains at Samaru Nigeria, *Hort. Res.*, 13(2/3), 97-104.

Richards, S. J., 1965. Porous black mulch for ornamental plantings, *Calif. Agric.*, **19**(12), 12-14.

Rowe-Dutton, P., 1957. The mulching of vegetables, *Commonwealth Agric. Bureau Comm. No. 24*, 167p.

Sales, P. M. J., 1966. Effect of petroleum mulch on the marketable yields of some vegetable crops, *Expl. Hort.*, **14**, 53-61.

Shadbolt, C. A., McCoy, O. D., and Whiting, F. L., 1962. The microclimate of plastic shelters used for vegetable production, *Hilgardia*, **32**, 251-256.

Van Doren, D. M., and Triplett, G. B., 1973. Mulch and tillage relationships in corn culture, *Soil Sci. Soc. America Proc.*, 37(5), 766-769.

Vick, F., 1961. The use of plastic mulches for forcing vegetables. I. The use of plastic sheets for covering unheated soil in a hangar glasshouse. *Vyz. Ustav. Zelin. Olomonc. Bull.*, 5, 53-72.

Willis, W. O., 1962. Effect of partial surface covers on evaporation from soil, *Soil Soc. America Proc.*, **26**, 598-601.

Zakharov, N. G., and Semikina, G. G., 1964. Investigations of the agrophysical effectiveness of mulching soil with film of polyethylene, *Dokl. Akad. S. kh. Nauk*, **12**, 14-16.

Cross-references: *Conservation; Management of Soils; Moisture Management; Soil Conditioners; Zero-tillage.*

N

NITRIFICATION—See NITROGEN CYCLE. VOL. IVA: NITROGEN CYCLE

NITROGEN CYCLE

Nitrogen occupies a unique position among the elements essential for plant growth because it is required in rather large amounts by most agricultural crops and because only trace quantities of available mineral forms are present in the soil at any one time. Nitrogen starvation by plants is shown by yellowing of the leaves and by slow and stunted growth. Other factors being favorable, a supply of available nitrogen in the soil promotes rapid plant growth and development of dark green color in the leaves.

The source of soil nitrogen is the atmosphere, where the strongly bonded molecule N_2 is the predominant gas (79.08% by volume of the gases). Only a few microorganisms have the ability to use molecular N_2; all other living organisms require combined nitrogen for carrying out their life activities.

An outline of the nitrogen cycle in soil is shown in Figure 1. In addition to interaction with the total ecosystem, an internal cycle is operative in which nitrogen passes from one system to another through a variety of biochemical and chemical processes. Gains in soil

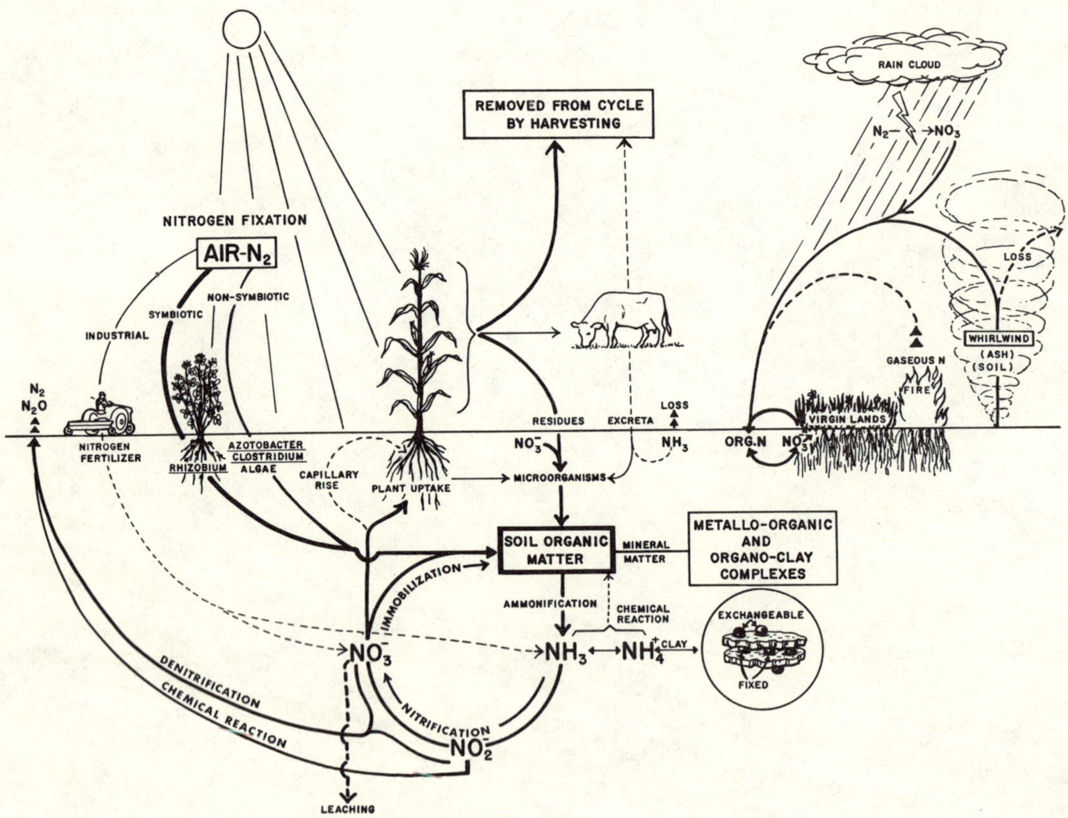

FIGURE 1. The nitrogen cycle in soil.

nitrogen occur through fixation of molecular N_2 by microorganisms and from the return of ammonia and nitrate in rain water; losses occur through crop removal, leaching, and volatilization. The conversion of molecular N_2 to combined forms occurs through *biological nitrogen fixation.* Organic forms of nitrogen in turn are converted to ammonia and nitrate by a process called *mineralization.* The conversion to ammonia is referred to as *ammonification;* the oxidation of this compound to nitrate is termed *nitrification.* The utilization of ammonia and nitrate by plants and soil organisms constitutes *assimilation* and *immobilization,* respectively. Combined nitrogen is ultimately returned to the atmosphere as molecular N_2 through either biological *denitrification* (q.v.) or chemical reactions involving nitrites, thereby completing the cycle. For a comprehensive discussion of agronomic aspects of the nitrogen cycle, refer to the book by Bartholomew and Clark (1965). Environmental considerations from the standpoint of air pollution have been covered by Robinson and Robbins (1970).

Ammonia and nitrite (NO_2^-), produced as products of the microbial decomposition of nitrogenous organic materials, are capable of undergoing chemical reactions with organic substances, in some cases leading to the evolution of nitrogen gases. Through the association of humic materials with mineral matter, organo-clay complexes are formed whereby the nitrogen compounds are protected against attack by microorganisms. The ammonium ion (NH_4^+) undergoes substitution reactions with other cations of the exchange complex, and it can be fixed by clay minerals.

The basic feature of nitrogen transformations centers on oxidation and reduction reactions. An entire sequence of events is involved, some of which take place in the cells of microorganisms and some in the tissues of higher plants. Included are gaseous, mineral, and organic forms of nitrogen.

Some of the oxidation levels, or valence states, of nitrogen are shown in Figure 2. They range from +5 for nitrate (NO_3^-) to -3 for ammonia (NH_3). In the oxidized (+) state, the outer electrons of nitrogen serve to complete the electron shells of other atoms; in the reduced (-) state, the two electrons required to fill the outer shell of nitrogen are supplied by other atoms.

A balance sheet for nitrogen, as devised by Delwiche (1970), is given in Figure 3. According to these data, nitrogen is being introduced into the biosphere at the rate of about 92 million metric tons per year while the amount being denitrified and returned to the atmosphere is about 83 million metric tons per year. The difference (9 million metric tons) represents the rate at which fixed nitrogen is building up in the biosphere and pedosphere.

VALENCE	COMPOUND	FORMULA	VALENCE ELECTRONS
+5	NITRATE ION	NO_3^-	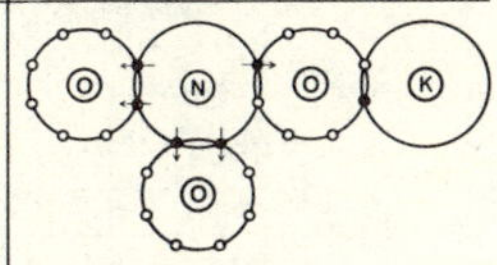
+3	NITRITE ION	NO_2^-	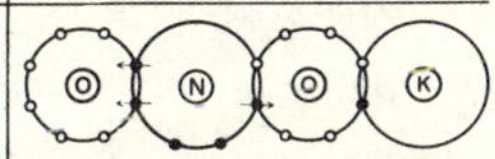
+1	NITROXYL	[HNO]	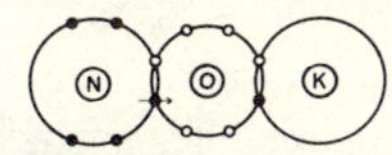
0	NITROGEN GAS	N_2	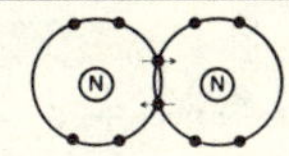
-1	HYDROXYLAMINE	$HONH_2$	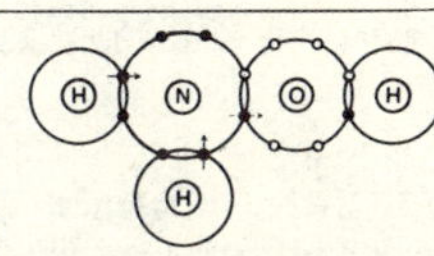
-3	AMMONIA	NH_3	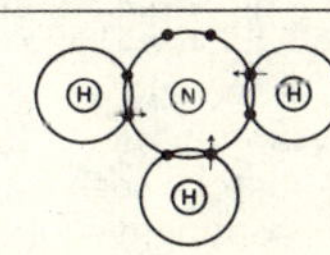

FIGURE 2. Oxidation levels, or valence states, of nitrogen. The schematic diagrams on the right show the disposition of electrons in the atom's outer shell. The ions are shown combined with potassium. Electrons of nitrogen are indicated with an arrow; those of other atoms as black dots or open circles. (From Delwiche, 1970; copyright © 1970 by Scientific American, Inc. All rights reserved.)

As Delwiche (1970) pointed out, the nitrogen balance of the earth has been upset by human activities. Fertilizer nitrogen production has been doubling about every six years, and some of this nitrogen is being carried into lakes and streams in runoff waters from agricultural lands. In some cases nitrate concentrations in water supplies are approaching the level (45 ppm) regarded by the U.S. Department of Health as unacceptable for human consumption. There is also the problem of disposal of nitrogenous organic wastes from an expanding human and domestic animal population. When possible, this waste nitrogen should be recycled back to the soil, but under present economic conditions, it is cheaper for the farmer to use chemical fertilizers. Research is now underway in attempts to develop efficient and adequate methods for utilization of organic wastes on soil.

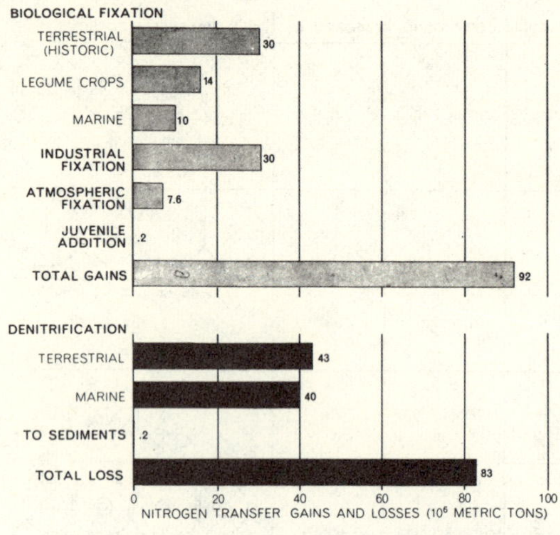

FIGURE 3. Balance sheet for the nitrogen cycle. Gains and losses are given in terms of metric tons per year while the figures for inventories are given in billions of metric tons. The value for the quantity of nitrogen contained in dead organic matter on land (760) may be high by a factor of 10. (From Delwiche, 1970; copyright © 1970 by Scientific American, Inc. All rights reserved.)

Gains in Soil Nitrogen

Several mechanisms have been proposed for the natural addition of nitrogen to soils, but thus far only fixation of molecular N_2 by microorganisms and accession of ammonia and nitrate in atmospheric precipitation are considered important. For many years Dhar (1960, 1961) has claimed that photochemical fixation of nitrogen occurs in soil, but this theory has yet to be substantiated.

Nitrogen in Atmospheric Precipitation. Combined nitrogen, consisting of ammonia, nitrite, nitrate, and organically bound nitrogen, is a common constituent of atmospheric precipitation. Nitrite occurs in trace amounts and is usually included with the nitrate value. The organically bound nitrogen is probably associated with cosmic dust and may not represent a new addition to land masses of the world.

The amount of nitrogen added to the soil each year in atmospheric precipitation is normally too small to be significant in crop production. For the United States and Europe, the estimates for ammonia- plus nitrate-nitrogen range from 0.7 to 19.6 lb/acre/yr, or from 0.78 to 22.0 kg/ha (Eriksson, 1952). Many of the higher values may represent analytical or sampling errors, although higher than normal amounts would be expected near highly industrial areas due to fossil-fuels burning.

Important sources of ammonia in the atmosphere include volatilization from land surfaces, combustion of fossil fuel, and natural fires. A small quantity of nitrogen may be fixed in the trail of meteorites (Hutchinson, 1944).

Sorption of Combined Nitrogen from the Atmosphere. Soils have the ability to absorb atmospheric ammonia and earlier investigators (notably Liebig) placed considerable emphasis on this process as a means of providing nitrogen to plants. The opinion of modern soil scientists is that the process is of little practical significance except under special circumstances such as downwind from a point source (e.g., a cattle feedlot) or in industrial areas where fossil fuels are burned.

Biological Nitrogen Fixation. Although a vast supply of nitrogen occurs in the earth's atmosphere (386×10^{13} metric tons), it is present as an inert gas and cannot be used by higher forms of plant and animal life. The covalent triple bond of the N_2 molecule ($N{\equiv}N$) is highly stable and can be broken chemically only at elevated temperatures and pressures. Nitrogen-fixing microorganisms, on the other hand, perform this seemingly impossible task at ordinary temperatures and pressures (see *Nitrogen Fixation*).

The ability of a few bacteria and blue green algae to fix molecular N_2 can be regarded as being equal in importance to photosynthesis for maintenance of life. Even with the tremendous expansion in facilities for producing fertilizer nitrogen since World War II, legumes are still to be regarded as a major source of fixed nitrogen for the majority of the world's soils (see Figure 3). For many years to come, the nitrogen necessary for crop production in underdeveloped countries will have to come from that which is indigenous to the soil or supplied by nitrogen-fixing microorganisms.

The total amount of atmospheric nitrogen returned to land surfaces of the earth each year through biological fixation is of the order of from 12 to 105×10^{12} g (Hutchinson, 1954). This is equivalent to from 0.7 to 6.2 lb/acre as an average or 0.78 to 6.9 kg/ha. As one might expect, this range is far too low for soils where legumes are grown and too high for most wastelands.

Mineralization and Immobilization of Nitrogen

An important feature of the internal nitrogen cycle in soil is the turnover of inorganic and organic forms of nitrogen through the opposing processes of mineralization and immobilization, as follows:

$$\text{Organic N} \xrightarrow{\hspace{4em}} NH_3 \xrightarrow{\textit{Nitrosomonas}} NO_2^- \xrightarrow{\textit{Nitrobacter}} NO_3^-$$

ammonification → | ← nitrification →

mineralization →

← immobilization

Mineralization encompasses two distinct microbiological processes, (i) ammonification, in which organic nitrogen is converted into ammonia; and (ii) nitrification, in which ammonia is oxidized to nitrate. Immobilization refers to the consumption of ammonium and nitrate by microorganisms and incorporation of the nitrogen into cellular components.

Both aerobic and anaerobic microorganisms are involved in the ammonification process whereas only aerobes oxidize ammonia to nitrate. The latter is the predominant available form of nitrogen in cultivated soils that are well aerated. Grassland soils often contain rather high levels of exchangeable ammonium but relatively low amounts of nitrate.

The oxidation of ammonia to nitrate is brought about largely by two groups of autotrophic bacteria, one deriving energy through oxidation of ammonia to nitrite, the other through oxidation of nitrite to nitrate. Five genera of ammonia oxiders and two genera of nitrite oxiders are recognized; however, only one genus of each is considered important, *Nitrosomonas* and *Nitrobacter*, respectively. A single species of *Nitrosomonas* (*N. euopaea*) is involved, whereas two species of *Nitrobacter* (*N. winogradskyi* and *N. agile*) are important.

Heterotrophic bacteria, actinomycetes, and fungi have all been shown to produce nitrite or nitrate from ammonia and organic nitrogen compounds in pure culture. However, the ecological importance of heterotrophic nitrification, has yet to be determined.

The initial stage of nitrification, catalyzed by *Nitrosomonas*, involves removal of three pairs of electrons from ammonia, with the formation of hydroxylamine, hyponitrite, and nitrite.

$$\underset{\substack{\text{Ammonia}\\-3}}{NH_3} \xrightarrow{\frac{1}{2}O_2} \underset{\substack{\text{Hydroxylamine}\\-1}}{HO{-}NH_2} \xrightarrow{-2H} \underset{\substack{\text{Hyponitrite}\\+1}}{\tfrac{1}{2}[HO{-}N{=}N{-}OH]} \xrightarrow{\frac{1}{2}O_2} \underset{\substack{\text{Nitrite}\\+3}}{HO{-}N{=}O}$$

Members of the genus *Nitrobacter* remove two electrons from nitrite. A dehydration reaction may characterize this reaction (Alexander, 1977).

$$\underset{\substack{\text{Nitrite}\\+3}}{HO{-}N{=}O} \xrightarrow{H_2O} \left[HO{-}N\begin{matrix}OH\\OH\end{matrix} \right] \xrightarrow{-2H} \underset{\substack{\text{Nitrate}\\+5}}{HO{-}N\begin{matrix}O\\O\end{matrix}}$$

In most soils the rate of oxidation of nitrite to nitrate is considerably faster than that of the formation of nitrite from ammonia; thus nitrites are seldom found. However, they often accumulate in alkaline soils, a result that may be due to ammonia toxicity of *Nitrobacter*.

The nitrogen content of organic residues, as reflected through the C/N ratio, is of primary importance in regulating the quantitites of ammonium and nitrate in the soil. Residues that have C/N ratios (q.v.) greater than about 30, equivalent to nitrogen contents of about 1.5% or less, result in lowering of mineral nitrogen reserves because of *net immobilization*. On the other hand, residues with C/N ratios below about 20, or nitrogen contents greater than about 2.5%, lead to an increase in mineral nitrogen levels through *net mineralization*.

Changes in nitrate levels attending the decay of low-nitrogen crop residues in soil are illustrated in Figure 4. Under conditions suitable for microbial activity, decomposition occurs with the concurrent liberation of carbon dioxide. To meet the nitrogen requirements of microorganisms, mineral nitrogen is consumed; that is, there is a *net immobilization* of nitrogen. However, when the C/N ratio of the decomposing material has been lowered to about 20, nitrate levels increase due to *net mineralization*.

Losses of Soil Nitrogen

Quantitative recovery of soil and fertilizer nitrogen by growing plants is seldom achieved under conditions existing in the field. Even under the best circumstances, no more than two thirds of the nitrogen added as fertilizer has been accounted for by crop removal; recoveries considerably less than one half are not uncommon. Numerous attempts have been made to account for the low recoveries, and it is now known that available forms of nitrogen (e.g., nitrate), whether added as fertilizer or produced through decay of organic matter, will not remain very long in the soil. Thus any mineral nitrogen not adsorbed by plants or incorporated into the organic matter will usually be lost by one or more of the processes discussed below.

Volatilization of Ammonia. Ammonia is often lost from soils by *volatilization.* Losses are of greatest importance on calcareous soils, especially when ammonium-containing fertilizers are used. Only slight losses occur in soils of pH 6 to 7, but losses increase markedly as the pH of the soil increases. Appreciable losses can also occur when nitrogenous organic wastes such as farmyard manures are permitted to decompose on the soil surface, even a soil that is acid, because of the localized increase in pH resulting from ammonia formation.

The use of anhydrous ammonia as fertilizer has become widespread in recent years. Losses through volatilization are not considered serious as long as the gas is injected well below the surface of the soil and the sorption capacity of the soil is not exceeded.

Bacterial Denitrification. Under suitable conditions nitrate can be lost rapidly from soil through denitrification. The ability to convert nitrate to nitrogen gases (N_2 and N_2O) is limited to a few organisms that are able to use the oxygen of nitrate as a substitute for molecular oxygen (O_2) in conventional metabolism. The organisms primarily involved are heterotrophic (live off organic matter) and belong to the genera *Pseudomonas, Micrococcus, Achromobacter,* and *Bacillus.* Several chemoautotrophs are also capable of using nitrate with the production of nitrogen gases, but they are of little importance in most soils.

The following pathway represents the probable mechanisms of bacterial denitrification:

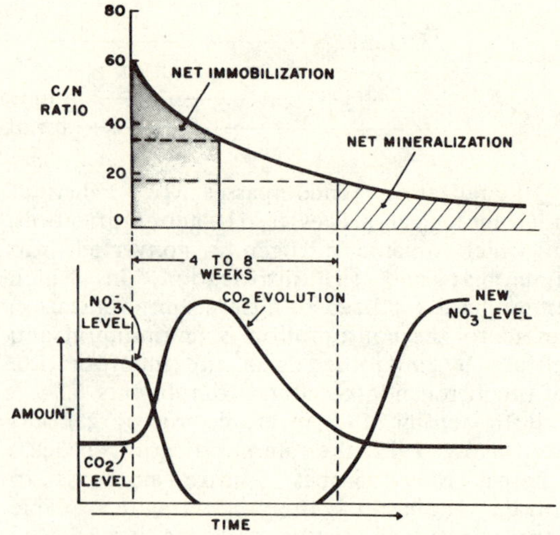

FIGURE 4. Changes in nitrate levels attending the decomposition of carbonaceous crop residues in soil.

nitrate content of groundwater. Denitrifying microorganisms are known to be present at considerable depths in soils, and it is possible that some of the nitrate leached into the subsoil may be volatilized before reaching the water table. Meek, Grass, and McKenzie (1969) concluded that much of the nitrate leached into the subsoil in irrigation waters was lost through denitrification. Stewart et al (1967) found that nitrate levels in soil under feedlots decreased sharply with depth and concluded that the decrease was due to denitrification.

Leaching. Nitrogen is lost by leaching mainly as nitrate, although ammonium may be

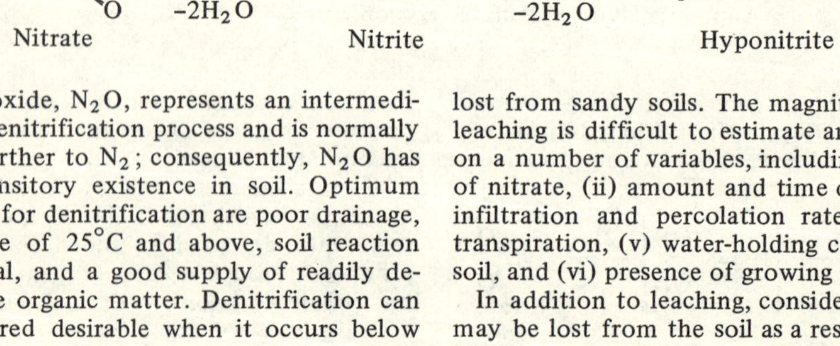

Nitrous oxide, N_2O, represents an intermediate in the denitrification process and is normally reduced further to N_2; consequently, N_2O has only a transitory existence in soil. Optimum conditions for denitrification are poor drainage, temperature of 25°C and above, soil reaction near netural, and a good supply of readily decomposable organic matter. Denitrification can be considered desirable when it occurs below the rooting zone because of reduction in the

lost from sandy soils. The magnitude of nitrate leaching is difficult to estimate and will depend on a number of variables, including (i) quantity of nitrate, (ii) amount and time of rainfall, (iii) infiltration and percolation rates, (iv) evapotranspiration, (v) water-holding capacity of the soil, and (vi) presence of growing plants.

In addition to leaching, considerable nitrogen may be lost from the soil as a result of erosion. Sheet erosion is highly selective in that the eroded fraction contains several times more nitrogen than the original soil.

Chemical Reactions of Nitrite. Nitrite is not usually present in detectable amounts in well-drained neutral or slightly acidic soils. Accumulations occur, however, in calcareous soils, and recent work indicates that this ion often persists, albeit temporarily, when ammonium or ammonium-type fertilizers are applied to soil. This nitrate accumulation has been attributed to inhibition of nitrification at the ammonium stage. Presumably, nitrate oxidizing organisms (*Nitrobacter*) are more sensitive to ammonia and an adverse soil reaction than ammonium oxidizers (*Nitrosomonas*).

Four reactions have been suggested for loss of nitrogen from soil as a result of the chemical degradation of nitrites. They are: (1) double decomposition of ammonium nitrite ($NH_4NO_2 \rightarrow 2H_2O + N_2$); (2) the Van Slyke reaction ($RNH_2 + HNO_2 \rightarrow ROH + H_2O + N_2$); (3) decomposition of nitrous acid ($3HNO_2 \rightarrow 2NO + HNO_3 + H_2O$; and (4) chemical reactions with humic and fulvic acids.

F. J. STEVENSON

References

Alexander, M., 1977. *Introduction to Soil Microbiology.* New York: Wiley, 467p.

Bartholomew, W. V., and Clark, F. E., eds., 1965. *Soil Nitrogen.* Madison: American Society of Agronomy, 615p.

Delwiche, C. C., 1970. The nitrogen cycle, *in* D. Flanagan, ed., *The Biosphere.* San Francisco: Freeman, 71-80.

Dhar, N. R., 1960. Role of organic matter in soil fertility, *7th Internat. Congr. Soil Sci. Trans.,* **3,** 314-320.

Dhar, N. R., 1961. Nitrogen problem, *Proc. 48th Indian Sci. Congr.,* 1-32.

Eriksson, E., 1952. Composition of atmospheric precipitation. 1. Nitrogen compounds, *Tellus,* **4,** 215-232.

Hauck, R. D., 1973. Nitrogen tracers in nitrogen cycle studies—Past use and future needs, *Jour. Environ. Quality,* **2,** 317-327.

Hauck, R. D., and Bremner, J. M., 1976. Use of tracers for soil and fertilizer nitrogen research. *Adv. Agron.,* **28,** 219-266.

Hutchinson, G. E., 1944. Nitrogen in the biogeochemistry of the atmosphere, *Am. Scient.,* **32,** 178-195.

Hutchinson, G. E., 1954. The biochemistry of the terrestrial atmosphere, *in* G. P. Kuiper, ed., *The Solar System, Vol. II. The Earth as a Planet.* Chicago: University of Chicago Press, 371-433.

Jansson, S. L., 1971. Use of ^{15}N in studies of soil nitrogen, *in* A. D. McLaren and J. J. Skujins, eds., *Soil Biochemistry,* Vol. 2. New York: Marcel Dekker, 129-165.

Keeney, D. R., 1973. The nitrogen cycle in sediment-water systems, *Jour. Environ. Quality,* **2,** 15-29.

Meek, D. B.; Grass, L. B.; MacKenzie, A. J., 1969. Applied nitrogen loss in relation to oxygen status of soils, *Soil Sci. Soc. America Proc.,* **33,** 575-578.

Nutman, P. S., ed., 1976. *Symbiotic Nitrogen Fixation in Plants.* London: Cambridge University Press, 584p.

Porter, L. K., 1974. Nitrogen transfer in ecosystems, *in* E. A. Paul and A. D. McLaren, eds., *Soil Biochemistry,* Vol. 4. New York: Marcel Dekker, 1-26.

Quispel, A., ed., 1974. *The Biology of Nitrogen Fixation.* New York: American Elsevier, 769p.

Robinson, E., and Robbins, R. C., 1970. Gaseous nitrogen compound pollutants from Urban and natural sources, *Jour. Air Pollution Control Assoc.,* **20,** 303-306.

Stewart, B. A.; Viets, F. G., Jr.; Hutchinson, G. L.; and Kemper, W. D., 1967. Nitrate and other water pollutants under fields and feedlots, *Environ. Sci. Tech.,* **1,** 736-739.

Stewart, W. D. P., ed., 1975. *Nitrogen Fixation by Free-living Microorganisms.* London: Cambridge University Press, 471p.

Cross-references: *Carbon Cycle; Carbon-Nitrogen Ratio; Clay-organic Reactions; Denitrification; Enzyme Activity; Fertilizers, Inorganic; Fertilizers, Organic; Humus; Metabolic Activity; Nitrogen Fixation; Nutrient Losses; Phosphorus Cycle; Plant Nutrients; Pollution; Soil Biochemistry; Soil Fertility. Vol. IVA: Cycles–Geochemical; Microorganisms; Nitrogen Cycle; Photosynthesis.*

NITROGEN FIXATION

Nitrogen (N_2) fixation in bacteria is commonly known as *biological N_2 fixation.* This biological process, which is distinguished from fixation by chemical or electrical means, involves the conversion of atmospheric N_2 to ammonia (NH_3) in organisms, which in turn combines with metabolic intermediates to form amino acids and proteins. Biological N_2 fixation indirectly depends on solar energy stored in plants as products of photosynthesis. This process may occur symbiotically or nonsymbiotically. Nonsymbiotic soil organisms that fix N_2 obtain energy for fixation through the decomposition of plant residues and soil organic matter. Symbiotic N_2 fixation occurs when soil organisms are associated with plant root tissue and obtain energy from the plant. Thus the N_2 fixing process requires energy that the N_2 fixing organisms must obtain either directly from plants in legumes or indirectly from the decomposition of plant residues in the soil. Exceptions include photosynthetic bacteria and blue green algae, which obtain their energy from sunlight.

A simplified version of the N_2 fixing reaction for all organisms is as follows:

$$N_2 \xrightarrow[\text{nitrogenase}]{6H + \text{energy}} 2NH_3$$

Three components in addition to N_2 are needed for the reaction: (1) a source of energy, which at the cellular level is supplied as Mg ATP (adenosine triphosphate, a high energy supplying molecule); (2) a reductant, designated as H, which is supplied as electrons from other organic compounds (ferredoxin and flavodoxin); and (3) an enzyme, nitrogenase, which catalyzes the reduction of N_2 to NH_3. This enzyme, characterized rather recently, is remarkably similar in many types of N_2 fixing bacteria.

Major Categories of N_2 Fixation

Organisms capable of N_2 fixation are limited to a small group of bacteria, blue-green algae, and possibly actinomycetes. They can be categorized in a number of ways (Figure 1). As shown, N_2 fixation can be divided into three major categories with subdivisions within.

Obligatory Symbiosis. *Symbiosis* describes the relationship between two organisms living together for mutual benefit. This type of association (Figure 1) occurs within root nodules (protrusions of the root) of either leguminous or nonleguminous angiosperms (a group of plants whose seeds are borne within a matured ovary or fruit). The organisms causing this symbiotic association are members of the genus *Rhizobium* in legumes and an actinomycete for the nonlegumes. The legumes will be considered first.

N_2 Fixation in Legumes. Nitrogen fixation in legumes is termed *obligatory* because both the host plant and the organism are required for fixation to occur. A classical example is the association between dicotyledonous plants (a subclass of angiosperms) of the family *leguminosae* and members of the bacterial genus *Rhizobium.* Plants in this category include legumes such as alfalfa, soybeans, peas, and beans. It has been estimated that 10,000 or more species of legumes exist, of which 200 or so are cultivated by man (Alexander, 1977). The actual N_2 fixation process occurs within nodules. The initial phase of nodulation occurs when a specific *Rhizobium* bacterium increases in number near the root environment and causes curling of the root hair (Dart, 1974). Actually the host plant functions in the nodulation process as specific root exudates encourage the multiplication of rhizobial numbers in the root zone (rhizosphere, q.v.). The bacteria then invade through the root hair usually by a thread-like material (infection thread) to the root cortex. Rapid multiplication of cells occurs, which eventually produces the protrusion of the root called *nodules.* One unique feature of the bacteria within nodules is that they increase in size, change shape, and become enclosed within a membrane envelope produced by the host cell (Bergerson, *in* Dart, 1974). These newly formed bacteria cells are called *bacteroids,* and it is here that N_2 fixation actively occurs. Because of the intimate contact between the nodules and the root vascular system, N_2 biologically fixed to NH_3 is used rapidly by the plant in the building of amino acids and proteins. Less efficient use of nitrogen by plants is usually experienced with the use of fertilizer nitrogen through which losses from leaching and gaseous evolution can occur.

Bacteria of the genus *Rhizobium* are gram-negative (a staining procedure), non-spore forming (a resting stage), aerobic (live in the presence of air) rods, 0.5 to 0.9 μm wide, and 1.2 to 3.0 μm long (Alexander, 1977). Many soil bacteria fit this description; however, only species of *Rhizobium* can produce nodules on legumes. Even more striking is the fact that certain species of *Rhizobium* produce nodules only on certain legumes. For example, sweet clover is nodulated only by bacteria called *R. meliloti.* The later term, the species, is derived from the genus of the host plant called *Melilotus.* Further descriptions are shown in Table 1 where the six species of *Rhizobium* are listed with their various cross-inoculation groups. A cross-inoculation group includes all host species that will form nodules by an individual bacterial strain. For example, *R. meliloti* in the alfalfa group will produce nodules on alfalfa, sweet clover, and fenugreek but not legumes outside this grouping. A more specific cross-inoculation group is soybeans, which are nodulated only by *R. japonicum.*

TABLE 1. Cross-Inoculation Groups and *Rhizobium*-Legume Associations.

Cross-inoculation Group	*Rhizobium* Species	Host Genera	Legumes Included
Alfalfa	*R. meliloti*	*Medicago* *Melilotus* *Trigonella*	Alfalfa Sweet clover Fenugreek
Clover	*R. trifolii*	*Trifolium*	Clovers
Pea	*R. leguminosarum*	*Pisum* *Vicia* *Lathyrus* *Lens*	Pea Vetch Sweetpea Lentil
Bean	*R. phaseoli*	*Phaseolus*	Beans
Lupine	*R. lupini*	*Lupinus* *Ornithopus*	Lupines Serradella
Soybean	*R. japonicum*	*Glycine*	Soybean
Cowpea		*Vigna* *Lespedeza* *Crotalaria* *Pueraria* *Arachis* *Phaseolus*	Cowpea Lespedeza Crotalaria Kudzu Peanut Lima bean

(After Alexander, 1977)

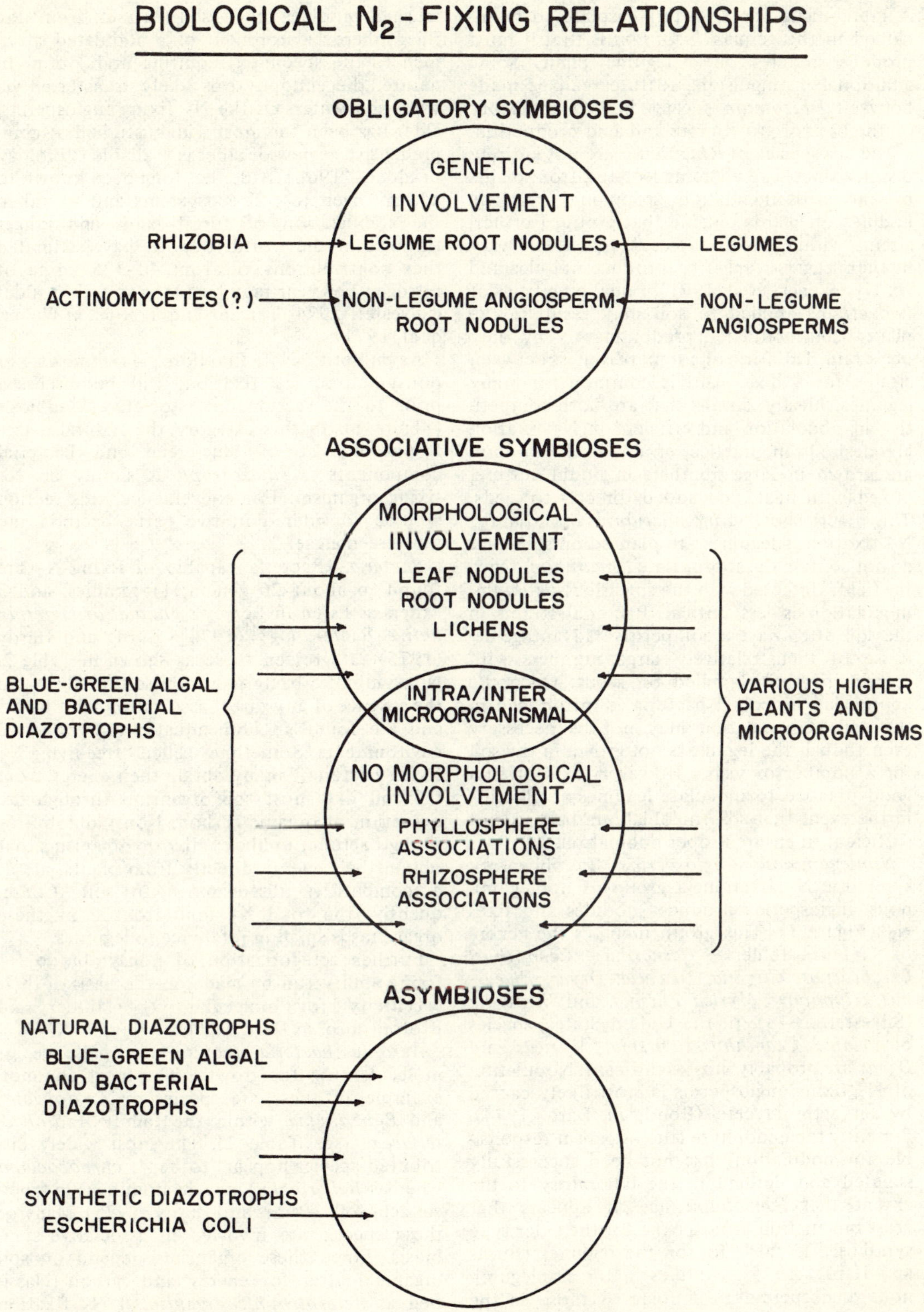

FIGURE 1. Biological N_2 fixing relationships. (From Burns and Hardy, 1975)

A prerequisite for a bacterial isolate to be included in the genus *Rhizobium* is that it must produce nodules on a legume plant. Some additional distinguishing features can be made between *Rhizobium* species in the laboratory on the basis of growth rate and acid production.

The six species of *Rhizobium* are not entirely distinct since a few strains isolated from plants of one cross-inoculation group may produce nodules on plants outside that group. Further, strains within, for example, *R. japonicum* vary in their aggressiveness to produce nodules and fix N_2 (Weber, et al, 1971). Equal quantities of two strains applied to soil may result in the plant being nodulated predominantly by only one strain. This form of competitiveness between strains for a host plant is common for many legumes. Ideally, strains that are both competitive in nodulation and efficient in N_2 fixation are desired. In practice, strains of *Rhizobium* are grown in large numbers in liquid culture, mixed with peat, and added directly to seeds. This practice, called *inoculation,* is essential for N_2 fixation if legumes are planted on soils that do not contain the appropriate *Rhizobium.* Once the field is infected with the specific *Rhizobium,* inoculation is less critical. Bacterial strains in the soil often have a competitive advantage because of their relatively large numbers with respect to those applied on seeds. Once the appropriate rhizobial bacteria is introduced in the soil, reinoculation may not be necessary even though the legume is not grown in the soil for a number of years. However, it is often a good practice to inoculate legumes at planting in the event that soil rhizobial numbers are not sufficient to ensure proper nodulation.

Nonleguminous Angiosperms. In obligatory symbiotic N_2 fixation, a group of nonleguminous angiosperms produces nodules and fixes N_2 (Figure 1). This group includes the genera of *Alnus* (alder), *Casuarina, Ceanothus, Cercocarpus, Coriaria, Discaria, Dryas, Elaeagnus, Hippophae, Myrica, Purshia,* and *Sheperdia* (Silvester, 1976). In the United States, species of *Alnus, Ceanothus, Purshia, Myrica,* and *Dryas* are probably most significant. Nodulation of N_2 fixing angiosperms is most likely caused by an actinomycete (Bond, *in* Dart, 1974). Presently the endophyte (the organism responsible for nodulation) has not been successfully isolated and defined in the laboratory to the extent that *Rhizobium* has. It appears that cross-inoculation groups exist for the alder-type symbiosis as they do for the legumes (Burns and Hardy, 1975). Nodules in the nonleguminous angiosperms are similar to those in the legumes, but they are different anatomically (Stewart, 1966). Inoculation is commonly achieved by transferring the endophyte via crushed nodules or a soil extract from the rhizosphere environment of a nodulated specimen to the specimen requiring nodulation. In nature, the endophyte is likely transferred via wind and water. Of the N_2 fixing angiosperms, alder has been the most widely studied. A comprehensive review of alder is available ("Biology of alder," 1968). Alder has long been known to add nitrogen to soil ecosystems and to aid in the stabilization of forest lands and ranges throughout the world. Reports have estimated that contributions of from 40–325 kg/ha of nitrogen per year may be added to soil by alder (Silvester, 1974; Tarrant and Trappe, *in* Weber, et al, 1971).

Asymbiotic N_2 Fixation. *Asymbiosis,* or nonsymbiotic N_2 fixation, will be discussed prior to the section on Associative Symbiosis (Figure 1). In this category the natural diazotrophs consist of blue-green and bacterial components. A *diazotroph* is simply an N_2 fixing organism. For convenience, this section will be considered in two parts, bacteria and blue-green algae.

Bacteria. Bacteria capable of fixing N_2 are found in about 26 genera, 11 families, and 3 orders as listed in Bergey's *Manual of Determinative Bacteriology* (1974). Burns and Hardy (1975) categorized these as shown in Table 2. Nonsymbiotic bacteria are those that fix N_2 in the absence of any direct association with plants and are found in both aquatic and terrestrial environments. Sometimes called "free-living N_2 fixing bacteria," many obtain their energy from the soil like most soil organisms through the oxidation of organic carbon. Nonsymbiotic N_2 fixing bacteria, unlike symbiotic organisms that obtain their energy directly from plants, are at a competitive disadvantage in soil. Consequently, the total N_2 input to soil by these organisms is small in reference to legumes.

Further categorization of nonsymbiotic N_2 fixing ability can be made on the basis of their specificity for molecular O_2 (Mulder and Brotonegoro, *in* Dart, 1974).

Aerobic bacteria require the presence of molecular O_2 for growth. The most common example of these are species of *Azotobacter* and *Beijerinckia* within the family *Azotobacteriaceae* (see Table 2). The most widely distributed species appears to be *A. chroococcum* whereas *Beijerinckia* may be dominant in tropical acid soils (Burns and Hardy, 1975). Many of these species are involved in associative symbiosis. Since these organisms depend on soil organic matter for energy and carbon (classified as *heterotrophic organisms*), N_2 fixation will likely occur in those environments in which the available carbon-nitrogen ratio (q.v.) is relatively high as is the situation for many

TABLE 2. N_2 Fixation Among the Bacteria.

Order	Family	Genus No.	Incidence of N_2 Fixation: Well-established	Probable or Possible
Pseudomonadales	Thiorhodaceae	14	*Chromatium* *Ectothiorhodospira*	All untested genera
	Athiorhodaceae	3	*Rhodopseudomonas* *Rhodospirillum* *Rhodomicrobium*	
	Chlorobacteriaceae	7	*Chlorobium* *Chloropseudomonas*	
	Thiobacteriaceae	5		*Thiobacillus*
	Pseudomonadacea	12		*Pseudomonas*
	Spirillaceae	11	*Desulfovibrio* *Desulfotomaculum*	*Spirillum* *Vibrio*
	4 other families			
Eubacteriales	Azotobacteriaceae	5	*Azotobacter* *Beijerinckia* *Derxia* *Azotococcus* *Azomonas*	
	Enterobacteriaceae	10	*Klebsiella*	
	Bacillaceae		*Bacillus* *Clostridia*	*Methanobacillus*
	9 other families	47		
Actinomycetales	Mycobacteriaceae	2	*Mycobacterium*	
	Actinomycetaceae	2		*Nocardia* *Actinomyces*
	2 other families			
7 other orders	20 families	50		

(From Burns and Hardy, 1975)

natural ecosystems. Under intensive agriculture by which chemical nitrogen is added, there is some doubt as to whether these organisms contribute substantial quantities of nitrogen. However, under pasture grasses and native rangelands, sufficient nitrogen may be fixed to maintain or even increase soil nitrogen. Research on plots at the Rothamstead Experiment Station in England has shown soil nitrogen accumulation in excess of 40 kg/ha/yr since 1881. These increases were due in part presumably to nonsymbiotic N_2 fixation (Harris and Dart, 1973).

Anaerobic bacteria fix N_2 only in the absence of O_2. The classic example is *Clostridium pasteurianum.* The clostridia tolerate a wide pH range and grow best in the presence of abundant organic matter. Although many soils are considered to be aerobic, microsites in soil aggregates and in the vicinity of plant roots (where O_2 may be low) are locations where *Clostridum* species appear to be in larger numbers. The contribution of N_2 fixed by these organisms appears to be low but not insignificant in those soil environments in which plants depend solely on nitrogen from natural sources. Species of *Bacillus* and *Klebsiella pneumoniae* comprise free-living diazotrophs that fix N_2 anaerobically even though they are classified as "facultative anaerobes" (that is, they grow in the presence or absence of O_2). These are widely distributed in both aqueous and terrestrial environments in which animal and plant residues are actively decomposing. Another possible anaerobic N_2 fixing organism was reported as *Methanobacillus omelianskii,* a methane-forming bacteria. Later work, however, shows that two bacteria may be involved (Bryant, et al, 1967, *in* Burns and Hardy, 1975). These organisms were studied in the proximity of natural gas lines where methane bacteria were believed to increase the soil nitrogen content. Another unique anaerobic N_2 fixing bacteria is within the genus *Desulfovibrio,* which has been shown to fix N_2 (Burns and Hardy, 1975).

Photosynthetic bacteria are aquatic, gram-negative bacteria that occupy fresh and marine-water environments such as mud flats, ponds, lakes, rivers, and ditches. The photosynthetic bacteria are of either the sulfur or nonsulfur type. Recent classification has shown the following (Arnon and Yoch, *in* Dart, 1974).

The purple sulfur bacteria: (family, *Chromatiaceae,* formerly *Thiorhodaceae*). Species of *Chromatium* are capable of N_2 fixation.

The purple nonsulfur bacteria: (family *Rhodospirillaceae,* formerly, *Athiorhodaceae*). An example is species of *Rhodospirillum.*

The green sulfur bacteria: (family *Chlorobiaceae,* formerly *Chlorobacteriaceae*). A common example is species of *Chlorobium.*

Bacterial photosynthesis is very similar to photosynthesis in higher plants with the exception that molecular oxygen (O_2) is not evolved. In green and purple sulfur bacteria, hydrogen sulfide (H_2S) can serve as a source of electrons in the same capacity as water in photosynthesis in higher plants. The nonsulfur photosynthetic bacteria require the presence of organic compounds in the nutrient medium and cannot use H_2S as the electron or hydrogen donor (Arnon and Yoch, *in* Dart, 1974). Further, N_2 fixation by the purple and green sulfur bacteria and the purple nonsulfur bacteria occurs in light only under anaerobic conditions. Aerobic growth of the purple nonsulfur bacteria can occur in the dark. Thus they are termed *facultative anaerobes and phototrophs.*

Blue-green algae, like higher plants, use sunlight for energy, fix carbon dioxide (CO_2) photosynthetically, and thus do not require an external source of energy for N_2 fixation. Blue-green algae are distributed widely on soil surfaces in tropical, temperate, and desert environments, in fresh-water lakes, and in the arctic. Blue-green algae are particularly important in rice fields where their N_2 fixation capacity has been studied extensively. Many of these are species of *Nostoc* and *Anabaena,* which have been reported to fix over 20 kg N/ha/yr (Stewart, 1973). Since blue-green algae require light for N_2 fixation, they are found on or near soils and water surfaces. They are often subject to periodic desiccation, freezing, and temperature extremes, but they remain extremely adaptable to a wide range of environmental conditions. In desert environments active N_2 fixation for even several months may add sufficient nitrogen to maintain the ecosystem.

Associative Symbioses. This category, shown in Figure 1, was discussed last because many of the asymbiotic diazotrophs are involved to some degree in associative symbioses. As suggested from the term, *associative symbiosis* can be considered a loose symbiotic relationship. Many of these relationships are based on the nondiazotroph contributing carbon compounds to the diazotroph as, for example, an exudate and the diazotroph contributing nitrogen to the host when these cells die (Burns and Hardy, 1975). Only recently have many of these relationships been recognized. Many, undoubtedly, remain yet to be discovered. Examples of these having some morphological involvement are leaf nodules that are formed on a tropical plant, *Psychotria* (Becking, *in* Weber, et al, 1971); blue green algae in association with liverworts (Stewart, 1966); associations of blue-green species of *Nostoc* with lichens (Burns and Hardy, 1975); and the use of *Azolla* for nitrogen fixation in rice fields (Stewart, 1973).

Many instances of associative symbiosis are known where no morphological involvement is present. Phyllospheric associations (those on leaf surfaces) involving species of *Beijerinckia* and others were reviewed by Ruinen (*in* Dart, 1974). *Rhizosphere associations* (those present on and near root surfaces) have been reported with great frequency as described in a review by Döbereiner (*in* Dart, 1974). Döbereiner and Day (1975) estimated N_2 fixation rates of up to 1 kg N/ha/day in associations in which a bacterium, *Azotobacter paspali,* forms a mucilagenous sheath on root surfaces of *Paspalum notatum,* a tropical grass. Other N_2 fixing associations have been reported between bacteria and the roots of rice (Dommergues, et al, 1973); sugar cane (Döbereiner, Day, and Dart, 1972); and maize (Raju, Evans, and Seidler, 1972). In another association a bacteria *Spirillum lipoferum* has been isolated from roots of a tropical plant *Digitaria decumbens,* which actually fixes N_2 within the root cortex (Döbereiner and Day, *in* Silvester, 1976). This association is more sophisticated than most since the bacteria actually fix N_2 within the root cortex, which likely involves a more specific symbiant and host. Another unique associative N_2 fixing phenomenon is the N_2 fixation by facultative anaerobic bacteria (species of *Enterobacter*) in rotting wood in association with decay fungi (Aho, et al, *in* Silvester, 1976). Such symbiosis is believed to speed the wood-rotting process and add nitrogen to the terrestrial environment. The preceding descriptions of associative N_2 fixation show the varied and unique features of these systems. Many remain to be identified and studied.

Genetics

Genes that are responsible for the process of N_2 fixation in organisms have been termed *nif*

TABLE 3. Reported N_2 Fixation Rates by Selected Systems.

N_2 System	N_2 Fixed kg/ha × yr	N_2 System	N_2 Fixed kg/ha × yr
Soybean	57, 84, 94	Soil-Canada	2–200
Peas, lentils	85	Natural grassland	1–2
Clovers	105–220	Pine forest	0–39
Lucerne	128, 158, 300	*Alnus*	56, 156
Rice paddy	10, 27, 45	*Ceanothus*	60
Wheat soil	4	Desert crust-Arizona	41

(After Burns and Hardy, 1975)

genes. Research on the possibility of transferring the *nif* genes from N_2 fixing organisms to those not capable of fixation is relatively recent. There exists the possibility that organisms not capable of fixing N_2 could fix N_2 if the *nif* gene were permanently established. In 1971 Streicher and coworkers transferred *nif* genes from one strain of *K. pneumoniae* to another with the use of a phage (virus), while more recently Dixon and Postgate were able to transfer the N_2 fixing gene from *K. pneumoniae* into *Escherichia coli*, creating the first intergeneric hybrid with N_2 fixing capacity (as cited in Brill, *in* Dart, 1974). These hybrids with N_2 fixing capacity can be termed *synthetic diazotrophs* as shown in Figure 1. More recent research has shown additional instances of transferring *nif* genes; however, considerable research is still needed in this relatively new area. A thorough review of the subject is available (Brill, *in* Dart, 1974).

Methodology

Many of the recent findings on N_2 fixation have been based on the acetylene reduction method. The history of this method is given in two reviews: one by Burris, as cited in Bond (1974), and one by Burns and Hardy (1975). This method is based on the finding that acetylene is reduced in the presence of nitrogenase, the enzyme involved in N_2 fixation.

In an enclosed container, the appropriate N_2 fixing sample, e.g., root nodules, soil, and seawater, is added and followed by the addition of acetylene (C_2H_2). Acetylene is reduced to ethylene in the following manner:

$$3C_2H_2 \text{ (acetylene)} \xrightarrow[\text{nitrogenase}]{6H + \text{energy}} 3C_2H_4 \text{ (ethylene)}$$

This equation is remarkably similar to the reaction shown previously. In this case the N_2 fixing system converts acetylene to ethylene rather than N_2 to NH_3. After incubation for a determined time (usually 1/2 to 1 hr), a gas sample is removed from the vessel and analyzed for ethylene. The quantity of ethylene determined with a gas chromatograph can provide a quantitative estimate on the N_2 fixation rate after appropriate controls and conversion factors are considered. This method has been used to discover N_2 fixation in environments where no other method would have been feasible. The method is simple, inexpensive, and much more sensitive than others. A discussion of its use and application is presented by Hardy, Burns, and Holsten (1973).

N_2 Fixation Estimates and Significance

Nitrogen is generally considered the limiting element for most plants in crop production. With the expanding world population and depletion of fossil fuels, efforts have intensified in N_2 fixation research in which methods are being developed to increase biologically fixed N_2 in forests, fields, and oceans. Table 3 provides some estimates on quantities of nitrogen fixed by various organisms. Legumes such as soybeans, clovers, and lucerne generally fix the largest quantities of N_2. However, *Alnus* (alder) has been reported to be as efficient in N_2 fixation as the legumes. Smaller, yet significant quantities of N_2 are fixed in rice paddy and desert crusts. Variable estimates have been reported in soil.

On a global scale, estimates have indicated that 175×10^6 metric tons of nitrogen are fixed from agricultural, forest and ocean sources (Table 4, p. 332). Of this total, 139×10^6 metric tons of nitrogen were estimated to be fixed on land. It is interesting to note that only 35×10^6 metric tons were fixed by legumes. Obviously, considerable N_2 fixation occurs in nonagricultural land areas such as in permanent grasslands and forests. Other estimates of global N_2 fixation exist in the literature. Because many estimates and assumptions are often used, the preceding values should be used with caution; however, the general magnitude given is reliable.

D. F. BEZDICEK

TABLE 4. N_2 Fixation Estimates in the Biosphere.

Land Use	ha $\times 10^{-6}$				kg N_2 Fixed Per ha $\times$ yr	Metric Tons Per yr ($\times 10^{-6}$)
Agricultural				4400		
Arable under crop			1400		140	35
Legumes		250				
Pulses	63					
Soybeans	34					
Groundnuts	18					
Other	135					
Nonlegumes		1150				
Rice	135				30	4
Other	1015				5	5
Permanent meadows, grasslands			3000		15	45
Forest and woodland				4100	10	40
Unused				4900	2	10
Ice covered				1500	0	0
Total land				14900		139
Sea				36100	1	36
Total				51000		175

(From Burns and Hardy, 1975)

References

Alexander, M., 1977. *Introduction to Soil Microbiology.* New York: Wiley, 467p.

Bergey's Manual of Determinative Bacteriology, 1974. eighth ed. Baltimore: Williams and Wilkins, 261-264.

Biology of alder, 1968. *In* J. M. Trappe; J. F. Franklin; R. F. Tarrant; and G. M. Hansen, eds., *Proceedings of a Symposium, Northwest Scientific Association, (Pullman, Washington).* Portland: Pacific N.W. Forest and Range Exp. Sta., 292p.

Burns, R. C., and Hardy, R. W. F., 1975. *Nitrogen Fixation in Bacteria and Higher Plants.* New York: Springer-Verlag, 189p.

Dart, P. J., 1974. The infection process, *in* A. Quispel, ed., *The Biology of Nitrogen Fixation.* New York: American Elsevier, 769p.

Döbereiner, J.; Day, J. M.; and Dart, P. J., 1972. Nitrogenase activity in the rhizosphere of sugar cane and some other tropical grasses, *Plant and Soil,* **37**, 191-196.

Döbereiner, J., and Day, J. M., 1975. Nitrogen fixation in the rhizosphere of tropical grasses, *in* W. D. P. Stewart, ed., *Proc. IBP Conference 6, Nitrogen Fixation by Free Living Microorganisms.* Edinburgh: Cambridge University Press, 39-56.

Dommergues, Y.; Balandreau, J.; Rinaudo, G.,; and Weinhard, P., 1973. Non symbiotic nitrogen fixation in the rhizospheres of rice, maize, and different tropical grasses, *Soil Biol. Biochem.,* **5**, 83-89.

Hardy, R. W. F.; Burns, R. C.; and Holsten, R. D., 1973. Applications of the acetylene-ethylene assay for measurement of nitrogen fixation, *Soil Biol. Biochem.,* **5**, 47-81.

Harris, D., and Dart, P. J., 1973. Nitrogenase activity in the rhizosphere of *Stachys sylvatica* and some other dicotyledenous plants, *Soil Biol. Biochem.,* **5**, 277-279.

Raju, P. N.; Evans, H. J.; and Seidler, R. J., 1972. An asymbiotic nitrogen fixing bacterium from the root environment of corn, *Proc. Natl. Acad. Sci., USA,* **69**, 3473-3478.

Silvester, W. B., 1976. Ecological and economic significance of the non-legume symbiosis, *in* W. E. Newton and C. J. Nyman, eds., *Proc. 1st Internat. Symposium on Nitrogen Fixation, vol. 2.* Pullman: Washington State University Press, 489-506.

Stewart, W. D. P., 1973. Nitrogen fixation by photosynthetic microorganisms, *Ann. Rev. Microbiol,* **27**, 283-316.

Stewart, W. D. P., 1966. *Nitrogen Fixation in Higher Plants.* London: Athlone Press, 168p.

Weber, D. F., Caldwell, B. E.; Sloger, C.; and Vest, H. G., 1971. Some USDA studies on the soybean–*Rhizobium* symbiosis, *Plant and Soil,* Special Volume 1971, 293-304.

Cross-references: *Algae, Denitrification; Enzyme Activity; Humus; Management of Soils; Metabolic Activity; Nutrient Losses; Soil Biochemistry; Soil Fertility.*

NUTRIENT CYCLES IN SOIL–See CARBON CYCLE; NITROGEN CYCLE; PHOSPHORUS CYCLE; SULFUR TRANSFORMATIONS

NUTRIENT LOSSES

Loss of plant nutrients and other chemical elements from soil is a natural and continuous phenomenon. Losses occur through leaching (q.v.), volatilization, surface runoff, and erosion (see *Water Erosion* and *Wind Erosion*), operating singly or in combination. (Nutrients removed in harvested crops are not considered losses because they have fulfilled their role in plant growth.) Soils that are part of a natural ecosystem are essentially nutrient conserving, but nutrients are lost to some extent from all soil-plant systems. Accurate quantitative measurements of nutrient losses from soils under natural field conditions are often difficult to obtain. Water is the principal transporting agent of most nutrients, so quantitative loss studies involve the measurement and continuous sampling of drainage waters. In the past leaching losses have been determined largely through the use of lysimeters (q.v.) and by the analysis of discharge waters from tile-drained fields. Usually field plots are used to measure runoff and erosion losses, while losses of nitrogen and sulfur to the atmosphere as gases have not yet been determined under field conditions due to the technical difficulties involved. Nevertheless, much information has been obtained about nutrient losses, particularly in the past century. In the past decade interest in plant nutrient losses from soils as a factor in water quality has greatly stimulated research on a watershed scale. The transport of nitrogen and phosphorus into surface and groundwaters is currently of primary concern. These nutrients are dominant in eutrophication, and elevated nitrate nitrogen levels in drinking water supplies pose potential health hazards. In addition, the loss of plant nutrients from soils represents an economic loss to farmers and gardeners.

Pathways of Nutrient Loss

Leaching. Deep percolation of water can occur when water added as precipitation or irrigation (q.v.) exceeds the water-holding capacity of soils. The probability of deep percolation (q.v.) can be estimated by subtracting water lost by evapotranspiration from precipitation. Deep percolation of water may range from none in certain western U.S. areas, for example, to around 50 cm in areas along the eastern U.S. Coastal Plain. There is a direct relationship between the volume of percolating water that passes through the soil and nutrient losses. Thus leaching losses of nutrients are minimized under an actively growing crop since the crop uses soil moisture and absorbs nutrients from the soil solution. In the humid regions of the United States, losses of nutrients by leaching are most apt to occur during the period of November to April.

The nutrients most vulnerable to the leaching process are nitrogen (N), potassium (K), calcium (Ca), magnesium (Mg), and sulfur (S). Nitrogen in the nitrate form is highly mobile and moves readily with water. Cations move largely as nitrates and soluble sulfates, although large amounts of sulfate sulfur can be absorbed by acidic soils high in exchangeable aluminum and iron. Very little phosphorus is lost by leaching because it forms relatively insoluble compounds (i.e., is fixed) with such soil constituents as calcium, aluminum, and iron, and thus its concentration in the soil solution (q.v.) is very low. Leaching losses of phosphorus, however, can be appreciable in organic soils and sands due to the low level of phosphorus fixation.

Volatilization. Losses of nutrients in gaseous forms are limited to nitrogen and possibly sulfur. Gaseous nitrogen may be lost from soil in forms of ammonia, elemental nitrogen, oxides of nitrogen, and in small amounts from plants as organic nitrogen; their relative importance under varying soil conditions is largely unknown. Ammonia losses are generally negligible in soils having pH values below 7 unless there is sufficient free ammonia present to raise the pH in localized areas. Ammonia losses increase as pH and temperature increase and are highest in soils having a low exchange capacity.

Molecular N_2 and N_2O can be formed by bacterial denitrification (q.v.) where nitrates or nitrites are present in quantity and the soil is very deficient in oxygen. The most common cause of oxygen deficiency in soils is flooding. Under certain conditions, NO may be formed, but losses of nitrogen in this form are thought to be minor, as are losses from plants as organic nitrogen.

Volatilization losses of sulfur as H_2S can be appreciable under flooded and waterlogged soil conditions. Gaseous losses of sulfur from cultivated and well-drained soils probably do not occur, although there is a possibility that such losses can occur from fresh organic materials when applied to soils. Scientists have given very little attention to this avenue of sulfur loss.

Runoff and Erosion. The transport of nutrients in solution by runoff water moving over a soil surface is generally of a low order of magnitude, particularly if the soil has a protective vegetative cover such as grass or trees. The highest solution losses occur in runoff following a surface application of fertilizer. Dead or dying vegetation can also contribute nutrients to runoff. Factors affecting the volume of runoff include rainfall characteristics, size of watershed, degree of slope, soil permeability, kind of vegetative cover, antecedent moisture, and others.

High amounts of potentially available nutrients are transported by soil erosion. It has been estimated that upwards of 4 billion tons of sediment is washed into tributary streams in the United States each year. Erosion tends to be a selective process for plant nutrients since eroded material frequently has a higher nutrient content than the soil from which it originated. Such differences may be expressed in terms of an *enrichment ratio* (ER):

$$ER = \frac{A}{B}$$

where A is the concentration of the nutrient in soil material in runoff, and B is the concentration of the element in soil from which the runoff originated (Barrows and Kilmer, 1963). Elevated enrichment ratios result from a higher proportion of finer soil mineral and organic particles in eroded materials. These finer materials are usually richer in plant nutrients.

Only a very small percentage of the total amount of plant nutrients (q.v.) in soils is available for plant growth at any given time; the release of nutrients from a solid soil particle to the soil solution is a very slow process. Early workers who studied plant nutrient losses by water erosion reported the total amount of nutrients in eroded material. Such data can be misleading unless one distinguishes between available and total amounts of plant nutrients found in soils. Only nutrients in solution are available to plants.

Nutrient Losses

Nitrogen. Nearly all the nitrogen found in soils is combined with organic matter. The plowed layer of most cultivated soils contains between 0.02 and 0.5% nitrogen, or approximately 450 to 11,000 kg of N/ha (Table 1). As organic matter decomposes, its nitrogen is released, and most is quickly converted to NO_3.

Assuming equivalent amounts, forms, and distribution of nitrogen in soils, losses of this element as NO_3 through leaching will depend entirely on the volume-rate relationships of water passing through the soil profile. Losses of nitrogen under various cropping systems are shown in Table 2. Losses are highest under corn and lowest under grass.

The leaching losses of nitrogen under legume-grass mixtures can be much greater than under grass alone. This is, of course, due to the symbiotic fixation of atmospheric nitrogen by legumes and its subsequent release to the soil. Lysimeter studies in Kentucky carried out over an 11-year period showed that nitrogen losses by leaching under legumes were 8 to 10 times the losses under bluegrass alone (Karraker, Bortner, and Fergus, 1950). Annual losses of nitrogen from steeply sloping fertilized grassed watersheds in western North Carolina were reported to be 3.3 kg/ha with optimum fertilization, but when excess amounts of nitrogen fertilizer were applied, the nitrogen loss averaged 12.1 kg/ha (Kilmer et al, 1974). Nitrogen losses from row crops such as corn or cotton tend to be higher than losses from grassed and forested soils.

TABLE 1. Total Amounts of Plant Nutrients Ordinarily Present in 15 cm of Surface Soil in the Humid Region.

Nutrient	Range (%)	Amount (kg/ha)
Nitrogen (N)	0.02-0.50	450-11,000
Phosphorus (P)	0.01-0.20	200-4500
Potassium (K)	0.17-3.30	3800-74,000
Calcium (Ca)	0.07-3.60	1600-81,000
Magnesium (mg)	0.12-1.50	2700-34,000
Sulfur (S)	0.01-0.20	200-4500
Iron (Fe)	0.50-5.00	11,000-112,000
Manganese (Mn)	0.20-1.00	4500-22,000
Zinc (Zn)	0.001-0.025	22-560
Boron (B)	0.0005-0.015	11-340
Copper (Cu)	0.0005-0.015	11-340
Chlorine (Cl)	0.001-0.10	22-2240
Cobalt (Co)	0.0001-0.005	2-112
Molybdenum (Mo)	0.00002-0.0005	0.5-11

(Adapted from Buckman and Brady, 1969)

Power (1970) reported that the possibility of leaching of NO_3-N under dry grassland agriculture at Mandan, North Dakota, is remote. This is a semiarid region having 40 cm of annual precipitation, and the soil is normally frozen for about 4 months during the winter period.

Terman and Allen (1970) studied leaching losses of soluble and slow-release nitrogen and potassium fertilizers under grass and fallow. They used 15- by 61-cm columns of Lakeland sand and obtained measurable amounts of leached nitrogen under ryegrass only when 10 cm of water was applied per week. Many other instances have been reported of greatly retarded leaching of soluble nutrients under an actively growing crop, particularly forages. This has been the usual finding of most workers both in the U.S. and abroad.

It is also evident that the rooting habits of plants influence nitrogen leaching to some degree. Studies in Florida have shown nitrogen losses to be higher under short-rooted plants (Volk, 1956). At the other extreme is alfalfa, which reportedly has the capacity to remove

TABLE 2. Average Annual Leaching Losses of Nitrogen (N) and Phosphorus (P) From Unfertilized and Fertilized Brookston Clay.

Cropping System	N Applied	N Loss *U**	N Loss *F**	P Applied	P Loss *U**	P Loss *F**
		(kg/ha/yr)				
Corn (1)†	129	5.6	15.1	30	0.13	0.24
Oats, alfalfa (2)†	17	4.3	5.7	30	0.13	0.13
Alfalfa (3)†	0	4.8	3.9	0	0.13	0.15
Alfalfa (4)†	0	4.7	8.6	0	0.08	0.22
Continuous corn	129	6.6	14.0	30	0.26	0.29
Continuous bluegrass	17	0.3	0.7	30	0.01	0.12

(Adapted from Bolton, Aylesworth, and Hore, 1970)
**U* = unfertilized; *F* = fertilized.
†4-year rotation.

NO_3 from alluvial soils to a depth of at least 6 m (Viets and Hageman, 1971).

Losses of nitrogen as a gas to the atmosphere can occur through denitrification (biochemical reduction of NO_3), resulting in the formation of N_2, NO, and N_2O. Losses by this route appear to account for rather large losses of nitrogen, amounting to 10 to 20% of the nitrogen applied in fertilizers. There is abundant evidence that such losses can occur in soils that are not necessarily poorly aerated, although losses can be progressively higher as soil moisture is increased. Evolution of nitrogen as a gas was 45 ppm from a soil incubated for 5 days at 8% moisture and 140 ppm at 79% moisture (McGarity, 1961). Such losses are largely unavoidable, although nitrification inhibitors may be effective under certain conditions.

Nitrogen can also be lost as NH_3 from alkaline soil solutions. If organic materials or fertilizers containing ammonia or urea are placed on the soil surface, nitrogen losses can be as much as 50% of the amount applied (Volk, 1959). Losses of nitrogen as NH_3 can be largely avoided if fertilizer nitrogen is mixed with the soil. Where nitrogen losses are related to NO_3 formation (leaching and volatilization), the use of nitrification inhibitors or slow-release nitrogen fertilizers may be effective.

Nitrogen can be lost in surface runoff even though soil erosion is prevented. Generally significant solution losses in runoff occur only when precipitation follows a surface application of soluble nitrogen fertilizer. Antecedent moisture is an important factor here. In Indiana rainfall on fertilized sod plots at 12.5% moisture resulted in an nitrogen loss of only 2% of that applied. Similar plots at 25.8% moisture lost 14% of the nitrogen applied (Moe, Mannering, and Johnson, 1967).

Dead or dying vegetation can also give rise to losses of nutrients in solution. In Minnesota Timmons, Holt, and Latterell (1970) determined nitrogen and phosphorus losses in plant leachates from alfalfa, Kentucky bluegrass, barley straw stubble, and oats straw stubble. They concluded that the leaching of alfalfa and bluegrass by surface runoff water could contribute substantial amounts of nutrients to lakes and streams. This mechanism of loss apparently accounts for the nitrogen loss in solution from a hay rotation. This loss was attributed to snow-melt leaching of the grain stubble.

Assuming that the total nitrogen content of surface soils is generally in the 0.02 to 0.50% range, each metric ton of surface soil lost by erosion carries with it up to 11 kg of potentially available nitrogen. This nitrogen is primarily in organic form, and from 2 to 10% of this is mineralized and becomes available to plants during a single growing season under aerobic conditions.

Stanford, England, and Taylor (1970) estimated that 5, 4, and 3 million tons of nitrogen were lost through erosion in 1930, 1947, and 1969, respectively, from U.S. harvested crop land. The authors attribute this decline to improved land use and management practices. Soil conservation practices can reduce or largely prevent nutrient losses by erosion. A comparison of losses in eroded material under a conservation versus a nonconservation system of management is shown in Table 3. Research in Wisconsin has shown that conservation practices do not increase nutrient losses through leaching (Kilmer, Hayes, and Muckenhirn, 1944).

Losses of nitrogen from forested land are very low in general since nutrient cycling in a forest ecosystem promotes nutrient conservation. However, operations such as clear-cutting, road building, and fertilization can temporarily increase nutrient loads in discharge waters from forested lands.

TABLE 3. Soil and Calculated Nitrogen (N), Phosphorus (P), Potassium (K), Calcium (Ca), and Magnesium (Mg)

Crop	Farming System*	Soil	Total N	Available P	Available K	Ca + Mg†
				(kg/ha)		
Corn	P	7100	19.2	2.8	4.40	107
Corn	C	1670	5.7	0.9	2.10	35
Meadow	P	210	0.6	0.1	0.13	3.4
Meadow	C	135	0.5	0.07	0.17	2.2

(Adapted from Stottenberg and White, 1953)
*P – prevailing system; C = conservation system.
†Calculated on a carbonate basis.

Regarding nitrogen losses from the soil system: Losses appear to be highest from fallow soils; cultivated lands lose more nitrogen than those in pasture and in forest; and leaching losses are highest during the winter months if precipitation exceeds evapotranspiration.

Phosphorus. More than half the phosphorus in soils can occur in organic forms. Leaching losses of phosphorus are very slight relative to the total amount present in soils. The formation of sparingly soluble phosphorus compounds in soils depends primarily on the presence of soluble iron, aluminum, calcium, and magnesium. Low concentrations of these elements are formed in sandy and organic soils, and increased mobility of phosphorus occurs in them. English scientists think it improbable that phosphorus will leach in soils containing 5% or more clay. Work in Florida has supported this thesis. Phosphorus moved to a depth of 210 cm in some sandy soils containing 1% or less clay (Spencer, 1957).

When soils are flooded, insoluble ferric phosphate is converted to more soluble ferrous phosphate through biological reduction of iron. The mobility of phosphorus is increased, and relatively large losses of this nutrient can occur under flooding. Normally, very little phosphorus in solution is lost from well-drained mineral soils. In North Carolina the highest annual loss of phosphorus from fertilized watersheds was 0.3 kg of P/ha (Kilmer et al, 1974). Typical leaching losses under various crops are shown in Table 2 (Bolton, Aylesworth, and Hore, 1970). Most workers report annual losses of phosphorus in solution to be less than 1 kg/ha.

Soil erosion is the principal avenue of phosphorus loss, probably accounting for as much as 90% (Kilmer, 1972). Surface soils in the humid regions of the United States contain an average of about 0.04% total phosphorus, although this is variable, ranging from 0.01 to 0.2% (Table 1). Thus each metric ton of soil lost by erosion carries up to 5 kg of phosphorus with it. Typical losses under corn and meadow are shown in Table 3. Note the relationship between phosphorus and soil losses. In Missouri where erosion was controlled by bluegrass sod, less than 0.1 kg/ha of total phosphorus was lost, but losses jumped to 15 kg/ha from a spaded, fallow soil (Duley and Miller, 1923). Incorporation of fertilizer phosphorus with the soil and erosion control can reduce phosphorus losses to an absolute minimum.

Potassium. Potassium is present in soils in relatively large amounts, ranging from about 3800 to 74,000 kg/ha in the plow layer (Table 1). Virtually all the potassium in soils is in inorganic form, and only a very small percentage of the total is available to plants at any given time. Potassium is lost from soils by leaching and erosion. Losses are of primary interest from a crop production standpoint since this element is seldom a limiting factor in the growth of algae and other aquatic plants.

Many studies of potassium losses by leaching were carried out during the first half of the present century. The classic work of Bizzell and Lyon (1927) at Cornell, as well as later investigations by Stauffer (1942) in Illinois and Kilmer, Hayes, and Muckenhirn (1944) in Wisconsin, are examples. Earlier lysimeter installations made no provisions for runoff, and hence high leaching losses of nutrients were often reported. For example, annual potassium losses from bare Dunkirk and Volusia soils in New York were 81 and 72 kg/ha/yr, respectively, where no runoff was permitted (Bizzell and Lyon, 1927). Similar losses in Illinois from bare Muscatine soil and Fayette in Wisconsin were 1.34 kg/ha where runoff was permitted (Stauffer, 1942; Kilmer, 1944). Bizzell and Lyon reported annual potassium losses by leaching of 70 kg/ha on Dunkirk silty clay loam under grass. This loss appears to be exceedingly high, particularly for sloping upland soils. Terman and Allen (1970) found that negligible amounts of potassium were leached from a sandy soil supporting a vigorous growth of bermudagrass. A study by Volk (1940) in Alabama demonstrated the influence of winter legumes on amounts of potassium leached. When no winter legumes were grown, the amount of potassium leached into the 20- to 61-cm soil layer and retained there over an 8-year period ranged from 9 to 32% of the amount added, as contrasted with a maximum of 5% leached when winter legumes were included in the rotation. Truog and Jones (1938) concluded that potassium leaching losses from cultivated soils would not exceed 11 to 17 kg/ha annually. Losses under grassland and meadow would likely be even lower.

Losses of potassium in surface runoff, particularly where pasture and meadow are involved, probably do not exceed 5 to 6 kg/ha annually. The data of Timmons, Burwell, and Holt (1968) showed an average 2-year loss of 3.4 kg of K/ha under a hay rotation. Stoltenberg and White (1953) showed around 0.5 kg of K/ha lost annually from meadow, but up to 8 times this amount in eroded material from corn (Table 3). Kilmer et al (1974) reported annual potassium losses by leaching and runoff in North Carolina to total about 4 to 6 kg/ha.

Calcium and Magnesium. The total content of soil calcium ranges from about 0.07 to 3.60% and of soil magnesium from 0.12 to 1.50%. Leaching losses of calcium and magnesium are usually higher than for any of the other plant nutrients. That both of these principal soil bases move primarily as nitrates has been known for nearly 100 years. In New York data from no-runoff lysimeters indicate annual calcium losses ranging from about 260 to 450 kg/ha; the highest losses reported were from bare soil. Magnesium losses under similar circumstances were reported to be from 30 to 70 kg/ha. Losses of calcium and magnesium under grass averaged 290 kg/ha and 43 kg/ha, respectively.

Leaching losses of calcium and magnesium from lysimeters permitting runoff are much lower. Stauffer (1942) obtained calcium losses ranging from 100 kg/ha on uncropped, well-drained Muscatine soil to 12.3 kg/ha on poorly drained Cowden. Magnesium losses from these two soils, respectively, were 52 and 3.7 kg/ha. Kilmer, Hayes, and Muckenhirn (1944) indicated annual calcium losses of 42.5 and 14.7 kg/ha from fallow and cropped Fayette silt loam. Magnesium losses were 20.8 and 6.5 kg/ha under similar circumstances.

Runoff losses of calcium and magnesium from pasture and meadow appear to be very low. Data obtained by Stoltenberg and White (1953) in Indiana indicated that losses of calcium plus magnesium calculated as carbonate ranged from around 2 to 3 kg/ha from meadow (Table 3). The downward movement of calcium and magnesium in soils is encouraged by the use of acid-forming nitrogen fertilizers. The problem of acid subsoils in humid regions is a difficult one, and care should be taken to include lime in soil-management programs.

Sulfur. Leaching losses of sulfur have been studied by many investigators. Haward and Reisenauer (1966) made a rather extensive review of the literature and concluded that sulfur losses might go as high as 320 kg/ha annually. Driebelbis (1947) reported annual leaching losses under meadow ranging from 3.9 kg/ha on Muskingham silt loam to 10 times that amount from Keene silt loam in Ohio. Lyon and Bizzell (1936) measured annual sulfur losses of 49 kg/ha from Dunkirk silty clay loam cropped to grass. Losses in Illinois and Wisconsin on fallow and cropped soils ranged from about 1 to 12 kg/ha, higher losses being obtained under fallow (Stauffer, 1942; Kilmer, Hayes, and Muckenhirn, 1944).

Movement of sulfur with eroded particles does not appear to be an important problem from permanent pastures and meadows. Erosion losses can account for large amounts of sulfur from bare or newly cultivated soils. Duley and Miller (1923) reported losses of 113 kg/ha total sulfur from uncultivated soils in Missouri. Kilmer et al (1974) reported sulfur losses by runoff and leaching to total about 3 kg/ha annually. Erosion losses of sulfur are likely to be less than 5 kg/ha from soils supporting a good stand of grass or meadow species (Kilmer, 1974).

Micronutrients. Studies of micronutrient losses by leaching and runoff from soils have not been made. This is a badly neglected area, and investigations should be initiated to assess the source and nature of micronutrients and other trace elements in drainage waters from soils. The range of micronutrient losses through erosion can be estimated by referring to Table 1, which gives the ranges of total amounts of micronutrients found in soils.

VICTOR J. KILMER

References

Barrows, H. L., and Kilmer, V. J., 1963. Plant nutrient losses from soils by water erosion, *Adv. Agron.*, **15**, 303-316.

Bizzell, J. A., and Lyon, T. L., 1927. Composition of drainage waters from lysimeters at Cornell University, *1st Internat. Congr. Soil Sci. Trans.*, **2**, 342-357.

Bolton, E. F.; Aylesworth, J. W.; and Hore, F. R., 1970. Nutrient losses through tile lines under three cropping systems and two fertility levels on a Brookston clay soil, *Canadian Jour. Soil Sci.*, **50**, 275-279.

Buckman, H. O., and Brady, N. C., 1969. *The Nature and Properties of Soils.* London: Macmillan, 653p.

Driebelbis, F. R., 1947. Some plant nutrient losses in gravitational water from monolith lysimeters at Coschocton, Ohio, *Soil Sci. Soc. America Proc.*, **11**, 182–188.

Duley, F. L., and Miller, M. F., 1923. Erosion and surface runoff under different soil conditions, *Mo. Agric. Exp. Sta. Res. Bull.*, **63**, 50p.

Engelstad, O. P., 1970. *Nutrient Mobility in Soils: Accumulation and Losses.* Madison: Soil Science Society of America, Special Publication No. 4, 81p.

Harward, M. E., and Reisenauer, H. M., 1966. Reactions and movement of inorganic soil sulfur, *Soil Sci.*, **101**, 326–335.

Karraker, P. E.; Bortner, C. E.; and Fergus, E. N., 1950. Nitrogen balance in lysimeters as affected by growing Kentucky bluegrass and certain legumes together and separately, *Ky. Agric. Exp. Sta. Bull.*, **557**, 1–16.

Kilmer, V. J., 1972. The relationship of soil and fertilizer phosphorus to water quality, *Soils Bull.*, **16**, Rome: FAO, 360p.

Kilmer, V. J., 1974. Nutrient losses from grasslands through leaching and runoff, *in* D. A. Mays, ed., *Forage Fertilization.* Madison: American Society of Agronomy, 341–360.

Kilmer, V. J.; Gilliam, J. W.; Lutz, J. F.; Joyce, R. T.; and Eklund, C. D., 1974. Nutrient losses from fertilized grassed watersheds in western North Carolina, *Jour. Environ. Qual.*, **3**, 214–219.

Kilmer, V. J.; Hayes, O. E.; and Muckenhirn, R. J., 1944. Plant nutrient and water losses from Fayette silt loam as measured by monolith lysimeters, *Jour. Am. Soc. Agron.*, **36**, 249–263.

Lyon, T. L., and Bizzell, J. A., 1936. *Lysimeter Experiment IV, Memoir 194.* Ithaca: Cornell University Agricultural Experiment Station, N.Y., 59p.

McGarity, J. W., 1961. Denitrification studies on some South Australian soils, *Plant and Soil*, **14**, 1–21.

Moe, P. G.; Mannering, J. V.; and Johnson, C. B., 1967. Loss of fertilizer nitrogen in surface runoff water, *Soil Sci.*, **104**, 389–394.

Power, J. R., 1970. Leaching of nitrate-nitrogen under dryland agriculture in the northern Great Plains, *in Relationship of Agriculture to Soil and Water Pollution.* Ithaca: Cornell University Conference on Agricultural Waste Management, January 19–21, 1970, Rochester, 111–122.

Spencer, W. F., 1957. Distribution and availability of phosphorus added to a Lakeland fine sand, *Soil Sci. Soc. America Proc.*, **21**, 141–144.

Stanford, G.; England, C. B.; and Taylor, A. W., 1970. *Fertilizer Use and Water Quality,* U.S. Department of Agriculture, ARS, ARS-41-168, 19p.

Stauffer, R. S., 1942. Runoff, percolate, and leaching losses from some Illinois soils, *Jour. Am. Soc. Agron.*, **34**, 830–835.

Stoltenberg, N. L., and White, J. L., 1953. Selective loss of plant nutrients by erosion, *Soil Sci. Soc. America Proc.*, **17**, 406–410.

Terman, G. L., and Allen, S. E., 1970. Leaching of soluble and slow-release N and K fertilizers from Lakeland sand under grass and fallow, *Soil and Crop Sci. Soc. of Fla. Proc.*, **30**, 130–140.

Timmons, D. R.; Burwell, R. E.; and Holt, R. F., 1968. Loss of crop nutrients through runoff, *Minnesota Sci.*, **24**, 16–18.

Timmons, D. R.; Holt, R. F.; and Latterell, J. J., 1970. Leaching of crop residues as a source of nutrients in surface runoff water, *Water Resour. Res.*, **6**, 1367–1375.

Truog, E., and Jones, R. J., 1938. Fate of soluble potash applied to soils, *Ind. and Eng. Chem.*, **30**, 882–885.

Viets, F. G., Jr., and Hageman, R. H., 1971. Factors Affecting the Accumulation of Nitrate in Soil, Water, and Plants, *U.S. Department of Agriculture,* ARS, *Agricultural Handbook No. 413,* 63p.

Volk, N. J., 1940. The effect of soil characteristics and winter legumes on the leaching of potassium below the 8-inch depth in some Alabama soils, *Jour. Am. Soc. Agron.*, **32**, 888–890.

Volk, G. M., 1956. Efficiency of various nitrogen sources for pasture grasses in large lysimeters on Lakeland fine sand, *Soil Sci. Soc. America Proc.*, **20**, 41–45.

Volk, G. M., 1959. Volatile losses of ammonia following surface application of urea to turf or bare soils, *Agron. Jour.*, **51**, 746–749.

Cross-references: *Carbon Cycle; Conservation; Denitrification: Fertilizers, Inorganic; Fertilizers, Organic; Leaching; Lysimeters; Nitrogen Cycle; Phosphorus Cycle; Plant Nutrients; Soil Fertility; Sulfur Transformations.*

NUTRIENT POTENTIAL

In thermodynamic terms, the chemical potential of a charged or uncharged chemical entity in a system represents its energy status therein; it is expressed with reference to its energy status in a convenient (though arbitrary) "standard" state. In single-phase, energetically homogeneous systems, its molar (or molal) concentration is related to the difference in its free-energy state from the "standard" state; when, however, its freedom to react is restricted (e.g., by chelation), the concentration or fugacity of the "free" species must be determined or inferred to evaluate its "free-energy" status. In multiphase systems, if the freedom of the chemical entity to move and react is restricted, the energy status of the free species *in the most mobile phase* (controlled by the partition coefficients, or equilibrium constants, between phases) defines its energy status in the system. Again, if the entity in question is charged (cation or anion), its chemical potential accounts for the influence of its charged environment (i.e., the concentrations and charges of other ions in the most mobile phase, the surface charge densities of the other phases in the system, etc.), and defines its free-energy state with reference to the standard state.

The soil systems used by plants are "soil:soil water:soil air" and "soil:soil water," plant nutrients (q.v.) pertaining to them being water, ammonia, oxygen and ionic forms of ammonium, nitrate, potassium, calcium, magnesium, and the trace elements (q.v.). Nutrient potential (or the free-energy state of the nutrient) in the first system is obtained from its partial pressure in the gaseous phase, and that in the "soil:soil water" system from its concentration (modified by its electrical environment, its activity, if a charged species) in the soil solution (q.v.). The essential feature of this definition is that the reversible simultaneous equilibria

Nutrient in soil matrix ⇌ nutrient adsorbed, held or precipitated in surface phase ⇌ nutrient in vapor or solution phase

must be established in the soil system as a whole when the concentration or partial pressure is measured or inferred from the measurements. The resultant nutrient potential defines its energy state in the *whole* system, and is a measure of the work required to remove the nutrient from the system *at that potential.* If, however, the nutrient potential of the system changes as a result, the new nutrient potential defines the work required to remove more nutrient.

Uncharged Nutrient Molecules in Soil

Water. The primary nutrient, water, is held mainly within expanding-type clay minerals, in organic matter, and in intra- and inter-particle pore space. The work required to remove water from a soil system can be most directly determined by measuring its vapor pressure. A characteristic property of a soil-water system is its vapor pressure as a function of its moisture content. The difference between the free energies of water in soil and pure bulk water at the same temperature and atmospheric pressure is a measure of this work, and is the height h in centimeters of a water column giving equivalent suction (i.e., the amount by which the hydrostatic pressure p of water in the soil pore space is less than the atmospheric pressure p_0), leading to a lowering of the relative vapor pressure. The consequent lowering in the free energy is given by

$$F = h = \frac{-RT}{981.\text{M}}\left(\ln \frac{p}{p_0}\right) \qquad (2)$$

where M is the gm mol wt of water, R the universal gas constant, and T the absolute temperature. In pure water

$$h = \int \frac{dS}{\rho} \qquad (3)$$

where S is the suction in dyn/cm^2 and ρ the density of water under tension. Incidentally, Schofield (1935) described the common logarithm of this difference as the pF value of a soil:water system, i.e.,

$$\text{pF} = \log_{10} h \qquad (4)$$

Figure 1 is reproduced from his paper with some additional captions illustrating some limiting pF values and the hysteresis in the ability of soil to hold and release water. This hysteresis indicates that at the same moisture content, more work needs to be done to withdraw water from the soil than that opposing the entry of an equivalent amount of water into the system.

The technical problem of measuring water potentials has been studied since the beginning of this century; the methods evolved vary in complexity considerably, the main difficulty being the time required for equilibrium to be attained between water in pores varying considerably in dimensions (diameter and length), and secondly, the problem of hysteresis in the "wetting" and "drying" relationships between vapor pressure and moisture content. However, the scope of this paper does not extend to technical details of measurement (see *Imbibition; Water Content and Retention;* and *Wetting Front*).

Another serious difficulty is that regions of varying water potential occur *within* soil profiles and crumbs, and equalization of potential between these regions is extremely slow, mainly because water flux (q.v.) is impeded by minute channels (relative to the size of water molecules) and because the escape of water molecules is accompanied by a temperature drop that further restricts the fugacity of water molecules. Such considerations must and do

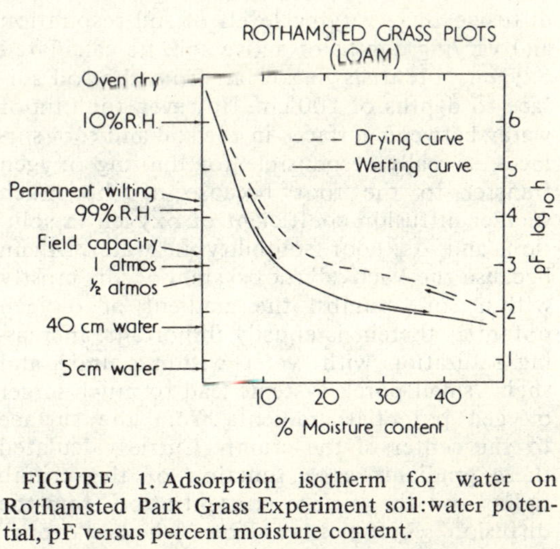

FIGURE 1. Adsorption isotherm for water on Rothamsted Park Grass Experiment soil:water potential, pF versus percent moisture content.

affect the potential of dissolved gases in soil water (q.v.) because their physical behavior is similar to that of water molecules.

All other soil nutrients, including gaseous nutrients (e.g., oxygen and ammonia) enter roots dissolved in water held in soil pores or in root membranes. Again, the thermodynamic activity of dissolved molecules of gaseous nutrients (and hence their potential in soil) is directly proportional to their partial vapor pressure in soil air.

Oxygen. Russell and Appleyard (1915) measured the mean concentration (by volume) in soil air on various plots of the Rothamsted Classical Experiments as 20.6 ± 0.2% compared with a value of 21.0 in atmospheric air. After removing soil air, they also pumped out small volumes of "adsorbed" air (from 5–10 cc per 100 g soil at a time successively) containing decreasing percentages of oxygen, the CO_2 concentrations being inversely related. Initial concentrations of oxygen varied from 0.0 in soils rich in organic matter to 15.1% in unmanured soils (and corresponding CO_2 concentrations from 99.3 to 6.3%). Clearly, measuring oxygen concentrations in soil air by suction without contamination by atmospheric air is difficult. The oxygen concentration in water sorbed in soil pores would seem much more relevant to the *oxygen potential* in the soil pertinent to uptake by plants.

The diffusion coefficient of a nutrient is its flux in unit time through unit cross-sectional area when the concentration or potential gradient is unity. In other words, if the appropriate parameters are known, such diffusive flux experiments can be used to estimate nutrient potentials in the soil.

Currie (1961, 1962) discussed the problems of oxygen concentrations in soil profiles and crumbs. In profiles of varying resistance to diffusive flux, varying levels of soil respiration and varying depths of active soil, he calculated oxygen potentials in soil air from the soil surface to depths of 100 cm. However, the film of water between surfaces in the soil and root surfaces is really responsible for limiting oxygen transfer to the root because of the much smaller diffusion coefficient of oxygen in solution and its poor solubility in water. Again because the "active" root system occurs mostly within soil crumbs, the gradient of oxygen potential therein is equally important. Increasing saturation with water within crumbs and their "small pore" systems lead to much larger oxygen potential gradients from the surface to the center of the crumb. Currie calculated these gradients as a function of the crumb radius and the surface concentration from the diffusion coefficients below which "death" occurs and an optimum crumb size above which anaerobic conditions become overwhelming.

Later Greenwood and Goodman (1967) showed that oxygen potentials in 1-cm-diameter soil spheres measured with a gold microelectrode agreed very well with those calculated by diffusion theory, showing that the oxygen potential fell from 6×10^{-3} ml of oxygen per ml of water (similar to that in air-saturated water) at the sphere surface to zero concentration at less than 2 mm from the surface.

Carbon Dioxide. Carbon dioxide (CO_2) concentrations in "adsorbed" air in soils are inversely proportional to oxygen concentrations (Russell and Appleyard, 1915). In soil water, CO_2 cannot be considered a plant nutrient, but indirectly the CO_2 potential in soil water buffers in the pH in the rhizosphere (q.v.) to around 6, thus mobilizing and increasing the chemical potential of nutrient phosphates, potassium, and trace elements in soils of higher pH or decreasing the activities of toxic nutrient elements, e.g., manganese, in more acid soils.

Ammonia. The increasing use of liquid ammonia and ammonia liquor as nitrogen fertilizers and the study of ammonifying bacteria engenders interest in measuring "ammonia" activities in soil. Pockets of NH_3 persisting in soil endanger root growth and microbial populations because of abnormal soil pHs. On the other hand, such pockets where NH_3 is "fixed" as NH_4-N and released on demand by the crop would act as sources of slow nitrogen release when soil reaction has neutralized the pH. Chemical methods and the use of ammonia electrodes do not, however, distinguish between ammonia and ammonium-nitrogen and certainly do not measure the thermodynamic activity or potential of ammonia (see *Electrochemistry*). The only method available at present is to measure the partial pressure of NH_3 in the equilibrium soil air and infer the NH_3 potential in soil water from the known solubility:vapor pressure relationship. High NH_3 potentials would then correspond with poor growth immediately following fertilization but in the long term would result in improved crop performance. Ashworth (1973) found by measuring partial vapor pressures of NH_3 in moist and dry soils systems that most of the NH_3 sorbed on his soils was by reactions other than cation exchange.

Nutrient Ions in Soil

Mineral or organic constituents of soils have charged surfaces, positive and negative, the positive charge occurring increasingly as the pH reaction of the soil system goes progressively more acid and/or the soil is increasingly ferruginous. Negatively charged surfaces usually predominate. These charges are compensated by counterions. In the soil:water system these ions dissociate to varying extents forming "parallel plate" and "diffuse-ion atmospheres" next to the charged surfaces and sites. The potential drop in the parallel plate condenser occurs across distances of molecular dimensions and linearly with distance from the surface. Its extent depends greatly on the specific attraction of the counterion for the charged surface, to a minor degree on the concentration of the counterion in the equilibrium and is independent of the ionic strength of the solution, virtually, the soil solution. The potential drop in the diffuse layer occurs exponentially across its thickness from the surface. This thickness x decreases with increasing ionic strength of the bathing solution according to the *Gouy relationship*

$$x = \left(\frac{\mathrm{DRT}}{8\mathrm{F}^2 z^2 c}\right)^{1/2} \quad (5)$$

and with increasing density of charge Γ on the soil surface and the valency of the counterion according to the *Schofield-Gouy relationship* (that includes the concentration c, in normality, of the bathing solution)

$$x = \left[\frac{1}{(z\beta)^{1/2}}\right]\left[\frac{q}{(c)^{1/2}} - \frac{4}{(z\beta)^{1/2}}\right]\left[\frac{1}{\Gamma}\right] \quad (6)$$

where z is the valency of the counterion and q a factor depending on the ratio of valencies of the counter and repelled ions (Schofield and Talibudeen, 1948).

One can infer from this that for a particular combination of density of charge on a soil surface and valency of the counterion, there is a critical ionic strength in solution at which the diffuse double layers around adjacent site interpenetrate and ions of like charge (to the sites) are excluded from the surface layer. We then have a nearly uniform volume distribution of charge on the dispersed soil phase akin to a Donnan system so that the cation activity ratio (or product when considering a cation-anion pair) in the diffuse double layer is the same as that in the bulk solution in equilibrium with the soil surface (Schofield, 1947). The potential of a nutrient ion in soil is then known from such an activity ratio (or product) by the relationship

$$\Delta F = -RT\left[\frac{1}{z_i}(\log_e A_i) - \frac{1}{z_s}(\log A_s)\right] \quad (7)$$

$$= RT\left\{\frac{1}{z_i}[p(A_i)] - \frac{1}{z_s}[p(A_s)]\right\} \quad (7)$$

where z_i, A_i, and z_s, A_s are the valencies and activities of the nutrient and "standard" ions. Where a cation-anion pair is concerned,

$$\Delta F = RT\left\{\frac{1}{z_i}\cdot[p(A_i)] + \frac{1}{z_s}\cdot[p(A_s)]\right\} \quad (8)$$

Three other characteristics of soil systems are relevant. If a sparingly soluble compound of the nutrient ion occurs in the soil, its constant thermodynamic solubility in the soil water controls the ionic equilibrium of the soil Donnan system. Second, if negatively *and* positively charged sites occur together in a soil, it will emulate a Donnan system only if one predominates over the other. Third, to evaluate a nutrient potential in soil, it must be referred to a "standard" ion dominating the surface ions because measuring single-ion activities is ambiguous.

The preceding discussion shows (equations 7 and 8) that, under certain conditions, in the bulk solution in equilibrium with soil of undisturbed surface ionic composition, the total anion concentration can be changed within wide limits while keeping the proportions of *activities* of nutrient and standard ions constant (Talibudeen, 1947). The range covering these limits is such that ion activities can be approximated to ion concentrations.

pH and Potentials of Cation Hydroxides. Schofield and Taylor (1955), working with soil from the Rothamsted Park Grass Experiment, evaluated these limiting ranges for the hydroxides of sodium, potassium, calcium, and aluminum by demonstrating that the preceding relationship did not hold above critical concentrations specific for each cation: Na 0.04*N*, K 0.10*N*, Ca 0.04*N*, and Al 0.005*N*. If solution concentrations exceed these critical values when any of these ions dominate the exchange equilibrium, the anion increasingly crowds, and the diffuse ion atmospheres are confined to the vicinity of each charged site (e.g., Figure 2 for the *lime potential* p(HCl) – ½p($CaCl_2$), (Schofield and Taylor, 1955).

These observations also indicate that the pH value of a soil in aqueous suspension will be very sensitive to ion concentrations less than the critical concentration in the soil solution derived from the soil or added by the experimenter. At greater concentrations the measured pH will decrease to a value unaffected by salt concentration, hence the modern soil chemist's practice of measuring soil pH values in *M* KCl. Alternatively, soil pH is measured in 0.02*N* $CaCl_2$ to give a known calcium concentration in the Donnan region.

In calcareous soils the equilibrium calcium activity, and hence the lime potential and soil

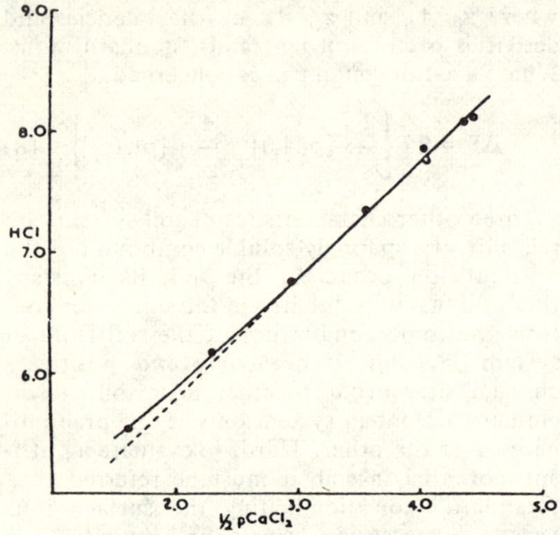

FIGURE 2. Lime potential of Rothamsted Sawyers I soil: $-\log(\mathrm{HCl})$ versus $-\frac{1}{2}\log(\mathrm{CaCl_2})$.

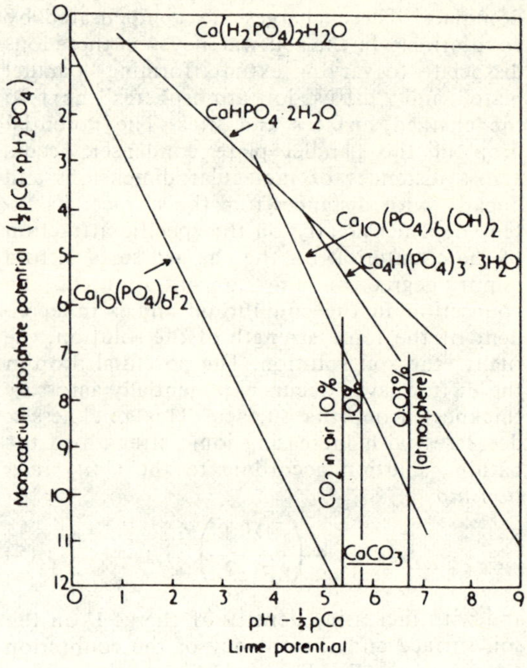

FIGURE 3. A phase diagram for the calcium phosphate system: phosphate potential versus lime potential.

pH, is controlled by the partial pressure of CO_2 in soil air. Hence soil pH in the rhizosphere is less than values in soil determined in the laboratory, and a knowledge of the partial pressure of CO_2 in soil air is essential (Russell and Appleyard, 1915, Tables I, II, and III).

Phosphate Potential. The soil phosphate equilibrium is basically governed by sparingly soluble phosphates, generally recognized to be alkaline earth and sesquioxidic phosphates. A solubility diagram (Figure 3) for the "calcium phosphate" system (condensed from papers by Bjerrum, Aslyng, and Schofield) relates the monocalcium phosphate potential (in soil in equilibrium with the soil solution containing calcium ions) with the lime potential. Stable and complex forms of the calcium phosphates (and sesquioxidic phosphates) take considerable time to crystallize out in soil systems where anion mobility is very restricted. The instantaneous sorptive equilibrium established, when soluble phosphate is added to moist soil, is with surfaces of its major mineral constituents modified by the adsorbed and soluble organic constituents in soil. This metastable equilibrium slowly transforms to a "solubility product" type of equilibrium at a speed dependent on water content, mixing of soil, temperature, disturbance by crop growth, soil characteristics, etc. So the soil phosphate equilibrium can be represented thus:

Phosphate added to soil solution ⇌ (Rapid) Surface adsorbed phosphate

Phosphate added to soil solution ⇌ Precipitated phosphate

Surface adsorbed phosphate ⇌ (Slow) Precipitated phosphate

Soluble phosphate added to the soil system is unlikely to change to precipitated forms rapidly and directly because (1) sparingly soluble "low-potential" phosphates (e.g., apatites and octocalcium phosphates) form *very* slowly, and (2) phosphate concentrations in solution do not reach levels high enough to precipitate more soluble "high-potential" phosphates (e.g., dicalcium phosphate), the "rapid" adsorption equilibrium lowering the phosphate concentration below the critical value for their precipitation.

Schofield's (1955) concept of a "monocalcium phosphate" potential is a measure of the work required to remove phosphate ions from the system *at that potential.* As with water, the *equilibrium* relationship between this potential and the "pool" of plant-available or labile phosphate characterizes precisely each soil phosphate system, and the kinetics of ion transfer from various sorptive phases in the pool to solution very often point to the nature of the pool phosphate in soil (Vaidyanathan and Talibudeen, 1968, 1970). Thus the fraction of the pool of soil phosphate extracted at various times by a chloride-saturated, strongly basic, anion exchange resin depends on the pH of the field soil (Figure 4). The more acid soil releases phosphate relatively much more slowly than the same soil at higher pH values. Between pH

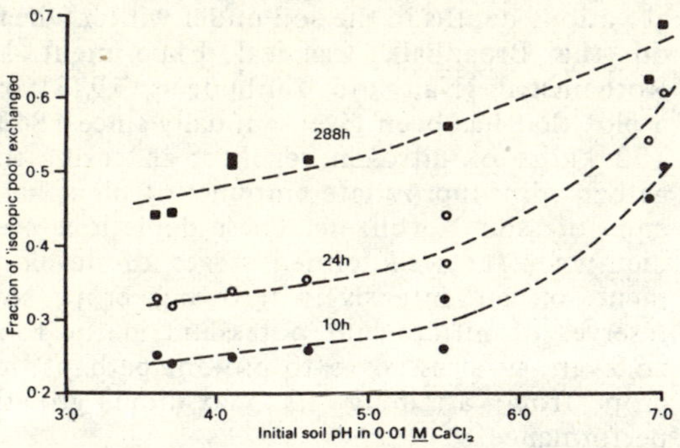

FIGURE 4. Fraction of isotopically exchangeable phosphate in soil extracted by chloride-saturated anion exchange resins as a function of soil pH and time of extraction.

values 6 and 7, the rate of release increases sharply.

Added phosphate increases soil phosphate potential quite independently of the precipitation of a sparingly soluble phosphate unless several months are allowed for equilibrium (Aslyng, 1954; Jensen, 1971; Webber and Mattingly, 1970). Cropping or ion exchange resins desorb phosphate likewise (Webber and Mattingly, 1970; Vaidyanathan and Talibudeen, 1968).

The lime potential of calcareous soils is directly related to the partial pressure of CO_2 in soil air. Thus the phosphate potentials of soils in the rhizosphere resulting from a sorptive mechanism or from sparingly soluble compounds must also depend on the CO_2 concentration in soil air (see Figure 3).

Potassium Potential. Equation 7 shows that the thermodynamic potential of potassium in soil is proportional to p(K) – ½p(Ca + Mg), p(K) –⅓p(Al), or p(K) – p(Na), depending on whether the exchange complex of the soil is dominated by Ca + Mg ions (pH > 5), aluminum ions (pH < 5), or sodium ions (saline soils). Taylor (1958), using an interpolation method on (Ca + Mg) rich soils at Rothamsted defined the constancy of the function p(K) – ½p(Ca + Mg) under various experimental variables, while Woodruff (1955) and Arnold (1962) related it to yields of maize and ryegrass, respectively. The potassium potential of soils of pH < 4.5 was evaluated by p(K) – ⅓p(Al) (Singh and Talibudeen, 1968). As with other nutrients, a most important characteristic of soil potassium is the relation between the change in its potential with the number of exchangeable potassium ions on the soil complex (Addiscott and Talibudeen, 1969; Matthews and Beckett, 1962).

Figures 5(a) and (b) illustrate the effect of

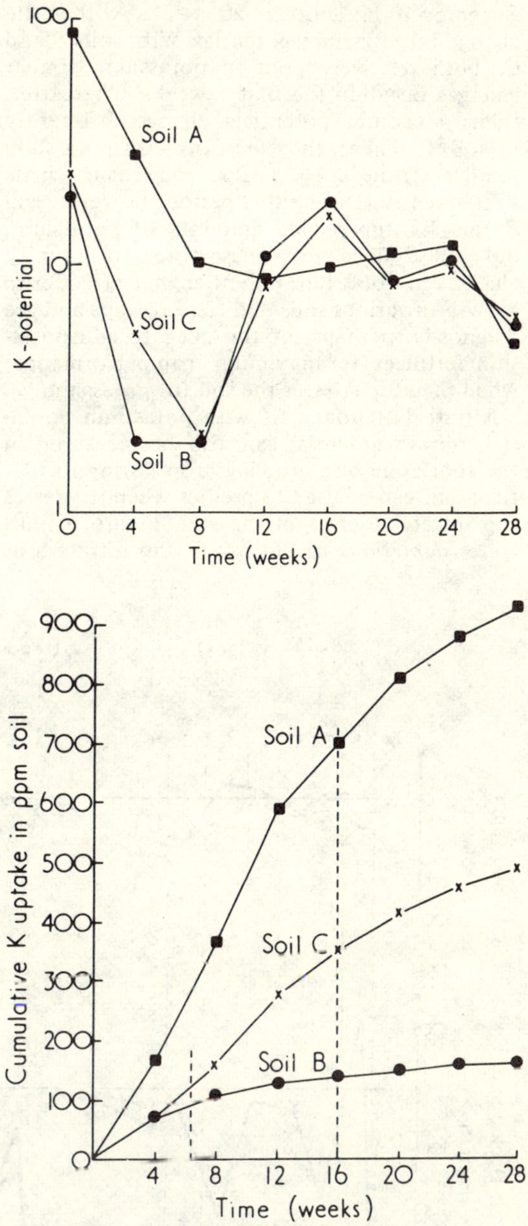

FIGURE 5. Effect of the period of cropping with ryegrass on the posassium status of three contrasting soils: Soil *A* rich in potassium; Soils *B* and *C* poor in potassium. (a) Change in potassium potential; (b) Potassium uptake.

removing increasing amounts of potassium from three soils [Figure 5(a)] by progressive cropping with ryegrass, on the potassium potential of the soil [Figure 5(b)]. The potassium potential of soil *A*, rich in potassium, drops sharply when large amounts of potassium are removed in 8 weeks. Thereafter, almost as much potassium is removed in another 20 weeks with little change in potassium potential. With soils *B* and *C*, both relatively poor in potassium, similar changes occur in the first 8 weeks. Thereafter, their potassium potentials change differently to soil *A*, that is, they *increase* with potassium removal from 8–16 weeks, and change little subsequently. During this period, however, soil *B* releases appreciable amounts of potassium, but soil *C* does not. These three soils typify changes in potassium potential that affect crop growth in various soils in different ways and are diagnostic, perhaps, of the need to add potassium fertilizer (for maximum crop performance) when the crop stresses the soil for potassium.

Nitrate Potential. As with potassium potential, nitrate potential can also be measured in the root zone of a growing crop during its lifetime and can be used to predict when it stresses the nitrate reserves in the soil. Figure 6 illustrates *depletions* of potassium and nitrate ions at various depths in the soil under winter wheat on the Broadbalk Classical Experiment at Rothamsted (Nair and Talibudeen, 1973) on a plot that has been given annually since 1843, 192 kg/ha of nitrogen fertilizer in spring together with appropriate amounts of phosphate and potassium fertilizers. These depletions predict when, at well-defined stages of development of an intensively growing crop, soil reserves of nitrate and potassium in the root zone are so stressed as to prevent perhaps the crop from attaining its maximum growth performance.

Using nutrient concentration maxima in the crop, corresponding with maximum soil nutrient potentials, to predict where maximum root activity occurs in the soil: When fertilizer is applied to soil, rain and/or irrigation and evaporation cause soluble nutrients to move to various depths in the soil depending on soil texture and structure and the reactions of the

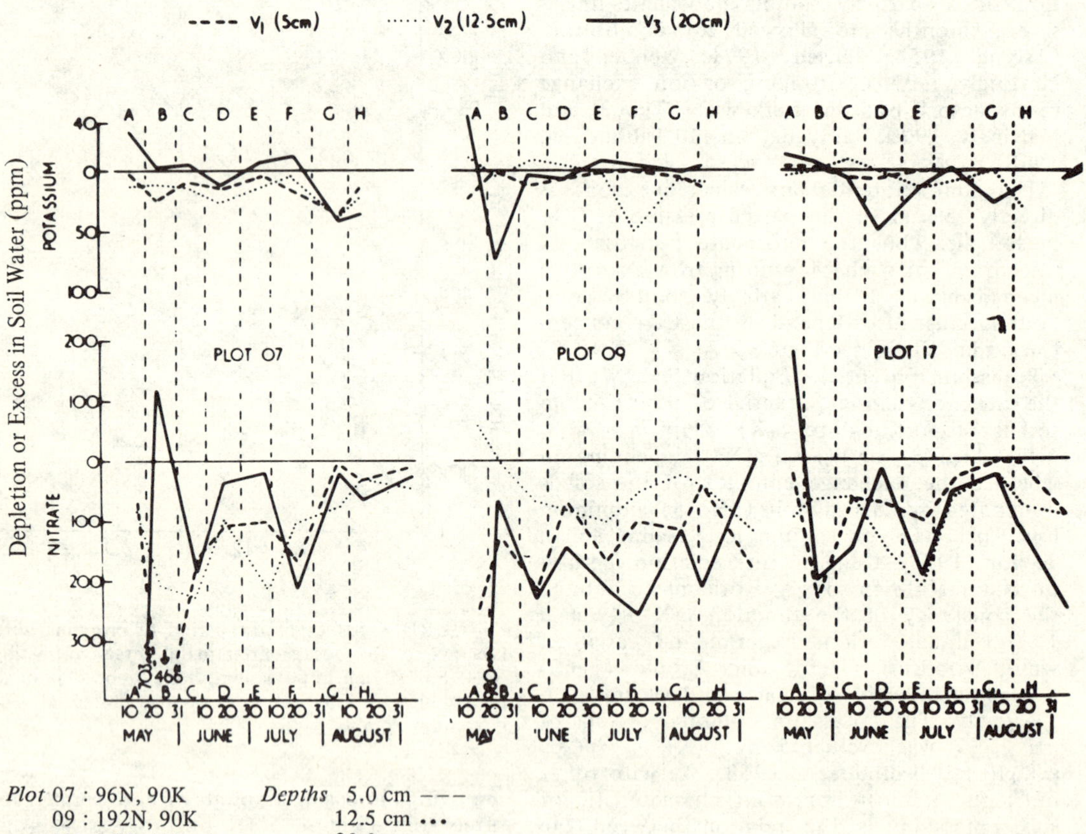

Plot 07 : 96N, 90K
09 : 192N, 90K
17 : 96N, 45K
in kg ha^{-1}

Depths 5.0 cm ‒ ‒ ‒
12.5 cm •••
20.0 cm ——

FIGURE 6. Depletion at various soil depths of potassium and nitrate ions in soil water at Rothamsted Broadbalk Experiment.

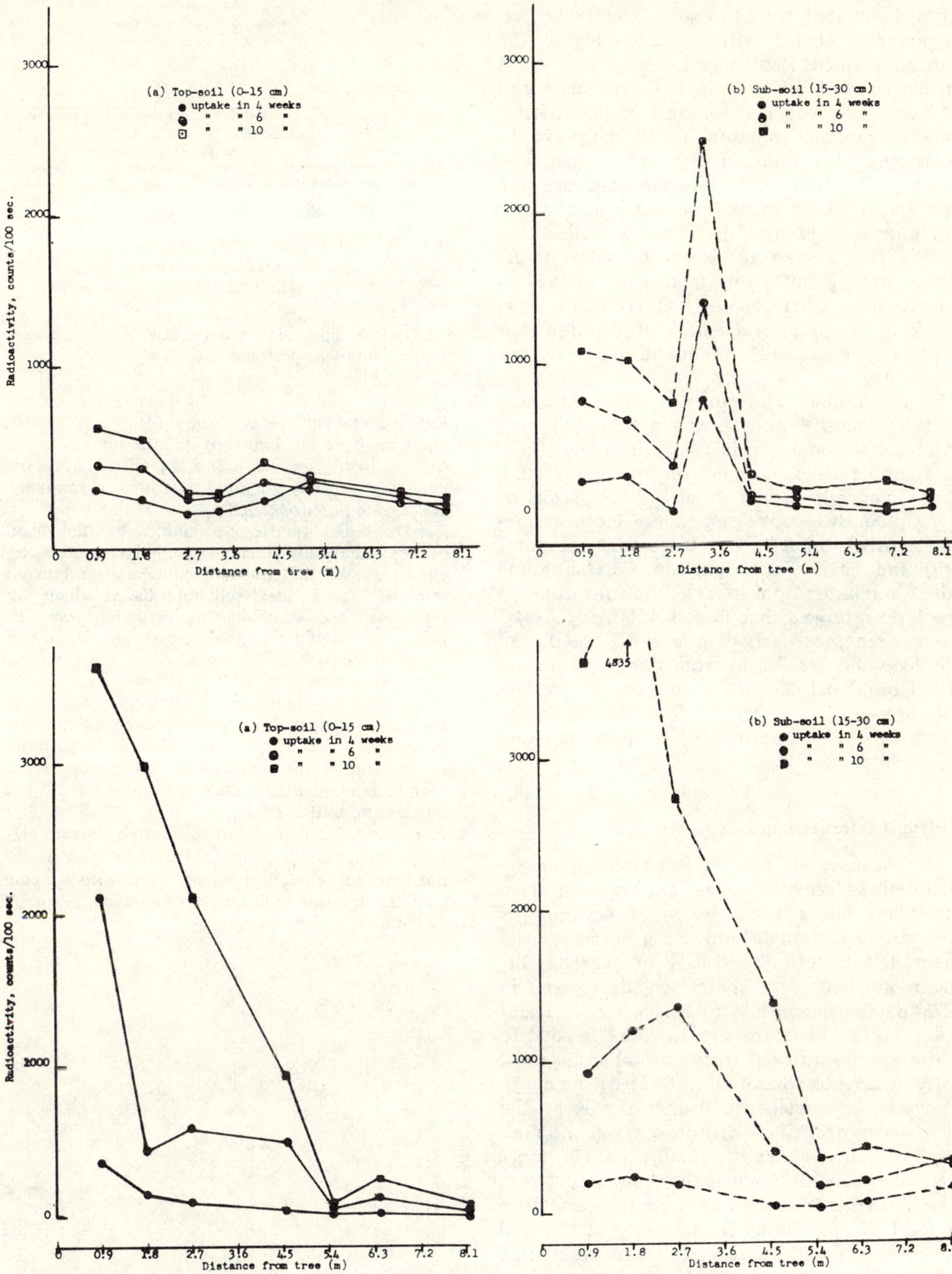

FIGURE 7. The effect of growth time on the uptake in latex of ^{32}P-labeled phosphate injected at various distances from the tree and at various depths. (a) Soil *A*; (b) Soil *B*.

applied nutrient with the soil. When such fertilizers are labeled with tracer atoms of the nutrient element (which have identical chemical properties and differ only in atom mass and perhaps the property to emit radioactivity), one can precisely measure fertilizer uptake by measuring this radioactivity in the growing plant. This method is effective because the radiotracer exists mostly in soil zones where the nutrient potential has been enhanced by adding the labeled fertilizer. In suitably designed experiments, this technique enables us also to trace where, in the soil, roots are most active in taking up nutrients–information that is valuable, practically, for manuring tree crops especially.

When soluble phosphate fertilizer labeled with radioactive phosphorus was placed in a three-dimensional grid in the root zones of mature rubber trees, growing on two contrasting Malaysian soils (Singh et al, 1972), peaks of ^{32}P-radioactivity appeared in the latex at predetermined times after placement [Figures 7(a) and (b)]. These maxima, corresponding with maximum phosphate concentration in the latex, showed that in soil *A* [Figure 7(a)], roots were most active in a soil zone 15–30 cm deep and 2.7–4.5 m from the trunk. In soil *B* [Figure 7(b)], however, the zone of greatest root activity was up to 1 m from the trunk to a depth of 15 cm, and up to 3 m from the trunk at 15–30 cm depth.

Nutrient Potential and Crop Performance

The lifetime of a crop is characterized by well-defined growth stages. During each stage the plant has a *work potential* defining its strength to assimilate and use a nutrient from the soil. Interactions, positive or negative, between nutrients play their recognized parts in defining the magnitude of this work. Nutrient potential in soil represents the work needed to withdraw the nutrient from the soil and is most aptly analogous (Schofield, 1955) to the depth of water in a well–the deeper the level, the more work needed to withdraw water, no matter how much water it contains. (Again, more water can be withdrawn at the same depth from a well of larger diameter to cause the same drop in level, thus defining the capacity of the well rigorously.) Considering all other nutrients in a plant:soil system to be optimum for growth, a study of the interplay of the work potential of the plant (at a particular growth stage) and the nutrient potential in the soil, leads to four possible diagnostics in the plant, predictable and recognizable in practice, at that stage of growth (Figures 8 and 9):

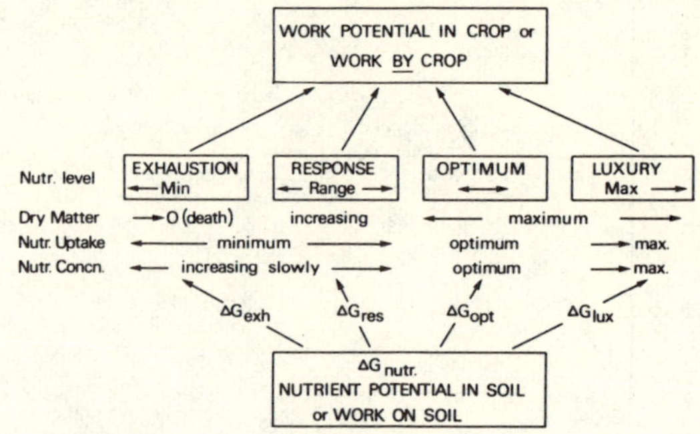

FIGURE 8. The relationship between soil nutrient potential and plant performance.

1. *Exhaustion.* The plant forms little dry matter, takes up little nutrient (but its nutrient *concentration* is high), and it eventually dies. The corresponding nutrient potential measured in soil is the *exhaustion potential,* ΔG_{exh}.
2. *Response.* Dry matter, produced by the plant photosynthetically, increases with increasing nutrient uptake so that nutrient concentration remains minimal. The smallest soil potential at which this crop response starts is the *response potential,* ΔG_{res}, or *uptake* potential –5600 and –4710 cal/equiv of potassium for ryegrass and potatoes, respectively (Addiscott, 1970).
3. *Optimum.* This indicates an optimum nutrient uptake, equivalent to minimum nutrient concentration in the plant that enables maximum photosynthesis, i.e., maximum dry matter produced, corresponding with maximum plant efficiency and a critical soil nutrient potential, ΔG_{opt}.
4. *Luxury.* At soil nutrient potentials increasingly more than ΔG_{opt}, nutrient uptake and hence nutrient concentration increase too without concomitant increase in dry matter produced.

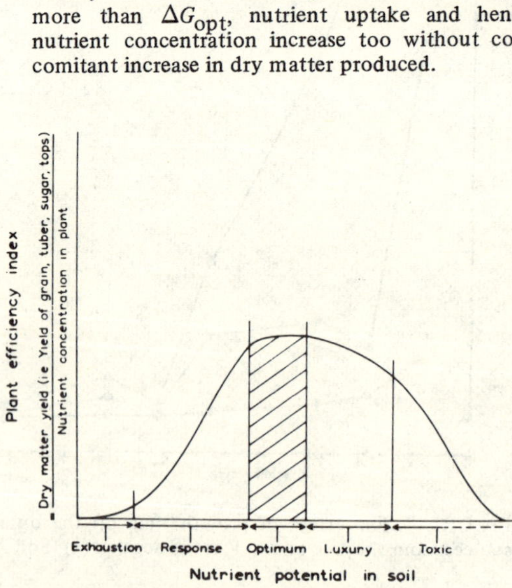

FIGURE 9. Diagrammatic representation of the effect of soil nutrient potential on crop performance.

Clearly, when manuring soils, soil nutrient potentials in regions 1 and 4 must be avoided; the ideal to aim for is region 3, progressively achieved by working through region 2. The efficiency of plant growth (see Figure 5) seems best defined by the yield of that part of the plant that is of interest (e.g., grain for cereals, sugar content for sugar beet, etc.) per unit of nutrient concentration in the vegetative parts.

The work potential of a plant is a function of many factors: soil water content, relative humidity of the air, CO_2 and oxygen contents in soil and atmospheric air, radiation input including soil and air temperature, and the correct combination of inorganic nutrients. To evaluate the effect of any one nutrient, all others would have to be optimum for the stage of growth of the plant under study.

O. TALIBUDEEN

References

Addiscott, T. M., 1970*a*, The uptake of initially available soil potassium by ryegrass, *Jour. Agric. Sci. Camb.*, **74**, 123-129.

Addiscott, T. M., 1970*b*, Potassium uptake by potatoes, *Jour. Agric. Sci. Camb.*, **74**, 495-500.

Addiscott, T. M., and Talibudeen, O., 1969. The buffering capacity of potassium reserves in soils, *Potash Rev.*, Subject 4, 45th Suite, 1-24.

Arnold, P. W., 1962. The potassium status of some English soils considered as a problem of energy relationships, *Proc. Fertil. Soc.*, **72**, 25-43.

Ashworth, J., 1973. Reactions of ammonia with soil. I. Adsorption isotherms and calorimetric heats of adsorption of ammonia gas on homo-ionic soil, *Jour. Soil Sci.*, **24**, 104-116.

Aslyng, H. C., 1954. The lime and phosphate potentials of soils. Solubility and availability of phosphates, *Yearbk. Roy. Vet. Agric. Coll., Copenhagen*, 1-50.

Currie, J. A., 1961. Gaseous diffusion in the aeration of aggregated soils, *Soil Sci.*, **92**, 40-45.

Currie, J. A., 1962. The importance of aeration in providing the right conditions for plant growth, *Jour. Sci. Fd. Agric.*, **13**, 380-385.

Greenwood, D. J., and Goodman, D., 1967. Direct measurements of the distribution of oxygen in soil aggregates and in columns of fine soil crumbs, *Jour. Soil Sci.*, **18**, 182-196.

Jensen, H. E., 1971. Phosphate solubility in Danish soils equilibrated with solutions of differing phosphate concentrations, *Jour. Soil Sci.*, **22**, 261-266.

Matthews, B. C., and Beckett, P. H. T., 1962. A new procedure for studying the release and fixation of potassium ions in soil, *Jour. Agric. Sci. Camb.*, **58**, 59-64.

Nair, P. K. R., and Talibudeen, O., 1973. Dynamics of K and NO_3 concentrations in the root zone of winter wheat at Broadbalk using specific-ion electrodes, *Jour. Agric. Sci. Camb.*, **81**, 327-337.

Russell, E. J., and Appleyard, A., 1915. The atmosphere of the soil: Its composition and the causes of variation, *Jour. Agric. Sci. Camb.*, **7**, 1-48.

Schofield, R. K., 1935. The pF of the water in soil, *3rd Internat. Cong. Soil Sci. Trans.*, **3**, 37-48.

Schofield, R. K., 1947. A ratio law governing the equilibrium of cations in the soil solution, *Proc. 11th Internat. Cong. Pure and Appl. Chem.*, **3**, 257-261.

Schofield, R. K.,1955. Can a precise meaning be given to "available" soil phosphorus?, *Soils Fertil.*, **18**, 373-375.

Schofield, R. K., and Talibudeen, O., 1948. Measurement of internal surface by negative adsorption, *Discussions of the Faraday Soc.*, **3**, 51-56.

Schofield, R. K., and Taylor, A. W., 1955. Measurements of the activities of bases in soils, *Jour. Soil Sci.*, **6**, 137-146.

Singh, M.; Pushparajah, E.; Soong, N. K.; and Talibudeen, O., 1972. Radiotracer studies on phosphorus uptake by *Hevea brasiliensis* from Malayan soils for determining "active" root distribution, *Proc. Conf. Isotopes and Radiation in Soil-Plant Relationships Including Forestry, International Atomic Energy Agency, Vienna*, 465-479.

Singh, M., and Talibudeen, O., 1968. Thermodynamic assessment of the nutrient status of rubber-growing soils, *Proc. Nat. Rubber Conf., Kuala Lumpur*, Part 1, 240-249.

Talibudeen, O., 1947. Negative adsorption and ionic concentrations at the internal surface of jute, Ph.D. Thesis, London.

Talibudeen, O., and Dey, S. K., 1968. Potassium reserves in British soils, *Jour. Agric. Sci. Camb.*, **71**, 405-411.

Taylor, A. W., 1958. Some equilibrium solution studies on Rothamsted soils, *Soil Sci. Soc. America Proc.*, **22**, 511-513.

Vaidyanathan, L. V., and Talibudeen, O., 1968. Rate-controlling processes in the release of soil phosphate, *Jour. Soil Sci.*, **19**, 342-353.

Vaidyanathan, L. V., and Talibudeen, O., 1970. Rate processes in the desorption of phosphate from soils by ion-exchange resins, *Jour. Soil Sci.*, **21**, 173-183.

Webber, M. S., and Mattingly, G. E. G., 1970. Inorganic soil phosphorus. I. Changes in monocalcium phosphate and lime potentials on mixing and liming soils, *Jour. Soil Sci.*, **21**, 111-126.

Woodruff, C. M., 1955*a*. Ionic equilibria between clay and dilute salt solutions, *Soil Sci. Soc. America Proc.*, **19**, 36-40.

Woodruff, C. M., 1955*b*. Cation activities in the soil solution and energies of cationic exchange, *Soil Sci. Soc. America Proc.*, **19**, 98-99.

Woodruff, C. M., 1955*c*. The energies of replacement of calcium by potassium in soils, *Soil Sci. Soc. America Proc.*, **19**, 167-171.

Cross-references: *Aeration, Respiration, Atmosphere; Activity Ratio; Adsorption Phenomena; Electrochemistry; Exchange Phenomena; Imbibition; Ionic Activities; Law of the Minimum; Lime, Liming; Nitrogen Cycle; Nutrient Losses; Phosphorus Cycle; Plant Nutrients; Soil Fertility; Soil Reaction; Soil Solution; Water Content and Retention.*

O

ORGANIC COMPLEXES–See METAL-COMPLEXING MACROMOLECULES

ORGANIC COMPOUNDS–See FERTILIZERS, ORGANIC; HUMUS; MULCH; ORGANIC MATTER; POLYSACCHARIDE; SOIL COMPONENTS, ORGANIC

ORGANIC MATTER

The term *soil organic matter* refers to all materials of vegetable and animal origin formed in or added to soils regardless of stage of decomposition. Soil organic matter thus includes the highly decomposed and colloidal soil fraction known as *humus* (q.v.) as well as organic residues that have not lost their anatomic structure (Kononova, 1961; Brady, 1974). It is the humus fraction that contributes most to soil properties that are of greatest importance in crop production.

Character and Importance

The nature of the organic material that accumulates in soil has been studied for well over a century (Waksman, 1938). Initial efforts focused on isolating definite chemical compounds from humus and establishing formulas for them. Recently, however, emphasis has shifted to investigations of the relationships between the main functional groups, to determining the causes for the chemical and physical properties of humus substances, and to assessments of organic matter transformations in soils (Allison, 1973). The complex nature of organic matter in soils is indicated by the great diversity of its components (see *Soil Components, Organic; Humus; Polysaccharide*). The plant and animal residues added to soils are composed of carbohydrates, proteins, lignins, fats, waxes, and resins. Products of microbial activity can also be carbohydrates and proteins. Intermediate products formed during the decomposition of organic matter consist of a great variety of compounds, as brought out in the article *Soil Components, Organic*. Humus consists of a complex group of compounds, as explained in the article on *Humus*.

The influence of organic matter on soil properties, and consequently on plant growth, is far greater than the low percentage (about 3 to 5 percent in a representative mineral soil) would indicate. Organic matter, for example, is largely responsible for the friable condition of productive soils and is also an important source of elements such as phosphorus, sulfur, and nitrogen. Through its effect on the physical conditions, organic matter increases the water-holding capacity of soils (see *Moisture Management*), facilitates the development of water-stable structure (see *Aggregation* and *Aggregate Stability*) and enhances the supply of nutrients to plants (see *Soil Fertility*). Organic matter is the main source of energy for soil microorganisms (see *Soil Microbiology*); without it, biochemical activity (see *Enzyme Activity; Metabolic Activity; Soil Biochemistry*) would be practically nil.

Organic materials also interact with other soil constituents (see *Clay-Organic Reactions*) and additives such as fertilizers (see *Fertilizers, Inorganic*), pesticides, and herbicides (see *Herbicide Diffusion*). Processes of accumulation and decomposition of organic matter play a significant role in soil management (q.v.) and are particularly relevant to methods of soil cultivation (see *Tillage* and *Zero-Tillage*) (Bartholomew, 1957). Organic matter (humus) content is a diagnostic feature in soil description and classification, and it is important to soil formation, including the weathering and decomposition of soil minerals (see Vol. XII, Pt. 2: *Weathering, Organic*).

Sources and Amounts

Most organic matter added to soils consists of dead parts and bodies of green plants. These add organic matter as litter on the surface and as roots within the soil. Another important addition comes from microorganisms in the soil. A minor part is contributed by the wastes and dead bodies of animals. Quantities of plant residues added as dry matter in kilograms per hectare per year (kg/ha/yr) range from 200 or less in wormwood deserts of Asia to 18,000 in the rainforests of western Africa (Rodin and Bazilevich, 1967; Nye, 1961). Intermediate amounts between 4000 and 5000 kg/ha/yr are added to the soil under prairie and deciduous forest vegetation in humid, cool-temperate regions. The comparative additions in the forms of litter and roots are about equal under prairie and shrub vegetation, but the first is roughly twice that of the second under forest.

Residence time of organic matter other than humus in soils is short (Bartholomew and Kirkham, 1960). Consequently, the bulk of the organic matter present is humus, as a rule. Quantities in soils to a depth of a meter expressed in kilograms per hectare range from as little as 1000 in deserts to 20 times that amount under tropical rainforest. Amounts in most soils of the world fall between those amounts, concentrating near the middle of the range.

Formation of humus is tied directly to conditions of soil formation. The small amounts of humus in desert soils (Aridisols) are due to the limited additions of organic residues under the sparse vegetation. Very large amounts of organic matter are added to soils under tropical rainforest to account for levels of humus in those soils. Annual additions are smaller, as are rates of decay, in Mollisols, the dark-colored soils of grasslands, to make those soils high in humus. Decomposition of organic matter in all of these examples proceeds under aerobic conditions. If decomposition is anaerobic, as occurs under wet soil conditions, organic matter can accumulate in the form of peat. Examples of large areas of peat are the Everglades of Florida and the large bodies of organic soils in the Northwest Territories of Canada.

Determination and Measurement

The amount of organic matter in a soil sample is measured indirectly by first determining the organic carbon and then calculating organic matter by use of a factor (Broadbent, 1965). Widely used factors are 1.724 and 1.73. Percent organic carbon is multiplied by the factor to give percent organic matter in the sample. The ratio of organic carbon to organic matter has been found to differ among soils, however, and a factor of 1.9 is therefore used in some instances, especially for samples of surface soils.

Organic carbon (see *Carbon Cycle*) is commonly determined by either dry or wet combustion, the former being the primary standard (Allison, 1965). The dry combustion method is used when precise data are required. Because the method is both faster and simpler, the Walkley-Black method, a wet-combustion procedure, is frequently used when semiquantitative organic carbon data are acceptable.

CHARLES W. FINKL, JNR.

References

Allison, F. E., 1973. *Soil Organic Matter and Its Role in Crop Production.* Amsterdam: Elsevier, 637p.

Allison, L. E., 1965. Organic carbon, *in* C. A. Black, ed., *Methods of Soil Analysis.* Madison: American Society of Agronomy, 1367-1378.

Bartholomew, W. V., 1957. Maintaining organic matter, *in Soil, The 1957 Yearbook of Agriculture.* Washington, D.C.: U.S. Government Printing Office, 245-252.

Bartholomew, W. V., and Kirkham, D., 1960. Mathematical descriptions and interpretations of culture-induced nitrogen changes, *7th Internat. Congr. Soil Science (Madison) Trans.*, 2, 471-477.

Brady, N. C., 1974. *The Nature and Properties of Soils.* New York: Macmillan, 639p.

Broadbent, F. E., 1953. The soil organic fraction, *Adv. Agron.*, 5, 153-183.

Broadbent, F. E., 1965. Organic matter, *in* C. A. Black, ed., *Methods of Soil Analysis.* Madison: American Society of Agronomy, 1397-1400.

Kononova, M. M., 1961. *Soil Organic Matter.* New York: Pergamon, 450p.

Kononova, M. M., 1975. Humus of virgin and cultivated soils *in* J. E. Gieseking, ed., *Soil Components: Vol. 1, Organic Components.* Berlin: Springer-Verlag, 475-526.

Nye, P. H., 1961. Organic matter and nutrient cycles under moist tropical forest, *Plant and Soil,* **13**, 333-346.

Rodin, L. E., and Bazilevich, N. E., 1967. *Production and Mineral Cycling in Terrestrial Vegetation.* London: Oliver & Boyd, 276p.

Waksman, S. A., 1938. *Humus. Origin, Chemical Composition and Importance in Nature.* Baltimore: Williams and Wilkins, 523p.

Cross-references: *Humus; Mulch; Soil Components, Organic.*

OXIDATION—See REDOX REACTIONS

OXIDES—See HYDROUS OXIDES. VOL. IVB: SOIL MINERALOGY. VOL. XII, Pt. 2: PEDOSESQUIOXIDES

P

PARTICLE DENSITY

Particle density is the density of the solid particles that collectively make up a soil sample. The value is commonly expressed in grams per cubic centimeter. The common range among soils is 2.55 to 2.70 g/cm^3. A value of 2.65 is therefore used generally, exceptions being made where great accuracy is required or where soils are known to depart from the common range.

Particle density of a soil sample is actually a weighted mean value for the various kinds of minerals and humus. The measured density depends on the relative proportions of constituent minerals and humus. Densities of widely occurring minerals in the sand and silt fractions are quartz 2.65, feldspars 2.5 to 2.8, micas 2.7 to 3.3, and apatite 3.1 to 3.3 g/cm^3 (Kretz, 1974–75). Densities of clay minerals range from 2 to 3, but many are near 2.65 g/cm^3. The density of humus is usually less than 1.5 g/cm^3. Particle densities of samples of surface soil containing humus in fair quantities are commonly between 2.5 and 2.6 g/cm^3.

Information on particle density is needed for estimates of porosity, air-filled voids, settling rates of particles in fluids, and transport of particles by wind or water. Particle density of a soil sample is measured by first determining its mass after drying to 105°C and then dividing that mass by the volume of the particles, excluding spaces among them. Rather precise measurements can be made with a good balance and a volumetric flask (Blake, 1965). A specially designed flask, called a *pycnometer,* can be filled to a precise volume because it has a ground-glass stopper pierced by a capillary opening through which water in excess of the volume overflows. Use of an ordinary volumetric flask results in a larger error of estimate.

Four weights are needed: weight of oven-dry flask, W_a; weight of flask filled with distilled water at room temperature, W_w; oven-dry weight of flask plus soil sample, W_s; and weight of flask plus soil plus distilled water needed to fill the flask when it contains the soil sample, W_{sw}. Temperature of the water must be known to find the density of the water ρ_w in standard tables. Particle density is then given by the equation:

$$\rho_p = \frac{\rho_w (W_s - W_a)}{(W_s - W_a) - (W_{sw} - W_w)}$$

Preboiled, distilled water stored in a stoppered bottle at room temperature is used in the determination. Any entrapped air can be eliminated by first placing the soil sample in the flask or pycnometer, partially filling it with water, gently boiling the contents with agitation, then stoppering and cooling to room temperature. The flask is then completely filled with water before determining W_{sw}.

About 10 g of dry soil are needed if a 50-ml pycnometer is used. When an ordinary volumetric flask is used instead, a larger sample should also be taken to help offset the reduced precision in filling that flask.

An error of 0.0003 g/cm^3 in the density results from a weighing error of 1 mg on a 10-g soil sample if there is no error in measuring water volume. Greater errors result from inexact measurements of liquid volume in volumetric flasks. If W_{sw} is based on a volume that exceeds the flask volume by 0.2 ml and W_w on a volume 0.2 ml less than the mark on the flask, the compounded particle density error is 0.07 g/cm^3 on a 40-g soil sample.

GEORGE R. BLAKE

References

Blake, G. R., 1965. Particle density, *in* C. A. Black, ed., *Methods of Soil Analysis, Part I.* Madison: American Society of Agronomy, 371–373.

Kretz, R., compiler, 1974–75. Physical constants of minerals, *in* R. C. Weast, ed., *Handbook of Chemistry and Physics,* 55th ed. Cleveland: Chemical Rubber Co., B193–197. (Any recent edition.)

Cross-references: *Bulk Density; Soil Mechanics.*

PARTICLE-SIZE DISTRIBUTION

Particle-size distribution of soils and sediments is important to many of their properties. Whether soils consist of sand, clay, or some mixture of those and silt, the size distribution affects the movement and retention of water, consistency and tilth, and capacity to shrink and swell—to name but a few properties. Moreover, particle-size distribution is influenced little by tillage or other manipulation unless it is drastic. Most soil classification systems therefore use particle-size distribution as one criterion.

Early Work in Particle-size Analysis

Attempts to measure particle-size distribution of soils date back to the seventeenth century. Keen (1931) mentions the investigations by Houghton reported in "A Collection of Letters for the Improvement of Husbandry and Trade" dated 1681. He mixed known weights of soil and water and then decanted the suspension into a second container. After materials in the two containers had dried, he weighed them and compared relative amounts of coarse and fine materials. Additional studies showed that the fine materials held more water than the coarse ones. Sir Humphrey Davy of England standardized Houghton's method and described it in "Elements of Agriculture," published in 1813 (Keen, 1931). Davy also realized that particle-size distribution affected the chemical properties of soils. Oddly enough, sieves were not used in these early investigations, although the Greeks and Romans had used them in mining operations from about 150 B.C. From these early beginnings of measuring particle-size distribution, many systems of size limits have been developed, eleven of which are shown in Figure 1.

Particle-size Limits in American Soil Science

The U.S. Department of Agriculture (USDA) system (Soil Survey Staff, 1951) is the standard for the National Cooperative Soil Survey and is used in all soil science related to agriculture in the United States. The system is based on the fine earth (less than 2.0 mm in diameter) and divides that into seven size classes. These classes are actually subdivisions of a continuum, and the changes in characteristics are therefore gradual across the full range. Subdividing the continuum is necessary, however, to permit the identification and description of soil materials. The current size limits in the USDA system are given as the first line in Figure 1.

The seven classes in the system evolved partly by accident. Thus, for example, the size limits for the subdivisions of sand were originally used by Osborne (1887) because they fit his eyepiece micrometer (Hopkins, 1899). The early upper limit of 0.005 mm for clay was set for a similar reason; measuring diameters of

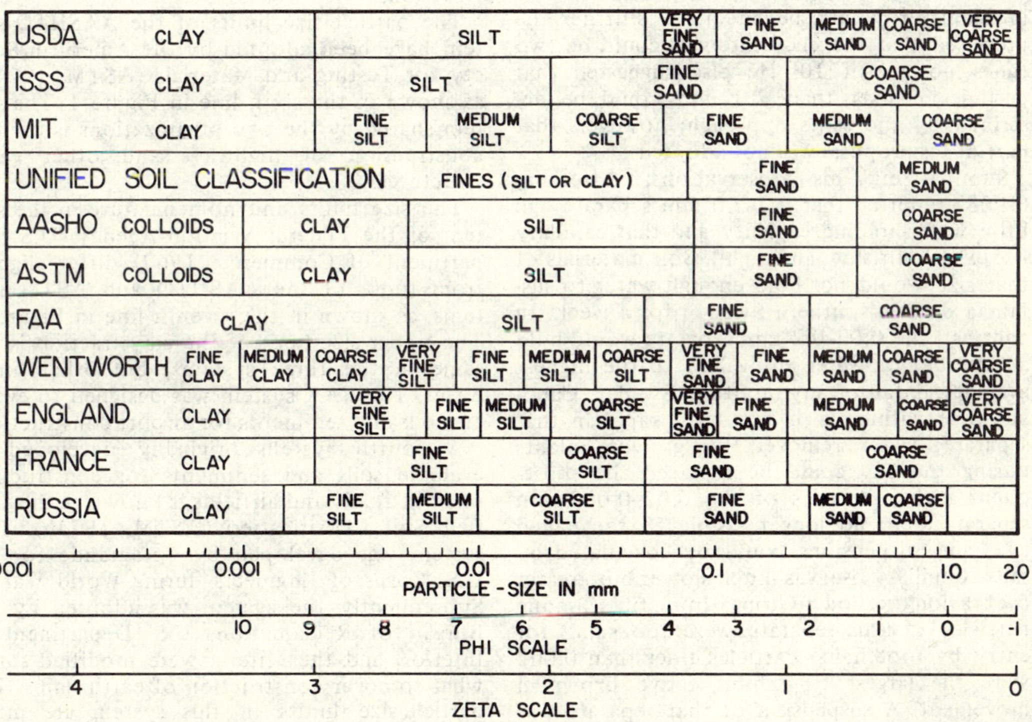

FIGURE 1. Particle-size class limits in several systems.

smaller particles was troublesome. The limit of 0.005 mm was replaced by 0.002 mm 40 years ago (Shaw and Alexander, 1936). Dispersing of samples was then more complete, and it had also been learned that proportions of primary minerals such as feldspars dropped off in soil materials finer than 0.002 mm. Moreover, that limit had been adopted earlier by the International Society of Soil Science. The agricultural significance of particle-size limits in the USDA system, as well as those in some other systems, had been discussed in a report of a Committee on Particle Size and Distribution (1967) of the Soil Science Society of America.

Particle-size Limits of the International Society of Soil Science

The size limits adopted by the International Society of Soil Science (ISSS) in 1913 are shown as the second line in Figure 1. These limits were proposed by Atterberg (1905) on the basis of studies in southern Sweden. He first examined the behavior of many soil separates with narrow size ranges, e.g., 1–2 mm, 0.1–0.2 mm, 0.02–0.05 mm, and 0.001–0.002 mm. Measurements were made of water-holding capacity, rate of capillary rise, tendency to coagulate in suspension, and Brownian movement. From these studies, Atterberg concluded that four size classes for fine earth less than 2.0 mm in diameter would be adequate. Further, he proposed that the class intervals could be two times powers of 10. He also suggested that particles coarser than 2.0 mm could be described by the same approach; however, that part of his proposal was not adopted.

Summarizing his observations, Atterberg (1908) reported that 0.2–2.0 mm separate had little water-holding capacity and that capillary rise was both slow and small. Soil materials of that size would not hold enough water to sustain a crop in southern Sweden for a week. In contrast, the 0.02–0.2 mm separate would hold an amount of water equivalent to the highest average rainfall of any month in Sweden. Capillary rise within 24 h was most rapid in that separate; water removed by growing plants during the day could be replaced the subsequent night. Particles of the 0.02–0.002 mm separate were no longer visible to the naked eye, and suspensions could be coagulated by salts. Capillary rise was much slower but greater over a long period of time. Interstices among particles of this separate were too small for entry by root hairs. Particles finer than 0.002 were the largest to exhibit active Brownian movement. A suspension of that separate was readily coagulated by salt solutions, and capillary rise was very slow.

Atterberg (1908) also proposed that each of the four size classes of soil particles could be subdivided if necessary, suggesting that limits of six as well as two times powers of 10 would be suitable. This proposal was not adopted by the International Society of Soil Science, but it has been revived within the last 30 years (Terzaghi and Peck, 1948) and is now part of the MIT classification of particle sizes. This is shown as the third line in Figure 1. The ISSS and MIT systems have size classes that can be represented conveniently in logarithmic graphs, which are useful for some purposes.

Particle-size Limits in Civil Engineering and Geology

The American Association of State Highway and Transportation Officials (AASHTO) had long used particle-size distribution in designing subgrades for roads and highways (Highway Research Board, 1945; AASHTO, 1974). Size limits in the AASHTO system, shown as the fifth line in Figure 1, were adapted from those of the USDA system many years ago. The upper size limit of 0.001 mm for colloids goes back to a subdivision of the clay fraction when its upper size limit was 0.005 mm in the USDA system. Several subdivisions of the gravel fraction (>2.0 mm) are made in the AASHTO system to aid in evaluating aggregate for construction.

The particle-size limits of the AASHTO system have been adopted by the American Society for Testing and Materials (ASTM, 1974*a*), as shown in the sixth line in Figure 1. The system shared by the two organizations is used in construction of highways and other earth structures.

The size limits and nomenclature in the system of the Federal Aviation Agency (U.S. Department of Commerce, 1962) differ slightly from those of the AASHTO and ASTM systems, as shown in the seventh line in Figure 1. The upper size limit of the clay fraction is the same in the three systems, but other limits differ. The FAA system was designed to evaluate soils and sediments for airport construction.

A fourth system originally developed to evaluate soils and sediments for construction of both roads and airfields is known as the Unified Soil Classification (ASTM, 1974*b*). This system was developed by Casagrande for the U.S. Corps of Engineers during World War II. Subsequently the system was adopted by the Bureau of Reclamation, U.S. Department of Interior, and the criteria were modified somewhat to cover construction of earth dams. The particle-size limits in this system are much the same as those of the AASHTO, ASTM, and FAA systems for the coarser fractions, but the finer fractions are lumped together, as shown in the fourth line in Figure 1. The silt and clay fractions are not separated. Furthermore, a diameter of 0.074 mm was set as the upper limit for fine-grained materials with undesirable engineering properties (ASTM, 1974*b*; U.S. Department of Interior, 1974).

The Wentworth particle-size classification is used primarily in sedimentology. The size limits are shown as the eighth line in Figure 1. Introduced originally by Udden and modified by Wentworth (Pettijohn, 1975), the system has a logarithmic scale with one limit set at 1 mm and others successively halved or doubled. A logarithmic scale facilitates certain analyses of data.

Krumbein (1936) represented the geometric intervals of the Wentworth scale on an arithmetic scale by means of the relationship:

$$\text{Phi scale number} = -\log_2 (\text{particle-size limit in mm})$$

The resulting scale is shown near the bottom of Figure 1. Krumbein also recommended a change in the ISSS system to what he called the Zeta scale to represent geometric intervals on an arithmetic scale:

$$\text{Zeta scale number} = 0.301 - \log_{10} (\text{particle-size limit in mm})$$

The Zeta scale is shown at the bottom of Figure 1. The transformations provide a better framework for statistical analysis.

Particle-size Limits Used in Europe

West-European soil scientists use several systems of particle-size limits. A discussion of these systems is given in a report compiled by the West-European Working Group on Soil Structure (1967), International Society of Soil Science. Although the report deals primarily with methods for measuring other physical and chemical and mineralogical properties of soils, particle-size limits are given in many cases. Figure 1 shows three sets of particle-size limits used in Europe to indicate differences from the USDA and ISSS systems.

The Soil Survey of England and Wales uses particle-size limits that include those of the ISSS system and provides additional subdivisions (Hodgson, 1974). A coarser upper limit is also used for the silt fraction (0.06 rather than 0.02 mm). These size limits are shown as the ninth line in Figure 1. The spans of particle-size classes in England and Wales provide almost equal divisions on a logarithmic scale.

Like the system in England and Wales, the Soil Science Laboratory in Versailles, France, uses particle-size limits that include those of the ISSS system but provide additional subdivisions, as shown in the tenth line in Figure 1. The upper size of silt is set at 0.05 rather than at 0.02 mm limit of the ISSS system. In this respect, the French and USDA systems are the same.

Several systems of particle-size limits used in the Soviet Union were reported by Kachinskii (1958). The system most commonly used between 1937 and 1957 is shown as the eleventh line in Figure 1. The total number of size classes in that system matches the one of the USDA, but the size limits themselves differ. Kachinskii (1958) considered size fractions arbitrary separations for convenience rather than direct reflections of geologic processes or direct indicators of other soil properties. He also thought that the ISSS system did not have enough classes to give a good picture of particle-size distribution in soils.

Soil Texture Classes in the United States

Most soils and sediments are mixtures of particles of various sizes. To facilitate the description of these mixtures, classes have been defined according to relative proportions of sand, silt, and clay. The most common set of classes used by American soil scientists are given by the USDA *texture triangle* (Soil Survey Staff, 1951), shown in Figure 2. These classes were defined for their practical value in agriculture. Thus, for example, sands and sandy loams are set apart because the latter retain more water and are less droughty than the former. Clays are more difficult to plow and prepare for crops than loams. Parallel distinc-

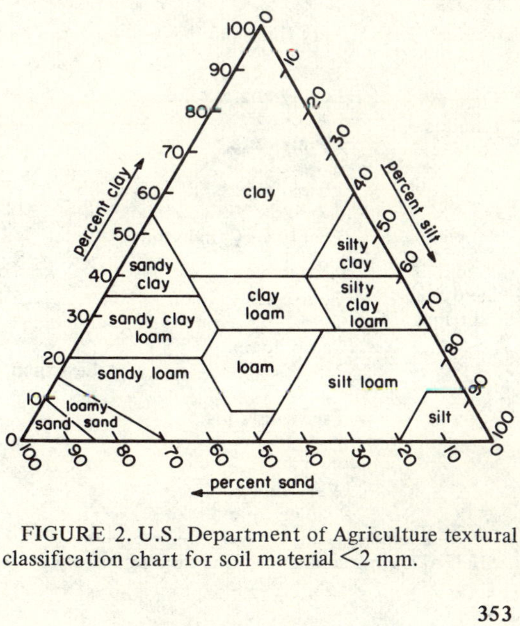

FIGURE 2. U.S. Department of Agriculture textural classification chart for soil material <2 mm.

tions apply generally. Names for the texture classes other than subdivisions of sands and sandy loams are given in Figure 2. One texture class in the system is called *loam,* which refers to soil material with certain proportions of sand, silt, and clay. That meaning is much different from the one that was long popular in English, namely, a dark, well-structured, and humus-rich soil. Texture class subdivisions have been devised for sands and sandy loams in addition to those that can be shown in the triangle. The names and definitions of these additional texture classes are given in Table 1.

Field estimates of texture are related to laboratory determinations of particle-size distribution by means of the texture triangle. Field soil scientists estimate the textures of soil materials by working them between their fingers. Accurate estimates require practice and periodic checks on samples of known texture. General descriptions of the field characteristics of soil materials of major texture classes, prepared many years ago by C. F. Shaw of California and reprinted in the *Soil Survey Manual* (Soil Survey Staff, 1951), are given below:

Sand: Sand is loose and single grained. The individual grains can be readily seen or felt. Squeezed in the hand when dry, it will fall apart when the pressure is released. Squeezed when moist, it will form a cast but will crumble when touched.

Sandy loam: A sandy loam is a soil containing much sand but which has enough silt and clay to make it somewhat coherent. The individual sand grains can readily be seen and felt. Squeezed when dry, it will form a cast that will readily fall apart, but squeezed when moist, it will form a cast that will bear careful handling without breaking.

Loam: A loam is a soil having a relatively even mixture of different grades of sand and of silt and clay. It is mellow with a somewhat gritty feel yet fairly smooth and slightly plastic. Squeezed when dry, it will form a cast that will bear careful handling, while the cast formed by

TABLE 1. Percentage of Sand Sizes in Subclasses of Sand, Loamy Sand, and Sandy Loam Basic Textural Classes as Defined by the U.S. Department of Agriculture.

		Soil Separates				
Basic Soil Class	Subclass	Very Coarse Sand, 2.0–1.0 mm	Coarse Sand, 1.0–0.5 mm	Medium Sand, 0.5–0.25 mm	Fine Sand, 0.25–0.1 mm	Very Fine Sand, 0.1–0.05 mm
Sands	Coarse sand	25% or more		Less than 50%	Less than 50%	Less than 50%
	Sand	25% or more			Less than 50%	Less than 50%
	Fine sand		Less than 25%	-or-	50% or more	Less than 50%
	Very fine sand					50% or more
Loamy Sands	Loamy coarse sand	25% or more		Less than 50%	Less than 50%	Less than 50%
	Loamy sand		25% or more	Less than 50%	Less than 50%	Less than 50%
	Loamy fine sand		Less than 25%	-or-	50% or more	Less than 50%
	Loamy very fine sand					50% or more
Sandy Loams	Coarse sandy loam	25% or more		Less than 50%	Less than 50%	Less than 50%
	Sandy loam	Less than 25%	30% or more	-and-	Less than 30%	Less than 30%
	Fine sandy loam		-or- Between 15 and 30%		30% or more	Less than 30%
	Very fine sandy loam		Less than 15%	-or-	More than 40%*	30% or more

*Half of fine sand and very fine sand must be very fine sand.

squeezing the moist soil can be handled quite freely without breaking.

Silt loam: A silt loam is a soil having a moderate amount of the fine grades of sand and only a small amount of clay, over half the particles being of the size called "silt." When dry, it may appear cloddy, but the lumps can be readily broken, and when pulverized it feels soft and floury. When wet, the soil readily runs together and puddles. Either dry or moist, it will form casts that can be handled freely without breaking, but when moistened and worked between thumb and finger, it will not "ribbon" and will give a broken appearance.

Clay loam: A clay loam is a fine-textured soil that usually breaks into clods or lumps that are hard when dry. When the moist soil is worked between the thumb and finger, it will form a thin "ribbon" that will break readily, barely sustaining its own weight. The moist soil is plastic and will form a cast that will bear much handling. When kneeded in the hand, it does not crumble readily but tends to work into a heavy compact mass.

Clay: A clay is a fine-textured soil that usually forms very hard lumps or clods when dry and is quite plastic and usually sticky when wet. When the moist soil is worked between the thumb and fingers, it will form a long, flexible "ribbon." Some fine clays very high in colloids are friable and lack plasticity in all conditions of moisture.

Not all the texture classes used in the United States are covered by the preceding descriptions. The nature of the undescribed classes can be inferred, however, from their names, e.g., sandy clay loam. That class is a clay loam containing more sand than normal for clay loams.

The boundaries between adjoining soil texture classes are limits in a continuum. Some soil materials consequently have textures that are at or near the shared limit of two classes. Practice plus continuing evaluation of field estimates are necessary for determination of the textures of such soil materials. Even then, errors are possible, but fortunately placing a given specimen in one or the other of adjacent classes is not a serious error.

Particle-size Classes at the Family Level in Soil Taxonomy

In the USDA soil classification system (Soil Survey Staff, 1975), efforts were made to define the family classes so that they would have meaning to both plant growth and engineering. The principal criteria for family classes are: particle size, mineralogy, reaction, soil temperature, and soil depth. Other characteristics are used as criteria in some subgroups.

The particle-size classes at the family level differ from soil texture classes for several reasons. The objectives to be served by the two systems are not the same. Furthermore, particle-size classes for families center on the deeper profile, a specified vertical thickness called the *control section,* rather than on any sample of soil material. Another distinction follows from consideration of particles of all sizes rather than the fine earth below a diameter of 2.0 mm, as is done in soil texture classes. The control section, as defined for the suborders and great groups in the classification system, is not uniform across the board. Most commonly, the control section is that part of the profile between depths of approximately 25 and 100 cm (10 and 40 in.).

A further difference between the particle-size classes for families and the soil texture classes is the disposition of the very fine sand fraction. Franzmeier et al (1960) found that soil materials with an upper limit of 0.1 mm, which include the very fine sand separate, would hold water at tensions usable by plants. Coarser separates would have pores too large to hold usable water. Consequently, the very fine sand fraction is included with coarser separates in the fine sand, loamy fine sand, and coarser textured classes. The very fine sand separate is included with silt in very fine sand, loamy very fine sand, sandy loam, silt loam, and finer textured classes. The application of these conventions makes the effective particle-size limit between sand and silt for the family classes much more nearly the same as that of 0.074 mm widely used by civil engineers. Definitions of the particle-size classes are as follows:

Fragmental: Stones, cobbles, gravel, and very coarse sand particles; too little fine earth to fill interstices larger than 1 mm in diameter.

Sandy-skeletal: Rock fragments 2 mm in diameter or larger make up 35% or more by volume; enough fine earth to fill interstices larger than 1 mm; the fraction finer than 2 mm is sandy as defined for the sandy particle-size class.

Loamy-skeletal: Rock fragments make up 35% or more by volume; enough fine earth to fill interstices larger than 1 mm; the fraction finer than 2 mm is loamy as defined for the loamy particle-size class.

Clayey-skeletal: Rock fragments make up 35% or more by volume; enough fine earth to fill interstices larger than 1 mm; the fraction finer than 2 mm is clayey as defined for the clayey particle-size class.

Sandy: The texture of the fine earth is sand or loamy sand but not loamy very fine sand or very fine sand; rock fragments make up <35% by volume.

Loamy: The texture of the fine earth is loamy very fine sand, very fine sand, or finer, but the amount of clay is <35%; rock fragments are <35% by volume.

Coarse-loamy: By weight, 15% or more of the particles are fine sand (diameter 0.25–0.1 mm) or coarser, including fragments up to 7.5 cm in diameter; <18% clay in the fine-earth fraction.

Fine-loamy: By weight, 15% or more of the particles are fine sand or coarser; including fragments up to 7.5 cm in diameter; 18 through 35% clay in the fine-earth fraction (<30% in Vertisols).

Coarse-silty: By weight, <15% of the particles are fine sand or coarser, including fragments up to 7.5 cm in diameter; <18% clay in the fine-earth fraction.

Fine-silty: By weight, <15% of the particles are fine sand or coarser, including fragments up to 7.5 cm in diameter; 18 through 34% clay in the fine-earth fraction (<30% in Vertisols).

Clayey: The fine earth contains 35% or more clay by weight, and rock fragments are <35% by volume.

Fine: A clayey particle-size class that has 35 through 59% clay in the fine-earth fraction (30 through 59% in Vertisols).

Very-fine: A clayey particle-size class that has 60% or more clay in the fine-earth fraction.

The class limits and names for the particle-size classes at the family level are shown by a triangular graph in Figure 3. The system provides for either seven or eleven classes, the choice depending on which better sets apart families that differ in usefulness to one or both of agriculture and engineering. Certain classes such as coarse-loamy, fine-loamy, coarse-silty, and fine-silty are optional in some subgroups. Full definitions and guides for use of the family classes are given in the monograph (Soil Survey Staff, 1975).

A set of modifiers can be used with the names of some great groups and subgroups to eliminate the need for particle-size classes at the family level. The mineralogy is also a criterion in classes above the family level for soils derived from volcanic ash.

Because of their effects on moisture relationships, strongly contrasting particle-size classes within the control section form bases for family distinctions. The break in materials introduces differences in pore sizes, which in turn alter

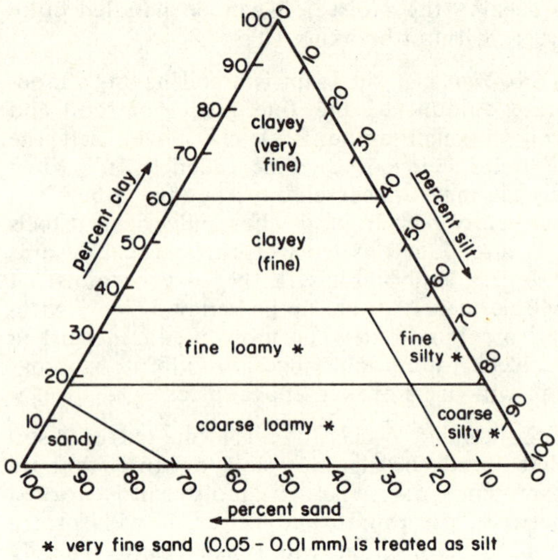

FIGURE 3. U.S. Department of Agriculture particle-size classes for the fine-earth fraction at the family level.

water movement and retention. At present, 40 combinations of contrasting materials within the control section are used in setting apart families within the United States. This total consists of combinations that have been found; others are possibilities.

Soil Texture Classes in Other Systems

Various texture classes differing from those used in soil science in the United States are used in engineering operations and in other countries. The AASHTO, ASTM, and FAA systems and the Unified Soil Classification, applied in the design and construction of one or more of roads, airports, and earth dams, group soils partly on the basis of particle-size distribution and partly on the basis of the Atterberg plastic and liquid limits.

Additional and different systems of texture classes are used in soil science in other countries. Names much like those in the United States are used in Australia, England and Wales, and New Zealand, but each country has its own set of definitions. Australia and New Zealand use the ISSS size limits in defining texture classes (Northcote, 1971; Taylor and Pohlen, 1962). The Soil Survey of England and Wales includes the ISSS size limits but adds others for defining texture classes (Hodgson, 1974). Several systems of texture classes have been used in the Soviet Union (Kachinskii, 1958). More examples of differing systems could be included, but these four indicate the current variety in soil texture classes throughout the world.

Particle-size Analyses

Particle-size distribution of soil samples is determined partly by sieving and partly by sedimentation. The sand fractions are separated from the silt and clay by sieving, which is also used to determine proportions of the size fractions of sand. The silt and clay fractions are determined by sedimentation, measurements being made with either a pipette or hydrometer (Day, 1965). Separation of the sand from the silt and clay may be done either before or after determination of the two latter fractions.

One or more pretreatments are normally used for particle-size analyses, also known as *mechanical analyses.* For samples containing organic matter, some treatment to destroy it is usually a first step. Samples must also be treated with compounds such as sodium metaphosphate to effect dispersion.

Dry sieving of sand to determine the proportions of the different size fractions is done with a nest of sieves having the proper openings. The National Bureau of Standards and the American Society for Testing and Materials have adopted specifications that ensure consistency among sieves, but careful measurement and periodic inspection of sieve openings are still desirable. Difficulties due to shapes of sieve openings and soil particles and to time of shaking cannot be eliminated. Consequently, a standard procedure must be followed to obtain comparable results. Specific techniques are given by Day (1965) and the Soil Conservation Service (1972).

Pipette and Hydrometer Methods

The pipette technique was introduced in the early 1920s by Jennings and Robinson (Day, 1965). The method involves extracting a known volume of a suspension and dispersing solution to measure density of the suspension at a specific depth after the critical particle-size fraction has settled to that depth according to *Stokes law.* Stokes law is an equation that relates the velocity at which a spherical particle of a particular diameter falls in a medium. The equation is given in this form:

$$V = \frac{2}{9} \frac{(d_p - d_l)gr^2}{\mu}$$

where V is velocity of fall in cm/s, d_p is the density of the particles (usually assumed to be about 2.6 g/cm^3 for soils), d_l is the density of the liquid, g is the constant of gravity, r is the radius of the particle, and μ is the viscosity of the liquid. The material extracted in the pipette is weighed, corrected for the weight of the dispersing reagent, and the total weight of clay is calculated from the amount in the aliquot. Specific techniques are given by Day (1965) and the Soil Conservation Service (1972). The pipette method is the standard technique used to characterize soils.

The hydrometer method is also an attempt to measure density of a suspension at a particular depth. To do this requires a hydrometer of a particular design and a controlled laboratory procedure that takes into account the temperature and time of sedimentation. This method also depends on Stokes law. Procedures for the hydrometer method are given by Day (1965) and by ASTM Standards (1974*a*). The hydrometer method is used mostly by engineers to determine particle-size distribution.

Neither method is completely free of difficulties, but they are the best developed so far to measure particle-size distribution. A comparison of the two methods by Liu et al (1966) showed that with few exceptions, similar data were obtained by the two methods. In general, the hydrometer analyses showed slightly higher clay contents than the pipette analyses.

A review of techniques used by soil scientists in Europe to measure particle-size distribution is found in a report from the West-European Working Group on Soil Structure (1967). Russian methods for particle-size analysis are reviewed by Kachinskii (1958).

GARY C. STEINHARDT

References

American Association of State Highway and Transportation Officials, 1974. Terms relating to subgrade, soil aggregate, and fill materials. AASHTO Designation: M146-70, *in Standard Specifications for Transportation Materials and Methods of Sampling and Testing,* Part 1, 685p.

American Society for Testing and Materials, 1974*a*. Standard method for particle-size analysis of soils. ASTM Standard D422-63, *in 1974 Annual Book of ASTM Standards,* Part 19, 464p.

American Society for Testing and Materials, 1974*b*. Method for classification of soils for engineering purposes. ASTM Standard D2487-69, *in 1974 Annual Book of ASTM Standards,* Part 19, 464p.

Atterberg, A., 1905. Die rationelle klassifikation der sande und kiese, *Chemiker Zeitung,* **29**, 195-198.

Atterberg, A., 1908. Studien auf dem Gebiete der Bodenkunde, *Landwirtschaftliche Versuchs-Stationen,* 69, 93-143.

Committee on Particle Size and Distribution, 1967. Considerations relative to a common particle-size scale of earthy materials, *Soil Sci. Soc. America Proc.,* **31**, 579-584.

Day, P. R., 1965. Particle fractionation and particle-size analysis, *in* C. E. Black, ed., *Methods of Soil Analysis, Part 1. Am. Soc. of Agron. Mon. No. 9,* 545-567.

Franmeier, D. P.; Whiteside, E. P.; and Erickson, A. E., 1960. Relationship of texture classes of fine earth to readily available water, *Proc. Seventh Internat. Cong. Soil Sci. (USA)*, **1**, 354–363.
Highway Research Board, 1945. Classification of soils and sub-grade materials for highway construction, *Proc. 25th Ann. Meeting Highway Research Board*, 376–382.
Hodgson, J. M., ed., 1974. *Soil Survey Field Handbook: Describing and Sampling Soil Profiles*, Soil Survey of England and Wales Tech. Mon. No. 5, 22–25.
Hopkins, C. G., 1899. A plea for scientific basis for the division of soil particles in mechanical analysis, *U.S. Depart. Agric., Div. Chemistry Bull. No. 56*, 64–66.
Kachinskii, N. A., 1958. *Mechanical and Microaggregate Composition of Soils: Methods of Investigation*, translated from Russian, 1966. Jerusalem: Israel Program for Scientific Translations, 134p.
Keen, B. A., 1931. *The Physical Properties of the Soil*. London: Longmans, 381p.
Krumbein, W. C., 1936. The application of logarithmic moments to size frequency distribution of sediments, *Jour. Sed. Petrology*, **6**, 35–47.
Liu, T. K.; Odell, R. T.; Etter, W. C.; and Thornburn, T. H., 1966. A comparison of clay contents determined by hydrometer and pipette methods using reduced major axis analysis, *Soil Sci. Soc. America Proc.*, **30**, 665–669.
Northcote, K. H., 1971. *A Factual Key for the Recognition of Australian Soils*. Glenside, S. A., Australia: Rellim Technical Publications, 26–29.
Osborne, T. B., 1887. The methods of mechanical soil analysis, *Conn. Agric. Exp. Sta. Ann. Rept.*, 1887, 141–158.
Pettijohn, F. J., 1975. *Sedimentary Rocks*, 3rd ed. New York: Harper, 27–31.
Shaw, T. M. and Alexander, L. T., 1936. A note on mechanical analysis and soil textures, *Soil Sci. Soc. America Proc.*, **1**, 303–304.
Soil Conservation Service, 1972. Soil survey laboratory methods and procedures for collecting soil samples, *U.S. Depart. Agric. Soil Survey Invest. Rept. No. 1*, 63p.
Soil Survey Staff, 1951. Soil Survey Manual, *U.S. Department of Agriculture Handbook No. 18*, 503p.
Soil Survey Staff, 1975. Soil taxonomy: A basic system of soil classification for making and interpreting soil surveys, *U.S. Department of Agriculture Handbook No. 436*, 754p.
Taylor, N. H., and Pohlen, I. J., 1962. "Soil survey method," *New Zealand Soil Bur. Bull. No. 25*, 82–89.
Terzaghi, K., and Peck, R. B., 1948. *Soil Mechanics in Engineering Practice*. New York: Wiley, 566p.
U.S. Department of Commerce, Federal Aviation Agency, 1962. *Airport Paving*. Washington, D.C.: U.S. Government Printing Office, 187p.
U.S. Department of Interior, Bureau of Reclamation, 1974. *Earth Manual*, 2d ed. Washington, D.C.: U.S. Government Printing Office, 810p.
West-European Working Group on Soil Structure, Internat. Soc. Soil Sci., 1967. *West-European Methods for Soil Structure Determination*. Ghent, Belgium: Faculty of Agricultural Sciences, State University, 529p.

Cross-references: *Mineral Analysis; Pore-size Distribution; Soil Mechanics. Vol. IVB: Soil Minerals. Vol. XII, Pt. 2: Field Texture. Vol. XIII: Unified Soil Classification System.*

PERCOLATION

Percolation has been qualitatively defined by soil scientists as the downward movement of water through soil in a saturated or nearly saturated state at hydraulic gradients of 1.0 or less (Soil Science Society of America, 1975). This definition implies that percolation of water is primarily controlled by the force of gravity at soil moisture contents greater than field capacity. Soil physicists would use *hydraulic conductivity* (q.v.) and soil water potential to describe percolation in more quantitative terms.

The speed at which water percolates depends mainly on the characteristics of the pore space (see *Pore Space, Drainable; Soil Pores*) in the soil through which it flows and to a lesser extent on the characteristics of the water itself (Kohnke and Bertrand, 1959). Water percolates by gravity through large pores and is retained by capillary forces in small pores. Pore-size distribution (q.v.) is therefore highly important. Pore sizes, however, can change in some soils due to the shrinking and swelling of soil clays and destruction of non-water-stable soil aggregates. To be an effective transmitter of water, the large pores must provide a continuous channel, however tortuous its path. Some soils have sufficiently large pore spaces, but collectively the channels they create are blind passages within the matrix of the structural units of the soil. Soils with layers of contrasting textures (fine over coarse) and structures (hardpan) also have contrasting pore sizes that markedly influence the percolation rate and spatial distribution of percolating water (Hill and Parlange, 1972).

Percolation can also be influenced by properties of the water itself. The effects of density, viscosity, and surface tension, although small by comparison to the effects of the porous medium, must be considered in very precise measurements of water movement. Viscosity effects are directly related to temperature and dissolved salts.

The most popular concept of percolation is that used by the sanitary engineer to evaluate the capacity of a soil to accept effluent from a septic tank drainfield. The percolation test has been devised to provide a quantitative measure of the rates of water movement (q.v.) in the soil from a test pit or auger hole. The percolation rate, expressed in terms of minutes per inch or inches per hour is a measure of the time for a standard head of water to drop 1 in. in a test hole after proper presoaking of the soil surrounding the hole. Contrary to definitions stated earlier, the flow of water from the percolation test hole is not confined to downward drainage but includes lateral and upward movement by capillary flow. Soil scientists would define the flow as a measure of three-dimensional infiltration (q.v.) under unsaturated flow where potential gradients are not restricted to values less than one (see *Flow Theory*). The percolation rate is first influenced by the condition of the wall of the test hole through which the water must infiltrate and second, the soil factors that control the speed of the wetting front as water moves through the soil (Hill, 1966). Moisture content is an important factor controlling the speed of the wetting front (q.v.). The front moves more slowly through the wet soil of early spring than the dry soil of summer. Seasonal variations in rates have been observed and can be influenced by cracks that develop in soils that shrink and swell with changing moisture content. Percolation rates measured in test holes may also vary according to the size and geometry of the test hole (Healy and Laak, 1972).

Thus the term *percolation* has both qualitative and quantitative connotations, and the use of the term by different disciplines may have different meanings. The manner in which water percolates through the soil is highly important, however, for it governs our ability to drain agricultural soils and recharge our water tables, and it controls the operation of a septic tank system.

DAVID E. HILL

References

Healy, K. A., and Laak, R., 1972. Factors affecting the percolation test, *Civil Engr. Dept. Report.*, Storrs: University of Connecticut, 26pp.

Hill, D. E., 1966. Percolation testing for septic tank drainage, *Conn. Agric. Exp. Sta. Bull. 678*, 25p.

Hill, D. E., and Parlange, J.-Y., 1972. Wetting front instability in layered soil, *Soil Sci. Soc. America Proc.*, 36, 697–702.

Kohnke, H., and Bertrand, A. R., 1959. *Soil Conservation.* New York: McGraw-Hill, 298p.

Soil Science Society of America 1975. *Glossary of Soil Science Terms.* Madison: Soil Science Society of America, 34p.

Cross-references: *Capillary Pressure; Conductivity, Hydraulic; Flow Theory; Infiltration; Permeability; Pore-size Distribution; Water Movement; Wetting Front.*

PERMEABILITY

Permeability is a property of a soil (or any porous medium) that permits the transmission of fluids. This property, denoted by k, is also called *intrinsic permeability* to stress that k is only a property of the soil and is independent of fluid properties such as viscosity or density. The *permeability* is related to the *hydraulic conductivity* (K) by the relation

$$K = \frac{k\rho g}{\mu} = \frac{kg}{\nu} \qquad (1)$$

where ρ is fluid specific mass, g is the acceleration of gravity, μ is the (dynamic) viscosity, and ν is the kinematic viscosity. Because K depends on both soil and fluid properties, it measures the mobility of a given fluid in a given porous medium. Both k and K are macroscopical properties distinctly defined by the two versions of Darcy's law:

$$q = -K\nabla\phi; \qquad q = -\frac{k}{\mu}\nabla(p + \gamma z) \qquad (2)$$

where q is the specific flux vector, $\nabla\phi$ the gradient of the total hydraulic head, z is the vertical coordinate oriented positive upward, and γ is the specific weight. The permeability k is a univalued scalar for an isotropic incompressible porous medium with stable structure. For anisotropic porous medium, equation 2 is valid if K and k are understood as second-order tensors.

In unsaturated soils, equation 2 is still generally used with ϕ defined for water by the relation:

$$\phi = \psi + z + \frac{P_a}{\gamma} \qquad (3)$$

where ψ is the *capillary head*, and P_a is the air pressure (defined equal to zero when atmospheric pressure prevails). In this case K can be considered a function of the *moisture content* (θ), or ψ, for a given soil. The generalized version of equation 2 for the unsaturated case is

$$q = -K(\theta)\nabla\left[\psi(\theta) + z + \frac{P_a}{\gamma}\right]$$

$$q = -\frac{k(\theta)}{\mu}\nabla\left[\gamma\psi(\theta) + \gamma z + P_a\right] \qquad (4)$$

Frequently the *degree of saturation* ($S = \theta/\theta_s$ where θ_s is the saturated values of θ) is used as the independent variable instead of the moisture content. It is commonly assumed that $k(S)$ relates to the saturated value k_s by a unique function, the so called *relative permeability* k_r

$$k_r = \frac{k(S)}{k_s} \tag{5}$$

It is generally assumed that k_r is a scalar function even for anisotropic soils whereas k is a tensor. Note also that from the relation of equivalence, equation 1, the *relative permeability* is equal to the *relative hydraulic conductivity*, namely, $k_r = K/K_s$.

Empirical Formulas

The $K(S)$ relationship constitutes a hydraulic property, which together with the corresponding $\psi(S)$ function, completely characterizes the flow system (see equation 4). Therefore, simple mathematical formulas adjustable to experimental data are sometimes very useful because they simplify computation. They may permit solution of unsaturated flow problems in analytical closed form and minimize the truncation error in numerical solutions. Several such empirical formulas are given in Table 1.

It is worthwhile to note that the $K(\psi)$ relationship is known to display a considerable hysteresis (see hysteresis of the hydraulic conductivity). All the empirical formulas for $K(\psi)$ can represent a continuous drainage or wetting process at best. For each process a different formula or different coefficient may be required. Also no single $K(\psi)$ relationship is valid for all types of soils. Moreover, even if a certain formula is suitable for a class of soils, the coefficients may vary considerably from soil to soil. Hence the empirical formulas do not permit in fact accurate predictions of K, and they can be used with confidence only after coefficients have been determined by a curve fitting of experimental measurements. It is generally preferable to use $K(S)$ rather than $K(\psi)$ because the relationship between K and S is less sensitive to hysteresis and in principle better adapted to unsaturated soils.

Quantitative Models

The experimental determination of the relative permeability relationship down to low water contents is laborious, time-consuming, and costly especially when field measurements are concerned. Often only a few data are available for computation thereby reducing considerably, the significance of the rigorous numerical procedures customarily used in solving unsaturated flow problems. Consequently, theoretical models that predict relative permeability are extremely important.

One may distinguish between two main groups of models used for relative permeability prediction. The first one is based on a generalization of Kozeny's approach for saturated porous media. The porous system is viewed as a conduit of irregular shape. The wetting fluid is thought to creep, more or less uniformly, along the entire solid surface while the nonwetting fluid occupies the internal portion. By introducing a "hydraulic radius" parameter, equal to θ/A_s, where A_s is the solid surface area, the relative permeability can be derived as a power function of the effective saturation (S_e),

$$k_r = S_e{}^n; \qquad S_e = \frac{\theta - \theta_r}{\theta_s - \theta_r} \tag{6}$$

where θ_r is the residual saturation (see *Capillary Pressure*). Averjanov (1950) and Yuster (1951) derived equation 6 with n equal to 3.5, 2.0, and 3.0, respectively. This power law formulation is very convenient for obtaining analytical solutions of partially saturated flows. It appears, however, that n is not a universal constant. While values of n in the range of 2.5 to 3.5 may

TABLE 1. Empirical Formulas for the Unsaturated Hydraulic Conductivity.

Reference	Wind (1955)	Gardner (1958)		Gardner's Type
Function	$K = a\lvert\psi\rvert^{-n}$	$K_r = e^{\alpha\psi}$	$K = \dfrac{a}{\lvert\psi\rvert^n + b}$	$K_r = \left[\left(\dfrac{\psi}{b}\right)^n + 1\right]^{-1}$
Reference	Brooks & Corey (1964)			Averjanov (1950)
	$\psi \geqslant \psi_{cr}$	$\psi \leqslant \psi_{cr}$		
Function	$K = K_s$	$K_r = \left(\dfrac{\psi}{\psi_{cr}}\right)^{-n}$		$K_r = S_e{}^n; \quad S_e = \dfrac{\theta - \theta_r}{\theta_s - \theta_r}$
Reference	Rijtema (1965)			King (1964)
	$\psi \geqslant \psi_{cr}$	$\psi_1 \leqslant \psi \leqslant \psi_{cr}$	$\psi < \psi_1$	
Function	$K = K_s$	$K_r = e^{\alpha(\psi - \psi_{cr})}$	$K = K_1\left(\dfrac{\psi}{\psi_1}\right)^{-n}$	$K_r = a\dfrac{\cosh[(\psi/\psi_1)^n] - 1}{\cosh[(\psi/\psi_1)^n] + 1}$

For detailed reference, see Mualem and Dagan (1976).

fit well sandy soils, higher values (up to 8) need to be used to accurately represent heavy soils (loamy and clay soils). Consequently, though equation 6 was derived on a relatively rigorous theoretical basis, the residual moisture content (θ_r) and the power n have to be determined by adjustment of the formula to measured data.

The second group, the "statistical models," includes those of Purcell (1949), Childs and Collis George (1950), Burdine (1953) and Wyllie and Gardner (1958), Farrel and Larson (1972), and Mualem (1976*a*). These models make use of the measured capillary head-moisture content $\psi(\theta)$ curve to derive the relative permeability in the unsaturated state. The basic assumption common to these models is that a porous medium can be characterized by a one-variable density distribution function $f(r)$, where r is length scale, geometrically interpreted as the radius of the pore. Using the mechanical modeling of a porous medium (see *Capillary Pressure*), r is uniquely related to ψ by the capillary law:

$$r = \frac{c}{\psi} \tag{7}$$

where c is a constant coefficient. On this basis the measured $S(\psi)$ curve is viewed as the pore water distribution function with

$$dS(r) = f(r)dr; \qquad S(R) = \int_0^R f(r)dr \tag{8}$$

Naturally, different assumptions regarding the geometrical configuration of the capillary model and the contribution of the elementary pore domain to the total permeability of the pore domain will yield different final formulas of k_r.

The first and the most simple model (Purcel, 1949) considered a porous medium to be analogous to a bundle of uniform, parallel capillaries whose radii distribution are given by $f(r)$. The unsaturated state is represented by assuming that for a given ψ, only tubes with a radius less than $R = \alpha/\psi$ are filled. Capillaries having a radius greater than R remain empty. According to the Hagen-Poiseuille equation, the specific discharge through the bundle of capillaries is

$$q(R) = \frac{g}{8\nu}\frac{d\phi}{dx}\int_{R_{min}}^R r^2 f(r)dr \tag{9}$$

The analogy of (9) and (2) indicate that the contribution of the filled tube to the hydraulic conductivity of the bundle is proportional to the integral in equation 9 leading to:

$$k_r = \frac{\int_{R_{min}}^R r^2 f(r)dr}{\int_{R_{min}}^{R_{max}} r^2 f(r)dr} \tag{10}$$

or as a direct function of ψ and S

$$k_r(S) = \frac{\int_{S_{min}}^S (dS/\psi^2)}{\int_{S_{min}}^1 (dS/\psi^2)} \tag{11}$$

where the subscripts min and max hold for the minimum and the maximum values. This approach has been modified by Fatt and Dykstra (1951) to account for the *tortuosity* of the flow path. They assumed that the tortuosity is inversely proportional to r^b and obtained the formula

$$k_r(S) = \frac{\int_{S_{min}}^S (dS/\psi^{2+b})}{\int_{S_{min}}^1 (dS/\psi^{2+b})} \tag{12a}$$

From comparing their formula with measured results, they found that b is close to 1. They indicated, however, that b may vary with the type of soil. Burdine (1953) accounted for tortuosity by applying a correction factor to equation 11 equal to S_e^2 leading to

$$k_r(S) = S_e^2 \frac{\int_{S_{min}}^S (dS/\psi^2)}{\int_{S_{min}}^1 (dS/\psi^2)} \tag{12b}$$

Considering two-phase flow, (e.g., water and air), a similar equation can be written for the nonwetting fluid:

$$k_{rn}(S) = (1 - S_e)^2 \frac{\int_S^1 (dS/\psi^2)}{\int_{S_{min}}^1 (dS/\psi^2)} \tag{13}$$

Using relative permeability data for both gas and oil, Burdine found that the suggested correction factors considerably improve the prediction of k_r for both the wetting and the nonwetting fluids (Figure 1).

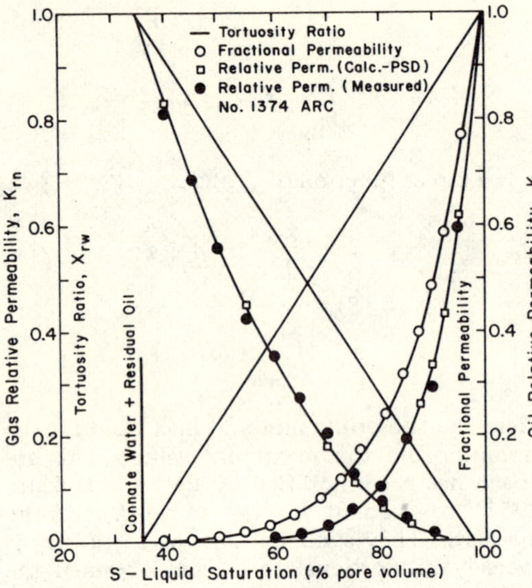

FIGURE 1. Measured and calculated relative permeabilities for liquid and gas as a function of liquid saturation. (After Burdine, 1953)

To take into account the random variation of the pore size, Childs and Collis George (1950) considered a porous column fractured at a plane normal to the direction of flow. The resulting identical faces are then randomly jointed together. The probability that a pore of radius r in one section should be connected to a pore of radius ρ in the other is

$$a(r,\rho) = f(r)f(\rho)dr\,d\rho \tag{14}$$

To simplify the matter, they assumed that the resistance to flow was dominated by the pore of smaller radius. Mualem (1974, 1976*a*) showed that this model leads to the final formula

$$k_r(S) = \frac{\int_{S_{\min}}^{S} [(S-s)/\psi^2]\, ds}{\int_{S_{\min}}^{1} [(S-s)/\psi^2]\, ds} \tag{15}$$

where s is a dummy variable of integration representing saturation as a function of ψ. This model has been discussed and experimentally tested by Marshall; Millington and Quirk; Jackson et al; Brutsaert; Kunze, Uehara, and Graham; Green and Corey; Bruce; Reisenauer; and Mualem and Dagan (see review and references in Mualem and Dagan, 1976). The attempt to use the model for a direct computation of the saturated hydraulic conductivity very often yields an error as large as several orders of magnitude. One would have to measure one value (probably K_s) to derive $K(S)$ with the aid of equation 15. Comparison between predicted and measured k_r values shows that better agreement is achieved by multiplying equation 15 with a correction factor $S_e{}^n$. Again no universal value of n has been found compatible to all types of soil. Millington and Quirk (1961) suggested $n = 4/3$ while Kunze, Uehara, and Graham found that using $n = 1$ improved the prediction of equation 15. Using a model similar to that featured by Childs and Collis George, Wylie and Gardner (1958) argued that as a result of the random rejoining of the cut porous slab, the probability of a pore of radius r at one slab to encounter pores of the other slab is $a_e = f(r)\theta/\lambda$, where λ is a constant correction factor. Hence the porosity of the contact interface reduces to θ^2/λ, which is interpreted as if each pore had an effective radius of $r_e = r\sqrt{\theta/\lambda}$. Substitution of a_e and r_e instead of $f(r)$ and r in equation 10 yields Burdine's formula (12b) analytically. Yet the idea of joining two cross-sections and considering a hydraulic conductivity of the intermediate interface is physically questionable because in flow there is no resistance at the cross-section.

Mualem (1976*a*) developed a conceptual modeling of porous media instead of the physical one of Wylie and Gardner. Consider a porous slab of thickness Δx ($x \to x + \Delta x$ along the axis). The pore area distribution at the two slab sides is identical. For $\Delta x \gg R_{\max}$, complete randomness of the relative position of the two slab faces is assumed. The probability of having pores of radii $r \to r + dr$ at x encountered with pores of radii $\rho \to \rho + d\rho$ at $x + \Delta x$ is given by equation 14. Here, no direct connection between the pores r and ρ exists along the x axis. The other extreme case occurs when $\Delta x \to 0$. Then the correlation between the two faces of the slab is complete. Since we are concerned with the effect of pore changes on the hydraulic conductivity, it is more relevant to consider Δx to be of the same order of magnitude as the pore radii. Then the probability of a pore of radius between r and $r + dr$ being connected to a pore of radius between ρ and $\rho + d\rho$ is

$$a(r, \rho) = G(R, r, \rho)f(r)f(\rho)dr \quad d\rho \tag{16}$$

$G(R, r, \rho)$ is a correction accounting for partial correlation between the pores r and ρ at a given water content $\theta(R)$.

The contribution of the actual flow configuration in the slab to the hydraulic conductivity cannot be accurately assessed. Therefore, two simplifying assumptions are used: (i) there is no by-pass flow between the slab pores, and (ii) the pore configuration may be replaced by a pair of capillary elements whose lengths (ℓ) are proportional to their radii $\ell_1/\ell_2 = r/\rho$. The hydraulic conductivity is then found to vary as $r_e^2 = r\rho$. A correction factor $T(R, r, \rho) < 1$ is used to account for eccentricity of the flow path (tortuosity factor). To simplify the computations, it is assumed that both correlation and tortuosity are power functions of the effective saturation, which leads to the final formula:

$$k_r(S) = S_e^{\,n} \left[\frac{\int_{S_{\min}}^{S} dS/\psi}{\int_{S_{\min}}^{1} dS/\psi} \right]^2 \qquad (17)$$

Obviously, n depends on the soil type. However, as a universal n is commonly desired, data of 45 various types of soils were used to determine n (by an analytical best-fitting procedure). The value $n = 0.5$ yields the minimum computed discrepancies between measured and predicted k_r.

All $K_r(S)$ formulas require the knowledge of the residual water content θ_r. However, $\theta(\psi)$ curves are very rarely measured up to this lower value of θ. Kunze, Uehara, and Graham (1968) investigated the influence of using partial and complete $\psi(\theta)$ curves on the computed hydraulic conductivity. It seems that complete $\psi(\theta)$ information improves the quality of the prediction mainly as a result of better fulfillment of the requirement that $K = 0$ for $\theta = \theta_r$. Mualem (1976*a, b*) suggested a computational method for the determination of θ_r and extrapolation of the partially measured $\psi(\theta)$ curve up to this value.

The accuracy and the deficiencies of the models of Averjanov, Childs and Collis George, Burdine and Mualem [equations 6(n = 3.5), 12, 15. (with $S_e^{4/3}$ as correction factor), and 17 (n = 0.5), respectively] were mutually tested against measured data of 45 soils (Mualem, 1976*a*). The deviations between the measured and predicted curves were quantitatively evaluated. It was found that different soils are best represented by different methods (Figure 2). However, it seems that on an overall basis, Mualem's formula yields the best results. Averjanov's formula yields good results for sands while often the agreement with experimental data of heavy soils is not satisfactory [see Figure 2(a)]. Among the other three statistical models, Burdine's formula yields the poorest results. For soils in which $d\theta/d\psi \neq 0$ as $\psi \to 0$, there is an abrupt fall of the computed hydraulic conductivity near saturation [Figure 2(b)], which only rarely is sustained by observations.

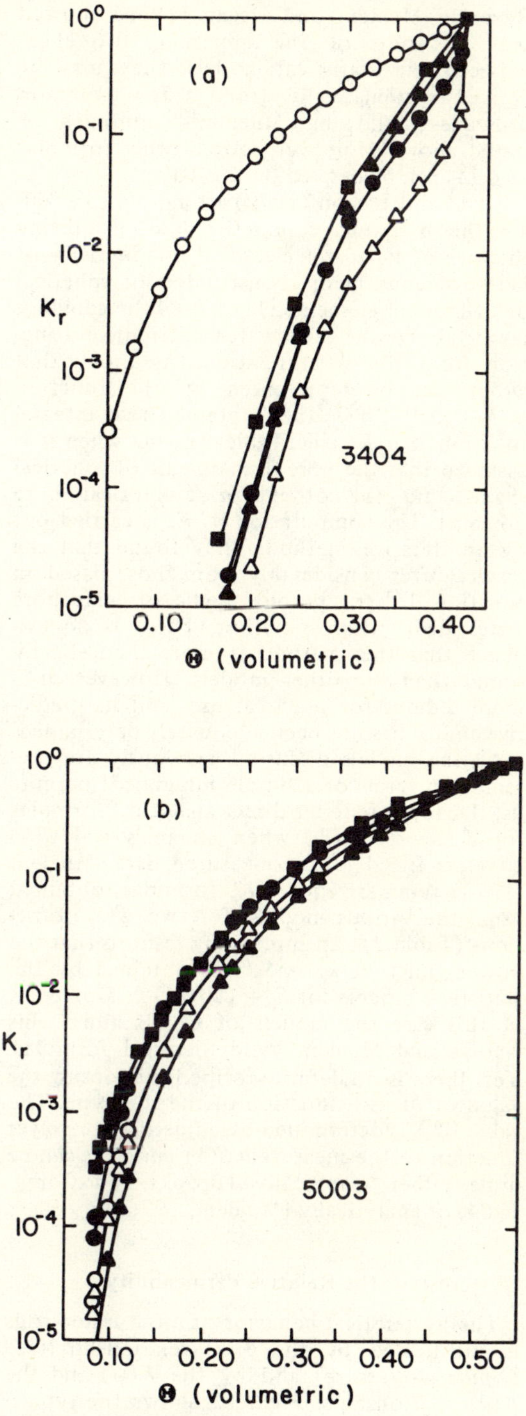

FIGURE 2. Computed relative hydraulic conductivities for two soils, using four models: Averjanov's (hollow circles); Wyllie and Gardner (hollow triangles); Childs and Collis George as modified by Millington and Quirk (triangles); and Mualem (circles). The measured curve is shown by squares. (After Mualem, 1976*a*)

Recently Mualem and Dagan (1976) showed that in spite of the apparent differences between the statistical models, they may be derived systematically from a few common principles following Mualem's approach as stated above. Moreover, several other formulas of $K_r(S)$ can be derived on this basis.

Farrel and Larson (1972) attempted to evaluate the hydraulic conductivity using existing theories of heat and electrical conductivity of heterogeneous media constructed of spherical or cylindrical elements. They used the capillary law to derive the pore water distribution function from the $\psi(\theta)$ relation. Assuming that $\psi(S_e)$ can be represented by the function $\psi = \psi_{cr}e^{\alpha(1-S_e)}$, they obtained two integral functions for K_r (see Table 2), one when it is assumed that the pore domains are of spherical shape, and the other for cylindrical pore domains. The computation of K_r is carried out by an iterative method. They found that the results differ considerably from those based on equation 15, for porous media of wide pore water distribution function. There is no evidence that this method is more theoretically sound than the other models. However, it is more tedious for practical use, and its predictive ability has not been adequately determined.

All the models discussed previously are presented in terms of a simple integral. It permits us, therfore, to derive direct analytical formulas for $K_r(S)$ or $K_r(\psi)$ when an analytical $\psi(S)$ curve is fitted to the measured data. Mualem (1974) summarized the K_r formulas, obtained using the various models, for two $\psi(S_e)$ functions (Table 2). An interesting result is that the power function $k_r = S_e{}^m$ is obtained by the statistical models for $S_e = (\psi/\psi_{cr})^{-\lambda}$. Moreover, in this case the models of Childs and Collis George and Mualem yield identical formulas. Yet, there is no definite recipe for choosing the value of m as a function of the soil type. The value of λ is determined by adjusting the power function to the measured $\psi(S)$ curve. It can be done either graphically (Brooks and Corey, 1964) or analytically (Mualem, 1976*a*).

Hysteresis of the Relative Permeability

The hysteretical behavior of unsaturated soils is manifested by the $\theta(\psi)$ relationship (see *Capillary Pressure*) and by the $K(\psi)$ and the $K(\theta)$ relationships. Figure 3 shows the typical three hysteretical loops measured by Poulovassilis (1970) in experiments carried out on a sand sample. Obviously the $K(\psi)$ hysteresis is magnified relative to the $\theta(\psi)$ hysteresis, while the magnitude of the $K(\theta)$ hysteresis is very limited. Moreover, for a soil sample that is rewetted after it has been drained to a low water content, hysteresis should be even more significant. For a given ψ, the relative permeability in a drying process may take a value a hundred times greater than the corresponding k_r in the wetting process. Yet the $k_r(\theta)$ may be almost a unique function. Consequently, for practical use in solution of problems of infiltration (q.v.) under intermittent irrigation or rainfall, where drying and wetting processes are involved, it is better to use the unique function $k_r(\theta)$ rather than a $k_r(\psi)$ function as suggested in Tables 1 and 2. Even though this typical phenomenon has been reported by several investigators such as Nielsen and Biggar, Topp and Miller, Topp, Talsma, Vachaud and Thony (see references in Topp, 1969) who found no significant $k_r(\theta)$ hysteresis, it is suspected that in certain cases it is not negligible. Staple (1965) and Dan and Wierenga (1975) found considerable $K(\theta)$ hysteresis, with an order of magnitude deviation between corresponding values of k_r on the main wetting and the main drying curve. Also for some soils the higher conductivity is measured on the drying branch while for other soils it is observed on the wetting branch.

Quantitative Models of the Relative Permeability Hysteresis

All the models of the relative permeability discussed in the preceding section do not take into account the capillary hysteresis. The statistical models that use an observed $\theta(\psi)$ curve for predicting $K(\theta)$ or $K(\psi)$ are based on the assumption that the pore distribution function $f(r)$ is characterized by only one variable, i.e., the pore is defined by only one geometrical length scale (r). Since r is related to ψ in a unique way, the relationships $\theta(\psi)$ and $K(\psi)$ are univalued. These models can be used for computation of $K(\psi)$ for a given $\theta(\psi)$ curve representing either drainage or imbibition (Kunze, Uehara, and Graham, 1968). Such an approach implies, however, that $f(r)$ is not determined by the soil structure only, but also depends on the process by which the porous medium is brought to a certain θ. One has, therefore, to introduce a new variable related to the process described by the $\theta(\psi)$ curve. One of the contradictory results of the existing methods is the derivation of different values of K_{sat} (or of the matching factor) depending on what branch of the $\theta(\psi)$ hysteresis is used in the computational procedure.

Recently Mualem (1974, 1976*b*) set forth a theory that permits the prediction of the $K(\psi)$ or the $K(\theta)$ hysteresis as well as the $\theta(\psi)$ hysteresis. In essence, the new models are an extension of the domain theory of hysteresis (see *Capillary Pressure*). A porous medium is charac-

TABLE 2. Prediction Methods of the Relative Hydraulic Conductivities and the Results for Two Analytic Approximations of the $\psi - s$ Relationship. (After Mualem, 1974)

Reference	$f(\psi, S_e)$ / $g(K, S_e, \psi)$	$S_e = (\psi/\psi_{cr})^{-\lambda}$	$\psi = \psi_{cr} e^{\alpha(1-S_e)}$
Mualem form of Childs and Collis George CCC, n=0 MQ, n=4/3 Kunze, n=1	$K_r(S_e) - S_e^{\,n} \dfrac{\int_0^{S_e}[(S_e-s)/\psi^2(s)]\,ds}{\int_0^1[(1-s)/\psi^2(s)]\,ds}$	$K_r(S_e) = S_e^{\,(2+2\lambda)/\lambda+n}$ $K_r(\psi) = \psi/\psi_{cr})^{-(2+2\lambda)-n\lambda}$	$K_r(S_e) = S_e^{\,n} \dfrac{[e^{2\alpha S_e} - 2\alpha S_e - 1]}{[e^{2\alpha} - 2\alpha - 1]}$
Burdine Wyllie and Gardner	$K_r(S_e) = S_e^{\,2} \dfrac{\int_0^{S_e} ds/\psi^2(s)}{\int_0^1 ds/\psi^2(s)}$	$K_r(S_e) = S_e^{\,(2+3\lambda)/\lambda}$ $K_r(\psi) = (\psi/\psi_{cr})^{-(2+3\lambda)}$	$K_r(S_e) = S_e^{\,2} \dfrac{[e^{2\alpha S_e} - 1]}{[e^{2\alpha} - 1]}$
Mualem n=0.5	$K_r(S_e) = S_e^{\,n} \left[\dfrac{\int_0^{S_e} ds/\psi(s)}{\int_0^1 ds/\psi(s)}\right]^2$	$K_r(S_e) = S_e^{\,(2+2\lambda)/\lambda+n}$ $K_r(\psi) = (\psi/\psi_{cr})^{-(2+2\lambda)-n\lambda}$	$K_r(S_e) = \dfrac{[e^{2\alpha S_e} - 2e^{\alpha S_e} + 1]}{[e^{2\alpha} - 2e^{\alpha} + 1]}$
Yuster m=2.0 Irmay m=3.0 Averjanov m=3.5 Corey m=4.0	$K_r(S_e) = S_e^{\,m}$	$K_r(\psi) = (\psi/\psi_{cr})^{-m\lambda}$	$K_r(\psi) = 1 - \frac{1}{\alpha}\ln(\psi/\psi_{cr})^{\,m}$
Farrel and Larson	$K_r(S_e) = \dfrac{\int_0^{S_e} [\nabla\phi(s)/\psi^2(s)]\,ds \,/\, \int_0^{S_e} \nabla\phi(s)\,ds}{\int_0^1 [\nabla\phi(s)/\psi^2(s)]\,ds \,/\, \int_0^1 \nabla\phi(s)\,ds}$ Spherical pores $\nabla\phi(s) = 3K/[2K + M/\psi^2(s)]$ Cylindrical pores $\nabla\phi(s) = 2K/[K + M/\psi^2(s)]$		$K_r = \left[\dfrac{\psi}{\psi_{cr}}\right]^2 \dfrac{[1-e^{-4\alpha S_e/3}]\,[e^{2\alpha/3} - 1]}{[1-e^{-4\alpha/3}]\,[e^{2\alpha S_e/3} - 1]}$ For spherical pores For cylindrical pores $K_r = \left[\dfrac{\psi}{\psi_{cr}}\right]^2 \dfrac{[1-e^{-\alpha S_e}]\,[e^{\alpha} - 1]}{[1-e^{-\alpha}]\,[e^{\alpha S_e} - 1]}$
Wind, Wesseling	$K_r = (\psi/\psi_{cr})^{-n}$; $\psi < \psi_{cr}$ $K_r = 1$ $\psi < \psi_{cr}$	$K_r(S_e) = S_e^{\,n/\lambda}$	$K_r(\psi) = e^{-\alpha n(1-S_e)}$
Gardner	$K = \dfrac{a}{\lvert\psi\rvert^n + b}$; $K_r = \dfrac{1}{(\psi/\psi_{cr})^n + 1}$	$K_r(S_e) = \dfrac{1}{1 + S_e^{\,-n/\lambda}}$	$K_r(\psi) = \dfrac{1}{e^{\alpha n(1-S_e)}}$
Gardner	$K_r = e^{a(\psi-\psi_{cr})}$; $\psi < \psi_{cr}$ $K_r = 1$; $\psi > \psi_{cr}$	$K_r = e^{a\psi_{cr}(S_e^{-1/\lambda}-1)}$	$K_r(S_e) = e^{a\psi_{cr}[e^{\alpha(1-S_e)}-1]}$

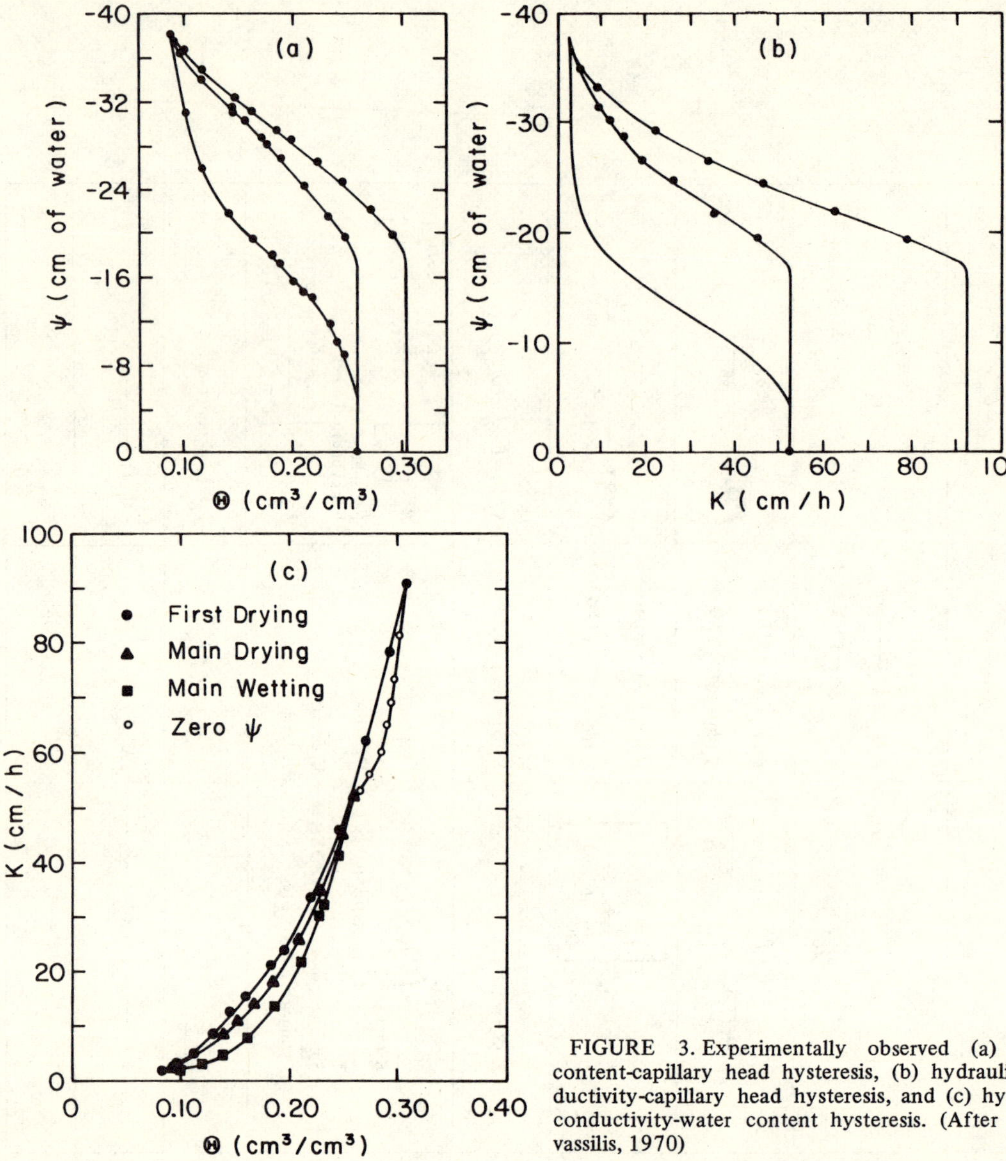

FIGURE 3. Experimentally observed (a) water content-capillary head hysteresis, (b) hydraulic conductivity-capillary head hysteresis, and (c) hydraulic conductivity-water content hysteresis. (After Poulovassilis, 1970)

terized in terms of elementary pore domains, each of which is defined by a two length scale r and ρ. The domains distribution function $f(r, \rho)$ for a medium can be quantitatively derived on the basis of some measured data. Moreover, the filled domains diagrams definitely determine the state of the system for a particular capillary head and a given process. Hence one can evaluate the corresponding contribution of an elementary pore domain to the hydraulic conductivity, and by integrating over the filled pore domains, obtain the actual hydraulic conductivity. Assuming that the hydraulic conductivity of such element is proportional to the product $r\rho$, the relative permeability can be expressed by the relation:

$$k_r(\theta) = \frac{\underset{\text{(filled pores)}}{\iint r\rho\, f(r, \rho)\, dr\, d\rho}}{\underset{\text{(all pores)}}{\iint r\rho\, f(r, \rho)\, dr\, d\rho}} \tag{18}$$

The domain of the integration in the r, ρ plane depends on the hysteretical process to which the porous medium is subjected, and therefore $k(\psi)$ or $k_r(\theta)$ are also hysteretic. In principle, then, if equation 18 is applied to a model of hysteresis, it will derive the relative permeabil-

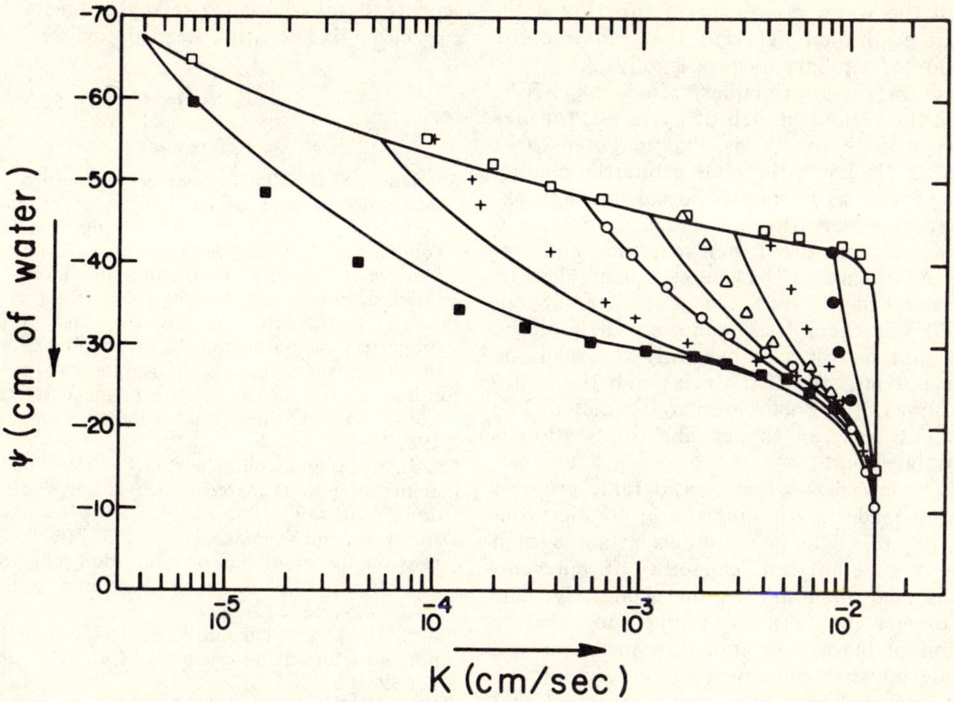

FIGURE 4. Predicted hydraulic conductivity-capillary head hysteresis (solid lines) compared with experimental results. (After Mualem, 1976*b*)

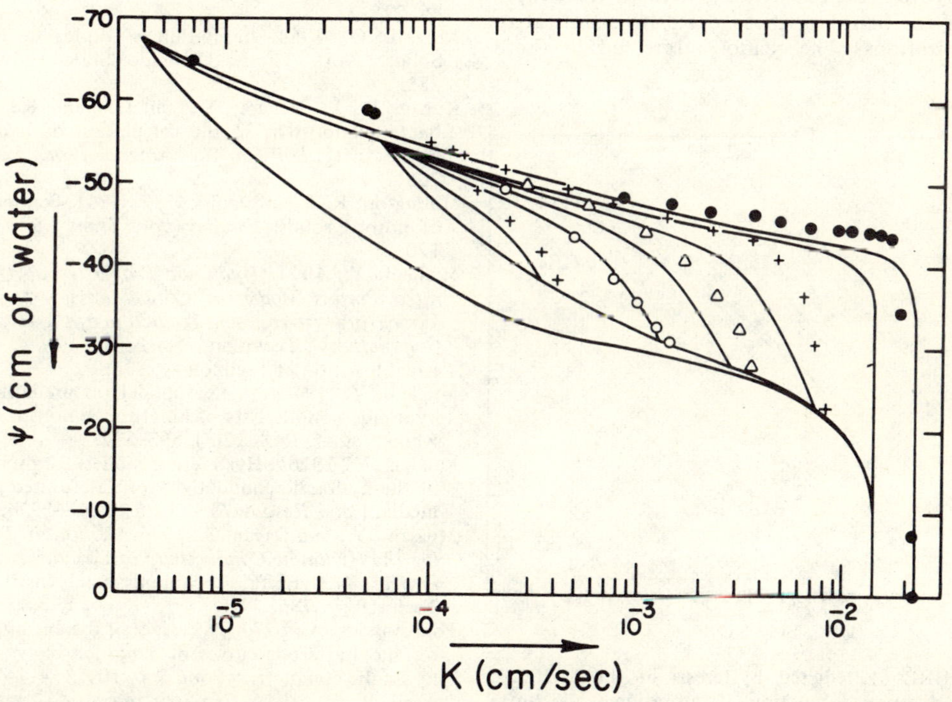

FIGURE 5. Predicted hydraulic conductivity-capillary head hysteresis including the first drying curve and the secondary scanning curves in drainage (solid lines) compared with the experimental results. (After Mualem, 1976*b*)

ity and the water content as a function of ψ without additional data (to that required for the study of capillary hysteresis only).

Based on these principles, Mualem (1976*b*) applied the various models of hysteresis for prediction of the k_r hysteresis. The predicted $k_r(\psi)$ hysteresis displayed the same properties characterizing the $\theta(\psi)$ hysteresis, as sustained by experimental observations. Figures 4, 5, and 6 show a comparison between measured and predicted $K(\psi)$ and $K(\theta)$ hysteresis, using data of Topp and Miller (1966). The generalized dependent domain theory has been used. It should be noted that in this case not only the scanning curves but also the first drying and the boundary curves of the main loop are predicted. The agreement between theory and observation is surprisingly good.

This review shows that considerable progress has been made in studying the hydraulic properties of soil. The new models attain a high degree of generality and capability of representing observed phenomena more accurately. One may expect that a more sound and accurate solution of the unsaturated flow problems will be achieved when these new models are used.

The reader should be aware of the fact that chemical solutions that interact with the soil matrix, swelling soils, and agricultural treatment are parameters that considerably affect the hydraulic properties. There is no simple recipe for handling these cases in quantitative computations. The reader interested in these

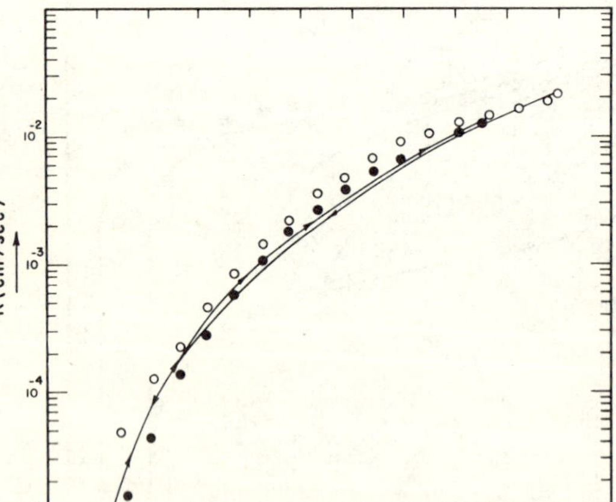

FIGURE 6. Predicted hysteresis of the hydraulic conductivity as a function of saturation (solid lines) compared with experimental results in drainage (open circles) and wetting process (solid circles). (After Mualem, 1976*b*)

aspects should look for related subjects in this encyclopedia and other scientific papers.

Y. MUALEM
H. J. MOREL-SEYTOUX

References

Averjanov, S. F., 1950. About permeability of sursurface coils in case of incomplete saturation, *Eng. Collection,* Vol. VII, (1950) as quoted by P. Ya Palubarinova, 1962, *The Theory of Ground Water Movement,* English translation by I. M. Roger DeWiest, Princeton University Press, 19–21.

Brooks, R. H., and Corey, A. T., 1964. Hydraulic properties of porous media, *Hydrology Paper 3,* Fort Collins: Colorado State University, 27p.

Burdine, N. T., 1953. Relative permeability calculation size distribution data, *Petroleum Trans. AIME,* **198**, 71–78.

Childs, E. C., and Collis George, N., 1950. The permeability of porous materials, *Royal Soc. Proc. (London),* **A201**, 392–405.

Dane, J. H., and Wierenga, P. J., 1975. Effect of hysteresis on the prediction of infiltration redistribution and drainage of water in a layered soil, *Jour. Hydrology,* **25**, 229–242.

Farrel, D. A., and Larson, W. E., 1972. Modeling the pore structure of porous media, *Water Resour. Res.,* **3**, 699–706.

Fatt, I., and Dykstra, H., 1951. Relative permeability studies, *Trans. AIME,* **192**, 249–255.

Gardner, W. R., 1958. Some steady state solutions of the unsaturated moisture flow equation with application to evaporation from a water table, *Soil Sci.,* **85**, 228–232.

King, L. G., 1964. Imbibition of Fluids by Porous Solids. Ph.D. Thesis, Colorado State University, 231p.

Kunze, R. J., Uehara, G.; and Graham, K., 1968. Factors important in the calculation of hydraulic conductivity, *Soil Sci. Soc. America Proc.,* **32**, 760–765.

Millington, R. J., and Quirk, J. P., 1961. Permeability of porous solids, *Faraday Soc. Trans.,* **57**, 1200–1206.

Mualem, Y., 1974. *Hydraulic Properties of Unsaturated Porous Media: A Critical Review and New Models of Hysteresis and Prediction of the Hydraulic Conductivity,* Technion, *Project No. 38/74,* Haifa: Israel Institute of Technology, 235p.

Mualem, Y., 1976*a*. A new model for predicting the hydraulic conductivity of unsaturated porous media, *Water Resour. Res.,* **12**(3), 593–522.

Mualem, Y., 1976*b*. Hysteretical models for prediction of the hydraulic conductivity of unsaturated porous media, *Water Resour. Res.,* **12**(6), 1248–1254.

Mualem, Y., and Dagan, G., 1976. *Methods of Predicting the Hydraulic Conductivity of Unsaturated Soils, Project 442.* Haifa: Technion, Israel Institute of Technology, 78p.

Poulovassilis, A., 1970. The effect of the entrapped air on the hysteresis curves of a porous body and its hydraulic conductivity, *Soil Sci.,* **109**, 154–162.

Purcell, W. R., 1949. Capillary pressures—Their measurement using mercury and the calculation of permeability therefrom, *Petroleum Trans. AIME,* **186**, 39–48.

Rijtema, P. E., 1965. An analysis of actual evapotranspiration, *Agric. Res. Rep. 659,* Center for Agricultural Publications and Documentation, Wageningen, Netherlands, 107p.

Staple, W. J., 1965. Moisture tension, diffusivity and conductivity of a loam soil during wetting and drying, *Canadian Jour. Soil Sci.,* **45,** 78–86.

Topp, C. C., 1969. Soil water hysteresis measured in a sandy loam compared with the hysteretic domain model, *Soil Sci. Soc. America Proc.,* **33,** 645–651.

Topp, G. C., and Miller, E. E., 1966. Hysteretic moisture characteristics and hydraulic conductivities for glass-bead media, *Soil Sci. Soc. America Proc.,* **30,** 156–162.

Wind, G. P., 1955. Field experiment concerning capillary rise of moisture in heavy clay soil, *Neth. Jour. Agric. Sci.,* **3,** 60–69.

Wyllie, M. R. J., and Gardner, G. H. F., 1958. The generalized Kozeny-Carman equation, *World Oil,* Houston, Texas, 210–228.

Yuster, S. T., 1951. Theoretical considerations of multiphase flow in idealized capillary systems, *Proc. 3rd World Petrol. Congr.,* **2,** 437–445.

Cross-references: *Capillary Pressure; Conductivity, Hydraulic; Flow Theory; Imbibition; Infiltration; Percolation; Pore Space, Drainable; Soil Water; Thermodynamics, Soil Water; Water Content and Retention; Water Movement; Wetting Front.*

PERMISELECTIVE PROPERTIES OF SOILS—See TRANSPORT PROCESSES

PESTICIDES—See POLLUTION

pH—See ACIDITY; pH, FIELD DETERMINATION; SOIL REACTION

pH, FIELD DETERMINATION

A highly important attribute of a soil is its pH. It indicates the reaction or degree of acidity or alkalinity (see *Soil Reaction*) of a soil horizon, which in turn affects solubilities of various compounds, the bonding of ions to exchange sites, and the activities of microorganisms (see *Acidity*). Thus, for example, a pH below 4 indicates the presence of free acids, generally formed by oxidation of sulfides; a pH between 4 and 5.5 commonly indicates much exchangeable aluminum; and a pH between 7.8 and 8.2 indicates the presence of calcium carbonate (Thomas, 1967). A pH above 8.2 indicates relatively high levels of monovalent bases such as sodium. Low pH values encourage the growth of fungi rather than bacteria and actinomycetes.

Because pH is an index to other important characteristics of soils, its value is determined in the field during soil mapping and in other studies such as those of weathered rock mantles. The two principal methods of field measurement rely on portable pH meters and on indicator dyes.

Portable pH Meters

Powered by batteries to permit mobility, most pH meters are a combination of glass or sensing electrode and an internal calomel or reference electrode. The construction of the glass electrode may be more sturdy in meters designed for field use than in those for laboratories. The principles of operation are the same for the meters used in the laboratory and in the field (McLean, 1973) (see *Electrochemistry*).

Measurements with a portable pH meter are normally made in soil-water suspensions. A supply of water must therefore be carried with the meter to prepare the suspensions and to wash the electrode and container.

Indicators

Dyes that change colors at different pH values are used widely in the field because of their convenience. The earliest known use of a chemical test for measuring soil acidity in the United States consisted of placing litmus paper on moist soil and noting any change in color (Wheeler, Tucker, and Hartwell, 1896). Today some commercial indicators consist of paper strips soaked in dyes, which can then be used in the same way as was the litmus paper. Generally, however, indicator dyes are used in solution.

Measurements are made by first adding a few drops of indicator solution to a pinch of soil in a small cavity in a spot plate. The mixture is stirred briefly, and the color of the indicator is then compared to a standard to estimate pH. Color changes of an indicator occur over a limited range in pH. For example, changes in the color of brom cresol purple occur between pH 5.0 and 6.7. Woodruff (1961) suggested that this indicator be used to measure the pH of soil samples in 0.01 M $CaCl_2$ solution. Another common indicator is brom thymol blue, which changes from yellow through green to blue over a pH range of roughly 5.3 to 8.7.

For pH measurements over the full range found in soils, a set of indicators is required. Kits consisting of solutions of several individual indicators to cover that range are therefore manufactured. To measure the pH of a given specimen, trials with more than one indicator may therefore be necessary. Kits with solutions containing a mixture of several indicators are also available, designed to cover the full pH range commonly encountered in soils. The two types of kits have their individual advantages.

Reliabilities of the Methods

All in all, measurements of pH with indicators are less accurate than those with pH meters. Furthermore, some soils adsorb dyes so rapidly that no indicator readings can be obtained. Indicators cannot be used to measure pH values of such soils. Indicator kits are easier to carry and use than portable pH meters, an important advantage in many situations. Furthermore, the approximate values so obtained are accurate enough for many purposes. Selection of a method is therefore made on the balance between needed accuracy and the facilities for transporting equipment.

E. O. McLEAN

[*Editor's Note:* Although measurement of pH in the field may be a useful practice under certain conditions, soil test kits in inexperienced hands can give anomalous results. As stressed here by Professor McLean, accurate results are best obtained by analyzing soils in well-equipped soil-testing laboratories.]

References

Graham, E. R., 1959. An explanation of theory and methods of soil testing, *Mo. Agric. Exp. Sta. Bull. 734,* 20p.

Mehlich, A., 1943. The significance of percentage base saturation and pH in relation to soil differences, *Soil Sci. Soc. America Proc.*, 7, 167–174.

McLean, E. O., 1973. Testing soils for pH and lime requirement, *in* L. Walsh and J. D. Beaton, eds., *Soil Testing and Plant Analysis.* Madison: Soil Science Society of America, 77–95.

Pearson, R. W. and Adams, F., eds., 1967. *Soil Acidity and Liming.* Madison: American Society of Agronomy, Monograph 12, 274p.

Schofield, R. K., and Taylor, A. W., 1955. The measurement of soil pH, *Soil Sci. America Proc.*, **19**, 164–167.

Snyder, G. H.; McLean, E. O.; and Franklin, R. E., 1969 Interactions of pH-dependent and permanent charges: II. Calcium and Rb bonding to bentonite and illite suspensions-clay-phase-retention, *Soil Sci. Soc. America Proc.*, **33**, 392–396.

Thomas, G. W., 1967. Problems encountered in soil testing methods, *in* G. W. Hardy, ed., *Soil Testing and Plant Analysis: Part 1, Soil Testing.* Soil Sci. Soc. America, Spec. Publ., **2**, 37–54.

Wheeler, H. J.; Tucker, G. M.; and Hartwell, B. L., 1896. On the neutralizing action of lime, *Rhode Island Agric. Exp. Sta. Annl. Rpt.*, **9**, 294–318.

Woodruff, C. M., 1961. Brom cresol purple as an indicator of soil pH, *Soil Sci.*, **91**, 272.

Cross-references: *Acidity; Electrochemistry; Management of Soils; Soil Chemistry; Soil Reaction; Soil Tests.*

PHOSPHORUS CYCLE

Phosphorus is found in soils, plants, and microorganisms in a number of organic (P_o) and inorganic (P_i) forms. It is second in amount only to nitrogen as a nutrient element required by plants and microorganisms (see *Macronutrients*). Phosphorus is considered an essential nutrient element because it is required or is favorable for (1) seed formation, (2) root development, (3) maturing of crops, especially cereals, (4) crop quality, (5) strength of straw in cereal crops, and (6) accumulation and release of energy during cellular metabolism (Tisdale and Nelson, 1964; Brady, 1974). Thus the element is critical in biological soil systems.

The phosphorus cycle differs somewhat in soils under natural conditions and in soils under cultivation. Under natural conditions the phosphorus taken up by growing plants is returned to soils in animal and vegetative residues and remains. Under cultivation, however, part of the element taken up by the crop is removed in harvests and only part is returned to the soils. To offset removals of phosphorus and to increase levels where they were naturally low, fertilizers are added to soils (see *Fertilizers, Inorganic; Fertilizers, Organic*). Thus the phosphorus cycle is important in soil management (q.v.), i.e., in the maintenance or improvement of fertility levels (Pierre, 1948; Halsted and McKercher, 1975).

Total Phosphorus in Soils

Amounts. Despite its importance to plant and animal life, phosphorus is a minor constituent of soils. Generally speaking, the element comprises roughly 0.04% of the soil mass. That proportion is appreciably less than potassium, which is approximately 1%, and nitrogen, which is approximately 0.2%, although these elements are also very important to plant growth (Black, 1968). If an average phosphorus content of 0.04% is assumed, the total amount in soils of the world to a depth of 20 cm would be 100 trillion tons (Larsen, 1967). Large as that number is, it is a tiny fraction of the soil-mantle mass.

In the United States, plow layers of arable land have an average content of approximately 0.062% phosphorus, and most soils fall within a range of 0.022 to 0.083% (Black, 1968). Extremes are 0.005% in St. Lucie sand, a Quartzipsamment in Florida, and as much as 1% in deeper horizons of Maury silt loam, a Paleudult derived in part from phosphatic limestone in central Tennessee (Kurtz, 1953).

The amounts of phosphorus in horizons below plow layers tend to be the same order of magnitude as those near the surface, although there are departures from this trend. To illustrate amounts in different soils and also show vertical distribution of total phosphorus, Table 1 gives quantities by horizons for a group of four profiles. Amounts are expressed in parts per million rather than in percent because some values are small. Values such as 0.04 and 0.062% become 400 and 620 ppm.

The four profiles represent markedly different kinds of soils. Furthermore, they also represent two appreciably different degrees of regolith weathering. Total quantities of phosphorus reflect the kind of soil, the degree of weathering of soil materials, and the original nature of soil parent materials. The Mahaska profile represents Mollisols, dark colored soils common to grasslands, that tend to be high in all nutrient elements including phosphorus. Amounts similar to those in the Mahaska soil of Iowa are values of 8–11 metric tons/ha-m in typical Chernozems of the USSR (Shkonde and Koroleva, 1968). The Aridisols, represented by the Desert soil profile in Table 1, also tend to be high in phosphorus. Such soils are restricted to dry regions where weathering and leaching are limited by moisture shortage. The Mollisols also occur in regions where weathering and leaching are slight to moderate but more moist than Aridisols.

TABLE 1. Total Phosphorus in Parts per Million by Horizons of Four Profiles.

	Horizon	Depth (cm)	P (ppm)
Mahaska sicl: Iowa[a] (Argiudoll)	*A*1p	0–20	691
	*A*12	20–33	631
	*A*3	33–43	547
	*B*1	43–54	504
	*B*21	54–75	425
	*B*22	75–88	464
	*B*31	88–105	598
	*B*32	105–128	645
	*C*1	128–138	766
	*C*2	138–156	747
	*C*3	156–168	796
	*C*4	168–188	800
	*C*5	188–206	856
Bonjema 1: Sierra Leone[b] (Plinthic Paleudult)	*A*1	0–10	152
	*A*3	10–41	75
	*B*1	41–63	76
	II*B*21	63–84	79
	III*B*22	84–145	78
	III*B*23	145–178	85
Orthic Podzol: Alberta[c] (Haplorthod?)	*A*2	0–7	18
	*B*ir	7–20	883
	C	At 32	306
Desert soil: California[d] (Profile *A*) (Haplargid?)	*A*	0–5	1047
	*B*21	5–12	655
	*B*22	12–27	655
	*B*3	27–34	1353
	*C*1	34–84	916

[a] Runge and Riecken, 1966
[b] McKenzie, Kurtz, and Melsted, 1977
[c] Alexander and Robertson, 1968
[d] Brown and Drosdoff, 1940

In contrast to the Mollisol and Aridisol profiles, Bonjema loam, representing the Ultisols common to warm-temperate, humid regions, is low in phosphorus. The soil has been formed in a strongly weathered regolith under a climate that promotes further weathering and leaching. Similar amounts of phosphorus, ranging from 192 down to 9 ppm, were reported earlier for a Red Latosol profile and a Yellow Latosol profile in Sudan (Middleton, 1954).

The effects of parent materials on phosphorus levels in soils show up especially well in the Ultisols. Soils such as Bonjema loam, derived from acidic rocks, are low in phosphorus. On the other hand, Ultisols derived from basic rocks are not nearly so low; levels are comparable to those of the Mahaska and Desert soils. Basic rocks tend to be higher than acidic rocks in phosphorus, which carries over into the soils derived from them. The effects are evident in all kinds of soils but less marked in those formed in less weathered regoliths than the Ultisols.

The total amount of phosphorus in the Orthic Podzol, illustrating the Spodosol order, is more like those of the Mahaska and Desert soils than like that of Bonjema loam. The Podzol formed in a slightly weathered regolith, and it occurs in a climate that limits weathering. To all intents and purposes, the soil is placed in cold storage when it is frozen for a part of each year.

Profile Distribution. Table 1 shows that the vertical distribution of total phosphorus tends to be similar for three profiles and much different for the fourth. The Mahaska profile has a distinct minimum in the *B* horizon with maxima in the *A* and *C* horizons. The minimum reflects upward transfers by plants and downward transfers by leaching (Runge and Riecken, 1966). A similar but less regular distribution holds for the Desert soil profile. It was formed in granitic residuum that could have been less uniform than the parent loess of Mahaska soils. The Bonjema profile also has a distinct maximum in the *A* horizon and the faintest suggestion of a possible second maximum at depth. A clear minimum is not really evident although the *A* horizon has almost twice as much phosphorus as do the deeper horizons.

The distribution curve of total phosphorus with depth in the Podzol profile is reversed from those of the other three. The minimum is in the *A*2 horizon, near the soil surface, the maximum is in the *B* horizon, and a second minimum marks the *C* horizon. The maximum in the *B* horizon is striking because the amount of phosphorus in that horizon is more than twice as large as the amount in the *C* horizon. The *B* horizons of Podzols, especially those *B* horizons high in iron as is the one in Alberta, retain both native and added phosphorus. The high capacity of such horizons to fix phosphorus was noted many years ago in Vermont (Weiser, 1933). The vertical distribution of total phosphorus in this Podzol profile is common but not universal for Spodosols.

Forms of Phosphorus

The nature of phosphorus compounds in soils is not well known. The relative amounts of phosphorus in organic and inorganic forms (see *Soil Minerals* in vol. IVB) have been determined, but all compounds in the two fractions cannot be identified. Because inorganic phosphorus is largely in the clay fraction, its compounds cannot be separated by physical means to facilitate identification (Larsen, 1967). Norrish (1968) for example, has concentrated inorganic phosphorus compounds by destroying other substances in soil samples in an initial processing stage prior to identification. Several indirect methods have also been developed and applied in efforts to identify compounds (Black, 1968). Indirect methods have been applied to both inorganic and organic forms of phosphorus.

Inorganic Forms and Amounts. Nearly all inorganic phosphorus exists in the form of orthophosphates, derivatives of phosphoric acid (H_3PO_4) (Black, 1968). The compounds are believed to be chiefly phosphates of calcium, aluminum, and iron with minor proportions of others. Some of the inorganic phosphorus is present in the lattices of silicate minerals and as inclusions in minerals, e.g., in quartz crystals. Those amounts are believed to be small.

The initial form of inorganic phosphorus is predominantly some form of the mineral apatite, perhaps most commonly calcium fluorapatite (Larsen, 1967). More than 60 years ago apatite was identified in the sand fractions of soil samples from many different parts of the United States (Black, 1968). Apatite has also been identified in the silt and sand fractions of all horizons of a Mollisol profile and a sandy Entisol profile in Nebraska (Shipp and Matelski, 1960). The mineral comprised less than 1% of the silt and sand in each horizon and could account for only part of the phosphorus present in those fractions. The amounts of the mineral did increase and decrease with the total quantities of phosphorus in the separates. Various forms of apatite are considered to be the principal phosphorus minerals in calcareous and alkaline soils of arid and semiarid regions (Olsen, 1953). Apatite could thus be a rather important form of inorganic phosphorus in Aridisols, Mollisols, and the associated soils of other orders.

Several phosphorus minerals other than varieties of apatite have been identified in soils. These include members of the plumbogummite series, vivianite, and wavellite (see *Soil Components, Inorganic*). Only those of the plumbogummite series have been found widely. Other members of that series are crandallite, florencite, and gorceixite. Plumbogummite is a lead-aluminum phosphate. The lead is replaced by calcium in crandallite, by rare earth metals in florencite, and by barium in gorceixite. Minerals of the series have been found in a considerable array of soils in Australia (Norrish, 1968). Florencite and gorceixite were found especially in Yellow Podzolic soils, most probably Ultisols, near Adelaide, and had apparently been formed in the soils. Crandallite was identified in samples of soils from the Cook Islands and in others over limestone. Between 12 and 60% of the total amounts of phosphorus in the soils could be accounted for by minerals of the plumbogummite series. The occurrence of these minerals in a large array of soils in Australia suggests that they may also be present elsewhere in the world.

Vivianite, a ferrous phosphate, has been identified in wet soils, as, for example, beneath peat, and wavellite, an aluminum phosphate, has been found in soils derived from phosphatic sediments in Florida (Black, 1968). How important the amounts of phosphorus in the forms of vivionite and wavellite may be in soils is not known.

Because only a small part of the inorganic phosphorus in soils can be attributed to identified minerals, efforts have been made to characterize the forms by indirect means (Black, 1968). The most widely used technique consists of successive extractions with a set of reagents, an approach initially proposed by Chang and Jackson (1957). The portions removed by four successive extractions are considered *calcium phosphates, aluminum phosphates, iron phosphates,* and *occluded phosphates.* A three-step extraction has also been used, and the portions removed are called *acid-extractable, nonoccluded,* and *occluded phosphates* (Walker and Syers, 1976). The nonoccluded phosphorus is believed to consist of phosphate ions sorbed on the surfaces of oxides and carbonates, whereas the occluded phosphorus is believed to be phosphate ions incorporated in the lattice structure or within concretions and coatings of hydrous oxides (Uzu, Juo, and Fayemi, 1975). Results from several laboratories indicate that most but not all of the inorganic phosphorus is removed by the successive extractions (see *Soil Tests*).

The relative amounts of inorganic phosphorus removed by the successive extractions are related to the degree of weathering of soil materials. If weathering has been slight, the first extraction removes large quantities of phosphorus, and little is left for the last extractions. If weathering has been strong, very little phosphorus is removed by the first extraction, and most comes out in the last. Thus, for example, 80% of the total extracted was removed by the first reagent from samples of the *A*1 horizon of a soil from North Dakota. In contrast, the first reagent removed very little phosphorus from samples of a soil from Hawaii, which was near the opposite end of the weathering scale, and the last extraction removed almost 80% (Black, 1968).

If soils of the world are considered collectively, the amounts of inorganic phosphorus form a much greater share of the total than does the organic form. The proportion of inorganic phosphorus may be as low as one third of the total in the *A*1 horizons of some Mollisols but is commonly twice that high even in those soils. Proportions of inorganic phosphorus are commonly three fourths or more of the total in the *A* horizons of soils other than the Mollisols. Below the *A*1 horizons, the proportions of inorganic phosphorus are generally 90% or more. Within profiles, the quantities of inorganic phosphorus tend to remain the same or decrease slightly in the uppermost half meter or so and then increase below that depth. Exceptions to this trend are an essentially uniform distribution or inorganic phosphorus within profiles, as occurs in some Oxisols, and distinct maxima in *B* horizons as occurs in Spodosols.

Organic Forms and Amounts. Efforts have been made to identify in soil organic matter three of the major groups of phosphorus compounds found in plants and animals (Black, 1968). The three groups are *inositol phosphates, nucleic acids,* and *phospholipids* or *phosphatides.* Only the first of the three has been found in soils in more than negligible quantities (Black and Goring, 1953; Anderson, 1964; Bourman and Cole, 1978).

Inositol phosphates consist of esters of inositol and phosphoric acid. Inositol is a saturated, cyclic, six-carbon alcohol with an alcohol group on each carbon atom. Four isomers of the phosphate have been found in soil organic matter, only one of which has been identified in plants. The other three isomers have been attributed to synthesis by microbial enzymes. As much as 35% of the organic phosphorus in soils has been estimated to be in the form of inositol phosphates, the most important being inositol hexaphosphate (Anderson, 1964).

Nucleic acids are major phosphorus compounds in plants and microorganisms but have not been positively identified in soil organic matter. Indirect indications of small amounts of nucleic acids have been obtained, however, and a range of 0.6–2.4% of the phosphorus in organic form has been ascribed to those acids.

Phospholipids are fatty compounds in plants. Fat solvents have therefore been tried as extractants of soil organic matter in efforts to find phospholipids. The quantities have been negligible, but there is some evidence that the compounds are insoluble or nearly so in the presence of silicate clays.

If it is assumed for the organic phosphorus in soils that 35% occurs as inositol phosphate, 2% in nucleic acids, and 1% in phospholipids, a maximum of 38% has been identified. The forms of 62%, or almost two thirds of the organic phosphorus, remain unknown.

Generally speaking, organic phosphorus is a smaller share of the total than the inorganic form. Moreover, the bulk of the organic phosphorus occurs in *A* horizons. For soils of the world collectively, organic phosphorus comprises between one fifth and one fourth of the total in *A*1 horizons. Extremes are 2 and 75%. Consequently, for a great majority of the *A*1 horizons of soils, amounts of organic phosphorus fall between 25 and 250 ppm. Common ranges, for example, are 100–250 for Mollisols and Alfisols and 25–100 ppm for Oxisols and Ultisols. Lower values are also found in Oxisols, e.g., 6 and 12 ppm in the *A*1 horizons of a pair of Haplustox profiles in Venezuela (Westin and Brito, 1969). The preceding estimates of ranges are based on data reported by Allaway and Rhoades (1951), Nye and Bertheux (1957), Olsen (1953), Pearson and Simonson (1939) and Schollenberger (1920).

As already indicated, the largest amounts of organic phosphorus within soil profiles are in the *A*1 horizons. Quantities drop off sharply below those horizons, as a rule, although there are some exceptions. The amounts of organic phosphorus were low in the *A*1 horizons and essentially the same in the *B*23 horizons of a pair of Hapluxtox (Oxisol) profiles in Venezuela (Westin and Brito, 1969). The latter horizons were below depths of 1.5 m. Distribution patterns of organic phosphorus such as those in the two Oxisol profiles are not common. The predominant vertical distribution consists of a maximum in the *A*1 horizon with a gradual but large change to the *B* or *C* horizon and continued decrease below. Very few soils have as many as 30 ppm of organic phosphorus below a depth of 1 m.

The Cycle

Phosphorus in the soil occurs in several "pools" (Figure 1). Considering soil phosphorus in terms of "pools" is a convenient means of summarizing the extremely complex transformations among different forms. Phosphorus transformations from one form of phosphorus to another take place at different rates depending on the processes involved.

The largest pool contains sorbed phosphorus, physically adsorbed phosphorus located mainly on colloid surfaces and immobilized (occluded and chemisorbed) or chemically "fixed" phosphorus (Ryden and Syers, 1977), and phosphorus precipitated directly from the soil solution into discrete and sparingly soluble forms (Mansell, Selim, and Fiskell, 1977). Organically bound phosphorus in living matter, i.e., the plant phosphorus and animal phosphorus constitutes the second major phosphorus pool. Decaying organic materials (humus and microbial phosphorus) contain a phosphorus pool of intermediate size. Water-soluble compounds in the soil solution (q.v.) make up a small, fluctuating phosphorus pool. Soluble forms are important because they are directly available for plant uptake.

Plants absorb phosphorus primarily in the ionic form (mainly $H_2PO_4^-$, $HPO_4^=$), though some organic forms may be used to a limited extent. Both organic and inorganic forms must be transformed into soluble phosphates, i.e., brought into the soil solution, before they become available to plants. The conversion of one form of phosphorus to another involves several complicated processes that are associated with the activities of microorganisms. The main microbiological transformations of phosphorus are the following processes: mineralization, solubilization, and immobilization.

Mineralization. The term *mineralization* refers to the conversion of organic forms of phosphorus into inorganic forms. Microorganisms (bacteria, fungi, and actinomycetes) break down organic matter in their life processes and in doing so, convert the constituent phosphorus into more soluble forms. Phosphorus release rates depend on microbial metabolism rates, higher rates being favored by temperatures in the thermophilic range, neutral pH and high contents of organic phosphorus. The release is also more rapid under conditions favoring ammonification (see *Nitrogen Cycle*). Through mineralization, the phosphorus that was tied up in organic matter is made available for new generations of plants and microorganisms.

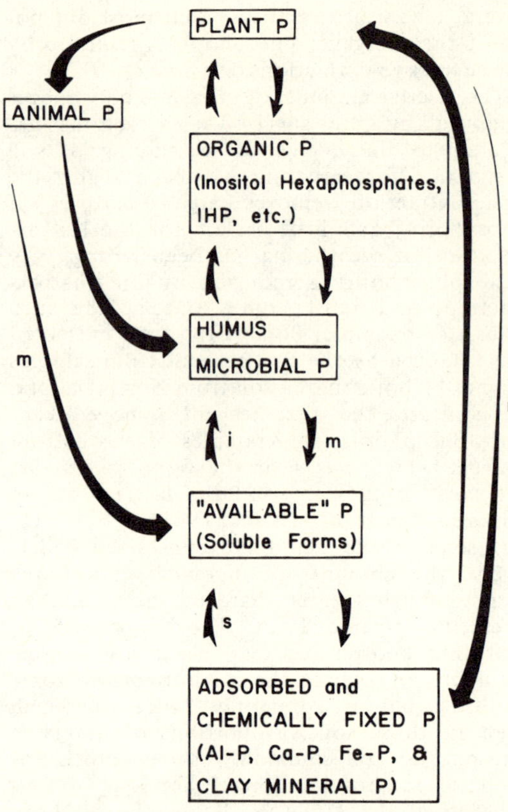

FIGURE 1. A simplified diagram of the phosphorus cycle showing the different phosphorus pools. The letters *m, i,* and *s* denote *mineralization, immobilization* (or chemisorption), and *solubilization* of phosphorus. Particularly important to the phosphorus cycle in soils are the microbial mineralization, immobilization, and solubilization reactions where one form of phosphorus is transformed (converted) into another. With the continual interplay of these processes, especially mineralization and immobilization, the availability of phosphorus to plants is enhanced or decreased. Most soil phosphorus (80–90%) is "fixed" in insoluble forms with only a small proportion (commonly less than 1%) being "readily available" for plant use. The addition of phosphate-bearing fertilizers and the fixation of this element in insoluble forms are important aspects of the phosphorus cycle in soils. (Modified after Brady, 1974; Alexander, 1977; and Armson, 1977)

Solubilization. The term *solubilization* refers to processes whereby a substance is put into solution or converted to more soluble forms. The agent responsible for the change in state is called a *solubilizer*. Most processes of solubilization of phosphorus in soils are mediated by microorganisms.

Many microorganisms produce organic acids through their metabolic processes, and the acids react with insoluble and sparingly soluble phosphorus compounds in the soil to make them more soluble (Alexander, 1977). Examples are species of *Pseudomonas, Myobacterium, Micrococcus, Bacillus, Flavobacterium, Penicillium, Sclerotium, Fusarium,* and *Aspergillus.* Sparingly soluble tri-calcium phosphate (Ca_3PO_4) is acidified to di-calcium or mono-calcium phosphate, both of which are more soluble and hence more available to plants. Part of the dissolved phosphorus is used by microorganisms in their own life functions, but part can be absorbed by plant roots. Amounts of the element brought into solution by the activities of *heterotrophs* (microorganisms able to derive energy only from decomposition of organic compounds) depend on the amounts of carbohydrates oxidized. Moreover, solubilization of phosphorus also requires that part of the carbohydrates be converted to organic acids.

Even though acid production by microorganisms is the main pathway by which phosphorus is solubilized, other mechanisms are also capable of converting phosphate. Under anaerobic conditions (e.g., in flooded soils), for example, ferric phosphates are reduced. The reduction processes lead to the formation of soluble iron and release of phosphate into solution. Phosphorus may also be made available by certain bacteria that liberate hydrogen sulfide, which reacts with ferric phosphate to yield ferrous sulfide and liberate phosphate (Alexander, 1977).

Immobilization. Phosphorus, like nitrogen, can be immobilized as well as mineralized. Which process will dominate depends on the percentage of phosphorus in decaying organic residues and the nutrient requirements of the microbial population. The assimilation of phosphorus into microbial nucleic acids or phospholipids, for example, puts a great demand on the available phosphorus reserve. Should the carbonaceous residue be deficient in phosphorus, the microbial assimilation of available phosphorus (humus/microbial phosphorus and "available" phosphorus pools) may lead to the accumulation of insoluble (unusable) forms of the element (Birch, 1961). The process by which phosphorus is converted from inorganic (P_i) to organic (P_o) forms in microbial tissues to make the element "not readily available" to other organisms or to plants is termed *immobilization.* Immobilization is different from *fixation* (a general expression), in that the latter refers to the loss of availability due to the formation of strong chemical bonds and hence conversion to less available inorganic combinations (Beckmann, Crack, and Prebble, 1976). Even though some soils have enormous phosphorus-fixing capacities, the phrase *phosphorus fixation* is an exaggeration because phosphorus reactions in soils are reversible to some extent (Kittrick and Jackson, 1956). Indeed, the conversion of phosphorus compounds from insoluble (unusable) to soluble (available) forms and vice versa is part of the cyclical behavior of the element.

Phosphorus Availability

Plants may suffer from phosphorus deficiencies even in the presence of relatively large quantities of this element. The problem is essentially one of *availability*. The availability of phosphorus in soils is largely determined by (1) pH; (2) soluble iron, aluminum, and manganese; (3) presence of iron-, aluminum-, and manganese-containing minerals; (4) available calcium and calcium minerals; (5) amount and stage of decomposition of organic matter, and (6) activities of microorganisms (Brady, 1974).

The availability of phosphorus to plants is governed by the ionic form of this element, which in turn is determined by the pH of the solution in which it is found. In highly acid solutions only H_3PO_4 ions are present. The H_2PO_4 ions are dominant under moderately alkaline conditions whereas PO_4 ions dominate under highly alkaline conditions. Maximum phosphorus availability tends to occur at intermediate pH levels between 6.0 and 7.0 when both H_2PO_4 and HPO_4 ions are present. The relationship is, however, greatly complicated by the presence or absence of other compounds or ions in soils. Chemical reactions occurring between soluble iron, aluminum, and manganese with H_2PO_4 ions, for example, may form hydroxy phosphates that are insoluble and hence unavailable to plants. Hydroxy phosphates are also formed by chemical reactions between H_2PO_4 ions and insoluble hydrous oxides of iron, aluminum, and manganese. The large quantities of hydrous iron and aluminum oxides in many soils contribute to the fixation of large total amounts of phosphorus. The availability of phosphorus may also be reduced in intermediate pH ranges by reactions with silicate clays through adsorption phenomena (q.v.) or anion exchange (see *Exchange Phenomena*). Phosphate availability at high pH values is restricted by its reactions with both the calcium and carbonate ions.

The availability of P_i and P_o compounds is commonly considered in terms of "very slowly available," "slowly available," and "readily available" phosphate groups. The "very slowly available" phosphates are mainly the Pi compounds in the chemically fixed pool (primary phosphorus-bearing minerals; aged aluminum, iron, and manganese phosphates) and the P_o compounds in the organic phosphorus pool (Figure 1). "Slowly available phosphates" are freshly formed aluminum, iron, and manganese (small crystals) phosphates, $Ca_3(PO_4)_2$, and mineralized organic phosphates. "Readily available phosphates" are mainly in the "available" phosphorus pool [ammonium phosphates, $Ca(H_2PO_4)_2$, $CaHPO_4$, $Ca(PO_3)_2$]. At any one time as much as 80–90% of the soil phosphorus may be in "very slowly available" forms with less than 1% being "readily available" (Brady, 1974).

The availability of P_o is related to the amount and degree of decomposition of soil organic matter. Though the equilibrium level depends on the nature of the soil environment, the amounts of available organic phosphorus are correlated with the contents of organic carbon and nitrogen. In general, P_o is more readily available in soils rich in organic matter. Moreover, the P_o level is directly proportional to the concentration of other humus constituents (Alexander, 1977).

Depletion and Replenishment of Available Phosphorus. The amount of phosphorus in the available or labile phosphorus pool (Figure 1) is of primary concern to soil scientists and agronomists because it provides soluble forms to plants. Soil-management (q.v.) practices are therefore geared to maintaining relatively high levels of available phosphorus. Because available phosphorus is so important to crop production, this phosphorus pool is a focal point in the phosphorus cycle. Two prime features of the phosphorus cycle are the addition of phosphate fertilizers and the fixation of this element in insoluble forms.

Available soil phosphorus is depleted first by chemical reactions that result in phosphorus fixation (conversion to insoluble forms) followed by crop removals and to a lesser extent by leaching and erosion losses (Ryden, Syers, and Harris, 1973; see also *Nutrient Losses*). Immobilization by microbial activity also results in depletion of the available phosphorus pool. As the available phosphorus pool is depleted, the chemically fixed, humus-microbial, plant, and animal phosphorus pools are enriched.

Replenishment of the available soil phosphorus pool is accomplished by adding inorganic and organic commercial fertilizers, through the mineralization of plant (and animal) residues or humus, and by solubilization of phosphorus-bearing soil minerals (Figure 1). The water-soluble phase represents the mobile portion of soil phosphorus ("available-phosphorus pool") that is immediately available to plant roots. Adsorption, immobilization, and precipitation are transformation reactions that remove phosphorus from the soil solution. Desorption, mobilization, and solubilization are reactions that convert phosphorus into soluble forms. Phosphorus adsorption and desorption reactions occur rapidly in soils and equilibrium may be reached in a few minutes or hours after fertilizer placement. Immobilization and mobilization reactions take place slowly so that days or months are required to approach equilibrium (true equilibrium may never be reached). Precipitation and solubilization reactions seem to occur at rates intermediate between those of adsorption-desorption and immobilization-mobilization (Mansell, Selim, and Fiskell, 1977).

CHARLES W. FINKL, JNR.
ROY W. SIMONSON

References

Alexander, M., 1977. *Introduction to Soil Microbiology.* New York: Wiley, 467p.

Alexander, T. G., and Robertson, J. A., 1968. Inorganic phosphorus forms in some Alberta soils as related to soil development, parent material and available phosphorus. *Canadian Jour. Soil Sci.,* 48, 289–295.

Allaway, W. H., and Rhoades, H. F., 1951. Forms and distribution of phosphorus in the horizons of some Nebraska soils in relation to profile development, *Soil Sci.,* **72,** 119–128.

Anderson, G., 1964. Investigations on the analysis of inositol hexaphosphate in soils, *8th Internat. Congr. Soil Sci. (Bucharest) Trans.,* **4,** 563–572.

Armson, K. A., 1977. *Forest Soils: Properties and Processes.* Toronto: University of Toronto Press, 390p.

Beckmann, G. G.; Crack, B. J.; and Prebble, R. E., 1976. Supplementary glossary of soil science terms as used in Australia, *Australian Soc. Soil Sci. Pub. No. 6,* 11p.

Birch, H. F., 1961. Phosphorus transformations during plant decomposition, *Plant Soil,* **15,** 347–366.

Black, C. A., 1968. *Soil-Plant Relationships.* New York: Wiley, 558–653.

Black, C. A., and Goring, C. A. I., 1953, Organic phosphorus in soils, *in* W. H. Pierre and A. G. Norman, eds., *Soil and Fertilizer Phosphorus in Crop Nutrition. Agron. Mon. No. 4.* New York: Academic, 123–152.

Bourman, R. A., and Cole, C. V., 1978. Transformations of organic phosphorus substrates in soils as evaluated by $NaHCO_3$ extraction, *Soil Sci.,* **125,** 49–54.

Brady, N. C., 1974. *The Nature and Properties of Soils.* New York: Macmillan, 639p.

Brown, I. C., and Drosdoff, M., 1940. Chemical and physical properties of soils and of their colloids developed from granitic materials in the Mojave desert, *Jour. Agric. Res.,* **61,** 335–352.

Chang, S. C., and Jackson, M. L., 1957. Fractionation of soil phosphorus, *Soil Sci.,* **84,** 133–144.

..sted, R. L., and McKercher, R. B., 1975. Biochemistry and cycling of phosphorus, *in* E. A. Paul and A. D. McLaren, eds., *Soil Biochemistry,* Vol. 4. New York: Marcel Dekker, 31-63.

Kittrick, J. A., and Jackson, M. L., 1956. Electron-microscope observations of the reaction of phosphate fixation in soils, *Jour. Soil Sci.,* 7, 81-90.

Kurtz, L. T., 1953. Inorganic phosphorus in acid and neutral soils, *in* W. H. Pierre and A. G. Norman, eds., *Soil and Fertilizer Phosphorus in Crop Nutrition. Agron. Mon. No. 4.* New York: Academic, 59-88.

Larsen, S., 1967. Soil phosphorus, *Adv. Agron.,* **19**, 151-210.

McKenzie, Jr., E.; Kurtz, L. T.; and Melsted, S. W., 1977. Phosphorus fertility of some tropical soils of Sierra Leone, *Soil Sci. Soc. America Jour.,* **41**, 373-381.

Mansell, R. S.; Selim, H. M.; and Fiskell, J. G. A., 1977. Simulated transformations and transport of phosphorus in soil, *Soil Sci.,* **124**, 102-109.

Middleton, K. R., 1954. Distribution of total phosphate in certain Sudan soils, *5th Internat. Congr. Soil Sci. (Belgian Congo) Trans.,* **3**, 218-222.

Norrish, K., 1968. Some phosphate minerals of soils, *9th Internat. Congr. Soil Sci. (Australia) Trans.,* **2**, 713-723.

Nye, P. H., and Bertheux, M. H., 1957. The distribution of phosphorus in forest and savanna soils of the Gold Coast and its agricultural significance, *Jour. Agric. Sci.,* **49**, 141-159.

Olsen, S. R., 1953. Inorganic phosphorus in alkaline and calcareous soils, *in* W. H. Pierre and A. G. Norman, Eds., *Soil and Fertilizer Phosphorus in Crop Nutrition. Agron. Mon. No. 4.* New York: Academic, 89-122.

Pearson, R. W., and Simonson, R. W., 1939. Organic phosphorus in seven Iowa soil profiles: Distribution and amounts as compared to organic carbon and nitrogen, *Soil Sci., Soc. America Proc.,* **4**, 162-167.

Pierre, W. H., 1948. The phosphorus cycle and soil fertility, *Jour. Am. Soc. Agron.,* **40**, 1-14

Runge, E. C. A., and Riecken, F. F., 1966. Influence of natural drainage on the distribution and forms of phosphorus in some Iowa prairie soils, *Soil Sci. Soc. America Proc.,* **30**, 624-630.

Ryden, J. C., and Syers, J. K., 1977. Origin of the labile phosphate pool in soils, *Soil Sci.,* **123**(6), 353-361.

Ryden, J. C.; Syers, J. K.; and Harris, D., 1973. Phosphorus in runoff and streams, *Adv. Agron.,* **25**, 1-45.

Schollenberger, C. J., 1920. Organic phosphorus content of Ohio soils, *Soil Sci.,* **10**, 127-141.

Shipp, R. F., and Matelski, R. P., 1960. A microscopic determination of apatite and a study of phosphorus in some Nebraska soil profiles. *Soil Sci. Soc. America Proc.,* **24**, 450-452.

Shkonde, E. I., and Koroleva, I. E., 1968. Mobility of nutrients in Chernozems, *9th Internat. Congr. Soil Sci. (Australia) Trans.,* **2**, 419-427.

Tisdale, S. L., and Nelson, W. L., 1964. *Soil Fertility and Fertilizers.* New York: Macmillan, 430p.

Uzu, F. O.; Juo, A. S. R.; and Fayemi, A. A. A., 1975. Forms of phosphorus in some important agricultural soils of Nigeria, *Soil Sci.* **120**, 212-218.

Walker, T. W., and Syers, J. K., 1976. The fate of phosphorus during pedogenesis, *Geoderma,* **15**, 1-19.

Weiser, V. L., 1933. Fixation and penetration of phosphates in Vermont soils, *Vermont Agric. Exp. Sta. Bull. No. 356.,* 31p.

Westin, F. C., and Brito, J. G. de, 1969. Phosphorus fractions of some Venezuelan soils as related to their stage of weathering, *Soil Sci.,* **107**, 194-202.

Cross-references: *Carbon Cycle; Macronutrients; Nitrogen Cycle; Nitrogen Fixation; Nutrient Losses; Nutrient Potential; Plant Nutrients; Soil Components, Inorganic; Soil Components, Organic; Sulfur Transformations.*

PHYLLOSILICATES—See CLAY MINERALS, SILICATES. VOL. IVA: COLLOIDS. VOL. IVB: SOIL MINERALS. VOL. VI: CLAY

PHYSICAL CHEMISTRY

Physical chemistry of soils concerns the relationships between physical and chemical phenomena of soils, especially borderline phenomena such as the common processes of adsorption (see *Adsorption Phenomena*) and ion exchange (see *Exchange Phenomena*). Like physical chemistry in general, physical chemistry of soils aims at predicting the pathways of chemical process over time and the final results under various conditions, beginning with the knowledge of structure and properties of the molecules involved in the reactions.

Historically development of physical chemistry of soils stems from the studies of Thompson and Way who discovered the ion exchange capacity of soils and formulated its main concepts from 1845-1854 (Russell, 1973). These studies provided the basis for investigation, development, and extensive application of various ion exchangers not only in soil science but also in modern chemistry and technology. Contributions have been made by many scientists in various countries, among them Antipov-Karataev, van Bemmelen, Ehrenberg, Gapon, Gedroiz, di Gleria, Gorbunov, Yarusov, Jenny, Kelley, Marshall, Mattson, Nikolskij, Oden, Ostwald, Remezov, Schofield, Tschernov, and Wiegner (See *Exchange Phenomena*).

The physical chemistry of soils involves the study of soil thermodynamics (q.v.), electrochemistry (q.v.) and photochemistry of soils, and kinetics of chemical soil processes as well as the investigation of soil solutions (q.v.), surface phenomena, and acid-base (see *Soil Reaction*) and reduction-oxidation (see *Redox Reactions*) equilibria. The scope of physical chemistry is broad because the resolution of problems in each subfield requires data from the entire body of knowledge pertaining to the chemical properties (see *Chemical Composition*) of soils.

Scope

Certainly, soil is one of the most complicated systems encountered in nature or technology, and it possesses a number of highly peculiar features. Such features include multiplicity of phases in soils (the number of phases always exceeds three), low natural humidity and large specific surface (which may be as high as 50–150 m^2/g for soil samples and up to 250–280 m^2/g for the fraction 0.001 mm). In such a system chemical interactions either occur at an interface or are directly affected by an interface. The chemical interactions may simultaneously involve clay minerals (montmorillonite, kaolinite, hydrous micas, etc.), organic substances (humic acids, fulvic acids, lipids, polypeptides, polycarbohydrates), ordinary and complex salts, and the products of reactions among the preceding components. Soil air (see *Aeration, Respiration, Atmosphere*) is typically rich in carbon dioxide, which affects all reactions occurring in the system. Finally, the natural or native soil is a changeable, dynamic formation, the processes in which are characterized by daily, seasonal, yearly, and secular dynamics (Rode, 1962). It is indeed not easy to apply the classical chemical laws derived for much simpler systems to such a complex multiphase and heterogenous system. However, the discrepancies often observed between the theoretical predictions and experimental results should be ascribed to inadequacy of theoretical frameworks, mathematical formulations, and methods of analyzing and processing results rather than to any essential impossibility of describing chemical processes in soils in terms of mathematical models and equations.

Thermodynamics

Focal points of soil physical chemistry are the application of thermodynamics in studies of soil formation and the theory of plant nutrition (see *Nutrient Potential).* Thermodynamics provides a basis for solving the problem of possible spontaneous chemical soil reactions, and it allows us to select the most plausible genetic hypotheses and the most probably pathways of transformations of minerals and organic substances in soils. The thermodynamics of soil moisture (see *Thermodynamics, Soil Moisture*) has so far been studied most, but it is traditionally classified as soil physics even though water is but one of the normal chemical components of soil.

Inasmuch as there is an exchange of matter and energy between a soil and its environment, soils are thermodynamically typical open systems. This factor limits applications of the equations derived for closed systems, but at the same time it provides new approaches to determining matter and energy balances in soil formation. Only in few model experiments can soil be treated as an equilibrium system; in many cases the soil state is close to being a stationary state so that the chemical reactions in the system have to be treated as irreversible ones. For such systems, the conventional thermodynamical potentials are characterized in terms of inequalities, for instance,

$$dH < TdS + Vdp \quad \text{and} \quad dZ < -SdT + Vdp$$

where H is enthalpy, S is entropy, Z is Gibbs free energy (isobaric potential), T is absolute temperature, V is volume, and p is pressure. The principal thermodynamic functions for the mineral component of soil have been estimated to have the following values: internal energy U about 4000–6000 kcal/100 g, enthalpy H 300–450 kcal/100 g, Gibbs free energy Z 300–400 kcal/100 g, and entropy S 15–30 cal/100 g. The combustion heats of soil humus fall in the 5000–6000 cal/g range and depend significantly on the humus type.

Soil scientists are interested in the processes that alter the composition of the system. The equilibrium state for such reactions may be characterized by the following equation, taking into account variations in chemical potentials:

$$dZ = Vdp - SdT + \Sigma\mu_j dn_j$$

The chemical potentials are assumed to be equal in all phases in a multiphase system in a state of equilibrium.

However, in the heterogenous soil system a potential drop is developed at each interface due to dissociation of the absorbed cations. Such a system has to be treated at equilibrium in terms of electrochemical rather than chemical potentials. For example, the distribution of sodium ions between the absorbing complex and the solution is therefore described by the following equation:

$$\mu'_{\text{Na}} + zF\psi' = \mu''_{\text{Na}} + zF\psi''$$

where prime and double prime denote different parts (or phases) of the system, ψ is the electrostatic potential, and F is the Faraday number.

Surface phenomena and the reactions at interfaces between soil phases are dealt with in the most general way by colloid chemistry of soils, which is traditionally classified as a part of physical chemistry. The study of soil colloids concerns the structure of the substances comprising them (primarily, the structure of clay minerals), the structure of the surface, development of the double electrical layer, and the theory of coagulation and peptization of colloidal particles (Marshall, 1964). Of essential importance are the turnovers of substances and the equilibria between the soil absorbing complex and the soil solution. Studies of the absorbing capacities of soils deal with these problems (Kelley, 1948).

Absorbing Capacity

Gedroiz (1931) suggested the following subdivision of the absorbing capacity of soils: (1) the mechanical absorbing capacity, i.e., the capacity to retain particles suspended in solution; (2) the physical absorbing capacity, which consists of adsorption and coagulation due to surface energy; (3) the physicochemical absorbing capacity or ion exchange capacity; (4) the chemical absorbing capacity due to formation and accumulation of difficulty soluble salts; and (5) the biological absorbing capacity.

Soil properties and plant growth are greatly affected by the capacity of soils to absorb and exchange cations, which depends on the pH value and the mechanical and mineral composition of the soil. A fundamental physicochemical characteristic of the soils is the absorption capacity for cations, which ranges from 5–10 meq/100 g in podzolic soils up to 50–80 meq/100 g chernozemic soils. The absorption capacity of the soil organic component may be as high as 500–700 meq/100 g, and it is highly dependent on pH value of the medium. The properties of soils depend not only on the total amounts of cations absorbed but also on their composition (Tucker, 1971). This introduces an important characteristic of acid soils, namely, the degree of saturation of soil with bases, the percent base saturation (PBS):

$$\mathrm{PBS} = \frac{\mathrm{TEB}\cdot 100}{\mathrm{CEC}} = \frac{S\cdot 100}{T}\,\%$$

where TEB = S, the total exchangable bases, and CEC = T, the cation exchange capacity (the absorption capacity). The corresponding expression for alkali and saline soils is ESP, the exchangeable sodium percentage.

The energy of cation absorption by soils depends on the charge of the cations, their radii, and their degree of hydration (Bolt and Bruggenwert, 1976). The absorption increases with increasing cation charge, and for the same charges it increases with increasing atomic weight according to the lyotropic series:

$$Li^+ < Na^+ < K^+ < Rb^+ < Cs^+$$

and

$$Mg^{2+} < Ca^{2+} < Sr^{2+} < Ba^{2+}$$

The order in this series may be changed somewhat according to the crystallochemical features of the sorbent.

The relationships between the ion composition of the solution and the composition of the exchange cations have been described by ion exchange isotherms. A number of equations were suggested as mathematical descriptions of exchange isotherms, as for instance, those derived by Freundlich, Jenny and Wiegner, Gans, Vanselow, Gapon, and Nikolskij (Marshall, 1964). The majority of these equations stem from the mass-action law and in the case of the exchange

$$S\mathrm{Ca} + 2\mathrm{Na}^+ \rightleftharpoons S\mathrm{Na}_2 + \mathrm{Ca}^{2+}$$

where S is the soil absorbing complex, we may write down the following relationship:

$$K = \frac{[S\mathrm{Na}_2]\cdot a_{\mathrm{CA}^{2+}}}{[S\mathrm{Ca}]\cdot a^2_{\mathrm{Na}^+}}$$

In this equation K is the exchange constant, $a_{Ca^{2+}}$ and a_{Na^+} are the activities of the calcium and sodium ions in the equilibrium solution, and $[S\mathrm{Na}_2]$ and $[S\mathrm{Ca}]$ are the concentrations of Na^+ and Ca^{2+} in the absorbed state; the activity coefficients of the absorbed cations are taken to be unity. None of the equations agree fully with the experimental data over wide ranges of concentrations and cation ratios. This lack of agreement is due to the multifunctional nature of the soil absorbing complex and to differences in the strength of bonding of the exchange cations with the adsorption centers. The latter fact may be taken into consideration if, when deriving the equation for equilibrium, we assume the electrochemical, rather than chemical, potentials in two parts of the system to be equal. The exchange isotherm is then most satisfactorily expressed as a log function.

The absorbed cations are capable of dissociation, which affects the composition of the soil solution. The highest degree of dissociation of univalent cations for weak dilutions ranges from 2–40% and for divalent cations, from 2–3%, frequently much less.

The degree of dissociation depends on the mineral composition, the humus content, and the strength of cation absorption by the absorbing complex. The dissociation and the Donnan equilibrium have been often related to the Wiegner effect (the suspension effect), but many modern results suggest that the potential difference observed between the suspension and the intermicellar solution is due to the diffusion potential. There are fewer data on absorption of anions even though that is also of great practical importance for evaluating the phosphate and nitrogen nutrition of plants.

Equilibria

Physicochemical analysis of equilibria in the heterogeneous system of solid phases and soil solution resulted in the development of the concepts of lime potential, phosphate potential, etc. (see *Nutriant Potential*). According to Schofield (1955), the phosphate potential PhP is given by the following expression:

$$\mathrm{PhP} = 0.5\mathrm{pCa} + \mathrm{pH_2PO_4}^-$$

where p is the negative logarithm of the activity of the respective component. The higher the PhP, the lower the activity and concentration of phosphates in the solution. If we make use of thermodynamical potentials, the decrease per mole of Gibbs free energy with transfer of phosphate from the solid phase into solution may be expressed through use of the chemical affinity of the reaction A as follows:

$$A = \frac{-2.3RT}{0.5p\mathrm{Ca} + \mathrm{pH_2PO_4}}$$

which differs from the phosphate potential only in sign and dimension. According to Schofield (1955), the lime potential LP is given by

$$\mathrm{LP} = \mathrm{pH} - 0.5\mathrm{pCa}$$

The potential LP does not depend on the ratio Ca:Mg; it is little affected by dilution, and many authors believe it to be a better characterization of soil acidity (q.v.) than pH value. A similar formula may be written for the potassium potential: pK – 0.5 p (Ca + Mg) (Woodruff, 1955). The preceding potentials are assumed to be stable characteristics of a soil though they may be used within only fairly narrow ranges. A similar concept underlies evaluation of provision of a certain element to plants using the quantity-intensity ratio according to Beckett (1971). For instance, to determine this parameter for potassium, the soil is stirred with a $KCl + CaCl_2$ solution, and in 12 hours the content of cations is measured in the filtrate; the difference yields the amount of the exchangeable potassium Δk in mg-equiv/100 g. The plot of Δk as a function of the ratio of the activities of K^+ and Ca^{2+} (or Ca + Mg) yields the activity of K^+ in the isotonic solution to characterize the state of the soil solution system at a given moment and the ability of soil to control the K^+ activity in the solution with changes in ambient conditions.

Soil Solution

Despite the importance of the heterogeneous nature of soils to equilibria, the soil solution (q.v.) is usually characterized independently of the solid phases, particularly in salinity studies. The best description of the soil solution may be obtained in terms of isotonic (isohydric for pH) solutions, that is, the solutions whose composition is not changed following interaction with the solid phases of soil. However, extraction and analysis of such solutions is a labor-consuming and complicated procedure, and simpler approximate methods are commonly used. The simplest, but at the same time the least satisfactory method, is to analyze water extracts of soil samples. The basic difficulty encountered in producing the native soil solution is that the diffuse ion layer of colloidal particles is in direct contact with the intermicellar medium, and therefore, separation of the phases inevitably results in an altered general state of the system. The primary thermodynamic function describing the soil solution is the activity a of the solution components, which is essentially different from the concentration c and may be formally calculated using the activity coefficient f from the formula $a = cf$. In the solutions from nonsaline mineral soils, the value of f for divalent ions ranges from 0.8 to 1.0, and for trivalent ions from 0.6 to 0.9. In the hydrocarbonate-sodium solutions with salt concentrations of 0.5–0.7 g/l, the activity coefficients for the respective ions are 0.6–0.7 and 0.3–0.4, whereas in chloride-sulphate solutions with salt concentrations of 2.5–5.0 g/l they are as low as 0.2–0.3 and 0.03–0.05. Thus the activity coefficients cannot be treated as small correction factors; they should be considered important characteristics because their values may have greater effects on the final results than the concentrations of components. Determinations of f from equations of the type

$$1nf = \frac{k \cdot \sqrt{J}}{1 + A\sqrt{J}}$$

where J is the ionic strength of the solution, and k and A are constants, are not always reliable and are labor-consuming, inasmuch as prior quantitative determinations of all components in solution are needed. A more promising method is the direct measurement of the activities of cations and anions using reversible ion-selective electrodes as is usual in electrometric determination of pH. Reliable enough electrodes have been suggested for direct determination in soils of the activities of K^+, Na^+, CA^{2+}, Cl^-, and some other ions (see *Electrochemistry*). One advantage of these electrodes is that they make possible characterization of soil without disturbing its native state.

The effects of the solid phases on the composition of the soil solution are especially strong in saline soils and in measuring pH values in a range from 3–5 to 8–9.

Acidity

The acid reaction (the *acidity*) of soils depends primarily on heterogeneous equilibria, the exchange cations H^+ and Al^{3+}, and the organic acids (see *Acidity*). An alkaline reaction is usually related to the presence of the alkali and alkali earth carbonates. For such systems, equations have been suggested taking into account the concentrations of CO_2 and carbonates. For instance,

$$pH = 6.00 - \tfrac{2}{3}\ 1\ gP_{CO_2} + \tfrac{1}{3}\ 1\ gf_{HCO_3^-} - \tfrac{1}{3}\ lgf_{Ca^{2+}}$$

or $$pH = \frac{pK_1 + \lg HCO_3^-}{[CO_2] + lgf_{H^+}}$$

where K_1 is the first-step dissociation constant for carbonic acid. As a rule, such equations yield results differing from the experimental pH values. Apparently, development of alkalinity also depends to some extent on substances such as various phosphates, sulphides, and silicates. As in other cases, evaluation of the soil reaction should take into consideration both the quantity factor and the intensity factor. The intensity factor is the measured pH value; the quantity factor comprises the potential (exchange and hydrolytic) acidity. The potential acidity is associated with the ability of soils to resist changes in the reaction. This *buffer capacity* of soil, (see *Buffers, Buffering*), as it may be called, is related not only to the reaction of the medium but also to many other characteristics such as reduction-oxidation potential and the C:N ratio (q.v.). In contrast to the buffer capacity of an isolated system, however, the buffer capacity of soils under natural conditions depends on the effects of soil forming factors and can be referred to as a *dynamic buffer capacity*. This is due to the fact that man-made effects such as liming (see *Lime, Liming*) do not eliminate the permanent climatic factors; the soil acidity reduced by liming can be restored by dissolution and leaching of carbonates.

Oxidation-Reduction

The reduction-oxidation processes in *automorphic* (well-drained) soils are characterized by reduction-oxidation potentials from +0.4 to +0.6 V, whereas in waterlogged and submerged soils in the presence of organic substances, reduction processes can decrease the potential down to +0.2 V and below. The principal factors controlling the potential level are aeration and the operation of biological processes (see *Redox Reactions*). A decrease in potential results in the increased mobility of certain elements through transformations such as Fe (III) → Fe(II) and Mn (IV) → Mn(II), thus affecting plant nutrition and intensifying migration of reduced substances. The oxidation regime, as a rule, contributes to accumulation of many difficultly soluble compounds in oxidized forms. According to the character of water regimes, soil areas may be classified as follows: zones of stable oxidation processes with potentials exceeding 0.40–0.45 V, zones of stable reduction processes with potentials below +0.40–0.45 V transitional zones, and zones of unstable reduction-oxidation regimes.

Photochemistry

We know relatively little about the photochemistry of soil processes. The electromagnetic waves in the form of solar radiation (direct or scattered) are a constant source of energy, especially great in equatorial zones. This energy is not only transformed into heat and thus affects the heat balance but it also has direct photochemical effects. The effect of radiant energy on organic substances has been studied more than has the effects on mineral substances. Under the effects of light, solutions of humic acids are more or less rapidly discovered, being completely bleached by large doses of radiation. The molecules of humic acids break down in the process into smaller units. On the other hand, other substances, for instance, some amino acids, are condensed under the effect of light to give rise to dark products. These processes involve primarily photochemical oxidation with formation of free radicals. Under natural conditions the photochemical reactions involve only the uppermost soil layer, but due to the transfers of matter among various layers, the products of photochemical reactions penetrate into the subsoil.

Although the thermodynamics of soils indicate that certain processes are possible in principle, their realization depends on kinetic parameters. The soil formation process as a whole, consisting of partial macro- and microprocess proceeds through individual and often numerous stages (Rode, 1962). By analogy with the processes in chemical technology, we can assume that the character and the final results of the soil formation process and its constituents depend on the slowest stage of the phenomenon as a whole. It should be stressed here that one specific feature of the soil chemical processes that affects their rate is the fact that they occur in the presence of solid phases exhibiting marked catalytic properties. It has been shown experimentally, for instance, that montmorillonite has a catalytic role in humification and that iron oxides, silicon oxides, and allophanes play a catalytic role in oxidation of phenols and autolyzates. It has been suggested that the principal active substances of soil are its organomineral complexes, which may be peculiar enzyme systems.

D. S. ORLOV

References

Bates, R. G., 1964. *Determination of pH: Theory and Practice.* New York: Wiley, 435p.

Beckett, P. H. T., 1971. Potassium potentials in soils, *Potash Review* (April/May 1971), Subject 5, *Soil Testing and Plant Analysis,* 30th Suite, 41p.

Bolt, G. H., and Bruggenwert, M. G. M., eds., 1976. *Soil Chemistry: A. Basic Elements.* Amsterdam: Elsevier, 281p.

Gedroiz, K. K., 1931. *Die Lehre vom Absorptionsvermögen der Böden.* Transl. from Russian by H. Kuron. Dresden: Steinkopf, 448p.

di Gleria, J.; Klimes-Szmix, A.; and Dvoracsen, M., 1962. *Bodenphysik and Bodencolloidik.* Budapest: Akedémiai Kiadó, 795p.

Hesse, R. P., 1972. *A Textbook of Soil Chemical Analysis.* New York: Chemistry Publications, 520p.

Kelley, W. P., 1948. *Cation Exchange in Soils.* New York: Rheinhold, 144p.

Marshall, C. E., 1964. *The Physical Chemistry and Mineralogy of Soils: Vol. 1. Soil Minerals.* New York: Wiley, 388p.

Rode, A. A., 1962. *Soil Science.* Transl. from Russian by A. Gourevich. Jerusalem: Israel Progressive Scientist Translations, 517p.

Russell, E. W., 1973. *Soil Conditions and Plant Growth,* 10th ed. London: Longmans, 849p.

Schofield, R. K., 1955. Can a precise meaning be given to "available" soil phosphorus? *Soils Fertil.* 18, 373-375.

Tucker, B. M., 1971. Basic exchangeable cations in soil. *CSIRO, (Australia) Div. Soils Tech. Paper, No. 8,* 22p.

Wiklander, L., 1964. Cation and anion exchange phenomena, *in* F. E. Bear, ed., *Chemistry of the Soil.* New York: Reinhold, 162-205.

Woodruff, C. M., 1955. Energies of replacement of calcium and potassium in soils, *Soil Sci. Soc. America Proc.,* 19, 167-171.

Cross-references: *Adsorption Phenomena; Electrochemistry; Electroosmosis; Exchange Phenomena; Flocculation; Ionic Activities; Nutrient Potential; Redox Reactions; Soil Biochemistry; Soil Chemistry; Soil Physics; Soil Solution; Thermodynanics, Soil Moisture; Thermal Regimes.*

PLANT NUTRIENTS

Substances plants require for successful growth and completion of their life cycles are referred to as *plant nutrients*. Examples are oxygen and carbon dioxide, water, and various mineral elements. Some nutrients are absorbed by plants in their elemental forms, and others as various compounds.

Essential Elements

The 16 elements listed in Table 1 have been shown to be essential for the growth of higher plants. Some elements (hydrogen, carbon, oxygen, nitrogen, potassium, phosphorus, calcium, magnesium, and sulfur) are required in rather large amounts and are therefore called *macronutrients* (q.v.). Others (chlorine, boron, maganese, iron, zinc, copper, and molybdenum) are required in relatively small amounts and are therefore called *micronutrients* (q.v.), *trace elements,* or *minor elements.* The optimum levels of individual nutrient elements in plants have a very wide range. On an equivalent basis, the ratio is about 1 for molybdenum and 60 million for hydrogen (Epstein, 1971; Gauch, 1972).

Some elements in addition to the 16 listed may be essential to certain species of plants. Thus, for example, sodium seems necessary for saltbush (*Atriplex vesicaria*) and silica for horsetail ferns (*Equisetum arvense*). Selenium may be essential for *Astragalus* spp. Rice does not grow normally when silicon is omitted from nutrient solutions, but the plants do survive. Cobalt seems to be required for *Rhizobium* bacteria to fix atmospheric nitrogen in nodules on the roots of legumes.

Much uncertainty remains about the nutritional needs of lower plants. Present evidence suggest that not all 16 elements required by higher plants are also required by lower forms. For example, it has not been possible to demonstrate that lower plants need boron. Moreover, elements like calcium that are

TABLE 1. Elements Essential for Plant Growth.

Element		Atomic Number	Atomic Weight	Principal Forms for Uptake	Plant Contents		
					%	Mole/g	Range %
Macronutrients							
Hydrogen	H	1	1	Water	6	60,000	
Carbon	C	6	12	Air CO_2 (soil)	45	40,000	
Oxygen	O	8	16	Water H_2O Air O_2	45	30,000	
Nitrogen	N	7	14	NO_3^- NH_4^{3+}	1.5	1,000	0.5–5
Potassium	K	19	39.1	K^+	1.0	250	0.5–5
Calcium	Ca	20	40.1	Ca^{2+}	0.5	125	0.05–5
Magnesium	Mg	12	24.3	Mg^{2+}	0.2	80	0.1–1
Phosphorus	P	15	31	$H_2PO_4^-$	0.2	60	0.1–0.5
Sulfur	S	16	32.1	$SO_4^=$	0.1	30	0.05–0.5
Micronutrients					*ppm*		*ppm*
Chlorine	Cl	17	35.5	Cl^-	100	3	100–10,000
Boron	B	5	10.8	H_3BO_3	20	2	2–100
Iron	Fe	26	55.9	Fe^{2+}, Fe^{3+}	100	2	50–1,000
Manganese	Mn	25	54.9	Mn^{2+}	50	1	20–200
Zinc	Zn	30	65.4	Zn^{2+}, Zincate	20	0.3	10–100
Copper	Cu	29	63.5	Cu^{2+}	6	0.1	2–20
Molybdenum	Mo	42	96.0	$MoO_4^=$	0.1	0.001	0.1–10

macronutrients for higher plants become micronutrients for many lower forms.

Historical Development of Current Knowledge

Ancient and Medieval Periods. Scholars of the classical period in Greece were much interested in and puzzled by plant nutrition. Theophrastus (372–287 B.C.), for example, devised a classification of plants and suggested the use of manures to improve poor land. Growing of legumes and the use of saltpeter (KNO_3) were also practiced. Existing knowledge of farm management, including that on plant nutrition, was summarized by Columella (ca. A.D. 60) in a book that was relied upon for several centuries and was a source of information for later writers as well. Much of the information assembled by Columella was restated and expanded by an Italian, Peter Crescentius (1230–1307) in a book that was also a prime guide for several centuries.

The Period of 1500 to 1810. During these three centuries numerous investigators searched for the "principle of vegetation." Discoveries were also made of the chemical and physiological bases for plant growth. As early as 1563 Palissy reported that salts were taken up by plants and returned to the soil in the form of ash, but his findings were not accepted. Water, potassium nitrate, and other substances were at one time or other considered to be the "principle of vegetation." Gradually enough experiments were conducted to demonstrate that plants required several substances, but the ideas widely held earlier disappeared slowly. In the early part of the nineteenth century, for example, Thaer, Davy, and Berzelius still held that plants fed on the humus in soils (Gauch, 1972; Russell, 1973).

Founding and Growth of Agricultural Chemistry. Boussingault (1802–1887) showed by pot experiments that the weight of plants grown was much larger than the loss of soil organic matter. Plants must use something other than humus. It remained for Liebig (1803–1873), however, through his trenchant and sarcastic writing to discredit the theory of humus as plant food and to gain acceptance for ideas formulated much earlier by de Saussure and others, namely, that carbon dioxide supplied carbon and oxygen, water supplied hydrogen and oxygen, and all other nutrients were derived from the soil. The needs of nonleguminous plants for a soil source of nitrogen were established in long-time experiments at Rothamsted in England after the time of Liebig. Fixation of gaseous nitrogen in root nodule bacteria was discovered as late as 1886. A few years earlier it had been learned that oxidation of ammonia to nitrite and then to nitrate was mediated by bacteria (Russell, 1973).

By the end of the nineteenth century, 10

elements (carbon, oxygen, hydrogen, nitrogen, phosphorus, sulfur, potassium, calcium, magnesium, and iron) were known to be essential for plant growth. Six more, all micronutrients, were added in the twentieth century (manganese, boron, zinc, copper, molybdenum, and chlorine). Knowledge of the physical chemistry of soils, including ion exchange and the fixation of nutrient elements, has increased greatly during the twentieth century. Mechanisms of nutrient transport in soils and plants and of nutrient absorption by plants have been subjects of intensive research in many places.

Plant Composition

The composition of plants differs with growing conditions, especially the nutrient supply. The composition also differs among plant parts, plant species and varieties, leaf position, and even time of day. For example, water contents range from less than 10% in some seeds to more than 90% in some roots and leaves. Ash contents of plants tend to go up and down with the supply of nutrients in the soil. Thus halophytes characteristically have high ash contents because they grow on saline soils. On a dry-weight basis, ash contents are relatively high in leaves; moderately low in seeds, tubers and roots; and very low in wood. Young leaves are usually high in most nutrients, whereas old leaves contain more of the immobile elements such as calcium, boron, and iron. Grains are high in nitrogen and phosphorus; roughly half of the amounts of those elements in the plant are present in the seeds.

Some plants accumulate very large amounts of specific elements, which can result in problems if the plants are being used by animals or humans. The content of fluorine, for example, is 0.2 to 2.0 ppm in most plants. Yet black tea may contain 100 to 200 ppm. Crop plants generally contain only 0.1 to 2 ppm of selenium, whereas species of certain wild plants such as *Astragalus racemosus* and *Stanleya bipinnata* may contain up to 15,000 ppm. These "converter plants" are able to use forms of selenium not available to other species. After such plants die and decay, the selenium is much more available to other species and can then cause "alkali disease" in cattle.

Plants take up solutes from the soil irrespective of actual needs. Hence plant composition does not reflect metabolic needs, nor does it reflect directly the composition of either the soil as a whole or of the solution at any given moment. Growing over a period time, a plant absorbs what is available in total during that interval.

Nutrient Transport and Uptake

Water and most nutrients are taken up by the roots of plants. Exceptions among nutrient elements are carbon and oxygen absorbed by the leaves. The efficiency of nutrient uptake depends on the aggregate root surface of a plant exposed in the soil. The entire root surface of a single plant can reach hundreds or even thousands of square meters, considering the major roots, rootlets, and root hairs. To be absorbed by a plant, nutrient elements in soils must reach its roots in a form suitable for uptake (Russell, 1973).

Transport to Plant Roots. Movement from the soil solution to the surface of a root is by diffusion and mass flow. Dominance of one mechanism or the other depends on the given situation, i.e., the ion, the rate of uptake, the rate of water flow, and various soil characteristics. Mass flow is dominant for ions that are present in relatively high concentrations (calcium and magnesium) and for those that are neither exchangeable nor otherwise stored (NO_3). In contrast, diffusion is dominant for ions present in low concentrations such as phosphates and iron. Because roots continue to grow and thus penetrate new regions in the soil, changes in the actual possibilities for absorption are continuous.

An alternative to the theory that nutrients are absorbed by plants from the soil solution is a theory of direct uptake of ions from either exchangeable form or from the solid phase. It has been shown that entry of nutrient elements into plants is enhanced by close proximity of roots to solids such as iron oxides and clays carrying exchangeable ions. These reactions can be explained, however, as due to increased nutrient concentrations in the aqueous boundary layer around roots because of the presence of a solid. Nutrient uptake would still be from the liquid phase.

Losses from Plants

These can occur through efflux from the plants, usually minor as compared to influx, and through rainwash, leaf drop, loss of guttation water, and secretions from special salt glands. Many of these losses return nutrients directly to the soil, a process of special importance in many tropical rain forests where relatively small amounts can be stored in the soils themselves.

Functions of Nutrients

Carbon and hydrogen are present in all organic compounds, whereas oxygen, nitrogen, sulfur, and phosphorus are constituents of important metabolites. The major functions of the remaining nutrient elements are as constituents or activators of enzymes. A few elements such as potassium play an important role in osmoregulation. Others have buffering effects on reaction and pH. The ionic environment also strongly affects the colloidal properties, especially membrane characteristics within the cells of plants. Major functions of the nutrient elements can therefore be summarized as follows (Epstein, 1971; Finck, 1969; Fried and Broeshart, 1967; Gauch, 1972):

Nitrogen. A constituent of a great number of important compounds such as proteins, nucleic acids, prophyrins (chlorophyll, cytochromes, etc.), and alkaloids. Nitrate induces the formation of the enzyme nitrate reductase, whereas ammonium seems to depress it.

Phosphorus. Plays a key role in energy metabolism. Many intermediates in the breakdown of sugars to water and carbon dioxide are phosphorylated. Adenosin triphosphate (ATP) functions universally in energy storage and transfer. Phosphorus is also a constituent of hydrogen carriers such as di- and triphosphopyridine nucleotides (NAD, NADP). It is a constituent of phospholipids that are contained in membranes. Phytic acid is a storage form of phosphorus in plants.

Sulfur. A constituent of the amino acids crystine, cysteine, methionine, homocysteine in proteins; of coenzymes thiamine and biotin and of coenzyme A. Ferredoxins, compounds that function in photosynthesis, contain sulfur. Some special compounds, often volatile and with disagreeable odors, contain sulfur, e.g., mercaptans in radish roots and onions, disulfides in onions, polysulfides and sulfoxides in garlic oil and mustard oils.

Potassium. Not known as a constituent; in any case, it contributes to activation of large numbers of enzymes, probably through effects on enzyme conformation. Examples are the synthesases· acetyl-CoA-synthesase, NAD-synthesase, and malic enzyme; the dehydrogenases: glycerindehydrogenase, malic acid dehydrogenase, and succinic acid dehydrogenase; and kinases such as pyruvate kinase, fructokinase, and phosphohexokinase. Osmotic effects include a role in operation of guard cells of stomata; a role in opening and closing of leaflets of some plants; and a role in osmotic adjustment to salinity. Potassium accelerates movement of photosynthates from leaves.

Calcium. A constituent of phytin and pectin (important for mechanical strength); present in the enzyme amylase and as precipitates (Ca-oxalates, -carbonates, and to a lesser extent as -sulfates and -phosphates). Calcium activates many enzymes and often acts as a competitive inhibitor for magnesium. It stabilizes membranes, inhibits cell elongation, regulates ion uptake mechanism, and is a chemotropic factor in pollen germination.

Magnesium. A constituent of chlorophyll and of phytin and pectin. The element is an activator of many enzymes, especially phosphorylating enzymes (connected with energy metabolism). It is antagonistic to potassium absorption.

Iron. A constituent of many compounds, viz., iron porphyrins (heme compounds) such as cytochromes, cytochrome oxidases, catalases, and peroxidases. The element has a predominant function in electron transport through valence change between di- and trivalent forms. Hemoglobin plays a role in nitrogen fixation by root nodule bacteria. Non-porphyrin compounds are ferredoxin and ferritin. Iron is an activator of enzymes such as aconitase.

Manganese. A constituent of the enzyme manganin and an activator of many enzymes, similar to magnesium for enzymes in energy metabolism (phosphorylation), arginase, in photosynthesis, formation of chlorophyll, and synthesis of vitamin C.

Zinc. A constituent of some dehydrogenases (alcohol dehydrogenase, lactic dehydrogenase), the element also functions in the formation of auxins, in carbonic anhydrase, and in peptidases.

Copper. A constituent of enzymes (oxydases, functions through change in valence) such as lactases, phenolases, ascorbic acid oxidases, and cytochrome oxidase. It is also a constituent of plastocyanin, a compound that functions in photosynthetic electron transport. Copper may also function in nitrite reductase.

Boron. Function is little understood. Participation in carbohydrate metabolism and transport has been suggested, as has interaction with growth hormones.

Molybdenum. A constituent of enzymes involved in nitrogen fixation and nitrogen reduction. It probably functions through change in valence.

Chlorine. This functions in photosynthesis and as an activator of enzymes involved in splitting water. It also functions in osmoregulation of plants growing on saline soils.

Cobalt. Required for nitrogen fixation and is a constituent of vitamin B_{12}.

J. J. OERTLI

References

Chapman, H. D., ed., 1966. *Diagnostic Criteria for Plants and Soils.* Berkeley: University of California, Division of Agricultural Sciences, 793p.

Epstein, E., 1971. *Mineral Nutrition of Plants: Principles and Perspectives.* New York: Wiley, 412p.
Finck, A., 1969. *Pflanzenerährung in Stichworten.* Kiel: Hirt, 200p.
Fried, M., and Broeshart, H., 1967. *The Soil-Plant System in Relation to Inorganic Nutrition.* New York: Academic, 358p.
Gauch, H. G. Jr., 1972. *Inorganic Plant Nutrition.* Stroudsburg, Pa.: Dowden, Hutchinson & Ross, 488p.
Richards, L. A., ed., 1953. Diagnosis and improvement of saline and alkali soils, *U.S. Department of Agriculture Handbook No. 60,* 160p.
Russell, E. W., 1973. *Soil Conditions and Plant Growth,* 10th ed. London: Longmans, 849p.
Sprague, H. B., ed., 1964. *Hunger Signs in Crops,* 3rd ed. New York: McKay, 461p.
Wallace, T., 1961. *The Diagnosis of Mineral Deficiencies in Plants,* 3rd ed. London: H.M. Sta. Office, 125p.

Cross-references: *Adsorption Phenomena; Diffusion Processes; Exchange Phenomena; Ionic Activity; Macronutrients; Metabolic Activity; Micronutrients; Nutrient Potential; Soil Fertility; Soil Solution.*

POLLUTION

Soil pollution is its contamination with various harmful compounds introduced by human activity. The definition thus hinges on the *identifiable presence* of a *contaminant* added by *human activity*. The last phrase excludes the presence in soil of large quantities of undesirable constituents due to natural causes. A typical example is the naturally high levels of nickel derived from serpentine in many soils. Such levels, though not considered as pollution, have the same harmful effects on plant growth and groundwater composition as does nickel added by land disposal of sewage sludge. Pollution forms one of a broader group of soil deterioration phenomena, which also includes erosion, salinization, and compaction.

Pollution can alter the functioning of soil (1) as the platform for a substantial part of life on the earth and (2) as the filter for water on its way into underground reservoirs or surface flow. Pollution may inhibit plant growth, or it may add toxic elements to the food chain through their uptake by plants. Pollution can result in the addition of undesirable elements to water passing through the soil or by failure of the soil filter to function properly. Emphasis is normally placed on the release of substances by the soil during passage of water. At the same time soil pollution may change its oxidation-reduction state or its microbial life so that undesirable substances are no longer removed from water during its passage.

The abiotic aspects of soil pollution have been discussed in a report by de Haan and Zwerman (1976). They trace the pathways through soils of a number of possible contaminants and provide numerous references to current literature. A survey of factors governing biotic aspects has been presented by Richards (1974).

Investigations of soil pollution have focused largely on transport and accumulation of contaminants. The principal carrier is water, but the gas phase is important in some instances as, for example, in the leakage of natural gas from distribution mains (Hoeks, 1972). The mobility of various compounds in soils is controlled chiefly by their reactions with the solid phase. Examples of approaches to characterize such reactions are the use of chromatography equations to describe displacement of cations (Reiniger and Bolt, 1972) and the movement of pesticides (Leistra, 1973; van Genuchten, Davidson, and Wierenga, 1974).

Presently known soil contaminants can be grouped into hydrocarbons, pesticides, and heavy metals. Each group is discussed in a subsequent section. In addition to these contaminants, substances normally present in soils and essential for plant growth may occur in high concentrations. Examples are nitrogen and phosphorus, which are discussed in the last section of this article.

Hydrocarbons

Examples of hydrocarbons are natural gas and petroleum. Leaks in gas mains (Hoeks, 1972) have damaged nearby vegetation through elimination of oxygen from the gas phase in soils following microbial oxidation of the methane. Displacement of the oxygen is complete under pavements. Sources of oil contamination are usually local spills or the disposal of oily waste. Water coming into contact with the oil acquires a disagreeable smell and taste through the transfer of small amounts of water-soluble components. Oil spreading can thus have undesirable effects because of its increased contact with water. On the other hand, oil spreading tends to favor its decomposition by microorganisms. Time required for decomposition depends on the nature of the hydrocarbon, temperature and moisture conditions, and the supply of oxygen and other elements essential to the microflora. A further factor is the time required by microflora to adapt to the foreign substance or substances.

Pesticides

A large number of compounds are applied on or in the soil as part of normal agricultural practice in many countries. The compounds are variously known as *fungicides, herbicides, insecticides,* and *nematicides,* depending on the target pest. Some pesticides containing heavy metals are still in use but the great majority now used are organic chemicals. Possible side effects of such chemicals on non-target organisms are a matter of public concern because of their widespread applications. Pesticides are a very diffuse source of soil pollution.

The rate of decomposition of pesticides in soils is of great importance because that rate largely governs chances for harmful accumulations. Transport and possible reactions with soil constituents can also affect concentration of pesticides. Although decomposition may be photolytic or chemical in nature, biodegradation is by far the most important process for the breakdown of the compounds.

The rate of decomposition is expressed as a "persistency value," the notation for which is DT 50 (disappearance time for 50% of the compound). The value given for DT 50 represents the days required for decomposition of 50% of a compound under the specified conditions. Table 1 presents the DT 50 values for a few pesticide species. Several tens of thousands of active compounds are now known, and the total is being expanded continuously.

The decomposition rate depends on conditions in the soil and on the bonding of a pesticide to soil constituents. Thus, for example, DDT breaks down 10 times faster under anaerobic than aerobic conditions. For most pesticides, however, the aerobic decomposition proceeds much faster than the anaerobic one. Bonding is predominantly governed by the organic fraction of the soil, but there is a wide range in the affinity of pesticides for the soil solid phase. Values of a distribution factor (indicating distribution in solid and liquid phases) range, for example, from about 2 for simazine to roughly 140,000 for DDT. These values indicate that simazine may be leached downward readily, whereas translocation of DDT in soils is almost restricted to mechanical displacement, the movement of soil materials by tillage or by fauna.

TABLE 1. Persistency Data (Disappearance Time 50%, Days) of Several Pesticides. (Adapted from Hamaker, 1972)

Compound	DT 50
Malathion	0.6
Aldicarb	16
Linuron	24
DDT (anaerobic)	35
Simazine	100
Terbacil	180
Lindane	570
Heptachlor	2000
DDT (aerobic)	3800
Endrin	4300

Heavy Metals

The phrase "heavy metals" is used here as a general name for metals with densities in excess of 5 g/cc. About 15 species are of practical concern, and some of these are discussed individually in alphabetical order later. Sources and general behavior of heavy metals in soils are considered next.

Heavy metals may be applied to soils deliberately to correct nutrient deficiencies or to kill pests. Very small amounts are needed to correct deficiencies, and these do not cause pollution. Repeated applications of inorganic pesticides containing heavy metals, as, for example in sprays applied to fruit trees, may add amounts to soils large enough to be harmful. In contrast to organic pesticides, heavy metals do not disappear through decomposition but remain in soils indefinitely. Additional sources of soil contamination by heavy metals are industrial and traffic exhausts, flooding of land by contaminated waters (e.g., foreland soils along rivers), sewage sludge applied to land, and disposal of other refuse.

Heavy metals participate in several kinds of reactions in soils, and these affect their concentrations and solubilities. The metal ions tend to be bonded to soil constituents through cation exchange (see *Exchange Phenomena*); this may amount to substantial quantities even though concentrations in the soil solution (q.v.) are usually low. Some soil constituents seem to have specific affinities for heavy metal ions, resulting in their preferential adsorption over more abundant cations (Lagerwerff and Brower, 1972; Harmsen, 1977). The concentrations of heavy metals in the soil solution are also affected by equilibria with hydroxyl, carbonate, and phosphate ions. Precipitation of heavy metals by these anions can limit concentrations even though fairly large amounts are added to soil. On the other hand, some heavy metal ions are strongly chelated by organic substances of low molecular weight, thereby altering their adsorption behavior and permitting rather high concentrations in the soil solution. The actual concentration in a soil is thus a function of reactions of heavy metals with a variety of soil constituents.

As a general rule, the concentrations in the soil solution rather than the total quantity in a soil are the primary concern. Even with low concentrations, however, there are possibilities for leaching of heavy metals from soils and subsequent accumulations in natural waters where they can be harmful.

Cadmium (Cd). At present cadmium is considered one of the most hazardous of the heavy metals because of its presumed effect on the development of vascular disease. Amounts of cadmium in soils are normally below 1 ppm, but values as high as 1700 ppm have been reported for surface samples collected near zinc-ore smelters. Cadmium is usually associated with zinc in nature, and the geochemical relationship between the two leads to their common occurrence with Zn/Cd ratios near 900. Cadmium is easily taken up by most plants. Some are quite sensitive to excess cadmium, and others are not.

Chromium (Cr). Often derived from refuse of the plating industry, this element may be toxic for plant growth but only at high concentrations. Chromium mobility within plants is extremely low. Soil pollution by chromium is seldom a problem because it is taken up by plant as chromate, a form that hardly occurs at prevailing pH values and redox potentials.

Cobalt (Co). This can be highly toxic to plants. Most plant species require solution concentrations of cobalt not exceeding 0.1 ppm. Usually cobalt contents of soil do not exceed 10 ppm. Preferential cobalt adsorption on soil constituents and fixation in clay mineral lattices have been described. Prospective health hazards due to cobalt soil pollution appear to be minor.

Copper (Cu). Copper is toxic to most plants at solution concentrations exceeding 0.1 ppm. Its concentration in drinking water for human consumption is considered safe when not exceeding 1.0 ppm. Concentrations above 20 ppm of copper in feed and forage are toxic to sheep. Normal copper contents of soils are around 20 ppm. Mobility and displacement of copper in soils are low because of its strong bonding with organic matter and clay minerals.

Lead (Pb). This may accumulate in soils along roads from traffic exhausts and in the vicinities of lead-zinc smelters. Roadside concentrations as high as 2400 ppm have been reported. While (excessive) intake of lead by humans and animals is considered a serious health hazard, the primary pathway of such intakes is probably via surface contamination of crops and grasses (eaten by grazing animals) rather than via plant uptake. The mobility of lead in soil and plants tends to be low, though a few cases are known where considerable uptake by plants was found without significant yield depression. Normal lead levels in plants range from 0.5–3 ppm. With respect to plant growth, lead toxicity levels in solution appear to differ considerably for different plant species.

Mercury (Hg). Extensive mercury poisonings were first reported at Minamata, Japan, in 1953. As a result of the strong interactions between mercury compounds and soil constituents, displacement of mercury in forms other than vapor is usually very low. Methylation of mercury, possibly occurring in nature under restricted conditions, constitutes one of the most serious hazards related to this element, because in this form mercury will accumulate easily in food chains. Because of this hazard, the use of alkylmercury fungicides for seed dressings has been outlawed in many countries.

Molybdenum (Mo). This element is best known for its deficiency in certain soils. Under normal conditions molybdenum predominates in anionic form (molybdate), subject to adsorption by iron oxides and hydroxides much like phosphate. While normal molybdenum content in plants is around 0.1 ppm, toxicity symptoms have been observed at levels above 200–300 ppm (dry matter).

Nickel (Ni). This element tends to be highly toxic to plants. As it is easily taken up by plants when present in soils, care must be exercised in disposal of waste containing nickel. Total nickel contents in soils range from 5–500 ppm, with 100 ppm as a rough mean value. The concentration in the soil solution is usually around 0.005–0.05 ppm, and contents in healthy plants do not exceed 1 ppm (dry matter).

Zinc (Zn). The use of this element in galvanized iron is widespread. Zinc commonly occurs in soils at levels of 10–300 ppm, with 30–50 ppm as a rough average range. Sewage sludges may have very high zinc contents, and the possible accumulation of zinc in soil after disposal of such wastes deserves attention. In plants zinc will become toxic at levels exceeding about 400 ppm (dry matter), where it probably interferes with the uptake of other essential elements. In soil zinc appears to be rather mobile.

Nitrogen (N) and Phosphorus (P)

Contamination of soil by hydrocarbons, pesticides, or heavy metals is easily recognized. The first two are not present naturally in soils, whereas the last occur in very small amounts. In contrast to these substances, others that are necessary for plant growth may occur in large enough amounts to lead to pollution in some circumstances. Examples are the major plant nutrient elements nitrogen and phosphorus. Modern agricultural practice requires higher levels of these elements than are found naturally in most soils.

Large dressings of nitrogen in various forms are applied to many soils to increase crop

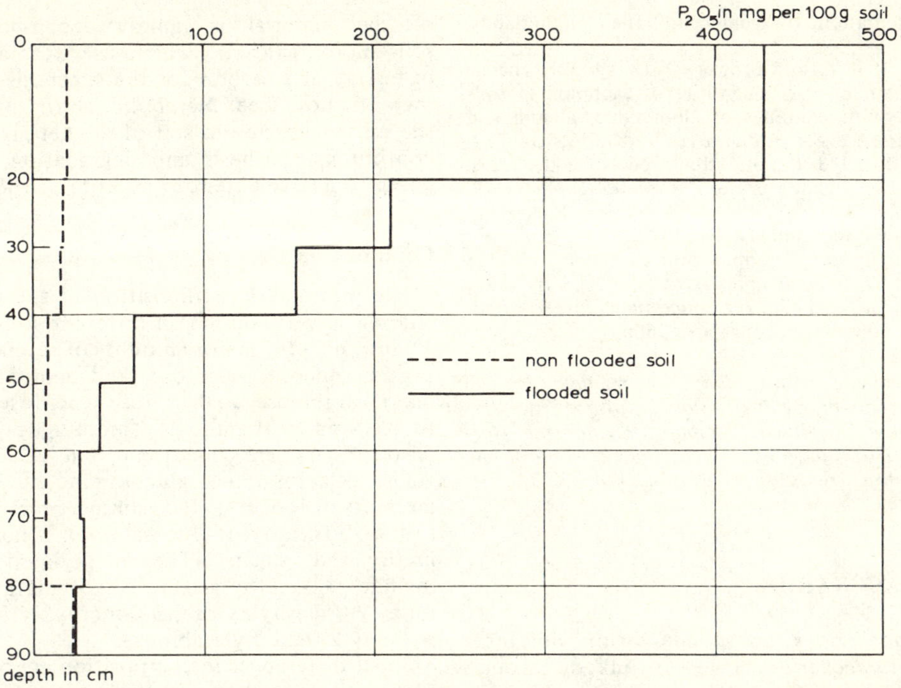

FIGURE 1. Accumulation of phosphate in soil following wastewater application. (Beck and de Haan, 1974)

yields and thereby expand production of food and fiber. Nitrate nitrogen is mobile in soils; amounts not taken up by plants or microorganisms can be leached and thus find their way into drainage waters. *Microbial denitrification* (see *Nitrogen Cycle* and *Denitrification*), the conversion of nitrate to nitrite and even nitrogen gas, is the major process eliminating surplus nitrate ions. Adjustment of nitrogen dressings to need insofar as possible is therefore good practice. Systems using soils to dispose of wastes high in nitrogen must be designed accordingly or nitrates will escape to surface and underground waters.

In contrast to nitrogen, phosphorus tends to be immobile in soils because of precipitation by calcium, aluminum, and iron and adsorption by aluminum and iron oxides and hydroxides (see *Phosphorus Cycle*). The very large capacity of soil to retain phosphorus is illustrated in Figure 1, which shows the effects of applications of sewage water for several decades. The amount of phosphorus retained was roughly 20,000 kg/ha as P_2O_5. Phosphorus can contribute to eutrophication of natural waters, but sources in populated areas are likely to be the direct entry of sewage or of effluent from treatment plants into open waters.

All in all, soil pollution follows from overtaxing a natural system with a generally large buffer capacity. That capacity reflects many different types of adsorption and precipitation reactions and the activities of immense numbers of microorganisms capable of decomposing many different compounds.

F. A. M. DE HAAN
G. H. BOLT

References

Beek, J., and Haan, F. A. M. de, 1974. Phosphate removal by soil in relation to waste disposal, *Proc. Int. Conf. on Land Use for Waste Management, Ottawa,* 77–87.

van Genuchten, M. Th.; Davidson, J. M; and Wierenga, P. J., 1974. An evaluation of kinetic and equilibrium equations for the prediction of pesticide movement through porous media, *Soil Sci. Soc. America Proc.,* 38, 29–35.

de Haan, F. A. M., and Zwerman, P. J., 1976. Pollution of soil, *in* G. H. Bolt and M. G. M. Bruggenwert, eds., *Soil Chemistry.* Amsterdam: Elsevier, 192–271.

Hamaker, J. W., 1972. Decomposition, quantitative aspects, *in* C. A. I. Goring and J. W. Hamaker, eds., *Organic Chemicals in the Soil Environment.* New York: Marcel Dekker, 253–340.

Harmsen, K., 1977. Behavior of heavy metals in soils, *Agric. Res. Rep. 866,* Center for Agricultural Publications and Documentation, Wageningen, The Netherlands, 171p.

Hoeks. J., 1972. Effect of leaking natural gas on soil and vegetation in urban areas, *Agric. Res. Rep. 778,* Center for Agricultural Publications and

Documentation, Wageningen, The Netherlands, 120p.

Lagerwerff, J. V., and Brower, D. L., 1972. Exchange adsorption of trace quantities of cadmium in soils treated with chlorides of aluminum, calcium and cadmium, *Soil Sci. Soc. America Proc.*, **36**, 734–737.

Leistra, M., 1973. Computation models for the transport of pesticides in soil, *Residue Reviews*, **49**, 87–130.

Reiniger, P., and Bolt, G. H., 1972. Theory of chromatography and its application to cation exchange in soils, *Netherlands Jour. Agric. Sci.*, **20**, 301–313.

Richards, B. N., 1974. *Introduction to the Soil Ecosystem.* New York: Longmans, 266p.

Cross-references: *Adsorption Phenomena; Clay-Organic Reactions; Denitrification; Diffusion Processes; Dispersion Phenomena; Exchange Phenomena; Herbicide Diffusion; Soil Microbiology; Soil-solvent Reactions; Transport Processes. Vol. XVI: Soil Pollution.*

POLYSACCHARIDE

Polysaccharides are condensation polymers of *monosaccharides* (sugars) linked through glycosidic bonds with the elimination of water (Finch, Hayes, and Stacy, 1971). They usually contain at least 10 sugar residues and fall into the class of substances known as *carbohydrates.*

The first evidence of the fact that soils contain carbohydrate was obtained towards the end of the last century when furfural derivatives were produced by boiling the soil with hydrochloric acid, while the first polysaccharide, thought to be a pentosan, was isolated from soil in 1910. Although there may well be many different polysaccharides in soil, they cannot be distinguished easily and for convenience they are referred to collectively as *soil polysaccharide.*

Polysaccharide usually constitutes between 5 and 25% of the organic matter (q.v.) of soils and may be as much as 40% of peaty soils. In any particular soil type the proportion is remarkably constant, usually being in the region of 10% in arable soils. Free *monosaccharides,* in contrast, account for less than 1% of the organic matter. Soil polysaccharide has several origins. During growth, plant exudates containing *oligosaccharides* as well as sugars are added to the soil through the roots. Mature plants may have up to 50% of their dry matter in the form of polysaccharide such as cellulose, hemicellulose, and pectin, and after death, plant remains decay in the soil. The amount of polysaccharide added to the soil in one year in this way may be as much as 20% of that present in the soil at a given time. Animals contribute much less. Much of the material is rapidly decomposed by soil microorganisms, which use it as a source of energy and carbon for the synthesis of their own tissues (see *Metabolic Activity*). Thus the polysaccharide in soil of the net product of continuous synthesis and degradation is composed to a large extent of microbial products.

Composition

The monomeric composition of the polysaccharide is very similar in different soils (Table 1) and includes major amounts of seven neutral sugars: glucose, galactose, and mannose (hexoses); arabinose and xylose (pentoses); and fucose and rhamnose (methyl pentoses). Also present are glucuronic and galacturonic acids, galactosamine, glucosamine, and minor amounts of N-acetyl glucosamine, muramic acid (3-0-carboxyethyl-D-glucosamine), ribose, and methylated sugars. The methylated sugars include 4-0-methylgalactose, 2-0-methylrhamnose, 2-0-methylxylose, 3-0-methylxylose, and possibly 2-0-methylarabinose.

Small differences in composition are observed from soil to soil particularly with respect to glucose, arabinose, and xylose, which form a greater proportion of the sugars in soils rich in plant remains. The composition in cultivated soils, however, does not appear to be affected by the type of crop. In uncultivated soils, there are often small variations with depth in the profile. These variations have been noted with arabinose, mannose, and rhamnose, but there are no consistent trends. The uronic acid content of soils ranges from less than 1% to about 6%. Five to 11% of the nitrogen of surface soils is present in amino sugar. The ratio of glucosamine to galactosamine ranges from 1.2 to 4.6, the highest ratios being observed in podzolic soils and mor litter, in which fungi are most active.

Part of the polysaccharide bears strong resemblance to constituents of plants. Thus 10–20% of the glucose in soil may be extracted with Schweitzer's cuprammonium hydroxide reagent in the form of cellulose. The only other polymer to be identified is a polygalacturonic acid, characterized by infrared spectroscopy, from the *H* horizon of a podzol. The presence of xylan in soil has been indicated by the identification of β (1–4) linked xylobiose and xylotriose.

These polymers aside, it has yet to be shown whether soil polysaccharide is a mixture of *homopolysaccharides,* each containing only one type of sugar in each molecule, or whether it consists of *heteropolysaccharides,* having many sugars within each molecule. Microbial polysaccharides that contain more than five

TABLE 1. Monosaccharide Composition of Soils Expressed as a Percentage of the Sum of Sugars Measured.

Soil	Horizon	Galactose	Glucose	Mannose	Arabinose	Xylose	Ribose	Rham-nose	Fucose	Total Sugars: As % of the Soil	Total Sugars: As % of the Soil Carbon	Total Soil Carbon (%)
A	0.5–1 m	6.6	80.2	6.2	1.1	3.3	-	2.6	-	25.5	17.0	60.0
B	*A*1	14.5	40.0	16.7	10.4	8.1	1.2	6.3	2.8	1.07	9.4	4.58
	*B*1	13.8	27.5	26.5	10.1	10.2	1.7	6.1	4.1	0.34	23.6	0.59
	*B*21	14.6	27.2	24.3	11.6	10.9	2.2	5.5	3.8	0.29	21.0	0.55
C	*S* 0–6 cm	12.2	39.3	19.4	11.7	7.5	1.9	4.4	3.6	0.290*		2.4
D	*S* 5–25 cm	11	39	12	13	14	-	8	3	1.52	9.7	5.9

*by Anthrone

Uncultivated Soils

A. Peat formed under *Calluna vulgaris, Eriophorum vaginatum* and *Sphagnum* spp. Scotland.

B. Typic Hapludalf, developed from cherty limestone under oak-hickory forest, Bardley series, U.S.A. (Folsom, Wagner, and Scrivner, 1974)

Cultivated Soils

C. Old pasture on Urrbrae fine sandy loam, Australia (Swincer, Oades, and Greenland, 1968)

D. Eight-course rotation on soil developed on basic igneous till. Insch series of the Insch Association, Scotland.

different sugar components, though rare, do occur, and the hemicellulose components of plants contain small quantities of complex polysaccharides. It has been thought that such complicated polysaccharides could gradually accumulate in soil; they would appear more stable than homopolysaccharides because a particular succession of enzymes of different specificities would be needed to hydrolize them, and this would be a very slow process.

Analysis

Analysis of polysaccharide usually involves a preliminary acid hydrolysis to convert it to monosaccharides, but conditions for obtaining the maximum hydrolysis of the polysaccharide with minimum destruction of the monosaccharides appear to differ from soil to soil. For complete hydrolysis a pretreatment with cold 12 M sulphuric acid to depolymerize the cellulosic fraction is necessary before treatment by dilute acid. Prolonged treatment with 12 M acid leads to destruction of some of the pentose components of the polysaccharide, and more accurate values may be obtained for some soils by a preliminary 20-min reflux with 2.5 MH_2SO_4, which releases most of the pentose, methyl pentose, galactose, and mannose. Hydrolysis of soil with 0.5 MH_2SO_4 at 100°C liberates only about three quarters of the sugars in the hemicellulose (noncellulose) fraction, but isolated water-soluble polysaccharide may be completely hydrolyzed by heating with dilute acid (0.5 MH_2SO_4, 100°C, 5 h). Fructose is very rapidly destroyed by acid hydrolysis and is rarely reported in hydrolysates but has been observed amongst the free sugars in soil. The isolated polysaccharide is sufficiently acidic to undergo autohydrolysis of its arabinose and part of its ribose components when heated in water. For maximum yields of uronic acids, the soil appears to need treatment with 12 M H_2SO_4. Maximum yields of hexosamines are obtained by hydrolysis with 6 M HC1 at 100°C for between 6 and 9 h, but losses caused by degradation occur.

Total quantities of reducing sugars in soil hydrolysates have been determined in colorimetric procedures using alkaline ferricyanide or alkaline copper salt solutions. Other methods are based on the formation of furfural derivatives in hot concentrated acid and their reaction with phenols. Reagents commonly used have been orcinol, anthrone, which is particularly sensitive for hexoses, and phenol, which gives high values because of interfering organic substances. Estimations of pentose have been obtained using strongly acidic reagents containing orcinol-ferric chloride or aniline acetate, although neither of these is specific.

Individual sugars can be determined after separation by paper chromatography. Alternatively, ion exchange chromatography of the borate complexes or gas liquid chromatography of the alditol-acetate derivatives can also be used. Glucose has been specifically determined in hydrolysates by glucose oxidase.

Uronic acids were originally measured in soils by the amounts of CO_2 formed on decarboxylation during treatment with 12% HC1, but it was subsequently shown that some humic substances decarboxylate even more readily than uronic acids. Subsequently colorimetric procedures using carbazole have been applied.

Hexosamines may be determined colorimetrically using N, N-dimethyl-p-aminobenzaldehyde or individually by ninhydrin after separation in an amino acid analyzer.

Isolation

Extraction. Polysaccharide, extracted from soil is a mixture of various neutral and acidic polysaccharides, some of which are chemically combined with proteins or peptides.

Part of the polysaccharide may be removed from the soil by extractants such as water, buffers, complexing agents, organic reagents, acids, and alkalis. Hot water at 70–80°C may extract as much as one quarter of the soil polysaccharide with comparatively little concomitant humic material, but behavior of the extract on gel columns indicates that some degradation occurs. Polysaccharide extracted with a buffer such as phosphate at pH 7.0 has a relatively high viscosity, indicating that less degradation occurs with this method than with others. Alkali is by far the most effective solvent, extracting as much as 50% of the soil polysaccharide and its efficiency may be increased by ultrasonic dispersion. Under alkaline conditions some shortening of poly saccharide chains occurs and in the presence of oxygen, some oxidation. To alleviate the oxidative effects, soils have been extracted under nitrogen or in the presence of stannous chloride. Extraction with dilute acid at 3°C has proved useful for peaty soils. Sequential extraction with acid, alkali, and acetic anhydride containing 2.5% sulfuric acid has given yields of more than 80% with some mineral soils, although acetolysis occurs in the last stage. Refluxing soil with 0.2M lithium bromide in anhydrous formic acid is also effective, but this may be because of hydrolysis. Dimethylformamide has given good yields of polysaccharide, but degradation may occur.

Purification. The extracted polysaccharide usually contains a large proportion of noncarbohydrate matter, which includes peptides, phenols, and humic substances. Acidification of alkaline extracts to pH 2 to 3 removes, by precipitation, a large fraction of the extracted noncarbohydrate organic matter (humic acid) and leaves polysaccharide in the acid-soluble (fulvic acid) fraction; however, as much as 12% of the polysaccharide may be coprecipitated with the humic acid. The adsorption of polysaccharide on humic acid, when the latter is precipitated by acidification, can be largely prevented by the pretreatment of the alkaline extract with the hydrogen form of a cation exchange resin, thus removing cations that probably form a linkage between the humic acid and the polysaccharide.

Adsorption on charcoal was one of the earliest and most effective methods of isolation of polysaccharide from fulvic acid. After adsorption of the fulvic fraction, the charcoal is eluted first with 0.1 M HCl and then with acetone containing 10% water. Water then elutes a colorless solution of the polysaccharide, but the yield is often less than a quarter of that in the fulvic acid applied to the column.

Alternatively phenolic and polyphenolic substances may be removed from the fulvic solution by selective adsorption using polyclar AT, a cross linked polyvinyl-pyrolidone, which does not absorb the polysaccharide. Various additional methods of purification have been applied. These include dialysis against water, precipitation by the addition of acetone, ethanol, or ethanol-acetic acid to an aqueous solution, treatment with HF and HCl, and Sephadex gel filtration. Filtration on Sephadex gels has been used to separate lower molecular weight humic substances from higher molecular weight polysaccharide.

Some of the nitrogen can be removed as protein by Sevag's emulsification technique using chloroform, or by precipitation using cadmium sulphate. Polysaccharide precipitated as a copper complex contains a lower percentage of nitrogen than the unprecipitated material. Gel filtration of the polysaccharide on Sephadex G-75 has given fractions containing different proportions of nitrogen.

Fractionation. Soil polysaccharide appears to consist of a mixture of substances, and attempts to separate the components have been only partially successful, judging by the complexity of the fractions obtained.

Those fractions obtained by fractional precipitation using ethanol or quarternary ammonium compounds show no difference in composition. Free boundary electrophoresis of the polysaccharide in borate or phosphate buffer usually gives a diffuse broad peak indicating a heterogeneous mixture of polysaccharides. The more successful of the methods involving electrophoresis have distinguished fractions rich in uronic acids but not differing markedly in the proportions of other sugars.

Gel chromatography has been used to obtain fractions comparatively rich in mannose. The most successful approach so far has been chromatography on charged gels or cellulose such as DEAE-Sephadex-A50 (Figure 1) or DEAE-cellulose-DE 52, where four or five fractions have been obtained by gradient elution with phosphate or chloride buffers. In studies on an organic soil, polysaccharides eluted early from the column had glucose as the principal component and are thought to have a plant-hemicellulose origin. Later fractions had galactose as the dominant sugar, were also richer in rhamnose, and were probably of microbial origin.

Properties

Isolated soil polysaccharide is a creamy white powder containing all the major kinds of neutral sugars present in the whole soil polysaccharide (Table 2), and these account for about one third of its weight. About 2–3% of each of the amino sugars, glucosamine and galactosamine, are present together with 2–3% of a mixture of amino acids in which alanine, glycine, aspartic acid, and glutamic acid are predominant. There is a small proportion of diaminopimelic acid. Much of the unidentified material in the polysaccharide is considered to be uronic acids, which are destroyed by hydrolysis during the analysis, but other organic acids are also present.

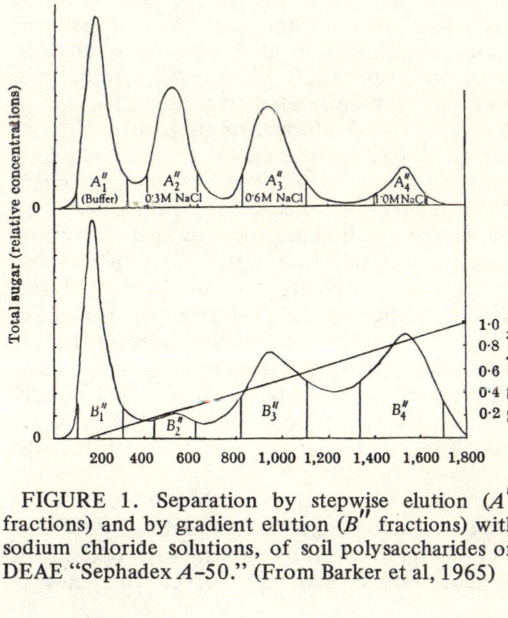

FIGURE 1. Separation by stepwise elution (A'' fractions) and by gradient elution (B'' fractions) with sodium chloride solutions, of soil polysaccharides on DEAE "Sephadex A-50." (From Barker et al, 1965)

TABLE 2. Monosaccharide Composition of Isolated Soil Polysaccharide from Different Soils Using Several Isolation Procedures Expressed as a Percentage of the Sum of the Sugars Measured.

				D				*E*		
								Molecular Weight of Fraction		
	A	B	C	F1a	F2a	F1b	F2b	<4,000	4,000–100,000	>100,000
Galactose	17	21	20	9.0	21.6	31.8	33.2	11.1	13.4	15.3
Clucose	21	24	38	68.9	32.3	19.5	18.1	27.4	26.8	32.9
Mannose	18	20	18	9.0	8.7	17.3	16.7	26.8	19.3	17.2
Arabinose	10	8.5	7.6	2.3	10.3	4.0	4.7	12.2	10.7	6.5
Ribose	+	+	–	–	–	1.5	0.7	4.7	2.8	0.4
Xylose	13	12	8.2	3.6	18.8	7.9	8.9	11.0	15.9	11.0
Fucose	0	6.5	–	1.9	3.3	3.5	5.2	7.5	3.8	7.5
Rhamnose	14	7	8.2	5.3	4.4	12.7	11.5	9.1	5.4	9.1
Galacturonic acid %	9*	2.2	17.9*	–	–	–	–	–	–	–
Glucuronic acid %		1.3								
Galactosamine %	+**	2.2	5.0**	–	–	–	–	–	–	–
Glucosamine %		1.7								

+ = present; * = uronic acid; ** = hexosamine.

A Soil Cultivated silty clay loam, Indiana. U.S.
Isolation procedure Extractant, water 85°C; acetone precipitation. (Whistler and Kirby, 1956)

B Soil Countesswells series under grass, Scotland.
Isolation procedure Extractant, 0.2 M NaOH; absorption of fulvic acid on charcoal, elution, precipitation with ethanol.

C Soil Calcareous meadow soil, England.
Isolation procedure Extractant, anhydrous formic acid containing LiBr; precipitation with cetavlon. (Parsons and Tinsley, 1961)

D Soil Highly organic soil, Norfolk fenland, England.
Isolation procedure Extractant, 2.5 M ethylenediamine; fractionation of the fulvic acid on Biogel P-150 (fractions F1 and F2) and DEAE-cellulose-DE 52 (fractions F1a and b, F2a and b). (Hayes, Stacey, and Swift, 1975)

E Soil Urrbrae fine sandy loam under old pasture, Australia.
Isolation procedure Extractant, 0.5 M NaOH, after 1 M HCl; treatment of the fulvic acid by Polyclar AT and Sephadex G-25, Biogel P100. (Swincer, Oades, and Greenland, 1968)

Infrared analysis of the polysaccharide shows absorption bands that are characteristic of acidic polysaccharide and peptide structures. Absorption at 1230–1250 cm^{-1}, sometimes observed, may be related to polymerized phosphate, and weak absorption at 1240 cm^{-1} and 820 cm^{-1} has been assigned to ester sulphate groups. Values for the mean molecular weight of extracted polysaccharide, determined by viscometry, sedimentation, or gel filtration, range from 10,000 to 450,000. The distribution of molecular weights for one soil polysaccharide found by gel filtration is shown in Table 3. Studies with the ultracentrifuge show that the polysaccharide is very polydisperse. Small negative values are usually obtained for the optical rotation, but there are exceptions, and reported values for $[\alpha]^D$ range from +36.8 to −80°.

Attempts to hydrolyze the polysaccharide using various enzymes including β (1–3), β (1–4), and α (1–6) glucanases, pectinases, β-glucuronidase, and the mixed enzymes in snail juice have not been successful in releasing mono- or oligocomponents, nor are amino acids released by proteinases. The failure of the proteinases suggests that the polysaccharide is protected from attack rather than that it is a heteropolysaccharide requiring a succession of enzymes of different specificities to degrade it. Nevertheless, an α-L-rhamnosidase released between 25 and 50% of the rhamnose, and it is suggested that this sugar is present in the polysaccharide as α-L-linked end groups. In

TABLE 3. Distribution of Molecular Weights for One Soil Polysaccharide.

Molecular Weight	Distribution (%)
4,000–20,000	29
20,000–40,000	25
40,000–100,000	9
100,000–200,000	11
>200,000	26

laboratory studies soil polysaccharide is strongly absorbed in large amounts by montmorillonite with which it forms an interlamellar complex. Such adsorption is typical of large linear uncharged molecules. It is thought that much of the polysaccharide in soil is associated with the clay fraction.

Complexing of Metal Ions

The polysaccharide fraction forms a complex with aluminum similar to that given by galacturonic acid. Iron is thought to form even stronger complexes than aluminum, and these elements usually account for a large proportion of the ash of the extracted soil polysaccharide. Infrared evidence suggests that copper forms complexes with the polysaccharide involving carboxyl and amino groups.

The separation on Sephadex gel of polysaccharide complexes, isolated from soil labeled with strontium 90 and yttrium 90, distinguished two negatively charged complexes with yttrium and a positively charged with strontium.

Metabolism of Polysaccharide in Soil

The rate of decomposition of isolated plant carbohydrates in soil is in the order glucose > starch > hemicellulose > pectin > cellulose. Polysaccharide in the intact plant tissue can show a much greater stability, however, as has been found with hemicellulose in mature cereal rye straw.

There are large differences in the rate of breakdown of microbial polysaccharides, some of them being very resistant. Breakdown rates have been studied by the measurement of CO_2, but more reliable methods involve the isolation and measurement of residues of ^{14}C-labeled material indistinguishable from the original. These studies and visual observations suggest an all-or-none pattern of degradation of intact plant material during which intermediate breakdown products are not easily detected. In comparison with added material, the polysaccharide already in the soil is decomposed only slowly. When soil polysaccharide is extracted, isolated, and returned to fresh soil, it too decomposes readily.

Therefore, it appears that the stability is related to the material with which the polysaccharide is associated rather than its complexity of structure or composition. In some instances, complexing of plant and microbial polysaccharides with metal ions reduces the breakdown rate, and large effects have been found with copper, zinc, iron, and aluminum. The effects are thought to relate to the spatial arrangement of active groups on the polysaccharide, which may reduce the ability of microbial enzymes to hydrolyze the polymers. Adsorption of a polysaccharide on clay reduces its rate of decomposition, the effect being greater with montmorillonite than with illite or kaolinite. Similarly, association of polysaccharide with humic substances also has a stabilizing effect.

The introduction of radioactively labeled substrates such as glucose to soil has shown that the microflora are capable of synthesizing all the major neutral sugar components, but the polysaccharides synthesized at normal laboratory temperatures (20 to 30°C) have invariably been deficient in the pentose components, xylose and arabinose, in comparison with soil polysaccharide. Under winter conditions, using remoistened soils, only arabinose is deficient, the xylose being synthesized by yeasts that compete more successfully with the bacteria for the glucose at these lower temperatures. Such synthesized polysaccharides are found to be less stable in soil than the native forms.

Significance

Since about 1945 when it was reported that gums produced by soil microorganisms could bind soil particles in stable aggregates, there has been an interest in the relationship between the physical condition of a soil and the nature and content of its polysaccharide.

It was soon shown that polysaccharides extracted from soils are able to stabilize existing soil aggregates. (see *Aggregation*). Significant correlations between the polysaccharide content of soil and extent of aggregation were subsequently shown for a number of soils. Correlations with other organic matter constituents have been observed in Australian soils, however, and it is evident that polysaccharides are not the only organic components capable of this action. Some such as plant cellulose have no effect, and gums produced by bacteria differ considerably in their aggregating ability.

Soils have been treated with periodate to oxidize and cleave sugars with vic-glycol groups to test the extent to which polysaccharides are involved in aggregation. It was found, for example, that young pastures and cultivated soils relatively low in organic matter show a marked decrease in aggregate stability. In soils under old pastures or forests, periodate treatment had little effect, and aggregation was considered to be caused by fungal hyphae or other organic matter components such as humic acid. With an American grassland soil, the stability of the aggregates was destroyed by periodate only after removal of a humic fraction by sodium pyrophosphate solution, showing that there was more than one type of binding agent in that soil.

Soils with high contents of organic matter have good moisture-retaining capacities, which has been attributed to the soil polysaccharide fraction and in particular to the highly viscous mucilaginous polysaccharides produced by soil microflora. The uronic acids in soil polysaccharide contribute to the cation exchange capacity of the soil. Similarly, in acidic soils, amino groups may contribute to the anion exchange capacity. The carbohydrate in soil seems to be the most important energy substrate for soil organisms.

MARTIN VEALE CHESHIRE

References

Barker, S. A.; Finch, P.; Hayes, M. H. B.; Simmonds, R. G.; and Stacey, M., 1965. Isolation and preliminary characterization of soil polysaccharides, *Nature,* **205**, 68–69.

Decau, J., 1968. Les polysaccharides du sol: Origine, évolution et role, *Ann. Agron.,* **19**, 65–82.

Finch, P.; Hayes, M. H. B.; Stacey, M., 1971. The biochemistry of soil polysaccharide, *in* A. D. McLaren and J. Skujins, eds., *Soil Biochemistry,* vol. 2. New York: Marcel Dekker, 257–319.

Folsom, B. L.; Wagner, G. J.; and Scrivner, C. L., 1974. Comparison of soil carbohydrate in several prairie and forest soils by gas-liquid chromatography, *Soil Sci. Soc. America Proc.,* **38**, 305–309.

Greenland, D. J., and Oades, J. M., 1974. Saccharides, *in* J. E. Gieseking, ed., *Soil Components,* Vol. 1: *Organic Components.* Berlin: Springer-Verlag, 213–261.

Gupta, U. C., 1967. Carbohydrates, *in* A. D. McLaren and G. H. Peterson, eds., *Soil Biochemistry,* Vol. 1. New York: Marcel Dekker, 91–118.

Hayes, M. H. B.; Stacey, M.; and Swift, R. S., 1975. Techniques for fractionating soil polysaccharides, *Supplementary volume, 10th Internat. Congr. Soil Sci., Moscow. Trans.,* **12**, 75–81.

Martin, J. P., 1971. Decomposition and binding action of polysaccharides in soil, *Soil Biol. Biochem.,* **3**, 33–41.

Mehta, N. C.; Dubach, P.; and Deuel, H., 1961. Carbohydrates in the soil, *Adv. Carbohydrate Chemistry,* **16**, 335–355.

Oades, J. M, 1972. Studies on soil polysaccharides. III. Composition of polysaccharides in some Australian soils, *Australian Jour. Soil Res.,* **10**(1), 113–126.

Parsons, J. W., and Tinsley, J., 1961. Chemical studies of polysaccharide material in soils and composts based on extraction with anhydrous formic acid, *Soil Sci.,* **92**, 46–53.

Swincer, G. D., Oades, J. M.; and Greenland, D. J., 1968. Studies on soil polysaccharides II. The composition and properties of polysaccharides in soils under pasture and under a fallow-wheat rotation, *Australian Jour. Soil Res.,* **6**, 225–235.

Swincer, G. D.; Oades, J. M.; and Greenland, D. J., 1969. The extraction, characterization and significance of soil polysaccharides, *Adv. Agron.,* **21**, 195–235.

Whistler, R. L., and Kirby, K. W., 1956. Composition and behaviour of soil polysaccharides, *Jour Am. Chem. Soc.,* **78**, 1755–1759.

PORE-SIZE DISTRIBUTION

Porosity and *pore-size distribution* are important soil properties in both the disposition of water falling on the earth and in the growth of plants. Soils with large and stable pores absorb rainfall and permit it to percolate (see *Percolation*) downward rather than flow away over the surface. Maximum plant growth depends on suitable distribution of large, intermediate, and small pores. Predominance of small pores favors waterlogged conditions and poor aeration, whereas too many large pores make soils droughty.

The nature of pores in soils, especially those in uppermost layers, can affect air temperatures (see *Thermal Regions*). A dry and fluffy surface layer with many large pores is a good insulator, causing incoming solar energy to be reflected and reradiated rather than absorbed. The reflected solar energy heats the air, which in turn helps promote the turbulence enjoyed

FIGURE 1. Soil pore: Electron micrograph showing a 10^{-3} mm pore with amorphous silica acid coating (arrow). (From Uehara and Jones, *in* Cary and Evans, 1975)

by glider pilots on warm summer days. In contrast, a moist and dense surface layer with predominantly small pores is a good conductor of heat and encourages downward transfer of the incoming solar energy. Moreover, the heat stored during the warmest part of the day is released to the air during cool nights, enough at times to prevent frost when temperature of the air above the ground is near freezing.

A loam soil has 40–60% pore space between the mineral particles which, in their great variety of sizes and shapes, form the matrix around the pores (Figure 1). The pores are irregular in shape and range in average diameter from a millimeter to a few nanometers. Clay-textured or "heavy" soils have many pores less than 0.001 mm in diameter, whereas sandy or "light" soils have more pores larger than 0.001 mm.

Pore Sizes and Soil Water Storage

Water rises into a glass capillary tube according to the relation

$$h = \frac{2\tau}{\rho g r} \quad \text{or} \quad P = \frac{2\tau}{r} \tag{1}$$

where h is the height of the water rise; τ, the surface tension; ρ, the density of the liquid; g, acceleration due to gravity; r, the radius of the tube; and P, the air pressure above ambient required in the capillary to prevent water from rising. This relation is used to measure pore-size distributions with the apparatus shown in Figure 2. The pores in the ceramic plate are so small that when they are filled with water, it cannot be displaced by the air pressure in the chamber. However, the water from larger pores in the soil can drain through the plate as the chamber pressure is increased. Measuring the incremental volumes of water outflow and using equation 1 leads to the release curves shown in Figure 3. These curves show both the water-storage capacities and the pore-size distributions of soils.

Pore Sizes and Soil Water Conductivity

Water flows through large soil pores rapidly and through small pores slowly. When the amount of water is not large enough to completely saturate the soil, the larger pores are not filled, and the water flows through the smaller ones, moving very slowly. A thin sheet of water standing on the surface during a rain is absorbed by a typical soil at a rate of about 1 cm/h, but 2 or 3 days after water disappears from the surface, flow rates in the soil decrease to less than 1 mm/day as the larger pores empty. Although infiltration rates of 1 cm/h are common, a change in the pore-size distribution at the surface may shift this value tenfold or more.

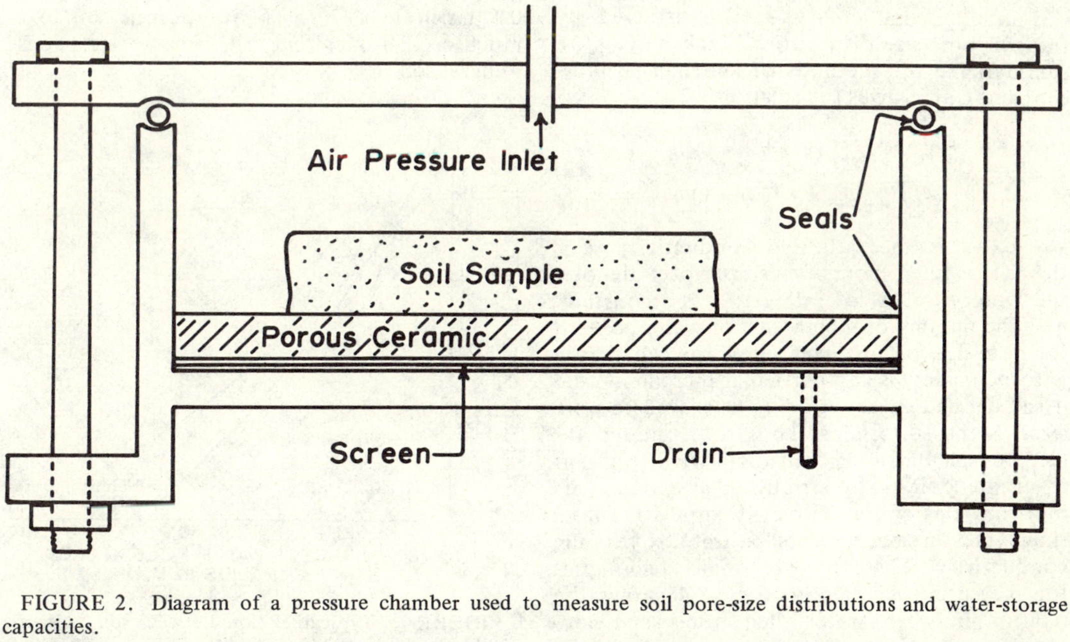

FIGURE 2. Diagram of a pressure chamber used to measure soil pore-size distributions and water-storage capacities.

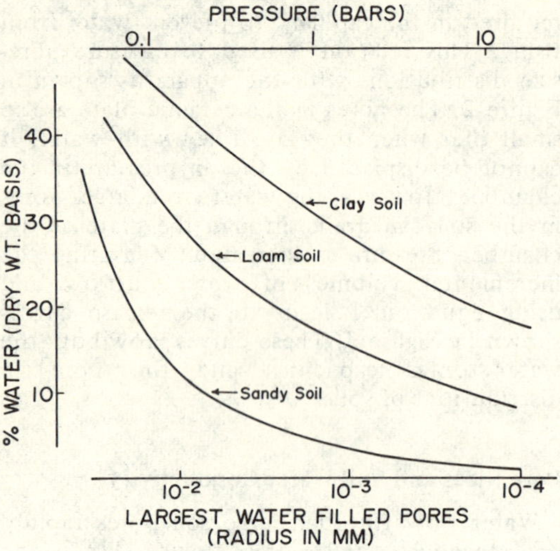

FIGURE 3. Typical relations between the soil water content, the largest water-filled pores, and the pressure in the apparatus shown in Figure 2.

The flow of water through soil is described by *Darcy's equation:*

$$J_w = -K\,\nabla\phi \tag{2}$$

where K is the hydraulic conductivity, and $\nabla\phi$ is the hydraulic gradient. Included in $\nabla\phi$ are both the gravitational and capillary pressure potentials. The capillary pressure P (as defined in equation 1) depends on the soil pore-size distribution.

Hydraulic conductivity is also affected by the soil pore-size distribution. Jackson (1972), summarizing the methods of Marshall, Millington, and Quirk, gives the relation

$$K_i = K_s \left(\frac{\epsilon_i}{\epsilon_1}\right)^n \frac{\Sigma_{j=i}^{m}\ [(2_j + 1 - 2_i)h_j^{-2}]}{\Sigma_{j=1}^{m}\ [(2_j - 1)h_j^{-2}]} \tag{3}$$

where K_s is the saturated conductivity; ϵ is the water-filled porosity; subscript i denotes the water content of interest; n is a constant; m is the number of increments of water content (or pore-size groups) over which the calculation is to be made; j is any particular increment; and h (as defined in equation 1) is related to pore sizes as the soil drains. Thus the conductivity can be calculated for any soil water content from the pore-size distribution as given in its moisture desorption curve (Figure 3) and a single measurement of its saturated hydraulic conductivity. There are also many techniques for measuring K directly. Figure 4 shows the relation between water-filled pores and conductivity. Conductivity decreases rapidly as the soil becomes unsaturated, causing water to flow very slowly, even though hydraulic gradients may increase. This means that the many tiny soil pores store water like a blotter, only transmitting it downward when the soil is near saturation and the larger pores are filled (see *Water Fluxes*).

Porosity Effects on Air and Heat Transfer

Heat and gas transfer, which occur primarily by diffusion, are both affected by the total pore space, whereas pore-size distribution is relatively less important. Transport by diffusion is proportional to the porosity, but viscous flow is proportional to the square of the pore radius. The pore-size distribution primarily affects the distribution of water in soil, which in turn affects heat and gas transfer. Viscous flow of air, which also transports heat, occurs principally in the surface few centimeters of soil where pore-size distribution is important.

Soil thermal conductivities, used in the *Fourier heat-flux equation,* are most often calculated by the method of deVries (1963):

$$\lambda = \frac{x_0\,\lambda_0 + k_1\,x_1\,\lambda_1}{x_0 + k_1\,x_1} \tag{4}$$

where λ is thermal conductivity; the subscript 1 denotes the solid phase, and 0 the air or water phase that predominantly fills the pores; x the volume fraction; and k a factor to account for microscopic thermal discontinuities between phases. Thermal conductivity may range from 0.5 mcal/cm s°C for a dry porous soil to as much as 10 mcal/cm s°C for a wet, dense mineral soil.

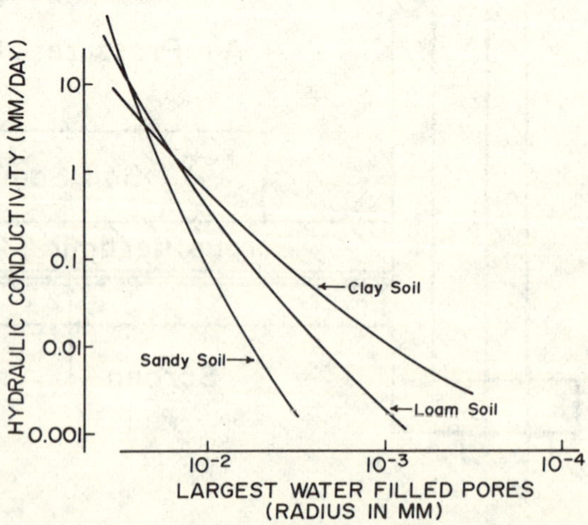

FIGURE 4. Typical relations between soil hydraulic conductivity and the largest water-filled pores.

Baver, Gardner, and Gardner (1972), in a review of the effects of porosity on soil gas diffusion, used the flux relation

$$J_g = D_0 S^n \nabla C \quad (5)$$

where D_0 is the diffusion constant of the gas in air; S is the air-filled soil pore space; n, a "constant" with values between 1 and 2; and ∇C, the concentration gradient.

Porosity Index and Plant Growth

A favorable soil pore-size distribution promotes vigorous plant growth. There must be enough large pores to allow rapid infiltration (q.v.) of surface water. These large pores must then drain quickly into smaller ones to allow aeration. Within a few hours after the surface is wetted, the soil must have at least 10% air-filled pores to dissipate carbon dioxide and provide oxygen for root respiration. There must be some pores with radii greater than 0.1mm to encourage root elongation, as well as a large group with radii in the range of 0.01–0.001 mm to store soil water at an energy level easily available to plants.

A porosity index I (Cary and Hayden, 1973), may be calculated from the pore-size distribution to judge how well a soil meets these criteria, as

$$I = 0.67M + 0.75 \quad (6)$$

where M can be evaluated from three points on the soil moisture release curve (Figure 3). The index is in effect the dimensionless arithmetic mean change in water content as the pressure is raised from 0–1.5 bars in the apparatus shown in Figure 2. Most soils have an index less than 10, but when it falls below 2.5, the pore-size distribution becomes unfavorable for optimum plant growth.

Stability of Pore-size Distributions

Pore-size distribution is a dynamic soil property. Surface traffic, the weight of overburden, tillage (q.v.), root growth, soil fauna (q.v.), wetting and drying, and freezing and thawing continually rearrange the pores. The changes are greatest near the surface and in the pore-size groups with radii greater than 0.001 mm. Surface traffic decreases the proportion of large pores and increases the smaller ones. The single passage of a tractor tire will cause 80% of the total reduction in porosity that could occur with multiple trips. This reduction may be significant to a 30-cm depth, although most will be concentrated in the first 5–10 cm beneath the tire. Tillage, of course, is often used to increase the number of large pores. Drying reduces the number of large pores, unless the soil is already severely compacted. In that case, wetting and drying and freezing and thawing tend to increase the proportion of pores with radii larger than 0.001 mm.

Soil particles are bound together by organic molecules, clay particles, the surface tension of adsorbed water films, and iron, aluminum, silicon, and lime cements. Consequently, increasing soil organic matter stabilizes soil particles and favors larger pore sizes. Increasing the amount of sodium, which disperses clay particles, and saturating soil with water, which eliminates surface tension forces, cause soil instability and a loss of the larger pore-size groups. Indeed, numerous factors influence soil pore-size distributions (Baver, Gardner, and Gardner, 1972; Gill and Vendenberg, 1967; and Cary and Evans, 1975), and many advances in soil management for maintenance of favorable porosity remain to be made.

JOHN W. CARY

References

Baver, L. D.; Gardner, W. H.; and Gardner, W. R., 1972. *Soil Physics,* 4th ed. New York: Wiley, 498p.

Cary, J. W., and Evans, D. D., eds., 1975. Soil crusts, *Ariz. Agric. Exp. Sta. Tech. Bull. No. 214,* 58p.

Cary, J. W., and Hayden, C. W., 1973. An index to pore size distribution, *Geoderma,* 9, 249–256.

Gill, W. R., and Vendenberg, G. E., 1967. Soil dynamics in tillage and traction, *U.S. Department of Agriculture Handbook, No. 316,* 511p.

Jackson, R. D., 1972. On the calculation of hydraulic conductivity, *Soil Sci. Soc. America Proc.,* 36, 380–382.

Uehara, G., and Jones, R. C., 1975. Bonding mechanisms for soil crusts: Part I. Particle surfaces and cementing agents, *in* J. W. Cary and D. D. Evans, Soil crusts, *Ariz. Agric. Exp. Sta. Tech. Bull. No. 214,* 17–28.

deVries, D. A., 1963. Thermal properties of soils, *in* W. R. van Wijk, ed., *Physics of Plant Environment.* Amsterdam: North Holland, 210–235.

Cross-references: *Aeration, Respiration, Atmosphere; Bulk Density; Capillary Pressure; Conductivity, Hydraulic; Conductivity, Thermal; Diffusion Phenomena; Flow Theory; Soil Drainage; Soil Pores; Soil Structure; Thermal Regimes; Water Content and Retention; Water Fluxes; Water Movement.*

PORE SPACE, DRAINABLE

The *drainable pore space* of a soil or similar porous material is the difference between its volumetric water content at saturation or near saturation (below a water table or after a rain or irrigation) and its volumetric water content after it has drained (following a falling water table or redistribution of rain or irrigation water in the soil). The water remaining in the soil after drainage is held against the force of gravity. The corresponding water content, sometimes called the *field capacity,* is not a constant because, even after most of the water has drained from the pores, the soil's water content will continue to decrease as soil-water pressure heads in the dewatered zone become more negative (e.g., by a falling water table).

The drainable pore space for a given soil mass and water-table drop is obtained by plotting the volumetric water content versus negative pressure head above the initial and final water-table position, and dividing the area between the resulting two curves (hatched area, Figure 1) by the distance of the water-table drop. Since, at equilibrium, the negative pressure head at each point above the water table is equal to the vertical distance of that point above the water table, the ordinate in Figure 1 directly represents distance above water table. Equilibrium water content profiles above water tables take a long time to develop and in practice may never be reached because other sources and sinks (e.g., rainfall, evaporation, and water up-take by plant roots) also affect the water content profile.

In clay soils with high water tables, like that found in agricultural land with subsurface drains, a water-table drop of only a few decimeters may produce little drainage of pore space because most of the soil above the water table may still be in the capillary fringe. Under these conditions drainable pore-space fractions often are only in the 0–5% range. Greater

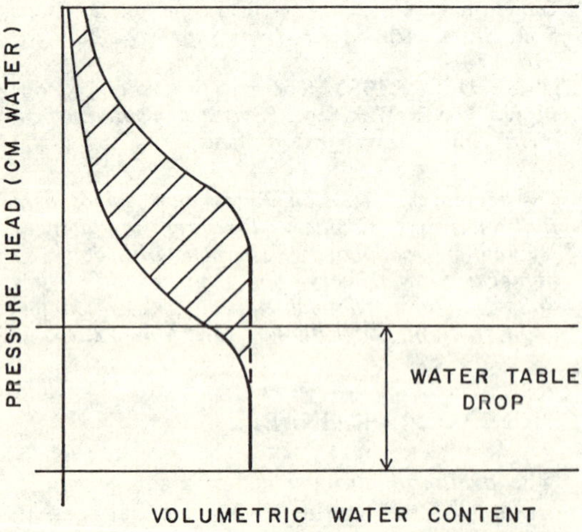

FIGURE 1. Water-content distribution curves above two successive water-table positions.

water-table drops will produce higher drainable pore space fractions, even for clay soils. For water-table drops in unconfined aquifers, consisting of fairly uniform sands or gravels, the drainable pore space may range from 20–30%. Consolidated materials have a drainable pore space that usually varies between 1 and 10% for fractured and cavernous rocks and may reach 20% or more for porous lava and similar materials. In groundwater hydrology, drainable pore space is called *specific yield,* which is defined as the amount of water released from storage in an unconfined aquifer per unit horizontal area and per unit drop of the water table. The specific yield is usually evaluated from the drawdown of the water table due to pumping a well (Lohman, 1972).

The rate of pore-space drainage is controlled by the material's hydraulic conductivity (q.v.). Pore-space drainage will lag behind the water-table drop if the rate of fall of the water table is not small in relation to the hydraulic conductivity of the soil (Bouwer and Jackson, 1974). For essentially instantaneous water-table drops, the drainage flow from the dewatered medium decreases with time according to an exponential, decay-type equation (Youngs, 1960; Jackson and Whisler, 1970). *Delayed yield* may affect the response of the water table in an unconfined aquifer during a pumping test to determine its transmissivity and specific yield. Procedures to account for this delayed yield were developed and used by Boulton (1963), Prickett, (1965), and Neuman (1975).

Because of hysteresis between water content and negative pressure head, the *fillable pore space* when the water table is rising is usually less than the drainable pore space of the medium (Bouwer and Jackson, 1974). Thus adding a given volume of water to an aquifer can cause the water table to rise more than it would fall if the same quantity of water were removed.

HERMAN BOUWER

References

Boulton, N. S. 1963. Analysis of data from non-equilibrium pumping tests allowing for delayed yield from storage, *Inst. Civil Eng. Proc.,* 26, 469–482.

Bouwer, H., and Jackson, R. D., 1974. Determining soil properties, *in* J. Van Schilfgaarde, ed., *Drainage for Agriculture.* Madison: American Society of Agronomy, 611–674.

Jackson, R. D., and Whisler, F. D., 1970. Equations for approximating vertical non-steady drainage of soil columns, *Soil Sci. Soc. America Proc.,* 34, 715–718.

Lohman, S. W., 1972. Ground-water hydraulics, *U.S. Geol. Surv. Professional Paper No. 708,* 70p.

Neuman, S. P., 1975. Analysis of pumping test data from anisotropic unconfined aquifers considering delayed gravity response, *Water Res. Res.*, **11**, 329–342.

Prickett, T. A., 1965. Type-curve solution to aquifer tests under water-table conditions, *Ground Water*, **3**, 5–14.

Youngs, E. G., 1960. The drainage of liquids from porous materials, *Jour. Geophys. Res.*, **65**, 4025–4030.

Cross-references: *Conductivity, Hydraulic; Field Water Cycle; Flow Theory; Water Content and Retention: Water Movement. Vol. XIII: Safe Yield.*

PROFILES, PHYSICAL MODIFICATION

Broadly speaking, physical properties are modified by all mechanical and chemical processes that operate in soils, whether natural or man-induced. The naturally occurring processes such as wetting and drying or freezing and thawing are not discussed here; instead, this article focuses on human modifications. Modifications at three different scales are considered, the first two briefly and the third at some length. As used here, *modification of physical properties* refers to changes in one or more characteristics affecting root growth in the soil profile, chiefly but not entirely below the plow layer or *Ap* horizon.

Microscale changes are alterations of soil properties over distances of millimeters, centimeters, or a few meters at most. Thus small changes in the soil occur when a person walks across a garden. Normally, the soil will be compacted slightly in each footprint. The compaction results in slightly higher bulk density (mass per unit volume) and in lower porosity, water-holding capacity, and hydraulic conductivity (capacity for water movement). Modifications at this scale are of minor importance and will not be considered further.

Tillage (q.v.) modifies the surface layers of soils in entire fields. Appreciable changes occur in the physical properties of those layers. Moldboard plows shear the uppermost horizons or layers, changing the total porosity and pore sizes. Secondary tillage such as disking or harrowing alters the uppermost part of the plow layer or *Ap* horizon to break down clods, to ensure adequate seed-soil contact, and on occasion to kill weeds. By altering the physical properties of all or part of the *Ap* horizon, tillage operations affect aeration, water movement, physical resistance to root penetration, and heat exchange. Although generally beneficial, tillage may also be harmful to plant growth. Plowing soils when they are wet can cause puddling (q.v.) and poor tilth, whereas disking can loosen the uppermost 10 cm (4 in) and compress the next deeper 10 cm. Tillage (q.v.) is normal practice in crop production; its effects have long been studied and are well known. It is therefore not considered further here. Attention is given instead to efforts to modify physical properties of all or much of the soil profile (a vertical section down through the soil to a depth of a meter or more—usually several feet). Such efforts hark back more than 50 years (Smith, 1925) but did not become widespread until powerful tractors and large machines were developed.

Some soils have one or more horizons (individual layers within profiles) with properties that hinder or completely inhibit root penetration and growth. Modification of such horizons can make some soil profiles more hospitable to plant roots. Primary attention is given in this article to efforts to modify physical properties of the deeper parts of soil profiles. Examples are mainly alterations below the *Ap* horizon (commonly 15 to 25 cm or 6 to 10 in thick). Modifications have been largely through mechanical means such as subsoiling and deep plowing. Chemical treatments such as deep placement of lime, fertilizer, or both have also been used in some cases (see *Soil Conditioners*).

Soil Profile Modification

Methods of modifying soil profiles range from chiseling and subsoiling to adding asphalt barriers at depth and even using dynamite. The methods follow, and some results of chiseling, subsoiling, deep plowing, asphalt barriers, dynamiting, and land forming are also discussed.

Chiseling and Subsoiling. *Chiseling and subsoiling* are used more widely than any other methods to modify the physical properties of soil profiles. The two are alike in that both are designed to disturb the soil to some depth without inverting it. Each is performed with a steel beam or shank, but subsoilers commonly have a shoe or sweep at the bottom and chisels do not. To distinguish the two in this article, chiseling is restricted to disturbance above and subsoiling below a depth of approximately 40 cm (Cooper, 1971). Each implement is set to penetrate to the desired depth and then drawn across a field at appropriate intervals. Both are meant to break up or shatter a subsurface layer or layers (see Vol. XII: *Soil Pans*), and both are most effective when a soil is dry. If a soil is wet enough to be plastic, neither operation will cause shattering. Cooper (1971) reports that very compact layers will shatter at water contents ranging from field capacity almost down to the wilting point (see *Field Water Cycle*).

Chisels are commonly used to break up a layer such as a plowpan immediately below the

Ap horizon. The chisel also loosens the deeper part of that horizon. Usually chiseling is done prior to ordinary tillage. Chiseling was markedly beneficial for long-cultivated Hapludalfs (Alfisols) of medium texture in the Delta region of Mississippi (Raney, Saveson, and Gill, 1956). Cotton yields were doubled in some instances. The practice was then extended to associated and more widespread heavy soils such as Sharkey clay without benefit. Those soils shrink and swell enough as they wet and dry to preclude formation of plowpans. A plow with chisels spaced about 30 cm (12 in) apart that penetrate to depths of 25 to 30 cm is gaining popularity in the Great Plains of the United States. In addition to the single blade or shaft, a subsoiler usually has a shoe or sweep at the bottom to increase the loosening of deeper horizons in profiles. Measurements made before and after subsoiling indicate that it reduced bulk density and soil strength and increased infiltration of water (Swain, 1975). These effects are not necessarily reflected in greater yields.

Subsoiling to depths of 40 or 50 cm did not by itself increase yields on an Albaqualf—a wet Alfisol—in central Missouri (Woodruff and Smith, 1947). If lime and fertilizer were added below the *Ap* horizon during subsoiling, yields increased slightly. The trials were made with a three-year rotation of corn, oats with lespedeza, and barley with sweet clover. However, the long-term responses to subsoiling and deep fertilization and liming were not large enough to offset the costs (Jamison, Smith, and Thornton, 1968). Indifferent results were also obtained from subsoiling six soil types representing the Entisol, Spodosol, and Ultisol orders in northwestern Florida (Robertson et al, 1957). Some yield increases were obtained, but they were not general enough to justify the expense of the operation.

Subsoiling in western Europe has produced results comparable to those reported in the United States (Swain, 1975). Appreciable yield increases have followed in some instances, small ones in others, and none in still others. The practice of subsoiling is tending to spread in western Europe, however, with the passing of time.

One exception to the generally indifferent results has followed from the ripping of soils with poorly or moderately expressed duripans (opal-cemented pans at depths near 60 cm—2 ft) in some soils of southern California. A massive machine with a chisel 2 m long is drawn and pushed across a field at intervals of about 2 m. The operation shatters the duripan, after which soils are suitable for a much wider range of crops than before. Thus soils previously suitable only for wheat and pasture can subsequently produce almonds with greatly increased returns (Simonson, R. W., personal communication).

Deep Plowing. Deep plowing has been tried in a number of places to improve unfavorable soil conditions. Mostly the efforts have been intended to mix part of a subsurface layer of clay accumulation such as a *Bt* horizon with a sandy *A* horizon. Some deep plowing has also been meant to mix and thus dilute soil materials with chemical characteristics unfavorable for plant growth such as horizons high in sodium.

Few efforts in deep plowing were made prior to the advent of tractors, although some had been made with steam-powered plows (Russell, 1956). In the United States the first efforts were to mix fresh sandy deposits with former plow layers after stream flooding. Deep plowing was also practiced rather widely in the middle forties to add clay to thick, sandy *A* horizons of Brownfield soils—Aridic Paleustalfs or Alfisols—and other similar soils in parts of western Texas and Oklahoma (Chepil et al, 1962). Giant moldboard plows have been used in most cases to invert and mix soils. Disk plows have also been tried but are generally less effective. A backhoe was used experimentally in one instance, and a ditching machine in another to mix the soil in the plots.

Most deep plowing has been to depths of 1 m (40 in) or less, but the maximum is around 2 m (7 ft). In practice, the degree of mixing ranges from poor to good. If horizons are distinct and the soil is not sandy, the sequence of horizons is not completely inverted but tends to be tipped over on its side. Multiple passes with a plow increase the amount of mixing. Effects of deep plowing have ranged from good to indifferent to poor.

The results of deep plowing of Pullman soils in northwestern Texas were governed by a combination of plow depth and irrigation practice. Pullman soils are Torrertic Paleustolls—Mollisols—with *A* horizons of moderately permeable clay loam 10 to 30 cm thick over slowly permeable *Bt* horizons extending down to 80 cm or more. Below is a horizon of carbonate accumulation, which may be but is not usually cemented. If a single preplant irrigation were applied, plowing to depths of both 90 and 150 cm increased the yield of sorghum (Eck and Taylor, 1969). In contrast, if two irrigations were applied during the growing season, sorghum yields were increased by plowing to a depth of 40 cm but not to depths of 60 and 80 cm. Infiltration and water storage were greater after plowing to a depth of 40 cm. Plowing to greater depths permitted loss of water through deep percolation (Musick and Dusek, 1971).

Deep plowing of Malheur soils in eastern Oregon has also been beneficial (Rasmussen, Moore, and Alban, 1972). These soils are Xerollic Nadurargids–Aridisols–with rather thin *A* and intractable *Bt* horizons and with duripans at some depth. The combination of *Bt* horizons high in sodium and the underlying duripan restrict root growth severely. Individual areas of Malheur soils range in size from a few meters in diameter up to a quarter of a hectare, occurring as small bodies in a matrix of Nyassa soils–Xerollic Durargids (Aridisols). The patches of Malheur soils, known locally as *slick spots* (q.v. in Vol. XII), produce low yields and interfere with timely cultivation.

An experimental area of the Nyassa-Malheur complex was plowed to a depth of 90 cm (3 ft), thus mixing the *A* and *Bt* horizons with part of the *C* horizon (Rasmussen, Moore, and Alban, 1972). Four years later water intake by the Malheur soils had doubled. The penetration by water and roots increased about fourfold, whereas small grain yields increased slightly more than threefold. Table 1 gives selected properties of individual horizons before and of corresponding layers of profiles after deep plowing of Malheur soils.

In contrast to the experience in Oregon, deep plowing (56 cm) of four soil types of the Black Solonetz group (Natriborolls–Mollisols) in eastern Alberta was not beneficial (Cairns, 1971). Deep plowing did increase the yields slightly on a fifth soil type with an especially prominent natric horizon.

Improvement of soils with one or more horizons high in sodium is feasible by deep plowing only if calcium is present in quantity within the depth reached by the plow. Mixing calcium-rich soil materials with those high in sodium can then facilitate removal of the latter element.

Results of deep plowing have been largely indifferent or poor in the southeastern United States. Plowing to depths of 18, 38, and 56 cm did not alter the yields of cotton, corn, and tobacco in a two-year trial in eastern North Carolina (Lutz, 1968). Soils in the plots are Ultisols, which have been strongly leached and are low in plant nutrient elements, especially in their very acid *Bt* horizons. Lutz (1968) concluded that it was not economic to improve the deeper soil layers if the top 20 or 30 cm were relatively homogeneous and otherwise favorable for plant growth.

Mixing of soil profiles to depths of 1 m (40 in) or more has been tried in a few experiments. A backhoe was used to mix the *A* and most of the *Bt* horizon of Freeman silt loam, a Haploxeralf (Alfisol), in the state of Washington. The *Bt* horizon is relatively high in bulk density and in clay. Mixing of the profile to a depth of 120 cm (48 in) nearly doubled the alfalfa yield and reduced frost heaving of the plants. Wheat yields also increased when nitrogen and phosphorus were applied heavily (Mech et al, 1967).

A ditching machine was used to mix Leon fine sand–a Haplaquod or wet Spodosol–to a depth of 1 m (40 in) in Florida (Robertson and Volk, 1968). Leon soils are very low in clay and are also low in organic matter except for a thin *A1* and somewhat thicker *Bh* horizon (the spodic horizon). Corn was grown on plots four years and each of bahia-grass and sorghum one year. Yields on the plots where the soils

TABLE 1. Selected Physical Properties of Malheur Silt Loam Before Deep Plowing and Four Years After Deep Plowing to a Depth of 90 cm.

Horizon	or	Soil Depth (cm)	Clay Content (%) Before	Clay Content (%) After	Bulk Density (g/cm^3) Before	Bulk Density (g/cm^3) After	Water-holding Capacity (% by wt.) ⅓ atm Before	⅓ atm After	15 atm Before	15 atm After
Ap	or	0–20	Sil (24.4%)	SiCl (28.7%)	1.24	1.22	31.1	32.7	13.1	11.7
*B*21	or	20–28	SiCl (34.6%)	SiCl (30.2%)	1.36	1.22	37.4	34.6	20.3	13.8
*B*22	or	28–40	SiCl (32.4%)	SiCl (30.7%)	1.41	1.25	34.7	34.0	14.1	12.4
*B*3ca	or	40–60	Sil (27.0%)	SiCl (29.2%)	1.38	1.12	38.7	33.0	11.8	10.4
*C*1ca	or	60–90	Sil	Sil	1.31	1.15	37.1	31.5	11.5	9.8

(From Rasmussen, Moore, and Alban, 1972)

had been mixed were no better than on plots that had received surface applications of fertilizer and lime.

Citrus growth, on the other hand, was better on Oldsmar fine sand—a Haplaquod or wet Spodosol—after mixing profile to a depth of 107 cm (43 in) plus deep incorporation of lime (Knipling and Hammond, 1972). Thickness of the sandy *A* horizons is twice as great (86 versus 40 cm), and the amount of clay in the *Bh* horizon is twice as large (8 versus 3%) in Oldsmar fine sand as in Leon fine sand. Furthermore, the relative proportions of aluminum and iron extractable with 0.1 N HCl from the spodic horizons of the two soils are reversed, aluminum being much higher in the Leon soil and iron in the Oldsmar soil (Gammon et al, 1953; Fiskell and Calvert, 1975). Although the reasons for the differing effects of mixing the Leon and Oldsmar soils are not known, some clay was added, and the water-holding capacity of the upper profile increased for the latter soil (Fiskell and Calvert, 1975).

On the whole, experience with deep plowing and mixing of soil horizons in the United States, Canada, and Europe thus indicates that benefits are obtained in some and not in other instances. For the most part, the practice has not been economic.

Asphalt Barriers. Experiments with asphalt layers placed at depth in sand profiles have been conducted during the last decade. The layer is continuous and forms a barrier to the movement of water, thus increasing the amount available for plant growth (Figure 1). Generally, the layer is placed at a depth near 60 cm (2 ft) by means of a subsoiler with a sweep and a spray attachment. The sweep is moved through the soil at a depth of about 60 cm, lifting the soil material so that an asphalt emulsion can be sprayed on the lower cut surface. Such a barrier roughly doubled the amount of water available for plant growth in some sands of Michigan (Erickson, Hansen, and Smucker, 1968). Yields of certain truck crops were also increased, e.g., cabbage 42–46%, cucumbers 33–107%, and green beans 24%.

FIGURE 1. Asphalt layer at about 60 cm depth.

Yield increases comparable to those in Michigan were obtained in Florida with a combination of an asphalt barrier, irrigation, and fertilization (Saxena, Hammond, and Lundy, 1968), as shown in Table 2. Crops were cabbage, potatoes, and tomatoes. The soil in the trial plots was a sand, a Quarzipsamment (an Entisol), that was very low in clay and organic matter (Robertson et al, 1973).

Dynamiting. Most dramatic among methods that have been tried to modify soil profiles has been the use of explosives. Dynamite was used with some success in southern California and without benefit in southern Illinois in the past.

Begun in the 1880s in southern California, the practice was to use about one fourth of a stick of dynamite to shatter the duripan (see Vol. XII: *Soil Pans*) wherever a fig was to be planted (Soil Survey Division, 1938). Duripans occur widely in the Durixeralfs (Alfisols) and Durargids (Aridisols) of southern California and parts of Arizona. Shattering of duripans with dynamite is no longer practiced. Dynamiting was also tried experimentally on the *Bt* horizons or claypans of Albaqualfs (wet Alfisols) and Argialbolls (wet Mollisols) in southern Illinois (Smith, 1925). The *Bt* horizons are very high in clay. Explosions did not shatter such horizons but formed "jugs" instead. As a result, physical properties were no better and more often worse after dynamiting. At present, dynamite is not being used in efforts to improve physical properties of soils. Its use in the past is an interesting historical note.

TABLE 2. Responses to Either or Both of Asphalt Layers and Irrigation for Some Vegetable Crops (cwt/acre*).

	Cabbage	Irish Potatoes	Vine ripened Tomatoes
Check	321.4	74	98.6
Irrigation	363.1	81	162.9
Mean	342.2	78	130.8
Asphalt layer (A.L.)	426.9	176	255.9
A.L. + irrigation	454.0	162	221.9
Mean	440.5	169	238.9

*Data are average of 3 fertilizer rates. Asphalt layer responses significant at 5% level of probability for all three crops. Irrigation response significant at 5% level for cabbage.

Land Forming

Grading or leveling a tract of land to shape the surface so that it is smooth and uniform, preferably with a gentle slope in one direction, is called *land forming*. The higher areas are cut down and soil materials moved into the lower ones. Part or all of the soil profile is thus modified wherever cuts and fills are made. Modification ranges from slight to drastic. Leveling or grading land has long been practiced in irrigated areas to facilitate uniform water distribution (see *Irrigation*). During the last 30 years, however, the practice has spread to humid regions as a water-control measure. Land forming ranges from no more than cutting off high spots and filling low ones at one extreme to precision grading at the other. After precision grading, a field has a uniform gentle slope.

Land forming has been most effective where *local relief*—differences in elevation between the high and low points in a field—is small and the soils have uniform profiles. Thus, for example, a field of slightly less than 50 ha (120 acres) in western Tennessee was graded when drainage was installed (Sewell and Hazlewood, 1966). The field lies within the floodplain of a stream, and the principal soil is Waverly silt loam (a Haplaquept or wet Inceptisol) that is silty throughout its profile. After grading, mechanized operations were possible for the first time so that tillage could be performed in more timely fashion. Delays were no longer caused by wet spots. Crop yields were also increased. Grading of sugarcane fields in Louisiana increased yields about 2 metric tons/ha (Saveson, 1963). Subsequent to grading, growth began earlier in the spring, water scald was eliminated, and weeds could be controlled more easily.

In the localities in Tennessee and Louisiana used as examples, the soils have rather uniform profiles and the local relief is small, both of which are important to the success of land forming. Much of the land in the Delta region of Mississippi has low relief, but the differences in elevation within a field are large enough in places to require substantial cuts and fills. Dubbs soils, Hapludalfs (Alfisols), have *Bt* horizons slightly higher in clay than the *A* horizons and much siltier and higher in clay than the deeper *C* horizons. If the *A* horizon and part of the *Bt* horizon are removed in land-forming operations, the truncated Dubbs soil can be productive again in a short time. Some extra care in soil management is required for a couple of years. On the other hand, if the *A*, *Bt*, and upper *C* horizons are all removed—as is necessary in some fields—a sandy deeper *C* horizon is exposed and must be cultivated. The sand cannot be made as productive as the Dubbs soil is prior to truncation. Lower production must be accepted on the cut areas. If land forming is then to be effective, the lower production on the cut areas of former Dubbs soils must be more than offset by increased production on the filled areas (Simonson, R. W., personal communication).

If the soils that must be truncated in land-forming operations have either or both *B* and *C* horizons that are unfavorable for plant growth because of physical or chemical characteristics, removal of the *A* horizons will result in reduced productivity of some part of a field. In extreme cases, the truncated soils cannot be made as productive as they were previously. Total production from a field as a whole may then be lowered by land forming. One possibility in such instances would be to stockpile the *A* horizon before grading is performed. Afterward the *A*-horizon materials could be redistributed. The cost would then be higher than otherwise, but so would production. Probable benefits would then have to be weighed against comparative costs of the two approaches.

Grading of occasional fields in the western part of the United States has included stockpiling of the uppermost several feet of soil prior to leveling. The gravelly sediments exposed by the stockpiling were then graded, after which the stockpiled soil materials were redistributed. The *A* and *B* horizons were mixed in the process of stockpiling and redistribution, but both were rather fine textured so that the soil materials after the operation were still much the same. The expense of such an operation is substantial.

The preceding examples indicate that the characteristics of soils of a tract where land forming is being contemplated are important to the success or failure of the operation, even as they are important to all methods for modifying the physical properties of soil profiles.

D. K. CASSEL
W. K. ROBERTSON

References

Cairns, R. R., 1971. Effect of deep plowing on the fertility status of black solonetz soils, *Canadian Jour. Soil Sci.*, 51, 411–414.

Chepil, W. S.; Hobbs, J. A.; Moldenhauer, W. C.; Nossaman, N. L.; and Taylor, H. M., 1962. Deep plowing of sandy soil, *U.S. Dept. Agric., Agric. Res. Serv. Production Rept., No. 64*, 14p.

Cooper, A. W., 1971. Effects of tillage on soil compaction, *in* K. K. Barnes et al, eds., *Compaction of Agricultural Soils.* St. Joseph Mich.: American Society of Agricultural Engineering, 315-364.

Eck, H. V., and Taylor, H. M., 1969. Profile modification of a slowly permeable soil, *Soil Sci. Soc. America Proc.*, **33**, 779–783.

Erickson, A. E.; Hansen, C. M.; and Smucker, A. J. M., 1968. The influence of subsurface asphalt barriers on the water properties and the productivity of sand soils, *9th Internat. Congr. Soil Sci. Trans. (Australia)*, **1**, 331–336.

Fiskell, J. G. A., and Calvert, D. V., 1975. Effects of deep tillage, lime incorporation and drainage on chemical properties of Spodosol profiles, *Soil Sci.*, **120**, 132–139.

Gammon, N., Jr.; Henderson, J. R.; Carrigan, R. A.; Caldwell, R. E.; Leighty, R. G.; and Smith, F. B., 1953. Physical, spectrographic and chemical analyses of some virgin Florida soils, *Fla. Agric. Exp. Sta. Bull. No. 524*, 130p.

Jamison, V. C.; Smith, D. D.; and Thornton, J. F., 1968. Soil and water research on a claypan soil, *U.S. Dept. Agric. Tech. Bull., No. 1379*, 111p.

Knipling, E. B., and Hammond, L. C., 1972. The SWAP study of soil water management for citrus on flatwood soils in Florida, *Soil and Crop Sci. Soc. Florida Proc.*, **31**, 208–210.

Lutz, J. F., 1968. Water movement-root relations in heterogeneous soils, *9th Internat. Congr. Soil Sci. Trans (Australia)*, **3**, 339–346.

Mech, S. J.; Horner, G. M.; Cox, L. M.; and Cary, E. E., 1967. Soil profile modification by backhoe mixing and deep plowing, *Trans. Am. Soc. Agric. Engineering*, **10**, 775–779.

Musick, J. T., and Dusek, D. A., 1971. Deep tillage of graded-furrowed-irrigated Pullman clay loam, *Trans. Am. Soc. Agric. Engineering*, **18**, 263–269.

Raney, W. A.; Saveson, I. L.; and Gill, W. R., 1956. Study of soil compaction in Mississippi River delta soils, *6th Internat. Congr. Soil Sci. Trans. (France)*, **D**, 521–524.

Rasmussen, W. W.; Moore, D. P.; and Alban, L. A., 1972. Improvement of a solonetzic (slick spot) soil by deep plowing, subsoiling and amendments, *Soil Sci. Soc. America Proc.*, **36**, 137–142.

Robertson, W. K.; Fiskell, J. G. A.; Hutton, C. E.; Thompson, L. G.; Lipscomb, R. W.; and Lundy, H. W., 1957. Results from subsoiling and deep fertilization of corn for two years, *Soil Sci. Soc. America Proc.*, **21**, 340–346.

Robertson, W. K.; Hammond, L. C.; Saxena, G. K.; and Lundy, H. W., 1973. Influence of water management through irrigation and a subsurface asphalt layer on seasonal growth and nutrient uptake of corn, *Agron. Jour.*, **65**, 866–868.

Robertson, W. K., and Volk, G. M., 1968. Effect of deep profile mixing and amendment additions on soil characteristics and crop production of a Spodosol, *9th Internat. Congr. Soil Sci. Trans. (Australia)*, **3**, 357–366.

Russell, E. W., 1956. The effects of very deep ploughing and subsoiling on crop yields, *Jour. Agric. Sci.*, **48**, 129–144.

Saveson, I. L., 1963. An efficient drainage system for sugarcane, *U.S. Dept. Agric., Agric. Res. Serv. Publ., No. 41-72*, 12p.

Saxena, G. K.; Hammond, L. C.; and Lundy, H. W., 1968. Effect of asphalt moisture barrier on soil salinity and yields of vegetables under variable irrigation and fertilization, *Soil and Crop Sci. Soc. Florida Proc.*, **28**, 310–318.

Sewell, J. I., and Hazlewood, B. P., 1966. Drainage and land grading on a bottomland field area in west Tennessee, *Tenn. Agric. Exp. Sta. Bull. No. 408*, 15p.

Smith, R. S., 1925. Experiments with subsoiling, deep tilling and dynamiting, *Ill. Agric. Exp. Sta. Bull. No. 258*, 158–170.

Soil Survey Division, Bureau of Chemicals and Soils, 1938. Soils of the United States, *in* Soils and Men, *U.S. Department of Agriculture Yearbook of Agriculture.* Washington, D.C.: U.S. Government Printing Office, 1105.

Swain, R. W., 1975. Subsoiling, *in Soil Physical Conditions and Crop Production, United Kingdom Ministry Agric., Fisheries, and Food Tech. Bull., No. 29*, 189–204.

Woodruff, C. M., and Smith, D. D., 1947. Subsoil shattering and subsoil liming for crop production on claypan soils, *Soil Sci. Soc. America Proc.*, **11**, 539–542.

Cross-references: *Irrigation; Management of Soils; Soil Conditioners; Soil Structure; Tillage. Vol. XII: Soil Pans.*

PUDDLING

The term *puddling* was defined by Buehrer and Rose (1943) as "the destruction of the aggregated condition of the soil by mechanical manipulation within a narrow range of moisture contents above and below field capacity (0.3 bars), so that soil aggregates lose their identity and the soil is converted into a structurally more or less homogeneous mass of ultimate particles." After puddling, a soil is called a *puddled soil,* to contrast it with a well-aggregated one.

In most cropping systems, puddling is an unintentional effect of tillage (q.v.) or traffic at the wrong moisture contents and usually results in severe yield decreases or delays in planting. In flooded rice culture, puddling is an important soil-management practice, conducted with great care for the purpose of destroying the structure of the surface layer. Puddling is almost synonymous with paddy rice culture in Asia (Sanchez, 1976).

Process

The process of puddling in rice culture is accomplished by a series of tillage operations beginning at soil moisture contents above saturation (i.e., flooded) and ending at moisture contents closer to field capacity (see Field Water Cycle). This process is best understood by considering the changes in soil strength *within* aggregates and *between* aggregates. According to Koenigs (1961), the cohesion *within* soil aggregates decreases with increasing soil moisture contents. The individual aggregates become soft and may or may not disintegrate depending on their stability. The cohesion *between* aggregates is very low at low moisture contents but increases rapidly with increasing moisture, peaking at about field capacity, and decreasing sharply as moisture contents approach saturation. The cohesion *between* aggregates depends primarily on the number of contact points between aggregates. The number of contact points are minimal in a dry soil and approach a maximum at about field capacity because of the increased thickness of water films and the swelling of the aggregates themselves. At higher moisture contents, the thick moisture films act as lubricants and decrease the number of contact points between aggregates. At approximately field capacity the cohesion within the aggregates is very low and the cohesion between the aggregates is maximum. When force is applied by a plow or a foot, the aggregates are easily destroyed because of the combined effects of high friction and low internal aggregate strength.

In practice, rice farmers start their puddling operations at moisture contents above saturation. In this range both the cohesion within and between the aggregates is minimal. Applying force while the soil is saturated does not cause maximum puddling, but it incorporates weeds and starts the aggregate destruction process. Farmers then continue to puddle the soil at progressively lower moisture contents until they reach maximum puddling at moisture tensions close to field capacity. The soil is converted into a uniform mud to a depth of 10 to 20 cm.

The degree of puddling attained differs with soil type and management. High clay contents facilitate puddling and permit more aggregate destruction. Nevertheless, sandy soils with low clay contents can be puddled. Montmorillonitic soils are puddled easier and more thoroughly than kaolinitic or oxidic soils. Sodium-saturated soils are the easiest to puddle because of their dispersed state. In general, the higher the contents of organic matter, iron oxides, or aluminum oxides, the more difficult it is to puddle a soil. Andepts and Oxisols are extremely difficult to puddle, and the degree of aggregate breakdown seems lower than in other soils.

Consequences of Puddling

Aggregate Destruction. The primary consequence of puddling is the destruction of soil aggregates. A puddled soil consists essentially of a two-phase or solid-liquid system. Individual clay particles or clusters thereof are oriented in parallel rows and are surrounded by capillary pores saturated with water. Sand and silt particles and some remaining aggregates are also part of the matrix. The degree of aggregate destruction is difficult to quantify because drying is necessary to measure aggregation. Kawaguchi, Kita, and Kyuma (1956) and others provide evidence of aggregate destruction after puddling and subsequent drying.

Changes in Porosity. Noncapillary pores are essentially eliminated in the process of puddling. Bodman and Rubin (1948) found that 91–100% of the volume occupied by such pores was destroyed by puddling a silt loam. Capillary porosity increases drastically. Because most of these pores are smaller than 0.2 μm in effective radii, water may move through pores as a liquid but can be lost only as vapor.

Bulk Density. Immediately after puddling a saturated soil, the apparent specific gravity or bulk density is less than that of the original soil because of the larger total pore volume occupied by water. With time, however, the bulk density of the flooded soils increases probably because of a slow settling of the clays. When dried, puddled soils shrink dramatically with resultant large increases in bulk density (Kawaguchi, Kita, and Kyuma, 1956; Sanchez, 1968).

Increased Soil Moisture Retention. As a consequence of the destruction of noncapillary pores, the increase in water-saturated capillary pores, and the decrease in initial bulk density, puddled soils hold more water than unpuddled soils at a given moisture tension. The effect is measurable within a range of 0 to 10 bars of soil moisture tension.

Decreased Moisture Losses. The changes in porosity and water retention result in sharply reduced soil moisture loss patterns in puddled soils. Table 1 shows the effects of puddling on drainage rates of six Philippine rice soils ranging widely in texture and mineralogy. Puddling decreased percolation losses by a factor of 1000 regardless of soil properties.

The benefits of puddling in decreasing the rate of soil water movement (q.v.) have been previously attributed to the formation of a relatively impermeable "plow pan" just below the puddled layer (see *Vol. XII: Pans, Soil*). These plow pans are found in loamy soils planted to rice for many years, e.g., in well-drained Oxisols and as concretionary materials in Andepts. Plow pans are absent in sandy and clayey soils, Vertisols, and young alluvial and calcareous soils (Sanchez, 1973). Puddling, however, is almost universally practiced in

TABLE 1. Effects of Puddling on the Drainage Rates of Six Flooded Philippine Soils in Pot Conditions.

			Drainage Rate (cm/day)	
Soil	Mineralogy	Clay (%)	Granulated Condi-tion	Puddled Condi-tion
Psamment	Siliceous	9	267	0.45
Fluvent	Mixed	24	215	0.17
Aquept	Montmorillonitic	30	183	0.05
Aqualf	Montmorillonitic	40	268	0.05
Ustox	Kaolinitic	64	155	0.05
Andept	Allophanic	46	214	0.31
Mean			217	0.18

(After Sanchez, 1973)

lowland rice farms. With the dikes surrounding the paddy fields and the puddled soil at the bottom, the rainfed fields hold the maximum amount of water possible for the rice crop growing during the rainy season. Another consequence of puddling is the slow rate of drying. The drying process may take several months in puddled clayey soils as opposed to much shorter periods in well-structured soils.

Reducing Conditions. Due to the absence of air, reduction processes can take place as soon as the soil is puddled (Breazeale and McGeorge, 1937). Puddled soils remain reduced regardless of whether they are flooded until cracks begin to form. Lack of oxygen in the soil pores inhibits the growth of most crops except rice and other anaerobic species. Nitrates are lost through denitrification (q.v.).

Organic Matter Decomposition. Puddling, like any other aggregate disruption process, temporarily hastens organic matter decomposition due to increased accessibility of the substrate by soil microorganisms. Puddling increases the mineralization of soil organic nitrogen during the first month after puddling and flooding (Harada, Hayashi, and Chikamoto, 1964), but the effect disappears at later stages (Briones, 1966).

Nutrient Availability. Puddling flooded soils does not *directly* increase the availability of nutrients to the rice plant (Sanchez 1968, 1973; Obermueller and Mikklesen, 1974). Small increases in iron and manganese availability have been recorded (Naphade and Ghildyal, 1971) but are not large enough to be of practical significance. Puddling often indirectly increases the availability of nutrients by decreasing leaching losses of cations such as Nh_4^+ (Sanchez, 1973; DeDatta and Kerim, 1974).

Regeneration of Structure

Puddling is not an irreversible process. The original structure can be regenerated through the processes of alternate wetting and drying or freezing and thawing. The puddled soil must be dried first, after which aggregates are reformed by these processes. Tillage at the appropriate moisture content facilitates regeneration of structure. This is accomplished most readily in soils high in organic matter or iron and aluminum oxides, (Koenigs, 1961).

Effects of Puddling on Crop Growth

The effects of puddling on crops other than rice are clearly detrimental (McGeorge and Breazeale, 1938). For rice, puddling is considered advantageous because it facilitates land leveling, permits the farmers to work the soil regardless of moisture status, reduces initial weed infestations, and, most important, decreases water and leaching losses.

The primary contribution of puddling to rice growth is to decrease water losses (Sanchez, 1973). Puddling is generally beneficial to rice grown on soils subject to leaching losses in irrigated paddy systems. This generalization is of particular relevance to sandy soils and well-aggregated Oxisols and Andepts, where often puddling is essential to flood such soils. At the other extreme, puddling is probably of little importance in soils having an impermeable layer in the subsoil, a constantly high water table, or in sodic soils that are naturally dispersed. Puddling is not practiced in soils having such properties in the rice-growing regions of the United States. In most rice soils puddling helps in decreasing percolation losses.

Puddling, however, is a double-edged sword in rainfed paddy systems. In most cases puddling attenuates the increases in soil moisture tension during temporary droughts and increases yields. But when intense droughts take place shortly after transplanting, the puddled soil may shrink, crack, and impede rice root development to a degree from which plants cannot recover afterwards (Sanchez, 1973; DeDatta and Kerim, 1974). Another potential detrimental effect of puddling is the time required for the soil to dry and be prepared for aerobic crops grown in rotation with rice. This time interval may be several months in clayey montmorillonitic soils but only several days in clayey kaolinitic, allophanic, or oxidic soils. In continuous paddy rice systems, this effect is irrelevant.

PEDRO A. SANCHEZ

References

Bodman, G. B., and Rubin, J. 1948. Soil puddling, *Soil Sci. Soc. America Proc.*, 13, 27–36.

Breazeale, J. F., and McGeorge, W. T., 1937. Studies on soil structure: Some nitrogen transformations in puddled soils, *Ariz. Agric. Exp. Sta. Tech. Bull. No. 69,* 225–252.

Briones, A. M., 1966. Nitrogen availability in five soils under different systems and its affect on growth and yield of rice. M.S. Thesis, University of the Philippines, Los Baños, 107p.

Buehrer, T. F., and Rose, M. S., 1943. Studies in soil structure. V. Bound water in normal and puddled soils, *Ariz. Agric. Exp. Sta. Tech. Bull, No. 100,* 155–218.

Croney, D., and Coleman, J. D., 1954. Soil structure in relation to soil suction (pF), *Jour. Soil Sci.,* **5**, 75–84.

DeDatta, S. K., and Kerim, M S. A. A., 1974. Water and nitrogen economy of rainfed rice on puddled and non-puddled soils, *Soil Sci. Soc. America Proc.,* **38**, 515–518.

Harada, T.; Hayashi, R.; and Chikamoto, A., 1964. Effect of physical pretreatment of the soils on the mineralization of native organic nitrogen in paddy soils, *Jour. Sci. Soil Manure,* **31**, 21–24.

Kawaguchi, K.; Kita, D.; and Kyuma, K., 1956. A soil core sample for paddy soils and some physical properties of the soil under waterlogged conditions, *Soil Plant Food,* **3**, 22–28.

Koenigs, F. F. R., 1961. *The Mechanical Stability of Clay Soils as Influenced by Moisture Conditions and Some Other Factors.* Wagnenigen: Verslag. Landbowk. Onderzoek. 67.7 Center for Agricultural Publications and Documentation, 171p.

McGeorge, W. T., and Breazeale, J. F., 1938. Studies on soil structure: Effects of puddled soils on plant growth, *Ariz. Agric. Exp. Sta. Tech. Bull. No. 72,* 413–447.

Naphade, J. D., and Ghildyal, B. P., 1971. Influence of puddling and water regimes on soil characteristics, ion uptake and rice growth, *Proc. Internat. Symposium Soil Fert. Evaluation (New Delhi)* **1**, 511–517.

Obermueller, A. J., and Mikklesen, D. S., 1974. Effects of water management and soil aggregation on the growth and nutrient uptake of rice, *Agron. Jour.* **66**, 627–632.

Sanchez, P. A., 1968. Rice performance under puddled and granulated soil cropping systems in Southeast Asia. Ph.D. Thesis, Cornell University, Ithaca, N.Y., 381p.

Sanchez, P.A., 1973. Puddling tropical rice soils. I. Growth and nutritional aspects. II. Effects of water losses, *Soil Sci.,* **115**, 149–158, 303–308.

Sanchez, P. A., 1976. Soil management in rice cultivation systems, *in Properties and Management of Soils in the Tropics.* New York: Wiley, 413–477.

Cross-references: *Aggregate Stability; Aggregation; Dispersion Phenomena; Management of Soils; Moisture Management; Pore Space, Drainable; Soil Drainage; Soil Structure; Surface-Soil Water Content; Tillage; Water Movement.*

R

RADIOCARBON DATING

Pedologists and geologists have used *radiocarbon dating* of organic-rich holocene soil horizons for some time. Quaternary geologists have used radiocarbon dating in their studies of the accumulation of authigenic carbonates (caliche, calcrete, etc.) in desert soils. Chronologies of changing environments, as reflected by pollen spectra in sediments and peats, have been elucidated with the aid of radiocarbon dating. These are a few examples of geoscientists' use of radiocarbon dating. The following sections review briefly typical problems entered in radiocarbon dating of soils, sediments, and peats. Because of variable and sometimes conflicting results, radiocarbon dating of these materials must be approached with caution.

Common problems associated with the radiocarbon dating of carbon-containing materials are: (1) obtaining materials not contaminated with modern ^{14}C or with old carbon; (2) misassociation of the sample with stratigraphic evidence; (3) establishing that the material has not been moved since the event of interest; and (4) location of substitute radiocarbon dating materials when charcoal, wood, etc., are not available. Variations in the long-term ^{14}C production rate (Ralph and Michael, 1974) and isotopic fractionation of ^{14}C (Craig, 1954) are separate problems that must also be considered, but they are not central to the problem of obtaining meaningful radiocarbon dates.

Radiocarbon dating of soils, sediments, and peats raises the problem of distinguishing between a mean residence time (MRT) of carbon and a discrete radiocarbon date. Soils are undesirable for radiocarbon dating because they have low carbon contents so that large samples must be processed. The chemical pretreatment is time-consuming and tedious. A more fundamental objection to the use of soil is that a radiocarbon date for this material is not a discrete date but is the mean residence time (MRT) for carbon in a specific soil horizon. The MRT is governed by a variety of redistribution factors. These factors include root growth, decay, animal transport, and downward percolation of water-soluble organic components. Thus there has been a constant search for materials that are less subject to these redistribution processes (Scharpenseel, 1972).

Radiocarbon dating procedures for peats and sediments are quite similar to those used for soils. Because peats and sediments are richer in organic matter, much smaller samples (20 to 50 g) are required for dating. More importantly, radiocarbon dates of sediments and peats are more likely to be discrete because the redistribution processes in sediments and peats are minimal, primarily because the moisture gradients in sediments and peats are small.

Potential radiocarbon-dating materials found in soils, sediments, and peats are: calcium carbonate, humic acid, fulvic acid, humine, lipids, and various mixtures of these materials. However, $CaCO_3$ is only occasionally used as dating material for soils, sediments, and peats because of variable results, especially for authigenic carbonates (caliche, calcrete, etc.) found in desert soils, (Williams and Polach, 1971).

Inasmuch as humic compounds have indefinite chemical compositions, they are defined in terms of their separation chemistry. Table 1 summarizes the separation chemistry of humic materials and gives an average order of their radiocarbon ages. If an "oldness" criterion is used, humic acid is the radiocarbon-dating material of choice; however, there are enough exceptions to the order given in Table 1 to raise questions. For example, Campbell et al (1967) found that humine was oldest, humic acids intermediate, and fulvic acid youngest.

Although sediments and peats contain the

TABLE 1. Radiocarbon-Dating Materials Derived from Soils.

Material	Chemistry	Relative MRTs in Soils*
Fulvic acid	Soluble in dilute NaOH and HCL	Youngest
Humic acid	Soluble in dilute NaOH, insoluble in HCl	Oldest
Hymatomelanic acid	Soluble in dilute NaOH, insoluble in HCl, soluble in ethanol	Variable
Humine (residue)	Insoluble in NaOH	Intermediate
Lipids	Soluble in CH_3Cl, CH_3OH benzene	Unknown

*Sequence derived from a large number of sources but mostly the journal *Radiocarbon*.

compounds listed in Table 1, their behavior is different. Most radiocarbon dates of humines (residues) and humic acids derived from peats and sediments are contemporaneous, but a few exceptions have been found, suggesting cautious interpretation of the peat and sediment data. Table 2 is a collection of representative soil, sediment, and peat data obtained by the Washington State University Radiocarbon Laboratory.

The large difference for the Lind Coulee dates is typical of radiocarbon dates of soils and emphasizes the point that such radiocarbon dates should be cautiously interpreted. The Hay radiocarbon dates show that charcoal can be used to determine soil ages and deposition rates. These data also indicate that radiocarbon dates of humic acid may give a rough or second-best date of a stratum since the humic acid MRT for the same stratum is slightly younger. More extensive research (Bailey et al, 1973) supports this observation. Other investigators also report that charcoal dates may be slightly younger than humic-acid MRTs. All these data suggest that humic-acid MRTs represent approximate radiocarbon ages of soil horizons as old as 12,000 and possibly out to 20,000 radiocarbon years. For MRTs less than 2000 radiocarbon years, the approximation is good to plus or minus 100 years. The difference increases to about 2000 years for MRTs of 20,000 radiocarbon years. But the MRT limitation must be remembered because large errors are possible. Careful use of stratigraphic evidence may eliminate some of these errors.

Humine (residue) and humic-acid dates of the Little Lake and Wildcat Lake sediments are very nearly contemporaneous and are consistent with other observations (Olson, 1972). The radiocarbon date of the Wildcat Lake sediment also gives a limiting date for a 4000-year-old volcanic ash layer from an eruption of Mount St. Helens, Washington (Crandell and Mullineaux, 1973). Agreement between these independently determined values engenders confidence in the radiocarbon dating of sediments. These observations also suggest that whole sediments can be dated by radiocarbon without separation of the humic acid, provided that the inorganic carbonates have been removed (Stuiver, 1975). But radiocarbon dating of the humic-acid and humine fractions eliminates questions about sediment dates when they are in close agreement.

The humine and humic acid dates for the Little Lake and Twin Lake peats also have excellent internal agreement, but it should be noted that two independent determinations on

TABLE 2. Representative Radiocarbon Dates for Soils, Sediments, and Peats.

Source Material	Depth (m)	Material Dated	Sample No.	Radiocarbon Age, Years, B.P.*
Lind Coulee Archeological Site, Washington				
Soil	1 (est.)	Whole soil	WSU-1422	8600 ± 70
		Humic acid	WSU-1500	9500 ± 120
Henley Site, Washington				
Soil	4.6 (est.)	Charcoal	WSU-1417	5120 ± 100
		Humic acid	WSU-1417H	4690 ± 120
Wildcat Lake, Washington				
Sediment	9.41–9.45	Humine	WSU-1451S	4180 ± 90
		Humic acid	WSU-1451H	3690 ± 90
Little Lake, California				
Sediment	1.26–1.34	Humine	WSU-1461	7160 ± 100
Peat	11.30–11.37	Humine	WSU-1464	5050 ± 140
		Humic acid	WSU-1464H	5000 ± 140
Twin Lakes, Colorado				
Peat	2.17–2.23	Lipid	WSU-1520	5990 ± 120
		Humic acid	WSU-1515H	5790 ± 90
		Humine	WSU-1515	5870 ± 120
Peat	2.52–2.58	Lipid	WSU-1529	6640 ± 140
		Humic acid	WSU-1516H	~6600 ± 120
		Humic acid	WSU-1529H	7200 ± 120
		Humine	WSU-1516	6760 ± 120

*Ages were calculated using the 5568 year ^{14}C half-life.

the 2.6m Twin Lake sample yielded humic-acid dates differing by 600 years. No reasonable explanation for this difference can be offered except that the humic acid was extracted by two investigators using the same method but under different conditions (time and temperature). Olson and Broecker (1958) and Dyck (1967) also found that radiocarbon dates of the humic-acid and humine fractions of peats usually agree. However, Runge, Goh, and Rafter (1973) have observed significant differences in the radiocarbon dates of humic acids and humines derived from peats, suggesting that care is required for the interpretation of radiocarbon dates of peats.

The Twin Lakes data indicate that lipids extracted from peats have radiocarbon dates that agree with those obtained for humine and humic acids and suggest that lipids were derived from either the humine or the humic acid. More research will have to be done to answer this question. However, it is worth noting that Bhandari, Maskina, and Randhawa (1969) found that a large fraction of the lipids are derived from humic acids.

The Little Lake data have an unusual relationship in that the peat date, obtained from the 11.30–11.37 m depth, is younger than the 1.26–1.34 m sediment sample. A likely explanation of these results is that the Little Lake sediment is derived from *Chara* and *Potamogeton,* which metabolize HCO_3^- derived from the reaction

$$Ca^{12}CO_3 + {}^{14}CO_2 + H_2O = Ca^{2+} + H^{12}CO_3^- + H^{14}CO_3^- \quad (1)$$

The effect of this reaction is to make sediments derived from *Chara* and *Potamogeton* appear to be about 5500 years older than they really are. Because the peat is derived from *Typha* and *Scirpus,* which metabolize atmospheric CO_2, the dilution of ^{14}C does not take place, and reasonable radiocarbon dates are obtained. Similar results have been obtained for Queechy Lake (Deevey et al, 1954). This effect, which is called the *hard-water* effect, emphasizes the importance of knowing the plants that produced the sediments, the chemical form of the carbon that the plants metabolized, and geological factors that produce hard water.

To summarize, radiocarbon dating of soils, sediments, and peats can yield useful ages of these materials. However, the complications and limitations discussed must be constantly kept in mind. If the limitations and complications are understood and accounted for, radiocarbon dates of soils, sediments, and peats can be very useful to the geoscientist.

J. C. SHEPPARD

References

Bailey, J. M.; Lee, R.; Rankin, P. C.; and Speir, T. W., 1973. Composition and properties of an age sequence of charcoals and associated humic acids, *New Zealand Jour. Geol. and Geophysics,* **16,** 703–15.

Bhandari, G. S.; Maskina, M. S.; and Randhawa, N. S., 1969. Characterization of lipids in some soils and their humic acids formed under different agroclimatic conditions, *Sci. and Culture,* **35,** 68–69.

Campbell, C. A.; Paul, E. A.; Rennie, D. A.; and McCallum, K. J., 1967. Applicability of the carbon-dating method of analysis to soil humus studies, *Soil Sci.,* **104,** 217–224.

Craig, H., 1954. Carbon-13 and Carbon-14 variations in nature, *Jour. Geol.,* **62,** 115–149.

Crandell, D. R., and Mullineaux, D. R., 1973. Pine Creek volcanic assemblage at Mount St. Helens, Washington, *U.S. Geol. Surv. Bull.,* **1383-A,** A1–A23.

Deevey, E. S., Jr.; Gross, M. S.; Hutchinson, G. E.; and Kraybill, H. L., 1954. The natural ^{14}C contents of materials from hard-water lakes, *Proc. Natl. Acad. Sci.,* **40,** 285–288.

Dyck, W., 1967. The geological survey of Canada Radiocarbon Dating Laboratory, *Geol. Survey of Canada Paper No. 66-45,* 33–37.

Olson, I., 1972. The pretreatment of samples and the interpretation of the results of ^{14}C determinations, *in* Y. Vasari, H. Hyvarinen, and S. Hicks, eds., *Climatic Changes in Arctic Areas During the Last Ten Thousand Years.* University of Oulu, Oulu, Finland, 9–37.

Olson, E. A., and Broecker, W. S., 1958. Sample contamination and reliability of radiocarbon dates, *Trans. N.Y. Acad. Sci.,* **20,** 593–604.

Ralph, E. K., and Michael, H. N., 1974. Radiocarbon-calendar age relationship, *Am. Sci.,* **62,** 553–560.

Runge, E. C. A.; Goh, K. M.; and Rafter, T. A., 1973. Radiocarbon chronology and problems in its interpretation for quaternary Loess deposits–South Canterbury, New Zealand, *Soil Sci. Soc. America Proc.,* **37,** 742–746.

Scharpenseel, H. W., 1972. Messung der naturlichen C-14 konzentration in der organischen substanz von rezenten boden, eine zwischenbelanz, *Archiv. für Acker-und Pflanzenban und Bodenkunde,* **133,** 241–263.

Stuiver, M., 1975. Climate versus changes in ^{13}C content of the organic component of lake sediments during the late quaternary, *Quat. Res.,* **5,** 251–262.

Williams, G. E., and Polach, H. A., 1971. Radiocarbon dating of arid-zone calcareous paleosols, *Geol. Soc. America Bull.,* **82,** 3069–3085.

Cross-references: *Carbon Cycle; Carbon-Nitrogen Ratio; Humus; Radioisotopes. Vol. IVA: Carbon-14 Dating. Vol. XIII: Geochronology.*

RADIOISOTOPES

Since radioactivity was discovered (Becquerel, 1896), the use of radioisotopes has continually increased. Radioisotopes have aided the understanding of soil in many diverse areas including ages of soil, origins of soils, chemical composition (q.v.) and mineral structure, the movement of ions through the soil, and the behavior of water in soil. In each of these areas, one or more of the many unusual properties that radioisotopes possess have been used. Some of the more useful properties are energy and type of particle emission, the half-life of a radioisotope, and ease of detection of these characteristics. Radioisotopes have their greatest utility when used in concert with the many other analytical and experimental methods available.

Types of Radioactivity

Most of the radiations emitted by radioactive substances arise from changes in an atomic nucleus, although a few of these transformations involve the emission of X-rays and electrons from the extranuclear electron levels (Friedlander, Kennedy, and Miller, 1964). Radiations from the nucleus are released if the ratio of neutrons to protons is unbalanced or if the atomic mass is above 140. Eventually these nuclei will undergo rearrangements to become more stable. The radiations of greatest interest to soil scientists include alpha particles, beta particles (positrons $\beta+$ and negatrons $\beta-$), gamma rays, neutrons, and X-rays.

Alpha particles are identical to helium nuclei and consist of two protons and two neutrons. Alpha radiation is emitted from a given radionuclide or radioisotope with a discrete kinetic energy. This discrete kinetic energy is one of the ways to identify a specific radioisotope. Common alpha-emitting radioisotopes are ^{235}U, ^{232}Th, ^{239}Pu, and ^{241}Am.

Beta-emitting ($\beta-$) isotopes occur when there is an excess of neutrons for the number of protons in the nucleus. To approach stability within the nucleus, one of the neutrons is converted to a proton, a neutrino, and an electron. The proton remains in the nucleus while the electron ($\beta-$) is ejected from the nucleus. Radioisotopes commonly used in soil research that are $\beta-$ emitters include tritium (^3H), ^{14}C, ^{32}P, and ^{35}S.

If the nucleus has a neutron-proton ratio lower than the stable forms of the element and the amount of energy available in a nucleus is at least 1.02 MeV, a proton may transform into a positron ($\beta+$), a neutron, and an antineutrino. The positron and antineutrino are emitted from the nucleus. Very few $\beta+$ isotopes are currently in use in soil research. If the energy is less than 1.02 MeV, the nucleus may capture an electron. As the K electrons in an atom are, on the average, closest to the nucleus, they are generally captured. In this process no radiation except a neutrino is usually emitted from the nucleus; however, X-rays are emitted from the atom when the electron captured by the nucleus is replaced by an electron from a higher energy level.

A nucleus that has more energy than is normally possessed by the stable nucleus may release this energy by emission of electromagnetic radiation. Such radiation is called *gamma* (γ) *radiation*. Gamma rays are characteristic of a particular radioisotope and are readily detected. Numerous gamma-emitting radioisotopes are used in soil research, including ^{60}Co, ^{137}Cs, and ^{198}Au. A useful radioisotope that spontaneously emits neutrons (n) is ^{252}Cf. This man-made radioisotope has been used for in situ neutron activation and soil moisture measurements.

The chart of the nuclides (Radiological Health Handbook, 1970) summarizes the various stable and radioactive isotopes for both the naturally occurring and man-made nuclides in readily understandable form. This chart provides a rapid way to determine the resulting isotope following nuclear processes involving γ, $\beta-$, $\beta+$, n, etc.,

Radioactivity Detection

One of the most useful properties of radioisotopes to the soil scientist is their ease of detection. Radioisotopes emit one or more of the foregoing types of radiation at characteristic energies. Specific detectors are available for alpha, beta, gamma, and neutron emissions. These detectors generally have the ability to determine the energy as well as the quantity of the radioactive events, thereby permitting the identification and quantification of the radioisotope of interest at extremely low concentrations.

Alpha Particles. The availability of semiconductor radiation detectors has greatly facilitated alpha pulse height analysis. As compared with the early techniques, the semiconductor radiation detectors have the advantages of better resolution, lower background, better stability, essentially "windowless" operation, and the ability to discriminate between particles of different energies. This type of detector is used extensively for determining ^{238}Pu and ^{239}Pu in environmental samples (Romney, Mork, and Larson, 1970; Krey and Hardy, 1970). The technique permits detection of materials to 10^{-15} curies per sample. An alternative method for alpha counting is a windowless, flow-type counter. This method provides gross alpha information and is useful in tracer research work with a single alpha-emitting radioisotope.

Photographic emulsions have been used with alpha-emitting radioisotopes where the location of the radioisotope is of interest. The positions of these radioisotopes are indicated by tracks in the emulsions following placement in contact with the system and subsequent development of the emulsion.

Beta Particles. Beta particles are counted efficiently with liquid scintillation counters because the sample is in intimate contact with the scintillator. Liquid scintillation methods are widely used in soil research to detect and measure ^{14}C-labeled pesticides, water containing ^{3}HHO, phosphate fertilizer tagged with ^{32}P, and sulfur compounds in soils containing ^{35}S. Other radioactive isotopes in soil research that are readily determined with liquid scintillation techniques include ^{36}Cl, ^{45}Ca, ^{59}Fe, ^{63}Ni, ^{64}Cu, and ^{65}Zn.

Gamma Radiation. Scintillation detectors are widely used as gamma-radiation detectors in conjunction with multichannel analyzers. There are four classes of scintillators: inorganic crystals, organic crystals, plastic phosphors, and liquid phosphors. The most commonly used scintillation material in soil research is sodium iodide activated with thallium [NaI(Tl)]. The resolution of the NaI(Tl) crystal is not as good as the semiconductor type of detector discussed below but is satisfactory for most soil research activities using gamma-emitting radioisotopes. Multichannel analyzers display the entire gamma spectrum simultaneously and provide greater flexibility than single-channel analyzers that count only the events within a specified energy range. The single-channel analyzers are adequate if tracer experiments are planned with only a single radioisotope.

The semiconductor in use for gamma-spectrum work is a lithium-drifted germanium detector. The efficiency of detection (number of counts recorded per gamma ray emitted) is less than that of the NaI(Tl) detector, but the semiconductor detector provides improved resolution permitting the differentiation of similar gamma energies. This improved resolution is particularly useful in neutron activation analysis and multielement tracer experiments (Murrmann, Winters, and Martin,1971).

Neutrons. Neutron detection in soil studies is limited to determining water content in soil. A slow-neutron detector is required for this purpose. The most common detector contains boron trifluoride gas. The alpha particles emitted when a neutron is captured by ^{10}B are detected in this method. Other available neutron detectors are thermoluminescent dosimeters and plastic scintillators. These latter detectors are not in general use in soil research at the present time.

Terms Used in Radioactivity

The subject of radioactivity has numerous terms with very specific meanings. A few that are of general importance to the soil scientist are listed.

Curie: One curie equals 3.7×10^{10} nuclear transformations per second.

Radioactive half-life: Time required for a radioactive element to lose 50% of its activity by decay.

Labeled compound: A chemical compound containing radioisotopes in certain positions. By observing radioactivity or isotopic composition, this compound or its fragments may be followed through physical, chemical, or biological processes.

Mass numbers: The number of protons and neutrons in the nucleus of an atom.

Radioactive series: A succession of nuclides, each of which transforms by radioactive disintegration into the next until a stable nuclide results. The first member is called the "parent," the intermediate members are called "daughters," and the final stable member is called the "end product."

Naturally Occurring Radioisotopes in Soils

A number of radionuclides exist in nature. These are used primarily for dating soil and rock formations. The more common radionuclides occurring in nature include ^{14}C, ^{40}K, and those in the thorium, neptunium, and uranium radioactive series.

Carbon-14. Carbon-14 (radiocarbon) has been used extensively to date organic matter that is 50,000 years or younger. Carbon-14 (half-life of 5730 years) is formed in nature when high-energy neutrons from cosmic rays enter the nucleus of a nitrogen atom in the outer atmosphere and release a proton (n,p reaction). The carbon-14 combines with oxygen to form $^{14}CO_2$, which circulates in the atmosphere. The abundance of radiocarbon is independent of longitude, latitude, altitude, and type of habitat. If the ^{14}C is incorporated in living matter, then the amount remaining decreases with time after the biological material dies. (Iwakura et al, 1966; Kerpen and Scharpenseel, 1967; Nakhla and Delibrias, 1967).

Potassium-40. In all soils the potassium consists of ^{39}K (93.08%), ^{40}K (0.01%), and ^{41}K (6.91%). The ^{40}K decays by β- emission with a half-life of 1.28×10^{9} years, but ^{39}K and ^{41}K are both stable. The amount of potassium in soils has been determined using ^{40}K (Benes, 1968), which is also useful as a tracer (Stewart, 1970).

Thorium and Uranium Series. The naturally occurring radioactive series (Bear, 1964) result in a number of long-lived radionuclides being present in soil (Table 1). These radionuclides are useful in dating older formations (greater

TABLE 1. Naturally Occurring Radioactive Series.

Element	Half-life	Type of Decay and Energy of Radiation (mev)
Radioactive Disintegration Series for Uranium-238		
${}_{92}U^{238}$	4.50×10^9	Alpha 4.21
${}_{90}Th^{234}$	24.1 days	Beta⁻ 0.13, gamma 0.09
${}_{91}Pa^{234}$	1.14 min	Beta⁻ 2.32, gamma 0.80
${}_{92}U^{234}$	2.35×10^5 yr	Alpha 4.76
${}_{90}Th^{230}$	8.0×10^4 yr	Alpha 4.66, gamma 0.68, 0.14–0.25
${}_{88}Ra^{226}$	1.6×10^3 yr	Alpha 4.79, gamma 0.19
${}_{86}Rn^{222}$	3.825 days	Alpha 5.49
${}_{84}Po^{218}$	3.05 min	Alpha 5.99
${}_{82}Pb^{214}$	26.8 min	Beta⁻ 0.65, gamma 0.29
${}_{33}Bi^{214}$	19.7 min	Alpha 5.50, beta⁻ 3.15, gamma 1.8
${}_{81}Tl^{210}$	1.32 min	Beta⁻ 1.80
${}_{84}Po^{214}$	1.5×10^{-4} s	Alpha 7.68
${}_{32}Pb^{210}$	22 yr	Beta⁻ 0.026, gamma 0.047
${}_{33}Bi^{210}$	4.97 days	Beta⁻ 1.17
${}_{34}Po^{210}$	138 days	Alpha 5.30, gamma 0.80
${}_{82}Pb^{206}$	S	

Note: ↘ denotes a parent-daughter relationship (that is, Sr^{90} decays by beta decay to Y^{90}, which is radioactive and decays by beta decay to stable Zr^{90}).

Element	Half-life	Type of Decay and Energy of Radiation (mev)
Radioactive Disintegration Series for Thorium-232		
${}_{90}Th^{232}$	1.39×10^{10} yr	Alpha 4.20
${}_{88}Ra^{228}$	6.7 yr	Beta⁻ 0.053
${}_{89}Ac^{228}$	6.13 hr	Beta⁻ 1.55, gamma 0.96
${}_{90}Th^{228}$	1.90 hr	Alpha 5.42
${}_{88}Ra^{224}$	3.64 days	Alpha 5.68
${}_{86}Rn^{220}$	54.5 s	Alpha 6.28
${}_{84}Po^{216}$	.158 s	Alpha 6.77
${}_{82}Pb^{212}$	10.6 hr	Beta⁻ 0.36
${}_{83}Bi^{212}$	60.5 min	Alpha 6.05, beta⁻ 2.20, gamma 0.040–2.20
${}_{81}Tl^{208}$	3.1 min	Beta 1.82, gamma 2.62
${}_{84}Po^{212}$	3×10^{-7}s	Alpha 8.95
${}_{82}Pb^{208}$ ${}_{82}Pb^{208}$	S	

TABLE 1. Continued

Element	Half-life	Type of Decay and Energy of Radiation (mev)
Radioactive Disintegration Series for Uranium-235		
$_{92}U^{235}$	7.07×10^{8} yr	Alpha 4.52
$_{90}Th^{231}$	24.6 hr	Beta$^-$ 0.2, gamma 0.03
$_{91}Pa^{231}$	3.2×10^{4} yr	Alpha 5.05, gamma 0.32
$_{89}Ac^{227}$	13.5 yr	Alpha 5.00, beta$^-$ 0.22
$_{90}Th^{227}$	18.2 days	Alpha 6.05, gamma 0.05
$_{87}Fr^{223}$	22 min	Beta$^-$ 1.15, gamma 0.05
$_{88}Ra^{223}$ $_{88}Ra^{223}$	11.7 days	Alpha 5.72, gamma 0.03
$_{86}Rn^{219}$	3.92 s	Alpha 6.82, gamma 0.27
$_{84}Po^{215}$	1.83×10^{-3}s	Alpha 7.36
$_{83}Bi^{211}$	2.16 min	Alpha 6.62, beta$^-$ 1.4, gamma 0.35
$_{84}Po^{211}$	0.5 s	Alpha 7.43
$_{81}Tl^{207}$	4.76 min	Beta$^-$ 1.47, gamma 0.89
$_{82}Pb^{207}$ $_{82}Pb^{207}$	S	

(From *Chemistry of the Soil,* 2nd ed., F. E. Bear, ACS Monograph Series No. 160, pp. 446–447. Washington, D.C.: American Chemical Society, 1964)

than 2,000,000 years) because the ratio of lead-206 to uranium-238 is a unique value depending on the length of time that decay has been occurring. Radon gas (^{222}Rn) produced during the decay of ^{226}Ra has been used to assess radioactive components of the soil and to study diffusion of gases through soils (Dios Vidal, 1964; Marquardt, 1967; Pearson, 1967; Pensko, Wardaszko, and Wochna, 1968). Other radioisotopes in these series used in soil and plant studies include ^{210}Pb, ^{210}Po, and ^{212}Bi (D'Souza and Mistry, 1970; Francis, 1967; Hansen, Watters, and Yaney, 1971; Holtzman, 1970; Iyengar and Markose, 1971).

Tracer Studies Using Artificially Produced Radioisotopes

Radiotracers have been used extensively in soil research to understand mechanisms responsible for plant growth, behavior of ions in soil solutions, and diffusion, exchange, and precipitation kinetics in the soil (Sokolov, et al, 1972). The use of radionuclides has also simplified the study of the behavior of organic compounds in soils and the role of microorganisms. In the following subsections, only a few examples where radioisotopes have played important roles are described to illustrate the versatility of the technique in specific areas of soil research.

Soil Fertility. Soil-fertility studies are concerned with the ability of a soil to supply essential mineral elements to plants. Elements essential to plant growth include carbon, hydrogen, oxygen, nitrogen, phosphorus, potassium, calcium, chlorine, sulfur, iron, boron, zinc, manganese, magnesium, copper, and molybdenum. Radioisotopes are commercially available for many of these elements. In other instances a stable element may be substituted for experimental purposes with an element (analogue) that has similar characteristics (e.g., cesium or rubidium have replaced potassium, and strontium has replaced calcium in plant nutrition studies). Table 2 lists some of the nuclear characteristics of the commercially available radioisotopes.

Extensive research has been conducted on the behavior of phosphorus fertilizer following its addition to the soil (Phillips and Peaslee, 1970; Levi, 1968; Wiebe and Kramer, 1954; Baker, 1964). The data show that commonly available phosphate fertilizers solubilize on contact with the soil to form orthophosphate ions. These

TABLE 2. Some of the Elements and Their Commercially Available Isotopes.

Nuclide	At. No.	Mass No.	Half-life	Gamma Energies (MeV)	Beta Energies (MeV)
Hydrogen	1	3	12.3 year		0.018
Carbon	6	14	5.73×10^3 year		0.155
Sodium	11	22	2.58 year	0.511	β+ 0.55
				1.27	Trace 1.8
Sodium		24	15.0 hr	1.37	Trace 4.2
				2.75	1.4
Magnesium	12	28	21.3 hr	1.35	
				3 others to	0.46
				0.95	
Aluminum	13	26	7.4×10^5 year	0.511	β+ 1.17
				1.81	
Silicon	14	31	2.62 hr	1.26	1.48
Phosphorus	15	32	14.3 day		1.70
Phosphorus		33	25 day		0.25
Sulfur	16	35	86.7 day		0.17
Chlorine	17	36	3.0×10^5 year		0.714
Chlorine		38	37.3 month	1.60	4.9
				2.17	
Potassium	19	42	12.4 hr	Trace to 2.4	2.0
				1.5	3.5
Potassium		43	22 hr		0.83
Calcium	20	45	165 day		0.25
Calcium		47	4.7 day	1.3	0.67
Manganese	25	52	5.7 day	1.4	0.6β+
				0.9	
				+0.5, 0.7	
Manganese		53	2.0×10^6 year	Cr X-rays	
Manganese		54	303 day	0.83	Trace 0.83 e^-
Manganese		56	2.6 hr	0.8	2.85
				1.8	
				2.1	
Iron	26	52	8.3 hr	0.16	0.8β+
				0.51	
Iron		55	2.7 year	~6 keV	
				Mn X-ray	
Iron		59	45 day	1.1	0.48
				1.3	1.57
Cobalt	27	56	77.3 day	0.85	1.5β+
				to 3.3	
Cobalt		57	267 day	0.12	4 e^-–0.13
				0.14	
Cobalt		58	71 day	0.81	0.47β+
				1.7	
Cobalt		60	5.26 year	1.17	0.31
				1.33	1.5
Nickel	28	63	92 year		0.067
Nickel		65	2.56 hr	1.1	Up to 2.1
				1.5	
Copper	29	61	3.3 hr	0.06–1.2	1.2β+
					Some 0.06 e^-
Copper		64	12.9 hr	0.51	0.57
				1.34	0.66β+

TABLE 2. continued

Nuclide	At. No.	Mass No.	Half-life	Gamma Energies (MeV)	Beta Energies (MeV)
Copper		67	61 hr	0.18 0.09	4 to 0.57 maximum
Zinc	30	65	245 day	1.12	Trace 1.1 e^- 0.33 [β+(2)]
Zinc		69 m	14 hr	0.44	0.4 e^-
Rubidium	37	83	83 day	0.53 0.79 Kr X-rays	Some e^- to 0.03
Rubidium		84	33 day	0.9 0.52	β+0.8 1.6 β–0.9
Rubidium		86	18.7 day	1.1	0.7 1.8
Strontium	38	85	64 day	0.51	0.5 e^-
Strontium		87 m	2.8 hr	Sr X-rays	0.4 e^-
Strontium		89	50.4 day		1.46
Strontium		90	28 year		0.6 2.2 From ^{90}Y
Molybdenum	42	99	66 hr	5–0.78	1.23 0.45
Ruthenium	44	97	2.9 day	0.22 0.32	Some 0.19 e^-
Ruthenium		103	40 day	0.50 0.61	0.70 0.21
Ruthenium		106	1.0 year		0.04
Iodine	53	123	13 hr	0.16; 3% others to 0.53	Some 0.13 e^-
Iodine		124	4.2 day	0.61 0.51 others to 2.3	2.1β+ 1.5β+
Iodine		125	60.2 day	0.027 X-rays 0.035	0.03 e^-
Iodine		126	13.2 day	0.67 0.39	Trace 1.1β+ 0.87 1.25
Iodine		129	1.6×10^7 year	0.04	0.15
Iodine		130	12.5 hr	0.54 0.67 0.74 others to 1.15	0.62 1.04
Iodine		131	8.05 day	0.36 others to 0.72	0.61 others to 0.81
Iodine		132	2.3 hr	0.67 0.77 others to 1.99	2.16 1.61 1.22 others to 2.1
Iodine		133	21 hr	0.53 others to 1.3	1.27
Cesium	55	131	9.7 day	Xe X-rays	
Cesium		132		0.67 trace to 1.3	0.4β+ 0.7β–

TABLE 2. continued

Nuclide	At. No.	Mass No.	Half-life	Gamma Energies (MeV)	Beta Energies (MeV)
Cesium		134	2.1 year	0.6 0.8 others to 1.4	0.09 0.66
Cesium		137	30 year	0.66 from ^{137m}Ba	1.18 0.51

(From *Chemistry of the Soil,* 2nd ed., F. E. Bear, ACS Monograph Series No. 160, pp. 446-447. Washington, D.C.: American Chemical Society, 1964)

ions are absorbed by soil colloids or precipitated very quickly. The radioisotopes ^{32}P and ^{33}P have been used to evaluate the chemical forms produced and to determine the very practical question of the effect of fertilizer placement on phosphorus availability to plants.

Because the ease with which plants obtain nutrients from the soil is related to the amount and nature of the cation exchange capacity, such exchange has received much attention from soil scientists. One approach has been to use radioactive ion pairs (Heald and Kinishi, 1965) for studying exchange reactions. Some useful ion pairs are ^{24}Na–^{66}Rb, ^{24}Na–^{85}Sr, ^{42}K–^{85}Rb, ^{42}K–^{85}Sr, and ^{47}Ca–^{85}Sr. Another approach has been to determine the isotopic exchange of ^{133}Ba with barium-saturated soils (Bache, 1970).

Fate of Materials. In an age concerned with the degradation of the environment and the effects of materials intentionally added to the soil to enhance agricultural production or as a convenient and inexpensive disposal technique, radioactive tracers have many roles. They provide unique tags for following the path of the material throughout the degradation or transport pathways. Many of the soil-solute-water interactions are still poorly defined, and prediction is extremely difficult; hence experimental information is required for various soil types, climatic conditions, and soil amendments. Compounds tagged with ^{14}C have been used extensively in research on the fate of herbicides and pesticides in the soil (Wagner and Chahal, 1966; Mills and Biggar, 1969). The ability to label specific locations within a compound has aided in measuring the decomposition rates of materials. Stable isotopic ratios that do not occur in nature have been used (enriched or depleted in ^{2}H, ^{15}N, ^{13}C, and ^{18}O) to tag compounds where radioactive isotopes are either not available or not suitable.

Water and Solute Movement. An understanding of water and solute behavior in soils is enhanced by the use of radionuclides. Radionuclides and in particular tritium (Blume, Zimmermann, and Muenich, 1967; Guizerix, et al, 1967; Csallany, 1967) have been used for tracing soil water movement (q.v.) as well as obtaining some of the parameters necessary to describe mathematically aquifer characteristics. Comparisons of the distribution in the soil of tritium, fertilizers, pesticides, etc., permit the relative flow rates of various compounds to be measured simultaneously. Tritium (Corey, 1970; Barry and Merritt, 1970) is also a useful tracer for following the evaporation of water under laboratory or specialized field conditions.

Fallout Isotopes. Radionuclides released to terrestrial ecosystems from fallout and from routine releases by nuclear facilities in various countries have permitted long-term studies of behavior of various radionuclides in diverse climatic regions and soil types. This research has principally examined ^{90}Sr, ^{137}Cs, ^{60}Co, ^{106}Ru, ^{125}Sb, and ^{65}Zn (Gulyakin, Dergunov, and Moroz, 1968; Romney, et al, 1960). Although the motivation was to define the health hazards to people if extensive fallout from nuclear weapons should occur, soil scientists have benefited from the extensive information developed in these studies.

Sealed Sources and Their Uses

In addition to tracer studies where the isotopes are placed directly in the soil, encapsulated (*sealed*) sources are used to study radiation effects on materials and living organisms. These sources are usually enclosed inside stainless steel capsules. This encapsulation prevents the radioactivity from reaching the environment. Encapsulated sources generally produce a higher radiation field than that produced by the tracer studies because larger quantities of radioactive materials are used. These sources are shielded by an appropriate material (gamma sources by lead or uranium, and neutron-emitting sources by water or hydrocarbons). Sealed sources are used for either soil physics and engineering studies or sterilization of soil.

Soil Physics and Engineering Studies

Methods for nondestructively measuring the water content (q.v.) and bulk density (q.v.) of soil under laboratory or field conditions include the use of a neutron meter or gamma probe (Gardner, 1965). The most common device for measuring the water content of soils under field conditions is called the *neutron meter.* The neutron meter measures the number of slow (thermal) neutrons produced from a source of fast neutrons. The number of thermal neutrons detected increases as the water content increases but is also affected by the density of the soil, the presence of significant amounts of chlorine, and the presence of hydrogen atoms in, for example, clay and organic matter. Table 3 summarizes the characteristics of some fast neutron sources. More recent developments have included the use of neutron-transmission and gamma-transmission methods for measuring water contents under field conditions.

Neutron-transmission Method. Wack (1967) has designed and patented a device for measuring the water content of surface soils based on the transmission of intermediate energy neutrons. A calibrated neutron source of intermediate energy is placed at known depths in a vertical tube in the soil, and the number of neutrons reaching the soil surface after being thermalized (slowed) by subsurface water is measured using a boron trifluoride counter surrounded by a paraffin cylinder 3 to 5 cm thick. The exterior of the paraffin cylinder is covered by a layer of cadmium to eliminate incidental thermal neutrons from reaching the detector. The number of neutrons reaching the detector decreases with increasing water content.

Gamma-transmission Method. Gamma-transmission techniques are in use under field conditions. This method uses a two-hole device: the source (^{137}Cs) is positioned in one hole, and the detector [a NaI(Tl) crystal] in the other hole. The fixed distance permits quantitative water content changes to be measured. The count rate at the detector is described by

$$I/I_0 = \exp - (\mu_s \rho_s + \mu_w \theta)\chi$$

where

I = the radiation intensity after passage through the soil
I_0 = the radiation intensity after passage through the air
μ_s = the mass attenuation coefficient for oven-dried soil
ρ_s = soil density
μ_w = the mass attenuation coefficient for water
θ = water content
χ = thickness of soil

Laboratory methods have been devised to measure soil density and water content simultaneously using a collimated beam of low- and high-energy gamma rays. Mansell, Hammond, and McCurdy (1973) have described the electronics and principles of operation. The technique is based on two equations identical to equation 1: the first for high-energy gamma rays, and the other for low-energy gamma rays. The only unknowns are ρ_s and θ, and solution

TABLE 3. Characteristics of Isotopic Neutron Sources (normalized to 5×10^{10} neutrons/s).

Source	Half-life	Gamma Dose Rate (rad/hr at 1 min)	Heat Generation[a] (W)	Volume[b] (cm^3)
^{124}Sb-Be	60 day	4.5×10^4	20	200
^{210}Po-Be	138 day	2.0×10^4	640	200
^{242}Cm-Be	163 day	0.3×10^4	600	2
^{252}Cf	2.65 year	2.9×10^4	0.8	<1
^{244}Cm-Be	18.1 year	0.2×10^4	600	70
^{238}Pu-Be	89 year	0.4×10^4	500	350
^{241}Am-Be	458 year	2.5×10^4	750	2.2×10^4

[a] 1 W = 1 J/s = 0.2319 g-cal/s.

[b] Not including void space for helium from alpha decay. ^{252}Cf sources need less void space for helium than the other sources because there are fewer alpha disintegrations per neutron emitted from ^{252}Cf sources.

(From *Chemistry of the Soil,* 2nd ed., F. E. Bear, ACS Monograph Series No. 160, pp. 446–447. Washington, D.C.: American Chemical Society, 1964) of the two equations for two unknowns is straightforward.

Soil Sterilization

Although autoclaving is the most common method for sterilizing soil, sterilization by gamma radiation has been used. Soil sterilization by gamma radiation is usually conducted in a shielded container containing kilocurie amounts of ^{60}Co or ^{137}Cs. The dose rate is usually in excess of 10^4 rads/min. Fungi (Jackson, et al, 1967) are generally unable to reproduce after receiving less than 10^6 rads, whereas bacteria and actinomycetes are generally unable to reproduce after 3×10^6 rads. Eno and Popenoe (1964) compared the effects of steam, methyl bromide, and gamma radiation on the exchangeable NH_4^+; extractable nitrogen, phosphorus, and sulfur; moisture equivalent; cation exchange capacity; and organic matter. None of the sterilization treatments significantly increased the amounts of nitrogen, phosphorus, and sulfur extracted or the organic matter content. Compared to the other two soil treatments, steam sterilization (autoclaving) significantly decreased the soil moisture equivalent percentage. Gamma-ray and methyl-bromide sterilization generally increased the release of nitrogen, phosphorus, and sulfur.

Neutron Activation

Whenever neutrons impinge on soil, some of the neutrons will be captured by some of the nuclei of elements in the soil. Capture of the neutrons results in an isotope of the same element as shown below:

$${}_b^a\chi + {}_0^1n \rightarrow {}^{a+1}{}_b\chi + \gamma$$

The amount of element ${}_b^a\chi$ in soil, if it is a gamma-emitting radioisotope, can be calculated as follows: the detection of gamma radiation using either a NaI(Tl) or Ge(Li) detector permits the amount of ${}^{a+1}{}_b\chi$ to be determined; the number of neutrons striking the target is known from the target position in the neutron flux; the affinity of ${}_b^a\chi$ for reaction with neutrons is known from published data; hence ${}_b^a\chi$ can be calculated.

Neutron activation is a useful analytical technique in soil investigations (Haskin and Ziege, 1971) permitting identification of 65 elements, for example: Al, Fe, K, Mg, Na, Cl, Sc, Ti, V, Cr, Mn, Co, Zn, As, Br, Sr, Ba, Sb, Cs, La, Ce, Sm, Yb, U, Eu, Ga, Dy, Se, F, Hg, Au, Tb, and Ta. The method is nondestructive and avoids detailed wet chemistry separations to bring the various elements into solution. Industrial and university facilities exist that permit these elements to be detected at the one-part-per-million level or less. Overall accuracies of ±10% are observed for most elements. The technique has been used to identify soils with similar parent materials (Borchardt and Harward, 1971) and as an analytical tool for trace metals.

JOHN C. COREY*

References

Bache, B. W., 1970. Barium isotope method for measuring cation-exchange capacity of soils and clays, *Jour. Sci. Food Agric.*, 21, 169–171.

Baker, D. E., 1964. A study of isotopic dilution as a method for relating phosphorus retention to availability of phosphorus in widely different soils, *Soil Sci. Soc. America Proc.*, 28, 511–517.

Barry, P. J., and Merritt, W. F., 1970. Perch Lake evaporation study. *Proc. Symposium on Use of Isotopes in Hydrology*, International Atomic Energy Agency, Vienna, 139–150.

Bear, F. E., ed., 1964. *Chemistry of the Soil*, New York: Reinhold, 515p.

Becquerel, H., 1896. *Compt. rend.*, 122, 422.

Benes, J., 1968. Natural radioactivity of soil and plants, *Jad. Energ.*, 14, 100–104.

Blume, H. P.; Zimmermann, U.; and Muennich, K. O., 1967. Tritium tagging of soil moisture: The water balance of forest soils, *Proc. Symposium on the Use of Isotope and Radiation Techniques in Soil Physics and Irrigation Studies*, International Atomic Energy Agency, Vienna, 315–331.

Borchardt, G. A., and Harward, M. E., 1971. Trace element correlation of volcanic ash soils, *Soil Sci. Soc. America Proc.*, 35, 626–631.

Corey, J. C., 1970. Contribution of tritiated water at various depths in soil to evaporation at the soil surface, *Proc. of a Symposium on Environmental Surveillance in the Vicinity of Nuclear Facilities* Springfield, Ill.: Charles C. Thomas, 309–319.

Csallany, S. C., 1967. Application of isotope techniques in groundwater hydrology, *in* M. A. Marino, ed., *Proc. of the Symposium on Groundwater Hydrology*, 351–357.

Dios Vidal, R., 1964. Radon-generating activity in different Galician soils in relation to other great world groups and the total lead content, *An. Edafol Agrobiol.*, 23, 53–63.

D'Souza, T. J., and Mistry, K. B., 1970. "Comparative uptake of thorium-230, radium-226, lead-210, and polonium-210 by plants, *Radiation Botany*, 10, 293–296.

Eno, C. F., and Popenoe, H., 1964. Gamma radiation compared with steam and methyl bromide as a soil sterilizing agent, *Soil Sci. Soc. America Proc.*, 28, 533–535.

Francis, C. W., 1967. *Source of Polonium-210 in Plants*, Ph.D. Thesis, University of Wisconsin, Madison, Wisconsin, 83p.

Friedlander, G.; Kennedy, J. W.; and Miller, J. M., 1964. *Nuclear and Radiochemistry*. New York: Wiley, 585p.

*This paper was prepared in connection with work under Contract No. AT(07-2)-1 with the U.S. Energy Research and Development Administration.

Gardner, W. H., 1965. Water content, *in* C. A. Black, ed., *Methods of Soil Analysis, Part 1. Physical and Mineralogical Properties, Including Statistics of Measurement and Sampling.* Madison: American Society of Agronomy, 82-127.

Guizerix, J.; Molinari, J.; Gaillard, B.; Santos, C. H.; Mornas, A.; and Corda, R., 1967. Localization of leakages from a large reservoir using radioactive tracers, *Proc. of a Conference on Isotopes in Hydrology,* International Atomic Energy Agency, Vienna, 587-599.

Gulyakin, I. V.; Dergunov, I. D.; and Moroz, V. D., 1968. Behavior of radioactive fission products in some soils of the Sierozem zone of Uzbekistan, *Izv. Timiryazev. Sel'skokhoz. Akad.,* 1, 84-93.

Hansen, W. R.; Watters, R. L.; and Yaney, N. D., 1971. Polonium-210 dioxide movement in a mountain watershed soil, *Health Phys.,* 20, 425-429.

Haskin, L. A., and Ziege, K. E., 1971. Neutron activation: Techniques and possible uses in soil and plant analysis, *In Instrumental Methods for Analysis of Soils and Plant Tissue.* Madison: Soil Science Society of America, 185-208.

Heald, W. R., and Kinishi, H. W., 1965. Use of radioisotopes in studying exchange reactions, *Proc. of Symposium on the Use of Isotopes and Radiation in Soil-Plant Nutrition Studies,* International Atomic Energy Agency, Vienna, 97-106.

Holtzman, R. B., 1970. Isotopic composition as a natural tracer of lead in the environment, *Environ. Sci. Technol.,* 4, 314-317.

**Isotope Techniques in Hydrology, Vol. II (1966-1971),* International Atomic Energy Agency, Vienna, 1973, 233p (814 references).

Iwakura, T.; Oda, A.; Miyahara, S.; Kasida, Y.; and Yamada, O., 1966. Recent radiocarbon in plants, *Radiation Res.,* 7, 134-141.

Iyengar, M. A. R., and Markose, P. M., 1971. An investigation into the distribution of uranium and daughters in the environment of a uranium ore processing facility, *Proc. of a Symposium on Radiation and Radioisotopes in Soil Studies and Plant Nutrition,* International Atomic Energy Agency, Vienna, 143-153.

Jackson, N. E., et al, 1967. Gamma irradiation and the microbial population of soils at two water contents, *Soil Sci. Soc. America Proc.,* 31, 491-494.

Kerpen, W., and Scharpenseel, H. W., 1967. Movement of ions and colloids in undisturbed soil and parent rock material columns, *Proc. of a Symposium on the Use of Isotope and Radiation Techniques in Soil Physics and Irrigation Studies,* International Atomic Energy Agency, Vienna, 213-226.

Krey, P. W., and Hardy, E. P., 1970. *Plutonium in Soil Around the Rocky Flats Plant. USAEC Report HASL-235,* 44p.

Levi, E., 1968. The distribution of mineral elements following leaf and root uptake, *Physiologia Plantarum,* 21, 213-226.

Mansell, R. S.; Hammond, L. C.; and McCurdy, R. M., 1973. Coincidence and interference corrections for dual-energy gamma ray measurements of soil density and water content, *Soil Sci. Soc. America Proc.,* 37, 500-504.

Marquardt, W. Z., 1967. Influences on natural radioactivity in the layer of air near the ground, *Meteorology,* 19, 222-231.

*Contains useful bibliographies.

Megumi, K., and Mamuro, T., 1977. Concentration of uranium series nuclides in soil particles in relation to their size, *Jour. Geophys. Res.,* 82(2), 353-356.

Mills, A. C., and Biggar, J. W., 1969. Solubility-temperature effect on the adsorption of gamma and beta-BHC from aqueous and hexane solutions by soil materials, *Soil Sci. Soc. America Proc.,* 33, 210-216.

Murrmann, R. P.; Winters, R. W.; and Martin, T. G., 1971. Determination of trace elements in soils and clay minerals by resonance neutron activation analysis, *Soil Sci. Soc. America Proc.,* 35, 647-652.

Nakhla, S. M., and Delibrias, G., 1967. Use of carbon-14 of thermonuclear origin to study the dynamics of carbon in the soil, *Proc. Symposium on Radioactive Dating and Methods of Low-level Counting,* International Atomic Energy Agency, Vienna, 167-176.

**Nuclear Science Abstracts,* 1977. U.S. Energy Research and Development Administration, Washington, D.C.

Pearson, J. E., 1967. *Natural Environmental Radioactivity from Radon-222, U.S. Public Health Service Report PHS-999-RH-26,* 31p.

Pensko, J.; Wardaszko, T.; and Wochna, M., 1968. Natural atmospheric radioactivity and its dependence on some geophysical factors, *Atompraxis,* 14, 255-258.

Phillips, R. E., and Peaslee, D. E., 1970. Effect of chemical pretreatment of kaolinite clay upon the apparent self-diffusion of phosphorus-32, *Soil Sci. Soc. America Proc.,* 34, 743-746.

Radiological Health Handbook, 1970. U.S. Department of Health, Education and Welfare, Rockville, Maryland, 458p.

**Radionuclide Movement in Soils and Uptake by Plants,* Oak Ridge National Laboratory, Oak Ridge, Tennessee, 1975, 367p (1397 references).

Romney, E. M.; Ehrler, W. L.; Lange, A. H.; and Larson, K. H., 1960. Some environmental factors influencing radiostrontium uptake by plants, *Plant and Soil,* 12, 41-48.

Romney, E. M.; Mork, H. M.; and Larson, K. H., 1970. Persistence of plutonium in soil, plants, and small mammals, *Health Phys.,* 19, 487-491.

Sokolov, A. V.; Andreeva, E. A.; Ginzburg, K. E.; Mel'nikova, M. K. Yu.; Polyakov, A.; Prokhorov, V.M.; Dedorovski, D. V.; and Shirshova, R. A., 1972. Use of radioactive and stable isotopes in agrochemical investigations, *Proc. of the 4th Internat. Conf. on the Peaceful Uses of Atomic Energy,* 12, 111-125, United Nations, New York.

Stewart, M. L., 1970. *Distribution and Movement of Radioactive Nuclides, Both Natural and Artificial, in Soils, Vegetation, and Wildlife,* Ph.D. Thesis. University of Utah, Salt Lake City, Utah, 181p.

Wack, B., 1967. Etude expérimentale de la mesure de l'humidité des sols par transmission de neutrons intermédiaires, *C.r. hebd. Séanc. Acad. Sci.,* Paris Ser. A., 264, 1098-1101.

Wagner, G. H., and Chahal, K. S., 1966. Decomposition of carbon-14 labeled atrazine in soil samples from Sanborn Field, *Soil Sci. Soc. America Proc.,* 30, 752-754.

Wiebe, H. H., and Kramer, P. J., 1954. Translocation of radioactive isotopes from various regions of roots of barley seedlings, *Plant Phys.*, 29, 342–348.

Cross-references: *Radiocarbon Dating; Soil Physics; Water Content and Retention. Vol. IVA: Radioactive Isotopes. Vol. VI: Radioactivity in Sediments.*

REDOX REACTIONS

Chemical reactions involving oxidation and reduction are referred to as *redox reactions*. These processes are important in soil formation, plant nutrition, and land use and management.

Early chemists thought oxidation and reduction involved the gain and loss of oxygen, but later studies showed that such reactions were merely examples of a much larger class among elements that can exist in more than one valence state, i.e., elements that can gain or lose electrons. In such reactions the electron donor (reducing agent) is oxidized and the electron acceptor (oxidizing agent) is reduced.

Redox Potentials

The intensity of oxidation-reduction reactions is a function of the electrical potential in the reacting system. The potential is given in volts with the potential of the standard hydrogen electrode,

$$2H^+ + 2e^- = H_2$$

taken as zero. Oxidizing systems are conventionally expressed as positive and reducing systems as negative potentials. In practice, the potential between an electrode that is relatively inert to most reactants (platinum) and a reference electrode (calomel) is measured with a *potentiometer*.

Quantitative treatment of redox potentials of natural systems, especially soils, is very complex, as the observed potentials usually represent nonequilibrium potentials of several redox systems simultaneously. As Bohn (1971) noted, oxidized transition metals (Fe^{3+}, Mn^{4+}, Cu^{2+}) are extremely insoluble in aerated soils and oxygen exchange currents at the electrode are very low. The system is poorly poised, i.e., poorly buffered against potential changes, so voltages drift and are highly variable. In waterlogged soils, larger quantities of reduced metals and organic reactants are dissolved, giving the system greater poise and much better redox potential stability.

Patrick and Mahapatra (1968) characterize soil redox potential ranges as:

1. Aerated soils, +700–+400 mv
2. Moderately reduced soils, +400–+100 mv
3. Reduced soils, +100– –100 mv
4. Highly reduced soils, –100– –300 mv

Their view on redox processes in soils is that as long as oxygen is available (reducible at +340–+320mv), other oxidized compounds are protected from biological and chemical reduction. Where the oxygen supply is depleted, nitrates are readily reduced (+225 mv) if an energy source is available, and manganese (+200 mv) and iron (+120 mv) tend to buffer redox potentials in the moderately reducing range. Sulfates are reduced at about –150 mv, and their presence can be suspected where highly reducing potentials are observed.

Well-drained Soils

The mineral matter of well-drained soils was largely accumulated in an environment rich in oxygen and is therefore relatively immune to further oxidation (Merkle, 1955). Major component minerals of acid igneous rocks such as quartz and feldspars were completely oxidized during formation of the original rocks. Moreover, these minerals lack elements readily subject to changes in valence states. Accessory minerals, present in small amounts, do contain elements such as iron, manganese, titanium, and copper that are subject to oxidation and reduction. Soil parent materials derived from basic igneous rocks contain larger amounts of elements, particularly iron, with more than one valence state. Half or more of the iron in such rock types is in ferrous or reduced form. It is subject to rapid oxidation when exposed to water carrying atmospheric oxygen. The oxidation results in a smaller size for the iron atom and thereby weakens the rock structure to make it more susceptible to further weathering.

In contrast to mineral matter in well-drained soils, organic substances are added as reduced forms of carbon, nitrogen, and sulfur derived from photosynthetic and other metabolic processes of plants. Forests, grasslands, and croplands in temperate regions annually contribute organic residues ranging from 2 to 7 metric tons of dry matter per hectare on or near the surface of the soil. That these organic deposits do not accumulate to any great extent in well-drained upland environments can be attributed to biological oxidation by various decomposers, the striking of a balance with annual production of organic residues. Thus in well-aerated soils, plant residues containing reduced forms of carbon, nitrogen, and sulfur in carbohydrates, fats, proteins, and lignin provide energy sources for respiration of soil organisms. In the respiratory process these reduced organic compounds are electron donors, and electron acceptance is mainly by respiratory reduction of atmospheric oxygen. When oxygen is abundant, the transition metals–iron, manganese, and copper–are present in the soil as oxidized compounds of very low solubility. Small amounts of these elements absorbed by soil organisms serve as catalysts in respiratory redox systems and thus are vitally involved in the oxidative decomposition of large quantities of soil organic matter.

Poorly Drained Soils

When oxygen is limited, as in soils subject to prolonged flooding or high water tables, and if organic substrates are available, aerobic organisms quickly deplete the supply of oxygen. They may then die or become dormant. Anaerobic microorganisms multiply rapidly and use oxidized soil compounds such as iron and manganese oxides, nitrate, sulfate, and their own metabolites as electron acceptors in respiration (Ponnamperuma et al, 1967, 1969) and also produce excretory products that are reducing agents in chemical reactions with oxidized soil constituents. Nitrates are rapidly converted to ammonia and nitrogen gas under moderately reducing conditions and may be lost from the soil by volatilization. Manganese is readily reduced to soluble manganous ions and may be leached from the soil, or more commonly, translocated to sites where it is precipitated as manganese carbonate, complex oxides with iron, or as manganic oxides where oxygen is available. As reducing environments become more intense, ferrous iron is brought into solution and may travel considerable distances in groundwater aquifers to emerge in the waters of chalybeate springs or perhaps to appear in well water as reddish brown colloidal suspensions of ferric hydroxide on exposure to the air. If retained in the soil profile under intense reducing conditions, iron precipitates as gray green ferrosoferric hydroxide [$Fe_3(OH)_8$], as pyrite (FeS_2) when sulfates are present, or less frequently as siderite ($FeCO_3$) if there is a large supply of CO_2.

Soil Formation Processes

The state of oxidation of soil materials during soil formation can be inferred broadly from the colors of horizons within a profile. Horizons of well-drained upland soils containing oxidized iron exhibit red, brown, and yellow hues as a result of coating of soil particles by oxides and hydrous oxides of ferric iron. Thus, for example, litter decomposition products in northern coniferous forests are highly acid and persist for some time. During wet periods ferrous iron is leached downward as organic complexes or chelates until adsorbed on ferric oxides in deeper parts of a profile. During dry periods these ferrous complexes become immobilized as oxidized ferric forms. This process for translocation of iron (and organic matter) can ultimately lead to formation of bleached upper horizons (*A*2) and reddish brown deeper horizons (*B*ir) in soils such as Haplorthods.

Where a soil remains wet for long periods because of flooding or restricted drainage, portions of the profile subject to waterlogging undergo marked reduction, provided organic matter is present. Such reduction has been called *gleying.* The appearance of a reduced horizon depends on the intensity of the reaction. A great majority of soil horizons subject to strong reduction are uniformly gray in color. Rarely, deeper parts of profiles, especially under peat, may be greenish gray or bluish gray, indicating ferroferrosic precipitates. Uniformly gray horizons or those with slightly greenish or bluish colors are wet continuously for long periods. If the water table fluctuates appreciably but the profile remains wet much of the time, the deeper horizons will have gray backgrounds spotted or streaked with yellowish brown, yellowish red, or red mottles. A fluctuating water table can also give rise to iron or iron-manganese concretions, which are commonly spherical in shape but may in some instances be elongated hollow cylinders around former root channels.

Redox Processes and Land Use

In most soils of the world used for crops, range, and forest, processes of oxidation and reduction play a passive but vital role in supplying small amounts of the mineral elements (iron, manganese, and copper) that catalyze respiratory processes of higher plants and soil organisms–processes in the cyclical production and decomposition of all members of living food chains.

Redox processes in soils tend to maintain balanced supplies of iron and manganese in well-drained soils having adequate annual additions of organic residues. Manganese is a powerful oxidizing agent, however, so an excess can immobilize iron in ferric form, thereby leading to iron deficiencies and stunted plant growth.

Redox processes of great consequence in land use take place when earthy materials containing much pyrite are exposed to atmospheric weathering. This and other sulfides are associated with coal beds and with permanently moist soils high in organic matter in brackish tidal marshes. Sulfides in exposed mine wastes and drained marshlands undergo rapid biological and chemical oxidation to form sulfuric acid, which may enter drainage waters or remain in place to give an acidity level highly detrimental to crop growth. Major areas of acid sulfate soils or potential acid sulfate soils (undrained wetlands high in pyrite and jarosite) are found in the tropics, often in coastal mangrove (*Rhizophora*) swamps (Bloomfield and Coulter, 1973). Total area of present and potential acid sulfate soils is more than 15 million hectares in tropical regions (Dost, 1972). These soils respond poorly to improvement of drainage but have been used with some success for rice culture in the tropics. Past experience provides an incentive for the research needed to use these soils effectively for food production.

CHARLES W. RALSTON

References

Bloomfield, C., and Coulter, J. K., 1973. Genesis and management of acid sulfate soils, *Adv. Agron.*, **25**, 265–326.

Bohn, H. L., 1971. Redox potentials, *Soil Sci.*, **112**, 39–45.

Dost, H., 1972. Acid sulfate soils: Advances in research, *Annl. Rept. Interntl. Inst. Land Reclam. and Imprvmnt*, 1972. Wageningen, 17–42.

Dost, H., ed., 1973. *Acid Sulfate Soils.* Wageningen: International Institute Land Reclamation and Improvement, vol. 1, 295p; vol. 2, 406p.

Merkle, F. G., 1955. Oxidation-reduction processes in soils, *in* F. E. Bear, ed., *Chemistry of the Soil.* New York: Reinhold, 200–218.

Patrick, W. H., Jr., and Mahapatra, I. C., 1968. Transformations and availability to rice of nitrogen and phosphorus in waterlogged soils, *Adv. Agron.*, **20**, 323–359.

Ponnamperuma, F. N.; Tianco, E. M.; and Loy, T. A., 1967. Redox equilibria in flooded soils: I. The iron hydroxide system, *Soil Sci.*, **103**, 374–382.

Ponnamperuma, F. N., and Tianco, E. M., 1969. Redox equilibria in flooded soils: II. The manganese oxide systems, *Soil Sc.*, **108**, 48–57.

Cross-references: *Aeration, Respiration, Atmosphere; Electrochemistry; Metabolic Activity; Soil Chemistry; Soil Reaction. Vol. IVA: Oxidation and Reduction. Vol. XII, Pt. 2; Gley Soils; Soil Color and Color Patterns; Soil Pans, Vol. XV: Coastal Soils.*

REMOTE SENSING—See SPECTRAL CHARACTERISTICS; SURFACE-SOIL WATER CONTENT; THERMOLUMINESCENCE

RESPIRATION—See AERATION, RESPIRATION, AND ATMOSPHERE

RHIZOSPHERE

The *rhizosphere* comprises the small zones immediately surrounding plant roots in the soil. Included are the microorganisms and the organic and inorganic materials released by the roots themselves. The rhizosphere is thus a poorly defined zone around any root, commonly extending out no more than 1 or 2 mm but occasionally as far as 20 mm. The name was originally proposed in 1904 by Hiltner for the volume of soil near legume roots in which the growth of heterotrophic bacteria might be stimulated by the nitrogen released from nodules (Neal, Atkinson, and Larson, 1970). The meaning has since been expanded to cover roots of all plants.

The root surface and adhering soil particles are called the *rhizoplane.* It is rather distinct as compared to the rhizosphere. Microorganisms invade and modify the viscous mucilaginous coatings on the epidermal surfaces of roots.

The microbial population of the rhizosphere differs both quantitatively and qualitatively from that outside these small zones. Biochemical and microscopic techniques and plate counts are used in the study of the microorganisms and various substances present in the rhizosphere. The numbers of viable organisms in the rhizosphere (R) are greater than those in the remainder of the soil (S). The *R:S ratio* for bacteria commonly ranges from 5 to 20 but may exceed 100. Ratios of 10 to 20 are common for fungi, based on estimates from numbers of fungal spores rather than active mycelia. Ratios for actinomycetes are lower than for bacteria, whereas reports on numbers of algae and protozoa are conflicting and therefore uncertain. The R:S ratio for free-living nematodes may be as high as 70. For the various organisms, the R:S ratios differ widely with differences in environmental conditions and in experimental techniques.

Qualitative effects are comparable to the quantitative ones. Within the rhizosphere, there is a selective stimulation of Gram-negative bacteria with fast growth rates and a reduction in the spore-forming *Bacillus* spp. and the Gram-positive cocci. Fungi become more abundant near the older parts of roots where specialized forms such as *Fusarium oxysporum, Penicillium* spp. and *Trichoderma viride* predominate. The detection of more responsive groups of microorganisms may therefore be more useful in defining the extent of the rhizosphere than total counts.

Substances within the rhizosphere also have evident nutritional effects on microorganisms. Bacteria that use amino acids directly are stimulated whereas those with more complex nutritional requirements are not. Selective stimulation of organisms with specific physiological activities also occurs. Examples are the ammonifiers, denitrifiers, and aerobic cellulose-decomposing microorganisms described by Rouatt, Katznelson, and coworkers and the nonsymbiotic nitrogen fixers observed by Dobereiner et al in the rhizospheres of tropical graminaceous plants (see references in Darbyshire and Greaves, 1973).

The basis of the rhizosphere phenomenon is apparently biochemical and nutritional. Rovira and McDougall (1967) list as excreted compounds: 10 sugars, 21 amino acids, 10 vitamins, 11 organic acids, 4 nucleotides, 16 fungal stimulators or attractors, 3 fungal inhibitors, 4 eelworm hatching factors, 2 eelworm attractors, and 11 miscellaneous compounds including enzymes. Root exudates and occasionally the balance of organisms may be modified by foliar sprays that incorporate urea, antibiotics, or plant growth regulators (Macura, 1967). Seed and root inoculation can also modify the population. More exciting prospects are indicated by the substitution of a chromosome pair from root-rot resistant wheat into a susceptible variety thereby developing for the latter rhizosphere microflora antagonistic toward the root-rot organism (Neal et al, 1970).

MICHAEL HERLIHY

References

Darbyshire, J. F., and Greaves, M. P., 1973. Bacteria and protozoa in the rhizosphere, *Pestic. Sci.*, **4**, 349–360.

Macura, J., 1967. Physiological studies of rhizosphere bacteria, *in* T. R. G. Gray and D. Parkinson, eds., *The Ecology of Soil Bacteria.* Liverpool: University Press, 379–395.

Neal, J. L., Jr.; Atkinson, T. G.; and Larson, R. I., 1970. Changes in the rhizosphere microflora of spring wheat induced by disomic substitution of a chromosome, *Canadian Jour. Microbiol.*, **16**, 153–158.

Rovira, A. D., and McDougall, B. M., 1967. Microbiological and biochemical aspects of the rhizosphere, *in* A. D. McLaren and G. H. Peterson, eds., *Soil Biochemistry, Vol. 1.* New York: Marcel Dekker, 417–463.

Cross-references: *Metabolic Activity; Microhabitats.*

ROOTS—See RHIZOSPHERE

S

SALT LEACHING

Salt-affected soils commonly occur in the arid and semiarid regions of the world. In humid areas, the soluble salts present in the soil or those released through weathering are carried downward through the soil profile and are ultimately transported by streams to the ocean. In arid regions, however, washing down of salts may be local, and the soluble salts may not be transported very far because of limited rainfall. Also the high evapotranspiration rates obtained in arid climates tend to decrease the limited amount of water available for removal and transportation of the soluble salts. As a result, high concentrations of soluble salts often build up in the root zone of crops, thereby rendering the soils less suitable for crop growth. The problem of excess salts in the rhizosphere (q.v.) is often accentuated when soils of arid and semiarid regions are brought under irrigation (q.v.). All waters used for irrigation contain some soluble salts, but if the soluble salt content of the irrigation water (q.v.) is low and the internal drainage of the soil is good, salt may not accumulate in the root zone. If the irrigation water contains soluble salts in appreciable quantities and/or the soil is poorly drained, continuous irrigation will enhance salt build-up in the root zone. Even though many soils are well drained and nonsaline, the drainage may not be adequate once irrigation is introduced. When water is used freely for irrigation, the groundwater table often rises from a considerable depth to within a few meters of the soil surface in a few years. Once the water table rises to 1 to 2 m below the soil surface, appreciable quantities of groundwater move upward into the root zone and are lost by evaporation (q.v.), leaving the dissolved salts in the root zone. Groundwater, as well as irrigation water, thus contributes to soil salinization.

Saline Soils

A *saline* soil contains neutral soluble salts sufficient to adversely affect crop production. The amount of soluble salts in a soil can be expressed by weight as a fraction of the total soil mass. The U.S. Salinity Laboratory (Richards, 1954) has defined a saline soil as one whose saturation extract has an electrical conductivity (q.v.) greater than 4 mmho/cm at 25°C. For a medium-textured soil, an electrical conductivity of 4 mmho/cm at 25°C corresponds to a salt content on a dry soil basis of about 0.2% by weight.

Salt Leaching-Reclamation

The presence of excess soluble salts in the root zone adversely affects the growth and final yield of most crops. The extent of yield reduction depends on such factors as the crop grown, the salt content of the soil, climatic conditions, etc. In extreme cases where there is a high salt concentration in the root zone, crop growth may be prevented. To improve crop growth in these soils, the excess salts must be removed from the root zone. The term *reclamation of saline soils* refers to the methods used to remove soluble salts from the root zone which render relatively unproductive soils more productive. The following methods have been adopted at one stage or another:

Scraping the salts that have accumulated on the soil surface by mechanical means has had a limited success. Although this might work in some cases, the ultimate disposal of salts still poses a major problem.

Flushing water over the surface is sometimes used to desalinize soils having surface salt crusts. Because the amount of salts that can be flushed off of a soil is rather small, this method does not have much practical significance.

Leaching is by far the most effective procedure for removing salts from the root zone of soils. It involves the removal of soluble salts by the passage of water through the soil. Leaching is most often accomplished by ponding fresh water on the soil surface and allowing it to infilterate (Figure 1). Leaching is effective when the salty drainage water is discharged through subsurface drains that carry the leached salts out of the area under reclamation. Leaching may reduce salinity levels in the absence of artificial drains when there is sufficient natural drainage, i.e., the ponded water drains without raising the water table.

Leaching should be conducted when the soil moisture content is low and the groundwater

FIGURE 1. Leveled and bunded fields ponded with water for salt leaching.

table is deep. Leaching during the summer months is, as a rule, less effective because large quantities of water are lost by evaporation. The actual choice, will, however, depend on the availability of water and other considerations. In some parts of India, for example, leaching is best accomplished during the summer months because this is the time when the water table is deepest and the soil is dry. This is also the only time when large quantities of fresh water can be diverted for reclamation purposes.

Some field experiments (Nielsen, Biggar, and Luthin, 1966) have shown that the quantity of salts removed per unit quantity of water leached can be increased appreciably by leaching at soil moisture contents less than saturation, i.e., under unsaturated conditions. In the field unsaturated conditions during leaching were obtained by adopting intermittent ponding or intermittent sprinkling at rates less than the infiltration rate of the soil. Nielsen, Biggar, and Luthin (1966), for example, showed that 25 cm of sprinkled water reduced the salinity of the upper 60 cm of soil to the same degree as 75 cm of ponded water. Although leaching under unsaturated conditions could result in saving water and consequently reduce pressure on drainage, the time required for leaching is considerably greater for soils with low permeability (Dahiya and Abrol, 1974).

Water Required for Leaching

It is important to have a reliable estimate of the quantity of water required to accomplish salt leaching. The initial salt content of the soil, desired level of soil salinity after the leaching, depth to which reclamation is desired, and soil characteristics are all factors that determine the amount of water needed for reclamation. A useful rule of thumb is that a unit depth of water will remove nearly 80% of the salts from a unit soil depth. Thus 30 cm of water passing through the soil will remove approximately 80% of salts present in the surface 30 cm soil. Similarly, to reduce the salt content of the surface 60 cm of soil to about 20% of the original value would require the passage of about 60 cm water through the soil. For more reliable estimates, however, it is desirable to conduct salt-leaching tests on a limited area and prepare *leaching curves.* Leaching curves (Figure 2) relate the ratio of actual salt content to initial salt content in the soil to the depth of leaching water per unit depth of soil. These curves provide more precise estimates of the amount of water required for complete leaching to desired levels (see Dieleman, 1963, for details).

The Leaching Process

Leaching of salts from the soil essentially involves the replacement of a saline soil solution by fresh water. During the displacement of a saline soil solution, the boundary between the fresh and saline water moves downward not as a sharp front but as a transition zone. Knowledge of the changes in salt concentration in the transition zone is essential to predict the amount of leaching required to reduce the salinity to a known value at any particular soil depth. Physical models describing the leaching process have been proposed by Molen (1956) and Gardner and Brooks (1957). The former assumed salt diffusion while the latter assumed dispersion due to distribution of flow velocities within the soil pores to be the major factors operative in salt displacement. Studies by Nielsen and coworkers (1961) have shown that the relative contribution of diffusion and dispersion mechanisms will vary with the flow velocity during leaching, pore geometry of the soil, as well as with exchange and adsorption phenomena (q.v.) during leaching.

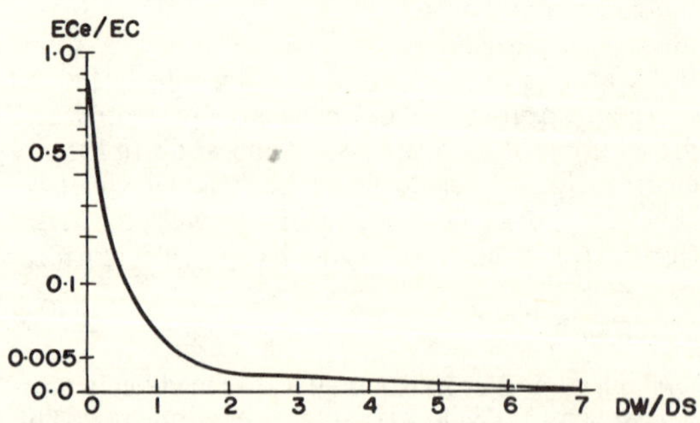

FIGURE 2. A typical leaching curve: DW/DS is the depth of leaching water per unit depth of soil; ECe/EC is the fraction of initial salt remaining in the soil. (After Dieleman, 1963)

Crops During Reclamation

Apart from the physical and chemical properties of soil that govern the choice of leaching methods, there are also economic and social factors that are intimately associated with reclamation programs. Provision of adequate soil drainage (q.v.), land preparation, and leaching are all costly inputs for the reclamation of saline soils. It is thus often desirable to grow crops during reclamation since some production, though small, is a welcome supplement to the farm income. The choice of crops during reclamation depends chiefly on the prevalent salinity levels and the climatic conditions. Crops differ widely in their tolerance to soil salinity (Table 1). By growing salt-tolerant crops and by adopting proper cultural, soil water management, and fertilization practices, it is often possible to obtain moderate yields of selected crops even under adverse conditions. Even though rice is considered to be only moderately tolerant to salinity (Table 1), the specific conditions of its growth make it a particularly suitable crop during the initial phases of reclamation. The shallow root system of rice, dilution of salts with applied water, and continuous downward water flux keep the salt concentration in the root zone relatively low, making rice a nearly ideal crop during reclamation.

I. P. ABROL

TABLE 1. Relative Tolerance of Crops to Salt.

Sensitive*	Moderately Tolerant	Tolerant
Fruit trees	Fruit trees	Date palm
Pear	Pomegranate	Barley
Apple	Fig	Sugarbeet
Orange	Olive	Cotton
Prune	Grape	Asparagus
Plum	Field crops and vegetables	Spinach
Almond	Wheat	
Apricot	Tomato	
Peach	Oats	
Other crops	Alfalfa	
Red clover	Rice	
Peas	Corn	
Beans	Flax	
Sugar cane	Potato	
	Carrot	
	Onion	
	Cucumber	

(From FAO/UNESCO, 1973)

*Crop yields are seriously affected if the electrical conductivity of the saturation extract of soil is greater than 10-16, 4-10, and 2-4 mmho/cm, respectively, for tolerant, moderately tolerant, and sensitive crops.

References

Dahiya, I. S., and Abrol, I. P., 1974. The redistribution of surface salts by transient and steady infiltration of water into dry soils, *Jour. Indian Soc. Soil Sci.*, 22, 209-216.

Dieleman, P. J., 1963. Reclamation of salt affected soils in Iraq, *in Soil Hydrological and Agricultural Studies.* Wageningen: Veenman and Zonen, N.V., 175p.

FAO/UNESCO, 1973. *Irrigation Drainage and Salinity–An International Sourcebook.* London: Hutchinson, 510p.

Gardner, W. R., and Brooks, R. H., 1957. A descriptive theory of leaching, *Soil Sci.*, 83, 295-304.

Molen, W. H. Van der, 1956. Desalinization of saline soils as a column process, *Soil Sci.*, 81, 19-27.

Nielsen, D. R., and Biggar, J. W., 1961. Miscible displacement in soils: I. Experimental information, *Soil Sci. Soc. America Proc.*, 25, 1-5.

Nielsen, D. R.; Biggar, J. W.; and Luthin, J. N., 1966. Desalinization of soils under controlled unsaturated flow conditions, *Internat. Comm. Irrig., Drainage, 6th Congr., New Delhi, India, Question 2,* 15-24.

Richards, L. A., ed., 1954. Diagnosis and improvement of saline and alkali soils, *U.S. Department of Agriculture Handbook 60,* 160p.

Cross-references: *Adsorption Phenomena; Conductivity, Electrical; Diffusion Processes; Evaporation; Exchange Phenomena; Infiltration; Irrigation; Irrigation Water; Leaching; Management of Soils; Permeability; Soil Drainage; Soil Solution.*

SILICATES—See CLAY MINERALS, SILICATES; SOIL COMPONENTS, INORGANIC

SLUDGE DISPOSAL

Sludge is the precipitated solid matter produced during water and wastewater treatment. The quantities of sludge resulting from water treatment are less than those from wastewater treatment, and disposal of water-treatment plant sludge is not a major problem (Hecht, Duvall, and Rashidi, 1975).

The disposal of sludge from wastewater-treatment plants is the most difficult and costly problem of sanitary district staffs (Kirkham, 1974). The traditional means of sludge handling and disposal are (1) ocean disposal through barging or pipeline transport, (2) dewatering and drying with disposal in a landfill, (3) incineration, and (4) lagooning. These means of disposal only displace the wastes and do not get rid of them. Dried sludge can sometimes be sold or given away as fertilizer, but drying is expensive. Land application of wet sludge is another method of disposal that is economical and solves the disposal problem in a beneficial way because nutrients and organic material are returned to the land. Land application of liquid sludge is considered in what follows.

The value of human and animal wastes has been recognized for a long time. "Night soil" has been used as a crop fertilizer and soil conditioner for centuries in the Orient. The extensive and successful sewage farms in Berlin and Paris were begun around 1850. Sludge has been applied to the land for many years in the United States. Most sludge, however, has been spread in the dried form. The disposal of liquid sludge rather than partially dewatered or dried sludge is becoming prevalent. In a survey of nearly 2000 treatment plants in the United States, 25% stated that they are spreading liquid sludge; only 3% have been spreading for 30 years or more (Carroll et al, 1975).

Terminology

Sludge is classified as (1) untreated (raw), and (2) treated (Grooms, 1975). *Primary* sludge results from the first major (sometimes the only) treatment in a sewage-treatment works. This treatment consists of sedimentation. Most suspended matter is removed, but colloidal and dissolved matter is not. In *secondary* treatment, which follows primary treatment, microorganisms are used to treat the sludge. *Activated* sludge is obtained in secondary treatment by settling flocculated bacterial cells feeding on soluble and suspended organic material in raw or settled sewage. Primary and activated sludges are treated to obtain a *stabilized* sludge in which organic matter has been decomposed into a stable material. During digestion, the usual method of sludge stabilization, anaerobic or aerobic microorganisms convert organic matter in sludge into less putrefiable matter. Anaerobic digestion is more common than aerobic digestion. Sludges can also be stabilized by pasteurization, chlorination, or lime treatment.

Characteristics of Sludge

Digested sludge contains 3–5% solids as finely divided and dispersed particles (Ewing and Dick, 1970). It looks like crude oil and can be easily transferred by pipes. Table 1 lists the nutrient and trace element content of anaerobically digested sludge (Keeney, Lee, and Walsh, 1975; Page, 1974).

TABLE 1. Concentration of Elements in Sludge.

Element	Range	Median	Maximum Amount Applied if Liquid Sludge (5% Solids) is spread (lb/acre-in)
	% of solids		
Total-N (moist)	3.4–9.5	...*	
Total-N (dried)	2.4–3.1	...	352
NH_4-N (moist)	0.8–4.1	...	
NH_4-N (dried)	0.02–0.26	...	29.5
Organic C	25.7–38.5	...	4,370
P	2.7–6.1	...	692
K	1.2–1.9	...	216
Ca	4.2–18.0	...	2,040
Mg	0.8–1.2	...	136
Na	0.6–2.2	...	250
Al	0.36–1.2	...	136
Fe	0.8–7.8	...	885
	mg/kg		
Ag	5–150	20	1.70
As	1–18	7.5	0.204
B	6–1,000	50	11.4
Ba	150–4,000	1,500	45.4
Cd	1–1,500	12	17.0
Co	2–260	12	2.95
Cr	20–40,615	380	461
Cu	52–11,700	700	133
Hg	0.1–56	3.0	0.636
Mn	60–3,860	400	43.8
Mo	2–1,000	5	11.4
Ni	10–5,300	52	60.2
Pb	15–26,000	480	295
Sn	40–700	120	7.95
Sr	52–7,810	...	88.8
V	20–400	60	4.54
Zn	72–49,000	2,200	556

(Data from Keeney, Lee, and Walsh, 1975; Page, 1974)
*Data not available.

Soil Physical Changes after Sludge Application

Once the sludge is applied to land, the most obvious physical change that takes place is the formation of a crust of sludge. The crust develops because liquid sewage sludge infiltrates slowly into soils due to the solids content of the crust. The rate of infiltration (q.v.) of sludge liquid depends on the initial soil moisture content and solids content of the sludge. The higher the soil moisture and sludge solids content, the lower the infiltration rate. Initially after a sludge application, the hydraulic conductivity of sand (or infiltration rate of sludge into sand) is greater than that of silt loam soils. After a few days of successive applications, the infiltration rate is about the same regardless of soil type. Therefore, the infiltration rate is eventually determined by the sludge cake and not by the soil surface. If successive sludge applications are made so that the sludge cake is not allowed to dry, infiltration rates decrease to low levels; but if the sludge cake is allowed to dry, the initial infiltration capacity is largely, if not entirely, recovered (Hinesly et al, 1974). If the sludge is plowed in, infiltration capacity is often maintained. Wet sludge applied to pasture or turf initially settles quickly into the soil.

Application of sludge increases soil aggregation (q.v.). Aggregation is influenced by microorganisms decomposing organic matter in sludge. Rapidly oxidizable organic matter can produce a desirable soil structure. As cementing agents decay, however, continual additions of organic matter are necessary to maintain good structure. If no further organic matter is added, the residual organic matter decomposes slowly.

Because addition of liquid digested sludge to land often has favorable effects on the physical characteristics of the soil, it has been used as a soil conditioner (q.v.). In clay soil sludge alleviates undesirable structure by changing pore spacing. In sandy soil it creates chemical-reaction sites for nutrient exchanges, improves soil aggregation, and makes a binder to hold sand from blowing away.

Soil Chemical Changes after Sludge Application

Nitrogen. Much of the nitrogen in sludge is in the ammonium-nitrogen form because digestion converts part of the organic nitrogen in raw sludge to ammonium-nitrogen (Table 1). Some ammonia is fixed by organic matter and clays and is protected from biological attack. Significant volatilization losses of ammonia do occur, however, from surface-applied sludge. Often about 50% of sludge ammonium-nitrogen is lost within a week after spreading liquid sludge.

Mineralization, the conversion of organic nitrogen to inorganic nitrogen as a result of microbial decomposition, proceeds at various rates depending on climate, soil, and type of organic matter. About 5–50% of the organic nitrogen in anaerobically digested sewage sludge can be mineralized to nitrate- plus nitrite-nitrogen in four months.

Even though the ammonium-nitrogen in sludge nitrifies readily in soils, some denitrification occurs. Denitrification is the transformation of nitrate- or nitrite-nitrogen to gaseous nitrogen. If free oxygen is absent or deficient and other conditions, including a supply of carbon, are favorable for biological activity, denitrification takes place. Some sludge nitrogen is also immobilized in the soil organic matter or remains as refractory organic nitrogen. Therefore, a variable amount of sludge-applied nitrogen is available for crop uptake or for leaching to the groundwater.

Under proper soil conditions (appropriate aeration and temperature), ammonium-nitrogen in anaerobically digested sludge is converted rapidly into mobile nitrate-nitrogen. The annual loading rate of sludge on land is limited by the amount of soluble nitrogen in the sludge-treated soil plus nitrogen from mineralization. If the applied nitrogen in the sludge greatly exceeds losses, concentrations of nitrogen in groundwater may be excessive. Also *eutrophication* results when surface water is enriched with nitrogen after runoff. Because high nitrate concentrations in drinking water are toxic to humans and livestock, large amounts of digested sludge on land can cause nitrate problems. About 2–3 in (5–8 cm) of sludge will satisfy the nitrogen needs of nonleguminous crops without producing excessive nitrates in percolated water. But, if higher loading rates are used, the nitrogen content of sludge should be reduced, for example, by flooding to induce denitrification or by collecting leachates for nitrogen removal.

Phosphorus. Many sewage-treatment plants remove phosphates from wastewater while concentrating them in the sludge as an insoluble residue. Phosphates in sludge, except for a small fraction bound in organic compounds, are present as calcium, iron, aluminum, or magnesium phosphate, depending on the process used. Calcium and magnesium phosphates are reasonably available to plants in neutral soils, while the phosphate in iron and alum (aluminum sulfate) precipitates is relatively unavailable. These precipitates can persist in the soil for long periods and slowly liberate phosphorus to plants. Recycling of sludge to the soil is the only economical way to conserve phosphates wasted in sewage.

In most soils, phosphorus is inactivated by adsorption and conversion to sparingly soluble precipitates. Therefore, phosphorus levels in drainage water from sludge-treated soils are usually less than 1 ppm. With high sludge-loading rates, movement of phosphorus to the groundwater may occur. Surface-water pollution with phosphorus is, however, a more serious problem because phosphorus-loaded surface soils can be eroded if soil conservation practices are not followed. If sludge application rates are based on the amount of nitrogen needed to meet crop requirements, phosphorus additions will be low and phosphorus will not be a eutrophication threat to water.

Potassium. Sludge is low in potassium compared to its nitrogen and phosphorus contents. Therefore, potassium may be in short supply for good crop growth. Inorganic potassium fertilizers sometimes are added to sludge-treated land.

Organic Material. Anaerobically digested sludge greatly increases soil organic matter and humus content. In Illinois, 136 dry tons per acre (309 metric tons per hectare) of anaerobically digested sludge incrementally applied during 4 years on a silt loam soil increased its organic carbon content from 1.2–2.4% in the surface 6 in (15 cm) (Hinesly et al, 1974). Six months after sludge addition to soil, about

20% of sludge carbon is evolved as carbon dioxide. The remainder is resistant to further breakdown.

Trace Elements. Sewage sludge contains an abundance of trace elements necessary for plant growth (boron, copper, iron, manganese, molybdenum, and zinc) as well as non-essential ones (cadmium, chromium, mercury, sodium, lead, and others) (Table 1).

The trace elements likely to cause toxicities to plants in soils treated with large amounts (for example, 400 metric tons per hectare) of domestic sludge for a number of years (15 yr) are cadmium, copper, zinc, boron, and possibly nickel (Page, 1974). As far as zinc, copper, and nickel are concerned, the British Ministry of Agriculture, Fisheries, and Food suggests that nickel is eight times as toxic to plants as zinc and that copper is twice as toxic as zinc (Kirkham, 1974). Work in the United States shows that nickel is only about four times as toxic to plants as zinc (Keeney, Lee, and Walsh, 1975). In industrialized sludges, in addition to the elements listed, concentrations of arsenic, barium, cobalt, chromium, manganese, lead, and vanadium may be sufficiently high to reduce crop yields when applied in large amounts. Plants can grow normally and yet contain concentrations of selenium, cadmium, molybdenum, and possibly lead that are toxic to man and animals. Cadmium is the element of most concern in sludge. If sludge is applied to soil in modest amounts (for example, 10 metric tons per hectare for 15 yr), the cadmium concentration in the soil will build up, and food obtained from plants grown on the soil could contain concentrations of cadmium toxic to man and animals.

Because tissues from different crop species and varieties vary widely in their concentrations of trace elements, the selection of crops provides a way to control the entrance of undesirable amounts of trace elements into food chains (Chaney, 1973). In addition, there are synergistic and antagonistic interactions between elements in sludge and soil that affect the absorption of elements by plant roots. For example, high phosphorus levels in soil often inhibit uptake of trace elements. Therefore, it is possible to decrease uptake of one element by supplying another element to the soil.

After application to soil, trace elements tend to remain in the plow layer with the sludge residue. Oxides of iron and manganese absorb trace elements. The distance the elements leach is determined by the oxidation-reduction potential and pH of the system. Practices that promote good soil aeration such as drainage and structure development lead to decreasing solubilities of trace elements. Also soils with high cation exchange capacities are desirable for sludge disposal because they have a great ability to hold and immobilize trace elements. Liming soils can control the uptake of trace elements of most concern in sludge (cadmium, copper, nickel, and zinc). Because of the nitrification reaction and the microbial production of carbon dioxide, sewage sludge usually lowers the pH of soils; thus liming is recommended at sludge-disposal sites. Trace elements can form inert and insoluble compounds with clays and organic compounds. Therefore, they are less available to plants than total concentrations of these elements in the soil indicate. Consequently, the availability rather than total amounts of sludge-derived trace elements in soils is of primary concern. Eventual movement of trace elements to groundwater is a possibility, especially because sludges contain metal-complexing agents. However, trace elements tend to move little with percolating water and remain at the point of application unless they are transported away on eroded sediments.

Pathogens. No incidence of disease transmission has been traced to the use of digested sludge on land. Most intestinal pathogenic bacteria are either destroyed or their populations greatly reduced by anaerobic digestion of sludge. The fate of viruses during anaerobic digestion is less certain. A 14-day anaerobic digestion period appears to destroy most viruses (Hinesly et al, 1974).

After sludge is applied on soil, the viability of pathogenic organisms varies, ranging from a few hours to several months. Spore-forming bacteria can remain viable for years in soil, and some microorganisms survive almost indefinitely; however, the death of many pathogenic organisms occurs rapidly. Bacteria and viruses seldom move more than 100 ft and usually only a few feet in the soil (see *Microorganism Transport*). Like many of the potential chemical pollutants, lateral movement of pathogenic organisms that survive digestion occurs only if soil erosion occurs at the sludge disposal site. Under these conditions, surface water pollution is a problem.

Contamination of groundwater with pathogens happens if undisinfected sludge is placed on shallow soils underlaid by porous material or on soils that crack when dry. Cracks and continuous channels that bypass the main soil mass can be more important than physical and chemical characteristics of soil that control movement of pathogens. Groundwater contamination is not a serious hazard in unfissured soil.

Field Application

Sludge can be applied on the surface of the soil or under the surface by injection into the soil (Keeney, Lee, and Walsh, 1975). Surface application of liquid sludge is done by spray, by ridge and furrow irrigation, or by tank truck or farm wagon. Portable spray irrigation systems using a single large-nozzle gun (2–5 cm orifice) are used. Sludge accumulating on plant leaves during spray irrigation can reduce photosynthetic rates. Ridge and furrow irrigation requires preparation of the land before spreading, and only level land can be used. It is suitable for row crops during the growing season. Tank trucks and farm tank wagons are most commonly used to apply sludge, especially in smaller communities. After surface application, sludge usually is plowed into the soil. Subsurface injection of liquid sludge has a number of advantages over surface application because odors and insects are not a problem, nitrogen is conserved because ammonia volatilization and runoff are minimized, and public acceptance is better.

M. B. KIRKHAM

References

Carroll, T. E.; Maase, D. L.; Genco, J. M.; and Ifeadi, C. N., 1975. *Review of Landspreading of Liquid Municipal Sewage Sludge, EPA-670/2-75-049.* Cincinnati, Ohio: Environmental Protection Agency, 96p.

Chaney, R. L., 1973. Crop and food chain effects of toxic elements in sludges and effluents, *in Recycling Municipal Sludges and Effluents on Land.* Washington, D.C.: National Association of State Universities and Land-grant Colleges, 129–141.

Ewing, B. B., and Dick, R. I., 1970. Disposal of sludge on land, *in* E. F. Gloyna and W. W. Eckenfelder, eds., *Water Quality Improvement by Physical and Chemical Processes.* Austin: University of Texas Press, 394–408.

Grooms, G. C., ed., 1975. Ohio guide for land application of sewage sludge, *Ohio State Univ. Bull. 598,* 12p.

Hecht, N. L.; Duval, D. S.; and Rashidi, A. S., 1975. *Characterization and Utilization of Municipal and Utility Sludges and Ashes,* Vol. II, EPA-670/2-75-033b. Cincinnati, Ohio: Environmental Protection Agency, 232p.

Hinesly, T. D.; Braids, O. C.; Dick, R. I.; Jones, R. L.; and Molina, J.-A. E., 1974. *Agricultural Benefits and Environmental Changes Resulting from the Use of Digested Sludge on Field Crops,* EPA/530/SW-30d.1. Washington, D. C.: Environmental Protection Agency, 375p.

Keeney, D. R.; Lee, K. W.; and Walsh, L. M., 1975. Guidelines for the application of wastewater sludge to agricultural land in Wisconsin, *Wisc. Depart. Nat. Resources Tech. Bull. 88,* 36p.

Kirkham, M. B., 1974. Disposal of sludge on land: Effect on soils, plants, and ground water, *Compost* Sci., **15**(2), 6–10.

Page, A. L., 1974. *Fate and Effects of Trace Elements in Sewage Sludge when Applied to Agricultural Lands, EPA-670/2-74-005.* Cincinnati, Ohio: Environmental Protection Agency, 98p.

Cross-references: *Pollution; Microorganism Transport; Soil Conditioners; Soils, Nonagricultural Uses; Trace Elements.*

SOIL

The noun *soil* originated from the Latin *solum,* which had the same general meaning as our modern word. The term "soil" has different shades of meaning and connotation, depending on the professional field in which it is being considered. Even within the purview of soil science (q.v.) the term "soil," like so many other old terms in common use, has several meanings (Simonson, 1968). According to Webster (Gove, 1971), soil is defined as "the loose surface material of the earth in which plants grow, usually consisting of disintegrated rock with an admixture of organic matter (q.v.) and soluble salts." This definition, nearly as old as the word itself, is still the common meaning and is thus of greatest interest.

In reviewing concepts of soils, Butler (1958) recognized three main categories, which he referred to as "edaphic," "pedologic," and "geographic" aspects. Though these headings are not entirely comprehensive, they nevertheless cover the greater part of the field. Another heading is that which treats soil as a material that has industrial or engineering uses.

The Edaphic Concept

The main sphere of interest here is restricted to those soil conditions that are important to the growth and well-being of plants (see *Edaphology*). Because the problems are to assess the plant growth potential of soils, to achieve that potential and to determine means for correcting any plant nutrient deficiency (see *Soil Fertility; Plant Nutrients*), the edaphologist or agronomist (see *Agronomy*) views soil as the natural medium for the growth of land plants. According to this view, the growth response of a crop is related to such factors as the soil's potential for supplying nutrients (see *Nutrient Potential*), its ability to provide a suitable medium for root development, and its response to cultural management practices (see *Management of Soils*). The edaphic concept thus considers soil in terms of the specific needs of plants and the significance of soil properties in meeting these needs.

The Pedologic Concept

The central theme of the pedologic concept (see Vol. XII, Pt. 2: *Pedology*) embraces the study of soils as independent natural bodies, each with a distinct morphology resulting from the integrated effect of climate and living matter, acting on parent materials, as conditioned by relief, over periods of time (Soil Survey Staff, 1951). According to this view, soil is the unconsolidated mineral matter on the surface of the earth that has been subjected to and influenced by genetic and environmental factors of parent material, climate, organisms, and topography, all acting over a period of time (see Vol. XII, Pt. 2: *Factors of Soil Formation*) resulting in a product (soil) that differs from the material from which it is derived in many physical, chemical, biological, and morphological porperties and characteristics (Soil Science Society of America, 1973). Pedologists thus use the term to refer to the upper portion of the lithosphere that has been altered into layers (*horizons*) that differ from each other and the underlying unaltered materials. Soil is a collective noun and refers to a natural universe of many different kinds of soils; it is a collection of natural bodies that occur in a continum at the earth's surface (Buol, Hole, and McCracken, 1973). *A soil* has a defined range in the properties of the profile and their horizontal variations, including those of each horizon (Soil Survey Staff, 1975). It usually occurs as a discontinuous entity in the natural landscape and is generally found wherever the soil-formation factors (Jenny, 1941) occur in the same combinations that are associated with the narrow range of soil properties defined as the range of that soil.

The Geographic Concept

The geographic concept of soil relates to the study of the areal distribution of soils. Its objective is the recognition of areal units and their delineation on maps, the study and rationalization of the distribution patterns, and the appraisal of the different units for use (Butler, 1958). Soil-mapping programs are conducted under the influence of the geographic concept, and the work has served a useful purpose in providing simplified models of complex landscapes. The soil surveyor's concept of soil is thus based on the soil mapping unit, which in turn is related to a classification unit or units.

The Engineering Concept

From the engineer's viewpoint, soil is a material with which and upon which he builds his structures (see Vol. XIII: *Engineering Soil Science*). Hence in civil engineering, the term "soil" includes not only the soil of the pedologist but also any unconsolidated material found between the ground surface and consolidated bedrock. It is generally understood that "soil" in the engineering sense means deposited derivatives from the rock crust of the earth in its natural undisturbed state or that which can be excavated and placed in an earthwork or what can be found in an earthwork already (Jumikis, 1962). In other words, soil includes all regolith (q.v. in Vol. XII, Pt. 2) materials, e.g., unconsolidated accumulations of rock fragments and any organic matter that is intermixed, or the entire vertical section down to consolidated rock (Lee, 1968; Peck, Hanson, and Thornburn, 1974).

Retrospect and Prospect

The soil mantle (q.v. in Vol. XII, Pt. 2) is part of the uniqueness of planet earth, for without soil, life could not exist as we know it. The soil is a dynamic natural entity that interacts with many aspects of the environment. The so-called pedosphere (q.v. in Vol. XII, Pt. 2) cuts across the other great natural spheres, e.g., atmosphere, biosphere, hydrosphere, and lithosphere, involving processes characteristic of all of them. Soil then acts as an interfacial unit; it is an unstable entity that is constantly changing, a turnstile through which passes an endless procession of atoms, a natural "synthograph" (Buol, Hole, and McCracken, 1973), which records a synthesis of much that has happened at the site. It is, as defined by Nikiforoff (1959), "the excited skin of the subaerial part of the earth's crust".

Soil is many things to many people, each defining soil within the scope of particular interests. Whether soil is regarded as an epidermal unit of a geologic body through which materials pass in erosion cycles, as a bank into which fertilizers are deposited by man to provide nutrients for plant growth, as part of the environment conditioned by plants, or as a record of the past is not critical to the central concept of soil. These different concepts of soil merely emphasize its diverse nature and underscore its universal importance. The concept of soil has been modified in the past (Cline, 1949; Nikiforoff, 1959, Simonson, 1968) and will undoubtedly change again in the future.

CHARLES W. FINKL, JNR.

References

Anon., 1973. *Glossary of Soil Science Terms.* Madison: Soil Science Society of America, 33p.

Buol, S. W.; Hole, F. D.; and McCracken, R. J., 1973. *Soil Genesis and Classification.* Ames: Iowa State University Press, 360p.

Butler, B. E., 1958. The diversity of concepts about soils, *Jour. Aust. Inst. Agric. Sci.*, **24**, 14–20.

Cline, M. G., 1949. The changing model of soil, *Soil Sci. Soc. America Proc.*, **25**, 442–446.

Gove, P. B., ed., 1971. *Webster's Third New International Dictionary of the English Language Unabridged.* Springfield: G. & C. Merriam, 2662p.

Jenny, H., 1941. *Factors of Soil Formation.* New York: McGraw-Hill, 281p.

Jumikis, A. R., 1962. *Soil Mechanics.* New York: Van Nostrand, 791p.

Lee, I. K., 1968. *Soil Mechanics.* Sydney: Butterworths, 625p.

Nikiforoff, G. C., 1959. Reappraisal of the soil, *Science*, **129**, 186–196.

Peck, R. B.; Hanson, W. E.; and Thornburn, T. H., 1974. *Foundation Engineering.* New York: Wiley, 514p.

Simonson, R. W., 1968. Concept of soil, *Adv. Agron.*, **20**, 1–47.

Soil Survey Staff, 1951. Soil Survey Manual, *U.S. Department of Agriculture Handbook No. 18*, 503p.

Soil Survey Staff, 1975. Soil taxonomy: A basic system of soil classification for making and interpreting soil surveys, *U.S. Department of Agriculture Handbook No. 436*, 754p.

SOIL CHEMISTRY

As the name implies, *soil chemistry* involves the application of chemical principles and techniques to the study of soils. It provides the basis for solutions to problems in such diverse areas as soil fertility (q.v.), land disposal of wastes (see *Sludge Disposal*), and corrosion of underground pipelines.

Soil is a complex, three-phase system containing solid, liquid, and gaseous components. It is also a dynamic system in which "equilibrium" conditions are achieved only if the reactions are very fast. The solid phase, which constitutes about 50% of the volume of an average soil, is an intimate mixture of mineral and organic matter. Except for peat and muck soils, which are composed dominantly of organic matter (q.v.), the mineral matter in the surface layer usually accounts for more than 95% of the solid phase. However, the organic component, both living and dead, compensates for its low abundance by being very reactive. The pore space (see *Soil Pores*) between the solid particles is shared by the liquid and gas phases in constantly changing proportions. The liquid phase contains dissolved substances, both inorganic and organic, and the gas phase includes the normal components of the atmosphere, but in different proportions. Because the wide variety of mineral and organic components present in soils constantly interacts in a series of physical, chemical, and biochemical reactions, the study of soil chemistry is very complex.

Mineral Composition

The soil chemist is concerned with the composition of this three-phase system and shares with the geologist an interest in the mineral composition of the solid phase and uses techniques such as X-ray diffraction, differential and integral thermal analysis, petrographic and electron microscopy, differential dissolution, and elemental analysis by various methods to identify the mineral constituents (see *Mineral Analysis*). Frequently the mineral portion will be fractionated into different particle-size groups to segregate some of the minerals. Primary minerals tend to segregate in the sand (0.05–2 mm) and silt (0.05–0.002 mm) fractions, secondary minerals in the clay (less than 0.002 mm). It is the clay fraction that, because of its large surface area and its surface charge characteristics, is the site of many important reactions in soils including the retention of plant nutrients (q.v.).

Organic Reactions

Identification of specific compounds making up the organic fraction of soils presents a particularly difficult challenge as there are very few techniques by which organic components can be identified without removal from the mineral matrix. As a result, soils are usually treated with a series of chemical extractants, and the extracted organic compounds are identified by various methods including elemental analysis, infrared and ultraviolet spectrophotometry, and gas, thin layer and paper chromatography. The compounds in the extracts are frequently only pieces of the original compounds because they may be altered during the extraction process. Putting all the pieces of this puzzle together to identify the original compounds is a difficult task, and a large part of the soil organic matter is still poorly characterized.

The living portion of the soil organic matter exerts a tremendous influence on the chemical reactions occurring in soils. Heterotrophic soil microorganisms obtain energy from oxidation of organic compounds, and they and their associated enzymes decompose dead organic materials, thus converting organically bound plant nutrients such as nitrogen, phosphorus, and sulfur to inorganic forms available to plants. Some autotrophic organisms, on the other hand, obtain energy from oxidation of reduced forms of nitrogen, sulfur, iron, and manganese and greatly influence chemical reactions involving these elements. Some of the organic compounds excreted by microorganisms and by plant roots and some of those formed as products of organic matter decomposition are very effective in forming stable complexes with transition metal cations and aluminum. Since iron, a transition metal cation, and aluminum are major constituents of many soil minerals, this complex formation accelerates weathering of those minerals and results in higher solution concentrations of the complexed cations and possibly of the other cations and anions present in those minerals.

Both higher plants and microorganisms affect soil chemical reactions by absorption of solutes from solution. Microbial competition for plant nutrients can sometimes result in nutrient deficiencies, most commonly of nitrogen, for higher plants (see *Nitrogen Cycle*).

As with all living matter, microorganisms and plant roots respire (see *Aeration, Respiration, Atmosphere*). In aerobic systems the microorganisms and plant roots use oxygen as the electron acceptor in the respiration process with production of carbon dioxide. If diffusion of oxygen from the atmosphere into the soil is restricted, as is the case with waterlogged soils, and if the microorganisms have a readily available energy source, they will deplete the oxygen in the soil air to negligible quantities, decreasing the redox potential. Facultative and anaerobic microorganisms then become dominant as they can use electron acceptors other than oxygen and in the process reduce nitrate ions to nitrogen or nitrogen oxides; manganese and iron oxides to soluble manganous and ferrous ions; sulfate to sulfur, hydrogen sulfide or ferrous sulfide; carbon dioxide to methane; and even water to hydrogen. This alters the chemical reactions associated with the mineral portion of the soil and drastically changes the plant root environment. Upland plants, which are unable to transport oxygen to root cells internally, die from lack of oxygen for root respiration.

Liquid Phase

Determining the composition of the liquid phase of soils also presents problems. Extraction of soil solution (q.v.) from relatively dry soils requires special techniques, and the concentrations of solutes vary with the proportion of the solution extracted because the ionic concentrations are not homogeneous throughout the solution phase in the presence of the charged surfaces of the solid phase. Once the solution is extracted, however, analysis of the solutes is relatively straightforward with established techniques such as flame emission and atomic absorption spectrophotometry, infrared, visible, and ultraviolet spectrophotometry, X-ray flourescence, and, for some trace constituents, isotope dilution and activation analysis. Electrical conductivity (q.v.) may be used to measure ionic concentrations in solution, particularly for diagnosing potential salinity problems.

Gas Phase

Sampling is also a major problem in analysis of the gas phase. In moist soils gas in small pores within structural aggregates may be effectively removed from contact with the gas in the large pores between aggregates and may have an entirely different composition. Most sampling techniques tend to be biased toward sampling the gas in the large pores. Once the sample of air has been obtained, however, the analysis is straightforward with instruments such as the oxygen analyzer, the gas chromatograph, and the infrared spectrophotometer.

Reactions across Phase Boundaries

Much of the research in soil chemistry is concerned with reactions that occur across phase boundaries. The processes of precipitation-dissolution and sorption-desorption across the solid-solution boundary are of particular interest. Solubility product approaches can be used to determine stability ranges of many soil minerals, but most of the more important short-term reactions are sorption-desorption reactions at particle surfaces.

The surfaces of the soil particles, particularly those of the clay-size fraction and the organic matter, exhibit an electric charge. The sign of this charge is usually negative except in strongly acid, highly weathered soils. The surfaces may be either of the constant charge type (planar surfaces of vermiculite or montmorillonite minerals) or constant potential (oxides and hydroxides of iron, aluminum, and manganese and edges of silicate minerals). Various modifications of classical adsorption and double-layer equations are used in attempting to describe both specific and nonspecific adsorption of cations and anions on these surfaces. These equations have been particularly useful in predicting problems of excessive levels of exchangeable sodium (sodic soils) and in describing phosphate sorption on particle surfaces.

Reactions occurring across the liquid-gas

interface include primarily condensation-evaporation of water and solution-volatilization of gaseous components. Evaporation has a marked effect on water content and thus on ionic concentrations in the surface layer of soils. The solution-volatilization reactions, under appropriate conditions, result in transfer of gaseous products of soil biological reactions such as carbon dioxide, oxides of nitrogen, nitrogen gas, ammonia, methane, mercaptans, and hydrogen sulfide to the atmosphere and absorption of atmospheric gases such as sulfur dioxide into the soil.

Future Research

One important application of research in soil chemistry is the development of analytical procedures for determining levels of available plant nutrients (see *Nutrient Potential*) and the fate of nutrients applied as fertilizers (q.v.). Closely allied to plant nutrient investigations are environmentally oriented studies of nitrogen and phosphorus concentrations in surface runoff and percolating waters, persistence of pesticide residues and recycling of various types of municipal and industrial wastes. The latter studies frequently involve determination of permissible loading limits due to the presence of relatively large quantities of nitrogen, phosphorus, soluble salts, carbonaceous materials or potentially toxic trace elements contained in the waste.

RICHARD B. COREY

References

Allison, F. E., 1973. *Soil Organic Matter and its Role in Crop Production.* New York: Elsevier, 637p.

Black, C. A., ed., 1965. *Methods of Soil Analysis.* Madison: American Society of Agronomy, 1–1572.

Bolt, G. H., and Bruggenwert, M. G. M., 1976. *Soil Chemistry. A. Basic Elements.* New York: Elsevier, 281p.

Garrels, R. M. and Christ, C. L., 1965. *Solutions, Minerals and Equilibria.* San Francisco: Freeman, 450p.

Gieseking, J. E., ed., 1975. *Soil Components: Vol. 1, Organic Components; Vol. II., Inorganic Components.* New York: Springer-Verlag, 1–534, 1–684.

Jackson, M. L., 1973. *Soil Chemical Analysis–Advanced Course,* 2nd ed. Madison: Published by Author, Department of Soil Science, University of Wisconsin, 895p.

Marshall, C. E., 1977. *The Physical Chemistry and Mineralogy of Soils II. Soils in Place.* New York: Wiley Interscience, 313p.

Ponnamperuma, F. N., 1972. The chemistry of submerged soils, *Adv. Agron.,* **24,** 29–96.

Walsh, L. M., ed., 1971. *Instrumental Methods for Analysis of Soils and Plant Tissue.* Madison: Soil Science Society of America, 222p.

Walsh, L. M., and Beaton, J. D., 1973. *Soil Testing and Plant Analysis.* Madison: Soil Science Society of America, 491p.

Cross-references: *Activity Ratio; Adsorption Phenomena; Base Saturation; Buffers, Buffering; Chemical Composition; Conductivity, Electrical; Dispersion Phenomena; Electrochemistry; Electro-osmosis; Exchange Phenomena; Ionic Activities; Mineral Analysis; Nutrient Potential; Redox Reactions; Soil Biochemistry; Soil Reaction.*

SOIL COMPONENTS, INORGANIC

The land portion of the earth's crust is extensively covered by soils. Because the earth's crust is roughly 50% oxygen, 25% silicon, and 15% sesquioxide-forming elements, inorganic soils are composed predominantly of minerals very high in silicates, especially aluminosilicates and/or oxides. Many elements substitute in varing degrees for aluminum, iron, and silicon in these silicate and oxide structures. These fortuitous isomorphous substitutions result in many complex series of minerals in soils with each member differing in composition possibly only slightly from the next member, or one end of a mineral crystal differing in composition from the other end.

Soils develop from the weathering of parent materials occurring in the surface of the earth's crust (1) by physical disintegration of minerals originally occurring at the site or disintegration of those deposited later by geological processes, (2) by chemical or biochemical degradation of less resistant minerals in the earth's crust leaving residues of the most resistant minerals, and (3) by recombinations and syntheses of new minerals from the moieties produced by 1 and 2. The soil components left after weathering are determined by the climatic and biologic environmental forces acting during the development of the soil and by the extent to which products of weathering have been removed by drainage.

Silicates

Table 1 gives the silicates expected in soils, according to Parfenova and Yarilova (1962) and MacKenzie (1975). The classes of silicates listed in the table and shown in Figure 1 are those defined by Strunz (1957), and described further by Bragg, Claringbull, and Taylor (1965). These classes include the following:

1. The nesosilicates, separate tetrahedra (SiO_4) [Figure 1(a)]
2. The sorosilicates, double tetrahedra (Si_2O_7) [Figure 1(b)]

TABLE 1. Silicates Commonly Found in Soils and Soil Parent Materials.[a]

Class	Group	Subgroup	Species	Approximate Formula
Nesosilicates	Andalusite			Al_2OSiO_4
	Chlorotoid			$(Fe^{++}, Mg, Mn)_2(AlFe^{3+})Al_3O_6SiO_4(OH)_4$
	Garnet			$*Fe^{++}{}_3Al_2(SiO_4)_3$
	Kyanite			Al_2OSiO_4
	Olivine			$MgFe^{++}SiO_4$
	Sillimanite			Al_2OSiO_4
	Sphene			$CaTiSiO_4(O, OH, F)$
	Topaz			$Al_2SiO_4(OH, F)_2$
	Zircon			$ZrSiO_4$
Sorosilicate	Epidote			$Ca_2Fe^{3+}Al_2(SiO_4)_3OH$
Cyclosilicates	Beryl			$Be_3Al_2(SiO_3)_6$
	Tourmaline			$(Na, Li, Mg, Mn, Al)_3Al_6(SiO_3)_6(BO_3)_3$
Inosilicates	Amphibole			$*Ca_2(Mg, Fe^{++})_5Si_8O_{22}(OH, F)_2$
	Pyroxene			$*(Mg, Fe^{++}, Ca)(Mg, Fe^{++})Si_2O_6$
	Wollastonite			$CaSiO_3$
Phyllosilicates	Pyrophyllite-talc	Pyrophyllites	Pyrophyllite	$Al_4Si_8O_{20}(OH)_4$
		Talcs	Talc	$Mg_6Si_8O_{20}(OH)_4$
			Minnesotaite	$(Fe, Mg)_6Si_8O_{20}(OH)_4$
	Montmorillonite-Saponite	Montmorillonites	Montmorillonite	$0.67M^+\cdot(Al_{3.33}Mg_{0.67})Si_8O_{20}(OH)_4$
			Beidellite	$0.67M^+\cdot Al_4(Si_{7.33}Al_{0.67})O_{20}(OH)_4$
			Nontronite	$0.67M^+\cdot Fe_4(Si_{7.33}Al_{0.67})O_{20}(OH)_4$
			Volkonskoite	$0.67M^+\cdot(Fe, Cr, Al)_4(Si, Al)_8O_{20}(OH)_4$
		Saponites	Saponite	$0.67M^+\cdot Mg_6(Si_{7.33}Al_{0.67})O_{20}(OH)_4$
			Sauconite	$0.67M^+\cdot(Mg, Zn)_6(Si_{7.33}Al_{0.67})O_{20}(OH)_4$
			Hectorite	$0.67M^+\cdot(Mg, Li)_6(Si, Al)_8O_{20}(OH)_4$
	Vermiculite	Dioctahedral vermiculites	Dioctahedral vermiculite	$1.73M^+\cdot(Al, Fe, etc.)_4(Si, Al)_8O_{20}(OH)_4$
		Trioctahedral vermiculites	Vermiculite	$1.73M^+\cdot(Mg, Fe, etc.)_6(Si, Al)_8O_{20}(OH)_4$
	Illite	Dioctahedral illites	Illite	$(K, H_3O)_2(Al, Fe, Mg)_4(Si, Al)_8O_{20}(OH)_4$
			Glauconite (?)	$(K, H_3O)_2(Fe, Al, Mg)_4(Si, Al)_8O_{20}(OH)_4$
		Trioctahedral illites	Ledikite	$(K, H_3O)_2(Mg, Fe, Al)_6(Si, Al)_8O_{20}(OH)_4$
	Mica	Dioctahedral micas	Muscovite	$K_2Al_4(Si_6Al_2)O_{20}(OH)_4$
			Pargonite	$Na_2Al_4(Si_6Al_2)O_{20}(OH)_4$
		Trioctahedral micas	Phlogopite	$K_2Mg_6(Si_6Al_2)O_{20}(OH)_4$
			Biotite	$K_2(Mg, Fe, Mn)_6(Si_6Al_2)O_{20}(OH)_4$
			Lepidolite	$K_2(Li, Al)_6(Si, Al)_8O_{20}(OH)_4$
			Zinnwaldite	$K_2(Li, Fe, Al)_6(Si, Al)_8O_{20}(OH)_4$

	Brittle mica	Dioctahedral brittle micas	Margarite	$Ca_2Al_4(Si_4Al_4)O_{20}(OH)_4$
		Trioctahedral brittle micas	Seybertite	$Ca_2(Mg_4Al_2)\ (Si_{2.5}Al_{5.5})O_{20}(OH)_4$
			Xanthophyllite	$Ca_2(Mg_4Al_2)\ (Si_{2.5}Al_{5.5})O_{20}(OH)_4$
	Chlorite	Dioctahedral chlorites	Sudoite	$Al_8(Si, Al)_8O_{20}(OH)_{16}$
		Trioctahedral chlorites		$(Mg, Fe, etc.)_{12}(Si, Al)_8O_{20}(OH)_{16}$
	Kaolinite-Serpentine	Kaolinites	Dickite	$Al_4Si_4O_{10}(OH)_8$
			Nacrite	$Al_4Si_4O_{10}(OH)_8$
			Kaolinite-Tc	$Al_4Si_4O_{10}(OH)_8$
			Kaolinite-Md	$Al_4Si_4O_{10}(OH)_8$
			Metahalloysite	$Al_4Si_4O_{10}(OH)_8$
			Halloysite	$Al_4Si_4O_{10}(OH)_8 \cdot 4H_2O$
			Anauxite (?)	$Al_4Si_6O_{14}(OH)_8$(?)
		Serpentines	Antigorite	$Mg_6Si_4O_{10}(OH)_8$
			Chrysotile	$Mg_6Si_4O_{10}(OH)_8$
			Amesite	$(Mg, Fe)_4Al_2(Si_2Al_2)O_{10}(OH)_8$
			Cronstedtite	$Fe_6(Si_2Fe_2)O_{10}(OH)_8$
			Berthierine	$(Fe, Mg, Al)_6(Si_3Al)O_{10}(OH)_8$
	Hormite	Mixoditrioctahedral hormites	Palygorskite	$(Al, Mg)_{3.5}Si_8O_{20}(OH)_2(H_2O)_4$
		Trioctahedral hormites	Sepiolite	$Mg_8Si_{12}O_{30}(OH)_4(H_2O)_4$
			Xylotile	$(Mg, Fe)_8Si_{12}O_{30}(OH)_4(H_2O)_4$
Tectosilicates	Analcime			$NaAlSi_2O_6 \cdot H_2O$
	Leucite			$KAlSi_2O_6$
	Nepheline			$Na_3KAl_4Si_4O_{16}$
	Sodalite			$Na_8Al_6Si_6O_{24}Cl_2$
	Feldspar			*$KAlSi_3O_8$

[a]Compiled from Mackenzie (1975), Kretz (1976), and Parfenova and Yarilova (1962).
*Formula of a typical member of an isomorphous series.

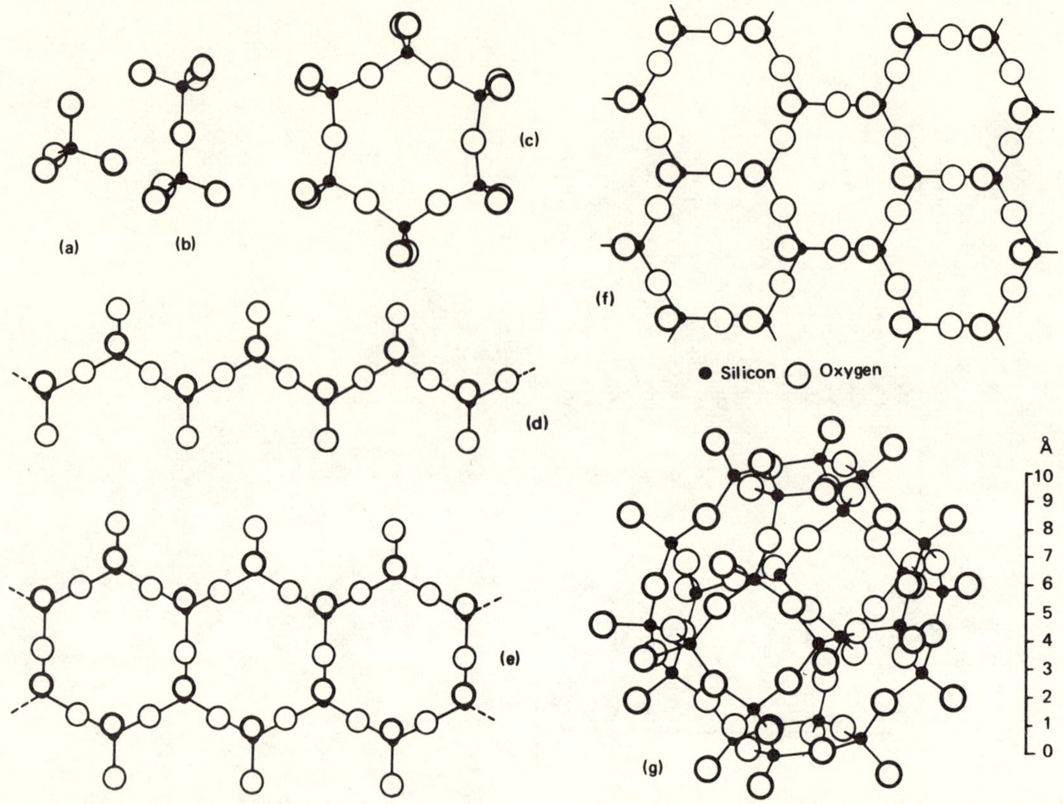

FIGURE 1. The linkage of SiO_4 tetrahedra in different classes of silicates: (a) Nesosilicates (separate tetrahedra); (b) sorosilicates (linked tetrahedra); (c) cyclosilicates (closed rings); (d) and (e) inosilicates (single and double chains); (f) phyllosilicates (sheets); (g) tectosilicates (framework). (From Mackenzie, 1975)

3. The cyclosilicates, rings of tetrahedra (SiO_3) [Figure 1(c)]
4. The inosilicates, chains of tetrahedra: single (SiO_3) double (Si_4O_{11}) [Figure 1(d) and (e)]
5. The phyllosilicates, sheets of tetrahedra (Si_2O_5) [Figure 1(f)]
6. The tectosilicates, framework of tetrahedra in corners common to vicinal cages (SiO_2) [Figure 1(g)]

Each formula given for the silicates in Table 1 should be considered a typical formula. These minerals are members of isomorphous series, and substitutions of ions by other ions having the same approximate size and chemical properties are so common that probably no two samples of the mineral could be found that would have exactly the same composition.

Oxides

Table 2 gives oxides that are formed in soils (MacKenzie, 1975; Mitchell, 1975). The formulas are only typical formulas because of the ease with which one sesquioxide-forming ion substitutes for another. These minerals always contain at least traces of other sesquioxide types of ions.

Phosphates

The phosphate minerals constitute less than 0.05% of the soil mass. Even though these phosphates are only a very small part of the inorganic part of the soil, they greatly influence the phosphorus cycles of plants and other organisms growing on and in the soil.

The amounts of phosphate minerals in most soils are too small to permit identification by the common mineralogical techniques. From total phosphorous determinations and from various fractional extractions of soil materials, it can be assumed that the phosphates compiled from Kretz (1976), given in Table 3, can reasonably be expected to occur in soils. Farmer and Palmieri (1975) give a bibliography of the literature related to the infrared spectra of most of the phosphate minerals. Future attempts

TABLE 2. Oxides Expected in Soils.

Mineral	Formula	Crystal System
Quartz	SiO_2	Hexagonal
Tridymite	SiO_2	Orthorhombic
Cristobalite	SiO_2	Tetragonal
Coesite	SiO_2	Monoclinic
Stishovite	SiO_2	Tetragonal
Amorphous silica		
Corundum	α-Al_2O_3	Hexagonal
Diaspore	α-AlOOH	Orthorhombic
Boehmite	γ-AlOOH	Orthorhombic
Bayerite	α-$Al(OH)_3$	Monoclinic
Gibbsite	γ-$Al(OH)_3$	Monoclinic
Nordstrandite	$Al(OH)_3$	Monoclinic
Spinel	$MgAl_2O_4$	Cubic
Hercynite	$FeAl_2O_4$	Cubic
Högbomite	$Mg(Al, Fe, Ti)_4O_7$	Hexagonal
Chrysoberyl	$BeAl_2O_4$	Orthorhombic
Brucite	$Mg(OH)_2$	Hexagonal
Hematite	α-Fe_2O_3	Hexagonal
Maghemite	γ-Fe_2O_3	Cubic, tetragonal
Goethite	α-FeOOH	Orthorhombic
Akaganéite	β-FeOOH	Tetragonal
Lepidocrocite	γ-FeOOH	Orthorhombic
Bixbyite	$(Mn, Fe)_2O_3$	Cubic
Groutite	α-MnOOH	Orthorhombic
Feitknechtite	β-MnOOH	? Hexagonal
Manganite	γ-MnOOH	Monoclinic
Magnetite	$FeFe_2O_4$	Cubic
Magnesioferrite	$MgFe_2O_4$	Cubic
Jacobsite	$MnFe_2O_4$	Cubic
Ulvöspinel	$TiFe_2O_4$	Cubic
Chromite	$FeCr_2O_4$	Cubic
Pyrolusite	β-MnO_2	Tetragonal
Nsutite	γ-MnO_2	? Orthorhombic
Birnessite	δ-MnO_2	Hexagonal
Ramsdellite	MnO_2	Orthorhombic
Psilomelane	$(Ba, Mn, R)_3Mn_8O_{16}(O, OH)_6$	Orthorhombic
Cryptomelane	KMn_8O_{16}	Tetragonal, monoclinic
Braunite	$(Mn, Si)Mn_2O_4$	Tetragonal
Hausmannite	$MnMn_2O_4$	Tetragonal
Rutile	TiO_2	Tetragonal
Anatase	TiO_2	Tetragonal
Brookite	TiO_2	Orthorhombic
Cassiterite	SnO_2	Tetragonal
Baddeleyite	ZrO_2	Monoclinic
Ilmenite	$FeTiO_3$	Hexagonal
Pseudobrookite	Fe_2TiO_5	Orthorhombic
Perovskite	$CaTiO_3$	Monoclinic

Compiled from Mackenzie (1975) and Mitchel (1975).

TABLE 3. Some Phosphates Expected to Occur in Soils in Trace Amounts.

Mineral	Formula
Chlorapatite	$Ca_5(PO_4)_3Cl$
Fluorapatite	$Ca_5(PO_4)_3F$
Hydroxyapatite	$Ca_5(PO_4)_3OH$
Isokite	$CaMgPO_4F$
Wagnerite	Mg_2PO_4F
Newberyite	$MgHPO_4 \cdot 3H_2O$
Struvite	$MgNH_4PO_4 \cdot 6H_2O$
Berlinite	$AlPO_4$
Phosphotridymite	$AlPO_4$
Variscite	$AlPO_4 \cdot 2H_2O$
Metavariscite	$AlPO_4 \cdot 2H_2O$
Wavellite	$Al_6(PO_4)_4(OH)_6 \cdot 5H_2O$
Augelite	$Al_2PO_4(OH)_3$
Taranakite	$Al_5(NH_4K)_3H_6(PO_4)_8 \cdot 18H_2O$
Amblygonite	$Al(LiNa)PO_4(F, OH)$
Montebrasite	$Al(Li, Na)PO_4(F, OH)$
Morinite	$Al_4NaCa_4H(PO_4)_4(OH)_4F_6$
Phosphosiderite	$Fe^{3+}PO_4 \cdot 2H_2O$
Strengite	$Fe^{3+}PO_4 \cdot 2H_2O$
Vivianite	$Fe^{3+}(PO_4)_2 \cdot 8H_2O$
Cacoxenite	$Fe^{3+}{}_4(PO_4)_3(OH)_3 \cdot 12H_2O$
Barrandite	$(Fe^{3+}, Al)PO_4 \cdot 2H_2O$
Clinobarrandite	$(Fe^{3+}, Al)PO_4 \cdot 2H_2O$
Lazulite	$(Fe^{2+}, Mg)Al_2(PO_4)_2(OH)_2$
Triplite	$(Fe^{++}{}_6Mn^{2+})_2PO_4F$
Zwieselite	$(Fe^{2+}, Mn^{2+})_2PO_4F$
Lithiophyllite	$(Fe^{2+}, Mn^{2+})LiPO_4$
Sicklerite	$(Fe^{2+}, Mn^{2+})LiPO_4$
Triphyllite	$(Fe^{2+}, Mn^{2+})LiPO_4$
Alluadite	$(Fe^{2+}, Mn^{2+})NaPO_4$
Arrojadite	$(Fe^{2+}, Mn^{2+})_5Na_2(PO_4)_4$
Natrophyllite	$Mn^{2+}NaPO_4$

Compiled from Strunz (1957) and Kretz (1976).

will no doubt be made to isolate soil phosphates in concentrated fractions so that they might be identified by infrared techniques.

Manganese, zinc, copper, molybdenum and boron are trace constituents or substituents of the minerals in all soils. Manganese is associated with the sesquioxides reported in Table 2, and along with zinc and copper, it can occur as an isomorphous replacement in the octahedral layers of the phyllosilicates. Molybdenum is a common impurity in olivine (Parfenova and Yarilova, 1962); boron, which occurs in tourmaline, can occur as impurities in phosphate minerals. Micro amounts of absorbable ions from these minerals are essential for plant growth in soils. Slight excesses of these elements in soluble form in soils are, however, toxic to plants.

Inorganic soils may be composed of gravel, sand, silt, and/or clay-size particles. The coarser fractions give mechanical stability to the soil. The clay fractions adsorb water and cationic plant nutrients with sufficient energy to prevent loss by leaching (q.v.) and drainage (q.v.). These substances are held loosely enough, however, to be available to plants. The minerals in clays are mostly phyllosilicates mixed with varying amounts of hydrated sesquioxide minerals. The thin sheet-like structures of the phyllosilicates give the clays enormous specific surface areas. These interfaces are responsible for high water-holding and cation exchange capacities. The M^+ ions in the montmorillonite and vermiculite formulas under phyllosilicates in Table 1 represent exchangeable cations in these minerals.

The foregoing reports only the most abundant inorganic constituents of soils and those trace inorganic constituents essential for plant growth. It should be emphasized, however, that any unconsolidated mineral or its weathering products can occur in soils.

JOHN E. GIESEKING

References

Bragg, L.; Claringbull, G. F.; and Taylor, W. H., 1965. *Crystal Structures of Minerals.* London: Bell, 166–382.

Farmer, V. C., and Palmieri, F., 1975. The Characterization of soil minerals by infrared spectroscopy, *in* J. E. Gieseking, ed., *Soil Components: Volume 2. Inorganic Components.* New York: Springer-Verlag, 573–670.

Kretz, R., 1976. *in* R. C. Weast, ed., *Handbook of Chemistry and Physics.* Cleveland: Chemical Rubber Company, B–196–201.

Mackenzie, R. C., 1975. The classification of soil silicates and oxides, *in* J. E. Gieseking, ed., *Soil Components: Volume 2, Inorganic Components.* New York: Springer-Verlag, 1–25.

Mitchell, B. D., 1975. Oxides and hydrous oxides of silicon, *in* J. E. Gieseking, ed., *Soil Components: Volume 2. Inorganic Components.* New York: Springer-Verlag, 395–432.

Parfenova, E. A., and Yarilova, E. A., 1962. *Mineralogicheskie issledovaniya v pochvovedenii.* Izdatel'stvo Akdemii Nauk SSSR (English translation: Mineralogical Investigations in Soil Science, Isreal Program for Scientific Translations, 1965).

Strunz, H., 1957. *Mineralogische Tabellen.* Leipzig: Akademische Verlagsgesellschaft, Geest & Portig, 57–74.

Cross-references- *Chemical Composition; Clay Minerals, Silicates; Hydrous Oxides; Soil Mineralogy. Vol. IVB: Soil Minerals; Also individual minerals.*

SOIL COMPONENTS, ORGANIC

Soil organic matter usually contains a range of substances from easily degradable biological products to extremely stable polymers. Polysaccharides, proteins, lignin, mercapto-compounds, nucleic acid derivatives, and other phospho-acids, added to soils, are partially decomposed quickly by enzyme systems of organisms in the soils to release the monomers or simpler polymers of the substrate plus large amounts of energy derived from the oxidation of many of the monomeric compounds. The remaining monomers and simple polymers are reused by plants or other organisms growing in or on the soil or they may re-co-polymerize with each other to form the extremely stable humic substances of soil organic matter.

Polysaccharides and Amino Compounds

The polysaccharides and their monomers added to soils and the amounts of saccharides isolated from many soils have been reported by Greenland and Oades (1975). Table 1 gives the polysaccharides and monomers occurring in soils. All the monomers in Table 1 have been identified in soils.

Stevenson (1954), Sowden (1955), Young and Mortensen (1958), Scharpenseel and Krause (1962), and Parsons and Tinsley (1975) have indicated the amino acids and sugars that are found in soils: glycine, α-alanine, β-alanine, serine, threonine, valine, leucine, isoleucine phenylalanine, tyrosine, cysteine, methionine, aspartic acid, glutamic acid, histidine, arginine, lysine, methionine, proline, hydroxyproline, α-aminobutyric acid, γ-aminobutyric acid, α, ϵ-diaminopimelic acid, methylhistidine, ornithine, glucosamine, and galactosamine. Some of the preceding amino compounds are abundant in crustacea, insects, and bacteria.

Anderson (1967) and Bremer (1967) have obtained the purine bases, adenine and guanine, and the prymidine bases, cytosine, thymine, and uracil, in soil organic matter hydrolysates. These bases are presumed to have been derived from ribonucleic acids and deoxyribonucleic acids.

Phosphorus and Sulfur

Anderson (1975) has studied the molecular structure and occurrence of organic phosphorus compounds in soils, which include derivatives of phytic acid nucleic acids, phospholipids, *myo-*, *scyllo-*, *chiro-* and *neo*-inisotol hexaphosphate and the penta-, tetra- and triphosphate forms of inisotol. Braids and Miller (1975) have investigated the presence of phosphatidyl choline in soils. Anderson (1975*a*) has also reported on the molecular structure and occurrence of the following organic sulfur compounds in soils. These compounds include: cystine, methionine, methionine sulfoxide, methionine sulfone, cysteic acid, cystine sulfinic acid, taurine, biotin, and thiamine.

Fats, Waxes, and Resins

Braids and Miller (1975) have reported on the occurrence, extraction, and molecular structures of fats, waxes, and resins that have been found in soils, dividing them into groups of hydrocarbons, alcohols, long-chain acids, sterols, terpenes, pigments, heterecyclic nitrogen compounds, and phospholipids.

Humic Substances

Many tons of crop residues are added per acre annually to highly productive soils. Some of these residues may contain as much as 20% lignin. A large part of the lignin is quickly destroyed by oxidation, and only small amounts of the degradation products are left in the soil to copolymerize with other organic components to form humic substances.

Flaig, Beutelspacher, and Rietz (1975) have extensively reviewed the results of work on the formation of humic substances in soils. Humic substances are interpolymerizations of phenolic compounds, peptides, amino acids, and sugars. Lignin is the source of the phenols in humic soil polymers. The nature of the phenolic compounds formed during the degradation of lignin in soils is described by Flaig, Beutelspacher, and Rietz (1975). They describe the procedures for the separation of soil organic matter polymers into humins, humic acids, fulvic acids, and hymatomelanic acids. They also give the properties of these humic fractions. The polymers in humic substances are very stable, being resistant to attack by the enzyme systems of microorganisms even though they contain large amounts of amino acid and amino sugar polymers in their structures.

Nitrogen

Flaig, Beutelspacher, and Rietz (1975) have found that only one half of the nitrogen in humic acids can be released by hydrolysis by boiling with 6N HCl. This hydrolysate is 40–80% amino acids and 20–50% amino sugars. They find that approximately 50% of the remaining nonhydrolysable nitrogen can be released by fusing with a sodium hydroxide-sodium acetate mixture. This explains why much of the nitrogen in highly organic soils is in a form that is only very slowly converted to forms available to plants.

Enzymes

According to Skujins (1967), only small quantities of a few enzymes have been extracted from soils, but activities of many enzymes have been reported in soil samples. These include: dehydrogenases, catalase, phenol

TABLE 1. Organic Soil Components.

Origin	Polysaccharide	Function	Component Sugars	General
Higher plants	*Cellulose*	Cell-wall material	Glucose	Forms 30 to 70% of tissue of higher plants. Linear polymer of β-(1→4)-D-glucopyranose units.
	Hemicelluloses, which include	Associated cell-wall material		Forms 2 to 5% of higher plants.
	glucuronoxylo-glycans		Glucuronic acid, xylose	Linear xylose polymer with attached uronic acid units.
	galacturono-arabino-xyloglycans		Galacturonic acid, arabinose, xylose	Linear xylose polymer with other units attached.
	xylans		Xylose	Linear polymers of β-(1→4)-D-anhydroxylopyranose units.
	glucomannans		Glucose, mannose, occasionally some galactose	Mannose usually predominates in β-(1→4)-D-pyranose chain.
	arabino galactans		Arabinose, galactose	Highly branched structures.
	Pectic substances, predominantly galacturonans associated with some arabans and galactans	In cell walls, some intercellular regions, and possibly in mucilages	Galacturonic acid	Linear polymer of α-(1→4)-D-pyranose units partially or fully methylated.
	Gums	Exudates from damaged tissue	Large number recorded	Generally complex, and frequently branched.
	Starches composed of	In storage organs		
	amylose and		Glucose	Linear α-(1→4)-D-glucopyranose polymer.
	amylopectin		Glucose	Branched glucopyranose structure.
Micro-organisms	*Exocellular,* include	Form protective capsules		
	dextrans		Glucose	α-(1→6)-D-polyglucosans
	levans		Fructose	β-(2→6)-D-polyfructosans.
	microbial gums		Large number recorded	
	Structural,			
	chitin	Cell-wall material in fungi	Glucosamine	Linear polymer of 2-N-acetamido-2-deoxy-β-D-glucopyranose units, linked (1→4).
	mucopeptides	Cell-wall material in bacteria	Amino sugars with amino acids	
	mucopoly-saccharides		Various sugars and amino sugars	

TABLE 1. continued

Origin	Polysaccharide	Function	Component Sugars	General
	teichoic acids		Glucose, glucosamine ribitol, and glycerol phosphates and alanine	
	Somatic	Energy storage materials		
	starches		Glucose	
	levans		Fructose	
	glycogens		Glucose	
Animals	chitin	Structural materials in lower animals	Glucosamine	Linear polymer of 2-N-acctamido-2-deoxy-β-D-glucopyranose units linked (1→4).
	hyaluronic acid	Intercellular cement	Glucosamine, glucuronic acid	
	glycogen	Energy storage material	Glucose	Branched chain.

(From Greenland and Oades, 1975)

oxidases, uricase, transaminase, transglycosylases, phosphatases, nucleases, lipases, amylase, invertase, saccharase, sucrase, maltase, lactase, cellobiase, cellulase, pepsin, trypsin, deaminases, decarboxylases, and urease.

Biomass

Live fauna and flora in macro- and microforms are an important part of the soil. This is called the *biomass* of the soil. The composition of the biomass of a Swiss meadow soil (Stöckli, 1946) is given as an example in Table 2. In addition to the biomass, there may be large quantities of living roots in the soil. The biomass is constantly being generated by growing organisms during temperate conditions as other organisms are dying. When these organisms die, their bodies are decomposed, compounds are oxidized and degraded, and the final products contribute to the group of compounds that polymerize to form humic substances. Under these conditions humic substances in soils are slowly but constantly being formed and decomposed at the same time.

TABLE 2. Organisms in the "Biomass" of a Swiss Meadow Soil to a Depth of 15 cm.

Organism	Weight to a Depth of 15 cm (g/m^2)
Bacteria, actinomycetes, mycelia, algae	2022
Earthworms	400
Mollusca	80
Protozoa	38
Arthropods	10
Nematodes	5
Enchytraeids	2
Mites, collembola, protura, diplura	2

(From Stöckli, 1946)

JOHN E. GIESEKING

References

Anderson, G., 1967. Nucleic acid derivatives and organic phosphates, *in* A. D. McLaren and G. H. Peterson, eds., *Soil Biochemistry, Vol. I.* New York: Marcel Dekker, 67–90.

Anderson, G., 1975*a*. Other organic phosphorus compounds, *in* J. E. Gieseking, ed., *Soil Components. Volume 1. Organic Components.* New York: Springer-Verlag, 305–331.

Anderson, G., 1975*b*. Sulfur in soil organic substances, *in* J. E. Gieseking, ed., *Soil Components. Volume 1. Organic Components.* New York: Springer-Verlag, 333–341.

Braids, O. C., and Miller, R. H., 1975. Fats, waxes and resins in soils, *in* J. E. Gieseking, ed., *Soil Components. Volume 1. Organic Components.* New York: Springer–Verlag, 343–368.

Bremner, J. M., 1967. Nitrogenous compounds, *in* A. D. McLaren and G. H. Peterson, eds., *Soil Biochemistry, Vol. I.* New York: Marcel Dekker, 19–66.

Flaig, W.; Beutelspacher, H.; and Rietz, E., 1975. Chemical composition and physical properties of humic substances, *in* J. E. Gieseking, ed., *Soil Components. Volume 1. Organic Components.* New York: Springer-Verlag, 1–211.

Greenland, D. J., and Oades, J. M., 1975. Saccharides, *in* J. E. Gieseking, ed., *Soil Components. Volume 1.*

Organic Components. New York: Springer-Verlag, 213-261.

Parsons, J. W., and Tinsley, J., 1975. Nitrogenous substances, *in* J. E. Gieseking, ed., *Soil Components. Volume 1. Organic Components.* New York: Springer-Verlag, 263-304.

Scharpenseel, H. W., and Krause, R., 1962. Aminosäureuntersuchungen an verschiedenen organischen Sedimenten, besonders, Grauund Braunhuminsäurefraktionen verschiedener Bodentypen (einschliesslich ^{14}C markierter Huminsäuren), *Z. Pfl. Ernähr. Dung.*, **96**, 11-34.

Skujins, J. J. 1967. Enzymes in soils, *in* A. D. McLaren and G. H. Peterson, eds., *Soil Biochemistry, Vol. I.* New York: Marcel Dekker, 371-414.

Sowden, F. J., 1955. Estimation of amino acids in soil hydrolysates by the Moore and Stein method, *Soil Sci.*, **80**, 181-183.

Stevenson, F. J., 1954. Ion exchange chromatography of the amino acids in soil hydrolysates, *Soil Sci. Soc. America Proc.*, **18**, 373-377.

Stöckli, A., 1946. Der Boden als Heimstäte des Lebens, *Schweiz Z. Forstw.*, **97**, 356-378.

Young, J. L., and Mortensen, J. L., 1958. Soil Nitrogen complexes: I. Chromatography of amino compounds in soil hydrolysates, *Ohio Agric. Exp. Sta. Research Circular.*, **61**, 1-18.

Cross-references: *Algae; Carbon Cycle; Enzyme Activity; Fauna; Humus; Metabolic Activity; Metal-complexing Macromolecules; Nitrogen Cycle; Organic Matter; Polysaccharide; Phosphorus Cycle; Sulfur Transformations.*

SOIL CONDITIONERS

Natural or synthetic materials that are applied to soils to modify certain physical and chemical properties are referred to as *soil conditioners.* Conditioners differ from *soil amendments* in that the latter term commonly refers to materials added as opposed to those used primarily as fertilizer (see *Fertilizers, Inorganic; Fertilizers, Organic*) that may contain appreciable amounts of nutrient elements. The general use of conditioners and the various application techniques are referred to as *soil conditioning* (De Boodt, 1972a), an aspect of ameliorative pedology (q.v. in Vol. XII, Pt. 2). There are many different types of soil conditioners and a wide variety of agricultural and non-agricultural uses. In general, soil conditioners are used to control or modify many soil properties such as volume stability, i.e., swelling and shrinking and strength (both shear and tensil). They also increase the permeability of the soil to air and water, increase its resistance to crusting when it dries out, and agglomerate (flocculate) soil colloids to improve tilth.

Historical Background

The benefits of an organic mulch, a natural soil conditioner, were recognized some 4000 years ago in the *Sumarian Georgica,* an early farmer's almanac (Kramer, 1958). Natural soil conditioners, however, are now frequently in short supply, difficult to locate, and have high handling costs. Even though synthetic soil conditioners were developed more than thirty years ago, they did not receive worldwide attention until the advent or urban and suburban development, which was brought about largely by highway construction. The intensification of soil tillage (q.v.) and associated slaking of soil particles with accelerated erosion has also encouraged research into methods of artificial soil aggregate stabilization (see *Aggregate Stability*).

Types of Conditioners and Applications

Soil conditioners include numerous types of materials that range from different kinds of natural and artificial mulches (see *Mulch*) to synthetic polyelectrolytes such as complex vinyl and acrylic compounds and certain cellulose and lignin derivatives. Such diverse products as calcium chloride, calcined clay, lime, composted garbage, fly ash, ground marble, Portland cement, molasses, resins, bentonite, and nonionic surfactants (wetting agents) have also been used because of their ameliorative effects in soil conditioning or as soil stabilizers (Blavia, Moldenhauer, and Low, 1971; Moldenhauer, 1975). These materials have many different uses. The following applications are indicative of the broad scope of soil conditioning: stabilization of embankments along highways, railways, and streams; prevention of evaporation losses (see *Evaporation*); control of water and wind erosion (see *Water Erosion; Wind Erosion*); reduction of pollution hazards associated with surface erosion and runoff; inhibition of leaching losses where rainfall is excessive (see *Leaching; Nutrient Losses*); improvement of soil structure (q.v.); and control of infiltration (q.v.) rates (Kijine, 1967; Kruger and Benksenstein, 1972; Gabriels, Moldenhauer, and Kirkham, 1973).

Synthetic soil conditioners, formerly in powder form, are now available in liquid form as solutions or emulsions (Emerson, 1965; Gardner, 1972). Liquid conditioners are less expensive and can be used in conjunction with pesticides. Polyvinyl alcohol (PVA), polyvinyl acetate (PVAc), polyacrylamide (PAM), hydrolyzed polyacrylonitrile, and salts of polyacrylic acid are water soluble. Rubber, bitumen, latexes, and copolymers of styrene are applied as emulsions. Emulsions may be applied in small amounts using a garden hand spray or by large commercial sprayers when large quantities are required (Toyoshima, 1973).

Mulches. Natural mulching materials such as straw, hay, woodchips, sawdust, shredded tree bark, peat moss, tobacco dust, ground corn cobs, peanut hulls, rice hulls, cocoa mulch, bagasse, and coir are frequently used for vegetable production and home gardening. Such organic mulches and even stone mulches have been used to control erosion and conserve moisture (Meyer and Mannering, 1963; Meyer, Johnson, and Foster, 1972; Roose, 1975). Petroleum by-products (e.g., bituminous emulsions, cutback asphalt (Chepil, 1955; Sale, 1966; Barnett, Diseker, and Richardson, 1967; Ahmed and Roblin, 1971), plastic papers (Linden, 1963), and light-reflecting materials (e.g., aluminum foil) are also used extensively as mulches.

Stabilizers. *Inorganic stabilizers* (e.g., lime, cement, phosphoric acids, etc.) promote both flocculation (q.v.) and chemical bonding (Moldenhauer and Gabriels, 1975). Flocculation results in an increase in the effective size of the clay crystals that are aggregated into domains. The chemical bonding effect differs according to the stabilizer used. The important difference between lime and cement stabilization, for example, is that the lime draws its silica from the soil clays ensuring a firmer bond to the clay than does cement. Cement, on the other hand, establishes its own strong lattice upon hydration and develops strength more rapidly than lime (Lee, 1968).

Organic stabilizers (for examples, PVAc, PAM) stabilize by various methods such as ion exchange flocculation or hydrogen bonding. These materials are subject to bacterial attack and hence are not permanent. In addition to their high cost, the mixing-in of small amounts of organic stabilizers is often difficult.

Whether hydrophilic or hydrophobic (see *Hydrophility, Hydrophobicity*) soil conditioners are used to stabilize the soil (see *Soil Stabilization*) depends on local soil conditions (Hemwall and Bozer, 1964). In some cases waterproofing agents such as long-chain organic cationic surfactants [e.g., ceytl trimethyl ammonium bromide (CTAB) and ceytl pyridinium bromide (CPB)] are used to reduce the ingress of water. In other soils a wetting agent is desirable because it can increase infiltration rates, especially into hydrophobic sandy soils (Pelishek, Osborn, and Letey, 1962). Polymers of PVA and PVAc are particularly effective in soil stabilization. For coarse-textured soils, emulsions of water-insoluble materials are most effective as aggregating agents (De Boodt, 1972b; Lehfeldt, Kullman, and Muller, 1973). However, emulsions are generally not as effective in improving the physical condition of finer-textured soils. Where PVAs can be readily incorporated by mechanical manipulation, they are an effective aggregate-stabilizing agent. When applied to cloddy dispersed sodic soils, PVAs tend to stabilize the dispersed condition (Carr and Greenland, 1975).

Physical methods of stabilization include the application of heat, which is directed toward fixation of the contact bonds between clay domains, and electrohardening, which cements the fine pores of the soil by the redeposition of hydrated alumina. Simple drying of soil is also frequently sufficient to ensure high strengths (Winterkorn, 1975).

Under certain conditions it is possible to obtain more favorable structural properties of soil by modifying the aggregate size, changing their spatial distributions and stability. Hoyle, Yamada, and Hoyle (1972) have described the technique of rototilling wet soil in a method called *aggresizing*. Better crop yields were obtained in aggresized soil due to greater nutrient uptake, more uniform distribution of water, and lack of surface crust formation (Hoyle and Yamada, 1975).

CHARLES W. FINKL, JNR.

References

Ahmed, N., and Roblin, A. J., 1971. Crusting of River Estate soil, Trinidad, and its effect on gaseous diffusion, percolation, and seedling emergence, *Jour. Soil Sci.,* 22, 23–31.

Barnett, A. P.; Diseker, E. G.; and Richardson, E. C., 1967. Evaluation of mulching methods for erosion control on newly prepared and seeded highway backslopes, *Agron. Jour.*, 59, 83–85.

Blavia, F. J.; Moldenhauer, W. C.; and Low, D. E., 1971. Materials for stabilizing surface clods of cropped soils, *Soil Sci. Soc. America Proc.*, 35, 119–121.

Carr, C. E., and Greenland, D. J., 1975. Potential applications of polyvinyl acetate and polyvinyl alcohol in the structural improvement of sodic soils, *in* W. C. Moldenhauer, ed., *Soil Conditioners.* Madison: Soil Science Society of America, 47–64.

Chepil, W. S., 1955. Effects of asphalt on some phases of soil structure and erodibility by wind, *Soil Sci. Soc. America Proc.,* 19(2), 125–128.

De Boodt, M., 1972*a*. Proceedings Symposium on the Fundementals of Soil Conditioning, *Medelelingen Fakulteit Landbouwwetenschappen, Rijksuniversiteit Gent,* 37(3), 897–1167.

De Boodt, M., 1972*b*. Improvement of soil structure by chemical means, *in* D. Hillel, ed., *Optimizing the Soil Physical Environment Towards Greater Crop Yield.* New York: Academic, 43–55.

Emerson, W. W., 1965. Synthetic soil conditioners, *Jour. Agric. Sci.,* 47, 117–121.

Gabriels, D. M.; Moldenhauer, W. C.; and Kirkham, D., 1973. Infiltration, hydraulic conductivity and resistance to water drop impact of clod beds as affected by chemical treatment, *Soil Sci. Soc. America Proc.,* 37(4), 634–637.

Gardner, W. H., 1972. Use of synthetic soil condi-

tioners in the 1950's and some implications to their further development, *Mededelingen Fakulteit Landbouwwetenschappen, Rijksuniversiteit Gent,* 37(3), 1046-1061.
Hemwall, J. B., and Bozer, K. B., 1964. Moisture and strength relationships of soils as affected by 4-tert. butyl-pyrocatechol, *Soil Sci.,* 98, 235-243.
Hoyle, B. J.; Yamada, H.; and Hoyle, T. D., 1972. Aggresizing—to eliminate objectionable soil clods, *Calif. Agric.,* 26(11), 3-5.
Hoyle, B. J., and Yamada, H., 1975. Seedbeds and aggregate stability improved by rototilling wet soil, *in* W. C. Moldenhauer, ed., *Soil Conditioners.* Madison: Soil Science Society of America, 111-130.
Kijine, J. W., 1967. Influence of soil conditioners on infiltration and water movement in soils, *Soil Sci. Soc. America Proc.,* 31, 8-13.
Kramer, S. N., 1958. *History Begins at Sumer.* London: Thames and Hudson, 333p.
Kruger, W., and Benkenstein, H., 1972. Effect of a layer of bitumen mulch on water infiltration, air composition, and condensation processes in soil, *Archiv. für Acker- und Pflanzenbau und Bodenkunde,* 16(4/5), 391-398.
Lee, I. K., 1968. *Soil Mechanics.* Sydney: Butterworths, 625p.
Lehfeldt, J.; Kullman, A.; and Muller, H. H., 1973. Relationship between emulsion properties, methods of application and the condition of the soil surface during liquid mulching, *Archiv. für Acker- und Pflanzenbau und Bodenkunde,* 17(7/8), 513-522.
Linden, R., 1963. Soil protection with plastic mulches, *Ann. Gembl.,* 69, 601-608.
Meyer, L. D.; Johnson, C. B.; and Foster, G. R., 1976. Stone and woodchip mulches for erosion control on construction sites, *Jour. Soil Water Conserv.,* 27, 264-269.
Meyer, L. D., and Mannering, J. V., 1963. Crop residues as surface mulches for controlling erosion on sloping land under intensive cropping, *Amer. Soc. Agric. Engin. Trans.,* 6, 322-323.
Moldenhauer, W. C., ed., 1975. *Soil Conditioners.* Madison: Soil Science Society of America, 186p.
Moldenhauer, W. C., and Gabriels, D. M., 1972. Some uses of soil stabilizers in the U.S.A., *Mededelingen Fakulteit Landbouwwetenschappen, Rijksuniversiteit Gent,* 37(3), 1014-1034.
Pelishek, R. E.; Osborn, J. F.; and Letey, J., 1962. Effect of wetting agents on infiltration, *Soil Sci. Soc. America Proc.,* 26, 595-598.
Roose, E. J., 1975. Natural mulch as chemical conditioner for reducing soil erosion in humid tropical areas, *in* W. C. Moldenhauer, ed., *Soil Conditioners.* Madison: Soil Science Society of America, 131-138.
Sale, P. M., 1966. Effect of petroleum mulch on seedling emergence, soil moisture and soil temperature, *Expl. Hort.,* 14, 43-52.
Toyoshima, K., 1973. General properties of poly (vinyl alcohol) in relation to its applications, *in* C. A. Finch, ed., *Poly (vinyl alcohol), Properties and Applications.* London: Wiley, 17-65.
Winterkorn, H. F., 1975. Soil stabilization, *in* H. F. Winterkorn, ed., *Foundation Engineering Handbook.* New York: Van Nostrand, 312-336.

Cross-references: *Aggregate Stability; Conservation; Mulch; Soil Stabilization; Tillage.*

SOIL CONSERVATION—See CONSERVATION; SOIL CONSERVATION SERVICE

SOIL CONSERVATION SERVICE

The Soil Conservation Act of 1935 established within the United States Department of Agriculture a "soil conservation service" to develop and prosecute a long-term program of soil and water conservation. Accordingly the Soil Erosion Service, which had been in existence for nineteen months as an emergency agency, was transferred from the Department of Interior to become a regular bureau, the Soil Conservation Service (SCS), within the Department of Agriculture. Through the years of its existence, the program of the SCS has been broadened and realigned from time to time.

Responsibilities

At present the major responsibilities of the Service are as follows: (1) the provision of technical assistance through conservation districts to landowners and operators who carry out locally adapted soil and water conservation programs; (2) the federal responsibility for the National Cooperative Soil Survey in conjunction with state agricultural experiment stations and other state and federal agencies; (3) the National Program of Land Inventory and Monitoring to provide soil, water, and related resource data and resource use trends necessary for planning and programing at all public agency levels; (4) administration of the flood-control and watershed and river investigation responsibilities of the Department of Agriculture; (5) provision of assistance to local sponsoring groups in planning and comprehensive development of land and water resources in multiple-county areas; and (6) general responsibilities for administration of the program designed to achieve greater agricultural stability in the ten Great Plain states.

Administration

The soil and water conservation program is carried on through technical assistance to locally organized and operated conservation districts, local sponsors of watershed-protection projects and resource conservation and development projects, and consultive assistance to other individuals and groups. More than 3000 conservation districts cover almost 2 billion acres in the 50 states, Puerto Rico, and the Virgin Islands.

The overall program is directed by the administrator and his staff from the central office in Washington, D. C. State offices administer state programs through area offices, which in turn coordinate work units offices. The latter essentially coincide with the conservation districts.

Soil and Water Conservation Needs Committee

By Memorandum No. 1396, April 10, 1956, the Secretary of Agriculture established the framework for a national inventory of soil and water conservation needs. The intent was to provide the Department of Agriculture with the information base necessary to effectively plan and carry out its responsibilities in soil and water conservation. The agencies of the department concerned with land use, soil and water conservation, and the management of land resources were designated as cooperators in the endeavor. The SCS was assigned responsibility for leadership. The goal for the completion of initial national inventory was three years, and it was provided that information be kept current with periodic review and revision.

The memorandum established the Department of Agriculture Soil and Water Conservation Needs Committee to plan, coordinate, and review the progress of the effort. The membership was defined as one representative of each of the relevant agencies. Provision was also made for a committee in each state and territory with like memberships from the agencies of the department. Additionally, representation was invited from the Land Grant Colleges, the State Foresters, and other appropriate agencies and groups who might provide assistance and useful data. The State Needs Committees were charged with the responsibility for developing inventory plans and providing guidance of the efforts within their respective states. The actual inventory was developed by the Needs Committees in each county.

The major thrust of the inventory to be conducted by the committees focused on non-federally owned land. The SCS, however, was given responsibility for collecting basic physical data on soil and water on non-federally owned land. The Forest Service was assigned the task of assuring adequate physical data on forestry on non-federally owned land.

In the three-year period of 1957–1960, the county committees studied and extrapolated soil data from randomly sampled areas and combined that with information from other sources to estimate land and water conditions for 1958 and the expected needs for 1975. In the early 1960s these data, with interpretations, were published in state reports. They were also summarized in two national reports, *Basic Statistics of the National Inventory of Soil and Water Conservation Needs* and a popularized digest, *Agricultural Land Resources, Capabilities, Uses, Conservation Needs (Agriculture Information Bulletin 263).*

RICHARD M. HIGHSMITH, JR.

References

Anon., 1957. *Policy and Procedure for Development of National Inventory of Soil and Water Needs.* Washington, D.C.: U.S. Department of Agriculture, 26p.

General Services Administration Staff, 1973/74. *United States Government Manual.* Washington, D.C.: U.S. Government Printing Office, 118–119.

SOIL DRAINAGE

Soil drainage is the removal of excess water from the land by artificial or man-made facilities. Most natural drainage occurs because of land forms and elevation differences that result from geologic erosion. All natural drainage is by gravity flow, even artesian water which moves upward to the land surface. *Runoff* is a more common term for drainage over the land surface, and *seepage* or *groundwater flow* are terms often applied to drainage below the land surface.

Man has improved on nature's drainage systems in order to construct buildings, highways, and other facilities; to improve the growth of more desirable vegetation; and to improve the environment for man, animals, and other organisms. Agricultural soil drainage enhances plant growth by increasing soil aeration and by removing toxic materials such as salts, which impair plant growth. Toxic salts accumulate near the surface in the more arid regions, and when such land is irrigated, either these salts must be removed or chemical amendments must be added to the soil. The most practical method of salt removal is irrigating and draining the excess water from the soil (Schwab et al, 1966). Some drainage method must eventually be provided, either natural or man-made, for leaching irrigated soil. The construction of a suitable drainage system may involve both extensive surface and subsurface drains.

The computed flow rate in an open channel or in a subsurface tube drain with gravity flow is the product of the average velocity and the cross-sectional flow area. The Manning equation is most commonly used to estimate average velocity. This velocity varies inversely with the roughness of the channel or tube and with the two-thirds power of the length of the wetted surface of the cross-section. It also varies directly with the two-thirds power of the cross-sectional area and with the square root of the hydraulic slope.

Surface Drainage

In agricultural land, surface drainage is accomplished by a combination of open ditches, shallow field drains, and land grading.

Open ditches should be built so that neither scouring nor sedimentation will occur, with sufficient flow capacity, with adequate depth, and with stable side slopes. They should have minimum initial cost and minimum long-term maintenance. Figure 1 shows a typical cross-section. Flow capacity will vary with different climatic regions and with the drainage needs of the crop to be grown.

Most big drainage ditches have been constructed under state laws enacted to facilitate drainage where several landowners are involved. Under such laws the construction and maintenance costs can be equitably distributed to the benefited landowners. Large ditches normally drain too much area and are too costly for a few owners to undertake individually. Regrowth of shrubs and trees in the channel causes reduced flow and is one of the major maintenance problems. In legal organized districts, the trend is toward annual maintenance assessments to assure continued proper functioning of the ditches.

Shallow field drains and land grading are often required in flat or undulating land to improve the flow to the open ditches. Small areas that are flooded even for a day or so may kill young plants or reduce crop yields appreciably. These fields drains may connect several low-lying areas in a random pattern or they may consist of parallel drains at uniform or variable spacing. Between these drains additional surface grading may be necessary to remove shallow water that accumulates on the surface. Such grading may also be required for removal of small ridges or furrows produced by normal tillage equipment. In irrigated areas, land grading is called *leveling,* but for some irrigation, the land is truely leveled as in a flooding system.

Shallow drains are often constructed with pan scrapers, blade graders, bulldozers, or similar equipment. Lazer-beam equipment for grade control is commercially available and is capable of greater accuracy than operator-controlled machines. The system of grade control is the same as for subsurface drains shown in Figure 2.

Subsurface Drainage

Subsurface drainage is required in agricultural land to lower the water table. Removing the excess groundwater permits oxygen and carbon dioxide gases to move into and out of the soil, increases the length of the crop growing season by early tillage and planting, removes toxic substances that may retard plant growth and increases the available water-storage capacity of the soil. Drainage of highways and buildings increase the structural bearing capacity of the soil by reducing the moisture content. Flow of water through the soil can be predicted by Darcy's law, which states that the flow rate is proportional to the soil hydraulic conductivity and the gradient of the flow. For one dimensional flow, this law can be easily applied, but for subsurface drains the flow is three-dimensional, and the flow region is not well defined. The geometry of the drains, varying soil characteristics, and complex boundary conditions greatly complicate mathematical flow equations and horizontal spacing equations for parallel drains.

Subsurface drains are contructed primarily with clay tile, concrete tile, and corrugated plastic tubing (Sisson, 1967). Figure 3 illustrates some common types of drain materials and openings. In the early days before such materials were available, drains were made by placing tree limbs, sand or gravel, rocks, and other natural porous material in the bottom of a covered trench. In some special soil and climatic conditions in humid areas *mole drains* (Ede, 1961) were successful. The mole channel was made by pulling a plug through the soil. It was attached to the bottom of a narrow vertical steel blade designed to keep the plow in the soil. Such drains rely entirely on the natural stability of soil to maintain its shape. Although they function for only a few years, mole drainage is still a common practice, primarily in Great Britain and New Zealand.

Clay drain tile (Alpers and Short, 1965) are made from surface clay or shale, which is pul-

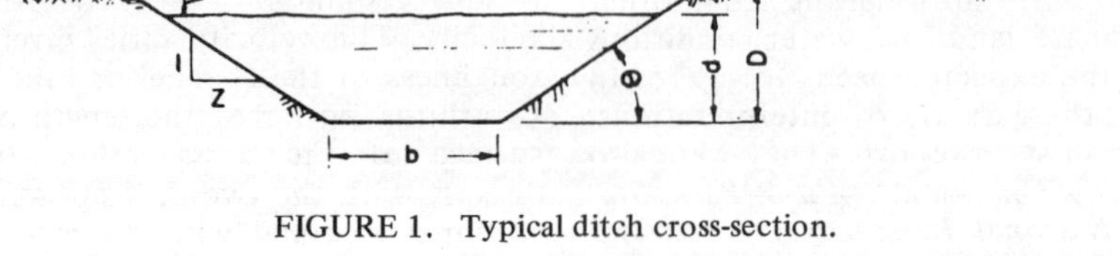

FIGURE 1. Typical ditch cross-section.

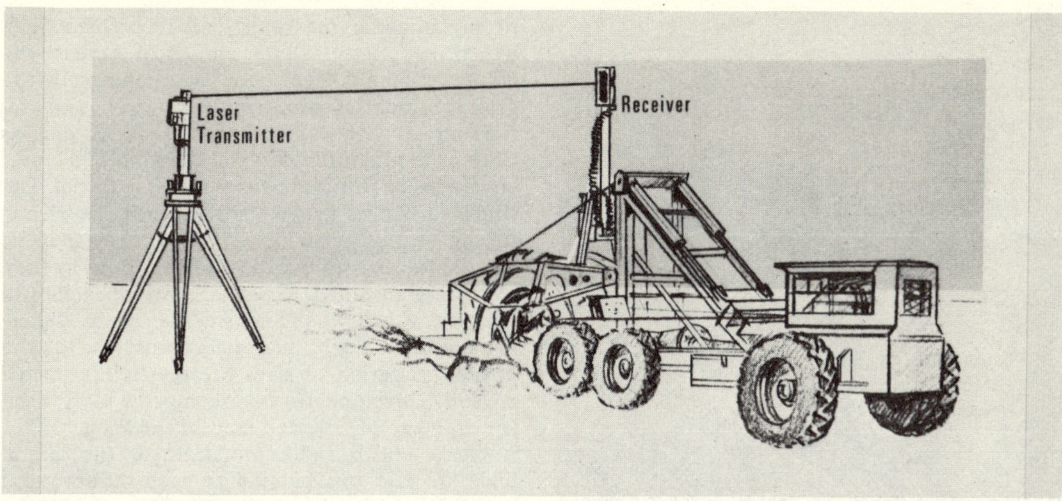

FIGURE 2. Laser-grade controlled trencher. (Photo: U.S. Department of Agriculture, Agricultural Research Service)

verized, extruded through a die, dried, and then burned in a kiln. Some early clay tile were made on a potter's wheel, while others were shaped from a flat slab of clay into a horseshoe shape. Clay tile are not affected by acid or alkaline soils, but those made from surface clay or poorly burned tile are subject to deterioration by freezing and thawing action. In general, tile installed more than 2 ft (60 cm) in depth are not affected by frost action. Many tile over

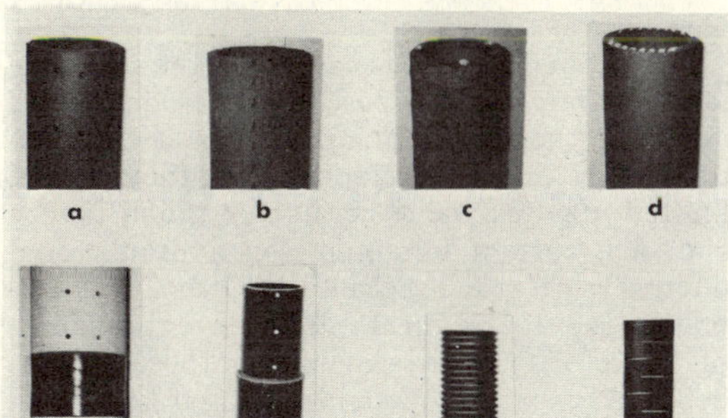

FIGURE 3. Common types of drain materials and openings: (a) clay tile with 16 5/16-in holes per foot in 4 rows; (b) clay tile with 36 3/16-in holes per foot in 3 rows; (c) concrete tile with 3 spacer projections on one end, 3/32-in joint spacing; (d) clay tile with corrugated ends, 1/16-in joint spacing; (e) asbestos cement pipe with 16 5/16-in holes per foot in 4 rows; (f) bituminous fiber pipe with 16 1/4-in holes per foot in 4 rows; (g) corrugated plastic pipe with 18 1/16-in by 1-in slits per foot in 3 rows; and (h) smooth plastic pipe with 15 1/32-in by 1 1/2-in slits per foot.

100 years old have been found. Good quality tile should last indefinitely in the soil.

Concrete tile, made from sand and gravel aggregate, are generally formed on a packerhead machine and then steam- or water-cured to obtain the desired strength (Manson, 1971). Concrete tile are resistant to freezing and thawing but may be subject to deterioration in acid and alkaline soils. For these soils the tile should be made with cement having special chemical characteristics.

Since the 1960s *thermoplastic tubing* (Figure 4) has become a common drain material (Fouss, 1974). The development of corrugating machines in Germany was largely responsible for the rapid expansion of plastic tubing. High-density polyethylene (PE) and polyvinylchloride (PVC) are the two most common plastics. PE tubing has a small amount of carbon black or other material added so that it will not deteriorate from ultraviolet rays in sunlight. In Europe PVC is the most widely used material, while PE is more common in the United States. The maximum size produced commerically in the United States is 457 mm (18 in) diameter. Corrugated plastic tubing, unlike concrete or clay tile, is a flexible material and will deflect vertically when soil is backfilled in the trench (Figure 5). As it deflects, the sides of the tubing move outward horizontally into the surrounding soil. The circular tubing changes to a slight oval shape, which becomes stabilized because the soil on the sides of the tubing resists the outward movement.

Corrugations in the tubing provide sufficient stiffness to resist the initial soil load. They also

FIGURE 4. Four-inch corrugated plastic tubing and modern transport method.

reduce the amount of plastic required to make the tubing as well as provide flexibility, which permits the smaller-size tubing to be coiled into a compact package.

Corrugated plastic tubing weighs only $\frac{1}{25}$ of concrete or clay tile, resists practically all soil chemicals, and can be installed in a continuous length (saves labor). Its stiffness decreases with increased temperature. Compared to concrete and clay tile, greater care is required in placing soil around plastic tubing because of deflection, and it is more difficult to locate the plastic lines with a steel probe.

Filter materials (Willardson, 1974) are sometimes placed around subsurface drains primarily to prevent the inflow of soil into the drains, which may cause failure, and/or to increase the effective diameter or area of openings in the drains, which increases inflow rate. These materials are of two types: (1) thin sheets such as fiber glass or spun nylon and (2) sand and gravel envelopes or other porous granular material. The thinsheet filters may be sealed onto the plastic tubing at the extruding plant, or they may be installed above and/or below the drains as they are being installed. Granular envelopes should be placed below and above the drains during installation. Such materials must have the proper gradation of sizes to prevent the inflow of soil. Envelope filters are most widely used for drainage of irrigated land in the West.

FIGURE 5. Modern trenching machine installing 4-in corrugated plastic tubing in Ohio.

Where entire fields are drained, the lateral lines are laid out parallel to each other, and a main drain is connected to each lateral at one end to carry the water into an outlet ditch. Where only low wet areas are drained, a system of random lines through the wet spots may suffice. Large systems may include a combination of random and parallel drains laid out so as to best fit the surface topography of the land. Where surface water is to be removed by the subsurface drain, vertical risers are installed to allow the water to enter the buried drain. In some areas where the level of water in the outlet ditch is too high, automatic pumping plants are required. On more sloping land, subsurface drains may be installed to intercept groundwater that causes springs or side-hill seeps. In such cases the drain is placed at the proper depth and location in relation to the seepage area.

For parallel subsurface systems, the principle factor affecting the cost is the spacing and depth of the drains. Many mathematical equations have been developed, but they are of limited practical value because of the difficulty and high cost of obtaining soil hydraulic conductivity data and because insufficient information is available on the drainage requirements of the crops to be grown. For these reasons depth and spacing of drains is based largely on experience and judgment (Schwab et al., 1966). Specific recommendations have been developed in each of the states where drainage is a major problem.

G. O. SCHWAB

References

Alpers, R. J., and Short, R. N., 1965. Clay draintile and its use, *in Drainage for Efficient Crop Production Conference (Chicago)*. St. Joseph, Mich.: American Society of Agricultural Engineering, 46–48.

Chow, V. T., 1959. *Open-channel Hydraulics.* New York: McGraw-Hill, 680p.

Ede, A. N., 1961. Floating-beam mole plough trails, *Farm. Mech.*, 13(144), 274–275.

Fouss, J. L., 1974. Drain tube materials and installation, *in* J. Van Schilfgaarde, ed., *Drainage for Agriculture.* Madison: American Society of Agronomy, Monograph 17, 147–177.

International Institute for Land Reclamation and Improvement, 1972–1975. *Drainage Principles and Applications,* Vol. I, II, III, IV. Wageningen: The Institute, 241, 374, 368, 476p.

Luthin, J. N., ed., 1957. *Drainage of Agricultural Lands.* Madison: American Society of Agronomy, Monograph 7, 620p.

Luthin, J. N., ed., 1973. *Drainage Engineering.* Huntington, N.Y.: Krieger, 250p.

Manson, P. W., 1971. Specifications for concrete drain tile, *Proc. Am. Soc. Agric. Eng. Conf. Nat. Drain. Symp.*, 29–31.

Richey, C. B.; Jacobson, P.; and Hall, C. W., eds., 1961. *Agricultural Engineers' Handbook.* New York: McGraw-Hill, 880p.

Schwab, G. O.; Frevert, R. K.; Edminster, T. W.; and Barnes, K. K., 1966. *Soil and Water Conservation Engineering.* New York: Wiley, 683p.

Sisson, D. R., ed., 1967. *Report on Drainage Materials, Soil and Water Division.* St. Joseph, Mich.: American Society of Agricultural Engineers, Special Publ. SP–01–67, 48p.

U.S. Soil Conservation Service Staff, 1973. *Drainage of Agricultural Land.* Port Washington, New York: Water Information Center, 430p.

Van Schilfgaarde, J., ed., 1974. *Drainage for Agriculture.* Madison: American Society of Agronomy, Monograph 17, 700p.

Willardson, L. S., 1974. Envelope materials, *in* J. Van Schilfgaarde, ed., *Drainage for Agriculture.* Madison: American Society of Agronomy. Monograph 17, 179–200.

Cross-references: *Conductivity, Hydraulic, Flow Theory; Irrigation; Leaching; Management of Soils; Moisture Management; Puddling; Salt Leaching Water Movement.*

SOIL DYNAMICS—See VOL. XIII: SOIL ENGINEERING

SOIL ENGINEERING—See VOL. XIII

SOIL EROSION—See WATER EROSION; WIND EROSION. VOL. IVA: SOIL EROSION

SOIL FERTILITY

The potential of a given site to produce is referred to as *productivity,* whereas the contribution of soil to productivity is called *fertility.* This is a comprehensive definition that covers physical, biological and chemical aspects of soils. A more restricted definition of soil fertility limits it to soil nutrient status. The more comprehensive definition is followed here, but discussions of aspects other than nutrient status will be brief. In accordance with this definition, the fertility of a virgin soil represents "its capacity to support the climax population of plants and animals" (Cooke, 1967), and the fertility of a soil under cultivation is its capacity to produce crops.

Soil Physics and Fertility

Texture. (See *Particle-size Distribution*) The *texture* of a soil affects productivity in several ways. Sandy soils are said to be "light" and clays "heavy." These terms came from the number of horses needed to pull a plow. Sandy soils are porous, have high infiltration rates, and retain little water. In contrast, clays have low infiltration rates, retain much water, and may be poorly drained. Specific surfaces and ion exchange capacities of sands are low as compared to clays. Aeration is good in sandy soils and fair to poor in clays, depending on their structure. Roots penetrate sands more easily than clays. Soils of intermediate textures such as loams are also intermediate in porosity, water retention, and drainage. The general tendency is for productivity to be better on medium-textured soils than on those that are either light or heavy (Foth and Turk, 1972).

Structure. The arrangement of elementary particles into aggregates is considered *structure,* which in combination with texture, governs porosity of the soil and thus affects water relations, aeration, root penetration, and the metabolic activities of soil flora and fauna. Modifications of structure are possible through soil management practices (see *Soil Structure*).

Water. A flow of water from the soil to the atmosphere through the plant is essential for its growth. The capacity of the soil to provide the necessary water is thus of major importance to fertility.

The state of water in soils is usually described by the water potential (which is equivalent to the chemical potential) relative to a standard state of pure free water at one bar pressure at a reference elevation. The potential is affected by the matrix, which gives a matric potential component, and by solutes, which give an osmotic component. The potential decreases as water is lost from the soil through transpiration by plants or by evaporation. The capacity of the soil to hold water gives a measure of how well the potential is buffered against reduction when water is lost (see *Water Content and Retention*). Soils such as sands have potentials readily subject to change and need frequent additions of water through rainfall or irrigation to be productive.

Aeration. The oxygen necessary for respiration of plant roots is largely provided by exchange of gases between the soil and the atmosphere. That exchange depends primarily on the volume-fraction of pores that drain rapidly after water is added by rainfall or irrigation. In addition to that exchange of gases, oxygen is also provided to roots through the interiors of many plants (see *Aeration, Respiration, Atmosphere*).

Whether aeration is good, indifferent, or poor greatly affects soil fertility. An adequate oxygen supply is necessary for respiration of roots. That affects plant metabolism which in turn affects the uptake of nutrients. Poor aeration can have harmful effects through the reducing conditions that follow. Solubilities of iron and manganese may reach toxic levels, nitrates can be reduced to nitrites or even to gaseous nitrogen and be lost completely, and sulfates can be reduced to hydrogen sulfide. Under certain conditions, decomposition of organic matter can be retarded sufficiently to produce peats and mucks. Conversely, when organic soils are drained and used for crops or pasture, aeration can be improved too much. If the water table is lowered too far, the organic residues are oxidized and disappear, as has happened in a number of places in the past (Russell, 1973).

Temperature. The effects of temperature on soil fertility are both direct and indirect. The metabolic rates of all organisms that use soil as a medium for growth are affected by temperatures. Low temperature reduces uptake of water and nutrients. Wilting of plants, for example, can be induced by irrigation with cold water. Germination of seeds, which has a temperature optimum, may be retarded by temperatures that are either too high or too low. Furthermore, low temperatures can also retard the growth of seedlings and thereby keep them susceptible for longer periods to attacks by pests. The maximum and minimum temperatures rather than the means tend to determine distribution of plants over the world through their effects on species survival. Release of nutrient elements by rock weathering affects fertility directly, release being faster under high than low temperatures.

Soil temperatures can be modified to some degree by management practices such as the use of mulches or the application of irrigation water.

Soil Biology and Fertility

Over and above the growth of plants themselves for crops, pasture or forest, biology has many effects on soil fertility. The plants being grown must compete with weeds and microorganisms. Various species of the latter participate in many reactions in the soil that can be either beneficial, neutral, or harmful to the plants being grown. Similarly, the fauna living in soils may also be harmful or beneficial to soil fertility.

Weeds interfere with the growth of desirable plants by competing for light, water, nutrients, and space. They may also damage crops by exuding toxic substances. They may harbor pests that attack crops and occasionally harbor insects that prey on the pests. Weeds may in some cases lead to unfavorable microclimatic conditions that result in frost damage.

Microorganisms (see *Soil Microbiology*) and soil fauna participate in transformation of many substances in soils. Decomposition of added organic residues is due in large measure to bacteria, fungi, actinomycetes, and soil fauna. The breaking down of the organic residues normally releases nutrients that can be used by plants, and it also gives rise to humus. In soils with high carbon-nitrogen ratios (q.v.), microorganisms compete with crop plants for nitrogen and other nutrients, the effects of which can be striking in semiarid regions. Quite a few species of bacteria are active in the nitrogen cycle (q.v.). Some fix atmospheric nitrogen and thereby incorporate it in the soil. Others convert ammonia to nitrite and nitrate and still others can reverse that reaction. Important fixers of atmospheric nitrogen are (a) free-living forms such as *Azotobacter* and some species of blue green algae and (b) bacteria of the *Rhizobium* genus that live in symbiosis with legumes. Still other forms of bacteria oxidize ferrous to ferric iron and convert sulfides to sulfates (see *Sulfur Transformations*).

Some microorganisms living in soils are pathogens and some microfauna are pests that feed on plants. The full set is large and includes bacteria, fungi, actinomycetes, nematodes, mites, and insects. Substantial losses of crops, forage plants, and forests are caused each year by these organisms.

Soil Chemistry and Fertility

Availability, Intensity-Extensity Relations, Potentials, and Capacities. The soil is a reservoir of essential mineral nutrients and occasionally of toxic chemicals, yet an elemental analysis has little indicative value for soil fertility. Soils may contain large quantities of a certain element that could fill the needs of vegetation for thousands of years, but plants growing on such soils may suffer from deficiencies of the nutrient element. Nutritional aspects of soil fertility are thus determined by the *status of nutrients* in soils, i.e., by their *availability* together with amount of nutrients present at a certain availability level, namely, the *capacity*. The status of nutrients is an *intensive* property; the expression for the capacity, an *extensive* property.

Schofield (1955) attempted to characterize the availability of nutrient elements by their chemical potentials. For phosphorus, he suggested the chemical potential of calcium phosphate in the p-form

$$\tfrac{1}{2}\ pCa + p\ H_2PO_4$$

where p is similar to the pH designation, the negative logarithm of the respective ion concentration or activity. Similarly, the availability of potassium was given by

$$pK - \tfrac{1}{2}\ pCa$$

These terms for the availabilities of phosphorus and potassium are useful only if soils of equal calcium levels are compared; otherwise, the differences in calcium will affect the quantitative expressions for the phosphorus and potassium availabilities.

Availability must refer to the nutrient uptake process. This makes availability a plant-oriented feature. Therefore, no soil property is likely to be a universal expression for the availability. For example, to maintain electroneutrality, potassium may be taken up in exchange for calcium, or in exchange for hydrogen, or together with NO_3 or together with SO_4, etc. Many or all of these processes occur simultaneously to various degrees. Each process, however, defines a different nutrient potential (q.v.). Hence it will hardly ever be possible to express availability by a single-valued term that depends only on soil properties. As an approximation, the availability can be taken as the concentration of nutrients in the soil solution. This criterion does not take into account nutrient interactions. Further problems arise from differing soil water contents.

The *extensity factor,* the capacity, is a measure of the quantity of nutrients present at a certain availability level. It can also be considered an expression of the rate of change of the intensity factor, ie., the potential, when nutrients are removed from the soil through plant uptake. Mathematically, it is the differential quotient

$$\left(\frac{\partial n_i}{\partial \mu_i}\right)_{I,\ P,\ nj,\ \ldots}$$

(*n* number of moles of component i, μ chemical potential of *i*). This tells how much of a nutrient can be removed from the soil before the potential suffers a decrease of one unit. The capacity concept, if applied to ionic nutrients, is as problematic as is the potential. Just as the nutrient concentration in the soil solution can serve as an approximation for the intensity factor, the sum of exchangeable and soluble ions can sometimes be used as an estimate for the nutrient capacity.

The difficulty of expressing availability of nutrients in quantitative terms is increased because it is usually not sufficient to describe soil fertility by intensity and extensity factors alone. Often consideration must be given to parameters affecting the transport of nutrients from the bulk of the soil solution (q.v.) to the root surface.

Soluble Nutrients. The soil is a three phase system consisting of solid, liquid, and gaseous phases. Nutrients in the liquid phase tend to be in equilibrium with those held in readily exchangeable form and with precipitates from which the dissolution process is rapid. Table 1 shows concentration ranges of ions in the soil solution. For some ions, in particular, nitrates (but not for total nitrogen), and for chlorides, the amounts in the soil solution represent the total reservoir available.

Sparingly Soluble Salts. Certain nutrients are present in the soil solution in such low concentrations that the total amount dissolved at any time will last for only a short period. For example, phosphorus concentrations are usually well below 1 ppm so that a crop contains 100 or more times the phosphorus content of the soil solution. In such cases, nutrients removed by plants from the liquid phase must be replaced continuously through dissolution from solid forms. For such sparingly soluble ions, the nutrient reservoir is thus present in a precipitated or solid form, and the rate of dissolution becomes an additional factor in soil fertility. Precipitates of sparingly soluble nutrients such

TABLE 1. Soil Solution Composition.
(After Fried and Broeshart 1967)

Element	Range	Acid Soils	Calcareous Soils
Ca	0.5–38*	3.4	14
Mg	0.7–100	1.9	7
K	0.2–10	0.7	1
Na	0.4–150	1.0	29
N	0.16–55	12.1	13
P	0.001–1	0.007	0.03
S	0.1–150	0.5	24
Cl	0.2–230	1.1	20

*Micromoles per liter

as phosphates usually occur in several different solid forms.

Exchangeable Cations. Cations are adsorbed on negatively charged surfaces of clay minerals and by negatively charged functional groups in soil organic matter. The negative charges on clay minerals originate from electrically unbalanced lattices and from the dissociation of H^+ ions at broken edges of minerals. A cation held at such a surface is called *exchangeable.* The total amount of charges that can be held in exchangeable form is called the *cation exchange capacity* (see *Exchange Phenomena*). It is often defined and determined at pH 7.

Many rules and laws have been stated and formulated to describe the ion exchange process (Wiklander, 1964). There is a tendency for ions to follow the Hofmeister series:

$$\mathrm{Li} < \mathrm{Na} < \mathrm{K} < \mathrm{Rb} < \mathrm{Cs}$$
$$\mathrm{Mg} < \mathrm{Ca} < \mathrm{Sr} < \mathrm{Ba}$$

There are exceptions to this rule depending on the exchanging substance. It is now assumed that cations are held as a diffuse ion swarm on clay surfaces. The distribution of positively charged ions may obey a Boltzman relation,

$$c = c_0 e^{-ze\Delta E/kT}$$

where c_0 is the equilibrium concentration in the unaffected soil solution, c the concentration where the electrical potential is ΔE with respect to the equilibrium solution, z the valency of ions, e the elementary charge, k the Boltzmann constant, and T the absolute temperature. Negatively charged surfaces attract cations and repel anions. Polyvalent cations are adsorbed more strongly than monovalent cations.

The *ratio law* (Schofield, 1955) states that the equilibrium between ions in the soil solution and exchangeable ions is not disturbed if in the soil solution all monovalent ions are changed in one ratio and all divalent ions in the square and all trivalent ions in the cube of that ratio (see *Activity Ratio*).

Numerous attempts have been made to formulate quantitative laws of ion exchange. *Donnan equilibrium* considerations for the exchange between mono- and divalent ions give

$$\frac{M_{ad}^{+}}{M^{+}_{solution}} = K_1 \sqrt{\frac{M_{ad}^{2+}}{M^{2+}_{solution}}}$$

where M stands for cations and ad for adsorbed, and K_1 is a constant. The difficulty is to find satisfactory expressions for the activities of adsorbed ions. Model considerations by Gapon (1933) and others give

$$\frac{M_{ad}^{+}}{M^{+}_{solution}} = K_2 \frac{M_{ad}^{2+}}{\sqrt{M^{2+}_{solution}}}$$

Based on the assumption that the activities of adsorbed ions can be expressed by their mole fractions, Vanselow (1932) and others derived the relation

$$\frac{M_{ad}^{+}}{M^{+}_{solution}} = K_3 \sqrt{\frac{M_{ad}^{2+}(M_{ad}^{2+} + M_{ad}^{+})}{M^{2+}_{solution}}}$$

Over limited ranges of concentrations, ion exchange equations are useful to predict soluble ion concentrations. In this way, they assume a role in describing fertility relations. The process of ion exchange is fundamental for soil fertility. Large quantities of nutrients can thereby be kept in soils in available form without causing salinity. The nature of exchangeable ions has other effects besides its role in plant nutrition. Clay platelets having much exchangeable sodium are often highly dispersed, causing poor soil structure, which in turn is a major cause of infertility in alkali soils.

Anion Exchange. Anions are repelled by negatively charged surfaces. This phenomenon is called *negative adsorption* (see *Exchange Phenomena*). However, soils also exhibit anion exchange capacities. Some anions (nitrates, chlorides) are adsorbed weakly or not at all, whereas phosphates interact strongly and sulfates take an intermediate position.

Anion exchange occurs at positively charged sites resulting from the dissociation of OH ions from clay minerals and from iron and aluminum hydroxides as well as from positively charged NH_2^+ and NH_3^+ groups of the soil organic matter. Anion exchange occurs predominantly at low pH values (Wiklander, 1964).

Immobilized, Fixed, and Difficulty Available Forms. In many soils large quantities of nutrient elements persist in forms originally present in soil parent materials. From these forms they become slowly available through decomposition of minerals. This process is too slow to furnish adequate nutrients for crop production.

Nutrients can become *immobilized* through incorporation in soil organic matter. Microorganisms, in the process of synthesizing new substances, use mineral nutrients that are then changed to organic forms. These nutrients can again become available only after renewed decomposition of the organic matter by microorganisms, a process called *mobilization* (see *Nitrogen Cycle*). Substantial quantities of nitrogen equivalent to crop needs for many years are immobilized in organic matter in many soils. Organic matter thus represents a valuable storage form for nitrogen from which little or no loss through leaching occurs (see *Nutrient Losses*). Whether a net immobilization of inorganic nitrogen occurs depends on the carbon-nitrogen ratio (C/N ratio) of the organic matter. With high ratios, >30, the nitrogen supplied to microorganisms directly by organic matter is inadequate; further nitrogen sources are tapped. This leads to significant depletions of inorganic nitrogen (NH_4^+, NO_3^-). Conversely, organic matter with a low C/N ratio, <20, contains more nitrogen than necessary for microorganisms, and some nitrogen is freed upon decomposition to become available to higher plants. Addition to soils of organic matter with low nitrogen contents (e.g., straw) can cause immobilization of nitrogen.

Some nutrients react with soil constituents so as to become less available. This process is referred to as *fixation* (the terms *fixation* and *immobilization* are often used interchangeably). Potassium and ammonium ions are fixed between layers of illitic and other clay minerals. Phosphates are fixed through reactions with calcium, aluminum, and iron compounds (see *Phosphorus Cycle*). Fixation is also of occasional importance with most micronutrients (q.v.). Fixation usually does not render the nutrients completely unavailable; transitions with various degrees of decreased availability are common.

pH-Fertility Relations

Plant communities on basic soils are strikingly different from those on acid soils. This association could be due to the different hydrogen and OH ion concentrations that interfere with plant metabolism, but most effects of differing acidity result from changes in physical, chemical, and biological conditions of soils. Consequently, pH is one of the most valuable indicators of soil fertility.

The pH is a measure of the hydrogen ion concentration (H^+) or more precisely, of hydrogen ion activity (H^+). It is defined as

$$pH = -\log(H^+)$$

Practically all soils fall in a pH range of very strongly acid (pH 3) to moderately basic (pH 9), with rare extremes as low as 2 and as high as 11 (see *Acidity; Soil Reaction*. Vol. IVB: *Coastal Soils*).

At extremely low pH values (below about pH 4), direct plant injury through interactions with hydrogen ions is suspected. For example, hydrogen ions could interfere by competing in plant uptake with essential nutrient elements such as potassium.

Levels of nutrient cations (calcium, magnesium, and potassium) are also characteristically low in acid soils. Most such soils have been strongly leached and are of low fertility generally. The degree of base saturation (q.v.) is also low, possibly below critical levels required for optimum crop production. Indications are that yields of sensitive crops such as alfalfa are reduced on some soils if base saturation is below 70%. Penetration of the soil by plant roots is also restricted by very low levels of calcium.

The availability (solubility) of many nutrients is a function of pH. In moderately to strongly acid soils, phosphates become less available through interactions with iron and aluminum hydroxides. At high pH values, phosphates become increasingly insoluble because of interactions with calcium to form apatites and other precipitates. Consequently, the highest solubility of phosphorus is found around pH 6. The solubility of many other nutrients is also affected by pH. Most become more soluble (available) under acid conditions (zinc, copper, manganese, iron, and boron). Molybdenum is an exception; its availability increases with increasing pH. In strongly acid soils concentrations of aluminum, manganese, and iron can reach toxic levels. In these situations aluminum is believed to be the dominating exchangeable cation. Aluminum causes injuries to roots, and relatively little is translocated to plant tops. It interferes with the nutrition of many other elements, in particular, phosphorus, and also induces wilting of intact plants. On the other hand, it improves the keeping quality of cut flowers, apparently by inducing partial closure of stomata. Occasionally concentrations of manganese reach toxic levels, mainly below pH 5. In contrast to aluminum, manganese is readily translocated into leaves. Symptoms of manganese toxicity often resemble those of iron deficiency. Because root growth is reduced through aluminum and manganese toxicity, unfavorable nutritional conditions prevail in acid soils not only because of reduced nutrient availabilities but also because plants exploit smaller volumes of soil.

The stability of soil aggregates is reduced in acid soils. This deterioration of structure, which is at least in part due to low calcium levels, results in poor permeabilities for water and gases. In sodic alkali soils—usually correlated with high pH—structural conditions are even less satisfactory. Alkali soils can become practically impervious to water.

Many biological processes depend on the soil pH. Nitrogen fixation (q.v.) and nitrification are largely inhibited at low pH values. Bacteria dominate the soil flora at neutral and basic pH values, whereas the relative importance of fungi increases in acid soils. Adjustment of soil pH can even be used to control plant diseases. Potato scab, caused by an actinomycete, is less severe under acid conditions.

Factors Favoring Acidity. Acid igneous parent material, humid climates with heavy leaching, cool climates, and detritus of certain plant genera (e.g., *Picea* and *Vaccinium*) favor the development of acid soils. Limestone as parent material and arid climates have the opposite effect. In modern agriculture fertility management greatly affects soil reaction. Fertilizer salts taken up predominantly as cations or nitrogen fertilizers that are subsequently nitrified to nitrate lower the pH. The fertility of strongly acid or basic soils can be improved through suitable corrective measures. Increases in pH are obtained through liming, decreases through application of sulfur, sulfuric acid, and suitable fertilizers.

Salinity and Alkalinity Factors

Salinity is a major factor in soil fertility of many irrigated areas in arid regions. In humid regions local problems of salinity may result from salt applications to ice- or snow-covered highways. Excessive local applications of fertilizers have also caused salt injuries to plants.

The salt contents of irrigation waters range from very low values to several thousand parts per million (ppm). Because annual water needs in irrigation can be as high as 1.5 m, many tons of salts may be added annually to each hectare. Such large additions combined with insufficient leaching can lead to rapid salinization, especially if the farmer does not realize the need for adequate drainage. The water table with its salts can soon rise into the root zone (see *Irrigation Water*).

Saline and nonsaline soils can be distinguished on the basis of electrical conductivity of the saturation extract. If conductivity is higher than 4 mmhos/cm, the soil is considered saline. If lower, the soil is nonsaline. In an *alkali* soil, more than 15% of the cation exchange capacity is saturated with alkali ions, most often sodium.

On the average, the electrical conductivity (EC) is related to the osmotic potential Ψ_π in bars (approximately equivalent to the negative of the osmotic pressure Π or to the negative of the solute suction) by

$$\Psi_\pi = -0.365 \text{ EC}$$

and to the solute concentration c in milligram-equivalents by

$$c = 10.37(\text{EC})^{1.065}$$

In reality, large deviations from this general rule can be expected.

Plant Responses. Salinity can first cause general growth reduction and then in more severe situations cause foliar necrosis starting from leaf tip and marginal areas. Based on yield reductions (by 10, 25, or 50%), tolerance classifications have been established by the U.S. Salinity Laboratory (Table 2). Among fruit crops, date palms are highly tolerant; pomegranate, fig, olive, grape and cantaloup have medium tolerance; pear, apple, orange, grapefruit, prune, plum, almond, apricot, peach, strawberry, lemon and avocado show low salt tolerance. (Bernstein, 1964)

At least part of the salinity injury can be attributed to unfavorable water relations. The hazard of alkali (sodic) soils often lies more in unfavorable chemical and physical conditions than in excess sodium. Usually growth is inhibited and plants are stunted, but the foliage is often dark green (although chlorotic leaves can also be observed). Sodic soils tend to be dispersed (deflocculated) so that primary soil particles are separated from each other and aggregation is poor. Water penetration is smaller by several orders of magnitude than in soils low in sodium.

The pH values of sodic soils are frequently high to give unfavorable chemical conditions as well. Concentrations of carbonate ions reach levels sufficiently high to induce calcium deficiency through precipitation of calcium carbonate. At pH values above 10.5, direct injuries from high OH-ion levels are suspected, and even aluminum toxicity is possible. Iron, zinc, and phosphorus deficiencies are not uncommon. The oxidation of nitrite to nitrate is inhibited preferentially to that of ammonium to nitrite, leading to temporary accumulations at toxic levels of nitrites after fertilization with ammonium-nitrogen.

Management. Management practices to correct salinity and alkalinity are of prime importance to irrigated agriculture in warm, arid regions. As a rule, excess water must be added above that required for consumptive use by plants. Formulas to estimate leaching requirements are available (see *Leaching*). The rise of the water table into the root zone must be prevented through adequate drainage.

Many existing saline soils can be improved by leaching (see *Leaching, Salt Leaching*). Special

TABLE 2. Salt Tolerance of Crops.

EC* with 50% Yield Reduction	Field Crops	Vegatable Crops	Forage Crops
19			
18			Crested and tall wheat grass and Bermuda grass
	Barley		
17			
16	Sugar beet, cotton		
15			
14	Safflower, wheat		Tall fescue
13			Barley hay
12	Sorghum		Perennial rye, Harding grass
		Beets	
11			Beardless wildrye
10	Soybean, sesbania		Birdsfoot trefoil
9			
	Rice		
8		Spinach, tomato, broccoli	Alfalfa, orchard grass
7	Corn	Cabbage	
6	Broadbean, flax	Potato, corn, sweetpotato	Meadow foxtail
5		Letticue, bell pepper	
4		Onion, carrot	Clover, alsike and red
3	Bean	Bean	
2			
1			

*Electrical conductivity.

care is required in saline-alkali soils, however, because removal of the soluble salts without reducing the exchangeable sodium percentage leads to highly undesirable soil structure. The sodium saturation percentage can be lowered before leaching through addition of calcium salts (gypsum) or through acid formers (sulfur, H_2SO_4, iron sulfate and organic matter).

Economic losses can sometimes be avoided or at least minimized by management practices in crop production on saline soils. Thus, for example, salt-tolerant crops can be grown. Planting of a crop can also avoid soils with the highest salinity. Saline conditions are usually patchy or spotty, as, for example, the concentrations of salt due to capillary rise at the tops of ridges between irrigation furrows. Planting can then be on the sides of the ridges or furrows.

Special attention must be paid to the quality of irrigation water (q.v.). Amendments of gypsum are necessary if the concentration of sodium relative to those of calcium and magnesium is too high. The sodium-adsorption ratio (SAR) is used as a measure of water quality with respect to the hazard of developing alkali soils (see *Irrigation Water; Salt-Leaching*). SAR values below 10 are adequate under nonsaline conditions, whereas SAR values above 18 are considered high enough so that chemical amendments are required. These tolerance limits decrease with increasing salinity (Richards, 1953). The salinity hazard is assessed by the electrical conductivity of irrigation water. A conductivity below 0.25 mmhos/cm is low and one above 750 mmhos/cm is high. In the latter case, only plants of high salt tolerance can be grown, and special management procedures to control salinity are also required.

Toxic Elements

Almost every element is toxic to plants if present in large enough amounts in the soil. Toxic levels of elements in soils are due to natural sources in some instances and to human activity in others (see *Pollution*).

Examples of more commonly occurring toxic elements are mentioned in this paragraph. Boron toxicity is a common local problem in arid and semiarid regions. Boron concentrations in the soil solution of 1 ppm can be harmful. Irrigation waters from wells that penetrate marine deposits are often high enough in boron to cause problems. One or more of aluminum, iron, and manganese may be toxic in strongly acid soils. Selenium can be toxic to animals in semiarid regions where the element has been concentrated in soils by certain plants, e.g., species of *Astragalus, Xylorrhiza, Oonopisi,* and *Stanleya,* to as much or more than 1%. Animals grazing on such soils suffer from what is called "alkali" disease. Concentrations of molybdenum are occasionally high enough to cause what is called "teart disease." Nickel toxicity, and less frequently chromium toxicity, contribute to the low productivity of soils derived from serpentines. Low phosphorus-molybdenum and high magnesium-calcium ratios also contribute to the low fertility of such soils.

Toxic concentrations of some elements originate in soils from human activity. Repeated applications of Bordeaux sprays as fungicides have resulted in copper toxicity in places. Similarly, arsenic toxities have followed continued application of arsenical sprays and dusts. Heavy metals such as lead, cadmium, and zinc have been added to soils near smelters, whereas fluorine toxicities have been reported in some instances near installations for processing aluminum, iron, and phosphate ores.

Nutrient Cycles

In an established, natural soil-plant system, nutrients move through cycles (see *Carbon Cycle; Nitrogen Cycle; Phosphorus Cycle; Sulfur Transformations*). The elements are removed from the soil by plant roots, a large fraction is transported to above-ground parts and deposited in leaves from whence some nutrients are remobilized and retranslocated to other parts of the plant, with some even going back to the soil at that stage. Additional nutrients are returned to the soil through leaching (by washing out of substances by rain and by falling guttation drops). Substantial quantities are returned to the soil surface in dead plant remains such as annual leaf fall, from which nutrients may be leached or liberated by decay. Through resynthesis of organic substances and through decomposition again, nutrient elements are kept in a never-ending secondary cycle (Figure 1). Nutrients are also moved about within the soil by water and by animals such as earthworms.

Contents of nutrient stores in different parts of such a cycle differ greatly with the type of vegetation. For example, the amount of nitrogen stored in a natural grassland soil (Mollisol-Udoll) exceeds many times the nitrogen content in plants growing on these sites whereas in a tropical rain forest the bulk of nutrients is contained in the living trees, and soil storage is minimal.

Soil-plant communities are open systems. Nutrients are lost permanently through runoff, leaching, and biochemical processes (see *Nutrient Losses*); gains result from decomposition of rocks, nitrification through microorganisms and lightning, and influx from other communities. In crop communities, some of these net gains and losses are much enhanced; for example, large quantities of nutrient elements are removed in harvested crops. If natural supply sources are insufficient, compensation through commercial fertilizers is necessary. Fried and Broeshart (1967) have collected extensive data about gains and losses of nturients. Crop removals of major elements (see *Fertilizers, Inorganic*) are 50 to several hundred kg/ha of nitrogen, 10 to 50 kg/ha of phosphorus, and 50 to several hundred kg/ha of potassium.

Leaching losses from soils depend, among other things, on climatic conditions, especially rainfall, and the forms of nutrients in soils. For example, annual losses of nitrogen range from essentially nil to a few hundred kilograms per hectare. Gains of nutrients also have a wide range; nitrogen fixation by root-nodule bacteria, for example, range from a few to many hundred kilograms per hectare per year. Additions of nitrogen in precipitation seem to be but a few kilograms per hectare per year, generally negligible as compared to other amounts in the nitrogen cycle (q.v.). In contrast, sulfur additions from the atmosphere may exceed crop needs in industrialized areas. Additions of nutrient elements in airborne dust are also important in some parts of the world such as the portion of Africa just south of the Sahara.

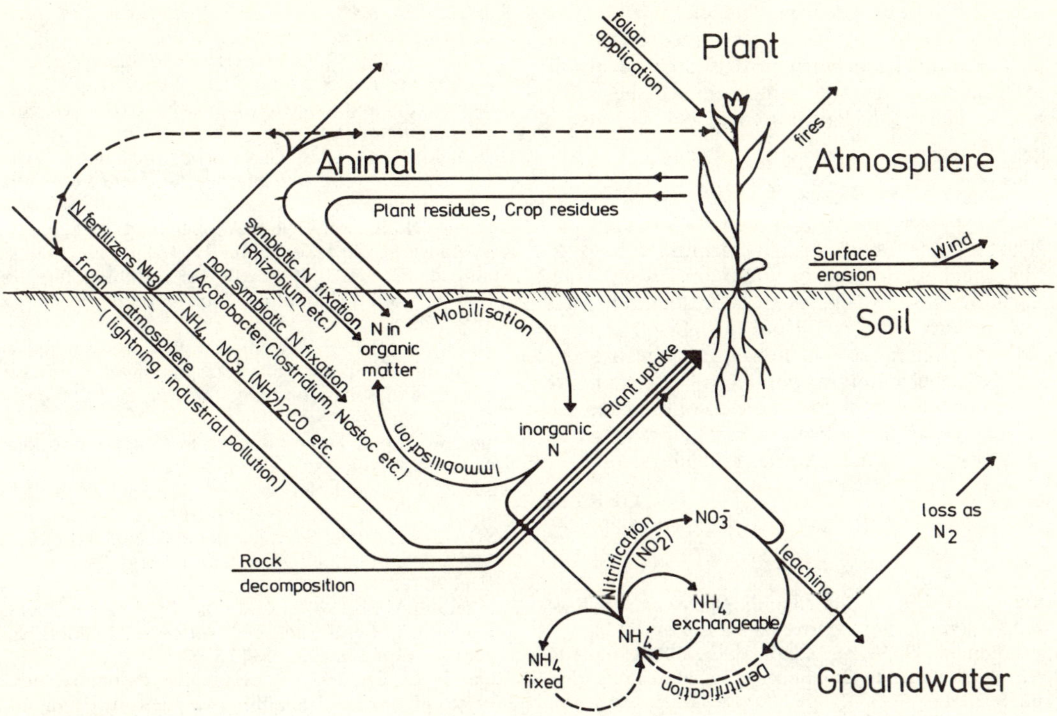

FIGURE 1. Nitrogen cycle.

Theoretical Aspects

Numerous attempts have been made to formulate laws and rules governing the dependence of yield (production) on various growth factors. Knowledge of such quantitative relations would be an immense help in fertility management.

Liebig's law of the minimum (anticipated by de Saussure) states that the rate of a process (such as growth) is limited by the factor at its minimum (see *Law of the Minimum*). The law was originally applied only to mineral nutrients, but it is equally useful for other growth factors such as light and atmospheric CO_2. The response to an improvement of a growth factor is not necessarily linear. A saturation response, e.g., can be expressed by a hyperbola

$$y = \frac{k_1 F}{1 + k_2 F}$$

where y represents the yield, F the growth factor at a minimum, and k_1 and k_2 constants, which may themselves be functions of other growth factors. (Russell, 1973)

Mitscherlich (also anticipated by another scientist, Wollny) states that improvement of a growth factor increases the yield in proportion to the distance from the maximum yield. In mathematical terms it is

$$\frac{dy}{dF} - k_3(y_{max} - y)$$

where y_{max} represents the maximum yield, and k_3 is a constant for which quantitative values are available.

Usually more than one factor is at a minimum, and the yield becomes

$$y = f(F_1, F_2, \ldots)$$

An attempt to arrive at a quantitative expression for such multifactorial effects has been made by *Baule* (Russell, 1973). A Baule unit is the amount of a factor required to produce 50% of the yield. Further yield increases per Baule unit are assumed to occur in a geometric fashion; i.e., a Baule unit will increase the yield by half the distance from the maximum yield. The simultaneous action of several factors is described by the Baule rule, which states that the fractional yields of each individual factor must be multiplied to give the total yield. Thus if the yield with only factor A at a minimum is, e.g., two thirds of the yield under the best conditions, the yield with only factor B at the minimum is three fourths of the maximum, then with both factors A and B simultaneously at the minimum at the same levels as above, the combined effect will give a yield of

$$Y = 2/3 \times 3/4 = 1/2$$

The success of efforts to formulate laws to show the dependence of yield on growth factors such as quantities of nutrient elements has been modest thus far. Some progress has been made, but far more will be required before any "law" can be widely applied. The efforts made in the past, however, do serve to emphasize the complexity of the soil-plant system, even when nutrient status alone is under consideration.

J. J. OERTLI

References

Allison, F. E., 1973. *Soil Organic Matter and Its Role in Crop Production.* Amsterdam: Elsevier, 637p.

Bartholomew, W. V., and Clark, F. E., eds., 1965. *Soil Nitrogen.* Madison: Soil Science Society of America, 615p.

Bernstein, L., 1964. Salt tolerance of plants, *U.S. Dept. Agric. Bull. No. 283,* 23p.

Black, C. A., 1968. *Soil-Plant Relationships,* New York: Wiley, 792p.

Cooke, G. W., 1967. *The Control of Soil Fertility.* New York: Hafner, 526p.

Finck, A., 1969. *Pflanzenernährung in Stichworten.* Kiel: Hirt, 200p.

Foth, H. D., and Turk, L. M., 1972. *Fundamentals of Soil Science.* New York: Wiley, 454p.

Fried, M., and Broeshart, H., 1967. *The Soil-Plant System in Relation to Inorganic Nutrition.* New York: Academic, 358p.

Gapon, E. N., 1933. Theory of exchange adsorption in soils, *Jour. General Chem. (USSR),* **3,** 144–152; **3,** 153–158. (In Russian)

Gauch, H. G., 1972. *Inorganic Plant Nutrition.* Stroudsburg, Pa.: Dowden, Hutchinson & Ross, 488p.

Kanwar, J. S.; Datta, N. P.; Bains, S. S.; Bhumbla, D. R.; and Biswar, T. D., eds., 1971. *Proc. Internat. Symposium on Soil Fertility Evaluation.* New Delhi: Indian Agricultural Research Institute, Vol. 1, 1117p.

Kilmer, V. J.; Younts, S. E.; and Brady, N. C., eds., 1968. *The Role of Potassium in Agriculture.* Madison: American Society of Agronomy, 509p.

Mengel, K., 1972. *Ernährung und Stoffwechsel der Pflanze.* Stuttgart: Fischer, 470p.

Mortvedt, J. J.; Giordano, P. M.; and Lindsay, W. L., eds., 1972. *Micronutrients in Agriculture.* Madison: Soil Science Society of America, 666p.

Pierre, W. H., and Norman, A. G., eds., 1953. *Soil and Fertilizer Phosphorus in Crop Nutrition.* New York: Academic, 492p.

Richards, L. A., ed., 1953. Diagnosis and improvement of saline and alkali soils, *U. S. Department of Agriculture Handbook No. 60,* 160p.

Russell, E. W., 1973. *Soil Conditions and Plant Growth.* London: Longmans, 849p.

Sanchez, P. A., 1976. *Properties and Management of Soils in the Tropics.* New York: Wiley, 618p.

Scheffer, F., and Schachtschabel, P., 1970. *Lehrbuch der Bodenkunde.* Stuttgart: Enke, 349p.

Schofield, R. K., 1955. Can a precise meaning be given to "available" soil phosphorus? *Soils and Fertilizers,* **18,** 373–375.

Sprague, H. B., ed., 1964. *Hunger Signs in Crops: A Symposium.* New York: McKay, 461p.

Stanford, G., 1973. Rationale for optimum fertilization in corn production, *Jour. Environ. Qual.,* **2,** 159–166.

Stewart, R., 1932. The Mitscherlich, Wiessman, and Neubauer methods of determining the nutrient contents of soils, *Commonwlth. Bur. Soil Sci., Tech. Comm. No. 25,* 46p.

Stiles, W., 1961. *Trace Elements in Plants.* Cambridge: Cambridge University. Press, 249p.

Tisdale, S. L., and Nelson, W. L., 1966. *Soil Fertility and Fertilizers.* New York: Macmillan, 694p.

USDA Staff, 1957. *Soil: The Yearbook of Agriculture 1957.* Washington, D.C.: Government Printing Office, 784p.

Vanselow, A. P., 1932. Equilibria of the base-exchange reactions of bentonites, permutites, soil colloids, and zeolites, *Soil Sci.,* **33,** 95–113.

Viets, F. G., Jr., 1977. A perspective on two centuries of progress in soil fertility and plant nutrition, *Soil Sci. Soc. America Jour.,* **41**(2), 242–249.

Walsh, L. M., and Beaton, J. D., 1973. *Soil Testing and Plant Analysis.* Madison: American Society of Agronomy, 621p.

Wiklander, L., 1964. Cation and anion exchange phenomena, *in* F. E. Bear, ed., *Chemistry of the Soil.* New York: Reinhold, 162–205.

Cross-references: *Agronomy; Carbon Cycle; Edophology; Fertilizers, Inorganic; Fertilizers, Organic; Law of the Minimum; Management of Soils; Macronutrients; Micronutrients; Phosphorus Cycle; Nitrogen Cycle; Nutrient Losses; Nutrient Potential; Plant Nutrients; Sulfur Transformations.*

SOIL MANAGEMENT—See MANAGEMENT OF SOILS

SOIL MECHANICS

That branch of applied mechanics that deals with the mechanical behavior of soils and other particulate matter in response to applied mechanical and hydraulic loads is referred to as *soil mechanics* (Lambe and Whitman, 1969).

Soil, as defined in civil engineering, includes all unconsolidated accumulations of rock fragments and any organic matter that may be intermixed. The equivalent geologic term is *regolith.* Boulders, gravel, sand, silt, clay, marl, mud, volcanic ash, marine ooze, and peat are all considered soils in the engineering sense. In pedology and geology the term *soil* is restricted to those unconsolidated materials at or near the surface that have been modified by mechanical, chemical, and biological agents to the extent that they will support rooted plants (see *Soil*).

Civil engineers have developed soil mechanics for the purpose of designing building foundations, retaining walls, earth dams, and other structures in which soils play a part; however, earth scientists in many other disciplines can make effective use of the principles of soil mechanics for the analysis of such problems as faulting, down-slope movements, and compaction of sediments.

A complete study of any problem in soil mechanics should ideally consist of four phases: subsoil exploration, determination of soil properties, theoretical prediction of soil behavior under applied loads, and observation of actual soil behavior. The thoroughness with which each phase is carried out in a particular investigation depends on the purpose for which the investigation is being made and the amount of money that can be reasonably spent on the investigation. Succeeding paragraphs treat the four phases in some detail.

Subsoil Exploration

Most investigations of problems in soil mechanics are complicated by the heterogeneity of the soil deposits involved. A first step in the analysis of a particular problem must therefore be exploration of the subsoil conditions at the site. Field geologic reconnaissance, airphoto study, and geophysical surveys often provide significant information about the general character of the soil deposits, but detailed information can be obtained only by making borings, test pits, and test shafts, from which soil samples can be recovered or in which the soil can be examined in situ.

Borings are the most common means of exploration. Their number and location should not be established in detail before the start of exploration. Study of a few preliminary borings will generally indicate where subsequent borings should be made to delineate most effectively the important heterogeneities in the subsoils. Because of the inevitable uncertainties, which even the most elaborate practical exploration program cannot diminish, subsoil conditions must be continuously studied as they may later be revealed (for example, by excavation during construction). Terzaghi and Peck (1967) and the U.S. Bureau of Reclamation (1974) have summarized the basic techniques of subsurface exploration.

Two types of samples can be taken, disturbed and undisturbed. *Disturbed samples* are usually taken by driving a thick-walled pipe into the soil at the bottom of a borehole. A qualitative classification of the soil recovered in such disturbed samples may suffice for the purposes of many investigations. In more important investigations it is necessary to measure the physical properties of the soil such as compressibility and shear strength, and for this purpose undisturbed samples are required. *Undisturbed samples* are usually taken in thin-walled tubes using special care to minimize disturbance of the soil structure that could result in a significant change of the physical properties.

Soil Properties

Soil properties can be divided into two broad groups, engineering properties and index properties. *Engineering properties* are the physical properties that determine a soil's response to applied load (Jumikis, 1962). They are:

1. *Permeability*, the resistance to flow of water through the void space.
2. *Compressibility*, the resistance to change in volume under applied load (which is called *consolidation* in soil mechanics and *compaction* in geology).
3. *Deformability and shear strength*, the resistance to change in shape and to rupture, with or without volume change, under applied shear load.

Index properties are the physical properties used to describe soil qualitatively and to classify it empirically into broad groups having similar engineering properties. They include:

1. *Water content*, weight of water in a given soil mass divided by the weight of solids.
2. *Dry unit weight*, weight of solids in a given soil mass divided by the total volume of the mass.
3. *Void ratio*, volume of voids in a given soil mass divided by the total volume of the solids.
4. *Porosity*, volume of voids in a given soil mass divided by the total volume of the mass.
5. *Grain-size distribution*, a graph of the percent by weight of grains smaller than a given diameter versus grain diameter (Figure 1).
6. *Atterberg limits:*
   Liquid limit, water content corresponding to an arbitrarily defined boundary between the semiliquid and plastic states.
   Plastic limit, water content corresponding to an arbitrarily defined boundary between the plastic and semisolid states.
   Plasticity index, liquid limit minus plastic limit.

Engineering properties are usually expensive and time-consuming to determine, especially since they must be measured in situ or else on undisturbed samples that are tested in the laboratory. It is impractical and uneconomical to attempt to measure the engineering proper-

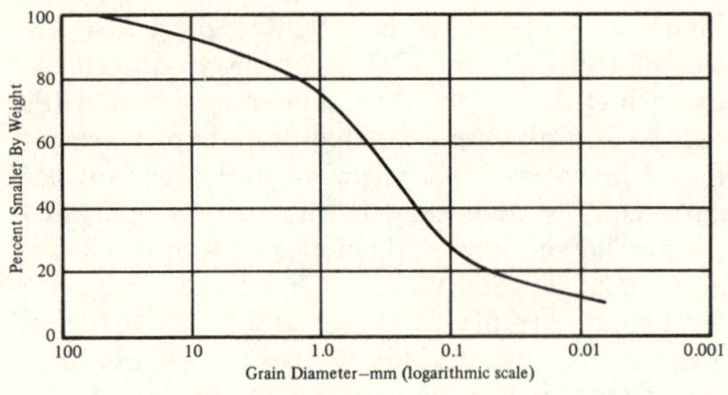

FIGURE 1. Grain-size distribution curve.

ties everywhere throughout a large mass of soil. Index properties, which can be measured more cheaply and easily, can be used for making empirical qualitative estimates of the engineering properties and for selecting the most critical subsoil zones for more intensive study of the engineering properties.

An essential part of any soil investigation is soil classification (q.v. in Vol. XII), which is based primarily on the index properties (Lee, 1968). Of the many soil classification systems used by engineers, geologists, and pedologists, the *Unified Soil Classification System* is best suited for conveying significant information about the engineering properties. In this system soils are divided into three groups: coarse-grained, fine-gained, and highly organic.

The *coarse-grained* soils are cobbles, gravel, and sand. They are cohesionless (see *Soil Structure*), i.e., when they are dried, the individual grains do not stick together. They are classified primarily on the basis of grain size, as follows:

1. *Cobbles,* retained on 3-in. sieve.
2. *Gravel,* passes 3-in. sieve and is retained on No. 4 sieve (4.76 mm).
3. *Sand,* passes No. 4 sieve (4.76 mm) and is retained on No. 200 sieve (0.074 mm).

The *fine-grained* soils are silt and clay. They are cohesive, i.e., when they are dried, the individual grains stick together in lumps. They are the soils that pass the No. 200 sieve, the distinction between silt and clay being made on the basis of the plasticity characteristics (Atterberg Limits). Fine-grained soils may be completely inorganic, or they may contain significant amounts of organic matter.

The *highly organic* soils such as peat consist primarily of partially decayed organic matter. Calcerous or siliceous plant and animal remains, although they are of organic origin, are not considered highly organic soils in this system. Organic silts and clays are classified as fine-grained rather than as highly organic, even though the organic matter they contain may have a predominant influence on their engineering properties.

Terzaghi and Peck (1967) and Sowers and Sowers (1970) give a thorough exposition of the use of index and engineering properties in soil mechanics; ASTM (1976) describes the laboratory tests for measuring the properties; and the U.S. Bureau of Reclamation (1974) describes the Unified Soil Classification System. A more detailed discussion of the engineering properties is given in the following paragraphs.

Permeability

Permeability (q.v.) is the resistance of a soil to the flow of water through its voids. A typical laboratory setup for measuring permeability is shown schematically in Figure 2. In this so-called constant-head permeameter, the heads (elevations of the free water surfaces) at the inlet and outlet ends of the specimen are maintained constant. Assuming laminar flow, which is almost always a reasonable assumption for the flow of water through soils of sand-size and smaller, the rate of flow can be expressed by the formula

$$q = kia$$

where

q = rate of flow (cc/s)
k = coefficient of permeability (cm/s)
$i = H/L$ = hydraulic gradient
a = cross-section area of soil specimen perpendicular to direction of flow (cm^2)

The *coefficient of permeability k* is the discharge velocity under unit hydraulic gradient, where discharge velocity is a fictitious velocity defined as the rate of flow divided by the gross cross-section area through which flow occurs.

Approximate ranges of the coefficient of permeability for some general soil types are: clean gravel (no silt or clay), 100–1 cm/s; clean sands (no silt or clay), 1–10^{-3} cm/s; very fine sands, silts, mixtures of sand, silt, and clay, 10^{-3}–10^{-7} cm/s; and clays, 10^{-7}–10^{-9} cm/s.

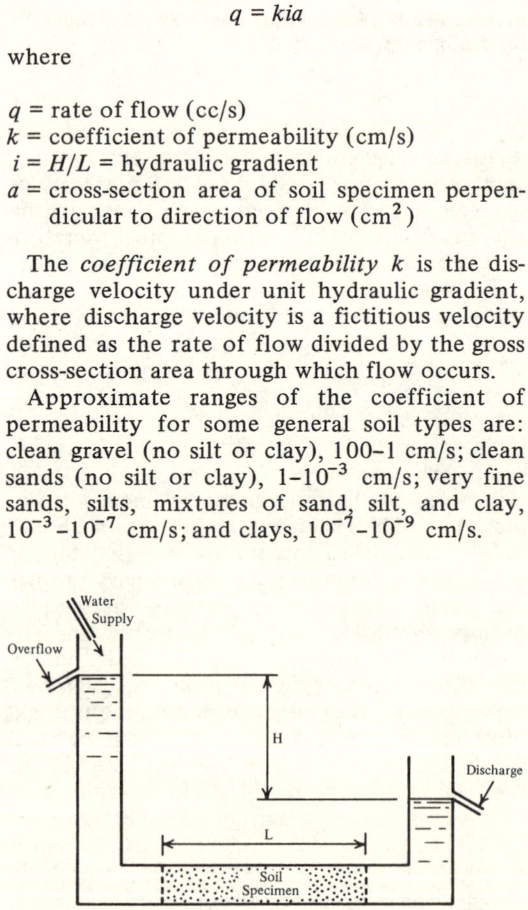

FIGURE 2. Laboratory permeability apparatus.

Compressibility

Consolidation (or *compaction* as it is called in geology) is the decrease in volume of a soil when the stresses carried by the soil skeleton are increased; *swelling* is the increase in volume when the stresses are decreased. Because the compressibility of water and individual soil grains is negligible compared to the compressibility of the soil skeleton, consolidation of a saturated soil can occur only if water is squeezed out of the soil.

The rate at which water is squeezed out of the soil is influenced by the permeability. Coarse-grained soils have a permeability that is high enough to permit water to be squeezed out practically as fast as load is applied. Fine-grained soils have much lower permeability, and water cannot usually be squeezed out as fast as load is applied. There is a consequent time lag between application of load to fine-grained soils and attainment of the equilibrium volume (or void ratio) corresponding to that load. This time lag is influenced by the thickness of the layer being loaded and the magnitude of volume decrease in addition to the permeability.

Figure 3 illustrates the general principles of the laboratory one-dimensional consolidation test. The soil specimen is contained in a rigid ring that prevents lateral deformation; drainage at the top and bottom faces of the specimen occurs through porous stones; and increments of consolidation load are applied in the vertical direction.

Two plots are used to summarize the results of the laboratory consolidation test:

1. *Time curve,* void ratio versus time, for each load increment.
2. *Pressure-void ratio curve,* equilibrium void ratio versus consolidation pressure.

Figure 4 is a typical semilogarithmic plot of the time curve for a single load increment on one

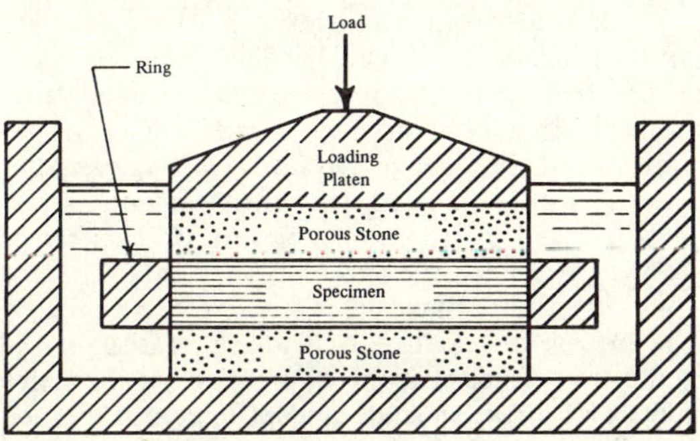

FIGURE 3. Laboratory consolidation apparatus.

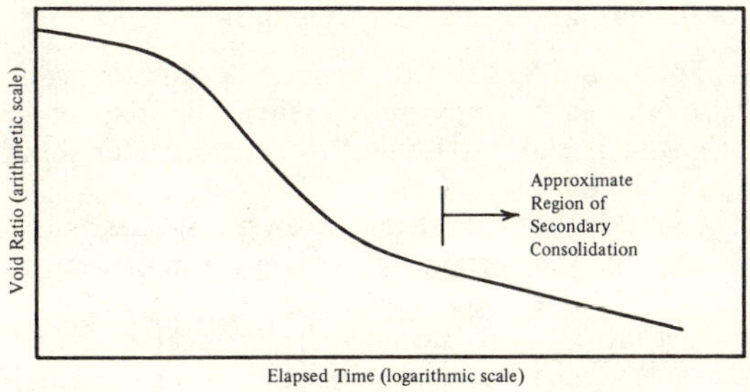

FIGURE 4. Time curve from consolidation test.

particular clay. In the analysis of consolidation, it is assumed that the time curve approaches a horizontal asymptote, but for practically all soils the time curve continues as an approximately straight line with a finite slope on the semilogarithmic plot. The latter slow continuing consolidation is referred to as secondary consolidation and may constitute a significant portion of the total consolidation, particularly for highly organic soils.

Figure 5 is a typical pressure-void ratio curve for an undisturbed sample of clay. This figure illustrates the procedure commonly used for finding the *preconsolidation pressure,* which is the maximum pressure that has ever been carried by the soil. A *normally consolidated clay* is one whose preconsolidation pressure is equal to the effective overburden pressure now acting on the clay in situ. An *overconsolidated clay* is one whose preconsolidation pressure is greater than the present effective overburden pressure because of past drying, loading by ice that has subsequently melted, or loading by a thick overburden of sediment that has been removed by erosion.

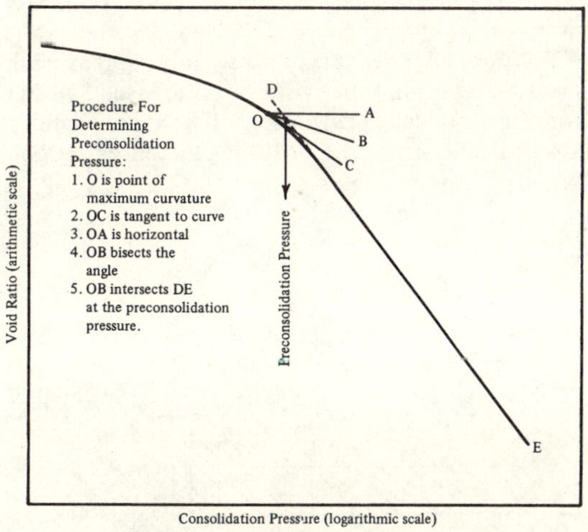

FIGURE 5. Pressure-void ratio curve.

Deformability and Shear Strength

Deformability is the resistance to deformation or distortion under shear loads; *shear strength* is the resistance to rupture under shear load.

The concept of shear strength is best visualized in the direct shear test illustrated in Figure 6. The soil specimen is confined in a box that is split along a horizontal plane. A normal force N is applied to the soil in the vertical direction; the shear force S required to produce shear failure along a horizontal plane is measured. This test is directly analogous to the simple physics experiment used to measure coefficients of friction. The normal stress σ is defined as the normal force N divided by the area of the specimen in the sliding plane, and the shear strength s (shear stress mobilized at failure) is defined as the shear load S divided by the area of the specimen in the sliding plane. Shear strength is usually plotted versus normal stress, as in Figure 7.

For dry, coarse-grained soils, the relation between shear strength and normal stress is usually curved as shown in Figure 7, but it can be approximated over a limited range of stress by a straight line through the origin. The inclination of this straight line is the *angle of internal friction, ϕ*. As illustrated in Table 1, which gives measured values of angle of internal friction at low normal stress (of the order of 1 kg/cm^2) for some soils, the angle of internal friction depends on the grain-size distribution, the shape of the grains, and the void ratio. In addition, the angle of internal friction is generally significantly smaller at high normal stresses than at low normal stresses. For example, an angular rockfill that has an angle of internal friction greater than 50° at a normal stress of 1 kg/cm^2 may have an angle of internal friction as low as 30° at a normal stress of 20 or 30 kg/cm^2.

Volume changes take place in a soil during shear. They may be volume increases, which commonly occur in dense soils sheared under low normal stress, or volume decreases, which commonly occur in loose soils sheared under high normal stress. If the soil is saturated, water must flow into or out of the voids to accommodate any volume changes. Coarse-grained soils usually have a sufficiently high permeability to permit the volume change and corresponding flow of water to take place as fast as the shear stress is applied (provided the shear stress is not applied very quickly, as may be the case during earthquake shaking). All the externally applied normal stress is carried by the soil skeleton, and the pressure in the porewater is practically zero at all times.

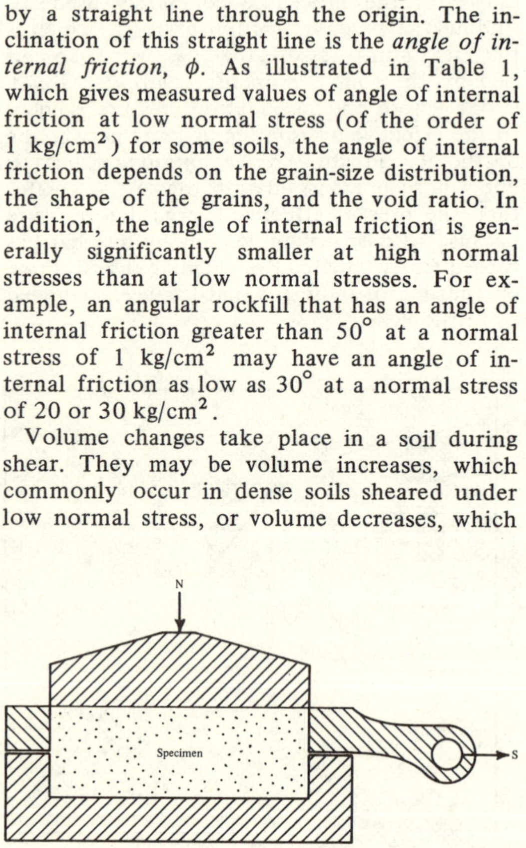

FIGURE 6. Direct-shear apparatus.

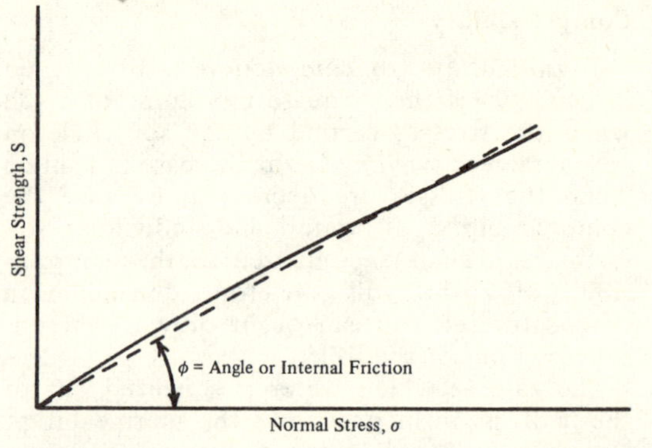

FIGURE 7. Relation between shear strength and normal stress.

The stress carried by the soil skeleton is called the *effective stress* $\bar{\sigma}$; the stress carried by the porewater is called the *pore pressure u;* and the total stress σ is the sum of the effective stress and the pore pressure: $\sigma = \bar{\sigma} + u$. Because water cannot resist static shear stress, the shear strength of a soil depends only on the magnitude of the effective stress.

Fine-grained soils behave much the same as coarse-grained soils except that they have much lower permeability. Volume changes during shear in fine-grained soils are constrained to occur more slowly because of the resistance of the soil to water outflow or inflow. Consequently, positive or negative pore pressures may build up in the water if the loads are applied quickly. If the induced pore pressure is positive, part of the externally applied total normal stress will then be carried by the porewater and the shear strength will be smaller than it would be if all of the total normal stress were carried by the soil skeleton. If the stresses acting on a coarse-grained soil are changed very quickly, there may be sufficient resistance to the inflow or outflow of water to result in the build-up of significantly positive or negative pore pressures. In such cases, the coarse-grained soil behaves much the same as the fine-grained soil.

TABLE 1. Measured Values of Angle of Internal Friction at Low Normal Stress.

Soil	Angle of Internal Friction	
	Loose State	Dense State
Well-rounded, uniform sand	28°	35°
Subrounded, silty sand	33°	37°
Subangular, silty sand	36°	40°
Well-graded, angular, sand-gravel mixture	42°	57°

Two limiting drainage conditions are used for the purpose of defining the strength characteristics of soils: drained and undrained loading. *Drained loading* is shear in which normal and shear stresses are changed so slowly that drainage is essentially complete at all times and there is no excess pressure in the porewater. The "drained" strength characteristics of both fine-grained and coarse-grained soils can be defined by the angle of internal friction. Angles of internal friction for fine-grained soils range from about 10° for highly plastic clays to 30° or more for silty clays and silts. At very large strains, there is a continuous reduction of strength for clays because of the continuing realignment of clay particles parallel to the direction of shear, and the ultimate strength reaches a minimum value that is termed the *residual strength*. Values of the "residual angle of internal friction" as low as about 6° have been measured.

Undrained loading is shear in which the normal and shear stresses are changed so quickly that no drainage can occur. In soft clays positive pore pressures are developed, and the "undrained" strength is smaller than the "drained" strength. In very stiff clays negative pore pressures are sometimes developed, and the undrained strength is larger than the drained strength.

Although the direct shear test is satisfactory for determining the drained strength of soils, it is not satisfactory for determining the undrained strength because it does not permit adequate control of drainage. Undrained tests are usually performed in the triaxial apparatus. Figure 8 is a schematic diagram of a triaxial apparatus. The cylindrical soil specimen is encased in a rubber membrane and is subjected to a hydrostatic pressure by the fluid in the triaxial chamber. Load can be applied to the specimen in the vertical direction to cause it to fail in compression. The pedestal is connected to a drainage burette in which the quantity of water entering or leaving the specimen during drained loading can be measured, or it may be connected to a device in which the pore pressure can be measured during undrained loading.

Deformation properties can be measured in the triaxial test the same as they are measured in conventional compression tests on steel and concrete. Defining the compressive stress as the vertical load P applied to the piston divided by the cross-section area of the cylindrical specimen, and the axial strain ϵ as the change in length of the specimen divided by the original length, one can plot a ***stress-strain curve,*** as shown in Figure 9. The ***strength envelope,*** determined by a series of triaxial tests, is shown in Figure 10. It is the envelope to the *Mohr strength circles,* whose centers lie on the normal-stress axis and which cut the normal stress axis at σ_3, the fluid pressure in the triaxial chamber, and σ_1, the sum of the chamber pressure and

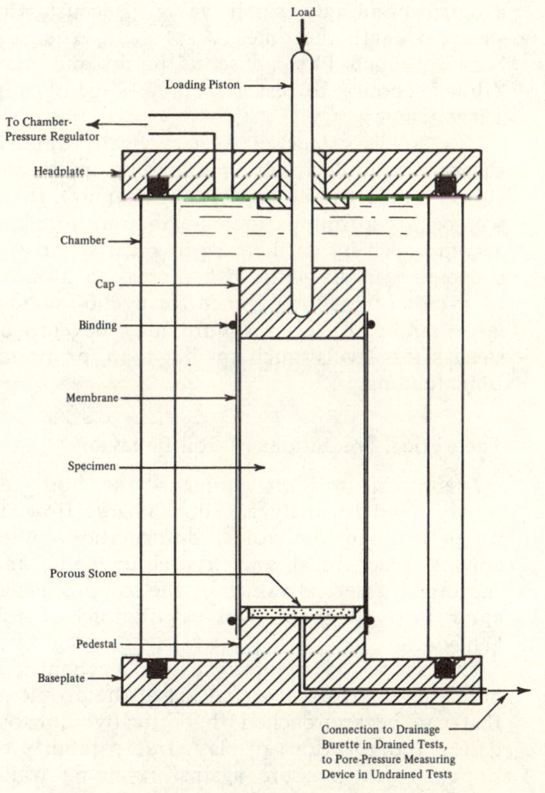

FIGURE 8. Triaxial apparatus.

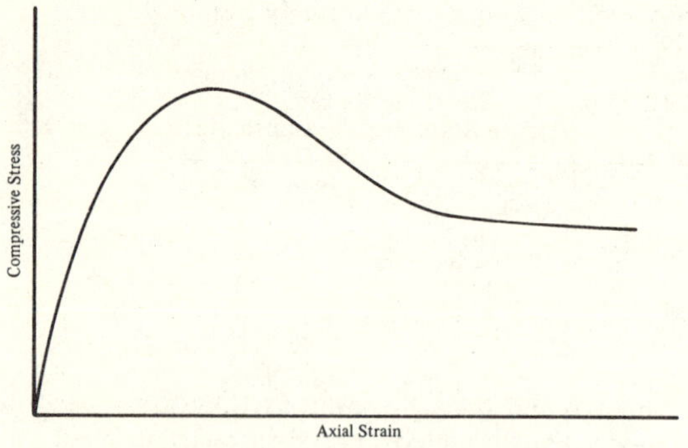

FIGURE 9. Stress-strain curve.

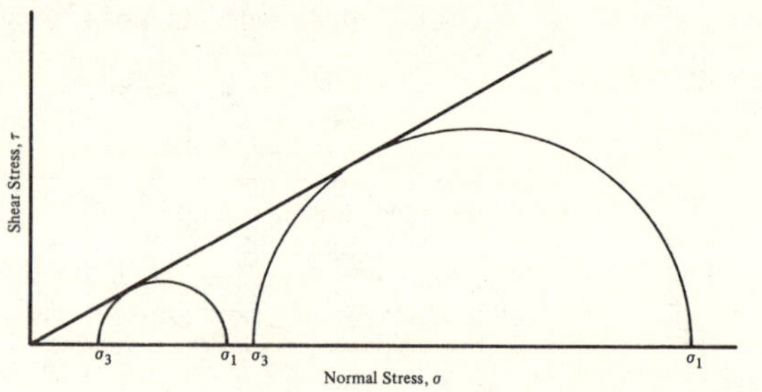

FIGURE 10. Mohr strength envelope.

the vertical stress produced by the piston load.

Some soils such as fine sands and clays may have unusually high porosity in situ. Such soils may undergo "liquefaction" or "flow" failures if the soil structure collapses due to dynamic or essentially static increases in shear stress. As a result of such a structural collapse, most of the total normal stresses acting on the soil are transferred from the soil structure to the pore-water, the effective stresses decrease to very small values, and the shear strength, which depends only on the effective stress, decreases to a correspondingly small value. Because the shear strength becomes very small, the soil behaves much like a viscous liquid and may "flow," coming to rest at a stable slope of only a few degrees.

If a soil is subjected to a cyclically applied shear stress under undrained loading conditions (for example, due to earthquake shaking), there will be a continuous increase in pore pressure accompanied by an increase in shear strain and a decrease in strength with increasing number of cycles of loading. Hence for cyclic loading, large strains or even failure may develop at shear stress levels much smaller than for monotonic loading.

Theoretical Predictions of Soil Behavior

Theoretical soil mechanics is the body of theory used to analyze soil behavior: flow of water through the voids, deformation under applied mechanical and hydraulic loads, and rupture. Numerical values of the soil properties enter into all theoretical calculations of soil behavior.

The scope of theoretical soil mechanics is best illustrated by a partial list of the problems that can be approached theoretically: consolidation (compaction) of clay strata; stability of slopes; earth pressure against retaining walls; bearing capacity of piles; and seepage through earth dams. An abridged presentation of theoretical soil mechanics will not be made because the abridgement would inevitably result in misleading oversimplification. Terzaghi (1943) should be consulted for a detailed exposition of theoretical soil mechanics that offers considerable insight into the limitations on applying theoretical soil mechanics to the solution of real problems. Poulos and Davis (1974) provide a complete summary of analytical solutions to problems that are encountered in soil mechanics.

Observation of Actual Soil Behavior

The final phase of any soil mechanics study should be to observe the actual soil behavior and to compare it to the predicted soil behavior. There are many reasons why the actual behavior may be different from the predicted behavior—subsoil details that escaped detection during the subsurface investigation; sample disturbance, which results in differences between the properties measured in the laboratory and the properties in situ; limitations of the laboratory or field tests used to measure the soil properties; and discrepancies between the assumptions made in the analysis of soil behavior and the actual conditions that prevail in the field.

Suffice it to say that no phase of applied soil mechanics can be handled in a routine manner—not the subsoil investigation nor the determination of the soil properties nor the theoretical analysis of soil behavior—and that observations of actual soil behavior constitute the final check on the effectiveness of all the preceding phases.

RONALD C. HIRSCHFELD

References

American Society for Testing and Materials, 1976. *Annual Book of ASTM Standards, Part 19: Natural Building Stones; Soil and Rock; Peats, Mosses, and Humus.* Philadelphia, Pa.: ASTM, 522p. (Annual editions are published.)

Jumikis, A. R., 1962. *Soil Mechanics.* New York: Van Nostrand, 791p.

Lambe, T. W., and Whitman, R. V., 1969. *Soil Mechanics.* New York: Wiley, 553p.
Lee, I. K., 1968. *Soil Mechanics.* Sydney: Butterworths, 625p.
Poulos, H. G., and Davis, E. H., 1974. *Elastic Solutions for Soil and Rock Mechanics.* New York: Wiley, 411p.
Sowers, G. B., and Sowers, G. F., 1970. *Introductory Soil Mechanics and Foundations.* New York: Macmillan, 556p.
Terzaghi, K., 1943. *Theoretical Soil Mechanics.* New York: Wiley, 510p.
Terzaghi, K., and Peck, R. B., 1967. *Soil Mechanics in Engineering Practice.* New York: Wiley, 729p.
U.S. Bureau of Reclamation, 1974. *Earth Manual.* Washington, D.C.: Government Printing Office, 810p.

Cross-references: *Aggregate Stability; Electroosmosis; Exchange Phenomena; Flow Theory; Particle Density; Particle-size Distribution; Permeability; Pore Space, Drainable; Soil Stabilization; Soil Structure; Thixotropy, Thixotropism; Water Content and Retention; Water Movement. Vol. XIII: Chemical Soil Tests; Clay, Engineering Geology; Consolidation Theory; Engineering Soil Science; Expansive Clays; Foundation Engineering; Pore Pressures; Quick Sand, Quick Clay; Rheology; Shear Strength; Soil Fabric.*

SOIL MICROBIOLOGY

The discipline of *soil microbiology* involves the study of the very small organisms in terrestrial ecosystems and of the biochemical processes that such organisms mediate. The subject includes not only the microorganisms in the soil but also those normally associated with plant litter and other organic debris and with the roots and shoots of live plants. The biochemical processes mediated variously contribute to soil formation, tilth and fertility, plant growth and welfare, and organic matter decay. Of especial interest are those microbial processes involved in elemental cycling (see *Carbon Cycle; Nitrogen Cycle; Nitrogen Fixation; Phosphorus Cycle*).

Soil microbiology is not concerned exclusively with the truly microscopic organisms and their activities. In the development of his discipline, the soil microbiologist devised cultural and observational techniques that were suitable not only for microscopic organisms but also for other single and multicelled organisms that are directly visible. Currently soil microbiology embraces not only the bacteria (including streptomycetes and actinomycetes), microfungi (including yeasts), and protozoa (these are motly microscopic) but also the algae (mostly macroscopic) and the macrofungi (including toadstools and mushrooms). The two last-named groups participate in many important soil processes. For example, the macrofungi are important agents of decay, and the blue green algae, of nitrogen fixation.

Although soil microbiology embraces some organisms that are directly visible, it does not necessarily embrace all microscopically sized organisms. As one example, the small roundworms known as *nematodes* are usually placed not in soil microbiology but in the separate discipline of nematology. Although many nematodes are microscopic in size, they are commonly studied by observations on their morphology and feeding habits rather than by observations on their cultural traits and the end-products they produce. Finally, it must be noted that soil microorganisms are often studied in other disciplines. Many microbes endemic to soil are intimately involved with man's activities or requirements and are studied as components of specific substrates, habitats, or processes rather than as components of terrestrial ecosystems. Their study then falls within other more circumscribed disciplines such as those of food, dairy, textile, industrial, or medical microbiology.

Historical Notes

Soil microbiology is a comparatively young science. Inasmuch as it deals primarily with single-celled organisms too small to be seen with the unaided eye, its growth as a science was in part dependent on the development of simple and compound lenses enabling magnifications of several to several hundredfold. The Dutch lensgrinder, Antonius van Leeuwenoek (1632–1723), is usually considered the father of microbiology. Inasmuch as some of his very small "animalcules" were seen "in water in which pepper had lain infused," soil microbiology must be ascribed the same longevity as the science of microbiology itself.

In the two centuries following Leeuwenhoek's observations, soil microbiology was almost completely ignored. Only during the last hundred or so years has it had a meaningful history. The classical experiments of Louis Pasteur (1822–1895) in disproving the hypothesis of spontaneous generation and in explaining the phenomena of fermentation and decay greatly stimulated the development of soil as well as general microbiology. Sergei Winogradsky (1856–1951) is usually considered the father of soil microbiology in its current context. Among his many contributions was elucidation of the phenomenon of nitrification. Although Pliny (23–79 A.D.) had written of the beneficial influence of legumes in crop rotations, it remained for Hellriegel and Wilfarth in 2888 to delineate the precise contribution made by legume-*Rhizobium* symbioses. Among other Europeans making noteworthy contributions were M. W. Beijerinck in the Netherlands; L. Hiltner and F. Löhnis in Germany; and E. J. Russell in the United Kingdom. In America, A. G. Lochhead did notable work in Canada; and Charles Thom, J. G. Lipman, E. B. Fred, and S. A. Waksman, in the United States. Lochhead (b. 1890) devised a nutritional grouping of soil bacteria and also contributed to rhizosphere microbiology. Thom (1872–1951) made monumental contributions to the taxonomy and physiology of the soil fungi. Lipman (1874–1939) studied the role of microorganisms in soil fertility; and Fred (b. 1887) was a leader in the development of legume bacteriology. Waksman (1888–1973) contributed to many aspects of soil microbiology; he became widely known for his role in the discovery of the antibiotic streptomycin and his reception of the 1953 Nobel Prize in Physiology and Medicine.

Limiting Factors for Soil Microorganisms

Substrate Availability. The principal factor limiting microbial growth in soil is scarcity or absence of a suitable and available source of energy. Nutritionally, the great majority of the soil microorganisms are heterotrophic. They use organic compounds synthesized by higher plants or by autotrophic microorganisms. Autotrophs may be either *photosynthetic,* as are the higher plants, or *chemosynthetic.* Chemosynthetic autotrophs use carbon dioxide as a source of cell carbon and secure their energy by oxidation of elements or inorganic compounds—for example, sulfur, iron, ammonia, and nitrite. The amount of organic matter supplied to soil heterotrophs by microbial autotrophs is very small in comparison to that supplied by higher plants.

However great the amount of organic matter produced annually in any given ecosystem, the number and voracity of the soil heterotrophs are such that in nearly all instances the annual amount of decomposition in an ecosystem balances out quite nicely with the annual productivity. Soil microbiologists have long observed that wherever available energy material is abundant in soil, microbes capable of using that material are also usually abundant. Thus more microorganisms are encountered in the presence of plant or animal residues than in their absence. Higher microbial populations are found in soil in the immediate vicinity of plant roots, the rhizosphere (q.v.), than in root-free soil.

That substrate availability is almost perennially inadequate or that there are many microorganisms in the soil and they are nearly always hungry can be shown by simple calculations. The respiratory output of carbon dixoide from soil is subject to quite precise measurement. The measured output falls far short of that calculable by using standard estimates of microbial numbers in soil and measured rates of carbon dioxide production by microbial cells in laboratory culture. Similar discrepancy is noted between calculated maintenance energy requirement of the measured microbial biomass in soil on the one hand and total ecosystem productivity on the other. The only plausible conclusion is that a good part of the soil microbiota must be in a dormant condition a good portion of the time simply because of inadequacy of substrate supply. This emphasis on the perennial shortage of available energy material in soil is not to affirm that microbial activity in soil faithfully mirrors the quantity of available substrate nor to deny that other factors may also be influential. Even in the presence of a suitable substrate, microorganisms may suffer restriction of their numbers or activities because of biotic and abiotic factors in the ambient microenvironment. Several of these factors are discussed in the following paragraphs.

Moisture. Either extremely wet or extremely dry soils are unfavorable for the growth of most soil microorganisms. Different mechanisms are involved for these two extremes. Wet soils are unfavorable for most aerobic organisms simply because filling up of the pore space with water diminishes soil aeration. The restrictive factor is lack of oxygen and not the excess of water in itself. In dry soils restriction or complete inhibition is imposed when the microorganisms can no longer obtain or hold water within their cells in amounts sufficient for their metabolism. Not all microbial subgroups are equally affected by low soil moisture content. Ammonification, for example, can proceed under more stringent drought than can nitrification. Degradation of organic residues occurs slowly at moisture tensions below that of the permanent wilting percentage. The rate, however, is severely restricted. At 15 bars, the rate is somewhat less than half that at field capacity; at 250 bars, the rate is barely measurable.

Although microbial activity is severely restricted in dry soil and the microbial populations therein markedly reduced, seldom are individual species eliminated entirely. When dry soils are rewetted, such diverse phenomena as nitrification, ammonification, cellulose decomposition, or biological nitrogen fixation almost invariably proceed without any soil reinoculation being required. One of course cannot conclude that specific bacteria are never entirely eliminated from soil during drought. In some field soils cropped to annual legumes, especially if such soils are poorly buffered and subject to severe drought during the nongrowing season, there may be insufficient carryover of viable rhizobia to effect nodulation in the following crop.

Aeration. Attention has already been called to the fact that very wet soils are unfavorable for many microorganisms simply because filling up of the pore space with water diminishes soil aeration (see *Aeration, Respiration, Atmosphere*). Usually in soils with moisture content lower than field capacity, oxygen content in the soil pores falls within the range of 10–20%. For pores not isolated by water barriers, such concentrations can supply oxygen by gaseous diffusion at a rate adequate for microbial oxygen demand. Inasmuch as the rate of oxygen diffusion through water is only about one ten-thousandth as fast as the rate through the air, the presence of water barriers can lead to oxygen concentrations limiting for many microbial species.

On a gross scale, such barriers can be established at the soil surface either by heavy rains or irrigation. If a water seal is established at the soil surface, the amount of oxygen in the soil atmosphere does not provide an inexhaustible reservoir for soil organisms. Disappearance of oxygen depends on the rate of its use by soil organisms and plant roots and on the amount of oxygen initially entrapped. This amount is determined by depth of the water table, the fraction of the soil volume occupied by air, and the initial oxygen content of the entrapped air. A much more difficult problem is that posed by water barriers in and around individual soil pores. One can be reasonably certain that these barriers are of little or no consequence at the moisture content of the permanent wilting percentage. As moisture content approaches field capacity, microsites of anaerobiosis develop. The observation that the populations of both strictly aerobic and strictly anaerobic bacteria are higher in the upper soil profile than in the deeper soil suggests that microsites of aerobiosis and anaerobiosis exist simultaneously or else that given sites fluctuate in their aeration status.

Temperature. The optimal temperature range for most soil microorganisms is about 25–35°C. Many grow quite well over the range of 10–40°, and a few thermophilic species are able to grow in the range of 60–80°. In the field high soil temperatures are seldom the primary factor limiting microbial growth. Limiting temperatures do occur in the surface layer of dry and barren soils, but in such soils lack of water is also restrictive. Moist tropical soils covered with vegetation usually provide near optimal temperatures for microorganisms. In temperate and Arctic soils, temperatures below the optimum may occur seasonally and be highly restrictive for microbial activity.

Reaction. The reaction range commonly tolerated by soil microbes is that between pH 4 and 10. For most species the optimum within this range is slightly on the alkaline side of neutrality. *Thiobacillus thioxidans,* an autotroph capable of oxidizing elemental sulfur, is perhaps the most acid tolerant of all living organisms. It tolerates an acidity of pH 0.6. Different pH tolerances often exist among species within the same genus. *Azotobacter chroococcum,* for example, is widely distributed in neutral and alkaline soils, but it is not found in soils below about pH 6. In contrast, *A. indicus* tolerates an acidity of pH 3. Burges and Raw (1967) offer detailed discussions of substrate preferences and environmental tolerances of the major subgroups of soil microorganisms.

Biotic Factors. Soil microorganisms exert pronounced restrictive effects on one another. These may be direct, as in predation and parasitism, or indirect such as occur when one species induces changes in the substrate or microenvironment that are unfavorable to a second species. Among microorganisms it is often difficult to distinguish between predation, parasitism, and lysis. A predator is usually stronger or more aggressive than its prey, whereas in parasitism, a smaller or weaker organism commonly feeds on a host over a period of time. Among the bacteria, some are spoken of as *micropredators* of other bacteria, but probably the relationship should be described as *lysis* rather than as predation. Among the fungi there are ameboid slime molds that are predatory on a wide variety of bacteria and nematophagous fungi that form specialized gelatinous or constrictive traps by which they capture nematodes. Many protozoa feed wholly by predation. Competition among microorganisms can best be defined as the injurious effects to them that result from consumption or diminution of their environmental resources. Parasitism among microorganisms is common, with either bacteria, fungi, or protozoa serving either as parasite or host. For an extended discussion of ecological relationships among microorganisms, see Clark (1969).

Injuries inflicted indirectly on microorganisms by means of deleterious changes in the substrate or environment are of two types. At one extreme are those that serve only to make growth conditions less optimal. Alcohols, organic acids, and carbon dioxide are products of microbial origin that may be restrictive to other microbes. Microbiologically produced nitrogen or oxygen deficiencies may also be restrictive. At the other extreme for restriction are those products of microbial origin that are quickly lethal. Antibiotics are usually defined as specific chemical compounds derived from or produced by living organisms and inhibitory or lethal to other organisms even in small quantities. Penicillin and streptomycin are examples. Although antibiotics may be dramatically effective when used *in vivo* in man or animals, their role in the soil is questionable, inasmuch as they themselves are decomposed in the presence of heterogeneous microorganisms. Nevertheless, either antibiosis or other biotic phenomena are sufficiently operative in soil that they are variously exploited by man in the control of soilborne plant pathogens. Early in this century it was observed that a host plant grown in the presence of a root-invading pathogen in sterilized soil suffered more damaging disease than did a second plant grown in nonsterilized soil containing an array of soil organisms in addition to the root pathogen. A logical inference was that if lack of saprophytic microorganisms increased disease severity, then stimulation of the soil microflora should be beneficial. This hypothesis was substantiated in field experiments. Some root diseases were controlled or ameliorated by additions to soil of organic materials such as green manures.

The mechanisms of biological control of soilborne parasites are now known to be many and varied. Diffusible antibiotics, even though not long-lasting in soil, are at times involved and likewise the phenomena of competition, predation, and hyperparasitism. Acceleration of decomposer activity by organic manures may either increase carbon dioxide or decrease oxygen concentration in the soil atmosphere to a level detrimental to the causal pathogen. Changes in the soil atmosphere may also be achieved by such management practices as compaction or flooding.

Determination of Soil Microbial Populations

Cultural Methods. Estimates of microbial populations are made either culturally or by direct microscopy. The cultural method most commonly used is the dilution plate technique. In this method successive serial dilution of soil is made in sterile water, and aliquots of the resultant suspensions are spread on or mixed into an appropriate agar substrate. After suitable incubation, the number of microbial colonies that develop are counted. A primary assumption of the technique is that each viable propagule in the soil suspension used as inoculant will develop into a visible colony during incubation. Any such assumption is highly idealistic. It is not possible to prepare any one combination of substrate ingredients and conditions of incubation that will permit growth of all propagules. This limitation can be overcome to some extent by tailoring the substrate to fit the growth requirements of different subgroups or species. Plate counts of microfungi are commonly made on different substrates than are those of bacteria, and dissimilar substrates are commonly used for different subgroups of bacteria.

A second cultural procedure for determining microbial populations is the extinction dilution or most-probable-number count. Again successive serial dilutions of soil are prepared, and from each suspension a number (usually five or ten) of replicate culture tubes of the chosen substrate (usually in liquid form) are inoculated. Turbidity or other macroscopic evidence of growth, or some change in substrate composition, is used to determine the number of positive growth responses for successive serial dilutions. Each individual culture tube is read only as either positive or negative for microbial growth. The readings thus obtained are converted into population estimates by reference to most-probable-number tables. For diverse plate count and dilution tube methodologies, see Black (1965).

Direct Counts. Direct counts are made by counting the number of cells or propagules in properly prepared microscopic fields. These preparations involve either very shallow aqueous layers or dried smears of soil suspensions on glass slides. In either case, addition of a staining agent is almost mandatory to permit differentiation between microbial cells and other similarly sized inorganic or organic particles. Direct counts are typically several to several hundred times higher than plate counts. This is to be expected. Ideally the direct method accounts for all microbial propagules, whereas in practice the plate method accounts for only those capable of growth on the specific substrate used. The widest ratios between direct and cultural counts are found in soils that have not recently received addition of easily decomposable organic residues. Cultural counts are especially useful for enumerating special subgroups of bacteria, and direct counts, for estimating the biomass of soil fungi. A long mycelial strand can be expected to produce only one to few colonies (some fragmentation may occur during plating) on cultural agar, whereas the same amount of biomass in the form of hundred of small fungal spores can be expected to produce hundreds of colonies.

Meaningful estimates of fungal biomass require measurement of the cumulative total length of all hyphal strands within a known soil volume and also of the average diameter of the hyphae.

Number of Microorganisms in Soil

Bacteria are the smallest and the most numerous of the free-living microorganisms in soil. Their number as determined by cultural methods ranges from a few thousand cells per gram for very poor or barren soils to as high as 500 million when conditions for their growth are especially favorable. For most arable soils, the bacterial population as shown by plate count ranges from 10 to 50 millions per gram. Bacterial counts made by direct microscopy range up to several billions per gram. Near the beginning of this century, some workers were optimistic that population counts might correlate directly with soil fertility, but such optimism was ill-founded.

Plate counting of soil fungi was done extensively during the first half of this century. Population estimates so obtained were commonly only about one-tenth the magnitude of the bacterial plate counts for the same soils. In recent years plate counting of soil fungi as a quantitative tool has lost much of its former popularity; cultural methods, however, remain extremely useful for qualitative and taxonomic studies of the soil fungi.

Protozoan numbers in soil are usually determined by the extinction dilution method and commonly range from several thousand to several hundred thousands per gram. Several workers have noted that the protozoan population in soil is negatively correlated with the bacterial population. Such negative correlation does not invariably occur. Inasmuch as protozoa feed heavily on bacteria, peak protozoan populations may engender depressed bacterial populations.

Figure 1 shows a common soil bacterium highly magnified. In actual size the cells average about 1 μm in length. If a photograph of a man 1.8 m (6 ft) tall could be similarly magnified, the resulting print, in relation to the actual height of the man, would be 45 km (28 mi) high. Bacterial cells possess no discrete nucleus but do contain disperse genetic material. The dark bodies occurring intracellularly in the micrograph (obtained by transmission electron microscopy) shown in Figure 1, can be interpreted as nutrient storage bodies. Some but not all bacteria possess whip-like appendages or flagellae by which they achieve mobility. Some species possess a single polar flagellum, as shown in Figure 2; others possess a tuft of polar flagellae; and still others, from several to many flagellae distributed over the cell exterior.

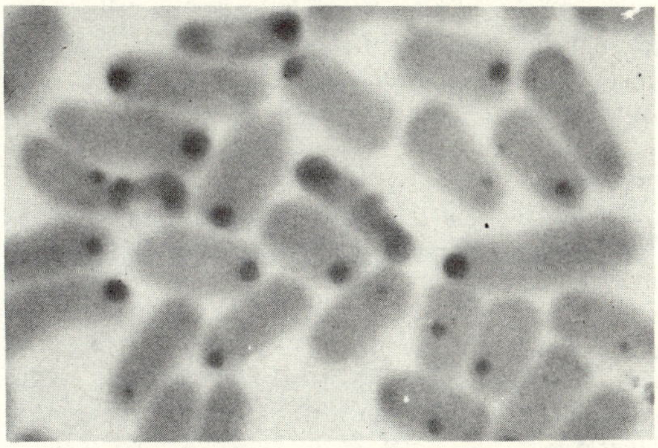

FIGURE 1. Electron microscope photograph of a common soil bacterium (X11,000).

Biomass of Soil Microorganisms

Of all the herbivores, carnivores, and decomposers in terrestrial ecosystems, the microorganisms are easily dominant in total biomass. Indeed, their biomass exceeds that of all the large and small mammals, birds, reptiles, insects, and all other invertebrates combined. At times the microbial biomass even approaches that of the plant component. The dry weight of the microbial component is on the order of 100–200 g/m^2 to a soil depth of 30 cm; that of large mammals, 1–5 g; of small vertebrates, 2 g or less; and of miscellaneous invertebrates, likewise no more than 2 g/m^2.

Of the total microbial biomass, the fungi constitute the major fraction. Their biomass has been measured in grassland soils as ranging from 50–300 g/m^2 to a soil depth of 30 cm. The bacterial biomass in the same soils was about one-third that of the fungi. Although in soil the biomass of bacteria tends to show positive correlation with that of fungi, this is by no

FIGURE 2. Electron microscope photograph of a soil bacterium bearing a polar flagellum (7,000).

means universally true. Differences in substrate, microclimate, and biotic factors dissimilarly affect the bacterial and fungal components. The biomass of the soil microfauna is sharply less than that of the microflora. In grassland, nematode biomass seldom exceeds 2 g/m^2 to a soil depth of 30 cm; that of the soil protozoa is generally less than 1 g/m^2. Collectively, the microbial biomass constitutes a major dynamic reservoir of available nutrients. In grassland the amount of nitrogen in microbial tissues is only slightly less than that in the vegetation, and the amount of phosphorus, twice that in the vegetation.

Soil Phenomena Mediated by Microorganisms

The role of microorganisms in nutrient and energy flows in terrestrial ecosystems is consistent with their high biomass, enzymic versatility, and rapid growth rates. Microbial involvement in carbon and nitrogen cycling is discussed under those headings elsewhere in this volume. That these two elements are accorded separate discussion is not to imply that microorganisms are not involved in the cycling of other elements—for example, they are intimately involved in the cycling of sulfur and phosphorus (see *Phosphorus Cycle; Sulfur Transformations*). In the cycling of some elements in soil-plant systems, microorganisms play only a minor role. The release of potassium from plant tissues, for example, is mediated primarily by physical processes known collectively as leaching (q.v.).

Microorganisms affect the availability of nutrients to plants by acting directly on the oxidation-reduction status of the elements themselves (e.g., nitrogen, sulfur, iron, and manganese) and by forming organic acids or other compounds that chelate with poorly soluble nutrients and thus increase their mobility and availability. Furthermore, microorganisms, particularly certain fungi, induce changes in root morphology and function that promote uptake of available soil nutrients. Fungal mantles on or hyphal penetrations of plant roots lead to modified root structures known as *mycorrhizae.* These fungus-root associations provide a greater surface area of live tissue for nutrient uptake by causing multibranching of colonized roots and by building hyphal bridges between root cells and root-adjacent or nearby soil. For woody plants that form ectomycorrhizae, the mantle of fungal tissue about the host rootlet and the multibranching of the roots are easily seen. For grasses and forbs, the mycobiont mycelijm on the root surface remains sparse, and the root morphology is not visibly changed. The most dramatic differences in nutrient uptake by mycorrhizal and nonmycorrhizal roots have been observed for woody plants.

Soil microorganisms are key agents in soil formation. They produce carbonic and organic acids, which in turn promote the weathering of soil parent materials and the release of their contained elements (see Vol. XII: *Weathering, Organic*). Organic acids also influence leaching phenomena and the differentiation of the soil profile into dissimilar layers. Microorganisms are primarily responsible for the formation of the long-lived organic matter component of soil commonly called humus (q.v.), a complex heteropolycondensate formed during microbial decomposition of organic residues. Humus influences the cation exchange and water-holding capacities of soil. Aggregate formation in soil is fostered by the binding action of fungal mycelia and by mucilaginous polysaccharides of microbial origin. However, not all aggregate formation in soil is attributable to the microbial component, nor are the aggregates of microbial origin necessarily long lived in soil. Both the mycelial strands and the polysaccharides (q.v.) are subject to microbial decomposition. Although aggregates of microbial origin may promote the water intake and storage properties of soils, other microbial products may hinder water intake (see *Hydrophility, Hydrophobicity*). Profuse growth of certain species of fungi makes the colonized soil highly water repellant and hence causes either lack of intake or else highly irregular patterns of water intake.

Conclusion

In summary, microorganisms are a major component of terrestrial ecosystems. Their biomass is less than that of green plants but far greater than that of all the invertebrate and vertebrate animals combined. Microorganisms influence soil-plant systems in many ways. Their influences may be either beneficial or harmful to plant growth and welfare and to the quality of the environment. Readers desiring book-length treatments of soil microbiology to extend the few remarks offered here should consult Alexander (1977), Gray and Williams (1971), or Dixon (1976).

FRANCIS E. CLARK

References

Alexander, M., 1977. *Introduction to Soil Microbiology.* New York: Wiley, 554p.

Black, C. A., ed., 1965. *Methods of Soil Analysis.* Madison: American Society of Agronomy, 1572p.

Burges, A., and Raw, F., eds., 1967. *Soil Biology.* London: Academic, 532p.
Clark, F. E., 1969. *Ecological Associations among Soil Microorganisms.* Paris: Unesco, 350p.
Clark, F. E., 1977. Soil microbiology–It's a small world, *Soil Sci. Soc. America Jour.,* 41 (2), 238-241.
Dixon, B., 1976. *Magnificent Microbes.* London: Atheneum, 251p.
Gray, T. R. G., and Williams, S. T., 1971. *Soil Microorganisms.* Edinburgh: Oliver & Boyd, 240p.

Cross-references: *Aeration, Respiration, Atmosphere; Carbon Cycle; Denitrification; Enzyme Activity; Fauna; Humus; Metabolic Activity; Microbial Ecology and Clay Minerals; Microorganism Transport; Nitrogen Cycle; Nitrogen Fixation; Microhabitats; Organic Matter; Phosphorus Cycle; Rhizosphere; Sulfur Transformations.*

SOIL MINERALOGY

Soils contain a rather limited number of minerals. The different mineral species in soils influence the release of isomorphously substituted structural ions and the chemical activity of exchangeable ions; hence the mineralogical composition of soils has an important bearing on ecology, soil productivity, and engineering properties (Stelly and Dinauer, 1968).

The minerals of soils can be classified into (a) *primary minerals,* which are formed at elevated temperatures and inherited from the igneous and metamorphic rocks, and (b) *secondary minerals,* those formed by low temperature reactions and inherited by soils from sedimentary rocks or formed in soils by weathering (Brewer, 1964; Dixon and Weed, 1977).

Primary Minerals of Soils

Primary minerals make up the main part of the sand (particles ranging from 1-0.05 mm in diameter) and silt (particles ranging from 0.05-0.002 mm in diameter) fractions of most soils. The combined percentage of sand and silt in a given soil mainly governs the percentages of primary minerals in that soil. Primary minerals occur in the clay fraction (particles less than 0.002 mm in diameter) of little weathered soils, e.g., those derived from fine-grained rock flour derived by glacial grinding of rocks. The most abundant primary minerals in soils the world over are *quartz* (SiO_2, silica) and *feldspars* ($MAlSi_3O_8$, silicate salts of basic cations, M^+), minerals which dominate in the rocks making up the earth's crust. Pyroxenes, amphiboles, olivine, and other accessory minerals of primary origin are important to the chemistry of soils, though usually present in smaller percentages.

Silica. Free silica occurs in soils mainly in the form of *quartz* (SiO_2). It also occurs as *opal* ($SiO_2 \cdot nH_2O$), including opal psuedomorphs after plant cells (phytoliths), *chalcedonite, agate, flint,* and *chert,* and in some soils as *cristobalite* (SiO_2). Quartz frequently makes up from 50-90% of the sand and coarser silts of many soils (Figure 1). Quartz occurs in the majority of soils, both alluvial and those developed from various types of sedimentary rocks and residual crystalline rocks. Free silica also occurs as amorphous sheets and as aggregates with amorphous iron and aluminum oxide gel (Mitchell, 1975). Quartz veins often introduce quartz into basic parent rock formations in which quartz would not normally be expected. Soils developed from quartz-free basic igneous rocks are occasionally found and have a quartz-free sand. Quartz-containing eolian dust has been carried by rainfall onto a belt of soils and ocean sediments 30° ± 15° N. and S. latitude, as established by oxygen isotopic ratios of quartz in upper horizons of soils developed on quartz-free basalt and in pelagic sediments (Jackson et al, 1973).

Feldspars. Feldspars commonly make up 5-25% of silt and sand of soils. They may be abundant (up to 40%) in the clay-size fraction of a few soils, derived from glacial rock flour, which have undergone little weathering.

Potassium feldspars ($KAlSi_3O_8$) of soils include *microcline, orthoclase, adularia,* and *sanidine.* Microcline is the most common in soils, derived from plutonic and metamorphic rocks; it is common in granites, syenites, gneiss, and some pegmatites. Sanidine is derived from volcanic rocks, particularly rhyolites and trachytes. Adularia is a low-temperature, lath-shaped form common in hydrothermal veins. Solid solution of $KAlSi_3O_8$ with $NaAlSi_3O_8$ occurs at high temperatures; the sodium-rich member is *anorthoclase,* and the potassium-rich member is included with *sanidine.*

The *plagioclase* series of feldspars of soils, a series consisting of a solid solution and exsolution mixtures of *albite* ($NaAlSi_3O_8$) and *anorthite* ($CaAl_2Si_2O_8$), provides an interesting and important relationship of the degree of weathering of soils. Given a full range of plagioclases initially, as the weathering intensity and time factors increase, the calcium is depleted progressively to give the more albitic plagioclase. In many well-weathered tropic and subtropical soils, the presence or absence of plagioclases in the parent rock makes the difference between presently productive and unproductive soils.

Accessory Minerals. The *accessory* minerals of soils include a wide variety of minerals occurring in small but significant amounts.

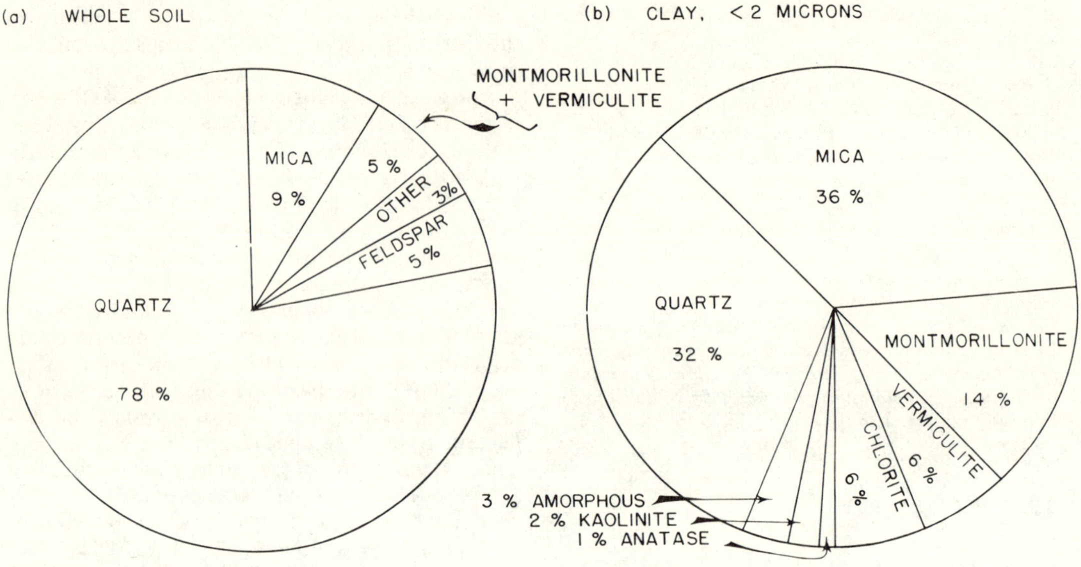

FIGURE 1. Mineralogical composition of a soil derived from weathering of limestone together with a little added glacial drift. (a) Quartz-dominated whole soil. (b) Clay fraction in which mica released by carbonate weathering from the limestone predominates.

The *pyroxene* compositions are represented by *enstatite* ($MgSiO_3$), *hypersthene* [$(Mg,Fe)SiO_3$], *diopside* ($CaMgSi_2O_6$), and *augite* [$Ca(Mg, Fe, Al) (Si,Al)_2O_6$], which is the most important in amounts as well as compositional significance to soils. Augite is a dark colored ferromagnesian mineral commonly derived from basalt and other igneous rocks and occurs as dark mineral grains in soils. *Amphiboles,* consisting of an isomorphous series between the endmembers, *tremolite* [$Ca_2Mg_5Si_8O_{22}(OH)_2$] and *actinolite* [$Ca_2Fe_5Si_8O_{22}(OH)_2$ or $Ca_2(Mg,Fe)_5Si_8O_{22}(OH)_2$], commonly make up 1% or less of soil. *Hornblende* [$(Na,Ca)_2(Mg,Fe,Al)_5(Si,Al)_8O_{22}(OH)_2$] is widely distributed in soils and is derived from igneous rocks such as syenite, granite, and gabbro and from metamorphic rocks such as gneiss, hornblende schist, and amphibolites. Hornblende weathers fairly rapidly, and its size distribution is a measure of the intensity and time of weathering of soils in cool, humid regions. *Olivines* [$(Mg,Fe)_2SiO_4$], olive green minerals forming an isomorphous series between the endmembers *forsterite* (Mg_2SiO_4) and *fayalite* (Fe_2SiO_4), make up a small percentage of easily weatherable minerals of soils. Other accessory minerals include *apatite* [$Ca_{10}(F,OH,Cl)_2(PO_4)_6$], *tourmaline* [$Na(Mg,Fe)_3Al_6(BO_3)_3Si_6O_{18}(OH)_4$ (of interest to soil chemistry because of its content of the plant-essential element, boron)], *epidote* [$Ca(Al,Fe^{3+})_3Si_3O_{12}(OH)$], *zircon* ($ZrSiO_4$), *topaz* [$Al_2SiO_4(OH,F)_2$], *garnets* [$X_3Y_2(SiO_4)_3$] in which X may be or Cr^{3+} and sometimes Ti or Mn^{3+}, *magnetite* (Fe_3O_4), *ilmenite* ($FeTiO_3$), *sphene* ($CaTiSiO_5$), *rutile* and *anatase* (TiO_2), and *kyanite, sillimanite,* and *andalusite* (Al_2SiO_5) and similarly structured *staurolite* [$2Al_2SiO_5 \cdot Fe(OH)_2$].

Carbonate and Sulfur-bearing Minerals in Soils

Among the secondary minerals in soils are found the carbonates, sulfur-bearing minerals, layer silicates, and oxide minerals (see Vol. IVB: *Soil Mineralogy*). The literature references on these minerals have been extensively reviewed (Brown, 1961; Farmer, 1974;, Gard, 1971; Grim, 1968; Jackson, 1964; Mackenzie, 1955; Thorez, 1975; van der Plas and van Reeuwijik, 1974; Warshaw and Roy, 1961).

Carbonates. The most abundant carbonate mineral of soils is *calcite* ($CaCO_3$), which accumulates in lower horizons of subsoils of subhumid and more arid regions. This horizon of accumulation occurs at variable depths that reflect the average depth of moisture penetration before the moisture is removed by roots. The calcite is commonly disseminated as silt crystals (2–50 μm in size) in the soil, but also occurs as in indurated forms, as nodules and/or as hard calcareous (travertinous) layers called *caliche* or calcrete, and mixed with the siliceous equivalent, *silcrete* (see Vol. VI: *Duricrust*).

Another hardened form of $CaCO_3$ is *rhizoconcretions* (or rhizomorphs) formed around roots as concentric layers of calcite (soil travertine). These reflect precipitation of carbonate on desiccation of the soil by roots or pseudomorphic filling of root channels. Rhizoconcretions have been noted in loess (northeastern Nebraska) and in semiarid belts of Africa, Western Australia, United States, Bermuda, Bahamas, and Hawaii. There are also instances of rhizoconcretions formed in acidic swamps, even mangrove swamps, but instead of the calcite cement there is silica. Silica is mobilized at very high or low pH (above 9 or below 5) and precipitates in the low intermediate range (below pH 8) in preference to $CaCO_3$, which is dissolved at this value and below.

Calcite in soil is also inherited from limestone, marl, and chalk rocks as well as from marble parent materials. *Dolomite* ($CaCO_3 \cdot MgCO_3$) is inherited in soils from dolomitic limestone and mantle rocks derived therefrom. *Siderite* ($FeCO_3$) is formed by decomposition of iron-bearing minerals in environments suitable for carbonate accumulation and is frequently associated with calcite. In rare instances the isomer of calcite, *aragonite* ($CaCO_3$), may be found in soils. Also rare in soils is *magnesite* ($MgCO_3$) because of its relatively high solubility in water.

Sulfur-bearing Minerals. *Gypsum* ($CaSO_4 \cdot 2H_2O$) accumulates in soils of semiarid and arid regions, frequently in a horizon situated somewhat beneath the horizon in which $CaCO_3$ accumulates. It sometimes occurs as indurated horizons and as crystals disseminated in surface soil horizons. Gypsum occurs in certain subsoils as a weathering product of *pyrite* (FeS_2), even in soils of humid regions. Drainage of some tidal marshes or exposure of acid-forming underclays of coal beds in outcrops causes H_2SO_4 to form from FeS_2, which results in extremely acid soils where pH 3.5 is quite common (see Vol. XV: *Coastal Soils*). Calcium is removed from calcium-bearing minerals by the action of the H_2SO_4, and gypsum crystals are formed. The reaction of the sulfuric acid with iron-bearing minerals in soils yields *jarosite* $[KFe_3(OH)_6(SO_4)_2]$ and with aluminum-bearing minerals often yields *alunite* $[KAl_3(OH)_6(SO_4)_2]$.

Layer Silicate Minerals of Soils

Layer silicates or *phyllosilicates* of soils (see also *Clay Minerals, Silicates*), of which *mica* is a representative familiar to the most casual observer, play a prominent role in most soils in determining the physical properties such as plasticity and structure and chemical properties such as cation exchange, ion release, and ion fixation when fertilizer is added to soil (Stelly and Dinauer, 1968). The unsubstituted, uncharged, idealized endmembers for the 2:1 layer silicates are *pyrophyllite* $[Si_8Al_4O_{20}(OH)_4]$, and *talc* $[Si_8Mg_6O_{20}(OH)_4]$. The deposition of significant amounts of mica, vermiculite, and other layer silicates as well as quartz and carbonates (above) occurs as aerosolic dusts. These are common from arid regions of both hemispheres, carried on circumglobal westerly winds and deposited in rainfall on lands and ocean bottoms (Jackson et al, 1973). Fixation of radionuclides such as ^{137}Cs in such dusts is important to rendering radioactive fallout less dangerous.

Micas. Micas occur almost universally in soil clays, primarily originating by inheritance from the parent rock from which the soil was formed (Figure 1). Well-ordered micas, inherited by soils mainly from igneous and metamorphic rocks, include dioctahedral *muscovite* $[K_2Al_2Si_6Al_4O_{20}(OH)_4]$ and its isomorphous analogue *paragonite* $[Na_2Al_2Si_6Al_4O_{20}(OH)_4]$, and trioctahedral *biotite* $[K_2Al_2Si_6(Fe^{2+}, Mg)_6O_{20}(OH)_4]$ and *phlogopite* $[K_2Al_2Si_6Mg_6O_{20}(OH)_4]$. A variety of additional ion substitutions occur, beyond those indicated for the endmembers. The layer charge is extremely high in *margarite* $[Ca_2Al_4Si_4Al_4O_{20}(OH)_4]$, which is representative of the "brittle" micas.

Imperfectly ordered micas of soils are both inherited from sedimentary rocks and formed by weathering of well-ordered micas. The imperfectly ordered micas from either of these sources contain less potassium and more water than the well-ordered micas. The extra hydrogen and deficiency of potassium is explained by the fact that the disordered micas of soils occur with mixed-layering with vermiculite, montmorillonite, and chlorite. The latter phases have hydrated exchangeable cations, water, and hydroxides intercalated between some of the expanded silicate layers, (Jackson, 1963), and these are distributed among mica sequences of layers, giving the extra H_2O and OH found on analysis. As the interlayer potassium is subjected to depletion through chemical weathering in soil, gain occurs in readily exchangeable ions and interlayer water (Figure 2) according to the equation of Jackson et al (1948):

$$\text{Mica} \rightleftharpoons \text{vermiculite} \rightleftharpoons \text{montmorillonite} \quad (1)$$

The levering action of hydrated sodium or calcium ions in curling back mica layers is revealed

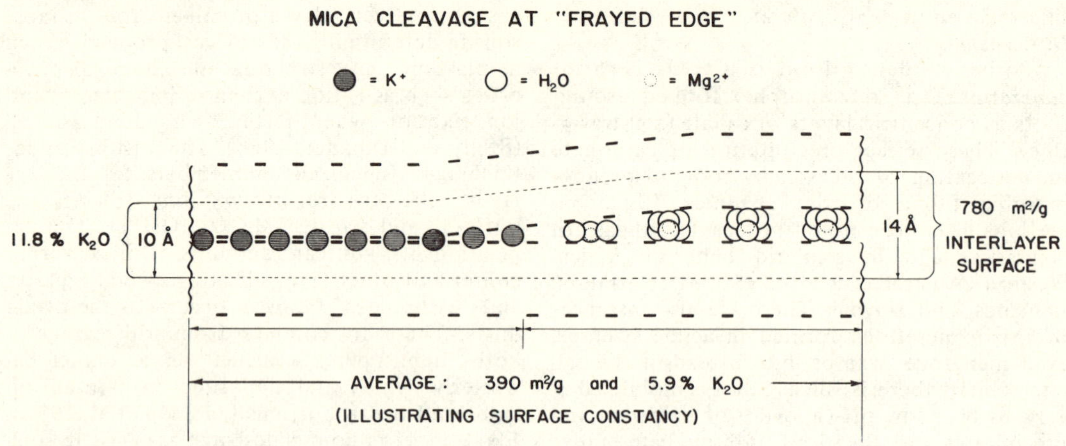

FIGURE 2. Wedge position at phase boundary between vermiculite (right) and mica (left) in weathering mica. The central potassium ion offers steric hindrance to rapid exchange by large, hydrated cations. (From Jackson, 1963)

by mica surface morphology (Figure 3). Addition of a high level of soluble potassium ions to soils or drying or freezing, by concentrating the potassium in solution, tends to shift equation 1 to the left; K or NH_4 ions bond strongly to the two crystallographic cavities kept aligned by the mica core (Figures 2 and 3) and close the layers together. Equation 1 has been amplified in equation 2 (Jackson, 1964) to include the chemical process of forming hydroxy sesquioxide interlayers, chlorite, kaolinite, and gibbsite:

Mica (Biotite, Muscovite) → Vermiculite → Montmorillonite ⇌ Pedogenic 2:1–2:2 swelling 18 Å intergrade (2)

(Fe, Mg, Al) chlorite → Secondary chlorite → Pedogenic 2:1–2:2 14 Å intergrade → Al chlorite → Kaolinite and halloysite → Gibbsite

(Mica, Vermiculite, Montmorillonite ↔ Pedogenic 2:1–2:2 14 Å intergrade; Pedogenic 2:1–2:2 swelling 18 Å intergrade → Kaolinite and halloysite; → Gibbsite)

Vermiculite. Vermiculite, which forms as a product of weathering of micas and possibly also of chlorites, occurs extensively in soils. The potassium of micas has been replaced predominantly by exchangeable calcium, aluminum, and magnesium and the interlayer space is expanded (Figure 2) to yield a unit cell c-spacing of 14Å or more so as to accommodate exchangeable interlayer cations and water (Ramon and Jackson, 1964). Vermiculite probably also forms from solution in soils. Dioctahedral vermiculite of soils may be represented by the formula, $X_{1.3}Al_2Si_6(Al_3Fe^{3+}_{1.1}Mg_{0.2})O_{20}(OH)_4 \cdot nH_2O$ in which X represents the exchangeable cations that counter the net negative charge (Jackson, 1965).

Montmorillonite. The minerals of the *montmorillonite* or *smectite* isomorphous series are freely expansible layer silicates responsible for soil plasticity. The spacing of the layers varies with the exchangeable cation species and the degree of interlayer solvation, and ranges from 12–18Å. Complete drying yields a spacing of less than 10Å. Full hydration can float the layers apart, independent of each other. A model soil montmorillonite formula is $X_{0.9}(Al_{0.4}Si_{7.6})(Al_{2.6}Fe^{3+}_{0.9}Mg_{0.5})O_{20}(OH)_4 \cdot nH_2O$, representative of several locations in the region of the United States between the Appalachian and Rocky Mountains (Figure 4), in India, in Queensland (Australia), and in Hawaii. *Nontronite* $[X_{0.7}(Al_{0.7}Si_{7.3})Fe_4^{3+}O_{20}(OH)_4 \cdot nH_2O]$ is the ferric endmember of the series. Montmorillonite with approximately half the negative charge in the octahedral layer and half in the tetrahedral layer appears to be common. *Beidellite* $[X_{0.9}(Al_{0.9}Si_{7.1})Al_4O_{20}(OH_4 \cdot nH_2O]$ is the ideal endmember of the moutmorillonite series with the entire negative charge arising in the tetrahedral layer.

An environment abundant in the exchangeable cations calcium and magnesium, silica-rich, moist, and not excessively leached favors the formation of montmorillonite. Mont-

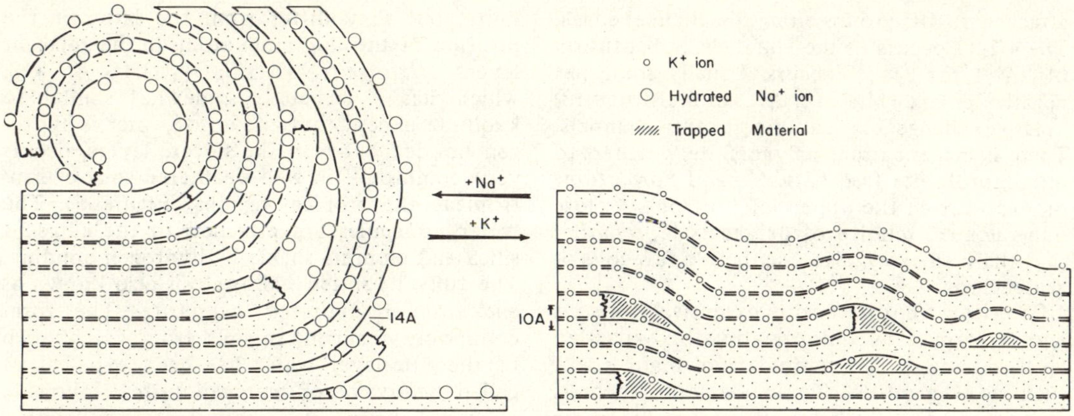

Chemistry of the Soil

FIGURE 3. Wedge position where curled vermiculite margins of mica cores are tangential to (001) cleavage plane of mica. Hydrated cations cause opening of the wedge and fixing cations cause unscrolling. (From Raman and Jackson, 1964)

morillonite structures may include lattice "mistakes" with some tetrahedra pointed toward the interlayer space. Excess hydroxyls occur in montmorillonite as positive hydroxy aluminum or sesquioxidic interlayers on the layer surfaces. When the hydroxy units adhere to only one layer, the clay remains swelling. Jackson (1965) proposed that these tactile hydroxy units combine with soluble silica–neutral $Si(OH)_4$ molecules in the interlayer–and form kaolinite or new montmorillonite layers, a clay nucleation mechanism. The hydroxy units may attach to adjacent layers to form chlorite areas of the structure.

Palygorskite. Palygorskite, which is also widely known as *attapulgite,* occurs abundantly in the clay fraction of soils in areas around the eastern Mediterranean Sea and occurs in other arid areas. The ideal endmember formula $[Si_8Mg_5(OH_2)_4O_{20}(OH)_2 \cdot 4H_2O]$ has protons

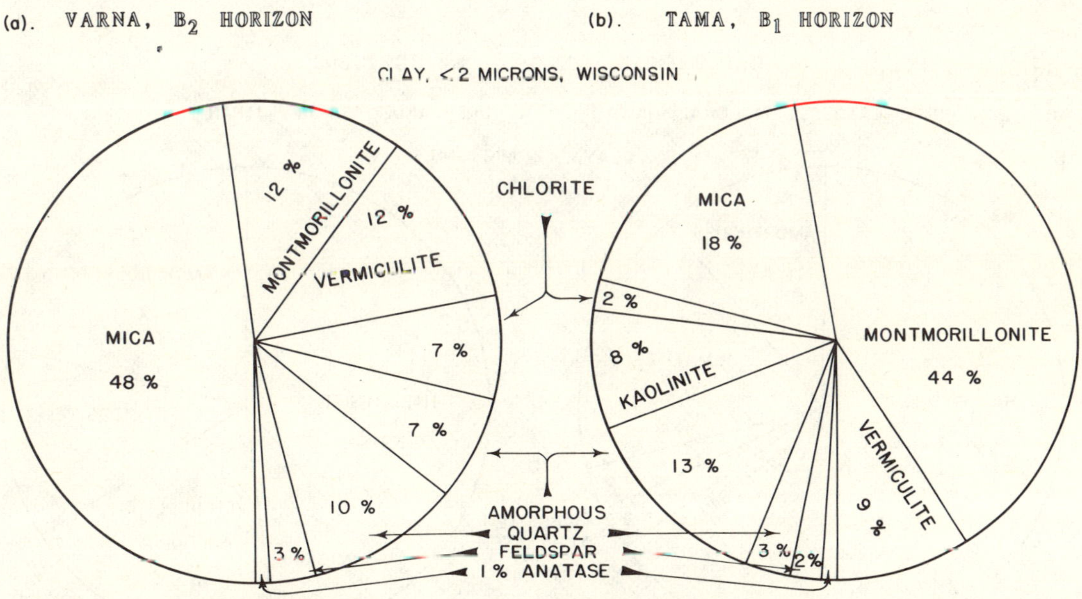

FIGURE 4. Mineralogical composition of clays from soils derived from glacial till and loess. (a) Varna soil of southeastern Wisconsin, derived from fine-textured till originating from shale; limited weathering has occurred. (b) Tama soil of southwestern Wisconsin, derived from loess originating from crystalline rocks, sandstones, and limestones; with the moderate permeability of the loess, weathering has been more extensive. (Percentages from Fanning and Jackson, 1965)

attached to OH groups along the channel edges. The $4H_2O$ occurs in the channels. Substitution of $2Al^{3+}$ for $3Mg^{2+}$ occurs. Usually some net negative charge (Mg^{2+} for Al^{3+} in the structure) holds exchangeable cations in the channels. There is evidence that palygorskite weathers to montmorillonite (see *Clay Mineral Formation*) on exposure in the upper soil horizons. A close mineralogical relative of palygorskite, *sepiolite* (a 3-chain silicate), also occurs in a few soils of arid regions.

Chlorite. *Chlorite* occurs extensively in soils and is formed by chloritization of expanded layer silicates and inherited from mafic rocks such as *serpentine.* Toxic elements such as chromium and nickel can occur in mafic chlorites. Serpentine-derived soils are sometimes outstanding for infertility and are known as "serpentine barrens" owing either to the presence of toxic elements or to deficiency of certain essential elements such as calcium. *Trioctahedral chlorite* commonly occurs in sedimentary rocks, and productive soils are derived therefrom (Figure 4). *Dioctahedral chlorite,* which may be represented by the formula $Al_{4.4}(OH)_{12} \cdot (Al_{1.2}Si_{6.8})Al_4O_{20}(OH)_4$, in which the $Al_{4.4}(OH)_{12}$ is the interlayer hydroxyl layer with 1.2 (+) charge, occurs in well-weathered, acid soils.

Kaolinite and Halloysite. *Kaolinite* [$Al_4Si_4O_{10}(OH)_8$] occurs commonly in well-weathered soils. *Dickite* and *nacrite* are isomers of kaolinite, but they differ from the latter in the position of the aluminum atoms in the adjacent layers. *Halloysite* [$Al_4Si_4O_{10}(OH)_8 \cdot 4H_2O$], which has a structure somewhat similar to kaolinite except that a layer of water is hydrogen bonded between the silicate layers, occurs very commonly in well-weathered soils of subtropical and tropical regions (Figure 5). The water molecules are attached to the adjacent silica and alumina sheets by hydrogen bonding. The fully hydrated halloysite is also known as *endellite* or *hydrated halloysite.* The form commonly occurring in soils, from which much of the interlayer water has been expelled, is called *halloysite* or *metahalloysite.* Halloysite of soils commonly forms by crystallization from extremely small particles of allophane and by the weathering of feldspathic rocks.

Oxide Minerals of Soils

Soils with free drainage are leached of silicon more rapidly than aluminum, iron, and titanium, and these elements accumulate as amorphous and crystalline oxides (Mackenzie, 1975).

Allophane. Allophane is a general term for amorphous aluminosilicate gels containing Al_2O_3, SiO_2, H_2O, and OH and usually some Fe_2O_3. Allophane of surface soils is associated with abundant humus, and such soil horizons are dark colored or Ando soils [Figure 5(b)].

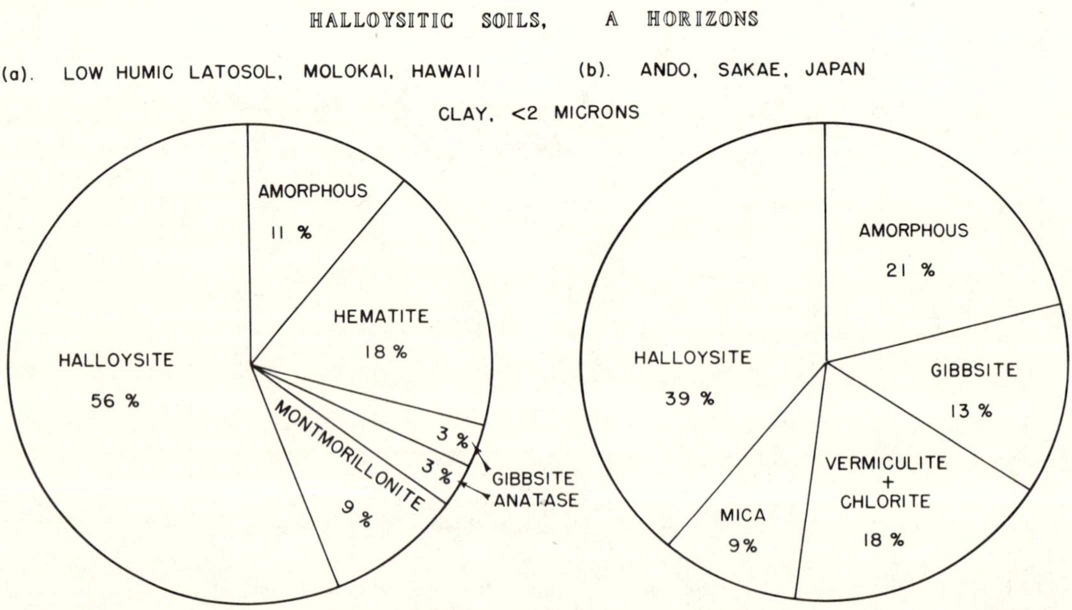

FIGURE 5. Mineralogical composition of clays of two halloysitic soils derived from volcanic rocks. (a) of Molokai soil from basalt and ash. (Tamura, Jackson and Sherman, 1953). (b) Sakae soil from volcanic ash (Aomine and Jackson, 1959).

Sesquioxidic allophanes, common (20–90%) in soils derived from volcanic ash, usually have a SiO_2/Al_2O_3 molar ratio range from 0.8 to 1.4 and the cation exchange capacity of about 1.0 meq/g. These allophanes are highly reactive and unstable. More siliceous, halloysitic allophanes of soils have a SiO_2/Al_2O_3 molar ratio of 1.5 to 2.3 and a cation exchange capacity of 0.1 to 0.2 meq/g. These allophanes are more stable as they are formed by weathering of halloysite and kaolinite or by silication of unstable allophane.

Aluminum Oxides. The crystalline mineral *gibbsite* [$Al(OH)_3$] is the most abundant (5–40% of the clay) free hydrous oxide of alumina in soils, occurring primarily in soils of tropical and subtropical regions that have undergone intensive leaching of silica from feldspars and other minerals (Figure 6). Amorphous hydrous oxides of aluminum occur in highly weathered wet soils as gels also containing iron (Millott, 1970).

The hydroxy aluminum oxide, *boehmite* (AlOOH), occurs in intensively leached, highly weathered soils, frequently in assocation with gibbsite. Soils that have been weathered for long time periods through unusual geomorphological circumstances that permit the weathering to extend for periods of several million years may develop appreciable amounts of boehmite along with the large amounts of gibbsite in deposits known as bauxite (Loughnan and Bayliss, 1961). The isomer of boehmite, *diaspore* (AlOOH), occurs in some aluminous fireclay deposits.

Iron Oxides. The most common iron oxides (see *Hydrous Oxides*) are *hematite* (Fe_2O_3), which gives the pink to bright red color of soils, and *goethite* [FeOOH (or $Fe_2O_3 \cdot H_2O$)], which gives the brown and dark reddish brown tints to soils [Figure 6(a)]. The iron oxide contents of soils vary widely, from less than 0.5% to as much as 80%. Hematite of coarsely crystalline form may occur in silt and sand fractions of soils; it then has a deep purple color. *Martite* (Fe_2O_3) has a structure similar to hematite. Dark iron oxide coatings are often associated with coatings of *pyrolusite* (MnO_2). When finely divided, the bright red color of hematite is intensified (as in the hematite streak) while the brownish color of goethite grades toward yellowish. Finely divided yellowish, hydrated goethite of soils is sometimes designated as *limonite* ($2Fe_2O_3 \cdot 3H_2O$).

Goethite is abundant (20–80% of the soil) in concretions in certain latosols (Oxisols), iron stone (Ortstein), and subsoils of podzols (Spodosols). Aluminum has been found in certain Australian soils to substitute for ferric iron ions in goethite structures up to approximately 20 atomic percent (Norrish and Taylor, 1961). In poorly drained organic soils, *lepido-*

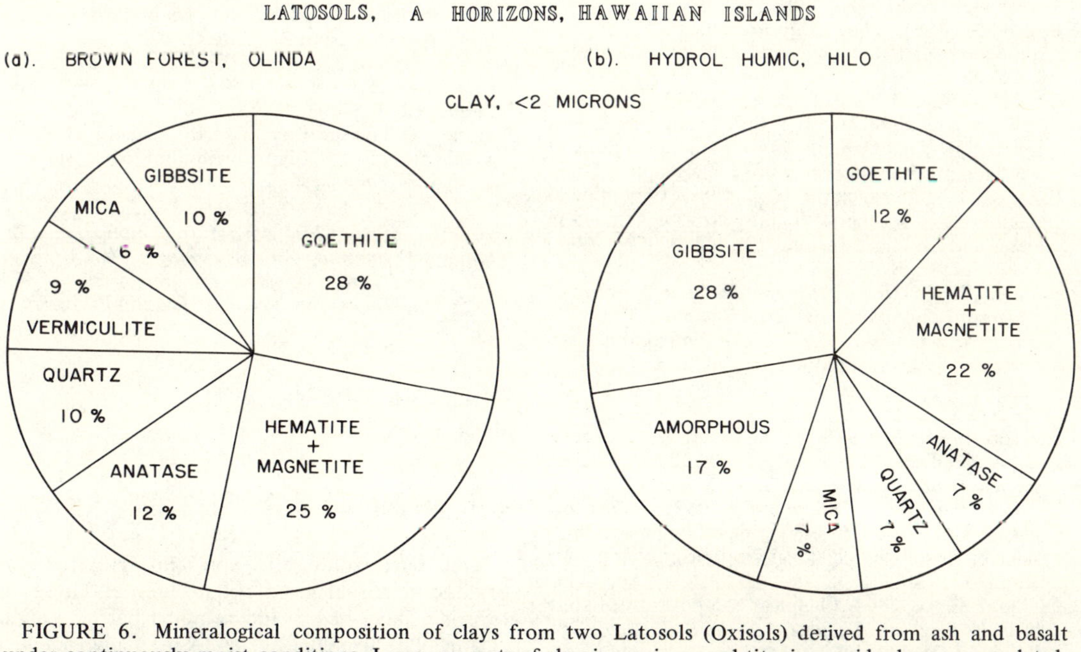

FIGURE 6. Mineralogical composition of clays from two Latosols (Oxisols) derived from ash and basalt under continuously moist conditions. Large amounts of aluminum, iron, and titanium oxides have accumulated. (a) Olinda soil with about 100 in annual rainfall. (b) Hilo soil with several hundred inches annual rainfall. (Percentages from Tamura, Jackson, and Sherman 1953, 1955)

crocite (FeOOH), an isomer of goethite, has frequently been reported. The magnetic iron oxide *magnetite* (Fe_3O_4) occurs in soils usually as a sand-size mineral. Oxidation of the iron in Fe_3O_4 yields *maghemite* (Fe_2O_3), which is still magnetic.

Titanium Oxides. *Rutile* (TiO_2) and *anatase* (TiO_2) commonly occur as 1% or so of soils and clay sediments. Rutile in soils may be inherited from rocks while anatase may be pedogenic and enriched during soil weathering (Figure 6). Although the titanium content of layer silicate samples is frequently placed in the octahedral layer in the calculation of formulas for soils and clay deposits, it is now recognized that titantium can usually be concentrated as euhedral rutile and anatase crystals that had been occluded in the layer silicate specimens. Amorphous hydrous titanium oxide is known as *leucoxene* ($TiO_2 \cdot nH_2O$). *Ilmenite* ($TiFeO_3$) is common in some tropical soils, inherited from mafic rocks. Substitution of titanium for iron in magnetite gives the *titanomagnetite* ($TiFe_2O_4$) to *titanomaghemite* ($TiFe_2O_5$) series, which occurs in Hawaiian soils.

M. L. JACKSON

References

Aomine, S., and Jackson, M. L., 1959. Allophane determination in Ando soils by cation-exchange capacity delta value, *Soil Sci. Soc. America Proc.*, **23**, 210–213.

Brewer, R., 1964. *Fabric and Mineral Analysis of Soils.* New York: Wiley, 470p.

Brown, G., 1961. *The X-Ray Identification and Crystal Structure of Clay Minerals.* London: Mineralogical Society, 544p.

Dixson, J. B., and Weed, S. B., eds., 1977. *Minerals in Soil Environments.* Madison: Soil Science Society of America, 948p. (Contains up-to-date discussions of most soil minerals and extensive bibliographies.)

Fanning, D. S., and Jackson, M. L., 1965. Clay mineral weathering in southern Wisconsin soils developed in loess and in shale-derived till, *Clays Clay Min.*, **13**, 175–191.

Farmer, V. C., ed., 1974. *The Infrared Spectra of Minerals.* London: Mineralogical Society, 539p.

Gard, J. A., ed., 1971. *The Electron-optical Investigation of Clay.* London: Mineralogical Society, 383p.

Grim R. E., 1968. *Clay Mineralogy.* New York: McGraw-Hill, 596p.

Jackson, M. L.; Tyler, S. A.; Willis, A. L.; Bourbeau, G. A.; and Pennington, R. P., 1948. Weathering sequence of clay-size minerals in soils and sediments, *Jour. Phys. Colloid Chem.*, **52**, 1237–1260.

Jackson, M. L., 1963. Interlayering of expansible layer silicates in soils by chemical weathering, *Clays Clay Min..* **11**, 29–46.

Jackson, M. L., 1964. Chemical composition of soils, *in* F. E. Bear, ed., *Chemistry of the Soil,* New York: Reinhold, 71-141.

Jackson, M. L., 1965. Clay transformations in soil genesis during the Quaternary, *Soil Sci.*, **99**, 15–22.

Jackson, M. L.; Gillette, D. A.; Danielsen, E. F.; Blifford, I. H.; Bryson, R. A.; and Syers, J. K., 1973. Global dustfall during the Quaternary as related to environments, *Soil Sci.*, **116**, 135-145.

Loughnan, F. C., and Bayliss, P., 1961. The mineralogy of bauxite deposits near Weipa, Queensland, *Am. Mineral,* **46**, 209–217.

Mackenzie, R. C., 1955. The classification and nomenclature of clay minerals, *Clay Min. Bull.*, 4, 52–66.

Mackenzie, R. C., 1975. The classification of soil silicates and oxides, in J. E. Gieseking, ed., *Soil Components; Volume 2, Inorganic Components.* New York: Springer-Verlag, 1-25.

Millot, G., 1970. *Geology of Clays.* New York: Springer-Verlag, 429p.

Mitchell, B. D., 1975. Oxides and hydrous oxides of silicon, *in* J. E. Gieseking, ed., *Soil Components; Vol. 2: Inorganic Components.* New York: Springer-Verlag, 395–432.

Norrish, K., and Taylor, R. M., 1961. The isomorphous replacement of iron by aluminum in soil goethites, *Jour. Soil Sci.*, **12**, 294–306.

Raman, K. V., and Jackson, M. L., 1964. Vermiculite surface morphology, *Clays Clay Min.*, **12**, 423–429.

Stelly, M., and Dinauer, R. C., eds., 1968. *Mineralogy in Soil Science and Engineering.* Madison: Soil Science Society of America, 106p.

Tamura, T.; Jackson, M. L.; and Sherman, G. D., 1953. Mineral content of low humic, humic and hydrol humic latosols of Hawaii, *Soil Sci. Soc. America Proc.*, **17**, 343–346.

Tamura, T.; Jackson, M. L.; and Sherman, G. D., 1955. Mineral content of a latosolic brown forest soil and a humic ferruginous latosol of Hawaii, *Soil Sci. Soc. America Proc.*, **19**, 435–439.

Thorez, J., 1975. *Phyllosilicates and Clay Minerals: A Laboratory Handbook for Their X-ray Diffraction Analysis.* Dison, Belgium: G. Lelotte, 582p.

van der Plas, C., and van Reeuwijk, L. P., 1974. From mutable compounds to soil minerals, *Geoderma,* **12**, 385–405.

Warshaw, C. M., and Roy, R., 1961. Classification and a scheme for the identification of layer silicates, *Geol. Soc. America Bull.*, **72**, 1455–1492.

Cross-references: *Clay Mineral Decomposition; Clay Mineral Formation; Clay Mineralogy; Clay Minerals, Silicates; Hydrous Oxides; Mineral Analysis. Vol. IVB: Soil Mineralogy. Vol. XII: Pedosesquioxides.*

SOIL MOISTURE—See SOIL SOLUTION; SOIL WATER

SOIL PHYSICS

Men have tilled the soil and irrigated and drained it for at least six millennia. These enterprises have been (and remain) basic to civilization. The use of physics in men's works with the soil had, necessarily, to await the emergence of physics itself, but precepts that make good physical sense had, of course, been abroad. The earliest known farmer's almanac, the *Sumerian Georgica,* composed more than 4000 years ago, includes instruction to the farmer to keep trampling oxen and other prowlers off newly irrigated soil (Kramer, 1958). And 2000 years ago, Virgil (*Georgics,* Book 1, lines 89–90) suggested that on some soils, the benefit of burning stubble is that "the heat opens up fresh ducts and hidden pores through which the juices of the soil may move to the growing plants."

The first natural scientist to give attention to the soil was probably John Evelyn, an Original Fellow of the Royal Society of London. During 1675 Evelyn gave two lectures to the Society titled "A Philosophical Discourse of Earth, Relating to the Culture and the Improvement of It for Vegetation and the Propoagation of Plants." The Society ordered the discourse printed, and it appeared the following year (Evelyn, 1676). This substantial treatise reports, among other things, his observations "with an indifferent Microscope" on various soils, unwashed and washed, unground and ground. Its content of physics is slight, though Evelyn, like Virgil, recognizes the vital role of pores and pore geometry. He writes, "Clay is . . . a cursed step-dame to almost all Vegetation, as having few or no Meatuses for the percolation of the alimental Showers, or expansion of the Roots."

Systematic scientific study of agriculture effectively began in the first half of the nineteenth century, and physical studies of the soil were a large part of the early work. In the period 1834–1850, however, there emerged the promise of a totally chemical theory of soil fertility, and soil studies became almost exclusively chemical. The turn of the present century saw the advent of a protracted period during which most soil physicists were preoccupied with the size distribution of the soil solid particles (see *Particle-size Distribution*). For a time the primary demand of physics was for methods of analysis and observation that would fit into procedures for surveying, naming, and classifying soils. Also there existed a fundamental perplexity as to how to proceed with the proper scientific study of the physical processes involving soil water. Nevertheless, in the context of engineering hydrology, the basis for quantitative prediction of fluid flow in *saturated* porous materials had been formally stated in 1856 by Henry Darcy, who demonstrated that the flow of water per unit area of "soil" is simply proportional to the gradient of water pressure (Darcy, 1856). Darcy's law (see *Flow Theory*), as conceived, did not pretend to describe flow in *unsaturated* materials, but it was to become a prime ingredient of a powerful conceptual basis for progress in soil-water studies enunciated in 1907 by Edgar Buckingham (Buckingham, 1907); in the succeeding two decades few soil physicists grasped its significance, and none built on it creatively.

Gradually, however, there came a recognition that it was the size and configuration of the pores (see *Soil Pores*) rather than of the particles that were the prime physical determinants of soil behavior. Schofield and Russell (1947) wrote, "The particle size distribution of soil has only limited significance for agricultural research work. It helps to place soils in categories such as sands, loams or clays, but this can be done almost equally well by the feel of the soil between the fingers and under foot . . . . The physical properties that are of importance in soil work are mainly concerned with the pore size distribution in the soil." And Childs and George (1948) observed that "physically a soil may be regarded as completely specified by the geometry of the interface between the solid component and the void space together with that of the air-water interface within the voids." These observations, which were by no means isolated, were symptomatic of a general return to a rational course of soil-physical study, the path of which had been charted forty years earlier.

The activities of contemporary soil physicists are reflected in the contents of ten 1975–76 issues of the Soil Science Society of America Journal. Of 66 soil physics papers, 58 are concerned either directly or indirectly with the behavior of the soil-water system. Of the 8 remaining papers, 3 deal with soil heat flux, 2 with the reflectance of the soil surface, and 1 each with apparatus, gaseous diffusion, and the effects of the size distribution of particles on the bulk density (q.v.) of soil.

In its natural state the soil (q.v.) is normally a three-component porous medium consisting of solid soil particles, water, and air; i.e., it is normally *unsaturated.* Much of the water involved in the hydrologic cycle (q.v. in vol. IVA) is located in unsaturated soil between the time of its arrival as rain at the soil surface and that of its return to the atmosphere. The processes of water movement (q.v.) in unsaturated soil therefore play a central part in the scientific study of the terrestrial sector of the hydrologic cycle and in the problems of dry-land and irrigated agriculture (see *Irrigation*), of plant ecology, and also of soil biology. They are, in addition, of great signficance in connection with the transport through the soil of materials in solution such as natural salts (see *Dispersion Phenomena*), fertilizers, and urban and industrial wastes. Scientific phenomena of great interest and importance include infiltration (q.v.); drainage (see *Soil Drainage*), and retention of water in the soil layers (see *Water Content and Retention*); extraction of water by plant roots; and the evaporation (q.v.) of water from the soil.

In situ soil can be considered as a matrix of discrete solid elements permeated by voids, or pore spaces (Figure 1). The solid elements vary in size and shape both within a particular soil and between constitutionally different soils. It follows, therefore, that the dimensional characteristics of pore spaces are also highly variable (see *Pore-size Distribution*).

The solid phase of the soil has mineral and principally plant-derived organic components (see *Soil Components*). The organic matter (q.v.), which is usually highly colloidal, seldom exceeds 5–10% by weight of soil. It does, however, together with clay mineral particles (see *Clay-Organic Reactions*), appear to play an important part in the genesis of soil structure, the arrangement of aggregates of ultimate individual soil particles, and voids (see *Aggregation* and *Soil Structure*). In an agricultural context, the main interest in soil structure is in terms of soil tilth. Tilth is a general term relating to the ability of aggregates to maintain their integrity when the soil is irrigated, tilled, or otherwise worked so that water retention and drainage and aeration are kept at favorable levels (see *Aggregate Stability; Tillage*).

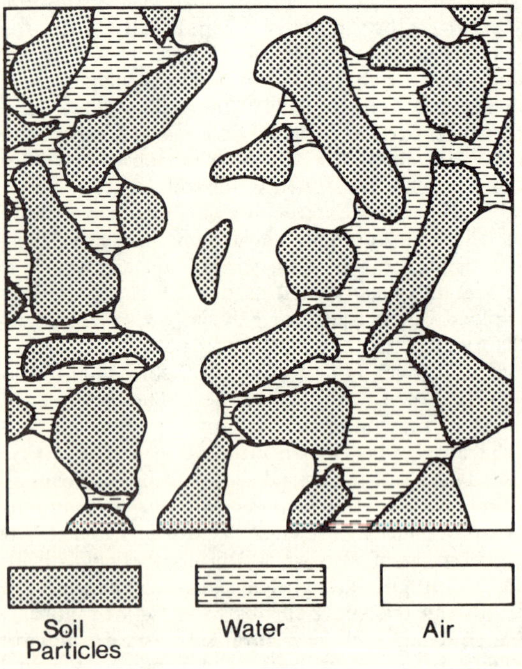

FIGURE 1. Schematic cross-section of unsaturated soil.

The interactions between the soil matrix and soil water (or more precisely, the soil solution, q.v.) are among the fundamental determinants of the physical behavior of water in unsaturated soil. The relative magnitudes of the interactive forces involved are to a degree dependent on the composition of the matrix, but more so on the geometric characteristics of the system as a whole. In other words, the physical behavior of water in a particular soil, over a range of moisture contents, is primarily dependent on the soil's architecture; the spatial arrangement of its solid elements and voids.

Despite the fact that the physical behavior of soil is so much a reflection of its structural state, it is often considered of use to have a measure of the size distribution of the ultimate individual particles making up a soil (see *Particle-size Distribution*). Methods of achieving this, which fall under the heading of particle-size analysis, are well established; see, for example, Day (1965). Figure 2 shows solid-void relationships for two soils, with total solids partitioned on the basis of particle-size analyses. The proportions of air- and water-filled pore space were determined a few days after heavy irrigation or rain when the characteristic initial period of rapid drainage had ceased (see *Field Water Cycle*).

Particle-size analysis is primarily a tool of the soil surveyors and classifiers; together with field estimates of soil texture (the feel of a soil when moistened and manipulated in the hand), it helps categorize soils as types of sands, loams, and clays (see Vol. XII; *Field Texture*). This information permits qualitative inference of many of the behavioral properties of a soil. Of particular interest are the relative proportions of the various clay minerals in soils because these constituents are physically and chemically the most active. It is well known, for example, that soils containing clay minerals with a high sodium saturation are likely to be physically unstable and intractable in irrigated agriculture, whereas the reverse is usually true for soils in which calcium is the dominant cation in the exchange complex (see *Exchange Phenomena*).

These and other important effects arise from the surface activity of clay mineral particles. An important and large share of the clays consists of a class of distinctly crystalline materials, the aluminosilicates (see *Clay Minerals, Silicates*). Their chemistry, which is a branch of colloid chemistry, is discussed by van Olphen (1963), for example. The occurrence of isomorphous substitutions and edge imperfections in the crystal lattices of clay minerals endow clay particle surfaces with a net negative charge. In

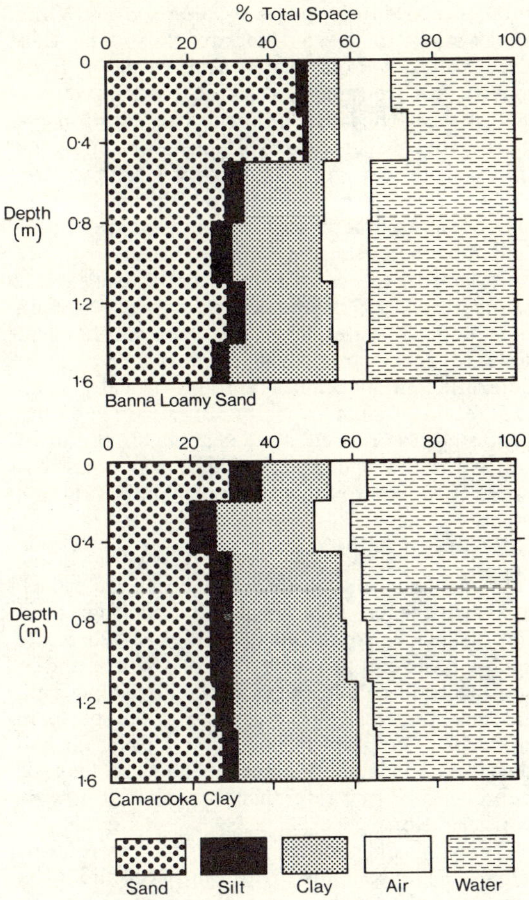

FIGURE 2. Solid-void relationships and particle-size distributions for two soils. *Particle-size analyses* (values are volumetric percentages of total solids): Banna loamy sand (0–0.4 m sample): sand (particles >20 μm diameter), 81.6; silt (2–20 μm), 4.5; clay (<2 μm), 13.9 Camarooka clay (0.75–1 m sample): sand, 42.8; silt, 8.1; clay, 49.1.

addition, and clearly not unrelatedly, since we are dealing with colloidal particles, there is an obvious correlation between particle size and specific surface (which simply means the total surface area of the particles in unit mass or volume of material). Though surface activity is displayed by all soil particles, the clay fraction especially with the inclusion of amorphous substances usually makes an overwhelming contribution to total specific surface.

Because they are negatively charged, clay mineral particles attract and bind cations from the surrounding soil solution (see *Clays, Physical Chemistry*). As a result, there is associated with the clay surface an electrical double layer, the extent of which depends on the cation type and concentration in the ambient equilibrium soil solution. The characteristics of the double layer affect the colloidal stability of clays, and since clay minerals (together with colloidal organic matter) are the main bridging agents linking the noncolloidal soil constituents, so also is soil structure affected.

A structural characteristic of all soils of high clay-colloid content is the swelling behavior they display on hydration (see *Soil Mechanics*). In other words, the soil solid matrix is not rigid, and the volume of a swelling soil system varies according to its moisture content. The swelling forces arising from hydration may, when severe, contribute to the disruption of aggregates, a first step in soil structural deterioration.

The reader seeking a more detailed knowledge of soil-physical topics is referred initially to the standard texts [e.g., Childs (1969); Hillel (1971); Baver et al (1972); Kirkham and Powers (1972)] and to relevant entries in this volume.

ED HIGHLEY

References

Baver, L. D.; Gardner, W. H.; and Gardner, W. R., 1972. *Soil Physics,* 4th ed. New York: Wiley, 498p.

Buckingham, E., 1907. Studies on the movement of soil moisture, *U.S. Dept. Agric. Bur. Soils Bull.*, 38, 61p.

Childs, E. C., 1969. *An Introduction to the Physical Basis of Soil Water Phenomena.* New York: Interscience, 493p.

Childs, E. C., and George, N. C., 1948. Soil geometry and soil water equilibrium, *Discussions of the Faraday Society,* **3,** 78–85.

Darcy, H., 1856. *Les Fontaines Publiques de la Ville de Dijon.* Paris: Victor Dalmont, 647p.

Day, P. R., 1965. Particle fractionation and particle-size analysis, *in* C. A. Black, et al., eds, *Methods of Soil Analysis: Part 1.* Madison: American Society of Agronomy, 454–567.

Evelyn, J., 1676. *A Philosophical Discourse of Earth, Relating to the Culture and the Improvement of It for Vegetation, and the Propagation of Plants.* London: The Royal Society, 182p.

Gardner, W. H., 1977. Historical highlights in American soil physics, *Soil Sci. Soc. America Jour.*, **41**(2), 221–229.

Hillel, D., 1971. *Soil and Water: Physical Principles and Processes.* New York: Academic, 288p.

Kirkham, D., and Powers, W. L., 1972. *Advanced Soil Physics.* New York: Interscience, 534p.

Kramer, S. N., 1958. *History Begins at Sumer.* London: Thames and Hudson, 333p.

van Olphen, H., 1963. *An Introduction to Clay Colloid Chemistry.* New York: Interscience, 301p.

Schofield, R. K., and Russell, E. W., 1947. The significance of particle size in relation to soils, *In Symposium on Particle Size Analysis,* Suppl. to *Trans. Inst. Chem. Eng. London,* **25,** 94–95.

Cross-references: *Aggregate Stability; Aggregation; Bulk Density; Clays, Physical Chemistry; Flow Theory; Imbibition; Infiltration; Particle Density;*

Percolation; Permeability; Physical Chemistry; Pore-size Distribution; Soil Drainage; Soil Pores; Soil Structure; Soil Water; Surface-Soil Water Content; Thermal Regimes; Thermodynamics, Soil Water; Thermoluminescence; Water Content and Retention; Water Fluxes; Water Movement; Wetting Front.

SOIL PORES

Soil pores are that portion of the bulk volume of soil not occupied by solid particles. The *pore space* of a soil thus consists of interstices and voids. Surface soils should ideally be about 50% pore space that is occupied by air and water in roughly equal proportions. This proportion is, however, subject to great fluctuations in soils. The volume composition of compact subsoils, for example, is characterized by higher percentages of small pores, minerals, and water and considerably lower contents of organic matter and air.

Amount of Pore Space

The amount of pore space is determined by the arrangement of solid particles and is measured by the *porosity*. If the solids are packed close together as in compact subsoils, the total porosity is low. When they are arranged in porous aggregates, the pore space per unit volume is high. The percentage of total pore space is given by the relation

$$\% \text{ pore space} = 100 - \left(\frac{\text{bulk density}}{\text{particle density}}\right) \times 100$$

The total pore space differs from soil to soil and depends on various factors such as the nature of the solid particles, presence of organic matter, granulation, drainage conditions, and biological activity. Sandy surface soils, for example, typically show a range from 35–50% whereas clayey soils may range from 40–60% or more. Some compact subsoils have porosities as low as 10%.

Size of Pores

Soil pores are classified according to size where measurements are standardized on the length of the shortest dimension (Brewer, 1964). The choice of size limits usually has some significance in soil genesis and for plant growth. The concept of *useful-pore-space* (De Boodt and De Leenheer, 1954), for example, is based on the notion that water is generally unavailable to plants from pores with a diameter smaller than 0.2 μm. The pore space between fabric units are larger than those between particles within fabric units. The larger interfabric pores are referred to as *macropores,* and the smaller intrafabric pores as *micropores* (Yong and Warkentin, 1975). Macropores so large that they will not hold water against the pull of gravity are frequently referred to as *noncapillary pores.* They are important in aeration because their aggregate volume represents the minimal air capacity of the soil (Daubenmire, 1964). *Mesopores,* which are intermediate in size between macro- and micropores, are important for the intensive renewal of air in fabric units and in transporting and distributing water (Jongerius, 1957). Smaller voids, which remain occupied by water after a wetted soil has had time to drain, are called *capillary pores.* They determine the maximum amount of water a soil can retain at field capacity. Brewer (1964) has proposed a comprehensive size classification of soil voids that recognizes several pore units that range from macrovoids to cryptovoids (Table 1).

Influence of Soil Behavior

In one way or another soil pores are associated with such diverse phenomena that their importance is sometimes overlooked. Pores are such a basic soil property that it is worth underscoring their role in the many biological, chemical and physical processes that take place in the soil. The importance of soil pores in soil development and in plant production may be appreciated by reference to the following considerations.

Capacity of soil for water flow (see *Flow Theory*) and soil distortion and consolidation, for example, require knowledge of characteristic water movement (q.v.) in pore spaces. The electroosmotic stabilization of soils (see *Soil Stabilization*) and the use of soil conditioners (q.v.) for engineering purposes is particularly important in this regard (Ingles and Metcalf, 1973). Quick conditions (see *Thixotropy, Thixotropism.* Vol. XIII: *Expansive Soils* and *Quick Sand-Quick Clay*) occur where there is a strong pressure grandient in the ground water and the porosity is such that grains are set free in their surrounding pore water to form a viscous liquid. This condition, which may be observed in exca-

TABLE 1. Size Classification of Voids.

Class Name	Subclass	Class Limits (μm)
Macrovoids		> 75
	Very fine macrovoids	75–1000
	Fine macrovoids	1000–2000
	Medium macrovoids	2000–5000
	Coarse macrovoids	5000
Mesovoids		30–75
Microvoids		5–30
Ultramicrovoids		< 5
Cryptovoids		< 0.1

From Brewer (1964)

vation pits or earth enbankments, can result in earth flows and flow slides. *Piping* (q.v. in Vol. XVI), which results from the selective removal of particles from macropores by moving water, can result in subsurface soil erosion.

Plowing when a soil is too wet or the continuous cropping of soils often results in a reduction of macropore spaces. The destruction of favorable soil structure (q.v.) resulting in a decrease in porosity and permeability (q.v.) may, for example, contribute to inferior aeration (see *Aeration, Respiration, Atmosphere*); poor drainage conditions (see *Pore Space, Drainable* and *Soil Drainage*); decreases in the activity of microorganisms (see *Metabolic Activity*) by destroying their living space (see *Microhabitats*); and possibly the development of denser, more compact (see *Bulk Density*) subsurface layers (see *Profiles, Physical Modification*. Vol. XII: *Soil Pans*). Thus one of the chief benefits of plowing, the production of porosity that will favor germination and seedling root growth, is lost. The continuity of pores (see *Pore-size Distribution*) is almost as important as the maintenance of their total volume because the network of voids provides conduits for the movement of water, the diffusion and dispersion of gases and solvents (see *Diffusion Phenomena* and *Dispersion Phenomena*), and the transport of microorganisms (see *Microorganism Transport*); it also provides pathways for the movement of nutrients in the soil solution (q.v.). Pores also play an important role in structural development (see *Aggregation*) as well as provide space for root development and plant anchorage.

Water retention is due partly to capillary forces (see *Capillary Pressure*), which are in turn related to the size of pores. Because water is retained in small pores, a changing water content results from a changing proportion of water and air in the pores or from changes in soil porosity. Thus the movement of soil water (or lack of it), which is related to the size, shape, and arrangement of pores, affects numerous attributes of soils such as their strength (as related to pore-water pressures in cohesive soils), tilth, pH, Eh, and fertility. In addition to influencing water movement, the shape and orientation of pores may contribute to the physical anistropy of soil materials, which may in turn affect the degree of leaching (q.v.) and the availability of water and nutrients to plant roots (see *Water Content and Retention*).

CHARLES W. FINKL, JNR.

References

Bouma, J., and Anderson, J. L., 1973. Relationships between soil structure characteristics and hydraulic conductivity, *in* R. R. Bruce, K. W. Flach, and H. M. Taylor, eds., *Field Soil Water Regime.* Madison: Soil Science Society of America, 77-105.

Brady, N. C., 1974. *The Nature and Properties of Soils.* New York: Macmillan, 639p.

Brewer, R., 1964. *Fabric and Mineral Analysis of Soils.* New York: Wiley, 470p.

Daubenmire, R. F., 1964. *Plants and Environment.* New York: Wiley, 422p.

De Boodt, M., and De Leenheer, L., 1954. The practical meaning of pore sizes with respect to the texture of soil, *5th Internat. Congr. Soil Sci. Trans. (Leopoldville)*, 2, 104-109.

Ingles, O. G., and Metcalf, J. B., 1973. *Soil Stabilization.* New York: Wiley, 374p.

Johnson, W. M.; McClelland, J. E.; McCaleb, S. A.; et al, 1960. Classification and description of soil pores, *Soil Sci.*, 89, 319-321.

Jongerius, A., 1957. Morphologic investigations of soil structure, *Bodemkundige Studies No. 2, Mededelinger van de Sticktung voor Bodemkartering (Wageningen).*

Jongerius, A.; Schoonderbeek, D.; Jager, A.; and Kowalinski, St., 1972. Electrooptical soil porosity investigation by means of quantimet-B equipment, *Geoderma*, 7(3/4), 177-198.

Katchinski, N. A., 1956. Soil structure and differential porosity, *6th Internat. Congr. Soil Sci. Trans. (Paris)*, 1, 35-52.

Maeda, T. and Warkentin, B. P., 1975. Void changes in allophane soils determining water retention and transmission, *Soil Sci. Soc. America Proc.*, 39(3), 398-403.

Marshall, T. J., 1968. A relation between permeability and size distribution of pores, *Jour. Soil Sci.*, 9(1), 1-8.

Yong, R. N., and Warkentin, B. P., 1975. *Soil Properties and Behaviour.* Amsterdam: Elsevier, 449p.

Cross-references: *Bulk Density; Pore-size Distribution; Pore Space, Drainable; Soil Structure.*

SOIL REACTION

The so-called reaction of soil indicates whether it is acid, neutral, or alkaline, and it is usually expressed in terms of a pH number. The neutral point is at pH 7, while values below 7 are acid and those above 7 are alkaline. Soil pH values generally lie between 4 and 9.

The pH is probably the most informative single chemical measurement that can be made on a soil because it relates closely to the genesis, classification, and composition of the soil, and it is "probably the most widely-operating factor influencing the form, amount and movement of plant nutrients in soil" (Place, Phillips, and Brown, 1968). However, when applied to soil, the pH concept loses the relatively precise meaning it has for solutions, and *soil reaction* is preferable as a general descriptive term.

The pH Concept

The biochemist Sørensen (1909) distinguished two categories of acidity in a weak-acid solution: the *total* acidity, which reacts when the solution is neutralized with a strong base such as sodium hydroxide; and the *degree* of acidity, which is related to the number of free hydrogen ions in solutions. Hydrogen ions exist in aqueous media as hydrated protons ($H^+ + H_2O \rightarrow H_3O^+$), sometimes referred to as *hydronium ions.* They may be produced by the dissociation of a weak acid such as acetic acid ($HOAc + H_2O \rightarrow OAc^- + H_3O^+$) and comprise only a small proportion of the total amount of hydrogen that can react with bases. The distinction between these two categories of acidity is also valid for soils, and the degree of acidity defines soil reaction.

The degree of acidity is best represented by the "quasi-thermodynamic" definition of the pH number:

$$pH = -\log_{10} a_H$$

where a_H represents the thermodynamic *activity* of the hydrogen ion H_3O^+ in solution and is related to the analytical *concentration m,* in moles per liter by the activity coefficient γ, that is $a = m\gamma$. Because of difficulties in evaluating the activity of a single ion species, chemists prefer an operational definition of pH:

$$pH_x = pH_R + \frac{(E_x - E_R)}{k}$$

where pH_R is the assigned pH of a reference solution, E_x and E_R are the values of the electromotive force (EMF) of a cell reversible to hydrogen ions with the electrodes immersed in the unknown solution X and the reference solution R, respectively, and k is a temperature-dependent constant, 0.059 at 25°C. The primary reference standard is 0.050 mol/l potassium hydrogen phthalate solution, which has been assigned the pH value of 4.005 at 25°C, and there are many subsidiary reference standards. The cell commonly used in pH measurements consists of a glass electrode and a calomel reference electrode with a saturated KCl salt bridge; the electrometer with which it is used measures the cell EMF, but is calibrated in pH units (see *Electrochemistry*).

Both hydrogen and hydroxyl ions are produced by the dissociation of water so that their activities are inversely proportional to each other and are given by the "ion product" of water K_w:

$$2H_2O \rightleftharpoons H_3O^+ + OH^-;$$
$$pK_w = pH + pOH = 14.00, \text{ at } 25°C$$

A detailed account of these topics is given by Bates (1964).

Problems with Soils

This pH concept applies only to solutions. Soils are porous solids that contain only a little solution in equilibrium with electrically charged surfaces of the colloidal clay and humus fractions. It is therefore impossible to define theoretically or to determine experimentally a unique pH value, and by its very nature a soil cannot be said to have one. For example, when measuring pH in a conventional soil:water suspension, the reading alters if the suspension is stirred and if the position of the reference electrode is changed. This suspension effect is caused by the electrically charged soil colloids altering the liquid junction potential of the reference electrode and can be overcome by adding sufficient salt to the solution to give ionic strength >0.005 (Clark, 1966). The reading also varies with the type and concentration of added salt and with the ratio of soil to solution in the suspension, reflecting the distribution of exchangeable cations between the charged surfaces and the solution. Differences of up to 2 pH units may be produced in extreme cases.

A mean value for the pH of the pore solution can be measured after obtaining sufficient free solution by displacement or by pressure extraction of moist soil. The reaction of the soil as a whole, however, is reflected by the balance between hydrogen ions and other mobile cations, and the only way of estimating this is to refer the H_3O^+ activity to the "reduced ratio" of the activity of the other cations. The appropriate reference ion concentration is the sum of the divalent cations, $D = (Ca, Mg)^{2+}$, because these cations dominate the cation exchange complex in agricultural soils. The required activity ratio is then $a_H/(a_D)^{1/2}$, or expressed as a negative logarithm, $pH - \frac{1}{2}pD$; the latter is often termed the *lime potential* of soil. This function does not vary when the salt concentration and soil: solution ratio vary within reasonable limits (Schofield and Taylor, 1955).

In strongly acid mineral soils the dominant cation is Al^{3+}, and the activity of H_3O^+ should ideally be referred to it; likewise in strongly alkaline soils it should be referred to Na^+; but even in these situations it is probably preferable to retain $(Ca, Mg)^{2+}$ as the reference cation in order to more readily compare the reaction of these soils with others.

Methods of Measurement

Although the necessity of expressing soil reaction in terms of lime potential has not been generally recognized, an adequate compromise can be achieved by equilibrating all soils with 0.01 mol/l calcium chloride solution at a soil: solution ratio of 1:1 or 1:2. The term ½pD is then approximately constant so that the recorded pH value equals the lime potential plus a constant. Moreover, the $(Ca, Mg)^{2+}$ concentration in the pore solution of many agricultural soils of temperate regions is close to 0.01 mol/l so that the pH value given by this procedure is close to the reaction of the pore solution. For strongly leached soils, it is more appropriate to use a lower $CaCl_2$ concentration.

At pH > 6, the pH of a soil suspension is very dependent on the partial pressure of carbon dioxide. Thus if precise results are required, all samples should be equilibrated with the same CO_2 concentration by bubbling air through the suspension for a few minutes before taking the reading.

Other methods for measuring soil reaction should be mentioned, apart from the unsatisfactory soil.water suspension referred to previously. A water-saturated soil paste is often prepared for the extraction of soluble salts, and pH is frequently measured on this paste. Although it may be difficult to get steady readings because of variable liquid junction potentials, the measured pH should be close to that of the pore liquid. A suspension of soil in 1.0 mol/l KCl solution has also been used for pH measurements; the high salt concentration desorbs most of the exchangeable H_3O^+ and Al^{3+} ions into solution, giving anomalously low pH values; the method does not therefore adequately differentiate the reaction of strongly acid mineral soils (pH < 5) because the aluminum in solution acts as an effective buffer system.

Buffering Mechanisms Controlling Soil Reaction

The small amount of free hydrogen ion normally present in soils arises from the dissociation of a complex system of acid-base pairs. The following are the more important components of this pH-regulating buffer system in soils.

Humic Acids. The proton-donating -OH groups occur mainly in the carboxyl and phenolic functional groups of humus. Titration curves of pH reveal a range of pK values for the dissociation constants of the carboxylic acids (Posner, 1964):

$$R.CO_2H + H_2O \rightleftharpoons R.COO^- + H_3O^+; \quad pK_d = 5.5 \pm 1.7$$

The dissociation of phenolic groups in humus has not been measured, but pure phenolic compounds similar to those that may exist in humus fractions have pK_d values between 7.8 and 9.8.

The range over which these functional groups exert buffering action is approximately one pH unit on either side of the pK_d value so that the humus fraction can provide effective buffering over the whole pH range found in soils.

Aluminum Ions. Aluminum ions occur as the dominant exchangeable cation in very acid mineral soils and are important at pH < 5.2 in all soils except peats. Their precipitation as $Al(OH)_3$ over the pH range 4.0–5.2 contributes to buffering in this region (Jackson, 1963), but simple solubility product equilibria do not apply because the solid phase does not consist of pure $Al(OH)_3$ and the solution contains hydroxy-aluminum polymers as well as Al^{3+}. The pH values of some acid mineral subsoils were found (Bache, 1974) to be close to that expected from the simply hydrolysis of aluminum ions according to the following shceme:

$$H_2O + [Al(H_2O)_6]^{3+} \rightleftharpoons [Al(H_2O)_5OH]^{2+} + H_3O^+$$

Amorphous Aluminosilicates. Acid subsoils, containing no humus and free of carbonate, frequently retain strong buffering action over the pH range 5–8 even when aluminum ions have been exhaustively extracted by neutral salts. The cause appears to be the dissociation of hydrogen from silanol (Si-OH) groups, which occur at the edges of clay crystals and in amorphous aluminosilicates such as allophane.

Carbon Dioxide. Carbon dioxide dissolves in water to give an acid solution containing H_3O^+, H_2CO_3, HCO_3^-, and CO_3^{2-}. A number of equilibria are involved, but the dominant reaction is

$$CO_2(g) + 2H_2O \rightleftharpoons HCO_3^- + H_3O^+$$

The pH of the resulting solution thus depends on the partial pressure of CO_2 in the air, as shown in Figure 1. The atmosphere contains about 0.03% CO_2, which gives an equilibrium pH of 5.63 in water; but at higher CO_2 partial pressures such as may occur in soil because of the respiration of soil fauna and flora, the pH of the CO_2-water system may drop to as low as 4.4. This is one of the principal mechanisms in the development of soil acidity, but the buffering action of other soil components usually prevents soil pH dropping to this low value.

Carbonates. Calcium carbonate (calcite) is a major constituent of some soils. A suspension of calcite in pure water gives a pH of 9.6, and the solubility reaction can be written

$$H_2O + CaCO_3 \rightleftharpoons Ca^{2+} + HCO_3^- + OH^-$$

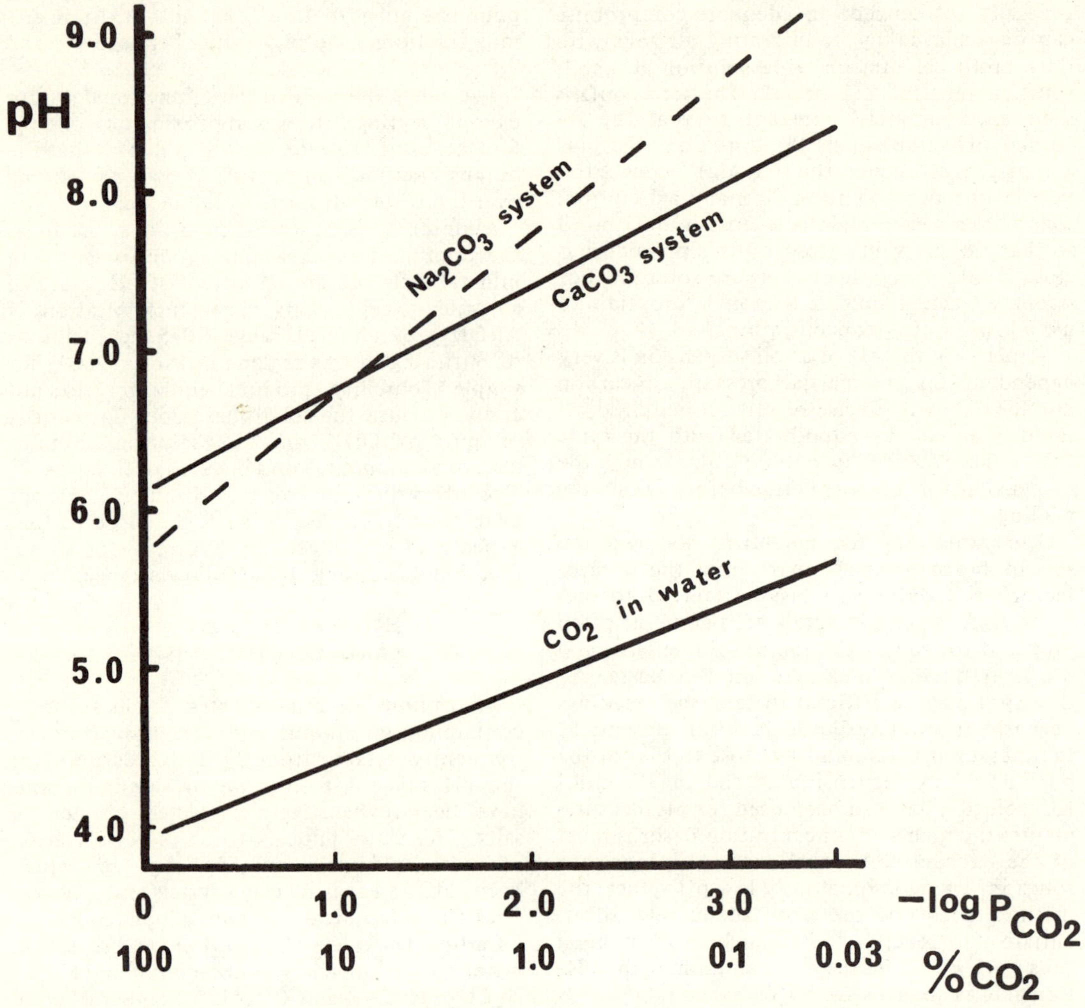

FIGURE 1. pH value as a function of the partial pressure (PCO_2) of carbon dioxide for water and for soil systems governed by $CaCO_3$ and Na_2CO_3.

However, the system is always in equilibrium with carbon dioxide, and the equilibrium pH drops to 8.35 for atmospheric CO_2 concentrations and to 6.7 when the gas phase contains 10% CO_2, as shown in Figure 1.

The saline-alkaline group of soils may contain considerable amounts of Na^+ and Mg^{2+}, which form carbonates; Figure 1 shows the pH variation with CO_2 content for Na_2CO_3 systems in 0.01 mol/l salt solution. Under waterlogged conditions, some soils contain Mn^{2+} and Fe^{2+}, which may form carbonates when the partial pressure of CO_2 is high; the pH values of such soils were found to be close to 6.1 when the soils were equilibrated with 100% CO_2, similar to the $CaCO_3$ system shown in Figure 1 (Ponnamperuma, Castro, Valencia, 1969). This may reflect the dominance of soil reactions by Ca^{2+} in the soil studied, in spite of the presence of manganese and iron.

Cation Exchange Equilibria. Buffering mechanisms that involve metal cations in solution are affected by reactions that determine the concentration of these cations. The extent of Al^{3+} hydrolysis clearly depends on the Al^{3+} concentration in solution, which in turn is governed by two further parameters, i.e., the proportion of Al^{3+} relative to other exchangeable cations, mainly $(Ca, Mg)^{2+}$, adsorbed by the soil surfaces, and the relative affinity of these cations for the surfaces as reflected by the cation exchange selectivity coefficient. These factors have been quantitatively related to the pH of subsoils (Bache, 1974).

The balance of Na^+ and Ca^{2+} in solution affects pH via the bicarbonate equilibria described previously. Thus the proportions of exchangeable Na^+ and Ca^{2+} and the exchange coefficient for this ion pair help to determine pH. A higher preference for Ca^{2+} by the exchanger lowers the Ca^{2+} concentration and increases pH (Cruz-Romero and Coleman, 1974). These reactions are relevant to saline and alkaline soils.

Soil Processes and Soil Reaction

The mineralogical composition of soils is related to soil reaction in a very general manner. For instance, calcareous soils and soils dominated by smectite clay (e.g., vertisols) normally give pH values above 7. If the prevailing weathering processes have allowed the dissolution of $CaCO_3$ and the decomposition of smectite, then soils typically give slightly acid pH values (7-5.5) while still containing 2:1 aluminosilicate clays of the vermiculite and illite type. More intense weathering, characteristic of leaching environments at higher temperatures, causes further decomposition of clay minerals giving soils containing mainly kaolinite clays and iron and aluminum hydrous-oxide minerals; these soils give pH values below 5.5.

The relation of soil reaction to soluble and exchangeable cations and its role in soil development are discussed in the entry *Base Saturation*.

The solution concentrations of specifically adsorbed anions (F^-, PO_4^{3-}, SO_4^{2-}, MoO_4^-, silicate), and their adsorption by soil minerals, is strongly pH dependent but in a rather complex manner. First the hydroxide ion competes with these anions so that they are desorbed from soil and thus give higher solution concentrations as pH rises. However, the nature of the anion-adsorbing minerals is also related to pH, as mentioned previously, so that phosphate, for example, is more strongly adsorbed and thus much less soluble in calcareous soils (pH > 7) than in soils at pH 6. Finally the adsorption of polybasic-acid anions has been related to the first dissociation constant of the acid (Hingston, Posner, and Quirk, 1972) so that a plateau in phosphate adsorption by iron and aluminum hydroxides occurs at pH 4 corresponding to pK_1 of H_3PO_4.

The concentration of heavy metal cations in solution varies widely over the pH range of soils. For a given soil, the concentrations would be expected to follow the solubility products of the hydroxides (that is, Fe^{3+}, < Al < Ca < Zn < Fe^{2+} < Co < Mn), and for a given metal, a lower solution concentration would be expected at a higher pH because of precipitation of the hydroxide by OH^-. However, this general guide is not necessarily true in practice because (a) the pure hydroxides rarely occur in soils, except iron and occasionally aluminum, and (b) the solubility may be decreased by specific adsorption on clays or hydroxide minerals, increased by the formation of soluble organic complexes, or altered by the oxidation-reduction potential prevailing in the soils.

The effects of soil reaction on plant growth are well documented by Jackson (1967) and Black (1968). They are not generally caused by specific effects of hydrogen or hydroxide ions themselves but by the ancillary chemical effects of soil reaction discussed previously. Plant species, and varieties within a species, differ greatly in their ability to adapt to these effects.

BRYON W. BACHE

References

Bache, B. W., 1974. Soluble aluminium and calcium-aluminium exchange in relation to the pH of dilute calcium chloride suspensions of acid soils, *Jour. Soil Sci.*, 25, 320–332.

*Bates, R. G., 1964. *Determination of pH, theory and practice.* New York: Wiley, 435p.

*Black, C. A., 1968. *Soil-Plant Relationships.* New York: Wiley, 792p.

Clark, J. S., 1966. The pH values of soils suspended in dilute salt solutions, *Soil Sci. Soc. America Proc.*, **30**, 11–14.

*Coleman, N. T., and Thomas, G. W., 1967. The basic chemistry of soil acidity, *in* R. W. Pearson and F. Adams, eds., *Soil Acidity and Liming.* Madison: American Society of Agronomy, 1–41.

Cruz-Romero, G., and Coleman, N. T., 1974. Reactions among $CaCO_3$, CO_2 and Na-adsorbents, *Soil Sci. Soc. America Proc.*, 38, 738–742.

Hingston, F. J.; Posner, A. M.; and Quirk, J. P., 1972. Anion adsorption by geothite and gibbsite. I. The role of the proton in determining adsorption envelopes, *Jour. Soil Sci.*, **23**, 177–192.

Jackson, M. L., 1963. Aluminium bonding in soils, *Soil Sci. Soc. America Proc.*, 27, 1–10.

*Jackson, W. A., 1967. Physiological effects of soil acidity, *in* R. W. Pearson and F. Adams, eds., *Soil Acidity and Liming,* Madison: American Society of Agronomy, 43–124.

Place, G. A.; Phillips, R. E.; and Brown, D. A., 1968. Self-diffusion of phsophorus in clays and soils: II. The effect of pH, *Soil Sci. Soc. America Proc.*, **32**, 657–660.

Ponnamperuma, F. N.; Castro, R. U.; and Valencia, C. M., 1969. Experimental study of the influence of the partial pressure of carbon dioxide on the pH values of aqueous carbonate systems, *Soil Sci. Soc. America Proc.*, **33**, 239–241.

Posner, A. M., 1964. Titration curves of humic acid, *8th Internat. Congr. Soil Sci. Trans.*, 3, 161–174.

Schofield, R. K., and Taylor, A. W., 1955. The mea-

*Additional bibliographic references may be found in these works.

surement of soil pH, *Soil Sci. Soc. America Proc.*, **19**, 164–167.

Søensen, S. P. L., 1909. Uber die Messung und die Bedeutung der Wasserstoffionenkonzentration bei enzymatischen Prozessen, *Biochem. Z.*, **21**, 131–301.

Cross-references: *Acidity; Activity Ratio; Base Saturation; Buffers, Buffering; Electrochemistry; Exchange Phenomena; Humus; Ionic Activities; Lime, Liming; pH, Field Determination; Redox Reactions; Soil Chemistry; Soil Solution; Soil Tests.*

SOIL SALINITY AND ALKALINITY—See VOL. IVA. VOL. XII, Pt. 2: SALT-AFFECTED SOILS; SLICK SPOTS

SOIL SCIENCE

Soil science refers to that body of knowledge pertaining to the soil system and its intersection with various aspects of the environment. Soil science studies specifically entail the characteristics, usefulness, management, and manipulation of the soil resource. *Soil systems* (Simonson, 1968) are generally described as (1) independent natural bodies on the earth's surface, (2) media for plant growth, or (3) as the surficial mantle of weathered rock. The characteristics of soil systems depend on the factors of soil formation (Jenny, 1941); these factors are climate, organisms, relief, geological material, and the influence of weathering time on geologic material. The length of time soils are influenced by weathering processes is vitally important in soil formation in that additions to the system (e.g., organic matter), horizonation, and other profile features and characteristics are strongly influenced by the time factor. Thus the study of soil science is broad in scope as a result of the variations and complex nature of soil systems throughout the world; the soil resource is also a vital aspect of human activity. This science is difficult to classify as a result of the physical and biological implications of soil systems.

Although historically the study of soil dealt primarily with agriculture and related problems, recent activity has enlarged the field to include a wide range of environmental problems. In addition to agricultural aspects, areas of study currently involving soil science include pollution of soils and water, waste disposal (e.g., sludge, industrial wastes, human wastes), reclamation, recreation, urban planning, and other nonagricultural activities. Most of the research in soil science occurs at universities and colleges, government agencies, industries, and numerous research institutes.

Soil science entails the application of numerous other sciences in the study of soil systems. Those sciences most extensively used in this study include biology, botany, plant physiology, ecology, chemistry, physics, mathematics, microbiology, mineralogy, geomorphology, and archeology. Many of the workers in soil science will minor, or in some cases major, in one of the preceding sciences in their college work; this is especially true of those working for M.Sc. and Ph.D. degrees. The strength of various curricula in soil science lies not only with the courses in soils but with the basic sciences such as mathematics, physics, chemistry, and geology. For example, a person interested in soil microbiology would take numerous background courses in biochemistry and microbiology, or a student pursuing the soil physics area would have a minor in some area of physics, engineering, or mathematics.

The discipline of soil science can be subdivided in numerous ways. The International Society of Soil Science, for example, lists the following seven commissions: soil physics; soil chemistry; soil biology, soil fertility and plant nutrition; soil genesis, classification and cartography; soil technology; and soil mineralogy. At many universities in the United States, soil science specialties are generally found in the chemistry, physics, microbiology, fertility, genesis and classification, and mineralogy areas. Another major area of emphasis in soil science deals with soil systems under forests. Major subdivisions of soil science previously mentioned (e.g., soil chemistry, soil classification, soil physics) also apply to scientists working specifically with forest soils. Soil science also plays an integral role in horticulture and range management.

Pedology, as defined by Dijkerman (1974), is the science of genesis, nature, distribution, and use potential of the soil resource. Thus this subdivision is generally considered synonymous with the soil classification, genesis, and cartography section in most institutions in the United States. A broader definition of pedology by Simonson (1970) states that "pedology is the science of the nature, properties, formation, distribution, and function of soils and of their use, management, and manipulation." In the USSR and a number of other countries, pedology and soil science are considered synonymous; thus the broader definition of pedology is used. Kovda and Dobrovolsky (1974), for example, mention several divisions of pedology in Russia such as forest pedology, ameliorative pedology, soil physics, and the chemistry and biochemistry of soils. A more detailed description of pedology is given by de Villiers in the *Encyclopedia of Soil Science, Part 2* (Volume XII).

Jacks (1966) summarized 35 years of world literature in soil science through study of relative numbers of research articles in specified divisions of soil science—i.e., chemistry, physics, biology, pedology, and analytical techniques. Jacks (1966) noted that the field of soil mineralogy actually began in the 1930s; that since then research on the chemistry of humus has doubled, and the largest and steadiest growth has been in soil biology. Yaalon (1964) studied the breakdown of soil research by area in the world literature on soils for 1941–62. He found the following breakdown of articles in soil science: soil chemistry 34.4.%, soil physics 19.6%, soil biology 21.8%, soil genesis and geography 17.6%, and soil technology 6.6% (see *Journals, Soil Science*).

At major universities in the United States, the soil science discipline is usually combined with plant science to form an agronomy or plant science department. In some cases soil science is a separate department and in others soil is mentioned in the title of the department (e.g., agronomy and soils, plant and soil science, or chemistry and soils). At other institutions soil science, or its subdivisions, is combined with agricultural engineering, geography, geology, earth science, or plant nutrition.

Subdivisions

Although subdivisions are recognized in soil science, e.g., the commissions mentioned earlier, the activities of this science are not clearly separated among the commissions because of the many research projects that overlap several subdivisions. For the sake of discussing the soil science disciplines, however, the subdivisions used by the International Soil Science Society will be used.

Soil physics (q.v.) is the study of the physical characteristics of the soil system. The major fields of study within soil physics include movement of water within soils (see *Water Movement*), moisture availability to plants (see *Moisture Management*), soil structure (q.v.), temperature (see *Thermal Regimes*), tillage (q.v.) methods, aeration (see *Atmosphere, Aeration and Respiration*) soil drainage (q.v.), irrigation (q.v.), and methods of measurement of physical parameters. Some of the major contributions in soil physics this century, as outlined by Kirkham (1961) and Philip (1974), were through studies of soil water (q.v.) (e.g., movement and moisture potential), work on soil aggregate stability (q.v.), developing techniques for soil moisture measurements (see *Water Content and Retention*), interpreting water-table and drainage characteristics of soils (see *Field Water Cycle*), and microclimatologic studies. Although studies in soil physics have focused largely on agricultural and forestry problems, in recent years soil physicists have studied pollution (q.v.) problems, waste disposal (industrial, agricultural and urban) (see *Sludge Disposal*), and other nonagricultural aspects (see *Soils, Nonagricultural Use*).

Soil chemistry (q.v.) is the study of chemical reactions associated with the soil system. As the soil system is so diverse, so the chemistry of soil involves a wide variety of reactions, both inorganic and organic in nature. As pointed out by Schuffelen (1974), the central point of most studies in soil chemistry has been the relationships between the solid and liquid phases of the soil. These relationships include ion exchange and solubility phenomena. The use of X-ray techniques has been an important factor in studying the clay minerals in soil, which with the organic colloids, provide the major materials responsible for the adsorption and exchange phenomena (q.v.). Other major areas of study in soil chemistry include soil acidity (see *Acidity; Soil Reaction*) (Jenny, 1961), fixation of potassium, nutrient availability (see *Plant Nutrients; Nutrient Potential*), clay-organic reactions (q.v.), pollution of groundwater, nutrient cycling, (see *Carbon Cycle; Nitrogen Cycle; Phosphorus Cycle; Sulfur Transformations*), and numerous methods of chemical analyses (see *Electrochemistry; Soil Tests*). Many activities in soil chemistry have close relationships to the commission on soil fertility (q.v.) and plant nutrition.

Soil biology is the study of macro- and microanimals and plants in soils and their associated reactions and processes influencing soil systems. Much of the work in soil biology has dealt with the microbiology of soil systems. Some of the major research areas, as described by Allison (1961) and Macura (1974), are soil organic matter (q.v.), nutrient mineralization and availability (see *Nutrient Potential*), nitrogen fixation (q.v.), population studies of organisms (see *Fauna*), nitrogen losses, (see *Nitrogen Fixation; Nutrient Losses*), microenvironment or habitat of microorganisms (see *Microhabits*), soil aggregation (q.v.), antibiotics, pesticides (see *Soil-Solvent Reactions*), rhizosphere (q.v.), enzymes and microbial transformations (see *Enzyme Activity; Metabolic Activity*), and mineral reactions (see *Clay-Organic Reactions; Microbial Ecology; and Clay Minerals*). Soil microbiology (q.v.) covers a wide range of topics, and as in other subdivisions, the work overlaps other areas of soil science, particularly soil chemistry, soil fertility, and mineralogy. Soil biology has also been concerned with environmental problems, especially with disposal of wastes of agricultural, industrial, and human origin (see *Sludge Disposal*).

Soil fertility and plant nutrition is concerned primarily with the interaction of plant and soil systems. As Bradfield (1961) has pointed out, soil fertility (q.v.) is a link connecting soil science and plant science, and a scientist working in this field must have a good background in the general principles of the other subdivisions of soil science. Some of the major areas of study in soil fertility and plant nutrition are plant responses to various soil systems and amendments, nutritive value of crops, trace element availability and toxicities, ion uptake, plant and soil testing, forest fertilization practices, soil moisture stress and nutrient availability, and organic matter influence on soil fertility (see *Nutrient Potential; Plant Nutrients, Trace Elements*). Soil and plant analyses are an important aspect of evaluating soil systems for plant growth. An interesting review of 50 years of soil testing has been made by Truog (1960). He noted several important aspects of soil testing such as the pH expression developed by the Danish biochemist Sørensen in 1909; the extractable nutrients tests (made possible by colorimetric and flame photometric methods); the study of micronutrients in the 1930s; and the making of reliable fertilizer recommendations. Recent studies have emphasized the pollution potential of various soil amendments and developing methods to control this hazard to our surface and groundwaters.

Soil genesis, classification, and *cartography* (q.v. in Vol. XII) involves the study of the formation of soils, the methods of classification of the various individual units in the landscape, and the methods of projecting these units on maps. As Kellogg (1974) has noted, Dokuchaev was one of the first, if not the first, scientist to recognize soils as discrete units in the environment. Kellogg (1974) also outlined other major contributors to the early work in this subdivision, namely Hilgard, Glinka, Coffey, Marbut, and Gedroitz. Examples of areas of study in soil genesis, classification, and cartography include soil geography, paleopedology, soil-geochemical landscapes, engineering properties of soils, delineation of soils by remote sensing, tropical soils, organic soils (Histosols), tundra soils, micromorphology, and soil interpretations. In the United States emphasis during the last decade has been given to soil interpretations for urban development, waste-disposal systems, urban planning, recreation areas, and tax assessment. Interest in using earth satellites for identification of soil resources and vegetative types has also increased drastically in the past few years.

Soil technology, as defined by Russell (1974), refers to the "practical management of the soil in the field." This rather diverse area includes the study of soil conservation (q.v.), soil cultivation (see *Microstructure, Manipulation; Tillage; Zero-Tillage*), water control (see *Soil Drainage; Irrigation; Water Erosion*), management of salt- and alkali-affected soils (see *Management of Soils*), soil reclamation (see *Salt Leaching*), wind erosion (q.v.), groundwater control, soil stabilization (q.v.), and management of organic soils. As noted in the subject areas, agricultural engineering plays an important role in this subdivision. Stoeckeler (1961) points out that fundamental to the improvement of forest and range management practices is a knowledge of site conditions. The site conditions include the various soil properties and their interaction with the growth of various tree and range species. Stoeckeler (1961) also suggests drainage of wetlands, fertilization, and water spreading to improve forest and range production.

Soil mineralogy (q.v.) involves the nature, properties, and reactions of the inorganic fractions of the soil system. Many of the study areas within soil mineralogy are closely related to soil chemistry (q.v.), both organic and inorganic phases. Much of the emphasis in soil mineralogy has been placed on the clay mineral fraction (see *Clay Minerals, Silicates*) and its identification, classification, quantitative estimation (see *Clay Mineralogy*), and the important reactions of this material (see *Clays, Physical Chemistry; Exchange Phenomena*). Other areas of investigation include the heavy and light minerals, acid sulphate soils, translocation and reactions of various key elements (e.g., iron), organic-clay mineral interactions (see *Clay-Organic Reactions*). formation and transformation of minerals, and various weathering reactions (see *Clay Mineral Decomposition; Clay Mineral Formation*). As mentioned by van der Plas and van Reeuwijk (1974), the study of clay minerals has been aided by the following methods of analysis: X-ray, electron microscopy; electron diffraction; scanning electron microscopy; and thermal and infrared analysis (see *Mineral Analysis*). They point out that soil mineralogy was not known prior to the First International Congress of Soil Science in 1927 but currently is an active subdivision of soil science.

Future of Soil Science

The history of soil science probably started in the 1800s, although as Simonson (1968) suggests, man first began thinking about soil as a medium for plant growth when he began to grow his food. In Russia soil science was officially recognized in the 1920s (Tyurin, 1957) with the research support from and membership in the Academy of Science of the USSR. Soil scientists in Russia number more than 5000 in the All-Union Society of Soil Science (Kovda and Dobrovolsky, 1974). Kellogg (1961) states that after the Soil Congress in 1927 and the one in the Soviet Union in 1930, soil science gained in interest and momentum in the United States. Today there are more than 3000 members of the Soil Science Society of America.

Soil science in the future will continue to focus strongly on the production of food and fiber. For example, Kovda and Dobrovolsky (1974) state that soil scientists will be working to double or triple the agricultural production in Russia in the next 20 to 30 years. In the United States, agricultural study will still be important, but more emphasis will be placed on agricultural pollution, production efficiency, maximum fertilizer efficiency, maximum production with minimum energy, preservation of some of our prime agricultural lands from development, and integration of the findings of research in soil science to other related areas of study. Consideration of soils information will also play a larger role in forest and range management in the future. The use of soils for purposes other than production of food and fiber have become important in industrial countries and will become even more important in the future.

JOHN E. FOSS

References

Allison, F. E., 1961. Twenty-five years of soil microbiology and a look to the future, *Soil Sci. Soc. America Proc.*, **25**, 432–439.

Bradfield, R., 1961. A quarter century in soil fertility research and a glimpse into the future, *Soil Sci. Soc. America Proc.*, **25**, 439–442.

Dijkerman, J. C., 1974. Pedology as a science: The role of data, models and theories in the study of natural systems, *Geoderma*, **11**, 73–93.

Jacks, G. V., 1966. The literature of soil science, *Soils Fertil.*, **29**, 227–230.

Jenny, H., 1941. *Factors of Soil Formation*, New York: McGraw-Hill, 281p.

Jenny, H., 1961. Reflections on the soil acidity merry-go-round, *Soil Sci. Soc. America Proc.*, **25**, 428–439.

Kellogg, C. E., 1961. A challenge to American soil scientists: On the occasion of the 25th anniversary of the Soil Science Society of America, *Soil Sci. Soc. America Proc.*, **25**, 419–423.

Kellogg, C. E., 1974. Soil genesis, classification, and cartography: 1924–1974, *Geoderma*, **12**, 347–362.

Kirkam, D., 1961. Soil physics 1936–61 and a look ahead, *Soil Sci. Soc. America Proc.*, **25**, 423–427.

Kovda, V. A., and Dobrovolsky, G. V., 1974. Soviet pedology to the 10th International Congress of Soil Science (the Centenary of Soil Science in Russia), *Geoderma*, **12**, 1–16.

Macura, J., 1974. Trends and advances in soil microbiology from 1924 to 1974, *Geoderma*, **12**, 311–329.

Philip, J. R., 1974. Fifty years of progress in soil physics, *Geoderma*, **12**, 265–280.

Russell, E. W., 1974. The role of Commission VI during the last 50 years, *Geoderma*, **12**, 363–382.

Schuffelen, A. C., 1974. A few aspects of 50 years of soil chemistry, *Geoderma*, **12**, 281–297.

Simonson, R. W., 1968. Concept of soil, *Adv. Agron.*, **21**, 1–47.

Simonson, R. W., 1970. Pedology, *McGraw-Hill Encyclopedia of Science and Technology*, **9**, 693.

Stoeckeler, J. H., 1961. Soil and water management for increased forest and range production, *Soil Sci. Soc. America Proc.*, **25**, 446–451.

Truog, E., 1950. Fifty years of soil testing, *7th Internat. Congr. Soil Sci. (Madison)*, **4**, 46–53.

Tyurin, I. V., 1957. The development of soil science in the USSR in forty years, *Pochvovedenie (Moscow)*, **11**, 1–13.

van der Plas, L., and van Reeuwijk, L. P., 1974. From mutable compounds to soil minerals, *Geoderma*, **12**, 385–405.

Yaalon, D. H., 1964. Has soil research national character?, *Soils Fertil.*, **27**, 89–93.

Cross-references: *Aeragronomy; Agronomy; Clay Mineralogy; Physical Chemistry; Soil Biochemistry; Soil Chemistry; Soil Fertility; Soil Mechanics; Soil Microbiology; Soil Mineralogy; Soil Physics; Soil Science and Society; Training in Soil Science. Vol. XII, Pt. 2: Pedology. Vol. XIII: Engineering Soil Science.*

SOIL SCIENCE AND SOCIETY

Civilization in the twentieth century faces the recurrent struggles of power, industrialization, progress, war, and rehabilitation. An increasing sense of man's concern for the future is commonly being expressed in terms of the "quality of life." This concept serves as a convergence or focal point for economic, social, and environmental considerations. A basic tenet in the quality-of-life concept is the ability to maintain a life-support system for human beings and other life forms. It is generally accepted that the life-support system of the earth is limited, and it is also being recognized that the values of society will determine the quality of its environment. The price that society is willing to pay will determine the level of quality obtained.

Insofar as the quality of life and the environment are intertwined, several social concerns can be identified. One is the continued support of man by the ecological systems as we know them today; another is the availability of suitable land, particularly for agricultural and recreational use; and a third is an adequate supply of air and water of suitable quality to support all life forms. These social concerns have been, and still are, vital concerns of soil science. Components of an ecosystem compete for the nutrient and other energy resources within their domain. Soil is a reservoir, a reactant, and a depository that, in part, sets limits for each delicately balanced system.

Use Concepts of Soil

Traditionally there have been two prevalent concepts about what soil means to man. On one hand, soil is a mysterious black-box commodity, which if disturbed on the surface and sowed with seeds, will produce useful plants. Lime, fertilizers, and water can be added to increase the amount of output. This concept treats soil like a bank account and is equivalent to saying that with *X* inputs, the soil will yield *Y* returns or outputs.

The other concept considers soil to be a dynamic part of the whole process and as such is also subject to manipulation and change. It may not be as flexible as some other inputs, but it is not a static, black-box unknown. In this case, society, or some segment of society, decides what it wants the output to be and then juggles the inputs to obtain the most favorable balance in keeping with their economic and social goals.

The challenge to locate land suitable for agriculture and recreation in today's world is greater than ever before in history (Kellogg, 1975). The frontiers tend to be more vertical than horizontal, and judgments about soil responses within various technological frameworks is beginning to tax the available store of such information. Assessing the suitability of land using the black-box concept of soil is far easier than assessing it using the concept of letting soil react as one of the inputs and deciding on appropriate interactions. Society is interested in the latter.

Integrating Soil Knowledge

Soil Science is, and has always been, a dynamic environmental science. Soil, by whatever name or concept, is essential to all societies. It has been stated that if an object does not fly or float, it depends on soil. A qualitative definition of soil indicates that soil is produced by the interaction of climatic and biotic processes, as modified by local topography, acting on a parent material through time. The many combinations of factors and rates of processes theoretically give rise to thousands of unique environmental units called *soils* (Soil Survey Staff, 1951). Fascinating as the historical development of soils and their landscapes may be, the major role of soil science is far more pragmatic and more demanding. Based on careful observation, measurement, and experimentation, soil scientists predict the expected behavior of soil when involved in specified uses. Projections into the past help us explain the life-support systems of our ancestors, and too often sadly follow their demise as the dynamic soil systems were pushed beyond their capabilities (Kellogg, 1966). Characterization of soil features and their correlation with responses of plants, buildings, and machines enable soil scientists to describe and evaluate soil qualities. The quality-of-soil concept is somewhat analogous to the quality-of-life concept in that it refers to suitabilities, capabilities, and potentialities of use depending on society's set of values, its degree of concern for "one world."

The chemistry and physics of pedological and biological processes that affect plant responses, environmental balances, and the direct use of soil as a material are not the sole propriety of soil science. These are aspects of most disciplines that deal with ecosystems and man's interactions with nature. Soil science focuses on the soil as an integrative entity of the environment that serves as a basis for understanding, explaining, and predicting outcomes. The concept of soil potential includes the cost, at least the relative cost, of tinkering with soil resources to produce the responses and behaviors that society desires. The history of society and of soil science will reflect on the wisdom of our present values of soil. Will the world continue to trade one form of pollution for another? Can we adequately and accurately measure the results of current and projected environmental policies such as clearing jungles, grazing high mountains, irrigating new lands, and altering cropping patterns? If we cannot, it is not possible to develop realistic plans for a harmony between nature and man's relationship to it. Without such a balance, it may well be that there can be no quality of life for human beings.

Changes and Challenges

Society is characterized by change. In America the percent of workers in agriculture has dramatically declined as those in manufacturing and services has become prominent. Almost as startling has been the rapid increase of effort devoted to attaining and disseminating knowledge. If the trend continues its spiral upward, then one consequence of the change inherent in the quality-of-life concept will be the need to provide for an unprecedented expansion of our knowledge. By what means will we handle the voluminous accumulation of new information we generate? Will society be able to develop classification schemes capable of permitting us to determine useful relationships among the fragmented bits of knowledge? The challenge appears to be one of integrating this explosion into comprehension and understanding relevant to the world's quality of life.

In its endeavors to learn more about the nature of soils, to understand their behavior, to locate and map the numerous kinds of soils, to interpret qualities and predict potentials, soil science has in the past assisted in what seems to be an agricultural panacea. Now soil science must move ahead to meet the challenges and demands of our ever-changing society whose values, goals, priorities, and policies will necessarily depend on the soil resources of the world. There is a responsibility to assure that, with the expansion of knowledge about soils, there is a commitment for an adequacy and accuracy of information about soil potentials and their relationship to the life-support system necessary for society to have "a quality of life."

RICHARD W. ARNOLD

References

Kellogg, C. E., 1966. *The Soils that Support Us.* New York: Macmillan, 370p.

Kellogg, C. E., 1975. *Agricultural Development: Soils, Food, People, Work.* Madison: Soil Science Society of America, 233p.

Soil Survey Staff, 1951. Soil Survey Manual, *U.S. Department of Agriculture Handbook No. 18,* 503p.

Cross-references: *Agronomy; Management of Soils; Soil Science; Training in Soil Science.*

SOIL SEPARATES—See PARTICLE-SIZE DISTRIBUTION

SOIL SOLUTION

Soil is a three-phase system—solid, liquid, and gas—although the interface between colloid surfaces and solution is so important that the "surface phase" should probably be considered a fourth phase. The main component of the liquid phase is water, but it contains a number of dissolved materials: gases, soluble organic compounds, and simple inorganic salts. Larger amounts of some of these materials are adsorbed at surfaces and act as a reserve from which the solution can be replenished. When soil is saturated, some of the water is in wide pores that will eventually drain; this water is unlikely to be in equilibrium with the bulk of the soil. The *soil solution* is therefore defined as "a quasi-equilibrium solution of electrolytes that occurs in the soil under unsaturated moisture conditions" (Pearson, 1971).

The main difficulty in the study of the soil solution is its separation, for some soils may contain as little as 10% by weight of water at field moisture capacity. The most reliable method is to displace the solution from a column of carefully packed soil by slowly percolating another liquid through it; the displacing liquid may be alcohol, an electrolyte solution, or pure water. This was first done by Schloesing in 1866 (Russell, 1973), and the classical work by Burd and Martin (1923) showed that the composition and electrical conductivity of successive fractions of leachate remain constant until the displacing solution breaks through the column. The alternative is to use air pressure or suction to displace the solution from moist soil supported on a porous membrane or ceramic filter. This practice can remove only solution held at relatively low free energies, and it gives inaccurate results where the soil solution is very dilute. Furthermore, serious losses of components such as phosphate, present in low concentration (of the order of 10^{-5} mol/l), can occur by adsorption on the filter.

The concentration of the soil solution varies widely between soils, depending on the leaching regime of the soil (see *Leaching*), fertilizer addition, crop uptake, and processes that affect mineralization of organic matter. The range of figures shown in Table 1 (p. 498), from 5–50 mequiv/l of cations and anions, is probably typical for nonsaline soils, but 100 mequiv/l is commonplace for saline soils. As the moisture content is reduced, the concentration of soluble salts increases proportionately, particularly for Ca^{2+}, Na^{+}, NO_3^- and Cl^-, but concentrations of K^+, SO_4^{2-}, and phosphate vary less because they are governed by adsorption or solubility equilibria.

The soil solution is important because it is the immediate source of plant nutrients (q.v.). However, the gross composition with respect to the more abundant ions, as shown in Table 1, represents only one experimental approach. Apart from more detailed chemical analysis, a thorough investigation of the amounts of each component present requires the use of activities rather than concentrations, and the consideration of ion pair formation, acid-base equilibria, hydrolysis equilibria, complex formation, and the interaction of gases (e.g., CO_2) with the species in solution. Adams (1971) lists the equilibrium constants of 53 such reactions. In addition, solubility product equilibria, cation exchange (see *Exchange Phenomena*), and specific adsorption equilibria (see *Adsorption Phenomena*) should be taken into account for a full understanding of the soil solution.

BRYON W. BACHE

References

Adams, F., 1971. Ionic concentrations and activities in soil solutions, *Soil Sci. Soc. America, Proc.,* 35, 420–421.

TABLE 1. Solution Composition of Some Nonsaline Soils, from Various Sources.

Soil Texture	Water (%)	pH	Cations (mequiv/l)						Anions (mequiv/l)			
			Ca^{2+}	Mg^{2+}	K^+	Na^+	NH_4^+	*Sum*	NO_3^-	Cl^-	HCO_3^-	SO_4^{2-}
Sandy loam	12	7.5	42.2	3.6	0.8	3.0	0.4	*50.0*	32.0	4.6	3.5	11.8
Loam	33	7.5	11.1	2.2	0.3	2.3	0.2	*16.1*	8.8	2.3	3.8	2.8
Loam*	20	3.9	1.0	1.4	0.4	0.4	0.1	*4.6**	3.8	0.2	–	0.8
Clay	18	5.7	47.8†		1.7	5.6	–	*55.1*	30.4	3.8	1.8	17.0
Clay	24	5.7	37.9†		1.5	4.3	–	*43.7*	22.1	2.6	1.6	14.4

*In addition to the cations listed here, this acid soil solution contained 0.7 mequiv Al^{3+} and 0.6 mequiv Mn^{2+}/l.

†This figure is for calcium and magnesium.

Burd, J. S., and Martin, J. C., 1923. Water displacement of soils and the soil solution, *Jour. Agric. Sci.*, 13, 265–295.

Pearson, R. W., 1971. Introduction to symposium—The soil solution, *Soil Sci. Soc. America Proc.*, 35, 417–420.

Russell, E. W., 1973. *Soil Conditions and Plant Growth*. London: Longmans, 849p.

Cross-references: *Electrochemistry; Ionic Activity; Soil Water.*

SOIL-SOLVENT INTERACTIONS

Soil consists mainly of mineral particles derived from crustal rocks by weathering, diagenesis, and differentiation. It contains an array of primary minerals (e.g., quartz), mineral residues (e.g., hydrated to anhydrous alumina), and secondary minerals (e.g., kaolinite) (see Vol. IVB, *Soil Mineralogy*). Soil also contains humus (q.v.) derived from plant and animal residues by microorganism digestion and synthesis. Soil particles range in size from the 2000 μm imposed by a 2-mm sieve to less than 0.02 μm (see *Particle-Size Distribution*), and these particles often interact to form coated particles and aggregates (*peds*) (see *Soil Structure*).

Salient soil characteristics from the standpoint of reactivity with molecules and ions are (1) specific surface area; (2) surface charge (kind, distribution, intensity); and (3) surface geometry (i.e., accessibility). The most reactive soil components are those of colloidal dimensions, e.g., aluminosilicate clays (see *Clay Minerals, Silicates*), aluminum and iron sesquioxides (see *Hydrous Oxides*), and humus (q.v.). Some soil colloids (e.g., montmorillonite) expand under certain conditions (e.g., hydration), creating additional surface area; others (e.g., kaolinite) are nonexpanding. Some soil colloids contain an assortment of micropores (see *Soil Pores*), that can accommodate small molecules (e.g., water) but exclude larger ones (e.g., benzene).

Inorganic soil colloids expose surfaces consisting of oxygen atoms or hydroxyl groups. Organic soil colloid (humus) exposes surfaces consisting of hydroxy, carbonyl, amino, methoxy, and hydrocarbon groups. Polarity generated by separation of opposing charges results in an array of van der Waals surface charges. Substitution of an atom by one of different valence, intrinsic ionic properties, and presence of assorted counterions result in an array of coulombic surface charges. The intensity of both partial and full electrostatic charges varies in accord with pH (which controls protonation-deprotonation) and distance from the particle surface.

Specific soil colloid characteristics are too numerous and entangling to discuss in detail (for some of these details, see Grim, 1968; Kononova, 1966). It is better, for clarity and progress, to visualize a simple model substrate incorporating the general characteristics described previously and use it as the basis for discussing reactivity.

Solvents

A spectrum of solvents can be assembled on the basis of differing molecular composition, size, shape, charge, and polarity. Water is foremost on the list, followed by the classical array of alcohol, aldehyde, ketone, ether, acid, ester, amide, amine, hydrocarbon, chlorinated hydrocarbon, and others. Among innumerable examples are ethanol, acetone, dioxane, dimethylformamide, pyridine, hexane, benzene, phenol, dibromoethane, acetonitrile, and dimethylsulfoxide. Even this abridged list is too long to deal with here. Use of selected hydrophilic and hydrophobic endmembers plus a single intermediate species might give appropriate "model" solvents from which a first approximation of soil-solvent behavior can be obtained. The abbreviated list, for example, can consist of water, ethanol, and benzene. These can be used as a starting point, to which others may be added as necessary for progress toward soil-solvent interaction principles.

The water molecule is polar (1.87 D*) because of its composition and structural geometry. Water molecules associate by hydrogen bonding and interact with oxygen-containing or OH-containing substrates by hydrogen bonding. They probably form hexagonal networks on planar surfaces to which they are attracted. Water molecules tend to encapsulate small ions through dipole-ion bonding. The area of largest cross-section of a water molecule is about 7.9 Å^2, but its effective diameter, imposed by its open-structure packing, is about 12.5 Å^2 (McClellan and Harnsberger, 1967).

The ethanol molecule is polar (1.70 D) because of separation of its opposite residual charges. Its MW 46, coupled with its BP 78°C, suggests moderate intermolecular association through hydrogen bonding. Ethanol can interact with O-containing and OH-containing substrates by hydrogen bonding and with ions by dipole-ion bonding. The area of largest cross-section of an ethanol molecule is about 18.1 Å^2. Loose packing gives it an apparent area about 28.3 Å^2 (McClellan and Harnsberger, 1967).

The benzene molecule is nonpolar (0 D) because of its symmetry. Its MW 78, coupled with its BP 80°C, suggests moderately low intermolecular association through van der Waals bonds. Benzene can interact with O-containing substrates and anions through weak hydrogen bonds. The area of largest cross-section of the benzene molecule is about 36.9 Å^2. Loose packing gives it an apparent area about 43.6 Å^2 (McClellan and Harnsberger, 1967).

Soil-solvent Interactions

Soil-solvent interactions fall into two interconnected categories: (1) primary interaction of a charged soil surface, including its counterions, and charged solvent molecules forming a monolayer envelope; and (2) interaction of excess solvent molecules, usually influenced by monolayer characteristics and always restricted by soil geometry, ultimately filling soil pore space (q.v.). The primary reaction at the soil-solvent interface is the *diagnostic reaction* and is of greatest interest here. Although the *secondary reaction* characterizes soil pore space, it is mostly intrasolvent in character and will not be considered in detail.

*D is charge asymmetry (dipole moment) expressed in Debye units. For simple solvents, D ranges from, for example, 0 for CCl_4, to for example, 3.9 for CH_3CN.

Adsorption Measurement

Interaction of soils and solvents cannot be studied easily using liquid solvents because monolayer responses are difficult to separate from overall responses. Adsorption of solvents from the vapor phase is more practicable and interpretable. In vapor-adsorption studies, the solvent (e.g., water) can be combined in varying proportions with a compatible nonvolatile liquid (e.g., sulfuric acid) and placed in a series of closed chambers, where the solvent volatilizes to some controllable equilibrium concentration. Exposed dry soil reacts with solvent vapor to a greater or lesser degree depending on system characteristics. The extent of interaction at equilibrium can be determined, for example, from soil weight gain. This gain can be plotted against equilibrium vapor pressure (P/P_0), and the plotted curve can be interpreted by inspection or by calculation based on a fitted equation. Actual procedures are usually more sophisticated (cf., e.g., Sharma, Uchara, and Mann, 1969), but the principles are the same.

Adsorption Curve Characteristics

A vapor-adsorption curve, in its simplest and most representative form, is composed of a linear segment that is completed at $P/P_0 \cong 0.05$ (Figure 1, segment 1); a convex segment (in

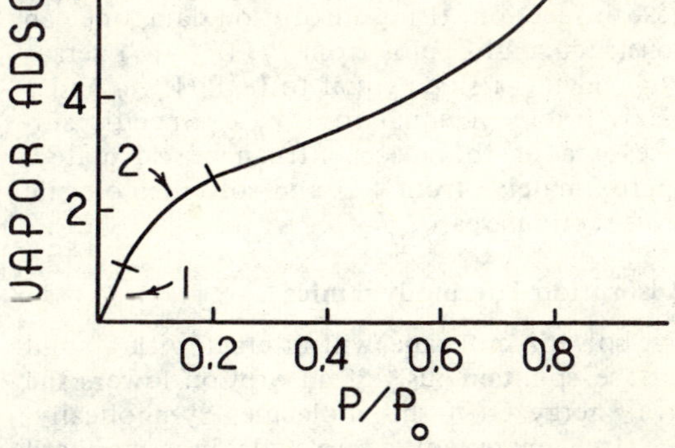

FIGURE 1. Vapor-adsorption curve (Type II) for a soil.

respect to the *x*-axis) completed at $P/P_0 \cong 0.20$ (Figure 1, segment 2); and a concave segment "completed" at $P/P_0 \to 1.0$ (Figure 1, segment 3). This curve is known as Type II (Brunauer et al, 1940). The linear segment probably represents reaction of solvent molecules with soil colloid counterions through dipole-ion bonds; the convex segment, reaction of solvent molecules with progressively decreasing soil colloid surface sites through dipole-dipole bonds, completing the monolayer; and the concave segment, interaction of solvent molecules through dipole-dipole bonds, forming multilayers.

A simple partition equation can be used to fit the linear portion of this compound curve, giving a slope that can be compared with others generated by different soil-solvent reactions:

$$\frac{V}{P/P_0} = s \qquad (1)$$

where V is volume (weight) of adsorbed vapor, and s is slope. The *Langmuir adsorption equation* (Langmuir, 1918) can be used to fit segments I and II of this sigmoid curve, defining its upper (i.e., monolayer) limit. The BET adsorption equation (Brunauer, Emmett, and Teller, 1938) can be used for the whole curve, defining its monolayer limit, describing accumulation of multilayers, and supplying heat of interaction. The Langmuir equation can be expressed as

$$\frac{P}{V} = \frac{P_0}{V_m c} + \frac{P}{V_m} \qquad (2)$$

and the BET equation as

$$\frac{P}{V(P_0 - P)} = \frac{1}{V_m c} + \frac{(c-1)P}{V_m c P_0} \qquad (3)$$

where V_m is monolayer volume (weight) of adsorbed vapor, and c is a constant related to heat of reaction. Using adsorption data, one can construct a BET plot from $P/V(P_0 - P)$ versus P/P_0, giving a slope equal to $(c/1)/V_m c$, and a y-axis intercept equal to $1/V_m c$. Specific surface area of the adsorbent can be calculated, approximately, from V_m and solvent molecule cross-section area.

Adsorption Thermodynamics

A solvent molecule will interact with a solid surface spontaneously if adsorption lowers the free energy G of this molecule. Symbolically, G° (free energy of a molecule in a standard state, e.g., liquid state) $-G_a$ (free energy of the molecule in the adsorbed state) $= -\Delta G_a$ (loss of free energy by the molecule during adsorption). In such spontaneous adsorption, heat H is released ($H^\circ - H_a = -\Delta H_a$). In addition, entropy of adsorption S decreases ($S^\circ - S_a = -\Delta S_a$). From the fundamental thermodynamic equation, $G = H - TS$ (T is temperature), we obtain $-\Delta G_a = -\Delta H_a + T\Delta S_a$ for spontaneous adsorption. The quantity ΔG_a is accessible from $G_a = -RT\ln K$, where K is the thermodynamic equilibrium constant for the adsorption reaction. The quantity ΔH_a is accessible from adsorption curves obtained at two temperatures and application of the *Clausius-Clapeyron equation;* and ΔS_a is obtained by difference. If we can assemble ΔG_a, ΔH_a, and ΔS_a values for a given solvent-adsorption reaction, we have captured its salient characteristics, which can then be compared with those of another adsorption reaction. The interpretation and value of differential entropy ΔS in vapor-adsorption study has been questioned (Hill, Emmett, and Joyner, 1951). Negative ΔS values, which were believed to indicate more orderly arrangement of adsorbed molecules (Low, 1961) may be misleading. It has been suggested that integral entropy, usually positive, is a more valid clue to degree of order-disorder, and that molecules (e.g., water) usually are less ordered in monolayer array than in free solvent (Sharma, Uehara, and Mann, 1969). Because thermodynamic functions vary continuously with equilibrium vapor pressure (Sharma, Uehara, and Mann, 1969), comparisons should be made at specific points, say, $P/P_0 = 0.05$ or $P/P_0 = 0.20$.

Solvent Comparison

Water, ethanol, and benzene vapors react with soil colloids spontaneously. Soil-water interactions have been examined extensively and intensively but are still incompletely characterized. Soil-alcohol and soil-hydrocarbon interactions have received much less attention, but perhaps enough is known to permit tentative comparisons.

Empirical soil-water reaction data give sigmoid (Type II) adsorption curves. Ion hydration is predominant at $P/P_0 < 0.05$. Completion of the first and second parts of the curve, marking the end of monolayer formation and beginning or multilayer formation, occurs at $P/P_0 \cong 0.20$ (Keenan, Mooney, and Wood, 1951). During monolayer adsorption, counterions and soil surface areas interact with water vapor forming ion-dipole bonds (e.g.,

$$\text{Soil}\text{—}Ca^{+}\text{- - -}\underset{\delta^-}{OH_2})$$

and dipole-dipole bonds (e.g.,

$$\text{Soil—Na}^{\delta^+}\text{---OH}_2^{\delta^-}, \quad \text{soil=O}^{\delta^-}\text{-----HOH}^{\delta^+}$$

$$\text{and} \quad \text{soil—OH}^{\delta^+}\text{---OH}_2^{\delta^-})$$

Water vapor invades the interlayer "space" of expandable species like smectites, hydrating interlayer cations, and interlayer surfaces. Hydration of counterions is a function of ion species and their environment. In general, trivalent ions (e.g., aluminum) bind more water than divalent ions (e.g., calcium), which bind more water than monovalent ions (e.g., potassium) (Jurinak, 1963). Water molecules, apparently, are not solidly packed within the monolayer. On interion surface areas, they may occur as a flat, hexagonal network of molecules, with hydrogen atoms directed toward surface oxygen atoms or with oxygen atoms directed to surface hydrogen atoms; around cations, they are clustered with oxygen atoms directed inwardly. The "monolayer" thus contains water molecules in assorted attitudes, and includes localized clusters of multilayer molecules. On organic soil colloids, water is probably adsorbed selectively by hydrogen bonding at hydrophilic sites (e.g., carbonyl, hydroxy, and amino sites). Hydrocarbon sites probably remain unhydrated. Approximate thermodynamic values for soil colloid-water vapor interaction at monolayer completion are: ΔG_a, -3300 cal/mol; ΔH_a, -10,000 cal/mol; $T\Delta S_a$, 6700 cal/mol (or ΔS_a, -22.5 cal/mol–deg) (calculations based on data from Alexander and Haring, 1936; Goates and Hatch, 1953; and Goates and Bennett, 1957). Useful supplementary details on soil-water interactions can be found in Rode (1969).

Empirical soil-ethanol reaction data give sigmoid (Type II) adsorption curves. On inorganic soil surfaces, interaction forms hydrogen bonds, and alkyl groups may be directed outward. For cations, interaction forms ion-dipole or dipole-dipole bonds, depending on whether the ion is multivalent or monovalent. Ethanol penetrates interlayer regions of expandable clays. On organic components, ethanol probably is adsorbed at hydrophobic and hydrophilic sites. The ethanol molecule is about twice as large as the water molecule so it has more restricted access to soil micropores. Comparable data for water and ethanol adsorption are available (Puri, 1949). The adsorption curves develop in an identical manner, but at different levels.

Other alcohols and other polar compounds react with soil colloids. Alkyl amines react with colloidal bentonite surfaces to a degree that depends on the length of the alkyl group (Slabaugh, 1954): CH_3NH_2 is not adsorbed spontaneously; for the C_4 amine, ΔG_a is -350 cal/mol; for the C_{14} amine, it is -2450 cal/mol. Corresponding ΔH_a values are -240 cal/mol and -2500 cal/mol. Protonated amines act as exchange cations; interaction ion-ion bonds are reinforced with dipole-dipole bonds. Acetone gives adsorption curves similar to those for alcohols (Puri, 1949). Glycol, glycerol, ketone, aldehyde, amide, phenol, and ether reactions with soils have been studied (Martin, 1955; Moore and Dixon, 1970; Parfit and Mortland, 1968; Larson and Sherman, 1964; Tahoun and Mortland, 1966; Luh and Baker, 1970; Bradley, 1945). Multilayer adsorption of solvents such as C_2H_5Cl and CF_2Cl_2 may be independent of soil surface characteristics (Pierce, 1960).

Empirical soil-benzene reaction data give sigmoid (Type II) adsorption curves (Barrer and Reay, 1957) similar to those for N_2 adsorption; both reagents form monolayers on external surfaces. The much larger benzene molecule cannot enter some of the soil colloid micropores accessible to N_2. Benzene does not penetrate between layers of expandable clays. Mikhail and Shebl (1972), working with silica gel, found that benzene gave a Type II BET adsorption curve, while cyclohexane and carbon tetrachloride gave Type V curves (i.e., segment 3 of the BET curve illustrated in Figure 1). Type V curves indicate negligible interaction between substrate surface and solvent. Cusumano and Low (1972), working with germania and benzene, calculated thermodynamic values ΔG_a (about -2 kcal/mol), ΔH_a (about -9 kcal/mol), and $T\Delta S_a$ (about -7 kcal/mol) for monolayer formation. Interaction probably took place via

$$\text{Benzene}^{\delta^-}\text{---HO}^{\delta^+}\text{—Ge} \quad \text{and} \quad \text{benzene}^{\delta^-}\text{---Ge}^{\delta^+}$$

bonds.

Ethylene dibromide adsorption curves are Type II; exchangeable cations strongly affect the degree of adsorption (Jurinak, 1957). Heat of reaction of this solvent with montmorillonite is 12–12.5 kcal/mol; and with kaolinite, 11.3–11.4 kcal/mol. Heat of wetting has been used as an approximation of solvent affinity for a soil. Expressed in calories per gram of soil, heat of wetting is relatively high for water, lower for alcohol, and low for hydrocarbons (Puri, 1949). Converted to calories per mole, adsorbed ethanol, benzene, and kerosene can have higher heats of wetting than water: bond strengths are additive, and larger molecules offer more bonding possibilities. Adsorption is restricted to external surfaces.

Some idea of the relative affinities of counterions and substrates for the selected solvents can be obtained by inspection of Figures 2–5.

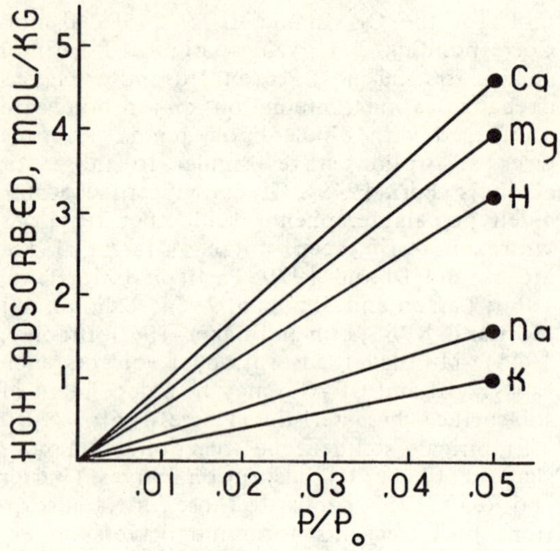

FIGURE 2. Initial adsorption of water vapor by homoionic montmorillonite. (Data from Orchiston, 1955)

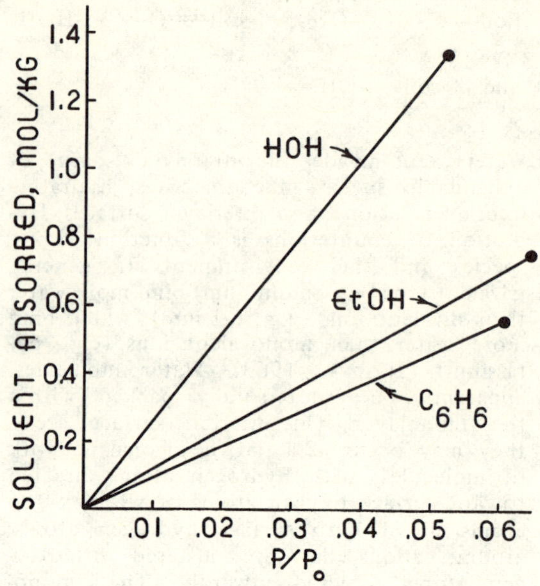

FIGURE 4. Initial solvent-vapor adsorption by active silica. (Data from Pigeon, 1935)

In Figure 2 the degree of hydration of cations in homoionic montmorillonite follows the order Ca > Mg > H > Na > K (Orchiston, 1955). In Figure 3 the degree of hydration of homoionic soil colloids follows the order Barnes > Carrington > Miami (mainly 2:1 clays) > Cecil (mainly 1:1 clay) (Alexander and Haring, 1936). In Figure 4 the order of reactivity of active silica with selected solvents is HOH > EtOH > C_6H_6 (Pigeon, 1955). In Figure 5 the order of reactivity of activated charcoal with solvents is EtOH > HOH (Puri, 1949).

Summary

It may be concluded that soil surfaces interact spontaneously with most solvent molecules,

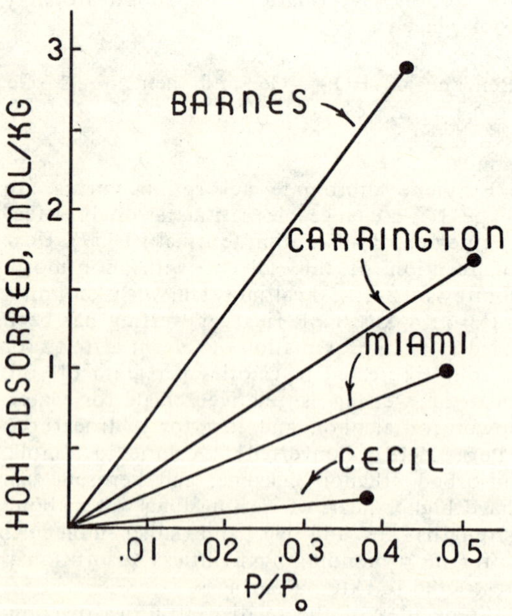

FIGURE 3. Initial water-vapor adsorption by various soil colloids. (Data from Alexander and Haring, 1936)

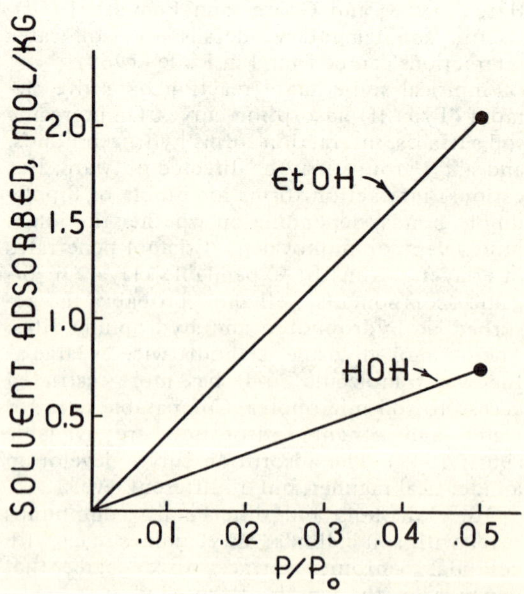

FIGURE 5. Initial adsorption of water and ethanol vapor by activated charcoal. (Data from Puri, 1949)

producing greater electron symmetry and stability. Ion-ion, ion-dipole, and dipole-dipole bonds can form during such interaction; for soil and nonionic molecules, ion-dipole, and hydrogen bonding are the principal mechanisms. Interaction tends to take place at specific soil sites. Soil-solvent interaction tends to reach a plateau at $P/P_0 \cong 0.2$, at which point a rough approximation to a molecule monolayer has accumulated. The closeness of approximation depends on the size, shape, charge, and attitude of solvent molecules at the interface. Solvent-solvent interaction occurs progressively thereafter, filling capillary pores to a degree depending on molecule size and intermolecular forces. Polar molecules can expose and interact with new sites after forcing interlayer entry. Strongly polar solvents such as acetonitrile (3.9 D), formamide (3.3 D), and nitromethane (3.4 D) do this very effectively. Nonpolar solvents are restricted to "external" surfaces.

Saturation with liquid solvent fills all accessible pores. Gravity drainage thereafter removes "free" solvent from large pores. Vacuum-suction drainage removes free solvent from pores of intermediate size, etc., extent of drainage being a function of intermolecular forces.

KERMIT STILLMAN LAFLEUR

References

Alexander, L. T., and Haring, M. M., 1936. Vapor pressure-water content relations for certain typical soil colloids, *Jour. Phys. Chem.*, **40**, 195–205.

Barrer, R. M., and Reay, J. S., 1957. Interlamellar sorption by montmorillonite, *Proc. 2nd Internat. Congr. Surface Activity*, **II**, 79–89.

Bradley, W. F., 1945. Molecular associations between montmorillonite and some polyfunctional organic liquids, *Jour. Am. Chem. Soc.*, **67**, 975–981.

Brunauer, S.; Emmett, P. H.; and Teller, E., 1938. The adsorption of gases in multimolecular layers, *Jour. Am. Chem. Soc.*, **60**, 309–316.

Brunauer, S.; Deming, L. S.; Deming, W. E.; and Teller, E., 1940. On a theory of van der Waals adsorption of gases, *Jour. Am. Chem. Soc.*, **62**, 1723–1732.

Cusumano, J. A., and Low, M. J. D., 1972. Interactions between surface hydroxyl groups and adsorbed molecules. IV. Benzene adsorption on germania, *Jour. Colloid Interface Sci.*, **38**, 245–255.

Goates, J. R., and Hatch, C. V., 1953. Standard adsorption potentials of water vapor on soil colloids, *Soil Sci.*, **75**, 275–278.

Goates, J. R., and Bennett, S. J., 1957. Thermodynamic properties of water adsorbed on soil minerals. II. Kaolinite, *Soil Sci.*, **83**, 325–330.

Grim, R. E., 1968. *Clay Mineralogy*. New York: McGraw-Hill, 596p.

Hill, T. L.; Emmett, P. H.; and Joyner, L. G., 1951. Calculations of thermodynamic functions of adsorbed molecules from adsorption isotherm measurements: Nitrogen on Graphon, *Jour. Am. Chem. Soc.*, **73**, 5102–5107.

Jurinak, J. J., 1957. The effect of clay minerals and exchangeable cations on the adsorption of ethylene dibromide vapor, *Soil Sci. Soc. America Proc.*, **21**, 599–602.

Jurinak, J. J., 1963. Multilayer adsorption of water by kaolinite, *Soil Sci. Soc. America Proc.*, **27**, 269–272.

Keenan, A. G.; Mooney, R. W.; and Wood, L. A., 1951. The relation between exchangeable cations and water adsorption on kaolinite, *Jour. Phys. Chemistry*, **55**, 1462–1474.

Kononova, M. M., 1966. *Soil Organic Matter*. New York: Pergamon, 450p.

Langmuir, I., 1918. The adsorption of gases on planes and surfaces of glass, mica and platinum, *Jour. Am. Chem. Soc.*, **40**, 1361–1403.

Larson, G. O., and Sherman, L. R., 1964. Infrared spectrophotometric analysis of some carbonyl compounds adsorbed on bentonite clay, *Soil Sci.*, **98**, 328–331.

Low, P. F., 1961. Physical chemistry of clay-water interaction, *Adv. Agron.*, **13**, 269–327.

Luh, M. D., and Baker, R. A., 1970. Vapor phase adsorption of phenol on selected clays, *Jour. Colloid Interface Sci.*, **33**, 539–547.

McClellan, A. L., and Harnsberger, H. F., 1967. Cross-sectional areas of molecules adsorbed on solid surfaces, *Jour. Colloid Interface Sci.*, **23**, 577–599.

Martin, R. T., 1955. Ethylene glycol retention by clays, *Soil Sci. Soc. America Proc.*, **19**, 160–164.

Mikhail, R. Sh., and Shebl, F. A., 1972. Adsorption in relation to pore structure of silica. III. Persorption versus specific interactions, *Jour. Colloid Interface Sci.*, **38**, 35–44.

Moore, D. E., and Dixon, J. B., 1970. Glycerol vapor adsorption on clay minerals and montmorillonite soil clays, *Soil Sci. Soc. America Proc.*, **34**, 816–822.

Orchiston, H. D., 1955. Adsorption of water vapor. III. Homoionic montmorillonites at 25°C, *Soil Sci.*, **79**, 71–78.

Parfit, R. L., and Mortland, M. M., 1968. Ketone adsorption on montmorillonite, *Soil Sci. Soc. America Proc.*, **32**, 355–363.

Pierce, C., 1960. The Frankel-Halsey-Hill adsorption isotherm and capillary condensation, *Jour. Phys. Chem.*, **64**, 1184–1187.

Pigeon, L. M., 1935. The sorption of vapors by active silica, *Canadian Jour. Res.*, **12**, 41–56.

Puri, A. N., 1949. Adsorption of vapors from organic liquids, Chapter X *in Soils: Their Physics and Chemistry*. New York: Reinhold, 447–454.

Rode, A. A., 1969. *Theory of Soil Moisture*. Springfield, Va.: U.S. Department of Commerce, 21–26, 49–91.

Sharma, M. L.; Uehara, G.; and Mann, J. A., Jr., 1969. Thermodynamic properties of water adsorbed on dry soil surfaces, *Soil Sci.*, **107**, 86–93.

Slabaugh, W. H., 1954. Cation exchange properties of bentonite, *Jour. Phys. Chemistry*, **58**, 162–165.

Tahoun, S. A., and Mortland, M. M., 1966. Complexes of montmorillonite with primary, secondary and tertiary amides. I. Protonation of amides on the surface of montmorillonite, *Soil Sci.*, **102**, 248–254.

Cross-references: *Adsorption Phenomena; Clay-Organic Reactions; Electrochemistry; Physical Chemistry; Soil Solution.*

SOIL STABILIZATION

Soil stabilization is the collective term for any physical, chemical, or biological method, or any combination of such methods, used to improve certain properties of a natural soil to make it serve adequately for an intended engineering purpose. Different soil uses require different mechanical strengths and resistances to environmental forces. To stabilize the slope of a cut or fill against erosion is vastly different from stabilizing a soil layer to successfully carry a heavy bomber in all kinds of weather. *Soil stabilization,* therefore, has a definite meaning only if both the nature of the soil and the engineering (including the environmental) requirements are properly identified. This article concentrates on soil stabilization as used in foundation engineering (q.v. in Vol. XIII). In this area soil stabilization is divided naturally into two parts: (1) increase in bearing power and/or decrease in permeability of deep foundation soils and other large soil masses, which because of their size or location, are difficult of access and must be treated by injections (Caron, Cattin, and Herbst, 1975); (2) the improvement of native soils for the construction of shallow foundations especially for highways, airfields, parking and feeding lots, and similar facilities. While fundamental soil considerations are the same in uses 1 and 2, injection stabilization involves many technological features that fall outside the legitimate scope of a treatise of applied soil science. Accordingly, emphasis here will be placed on uses in area (2).

In these uses soils and foundations built therefrom are subjected not only to mechanical stresses incidental to their usage but also to dynamic interaction with their environment, an important aspect of which is the local or microclimate. Consequently, soil stabilization involves more than a mere increase in compressive strength or shear resistance or the improvement of any other physical property; it must also supply a defense mechanism against the continuously acting dynamic forces associated with the subjection of the soil system to daily and seasonal temperature and moisture changes and to microbial and other biologic activity, all of which tend to destroy its integrity. The stabilization requirements thus increase with increasing proximity of the soil structure (q.v.) to the pavement surface where the normal temperature and moisture variations exhibit their greatest amplitudes. Also a highway base course of the same soil type and built for the same traffic loads and frequencies has considerably higher stabilization requirements in severe than in mild climates.

The main established uses of soil stabilization include the following:

1. Lifting a country or a region out of the mud or out of sand for better economic development, now especially important for developing areas
2. Providing bases and surfaces for secondary and farm-to-market roads where good primary roads are already in existence
3. Providing bases for high-type pavements where high-type rock and crushed gravel normally used for such bases are not economically available
4. For city and suburban streets where the noise absorbing and elastic properties of certain stabilized soil systems possess definite advantages over other construction materials
5. For military and other emergencies where an area must be made trafficable within a short period of time.

Soil stabilization involves:

1. Diagnosis of the resistance properties of a given soil and of their required supplementation for the intended use in physicsl, physicochemical, and/or chemical terms
2. Translation of the supplementary requirements into available materials and processes and decision on specific method to be used on the basis of economy, practical feasibility, or special (military or other emergency) considerations
3. Construction, consisting normally of comminution, mixing with stabilizing material, and densification followed by curing
4. Economic considerations relating to total cost composed of cost of materials, construction, and maintenance for the expected service life of the structure (Winterkorn, 1946).

The three pillars of the science of soil stabilization are shown in Table 1.

Importance of a Granular Bearing Skeleton

For stabilization purposes soils are conveniently subdivided into those that in a compacted state possess a granular bearing skeleton composed of contacting sand and gravel particles, and soils without such a bearing skeleton (see *Engineering Soil Classifications;* and *Soil Mineralogy, Engineering Applications* in Vol. XIII; also Winterkorn, 1975). The wider the range from maximum to minimum size of the skeleton-forming particles in a continuous gradation, the smaller is the space left for the accommodation of silt-clay particles while maintaining the integrity of the skeleton.

Soils with a granular bearing skeleton normally possess good volume stability and fric-

TABLE 1. The Three Pillars of the Science of Soil Stabilization.

Pedology	Physical and Chemical Sciences	Body of Knowledge on Traditional Construction Materials
The science of soil formation and of the physical and chemical soil characteristics as function of 1. parent material 2. climate 3. topography 4. organisms 5. human activity 6. time	These allow judgment of soil deficiencies and indicate supplementation in physical and chemical terms that can be translated into economically available materials such as 1. inorganic cements 2. organic cements 3. waterproofing agents with or without improvement of soil granulometry, etc.	A stabilized soil, being a construction material, must resemble traditional construction materials in its essential mechanical and chemical resistance properties and structure. Especially important are granulometric considerations. Most nonmetallic construction materials and most stabilized soils fall into the category of collameritic (cemented particle) systems.

(After Winterkorn, 1955)

tional resistance properties. They may require bonding or cementation and also decrease in permeability or water-storage capacity if deficient in fines. As seen from Table 2, stabilized granular soils belong to the vast class of *collameritic systems* (colla=glue; meros=particle). Realization of this fact renders available for soil stabilization the large amount of information already collected and organized for other members of this important category of construction materials. In the terminology of materials science, such bonded stabilized soils belong to the class of "concretes" if their maximum particle size is larger than 4.76 mm, and to the "mortars" if their largest particles are of sand size.

Soils with enough granular constituents to form a bearing skeleton and enough silt-clay to provide effective cohesion under both moist and dry conditions, are often found in nature and require only densification at their optimum moisture content to yield structures of good mechanical stability. They are commonly called *sand-clay* and *stabilized gravel,* respectively. They can be artifically produced by combining granular soils that lack binder fractions with clay soils deficient in granular fractions. This is the oldest method of stabilization. Sometimes natural soils have a combination of amount and activity of silt-clay constituents that the tolerances for a stable bearing skeleton are slightly exceeded. In such cases treatment with small amounts of waterproofing or cementing agents may decrease the swelling power of the silt-clay fraction sufficiently to avert danger of instability under wet conditions. On the other

TABLE 2. Collameritics: The Science of Composition and Properties of Nonmetallic Construction Materials.

Properties of the Particles	Properties of the Cementing Agents	Examples of Cemented Systems
A. Physical I. Granulometry Laws of arrangement and packing as functions of size, gradation, and shape factors. II. Mechanical Strength, toughness abrasion resistance. B. Physicochemical and Chemical I. Interaction and bonding with cementing agents. II. Reactivity with deleterious substances in environment.	A. Inorganic I. Simple Gypsum and lime plasters. II. Complex Sorel-, hydraulic, and other cements. III. Clay and binder soil B. Organic I. Bituminous Asphalt, pitches, tars. II. Natural and synthetic resins and other polymers. III. Gums, glues of various types, etc.	1. Mortars with inorganic and organic cements including natural and artificial sand stones. 2. Concretes Portland cement-, bituminous, resinous-, clay-, etc., including naturally cemented conglomerates. 3. Plastics Powder-, paper-, cloth-, and fiber-filled; also natural wood in which cellulose fibers are bonded together by lignin.

(After Winterkorn, 1955)

hand, if there is too small a binder fraction for the desired cohesion and low permeability (q.v.), this can be remedied by the addition of silt-clay or other filler-bonding substances.

Stabilization of silt-clay soils when sand and gravel for the formation of a granular skeleton are not available involves the following two steps: (1) reinforcement of the existing natural or artificial soil crumbs or secondary aggregations through the action of added cementing or waterproofing agents with which they become covered in the mixing process and which penetrate for a certain distance into the crumbs; and (2) agglomeration of these surface-waterproofed or case-hardened crumbs and their cementation by means of the remaining stabilizer or the richer mixtures of the stabilizer with the smaller secondary particles produced by the comminution, mixing, and compaction processes. The extent and rate of penetration of the stabilizing agents into the crumbs and the concomitant case-hardening or waterproofing depend on such factors as the size, permeability, and moisture content of the crumb, the type of stabilizer, and the mechanism of its reaction with the solid, liquid, and gaseous soil constituents as well as on the temperature and pressure conditions of the system. Figure 1 shows grain-size composition ranges for sand-clay and clay-bonded coarser mixtures recommended for various uses, as well as suggested stabilizing treatments for fine-grained soils without granular bearing skeleton.

Nomenclature and Methods of Stabilization

Of the innumerable theoretical and practical possibilities of stabilizing soils, the following systems and methods have crystallized as practical and economical solutions (Winterkorn, 1946, 1975):

1. Combination of granular-bearing skeleton with pore filling and/or cementing natural and extraneous materials (clay- and other concretes and mortars) (Winterkorn, 1955)

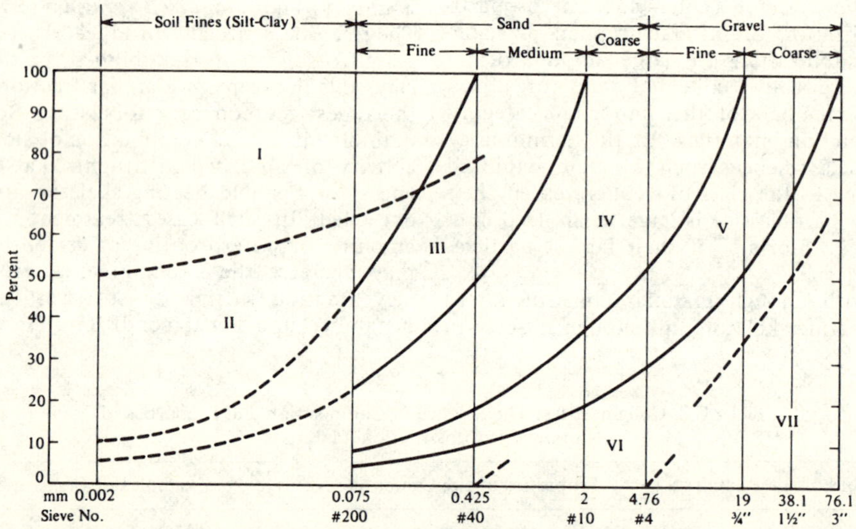

Area	Properties and Suggested Treatment
I	Heavy clay soils of high plasticity; treatment with quick or hydrated lime by itself or for rendering soil susceptible to stabilization with hydraulic cements or bituminous materials (asphalts, tars, etc).
II	Intermediate clay contents; hydraulic cements if LL of −#40 sieve fraction < 40 and PI < 15, and bituminous binders if PI < 10; in borderline cases toward I combined with use of hydrated lime.
III	Well-graded sandy soil, easily stabilized with hydraulic cement or bitumen.
IV	Well-graded soil in the sand-clay (clay-mortar) range easily shaped and compacted at optimum moisture content to form light traffic surfaces if PI = 4–12 and bases if PI = 0–6. Higher type mortars are obtained by partial or complete substitution of soil-binder by inorganic or organic cementing agents.
V	Stabilized gravel or soil concrete suitable for heavy traffic surfaces if PI = 4–9 and bases if PI = 0–6. Higher type concretes are obtained by partial or complete substitution of soil binder by inorganic or organic cementing agents.
VI	If minimum sizes fall between #40 and #4 sieve, material has open texture and is subject to ravelling under traffic.
VII	Harsh mixtures, difficult to compact and shape during construction.

FIGURE 1. Grain-size distribution and plasticity properties as related to various methods of stabilization. (From Winterkorn 1975; copyright © 1975 by Litton Educational Publishing, Inc. Reprinted by permission of Van Nostrand Reinhold Company.)

2. Soil-(Portland) cement
   a. compacted soil cement
   b. cement-modified soil
   c. plastic soil cement (Portland Cement Association, 1956, 1971)
3. Soil-lime for the stabilization or amelioration of heavy clay soils and extension into sandy and silty soil range (National Lime Association, 1954, 1972; Winterkorn, 1975)
4. Soil-bitumen
   a. soil-bitumen proper (absence of granular-bearing skeleton)
   b. sand-bitumen
   c. waterproofed granular stabilization
   d. oiled earth (Highway Research Board, 1946)
5. Soil-resin
   a. waterproofing of cohesive soils with small (less than 1%) amounts of completely or partially neutralized natural resins, or with certain organic cations including aniline-furfural condensation products
   b. waterproofing and cementing cohesive or noncohesive natural soils by means of natural or artificial resins (Mainfort, 1955)
6. Chemical stabilization, a general term for either all those methods in which chemical rather than physical or physicochemical reactions play a dominant role and would cover soil-cement, soil-lime, addition-, and condensation polymerizations (e.g., aniline-furfural, and calcium-acrylate), and waterproofing with amines and other chemically-acting agents, or the term should be reserved for cases in which one or more relatively pure chemicals are added to a soil and react with it (and/or with each other in a manner that can be described by one or more simple chemical equations (Winterkorn, 1946, 1975)
7. Thermal methods
   a. freezing
      1. temporary soil refrigeration in troublesome foundation soils during construction work
      2. continuous refrigeration to keep foundation soils in permafrost regions from thawing
   b. heating
      1. in situ for road bases and surfaces by burning piled-on combustible material or by the use of traveling road burners
      2. burning of cohesive soils in kilns to produce artificial granular aggregate to be used as natural sand and gravel
      3. stabilization of collapsible loessial and similar foundation soils by burning liquid or gaseous fuels in externally sealed bore holes
8. Electroosmotic and electrochemical stabilization
   a. electroosmotic drainage and also detouring of natural seepage by imposing strong electric potentials on the soil system
   b. electrochemical hardening, which combines water removal by electroosmosis (q.v.) with the introduction of ions or ion pairs from the electrodes that form pore-filling and cementing substances in place of the water that is removed by electroosmosis

Auxiliary or supplementary to the previously described methods and materials are the following:

1. Hygroscopic salts used as dust palliatives and compaction aids
2. Resins and resin derivatives used as construction aids for temporary waterproofing of soil access roads to construction projects
3. Use of quick- and hydrated lime for "drying out" soils on construction projects when more soil than could be treated or protected had been torn up, soaked by heavy precipitation, and turned into a morass by construction traffic
4. Dispersion agents for either or both soil and stabilizing materials, to modify the mixing, compaction, and strength characteristics of the system and to decrease its permeability

For details on all the listed methods including construction procedures, refer to Winterkorn (1975) and to the pertinent literature cited in that publication.

HANS F. WINTERKORN

References

Caron, C.; Cattin, P.; and Herbst, T. F., 1975. Injections, *in* H. F. Winterkorn and F. Y. Fang, eds., *Foundation Engineering Handbook.* New York: Van Nostrand Reinhold, 337–353.

Highway Research Board Staff, 1946. Soil bituminous roads, *Current Road Problems Bull. No. 12,* 52p.

Mainfort, R. C., 1955. Soil stabilization with resins and chemicals, *Highway Res. Board Bull. 108,* 112–119.

National Lime Association Staff, 1976. Lime stabilization construction manual, *Nat. Lime Association Bull. 326,* 49p.

Portland Cement Association Staff, 1956. *Soil Cement Construction Handbook.* Skokie, Ill.: Portland Cement Association, 99p.

Portland Cement Association Staff, 1971. *Soil Cement Laboratory Handbook.* Skokie, Ill.: Portland Cement Association, 61p.

Winterkorn, H. F., 1946. Principles and practice of soil stabilization, *in* J. Alexander, ed., *Colloid Chemistry.* New York: Reinhold, 454–492.

Winterkorn, H. F., 1955. The science of soil stabilization, *Highway Res. Board Bull. 108,* 1–24.

Winterkorn, H. F., 1975. Soil stabilization, *in* H. F. Winterkorn and F. Y. Fang, eds., *Foundation Engineering Handbook.* New York: Van Nostrand Reinhold, 312–336.

Cross-references: *Aggregate Stability; Electroosmosis; Lime, Liming; Soil Conditioners; Soil Mechanics; Soil Structure; Soils, Nonagricultural Uses. Vol. XIII: Engineering Soil Science; Soil Mechanics; Soil Mineralogy, Engineering Applications.*

SOIL STERILIZATION—See RADIOISOTOPES

SOIL STRUCTURE

The term *soil structure* refers to the physical constitution of soil material as expressed by size, shape, arrangement, and degree of development of the primary soil particles and voids into naturally or artificially formed structural units (Brewer, 1964). This broad concept of structure applies to naturally occurring or fabricated materials at any level of organization. *Macrostructures* may be studied in the field by using the senses of touch and sight. *Microstructures,* on the other hand, are investigated in the laboratory because their small sizes require the use of a microscope. Moreover, they are intimately related to investigations of soil life (see *Microhabitats; Soil Microbiology*) and fertility (see *Soil Fertility*) (Clarke and Beckett, 1971; Low, 1973). Detailed considerations of soil microstructure are discussed in Vol. XII, Pt. 2 in the articles dealing with micromorphology, e.g., *Micromorphological Structure Analysis, Micromorphometry, Micromorphometric Analysis,* and *Microvariability.*

A single natural soil aggregate is called a *ped.* Peds are relatively permanent aggregates that are separated from each other by voids or natural surfaces of weakness. *Primary peds* are the simplest forms in soil materials. They cannot be divided into smaller natural units but may be grouped to form compound peds at higher levels of organization (Brewer, 1964). Primary peds may, for example, be grouped to form *secondary peds,* which in turn may be grouped to form larger *tertiary peds.* Soil materials in which there is no observable aggregation, i.e., the materials lack peds, are termed *apedal.*

Peds are distinguished from less permanent aggregates such as clods or fragments (Soil Survey Staff, 1951). Clods and fragments are usually formed at or near the soil surface by cultivation or other manipulation. Besides being more persistent, peds commonly have coatings of organic matter or clay (*cutans*) and consequently undergo a change in color when crushed. Clods and fragments lack such coatings on their surfaces. Plowing of a soil or digging in it can form clods by molding small quantities of material into transient masses that do not persist through wetting and drying cycles. Fragments are formed by breaking large peds or by rupturing a massive soil horizon. Such units are then marked by fracture faces rather than by surfaces of natural weakness. Clods are often in the size range of coarse blocky structure, whereas fragments have no common size.

Structure Formation

The development of structure in soils is a complicated process that is still poorly understood. Primary factors in development seem to be shrink and swell phenomena during wetting and drying, pressures exerted by plant roots and the bonding together of particles by clays, iron and aluminum compounds, and organic substances such as humus, polyuronides, polysaccharides, and proteins (Baver, Gardner, and Gardner, 1972; Greenland, Lindstrom, and Quirk, 1962; Hénin, 1939; Marshall, 1962; Mehta et al, 1960; Myers, 1937; Russell, 1971; Swaby 1950). Further information on possible mechanisms and bonding agents is given in the articles on *Aggregation; Aggregate Stability; Microstructure;* and *Polysaccharide in Soil.*

Efforts have been made to name certain structures so as to suggest their mode of origin. Thus, for example, Jongerius (1957, 1958) proposed that the terms *active* and *passive,* respectively, be applied to structures due to biologic activity and those due to physical processes alone. He also proposed the term *geogenic* for structures persisting unchanged from the parent materials. Slager (1966) used the terms *biogenic* and *physiogenic* rather than *active* and *passive.*

Classification

Following the early work of Zakharov (1927), various sytems of soil structure classification have been devised (Butler, 1955; Jongerius, 1958; Nikiforoff, 1941). Most current systems for describing and classifying structure in the field (FAO Staff; Hodgson, 1974; Jongerius, 1958; National Soil Survey Committee of Canada, 1974; Taylor and Pohlen, 1970) are, however, adaptations of the system outlined in the U.S. Department of Agriculture's *Soil Survey Manual* (Soil Survey Staff, 1951).

Field description and classification of structure is based on (1) the *shape* and arrangement, (2) the *size,* and (3) the *distinctness* and durability of peds. The terminology for structure consists of separate sets of terms that designate each of these three characteristics. Shape and arrangement of peds is designated as type of soil structure. Type, in turn, is subdivided into *kinds* on the basis of the character of ped faces and edges. The *class* of soil structure refers to the size of the peds, whereas the degree of distinctness is expressed by *grades.*

Table 1 gives the four principal types of structure. *Blocklike* (the blocky and subangular blocky structures) (Figure 1), or *polyhedron-like* and *spheroidal* (the granular or crumb structures) (Figure 2), structural types include those peds arranged around a point and bounded by flat or rounded surfaces. *Peds* that are arranged around a horizontal plane and generally bounded by relatively flat horizontal surfaces are referred to as *platelike* (the platy

TABLE 1. Types and Classes of Soil Structure.

Class	Type (Shape and Arrangement of Peds)						
				Blocklike; polyhedronlike, or spheroidal, with three dimensions of the same order of magnitude, arranged around a point			
	Platelike, with one dimension (the vertical) limited and greatly less than the other two; arranged around a horizontal plane; faces mostly horizontal	Prismlike with two dimensions (the horizontal) limited and considerably less than the vertical; arranged around a vertical line; vertical faces well defined; vertices angular		Blocklike; blocks or polyhedrons having plane or curved surfaces that are casts of the molds formed by the faces of the surrounding peds		Spheroids or polyhedrons having plane or curved surfaces which have slight or no accommodation to the faces of surrounding peds	
		Without Rounded Caps	With Rounded Caps	Faces Flattened; Most Verticles Sharply Angular	Mixed Rounded and and Flattened Faces with Many Rounded Vertices	Relatively Nonporous Peds	Porous Peds
	Platy	Prismatic	Columnar	(Angular) Blocky[1]	Subangular Blocky[2]	Granular	Crumb
Very fine or very thin	Very thin platy; <1 mm	Very fine prismatic; <10 mm	Very fine columnar; <10 mm	Very fine angular blocky; <5 mm	Very fine subangular blocky; <5 mm	Very fine granular; <1 mm	Very fine crumb; <1 mm
Fine or thin	Thin platy; 1 to 2 mm	Fine prismatic; 10–20 mm	Fine columnar; 10–20 mm	Fine angular blocky; 5–10 mm	Fine subangular blocky; 5–10 mm	Fine granular; 1–2 mm	Fine crumb; 1–2mm
Medium	Medium platy; 2–5 mm	Medium prismatic; 20–50 mm	Medium columnar; 20–50 mm	Medium angular blocky; 10–20 mm	Medium subangular blocky; 10–20 mm	Medium granular; 2–5 mm	Medium crumb; 2–5 mm
Coarse or thick	Thick platy; 5–10 mm	Coarse prismatic; 50–100 mm	Coarse columnar; 50–100 mm	Coarse angular blocky; 20–50 mm	Coarse subangular blocky; 20–50 mm	Coarse granular; 5–10 mm	
Very coarse or very thick	Very thick platy; >10 mm	Very coarse prismatic; >100 mm	Very coarse columnar; >100 mm	Very coarse angular blocky; >50 mm	Very coarse subantular blocky; >50 mm	Very coarse granular; >10 mm	

After Soil Survey Staff, (1951)

[1](a) Sometimes called *nut* (b) The word "angular" in the name can ordinarily be omitted.

[2]Sometimes called *nuciform, nut,* or *subangular nut.* Since the size connotation of these terms is a source of great confusion to many, they are not recommended.

FIGURE 1. Blocky and subangular blocky (blocklike) types of structure. (Photo by Roy W. Simonson)

FIGURE 3. Platy (platelike) type of structure. (Photo by Roy W. Simonson)

structures) (Figure 3). *Prismlike* types (Figure 4) (the prismatic or columnar structures) consist of types arranged around a vertical axis and bounded by relatively flat vertical surfaces.

The different *classes* are indicated by their names and size limits (Table 1). The five size classes recognized in each of the primary types are (1) very fine or very thin, (2) fine or thin, (3) medium, (4) coarse or thick, and (5) very coarse or very thick.

Grade of structure expresses the differential between cohesion within peds and adhesion between neighboring peds. It is determined mainly by noting the durability of the aggregates and the proportions of aggregated and unaggregated material when the peds are displaced or gently crushed. Ped grade is assessed by the ease with which a soil horizon separates into peds and their durability when separated. Grade depends in part on moisture status but is normally described at the soil water content in which the soil is found.

Table 1 gives the terms for describing the degree of ped expression (grade of structure).

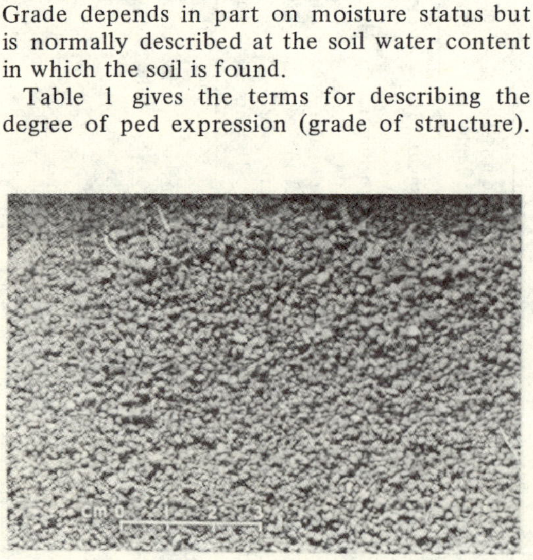

FIGURE 2. Granular or crumb (spheroidal) type of structure. (Photo by C. C. Nikiforoff)

Apedal or *structureless* soil materials lack a definite orderly arrangement of natural lines of weakness and do not exhibit observable aggregation. *Massive,* if coherent, and *single grain,* if loose, grades are recognized. The modifiers *weak, moderate,* and *strong* are used to indicate the degree of expression and durability of peds in pedal soils.

The sequence followed in combining terms to characterize soil structure is (1) grade (distinctness), (2) class (size), and (3) type (shape). Thus the designation for the soil structure in which the peds are loosely packed and roundish, dominantly below 2 mm in diameter, and quite distinct is *strong fine granular.* The designation of soil structure by grade, class, and type can be modified with other appropriate terms to describe additional characteristics of the peds.

Agricultural Importance

The importance of soil structure to plant growth and consequent agricultural production is substantial. Soils must have favorable structures for maximum crop yields (Low, 1972, 1973). Most crops require soils in which the structure (1) is amenable to root development, (2) permits adequate aeration and drainage, (3) has sufficient water-storage capacity, (4) provides access to a nutrient supply, and (5) allows good anchorage of roots (Russell, 1973). No one structure is completely ideal, however, because requirements differ among crop plants. Moreover, crops will generally grow satisfactorily over a range of structural conditions. A notable exception to the general requirements is that of paddy rice. For that crop, soils are puddled (see *Puddling*) so as to reduce a surface layer to a homogeneous mass of the ultimate particles. Each of the general requirements of crops for favorable structure is discussed briefly in one of the following subsections.

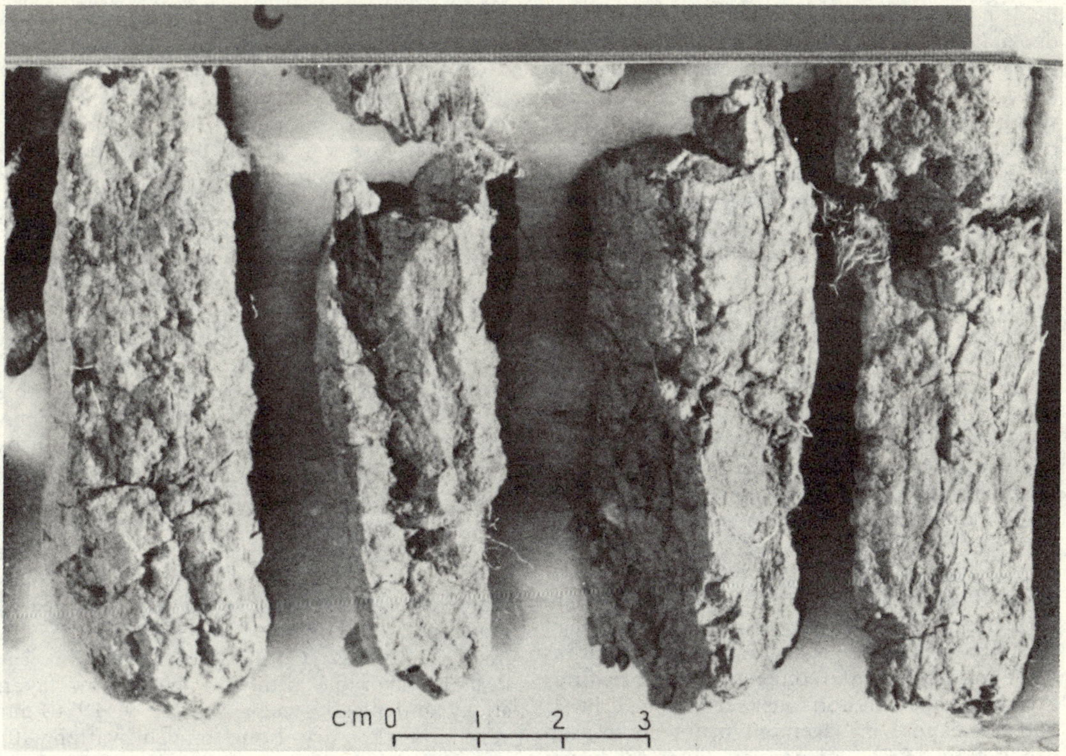

FIGURE 4. Prismatic (prismlike) type of structure. (Photo by C. C. Nikiforoff)

Root Development. The size distribution of voids in soils (see *Pore-size Distribution*) is important to root growth. Seedlings in particular require an adequate network of interconnecting voids or channels to permit rapid proliferation of their roots. An open network of voids or channels is also needed for the root systems of larger plants to develop properly (Stolzy and Barley, 1968). Otherwise, root systems will not be adequate. For example, where the soil structure inhibits root development of cereals, thickening and suberization of the seminal roots occur (Wiersum, 1957). The minimum pore sizes required for plant growth are different for different crops. Measurements of winter wheat roots, for example, indicate that the range of pore sizes required for free development of the primary seminal roots is 390–450 μm whereas first pairs of laterals require 300–350 μm and root hairs 8–12 μm (Finney and Knight, 1973).

Aeration and Drainage. Most plants begin to die after 24 hr of waterlogging (Van't Woudt and Hagan, 1957). Thus for the majority of plants, it is essential that soil structure permits rapid restoration of aerobic conditions after heavy rains. At field capacity (see *Soil Water*), the air-filled porosity should be at least 10% of the total soil volume (Marshall, 1962). When the oxygen concentration in a soil falls below about 1%, ethylene is produced by microorganisms from organic compounds within 24 hr. Ethylene has a marked deleterious effect on root development and may persist in a poorly drained soil for weeks (Dowdell et al, 1972). Under even lower oxygen concentrations, biological reduction of nitrates may occur (see *Denitrification*). Greenwood (1962) found that the concentration of dissolved oxygen in the solution bathing bacteria must fall to less than about 5×10^{-6} M before active reduction begins (this corresponds to the concentration that is in equilibrium with air containing 0.3% oxygen).

Water moves easily in pores 300 μm or more in diameter that are emptied at a suction or soil water pressure (Childs, 1969) of about 10 mbar (see *Flow Theory*). To obtain an air-filled porosity of 10% within 24 hr, rapid drainage of a saturated soil requires an adequate number of pores 300 μm or greater in diameter. At field capacity the suction generally ranges between 50 and 350 mbar. Pores greater than 60 μm will be emptied in about 48 hr at 50 mbar.

Water Storage. Water available for plant growth is stored in the soil pores. This water is held in pores mainly between 10 and 60 μm, depending on the suction at field capacity. Plants remove water from the soil primarily to replace losses due to transpiration. When the suction is about 15 bar, the rate of movement of water into plants is so slow that transpiration losses result in wilting or even death.

Nutrient Uptake. Plant nutrients (q.v.) enter plants through the root surfaces (see *Rhizosphere*). The nutrients move down a concentration gradient with the soil solution (q.v.) in the interconnected pores, both with mass flow of water and by diffusion (see *Diffusion Processes*). *Mass flow* comprises movement of soil water with its content of nutrients toward the roots to meet needs of evapotranspiration, downward because of gravity, and upward because of evaporation at the soil surface. The rate of nutrient transport (see *Transport Processes*) by mass flow is given by the product of its concentration in the soil solution and the rate of water movement (Graham-Bryce, 1967).

Anchorage. In order that plants may be anchored securely, they require a well-developed root system. Because root development requires a deep system of interconnected pores, features such as compacted horizons (see Vol. XII, Pt. 2: *Soil Pans*) pose management problems in crop production. Soil compaction due to heavy stocking or excessive rolling of pastures, for example, may result in an inadequate root system allowing the plants to be easily uprooted by grazing animals (Low, 1972). Unfavorable structural features such as compacted horizons may, however, be amenable to modification (see *Profiles, Physical Modification*).

Management of Soil Structure

Soil structure is described as *good* if it meets, at any given moment, all the criteria for optimum plant growth (Russell, 1973). A soil is said to have a *poor* structure when there is inadequate pore space for root development, drainage, or gas exchange. Soils with undesirable structural features can frequently be made productive with appropriate management practices (see *Soil Management*). Sound management practices are also essential to the stabilization and preservation of desirable soil structures (Russell, 1971).

Modification. Changes in soil structure are to a large extent associated with type of land use. For example, when old grasslands are plowed, a general degradation of structure occurs along with a decrease in organic matter (q.v.) content. Within a year or two of plowing, many soils that were previously under permanent grass exhibit a decrease in aggregate stability (q.v.), loss of resistance to rapid wetting, and increased apparent density (see *Bulk Density*) (Low, 1976). When old arable land is sown down to grass, these changes tend to be reversed, and structure is improved (Low, 1955). The development of favorable soil structure, however, takes three to four years under optimum conditions, whereas it can be destroyed in as few as one or two.

The effects of tillage on soil structure can be controlled to a large extent by the nature and timeliness of operations (Slager, 1966). The consequent modifications of structure will then alter other attributes so that they are also more favorable to plant growth. Thus tillage (q.v.) loosens surface layers so that they are more porous and open, which in turn reduces mechanical strength and facilitates penetration by plant roots and permits water to move more freely and so promotes drainage.

A soil is said to be *mellow* or to be in *good tilth* when aggregates in the surface layer are soft and crumbly. Plow layers of intermeodiate texture such as loams are most easily made and kept mellow, especially if they are high in humus. Continuous cultivation can destroy favorable tilth and leave plow layers sticky and plastic when wet (Low, 1976) and very hard when dry. Long-time cultivation without offsetting management practices reduces porosity of the plow layer, which seems to be a prime factor in mellowness. Numerous large pores seem to be characteristic of mellow soil layers. The sizes of pores or voids among aggregates are also important to the energy required in plowing soils (Low and Piper, 1973). If numbers of large pores are small, much more energy is required to move the plow through the soil.

Stabilization. Soil structure may be stabilized by natural or artificial means. Because organic materials appear to be intimately associated with the formation and maintenance of desirable structural properties, natural methods generally involve the incorporation of organic materials such as organic mulches (see *Mulch*) or the establishment of cover crops. Chemical, thermal, electroosmotic, and electrochemical methods have also been employed in soil stabilization (q.v.). These methods are, however, applicable mainly to engineering uses of soils (see *Soils, Nonagricultural Uses* and also Vol. XIII: *Engineering Soil Science*). Various types of soil conditioners (q.v.) are used to stabilize aggregates, especially at soil surfaces. Inorganic stabilizers (such as lime, cement, or phosphoric acids) and synthetic organic stabilizers (polyvinyl acetate, PVAc; polyacrylamide, PAM; vinylacetate-maleic anhydride polymer, VAMA; and hydrated polyacrylonitrile, HPAN) have, for example, been used to promote flocculation (q.v.) and chemical bonding.

Preservation. A fundamental problem of soil management is the maintenance of stable interconnected networks of pores. Because the development of soil structure is closely tied to the direct and indirect effects of roots, the activities of soil fauna (q.v.) and the production of bacterial "glues," vigorous plant growth must be maintained to supply the organic matter needed by macro- and microorganisms. To obtain optimum plant growth, there must be an adequate supply of nutrients in the soil (see *Plant Nutrients*). Thus maintaining a high level of soil fertility (q.v.) does, in effect, promote conditions favorable for desirable soil structure.

Another way to preserve good soil structure is to minimize field traffic of heavy machinery. Continued compaction by farm implements eventually destroys pore systems and degrades structural conditions, forming traffic pans (see Vol. XII, Pt. 2: *Soil Pans*) that are not conducive to crop production. Zero-tillage (q.v.), for example, keeps the traffic flow over fields to a minimum because weeds are controlled by herbicides instead of cultivation. The pores developed by roots, particularly cereals, biopores created by worms, and voids that link surface and subsoil pores are thus preserved (Ehlers, 1975).

A. J. LOW

References

Baver, L. D.; Gardner, W. H.; and Gardner, W. R., 1972. *Soil Physics.* New York: Wiley, 498p.

Boekel, P., 1963. Soil structure and plant growth, *Netherlands Jour. Agric. Sci.*, **11**, 120–127.

Bradfield, R., 1950. Soil structure, *4th Internat. Congr. Soil Sci. Trans. (Amsterdam)*, **2**, 9–19.

Brewer, R., 1964. *Fabric and Mineral Analysis of Soils.* New York: Wiley, 470p.

Butler, B. E., 1955. A system for the description of soil structure and consistence in the field, *Jour. Australian Inst. Agric. Sci.*, 239–249.

Childs, E. C., 1969. *An Introduction to the Physical Basis of Soil Water Phenomena.* London: Wiley, 347p.

Clarke, G. R., and Beckett, P. H. T., 1971. *The Study of Soil in the Field.* London: Oxford, 155p.

Dowdell, R. J.; Smith, K. A.; Crees, R.; and Restall, S. W. F., 1972. Field studies of ethylene in the soil atmosphere–equipment and preliminary work, *Soil Biol. Biochem.*, **4**, 325–332.

Ehlers, W., 1975. Observations on earthworm channels and infiltration on tilled and untilled soils, *Soil Sci.*, **119**, 242–249.

FAO Staff, no date. *Guidelines for Soil Profile Description.* Rome: Multigrafica, 53p.

Finney, J., and Knight, B. A. G., 1973. The effect of soil physical conditions produced by various cultivation systems on the root development of winter wheat, *Jour. Agric. Sci.*, **80**, 435–442.

Gardner, W. H., 1972. Use of synthetic soil conditioners in the 1950's and some implications to their further development, *Mededelingen Fakulteit Landbouwwetenschappen, Rijksuniversiteit Gent*, **37**(3), 1046–1061.

Graham-Bryce, R. A., 1967. The movement of potassium and magnesium ions in soil in relation to their availability, *in HSMO (UK) Tech. Bull. No. 14, Soil Potassium and Magnesium*, 20–31.

Greenland, D. J.; Lindstrom, G. R.; and Quirk, J. P., 1962. Organic materials which stabilize natural soil aggregates, *Soil Sci. Soc. America Proc.*, **26**, 366–371.

Greenwood, D. J., 1962. Nitrification and nitrate dissimilation in soil. 2. Effect of oxygen concentration, *Plant and Soil*, **17**, 378–391.

Hénin, S., 1939. L'Influence des facteurs climatiques sur la stabilité structurale des sols de limon, *Ann. Agron. (Paris)*, **9**, 300–311.

Hodgson, J. M., ed., 1974. Soil Survey Field Handbook, *Soil Survey of England and Wales Tech. Mon. No. 5*, 99p.

Jongerius, A., 1957. Morfologische onderzoekingen over de bodemstructuur Mededelingen van de Stichting voor Bodemkartering, *Bodemkundige Studie No. 2*, Wageningen, 93p.

Jongerius, A., 1958. The morphological soil structure classification of the Dutch Soil Survey Institute, *Mededelingen van de Lanbouwhogeschool en de Opzoekingsstations van de Staat Universitet te Gent (Belgium)*, **24**(1), 206–215.

Low, A. J., 1955. Improvements in the structural state of soils under leys, *Jour. Soil Sci.*, **6**, 179–199.

Low, A. J., 1972. Some aspects of soil structure, *Chem. Ind.*, 373–378.

Low. A. J., 1973. Soil structure and crop yield, *Jour. Soil Sci.*, **24**, 249–259.

Low, A. J., 1976. Effect of long periods under grass on soils under British conditions, *Jour. Sci. Food Agric.*, **27**, 571–582.

Low, A. J., and Piper, F. J., 1973. Some examples of the effect of soil physical conditions on the tractive effort required during ploughing, *Jour. Sci. Food Agric.*, **24**, 1011–1020.

Low, A. J.; Piper, F. J.; and Roberts, D., 1963. Soil changes in ley-arable experiments, *J. Agric. Sci.*, **60**, 229–238.

Marshall, T. J., 1962. The nature, development and significance of soil structure, *Joint Mtg. Com IV and V, Internat. Soc. Soil Sci. Trans. (New Zealand)*, 2–15.

Mehta, N. C.; Streuli, H.; Muller, M.; and Deuel, H., 1960. Role of polysaccharides in soil aggregates, *Jour. Sci. Food Agric.*, **11**, 40–47.

Myers, H. E., 1937. Physicochemical reactions between organic and inorganic soil colloids as related to aggregate formation, *Soil Sci.*, **44**, 331–359.

National Soil Survey Committee of Canada, 1974. The System of Soil Classification for Canada, *Department of Agriculture Publ. No. 1455*, 255p.

Nikiforoff, C. C., 1941. Morphological classification of soil structure, *Soil Sci.*, **52**, 193–212.

Russell, E. W., 1971. Soil structure, its maintenance and improvement, *Jour. Soil Sci.*, **22**, 137–151.

Russell, E. W., 1973. *Soil Conditions and Plant Growth*, 10th ed. London: Longmans, 849p.

Slager, S., 1966. Morphological studies of some cultivated soils, *Agricultural University Agric. Res. Report 670*, Wageningen, 106p.

Soil Survey Staff, 1951. Soil Survey Manual, *U.S. Department of Agriculture Handbook No. 18,* 503p.

Stolzy, L. H., and Barley, K. P., 1968. Mechanical resistance encountered by roots entering a compact soil, *Soil Sci.,* **105,** 297–301.

Swaby, R. J., 1950. The influence of humus on soil aggregation, *Jour. Soil Sci.,* **1,** 182–194.

Taylor, N. H., and Pohlen, I. J., 1970. Soil Survey Method, *New Zealand Soil Bur. Bull. No. 25,* 242p.

Van't Woudt, B. D., and Hagan, R. M., 1957. Crop responses to excessively high soil moisture levels, *in* J. N. Luthin, ed., *Drainage of Agricultural Lands.* Madison: American Society of Agronomy, 514–578.

West-European Working Group on Soil Structure, Internat. Soc. Soil Sci., 1967. *West-European Methods for Soil Structure Determination.* Ghent (Belgium): State University, Faculty of Agricultural Sciences, 529p.

Wiersum, L. K., 1957. The relationship of the size and structural rigidity of pores to their penetration by roots, *Plant and Soil,* **9,** 75–85.

Zakharov, S. A., 1927. Achievements of Russian science in the morphology of soils, *Acad. Sci. USSR (Leningrad) Russ. Pedological Investig. No. 2,* 47p.

Cross-references: *Aggregate Stability; Aggregation; Microstructure, Manipulation; Profiles, Physical Modification; Soil Pores, Vol. XII, Pt. 2: Micromorphological Structure Analysis.*

SOIL TECHNOLOGY–See CONSERVATION; IRRIGATION; LIME, LIMING; MANAGEMENT OF SOILS; PROFILES, PHYSICAL MODIFICATION; PUDDLING; SALT-LEACHING; SOIL CONDITIONERS; SOIL DRAINAGE; SOIL STABILIZATION; SOIL TESTS; TILLAGE; ZERO-TILLAGE

SOIL TESTS

Soil testing for determining lime and fertilizer needs has been used widely in the country for about thirty years. Most of the test procedures being used by state and commercial soil-testing laboratories had their origin in the late 1940s and early 1950s. In a survey conducted in 1951 by the Soil Test Work Group of the National Soil and Fertilizer Research Committee (Nelson et al, 1953), the 48 state soil-testing laboratories used 28 different extractants for determining phosphorus and 19 different extractants for potassium. Most procedures were locally developed methods and techniques. Only the Morgan method had wide use (Lunt, Swanson, and Jacobson, 1950), particularly in the northeastern United States and in several western states. In a 1973 study of the same laboratories regarding soil test methodology in current use, Jones (1973) found that essentially three extraction methods are used to determine extractable phosphorus (Bray Pl, Double Acid, and Olsen), and three for extractable potassium (Neutral Normal Ammonium Acetate, Double Acid, and Morgan). This is a remarkable change from the earlier survey. One might conclude that considerable standardization in methodology had indeed taken place. However, although the number of extractants used for determining soil phosphorus and soil potassium, respectively, are essentially three in number, there remains considerable variation in laboratory technique. A handbook describing reference methods for soil testing was recently published by the Council on Soil Testing and Plant Analysis that describes in detail 18 soil test methods (Jones, 1974) including a newer technique by Baker (1973). The handbook presents standard procedures for the various test methods.

Changing Role and Needs

Soil testing and soil test recommendations are essential if lime and fertilizer materials are to be correctly and economically used by farmers. Soil testing is also big business for state, private, and fertilizer company laboratories. There is a growing interest among farmers and growers to test soil. This interest is due in part to the increasing cost of fertilizer materials (see *Fertilizers, Inorganic and Organic*) and the desire of some to be environmentally correct in their use of fertilizers. However, if soil testing is to be an effective means of evaluating the fertility status of soils, correct methodology is essential. There is no such thing as A soil test, for there are a number of procedures for determining soil pH, lime requirement, and level of extractable nutrient elements. The use of widely varying laboratory techniques has caused concern among those who see an increasingly significant role for soil testing as a monitoring tool and as a means of estimating lime and fertilizer needs on a regional or national scale.

The variance in methodology introduces differences in test results that create confusion among farmers and farmer advisors. Failing to realize that there is no such thing as *A* soil test, they may unknowingly send soil samples to a laboratory that does not use tests applicable to their particular soil type. Also some laboratories may fail to realize that a particular soil-testing procedure may not be applicable to all soils received. The result is a faulty interpretation and recommendation that may produce a deficiency or imbalance problem for the farmer.

Although the number of extractants used has

been narrowed markedly in the last twenty years, there is considerable variance in technique. For example, among those who use the Bray P1 extraction procedure for determining phosphorus, soil to solution ratios were found to vary from 1:6.7 to 1:10 and shaking times from 40 to 5 min (Jones, 1973). Bray's original procedure called for a 1:7 soil to solution weight ratio with a 40-s shaking time.

Similar differences exist for other test methods. Potassium, extracted by neutral normal ammonium acetate, is being determined with soil to solution ratios varying from 1:3 to 1:10 with various shaking times. Soil pH is determined in either 1:1 or 1:2 soil-water mixes, in saturated pastes, or in dilute salt solutions (Jones, 1973). Some laboratories weigh samples, others scoop to known volume or estimated weight. Therefore the amount of soil extracted can determine the test result, which affects the interpretation. For those who rely on more than one laboratory to provide soil tests for farmers in various regions of the United States, these differences in methodolgy create an almost intolerable situation.

The various ways of expressing results cause confusion too, particularly for those who work across state lines or submit samples to more than one laboratory. The volume measurement (pounds per acre), and the weight measurement (parts per million), are the two most commonly used expressions denoting the level for an extracted element. Few realize an error is introduced into the soil test result when the volume measurement is used for weighed laboratory samples or when volume measured samples are expressed in the final result as parts per million. To the unsuspecting, differing results for the same soil may be due only to a faulty mathematical calculation. A useful discussion of the volume and weight concepts has recently been given by Mehlich (1972).

A soil test result itself is of relatively less importance than its interpretation. For the farmer who tests soil, the lime and fertilizer recommendation is of primary importance. Therefore if the laboratory test result is well correlated with crop response or yield, then how the test result was obtained is of minor concern. In too many instances, however, laboratory modifications of an already calibrated soil test are made without recalibration. For example, many laboratories use Bray's P1 phosphorus test calibration and interpretation data but have modified Bray's original laboratory procedure. In these instances the modification alters the test result sufficiently to invalidate the interpretation as originally proposed. The result is a faulty interpretation and an improper phosphorus fertilizer recommendation.

Reference Methods

The need to standardize soil-testing procedures and methods is being more than talked about today, although there is little unanimity on the subject. Several groups are evaluating various testing methods and making efforts to set parameters for each laboratory procedure. Much of this work is being done by regional research committees on soil testing. In 1970 the Council on Soil Testing and Plant Analysis was organized with one of its primary goals to research and prepare reference methods for soil testing. In 1974 the council published the *Handbook on Reference Methods for Soil Testing.*

Brief summaries of the most commonly used soil-testing procedures included in the Handbook (Jones, 1974) are given in Tables 1 through 7 (pages 516–521). This handbook, which is the most authoritative work on standard methods for soil testing, describes in detail eighteen methods of soil testing.

J. BENTON JONES, JR.

References

Baker, D. E., 1973. A new approach to soil testing. II. Ionic equilibrium involving H, K, Ca, Mg, Mn, Fe, Cu, Zn, Na, P, and S, *Soil Sci. Soc. America Proc.*, 37, 537–541.

Jones, J. B., 1973. Soil testing in the United States, *Comm. Soil Sci. Plant Anal.*, 4, 307–322.

Jones, J. B., ed., 1974. *Handbook on Reference Methods for Soil Testing.* Athens: Council on Soil Testing and Plant Analysis, University of Georgia Press, 101p.

Lunt, H. A., 1973. Swanson, C. L. W., and Jacobson, H. G. M., 1950. The Morgan soil testing system, *Conn. Agric. Exp. Sta. Bull. 541,* 1–60.

Mehlich, A., 1972. Uniformity of expressing soil test results: A case of calculating results on a volume basis, *Comm. Soil Sci. Plant Anal.*, 3, 417–424.

Nelson, W. L.; Fitts, J. W.; Kardos, L. D.; McGeorge, W. T.; Parks, R. G.; and Reed, J. F., 1953. *Soil Testing in the United States.* Washington, D.C.: National Soil and Fertilizer Research Committee, U.S. Government Printing Office, 0-979953, 50p.

Cross-references: *Lime, Liming; Management of Soils; PH, Field Determination; Soil Chemistry; Soil Fertility.*

SOIL TEXTURE—See PARTICLE-SIZE DISTRIBUTION

SOIL USE—See AGRONOMY; EDAPHOLOGY; MANAGEMENT OF SOILS; SOILS, NON-AGRICULTURAL USES

TABLE 1. Soil pH in Water, 0.01m Calcium Chloride, SMP and Adams-Evans Buffers, and Soil Paste.

Soil Test	Sample Size	Extraction Reagent	Volume Used	Interaction or Shaking Time	Method of Determination	References
Soil water pH (for SMP buffer)	4.25 cm^3 (5.0 g)	Water	5 cc	10 min. with intermittent stirring	Read pH to nearest 0.1 with glass electrode pH meter calibrated with buffers pH 4.0 and 7.0	Methods of Soil Analysis, Amer. Soc. Agron. 1965. Chapter 60, pp 914–926
Soil water pH (for Adams-Evans buffer)	10 cm^3	Water	10 cc	10 min. with intermittent stirring	Read pH to nearest 0.1 with glass electrode pH meter calibrated with buffers pH 4.0 and 7.0	Soil Testing and Plant Analysis, Soil Sci. Soc. Amer. 1973, Chapter 7, pp. 77–95
Soil pH in 0.01M $CaCl_2$	4.25 cm^3 (5.0 g)	0.01m $CaCl_2$	5 cc	30 min. with intermittent stirring	Read pH to nearest 0.1 with glass electrode pH meter calibrated with buffers pH 4.0 and 7.0	Soil Sci. Soc. Amer. Proc. 19: 164–1967. 1955
SMP buffer pH	4.25 cm^3 (5.0 g) soil in 5 cc water	SMP buffer	10 cc	Shake 10 min., intermittent stirring for 20 minutes	Read pH to nearest 0.1 with glass electrode pH meter calibrated with buffers pH 4.0 and 7.0	Soil Sci. Soc. Amer. Proc. 24: 274–277. 1961
Adams-Evans buffer pH	10 cm^3 soil in 10 cc water	Adams-Evans buffer	10 cc	Shake 10 min., intermittent stirring for 20 minutes	Read pH to nearest 0.5 with glass electrode pH meter calibrated to read pH 8.00 in mixture of 10 cc buffer reagent and 10 cc water	Soil Sci. Soc. Amer. Proc. 26: 355–357. 1962
Soil paste pH	About 150 g	Water	Varies with texture	1 hour	Read pH to nearest 0.1 with glass electrode pH meter calibrated with buffers pH 4.0 and 7.0	Diagnoses and improvement of saline and alkali soils. USDA

TABLE 2. Soil Test Phosphorus Methods.

Parameter	Name of Method Double Acid	Bray P_1	Morgan	Olsen
Adaptability limits	Sandy soils, acid, low in CEC	Acid soils with moderate CEC	Acid soils with moderate CEC	Alkaline soils
Sample size, volume (weight)	5 cm^3	2.5 cm^3	5 cm^3	2 cm^3
Volume extractant, cc	25	25	25	50
Extracting solution	0.05N HCl in 0.025N H_2SO_4	0.03N NH_4F in 0.025N HCl	lN NaOAc at pH 4.8	0.5N $NaHCO_3$ at pH 8.5
Shaking time, minutes	5	5	30	30
Shaking action and speed	Reciprocating 180+ oscillations/min.	Reciprocating 180+ oscillations/min.	Reciprocating 180+ oscillations/min.	Reciprocatting 180+ oscillations/ min.
Method of P determination in extract	Molybdenum Blue	Molybdenum Blue	Molybdenum Blue	Molybdonum Blue
Range in soil P conc. without dilution, kg/ha	2–100	10–250	10–250	10–200
Sensitivity, kg/ha	1	1	1	1
Primary reference	North Caroline Soil Test Div. Mimeo (1953)	Soil Sci. 59:39 (1945)	Cornell Agric. Exp. Sta. Bull. 960 (1965)	USDA Circular 939 (1954)

TABLE 3. Soil Test Potassium Methods.

	Name of Method			
Parameter	Double Acid	1N NH_4OAc pH 7.0	Morgan	Water
Adaptability limits	Sandy soils, acid, low in CEC	Wide range of soils	Wide range of soils	Wide range of soils
Sample size	5 cm^3	5 cm^3	5 cm^3	5 cm^3
Volume of extractant, cc	25	25	25	25
Extracting solution	0.05N HCl in 0.025N H_2SO_4	1N NH_4OAc at pH 7.0	1N NaOAc at pH 4.8	Pure water
Shaking time, minutes	5	5	30	30
Shaking action and speed	Reciprocating 180+ oscillations/min.	Reciprocating 180+ oscillations/min.	Reciprocating 180+ oscillations/min.	Reciprocating 180+ oscillations/min.
Method of K determination extract	Flame emission spectroscopy	Flame emission spectroscopy	Flame emission spectroscopy	Flame emission spectroscopy
Range in soil K conc. without dilution, kg/ha	50–500	50–1000	50–1000	50–500
Sensitivity, kg/ha	5	5	5	5
Primary reference	North Carolina Soil Test Div. Mimeo (1963)	Soil Sci. 59:13 (1945)	Cornell Agric. Exp. Sta. Bull. 960 (1965)	Amer. Soc. Agron. Pub. No. 9, p 935 (1965)

TABLE 4. Soil Test Calcium Methods.

	Name of Method			
Parameter	Double Acid	1N NH_4OAc pH 7.0	Morgan	Water
Adaptability limits	Sandy soils, acid, low in CEC	Wide range of soils	Wide range of soils	Wide range of soils
Sample size	5 cm^3	5 cm^3	5 cm^3	5 cm^3
Volume of extractant, cc	25	25	25	25
Extracting solution	0.05N HCl in 0.025N H_2SO_4	1N NH_4OAc at pH 7.0	1N NaOAc at pH 4.8	Pure water
Shaking time, minutes	5	5	30	30
Shaking action and speed	Reciprocating 180+ oscillations/min.	Reciprocating 180+ oscillations/min.	Reciprocating 180+ oscillations/min.	Reciprocating 180+ oscillations/min.
Method of Ca determination in extract	Atomic absorption spectroscopy	Atomic absorption spectroscopy	Atomic absorption spectroscopy	Atomic absorption spectroscopy
Range of soil Ca conc. without dilution, kg/ha	150–1500	500–2000	500–2000	150–1000
Sensitivity, kg/ha	10	10	10	10
Primary reference	North Carolina Soil Test Div. Mimeo (1953)	Soil Sci. 59:13 (1945)	Cornell Agric. Exp. Sta. Bull. 960 (1945)	Am. Soc. Agron. Pub. No. 9, p. 935 (1965)

TABLE 5. Soil Test Magnesium Methods.

	Name of Method			
Parameter	Double Acid	lN NH_4OAc pH 7.0	Morgan	Water
Adaptability limits	Sandy soils, acid, low in CEC	Wide range of soils	Wide range of soils	Wide range of soils
Sample size	5 cm^3	5 cm^3	5 cm^3	5 cm^3
Volume of extractant, cc	25	25	25	25
Extracting solution	0.05N HCl in 0.025N H_2SO_4	lN NH_4OAc at pH 7.0	lN NaOAc at pH 4.8	Pure water
Shaking time, minutes	5	5	30	30
Shaking action and speed	Reciprocating 180+ oscillations/min.	Reciprocating 180+ oscillations/min.	Reciprocating 180+ oscillations/min.	Reciprocating 180+ oscillations/min.
Method of Mg determination in extract	Atomic absorption spectroscopy	Atomic absorption spectroscopy	Atomic absorption spectroscopy	Atomic absorption spectroscopy
Range in soil Mg conc. without dilution, kg/ha	50–500	50–500	50–500	50–500
Sensitivity	5	5	5	5
Primary reference	North Carolina Soil Test Div. Mimeo (1953)	Soil Sci. 59:13 (1945)	Cornell Agric. Exp. Sta. Bull. 960 (1945)	Am. Soc. Agron. Pub. No. 9, p 935 (1965)

TABLE 6. Soil Test Sodium Methods.

Parameter	Name of Method: Double Acid	1N NH_4OAc pH 7.0	Water
Adaptability limits	Sandy soils, acid, low in CEC	Wide range of soils	Wide range of soils
Sample size	5 cm^3	5 cm^3	5 cm^3
Volume of extractant, cc	25	25	25
Extracting solution	0.05N HCl in 0.025N H_2SO_4	1N NH_4OAc at pH 7.0	Pure water
Shaking time, minutes	5	5	30
Shaking action and speed	Reciprocating 180+ oscillations/min.	Reciprocating 180+ oscillations/min.	Reciprocating 180+ oscillations/min.
Method of Na determination in extract	Flame emission spectroscopy	Flame emission spectroscopy	Flame emission spectroscopy
Range in soil Na conc. without dilution, kg/ha	10–200	10–250	10–200
Sensitivity, kg/ha	2	2	2
Primary reference	North Carolina Soil Test Div. Mimeo (1953)	Soil Sci. 59:13 (1945)	Am. Soc. Agron. Pub. No. 9, p 935 (1965)

TABLE 7. Soil Test Zinc Methods.

Parameter	Name of Method: Double Acid	0.1N HCl	DTPA
Adaptability limits	Sandy soils, acid, low in pH	Acid soils	Wide range of soils
Sample size	5 cm^3	5 cm^5	10 cm^3
Volume of extractant, cc	20	20	20
Extracting solution	0.05N HCl in 0.025N H_2SO_4	0.1N HCl	0.005M DTPA 0.1M TEA 0.01M $CaCl_2$
Shaking action and speed	Reciprocating 180+ oscillations/min.	Reciprocating 180+ oscillations/min.	Reciprocating 180+ oscillations/min.
Method for Zn determination in extract	Atomic absorption spectroscopy	Atomic absorption spectroscopy	Atomic absorption spectroscopy
Range in soil Zn conc. without dilution, kg/ha	2–30	2–30	1–20
Sensitivity, kg/ha	1	1	0.5
Primary reference	Comm. Soil Sci. Plant Anal. 1:35 (1970)	Soil Sci. Soc. Amer. Proc. 32:543 (1968)	Soil Sci. Soc. Amer. Proc. 33:62 (1969)

SOIL WATER

Water is a major component of soil and affects almost all the physical and chemical properties that are associated with plant and animal life in and on soil. In addition to its direct role as a plant growth substance, it carries nutrients into plants in the transpiration stream. Soil water has dissolved in it the growth substances required by plants and microflora and fauna that occupy the soil mass (see *Soil Solution*). Movement of such gases as oxygen and carbon dixoide (see *Diffusion Processes*), both involved in growth processes, are inversely related to soil water content. Heat flow and retention are directly related. Also water materially affects mechanical properties of soil. At high water content it acts as a lubricant, facilitating sliding of particle on particle surface. Yet at lower water content it binds particles together by means of surface-tension forces (Rose, 1966).

Water is a dominant material in physical and chemical weathering (q.v. in vol. XII) of soil. Forces exerted by water as it freezes in cracks or pores help to break up rocks or aggregates of soil particles. Water carries suspended particles that abrade rocks or minerals and chemical agents that react with soil materials changing their chemical nature. Additionally, latent heat, which is given up or absorbed as water changes state back and forth from liquid to gas or liquid to solid, strongly affects soil temperature and the many processes dependent on temperature (see *Thermal Regimes*). Transport of substances in liquid water, with their deposition at the point of evaporation or precipitation, is particularly important in concentrating chemical compounds at the soil surface or within the soil profile.

Water contents range from saturation to extremely low values in dry atmospheres where water may be present in discontinuous films as little as one molecule thick (Gardner, 1965). Water vapor exists in nonfilled pores with relative humidities varying from 100% at near saturation to values approaching those of the atmosphere in air-dry soils. Liquid water is held by strong adsorptive forces on particle surfaces and in pores between particles by surface tension in much the same way as water is held in capillary tubes. In moist soil the force with which water is held is indirectly indicated by the total radius of curvature R of air-water interfaces as given by the equation

$$\frac{1}{R} = \frac{1}{r_1} + \frac{1}{r_2}$$

where, as shown in Figure 1, r_1 is the concave radius (negative) of the air-water interface viewed between the particles, and r_2 is the convex radius (positive) at right angles to r_1, but which could be observed directly only if a particle could be removed without changing the geometric configuration of the water (Baver, Gardner, and Gardner, 1972). Where $r_1 = r_2$, the total radius of curvature R is infinite, and the relative humidity immediately adjacent to this surface is 100%. The vapor pressure in moist soil air is given by the equation for pressure across a liquid-air interface, $p = 2\sigma/R$, where σ is the surface tension and R is the total radius of curvature. But when most of the water is associated with particle surfaces rather than held in the pores, as can be imagined in the dry soil at the right in the figure, the capillary model no longer holds. Here adsorptive forces between water and the particle surface dominates over surface tension and water is bound so tightly that the relative humidity in the soil atmosphere has much lower values than in moist soil (Nielsen et al, 1972).

The two most important physical properties of soil water are surface tension, which affects water retention, and viscosity, which affects water flow (Childs, 1969; Baver, Gardner, and Gardner, 1972; Hillel, 1971). Surface (force per unit length) that arises from an unbalance of forces acting on polar water molecules at the surface. Water molecules are tetrahedrons with two positive and two negative poles. They are randomly oriented in the deep water but become oriented at the surface as a consequence of unequal forces exerted by water molecules on one side and air molecules on the other. Surface tension acts something like the stretched surface in a rubber membrane, but the force does not increase as the surface expands as would be the case with a rubber membrane. Viscosity is the property

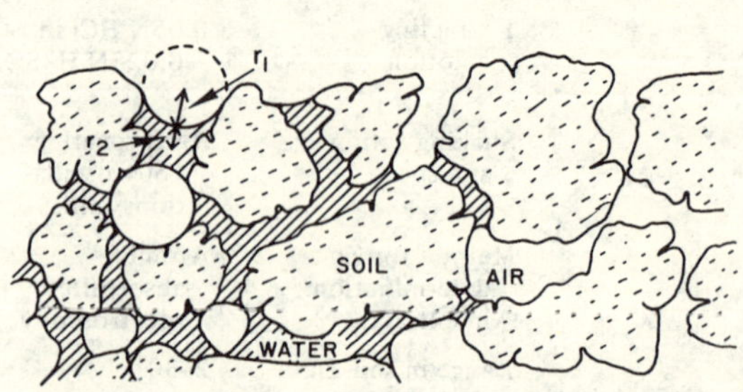

FIGURE 1. Partially wetted soil showing air-water interfaces. The water surface has a negative radius r_1 in the plane of the paper and a positive curvature r_2 at right angles.

that relates the shear force created when water "layers" slide on other water "layers." These shear forces are proportional to the change in velocity from point to point in a flowing fluid (gradient of the velocity), and viscosity is the proportionality factor—the higher the viscosity, the greater the shear. The velocity of flow is a maximum at the middle of a water-filled flow channel and zero at the particle surface. Because of shear (viscosity), the kinetic energy associated with flow is lost to friction with water being warmed by such flow (Kirkham and Powers, 1972). The effect of viscosity on water flow in soil may be seen by analogy with flow through a capillary tube. Rate of flow in a cylindrical capillary tube is proportional to the fourth power of the radius a of the tube as given by the *Hagen-Poiseuille equation,* Flux = $(\pi a^4/8\eta)(\Delta p/S)$ (flux per unit cross-section of the channel is $1/\pi a^2$ of this) where η is the viscosity and Δp is the pressure difference across the tube of length S. Thus because of the influence of viscosity, the smaller the flow channel, the greater flow is affected.

WALTER H. GARDNER

References

Baver, L. D.; Gardner, W. H.; and Gardner, W. R., 1972. *Soil Physics,* 4th ed. New York: Wiley, 498p.

Childs, E. C., 1969. *An Introduction to the Physical Basis of Soil Water Phenomena.* New York: Wiley, 493p.

Gardner, W. H., 1965. Water content, *in* C. A. Black, et al, eds. *Methods of Soil Analysis.* Madison: American Society of Agronomy, 82–127.

Hillel, D., 1971. *Soil and Water: Physical Principles and Processes.* New York: Academic, 288p.

Kirkham, D., and Powers, W. C., 1972. *Advanced Soil Physics.* New York: Wiley, 534p.

Nielsen, D. R., et al, eds., 1972. *Soil Water,* Madison: American Society of Agronomy, 176p.

Rose, C. W., 1966. *Agricultural Physics.* Oxford: Pergamon.

Cross-references: *Capillary Pressure; Conductivity, Hydraulic; Evaporation; Field Water Cycle; Flow Theory; Imbibition; Infiltration; Irrigation Water; Leaching; Lysimeters; Moisture Management; Percolation; Soil Drainage; Soil Solution; Surface-Soil Water Content; Thermodynamics; Water Content and Retention; Water Fluxes; Water Movement; Wetting Front.*

SOILS, NONAGRICULTURAL USES

Soils differ in their capacity for plant growth and agricultural crop production. This soil variation also manifests itself in the nonagricultural uses of soils. The ever-increasing use of land and soils in, for example, construction sites, highways, airports, waste disposal, recreation, and underground utilities increases the danger of costly mistakes because the uses of the soil may not correspond with its capability. In allocating a parcel of land to a specific use, the decision maker should have some idea about, for example, the soil's corrosion potential, bearing strength, shrink-swell ratio, permeability, effluent renovation capacity, susceptibility to failure from earthquake shock waves, flooding, erosivity, and energy and cost requirements for alteration and management. Much information and data gathered by soil scientists over the decades has increasingly found application in the nonagricultural uses of soils. Several references (Bartelli, et al, 1966; Buol, Hole, and McCracken, 1973; McHarg, 1969; Simonson, 1974) have summarized the application of soil information to the planning of nonagricultural land uses.

Early Nonagricultural Uses

Simonson (1968) has reviewed the early concepts and recognition of soils as components of the natural resource base. Early civilizations, while often located near major transportation routes, were also conscious of soils suitable for agriculture. The pattern of prehistoric occupation of soils in western Europe suggests that Neolithic people recognized distinctions between soils (Clark and Evans, cited by Simonson, 1968). Early Chinese, Egyptian, Messopotamian, Indus, and Roman civilizations recorded soil classification systems that were based on productivity (Simonson, 1968).

Early nonagricultural soil uses while not well documented nevertheless recognized the importance of soil. The early brickmakers and potters must have appreciated the amount and types of clay contained in certain soils and deposits. Examination of many American Indian camps reveals that they were often located on the higher and better-drained soils.

Soil Survey Interpretations for Nonagricultural Land Uses

Organized efforts for soil classification and mapping in the United States began in 1899 as a cooperative research effort of the U.S. Department of Agriculture (USDA) and the state land-grant universities. The original objective of this program was to guide people in selecting soils that were responsive to current farm management systems. According to Kellogg (cited in Buol, Hole, and McCracken, 1973) the primary purpose of the soil classification program of the USDA has been its use in soil surveys. Furthermore, the principal uses of the soil survey in the USDA have been oriented toward interpretations for technical assistance in soil conservation programs, (see *Soil Conservation*), for planning agricultural programs, and as a basis for credit (Smith as cited by Buol, Hole, and McCracken, 1973). As the agricultural uses of the land intensified and grew more complex, it became obvious that the combination of soil properties responsible for a specific soil having a characteristic crop production and management potential also determined its suitability for a variety of nonagricultural uses.

The early attempts in the United States at applying the basic soil classification system to nonagricultural land uses began in the mid and late 1920s. By 1930 the system began to guide highway construction and was later applied in defense planning (Kellogg, in Bartelli, et al, 1966). As nonagricultural land use intensified after World War II, soil survey interpretations accelerated in the areas of community planning, engineering, transportation, and waste-disposal technology.

Aitchison et al (1954) began a study in 1945 on the urban-related soil problems of Adelaide, South Australia, indicating the control exerted on the city's expansion pattern by the nature of the soil. Of particular interest to Aitchison in the 1954 report and subsequent applications of soil survey principles to urban planning uses was the correlation of soil types with building foundation behavior and design. Aitchison concluded "that the soil profile could serve as the basis for the quantitative extrapolation of knowledge basic to foundation design" (in Simonson, 1974). Such information and its extrapolation to similar soils has proved valuable to engineers and architects and aided in the development of building codes and practices to overcome soil limitations.

One of the earliest soil surveys made specifically for planning the expanding nonagricultural land uses on the fringes of suburban areas in the United States was in Fairfax County, Virginia (Kellogg, in Bartelli, et al, 1966). The soil mapping, begun in 1953 at the request of the county, was completed in 1955 with the final report published by the Soil Conservation Service (SCS) in 1963. This county was the first in the United States to hire a full-time soil scientist for multipurpose interpretations. The Fairfax County experience has been recently reviewed by Pettry and Coleman (in Simonson, 1974).

The use of soil surveys and soil data for regional planning in a seven-county area of southeast Wisconsin has been reviewed in detail by Bauer (in Simonson, 1974). Bauer states that, to help avoid further abuse and misuse of the soil element of the natural resource base, a need exists in any comprehensive regional planning program to examine not only how land and soils are presently used but how they can *best* be used and managed. This effort requires definitive data and knowledge about the geographic location and extent of the various soils; about their physical, chemical, mineralogical, and biological properties; and about the capability of these soils to support various kinds of both rural and urban uses.

Modern soil surveys in the United States provide a variety of interpretive tables for many nonagricultural soil uses. These interpretations usually take the form of indicating the suitability (good, fair, poor) or limitations (slight, moderate, severe) of the soils for the use being considered.

Nonagricultural Uses of Soil Information

Floodplains. Because urban development often encroaches on floodplains of streams and rivers, it is necessary to develop restrictions preventing such unwise land use in the interest of public welfare and safety. To delineate the various floodplain boundaries with respect to maximum probable flood events, expensive hydrologic surveys are required to determine the various limits of the floodplain. Because soils in a floodplain are unique and distinct from other upland soils, a floodplain can be delineated from soil maps. Local jurisdictions have used soil maps as tools in delineating floodplain zones. In the humid regions of the United States, most floodplains delineated from soil maps would correspond to the 25- to 40-year frequency flood event.

Waste Disposal. There are about 15 to 20 million homes and other buildings in the United States using on-site liquid waste disposal systems in which the soil medium is used as the recipient and renovator of the effluent. Septic tank effluents contain suspended solids, BOD (biochemical oxygen demand) components, bacteria, virus and salts, including nitrogen and phosphorous.

The major criteria in evaluating soils for their suitability as a medium for effluent renovation are: (1) hydraulic conductivity; (2) adequate unsaturated soil volume or depth beneath the discharge point; and (3) lack of constraints such as flooding, stoniness, and concave landscape position.

Because soils differ in their ability to transmit liquids (hydraulic conductivity) and their depth as well as drainage condition under a particular moisture (climate) regime, their suitability or potential as a medium for waste disposal and renovation also varies. A septic system with a properly operating soil absorption field in a deep, medium-textured soil will result in a highly purified effluent by the time the effluent reaches the water table. Virus, bacteria, solids, phosphates and the BOD load are effectively absorbed or deactivated by the soil. Only a portion of the nitrate and salts eventually reach the groundwater—a condition of little concern unless there is a high density of such systems.

For many years soils have been interpreted in soil surveys for their suitability for on-site waste disposal. This use of soil information has been an important tool in town and regional planning (Bauer, in Simonson, 1974). A review of site selection, soil behavior, and effluent renovation under septic system discharge can be found in Beatty and Bouma (in Simonson, 1974) and in Miller and Wolf (1976).

Soils are also being used as a medium for other wastes and waste-disposal systems. More communities are going to land disposal of the centrally collected effluent, either spraying it on the land or flooding certain soil areas for infiltration and renovation. This practice has been used in Europe and Australia since the 1800s. The environmental constraints on ocean dumping and incineration are forcing sewage utilities in metropolitan areas to look at the land as a medium for disposing of sewage sludge. This sludge contains primarily the organic residues of the treatment process as well as nutrients such as nitrogen, phosphorous, and small amounts of potash and some micronutrients. This waste product can be beneficial to agricultural crops. But the sludge also contains heavy metals such as zinc, copper, nickle, cadmium, and lead that can be toxic to plants and the food chain if certain concentrations are exceeded (see *Pollution*). Again soils differ in their ability to accommodate such sludges, and those responsible for using sludges on agricultural or other lands must be mindful of the limitations of the soil medium in degrading this product.

The soil medium is also used as a receptacle for solid waste. Where large volumes of wastes are buried, these sites are called *landfills* (see vol. XIII: *Sanitary Landfills*). Pesticide containers, radioactive wastes, organic compounds, and other wastes are often discharged to the soil. Soil scientists are called on to interpret soil data and evaluate soils for their capability to handle such wastes without contaminating adjacent ecosystems. As soils are disturbed and mixed with these solid wastes, new or manmade soils result, requiring special management and care when used for parks or recreation areas after the landfill operation is complete.

Engineering. Soils differ not only in their capacity to grow crops and accept wastes but also in their ability to support loads and remain stable when disturbed or graded into embankments and earthen dams. Engineers run complex soil tests to determine suitability as a foundation or building material for a large building or other major project. But for most nonagricultural land uses, there is no on-site engineering determination of the soil's suitability for the specific use.

Many residential streets, houses, small or low-load buildings, and other urban-oriented land uses are placed on soils without on-site soil sampling and testing. These types of land uses are protected to a degree by building codes and other ordinances that require minimum foundation and subbase standards to minimize the dangers of soil failure.

Over the years soil scientists and engineers have learned a great deal about the soil properties responsible for a soil's stability and shear strength under various loads and slopes. These properties include the soil's texture or particle-size distribution. Texture in turn is related to the amount and size of the pores within the soil matrix (see *Soil Pores*). The amount, size, and shape of the pore space determines the soil's permeability (q.v.) and the amount of water the soil can hold against gravity. The amount and type of clay mineral in the soil will determine the soil's potential to swell when wetted and shrink when dried. Furthermore, the texture and permeability of the various horizons or layers within the soil profile affect the soil's behavior (see vol. XIII: *Expansive Soils*).

Engineers have developed several soil classification schemes that relate the soil's behavior to its texture under various moisture levels. These include the Unified System (U.S. Army Corps of Engineers), the AASHTO system (American Association of State Highway and Transportation Officials), and several others (q.v. in Vol. XII, Pt. 2). The Unified and AASHTO systems are published in the U.S. soil surveys for the major soil series horizons, along with the USDA textural classes so that both engineers and soil scientists are familiar with the characteristics of the soil horizons.

Knowing the properties then of the various horizons in each type of soil, one can predict to a degree how the soil will respond or behave when loaded or used as a material for a road subbase or a dam, for example. It is this technique that soil scientists and engineers use when considering an area for a nonagricultural use when on-site testing is either impractical or too costly for the proposed use.

Interpretive maps can be made of an area,

showing the soil depth, water-table limits, permeability classes, shrink-swell classes, and engineering classes. These and other data can be used to predict the soil's corrosion potential and depth constraints for pipelines, suitability for road fill and foundation loads, reservoir site suitability, grading limitations and constraints for other uses. These engineering soil properties can be and have been used to predict a soil's behavior and susceptibility to damage under earthquake shocks. Soil maps and data have been used for military purposes to determine trafficability and potentials and limitations of lands for certain types of equipment.

Again the principle of extending the application of knowledge about one kind of soil to other areas with similar soils is used. Since each kind of soil can occur over a large area, there is a wide potential application for all information and data relevant to the specific behavior of each soil. Building codes, subdivision specifications, urban zoning ordinances, and highway engineering standards have incorporated soil maps and information as a bisis for avoiding and/or correcting problem soils (Baker, 1973; Thornburn 1969; Beatty and Yanggen, 1966, cited in Bartelli, et al, 1966).

Other Uses. Numerous other nonagricultural uses of soil information and surveys can be cited. Soil information and data are used as criteria in park and recreation area planning, wildlife habitat management, identification of sand and gravel sources, analyzing the pattern of insurance claims from earthquake damage, and for environmental-impact statements and site analysis.The disciplines of geology and archeology use soil information in determining the age relationships of specific sites and layers such as the alluvial infilling of basins, migration of shorelines, ash accumulation from volcanic activity, and history of vegetation in swamps. Old geologic faults penetrating into the subsoil horizons can assist the geologist in dating the fault's occurrence, a critical measurement now required in the location of nuclear power plants. Soil science has also been applied in criminology and in forensic science to verify or link an act or individual to a specific area via the soil characteristics unique to the region.

The Future

The future of soil survey interpretation and utilization of soil data seems pointed in the direction of not just describing soil conditions and indicating soil suitability but also attempting to predict the potentials of soils for alternative land uses including the measures and costs involved in making the soils more suitable for a particular use. The Dutch soil scientists have been forced into this direction for years because of the high intensity of land use in that small country (Westerveld and Van Den Hurk, in Simonson, 1974). Similar attempts in the United States have been made in local areas and now there is a national movement in this direction.

The costly land-use mistakes of the past, the increasing concern for preserving prime agricultural lands in the United States and elsewhere, and the desire among planners to expand urban areas onto known soil conditions have all played a role in forcing the soil scientist to be more quantitative and precise in his predictions and more interdisciplinary in considering alternatives. Soil properties will be weighed as criteria for different use potentials. As more soil data are obtained, computers will be able to digest the information, assigning different, or perhaps relative, weights for each property. Eventually information will be available on which planners and others can decide which soil to use for what purpose and how much energy and cost will be required to accommodate that soil for the particular use. Furthermore, the alternative soil use decisions can be made on cost-benefit analysis with the potential costs and trade-offs known before the soil is disturbed.

FRED P. MILLER

References

Aitchison, G. D.; Sprigg, R. C.; and Cochrane, G. W., 1954. The soils and geology of Adelaide and suburbs, *Geol. Surv. So. Aust. Bull. 32,* 130p.

Baker, W. B., 1973. Soil surveys–A base for building codes, *Soil Conserv.,* **38,** 264–265.

Bartelli, L. J.; Klingebiel, A. A.; Baird, J. V.; and Heddleson, M. R., 1966. *Soil Surveys and Land Use Planning.* Madison: Soil Science Society of America and American Society of Agronomy, 196p.

Buol, S. W.; Hole, F. D.; McCracken, R. J., 1973. *Soil Genesis and Classification.* Ames: Iowa State University Press, 360p.

McHarg, I. L., 1969. *Design with Nature.* Garden City, New Jersey: Natural History Press.

Miller, F. P., and Wolf, D. C., 1976. Renovation of sewage effluents by the soil, *in Second National Conference on Individual Onsite Wastewater Systems.* Ann Arbor: National Sanitation Foundation, 87–101.

Simonson, R. W., ed. 1974. *Non-Agricultural Applications of Soil Surveys.* New York: American Elsevier, 178p.

Simonson, R. W., 1968. Concept of soil, *Adv. Agron.,* **20,** 1–47.

Thornburn, T. H., 1969. Geology and pedology in highway soil engineering, *in* D. J. Varnes and G. A. Kiersch, eds. *Reviews in Engineering Geology, Vol. II.* Boulder, Co.: Geological Society of America, 17–57.

Cross-references: *Sludge Disposal; Soil Mechanics; Soil Stabilization. Vol. XIII: Expansive Soils; Geomechanics; Geotechnical Engineering; Highway Engineering; Lime Stabilization; Pedogeochemical Prospecting; Soil Mechanics.*

SOIL ZOOLOGY–See FAUNA; SOIL MICROBIOLOGY

SPECTRAL CHARACTERISTICS

Surface soils may be characterized, identified, and mapped according to their spectral properties. Remote sensors have been used to identify moisture content, amount of organic carbon, iron content, clay types, and other soil parameters. The *electromagnetic spectrum,* remote sensors used to measure reflected energy, and some of the spectral properties of soils, is discussed here.

Electromagnetic Spectrum

Soils have been studied in the laboratory and field with remote sensors that operate in the electromagnetic spectrum (Figure 1) from the ultraviolet to the microwave region (0.2–1000 μm). The specific names for the various regions of the spectrum as listed by Stewart (1970) are: ultraviolet 0.1–0.4 μm, visible 0.4–0.8 μm, photographic infrared 0.8–1.0 μm, near infrared 1.0–5.0 μm, midinfrared 5.0–50.0 μm, far-infrared, 50.0–1000 μm, and microwave 1×10^3 to 3×10^5 μm. Most portions of the spectrum listed have been used in the study of soils in the laboratory, but the majority of outdoor studies have concentrated on the reflective region, 0.4–2.6 μm (Figure 1). The energy radiated by the sun reaching the earth's surface is greater in the short wavelengths (Figure 2) and explains why most outdoor studies have been concentrated in this region.

Remote Sensors

Remote sensing is the acquisition of information about an object (or phenomenon) that is not in intimate contact with the information-gathering device (Carroll, 1973*a*). Black and white film sensitive to the visible portion of the electromagnetic spectrum (0.4–0.8 μm) was the

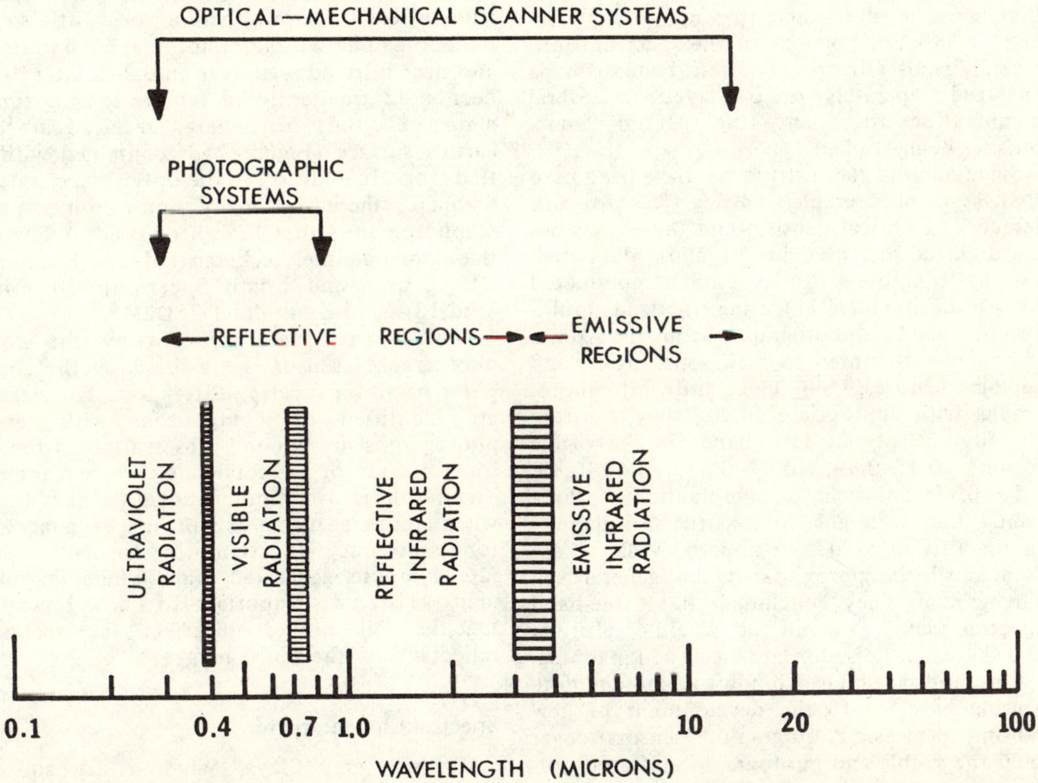

FIGURE 1. A portion of the electromagnetic spectrum. (After Hoffer and Johannsen, 1969) [Correction: For micron, μ, read micrometer, μm, the modern S.I. unit]

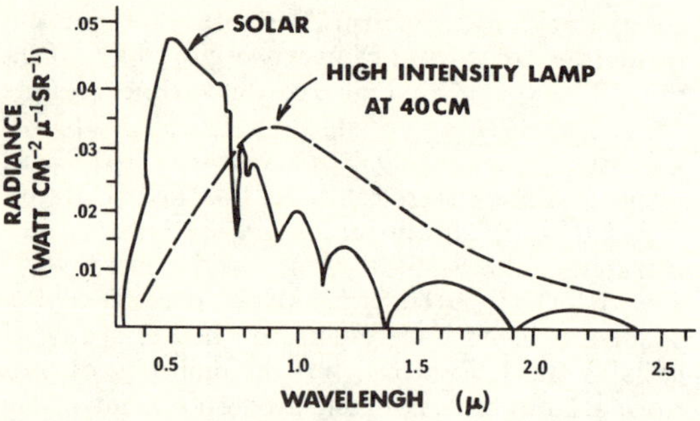

FIGURE 2. A comparison between solar radiation at sea level and the General Electric DXW lamp. (After DeWitt and Robinson, unpublished) [Correction: For micron, μ, read micrometer, μm, the modern S.I. unit]

first remote-sensing data recorder. The development of color film offers the photointerpreter more information since the photo, like the landscape, is usually seen in color by the interpreter. Photographic remote sensors have been developed to the point that through the use of special films and filter combinations, reflected solar radiation measurement has been extended to 0.3–1.0 μm. Several systems of multispectral photography have been developed that separate the visible through the photographic infrared portions of the spectrum into several bands (Figure 1). These bands can be analyzed separately or overlayed in several combinations to enhance the spectral characteristics being studied.

Since about 1965 instruments have been used that scan photographs or negatives for differences in optical density, and these densities are digitized and used in automatic data analysis by computers. An experiment conducted by Anuta et al (1971) using digitized, multispectral, satellite photography from the Apollo 9 mission attempted to map soils, crops, and geologic features. Using black and white photographs with appropriate filters, they digitized the film density of four bands in the visible region; 0.47–0.89, 0.47–0.61, 0.68–0.89, 0.59–0.71 μm. Using a combination of three bands, they were able to construct a soils map of the El Centro area in Imperial Valley, California, which approximated the general soil survey map. They concluded that three-band spectral data gathered at satellite altitude shows great promise for large-scale soil mapping.

Certain disadvantages in photographic remote sensing have led to the development of new sensing devices. Photographic sensors cover only the visible and photographic infrared portion of the electromagnetic spectrum. Much additional information is available from the infrared portions of spectrum beyond that recorded by photographic sensors (Figure 1). This is because solar energy is still being reflected (0.9–2.5 μm), and there is emissive radiation in the thermal infrared region. Carroll (1973*a*) concludes, "it would seem that the well proven techniques of photographic interpretation will continue to be an important aid to the soil scientist, but the newer sensors may prove of equal value if they are carefully developed. These sensors are complimentary to, rather than competing with, conventional photographic sensors."

Nonphotographic Sensors. The nonphotographic sensors generally record the data on magnetic tape rather than photographic film (Carroll, 1973*b*). Storage of data on magnetic tape lends itself to automatic data processing with computers.

The optical-mechanical scanner consists of a multiband spectrometer with a motor-driven mirror used to scan across the field of view. All the energy from a given scene passes through the same optical aperture and is dispersed, measured, and recorded on magnetic tape according to its position in the spectrum. These multispectral sensors scan discrete bands, selected at wavelengths that allow collection of data in a portion of the electromagnetic spectrum where the atmosphere absorbs little solar radiation. The visible, photographic infrared, and near infrared regions from 0.3–2.5 μm have been used frequently in remote sensing since almost all the solar energy received at the earth's surface (Figure 2) is contained within that range. In addition to the optical-mechanical scanners, there are many nonscanning spectrophotometers that have been used. A few of these include the Beckman DK–2A, Beckman DU, Bausch and Lomb Spectronic 20, Cary Model 14, and Hunterlab D25D2M.

The digitized data gathered by the nonphotographic sensor is classified by the computer based on spectral differences. The classes are identified by comparisons with aerial photographs or ground observations from a small portion of the total land surface scanned. The result is a computer-generated map that, when compared to an existing map, is analyzed for its accuracy. In attempting to make accurate computer-generated maps where no soils maps exist, it is important to know how the specific soil properties affect the spectral reflectance of the sun's energy.

Spectral Characteristics

Moisture and Clays. When a dry soil is wetted, it appears darker. This phenomenon has been the primary subject of investigation using remote sensors. Angstrom, reported by

Planet (1970), concluded that the decrease in reflectance of soils was due to the total internal reflections in the water film of the energy reflected from the soil surface itself. Planet's work (1970) supports Angstrom's conclusion that the darkening of soil on wetting is due to the optical effects of a thin layer of liquid on the surface of the soil. Since the amount of water held by mineral soils is greatly influenced by the amount of clay present, it is difficult to study the effects of one without considering at the same time the effects of the other.

With this in mind, work was conducted by Bowers and Hanks (1965) using a Beckman DK-2A spectrophotometer (0.185–2.500 μm), and a Beckman DU spectrophotometer (0.21–1.00 μm). Three soils were studied to evaluate the influences of moisture content, organic matter, and particle size on the reflected spectra. They noticed that as moisture increased, the percent reflectance of the energy at all wavelengths decreased and that the curves of percent reflectance versus wavelength had similar shapes at all moisture levels. For the Newtonia silt loam (Typic Paleudoll) (Figure 3), they noticed two distinct absorption bands located at 1.4 and 1.9 μm. These are the wavelengths where solar radiation is strongly absorbed by water and specifically represent overtones of the fundamental frequencies (3.106, 6.08, and 2.903 μm) at which water molecules vibrate. These same types of results were observed by Beck (1975) for 15 surface soils from northern Indiana. One example is the Brookston (Typic Argiaquoll) (Figure 4), which shows the decrease in spectral reflectance as measured with the Exotech 20-C spectroradiometer as the moisture tension decreased.

Hunt and Salisbury (1970) studied montmorillonite and kaolinite clays with a Cary Model 14 spectrophotometer covering the spectrum from 0.3–2.5 μm and reported the spectral reflectance curve of montmorillonite (Figure 5), which looks similar to that of the Newtonia

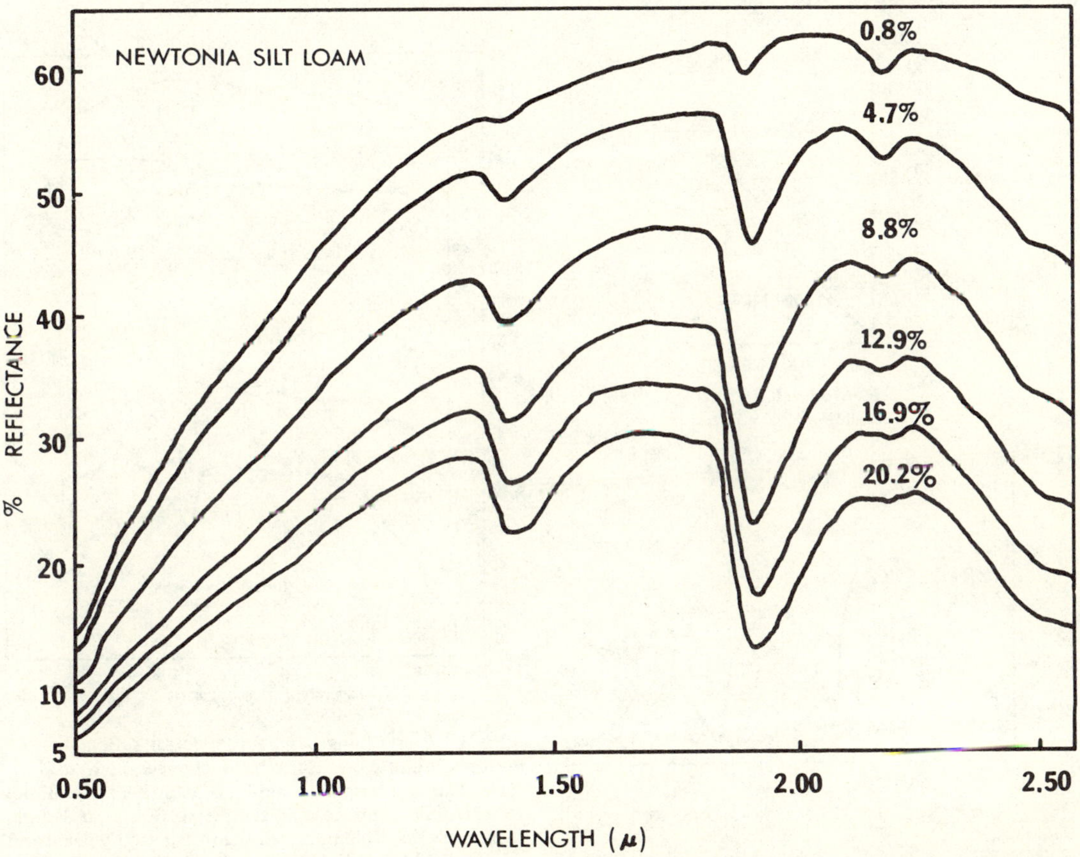

FIGURE 3. Percent reflectance versus wavelength of incident radiation for the Newtonia silt loam at various moisture contents. (After Bowers and Hanks, 1965) [Correction: For micron, μ, read micrometer, μm, the modern S.I. unit]

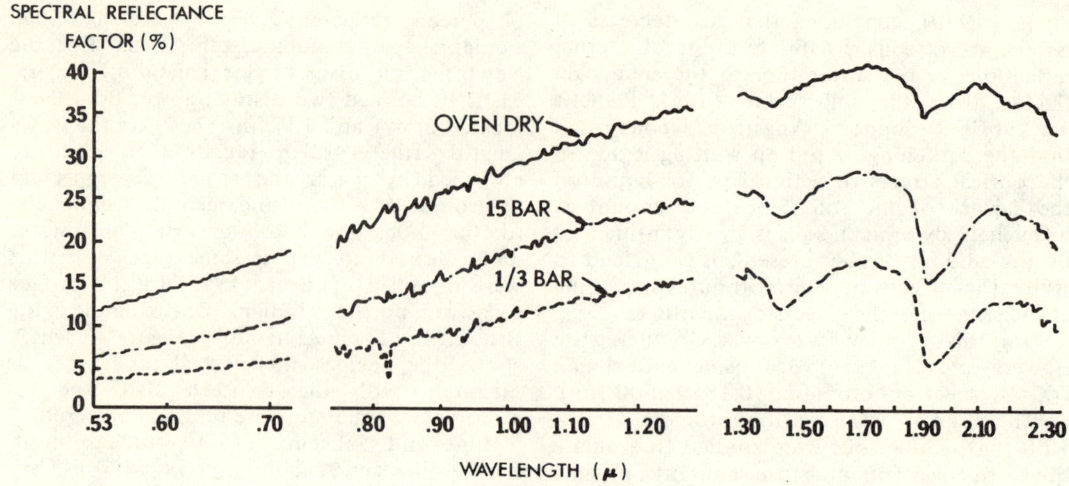

FIGURE 4. Percent reflectance versus wavelength of incident radiation for Brookston silty clay loam. [Correction: For micron, μ, read micrometer, μm, the modern S.I. unit]

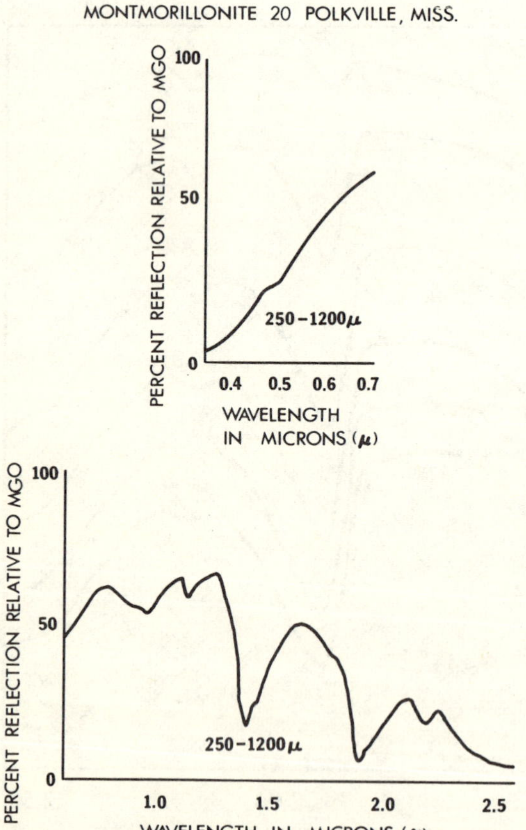

FIGURE 5. Percent reflectance versus wavelength of incident radiation for montmorillonite. (After Hunt and Salisbury, 1970) [Correction: For micron, μ, read micrometer, μm, the modern S.I. unit]

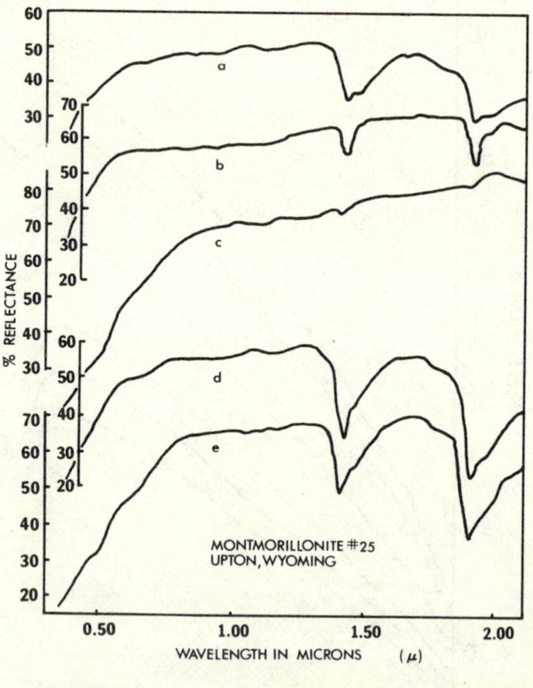

FIGURE 6. Changes in the spectral reflectance of montmorillonite as a result of changes in hydration state. Curve *a* original sample; *b* after heating sample to 120°C; *c* after heating sample to 600°C; *d* sample from curve *b* after humidification for two weeks; and *e* sample from curve *c* after humidification for two weeks. (After Lindberg and Snyder, 1972) [Correction: For micron, μ, read micrometer, μm, the modern S.I. unit]

silt loam and the Brookston silty clay loam. The montmorillonite spectral reflectance curve (Figure 6) is dominated by very strong absorption bands at 1.4 and 1.9 μm due to bound water typical of montmorillonite and usually a weaker absorption band at 1.16 μm possibly due to absorbed water (Lindberg and Snyder, 1972).

Kaolinite's major spectral reflectance features (Figure 7) are several very strong hydroxyl bands in the near infrared centered near 1.4 and 2.2 μm. Lack of appreciable bound water is typical of kaolinite and is indicated by the weakness of the band at 1.9 μm (Figure 8). Lindberg and Snyder's (1972) explanation of the absorption bands states that no bands due to the fundamental vibration modes occur in the range less than 2.5 μm, so that all features observed in the spectral curve are to be considered exclusively overtones or combination tones of fundamental frequencies that occur in the mid- and far-infrared regions.

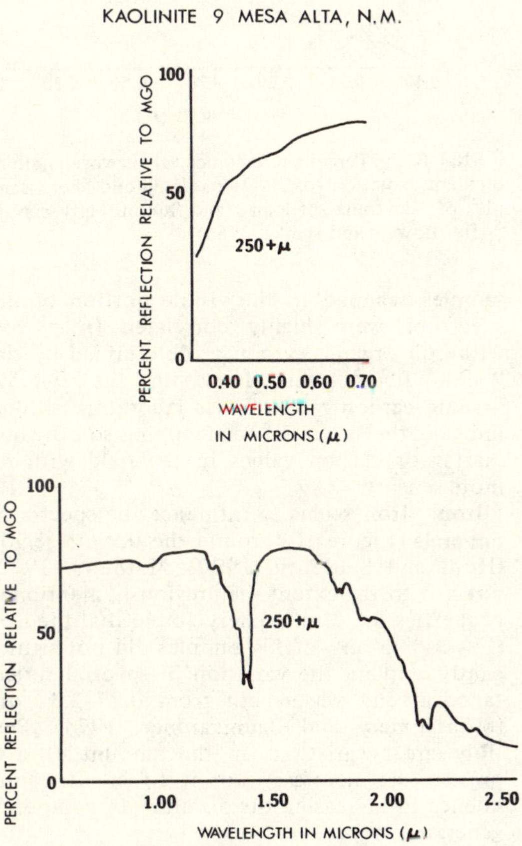

FIGURE 7. Percent reflectance versus wavelength of incident radiation for kaolinite. (After Hunt and Salisbury, 1970) [Correction: For micron, μ, read micrometer, μm, the modern S.I. unit]

In the visible range the color imparted to a material may be due to the occurrence of specific absorptions in this region, or the color may be caused by intense absorptions in either or both the ultraviolet and infrared, and their shoulders may extend forward and backward into the visible. This will impart a color to the material without its having any specific chromophoric group absorbing energy in the visible region (Hunt and Salisbury, 1970).

The water molecule is very commonly associated with rocks and minerals. According to Hunt and Salisbury (1970), water molecules have three fundamental vibration modes, and all are infrared active. They are γ_1, the symmetric OH stretch; γ_2, the H-O-H bend; and γ_3, the asymmetric OH stretch. In the liquid phase these bands appear at 3.106 μm (3219 cm^{-1}), 6.08 μm (1645 cm^{-1}), and 2.903 μm (3445 cm^{-1}). In the spectra of minerals and rocks, whenever water is present, two characteristic bands appear at 1.4 μm due to $2\gamma_3$ overtones and at 1.9 μm due to $\gamma_2 + \gamma_3$ overtones. Well-defined bands usually indicate that the water molecules are located in well-defined, ordered similar environments. Broad bands indicate that they are relatively unordered and are not all found in the same environment. The presence of both the 1.4 and 1.9 μm bands together is diagnostic of undissociated water molecules in the structure, i.e., water of hydration. The appearance of the 1.4 μm band without the 1.9 μm band indicates that OH groups

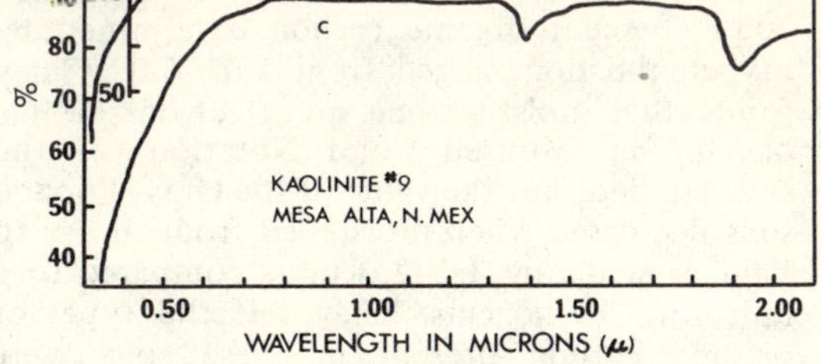

FIGURE 8. Changes in the spectral reflectance of kaolinite as a result of changes in hydration state. Curve *a* original sample; *b* after heating to 600°C; *c* sample from "curve" *b* after humidification for two weeks. (After Lindberg and Snyder, 1972) [Correction: For micron, μ, read micrometer, μm, the modern S.I. unit]

other than those in water are present in the material, i.e., hydroxyls (Hunt and Salisbury, 1970). The 1.4 μm band appears with the presence of any hydroxyl, but the 1.9 μm band appears only when water is present. This is due to the presence of the vibration mode that is found only in water.

The previously mentioned studies dealt with the spectrum from 0.185–2.5 μm and could detect water to a maximum depth of only 15 mm (Montgomery and Baumgardner, 1974). Idso, Jackson, and Reginato (1975) have studied soil moisture contents using microwave brightness temperature techniques and have been able to determine the integral moisture content of soil layers to a depth of 10 cm. There are problems with these techniques such as salinity and surface temperatures that must be worked out, but the possibility of noninterference of clouds suggests that these microwave techniques have potential in the detection of soil moisture by remote sensors.

Organic Carbon. Organic carbon (organic matter) has been demonstrated to affect the spectral reflectance of soils (Bowers and Hanks, 1965; Mathews, 1972) by scanning them before and after treatment with hydrogen peroxide (Figures 9 and 10). Regardless of the type of clay present, the organic carbon seemed to lower the spectral reflectance at all wavelengths with no well-defined absorption bands. Beck (1975) found that organic carbon accounts for most of the variability in spectral reflectance of soils at the same moisture tension in the region of 0.90–1.22 μm (Figure 11). The greatest difference in the amount of energy reflected between the oxidized samples of Bowers and Hanks (1965) and Mathews (1972) is in this same region (Figure 9 and 10).

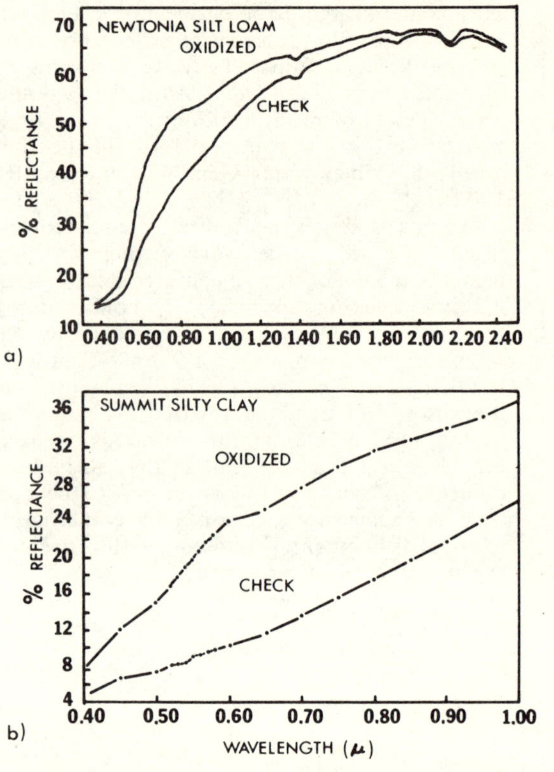

FIGURE 9. Percent reflectance versus wavelength of incident radiation for H_2O_2 oxidized and check samples of Newtonia silt loan *a* and Summit silty clay *b*. (After Bowers and Hanks, 1965)

Shields et al (1968) used a Bausch and Lomb Spectronic 600 with an ultraviolet attachment to distinguish between the Dark Gray Chernozemic and Dark Gray Wooded Great Groups (Canadian Soil Survey) based on their soil color. Percent organic carbon determined by dry combustion ranged from 1.01–6.02. They found that moisture had no effect on the hue or chroma (Munsell Color Notation) of the soils studied, but the value of the Gray Wooded soils decreased when moistened from air dry to field capacity by 1.5–2.0 units compared to 1 unit for Chernozems. They felt the types of organic carbon present in these soils were different and accounted for the variation in the color upon wetting. Another study, Page (1974), estimated organic carbon percentage in 96 coastal plain soils using a Hunterlab Model D25D2M spectrophotometer calibrated with two reflectance standards (94.5% and 21.3% reflectance). Reflectance values of air-dried samples scanned in the visible portion of the spectrum were highly correlated (r = .89) with soil organic carbon as determined by the Walkley-Black method within the 0–2.5% organic carbon range. These laboratory studies indicate the potential for mapping soil organic matter or carbon values in the field with remote sensors.

Iron. Iron seems to influence the spectra of minerals (Figure 10) around the 0.9 μm region (Hunt and Salisbury, 1970); Mathews, 1972) but not to the extent the previously mentioned properties do. Montgomery found that the iron (Fe_2O_3) in any of his samples did not significantly explain the variation in spectral reflectance at any wavelength from 0.37–2.32 μm (Montgomery and Baumgardner, 1974). Possibly great variations in the amount of iron present are necessary for it to be of consequence in increasing the accuracy of computer-generated maps.

Conclusion

Study of individual properties previously mentioned and their effects on the spectral

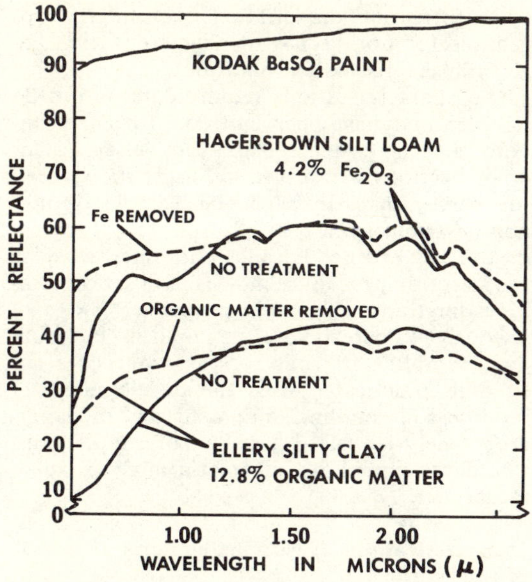

FIGURE 10. Percent reflectance versus wavelength of incident radiation for H_2O_2 oxidized and check samples of Ellery silty clay and iron oxide removed and check samples of Hagerstown silt loam. (After Mathews, 1972)

reflectance of soils has not preceded field application of remote sensing by many years, and sometimes both have been studied simultaneously. Most field work has been conducted using spectral bands in the visible region and not much emphasis has been placed on the infrared region. Mapping soil by remote sensing

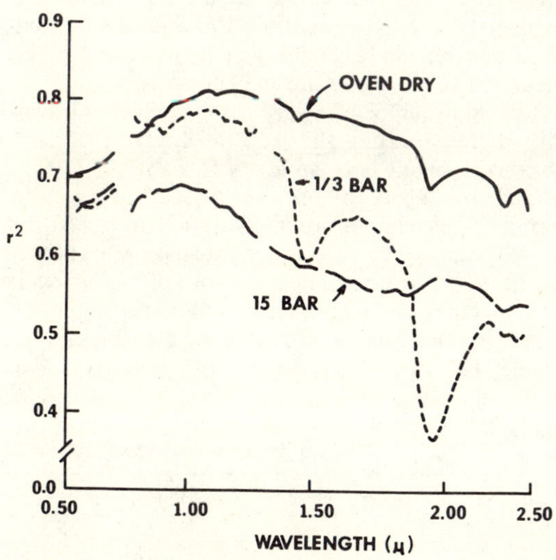

FIGURE 11. Coefficients of determination versus wavelength for organic carbon and spectral reflectance at each moisture level.

has been somewhat successful (Cipra et al, 1972; Zachary et al, 1972; Kristoff and Zachary, 1974). Possibly the information reported about spectral characteristics of individual soil properties may lead to more success in attempting to map surface soils by remote sensing.

ROBERT H. BECK

References

Anuta, P. E.; Kristof, S. J.; Levandowski, D. W.; Phillips, T. L.; and MacDonald, R. B., 1971. Crop, soil, and geological mapping from digitized multispectral satellite photography, *Proc. 7th Internat. Symp. on Remote Sensing of Env.; Env. Res. Inst. of Mich., Ann Arbor, Mich.*, e:1983–2016.

Beck, R. H., 1975. Spectral Characteristics of Soils Related to the Interaction of Soil Moisture, Organic Carbon, and Clay Content. M.S. Thesis, Purdue University, 95p.

Bowers, S. A., and Hanks, R. J., 1965. Reflection of radiant energy from soils, *Soil Sci.*, **100**, 130–138.

Carroll, D. M., 1973*a*. Remote sensing techniques and their application to soil science. Part 1. The photographic sensors, *Soils Fert.*, **36**(7), 259–266.

Carroll, D. M., 1973*b*. Remote sensing techniques and their application to soil science. Part 2. The nonphotographic sensors, *Soils Fert.*, **36**(8), 313–320.

Cipra, J. E.; Swain, P. H.; Gill, J. H.; Baumgardner, M. F.; and Kristof, S. J., 1972. *Definition of Spectrally Separable Classes for Soil Survey Research.* West Lafayette, Indiana: Laboratory for Applications of Remote Sensing Print 100372, 6p.

Hoffer, R. M., and Johannsen, C. J., 1969. Ecological potentials in spectral signature analysis, *in* P. L. Johnson, ed., *Remote Sensing in Ecology.* Athens, Georgia: University of Georgia Press, 1–16.

Hunt, G. R., and Salisbury, J. W., 1970. Visible and near-infrared spectra of minerals and rocks: I. Silicate minerals, *Modern Geol.*, **1**, 283–300.

Idso, S. B., Jackson, R. D., and Reginato, R. J., 1975. Detection of soil moisture by remote surveillance, *Am. Sci.*, **63**(5), 549–557.

Kristof, S. J., and Zachary, A. L., 1974. Mapping soil types from multispectral scanner data, *Photogramm. Eng.*, **40**(12), 1427–1434.

Lindberg, J. D., and Snyder, D. G., 1972. Diffuse reflectance spectra of several clay minerals, *Am. Mineral.*, **57**, 485–493.

Mathews, H. L., 1972. Application of Multispectral Remote Sensing and Spectral Reflectance Patterns to Soil Survey Research. Ph.D. Thesis, The Pennsylvania State University, 110p.

Montgomery, O. L., and Baumgardner, M. F., 1974. *The Effects of the Physical and Chemical Properties of Soil on the Spectral Reflectance of Soils.* West Lafayette, Indiana: Laboratory for Applications of Remote Sensing Information Note 112674, 110p.

Page, N. R., 1974. Estimation of organic matter in Atlantic Coastal Plain soils with a color-difference meter, *Agron. Jour.*, **66**, 652–653.

Planet, W. G., 1970. Some comments on reflection measurements of wet soils, *Remote Sensing Environment*, **1**, 127–129.

Shields, J. A.; Paul, E. A.; St. Arnaud, R. J.; and Head,

W. K., 1968. Spectrophotometric measurement of soil color and its relationship to moisture and organic matter, *Canadian Jour. Soil Sci.*, 48, 271–280.

Stewart, J. E., 1970. Infrared spectroscopy, *Experimental Methods and Techniques.* New York: Marcel Dekker, 636p.

Zachary, A. L.; Cipra, J. E.; Diderickson, R. I.; Kristof, S. J.; and Baumgardner, M. F., 1972. *Application of multispectral remote sensing to soil survey research in Indiana.* West Lafayette, Indiana: Laboratory for Applications of Remote Sensing Print 110972, 12p.

Cross-references: *Soil Physics; Surface-soil Water Content; Thermoluminescence. Vol. XII, Pt. 2: Soil Survey, Aerospace Remote Sensing. Vol. XIII: Remote Sensing.*

STATISTICAL-CHEMICAL CHARACTERIZATIONS

Soils are defined and characterized on the basis of their morphological profiles because the assemblage of obvious physical features represented by these units are often related to the less obvious features of their chemical composition, chemical properties, and fertility. Accordingly, morphological units of soil are usually characterized by chemical analyses, giving so-called chemical characterizations. However, the closeness of the relationship between chemical features and morphological profile type varies both with the chemical features and morphological type. This is because soil-forming processes may have quite different effects on the development of chemical features than on physical features. Also chemical features may be altered considerably by land use, especially with exploitative cropping, irrigation, and fertilizer applications. For these reasons it is important to indicate the specificity and variability of individual properties in any characterization by applying statistical procedures to obtain *statistical-chemical characterizations.*

Statistical characterizations of soils are made by procedures developed for studying trends in individual properties within soil profiles, corresponding in concept to those used for characterizing simple variables. Thus the general levels of a chemical property within the profiles of a soil are indicated by estimating the mean profile trend of the property; variability is indicated by estimating the variance of the trend; and standard-error estimates are used to test the significance of differences in corresponding trends for different soils. Simpler characterizations can also be made of the features of individual depths or horizons in the soil profile, but these are less informative because they do not use information on variations with depth. Procedures for both types of characterization are given in the discussion that follows.

The characterizations require data on profile samples that have been collected by some random sampling scheme, as on samples from random locations within the soil units shown on a soil survey map. In particular, samples should not be taken from carefully selected "typical" or "type" profiles because with such samples the selection procedure may cause a serious bias in estimations of general properties and may seriously underestimate variability. The random sampling of soils for these purposes has become practicable with the development of modern soil sampling equipment, and the laboratory analysis of large numbers of samples with the development of automatic analysis apparatus.

Characterizations of Nominated Parts of the Profile

Characterizations of particular parts of the profile are conducted by sampling the nominated depth or horizon at different locations and ignoring as negligible any vertical variations within the samples. Analyses on these samples are then treated as random observations on a variable population (the soil) and characterized by standard statistical procedures. Thus if n soil samples are taken from random sites and analyzed for a particular chemical property Y, then the n analyses, denoted by Y_i, $i = 1, 2, \ldots n$, are used to calculate the mean $[\bar{Y} = (1/n) \Sigma_{i=1}^{n} Y_i]$ and the variance $[V = (1/n-1) \Sigma_{i=1}^{n} (Y_i - \bar{Y})^2]$. Given such data for several soils and given that the soil variances do not differ significantly, a single-classification analysis of variance can be made to test for differences in the level of the property among the soils.

The number of analyses for each soil will usually differ so that the general analysis of variance will have the form given in Table 1.

In Table 1 g is the number of soils being compared, n_i is the number of data for the soil i ($i = 1, 2, \ldots g$), E denotes statistical expectations, σ_w^2 is the variance within soils, and σ_b^2 is the variance between soils. The ratio of calculated mean squares provides a test for significance (F ratio), and the within soils mean

TABLE 1. General Analysis of Variance.

Source	Degrees of Freedom	E (Mean Squares)
Between soils	$(g-1)$	$\sigma_w^2 + \frac{1}{(g-1)}\left[\Sigma n_i - \frac{\Sigma n_i^2}{\Sigma n_i}\right]\sigma_b^2$
Within soils	$(\Sigma n_i - g)$	σ_w^2
Total	$(\Sigma n_i - 1)$	

square provides an estimate of the standard error for testing differences between individual soils.

If the variances for different soils are not the same (*nonhomogeneity of variances*), then tests of significance can still be obtained, using now a weighted t test,

$$t' = (\bar{Y}_a - \bar{Y}_b) \Big/ \sqrt{\frac{V_a}{n_a} + \frac{V_b}{n_b}}$$

where Y is the mean, V is the variance, n is the number of samples, and subscripts indicate values for the different soils, soil a and soil b. Procedures for the preceding calculations are detailed in standard texts (e.g., Snedecor and Cochran, 1967).

Profile Characterizations

Characterizations corresponding to the previous ones can be obtained for a quantity that varies down the soil profile by representing its vertical variation by orthogonal polynomial trends. The profile analyses are represented by a depth function Y_d, the subscript d indicating that the analysis varies with depth.

$$Y_d = p_0\xi_0 + p_1\xi_1 + p_2\xi_2 + p_3\xi_3 + \cdots \quad (1)$$

The ξ_j denotes orthogonal polynomial trends of degree j, ξ_0 is zero or mean trend, ξ_1 is linear trend, ξ_2 is quadratic trend, ξ_3 is cubic trend, and so on to higher degrees, the highest degree being selected to give a desired smoothing of the profile variations with depth. The coefficients p_j provide quantitative measures of the size of the respective trends in the depth function Y_d. The advantage of this representation of the vertical variations lies in the fact that the trends ξ_j are discrete and independent of each other (hence the name "orthogonal") so that each of the coefficients can be treated as an independent soil feature and used to characterize the profile variations of a soil in the landscape.

If n profiles are sampled to represent a soil, then the analyses on these profiles may be denoted by Y_{di}, $i = 1, 2, \ldots n$, and the orthogonal polynomial model written,

$$Y_{di} = P_{0i}\xi_0 + p_{1i}\xi_2 + p_{3i}\xi_3 + \cdots$$

the subscript i being added to denote individual samples of profile. The means and variances of the trend coefficients are calculated as before by

$$\bar{p}_j = \frac{1}{n}\sum_{i=1}^{n} p_{ji} \quad \text{and} \quad V_j = \frac{1}{(n-1)}\sum_{i=1}^{n} (p_{ji} - \bar{p}_j)^2$$

and the mean profile trend is obtained simply by the sum of the mean orthogonal trends,

$$\bar{Y}_d = \bar{p}_0\xi_0 + \bar{p}_1\xi_1 + \bar{p}_2\xi_2 + \bar{p}_3\xi_3 + \cdots \quad (2)$$

It follows from the procedure for calculating the variance of a sum that the profile variance is

$$V(Y_d) = \frac{1}{n-1}\sum_{i=1}^{n} (Y_{di} - \bar{Y}_d)^2$$

$$= \frac{1}{n-1}\sum_{i=1}^{n}\left[\sum_{j=0}^{t} (p_{ji} - \bar{p}_j)^2 \xi^2_{jd} + 2\sum_{j=0}^{t}\sum_{k=j+1}^{t} (p_{ji} - \bar{p}_j)(p_{ki} - \bar{p}_k)\xi_{jd}\xi_{kd}\right] \cdots \quad (3)$$

for t trends. Note that computed this way both the mean analysis and the variance vary with depth.

The orthogonal depth function (equation 1) can be converted to the nonorthogonal polynomial function

$$Y_d = b_0 + b_1 D + b_2 D^2 + b_3 D^3 + \cdots$$

where D is the depth, and the coefficients b_j are calculated from the coefficients p_j of equation 1. These coefficients b_j are the coefficients obtained by the least-squares regression procedure, and formulas for the conversion and computing programs are described by Colwell (1978). Similarly, the mean depth function

$$\bar{Y}_d = \bar{b}_0 + \bar{b}_1 D + \bar{b}_2 D^2 + \bar{b}_3 D^3 + \cdots$$

can be obtained from the mean function (equation 2), and standard errors for the mean from the variances (equation 3). The variation of an analysis with depth for a soil thus characterized can be represented graphically, as in Figure 1, the mean profile function being indicated by a center curve, and the variability of the trends in the landscape for a particular soil by confidence intervals (broken lines). The trends illustrated are for pH values in the profiles of four types of soil in a region of New South Wales, Australia.

Differences between the profile properties of different soils can be tested, statistically, as for the properties of a particular depth previously described, using the trend coefficients in place of the sample analyses. Again if variances can be assumed to be the same for the different soils, the test is carried out by a single classification analysis of variance, and for differing vari-

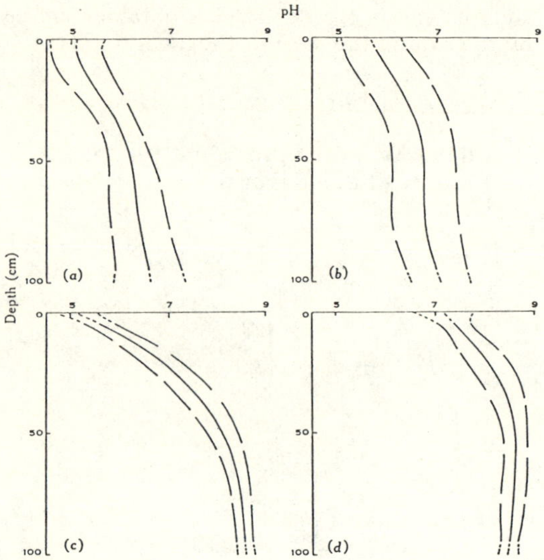

FIGURE 1. Variability of pH profiles. The mean profile function is indicated by the solid line, and confidence intervals by the broken lines. The four types of soil represented are (a) Podzolic, (b) Red Earth, (c) Red-Brown Earth, and (d) Grey Clay. (After Colwell, 1970)

ances a weighted t test is used. For example, the four soils represented in Figure 1 show significant differences in the zero, linear, and quadratic trend coefficients, and these differences can be seen by inspection of the graphs. Zero trend differences are shown by the differing overall displacement of the curves, left to right, the linear trend differences by differing rates of increase with depth, and the quadratic differences by differing curvatures.

The statistical characterization of profiles requires that all profiles be sampled in the same consistent manner. Samples may be taken at fixed depths, not necessarily evenly spaced, which provides representations independent of physical horizon distribution with the profiles. Alternatively, if it is known that analysis variations are closely associated with horizons, then a set of nominated horizons may be sampled, each horizon being coded by numerals 0, 1, 2, . . . In the latter case, vertical axes of a graphical representation would be labeled by horizon rather than by depth, and variations from horizon to horizon would be represented by smooth curves through the central depth of each horizon. Polynomial models are not suitable to represent discontinuities such as may sometimes occur at horizon boundaries.

J. D. COLWELL

References

Colwell, J. D., 1970. A statistical-chemical characterization of four great soil groups in southern New South Wales based on orthogonal polynomials, *Aust. Jour. Soil Res.*, 8(3), 221-38.

Colwell, J. D., 1978. *Computations for Studies of Soil Fertility and Fertilizer Requirements.* Farnham Royal Slough: Commonwealth Agricultural Bureau, and Melbourne: Commonwealth Scientific and Industrial Research Organization (in press).

Snedecor, G. W., and Cochran, W. G., 1967. *Statistical Methods.* Ames: Iowa University Press, 593p.

Cross-references: *Chemical Composition; Soil Chemistry. Vol. XII, Pt. 2: Depth Function.*

SULFUR TRANSFORMATIONS

The transformations of sulfur in soil are many and varied, and the changes are often cyclic; the element changes from inorganic to organic forms and back again. These transformations of sulfur are mainly biological, although nonbiological reactions can be involved (e.g., the oxidation of iron sulfides). A great diversity of reactions is possible because sulfur occurs in many organic and inorganic combinations in various states of oxidation, e.g., sulfate, +6; sulfite, +4; thiosulfate, +6 and −2; elemental sulfur, 0; disulfide, −1, and sulfice, −2, (Starkey, 1966).

Sulfur is required for the metabolism of all organisms. Some forms of life can use simple forms of sulfur such as elemental sulfur, sulfide, or sulfate for the synthesis of organic compounds for metabolism and growth. The organic sulfur so created can be used as sulfur sources by other living organisms. These organic sulfur compounds can be broken down by other organisms to regenerate the simple inorganic forms of sulfur.

Forms of Sulfur in Soil

Sulfur occurs in soils in both organic and inorganic forms, and the amounts of these two forms may differ widely depending on climatic conditions, soil types and depths in the profile. While it has been possible to identify the chemical nature of most of the inorganic forms of sulfur occurring in soils, little is known of the identity of the organic forms of sulfur.

Inorganic Sulfur

Inorganic sulfur may occur in soils as sulfate and as compounds of lower oxidation state such as sulfide, polysulfide, sulfite, thiosulfate, and elemental sulfur. In well-drained, well-aerated soils, most of the inorganic sulfur occurs as sulfate. Sulfur in reduced forms is normally less than 1 ppm.

Sulfides. Considerable amounts of sulfides can be formed under anaerobic conditions, particularly in tidal swamps and waterlogged soils. In waterlogged soils, ferrous sulfide occurs as a coating on clay particles, sometimes uniformly distributed throughout the soil but often as discrete black patches around decaying plant remains. Pyrite (FeS_2) is usually the most important sulfur compound in permanently waterlogged soils in which the monosulfide is a minor constituent. Pyrite and marcasite (FeS_2) are frequently found in shales and in other sedimentary rocks including limestones.

Water-Soluble Sulfate. The amount of soluble sulfate in soil ranges widely among soil types and within soil profiles, but in the surface horizons of most well-drained soils, the contents are usually very low. In dry areas where there is insufficient rain to leach sulfate from the soil profile, gypsum ($CaSO_4 \cdot 2H_2O$) and epsomite ($MgSO_4 \cdot 7H_2O$) can accumulate.

Seasonal fluctuations in the amounts of soluble sulfate in the surface soil occur that result from the interactions of seasonal conditions on mineralization of organic sulfur, leaching of the soluble sulfate, application of fertilizers, and plant uptake (Williams, 1968).

Adsorbed Sulfate. Sulfate can be retained in some soils by adsorption on inorganic colloids such as kaolinite, aluminum oxides, and iron oxides. The capacity of soils to adsorb sulfate depends on a number of factors; the most important of these are pH, type of clay, clay content, and exchangeable cations.

Little adsorption of sulfate occurs above pH 6.5 whereas adsorption increases with decreasing pH below this value. Kaolinite minerals retain more sulfate than montmorillonite-type minerals, and aluminum oxide appears to have a greater capacity for adsorption of sulfate than iron oxide. Sulfate adsorbed on kaolinite clay is weakly held whereas that adsorbed on iron and aluminum oxides is more tightly bound and thus more resistant to leaching (for a full discussion, see Barrow, 1975, and Harward and Reisenauer, 1966).

Insoluble Sulfates. The most common form of insoluble sulfate occurs as a cocrystallized impurity in calcium carbonate in calcareous soils. In some soils it may acount for 95% of the total sulfur, and in subsoil horizons of calcareous soils it frequently accounts for 40–50% of the total sulfur (Williams and Steinbergs, 1962).

Basic iron sulfate, jarosite [$KFe_3(OH)_6(SO_4)_2$], and coquinbite [$Fe_2(SO_4)_3 \cdot 5H_2O$] have been identified in material taken from tidal marsh soils; basalunite [$Al_4(OH)_{10}SO_4 \cdot 5H_2O$] has been isolated from weathered sandstone deposits; and barytes [$BaSO_4$] has been identified as small concretions in certain soils (Lynn, Tu, and Franzmeier, 1971, and Williams, 1975).

Organic Sulfur

Even though it is accepted that practically all the sulfur is organic in many soils, little is known of the chemical nature of this organic complex. A great variety of organic sulfur compounds is added to, or produced in, soil by higher plants, animals, and microorganisms, but because the compounds are susceptible to decomposition, they do not accumulate in soil and are not readily detected. Thus the concentration of these simple sulfur compounds in soil has been found to be very low and also variable.

For example, only small amounts of the sulfur-containing amino acids (cysteine, cystine, cysteic acid, methionine, methionine sulfoxide, and taurine) have been isolated in uncombined form from soil (Putnam and Schmidt, 1959). The concentration of some of these amino acids may be slightly higher in the vicinity of plants as it is known that they are exuded from the roots of certain species, but even here the concentration is very low. Most of the sulfur-containing amino acids in soil appear to be bound in some manner to the mineral and humus fractions and are released only on treatment of the soil with boiling acid (e.g., 6NHCl).

Many workers have found a close relationship between organic carbon, organic nitrogen, and total sulfur in surface soils, and thus it is inferred that most of the sulfur in the surface horizons of well-drained agricultural soils is in the organic form. While there is considerable variation in the C:N:S ratio of individual soils and smaller variations between groups of soils, the mean ratios for soils from different regions of the world are remarkably similar (C:N:S: = 140:10:1.3) (Williams, 1975).

Except for soils over buried *A* horizons and soils with organic matter accumulations in the *B* horizons, organic sulfur decreases with depth in the profile.

Attempts to characterize the organic sulfur complexes in soil have been hindered by the lack of satisfactory methods of separating organic and inorganic soil constituents. The classical method of separation is to have soil react with sodium hydroxide. This reagent dissolves a substantial amount of organic matter, but it degrades some of the sulfur fractions during extraction. High molecular weight compounds are broken down to low molecular weight compounds, and some organic sulfur is converted to inorganic sulfate by the alkaline reagent.

Much of the information on the chemical nature of organic sulfur in soil has been derived from the comparative reactivity of soil sulfur with reducing agents (see reviews by Freney, 1967, and Williams, 1975). Inorganic sulfate is first removed by leaching with a phosphate solution, and the organic sulfur remaining then reacts with a reducing mixture containing hydriodic acid. This separates organic sulfur into HI reducible and HI nonreducible fractions. The HI nonreducible fraction can then be further fractionated by reaction with Raney nickel to give Raney nickel reducible and non-reducible fractions (see Figure 1) (Freney, Melville and Williams, 1975).

Hydriodic-Acid Reducible Sulfur. This fraction accounts for approximately 50% of the total organic sulfur although analyses covering wide ranges of soils have shown that the percentage may range from 30–70%. Hydriodic acid will reduce, to hydrogen sulfide, only that organic sulfur which is not directly bonded to carbon. For example, it will reduce to hydrogen sulfide, the sulfur in ester sulfates (-C-O-S-), sulfamic acid (-C-N-S), and the second sulfur in S-sulfocysteine (-C-S-S-) where the sulfur and carbon atoms are separated by oxygen, nitrogen, and sulfur atoms, respectively. Nor will it reduce, to hydrogen sulfide, when the sulfur is directly bonded to carbon, e.g., the sulfur in cysteine, methionine, or the first sulfur atom in S-sulfocysteine. The sulfur in this fraction is believed to be in the form of organic sulfates because of its reactivity with hydriodic acid, its infrared spectra, and because it is readily converted to inorganic sulfate by reaction with acid and alkali (Freney, 1961).

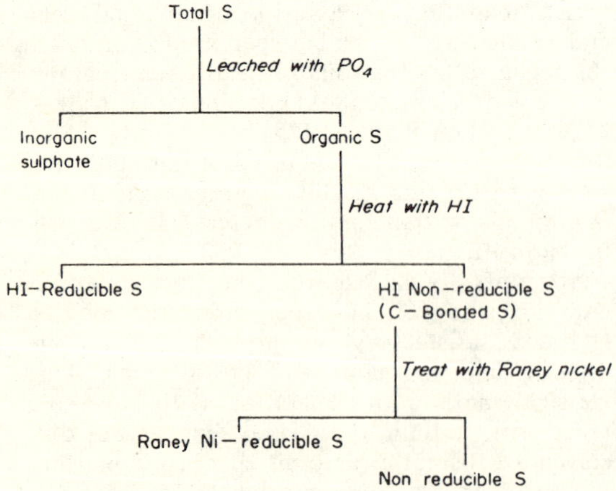

FIGURE 1. Flow sheet for the fractionation of soil sulfur. (From Freney, Melville, and Williams, 1975)

Carbon-Bonded Sulfur. This fraction represents the difference between total organic sulfur and hydriodic acid reducible sulfur and is so designated because it does not react with hydriodic acid. In general, it represents half of the total organic sulfur. This fraction is not converted to inorganic sulfate by acids or alkalis and would include the sulfur-containing amino acids, mercaptans, disulfides, etc.

Raney Nickel Reducible Sulfur. Attempts were made to measure carbon-bonded sulfur directly by reaction of the soil with Raney nickel (Lowe and De Long 1963; Freney, Melville, and Williams, 1970). This reagent reacts with all reduced sulfur bonded to carbon, for example, mercaptans, R-SH, and oxidized sulfur in the form of sulfoxides R-SO-CH_3, sulfinic acids R-SO-OH, or sulfonic acids attached to an aromatic nucleus, for example, sulfanilic acid H_2N-C_6H_4-SO_2OH. However, it does not react with sulfur in the form of aliphatic sulfones, for example, methionine sulfone HOOC-CH(NH_2)-CH_2-CH_2-SO_2-CH_3 or aliphatic sulfonic acids, for example, cysteic acid HOOC-CH(NH_2)-CH_2-SO_2OH.

In addition, iron and manganese in soils interfere with this reduction procedure, and the result depends on the amount of alloy, the reduction time, and the alkali concentration. Even when all these variables are adjusted to the optimum, the method will reduce only about 30% of the total sulfur in soil, whereas carbon-bonded sulfur should account for approximately 50% of the total.

Although only a few soils have been studied, there was good agreement between the amounts of Raney nickel reducible sulfur and amino acid sulfur, and because of this agreement, it has been suggested that the Raney nickel method determines mainly amino acid sulfur (Freney, Melville, and Williams, 1975).

Nonreducible Sulfur. The remaining carbon-bonded sulfur that is not reduced by Raney nickel may be in the form of aliphatic sulfones or aliphatic sulfonic acids that are not reduced to sulfide by either hydriodic acid or Raney nickel. It may also be sulfur that is protected from reacting with these reducing agents by the mineral or organic matter in the soil. Approximately 20% of the total sulfur in soil is not reduced to sulfide by either hydriodic acid or Raney nickel.

Transformations of Sulfur in Soil

Transformations of sulfur are frequently represented by a cycle that depicts an over-simplified outline of the processes that occur.

In reality, there is an interwoven network of reactions rather than a simple cycle. Diverse chemical reactions and products occur in the decomposition of organic sulfur compounds, but they may not be related by a simple sequence of changes from one compound to another; side reactions and alternative pathways may be involved. Even with the microbial transformations of the inorganic sulfur compounds that are fewer in number and kind than organic forms, there may not be a simple order of events. Elemental sulfur and practically all the inorganic sulfur compounds can be either oxidized or reduced, often by diverse reactions (Starkey, 1966).

The microbial transformations of sulfur in soil (see Figure 2) can be grouped into four main categories: (1) *immobilization* (i.e., assimilation of inorganic sulfur into organic compounds); (2) *mineralization* (decomposition of organic sulfur compounds with release of inorganic sulfur); (3) *oxidation* of sulfur and inorganic sulfur compounds; and (4) *reduction* of sulfate and incompletely oxidized inorganic sulfur compounds. The types of transformations are affected by the oxidation state of the sulfur and the environmental conditions, in particular, the availability of oxygen (Starkey, 1966).

The term *immobilization* usually denotes the conversion of inorganic compounds to the organic form by microorganisms. In the case of sulfur, microorganisms take in inorganic sulfate and convert it into organic forms for the syn-

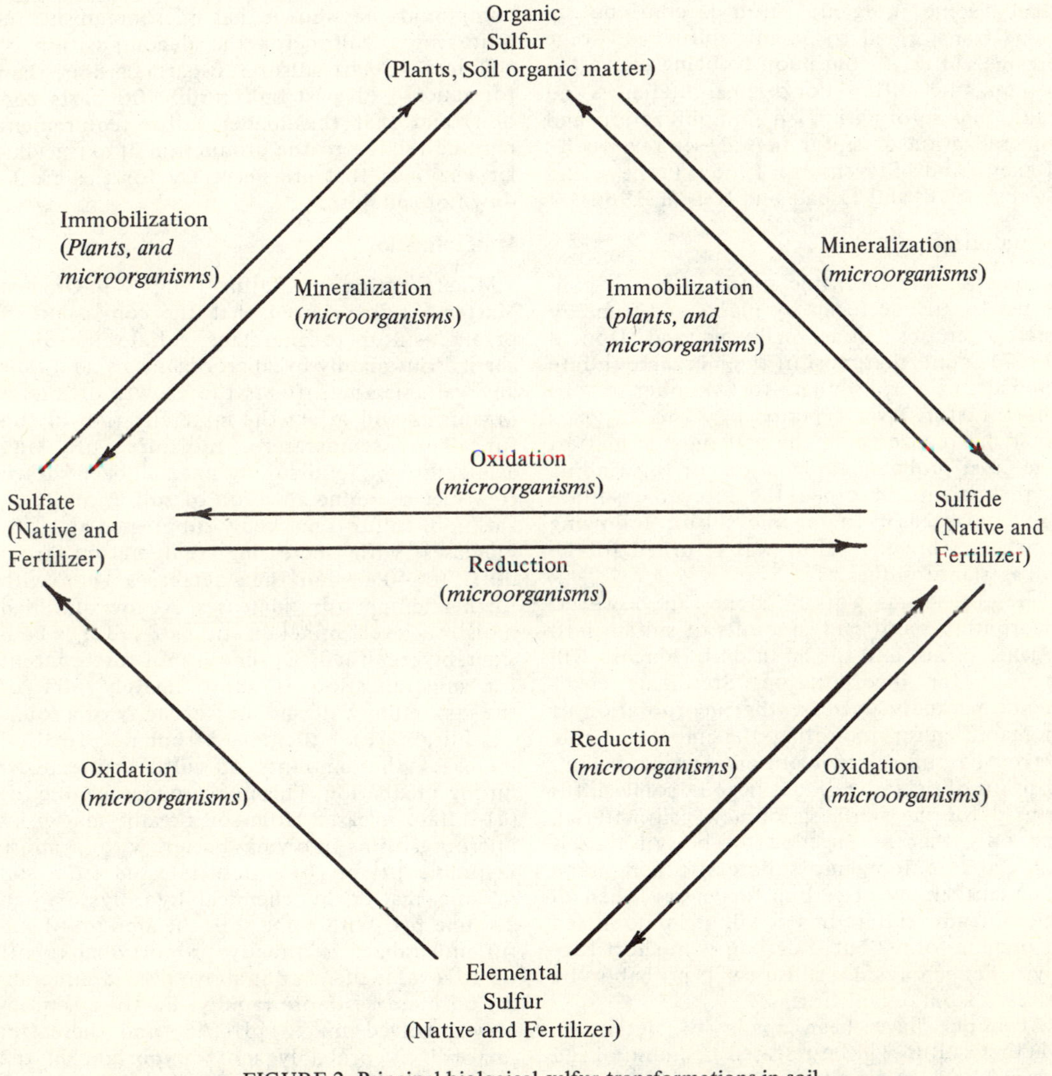

FIGURE 2. Principal biological sulfur transformations in soil.

thesis of microbial tissue and cell constituents. The microorganisms involved in this reaction obtain their energy from the oxidation of carbonaceous materials in soil. An alternative explanation is that when microorganisms decompose natural organic residues or added carbon compounds in soil, they need sulfur for metabolism and growth and in the process will convert inorganic sulfate to organic forms and so render the sulfur unavailable for the growth of plants.

Mineralization, which is the reverse of immobilization, is the conversion of an element in organic combination to its inorganic form (e.g., organic sulfur to inorganic sulfate). The two opposing processes–immobilization and mineralization–occur simultaneously in various locations in a soil where organic debris is undergoing microbiological decomposition. As a result, some inorganic sulfur is continuously being transformed to organic sulfur, and some organic sulfur is continuously being converted to inorganic sulfur. (For original references and additional information on immobilization and mineralization of sulfur in soil, see reviews by Freney and Stevenson, 1966; Freney and Swaby, 1975; and Tisdale and Nelson, 1966).

Immobilization

Some of the inorganic sulfur in soil is converted to organic forms by plants and some by microorganisms alone. Williams and Donald (1957) found that most of the inorganic sulfate applied as superphosphate to a number of soils under pasture over a period of 15 to 25 years could be recovered in the soil organic matter. The most probable explanation for this finding is that plants absorbed the inorganic sulfate and converted it to organic sulfur, following which the organic sulfur was returned to the soil as plant residues.

Microorganisms working alone can, however, incorporate significant amounts of sulfate into organic forms, and the addition of glucose will increase the incorporation. Sterilizing treatments markedly decrease the incorporation of inorganic sulfur indicating the important role played by soil microorganisms. During decomposition of organic residues in soils, some sulfur is used for the synthesis of new cell material, and only that not needed for biosynthesis is converted to inorganic sulfate. If there is an abundant supply of carbon for energy, then all the inorganic sulfate in soil will be immobilized in organic forms; but if carbon is in short supply, then inorganic sulfur will probably be released from organic forms.

Attempts have been made to determine whether sulfur will be released or immobilized during the decomposition of organic residues in soil by considering the carbon:sulfur ratio of the added residue. Model studies using glucose as a carbon source suggested that microbiol activity would be maximal if sufficient sulfur were present to give a carbon:sulfur ratio of 900:1 or less. As most plant residues have carbon to sulfur ratios less than 900:1, sulfur should not be limiting for the decomposition of residues under natural conditions. However, in practice, factors other than the carbon:sulfur ratio are involved, and it has been shown that immobilization of sulfur will occur during the decomposition of plant residues with carbon: sulfur ratios well below 900:1. This suggests that all the organic sulfur in plant residues is not readily available for the needs of microorganisms, or that the organisms have a higher sulfur requirement for the decomposition of plant residues than glucose. Work with pure compounds has shown that microorganisms require more sulfur for the decomposition of cellulose (1 part sulfur:300 parts carbon) than for glucose (1 part sulfur:600–900 parts carbon) and that this higher sulfur requirement may be related to the production of extracellular enzymes that are necessary for the breakdown of cellulose.

Mineralization

Mineralization of Sulfur from Soil Organic Matter. It is believed that the conversion of organic sulfur to inorganic sulfate in soil is carried out mainly by microorganisms, and thus any variable that affects the growth of microorganisms will affect the mineralization of sulfur. Thus temperature, moisture, pH, and availability of food supply have all been shown to affect the mineralization of sulfur. Mineralization of sulfur is markedly suppressed at 10°C, increases with increasing temperature from 20°C to 40°C, but then decreases again with further temperature increases. At low and high moisture levels mineralization of sulfur is considerably retarded. Optimum moisture content for mineralization is approximately 60% of the maximum water-holding capacity of a soil.

Addition of calcium carbonate to soil increases the amounts of sulfur mineralized during incubation. This increase may be due to: (a) sulfate released from soil organic matter by microorganisms growing better with a more favorable pH or (b) sulfate released from soil organic matter by chemical hydrolysis at an alkaline pH. With some soils the amount of sulfur mineralized is directly proportional to pH up to a value of 7.5, but above pH 7.5 mineralization increases more rapidly. Factor a is probably involved up to pH 7.5, and thereafter factor b is probably more important in the release of inorganic sulfur. Sterilizing treatments with toluene and formaldehyde enable some distinction to be made between microbial and chemical hydrolysis (Williams, 1967).

When soils are dried and then remoistened prior to incubation, they release appreciably more sulfate than they do when incubated without prior drying. In general, the amount of sulfur mineralized from unamended soils does not appear to be directly related to soil type, to total amounts of carbon, nitrogen or sulphur, to C:S, N:S, or C:N ratios (q.v.), to soil pH, or to mineralizable nitrogen. Some workers using a restricted range of soils have found strong relationships between some of these factors and mineralizable sulfur.

The growth of plants in soils has been found to bring about increased mineralization of sulfur from soil organic matter. This increase may be due to the greater proliferation of organisms under plants or to the excretion of enzymes from plant roots that catalyze the decomposition of organic matter (see *Enzyme Activity*).

The mineralization of sulfur from soil organic matter follows a number of patterns: (a) Sulfur can be immobilized during the initial stages of incubation but is mineralized in the later stages; (b) a steady, linear release with time over the whole period of incubation; (c) a rapid release of sulfate during the first few days followed by a slower linear release; and (d) a rate of release that decreases with time (Williams, 1967). The release pattern does not appear to be related to any soil property but is probably due to the chemical nature of the decomposing fraction of the soil organic matter. Better predictions of mineralization may be obtained when the main sulfur compounds in soil are identified and their turnover rates determined.

Seasonal Fluctuations in Mineral Sulfur. Seasonal changes in sulfate sulfur in either bare soil, cultivated fields, or soil under pasture have been studied in a number of locations. In soil under subterranean clover pasture, sulfate was accumulated in the surface soil during summer immediately after senescence of the pasture. High concentrations were maintained throughout the summer-autumn period, and these decreased to low values in winter and spring. The higher values in summer probably resulted from mineralization of soil organic matter under favorable moisture and temperature conditions and lack of plant uptake. Desiccation of the soil organic matter may also affect this release. The low values in winter and spring are probably due to leaching, to plant uptake, and to low rates of mineralization at low soil temperatures.

Source of Mineralizable Sulfur. Few attempts have been made to study the dynamics of sulfur in soil, i.e., to evaluate the amount of plant available sulfur in soil, the amount of the biologically and chemically active reserve forms, and the rate of turnover that regulates the transition from one form to the other. Fortini, Cardas, and Panella (1966) attempted to do this with ^{35}S-labeled sulfur and calculated that the daily rate of regeneration of available sulfate in a volcanic soil was 2.7 μg S and in an organic soil was 4.6 μg S. In the volcanic soil the physicochemical and biological transformations occurring daily in total sulfur affected 7.7% of the soluble sulfate, 7.3% of the adsorbed sulfate, and 9.4% of the reserve sulfur.

Other work with ^{35}S-labeled soil organic matter attempted to identify the organic fractions of soil sulfur that decompose during the growing season to provide plant available sulfur (Freney, Melville, and Williams, 1975). All indigenous organic sulfur fractions (which represent the older, more stable organic matter) decreased during plant growth and the carbon-bonded sulfur decreased more than the hydriodic acid reducible sulfur. The net results suggested that the plants derived 60% of their sulfur requirements from the carbon-bonded sulfur fraction and the remaining 40% from the hydriodic acid reducible fraction. The changes in carbon-bonded sulfur involved changes in both Raney nickel reducible and nonreducible sulfur fractions. Within the labeled sulfur fractions (which represent the recently formed organic matter), the hydriodic acid reducible fraction provided most of the plant available sulfur, and there was little measurable change in the other fractions. Thus there is a considerable difference between the reactivity of the old and the newly formed organic matter.

Aryl Sulfatase Activity. As hydriodic acid reducible sulfur (which is most likely ester sulfate) accounts for approximately 50% of the total sulfur in surface soils, it has been suggested that sulfatases may play an important role in the mineralization of organic sulfur in soil.

Soils have been shown to contain enzymes that hydrolyze aryl sulfates, and a method has been developed for the determination of aryl sulfatase activity in soil. In general, aryl sulfatase activity was found to be correlated with organic matter content, and activity decreased with depth in the profile as organic matter decreased (Tabatabai and Bremner, 1970*a*, 1970*b*). However, sulfur mineralized from a range of soils in Iowa was not related to aryl sulfatase activity (Tabatabai and Bremner, 1972).

Oxidation of Sulfur and Inorganic Sulfur Compounds

Although elemental sulfur is formed in soil from the oxidation of ferrous and ferric sulfides, it seldom accumulates. There have been cases, however, where sulfur has been found to collect. In southern India, for example, delta soils are periodically flooded by the sea. They are wet for 6 months and dry for approximately 3 months. In some areas the sulfur content was as high as 35%. Apparently sulfate is reduced to sulfide when the soil is waterlogged, and then oxidized when the soil is dried. Sulfur can also occur in soils as a result of oxidation of hydrogen sulfide and metallic sulfides by atmospheric oxygen, and it is also added to soils as a fertilizer (see *Fertilizers, Inorganic*).

Sulfur-Oxidizing Organisms. A host of organisms can take part in sulfur oxidation, some of which carry the process only a step or two, but the overall reaction results in the production of sulfuric acid (Tisdale and Nelson, 1966)

$$2S + 3O_2 + 2H_2O \rightarrow 2H_2SO_4$$

The best known, and usually considered to be the most important, group of sulfur-oxidizing microorganisms are the autotrophic bacteria belonging to the genus *Thiobacillus*. This group can use elemental sulfur and incompletely oxidized inorganic sulfur compounds as specific hydrogen donors for the assimilation of carbon dioxide. Carbon dioxide can supply all their carbon requirements.

The *Thiobacillus* group of bacteria has been more closely studied than the heterotrophic bacteria, fungi and actinomycetes, whose importance in the oxidation of sulfur in soil has often been overlooked. Vitolins and Swaby (1969) found that heterotrophic yeasts and several genera of heterotrophic and facultative autotrophic bacteria were far more numerous than the strict autotrophs and could play an important role in the oxidation of sulfur in many soils.

The mechanism of oxidation of sulfur by autotrophic bacteria is obscure despite much investigation. Several pathways have been suggested, the one given below being proposed by Vishniac and Santer (1957).

$$\underset{\text{Sulfur}}{S^{\circ}} \rightarrow \underset{\text{Thiosulphate}}{S_2O_3^{=}} \rightarrow \underset{\text{Tetrathionate}}{S_4O_6^{=}} \rightarrow \underset{\text{Trithionite}}{S_3O_6}$$
$$\rightarrow \underset{\text{Sulfite}}{SO_3^{=}} \rightarrow \underset{\text{Sulfate.}}{SO_4^{=}}$$

Factors Affecting the Oxidation of Elemental Sulfur in Soil

The oxidation of elemental sulfur in soil is affected by (a) the population of microorganisms, (b) soil moisture and aeration, (c) soil temperature, (d) soil pH, and (e) particle size. (Original references are cited in reviews by Attoe and Olson, 1966; Tisdale and Nelson, 1966; and Weir, 1975).

Microbial Population. Soils differ markedly in their ability to oxidize sulfur, which is commonly explained by differences in microbial populations. The rate of oxidation of elemental sulfur in soil can often be increased markedly by addition of a suitable inoculum. Inoculation either with *Thiobacillus thiooxidans* or heterotrophic organisms produce a marked improvement in the initial rate of oxidation in pot and incubation studies.

Soil Moisture and Aeration. The most rapid oxidation of elemental sulfur takes place at a moisture level near field capacity, i.e., in the soil moisture range optimal for the growth of agricultural plants. The rate of oxidation is markedly depressed at very low or very high soil moisture levels. Soil aeration is intimately associated with soil moisture, and thus at high moisture levels oxygen will be limiting for sulfur oxidation.

Soil Temperature. Microbial oxidation of sulfur will occur at temperatures as low as 4°C, but the process is slow below 10°C. Increasing the temperature to 40°C increases the rate of sulfur oxidation. At temperatures between 55° and 60°C, the organisms responsible for this oxidation are inactivated. Optimal temperatures are not the same for all microorganisms active in the oxidation of sulfur, but temperatures between 25° and 40°C seem to be optimum for most organisms.

pH. *Thiobacillus thioparus* and *T. denitrificans* grow best at reactions close to neutrality whereas *T. thiooxidans* and *T. ferrooxidans* develop best in the pH ranges 2.0–3.0 and 2.2–4.7, respectively (Starkey, 1966). Addition of lime to soil can increase sulfur oxidation, have no effect, or inhibit oxidation. These conflicting results may be due to a direct effect on pH or to an indirect effect through a reduction in solubility of other elements.

Particle Size. Because elemental sulfur is insoluble in water, it appears that its oxidation rate would be a function of the surface area exposed to microorganisms. Thus the smaller the particle size, the greater the surface area and the greater the rate of oxidation. This relationship has been frequently demonstrated in incubation studies, pot, and field trials.

Oxidation of Sulfide

The oxidation of iron sulfides in soil is more complex than the oxidation of elemental sulfur. The oxidation appears to consist of chemical as well as biological processes. The oxidation of both ferrous sulfide, FeS, and pyrite, FeS_2 appear to be basically similar, consisting of an initial chemical splitting of the sulfide followed by oxidation of the iron to the ferric state. This is then followed by oxidation of elemental sulfur to sulfate.

The following sequence has been proposed for the oxidation of pyrite (Temple and Delchamps, 1953). The first step in the oxidation is nonbiological, and ferrous sulfate is formed.

$$2FeS_2 + 2H_2O + 7O_2 \rightarrow 2FeSO_4 + 2H_2SO_4 \quad (1)$$

This reaction is then followed by the bacterial oxidation of ferrous sulfate, a reaction normally carried out by *T. ferrooxidans.*

$$4FeSO_4 + O_2 + 2H_2SO_4 \rightarrow 2Fe_2(SO_4)_3 + 2H_2O \quad (2)$$

Subsequently ferric sulfate is reduced, and pyrite is oxidized by what appears to be a strictly chemical reaction.

$$Fe_2(SO_4)_3 + FeS_2 \rightarrow 3FeSO_4 + 2S \quad (3)$$

Elemental sulfur so produced may then be oxidized by *T. thiooxidans* (equation 4), and the acidity generated favors the continuation of the process.

$$2S + 3O_2 + 2H_2O \rightarrow 2H_2SO_4 \quad (4)$$

The main difference between the oxidation of iron sulfide FeS and pyrite FeS_2 is the very much greater resistance of the latter to attack by atmospheric oxygen (Weir, 1975).

Reduction of Sulfate

Sulfate is unstable in anaerobic environments where conditions are favorable for bacterial development. It serves as a hydrogen acceptor for sulfate-reducing bacteria and is reduced to sulfide. This dissimilative sulfate reduction is the distinctive property of a small number of bacteria that are strict anaerobes. Two principal genera are recognized: *Desulfovibrio* and *Desulfotomaculum* (Roy and Trudinger, 1970). These bacteria use a variety of fermentation products and molecular hydrogen to reduce sulfate, they tolerate high concentrations of salt and hydrogen sulfide, and they function best in the pH range 5.5–9.0. Nitrate or low temperature retards the reaction (Ponnamperuma, 1972; Starkey, 1966).

Whereas oxidation of sulfur results in an increase in hydrogen-ion concentration, reduction of sulfate leads to a decrease.

$$Na_2SO_4 + Fe(OH)_3 + 9H(\text{organic matter}) \rightarrow FeS + 2NaOH + 5H_2O$$

The sodium hydroxide produced reacts with carbon dioxide from the decomposition of organic matter to produce carbonate and bicarbonate.

Most of the sulfide that accumulates as iron sulfide in waterlogged soils, bogs, ditches, and marine sediments is produced by sulfate-reducing bacteria. Hydrogen sulfide formed in the soil reacts with iron compounds to form FeS; under certain conditions the disulfide, usually pyrite, is formed more slowly. Partly reduced sulfur such as that in sulfite, polythionate, thiosulfate, and elemental sulfur is reduced more readily than sulfate. Many bacteria and other microorganisms, in addition to the sulfate-reducing bacteria, produce sulfide from these substances (Starkey, 1966). The pathway of reduction of sulfate to sulfide by *Desulfovibrio* has been discussed recently by Kelly (1972) and Roy and Trudinger (1970) and in simplified form is

$$SO_4^{=} \xrightarrow{ATP} APS \xrightarrow{2e} SO_3^{=} \xrightarrow{6e} S^{=}$$

Conclusions

There is relatively little information on the transformations of specific organic sulfur compounds in soil. The little information available suggests that the end products of the transformation will depend on the organisms present and on the environmental conditions. Much still needs to be done to provide a complete understanding of sulfur transformations in soils.

J. R. FRENEY

References

Attoe, O. J., and Olson, R. A., 1966. Factors affecting rate of oxidation in soils of elemental sulfur and that added in rock phosphate-sulfur fusions, *Soil Sci.,* **101**, 317–325.

Barrow, N. J., 1975. Reactions of fertilizer sulphate in soils, *in* K. D. McLachlan, ed., *Sulphur in Australasian Agriculture.* Sydney, N.S.W.: Sydney University Press, 50–57.

Fortini, S.; Cardas, G.; and Panella, S., 1966. Quantitative evaluation of the available sulfur fractions and soil reserves, and of their rate of regeneration, *Agrochimica,* **10,** 281–300.

Freney, J. R., 1961. Some observations on the nature of organic sulphur compounds in soil, *Aust. Jour. Agric. Res.,* **12,** 424–432.

Freney, J. R., 1967. Sulfur containing organics, *in* A. D. McLaren and G. H. Peterson, eds., *Soil Biochemistry.* New York: Marcel Dekker, 229–259.

Freney, J. R.; Melville, G. E.; and Williams, C. H., 1970. The determination of carbon bonded sulfur in soil, *Soil Sci.,* **109,** 310–318.

Freney, J. R.; Melville, G. E.; and Williams, C. H., 1975. Soil organic matter fractions as sources of plant-available sulphur, *Soil Biol. Biochem.,* **7,** 217–221.

Freney, J. R., and Stevenson, F. J., 1966. Organic sulfur transformations in soils, *Soil Sci.*, **101**, 307–316.
Freney, J. R., and Swaby, R. J., 1975. Sulphur transformations in soils, *in* K. D. McLachlan, ed., *Sulphur in Australasian Agriculture.* Sydney, N.S.W.: Sydney University Press, 31–39.
Harward, M. E., and Reisenauer, H. M., 1966. Reactions and movement of inorganic soil sulfur, *Soil Sci.*, **101**, 326–335.
Kelly, D. P., 1972. Transformations of sulphur and its compounds in soils, *Symp. Internat. sur le Soufre in Agric. Versailles (1970)*, 217–232.
Lowe, L. E., and De Long, W. A., 1963. Carbon bonded sulphur in selected Quebec soils, *Canadian Jour. Soil Sci.*, **43**, 151–155.
Lynn, W. C.; Tu, H. Y.; and Franzmeier, D. P., 1971. Authigenic barite in soils, *Soil Sci. Soc. America Proc.*, **35**, 160–161.
Ponnamperuma, F. N., 1972. The chemistry of submerged soils., *Adv. Agron.*, **24**, 29–96.
Putnam, H. D., and Schmidt, E. L., 1959. Studies on the free amino acid fractions of soils., *Soil Sci.*, **87**, 22–27.
Roy, A. B., and Trudinger, P. A., 1970. *The Biochemistry of Inorganic Compounds of Sulphur.* Cambridge, U.K.: Cambridge University Press, 400p.
Starkey, R. L., 1966. Oxidation and reduction of sulfur compounds in soils, *Soil Sci.*, **101**, 297–306.
Tabatabai, M. A., and Bremner, J. M., 1970*a*. Arylsulfatase activity of soils, *Soil Sci. Soc. America Proc.*, **34**, 225–229.
Tabatabai, M. A., and Bremner, J. M., 1970*b*. Factors affecting soil arylsulfatase activity. *Soil Sci. Soc. America, Proc.*, **34**, 427–429.
Tabatabai, M. A., and Bremner, J. M., 1972. Distribution of total and available sulfur in selected soils and soil profiles, *Agron. Jour.*, **64**, 40–44.
Temple, K. L., and Delchamps, E. W., 1953. Autotrophic bacteria and the formation of acid in bituminous coal mines, *Applied Microbiol.*, **1**, 255–258.
Tisdale, S. L., and Nelson, W. L., 1966. *Soil Fertility and Fertilizers.* New York: Macmillan, 694.
Vishniac, W., and Santer, M., 1957. The thiobacilli, *Bact. Rev.*, **21**, 195–213.
Vitolins, M. I., and Swaby, R. J., 1969. Activity of sulphur-oxidizing microorganisms in some Australian Soils, *Aust. Jour. Soil Res.*, **7**, 171–183.
Weir, R. G., 1975. The oxidation of elemental sulphur and sulphides in soil, *in* K. D. McLachlan, ed., *Sulphur in Australasian Agriculture.* Sydney, N.S.W.: Sydney University Press, 40–49.
Williams, C. H., 1967. Some factors affecting the mineralization of organic sulphur in soils, *Plant and Soil*, **26**, 205–223.
Williams, C. H., 1968. Seasonal fluctuations in mineral sulphur under subterranean clover pasture in southern New South Wales, *Aust. Jour. Soil Res.*, **6**, 131–139.
Williams, C. H. 1975. The chemical nature of sulphur compounds in soils, *in* K. D. McLachlan, ed., *Sulphur in Australasian Agriculture.* Sydney, N.S.W.: Sydney University Press, 21–30.
Williams, C. H., and Donald, C. M., 1957. Changes in organic matter and pH in a podzolic soil as influenced by subterranean clover and superphosphate, *Aust. Jour. Agric. Res.*, **8**, 179–189.
Williams, C. H., and Steinbergs, A., 1962. The chemical nature of sulphate in some Australian soils, *Plant and Soil*, **17**, 279–294.

Cross-references: *Carbon Cycle; Enzyme Activity; Fertilizers, Inorganic and Organic; Metabolic Activity; Nitrogen Cycle; Nutrient Potential; Phosphorus Cycle. Vol. IVA: Sulfur Cycle; Sulfur Oxidation: Bacterial; Sulfur Reduction: Microbial; Sulfur in Sediments. Vol. IVB: Soil Minerals. Vol. XV: Coastal Soils.*

SURFACE SOIL WATER CONTENT

Water contained in the surface soil is of great practical importance for agriculture, forestry, and rangeland management (Skidmore, Dickerson, and Schimmelpfennig, 1975; Ulaby, Cihlar, and Moore, 1974). Idso et al (1975*b*) emphasized the importance as follows:

> The presence of an adequate water supply in the upper-most few centimeters of the soil is essential for proper seed germination and crucial to early development of emerging crops. Its presence is also a factor in partitioning water income via rainfall and irrigation into runoff, deep percolation and storage. Disposition of stored water through evaporation is further dependent upon soil water content, as is erosion of the soil itself by wind action. Even populations of many economically important insect pests rise and fall with variations in the water content near the soil surface, since it is here that their eggs are deposited. Also, surface water content is a valuable boundary condition for meteorologists modeling the general circulation of the atmosphere.

Referring to the significant role of water in determining the nature and properties of soils, Low (1973) reported that in its absence, life in the soil comes to a virtual standstill and that all the physical, chemical, and biological reactions and processes that occur in the soil depend on its presence. Surface soil water has not only great practical importance but it also has theoretical implications. It enters as a basic component into the hydrological cycle, into energy exchanged near the surface, and into modeling of various ecosystem processes. It is also highly dynamic.

The water content at the interface of the soil and the atmosphere is constantly changing. It may change rapidly because of precipitation, irrigation (q.v.), soil drainage (q.v.), and/or evaporation (q.v.). Precipitation may occur violently in the form of wind-driven rain and/or hail, or gently in the form of dew or soft-landing snowflakes. The water thus received at the soil surface may run off or infiltrate into the soil and be distributed in the soil volume according to the laws that govern saturated and unsaturated moisture flow (see *Flow Theory*), or it may evaporate from the surface.

Surface Soil Water Content and Soil Detachment

The amount of water in surface soil particles greatly influences their detachment and transport by wind (see *Wind Erosion*). Chepil (1956) found that cohesion of water films between erodible-size soil particles varies directly with water content. He found that resistance of soil particles was equal to $6W^2$ where W, equivalent moisture, is a ratio of the water content in question to water content at -15 bar pressure potential. When the equivalent moisture is 1.0 (-15 bar pressure potential), the force of attraction between soil particles is about 6 dyn/cm^2. A friction velocity of 70 cm/s is required to produce that much surface drag.

Most natural winds are not strong enough to exert that much force to overcome the cohesive force of water at -15 bar pressure potential. However, as noted earlier, water at the surface is subject to drying rapidly. The pressure potential of the water in the atmosphere above the soil at 30°C and 50% relative humidity would be only -970 bars. Therefore, there is usually a steep water-potential gradient (conducive to rapid drying) between a moist soil and the atmosphere above it.

Water and Crust Formation

Water content and the wetting method affect the physical conditions of the soil surface. Surface flooding (Kemper, Evans, and Hough, 1974) and driving rain (Lyles et al, 1969) both destroy soil aggregates. Breakdown and dispersion of soil aggregates create a condition conducive to crust formation.

A saturated or near-saturated state facilitates particles' coming close to each other, a condition necessary for them to become cemented together (Uehara and Jones, 1974). Then, as stated by Uehara and Jones (1974):

> As water is removed by drainage and/or evaporation, pore water pressure increases negatively. If the surface layer consists of water-stable aggregates, water drains rapidly from the large inter-aggregates' pores, and subsequent evaporation results in increased negative pore water pressure within the aggregate, which in turn brings particles in aggregates closer together. If, on the other hand, the surface layer consists of aggregates which slake in water, large pores disintegrate upon wetting and pore size distribution is narrowed and shifted to the fine range.

The effects of reduced pore size (bringing the particles closer to each other) and therefore higher attainable negative pore-water pressure enhance bond formation or crusting.

As the soil drys, bonds form between soil particles, and crust develops. Cracks develop in the crust as the soil shrinks from drying.

If the soil is wetted gently, the aggregates may remain intact through wetting. However, when they are subsequently dried by evaporation, uneven forces develop. The aggregates usually dehydrate nonuniformly so that unequal strains develop. Strains and stresses that develop from wetting and drying break down soil aggregates.

Changeableness of Surface Soil Water Content

The surface or thin layer of porous soil, bounded by the atmosphere above and more soil below, is subject to extreme variations in several parameters, including water content. How many of these parameters vary depends on water content; thus various methods have been devised to detect and measure water content (see *Water Content and Retention*).

A unit volume of most mineral soils is about half solid and half pore space with the pores nonuniform and ranging in diameter from <0.1 to >1000 μm. The total pore volume is the potential volume of water that can be stored in the soil. When the soil is dry, the pores contain no water and are filled with air.

Water received at the soil surface displaces the air in the soil pores. Gravity causes some water to drain from the large pores. The distribution of water in soil depends on a balance between forces that pull the water toward the soil particles and surface-tension forces at the liquid-vapor interface, which keep the surface area of the liquid water at a minimum. The effective diameter of the largest water-filled or smallest air-filled pore in a system can be calculated from

$$D = \frac{\sigma}{RT \ln P/P_0} \quad (1)$$

where σ is water-surface tension; R is universal gas constant; and P/P_0 is relative humidity. The denominator on the right side of equation 1 represents free energy or water potential. At water potentials of 10^4 (0.01 bar), 10^5, 10^6, 10^7, 10^8 ergs/g, the diameters of the largest water-filled pores are 288, 28.8, 2.88, 0.288, and 0.0288 μm, respectively.

As the diameter of the largest water-filled pore decreases with lesser water contents, the mobility of the water also decreases (Table 1). Hydraulic conductivity (q.v.) may change over

TABLE 1. Some Soil Properties Influenced by Amount of Water Contained in the Soil.

Parameter	Dimension	Soil Material	Wet	Dry	Reference
Albedo	–	Avondale loam	0.14	0.30	Idso et al, 1975*c*
Thermal conductivity	Cal/cm/s/C	Fairbanks sand	5.49×10^{-3}	0.79×10^{-3}	deVries, 1966
Thermal conductivity	Cal/cm/s/C	Healy clay	3.68×10^{-3}	0.38×10^{-3}	deVries, 1966
Heat capacity	$Cal/cm^3/C$	Mineral soil	0.60	0.21	deVries, 1966
Soil temperature*	C	Silty clay	22	54	Idso et al, 1975*b*
Water potential	ergs/g	–	1.4×10^6	960×10^6	Calculated**
Hydraulic conductivity	cm/day	Fine sand	188 @ sat.	0.01 @ θv = 0.10	Bruce, 1972
Hydraulic conductivity	cm/day	Yolo loam	50 @ sat.	0.01 @ θv = 0.30	Davidson et al, 1969
Hydraulic conductivity	cm/day	Slate dust	86 @ θv = 0.10	0.086 @ θv = 0.40	Youngs, 1964
Polarized light	–	–	0.89	0.16	Doll, 1973
Microwave emissivities	–	Avondale loam	0.50	0.90	Poe et al, 1971
Dielectric constant	–	Vernon clay loam	30 @ 21 cm	3 @ 21 cm	Lundien, 1971
Dielectric constant	–	Silty clay loam	16 @ 1.55 cm	3 @ 1.55 cm	Schmugge, 1974

*Diurnal amplitude of surface-soil temperature wave.

**Water potential equals RT 1n P/P_0; where R is universal gas constant 0.462 ergs/°K/g; T is temperature in degrees Kelvin; and P/P_0 is relative vapor pressure or humidity. P/P_0 was 0.999 and 0.50 for wet and dry, respectively.

several orders of magnitude with change in water content.

When the evaporation rate from the soil exceeds the soil's ability to transmit water to that surface, the amount of water in the surface soil can change rapidly. Idso et al (1975*c*), using albedo measurements, found that surface volumetric water of a bare Avondale loam soil at Phoenix, Arizona, decreased from 0.20 to 0.07 cm^3/cm^3 between 1300 and 1500 h 2 days after the soil was irrigated in July. This process was slower in December when the surface soil did not dry until 10 days after irrigation; the volumetric water content decreased from 0.16 to 0.06 cm^3/cm^3 between 1000 and 1500 h.

Jackson (1973), who gravimetrically measured soil water in samples taken at 0- to 0.5-cm increment at 0.5-h intervals, showed that the amount of water in that layer varied widely during a day (Figure 1). He found that water content decreased from 0.23 cm^3/cm^3 at sunrise to 0.10 cm^3/cm^3 at 1400 h, five days after irrigation; that during the night the volumetric water content increased to 0.19 cm^3/cm^3, then decreased rapidly again during the day; and that water content and amplitude of change in content decreased gradually with time after irrigation. This diurnal pattern continued for all days of his experiment.

Measurement

The standard method for measuring soil water content is gravimetric. A soil sample is obtained, weighed, dried at about 105°C, and then reweighed. Then gravimetric water content is calculated by first subtracting the dry-soil weight from the wet-soil weight and dividing the result by the dry-soil weight. To convert that gravimetric or mass ratio (mass of water per unit mass of soil) to volumetric water content, multiply it by the soil bulk density. When the soil bulk density is not known, a separate bulk-density determination is required. (Usually for mineral soils the bulk density (q.v.) is between 1.0 and 1.4 g/cm^3.)

Soil samples to be used in determining gravimetric and volumetric water content must be obtained over some finite depth, usually several centimeters. For detailed measurements, samples have been carefully obtained at 1-cm intervals (Jackson, 1973). Of the several methods described in recent literature, some are for sampling at the near surface only; others for sampling a thin layer of soil. Several involve noncontact remote sensing, which with further refinement and evaluation, may be useful in detecting water amounts in soil by satellite or airplane overflight (Idso, Jackson, and Reginato, 1975*a*).

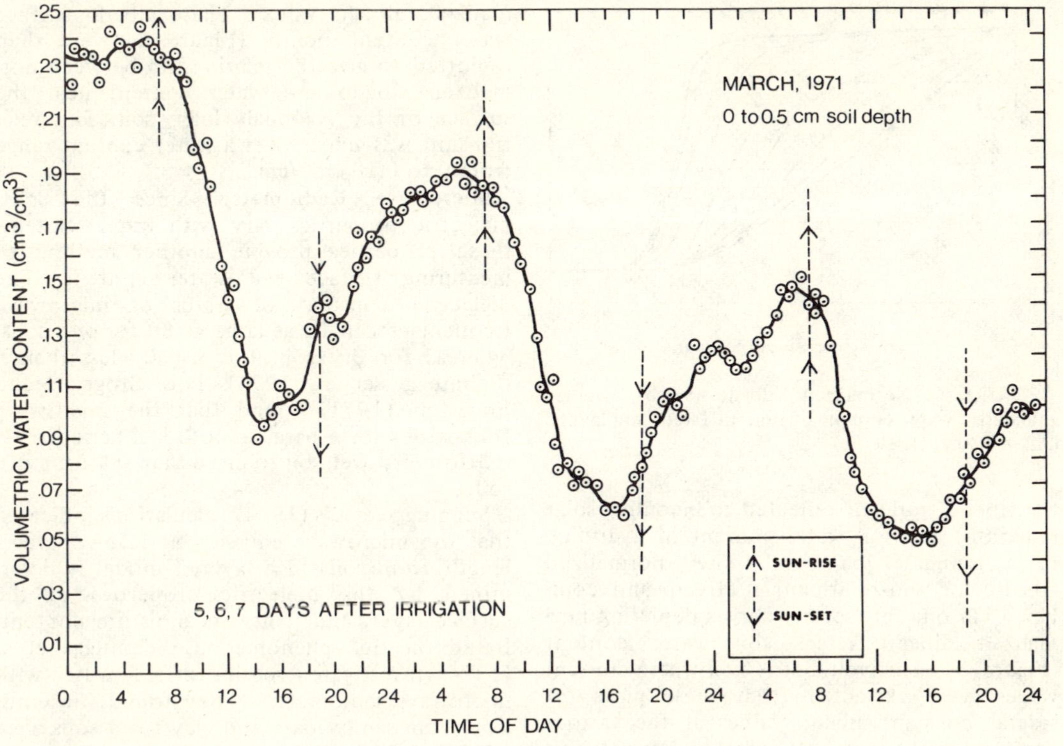

FIGURE 1. Volumetric water contained in the 0- to 0.5-cm increment versus time for 3 days during March 1971. (After Jackson, 1973)

Infrared Reflectance. Using water's property to absorb certain wavelengths in the near infrared as a basis, Skidmore, Dickerson, and Schimmelpfennig (1975) evaluated surface soil water content by measuring reflectance of near-infrared radiation from a soil surface. Previously Bowers and Hanks (1965) had demonstrated the potential of using reflectance measurements to determine soil water content by showing that reflectance at a 1.9-μm wavelength, for a Newtonia silt loam, decreased as water increased. Initial testing of the reflectometer indicated that water 10 μm thick (10 mg/cm^2) absorbed most electromagnetic radiation at the 1.95 μm wavelength. With as little as 1.0 mg/cm^2 of water, the reflectance attenuated >25%. Since infrared radiation does not penetrate deeply, the amount of IR energy reflected is mainly a surface phenomenon, thus indicating the water content only very near the surface.

The shape of the reflectance versus soil water content curve (Figure 2) closely resembles a log-linear relationship for water contents between oven-dried soil and the point of small change in reflectance as water content is increased. The least-squares fit for a log-linear relationship for soil water contents between oven-dried and 0.3 bar tension gave R^2 values of 0.99, 0.96, 0.99 for Carr sandy loam, Farnum sandy clay loam, and Smolan silty clay loam, respectively. At low water contents soil properties (other than water) strongly influenced soil reflectance.

Albedo. In a series of experiments Idso et al (1975*c*) intensively and concurrently measured

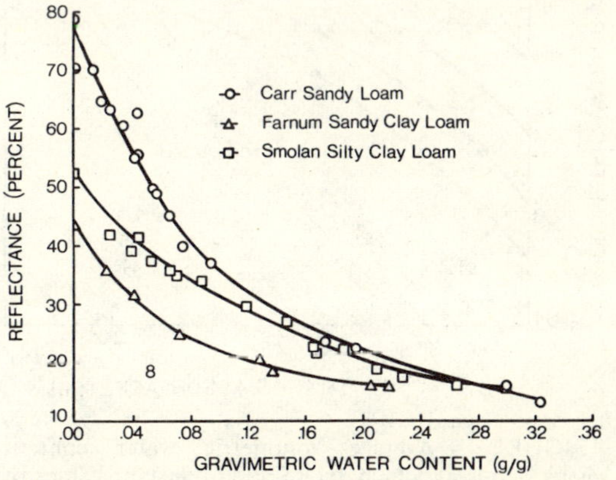

FIGURE 2. Reflectance at 1.95 μm of three soils as influenced by their water content. (After Skidmore, Dickerson, and Schimmelpfennig, 1975)

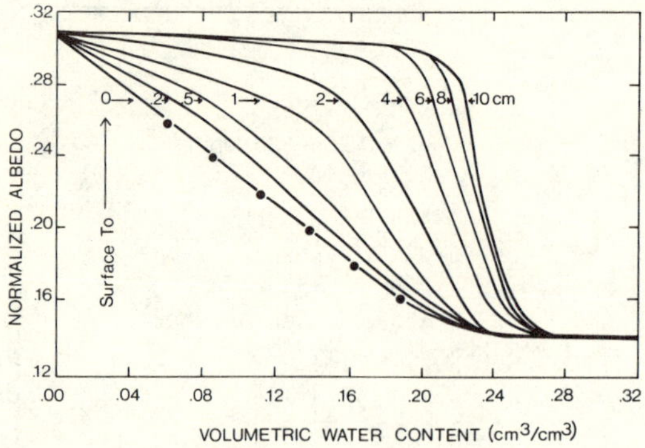

FIGURE 3. Normalized albedo versus average volumetric water content of nine different soil layers. (After Idso et al, 1975c)

the albedo (ratio of reflected to incoming solar radiation) and soil water content of a drying, bare Avondale loam soil. They normalized albedo for sun-zenith-angle effects and combined (in one set) several curves depicting normalized albedo versus soil water content (Figure 3). The volumetric soil-water content values were extracted from these plots for several constant albedo values at the various depth-interval lines and plotted against this latter parameter (Figure 4). Smooth lines were drawn connecting those points and extrapolated to yield the volumetric soil water contents at the soil surface for the chosen constant, normalized, albedo values. These albedo surface water content points (Figure 3) were then replotted to give the relationship between normalized albedo and water content near the surface of the Avondale loam soil. This relationship was linear over a water content range from 0 to 0.18 cm^3/cm^3.

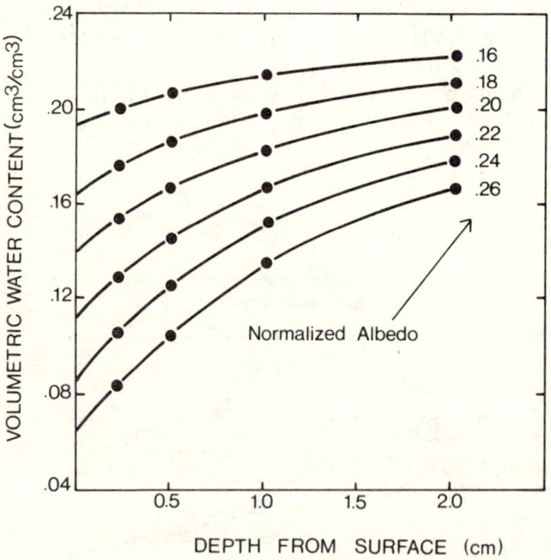

FIGURE 4. Average volumetric water content versus depth at which six selected constant values of normalized albedo were averaged. (After Idso et al, 1975c)

Microwave Radiometry. Since the soil's dielectric properties vary with water content, these properties provide another method of measuring surface soil water content. The dielectric constant of water at microwave frequencies can be as large as 80 for wet soils, whereas for dry soil it is usually less than 5 (Schmugge et al, 1974). Poe, Stogryn, and Edgerton (1971) found that the microwave emissivities for a bare, smooth soil varied from 0.5 for very wet soil to more than 0.9 for a dry soil.

Schmugge et al's (1974) calculations indicated that the microwave emission at 1.55-cm wavelength from soils in a layered model is determined by the dielectric properties of the surface layer, and hence its moisture content. Based on this phenomenon, Schmugge et al (1974) investigated sensing of soil water with microwave radiometers. They found the emission from sandy loam and clay loam soils were a linear function of soil water content over the 0–35% range at a wavelength of 21-cm (Figure 5).

Radar. Radar or active microwave is also used to detect soil water content. Water in the surface soil strongly affects the microwave return, which is not directly influenced by the soil's chemical and mineralogical properties (Lundien, 1971; Edgerton et al, 1971). However, Ulaby's (1974) results indicated that the radar response to soil moisture content depends highly on surface roughness, microwave frequency, and look angle. In a subsequent study, on active microwave response to soil moisture based on the soil skin-depth (attenuation) concept, Ulaby, Cihlar, and Moore (1974) reasoned that because the soil thickness interacting with the electromagnetic radiation varied with moisture content and incident angle, microwave response to soil moisture should include these variations. Hence, they proposed a skin-depth model for active microwave response to soil moisture, defining skin depth as the reciprocal of attenuation coefficient (in terms of wavelength and dielectric constant). They also indicated that presently moisture dependence of soil dielectric properties has not been well defined.

Skin depth ranges from less than 1 cm for wet soils to several centimeters for dry soil (Figure 6). Ulaby, Cihlar, and Moore (1974) found reasonable correlation between radar backscat-

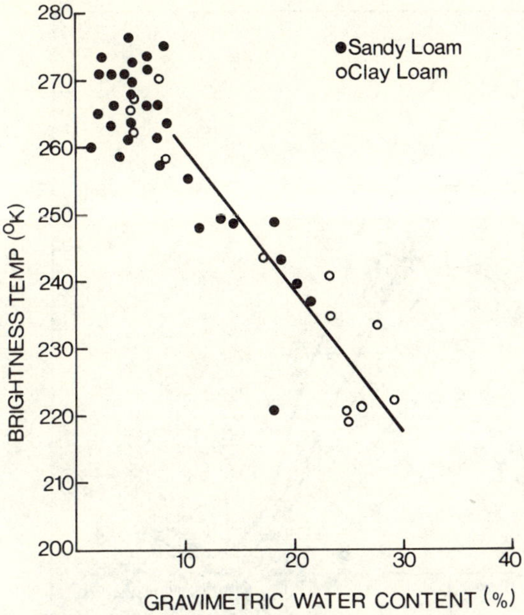

FIGURE 5. Plot of the 21-cm wavelength brightness temperatures versus soil water content. (The brightness temperature is the product of the temperature and emissivity of the surface.) (After Schmugge et al, 1974)

ter from bare soil and moisture content expressed in terms of the mean attenuation coefficient over a distance of one skin depth below the surface. Their results indicated that using radar techniques for determining soil moisture are promising but far from proven.

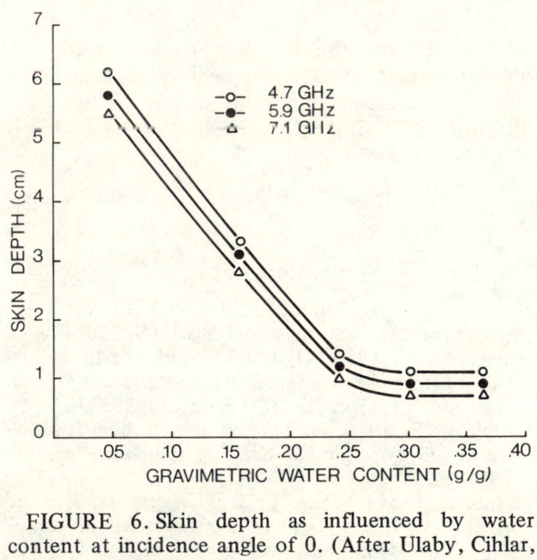

FIGURE 6. Skin depth as influenced by water content at incidence angle of 0. (After Ulaby, Cihlar, and Moore, 1974)

Radar techniques also have the potential for much better spatial resolutions, but unlike the capabilities of radiometric techniques, they have not been demonstrated from an aircraft platform. Also radar techniques would be independent of soil temperature, while the radiometric techniques may yield information on soil temperatures. However, both techniques respond to moisture variations in surface layers of the same thickness.

Thermal Inertia. The thermal inertia concept of measuring soil moisture, which relates the amount of soil water to the amplitude of the diurnal surface soil temperature wave, was evaluated by Idso et al (1975*b*). They found on an Avondale loam soil at Phoenix, Arizona, that the volume of water contained in the surface soil layer (2 to 4 cm thick) was a linear function of the amplitude of the diurnal surface-soil temperature wave for clear day-night periods and also of the daily maximum value of the surface soil-air temperature differential. Diurnal plots of measured surface soil (minus air-temperature differential) on selected clear day-night periods after the plots had been irrigated (Figure 7) showed that the amplitude of the temperature wave increased as the soil water decreased. Figure 8 shows the maximum value of the surface soil minus air-temperature differential as a function of volumetric soil water content.

With the thermal inertia concept, like some other methods, measuring the amount of water in the soil is difficult because different soils have unique relationships between their water content and the measured parameter. Thus different soils have different relationships between amplitude of diurnal temperature and volumetric water content. However, Idso et al (1975*b*) found that soils had a closer relationship between amplitude of diurnal surface soil temperature and pressure potential. They did not evaluate the dependence on climate of the amplitude of diurnal surface soil temperature wave.

Light Polarization. Another method for determining surface-soil water content is to measure light polarization when it is reflected from the soil surface. Any natural surface alters the polarizing properties of light reflected from it. Most dry surfaces reflect weakly polarized light, whereas light reflected from a wet surface may be nearly 100% polarized (Doll, 1973). Thus as water is added to dry soil, the percentage of light polarization increases steadily as it increasingly adopts the reflecting properties of the water. Doll (1973) found that at the optimum reflection angle (about 60°), polarization of the reflected light ranged from

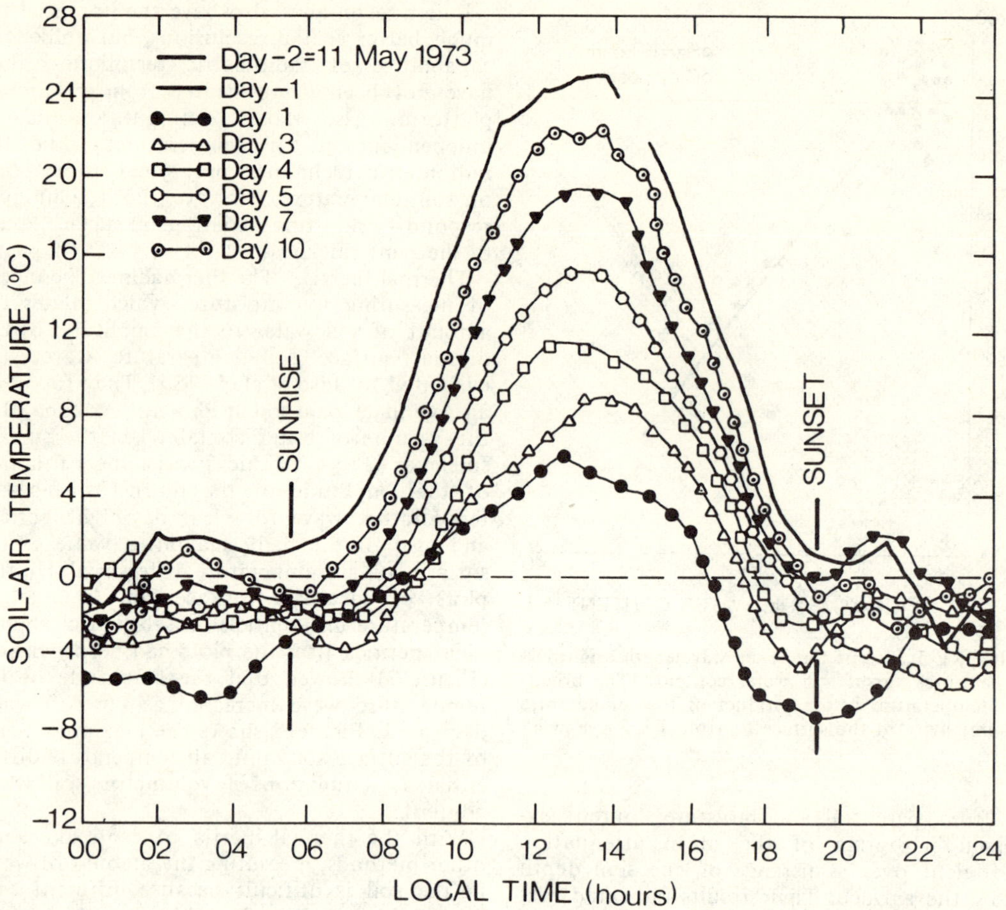

FIGURE 7. Diurnal plots of measured surface soil minus air-temperature differential on selected clear day-night periods after plots had been irrigated with 10 cm of water on day zero. (After Idso et al, 1975b)

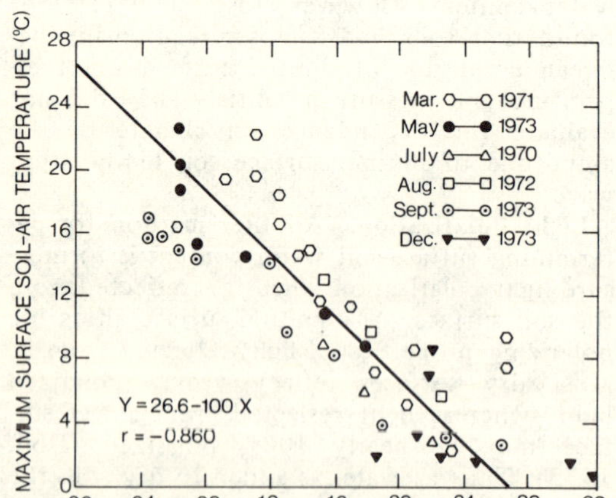

FIGURE 8. The maximum value of the surface soil minus air-temperature differential versus the mean daylight volumetric soil water content of 0- to 2-cm increments. (After Idso et al, 1975b)

15.5% for dry soil to 89% for saturated soil, and increased nearly linearly. Doll suggested that by using range-polarization values, the amount of water in the soil could be determined accurately.

EDWARD L. SKIDMORE

References

Bowers, S. A., and Hanks, R. J., 1965. Reflection of radiant energy from soils, *Soil Sci.,* **100,** 130–138.

Bruce, R. R., 1972. Hydraulic conductivity evaluation of the soil profile from soil water retention relations, *Soil Sci. Soc. America Proc.,* **36,** 555–561.

Chepil, W. S., 1956. Influence of moisture on erodibility of soil by wind, *Soil Sci. Soc. America Proc.,* **20,** 288–292.

Davidson, J. M.; Stone, L. R.; Nielsen, D. R.; and LaRue, M. E., 1969. Field measurement and use of soil-water properties, *Water Resources Res.,* **5,** 1312–1321.

deVries, D. A., 1966. Thermal properties of soils, *in*

W. R. Van Wijk, ed., *Physics of Plant Environment.* Amsterdam: North-Holland, 210–235.

Doll, B., 1973. The potential use of polarized reflected light in the remote sensing of soil moisture, *Report No. ECOM-5501, U.S. Army Electronics Command, Fort Monmouth, New Jersey,* 1–19.

Edgerton, A. T.; Ruskey, F.; Williams, D.; Stogryn, A.; Poe, G.; Meeks, D.; and Russel, O., 1971. Microwave emission characteristics of natural materials and the environment, *Final Technical Report 9016R-8, Aerojet General Corporation, El Monte, Calif.,* 284p.

Idso, S. B.; Jackson, R. D.; and Reginato, R. J., 1975*a*. Detection of soil moisture by remote surveillance, *Amer. Sci.,* **63,** 549–557.

Idso, S. B.; Schmugge, T. J.; Jackson, R. D.; and Reginato, R. J., 1975*b*. The utility of surface temperature measurements for the remote sensing of surface soil water status, *Jour. Geophys. Res.,* **80,** 3044–3049.

Idso, S. B.; Jackson, R. D.; Reginato, R. J.; Kimball, B. A.; and Nakayama, F. S., 1975*c*. The dependence of bare soil albedo on soil water content, *Jour. Applied Meteor.,* **14,** 109–113.

Jackson, R. D., 1973. Diurnal changes in soil water content during drying, *in* R. R. Bruce; K. W. Flach; and H. M. Tayler, eds., *Field Soil Water Regime.* Madison: Soil Science Society of America Special Publication No. 5, 37–55.

Kemper, W. D.; Evans, D. D.; and Hough, H. W., 1974. Crust strength and cracking. Part 1: Strength, *in* J. W. Cary and D. D. Evans, eds., *Soil Crusts.* A Western Regional Research Publication, Tech. Bulletin 214. Agricultural Experiment Station, University of Arizona, 31–38.

Low, P. F., 1973. Foreword, *in* R. R. Bruce; K. W. Flach; and H. M. Taylor, eds., *Field Soil Water Regime.* Madison: Soil Science Society of America Special Publication No. 5, 212p.

Lundien, J. R., 1971. Terrain analysis by electromagnetic means, *Technical Report No. 3-393, Report 5, U.S. Army Engineer Waterways Experiment Station, Vicksburg, Mississippi,* 34p.

Lyles, L.; Ambrust, D. V.; Dickerson, J. D.; and Woodruff, N. P., 1969. Spray-on adhesives for temporary wind erosion control, *Jour. Soil Water Conserv.,* **24,** 190–193.

Poe, G.; Stogryn, A.; and Edgerton, A. T., 1971. Determination of soil moisture content using microwave radiometry, *Final Report, 1684FR-1 DOC Contract 0-35239, Aerojet-General Corp., Microwave Division, El Monte, Calif.,* 43p.

Schmugge, T.; Gloersen, P.; Wilheit, T.; and Geiger, F., 1974. Remote sensing of soil moisture with microwave radiometers, *Jour. Geophys. Res.,* **79,** 317–323.

Skidmore, E. L.; Dickerson, J. D.; and Schimmelpfennig, H., 1975. Evaluating surface-soil water content by measuring reflectance, *Soil Sci. Soc. America Proc.,* **39,** 238–242.

Uehara, G., and Jones, R. C., 1974. Bonding mechanisms for soil crusts: Part 1. Particle surfaces and cementing agents, *in* J. W. Cary and D. D. Evans, eds., *Soil Crusts.* A Western Regional Research Publication, Technical Bulletin 214. Agricultural Experiment Station, University of Arizona, 17–28.

Ulaby, F. T., 1974. Radar measurement of soil moisture content, *Trans. Antennas Propagation,* **AP-22** (2), 257–265.

Ulaby, F. T.; Cihlar, J.; and Moore, R. K., 1974. Active microwave measurement of soil water content, *Remote Sensing Environment,* **3,** 185–203.

Youngs, E. G., 1964. An infiltration method of measuring the hydraulic conductivity of unsaturated porous materials, *Soil Sci.,* **97,** 307–311.

Cross-references: *Aggregate Stability; Conductivity, Hydraulic; Energy Balance; Evaporation; Flow Theory; Heat Capacity; Infiltration; Pore Space; Spectral Characteristics; Thermal Regimes; Thermoluminescence; Water Content and Retention.*

T

TESTS AND TESTING–See ELECTROCHEMISTRY; MINERAL ANALYSIS; PARTICLE-SIZE DISTRIBUTION; pH, FIELD DETERMINATION; SOIL MECHANICS; SOIL TESTS

TEXTURE AND TEXTURAL PARAMETERS–See PARTICLE-SIZE DISTRIBUTION. VOL. XII, Pt. 2: FIELD TEXTURE

THERMAL REGIMES

Ordinarily the *thermal regime* of the soil is referred to as the regular pattern of soil temperature fluctuations in the soil, without relating it to the source of energy that controls that regime. To describe better the thermal regime of the soil, it is helpful to introduce the concept of the soil surface as an active surface with respect to radiation, heat, and mass exchange with the soil proper underneath it and the atmosphere above it.

At the soil surface, solar radiation is intercepted, transformed, and redistributed between the atmosphere and the soil. The energy balance for the soil surface is presented schematically in Figure 1. Most of the incoming radiation is dissipated through reflection, long wave back radiation, soil water evaporation, sensible heat transfer to the atmosphere as convection and conduction, while only a relatively small amount of energy dissipates into the soil proper as the soil heat flux.

The flux of heat into and out of the soil is a process of thermal energy transfer in the soil, which determines the thermal regime of the soil, of which the fundamental index is the soil temperature. Thus the thermal regime of a soil is characterized by giving the temperature distribution with respect to depth, time, and season for a given soil at a given location.

Soil Temperature Profiles

Soil temperature profiles depend on the heat flux density and the soil thermal properties, namely, the thermal conductivity (q.v.) and heat capacity (q.v.) of the soil.

Global Pattern. Data presented by Houghton (1954) show the global radiation to be the highest and the albedo the lowest for lower latitudes as compared to corresponding values for higher latitudes (Figure 2). These differences in solar energy intercepted at the soil surface are responsible for most of the differences between soil temperature profiles and thus to the different thermal regimes at different locations. Baver, Gardner, and Gardner (1972) have presented approximations of soil temperatures for summer and winter periods for the northern hemisphere for different soil depths and latitudes (Figure 3). It is interesting to note that (a) soil temperatures below 10 cm depth are about the same for all soil depths for 20°N or S during the summer, but at 10 and 30 cm depth are somewhat higher at 40°N than at 40°S and (b) soil temperatures increase at all depths from the higher latitudes toward the equator, yet are somewhat lower at 40°N than at 40°S. These

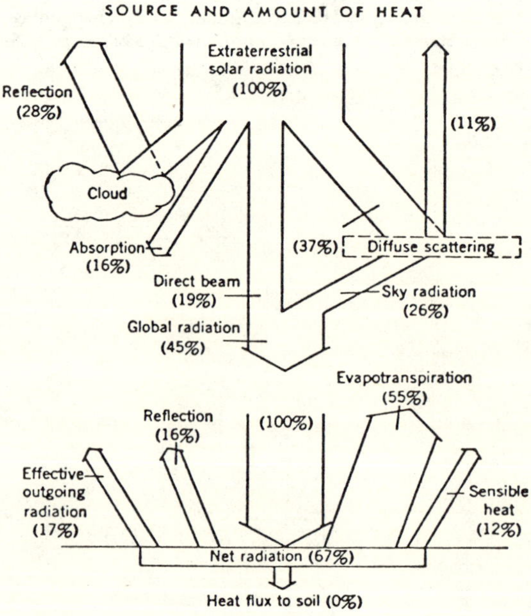

FIGURE 1. Daytime heat balances. (After Baver, Gardner, and Gardner, 1972)

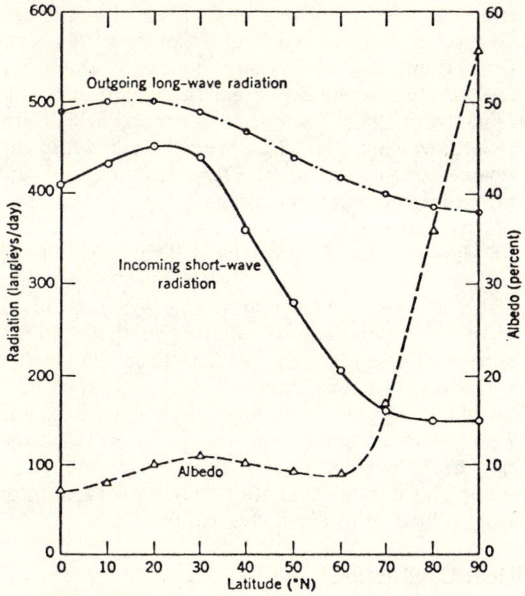

FIGURE 2. Radiation distribution in the northern hemisphere in relation to latitude. (After Houghton, 1954, and Baver, Gardner, and Gardner, 1972).

observations are very general, yet they serve well to show the global pattern.

Local Pattern of Temperature Profiles. It is obvious that on the global pattern soil temperatures taken by averaging the pattern over large areas of different exposures, coverage and thermal properties do not show the diversity in soil temperature profiles caused by the local conditions. The local pattern of soil temperature profiles is divided into the diurnal pattern and the seasonal or annual pattern.

The diurnal patterns of soil temperature are affected by the soil thermal properties, the soil cover, and the energy balance at the soil surface. Data collected by Yakuwa (1945), and given in Figure 4 show that in early morning the temperature is lowest at the soil surface and higher at deeper zones of the soil. After sunrise the soil surface starts to warm up, reaches a maximum when the sun reaches its highest point (local noontime), and subsequently drops down once again. The pattern of the temperature that changes at the surface with respect to time has a wave form, similar to a sinusoidal curve. Below the soil surface the temperature waves of different soil levels show that as depth increases, the differences between maximal and minimal temperatures decrease, yet this occurrence is delayed the deeper the level of measurement is. If we define the temperature wave amplitude as the half span of temperature change between the maximum and minimum at a given soil level, then the temperature amplitudes decrease with soil depth. Moreover, these temperature amplitudes are different for different soils, as can be seen from data presented by Baver, Gardner, and Gardner (1972), and given in Table 1. The temperature amplitudes at the soil surfaces and at deeper layers depend on the thermal balance of the soil surface and the soil thermal properties, which affect the heat flux dissipation at the soil surface into the soil profile.

The seasonal soil temperature profile varies similarly to the diurnal variations. For the northern hemisphere the maximum temperature will be reached during the summer months of June and July (January and February in the southern hemisphere) in similarity to midday, and the minimum temperature will be reached during the winter months of December and

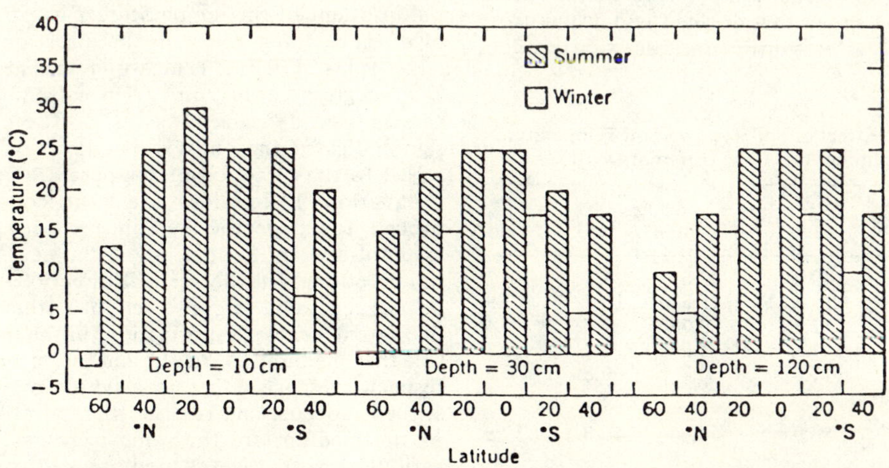

FIGURE 3. Variations in soil temperature with latitude in the Western hemisphere. (After Baver, Gardner, and Gardner, 1972)

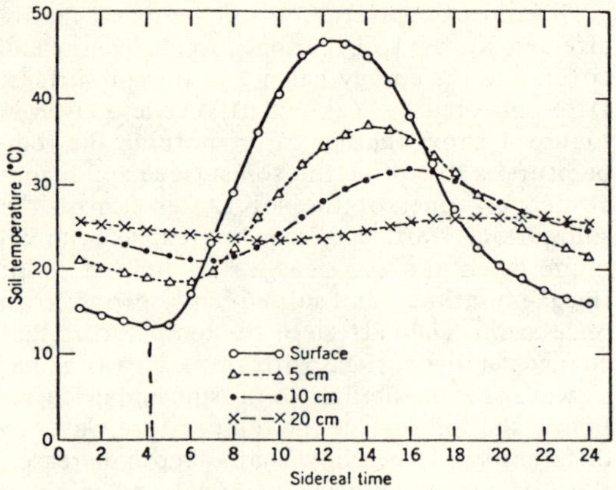

FIGURE 4. Diurnal variations in temperature in a loam soil. (After Yakuwa, 1945)

January (June and July in the southern hemisphere) in similarity to the nocturnal temperature in the diurnal pattern. Baver, Gardner, and Gardner (1972) used temperature data published by Smith (1932) to show the monthly variation of temperatures in relation to depth. These data also show clearly that the annual temperature amplitudes decrease with depth, and their maximum is delayed the deeper the level of measurement, in resemblance to the diurnal temperature wave changes.

Soil Heat Flux

The diurnal and annual patterns of soil temperature distributions illustrate the fact that heat is continuously moving into or out of the soil and that thermal energy is continuously redistributed within the soil. Heat is flowing from the warmer regions into the cooler regions of the soil. The diurnal temperature pattern given in Figure 4 shows that during the daytime from some time after sunrise until early afternoon, there is a warming tendency from the soil surface to a depth of 10 cm, yet during the nighttime there is a cooling tendency of the same layers. During the day, heat flows from the surface toward the deeper layers as long as the surface is warmer than the deeper layers, while the reverse occurs at night. The amount and the rate of heat being redistributed within the soil depend on the thermal energy balance, namely, on the soil heat flux at the soil surface and the soil heat capacity and its thermal properties. Table 2 gives examples of annual soil heat fluxes at different latitudes of the northern hemisphere. These data show that there are smaller variations between maximal gain or loss of heat as given by the soil heat flux densities, the nearer the site is to the equator, and that the net heat balance is not zero; there are years when the annual heat flux shows a total gain or loss of heat in or from the soil profile.

TABLE 1. Effect of Soil Type on Soil Temperature Amplitudes (in °C) at Different Depths.

Soil Depth (cm)	Sand	Loam	Clay	Peat
0	40.0	33.6	21.5	23.2
5	19.4	18.5	13.7	13.9
10	12.3	10.7	7.7	5.4
20	4.8	3.0	2.2	0.7
30	1.6	0.7	0.6	0.3
Depth at which temperature amplitude is 0.1°C [cm]	56.9	46.6	46.8	39.5

(After Bauer, Gardner, and Gardner, 1972)

Heat Conduction

To analyze or estimate the thermal regime of a soil, the soil temperature profiles and the heat flux densities should be analyzed mathematically as a process of heat conduction. This process can be described by *Fourier's law* of heat conduction given in equation 1:

$$G = -\lambda \frac{\partial T}{\partial Z} \tag{1}$$

where G is the heat flux density defined as the quantity of heat transferred during a unit of time across a unit of surface area perpendicular to axis Z. The proportionality factor λ depends on the soil and is termed the *thermal conductivity* (q.v.) of the soil. The term $\partial T/\partial Z$ is called the *temperature gradient,* which is the differential quotient of the temperature T as a function of distance along axis Z in the direction of the larger value of T. If Z is measured in centimeters T in degrees celsius, and G in cal $cm^{-2} s^{-1}$, λ is expressed by cal $cm^{-1} s^{-1} {}^\circ C^{-1}$. The negative sign indicates that heat flow direction is opposed to the direction of temperature increase.

Equation 1 describes the heat conduction process for a specific medium provided several assumptions are kept, e.g., (a) the heat flow is unidirectional (in the vertical direction); (b) no heat is generated or converted into other forms, like latent heat of vaporization, in the medium; (c) the medium is rigid, and no changes in respect to space either as a whole or its constituents occur; and (d) the thermal properties of the medium are the same in respect to the various axes, that is, the medium is isotropic.

Since soil is a composite medium, strictly speaking, equation 1 cannot be applied to soils,

TABLE 2. Annual Soil Heat Flux Density at the Soil Surface (cal cm^{-2} $month^{-1}$) for Four Different Stations.

Station	Jan.	Feb.	March	April	May	June	July	Aug.	Sept.	Oct.	Nov.	Dec.	Total
Sodankila (67°N 27°E) 1915–1916	−119	−132	−210	+664	1716	1209	1018	80	−338	−808	−742	−649	+1689
Potsdam (52°N 13°E) 1903	−248	−96	+88	321	536	568	384	112	−120	−332	−504	−528	+181
Djakarta (6°S 107°E) 1922	10	38	−5	+4	+4	−35	−20	+43	33	10	10	−17	+75
Ikengueng (42°N 108°E) 1931–1932	−224	−119	−20	+310	595	606	494	130	−190	−494	−639	−628	−249

(After Van Wijk and deVries, 1963)
Negative values show heat loss from the soil profile.

but for all practical purposes if an appropriate mean value of λ can be specified, it can be used. The average property of λ depends on the soil constituents, i.e., its mineralogical composition, bulk density, texture, structure, water and air contents, organic matter, and temperature. The λ dependence on these factors will be considered in a subsequent section dealing with thermal conductivity.

Usually the soil's water contents, air contents, and bulk densities change with depth and time; thus the thermal conductivity changes accordingly with depth and time. If the temperature distribution around a soil volume of given λ does not vary with time, then the heat flux G is independent of time; such a situation is called a *stationary state,* or sometimes a *steady-state* condition. If the temperature varies with respect to position within the soil and time, then we have a *nonsteady state condition,* or *transitory state.* This is the common case under natural conditions; thus since equation 1 cannot describe a transitory state, another equation should be derived to account for the change of temperature with time and depth.

According to the principle of energy conservation, heat should also be conserved. Taking a slab of soil of unit cross-section at depth Z and height dz, a balance of heat flux can be made. Let us denote G_z as the heat flux density at depth z entering the slab, and G_{z+dz} the heat flux at depth $z + dz$ leaving the slab; then the difference between G_z and G_{z+dz} will equal the differential $-\partial G/\partial z \cdot dz$ of heat flux density for the slab or the amount of heat stored in the slab. This is given in equation 2, which is called the *equation of continuity,* and is an expression for the principle of energy conservation. Equation 2 equates the differential of the heat flux density and the heat storage as expressed by the temperature change, where C represents the *volumetric heat capacity,* defined as the amount required to raise the temperature of a unit volume of the soil by one degree [*i.e.,* it is a unit that equals

$$-\frac{\partial G}{\partial z} = C\frac{\partial T}{\partial t} \tag{2}$$

(cal cm^{-3} of soil)], and t is time. Substituting equation 1 into equation 2 gives the generalized heat flow equation

$$C\frac{\partial T}{\partial t} = \frac{\partial}{\partial z}\left(\lambda\frac{\partial T}{\partial z}\right) \tag{3}$$

In case the soil is homogeneous and isotropic, λ and C are independent of z, and equation 3 becomes equation 4, where $\alpha = \lambda/c$ is called the *thermal diffusivity* of the soil and has the dimensions of unit area per unit time [$cm^2 s^{-1}$].

$$\frac{\partial T}{\partial t} = \frac{\lambda}{c}\frac{\partial^2 T}{\partial_z{}^2} = \alpha\frac{\partial^2 T}{\partial_z{}^2} \tag{4}$$

For a given soil, of which the thermal properties are known, equation 4 will yield the temperature profiles at any time provided conditions of soil heat flux density at the boundaries of the soil or temperature variations of time dependent are given or assumed.

Since the global radiation fluxes intercepted at the soil surface are of a periodic nature, the soil surface temperature and the soil heat flux density fluctuations have a periodic nature. Their periodic nature cannot, however, be described by simple sine or cosine functions. The soil temperature distribution with respect to depth and time can be described by a general periodic function given by equation 5 where

$$T(z, t) = T_A(z) + \sum_{k=1}^{\infty} T_{0,k}^* \exp\left\{-\frac{zk}{\sqrt{2\alpha/\omega}}\right\}$$

$$\sin\left(k\omega t - z\sqrt{k\frac{\omega}{2\alpha}} + \psi_k\right) \qquad (5)$$

$T_A(z)$ and T_0^* are the average mean temperature as a function of depth and the temperature amplitude at the soil surface ($z = 0$), respectively, ψ_k is the phase constant, and ω is the radial frequency. The radial frequency is defined as $\omega = 2\Pi/P$, where P is the period duration [for example, ω equals $2\Pi/86400 = 7.27 \times 10^{-5}$ (s^{-1}) for a diurnal variation and 1.99×10^{-7} (s^{-1}) for an annual variation, and k signifies the harmonics of the variations. Equation 5 is simplified if k is assumed to equal 1 and a value called the *damping depth*, $D = \sqrt{2\alpha/\omega} = \sqrt{2\lambda/c\omega}$, is introduced, as given in equation 6 where ψ is the phase constant.

$$T(z, t) = T_a(z) + T_0^* \exp\left\{-z/D\right\}$$

$$\sin(\omega t - z/D + \psi) \qquad (6)$$

The decrease in temperature amplitude with depth and the shift of phase when it occurs, by a factor $-z/D$ are typical of the propagation of periodic temperature variations in a soil and are found also when λ and C change with depth, but then the damping depth D changes with depth. The damping depth is the depth at which the amplitude decreases to $1/e$ or 0.37 of its initial value (e is the base of natural logarithms). The physical cause for the damping depth and the reduction of the temperature wave variation with depth is the fact that the soil layers have heat capacity, and thus some heat is either stored or released in the layers when the temperature increases or decreases, respectively. The damping depth is also a function of the temperature variation period, by definition, since it includes the radial frequency. If the diurnal damping depth of the soil is taken as a unit, then the annual damping depth is larger by $\sqrt{365/1} \approx 19.1$ times than the diurnal damping depth.

Heat Flux Densities

The harmonic heat flux density $G(z, t)$ can now be obtained by using the derivative of equation 6 with respect to z and substituting it into equation 1; the result is given in equation 7. Usually the term

$$G(z, t) = -\lambda \frac{\partial}{\partial z}[T_0(z)] + T_0^* \sqrt{\lambda c\omega} \cdot \exp$$

$$\left\{-z/D\right\} \sin(\omega t - z/D + \psi + \Pi/4) \qquad (7)$$

is small and can be neglected. The sine function argument in equation 7 when compared with that in equation 6 is advanced by $\Pi/4$, which means that the heat flux wave precedes the temperature wave by $\Pi/4$, which is 3 h for a diurnal fluctuation and about 6 weeks for the annual variation (i.e., the time of maximum temperature at a certain depth will occur about 3 h later than that at which the heat flux density will reach its maximum at the same depth). The amount of heat released or stored in the soil or at a given depth is obtained directly by integrating equation 7 with respect to time intervals and the soil layer depths under consideration.

Modification of the Soil Thermal Regime

The thermal regime of a soil can be modified through changing the energy balance at the soil surfaces by one of the following ways: vegetative cover, various mulches, or changing the thermal properties of the soil surface through tillage (q.v.) or irrigation (q.v.).

Vegetative covers change the energy balance by changing the reflectivity, decreasing the amount of global radiation to the soil surface beneath it, and decreasing the heat losses or gains by acting as an insulator. A vegetative cover acts as an insulator above the base soil, which during summer reduces the energy interception at the soil surface and consequently the soil heat flux, as compared with bare soil, thus reducing the soil surface temperature amplitudes and the rate at which the soil warms up. During winter the same cover insulates the soil and decreases heat loss from the cover as compared with bare soil. The vegetative cover also reduces the diurnal variations for a covered soil.

Mulches can affect the thermal regime of the soil by their color, transparency, or thermal conductance. Light colored or transparent mulches will increase the reflection of the global radiation during the periods of low interception angle (i.e., morning or late afternoon hours), while at midday they will transmit the solar energy through them to the soil surface underneath and intercept the outgoing long-wave radiation, thus producing a greenhouse effect of heating up the soil. Mulches having a low thermal conductivity will act as insulators and thus will consequently decrease the heat gain or loss from the soil.

TABLE 3. Thermal Properties of Tilled and Untilled Soils.

Soil Type	Volumetric Water Content (cm^3 H_2O/cm^3 soil)	Thermal Conductivity (λ) (cal^{-1} s^{-1} $°C^{-1}$)	Volumetric Heat Capacity (cal cm^{-3} $°C^{-1}$)	Damping Depth (cm)	Source
Sand	0.0	0.7×10^{-3}	0.3	8.0	Van Wijk and Derksen, 1963
	0.2	4.2×10^{-3}	0.5	15.2	Table 4.2
Clay	0.2	0.7×10^{-3}	0.3	7.4	Van Wijk and Derksen, 1963
	0.4	3.8×10^{-3}	0.7	12.2	Table 4.2
Nontilled sandy soil	0.175	3.4×10^{-3}	0.44	14.6	Van Wijk and Derksen, 1963 Table 6.2
Tilled sandy soil	0.125	1.7×10^{-3}	0.31	12.3	Van Wijk and Derksen, 1963 Table 6.2

Changes in Soil Thermal Properties

Changing the soil thermal properties can be achieved by changing the soil's water content, its bulk density, or both. Changes such as an increase in soil water content increase the soil's heat capacity and its thermal conductance (i.e., after irrigation or after the winter); therefore the diurnal and annual variations are reduced. That is, the rate of the soil warming is reduced during the morning on a daily basis, or during the spring on an annual basis (Table 3). Decreasing the soil water content, as by soil drainage (q.v.) or evaporation (q.v.), will decrease the soil's heat capacity and will raise the soil surface temperature.

Tillage operations usually produce a loose surface layer that acts as a mulch having a lower heat capacity and thermal conductivity as compared to a nontilled soil under equal conditions. Because of its thermal properties, the tilled layer is warmer by day and colder by night, which makes it more vulnerable to radiative frost. The diurnal and annual temperature amplitudes in the tilled layer are larger than those for a nontilled soil. Table 3 gives thermal properties of various soils under different conditions. Tilled soils may be looked upon as being nonhomogeneous soils from the soil structural point of view, and as such it is possible to treat them theoretically as soils made up of two distinct layers as reported in a review by Van Wijk and Derksen (1963). However, soils are normally nonhomogeneous in their texture, structure, and water content distributions, and hence their thermal properties will vary with depth and time. Under these circumstances simple analysis cannot be performed, and theoretical models become very cumbersome and complicated. Such analytical attempts were made by Lettau (1954, 1962) and more recently, using digital computers, by Wierenga and de Wit (1970).

AMOS HADAS

References

Baver, L. D.; Gardner, H. R.; and Gardner, W. R., 1972. *Soil Physics*, 4th ed. New York: Wiley, 498pp.

Houghton, H. G., 1954. On the annual heat balance of the northern hemisphere, *Jour. Meteorol.*, **11**, 1-9.

Lettau, H. H., 1954. Improved models of thermal diffusion in the soil, *Trans. Am. Geophys. Un.*, **35**, 121-132.

Lettau, H. H., 1962. A theoretical model of thermal diffusion in nonhomogeneous conductors, *Gerland. Betr. Geophys.*, **71**, 257-271.

Smith, A., 1932. Seasonal subsoil temperature variations, *Jour. Agric. Res.*, **44**, 421-428.

Wierenga, P. J., and de Wit, C. T., 1970. Simulation of heat transfer in soils, *Soil Sci. Soc. America Proc.*, **32**, 326-328.

Van Wijk, W. R., and Derksen, W. J., 1963. Sinusoidal temperature variation in a layered soil, *in* ed. Van Wijk, *Physics of Plant Environment.* Amsterdam: North Holland, 171-209.

Van Wijk, W. R., and de Vries, D. A., 1963. Periodic temperature variation in a layered soil, *in* ed. W. R. Van Wijk, *Physics of Plant Environment.* Amsterdam: North Holland, 171-209.

Yakuwa, R., 1945. Uber die Bodentemperaturen in dem verschiedenen Bodenarten in Hokkaido, *Geophys Mag. (Tokyo)*, **14**, 1-12.

Cross-references: *Conductivity, Thermal; Energy Balance; Evaporation; Heat Capacity; Spectral Characteristics; Thermodynamics; Soil Water; Thermoluminescence.* *xmxmxmxmxmxmxmxm*

THERMODIFFUSIVE AND THERMO-ELECTRIC PHENOMENA IN SOILS—See **DIFFUSION PROCESSES; TRANSPORT PROCESSES**

THERMODYNAMICS, SOIL MOISTURE

The thermodynamics of soil moisture concerns the equilibrium properties of the aqueous content of a soil matrix and the interrelationships of these properties. The normally stated requirement for equilibrium is that the energy of a system be minimized. The internal energy U is usually expressed as the sum of products of extensive and intensive variables.

$$U = TS - PV + \Sigma\mu_i n_i + \Sigma\sigma_j A_j \qquad (1)$$

where the intensive variables are temperature T, pressure P, chemical potential μ_i, and interfacial tension, σ_j, and the extensive variables are entropy S, volume V, number of moles of ith component n_i, and surface area of the jth surface A_j. In partially saturated soil systems, geometric variables and surface properties largely determine the equilibrium distribution of soil water. The energy associated with boundaries between phases, which can often be safely neglected in thermodynamic treatments of multiphase systems, assumes particular importance with respect to the behavior of soil moisture. Constraints on reversible variations in positions of phase boundaries are such that the energy of the soil system does not represent a global minimum.

Energy State of Soil Water

The total energy of soil moisture Ψ is given by the sum of the internal energy and the potential energy Φ due to position in the gravitational field, with respect to some reference plane, usually taken as that of the free-water surface.

$$\Psi = U + \Phi \qquad (2)$$

The contribution of kinetic energy to the total energy of soil water can usually be neglected. The condition for equilibrium is that the total energy of the system is minimized and that for any possible reversible variation in the total energy of the system

$$d\Psi \geqslant 0 \qquad (3)$$

This condition is satisfied when the total potential is everywhere equal. Equations of condition for thermal, mechanical, and chemical equilibrium can be separated from this general equation (Gibbs, 1961).

For systems in thermal equilibrium, minimization of the Helmholtz free energy A, defined by

$$A = U - TS \qquad (4)$$

is commonly used as a criterion of equilibrium. When the energy of the system also depends on position in a gravitational field, as is often the case for soil moisture, change in total free energy is given by

$$dA + d\Phi = 0 \qquad (5)$$

Edlefsen and Anderson (1943) have given an extensive discussion of equilibrium thermodynamics, in terms of changes in free energy, that pays particular attention to situations of relevance to the behavior of soil moisture.

Many factors must be considered in determining the energy state of soil water and the details and nature of equilibrium conditions for soil water (Childs, 1969; Baver, Gardner, and Gardner, 1972). Soil scientists have adopted the concept of a total water potential that gives the difference in chemical potential of water with respect to some reference state (see *Water Content and Retention*). The three commonly considered components of the water potential are the *pressure potential,* given by pressure differences with respect to the reference pressure; *matrix potential,* due to the presence of the soil matrix; and *solute* or *osmotic potential* that arises from differences in concentration and type of dissolved substances in the soil water. The total potential is given by the sum of the water potential and the gravitational potential (Taylor and Ashcroft, 1972). There can be other contributions to total potential such as electrical and overburden potentials. A number of other potentials or groups of potentials have been defined to suit specific applications, and there is duplication and overlap in the definitions of the various types of soil water potentials. In some respects this inconsistency may have contributed to the confusion that has arisen in the concepts of soil water potentials and their thermodynamic basis (Low, 1951; Iwata, 1973; Corey and Kemper, 1961; Low, 1961).

Soil Water Systems

Soil and its water content present a system that is highly complex, both chemically and physically. The behavior and properties of soil moisture are intimately related to the properties of the soil matrix in which the water is retained. The soil matrix is made up of mineral particles of various shapes, sizes, and chemical composition, and both living and decaying organic materials. Soil particles of colloidal size may occur either as aggregates or as dispersions in water contained within the pore spaces of the soil (see *Soil Solution*).

Water that contacts the soil through rainfall or melting snow is in a relatively pure state.

Within the soil, this water (see *Soil Water*) takes salts into solution. Plants that take up soil moisture and return it to the atmosphere through transpiration often form an integral part of the soil system (see *Evaporation*). Continued changes in soil moisture content and distribution are such that assumption of equilibrium must necessarily be approximate. Furthermore, many of the soil processes are inherently irreversible; idealized control of external constraints does not ensure that changes within the soil from one state to another would be achieved by passing through a series of continuously reversible equilibrium states. Examples of such processes are the relationship between pressure and bulk volume for a swelling soil; dehydration of ferric hydroxide, a process that is of importance in the formation of stable soil aggregates; and hysteresis in relationships between capillary pressure (q.v.) and water content.

In considering the properties of soil moisture, it is convenient to distinguish between water that has properties that are materially affected by the presence of the soil particles and the remaining water that is not directly affected by the solid. Estimates of the thickness of the physically adsorbed layer at the solid surface range upward from about four molecular layers (~10 Å). Longer range effects can arise because of the effects of surface charge on the distribution of ions in solution (Taylor and Ashcroft, 1972). In the vicinity of the solid surface, there is a nonhomogeneous region where the thermodynamic properties of soil water change with distance from the surface.

Swelling and Nonswelling Soils

The behavior of soil moisture is often discussed in terms of two extreme types of soil: nonswelling and swelling. In sandy soils there is essentially direct contact between soil particles, and the bulk volume is fixed. Change in moisture content is achieved by capillary processes involving movement of air and water within the pore space of the soil (see *Pore Space, Drainable*). For the swelling soil (q.v. in Vol. XIII), the particles are finely divided and separated from their neighbors by water that has properties that are influenced by the presence of the solid particles. Changes in moisture content are achieved through change in spacing between the particles with corresponding change in bulk volume of the soil. In the initial stages of drying, air does not penetrate the interparticle pore space, although the soil may become fissured. The behavior of most soils falls somewhere between the two extremes of swelling and nonswelling soils.

Clay soils generally exhibit behavior that is close to that of swelling soils. The bulk volume of the soil can be varied through application of overburden pressure or through suction applied to the interstitial water via a semipermeable membrane. The distance of separation between clay platelets varies with hydrostatic pressure and a considerable volume of water may be removed before air penetrates the interstices of the soil. Observed hysteresis in relationships between swelling and shrinkage has been ascribed to reorientation of the soil particles (Childs, 1969). The changes in mechanical properties with moisture content of these soils is of special concern in soils engineering (Spangler, 1963).

In contrast to swelling soils, the pressure relative to that of the air phase at which water is held within a sandy soil is directly related to the soil's capillary properties. This pressure is referred to as the *matric* or *capillary potential.* Simple soil models are often used to investigate particular aspects of moisture rentention in sandy soils. Slichter (1899) introduced the term *ideal soil* for regular packings of equal spheres. The term was later extended to random packings of equal spheres, which are often used as a laboratory model of sandy soils. Ideal soil models have been used extensively in both theoretical and experimental studies of the thermodynamic properties of soil water and the more general problem of the behavior of fluid interfaces within a porous media (Green and Ampt, 1912; Hackett and Stretton, 1928; Haines, 1930; Smith, 1933; Mason, 1971). The pendular ring of fluid held by capillary forces between two equal spheres is an important feature of the ideal soil model because this capillary structure lends itself to precise thermodynamic analysis (Melrose, 1966).

Capillary Equilibrium

For equilibrium of capillary-held water, a number of conditions must be satisfied. Thermal equilibrium requires that the temperature be constant throughout the system. The condition for chemical equilibrium at a given height for all phases into which soil water can freely pass is that its chemical potential be equal

$$\mu_l = \mu_v = \mu_a \tag{6}$$

where subscripts *l*, *v*, and *a* refer to liquid and vapor phases, and *a* to adsorbed water that differs in its properties from bulk water because of its proximity to an interface. Thus although the composition of soil water may be quite complex, its chemical potential can be determined from vapor pressure.

The chemical potential of the adsorbed phase as defined by equation 6 has also been referred to as the total potential of the constituent (Low, 1961). In this approach contributions to free energy that depend on the intrinsic nature and state of water are distinguished from those due to position in the force field in the vicinity of the solid surface. However, these two contributions to the state of soil water cannot be distinguished experimentally, and it is more usual to treat the whole of the energy change as change in the internal energy of the system.

When considering equilibrium with respect to position in a gravitational field, the contribution of potential energy to the total energy of the system is clearly distinguishable from the internal energy. The condition for chemical equilibrium is modified:

$$\mu + Mgh = \text{constant} \tag{7}$$

where M is the molecular weight of water, h is some height above a reference plane, and g is the acceleration due to gravity (Gibbs, 1961). Hydrostatic equilibrium implies that the fluid pressure P, with respect to some reference level h, is given by the hydrostatic head:

$$P = \Delta\rho gh \tag{8}$$

The pressure differences across a curved interface, often described as the *capillary pressure* or *capillary potential* P_c, is given by Laplace's (1806) equation:

$$P_c = \sigma_{wa} C \tag{9}$$

where C is the surface curvature and σ_{wa} is the air-water interfacial tension. A further requirement for equilibrium is

$$P_c = \Delta\rho gh' \tag{10}$$

h' being the height of the point in the air-water surface at which the surface curvature is C. For equilibrium at a point on a soil-water-air, three-phase line of contact, at the surface of the soil particles where they are smoothly continuous, *Young's equation* (Young, 1805) must be satisfied

$$\sigma_{sa} - \sigma_{sw} = \sigma_{wa} \cos\theta \tag{11}$$

where σ_{sa} and σ_{sw} are the solid-air and solid-water interfacial tensions, respectively, and θ is the angle of contact. Most soils appear to give spreading of water against air, which corresponds to a contact angle of zero. When θ is finite, conditions under which θ can be observed to be a single valued property independent of the direction of movement of a three-phase line of contact are highly restrictive. Most soils are believed to be hydrophilic (see *Hydrophility, Hydrophobicity*), but soils that are not strongly wetted by water against air are known to exist (DeBano and Letey, 1969). Such soils will probably have nonuniform wetting properties because of surface heterogeneity.

A further local condition of eqilibrium for capillary systems is given by the *Gauss equation* (Gauss, 1830), which relates the change in interfacial areas of the system to the displaced volume V_D, given by movement of the air-water interface

$$d(A_{wa} - A_{ws} \cos\theta) = C\,dV_d \tag{12}$$

where A_{wa} and A_{ws} are the water-air and water-solid interfacial areas, respectively.

Movement of air-water interfaces within the interstices of porous media inevitably results in the formation of unstable interfaces that undergo spontaneous movement to new positions of stability (Haines, 1930). Such transitions are known as *Haines' jumps.* In general, the constraints that pore geometry places on stable interface configurations prevent the soil system from spontaneously reaching a global minimum free energy. These constraints also lead to time invariant hysteresis, which is observed in relationships between capillary pressure (q.v.) and soil moisture content (Miller and Miller, 1956). The state of the soil-water system generally depends on its history. Uniformity of pressure is a necessary condition of equilibrium, but uniformity of moisture content is not. Capillary pressure relationships cannot be treated as being representative of quasistatic processes, analogous to pressure-volume relationships for a gas. A basis for using these relationships to determine limits of change in free energy with moisture content has, however, been proposed (Morrow, 1970).

Lowering of vapor pressure ($p_0 - p$) at a curved surface, for vapor of molar volume V, is given by the *Kelvin equation*

$$\frac{p}{p_0} = \exp - \frac{P_c V}{RT} \tag{13}$$

More precise formulation of this equation has been given, which takes into account the increase in volume of capillary-held liquid with reduction in pressure and conditions for the stability of water in liquid form when under the high tensile stresses that develop in systems of very high interfacial curvature (Melrose, 1972).

Attainment of Equilibrium

Changes in moisture distribution will occur within the soil through both liquid flow and diffusion of soil moisture. Spontaneous changes in interface configuration can arise through either process and have been described as *Laplace unstable* or *Kelvin unstable* according to the mechanism by which the instability arises (Everett and Haynes, 1972; Everett, 1975).

In accounting for the behavior of soil moisture, it is important to distinguish between those conditions for equilibrium that are likely to be met rapidly or within a reasonable time period and those that will be met only very slowly, if at all. The equilibrium condition expressed by equation 8 will be achieved almost instantaneously by local adjustment of the curvature of an air-water interface. For spreading (hydrophilic) systems ($\theta = 0^\circ$), equilibrium at the three-phase line of contact is reached rapidly whereas hydrophobic systems ($\theta > 0^\circ$), which are not readily wetted, may take considerable time to reach a condition that is observably stable. Balance of gravity and capillary forces (equation 9) achieved by hydraulic flow will be much slower than the time required to achieve local equilibrium expressed by equation 8 but will usually be rapid compared to diffusional processes when they are governed by small differences in vapor pressure as given by equation 13. As an example, drainage of a column of soil results in retention of a significant fraction of water that essentially loses continuity with respect to hydraulic flow (see *Flow Theory*). Studies of drainage in sphere packings have shown that the discontinuous water does not achieve equilibrium within a reasonable time period (Morrow, 1970). Similar loss of hydraulic continuity during drainage of soils results in a characteristic fraction of water, known as the *field capacity*, being left in the soil. Retention of water is highly important to plant growth. For such systems it can be assumed that equation 8 is satisfied locally whereas equation 9 is not likely to be satisfied if the soil water has lost hydraulic continuity with the free-water surface. One approach to thermodynamic treatment of such behavior is to assume that while local equilibrium (equation 9) holds, changes due to diffusion can be neglected (Morrow and Harris, 1965). An indication of the circumstances under which equation 10 is likely to become satisfied through diffusion may be obtained by comparing the lowering of vapor pressure with curvature given by equation 13, with changes in vapor pressure with temperature. Small variations in temperature will have a dominant effect on transport through the vapor phase, and temperature differences between the roots and leaves of a plant may have a dominant effect on the potential gradient within the plant. Variations in salinity may also play an important role in determining the direction of moisture transport.

Soil Water in Plants

The soil water potential concept has been extended to moisture within plant systems in which water uptake is related to the difference in potential of water available to the root system, and the turgor pressure of water contained within the cells of plant leaves through which soil moisture is returned to the atmosphere by transpiration (Taylor and Ashcroft, 1972).

The matrix potential provides a useful measure of the availability of soil moisture to plants. In this respect it has been shown that soil water potential can be considerably more significant to plant growth than the actual amount of moisture the soil contains. Most mesophytic plants exhibit wilting when the matric potential is about −1500J/kg. This potential corresponds to a relative humidity of about 99% and interface curvature significantly less than $10^{-4}\,cm^{-1}$. Further reduction in moisture content involves removal of both capillary-condensed and surface-adsorbed water and is usually marked by significant lowering of vapor pressure. In this range of moisture content, measurement of vapor pressure, a method common to study of adsorption behavior, can be used to determine relationships between soil water potential and water content (see *Water Content and Retention*).

Freezing of Soil Moisture

Equilibrium thermodynamics has been applied to the problem of freezing of soil moisture (Everett, 1961; Koopmans and Miller, 1966). The development of a two-phase system within a confined soil gives rise to capillary forces analogous to those given by invasion of air. Changes in capillary pressure across the ice-water interface are directly related to lowering of freezing point with increased ice content. For soils in which ice grows within the pore structure of the soil, thermodynamic considerations show that ice will grow preferentially in the larger pore spaces. If the confining pressure exerted by the pore walls is not sufficient to contain the growing ice crystal, frost heaving will occur (see Vol. XIII: *Soil Water Problems, Cold Regions*). Growth of ice lenses can be prevented by application of sufficient overburden pressure. In swelling soils, the mechanism of freezing and heaving is related to the formation of ice lenses accompanied by decrease in distance of separation between soil particles.

Steady-state Processes

Although equilibrium thermodynamics can be used to predict the direction of change in moisture content, the methods of irreversible thermodynamics are needed to account for fluxes of energy and matter (de Groot and Mazur, 1962). This more recently developed branch of thermodynamics deals with entropy production of steady-state transport processes. Generally, each gradient is related by a phenomenological coefficient to each flux such that fluxes and forces are coupled. In applications to the transport of soil moisture, the commonly considered gradients are temperature, concentration, and hydraulic potential (Taylor and Ashcroft, 1972; Groenevelt and Bolt, 1969).

NORMAN R. MORROW

References

Baver, L. D.; Gardner, Walter H.; and Gardner, Wilford R. 1972. *Soil Physics,* 4th ed. New York: Wiley, 498p.

Childs, E. C., 1969. *An Introduction to the Physical Basis of Soil Water Phenomena.* London: Wiley, 493p.

Corey, A. T., and Kemper, W. D., 1961. Concept of total potential in water and its limitations, *Soil Sci.,* **91,** 299.

DeBano, L. F., and Letey, J. L., eds., 1969. *Proc. Symposium on Water-repellant Soils,* University of California (March), 354p.

Denbigh, Kenneth, 1966. *The Principles of Chemical Equilibrium.* Cambridge: Cambridge University Press, 494p.

Edlefsen, N. E., and Anderson, Alfred B. C., 1943. Thermodynamics of soil moisture, *Hilgardia,* **15,** 31.

Everett, D. H., 1961. The thermodynamics of frost damage to porous solids, *Trans. Faraday Soc.,* **57,** 1541.

Everett, D. H., and Haynes, J. M., 1972. Model studies of capillary condensation, *Jour. Colloid Int. Sci,* **38,** 125.

Everett, D. H., 1975. Thermodynamics of multiphase fluids in porous media, *Jour. Colloid Int. Sci.,* **52,** 189.

Gauss, C. F., 1830. In comment, *Soc. Reg. Scient. (Gottingen),* recent 7.

Gibbs, J. Willard, 1961. *The Scientific Papers of J. Willard Gibbs, Vol. 1, Thermodynamics.* New York: Dover, 434p.

Green, H., and Ampt, G. A., 1912. The permeability of an ideal soil to air and water, *Jour. Agric. Sci.,* **5,** 1.

Groenevelt, P. H., and Bolt, G. H., 1969. Non-equilibrium thermodynamics of the soil-water system, *Jour. Hydrology,* **7,** 358.

Groot, S. R. de, and Mazur, P., 1962. *Non-equilibrium Thermodynamics.* Amsterdam: North-Holland, 510p.

Hackett, F. E., and Stretton, J. S. 1928. The capillary pull of an ideal soil, *Jour. Agric. Sci.,* **18,** 671.

Haines, W. B. 1930. Studies in the physical properties of soil, Part V., *Jour. Agric. Sci.,* **20,** 97.

Iwata, S., 1973. On the energy concept of soil water, *in* E. Hadas, Schwartzendruber, D.; Rijtema, P. E., Fuchs, M., and Yaron, B., eds., *Physical Aspects of Soil Water and Salts in Ecosystems.* Berlin: Springer-Verlag, 460p.

Koopmans, R. W., and Miller, R. D., 1966. Soil freezing and soil water characteristic curves, *Soil Sci. Soc. America Proc.,* **30,** 680.

Laplace, P. S., 1806. Supplement "On Capillary Attraction," *Mécanique celéste,* **10.**

Low, P. F., 1951. Force fields and chemical equilibrium in heterogeneous systems with special reference to soils, *Soil Sci.,* **71,** 409.

——, 1961. Concept of total potential in water and its limitations: A critique, *Soil Sci,* **91,** 303.

Mason, G., 1971. A model of the pore space in a random packing of equal spheres, *Jour. Colloid Int. Sci.,* **35,** 279.

Melrose, J. C., 1966. Model calculations for capillary condensation, *AIChe Jour.,* **12,** 986.

——, 1972. Chemical potential changes in capillary condensation, *Jour. Colloid Int. Sci.,* **38,** 312.

Miller, E. E., and Miller, R. D., 1956. Physical theory for capillary flow phenomena, *Jour Appl. Phys.,* **27,** 324.

Morrow, N. R., and Harris, C. C., 1965. Capillary equilibrium in porous materials, *Soc. Petroleum Engineers Jour.,* **5,** 15.

——, 1970. Physics and thermodynamics of capillary action in porous media, *Ind. Eng. Chem.,* **62,** 32.

——, 1970. Irreducible wetting-phase saturations in porous media, *Chem. Eng. Sci.,* **25,** 1799.

Slichter, C. S., 1899. *U.S. Geol. Survey, 19th Annual Report, Part 2,* 295.

Smith, W. O., 1933. Minimum capillary rise in an ideal uniform soil, *Physics,* **4,** 184.

Spangler, M. G., 1963. *Soil Engineering.* Scranton, Pa.: International Textbook, 483p.

Taylor, Sterling A., and Ashcroft, Gaylen L., 1972. Energy relations of soil water, *Physical Edaphology.* San Francisco: Freeman, 533p.

Thomson, W., 1871. On the equilibrium of vapor at curved surface of liquid, *Phil. Mag.,* **42,** 448.

Young, T., 1805. *Phil. Trans. Roy. Soc. (London),* **95,** 65.

Cross-references: *Capillary Pressure; Flow Theory; Soil Water; Surface-Soil Water Content; Water Content and Retention; Water Fluxes. Vol. IVA: Free Energy and Free Enthalpy. Vol. XIII: Soil Water Problems, Cold Regions; Swelling Soils.*

THERMOLUMINESCENCE

To exhibit thermoluminescence, the emission of light produced by heating a solid to a temperature below that of incandescence, a substance must have an ordered structure such as is found in crystals or a semiordered structure as in glasses. In addition, it must be electrically an insulator or a semiconductor. Briefly, the heating of the sample provides the activation energy to release the charge carriers trapped in metastable energy states. The released charge carriers may be retrapped or drop to the ground state through an electron-hole recombination process. When the latter occurs, the energy in excess of that of the ground state is released as heat and/or light. The nature of thermoluminescence has been discussed in some detail by several authors (e.g., Fonda and Seitz, 1948; Garlick 1949; Leverenz, 1950; Schulman and Compton, 1962; Curie, 1963; Attix, 1967, McDougall, 1968).

Soils show a variety of thermoluminescent characteristics (Nishita and Hamilton, 1968, 1973, 1974). In general, the soils that are high in lime content show high thermoluminescence, while those very high in free-iron oxides or organic matter show only little or no thermoluminescence. Examples of these are shown in Figure 1, which gives the thermoluminescent glow curves (plots of glow intensity as a function of sample temperature) for several different soil types. Of the soils represented, Egbert muck was high in organic matter content (32.9%). Soil 4FF was highly calcareous (33% lime). Aiken clay loam, which showed no natural thermoluminescence, was relatively high in free-iron oxide content (15.4%). The other soils were noncalcareous, nonsaline soils with 1.5–2.0% organic matter and less than 2.0% free-iron oxides. The area under the glow curve for each soil represents its total light output. This output depends on the amounts and the kinds of different crystalline substances composing the soil because soil thermoluminescence is a composite light output of all substances in it. The shapes of the glow curves are determined by the distribution of *trap depths* (thermal activation energies required to liberate the charge carriers trapped at different metastable states) within the crystalline substances. The shallower the trap (lower the metastable state), the lower the temperature at which light is emitted.

Most soils show only little, if any, natural thermoluminescence below the threshold temperature of around 200°C (Nishita, Hamilton, and Haug, 1974). Natural thermoluminescence results from the trapping of charge carriers at lattice imperfections in the crystalline substances in the soil as well as in other geologic materials such as minerals, clays, and rocks. The crystal lattice imperfections were formed during the crystallization process and by the continuous low-level radiation by the penetrating cosmic radiation from space and the radioactive impurity elements (particularly K^{40} and the uranium and thorium decay series) present (Johnson and Blanchard, 1967; Aitken, Zimmerman, and Fleming, 1968; Medlin, 1968; Bothner and Johnson 1969; Fleming, 1971). Natural thermoluminescence also depends on the temperature regime to which the soils were exposed over geologic time.

The exposure of soils to high energy ionizing radiation (Figure 2) increases their thermoluminescence (Nishita and Hamilton, 1968, 1972). The exposure of soils to ionizing radiation de-

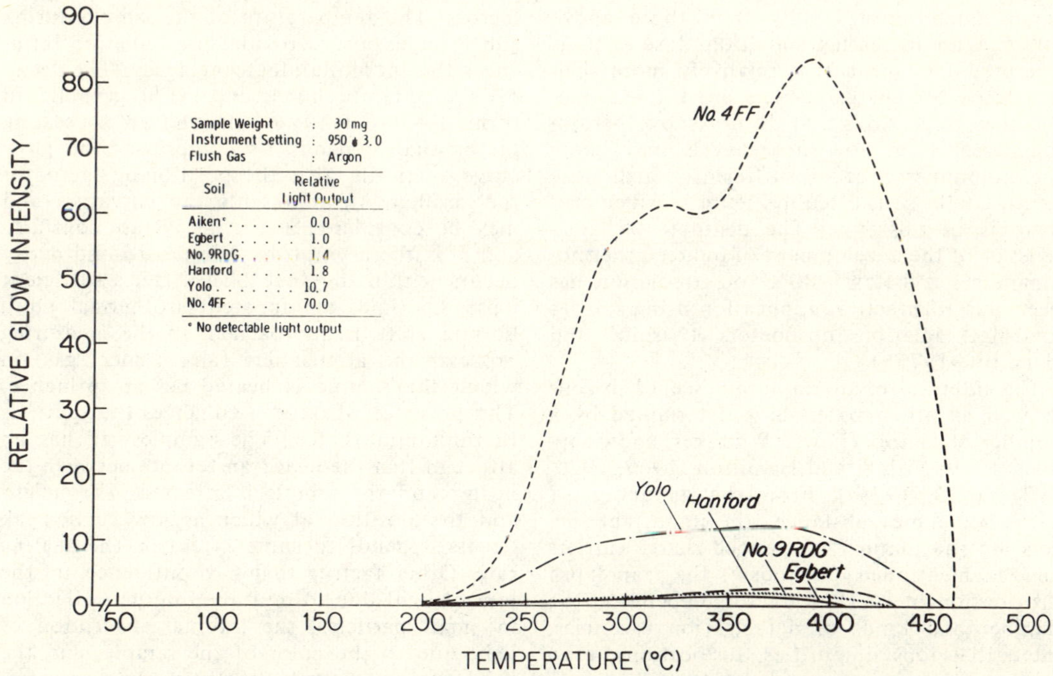

FIGURE 1. Natural thermoluminescence of different types of soils.

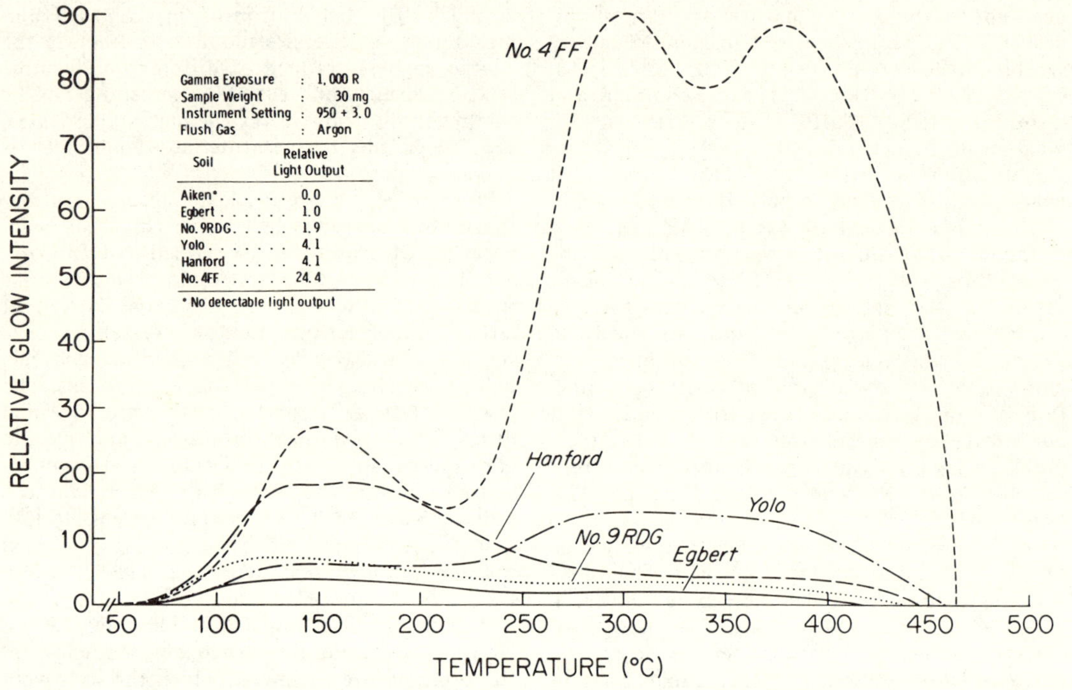

FIGURE 2. Thermoluminescence of different types of soils measured 30 minutes after their exposure to 1000 roentgens of Co^{60} photons.

velops glow curve peaks below 200°C and increases the peak heights above 200°C (compare Figures 1 and 2). The peaks below 200°C develop much more rapidly than those above 200°C with increasing radiation dose. This is interpreted to mean that relatively more shallow traps for charge carriers are formed than the deep traps. Soils that do not show thermoluminescence at low dose levels may show thermoluminescence at sufficiently high dose levels. Of the soils listed in Figure 1, Aiken clay loam is of this type. The distinctive characteristics of the development of induced thermoluminescence below 200°C on irradiation has been put to practical application using soils as ionizing radiation dosimeters (Nishita and Hamilton, 1975*b*).

The intensity of thermoluminescence in soils as well as other materials is determined by a number of factors (Lewis, Whitaker, and Chapman, 1959; Nishita and Hamilton, 1969, 1970, 1971, 1973, 1975*a*). From the standpoint of each crystalline substance, the important factors are the number of trapped-charge carriers and as mentioned previously, the trap-depth distribution in it. These factors depend on the number and kinds of defects (ion vacancies, interstitial ions, impurities, dislocations, etc.) in the lattice structure of the crystalline substance. From the radiation source aspect, the dose level and the kind and energy of the radiation source (X-rays, γ-rays, β-particles, α-particles, neutrons, etc.) are the determining factors. The temperature of the sample during and after exposure to ionizing radiation influences the thermoluminescent decay. The decay of soil thermoluminescence is hyperbolic in form, $Y = \alpha(\beta + X)\gamma + \delta$, in which α is a scaling factor that expands or compresses the glow curve vertically, β is the positioning factor of the ordinate, γ represents the curvature and may be considered as the decay rate constant, and δ is the asymptote. The most rapid decay occurs within the first 24 h. After a sufficient lapse of time, an apparent isothermal equilibrium state is approached. In the measuring process, the atmosphere (air or inert gas) in which the sample is heated has an influence. The presence of oxygen enhances the spurious thermoluminescence. The sample size has an affect in that the heat transfer rate between the heater and the sample is affected. The height and the position at which a glow curve peak occurs depends to some extent on the heating rate. Other factors that have influence are the loss of light due to multiple internal reflection in small particles, the internal absorption of light due to the color of the sample, and the solid angle of light collection in the apparatus.

H. NISHITA

References

Aitken, M. J.; Zimmerman, D. W.; and Fleming, S. J., 1968. Thermoluminescent dating of ancient pottery, *Nature,* **219**, 442–445.

Attix, F. H., ed., 1967. Luminescence dosimetry, *Proc. Internat. Conf. on Luminescence Dosimetry.* Stanford University, Stanford, California, June 21–23, 1965, CONF-650637, 514p.

Bothner, M. H., and Johnson, N. M., 1969. Natural thermoluminescent dosimetry in late Pleistocene pelagic sediments, *Jour. Geophys. Res.,* **74**, 5331–5338.

Curie, D., 1963. *Luminescence in Crystals.* London: Methuen, 332p.

Fleming, S. J., 1971. Thermoluminescent dating principles and application, *Naturwissenschaften,* **58**, 333–338.

Fonda, G. R., and Seitz, F., eds., 1948. *Solid Luminescent Materials.* New York: Wiley, 459p.

Garlick, G. F. J., 1949. *Luminescent Materials.* Oxford: Clarendon Press, 254p.

Johnson, N. M., and Blanchard, R. L., 1967. Radiation dosimetry from the natural thermoluminescence of fossil shells, *Am. Mineralogist,* **52**, 1297–1310.

Leverenz, H. W., 1950. *An Introduction to the Luminescence of Solids.* New York: Wiley, 569p.

Lewis, D. R.; Whitaker, T. N.; and Chapman, C. W., 1959. Thermoluminescence of rocks and minerals, *Am. Mineralogist,* **44**, 1121–1140.

McDougall, D. J., ed., 1968. *Thermoluminescence of Geological Materials.* New York: Academic, 678p.

Medlin, W. L., 1968. Natural thermoluminescence of Cretaceous limestones in terms of alpha activity, *Geochim. et Cosmochim. Acta,* **32**, 1165–1176.

Nishita, H., and Hamilton, M., 1968. Some thermoluminescent characteristics of gamma irradiated soils, *Soil Sci.,* **106**, 76–84.

Nishita, H., and Hamilton, M., 1969. Heating effects on the thermoluminescence of gamma irradiated soils, *Soil Sci.,* **108**, 1–10.

Nishita, H., and Hamilton, M., 1970. Spurious thermoluminescence of soils, *Soil Sci.,* **110**, 371–378.

Nishita, H., and Hamilton, M., 1971. Influence of temperature during the gamma irradiation period on the subsequent thermoluminescence of soils and LiF and CaF_2 dosimeters, *Soil Sci.,* **111**, 393–398.

Nishita, H., and Hamilton, M., 1972. Soil thermoluminescence in relation to radiation exposure under field conditions, *Soil Sci.,* **114**, 430–439.

Nishita, H., and Hamilton, M., 1973. Effects of some chemical treatments of soils on their thermoluminescence, *Soil Sci.,* **115**, 30–39.

Nishita, H.; Hamilton, M.; and Haug, R. M., 1974. Natural thermoluminescence of soils, minerals, and certain rocks, *Soil Sci.,* **117**, 211–219.

Nishita, H.; Moore, R. H.; Beckman, R. J.; and Hamilton, M., 1975*a*. Decay characteristics of soil thermoluminescence, *Soil Sci.,* **119**, 259–268.

Nishita, H., and Hamilton, M., 1975*b*. Empirical methods of using soils as radiation dosimeters, *Soil Sci.,* **120**, 96–106.

Schulman, J. H., and Compton, W. D., 1962. *Color Centers in Solids.* New York: Macmillan, 368p.

Cross-references: *Spectral Characteristics; Thermal Regimes. Vol. IVB: Thermoluminescence. Vol. XIII: Remote Sensing, Engineering Geology.*

THIXOTROPY, THIXOTROPISM

Thixotropy is an isothermal, reversible, time-dependent process whereby a material stiffens under conditions of constant composition and volume while at rest and softens or liquifies under transitory stress (Freundlich, 1935; Mitchell, 1961; Gaucher, 1968). This process is characteristic of gels that liquify under vibration and then solidify again when left standing. It is displayed strongly by clays such as bentonites when holding large amounts of exchangeable sodium, a hydrophyllic ion (Russell, 1973).

Strength Properties

Thixotropic behavior is exhibited by some soils high in crystalline clay minerals and by others in which the clay fraction is amorphous. The clay minerals, responsible for the plasticity and cohesion of soils, occur as very small, thin, plate-like particles (less than 2 μm in size). Among the three major groups of crystalline clay minerals, montmorillonite exhibits a high strength gain due to its thixotropic properties. Kaolinite, on the other hand, gains little or no strength from thixotropy, whereas illite is intermediate between the other two. Kaolinite can be made more thixotropic by the addition of a dispersing agent (see *Soil Conditioners*) to prevent the flocculation (q.v.) that commonly prevails in natural specimens. Amorphous clay fractions can be similar in chemical composition to crystalline clay minerals, although some are essentially gels of hydrous iron and aluminum oxides. Soils that are high in amorphous clays also exhibit thixotropy to a high degree.

Although dilute suspensions of colloidal particles may exhibit thixotropism, the agronomist is concerned primarily with the age hardening that occurs in concentrated suspensions such as very soft clays between the plastic limit and the liquid limit (see *Soil Mechanics*). Some very soft clays stiffen and gain strength when allowed to remain undisturbed after being remolded while in a concentrated suspension. Other clays such as montmorillonites, however, gain considerable strength with time after being compacted at water contents somewhat less than 100%. This age hardening after remolding is quite different from the gain in strength produced by changing the volume of the clay by driving off a part of the pore water during consolidation (Lee, 1968). The gain in strength due to thixotropy occurs without removing any of the pore water and without volume change.

Thixotropic Soils

Thixotropic clays are soils that tend to liquify when disturbed. The sensitivity of a clay, a measure of its tendency toward liquifaction, is determined by the relation

$$\text{Sensitivity} = \frac{\text{strength of undisturbed material}}{\text{strength of remolded material}}$$

where the sensitivity is the ratio of undisturbed to remolded strength (Peck, Hanson, and Thornburn, 1974). Sensitivity values for most soils range between 1 and 8. Natural soils having values greater than 4 are known as *sensitive clays*. *Extrasensitive clays* have values greater than 8. If the sensitivity is greater than 16, such clays are described as *quick clays* (Skempton and Northey, 1952). Most clay soils are naturally sensitive to some extent because their strength decreases when they are remolded. Extrasensitive or thixotropic clays become quick when there is a loss of strength (see Vol. XIII: *Quick Sand-Quick Clay*). These clays have a rather open structure with a random orientation (see *Soil Structure*), but because the clay minerals are plate-shaped rather than rounded, the volume of voids is usually much greater than the volume of solid material. When the structure is broken by disturbance, the particles are set free in their surrounding pore water and the mass becomes a viscous liquid because the water cannot escape from the mixture (Fahn, Weiss, and Hofman, 1953; Seed and Chan, 1957). Once the structure is destroyed by shock or distortion, it cannot be reformed artificially, although the strength of the remolded material may be increased by drying or by chemical changes (Ingles and Metcalf, 1973) (see *Soil Stabilization*).

Hazards

Thixotropic soils can be very troublesome in agricultural operations or in construction of roads and other works. They also create natural hazards on hillslopes because they are subject to mass movements in the form of earth flows (Sharpe, 1960). Substantial downslope movement over time is a normal feature of landscapes such as the Appalachian region of Pennsylvania and West Virginia. Earth movements may be rapid in the form of slides as well as the slower flows. The bearing capacities of thixotropic soils is low and deteriorates rapidly under traffic. Thus, for example, the soils will support a tractor on the first pass, may on the second, but not on the third. Because they can liquify when subject to vibration, thixotropic soils pose major problems in construction of highways and other works. They had best be avoided or removed, if either is feasible. Because they do pose serious problems, identification of bodies of thixotropic soils in early stages of planning construction projects or land-use programs is necessary.

CHARLES W. FINKL, JNR.

References

Fahn, R.; Weiss, A.; and Hofman, H., 1953. Thixotropy in clays, *Berichte der Deutsche Keramischen Gessellschaft*, **30**, 21.

Freundlich, H., 1935. *Thixotropy*. Paris: Hermann et Cie, 50p.

Gaucher, G., 1968. *Traité de Pedologie: Agricole Le Sol*. Paris: Dunod, 578p.

Ingles, O. G., and Metcalf, J. B., 1973. *Soil Stabilization*. New York: Wiley, 374p.

Kerr, P. F., 1963. Quick clay, *Sci. Am.*, **209**(5), 132–142.

Knight, K., 1963. The origin and occurrence of collapsing soils, *Proc. 3rd Reg. Conf. for Afr. on Soil Mech. and Found. Eng.*, **1**, 127–130.

Lee, I. K., 1968. *Soil Mechanics*. Sydney: Butterworths, 625p.

Mitchell, J. K., 1961. Fundamental aspects of thixotropy in soils, *Trans. Am. Soc. Civil Engineers*, **126**, 1586–1620.

Peck, R. B.; Hanson, W. E.; and Thornburn, T. H., eds., 1974. *Foundation Engineering*. New York: Wiley, 514p.

Rosenqvist, Th., 1959. Physico-chemical properties of soils: Soil-water systems, *Jour. Soil Mech. Foundation Div., ASCE*, 85(SM2), 31–53.

Russell, E. J., 1973. *Soil Conditions and Plant Growth*. London: Longmans, 849p.

Seed, H. B., 1968. Landslides during earthquakes due to soil liquefaction, *Proc. Am. Soc. Civil Engineers*, 94(SM5), Proc. Paper 6110, 1053–1122.

Seed, H. B., and Chan, C. K., 1957. Thixotropic characteristics of compacted clays, *Proc. Am. Soc. Civil Engineers*, 83(SM4), Proc. Paper 1427, 1–34.

Sharpe, C. F. S., 1960. *Landslides and Related Phenomena*. Paterson, N.J.: Pageant Books, 137p.

Skempton, A. W., and Northey, R. D., 1952. The sensitivity of clays, *Geotechnique*, **3**(1), 30–52.

Cross-references: *Aggregate Stability; Flocculation; Moisture Management; Soil Conditioners; Soil Mechanics; Soil Stabilization; Soil Structure. Vol. XIII: Quick Sand–Quick Clay.*

TILLAGE

In modern agriculture, *tillage* is the means of implementing the soil-handling management practices that are necessary for the production of food, timber, fiber, and oil crops. When "tillage" is defined as "the mechanical manipulation of soil," the theory and principles include all soil-handling operations in agriculture, earthmoving, mining, and construction.

Agricultural management practices vary widely, depending on the soil, climatic regions, and crops (Donahue, Shickluna, and Robertson, 1971). Areas of low rainfall are managed for the maximum use of natural precipitation: Drought-tolerant crops are used, plant residues are retained on the surface of the soil to prevent evaporation (q.v.) and to control wind erosion (q.v.), and areas may be fallowed in some years to accumulate sufficient moisture to produce a crop. Tillage tools such as chisels, rod weeders, and sweeps loosen the soil without inverting it and causing the loss of soil moisture. Windy areas require protection that is provided by maintaining a rough soil surface and using windbreaks or surface mulches to reduce wind erosion. Chisels, rod weeders, and sweeps are used to control wind erosion. Hilly areas subject to high intensity rainfall are contoured, mulched with plant residue, or double-cropped to prevent soil loss by erosion. The use of cover crops in the winter and the interplanting of crops during the summer ensure continuous soil cover. Arid regions must be irrigated for successful agriculture. Tillage is integrated closely with irrigation water management to produce the desired soil conditions for crop production and to reduce energy requirements.

Tillage is one of the most expensive operations of crop production; new emphasis has been placed on reducing the energy required for tillage. Energy has been reduced by applying principles of so-called minimum tillage systems (see *Zero-tillage*). Operations are combined so that fewer trips are made through the fields, and some operations (especially plowing) are eliminated. The concept is developed to the extent that "no-tillage" is possible in some instances. As described by Phillips and Young (1973), no-tillage involves planting the crop with little or no soil preparation. Seeding from the air is practiced for some situations. Weed control is accomplished by chemicals so there is no need for mechanical cultivations for weed control. It has been found that soil may naturally have suitable tilth without excessive tillage and that crops can be produced economically without intensively tilling the soil.

Tillage Machines

Plows used for primary tillage break out, invert, and crumble furrows of soil (Kepner, Bainer, and Barger, 1972). Disk harrows, chisels, drags, and other implements further refine the coarse soil conditions during secondary operations. Rotary hoes, spring-tooth harrows, spike-tooth harrows, and other weeding cultivators are used to control small weeds and disrupt soil crusts. Cultivator sweeps and blades are used to cut the roots and destroy larger weeds. A series of special-purpose tools, including hillers, ridgers, subsoilers, and openers, perform specific soil-handling functions. Tillage tools are the machine components that are used to apply forces to the soil to cause desired effects such as pulverization, cutting, inversion, or movement of the soil. Tillage tools usually produce several effects simultaneously. The ultimate aim of tillage is to manipulate a soil from a known condition into a different, desired condition by mechanical means. Some form of these tools has been used since antiquity.

The physical properties of the soil determine to a large extent the nature of soil reaction to a tillage tool. Water, the most important of these properties, can cause the soil to vary from a hard condition to a plastic mass. Sandy soils are easily manipulated, whereas dry clays may be extremely difficult to manipulate—possessing mechanical strengths in the range of building materials. The strength of machines, the rate of wear, and the power requirements are direct functions of the soil conditions during operation. The design of machines is based on the knowledge of the soil-machine system and a knowledge of the desired soil conditions. The improper design and use of machines can cause harmful soil compaction, which requires correction by tillage (American Society of Agricultural Engineers, 1971). Normally conditions of interest are those associated with plant growth. According to Gill and Vanden Berg (1967), some of the common reasons for tilling soil are:

1. Conditioning of Soil: Tillage is performed to loosen, granulate, or otherwise condition soils. Often this is done to promote plant growth. Soil may be cut, loosened, or broken up so that it may be moved or so that water and air may move more freely through it. Soil may be loosened to promote drying or to reduce the mechanical strength of the soil mass.

2. Eradication or Control of Plants or Plant Materials: Tillage is frequently performed to control weeds during the production of crops. Sometimes weeds are controlled by chemical means. Other times, the complete removal of plant materials—as in the creation of a firebreak—may be required. Plant thinners that operate in the soil control plant growth. Not all mulching operations involve soil tillage; some operations merely process plant residue.

3. Establishment of Soil Boundaries and Surface Configurations: In planting (listing and bedding), irrigation, drainage, and landscaping, the soil often is tilled to create some special soil configuration on or in the soil profile. These tillage operations would generally fall in the class of land forming, but this would not preclude the inclusion of a ditch cleaner since its action would be to restore a desirable boundary by moving the soil. Transient boundaries may be used to influence soil-handling operations. For example, a coulter or jointer may be used to establish a clean or straight boundary to promote good plow action. The boundary is transient since it may be removed within a few seconds after it is established.

4. Incorporation, Covering, or Handling of Foreign Materials in Soil: Fertilizers, plant residues, mulch papers, soil amendments, flexible pipes, communication wires, military mines, and many other materials may be inserted into the soil by tillage. Often these materials are distributed in specific portions of the soil profile, so the designs of tillage tools and the principles of their use may vary considerably. The inversion of soil by tillage to cover materials such as crop residues or radioactive dust may serve no other useful purpose, although granulation and other desirable effects result from the operation as a byproduct of the primary action.

5. Segregation: Tillage may be used to move soil or other materials from one layer of soil to another such as when deep tillage is used to create clods and to move these clods to the surface to minimize wind erosion. Segregation may also involve the removal of materials from the soil. Rock picking and root harvesting are included in this category.

6. Mixing: When materials are incorporated into the soil, some type of mixing is frequently needed. This may be done to promote drying of the soil, to mix moisture throughout a depth of soil, or to distribute amendments such as soil conditioners, fertilizers, or soil-stabilizing materials. In special cases several layers of soil may be mixed to form a different, more desirable soil texture.

7. Compaction or Firming: The strength of the soil must be increased by compaction in several situations such as in the construction of foundations. In addition, compaction may be used to decrease the permeability of the soil to water to make water-retaining structures such as ditches, terraces, or dams. The manipulation of soil by means of a roller or tamper is a widely used form of tillage. Firming operations are needed to gain good contact between seeds and the soil in seedbed preparation. However, although the added contact is desirable, the mechanical strength increase in the seedbed may be detrimental.

8. Excavation and Transportation: When construction or mining operations require the movement of earth, soil must be excavated, transported, and frequently recompacted into some desirable condition.

Soil-Machine Mechanics

The key area of tillage as a science is the establishment of a soil-machine mechanics that can be used to describe and predict the action of a tillage tool on the soil (Gill and Vanden Berg, 1967; Bernacki, Haman, and Kanafojski, 1967; Zelenin, Balovnev, and Kerov, 1975). Given an initial soil condition, tillage transforms the soil into a new, desired condition. The duration of a tillage action is very short; hence, changes are effected only during this application of forces. The soil conditions as shown by Baver, Gardner, and Gardner (1972) can be further altered by natural forces (e.g., wetting, drying, freezing, and warming) and by biological forces (plants and animals).

The objective of a mechanics of tillage tools is to provide a method for describing the application of forces to the soil and for describing the soil's reaction to the forces. An accurate mechanics would provide a method by which the effects could be predicted and then controlled by the design of an appropriate tillage tool or by the use of a sequence of tillage tools. Furthermore, the efficiency and economy of the tillage operation could be evaluated from the mechanics. A thorough knowledge of the basic forces and reactions is required to develop the mechanics. Such knowledge does not exist at present, and soil reactions cannot be predicted or controlled. As a result, an operation is performed, the conditions are arbitrarily evaluated, and additional operations are performed in sequence until the conditions are adjudged to be adequate. Thus, today, tillage is more an art than a science.

Progress has been made, however, in developing mechanics in which simple tools or simple actions are involved and in which forces and reactions can be described. Several approaches have been used toward the development of a simple form of soil-tillage tool mechanics. The simplest approach, suggested by Soehne (1956), involves the use of an inclined wedge as the tillage tool. Movement of the wedge into the soil causes a repetitive compression and shearing of soil blocks (Figure 1), and the equilibrium forces on the soil are:

$$W = N_0 \sin \delta + \mu' N_0 \cos \delta + kb$$

where

W = draft force
N_0 = normal load on the inclined tool
δ = lift angle of the tool
μ' = coefficient of soil-metal friction
k = pure cutting resistance of soil per unit width
b = width of tool

Attempts to include the influence of velocity into such a relationship have been somewhat successful (Luth and Wismer, 1971). The most universally used draft equation, developed by Goryachkin (1968), is:

$$D = D_0 + \epsilon V^2$$

where

D = draft at velocity
D_0 = basic draft independent of velocity
ϵ = a constant
V = actual velocity

A simple mechanics has been developed with some success for plows, harrows, bulldozers, scrapers, and other machines. In all instances the limiting factor is the lack of realistic soil parameters. Unlike processed materials such as steel the magnitude of soil parameters changes with soil conditions, particularly moisture. A concept of "soil constants" is unrealistic, and prediction of performance is impossible.

Historically, tillage is accomplished by drawing a tool through the soil with a tractive unit. More complex methods of applying forces to the soil may help control the soil reaction and make better use of available power. The dynamic properties of the soil govern the performance of the tractive unit, as well as that of the tillage tool. The use of dynamic tools that rotate or vibrate to impart motion to the soil have a reaction that may be designed to help propel the tillage machine. The tractive loss between traction devices and the soil is large, and contributions to the propulsion of a tillage machine by the tool are important. Tractors can provide power in the form of power take-offs and hydraulic systems to operate dynamic tillage tools. A series of multipowered tools is being explored as a means of improving the performance of tillage tools. (Dalin and Pavlov, 1950; Ul'yanov, 1969; Yatsuk, 1971). The soil-machine relations of such tools are quite complex. Electrical power used in electro-osmotic action can reduce the components of forces originating from soil-metal friction and improve performance.

Models

Modeling, or similitude, offers possibilities for the design and use of machines without having a detailed knowledge of the soil-machine dynamic relation (Wismer, Freitag, and Schafer, 1975). A system consisting of n variables and with r functional dimensions may be defined by a function in the form

$$g(\pi_1, \pi_2, \pi_3, \ldots, \pi_{n-r}) = 0$$

where each Pi term is an independent dimensionless ratio. The function can be arranged to relate dependent and independent terms as:

$$\pi_1 = f(\pi_2, \pi_3, \ldots, \pi_{n-r})$$

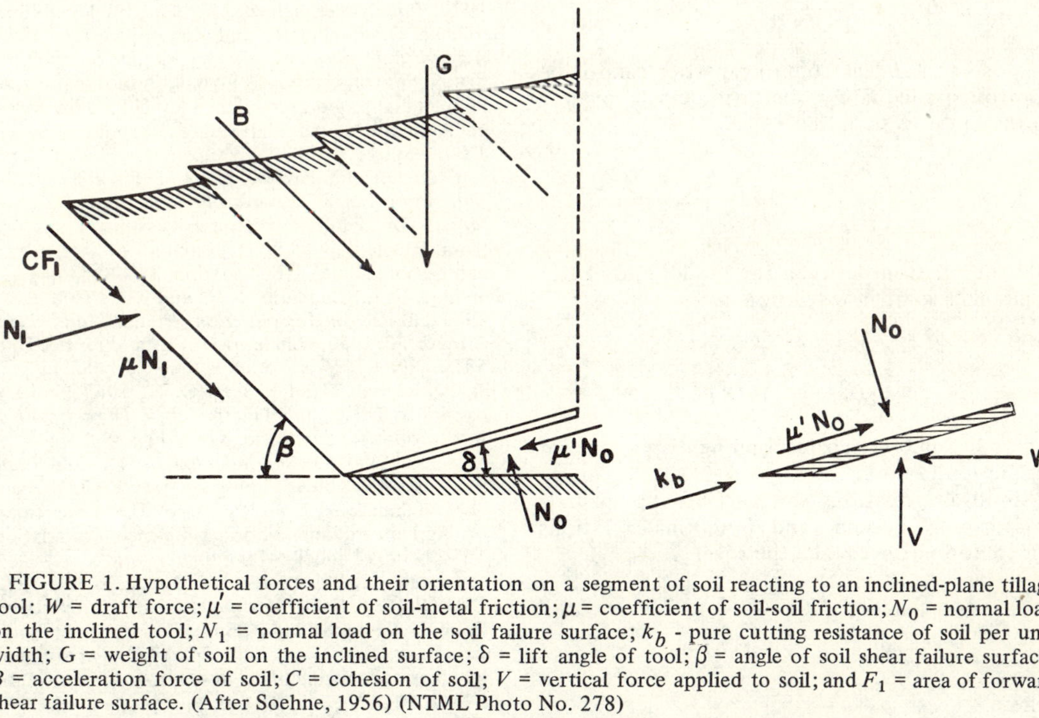

FIGURE 1. Hypothetical forces and their orientation on a segment of soil reacting to an inclined-plane tillage tool: W = draft force; μ' = coefficient of soil-metal friction; μ = coefficient of soil-soil friction; N_0 = normal load on the inclined tool; N_1 = normal load on the soil failure surface; k_b - pure cutting resistance of soil per unit width; G = weight of soil on the inclined surface; δ = lift angle of tool; β = angle of soil shear failure surface; B = acceleration force of soil; C = cohesion of soil; V = vertical force applied to soil; and F_1 = area of forward shear failure surface. (After Soehne, 1956) (NTML Photo No. 278)

These relations can be developed both for a model and a prototype so that

$$\text{Model} = \pi_{1m} = f[\pi_{2m}, \pi_{3m}, \ldots, \pi_{(n-r)m}]$$

$$\text{Prototype} = \pi_1 = f(\pi_2, \pi_3, \ldots, \pi_{n-r})$$

In principle, the objective is to predict the performance of the prototype from that of the model

$$\pi_1 = \pi_{1m}$$

For a simple disk plow, the following Pi terms have been suggested:

$$\frac{R}{\gamma D^3} = f\left(\frac{\lambda}{D}, \frac{V^2}{gD}, \alpha, \beta, \phi, \frac{C}{\gamma D}, \frac{\mu}{\gamma D}\right)$$

where

R = dependent draft force
λ = other pertinent disk lengths
D = disk diameter
V = tool velocity
g = acceleration due to gravity
α = angle of disk inclination
β = disk approach angle
ϕ = soil-friction angle
C = soil cohesion
γ = soil specific wet weight
μ = angle of soil-disk friction

Because soil and machine properties cannot be controlled adequately, distortion exists in some of the Pi terms such that

$$\left(\frac{C}{\gamma D}\right)_m = \psi \left(\frac{C}{\gamma D}\right)$$

and the relation between the model and prototype takes some new relation, as

$$\left(\frac{R}{\gamma D^3}\right) = \delta \left(\frac{R}{\gamma D^3}\right)$$

where ψ and δ are corresponding distortion and prediction factors.

Similitude relations can also be applied to a tractor-plow system, and functional relations for traction devices take the form

$$\frac{P}{W} = f\left(\frac{Cd^2}{W}, \frac{b}{d}, \frac{\delta}{h}\right)$$

where

P = pull of tire (or track)
W = load on tire
C = cohesion
b = tire section width
d = tire diameter
δ = tire deflection
h = tire section height

When the total soil-machine system includes both the tractor and the tillage machine, the contributions of both components can be expressed in Pi terms and a total performance evaluated.

Future Requirements

The maximum use of soil and water resources is necessary to meet world food needs to the extent consistent with conserving natural resources. Tillage can be optimized in support of this goal through a complete knowledge of the soil-air-plant-machine system. Research toward this goal will be an important part of future agricultural research.

WILLIAM R. GILL

References

American Society of Agricultural Engineers, 1971. "Compaction of agricultural soils," *ASAE Monograph.* St. Joseph, Michigan: American Society of Agricultural Engineers, 471p.

Baver, L. D.; Gardner, W. H.; and Gardner, W. R., 1972. *Soil Physics,* 4th ed. New York: Wiley, 498p.

Bernacki, H.; Haman, J.; and Kanafojski, Cz., 1967. *Agricultural Machines, Theory, and Construction, Vol. I.* Warsaw, Poland: PWRIL Publishing House, 883p. (Translated from Polish by U.S. Department of Agriculture and National Science Foundation, TT-69-50019.)

Dalin, A. D., and Pavlov, P. V., 1950. *Rotary Soil Cultivating and Excavating Machines.* USSR: MASHGIZ Publishing House, 248p. (Translated from Russian by U.S. Department of Agriculture and National Science Foundation, TT-70-50041.)

Donahue, R. L.; Shickluna, J. C.; and Robertson, L. S., 1971. *Soils–An Introduction to Soils and Plant Growth,* 3rd ed. Englewood Cliffs: Prentice-Hall, 587p.

Gill, W. R., and Vanden Berg, G. E., 1967. Soil Dynamics in Tillage and Traction, *U.S. Department of Agriculture Handbook No. 316,* 511p.

Goryachkin, W. P., 1968. *Collected Works, Vols. 1 and 2.* Moscow: Kolos Publishing House, 737p. and 465p. (Translated from Russian by U.S. Department of Agriculture and National Science Foundation, TT-72-50023 and TT-71-50087.)

Kepner, R. A.; Bainer, R.; and Barger, E. L., 1972. *Principles of Farm Machinery,* 2nd ed. Westport, Conn.: AVI Publishing, 486p.

Luth, H. A., and Wismer, R. D., 1971. Performance of plane soil cutting blade in sand, *Trans. ASAE,* **14**, 225-259, 262.

Phillips, S. H., and Young, H. M., Jr., 1973. *No-tillage Farming.* Milwaukee: Reiman Associates, 224p.

Soehne, W., 1956. Some principles of soil mechanics as applied to agricultural engineering, *Grundlagen der Landtechnik,* 7, 11-27. (National Institute of Agricultural Engineering English Translation 53, Wrest Park, Silsoe, Bedford, England.)

Ul'yanov, N. A., 1969. *Theory of Self-propelled Wheeled Earth Working Transport Machines* [Teoriya samokhodnykh holesnykh zemleroinotransportnykh mashin]. Moscow, USSR: Machine Construction Publishing House, 520p.

Wismer, R. D.; Freitag, D. R.; and Schafer, R. L., 1975. *Application of Similitude to Soil-Machine Systems.* St. Joseph, Michigan: American Society of Agricultural Engineers, 37p.

Yatsuk, E. P., e dr., 1971. *Rotary Soil Working Machines* (Rotatsionnye pochvoobrabatyvayushchie mashiny). Moscow, USSR: Machine Construction Publishing House, 255p.

Zelenin, A. N.; Balovnev, V. I.; and Kerov, I. P., 1975. *Machines for Earthmoving Work (Mashiny dlya Zernlyanteykh Rabot).* Moscow: Machine Construction Publishing House, 423p.

Cross-references: *Agronomy; Management of Soils; Moisture Management; Profiles, Physical Modification; Soil Mechanics; Soil Physics; Soil Structure; Zero-tillage.*

TOXIC ELEMENTS IN PLANTS—See VOL. IVA

TOXIC ELEMENTS IN SOILS—See POLLUTION. VOL. XVI: SOIL POLLUTION

TRACE ELEMENTS

Fourteen elements in the crust of the earth (aluminum, carbon, calcium, iron, hydrogen, potassium, magnesium, nitrogen, sodium, oxygen, phosphorus, sulfur, silicon, and titanium) constitute over 99% of the total elemental content. The remaining elements are the so-called trace elements. The words "trace elements" are generally used for those elements occurring in the soil in minute amounts without regard to their requirement by organisms. The occurrence of trace elements in soils as well as their reaction with soil constituents is of interest to the agronomist, soil scientist, and even the exploration geochemist. Accounts detailing the geographic distribution, forms (inorganic/organic, adsorbed, complexed, fixed, etc.), and elemental content of soils are given in the following noteworthy works by Swaine (1955), Vinogradov (1959), Mitchell (1965), Keeney and Jacobs (1971), Ermolenko (1972), and Aubert and Pinta (1977). The following list of trace elements is mentioned because information regarding their concentration in the soil is generally available: silver, arsenic, gold, boron, barium, beryllium, bromine, cadmium, cerium, chlorine, cobalt, chromium, cesium, copper, fluorine, gallium, germanium, hafnium, mercury, iodine, indium, lanthanum, lithium, manganese, molybdenum, niobium, nickel, lead, radium, rubidium, antimony, scandium, selenium, tin, strontium, tantalum, thorium, thallium, uranium, vanadium, wolfram, yttrium, zinc, and zirconium.

The term *micronutrients* (q.v.) is used for trace elements in the soil that are essential for healthy development of plants, animals, or microorganisms (boron, chlorine, cobalt, chromium, copper, fluorine, iodine, manganese, molybdenum, nickel, selenium, vanadium, and zinc and possibly barium, bromine, strontium, and wolfram).

Trace Element Content

Trace element composition of soils is determined in terms of total and extractable amounts present per unit weight of dry soil (Table 1). Total amount is commonly measured by dry ashing (500 to 550°C for 3 to 4 h) or by wet ashing with a mixture of perchloric and nitric or sulfuric acids. Extractable amounts, obtained to evaluate availability to higher plants, are determined by using mineral acids (for example, 0.1 *N* HCl), organic acids (for example, acetic, citric), and chelating agents (for example, EDTA, DTPA) Hot water extraction is also done, especially for trace elements that occur as anions (boron, molybdenum, and selenium).

Distribution in Soil Profiles

The pedological factors involved in profile differentiation of the trace elements include the following (Mitchell, 1965; Korte et al, 1976; Aubert and Pinta, 1977):

1. Surface enrichment because of trace elements taken up by plants. This is especially obvious in profiles with surface horizons rich in organic matter (q.v.).
2. Leaching (q.v.) of mobilized constituents such as boron, lithium, manganese, or selenium either completely out of the profile or to zones of accumulation. This is particularly important in tropical soils of humid regions.
3. Translocation, in the course of soil-forming processes such as podzolization, of trace elements together with iron and aluminum. Organic and organic mineral complexes are probably involved.
4. Mobilization of trace elements through breakdown of soil minerals as a result of alternate wetting and drying.
5. Mechanical translocation of clay, which increases in the layers of the clay accumulation the contents

TABLE 1. Trace Element Content of Soils.

Element	ppm* Total	Extractable	Comment
Ag	0.1(0.1–5)	0.02(0.01–0.05)	Tightly bonded by organic and inorganic soil constituents. Usually less than 0.01 ppm in plants.
As	6(0.1–40)	2.7(0.03–11)	Accumulates in clay soils. Leached in acid sandy soils. Closely resembles P chemically. Soils containing enough As to be toxic to mammals occur in Argentina and New Zealand.
Au	<1	–**	Stays in oxidized zone in soil formation. Accumulates in soils through solubilization and reprecipitation.
B	10(2–130)	1.9(0.01–130)	Highest in saline and alkaline soils. In parts of California plants and animals may be poisoned by excessive B in the soil. Deficiencies in B lead to poor crops in many other areas. Present in soil largely as H_3BO_3. Can be readily leached in acid soil. Lime promotes fixation.
Ba	500(100–3500)	138(4–3500)	More common in soils near mines where Ba toxicity to plants may occur. Can be fixed by clay minerals such as montmorillonite.
Be	6(0.1–40)	–	Acts in soil similarly to Al. Not accumulated by plants.
Br	5(1–10)	–	Enriched in soil organic matter. Considered nonessential and nontoxic to plants.
Cd	0.5(0.01–0.7)	0.06(0.01–0.5)	Acts in soil similarly to Zn. High levels of Zn in soil can reduce Cd uptake by plants.
Ce	50(30–50)	–	Concentrates in soil clays. Humus reduces Ce uptake. Chelates increase solubility and leaching. Some mobility in soils. Much remains within a few centimeters of surface.
Cl	100	10(7–50)	Much higher in alkaline soils, near the sea, and in salt deserts. A major exchangeable anion in many soils.
Co	8(1–40)	1.1(0.001–15.4)	Higher in soils derived from basalt or serpentine. Extensive areas of soil deficiency are known, which can cause diseases of ruminant mammals.
Cr	100(5–3000)	0.3(0.01–3.9)	Cr(VI) more mobile in soils than Cr(III). Accumulates in roots and is not translocated. Usually less than 1 ppm in plants. Highest in soils derived from basalt or serpentine. Such soils may have an adapted flora.
Cs	6(0.3–25)	–	Can be fixed on montmorillonite.
Cu	20(2–100)	2.9(0.002–19.2)	Strongly adsorbed by humus. Extensive areas of soil deficiency, as in northern Germany, cause plant and animal diseases. Toxicities rare. Soils able to fix large amounts of Cu.
F	200(30–300)	–	Fixed in many clay minerals and in apatite. Vegetation from F-rich soils in Madras and South Africa toxic to grazing mammals. F deficiency is associated with dental decay.
Ga	30(0.4–300)	–	Accumulates in humus layer and in illuvial horizons. Congener of Al. Seldom occurs in concentrations over 1 ppm in plants.
Ge	1(1–50)	–	Adsorbed by humus, especially in alkaline soils. More reactive than Si. Most Ge salts have low toxicity to man and plants. Reacts similarly to Si in soils and waters.
Hf	6	–	Chemically similar to its congener Zr. Low toxicity to plants and mammals.
Hg	0.03(0.01–0.8)	–	Lowest in upper layer of soil since it volatilizes.
I	5	0.01	Strongly adsorbed by humus. Extensive areas of soil deficiency, resulting in mammalian goiter.
In	0.1	–	Congener of Al.

TABLE 1. continued

Element	ppm* Total	Extractable	Comment
La	30(1–5000)	–	Concentrates in clay fraction. Addition of chelating agent increases soil mobility.
Li	30(7–200)	–	Li^+ freely mobile in soils.
Mn	850(100–4000)	110(15–170)	May be a major exchangeable cation in very acid soils. Oxidation state (Mn^{2+} or MnO_2) depends on microbial action. Available form: Mn^{2+}. Toxicities occur on poorly drained and acid soils. Remedied by liming, drainage. Mn toxicity sometimes found in steamed soils. Deficiencies chiefly on neutral or alkaline soils, particularly those high in organic matter. Mn deficient soils lead to infertility in mammals.
Mo	2(0.2–5)	0.9(0.001–4.8)	Adsorbed by humus, especially in alkaline soils. A few soils are rich enough in Mo to yield herbage toxic to animals, while others give poor crops owing to Mo deficiency.
Nb	24	–	Essentially unavailable in soils. Little leaching. Uptake of Nb by plants limited.
Ni	40(5–5000)	18(0.01–403)	Higher in soils derived from serpentine where it may limit or prevent the growth of woody plants, although an adapted flora survives. Ni contents increase with depth in soils.
Pb	10(2–200)	4.4(0.05–46)	Low soluble amounts in most soils. Higher in some limestone soils. Strongly adsorbed by humus.
Ra	0.8×10^{-6} $(0.3–2 \times 10^{-6}$	–	Chemistry of Ra similar to Ca. Leaches readily in many soils. Not highly mobile in plants. All isotopes radioactive.
Rb	100(20–600)	–	Can be fixed by some clay minerals.
Sb	2–10(?)	–	Decreaseing pH increases sorption of Sb on soil. Moves in neutral and alkaline soils, not acid soils. Sb taken up by plants, fixed in root, but some accumulates in leaves. Moderately toxic to plants
Sc	7(10–25)	–	Congener of Al. Sc content related to clay content of soil. Is translocated downward. Plants contain less than 0.01 ppm.
Se	0.2(0.01–38)	0.1(0.005–9.1)	Chemistry similar to sulfate. Adsorbed by humus, especially in alkaline soils. Vegetation from Se-rich soils may be toxic to mammals, as in Queensland, Ireland, South America, western United States, and other areas of low rainfall.
Sn	10(2–200)	1.4(0.04–6.8)	Strongly adsorbed by humus. Usually less than 2 ppm in plants. Sn in plant not related to Sn in soil.
Sr	300(50–1000)	–	Chemical properties similar to those of Ba and Ca. Concentrations as high as 17,500 in plants. Sr^{2+} may partially replace Ca^{2+}.
Ta	2?	–	Essentially unavailable in soils. Little Ta leaches with water in soil. Little Ta uptake by plants, and Ta concentrates in roots.
Th	5(0.1–12)	–	Accumulates in silt or clay fraction. Not available for plant uptake.
Tl	0.1	–	Congener of Al. Strongly fixed by soils. In neutral soil, not phytotoxic except at high levels (100 ppm).
U	1(0.9–9)	–	Accumulates in top soil due to tight bonding to organic matter. Accumulates also in silt or clay fractions. Can be oxidized from +4 to soluble +6 oxidation state, where it can be readily leached by dilute acids or complexing agents.

TABLE 1. continued

Element	ppm* Total	Extractable	Comment
V	100(20–500)	0.17(0.01–8.8)	Adsorbed by humus, especially in alkaline soils.
W	1	–	Related to Mo. More strongly sorbed onto acid than alkaline soils and more tightly bound to kaolinite and illite than bentonite. Moves readily in alkaline soils when leached with water. Moderately toxic to plants.
Y	50(10–500)	–	Concentrated in clay fraction. High concentrations in phosphate rock. Strongly adsorbed by soils. Unavailable for plant uptake.
Zn	50(10–300)	9.6(0.01–200)	Some soils yield poor crops owing to Zn deficiency. Zn toxicity rare and only found near zinc ores. Can be lost by leaching in acid soils. Organic matter complexes relatively unavailable.
Zr	300(60–2000)	–	Congener of Hf. Hf found distributed in nature in all Zr materials at less than 2% of their Zr content. Low solubility in soil. Generally not detectable in plants.

(Compiled from Swaine, 1955; Vinogradov, 1959; Bowen, 1966; Keeney and Jacobs, 1971; Ermolenko, 1972)
*The number preceding parentheses is the mean value found in the references cited.
**No data available.

of trace elements more abundant in clay than in other soil fractions (see Vol. XII, Pt. 2: *Pedotranslocation*).

6. Surface accumulation of soluble salts such as borates in arid regions (see *Salt Leaching*. Vol. XII, Pt. 2: *Salt-Affected Soils*).
7. Mobilization or fixation arising from microbiological activity (see *Enzyme Activity; Metabolic Activity*).

Trace Element Availability

The availability to plants of the micronutrient trace elements is influenced by many factors (Allaway, 1968; Hodgson, 1963; Sauchelli, 1969; Mortvedt, Giordano, and Lindsay, 1972). The important role of some soil properties that may enhance or retard micronutrient availability is briefly indicated by the factors discussed below. For a more comprehensive consideration of the role of trace elements in plant nutrition and of the factors that influence the mobility of trace elements in soil systems see also *Law of the Minimum, Micronutrients, Nutrient Potential, Plant Nutrients,* and *Soil Fertility*.

pH. The availability of boron, cobalt, copper, manganese, and zinc decreases with increasing pH; the availability of molybdenum and selenium, however, increases with increasing pH. The application of lime is the most widespread method used to control trace element movement from the soil to the plant. In general, liming tends to reduce the availability of the metallic cations and to increase the availability of trace elements taken into the plant as anions (see *Lime, Liming*). Changes in soil reaction (q.v.) resulting from liming can affect the solubility of trace elements by changing the oxidation state of the element.

Organic Matter. Applications of organic matter to the soil are used to increase or decrease trace element concentrations in plants. The presence of organic matter can promote the availability of trace elements, presumably by supplying soluble complexing agents that interfere with their fixation (Nicholas and Egan, 1975). However, soils that are commonly deficient in certain metallic trace elements, especially copper, are organic in nature. Organic matter chelates the metals and makes them less available to plants (see *Metal-complexing Macromolecules*). Chelation appears to be more important than cation exchange in binding the elements to the organic matter (Bowen, 1966). At low pH, organic matter in soil reduces metal availability. At high pH, organic matter appears to increase availability, at least for zinc.

Cation Exchange Capacity. The cation exchange capacity of the soil is important in binding the cationic trace elements (see *Exchange Phenomena*). Therefore soil horizons rich in clays or organic matter generally have higher contents of trace elements than silty or sandy horizons (Ermolenko, 1972).

Oxidation-Reduction. Many trace elements (cobalt, copper, manganese, molybdenum, nickel, lead, vanadium, and zinc) are mobilized in poorly drained soils. The increased movement of trace elements under conditions of even slightly impeded drainage is often so marked that changes in extractable trace element contents due to other causes are obscured. Drainage of wet soils can affect the availability of trace elements because more highly oxidized forms of these elements are formed (see *Redox Reactions*). For example, toxicities of manganese that occur on poorly drained soils are remedied by drainage. In some instances, chelating agents like EDTA and DTPA appear to increase trace element uptake because they reduce oxidized forms of cations.

Microorganisms. Deficiency of trace elements is aggravated by microorganisms because they either compete directly for nutrients or decompose organically bound forms. They can oxidize an element, usually to a less available form, or reduce an oxidized form of an element under conditions where oxygen is limited. They can also make trace elements more available for plant uptake by releasing inorganic ions or soluble organic complexes during decomposition of organic materials (see *Metabolic Activity*).

Phosphorus. Phosphate is well known for decreasing injury caused by excessive levels of trace elements. In agronomic practice, heavy applications of phosphorus fertilizers are used to reduce the availability of the metallic cation trace elements (see *Fertilizers, Inorganic*).

Ion Competition. Competing elements are used to reduce the uptake and accumulation by plants of specific trace elements. For example, plants generally take up more cadmium if the zinc content of the soil is low. However, one of the most celebrated attempts to use ions that compete resulted in failure. In the western United States sulfate was added to the soil to compete with undesired selenates in the plant-uptake process. The seleniferous soils, however, contained an abundance of native sulfates, and the experiment failed.

Plant Type. Plant species and varieties differ widely in tolerance to trace elements (Sauchelli, 1969). Vegetable crops sensitive to trace elements are beets, chard, kale, mustard, spinach, tomatoes, and turnip. Beans, cabbage, collards, and other vegetables are less sensitive. Many farm crops such as corn, small grains, and soybeans are moderately tolerant. Most grasses are tolerant. Highly tolerant ecotypes of grasses are found on ore outcrops or near mines that contain high concentrations of trace elements. Plants also vary in amounts of trace elements accumulated in roots, leaves, stems, flowers, and fruits.

Seasonal Variation. The availability of many elements in soils changes during the year. Manganese exhibits the most pronounced seasonal variation in availability, probably due to microbially induced oxidation and reduction. It is more available in the summer than other times of the year. Cobalt content of forage is high in spring. Copper and zinc are high in rainy, cool weather. Molybdenum is high in autumn. Variations in boron uptake by plants are more striking from year to year or within a season than from season to season. Outbreaks of boron deficiency are commonly associated with dry years. Boron deficiency also occurs in wet seasons because it is lost from the soil by leaching.

Rhizosphere Effects. Plant roots (see *Rhizosphere*) exude a variety of compounds in quantities sufficient to change the availability of nutrients in their environment. Root exudates can alter the chemical environment of the root either directly through an interaction with trace elements in the soil or indirectly through their influence on microorganisms.

M. B. KIRKHAM

References

Allaway, W. H., 1968. Agronomic controls over the environmental cycling of trace elements, *Adv. Agron.*, **20**, 235–274.

Aubert, H., and Pinta, M., 1977. *Trace Elements in Soils.* New York: Elsevier, 394p.

Bowen, H. J. M., 1966. *Trace Elements in Biochemistry.* London: Academic, 241p.

Ermolenko, N. F., 1972. *Trace Elements and Colloids in Soils.* Jerusalem: Israel Program for Scientific Translations Ltd., 259p (Translated from the Russian).

Hodgson, J. F., 1963. Chemistry of the micronutrient elements in soils, *Adv. Agron.*, **15**, 119–159.

Keeney, D. R., and Jacobs, L. W., 1971. The remaining seventy elements, *in* G. Chesters, and J. M. Bremner, eds., *Soil Chemistry.* New York: Marcel Dekker, 81p.

Korte, N. E.; Skopp, J.; Fuller, W. H.; Niebla, E. E.; and Alesii, B. A., 1976. Trace element movement in soils: Influence of soil physical and chemical properties, *Soil Sci.*, **122(6)**, 350–359.

Mitchell, R. L., 1965. Trace elements in soils, in F. E. Bear, ed. *Chemistry of the Soil.* New York: Reinhold, 320–368.

Mortvedt, J. J.; Giordano, P. M.; and Lindsay, W. L., eds., 1972. *Micronutrients in Agriculture.* Madison: Soil Science Society of America, 666p.

Nicholas, D. J. D., and Egan, A. R., eds., 1975. *Trace Elements in Soil-Plant-Animal Systems.* New York: Academic, 417p.

Sauchelli, V., 1969. *Trace Elements in Agriculture.* New York: Van Nostrand Reinhold, 248p.

Swaine, D. J., 1955. Trace element content of soils, *Commonwealth Agr. Bur. Tech. Commun. 48*, 157p.

Vinogradov, A. P., 1959. *The Geochemistry of Rare and Dispersed Chemical Elements in Soils.* New York: Consultants Bur., 209p (Translated from the Russian).

Cross-references: *Metal-complexing Macromolecules; Micronutrients; Pollution; Plant Nutrients; Soil Fertility.*

TRAINING IN SOIL SCIENCE

Instruction in any discipline includes two components, subject matter content and teaching procedure. The aim in teaching soil science is to develop an appreciation and understanding of soil as a natural resource, of how and why soils differ, and of the consequences of soil differences as man manipulates it to produce food and fiber. Knowledge of soils is also useful, and indeed sometimes imperative, when making plans for industrial and residential sites, parks, and recreation areas. The student of soil science learns to observe, classify, interpret, predict, and gain experience in the scientific method. In the application of both the physical and biological sciences, the instructor develops concepts and principles necessary in problem solving. However, teachers recognize that technical or scientific soil problems cannot be considered in isolation from environmental and socioeconomic factors. Their teaching reflects the concern that not only could the history of man be written in terms of soil and water resources but his future may well depend on how he uses them.

Today man is more conscious than ever that life on our planet exists as a complex ecosystem in sensitive equilibrium with the total environment. In the teaching of soil science, it is essential to view soil as an intermediary for the fundamental steps in the operation of the earth's ecosystem. These steps include the reception of energy, a reservoir for water, production of organic matter by producers, consumption and elaboration of organic consumers, decomposition of rocks and minerals and their transformation to forms useful in plant nutrition.

Man's restricted approaches designed to maximize food, fiber, power, and industrial outputs that have dominated our management of natural resources here invariably led to imbalances and instabilities in ecosystems. Instruction in soil science takes into account that soil management like all resource management enterprises will be based increasingly on an understanding of the earth's ecosystems, interconnections, and interactions. The teaching of soil science stresses the dynamic nature of soil, that is, soil must be considered as a flowing and open system. As such, soil plays a pivotal role both in fundamental aspects within the ecosystem involving circulation transformation and accumulation of energy (photosynthesis and decomposition) and determining the kind of land use such as urban environmental planning. Indeed, instruction in soil science will increasingly identify the important role of soils by perceiving the subject in context with man's relevant concerns such as those identified by the Environmental Studies Board of the National Academies of Sciences and Engineers (NASE) (Table 1).

No teaching would be complete without the consideration of employment opportunities. The burgeoning needs and desires of people for new and different uses of land strain the capabilities of the soil resource. From soil scientists and technologists come the knowledge and skills that help conserve these precious resources we call "soil" and "water" and that find new ways of manipulating and using them for the benefit of mankind. Indeed, a greater problem today than man's reaching the moon is man's reaching mankind with food and humanity. The teaching of soil science will inevitably stress the many jobs that require knowledge and understanding about soils.

Building Blocks

Chemistry, physics, biology, and earth science are the building blocks of soil science. These disciplines provide the basis for understanding soil—its genesis, characterization, improvement, and wise use. Such understanding enables man to wisely manipulate soil not only as a resource for food and fiber but as a resource on which he must plan his industries, highways, cities, and recreational areas. Advanced instruction in soil science tends to specialize in such areas as soil fertility, soil chemistry, soil physics, soil biology, mineralogy, soil genesis and classification, and soil survey and interpretation.

TABLE 1. Significance of Soil in an Agenda of Concern for our Ecosystem.

Urgent Studies Suggested by NASE to Monitor Our Ecosystem	Significance of Soil to Suggested Item
1. Physical and chemical properties of land, air, and water	The use of fertilizers and pesticides in agriculture and forestry
2. Distribution of plants and animals in land, air, and water	Sediments from eroded soil, municipal, industrial, and livestock wastes
3. Land use, including diversity of purpose	Urban environmental planning, soil survey interpretation, zoning land for specific uses
4. Construction	Fertile and productive soil produces healthy plants and animals
5. Epidemiology of man, animals, and plants	
6. Aesthetic	In addition to assuring soil for food and fiber, need exists for outdoor recreation, wildlife, and natural scenic areas
7. Evidence of environmental stress such as tranquilizer consumption and social behavior	
8. Noise	

Teaching Methods

Soil science teachers commonly use a multisensory approach in creating teacher-learning situations. Audio-programed instruction units with slides and synchronized tape are often provided for the student who can effectively study at a rate commensurate with his ability to learn (Figure 1). Color slides provide an excellent means by which soils of the world can be observed, their differences noted, and their use and properties recorded. Two principle methods of examining soils are field examination and laboratory analysis. Opportunity is provided for the student to examine and describe soils in the field and to map their boundaries (Figure 2). The audio-programed laboratories are also generally used to introduce the student to laboratory equipment and methods of analysis that he can subsequently carry out with facilities common to most chemistry or plant science laboratories.

Instruction in soil science gives the student an opportunity to identify, analyze, and offer solutions to soil-use problems in the field. Selected field trips are made to locations where soil erosion is common or where poor use of a soil resource has occurred. Students frequently sample soils in the field for laboratory analyses to determine their fertilizer needs or to identify polluted areas where eutrophication is suspected.

Mode of Thought Experience

Films, guest speakers, and seminars designed to examine the relationships between mode of thought in several fields of inquiry essential to the understanding of interrelationship of soils with interface disciplines are presented. Indeed, the teaching of soil science is aimed at developing imagination and flexibility of mind sufficient to deal with the present and future, a readiness to think for oneself yet be tolerant to the views of others coupled with enough knowledge and shrewdness to judge the worth of these opinions, a generosity of spirit to appreciate what others are doing in every field of creative endeavor, and a sense of humility so that one never stops learning.

FIGURE 1. Soil science class using audio-programed instruction at Oregon State University. Students study at their own learning rate. Color slides are synchronized with an audio tape.

FIGURE 2. Students' studying soils and landscape on a class field trip.

MURRAY D. DAWSON

References

Barbarick, K. A., and Post, D. F., 1974. Use of undergraduates as teaching assistants for basic soils instruction, *Jour. Agron. Educ.*, **3**, 101-103.

Black, C. A. 1972. A perspective of graduate education in soil science: The future, *Jour. Agron. Educ.*, **1**, 2-6.

Foth, H. D., 1972. Effective teacher evaluation, *Jour. Agron. Educ.*, **1**, 21-24.

Foth, H. D., 1973. A mastery learning program in soil science, *Jour. Agron. Educ.*, **2**, 65-68.

Huddleston, J. H., and McIntosh, T. H., 1973. Soil science teaching in an environmental science curriculum, *Jour. Agron. Educ.*, **2**, 44-49.

Kellogg, C. E., 1974. Education of soil scientists, *Jour. Agron. Educ.*, **3**, 98-101.

Postlethwait, S. W., 1965. Audio-tutoring–A practical solution for independent study, *Med. Biol. Illus.*, **15**, 183-187.

Postlethwait, S. N.; Novak, W. J.; Murray, H. T., Jr., 1969. *The Audio-tutorial Approach to Learning.* Minneapolis: Burgess, 149p.

Scholles, W. H., 1972. Students can be motivated, *Jour. Agron. Educ.*, **1**, 43-46.

Cross-references: *Soil Science; Soil Science and Society.*

TRANSPORT—See DIFFUSION PROCESSES; DISPERSION PHENOMENA; MICROORGANISM TRANSPORT; TRANSPORT PROCESSES; WATER FLUXES; WATER MOVEMENT

TRANSPORT PROCESSES

Soil is a three-phase system, and transport processes occur in each of the phases. The main entities of concern are matter (water, salt, nutrients, pollutants, oxygen, carbon dioxide), heat, and electric charge. For some processes, e.g., the transport of salt, it is sufficient to consider only one of the phases of the system, i.e., the liquid phase. For other processes, e.g., the transport of heat, it is mandatory to consider the mixture as a whole. In the case of heat transport, one has to consider both the conductive and the convective transport. Heat is conducted through the solid and the liquid phase. Because of the intricate geometry of both phases and their intimate contact, the conduction of heat through both phases is almost always inseparable. Convective transport of heat occurs when heat is carried along, e.g., by the liquid. Other transport processes may also consist of a conductive and a convective part. In case of the dispersion of solutes in soils, the conduction (diffusion) and convection (drag by the liquid) are usually highly interwoven. The most important transport processes occur in the liquid phase. However, the transport of oxygen, carbon dioxide, and water through the vapor phase are of great importance to plant growth. On the other hand, the solid phase itself might be moving as happens in swelling and frozen soils. The formation of cracks and ice lenses has important consequences for the physical properties of the soil.

Many books and articles have been published on transport processes in soils. The following references constitute only a small selection: Baver, Gardner, and Gardner (1972); Childs (1969); Hillel (1971) Kirkham and Powers (1972); Rose (1966); De Wiest (1969); de Vries and Afgan (1975).

Transport occurs due to the presence of a driving force or to convection. The process is then formulated by relating the flux of the entity of concern to the appropriate driving force. The contribution due to convection is then added as a separate term. The resultant constitutive equation is called the *transport equation.* To find the distribution of the entity in the space of concern as a function of time, the appropriate continuity equation has to be fomulated. This equation contains the divergence of the flux. By substituting this flux by the one as formulated in the transport equation, one obtains the *flow equation,* which is usually a second-order, partial-differential equation. This flow equation has then to be solved for the appropriate boundary conditions. There is extensive literature on this part of the subject [cf. Carslaw and Jaeger (1959); Crank (1957); Bird, Stewart, and Lightfoot (1960)].

To limit the discussion on transport processes in soils, the following will be concerned with only the transport equations. Inspection of the variety of transport processes in soils and all the possible interactions reveals that an attempt to formulate a general transport equation including all causes and effects will lead to an intractable result. Therefore, we will first describe the simple processes and from there expand into somewhat more complicated processes.

Darcy's Equation

The liquid present in the soil (i.e., the water plus the dissolved salts and other solvents) moves due to a gradient of the hydraulic pressure $p^{\dagger}$. This hydraulic pressure is the sum of the equilibrium liquid pressure (as measured, e.g., by a tensiometer) and the pressure equivalent of the gravimetric potential. The movement of the liquid is expressed as a volume flux, e.g., in cubic centimeters of liquid per square centimeters of soil per second.

Darcy's equation then states that a linear relationship exists between the liquid flux j^V and the gradient $\nabla P^{\dagger}$,

$$j^V = -L_V \nabla P^{\dagger}$$

where L_V is the hydraulic conductivity. The magnitude of L_V depends strongly on the type of soil and the degree of saturation. This relationship holds for almost all liquid transport processes occurring in soils.

Fick's Equivalent Equation

Salt moves through the liquid phase in the soil due to a concentration gradient or, in case of the presence of a single salt, due to the gradient of the osmotic pressure of the equilibrium liquid $\nabla\pi$. The movement of the salt may be expressed as a macroscopic diffusion flux, e.g., in grams of salt per square centimeter of soil per second. Dividing by the equilibrium salt concentration, the preceding flux turns into a linear diffusion flux j^D in centimeters per second. Fick's equation then suggests that a linear relationship exists according to:

$$j^D = -L_D \nabla\pi$$

Fourier's Equivalent Equation

A temperature gradient in the soil causes a heat flux. Although the system of concern is a mixture of phases, one may assume that the heat flux j_q' can be related to the gradient of the temperature according to

$$j_q' = -L_T \frac{\nabla T}{T}$$

Ohm's Equivalent Equation

Soil is an electric conductor. The application of an electrostatic potential gradient ∇E causes an electric current I. The relationship between the two appears again to be linear, viz.

$$I = -L_E \nabla E$$

The preceding four equations formulate the main transport processes in soils. The parameters are characteristic of the soil. They depend on the type of soil, the packing density, the salt concentration, the temperature, etc. However, the relationships are linear for almost all practical purposes, provided that the "other" driving forces are absent. The possibility that a certain driving force contributes to a nonconjugated flux might result into the unwarranted conclusion that the flux-force relationships are nonlinear. In unsaturated soils, the interaction between the transport of heat and water is an important phenomenon. In particular, in clay soils the interaction of the movement of water and salts is of great practical interest. This interaction is directly related to the *permiselectivity* of the soil, being a property by which the soil can select between different molecules or ions during the transport process.

In the field where day and night, rainfall and evaporation, fertilizing, and water uptake by plants cause the simultaneous presence of temperature gradients, hydraulic pressure gradients, and salt concentration gradients, it is mandatory to consider the possibility of interaction between these processes. An attempt will now be made to formulate the flux-force relationships such that this interaction is taken into account. In principle, one first has to find an expression for the total energy dissipation in the system. The flux-force pairs in that equation then provide the lead, which together with the postulate that every flux is a linear function of every force appearing in the equation of energy dissipation, results into an appropriate set of flux-force relationships. The following development is taken from Groenevelt and Bolt (1969).

Expressing the thermodynamic forces F per unit macroscopic length of the porous system, the equations relating the chosen fluxes and forces in the present single salt system may be written as

$$j_q' = L_T F_T + L_{TV} F_V + L_{TD} F_D + L_{TE} F_E$$

$$j^V = L_{VT} F_T + L_V F_V + L_{VD} F_D + L_{VE} F_E$$

$$j^D = L_{DT} F_T + L_{DV} F_V + L_D F_D + L_{DE} F_E$$

$$I = L_{ET} F_T + L_{EV} F_V + L_{ED} F_D + L_E F_E$$

in which

$$F_T = \frac{(-\Delta T)}{T \Delta X}; \quad F_V = \frac{(-\Delta P^\dagger)}{\Delta X};$$

$$F_D = \frac{(-\Delta \pi)}{\Delta X}; \quad F_E = \frac{(-\Delta E)}{\Delta X}$$

The four straight coefficients are all positive and describe the above-mentioned primary transport phenomena, i.e., those of Fourier, Darcy, Fick, and Ohm. Comparison with these equations gives: $L_T/T = \lambda$ (soil), the thermal conductivity of the soil, and $L_V = K_h$ (hydraulic conductivity). $L_D = D$ (soil)$/RT\rho_s/M_s$, with D (soil) equal to the effective salt diffusivity in the soil system. Of course D (soil) should be related to D_0 (diffusion in aqueous solutions), inasfar as the frictional resistance of the salt against the water will be roughly the same as in solution (aside from the reduction of the cross-section available for diffusion θ and a tortuosity effect). In addition, however, D (soil) contains the effect of the volume flux induced by the very gradient of the osmotic pressure (cf. the coupling phenomenon discussed below). M_s indicates the molecular weight of the salt.

$L_E = \kappa$ (soil), the effective electrical conductivity of the soil system. There exists a relationship between κ (soil) and κ_0 (the conductivity of the equilibrium solution) comparable to the relationship between D (soil) and D_0. Obviously, all straight coefficients are defined as the ratio of a flux to its conjugated force at zero value of all other forces.

The twelve coupling coefficients, which are defined as flux-force ratios at zero value of the other three forces, are equal in pairs carrying the same two subscripts (*Onsager relations*). In the following, these pairs of coefficients will be discussed qualitatively by describing the mechanisms causing a flux j_i to arise in the presence of a nonconjugated force F_k. In certain cases the existence of coupling is easier demonstrated by a complementary phenomenon, i.e., the fact that a finite value of the conjugated force F_i arises if a zero flux j_i is maintained in the presence of an applied nonconjugated force F_k. It should be pointed out that the linearity of the relations between a flux and the nonconjugated forces is an intrinsic consequence of the linearity of the main relations. As long as the straight relations are linear, the coupling processes are adequately formulated by linear relationships. In general, it may be stated that coupling arises as a result of "selective" properties of the transporting medium with respect to the transported matter. The coupling coefficients may be positive or negative, and their absolute value must lie between zero (no coupling) and the geometric mean value of the corresponding straight coefficients.

L_{DV} and L_{VD}, Filtration (Salt-sieving Effect) and Osmosis

The simplest example of the mentioned selection mechanism is found in systems with very fine pores (in the Å range), where the resistance against passage of water molecules is much smaller than that for dissolved larger molecules. A pressure gradient will then cause preferred transport of the water, giving rise to a net movement of the solute with respect to the solvent in a direction opposite to the solvent flow. Thus L_{DV} is negative and finite (salt-sieving effect). Alternatively, a gradient of the osmotic pressure (in the absence of a pressure gradient) across such a barrier will lead to a net force on the solution as a whole (a relatively larger share of the diffusive "force" on the solute molecules being used to overcome the friction of the membrane), and j^V becomes finite. Accordingly, L_{VD} is again negative and finite. The two extremes of this situation are found in the perfectly semipermeable barrier where $-L_{DV} = -L_{VD} = L_V = L_D$ and a non-selective solutioncolumn where $L_{DV} = L_{VD} = 0$. The permiselectivity of soil systems may be substantial even for much thicker water layers as a result of the electric double layer formed on the pore walls.

L_{EV} and L_{VE}, Streaming Current and Electroosmosis

Another familiar example of coupling coefficients of finite magnitude is found in the electrokinetic phenomena. These arise if the solution phase carries a net volume charge as a result of an electric field emanating from charges present on the solid matrix. This field exerts a selectivity with regard to the passage of cations and anions. A pressure gradient will then cause preferred movement of one of the two species, giving rise to a streaming current, which shows that L_{EV} has a finite value. This streaming current requires a possibility for return flow of, e.g., electrons outside the system as may be provided by, e.g., inserting a set of shorted electrodes (condition of macroscopic electroneutrality of the system!).

If such an "external connection" is not present, a macroscopic gradient of the electric potential will arise instantaneously, making the net current zero. The existence of this potential gradient (termed *streaming potential*) proves again the existence of the coupling coefficient L_{EV}. Alternatively, an imposed gradient of the electric potential will lead to a net force on the solution phase (because of the presence of non-equivalent amounts of cations and anions). The ensuing volume flux is termed *electroosmosis* (q.v.), its coefficient being L_{VE}. The coupling coefficients vanish in the absence of the selective action of the electric field emanating from the pore wall.

L_{ED} and L_{DE}, "Diffusion" Current and Electrophoresis

The coupling between diffusion of salts and an electric potential gradient is of course almost always present in any aqueous solution. Here the selectivity mechanism lies in the (usually) unequal resistance exerted by water on diffusing ions. A concentration gradient of salts will thus induce a "diffusion" current provided an external return flux of electrons is possible. Usually such an external connection (via shorted electrodes) is absent; then a diffusion potential arises.

The complementary phenomenon, i.e., the transport of salt under influence of a gradient of the electric potential, is well known from, e.g., the determination of transport numbers with the *Hittorf method.* The effects vanish for equal mobilities, e.g., in KCl solution. The magnitude of these effects in porous systems may be substantially larger than in aqueous solutions if the transport number of either one of the ion species is influenced substantially. This is almost always the case in soil systems where the mentioned electric field in the liquid layers against the solid matrix induces high or sometimes very high values for the transport number of the cation. The potential gradient then arising as a result of concentration gradients (even of KCl) has been discussed extensively in soil science literature in relation to potentiometric measurements in soils involving salt bridges and was then referred to as "liquid junction potential."

L_{TV} and L_{VT}, "Thermofiltration" and Thermoosmosis

The coupling between heat and mass fluxes is less obvious on first sight. In line with the preceding reasoning, coupling will arise only if the system exerts a selectivity with respect to "hot" and "cold" molecules ["hot" molecules indicating the molecules with a high value of $(\bar{H} + \psi)$ or total specific enthalpy, or specific heat content]. In that case, an applied pressure gradient causes, e.g., preferred transport of the "hot" ones, thus increasing the heat content of the system at the low pressure end, which amounts to a heat flux in addition to the mass-flux ($L_{TV} > 0$). In a liquid saturated soil system, not much can be expected of this effect, unless the liquid layers are extremely thin (which is difficult to effectuate in the saturated system). As follows from the observed heat of wetting of dry soil, the heat content is indeed decreased for the first few layers of water due to the presence of strong adsorption fields. These layers are presumably also less mobile than further layers. For thick water films, the difference in the "average" heat content of the mobile layers as compared to that of the entire film becomes vanishingly small. Alternatively, a temperature gradient applied to a water film subject to an adsorption field will induce a mass flux as a temperature rise will "lift" a larger part of the molecules from the adsorbed (immobile) layer to the mobile one. Again this effect should be expected to be quite small for the saturated soil system because of the width of pores, which is usually of concern.

For unsaturated and frozen soils, the phenomena of thermofiltration and thermoosmosis are quite important (cf. Kay and Groenevelt, 1974; and Groenevelt and Kay, 1974).

L_{TD} and L_{DT} and L_{TE} and L_{ET}, Thermodiffusive and Thermoelectric Phenomena

In these phenomena coupling will arise if the selection mechanism operates on the salt molecules and the countercharge, distinguishing between "hot" and "cold" specimen. For the salts present in charged pores, the effect must be extremely small. The salt being repelled from the immobile layer close to the liquid-solid interface, only a tiny fraction of the salt (this time the fraction with high heat content) is situated in this layer. The difference in average heat content of the mobile part compared with that of all salt molecules will be vanishingly small. A locally applied temperature increase will likewise not contribute measurably to a decrease of mobile salt molecules. The coefficients L_{TD} and L_{DT} will thus be negative in principle and presumably zero in practice. In contrast the countercharge is accumulated in the immobile layer, so L_{TE} and L_{ET} should be positive (for a positive countercharge). As to the practical significance of these coupling phenomena, it appears that L_{ET} might indeed be easily detectable. As was indicated earlier in the discussion of L_{EV}, this effect is usually met in the form of a gradient of the electric potential arising here when a temperature gradient is applied under the condition of zero current. In systems with low electrolyte concentration, the majority of the "relatively few" conducting ions is situated close to the solid surface, where the ionic mobility may be impaired considerably. A temperature increase could then increase substantially the number of actually conducting ions. A considerable gradient of the electric potential might thus be necessary to maintain a zero current when a temperature gradient is applied.

P. H. GROENEVELT

References

Baver, L. D.; Gardner, W. H.; and Gardner, W. R., 1972. *Soil Physics.* New York: Wiley, 498p.

Bird, R. B.; Stewart, W. E.; and Lightfoot, E. N., 1960. *Transport Phenomena.* New York: Wiley, 780p.

Carslaw, H. S., and Jaeger, J. C., 1959. *Conduction of Heat in Solids.* Oxford: Clarendon Press, 510p.

Childs, E. C., 1969. *The Physical Basis of Soil Water Phenomena.* New York: Wiley, 493p.

Crank, J., 1957. *The Mathematics of Diffusion.* Oxford: Clarendon Press, 347p.

de Vries, D. A., and Afgan, N. H., 1975. *Heat and Mass Transfer in the Biosphere.* New York: Wiley, 594p.

De Wiest, R. J. M., 1969. *Flow Through Porous Media.* New York: Academic Press, 530p.

Groenevelt, P. H., and Bolt, G. H., 1969. Non-equilibrium thermodynamics of the soil-water system, *Jour. Hydrology,* 7, 358–388.

Groenevelt, P. H., and Kay, B. D., 1974. On the interaction of water and heat transport in frozen and unfrozen soils: II. The liquid phase, *Soil Sci. Soc. America Proc.*, 38, 400–404.

Hillel, D., 1971. *Soil and Water.* New York: Academic Press, 288p.

Kay, B. D., and Groenevelt, P. H., 1974. On the interaction of water and heat transport in frozen and unfrozen soils: I. Basic theory: The vapor phase, *Soil Sci. Soc. America Proc.*, 38, 395–400.

Kirkham, D., and Powers, W. L., 1972. *Advanced Soil Physics.* New York: Wiley, 534p.

Rose, C. W., 1966. *Agricultural Physics.* Oxford: Pergamon Press, 226p.

Cross-references: *Aeration, Respiration, Atmosphere; Conductivity, Electrical; Conductivity, Hydraulic; Conductivity, Thermal; Diffusion Processes; Energy Balance; Electroosmosis; Flow Theory; Microorganism Transport; Thermal Regimes; Thermodynamics, Soil Water; Water Fluxes.*

VAN DER WAALS FORCE—See VOL. IVA

W

WASTE DISPOSAL—See MULCH; SLUDGE DISPOSAL; SOILS, NON-AGRICULTURAL USES

WATER CONTENT AND RETENTION

The volumetric water content of a soil when saturated is the same as the volume of the pores. As the water content is reduced, the largest pores empty first, water being retained most forcefully in small pores, in the interstices between particles, and on particle surfaces. Additional structural water is present internally in the clay lattices and as water of crystalization. This water ordinarily is not regarded as a part of the soil water (q.v.). However, water content quantification requires a definition of a dry soil, and it is not always possible to distinguish clearly between structural water and absorbed water. Hence a dry state cannot be unequivocally defined, and where high accuracy in measurement is required, the method of removing water must be carefully specified. In soil science as well as in many other areas where water content determinations are made, the *dry state* is specified as that which exists when a sample is dried in an oven to constant weight at a temperature between 100 and 110°C. The implications of such a temperature choice with respect to soil minerals are shown in Figure 1 where weight loss at increasing temperature for a variety of constituent minerals is plotted. It is evident from this figure that depending on the soil mineral, small errors in temperature measurement or control can easily lead to large errors in water content measurements. Data in Figure 1 suggest that a choice of, say, 165–175°C would reduce the dependence on precise temperature control and measurement. However, at such high temperatures organic constituents would be lost rapidly through oxidation. Even at temperatures of 100–110°C, organic materials oxidize, and a soil sample high in organic matter will continue to lose weight over a period of time in oven drying. Because of the decomposition problem, organic materials are often dried at 50–70°C in the presence of drying agents or in a vacuum oven. Hence the choice of 100–110°C for drying soil containing both minerals and organic matter may be seen to be a compromise. Where high precision and accuracy in water content measurements are required, careful control of both temperature and temperature distribution in an oven and specification of drying time become essential.

Expression and Measurement of Water Content

Water content may be expressed in several ways. The most common, because of the fact that samples are weighed, dried, and reweighed to determine water loss, is the expression of mass of water per unit mass of oven-dried soil, referred to as "mass basis" or "dry mass basis" water content. Where a soil sample is taken in

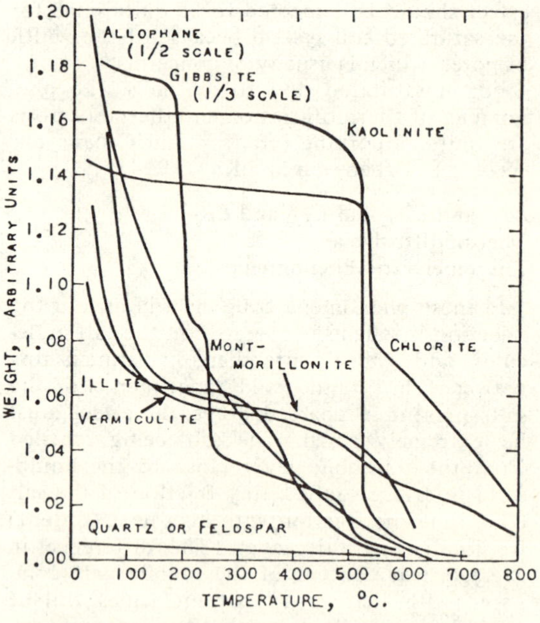

FIGURE 1. Weight loss of clay minerals as a function of temperature. The weight of the sample at a temperature of 800°C or greater is taken to be unity. Water content percentages on a mass basis may be obtained by subtracting 1 from the ordinate figure and multiplying by 100. (The numbers to the right of the decimal thus may be regarded as percentages.) (Nutting, 1943). (From Gardner, Walter H., Water content, *in* C. A. Black et al, eds., *Part 1: Methods of Soil Analysis.* Madison: American Society of Agronomy, 1965. Reprinted by permission of the publisher.)

such a way that its in situ volume is known, the expression of mass or volume of water per unit of bulk volume becomes possible. In a great many situations the volume of water associated with unit volume of bulk soil is the most useful value. This value may be readily converted to the depth of water associated with a sample if it were accumulated in a layer. Thus, this value is obtained by multiplying the volumetric water content by the depth of soil to which it pertains to get the volume of water per unit area for the specified depth increment, or merely centimeters or inches of water, in the same way as rainfall is specified in length units. Water content on a mass basis may be converted to volumetric water content through multiplication by the ratio of the dry bulk density of the soil to the density of water.

Direct Methods. The common and generally accepted standard for water content measurement in soil involves measuring the mass of the wet soil, oven drying to constant mass, and then computing the water content as the mass of water removed per unit mass of oven-dry soil. Problems of defining an unequivocal dry mass and problems of assuring that the temperature and drying time are appropriately controlled necessitate careful attention to technique and explicit description of the method used wherever high precision and accuracy are required. Because of such problems, it is more realistic to consider reproducibility rather than accuracy of measurement. Under less demanding conditions, it is possible to follow somewhat standard procedures using ordinary laboratory ovens and balances and to achieve a precision and reproducibility of roughly half of a water content percentage with most soils. Temperature must be controlled within a few degrees of 105°C, and samples must be dried for a uniform period of at least 10 h where a forced-draft oven is used and for at least 24 h in a convection type oven. Overloading an oven or adding wet samples prior to removal of almost dry samples will increase the size of errors. Considering weighing errors only, the size of oven-dry sample required to achieve a particular precision in water content determination is given in terms of standard deviations by the approximate equation

$$\text{Oven-dry sample size} = \frac{2\sigma_b}{\sigma_\theta}\sqrt{\theta + 1}$$

where σ_b is the standard deviation for the balance used, and σ_θ is the standard deviation in mass basis water content θ associated with weighing.

Other methods for drying samples may be used. Use of a microwave oven is particularly attractive for rapid determination of water contents, it being possible to complete the drying process is as little as 15 min. Temperatures achieved in samples are as much as 200°C, considerably higher than those usually specified for oven drying. Whether or not measurements made at such temperatures are useful depends on the purpose for which used and the precision required. A drying temperature of 200°C would be as good or better than 105°C for most soil minerals. However, such high temperatures would result in badly oxidized organic components and would introduce difficulties with some soil minerals.

Indirect Methods. Several indirect methods for measuring water content of soils are available. Two of the most common are the gypsum block method and the neutron-scattering method.

Gypsum Block Method. Porous blocks containing electrodes that permit measurement of electrical conductivity can be calibrated against water content. When these blocks are buried in soil, their water content comes to potential equilibrium with that of the soil. Measurement of electrical conductivity is usually made with an alternating current at about 1000 cps to avoid polarization at the electrodes. Also instead of conductivity, some blocks and measuring units are made to read in electrical capacitance. Since electrical conductivity (q.v.) depends on ion content of the soil solution (q.v.), inert porous blocks are sensitive to solutes present in the soil solution. Use of slightly soluble substances like gypsum for the blocks reduces sensitivity to solutes through a buffering action. Many of these electrical conductivity blocks are insensitive in the very wet range, but by increase in pore size, the sensitivity can be extended somewhat into the wet range. These moisture blocks may also be calibrated to read matric potential rather than water content. Although useful in many applications, such blocks do not have the precision ordinarily obtained using oven-drying techniques. On the other hand, they have certain advantages in following water content changes with time inasmuch as they are not confounded with spatial variability as are destructive-sampling techniques.

Neutron-scattering Method. Neutron-scattering techniques for water content measurements are based on the fact that fast neutrons emitted from a neutron source are slowed down to thermal energies to a considerably greater extent by light elements like hydrogen than by the heavier elements most common in soil. A neutron source (beryllium bombarded by gamma rays from Americium-241) is fitted to a thermal neutron-detector probe and lowered into soil through a thin-walled metal access tube. The detector tube is essentially transparent to the fast neutrons emitted so these fast neutrons move outward into the soil in a solid angle. In the soil they are slowed down to thermal velocities, largely by interaction with water due to collisions with hydrogen and oxygen nuclei. The more water present in the soil, the less the distance of travel outward into the soil. A proportion of the thermalized neutrons interact with or are captured by heavier atoms, and a balance between neutrons emitted and thermalized and neutrons captured is almost instantly achieved with the density of the thermal neutrons in the vicinity of the thermal neutron detector being linearly related to the water content. Certain compensating factors exist between the processes of thermalization and capture in soil materials such that the relationship between thermal density and water content is to some degree independent of variation in soil materials, including the presence of hydrogen-containing organic matter. This factor is highly advantageous in practical application of the technique. However, in some soils calibration is desirable.

Apart from electronic problems that have been greatly minimized in recent years, the most important limitations of the method are lack of spatial resolution and precision errors associated with random emission of neutrons at the neutron source. Emission of neutrons follows a Poisson distribution for which the standard deviation is the square root of the number of neutrons emitted. Thus it is necessary to take into account the random nature of neutron emission. The greater the number of counts, the more reliable the count rate average from which the water content is inferred. Since the size of the neutron source is generally fixed in the design of the instrument, increased precision requires larger counts that are obtainable only by increasing counting time intervals. Because the standard deviation in neutron emission is a square root function, water content precision also depends on the square root of counting time. Hence reduction of standard deviation in water content by a factor of 2 requires counting 4 times as long. Commonly used counting times for equipment generally available varies from 20 s to 1 min, although much longer time intervals are used in special situations. The standard deviation associated with 10^5 counts in the reference standard is about 0.5% water content with good electronic equipment.

Spatial resolution of nuclear water content measurements is inherently limited by the fact that once in the soil, the path of a neutron is determined only by its collisions, which means that the region of influence cannot be limited to a fixed layer. Use of fast neutron absorbers to confine initial emission to a layer of soil rather than the solid angle achieved otherwise can increase the spatial resolution somewhat but not extensively. Water content-depth curves produced using nuclear scattering are generally smooth average curves, even under circumstances in which sharp breaks exist such as might be produced by soil stratification.

Water Potential

Although indispensable in many situations, water content of a soil often is less useful than knowledge of how tightly water is held. Supply of water for plant growth depends on two soil properties: (1) the forces with which water is held and (2) the water conductivity of the soil adjacent to absorbing roots. Many engineering properties of soil are more clearly related to forces of retention than to water content. For example, yield of water from a watershed and structural properties of soil at the foundation of buildings or bridges depend more directly on water-retention forces than on water content itself. Force of retention of water held in pores is generally a capillary force, i.e., the force associated with surface tension in the air-water interface. Forces holding water close to particle surfaces and in narrow interstices between particles are strong absorptive forces, generally called *van der Waal-London forces.* These vary with the 6th or 7th inverse power of the distance to the surface. Other forces retain water such as the force exerted by solutes on nearby water molecules, *osmotic forces.* Such forces are in evidence during evaporation and in the presence of membranes that limit the passing of dissolved ions. The complicated geometry of soil surfaces and the nature of the many forces associated with water in soil makes quantification of the retentive forces nearly impossible. Hence a unifying principle is used to deal with these forces. Rather than to specify the several forces explicitly, the energy required to move water from the soil to a reservoir of water at a defined reference state is used. By specifying only the work required, the nature and direction of forces need not be indicated. This work is generally specified on a unit mass or unit volume basis and is known as the *water potential. Kinetic energy,* or energy of motion, usually need not be specified in slow-moving steady-state or quasi-steady-state water flow situations ordinarily encountered in soil. Use of potential terminology makes it possible to add the effects of the several types of forces acting on water in soil.

Terminology. Water potential is made up of several components, the first two of which are closely related to water content: (1) *matric potential,* the work required to remove water against surface tension and particle surface forces, a negative value; (2) *osmotic potential,* the work necessary to remove water from ions in the solution, also a negative value; (3) *gravity potential,* the work involved in moving water from soil at a particular elevation to a reference elevation—this may be either positive or negative, depending on the choice of reference elevation; (4) *pneumatic potential* (or *gas pressure potential*), involved where a gas pressure gradient exists; and (5) *overburden potential,* involving the weight of those soil particles that are somewhat free to move and to exert a pressure on water below (somewhat comparable to the contribution of suspended materials to total pressure of a liquid column). For soil beneath the water table, the *matric potential* is replaced by a positive pressure or *submergence potential* that exists because of the weight of the water. Where a free water surface coincides with the soil surface, the *submergence potential* is exactly balanced by the *gravity potential.*

Water content and matric potential are usefully related by curves known as "sorption" or "desorption" curves or sometimes referred to as "characteristic curves." Figure 2 shows such curves covering a wide variety of soil materials. These curves are desorption curves, having been made by drying the soil step-wise down from saturation. Curves relating water content and matric potential are not unique, but depend on wetting history. Soils undergoing wetting are somewhat dryer at the same matric potential than soils undergoing drying. Reversal of the wetting or drying process at intermediate values between completely dry and saturation produces intermediate curves within what is known as the *hysteresis loop.* These are referred to as "scanning curves." This hysteretic phenomenon results from the fact that matric potential depends on the energy state of water at the entry or entries to larger soil pores and not on the water content of isolated large pores. The larger cross-sectional space may be either wet or dry depending on wetting history.

Matric potential and osmotic potential together determine the vapor pressure or relative humidity in the soil as given by

$$(\text{Matric} + \text{osmotic potential}) = \frac{RT}{V_m} \ln \text{RH}$$

where R is the universal gas constant, T is the

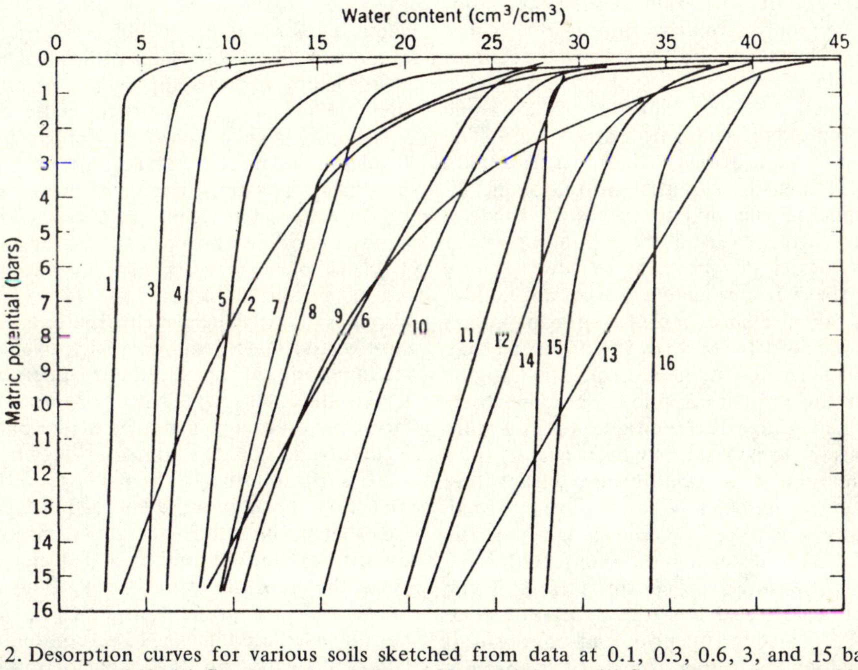

FIGURE 2. Desorption curves for various soils sketched from data at 0.1, 0.3, 0.6, 3, and 15 bars tension given by Holtan el at, 1968. (1) Continental gravelly sand loam, Arizona. (2) Sassafras sandy loam, Mayland. (3) Progresso fine sandy loam, New Mexico. (4) Vaucluse sandy loam, Georgia. (5) Albion loam, Oklahoma. (6) Abilene clay loam, Texas. (7) Hartsells loam, Ohio. (8) Palouse silt loam, Washington. (9) Fayette silt loam, Wisconsin. (10) Nellis gravelly loam, New York. (11) Larb-like silt clay loam, South Dakota. (12) Memphis silt loam, Mississippi. (13) Drummer silty clay loam, Illinois. (14) Auston silty clay, Texas. (15) Marshall silty clay loam, Iowa. (16) Bascom-like clay, South Dakota. (From Baver, Gardner, and Gardner, 1972)

absolute temperature, and V_m is the molecular volume of liquid water. In the water content range where air-liquid interfaces may be observed, the matric potential is a function of the radius of the air-water interface; thus

$$\text{Matric potential} = h\rho g = \frac{2\sigma}{r}$$

where h is the height of a hanging water column in equilibrium with the soil, ρ is the density of liquid water, and σ is the surface tension of water.

In the absence of significant temperature gradients, soil temperature is usually neglected in dealing with water except that it does influence the size of the matric potential and the viscosity, which in turn affect water flow. However, if an appreciable temperature gradient exists, as is possible, say, due to solar heating at the soil surface, water flow is affected, particularly in the vapor phase. Under such conditions appreciable water vapor can move from regions of high to regions of low temperature. Such flow can give rise to return liquid flow, and in many irrigated areas, salt is transported in the liquid water toward warm regions and then evaporated, leaving salt behind. Apart from involvement with salt deposition, vapor flow has its greatest relative importance in the dry range where liquid flow becomes inappreciable.

Measurement. Several methods are available for measurement of water potential in soil. The most common and probably the most useful are methods for measuring matric potential and involve balancing the matric forces in the soil against the force exerted by a hanging water column across a porous cup, or alternatively balancing these forces against forces applied by means of air pressure across a porous plate. When equilibrium is reached, the potential may be obtained from the height of the hanging column or the pressure applied. These methods are shown diagramatically in Figure 3. Figure 3(a) shows a tensiometer where water in a porous cup comes to equilibrium with forces exerted by the hanging water column through the cup as indicated by the height of the column h. The potential is $-h\rho g/1020$ bars, where h is measured in centimeters, ρ is the density of water, and g is the acceleration of gravity. (Potential is the potential energy/unit volume, which is dyn-cm/cm^3 or dyn/cm^2, which is a pressure, hence the use of a pressure unit; 1020 cm water is equivalent to 1 bar of pressure. Potential may also be given as potential energy per unit mass, but where the density of water may be taken as unity, the numerical

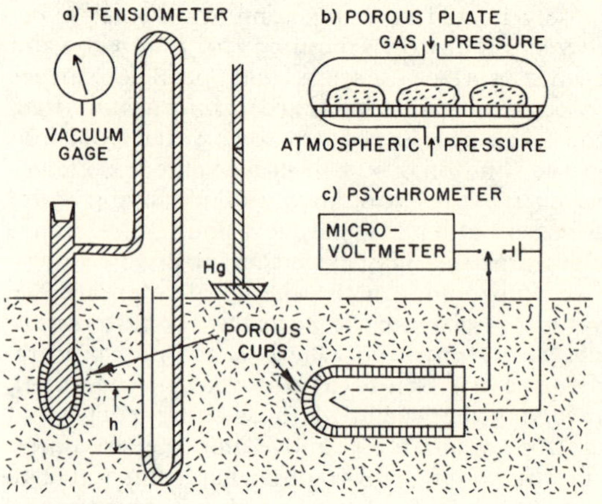

FIGURE 3. Tensiometer and pressure plate for measurement of matric potential in field soil and in the laboratory and thermocouple psychrometer for measurement of matric and osmotic potential. In practice, either the mercury manometer or the vacuum gage replaces the hanging water column shown attached to the tensiometer.

value remains the same. The hanging water column is shown only to illustrate the principle of the measurement. In practice, this column is replaced by a mercury manometer, which may then be located above the soil surface, or by a vacuum gage. Figure 3(b) shows a pressure plate apparatus in which soil samples are brought to equilibrium with an applied gas pressure, which may be measured directly in bars. This equipment may also be used to determine the relationship between water content and matric potential. The tensiometer is limited practically to the wet range from 0 to about –0.85 bars by problems of cavitation in the water column.

A *soil psychrometer* may be used to obtain water potential measurements in soil below about –½ bar. The psychrometer measures the sum of the matric and osmotic potential so that an independent measurement of osmotic potential is necessary if matric potential is desired. However, in many instances where plant growth is involved, it is the sum of the two potentials that is often desired. The psychrometer consists of a tiny thermocouple encased in a porous cup together with a microvoltmeter and an electrical circuit for application of a 3–8 ma current to cool the thermocouple by Peltier cooling to below the dew point. In operation, the thermocouple is cooled for 5–15 s, condensing a thin layer of water on the thermocouple surface, after which the temperature of the thermocouple quickly comes to a value determined by rate of evaporation of the water back into the cup atmosphere. From the temperature difference between the thermocouple before cooling and that during evaporation afterward, the relative humidity may be determined. It is related to the sum of the matric and osmotic potential by the equation given earlier. Imperfect boundary conditions for the psychrometer prevent direct association of temperature measurements with relative humidity. Hence the instrument is calibrated against relative humidities established by osmotic solutions. Calibration is usually carried out to give the potential directly as a function of the microvoltage reading. The relative humidities in soil in the range of interest where plant growth is concerned are in the difficult range of about 99% and above. Hence temperature is an important factor. However, design and techniques have reduced the influence of temperature to the point at which practical measurements can easily be made. The psychrometer may also be applied to measurements on plants and to osmotic potentials of plant saps.

WALTER H. GARDNER

References

Baver, L. D.; Gardner, W. H.; and Gardner, W. R., 1972. *Soil Physics.* New York: Wiley, 498p.

Childs, E. C., 1969. *An Introduction to the Physical Basis of Soil Water Phenomena.* New York: Wiley, 493p.

Gardner, W. H., 1965. Water content, *in* C. A. Black, ed., *Methods of Soil Analysis.* Madison: American Society of Agronomy, 82-127.

Hillel, D., 1971. *Soil and Water: Physical Principles and Processes.* New York: Academic, 288p.

Holtan, H. N.; England, C. B.; Lawless, G. P.; and Shumaker, 1968. Moisture-tension data for selected soils on experimental watershed. U.S. Department of Agriculture, *Agric. Res. Serv.* ARS 41-144, 609p.

Kirkham, D., and Powers, W. L., 1972. *Advanced Soil Physics.* New York: Wiley, 534p.

Nielson, D. R.; Jackson, D. R. Cary, J. W.; and Evans, D. D., eds, 1972. *Soil Water.* Madison: American Society of Agronomy, 175p.

Rose, C. W., 1966. *Agricultural Physics.* Oxford: Pergamon Press, 226p.

Taylor, S. A., and Ashcroft, G. L., 1972. *Physical Edaphology.* San Francisco: Freeman, 533p.

Cross-references: *Capillary Pressure; Conductivity, Hydraulic; Flow Theory; Soil Solution; Soil Water; Surface-Soil Water Content.*

WATER EROSION

Soil erosion by water is the detachment and transportation of soil particles by rainfall and runoff. It is a serious problem on more than half the 180 million hectares of cropland in the United States (USDA, 1965) and on a similar area of forest, pasture, and rangeland. Excessive erosion has historically caused disastrous consequences to the once-productive lands of many nations (Bennett, 1939; Lowdermilk, 1950). Today major shifts in farming systems throughout the world must deal with the threat of erosion to future soil productivity (Lal, 1974; Greenland, 1975).

Some erosion occurs naturally under even pristine conditions (geological erosion), and such erosion has been a tremendous factor in forming the earth's landscape. However, man's activities often cause much additional erosion (accelerated erosion) by upsetting the equilibrium established during natural land uses. Soil losses that are appreciably in excess of about 12 t/ha deplete the potential of upland soils for agricultural production (Wischmeier and Smith, 1965). The 4 billion people of our world rely on a productive agriculture, so wise management of agricultural land, using cropping systems that include adequate erosion-control practices, is essential in meeting present world food needs and maintaining future soil-productivity potentials.

Soil erosion usually occurs at much greater rates on land that is denuded and reshaped for urban development, roadways, reservoirs, and mining than on agricultural land. Approximately one million hectares of such nonagricultural land are in this condition each year in the United States alone. Although erosion is less harmful to these sites themselves than to agricultural land, the eroded soil or sediment from both nonagricultural and agricultural sources often causes major problems downstream. There it may muddy streams and lakes, clog ditches and rivers, cover roads and bottomland, and fill water-supply reservoirs and harbors. Thus eroded soil may cause triple damage: first to the land where it originates; second to the water that transports it; and third to the site where it is deposited. *Sediment,* a common term for eroded soil, is the world's greatest pollutant of surface waters (Robinson, 1971). About 20 billion tons or cubic meters of sediment are transported to the oceans of the earth each year (Holeman, 1968; Jansen and Painter, 1974).

The Soil Erosion-Sedimentation Process

The companion processes of soil erosion and sedimentation involve complex interrelationships among the many factors that influence the detachment, transportation, and deposition of soil particles by wind or water. This entry will consider the water (*fluvial*) aspects, while the topic **Wind Erosion** will cover *aeolian* aspects of soil erosion and sedimentation.

Soil erosion by water involves the detachment of soil particles from the soil mass by raindrop impact and/or runoff shear and their transport by raindrop splash and/or flowing runoff (often termed *overland flow*). Eroded soil in transport is sediment, and *sedimentation* occurs when sediment is deposited. Sediment movement through the watershed system is often intermittent, with detachment, transport, and deposition of the eroding material recurring repeatedly. The rate of erosion depends on the climate, soil, topography, and plant cover (Smith and Wischmeier, 1962).

The principles and processes involved in soil erosion by water have been described by Bennett (1939), Ellison (1947), Stallings (1953 and 1957), Smith and Wischmeier (1962), Robins and Neff (1963), Leopold, et al (1964), Schumm (1969), Guy (1970), Hudson (1971), Partheniades (1971), and ASCE (1975). These knowledgeable writings and their bibliographies are the basis for much of the following material.

Erosion by Raindrop Impact. The water-erosion process usually begins when raindrops strike unprotected soil on the earth's surface and detach soil particles (Ekern, 1950; Mihara, 1951). The erosive potential of rain depends on the raindrop size distribution, fall velocities and total mass at impact (Hudson, 1971). In areas where the annual rainfall is about 1 m, several trillion raindrops, falling at speeds up to 9 m/s, annually bombard each hectare of land. This amount of water has a volume of 10,000 m^3, a mass of about 10 million kg, and falls with an impact energy of 200 to 300 million J. Unless the soil surface is protected by vegetation, mulch (q.v.), or other cover, these raindrops detach tremendous quantities of soil from the soil mass. In addition, the outward splash occurring when the drops strike exposed soil will cause net movement downslope. The rate of soil detachment and net downslope movement by rainfall depend on soil characteristics, surface condition, slope steepness, and rainfall characteristics (Ellison, 1947; Ekern, 1950).

Erosion by Runoff. Most soil eroded by water is transported downslope by surface runoff. Runoff does not begin, however, until the rainfall intensity exceeds the infiltration rate of the soil and the surface storage capacity of the land has been satisfied. Thus soil conditions having high intake rates or large surface-ponding capacities may appreciably delay runoff and reduce subsequent runoff rates.

Once runoff begins, the quantity and size of material that it can transport depend on runoff velocity and turbulence (ASCE, 1975), and these increase as the slope steepens and the flow increases. The larger and denser the eroding material, the greater must be the flow velocity and turbulence to transport it. Major rates of runoff can generally transport all rainfall-detached soil and in addition, detach soil by hydraulic shear and transport it downslope.

Interrill Erosion. Soil eroded from upland slopes comes from (a) *interrill areas,* that part of the land surface between runoff channels, (b) *rills,* the eroded portions of runoff channels that can be obliterated by subsequent tillage, and (c) *gullies,* eroded channels that are too large to be readily traversed by farming equipment. Another term, "sheet erosion," has often been used to include all erosion that can be obliterated by tillage, thus encompassing both interrill and rill erosion. However, true sheet erosion occurs only before runoff has sufficient detaching capacity to cause rill erosion. Use of the terms "interrill" and "rill" erosion eliminates the need for the confusing term "sheet" erosion because interrill erosion is literally sheet erosion, and rill erosion implies the nonsheet, locationally selective removal caused by runoff detachment (Meyer, Foster, and Romkens, 1975).

Interrill erosion is a relatively uniform removal of soil, so it is not so visually apparent as rill or gully erosion. It results primarily from the effects of raindrop impact, and it includes both movement by splash and transport of raindrop-detached soil by thin-film runoff, enhanced by raindrop-impact turbulence. On land where the soil surface is not protected by cover, soil is lost from most of the land surface by interrill erosion. The interrill erosion rate is not greatly affected by the steepness of the interrill surface or the location on the land slope since rainfall impact is relatively uniform all over an area of land (Lattanzi, Meyer, and Baumgardner, 1974; Meyer, Foster, and Romkens, 1975).

Rill Erosion. Rill erosion results primarily from soil detachment by concentrated runoff. It usually occurs on only a very limited percentage of the land surface. Rill erosion is much more intensive and noticeable than interrill erosion (Figure 1). Rills may develop where runoff channelizes due to topographic variations, tillage marks, or random irregularities on the land surface. However, channelized flow does not cause rill erosion until the flow's shear characteristics exceed the soil's resistance to them. Thus concentrated runoff may flow for a considerable distance downrow or downslope before rilling starts (Meyer and Monke, 1965). Once rilling begins, it increases rapidly with greater flow accumulations, so rill erosion increases with the length of land slope. It also increases with slope steepness. Rilling typically progresses upslope by a series of intensely eroding headcuts or knickpoints (Meyer, Foster, and Nikolov, 1975).

FIGURE 1. Rill erosion between rows in a field of young soybeans. Interrill erosion occurred between rills, although it is not apparent. Note the deposition of the coarser sediment at the toe of the slope where the land steepness decreased.

Gully Erosion. Gullies may be considered as large rills that cannot be readily crossed or obliterated by subsequent tillage. They develop through massive soil removal by concentrated runoff. Slumping of gully sidewalls from gravitational forces and excess soil moisture provide much of the soil debris removed by the subsequent runoff. Gully erosion characteristically proceeds upslope as a series of large headcuts (Piest and Bowie, 1974; Leopold, Wolman, and Miller, 1964).

The sediment from an area of sloping land may include soil eroded by interrill erosion, rill erosion, and gully erosion. Figure 2 illustrates the contributions of each. Only interrill erosion occurs near the top of the slope, and it continues at a relatively uniform rate all along the slope. Rill erosion begins a short distance downslope and increases rapidly with slope length. Gully erosion begins at a considerable distance downslope. The relative contributions from interrill areas, rills, and gullies to the total erosion depend on the slope length and vary with the climatic and soil conditions (Meyer, DeCoursey, and Romkens, 1976).

FIGURE 2. Typical contributions of interrill erosion, rill erosion, and gully erosion to total erosion from sloping land. Note that rilling begins a short distance downslope and that gullying begins farther downslope.

Soil Erodibility. Soil erodibility is the susceptibility of a given soil condition to erosion. A soil's resistance to detachment and transport by rainfall and runoff depends on soil properties such as particle size, shape and density, cohesiveness, and aggregate strength. The physical and chemical properties of cohesive soils greatly affect their erodibility (Partheniades, 1971). Since soil is often eroded as aggregates rather than as primary (textural) particles, the larger size and lesser density of aggregates, as compared to their component separates, are important characteristics affecting erodibility. Sediment that is eroded as aggregates may have the transportability characteristics of sand or even gravel although consisting mostly of silt and clay material.

Generally, cohesive or fine-textured soils are less easily detached from the soil mass than noncohesive soils, but once detached (unless as sizeable aggregates), the finer materials are more easily transported. Coarse-textured soils are not held to the soil mass by cohesion, but they usually have greater rain-intake rates and thus less runoff. The net result is that medium-textured soils such as loams and silt loams are generally more erodible than soils with high clay or high sand contents because they are susceptible to surface sealing and are also more readily detached and transported (Wischmeier and Mannering, 1969).

The characteristics of sediment from interrill, rill, or gully erosion may differ according to the sediment source. Raindrop impact on interrill areas tends to break soil into smaller aggregates and primary particles, and the very thin film of runoff may be unable to transport the larger material, so sediment from interrill areas may be much finer than that resulting from the more massive removal in rills and gullies (Stallings, 1953; Meyer, Foster, and Romkens, 1975). In addition to such physical differences, the chemical-pollution potential may be different. Interrill erosion removes a relatively shallow depth of soil from a large percentage of the area, so sediment from interrill areas may be quite rich in chemical pollutants applied to the land surface. However, sediment from rill erosion may have a much lower content of surface-applied chemicals, but a much higher content of subsurface chemicals than interrill sediment. Sediment from gullies may be relatively free of chemical pollutants since much of this sediment comes from sterile soil horizons of land that is seldom treated with agricultural chemicals.

Soil-surface Cover. The extent to which soil surfaces are protected by vegetation, mulches, or other cover greatly influences their susceptibility to erosion by rainfall and runoff. Such covers dissipate raindrop impact energy and slow the velocity of runoff. When the soil surface is not well protected, impacting raindrops promote soil erosion by (a) detaching soil particles from the soil mass, (b) breaking soil crumbs and aggregates into smaller, more easily transported sizes, (c) reducing infiltration rates through soil surface sealing, (d) splashing soil particles dominantly downslope or into rills where they can be transported downslope by runoff, and (e) increasing the sediment-carrying capacity of runoff. When the soil surface is well protected with surface-contact or canopy cover, interrill erosion may be essentially eliminated (Lattanzi, Meyer, and Baumgardner, 1974), and rill erosion may be appreciably decreased (Meyer, Foster, and Nikolov, 1975).

Surface cover often affects erosion even more by slowing runoff velocity than by dissipating raindrop impact energy, particularly on the longer slopes. In addition to reducing runoff amount by maintaining higher intake rates and decreasing runoff turbulence by intercepting raindrops before they strike the overland flow, vegetation and mulches dissipate runoff energy by (a) absorbing flow shear forces that would otherwise be expended on the soil, (b) reducing effective slope steepness by forcing the flow to meander as it moves downslope, and (c) increasing the hydraulic roughness of the flow surface. Growing vegetation also acts as a pump that removes soil water from the root profile and thus provides a greater storage potential for subsequent precipitation.

Deposition of Sediment. Eroded soil (that detached and transported from its original location) is seldom all lost from an area of sloping land because some of it usually deposits within the area. Deposition occurs when the runoff can no longer carry all its sediment load, usually because of decreased slope steepness or increased density of vegetative cover. Deltas of sediment at the bases of erodible slopes (see Figure 1) and in filter strips of vegetation are examples. Such deposition is a selective process, with the largest and densest materials settling out first and finer materials carried farther with the runoff (Stallings, 1953). Therefore the size distribution of the eroding sediment is important in determining the portion of the sediment load that is deposited and the size distribution of the remaining sediment load.

The sediment yield from a land area or watershed is usually less than the total gross erosion from all sediment sources because of subsequent deposition within the watershed. The ratio of sediment yield to gross erosion for an area is known as the "sediment-delivery ratio." Clearly, the sediment-delivery ratio for a given watershed depends on sediment transportability characteristics such as size and density and on the opportunities for sediment deposition within the watershed.

Erosion Modeling

A mathematical model to simulate the soil erosion-sedimentation process must incorporate the rather complex interrelationships among the various components that influence soil detachment, transport, and deposition. The basis for such a model is the useful definition of soil erosion, "A process of detachment and transportation of soil materials by erosive agents" (Ellison, 1947). Based on this definition and knowledge that the erosive agents are rainfall and runoff, the process of soil erosion by water may be divided into four subprocesses: (a) detachment by rainfall, (b) detachment by runoff, (c) transport by rainfall, and (d) transport by runoff (Meyer and Wischmeier, 1969; Rowlison and Martin, 1971). The dynamics of each subprocess may then be evaluated using fundamental hydraulic, hydrologic, meteorologic, or other physical relationships plus parameters to characterize the soil's susceptibility to each of these four aspects of the erosion process.

The interrelationships of these subprocesses are then considered, as illustrated in Figure 3, to describe soil movement at all locations along the slope. On each successive length segment, the soil available for transport is that detached on the segment by rainfall and runoff plus that carried to it from the segment upslope. This total is compared with the total of the rainfall and runoff transport capabilities at the end of the segment. If available detached soil is less than the total transport capability, available soil is the limiting factor on the segment, and the sediment load carried to the next segment equals the amount of the available material. However, if instead the total transport capability is less than the available detached soil, transportation is the limiting factor, and the sediment load equals the transport capability. The important concept to recognize is that the limiting factor determines the soil-erosion rate, just as the weakest link in a chain determines its strength. Successive segments of the slope may

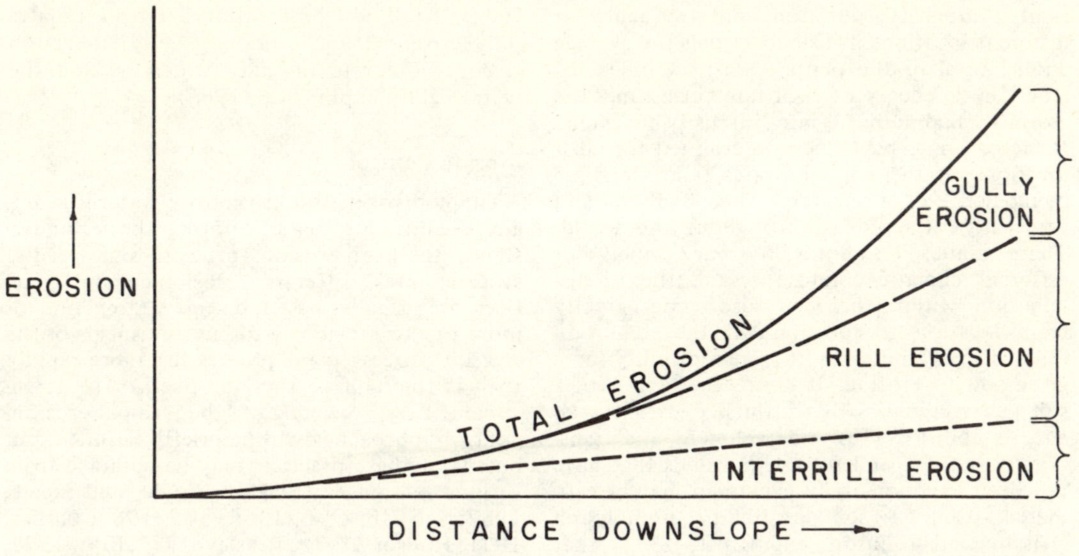

FIGURE 3. Approach used to simulate the process of soil erosion by water on each successive slope-length segment. The four subprocesses are evaluated for each segment, and sediment movement is routed downslope as illustrated. (From Meyer and Wischmeier, 1969)

differ in cover, steepness, runoff rate, and soil characteristics, and these differences may be considered in evaluating each segment. The sediment and runoff are routed from the top to the bottom of the slope using conservation-of-mass techniques, and soil erosion can be simulated as a dynamic process with soil movement described at locations all along the slope at any time.

Knowing whether available detached soil or the transport capability is limiting is important not only for evaluating erosion rates but also for developing erosion-control practices. For example, when available detached soil is limiting, erosion-control practices that reduce raindrop impact effects and shear forces of runoff would be effective. However, when transportation capability is limiting, practices that decrease runoff transport capability would be most beneficial.

An erosion model such as described provides a basic framework for a more complete mathematical simulation of the soil erosion-sedimentation processes. Subsequent developments in erosion modeling are described by Foster and Meyer (1972), Bennett (1974), Foster and Meyer (1975), Fleming (1975), Onstad and Foster (1975), David and Beer (1975), Frere, Onstad, and Holtan, (1975), and Smith (1976).

Erosion Prediction

Accurate predictions of soil-erosion rates for specific conditions and land uses are important for determining sediment losses and designing erosion-reducing practices. The *Universal Soil Loss Equation* (Wischmeier and Smith, 1965; Wischmeier, 1973*a*) is the most comprehensive technique currently available for estimating upland erosion. It incorporates the six major factors that affect upland soil erosion by water: rainfall erosiveness, soil erodibility, slope length, slope steepness, cropping and management techniques, and supporting conservation practices. It is a methodical procedure developed from statistical analyses of more than 10,000 plot-years of data from runoff plots and small watersheds at 47 locations in 24 states. It enables land-management planners to estimate erosion rates for a wide range of rainfall, soil, slope, crop, and management conditions and to select alternative land-use and practice combinations that will limit erosion rates to acceptable levels.

The universal soil loss equation is

$$A = R\,K\,L\,S\,C\,P$$

where A, the computed soil loss per unit of area, is the product of the other six factors. It is an estimate of the average annual interrill plus rill erosion from rainstorms for field-sized upland areas. It must be increased to account for snowmelt erosion and for erosion from gullies and streambanks along the flow route and/or decreased to account for deposition of eroded material within the watershed. R, the rainfall factor, accounts for effects of differences in rainfall intensity, duration, and frequency at different locations. It usually equals the average annual total of the rainfall erosivity index, EI (the kinetic energy of each rainstorm times the storm's maximum 30-min intensity). Annual *R* factor values have been determined for most locations in the United States, and they are being derived for the remaining locations and for many other areas throughout the world. These annual *R* factor values vary widely for different climatic conditions, and the EI distribution within the year also varies greatly from location to location. *K*, the soil-erodibility factor, indicates the susceptibility of a given soil to erosion. It lumps the effects of a soil's characteristics that influence water intake and its ability to resist detachment and transport by rainfall and runoff. *K* values that have not been experimentally evaluated may be estimated using a soil erodibility nomograph (Wischmeier, Johnson, and Cross, 1971) that requires knowledge of only the soil particle-size distribution, organic matter content, structural class, and permeability group. *L*, the slope-length factor, accounts for the effects of greater accumulation of runoff with distance downslope. *S*, the slope-steepness factor, accounts for the increased erosiveness of runoff and ease of sediment movement as slope steepens. *L* and *S* are commonly combined into a topographic *LS* factor. Available *LS* values based on data from relatively short slopes of moderate steepnesses have been extrapolated to estimate *LS* factor values for steep, long, and very flat slopes (SCS, 1975*a*). *C*, the cropping-management factor, accounts for the cropping system used, tillage method, productivity level, and other cultural variables. The annual *C* factor is a weighted composite based on the land condition and seasonal distribution of the *R* factor for each stage of crop growth for all crops in the system. *P*, the erosion-control practice factor, accounts for the effect of conservation practices such as contouring, strip cropping, and terracing on the resulting erosion.

Values for the six independent variables in this equation are available for many agricultural and nonagricultural conditions (Wischmeier and Smith, 1965; Wischmeier, Johnson, and Cross, 1971; Wischmeier, 1973*b;* Wischmeier, 1974; Foster and Wischmeier, 1974; Chen, 1974; Wischmeier, 1975; SCS, 1975*a;* Ports, 1975). Handbooks with local values of these factors have been developed by agencies such as the USDA Soil Conservation Service. Research is continuing toward improved estimates of erosion rates for conditions that have not been experimentally evaluated.

Reliable procedures for predicting gully-erosion rates are unavailable (Piest and Bowie, 1974). The Soil Conservation Service estimates gully erosion from the rate of gully elongation upslope, the rate of gully widening, and the average gully depth (SCS, 1966).

Erosion Control

The control of soil erosion by water is generally the ultimate goal of efforts to better understand the soil-erosion process and predict erosion rates. Effective erosion-control practices are usually based on preventing one or more of the previously discussed aspects of the erosion process from progressing more rapidly than is tolerable to the land itself or to downstream sites. Several of the major erosion-control approaches will be briefly outlined, but details of these practices may be obtained from various references (Blakely, Coyle, and Steele, 1957; Schwab, et al, 1966; SCS, 1969; Barnes, 1971; Hudson, 1971; Beasley, 1972; EPA, 1972; Chen, 1974; SCS, 1975*b*; Stewart et al, 1975; Meyer and Romkens, 1976) or from publications and consultations with agencies such as the Soil Conservation Service in the United States or similar organizations in other countries.

Vegetative Control. Growing vegetation is nature's major way of controlling erosion, and it is also one of man's best ways. Potentially erodible pasture, range, and forest lands have low erosion rates when good vegetative cover is maintained (Wischmeier, 1975). Cropland erosion is reduced during periods when dense meadow crops are grown in a rotation and when grain or fiber crops provide good soil cover (Wischmeier and Smith, 1965). During other periods cropping practices such as no-till planting that keep residues of past vegetation on the surface also greatly decrease erosion (Wischmeier, 1973*b*). Even when worked into the soil profile, plant residues from heavy vegetative growth decrease erosion during several succeeding years (Wischmeier and Smith, 1965). On construction sites and urbanizing areas, vegetative mulches and rapid revegetation of denuded areas are important erosion-control practices (SCS, 1975*b;* EPA, 1972).

Conservation Tillage. Tillage (q.v.) of land is often essential or at least desirable to prepare the soil for crop production, yet tillage usually destroys vegetative cover. Tillage also breaks surface soil into erodible sizes, just as rainfall and runoff do. However, certain tillage methods leave the land in a less erodible condition by (a) reducing the amount of surface disturbance and traffic, (b) leaving the surface rough and cloddy, (c) using existing vegetation and plant residues for soil-surface protection and (d) maximizing the permeability of the subsurface profile. These conservation tillage methods include no-till or zero tillage, strip tillage, minimum or reduced tillage, mulch tillage, contour listing and ridging, and deep tillage. Such tillage methods offer a great potential for erosion reduction on sloping land when adapted for major cropping and land-use systems (Meyer and Mannering, 1967; Wischmeier, 1973*b;* Stewart, et al, 1975).

Supporting Conservation Practices. Erosion control for intensively cropped land and construction sites often cannot be achieved by improved vegetation and conservation tillage practices only. For such conditions, supporting practices that increase surface storage for runoff, reduce runoff velocity, and trap detached sediment may provide the desired erosion control. Some supporting practices are applied annually, and others require more permanent land modification.

Contouring and controlled row grades are supporting practices that are applied by farming across the slope at a small grade rather than with the field boundary or up-and-down slope. Average annual soil loss will be reduced to about half on slopes of moderate steepnesses and lengths (Wischmeier and Smith, 1965); however, these practices are less effective on other steepnesses, and other practices are more dependable for long slopes. Contouring and graded rows pond runoff and carry it around the slope during small to moderate rainstorms, but they may lose their effectiveness for high-intensity storms. They are more effective with tillage methods that produce a large ridge such as listing than for methods with only small relief.

Contour strip cropping is a practice by which alternate strips of close-growing vegetation are used between strips of other crops. These vegetative strips act as filters to slow the downslope runoff and reduce its sediment-carrying capacity. Contour strip cropping reduces erosion to approximately half that occurring with contouring alone when alternate strips are in meadow and to three-fourths when only one strip in four is in meadow (Wischmeier and Smith, 1965).

Terracing is a practice by which large permanent ridges are built cross-slope to intercept downslope flow and impound it or carry it around the slope to stabilized outlets. Terracing basically decreases the length of slope, thereby changing the runoff patterns and reducing flow velocity. Diversion terraces are used to collect runoff from upslope and prevent it from crossing erodible land downslope of the terrace. A system of several terraces on an area of sloping land is usually a very effective supporting practice for erosion control, but such terraces are often relatively expensive and may cause difficulty in farming operations. Systems of parallel terraces and other terrace designs have been developed to improve the farmability of these systems with modern farming machinery (Phillips and Beauchamp, 1966; Beasley, 1972).

Erosion rates, as predicted by the universal soil loss equation (Wischmeier and Smith, 1965) for cropland with no supporting practices and with the three major supporting practices, are given in Table 1 for two tillage methods, two crop systems, and three slope steepnesses. Rainfall and soil conditions that are typical in the central Cornbelt of the United States were assumed. The tolerable erosion rates for these soils are 9 to 12 t/ha. As the erosion hazard increases due to steeper topography, adequate erosion control requires more effective supporting practices, conservation tillage, and/or a meadow-based rotation.

Sediment basins are runoff-ponding structures that are commonly used to collect runoff and temporarily detain it so that much of the sediment may be trapped and retained in the pool. They usually consist of a temporary barrier or dam (although an excavated basin may serve the same purpose), a pipe outlet, and an emergency spillway. They are usually situated in the natural drainageways at the low part of rapidly eroding areas or land undergoing development, construction, or mining. The sediment-trapping efficiency of sediment basins depends on their volume and shape in relation to the rate of inflow and the settling characteristics of the sedi-

TABLE 1. Average Annual Erosion Rates (t/ha) for Several Tillage Methods and Supporting Practices (Central Cornbelt States).

		Crop Rotation			
		Continuous Corn		Corn-Corn-Wheat-Meadow	
Slope Length and Steepness	Supporting Practice*	Conventional Tillage	Minimum Tillage	Conventional Tillage	Minimum Tillage
90 m, 4%	None	31	22	10	6.5
	Cntr	16	11	4.9	3.2
	CSC	*	*	3.6	2.5
	Terr	5.6	4.0	1.8	1.1
90 m, 8%	None	76	54	25	16
	CSC	*	*	11	6.9
	Terr	13	9.4	4.3	2.7
90 m, 12%	None	140	99	45	29
	CSC	*	*	20	13
	Terr	22	15	6.7	4.3

*None: Up-and-down slope or field boundary row direction.
Cntr: Contouring (not recommended for lengths of 90 m on 8% and steeper slopes).
CSC: Contour strip cropping (not applicable for continuous corn).
Terr: Terracing at intervals of 45 m at 4%, 30 m at 8% and 23 m at 12%.

ment. Sediment basins are usually a final effort to trap sediment that was not controlled by other upland practices (SCS, 1975*b*).

Gully stabilization practices usually involve reshaping to establish vegetation or a series of brush, wood, or stone dams to trap sediment and reduce the gradient of the erodible gully channel (Francis, 1957; Hudson, 1971). These practices plus fertilization help to develop an environment conducive to revegetation.

Other supporting practices such as *grassed waterways, nonvegetative mulches,* and *riprap* are useful for erosion control on certain conditions. The selection of effective practices for erosion control requires that erosion-control plans be individually designed to fit specific situations by people experienced in such planning and knowledgeable concerning the available alternatives. Qualified planning for such conditions is usually available through local governmental agencies and consulting firms.

The control of soil erosion by water is essential for continued productivity on agricultural land and for environmental protection of downstream lands and waters from sediment and its associated pollutants. The future development of better erosion-control practices will require (a) critically assessing the strengths of current practices, (b) intensively investigating the basic processes involved in soil erosion by water, and (c) actively pursuing promising new erosion-control approaches. The current research and development effort toward these goals is very small compared to the present activity in field erosion-control programs and the potential future damages from inadequate practices.

L. DONALD MEYER

References

American Society of Civil Engineers, 1975. Sedimentation engineering, *ASCE Manual No. 54,* 745p.

Barnes, R. C., Jr., 1971. Erosion control structures, *in* H. W. Shen, ed., *River Mechanics,* Ft. Collins, Colorado: H. W. Shen, 28–1 to 28–26.

Beasley, R. P., 1972. *Erosion and Sediment Pollution Control.* Ames: Iowa State University Press, 320p.

Bennett, J. P., 1974. Concepts of mathematical modeling of sediment yield, *Water Resources Res.,* **10,** 485–492.

Bennett, H. H., 1939. *Soil Conservation.* New York: McGraw-Hill, 993p.

Blakely, B. D.; Coyle, J. J.; and Steele, J. G., 1957. Erosion on cultivated land, *in Soil, the 1957 USDA Yearbook of Agriculture,* 290–307.

Chen, Chang-Ning, 1974. Evaluation and Control of soil erosion in urbanizing watersheds, *Proc. Nat. Symp. on Urban Rainfall and Runoff and Sediment Control,* **UKY BU106,** 161–173.

David, W. P., and Beer, C. E., 1975. Simulation of soil erosion, *Trans. ASAE,* **18,** 126–133.

Ekern, P. C., 1950. Raindrop impact as the force initiating soil erosion, *Soil Sci. Soc. America Proc.,* **15,** 7–10.

Ellison, W. D., 1947. Soil erosion studies, *Agric. Engin.,* 28, 145–146, 197–201, 245–248, 297–300, 349–351, 402–405, 442–444.

Environmental Protection Agency, 1972. Guidelines for Erosion and sediment control planning and implementation, *Environmental Protection Technology Series EPA-R2-72-015,* 228p.

Fleming, G., 1975. Sediment erosion-transport-deposition simulation: State-of-the-art, *Proc. 1972 Sediment-Yield Workshop, Oxford, Mississippi,* **ARS-S-40,** 274–285.

Foster, G. R., and Meyer, L. D., 1972. A closed-form soil erosion equation for upland areas, *in* H. W. Shen, ed., *Sedimentation: Einstein Symposium.* Ft. Collins, Colorado: H. W. Shen, 12–1 to 12–19.

Foster, G. R., and Meyer, L. D., 1975. Mathematical simulation of upland erosion by fundamental erosion mechanics, *Proc. 1972 Sediment-Yield Workshop, Oxford, Mississippi,* **ARS-S-40,** 190–207.

Foster, G. R., and Wischmeier, W. H., 1974. Evaluating irregular slopes for soil loss prediction, *Trans. ASAE,* **17,** 305–309.

Francis, C. J., 1957. How to control a gully, *in Soil, the 1975 USDA Yearbook of Agriculture,* 315–320.

Frere, M. H.; Onstad, C. A.; and Holtan, H. N., 1975. *ACTMO, an Agricultural Chemical Transport Model,* **ARS-H-3,** 54p.

Greenland, D. J., 1975. Bringing the green revolution to the shifting cultivator, *Science,* **190,** 841–844.

Guy, H. P., 1970. Fluvial sediment concepts, *in U.S. Geological Survey Techniques of Water Resources Investigations, Book 3,* 55p.

Holeman, J. N., 1968. The sediment yield of major rivers of the world, *Water Resources Res.,* **4,** 737–747.

Hudson, N., 1971. *Soil Conservation.* Ithaca: Cornell University Press, 320p.

Jansen, J. M. L., and Painter, R. B., 1974. Predicting sediment yield from climate and topography, *Jour. Hydrology,* **21,** 371–380.

Lal, R., 1974. Soil erosion in shifting agriculture. *FAO Soils Bull. No. 24,* 48–71.

Lattanzi, A. R.; Meyer, L. D.; and Baumgardner, M. F., 1974. Influences of mulch rate and slope steepness on interrill erosion, *Soil Sci. Soc. America Proc.,* **38,** 946–950.

Leopold, L. B.; Wolman, M. G.; and Milier, J. P., 1964. *Fluvial Processes in Geomorphology.* San Francisco: Freeman, 522p.

Lowdermilk, W. C., 1950. Conquest of the land through seven thousand years, **SCS MP-32,** 38p.

Meyer, L. D.; DeCoursey, D. G.; and Romkens, M. J. M., 1976. Soil erosion concepts and misconceptions, *Proc. Third Federal Inter-Agency Sedium. Conf., Denver, Colo., Symposium,* **2,** 1–12.

Meyer, L. D.; Foster, G. R.; and Nikolov, S., 1975. Effect of flow rate and canopy on rill erosion. *Trans. ASAE,* **18,** 905–911.

Meyer, L. D ; Foster, G. R.; and Romkens, M. J. M., 1975. Source of soil eroded by water from upland slopes, *Proc. of the 1972 Sediment-Yield Workshop, Oxford, Mississippi,* **ARS-S-40,** 177–189.

Meyer, L. D., and Mannering, J. V., 1967. Tillage and land modification for water erosion control, *Tillage for Greater Crop Production (Conf. Proc., ASAE-ASA-SCSA Tillage Conference), ASAE Pub. Proc.,* 168, 58–62.

Meyer, L. D., and Monke, E. J., 1965. Mechanics of soil erosion by rainfall and overland flow, *Trans. ASAE,* 8, 572–577, 580.

Meyer, L. D., and Romkens, M. J. M., 1976. Erosion and sediment control on reshaped land, *Proc. Third Federal Inter-Agency Sedim. Conf., Denver, Colo., Symposium,* 2, 65–76.

Meyer, L. D., and Wischmeier, W. H., 1969. Mathematical simulation of the process of soil erosion by water, *Trans. ASAE,* 12, 754–758, 762.

Mihara, Y., 1951. Raindrops and soil erosion, *Natl. Inst. Agric. Sci., Tokyo, Japan, Bull.* Ser. A., 76p.

Onstad, C. A., and Foster, G. R., 1975. Erosion modeling on a watershed, *Trans. ASAE,* 18, 288–292.

Partheniadas, E., 1971. Erosion and deposition of cohesive materials, *in* H. W. Shen, ed., *River Mechanics.* Ft. Collins, Colorado: H. W. Shen, 25-1 to 25-19.

Phillips, R. L., and Beauchamp, K. W., 1966. Parallel terraces, Lincoln, Nebraska: Soil Conservation Service, E&WP Unit, 78p.

Piest, R. F., and Bowie, A. J., 1974. Gully and streambank erosion, *Proc. 29th Annual Meeting, Soil Conserv. Soc. of America, Syracuse, N.Y.,* 188–196.

Ports, M. A., 1975. Urban sediment control design criteria and procedures, *ASAE Paper No. 75-2567* (mimeographed), ASAE Winter Meeting, Chicago, Illinois, 28p.

Robins, J. S., and Neff, E. L., 1963. Principles of the erosion process, *Forest Watershed Management Symposium, Oregon State University,* 127–139.

Robinson, A. R., 1971. Sediment, our greatest pollutant? *Agric. Engin.,* 53, 406–408.

Rowlison, D. L., and Martin, G. L., 1971. Rational method for describing slope erosion, *Jour. of Irrigation and Drainage Div., ASCE,* 97, 39–50.

Schumm, S. A., 1969. A geomorphic approach to erosion control in semiarid regions, *Trans. ASAE,* 12, 60–68.

Schwab, G. O.; Frevert, R. K.; Edminster, T. W.; and Barnes, K. K., 1966. *Soil and Water Conservation Engineering.* New York: Wiley, 683p.

Smith, D. D., and Wischmeier, W. H., 1962. Rainfall erosion, *Adv. Agron.,* 14, 109–148.

Smith, R. E., 1976. Simulating erosion dynamics with a deterministic distributed watershed model, *Proc. Third Federal Inter-Agency Sedim. Conf., Denver, Colo., Symposium,* 1, 163–173.

Soil Conservation Service, 1966. Procedure for determining rates of land damage, land depreciation and volume of sediment produced by gully erosion, *SCS Technical Release No. 32,* 18p.

Soil Conservation Service, 1969. *Engineering Field Manual for Conservation Practices.* (available at every local SCS district office).

Soil Conservation Service, 1975*a*. Procedure for computing sheet and rill erosion on project areas, *SCS Technical Release No. 51* (rev.), 16p.

Soil Conservation Service, 1975*b*. *Standards and Specifications for Soil Erosion and Sediment Control in Developing Areas.* College Park: Maryland State SCS Office, 200p.

Stallings, J. H., 1953. Mechanics of water erosion, *SCS TP-118,* 26p.

Stallings, J. H., 1957. *Soil Conservation.* Englewood Cliffs: Prentice-Hall, 575p.

Stewart, B. A.; Woolhiser, D. A.; Wischmeier, W. H.; Caro, J. H.; and Frere, M. H., 1975. Control of water pollution from cropland, vol. I, *Cooperative ARS-EPA Manual: ARS-H-S-1 and EPA 600/2-75-026a,* 111p.

U.S. Department of Agriculture, 1965. Soil and water conservation needs–A national inventory, *USDA Misc. Pub. 971,* 94p.

Wischmeier, W. H., 1973*a*. Upslope erosion analysis, *in* H. W. Shen, ed., *Environmental Impact on Rivers.* Ft. Collins, Colorado: H. W. Shen, 15-1 to 15-26.

Wischmeier, W. H., 1973*b*. Conservation tillage to control water erosion, *Proc. National Conserv. Tillage Conf., Des. Moines, Iowa,* 133–141.

Wischmeier, W. H., 1974. New developments in estimating water erosion, *Proc. 29th Annual Meeting, Soil Conserv. Soc. of America, Syracuse, N. Y.,* 179–186.

Wischmeier, W. H., 1975. Estimating the soil loss equation's cover and management factor for undisturbed areas, *Proc. 1972 Sediment-Yield Workshop Oxford, Mississippi,* ARS-S-40, 118–124.

Wischmeier, W. H.; Johnson, C. B.; and Cross, B. V., 1971. A soil erodibility nomograph for farmland and construction sites, *Jour. Soil Water Conserv.,* 26, 189–193.

Wischmeier, W. H., and Mannering, J. V., 1969. Relation of soil properties to its erodibility, *Soil Sci. Soc. America Proc.,* 23, 131–137.

Wischmeier, W. H., and Smith, D. D., 1965. Predicting rainfall-erosion losses from cropland east of the Rocky Mountains, *U.S. Department of Agriculture Handbook 282,* 47p.

Cross-references: *Conservation; Management of Soils; Soil Drainage. Vol. III: Erosion; Gully Erosion; Sediment Transport; Sheet Erosion. Vol. IVA: Soil Erosion. Vol. XVI: Erosion Hazards; Erosion and Sediment Pollution Control; Gullies and Gully Control; Watershed Management.*

WATER FLUXES

Water is seldom at rest in soil. The amount of water and its rate and direction of movement is of fundamental importance to the many physical, chemical, and biological processes that occur within the soil profile, especially within the rooting zone of crops. The rate of water movement in soil, or the *soil-water flux,* is the quantity of water that passes through a unit cross-sectional area of soil normal to the flow direction in a unit time. Although water moves in all directions within soils, certain directional fluxes are referred to by specific names. The *evaporation rate* is the upward soil-water flux at the soil surface (see *Evaporation*) The *infiltration rate* is the downward soil-water flux at the soil surface (see *Infiltration*). *Deep percolation* is the downward soil-water flux below the rooting zone of a crop (see *Percolation*). Other fluxes of interest are: the flux of water to drainage tubes, the flux to a plant root, the flux to a germinating seed, the flux to a fertilizer granule, and the flux to insect larvae that develop in the soil.

Flux Equations

The flow of water in soil is a special case of fluid flow in porous media (DeWiest, 1969; Baver, Gardner, and Gardner, 1972; Nielsen et al, 1972). The *Darcy equation* may be written as

$$J = K\nabla\Psi \qquad (1)$$

where J is the soil-water flux, K the hydraulic conductivity, and $\nabla\Psi$ the total potential (Baver, Gardner, and Gardner, 1972; Klute, 1973; Bouwer and Jackson, 1974). The complexity of measuring soil-water flux becomes apparent when K and Ψ are examined. The hydraulic conductivity is temperature dependent, and very markedly dependent on the soil-water content. A loam soil at a water content one-half that of saturation may have a hydraulic conductivity 1/10,000 of that when the soil is saturated. The total potential Ψ is the sum of five potentials: matric, gravity, pressure, osmotic, and overburden (for definitions and discussions of these potentials, see *Water Content and Retention*). Generally, the matrix and gravity potentials are dominant. Above the water table the matrix potential is always negative. The absolute value of the matrix potential is frequently referred to as *tension* or *suction*. For a particular soil a relation between the matrix potential and the water content (known as the *water-retention relation*) can be obtained experimentally.

Equation 1 indicates that soil-water flux is directly proportional to the potential gradient. Some experiments have shown that the assumed proportionality does not always hold for water flow in soils. Swartzendruber (see Chapter 6 in DeWiest, 1969) discussed some examples of nonproportionality and proposed some possible explanations of its cause. Usually, however, proportionality is assumed, and equation 1 is used to describe the flux of water in soils.

If the matrix potential is dominant, then the derivation of the water-retention relation ($d\psi_m/d\theta$, where θ is the volumetric water content) can be used to translate equation 1 to a water-content basis, that is,

$$J = D\nabla\theta \qquad (2)$$

where D is equal to $K\partial\Psi/d\theta$ and is the soil water diffusivity (Philip, 1955; Klute, 1973).

In addition to the various potential gradients that cause soil-water flow, water in soil moves in response to temperature gradients. Philip and deVries (1957) have outlined a theory to describe soil-water flux under both water-content and temperature gradients. Under these conditions four diffusion coefficients are required. The flux is given by

$$J = -D_{\theta l}\nabla\theta - D_{\theta v}\nabla\theta - D_{Tl}\nabla T - D_{Tv}\nabla T \qquad (3)$$

where $D_{\theta l}$ is the "isothermal" (but temperature-dependent) liquid water diffusivity; $D_{\theta v}$ the "isothermal" (and temperature-dependent) water-vapor diffusivity; D_{Tl} is the "thermal" (but water-content dependent) liquid diffusivity; and D_{Tv} is the "thermal" (and water content dependent) water-vapor diffusivity (Jackson et al, 1974).

Magnitudes of Fluxes

Figure 1 gives an example of soil-water flux at the soil surface. The figure depicts the diurnal trends in evaporation of water from bare soil one day after irrigation (curve A) and five days after irrigation (curve B). One day after irrigation, the evaporation rate of 2.2×10^{-5} cm s^{-1} peaked at 1430 h, mountain standard time. For the fifth day, the peak rate was 0.55×10^{-5} cm s^{-1}, at about 1300 h. As time proceeds with no further additions of water, the peak rate continues to decrease. However, the diurnal trend of soil-water flux as shown in Figure 1 is evident for many days after irrigation.

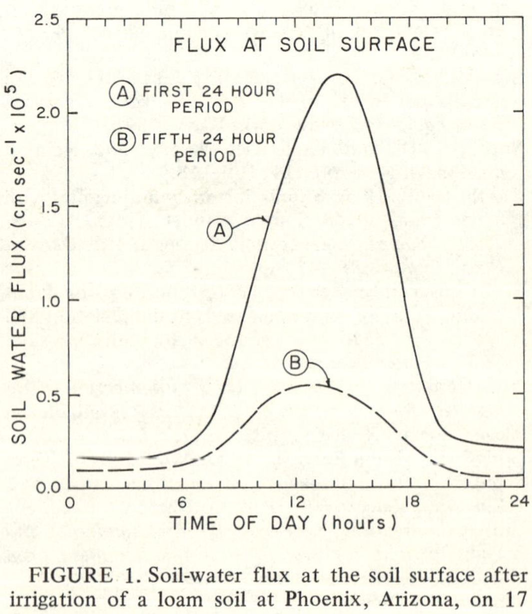

FIGURE 1. Soil-water flux at the soil surface after irrigation of a loam soil at Phoenix, Arizona, on 17 September 1973.

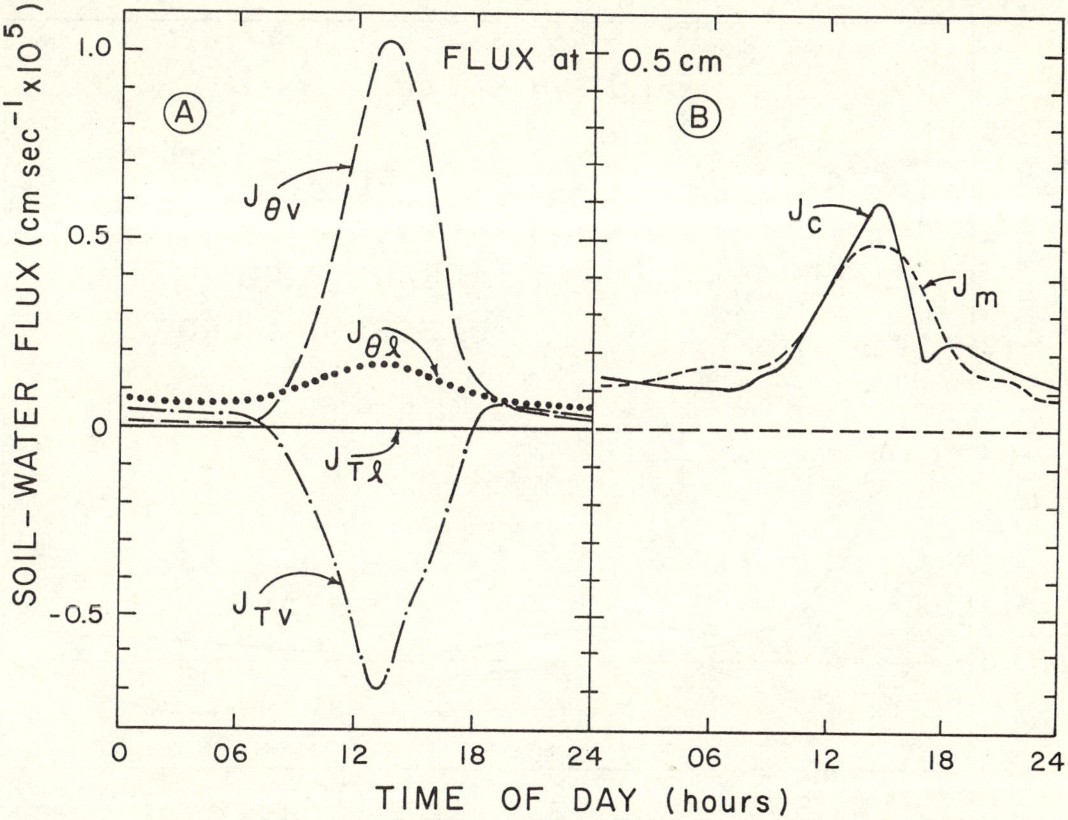

FIGURE 2. Soil-water flux at 0.5 cm depth in a loam soil at Phoenix, Arizona on 12 March 1971, 10 days after irrigation. Part *A* shows the various components of soil-water flux: $J_{\theta v}$ is the flux of water vapor, and $J_{\theta l}$ is the flux of liquid water due to a water-content gradient; J_{Tv} and J_{Tl} are the fluxes of water vapor and liquid water due to a temperature gradient. Part *B* shows the total flux calculated using equation 3, J_c, and the measured flux J_m.

Figure 2 shows some factors that determine the total soil-water flux. Part *A* shows four components of the total flux, comprising the four terms on the right-hand side of equation 3. They are: $J_{\theta l} = -D_{\theta l} \nabla \theta$, $J_{\theta v} = -D_{\theta v} \nabla \theta$, $J_{Tl} = -D_{Tl} \nabla T$, and $J_{Tv} = -D_{Tv} \nabla T$. The sum of the four calculated fluxes J_c is shown in part *B* along with the measured total flux for the 0.5 cm depth (see Jackson et al, 1974 for additional details). In a drying soil, the water-content gradient tends to move water upward toward the soil surface in both the liquid and the vapor phases. During night-time hours the temperature gradient moves water upward also, but during the daytime the soil surface becomes much warmer than the soil beneath it, and the temperature gradient tends to move water downward. Thus during the daytime the total soil-water flux is the net result of fluxes whose vectors are in opposite directions. The data in Figure 2 were taken 10 days after irrigation. Since the soil was quite dry, soil-water flux was largely in the vapor phase. The thermal liquid flux J_{Tl} was essentially zero.

A third example is the flux as 170 cm depth as shown in Figure 3 (Van Bavel, Brust, and Stirk, 1968). The flux at this depth decreases rapidly with time after irrigation. Curve *A* shows the flux for a bare soil. Curve *B* is for a soil containing a sorghum crop. Curve *A* approaches zero but is always in a downward direction. For curve *B*, after about 160 h the flux becomes negative, indicating upward flow. The sorghum roots apparently dried the soil profile sufficiently in 160 h so that the potential gradient upward became greater than the downward gravity gradient.

If the amount of water that enters the soil surface as rainfall or irrigation exceeds the amount lost to the atmosphere by evaporation (q.v.) and transpiration, then the flux below the root zone will always be downward. This water

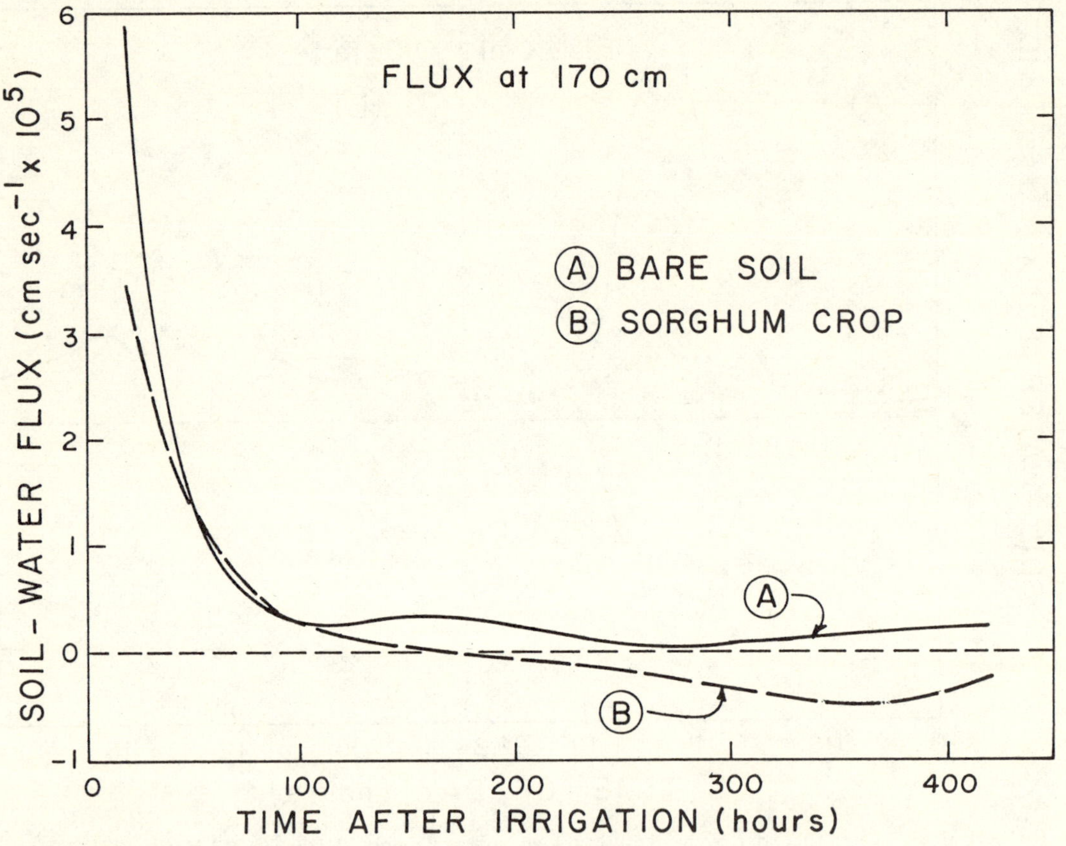

FIGURE 3. Soil-water flux at 170 cm as a function of time after irrigation for soil in which a sorghum crop was growing, at Phoenix, Arizona, April–July 1966. (Adapted from Van Bavel, Brust, and Strik, 1968)

will eventually reach the groundwater table and may then flow in a near-horizontal direction to a stream or spring.

Measurement of Soil-Water Fluxes

The magnitude of soil-water flux within a soil profile can be obtained by directly measuring flux at one plane and measuring water content changes with time within the profile. Van Bavel, Brust, and Stirk (1968) and Jackson et al (1974) measured the flux at the soil surface J_0, using weighing lysimeters (q.v.), and sampled increments of soil for water content. The flux at a particular depth z was given by

$$J_z = J_0 - \int_0^z \frac{\partial \theta}{\partial t} dz \qquad (4)$$

where t is time, and z is depth below the soil surface.

If the hydraulic conductivity (q.v.) is known and the potential gradient measured, the flux can be calculated using equation 1. However, the hydraulic conductivity is difficult to measure precisely in a field soil. Cary (1973) has developed a fluxometer that can be inverted in the soil profile to directly measure soil-water flux. His device uses porous glass bead plates of known hydraulic characteristics. The device works on a principle similar to a heat flux plate, i.e., the gradient is measured across a material whose conductivity is known and the flux is calculated using equation 1.

In soil profiles where the flux of water is predominately downward, the vacuum extraction device of Duke and Haise (1973) can be used. Their device uses a porous ceramic tube within which the water pressure (negative) is kept the same as the soil-water pressure at the same plane as the ceramic tube. In this manner the magnitude of flux can be measured, and a sample of soil-water can be collected for chemical and biological analysis.

Conclusion

The three examples demonstrate the dynamic aspects of soil-water flux. If the flux at the soil surface and below the rooting zone are measured, the water available to crops can be determined. A difficult but important research area is the measurement and modeling of soil-water flux to a growing plant root. Since water carries nutrients with it, the entire well-being of a plant depends on the flow of water to its roots.

RAY D. JACKSON

References

Baver, L. D.; Gardner, W. H.; and Gardner, W. R., 1972. *Soil Physics.* New York: Wiley, 498p.

Bouwer, H., and Jackson, R. D., 1974. Determining soil properties, *in* J. Van Schilfgaard, ed., *Drainage for Agriculture.* Madison: American Society of Agronomy, 611–672.

Cary, J. W., 1973. Soil water flowmeters with thermocouple outputs, *Soil Sci. Soc. America Proc.*, **37**, 176–181.

DeWiest, R. J. M., 1969. *Flow Through Porous Media.* New York: Academic, 530p.

Duke, H. R., and Haise, H. R., 1973. Vacuum extractors to assess deep percolation losses and chemical constituents of soil water, *Soil Sci. Soc. America Proc.*, **37**, 963–964.

Jackson, R. D.; Reginato, R. J.; Kimball, B. A.; and Nakayama, F. S., 1974. Diurnal soil water evaporation: Comparison of measured and calculated soil-water fluxes, *Soil Sci. Soc. America Proc.*, **38**, 861–866.

Klute, A., 1973. Soil water flow theory and its application in field situations, *in* R. R. Bruce, K. W. Flach, and H. M. Taylor, eds., *Field Soil Water Regime.* Madison: Soil Science Society of America, 9–35.

Nielsen, D. R.; Jackson, R. D.; Evans, D. D.; and Cary, J. W., 1972. *Soil Water.* Madison: American Society of Agronomy and Soil Science Society of America, 175p.

Philip, J. R., 1955. The concept of diffusion applied to soil water, *Natl. Acad. Sci., India Proc.*, **24A**, 93–104.

Philip, J. R., and deVries, D. A., 1957. Moisture movement in porous materials under temperature gradients, *Trans. Am. Geophys. Union*, **38**, 222–232.

Van Bavel, C. H. M.; Brust, K. J.; and Stirk, G. B., 1968. Hydraulic properties of a clay loam soil and the field measurement of water uptake by roots: II. The water balance of the root zone, *Soil Sci. Soc. America Proc.*, **32**, 317–321.

Cross-references: *Conductivity, Hydraulic; Evaporation; Flow Theory; Imbibition; Infiltration; Percolation; Thermal Regimes; Transport Processes; Water Content and Retention; Water Movement; Wetting Front.*

WATER MOVEMENT

Water movement into and through the soil is essential to the growth of natural vegetation and agricultural crops and essential in the replenishment of the groundwater aquifer, which is used increasingly as a water source for domestic and industrial purposes. Liquid entering the soil can have a natural origin such as rain but may also be added by man as irrigation water (q.v.) or as a liquid waste to be purified during soil percolation by processes of filtration and adsorption (see *Sludge Disposal*). Inadequate infiltration of liquid into the soil may result in surface runoff, which can have very serious environmental effects due to sediment and nutrient pollution of lakes and streams. Such processes may be caused by crusting or compaction of the infiltrative surface, which results in a sharp decline of the infiltrative capacity. Soils with a natural vegetation and a permanent vegetational cover have generally higher rates of infiltration (q.v.) and less erosion as compared with those that are used for agricultural practices that involve use of heavy machinery (see *Tillage*) and that imply the periodic exposure of the soil surface to the mechanical impact of raindrops. Water movement often implies the occurrence of flow through unsaturated soil, in which not all but only the finer pores are filled with liquid, down to the level of the groundwater, where all pores are filled with water. Major advances have been made in recent years in soil physics and hydrology in describing processes of water movement in soils (see references).

Physical Theory of Water Movement

Soil Water Potentials. Movement of water occurs from locations in the soil where the water in the soil has a high energy to locations where that energy is lower. The energy of soil water (q.v.) must be expressed in terms of the total potential, which is energy per unit quantity. The *total potential* of water Ψ is defined as the mechanical work required to transfer unit quantity of water reversibly and isothermally from a standard reference state, where Ψ is taken as zero, to the situation where the potential has the defined value. A pool of pure water at an elevation that can be arbitrarily specified and that experiences a gas pressure of one standard atmosphere has been adopted as this standard reference state by the ISSS (1963). Many forces may affect the total potential of water in the soil. The *gravitational potential Z* is due to the attraction by the earth and increases with height from the earth's surface. The pressure in soil water at a particular point will be greater than one atmosphere if that point is submerged beneath a free-water surface, which can be the level of the groundwater. The potential due to this cause is the *submergence potential S.* If the soil is unsaturated, the pressure in the soil water is less than that of the local atmosphere. It is convenient to refer to a pressure less than atmospheric as a "suction" or "tension" (see *Wetting Front*). The potential due to this cause is the *matric* or *capillary potential M.* From their character it follows that the *S* and *M* potential are mutually exclusive possibilities. Another possible cause of pressure change in soil water is a change in the pressure of the air adjacent to it. This may be expressed by the term *pneumatic potential G.* It is convenient to define the *pressure potential P*: $P = M$ (or S) $+ G$, which sums the effect of all pressure changes on the potential of soil water, whatever the cause of these pressure changes may be. In addition, osmotic effects of solutes in soil water must be considered because they affect the thermodynamic properties of the soil water. The effect of solutes on the total potential of soil water, expressed as the *osmotic potential O,* becomes of primary significance if the water is separated by a membrane whose permeability to water molecules differs from that to solute molecules. This aspect is relevant for water movement into and through plant roots. The total potential Ψ of soil water is the sum of all the component potentials:

$$\Psi = Z + P + O$$

It is often convenient to define potentials on a unit weight basis rather than on a unit mass basis. For example, for the gravitational potential Z, it follows that $Z_w = W/\rho g(dv) = z$ (cm) where W equals $\rho(dv)\,gz$; representing the work needed to transfer the quantity dv over a distance z (ρ equals density of water; g equals acceleration of gravity). For many flow problems, only the Z and P potentials are relevant. Therefore a *hydraulic potential* ϕ has been defined as follows: $\phi = Z + P$. When potentials are expressed per unit weight, it follows that $\phi = H$ (cm). The height H is called the *hydraulic head.*

Soil Water Movement and the Law of Darcy. Whatever the state of saturation, the volume flux, or the volume of water crossing unit area per unit time, is proportional to the gradient of the hydraulic potential in the direction of the flux (see *Water Fluxes*). Thus denoting distance as *s*, we describe the law of Darcy

$$v = -K\frac{dH}{ds} \quad \text{or} \quad v = -K\left(\frac{dP}{dz} + 1\right) \qquad (1)$$

Clearly, v has to be considered as a bulk flow velocity or rather as a hypothetical velocity that the water would have if flowing through the given cross-section quite unobstructed by solid particles. *Hydraulic conductivity K* is strongly dependent both on the detailed pore geometry of the soil and on its water content (see *Water Content and Retention*). Thus K is the volume flux resulting from unit gradient in hydraulic potential in the particular soil and water situation under consideration (ISSS, 1963). Values of K are characteristic for different soils at a given moisture content. Flow rates, on the contrary, may vary greatly as a function of varying environmental conditions that determine the hydraulic gradients in the flow system. The analysis of vertical entry of water or drainage in the soil requires the addition of a term to satisfy the continuity requirement. Moreover, what is usually measured as v is not a real velocity but a change in volumetric moisture content θ in time at various depths. An expression for $\partial\theta/\partial t$ can be obtained by differentiating equation 1 with respect to z and by substituting

$$D = -K\frac{\partial P}{\partial \theta}$$

where D is the *soil water diffusivity.*

$$\frac{\partial v}{\partial z} = -\frac{\partial}{\partial z}\left(D\frac{\partial \theta}{\partial z}\right) - \frac{\partial K}{\partial z} \qquad (2)$$

Considering the continuity requirement

$$\frac{\partial \theta}{\partial t} = \frac{-\partial v}{\partial z}$$

this changes to

$$\frac{\partial \theta}{\partial t} = \frac{\partial}{\partial z}\left(D\frac{\partial \theta}{\partial z}\right) + \frac{\partial K}{\partial z} \qquad (3)$$

Sometimes it is advantageous to use the pressure-head form of the equation by using the *water capacity C,* which is defined as $d\theta/dP$, as follows:

$$C\frac{\partial p}{\partial t} = \frac{\partial}{\partial z}\left(K\frac{\partial p}{\partial z}\right) + \frac{\partial K}{\partial z} \qquad (4)$$

Flow equations 3 and 4 can be solved analytically or numerically. Many examples are presented in the literature (see review by Klute, 1973). For many field applications, approximate methods can be used that are much easier to work with but physically still quite acceptable in view of the variability found in the field (Bouma, 1973; Klute, 1972).

Methods to Determine *K*

Hydraulic conductivity *K* values are needed to solve flow problems (see *Flow Theory*). Many methods, both for laboratory and field use, are available (see review by Klute, 1972). All analytical methods require the measurement of soil-moisture potentials with tensiometers at at least two locations and an estimate of a flow rate, occurring at that time of measurement. The flow rate, which may be upward or downward, is usually derived from successive measurements of moisture contents. Values of *K* are calculated with equation 1. Generally, in situ measurements are preferable to those using "undisturbed" samples in the laboratory. The sampling procedure, transport, and laboratory conditions may induce changes in the sample that cause deviations from conditions existing in the field. Figure 1 shows a series of representative *K* curves for a sand, a sandy loam, a silt loam, and a clay soil. These curves, showing *K* as a function of the (negative) matric or capillary potential *M*, which can also be described as a positive soil-moisture tension in centimeters or millibar, express the effect of the pore-size distribution of the soil on the value of *K*. Sands, with only relatively coarse pores, have a relatively high value of *K* at saturation, but *K* drops strongly with increasing tension. Clays, with only relatively fine pores, have a low *K* at saturation, but *K* decreases more slowly on increasing tension. The curve for the silt loam demonstrates the physical effect of the occurrence of relatively large cracks between aggregates, and root and worm channels. The aggregates are fine-porous, and these fine pores hardly contribute to flow. The large structural pores result in high *K* values at saturation (140 cm/day for the silt loam), but *K* values drop very strongly between saturation and a tension of only 20 cm (1.5 cm/day) because then these larger voids are filled with air and do not contribute to flow, and flow can occur only through the peds. The functional relationships between pore sizes and permeability (q.v.) has been used in attempts to calculate *K* curves (see review by Klute, 1972; Anderson and Bouma, 1973). The strong effect of pore size (see *Soil Pores*) on measured permeabilities can be explained by considering that the flow rate through a tubular pore is proportional to r^4 (r equals radius pore), whereas flow through a plane slit is proportional to D^3 (D is the width plane slit) (Childs, 1969). Values of *K* are increasingly used in simulation models that attempt to predict soil hydrology as a function of climatic conditions and growth of certain crops (Amerman, 1973; De Laat, van den Akker, and van de Nes, 1975). These models require *K* curves that have to apply to certain areas in a landscape rather than to just a point of measurement. Studies on variability of *K* in the field are limited, but a considerable variation is usually found (Nielsen, Biggar, and Erh, 1973). Use of soil maps, prepared in the context of the soil survey program, which is in progress in many countries, may be quite valuable to identify specific areas that may be expected to have particular types of *K* "bands." The soil survey program also defines certain soil-moisture regimes for the types of soil shown on the maps. Criteria used to define such regimes become increasingly specific although they still express a significant amount of practical experience and know-how of soil surveyers and farmers (Soil Survey Staff, 1951, 1975).

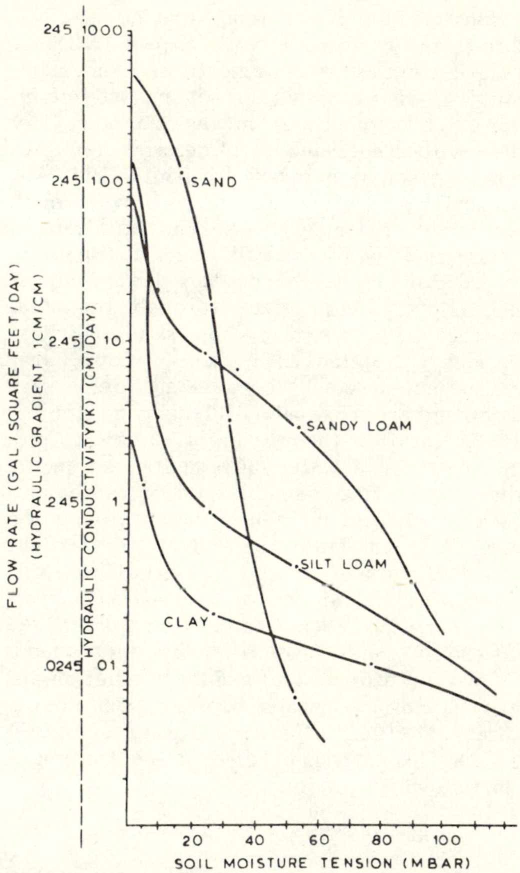

FIGURE 1. Hydraulic conductivity *K* as a function of soil moisture tension measured *in situ* with the crust-test procedure.

Complications Associated with Flow Theory

Flow theory, as expressed by equations 2, 3, and 4, really applies only to a very idealized porous medium that is rigid, does not swell or shrink, and is isotropic with respect to Darcy flow. Natural soils are anisotropic, and many soils with clay do swell on wetting (see *Imbibition*). This sometimes means that the flow theory does not adequately describe flow processes in soil (see review by Klute, 1973). A particular problem occurs during flow through natural heterogeneous soils. The flow rate in the Darcy equation is a bulk rate. The real velocity of flow inside the pores is always higher since flow cannot occur through the solid particles in the soil. When saturated flow through aggregated clayey soils follows larger voids such as cracks between aggregates and worm and root channels (Anderson and Bouma, 1973), the flow theory cannot be used to predict patterns of water movement. Instead, the theory of hydrodynamic dispersion can be useful to predict the moment of breakthrough of a newly applied liquid, as well as the moment when all initially present liquid has been displaced. The calculation of an *apparent dispersion coefficient D* (not to be confused with the diffusivity) can be made from data derived from a breakthrough curve of a chloride solution displacing untraced water from a soil column (Brenner, 1962; Nielsen, Biggar, and Erh, 1972). This analysis is based on the solution of the following equation:

$$\frac{\partial C}{\partial T} + U\frac{\partial C}{\partial X} = D\frac{\partial^2 C}{\partial X^2} \qquad (5)$$

where T is the time from the commencement of the displacement, X is the distance from the point of introduction of the displacing fluid, $C = C\ (X,\ T)$ is the solute concentration, D is the axial dispersion coefficient, and U is the average interstitial velocity of the fluid. Equation 5 can be solved for specific boundary conditions (Brenner, 1962). Additional terms have to be added to express adsorption or precipitation of percolating compounds.

JOHANNES BOUMA

References

Amerman, C. R., 1973. Hydrology and soil science. *in* R. R. Bruce, K. W. Flach, and H. M. Taylor, eds., *Field Soil Moisture Regime.* Madison: Soil Science Society of America, Special Publication No. 5, 167–180.

Anderson, J. L., and J. Bouma, 1973. Relationships between saturated hydraulic conductivity and morphometric data of an argillic horizon, *Soil Sci. Soc. America Proc.*, **37**, 408–413.

Baver, L. D.; Gardner, W. H.; and Gardner, W. R., 1972. *Soil Physics.* New York: Wiley, 498p.

Bouma, J., 1973. Use of physical methods to expand soil survey interpretations of soil drainage conditions, *Soil Sci. Soc. America Proc.*, **37**, 413–421.

Bouma, J.; Baker, and Veneman, 1974. Measurement of water movement in soil pedems above the water table, *Information Circular No. 27,* Geol. and Nat. Hist. Survey. Madison Wi.: University of Wisconsin.

Brenner, H., 1962. The diffusion model of longitudinal mixing in beds of finite length, *Chem. Eng. Sci.*, **17**, 229–243.

Childs, E. C., 1969. *The Physical Basis of Soil Water Phenomena.* New York: Wiley, 439p.

De Laat, P. J. M.; Akker, C. van den, and Nes, Th. J. van de., 1975. Consequences of groundwater extraction on evapotransporation and saturated-unsaturated flow, *Proc. Symp. Application of Mathematical Models in Hydrology and Water Resources Systems (Bratislava). IAHS-AISH Publ. No. 115.*, 67–76.

Hillel, D. I., 1971. *Soil and Water: Physical Principles and Processes.* New York: Academic, 288p.

International Society of Soil Science (ISSS), 1963. Soil physics terminology, *Bulleton No. 22,* 1–5.

Klute, A., 1972. The determination of the hydraulic conductivity and diffusivity of unsaturated soils, *Soil Sci.*, **113**(4), 264–277.

Klute, A., 1973. Soil water flow theory and its application in field situations, *in* R. R. Bruce, K. W. Floch, and H. M. Taylor, eds., *Field Soil Moisture Regime.* Madison: Soil Science Society of America Special Publication No. 5, 9–35.

Nielsen, D. R., 1972. *Soil Water.* Madison: American Society of Agronomy, 175p.

Nielsen, D. R.; Biggar, J. W., and Erh, K. T., 1973. Spatial variability of field measured soil water properties, *Hilgardia*, **42**(7), 215–260.

Rose, C. W., 1966. *Agricultural Physics.* Oxford: Pergamon, 226p.

Soil Survey Staff, 1951. Soil Survey Manual, *U.S. Department of Agriculture Handbook No. 18,* 503p.

Soil Survey Staff, 1975. Soil Taxonomy: A Basic System of Soil Classification for Making and Interpreting Soil Surveys, *U.S. Department of Agriculture Handbook No. 436,* 754p.

Cross-references: *Capillary Pressure; Conductivity, Hydraulic; Field Water Cycle; Flow Theory; Imbibition; Leaching; Percolation; Permeability; Soil Drainage; Soil Solution; Soil Water; Water Content and Retention; Water Fluxes; Wetting Front.*

WETTING FRONT

As water penetrates a dry soil, the wetted soil takes a darker color in contrast to its original light appearance. Because of this change in color, the transition zone between the dry soil and the moistened soil can be appreciated visually. For a number of soils the transition is quite sharp, and the propagation of this color change, as more water moves into the soil, can be followed by eye readily. The region of rapid change in color resulting from an increased water content is called the *wetting front.*

From this clear visual separation between the light dry soil and the dark moist soil, one might

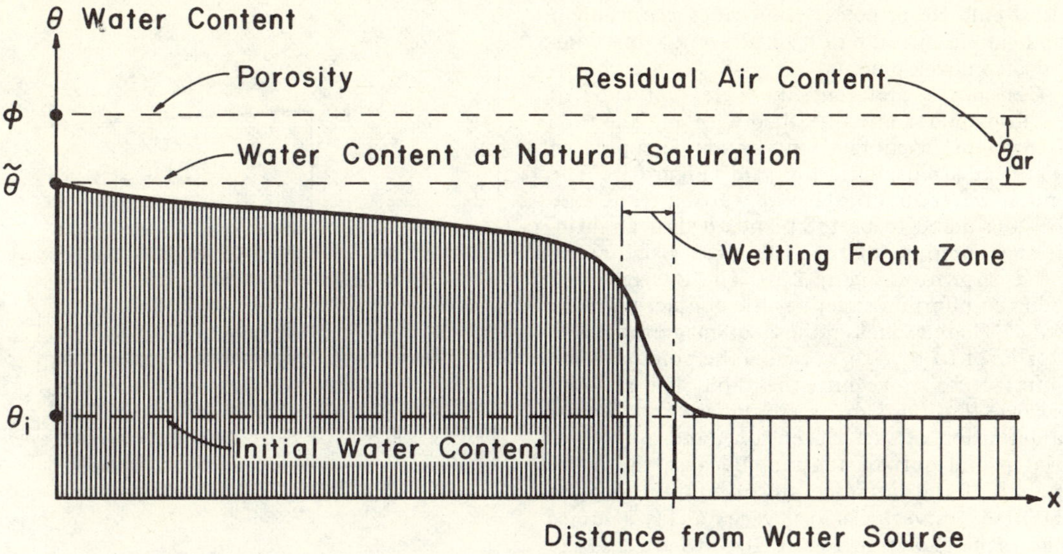

FIGURE 1. Variation of water content and soil color with distance from water source.

infer that ahead of the wetting front, the soil is in its original state of dryness and that behind the wetting front, water now fully occupies the pore space vacated by the displaced air. But this is not the case. If one were to measure water content as a function of distance away from the water source, one would obtain a profile as shown in Figure 1. Though there is no discernible difference in color behind the wetting front, the water content behind it is far from uniform. In fact, it reaches its maximum value $\tilde{\theta}$, that is, a state of full saturation except for a certain amount of trapped residual air, only close to the source of water.

Whereas the phenomenon of propagation of the wetting front must have been known to primitive man as a matter of daily observation, the mechanism by which water actually invades the soil was not elucidated until the seventeenth century. Certainly if water is poured on top of a soil column, one expects water to percolate through the soil by gravity. On the other hand, if a thin layer of water and a soil sample lie side by side as shown in Figure 2(a), in contact, on top of a horizontal tray, why would water, at atmospheric pressure, penetrate the porous medium and displace air, also at atmospheric pressure? Yet, apparently for no reason, it does. It is as though it were sucked in!

Propagation in a Tube

The thorough understanding of the reason why water, at atmospheric pressure, will penetrate a porous medium and displace another fluid such as air also at atmospheric pressure, is of capital importance for a sound physical understanding of water movement (q.v.) in soils. To the extent that a porous medium, as a first rough approximation, can be visualized as a bundle of unconnected cylindric tubes, some small, some large, there is merit in the study of the wetting of a tube. This highly idealized model of the porous medium is an obviously inaccurate picture of an extremely complex reality. Its merits as a model of reality are al-

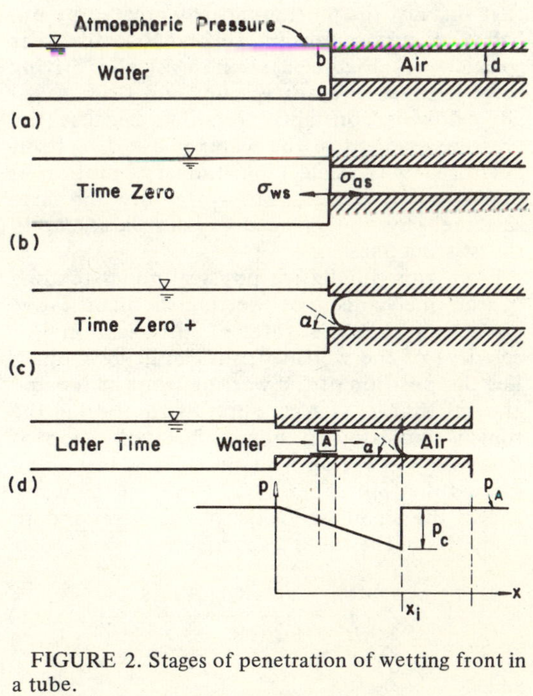

FIGURE 2. Stages of penetration of wetting front in a tube.

most null. Nevertheless, it provides a convenient and simple introduction to ulterior, more complicated developments.

Consider a preferentially water-wet tube of small diameter initially filled with air. The tube is brought in contact with an ample supply of water at atmospheric pressure (Figure 2). The initial water-air interface or wetting front may be considered to be the plane *ab*. For an infinitesimal amount of time the force diagram will be as shown on Figure 2(b). The interface will be in equilibrium except at the contact with the solid. There, a net impulse of magnitude proportional to $\sigma_{as} - \sigma_{ws}$ causes the point *a* of the interface to move into the tube. The air-solid surface tension is σ_{as} (see *Capillary Pressure* for a definition of surface tension), and σ_{ws} is the water-solid surface tension. Because the other points of the interface are in equilibrium, the interface curves in, and it very quickly assumes the shape shown in Figure 2(c), where α is the equilibrium wetting angle. The interface is now in equilibrium because the surface tension force and the capillary pressure force balance one another. Because the pressure drop in the moving air phase is essentially negligible compared to that in water, the pressure in the air just ahead of the interface is essentially atmospheric p_A. Just behind the interface, the pressure in the water p_w is less than atmospheric by an amount equal to the *capillary pressure* p_c. At the inlet the pressure is atmospheric. As a result, there exists an unbalanced pressure gradient in the water phase that causes the water to flow and that sustains the motion initially created by unbalanced surface-tension forces. Viscous shear (particle *A*, Figure 2) will resist the driving force tending to enforce a uniform flow. Actually, flow is constantly retarded because the pressure gradient in the water phase p_c/x_i itself decreases with time. In summary, motion is originated by preferential wettability and later sustained by the induced pressure gradient in the wetting phase.

From this qualitative physical understanding of the mechanism of wetting, a quantitative prediction for the rate of propagation (or celerity) of the wetting front can be developed. Let the position of the wetting front be located at x_f at time *t*. The pressure distribution in the tube is indicated in Figure 3. Application of *Poiseuille's equation* for laminar flow in a cylindrical tube (Morel-Seytoux, 1969) in the water and in the air phase yields for the water and air velocities in the tube, respectively,

$$u_w = \frac{dx_f}{dt} = \frac{R^2}{8\mu_w} \frac{(p_A - p_w)}{x_f} \qquad (1)$$

and

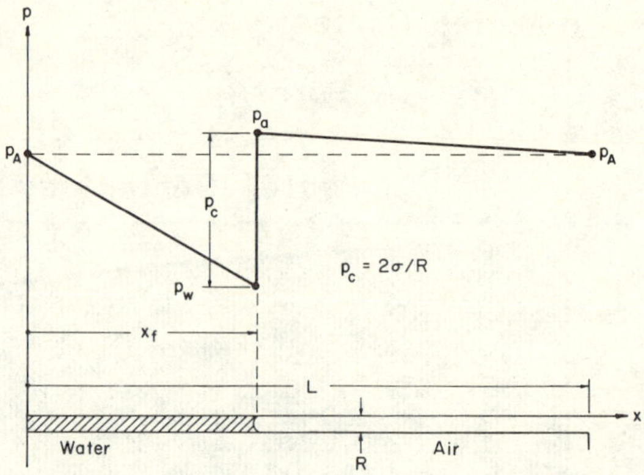

FIGURE 3. Fluid pressure distribution in a wetting tube.

$$u_a = \frac{dx_f}{dt} = \frac{R^2}{8\mu_a} \frac{(p_a - p_A)}{(L - x_f)} \qquad (2)$$

where L is the length of the tube and μ_w and μ_a are the viscosities of water and air. As the two velocities must be equal, it follows by equating u_w and u_a, that the unknowns p_w and p_a must be related by the equalities,

$$\frac{p_a - p_A}{\mu_a (L - x_f)} = \frac{p_A - p_w}{\mu_w x_f} = \frac{p_c}{\mu_a (L - x_f) + \mu_w x_f} \qquad (3)$$

from which the velocity dx_f/dt can be obtained as a function of the wetting-front position, namely,

$$\frac{dx_f}{dt} = \frac{R^2}{8} \left[\frac{p_c}{x_f (\mu_w - \mu_a) + \mu_a L} \right] \qquad (4)$$

If the viscosity of air is neglected compared to that of water and if one expresses the capillary pressure as an equivalent water height h_c, equation 4 takes the simpler form

$$\frac{dx_f}{dt} = \frac{R^2 \rho_w g}{8\mu_w} \cdot \frac{h_c}{x_f} \qquad (5)$$

After integration one obtains

$$x_f = \left(\frac{R^2 \rho_w g h_c}{4\mu_w} \right)^{1/2} (t^{1/2}) \qquad (6)$$

from which the imbibition rate, $I = (dx_f/dt)$, (that is, volumetric water inflow rate per unit cross-section area) can be deduced as an explicit function of time:

$$I = \frac{1}{2} \frac{S_p}{\sqrt{t}} \quad (7)$$

where

$$S_p = \left(\frac{R^2 \rho_w g h_c}{4\mu_w} \right)^{1/2} \quad (8)$$

can be called the *Poiseuille sorptivity.* This sorptivity characterizes the ability of the tube to imbibe water. Alternately the imbibition rate can be expressed as a function of position of the wetting front, namely,

$$I = \frac{1}{2} \frac{S_p^2}{x_f} \quad (9)$$

Equations 9 and 7 are equivalent since $x_f = S_p \sqrt{t}$. Assuming that the interface has the shape of a spherical cap, the capillary pressure at the wetting front is

$$h_c = \frac{2\sigma \cos \alpha}{R \rho_w g} \quad (10)$$

from which a more explicit form of the Poiseuille sorptivity is obtained

$$S_p = \left(\frac{R \sigma \cos \alpha}{2\mu_w} \right)^{1/2} \quad (11)$$

Propagation in a Soil

If one assumes that in the soil, as in the capillary tube, the wetting front separates the dry soil from a fully saturated soil (Figure 4), then the imbibition rate in a horizontal soil is given by the application of Darcy's law for saturated soil between the soil-entry face and the wetting front, namely,

$$I = \tilde{K} \frac{[H_A - (H_A - H_f)]}{x_f - 0} = \tilde{K} \frac{H_f}{x_f} \quad (12)$$

since the water is at atmospheric pressure at the entry face and is below atmospheric pressure just behind the wetting front. The term $\tilde{K}$ is the hydraulic conductivity (q.v.) at natural saturation. The term H_f is the so-called *wetting-front suction.* Whereas the capillary pressure across a meniscus in a capillary tube is well defined by equation 10, the wetting-front suction is not. For, clearly, there are as many capillary-suction values as there are menisci in all the complex-shaped interconnected capillaries that constitute the real soil. Proceeding by analogy, one can, at best, say that the wetting-front suction is an average capillary suction in some sense across a representative pore radius.

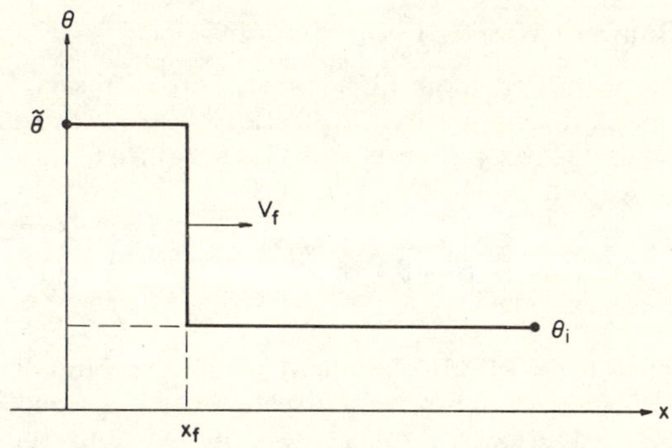

FIGURE 4. The piston displacement model of Green and Ampt.

For flow in the vertical direction, the imbibition (or infiltration) rate is given, again by application of Darcy's law, by the slightly different formula

$$I = \tilde{K} \frac{(H_f + z_f)}{z_f} = \tilde{K} + \tilde{K} \frac{H_f}{z_f} \quad (13)$$

which shows the additive effects of the gravity and capillarity forces on the infiltration rate. The celerity of the wetting front is related to the infiltration rate by the equation

$$I = (\tilde{\theta} - \theta_i) \frac{dz_f}{dt} \quad (14)$$

since only the volume change from θ_i to $\tilde{\theta}$ is available to water rather than the total bulk volume of soil. Combining equations 13 and 14 and integrating leads to a relation between the wetting-front position and time in the form

$$\frac{\tilde{K} t}{\tilde{\theta} - \theta_i} = z_f - H_f \ln \left(1 + \frac{z_f}{H_f} \right) \quad (15)$$

known as the *Green and Ampt equation* (Green and Ampt, 1911). The Green and Ampt equation cannot be used for prediction purposes since the wetting-front suction is not known. However, if an experiment is performed in which the various positions of the wetting front are visually recorded as a function of time, one can deduce indirectly from the observations the value of H_f. The best value of H_f is that value which when used in equation 15 gives predicted positions of the wetting front at given times closest to the actual observations. The wetting-front suction, in spite of its impressive name, has been reduced to the low status of an empirical constant (Childs, 1969).

Bouwer's Wetting-Front Suction

The rather good fit between actual observations of cumulative infiltration W and those predicted by a slightly modified form of equation 15, namely,

$$\tilde{K}t = W - H_f(\tilde{\theta} - \theta_i) \ln \left(1 + \frac{W}{H_f(\tilde{\theta} - \theta_i)}\right) \quad (16)$$

since for a full displacement of air by water $W = z_f(\tilde{\theta} - \theta_i)$, led several investigators to suspect that equation 16 might be valid without the assumption of complete saturation (*piston* displacement) behind the wetting front. The undiscovered relation between the wetting-front suction and the soil characteristic curves such as capillary pressure (see *Capillary Pressure*) and water relative permeabilities (see *Permeability*) kept teasing the brains of various scientists.

Bouwer (1964, p. 142) suggested an approximate determination of H_f by substituting for it what he called the *critical pressure head,* defined by the relation

$$H_b = \int_0^\infty k_{rw}\, dh_c \quad (17)$$

where k_{rw} is the relative permeability to water (that is, relative to the wetting permeability at natural saturation), and h_c is the capillary pressure (expressed as a water height). Intuitively, viewing the porous medium as an ensemble of capillaries, the use of a weighted average of capillary pressure across the wetting front is appealing. The problem is of course the choice of the weight. The treatment of a different problem, that of unsaturated flow above mildly sloping water tables (Bouwer, 1964, p. 130) suggested naturally k_{rw} as the weighting function. By comparison with observations, it was found that the use of H_b yielded accurate predictions of infiltration rates from a canal in sand.

A Two-fluid Formulation

As water penetrates the soil, it displaces the air. The problem of water movement is therefore also a problem of air movement. Traditionally, the description of the air movement has been neglected. The resulting mathematical simplification is secured at the expense of physical significance.

Darcy's law for two immiscible fluids (Morel-Seytoux, 1969) such as air and water can be written in the form

$$v_w = -\frac{kk_{rw}}{\mu_w}\left(\frac{\partial p_w}{\partial z} - \rho_w g\right) \quad (18)$$

$$v_a = -\frac{kk_{ra}}{\mu_a}\left(\frac{\partial p_a}{\partial z} - \rho_a g\right) \quad (19)$$

where v_w and v_a are the fluid velocities (in the Darcy sense), k is the intrinsic permeability, k_{rw} and k_{ra} are the fluid relative permeabilities, ρ_w and ρ_a their specific masses, g is the acceleration of gravity, and z is the vertical coordinate oriented positive downward. The law of conservation for both fluids yields

$$\frac{\partial \theta}{\partial t} + \frac{\partial}{\partial z}(v_w) = 0 \quad (20)$$

and

$$\frac{\partial \theta_a}{\partial t} + \frac{\partial v_a}{\partial z} = 0 \quad (21)$$

where θ is water content, and θ_a is air content. Adding equations 18 and 19 gives

$$v_w + v_a = v = -(\lambda_w + \lambda_a)\frac{\partial p_w}{\partial z} - \lambda_a \frac{\partial p_c}{\partial z} + \lambda_w \rho_w g + \lambda_a \rho_a g \quad (22)$$

where v (the sum of v_w and v_a) is defined as the *total* velocity and λ_w and λ_a (short notations for kk_{rw}/μ_w and kk_{ra}/μ_a) are the fluid *mobilities.* Equation 22 can be rewritten as

$$\frac{v}{\Lambda} = -\frac{\partial p_a}{\partial z} + f_w \rho_w g \frac{\partial h_c}{\partial z} + \rho_w g f_w + \rho_a g \quad (23)$$

where Λ (sum of the mobilities) is the total mobility, and f_w is a function of water content θ defined as

$$f_w = \frac{\lambda_w}{\Lambda} \quad (24)$$

Integration of equation 24 between the water-entry face and just below the wetting front yields

$$\int_0^{z_f} \frac{dz}{\Lambda} = \rho_w g \int_0^{h_{ci}} f_w dh_c + \rho_w g \int_0^{z_f} f_w dz + \rho_a g z_f \quad (25)$$

Adding similarly equations 20 and 21, one obtains

$$\frac{\partial(\theta_w + \theta_a)}{\partial t} + \frac{\partial v}{\partial z} = \frac{\partial \phi}{\partial t} + \frac{\partial v}{\partial z} = \frac{\partial v}{\partial z} = 0 \quad (26)$$

since the porosity does not vary with time if the medium is undeforming and stable. Since v

does not vary with z, it can be taken outside the integral sign in equation 25 and one obtains for v the expression

$$v = \frac{\rho_w g \left[\int_0^{h_{ci}} f_w dh_c + \int_0^{z_f} f_w dz\right]}{\int_0^{z_f} \frac{dz}{\mu_w \Lambda_r}} \tag{27}$$

having neglected the weight of air. Or with minor changes in notations

$$v = \frac{\bar{K}\left(H_c + \int_0^{z_f} f_w dz\right)}{\int_0^{z_f} \frac{dz}{\mu_w \Lambda_r}} \tag{28}$$

where Λ_r is the relative total mobility (i.e., the sum of $(k_{rw}/\mu_w) + (k_{ra}/\mu_a)$, and H_c is the *effective capillary drive* defined by the relation

$$H_c = \int_0^{h_{ci}} f_w dh_c \tag{29}$$

The relevance of equation 28 to the problem of Green and Ampt is clearly understood if one rewrites equation 13 in the form

$$I = \frac{\bar{K}\left(H_f + \int_0^{z_f} dz\right)}{\int_0^{z_f} \frac{dz}{\mu_w(1/\mu_w + 0)}} \tag{30}$$

and if one recognizes that at the water-entry face $v = I$. Indeed, if one assumes full saturation behind the wetting front in equation 27, it transforms into

$$I = \bar{K}\left[\frac{H_c + z_f}{z_f}\right] \tag{31}$$

which by comparison with equation 13 yields the result that the *wetting-front suction* H_f is identical to the *effective capillary drive* H_c. The empirical wetting-front suction turns out to be a well-defined theoretically derivable quantity, namely,

$$H_f = H_c = \int_0^{h_{ci}} f_w dh_c \tag{32}$$

Figure 5 shows a typical f_w versus h_c curve for a sand. For h_{ci} = 66 cm of water, H_c (surface of crosshatched area) = 24 cm.

H. J. MOREL-SEYTOUX

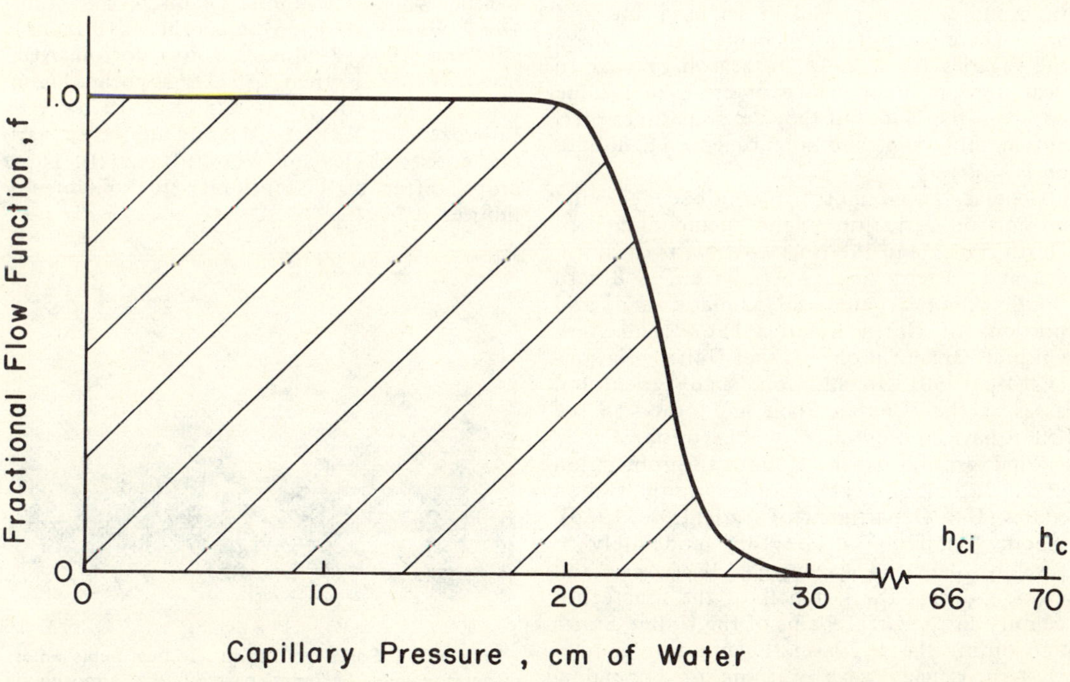

FIGURE 5. Fractional flow function versus capillary pressure (expressed as a water height). Crosshatched area expresses the effective capillary drive in centimeters of water.

References

Bouwer, H., 1964. Unsaturated flow in ground-water hydraulics, *Jour. Hydraul. Div., Am. Soc. Civil Eng.*, 90(HY5), 121–144.

Childs, E. C., 1969. *The Physical Basis of Soil Water Phenomena.* New York: Wiley, 439p.

Green, W. H., and Ampt, C. A., 1911. Studies on soil physics. 1. Flow of air and water through soils, *Jour. Agric. Sci.* **4**, 1–24.

Morel-Seytoux, H. J., 1969. Introduction to flow of immiscible liquids in porous media, *in* R. de Wiest, ed., *Flow Through Porous Meida,* New York: Academic, 455–516.

Morel-Seytoux, H. J., and Khanji, J., 1974. Derivation of an equation of infiltration, *Water Resources Res. Jour.*, **10**(4), 795–800.

Cross-references: *Capillary Pressure; Conductivity, Hydraulic; Flow Theory; Imbibition; Infiltration; Soil Water; Water Fluxes; Water Movement.*

WIND EROSION

Although soil erosion by wind is generally believed to be of consequence only in semiarid and arid areas, it can be a problem wherever soil, vegetative, and climatic conditions are conducive (FAO, 1960). These conditions exist when: (1) The soil is loose, dry, and reasonably finely divided; (2) the soil surface is smooth and vegetative cover is absent or sparse; (3) the field is sufficiently large; and (4) the wind is sufficiently strong to move soil. Those conditions more often prevail in semiarid and arid areas where precipitation is inadequate or where the vagaries from season to season or year to year prevent maintenance of crops or residue cover on the land, but they are sometimes present in subhumid, and sometimes even humid, areas.

General areas most susceptible to wind erosion on agricultural land include much of North Africa and the Near East, parts of southern and eastern Asia, Australia and southern South America, and the semiarid and arid portions of North America [Food and Agricultural Organization of the United Nations (FAO), 1960]. In addition, such agricultural areas as the Siberian Plain and others in the USSR have a potential for wind erosion.

Wind erosion is the dominant problem on about 30 million hectares of land in the United States (U.S. Department of Agriculture, 1965). About 2 million hectares are moderately to severely damaged each year. Extensive soil erosion during the last half of the nineteenth century in the Great Plains of the United States and during the 1920s in the prairie region of western Canada gave warning of impending disaster; during the 1930s a prolonged dry spell culminated in dust storms and soil destruction of disastrous proportions in the prairie regions of western Canada and the Great Plains of the United States (Anderson, 1975; Johnson, 1947; Malin, 1946*a,b,c;* Svobida, 1940).

Wind erosion physically removes from the field the most fertile portion of the soil and therefore lowers productivity (Daniel and Langham, 1936). By relating crop yield to soil thickness and determining potential annual soil loss, Lyles (1975) estimated annual yield reductions of 9.2 thousand tons (metric) of wheat and of 13.8 thousand tons (metric) of grain sorghum on 0.5 million hectares of sandy surface soils in southwestern Kansas.

Some soil from damaged lands enters suspension and becomes part of the atmospheric dustload (Figure 1). Hagen and Woodruff (1973) estimated that eroding lands of the Great Plains contributed 244 and 77 million tons of dust per year to the atmosphere in the 1950s and 1960s, respectively. Dust obscures visibility and pollutes the air, causes traffic hazards, fouls machinery, and imperils animal and human health. Blowing soil fills road ditches; reduces seedling survival and growth; lowers the marketability of vegetable crops like asparagus, green beans, and lettuce; and increases the susceptibility to and transmission of some diseases (Claflin, Stuteville, and Armbrust, 1973; Hayes, 1965, 1966). Extent of damage and plant response to windblown soil-abrasive injuries have been evaluated for several crops: winter wheat (Armbrust, Paulsen, and Ellis, 1974; Woodruff, 1956); green beans (Bubenzer and Weis, 1974; Skidmore, 1966); cotton (Armbrust, 1968; Fryrear, 1971); soybean (Armbrust, 1972); tomato (Greig et al, 1974); peas Bubenzer and Weis, 1974); and alfalfa and various grasses (Lyles and Woodruff, 1960). These crops differ in susceptibility to soil-abrasive injuries.

FIGURE 1. Some soil from eroding fields enters suspension and becomes part of the atmospheric dustload. The suspended dust obscures visibility and pollutes the air.

Erosion Mechanics

Surface Wind. Movement of soil particles is caused by wind forces exerted against the surface of the ground. The average forward velocity of the wind near the ground increases logarithmically with height above the ground surface (Figure 2). The change in velocity with height is known as the *velocity gradient.* This gradient determines the shear stress of drag force exerted on the ground surface.

The velocity gradient or the shape of the adiabatic windspeed profile is given by

$$\frac{\partial u}{\partial z} = \frac{u^*}{kz} \tag{1}$$

where u is mean windspeed at height z above the mean ground surface, k is the von Karman constant (0.4); and u^* is friction velocity further defined as $(\tau/\rho)^{1/2}$ where τ is surface shear (force per unit area), and ρ is fluid density. The surface shear then is

$$\tau = \rho u^{*2} \tag{2}$$

The surface shear associated with the decrease in wind velocity near the surface is a vertical transfer of horizontal momentum. Momentum (mass times velocity) decreases as the surface is approached. The eddy diffusion equation for steady-state, one-dimensional momentum transport is

$$\tau = \rho\, Km \frac{\partial u}{\partial z} \tag{3}$$

where Km is momentum-transfer coefficient.

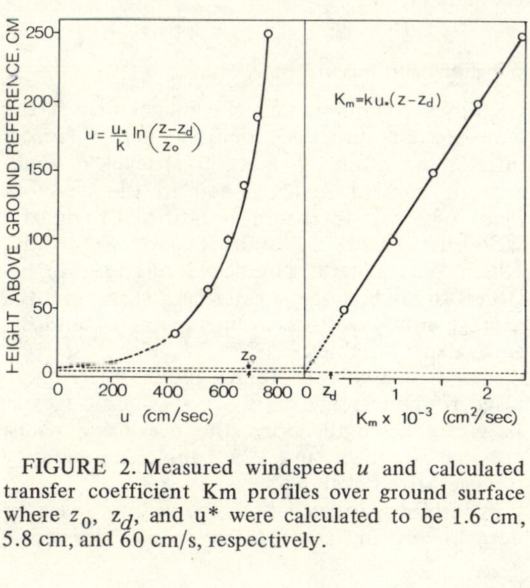

FIGURE 2. Measured windspeed u and calculated transfer coefficient Km profiles over ground surface where z_0, z_d, and u* were calculated to be 1.6 cm, 5.8 cm, and 60 cm/s, respectively.

The integrated form of equation 1 over a rough surface becomes

$$u = \frac{u^*}{k} \ln\left(\frac{z - z_d}{z_0}\right) \tag{4}$$

which is the well-known logarithmic law. The parameter z_d, the effective displacement height, is the distance from the ground surface to the plane at which the momentum-transfer coefficient extrapolates to zero (Figure 2). The roughness parameter z_0 is the distance from the displaced reference plane to the surface at which the wind profile extrapolates to zero.

Equilibrium Forces. In addition to surface shear, another force tending to dislodge a soil grain is a negative pressure on the top as compared with the bottom of the grain. This Bernoulli effect causes lift on the grain (Chepil, 1959*b*). Chepil (1959*b*) analyzed the drag, lift, and gravity forces on soil grains at the threshold of their movement by wind. Equilibrium between those forces and the soil grains was influenced by the diameter, shape, and density of the grains; the angle of repose of the grains with respect to the mean drag level of the fluid; the closeness of packing of top grains, and the impulses of fluid turbulence associated with drag and lift. The relationship was

$$\bar{\tau}_c = \frac{0.66\, g d \rho' \tan \phi \eta}{(1 + 0.85 \tan \phi)\, T} \tag{5}$$

where $\bar{\tau}_c$ is the mean critical drag per unit horizontal area of the whole bed; g is acceleration of gravity; d is diameter of the spherical grain; ρ' is difference in density of grain and fluid; ϕ is angle of repose of the grain with respect to the mean drag level of the fluid; η is ratio of mean drag and lift per unit area on the whole bed to mean drag and lift per unit area on the top grain moved by the fluid; and T is the ratio of maximum to mean drag and lift on the soil grain. Chepil (1959*b*) experimentally determined these values for the constants of equation (5): $T = 2.5$, $\tan\phi = 0.45$, and $\eta = 0.21$. Lyles and Krauss (1971) found, however, that T and η varied with surface roughness.

When the mean critical drag on a particle is exceeded, the particle is dislodged and transported by the wind. This happens for loose grains 0.25 mm in diameter when the friction velocity u^* is 20 to 44 cm/s (Bagnold, 1943; Chepil, 1959*b*; Lyles and Krauss, 1971; Zingg, 1953*a*), which corresponds to the surface drag of 0.48 to 1.94 dyn/cm^2. The windspeed at initial particle movement is from 4.0–5.8 m/s at 30 cm (Chepil, 1945*b*, *c;* Malina, 1941).

Initiation of Particle Motion. The windspeed at which sand movement starts as a result of direct pressure of the fluid was called "fluid threshold" by Bagnold (1943). Bagnold described the initial motion as "surface grains, previously at rest, began to be rolled along the surface by the direct pressure of the wind. . . . A foot or so downwind of the point at which the rolling began, the grains could be seen to have gathered sufficient speed to start bouncing off the ground." Others (Bisal and Nielsen, 1962; Lyles and Krauss, 1971) observed that as the fluid threshold was approached, some particles began to vibrate, or rock back and forth. Erosive particles vibrated with increasing intensity as windspeed increased and then left the surface instantaneously as if ejected. Evidence supported the hypothesis that the particle-vibration frequency is related to the frequency band containing the maximum energy of the turbulent motion.

Saltation. The eroding particle's bouncing or ejecting off the surface bed into the airstream and moving forward is referred to as *saltation.* Almost 50–75% of the movement of soil particles is through saltation (Chepil, 1945*a*). In saltation the particles rise almost vertically, rotating from 20–1000 revolutions per second, they travel 10 times their height of rise, then return to the surface with an angle of descent of about 6–12° from the horizontal (Chepil and Woodruff, 1963; Vanoni, 1975; Zingg, 1953*b*). On striking the surface, they either rebound and continue their movement in saltation or impart most of their energy by striking other grains, causing those grains to rise upward or roll along the surface. Most of the saltating particles range from 0.1–0.5 mm in diameter.

Creep. The rolling or sliding of larger particles with energy derived from saltating particles is called *creep.* Individual grains are knocked onward by the blow they receive from behind. Bagnold (1943) observed that at low windspeeds, the grains move in jerks, a few millimeters at a time; but as the windspeed is increased, the distance moved increases and more grains are set in motion until, in high winds, the whole surface appears to be creeping forward.

Suspension. Particles smaller than about 0.1 mm may enter suspension and be carried to great heights by the eddies of the erosive winds. Impact of particles in saltation usually starts movement of these fine particles. Although most soil is moved by saltation and surface creep, that moved by suspension is the most spectacular and easily recognized from a distance (Figure 1).

Sorting. An eroding wind has been said to act on the soil like a fanning mill acts on grain, removing the fine, porous particles and leaving the coarser and denser behind (Chepil, 1957*a;* Daniel, 1936; Moss, 1935). Of that being removed, the coarsest particles usually end up in a soil drift, and the rest enter suspension to be transported oftentimes great distances before being deposited (Malina, 1941). Chepil (1957*a*) observed that the most distinct feature in the sorting process is the peak (predominant) diameter of saltating grains. Fractions larger than the peak diameter tend to remain in the wind-eroded fields; particles smaller than that diameter tend to be carried, in suspension, far through the atmosphere.

Peak diameter of drifted material derived from sand and loamy sand fields was about 0.4 mm, and that of drifted material from the finer textured soils was about 0.6 mm. The drifted materials derived from sand and loamy sand fields were composed of principally discrete, nonporous grains having an average bulk density of 2.37 g/cm^3. But the materials drifted from the finer textured soils were predominantly aggregates with a distinct degree of porosity and an average bulk density of 1.70 g/cm^3 (Chepil, 1957*a*). By applying equation 5 for peak diameters and average bulk densities, one finds that critical mean drag is about the same for both conditions: 1.7 for the single grain and 1.8 dyn/cm^2 for porous grains.

Very little sorting occurs on some fine-textured soil derived from loess. Moss (1935) found that clay soils and the corresponding drifted materials were practically identical in composition. Wind erosion sometimes virtually removes the surface soil (Chepil, 1957*a, b;* Zingg, 1954). This nonselective removal by wind is associated primarily with loess sorted and deposited from the atmosphere during past geologic eras.

Wind-erosion equation

Studies to understand the mechanics of the wind-erosion process, identify major factors influencing wind erosion, and develop wind-erosion control methods led to the development of a wind-wrosion equation (Chepil and Woodruff, 1963; Woodruff and Siddoway, 1965). The general functional relationship between the independent variable E (the potential average annual soil loss in tons per acre), and the equivalent variables is: $E = f(I', K', C', L', V)$, where I' is a soil-erodibility index; K' is a soil-ridge roughness factor; C' is a climatic factor; L' is field length along the prevailing wind-erosion direction; and V is equivalent quantity of vegetative cover.

Erodibility Index. Soil erodibility (ease of detachment and transport by wind) was recognized early as a primary variable affecting wind erosion. From wind-tunnel tests, Chepil (1950) determined relative erodibilities of soils reasonably free from organic residues as a function of apparent specific gravity and of proportions of dry soil aggregates in various sizes. Clods larger than 0.84 mm in diameter were nonerodible. Since then, the nonerodible soil fraction greater than 0.84 mm, as determined by dry sieving, has been used to indicate erodibility of soil by wind. In an early version of the wind-erosion equation (Chepil and Woodruff, 1954), it was one of the three major factors developed from results obtained principally with a portable wind tunnel (Zingg, 1951*a, b;* Zingg and Woodruff, 1951). During the severe wind-erosion season, January 1 through April 1, 1954, 1955, and 1956, Chepil (1960) studied 69 fields in western Kansas and eastern Colorado to determine the quantity of soil loss (tons per acre per year) for any field erodibility as determined from various field conditions. Seasonal loss was converted to annual soil loss, and relative field erodibility for each field was determined (Chepil, 1959*a;* Chepil and Woodruff, 1954).

Ridge Roughness Factor. Chepil and Milne (1941*a*), investigating the influence of surface roughness on intensity of drifting dune materials and cultivated soils, found that the initial intensity of drifting was always much less over a ridged than over a smooth surface. Ridging cultivated soils reduced the severity of drifting, but ridging highly erosive dune materials was less effective because the ridges disappeared rapidly. The rate of flow varied inversely with surface roughness.

Early versions of the wind-erosion equation (Chepil and Woodruff, 1954; Chepil and Woodruff, 1959; Chepil, Woodruff, and Zingg, 1955) contained a ridge roughness equivalent derived as the product of residue and ridge roughness. As the wind-erosion equation evolved, the influences of soil ridge roughness and vegetative cover were evaluated independently. Armbrust, Chepil, and Siddoway (1964) studied the effects of ridge roughness equivalent on total quantity of eroded material from three simulated, cultivated soils exposed to different friction velocities. From their data a curve can be constructed showing the relationship between relative quantity of eroded material and ridge roughness equivalent.

Climatic Factor. To determine average annual soil loss for climatic conditions other than those prevailing when the relationship between wind tunnel and field erodibility was obtained, Chepil, Siddoway, and Armbrust (1962) proposed a climatic factor. It is an index of average rate of soil movement by wind as influenced by moisture content in surface soil particles and average windspeed. The soil moisture term of the climatic factor of the wind-erosion equation was developed on the basis that erodibility of a soil varies inversely with the equivalent moisture in surface soil particles (Chepil, 1956). The windspeed term of the climatic factor is based on the rate of soil movement being proportional to windspeed cubed. Several researchers (Bagnold, 1943; Chepil, 1945*a;* Zingg, 1953*a*) have reported that when windspeed exceeds those speeds required barely to move the soil, the soil-movement rate is directly proportional to friction velocity cubed. Over a specified surface, windspeed and friction velocity are proportional.

Field Width. Chepil and Milne (1941*b*) reported that rate of soil movement began with zero on the windward side (side from which the wind blows) of fields or field strips and increased with distance downwind. Later Chepil (1946) found that the cumulative rate of soil movement with distance away from the windward edge of eroding fields was the main cause of increasing abrasion and gradual decrease in surface roughness along the direction of wind. He called the increase in rate of soil flow with distance downwind "avalanching."

"Rate of soil flow increased with distance downwind across an eroding field until, if the field was large enough, it reached a maximum that a wind of a given velocity can carry. Beyond that point the rate of flow remained essentially constant" (Chepil, 1957*c*). Relationships between field erodibility and field width (Chepil, 1946, 1957*c;* Chepil and Milne, 1941*b*), considering the many associated factors, gave rise to the field-length term of the wind-erosion equation (Chepil and Woodruff, 1963; Woodruff and Siddoway, 1965). Field width was initially considered as the distance across a field in the prevailing wind-erosion direction. However, sometimes there is essentially no prevailing wind-erosion direction. Therefore preponderance of wind-erosion forces in the prevailing wind-erosion direction is now used to assess equivalent field width (Skidmore, 1965; Skidmore and Woodruff, 1968).

Vegetative Factor. Value of crop residue for controlling wind erosion was recognized early, and quantitative relationships were reported (Chepil, 1944). Amounts of wheat straw needed to protect most erodible dune sand and less erodible soils against strong winds were established (Chepil et al, 1960; Chepil et al, 1963). Standing stubble was much more effective than flattened stubble (Chepil, Woodruff, and Zingg, 1955). Standing sorghum stubble with rows perpendicular to wind direction controlled wind erosion much more effectively than did rows parallel to wind direction (Englehorn,

Zingg, and Woodruff, 1952; Skidmore, Nossaman, and Woodruff, 1966).

Siddoway, Chepil and Armbrust (1965) quantified the specific properties of vegetative covers influencing soil erodibility and developed regression equations relating soil loss by wind to selected amounts, kinds, and orientation of vegetative covers, wind velocity, and soil cloddiness. Those studies led to the relationship developed by Woodruff and Siddoway (1965) showing the influence of an equivalent vegetative cover of small grain and sorghum stubble for various orientations (flat, standing, height), then relating soil loss to equivalent vegetative cover.

Craig and Turelle (1964) presented equivalent vegetative cover for additional crops, including a figure for converting quantity of various crop residues (peanuts, soybeans, shredded cotton, quar, sesame, standing cotton stalks) to quantity of equivalent flat, small-grain residue. Woodruff et al (1974) developed an equation for converting quantity of cattle feedlot manure to flat, small grain residue, wind-erosion-control equivalent. Recent research (Lyles, Schrandt, and Schmeidler, 1974) indicates that if residue is standing and is equidistantly spaced, much less residue is needed to control wind erosion than has been previously reported.

Application. Relations among variables are complex, and a single equation that expresses *E* as a function of the dependent variables has not been devised. The equation was solved in a stepwise procedure involving graphical solutions.

Because many tables and figures are required to solve the functional relationships of the equation, a computer solution has been developed to simplify the procedure (Fisher and Skidmore, 1970; Skidmore, Fisher, and Woodruff, 1970). Other efforts to implement the use and to improve the accuracy of the wind-erosion equation include: evaluating monthly climatic factor (Skidmore and Woodruff, 1968; Woodruff and Armbrust, 1968); assessing wind-erosion forces, prevailing wind-erosion direction (Skidmore, 1965; Skidmore and Woodruff, 1968); evaluating the erodibility of organic soils (Woodruff and Dickerson, *Wind Erosion Research Laboratory Annual Reports,* 1971 and 1972); correlating feedlot solids with other types of vegetative cover (Woodruff et al, 1974); and improving trap-strip design (Hagen, Skidmore, and Dickerson, 1972).

The equation was designed as a tool to determine the potential erosion from a particular field and the field conditions (soil cloddiness, roughness, vegetative cover, sheltering by barriers, or width and orientation of field) necessary to reduce potential erosion to a tolerable amount (Woodruff and Siddoway, 1965). The information needed to assess potential soil loss from a field is: (1) percentage of soil aggregates exceeding 0.84 mm; (2) windward knoll slope; (3) ridge height and spacing; (4) climatic factor; (5) angle of deviation of prevailing wind-erosion direction from right angles to field strip; (6) preponderance of wind-erosion forces in prevailing wind-erosion direction; (7) height of wind barrier, if any; (8) field width; (9) quantity of vegetative cover; and (10) type of vegetative cover. Information for items 4 and 6 and for determining item 5 can be obtained from the literature (Skidmore and Woodruff, 1968) by month for many U.S. locations. The percentage of soil aggregates exceeding 0.84 mm (item 1) can best be obtained by dry sieving; however, in practice, the percentage is often determined from wind-erodibility groups (Hayes, 1972), based on soil type or predominant soil textural class. Other factors can be measured in the field or estimated by comparing field conditions with similar field conditions for which the factors have been measured.

The solution of the wind-erosion equation gives the amount of erosion expected, in tons per acre per year, from a given agricultural field. The second application of the equation is to specify the amount of erosion that can be tolerated, substitute it for *E,* and then solve the equation to determine conditions—such as amount of residue and field width—needed to limit soil loss to the tolerance level. The equation has been used widely for both purposes (Hayes, 1975; Skidmore, Fisher, and Woodruff, 1970; Skidmore and Woodruff, 1968; Woodruff and Siddoway, 1965). The U.S. Soil Conservation Service has used the equation extensively to plan wind-erosion-control practices and to determine crop tolerance to wind erosion (Hayes, 1965, 1966, 1972, 1975). The equation is also a useful guide to wind-erosion-control principles (Carreker, 1966; Moldenhauer and Duncan, 1969; Woodruff et al 1972). Other uses of the equation include: (1) determining spacing for barriers in narrow strip-barrier systems (Hagen, Skidmore, and Dickerson, 1972); (2) estimating fugitive dust emissions from agricultural and subdivision lands (PEDCO-Environmental Specialists, Inc., 1973; Wilson, 1975); (3) predicting horizontal soil fluxes to compare with vertical aerosol fluxes (Gillette, Blifford, and Fenster, 1972); (4) estimating effects of wind erosion on productivity (Lyles, 1974, 1975); and (5) evaluating stubble requirements in field strips to trap windblown soil (Lyles, Schmeidler, and Woodruff, 1973.)

The principles of wind-erosion control that have evolved through the development of a wind-erosion predictive model are: pro-

FIGURE 3. Establishing and maintaining vegatative cover is the "cardinal rule" of wind-erosion control.

duce greater percentage of clods greater than 0.84 mm, roughen the surface, reduce field length, and establish and maintain vegetative cover. This last item has been called the "cardinal rule" of wind-erosion control (Figure 3).

EDWARD L. SKIDMORE

References

Anderson, C. H., 1975. A history of soil erosion by wind in the Palliser Triangle of Western Canada, *Historical Series No. 8,* Research Branch, Canada Department of Agriculture, 25p.

Armbrust, D. V., 1968. Windblown soil abrasive injury to cotton plants, *Agron. Jour.,* **60,** 622–625.

Armbrust, D. V., 1972. Recovery and nutrient content of sandblasted soybean seedlings, *Agron. Jour.,* **64,** 707–708.

Armbrust, D. V.; Chepil, W. S.; and Siddoway, F. H., 1964. Effects of ridges on erosion of soil by wind, *Soil Sci. Soc. America Proc.,* **28,** 557–560.

Armbrust, D. V.; Paulsen, G. M.; and Ellis, R., Jr., 1974. Physiological responses to wind and sandblast damaged winter wheat plants, *Agron. Jour.,* **66,** 421–423.

Bagnold, R. A., 1943. *The Physics of Blown Sand and Desert Dunes.* New York: William Morrow, 265p.

Bisal, F., and Nielsen, K. F., 1962. Movement of soil particles in saltation, *Canadian Jour. Soil Sci.,* **42,** 81–86.

Bubenzer, G. D., and Weis, G. G., 1974. Effect of wind erosion on production of snap beans and peas, *Jour. Am. Soc. Hort. Sci.,* **99,** 527–529.

Carreker, J. R., 1966. Wind erosion in the Southeast, *Jour. Soil and Water Conserv.,* **21,** 86–88.

Chepil, W. S., 1944. Utilization of crop residues for wind erosion control, *Sci. Agric.,* **24,** 307–319.

Chepil, W. S., 1945*a*. Dynamics of wind erosion: I. Nature of movement of soil by wind, *Soil Sci.,* **60,** 305–320.

Chepil, W. S., 1945*b*. Dynamics of wind erosion: II. Initiation of soil movement, *Soil Sci.,* **60,** 397–411.

Chepil, W. S., 1945*c*. Dynamics of wind erosion: III. The transport capacity of the wind, *Soil Sci.,* **60,** 475–480.

Chepil, W. S., 1946. Dynamics of wind erosion: V. Cumulative intensity of soil drifting across eroding fields, *Soil Sci.,* **61,** 257–263.

Chepil, W. S., 1950. Properties of soil which influence wind erosion: II. Dry aggregate structure as an index of erodibility, *Soil Sci.,* **69,** 403–414.

Chepil, W. S., 1956. Influence of moisture on erodibility of soil by wind, *Soil Sci. Soc. America Proc.,* **20,** 288–292.

Chepil, W. S., 1957*a*. Sedimentary characteristics of dust storms: I. Sorting of wind-eroded soil material, *Am. Jour. Sci.,* **255,** 12–22.

Chepil, W. S., 1957*b*. Sedimentary characteristics of dust storms: III. Composition of suspended dust, *Am. Jour. Sci.,* **255,** 206–213.

Chepil, W. S., 1957*c*. Width of field strip to control wind erosion, *Kansas Agric. Expt. Sta. Tech. Bull. 92.,* 16p.

Chepil, W. S., 1959*a*. Wind erodibility of farm fields, *Jour. Soil and Water Conserv.,* **14,** 214–219.

Chepil, W. S., 1959*b*. Equilibrium of soil grains at the threshold of movement by wind, *Soil Sci. Soc. America Proc.,* **23,** 422–428.

Chepil, W. S., 1960. Conversion of relative field erodibility to annual soil loss by wind, *Soil Sci. Soc. America Proc.,* **24,** 143–145.

Chepil, W. S., and Milne, R. A., 1941*a*. Wind erosion of soil in relation to roughness of surface, *Soil Sci.,* **52,** 417–433.

Chepil, W. S., and Milne, R. A., 1941*b*. Wind erosion of soils in relation to size and nature of the exposed area, *Sci. Agric.,* **21,** 479–487.

Chepil, W. S.; Siddoway, F. H.; and Armbrust, D. V., 1962. Climatic factor for estimating wind erodibility of farm fields, *Jour. Soil and Water Conserv.,* **17,** 162–165.

Chepil, W. S., and Woodruff, N. P., 1954. Estimations of wind erodibility of field surfaces, *Jour. Soil and Water Conserv.,* **9,** 257–265, 285.

Chepil, W. S., and Woodruff, N. P., 1959. Estimations of wind erodibility of farm fields, *U.S. Dept. Agric. Prod. Res. Rpt. No. 25,* 21p.

Chepil, W. S., and Woodruff, N. P., 1963. The physics of wind erosion and its control, *Adv. Agron.,* **15,** 211–302.

Chepil, W. S.; Woodruff, N. P.; Siddoway, F. H.; Fryrear, D. W.; and Armbrust, D. V., 1963. Vegetative and nonvegetative materials to control wind and water erosion, *Soil Sci. Soc. America Proc.,* **27,** 86–89.

Chepil, W. S.; Woodruff, N. P.; Siddoway, F. H.; and Lyles, L., 1960. Anchoring vegetative mulches, *Agric. Engin.,* **41,** 754–755, 759.

Chepil, W. S.; Woodruff, N. P.; and Zingg, A. W., 1955. Field study of wind erosion in western Texas, *U.S. Dept. Agric., SCS-TP-125,* 60p

Claflin, L. E.; Stuteville, D. L.; and Armbrust, D. V., 1973. Windblown soil in the epidemiology of bacterial leaf spot of alfalfa and common blight of beans, *Phytopathology,* **63,** 1417–1419.

Craig, D. G., and Turelle, J. W., 1964. Guide for wind erosion control on cropland in the Great Plains states, Washington, D.C.: *U.S. Depart. Agric., SCS,* 104p.

Daniel, H. A., 1936. Physical changes in soils of the Southern High Plains due to cropping and wind erosion and the relation between the (sand and silt)/clay ratios in these soils, *Jour. Am. Soc. Agron.,* **28,** 570–580.

Daniel, H. A., and Langham, W. H., 1936. The effect of wind erosion and cultivation on the total nitrogen and organic matter content of soils in the Southern High Plains, *Jour. Am. Soc. Agron.*, **28**, 587–596.

Englehorn, C. L.; Zingg, A. W.; and Woodruff, N. P., 1952. The effects of plant residue cover and clod structure on soil losses by wind, *Soil Sci. Soc. America Proc.*, **16**, 29–33.

Fisher, P. S., and Skidmore, E. L., 1970. WEROS: A Fortran IV program to solve the wind erosion equation, *U.S. Dept. Agric., ARS 41-174*, 13p.

Food and Agricultural Organization of the United Nations (FAO), 1960. Soil erosion by wind and measures for its control on agricultural lands, *FAO Agricultural Development Paper No. 71*, 88p.

Fryrear, D. W., 1971. Survival and growth of cotton plants damaged by windblown sand, *Agron. Jour.*, **63**, 638–642.

Gillette, D. A.; Blifford, I. H., Jr.; and Fenster, C. R., 1972. Measurements of aerosol size distribution and vertical fluxes of aerosols on land subject to wind erosion, *Jour. Appl. Meteorology*, **11**, 977–987.

Greig, J. K.; Bokhari, N.; Armbrust, D. V.; and Anderson, L. C., 1974. Residual effects of wind and sandblast-damage on tomato plants at different stages of development, *Jour. Am. Soc. Hort. Sci.*, **99**, 530–534.

Hagen, L. J., and Woodruff, N. P., 1973. Air pollution from duststorms in the Great Plains, *Atm. Environ.*, **7**, 323–332.

Hagen, L. J.; Skidmore, E. L.; and Dickerson, J. D., 1972. Designing narrow strip barrier systems to control wind erosion, *Jour. Soil and Water Conserv.*, **27**, 269–270.

Hayes, W. A., 1965. Wind erosion equation useful in designing northeastern crop protection, *Jour. Soil and Water Conserv.*, **20**, 153–155.

Hayes, W. A., 1966. Guide for wind erosion control in the northeastern states, *U.S. Dept. Agric. SCS*, Upper Darby, Pennsylvania, 26p.

Hayes, W. A., 1972. Designing wind erosion control systems in the Midwest Region, *RTSC–Technical Note–Agronomy–LI-9*, 23p.

Hayes, W. A., 1975. Estimating wind erosion in the field, *in Land Use: Food & Living. Proc. 30th Ann. Mtg., SCSA, August 10-13, San Antonio, Texas*, 139–143.

Johnson, V., 1947. *Heaven's Tableland–The Dust Bowl Story*. New York: Farrar, Straus, 155, 156, 157.

Lyles, L., 1974. Speculation on the effect of wind erosion on productivity. *Special Rpt. to USDA Task Force on Wind Erosion Damage Estimates*, 17p.

Lyles L., 1975. Possible effects of wind erosion on soil productivity, *Jour. Soil and Water Conserv.*, **30**, 279–283.

Lyles, L., and Krauss, R. K., 1971. Threshold velocities and initial particle motion as influenced by air turbulence, *Trans. ASAE*, **14**, 563–566.

Lyles, L., and Woodruff, N. P., 1960. Abrasive action of windblown soil on plant seedlings, *Agron. Jour.*, **52**, 533–536.

Lyles, L.; Schmeidler, N. F.; and Woodruff, N. P., 1973. Stubble requirements in field strips to trap windblown soil, *Kansas Agric. Expt. Sta. Res. Pub. 164*, 22p.

Lyles, L.; Schrandt, R. L.; and Schmeidler, N. F., 1974. How aerodynamic roughness elements control sand movement, *Trans. Am. Soc. Agric. Engineers*, **17**, 134–139.

Malin, J. C., 1946*a*. Dust storms–Part one, 1850–1860, *Kansas Histor. Quart.*, **14**, 129–144.

Malin, J. C., 1946*b*. Dust storms–Part two, 1861–1880, *Kansas Histor. Quart.*, **14**, 265–296.

Malin, J. C., 1946*c*. Dust storms–Part three, 1881–1900, *Kansas Histor. Quart.*, **14**, 391–413.

Malina, F. J., 1941. Recent developments in the dynamics of wind erosion, *Am. Geophys. Union Trans.*, 262–284.

Moldenhauer, W. C., and Duncan, E. R., 1969. Principles and methods of wind erosion control in Iowa, *Special Rpt. No. 62, Iowa State University*, 19p.

Moss, H. C., 1935. Some field and laboratory studies of soil drifting in Saskatchewan, *Sci. Agric.*, **15**, 665–675.

PEDCO-Environmental Specialists, Inc., Cincinnati, Ohio, 1973. *Investigations of fugitive dust-sources, emissions, and control.* Report prepared under Contract No. 68-02-0044, Task Order No. 9, U.S. Environmental Protection Agency, Office of Air Quality Planning and Standards, Research Triangle Park, North Carolina, 151p.

Siddoway, F. H.; Chepil, W. S.; and Armbrust, D. V., 1965. Effect of kind, amount and placement of residue on wind erosion control, *Trans. Amer. Soc. Agric. Engineers*, **8**, 327–331.

Skidmore, E. L., 1965. Assessing wind erosion forces: Directions and relative magnitudes, *Soil Sci. Soc. America Proc.*, **29**, 587–590.

Skidmore, E. L., 1966. Wind and sandblast injury to seedling green beans, *Agron. Jour.* **58**, 314–315.

Skidmore, E. L.; Fisher, P. S.; and Woodruff, N. P., 1970. Wind erosion equation: Computer solution and application, *Soil Sci. Soc. America Proc.*, **34**, 931–935.

Skidmore, E. L.; Nossaman, N. L.; and Woodruff, N. P., 1966. Wind erosion as influenced by row spacing, row direction, and grain sorghum population, *Soil Sci. Soc. America Proc.*, **39**, 505–509.

Skidmore, E. L., and Woodruff, N. P., 1968. Wind erosion forces in the United States and their use in predicting soil loss, *USDA, ARS Agric. Handbook No. 346*, 42p.

Svobida, L., 1940. *An Empire of Dust.* Caldwell, Idaho: Caxton Printers, 203p.

U.S. Department of Agriculture, 1965. *Soil & Water Conservation Needs–A National Inventory.* Misc. Pub. No. 971, 94p.

Vanoni, V. A., 1975. Sedimentation engineering. *ASCE–Manuals and Reports on Engineering Practice, No. 54*, pp. 230–245.

Wilson, L., 1975. Application of the wind erosion equation to predict fugitive dust emissions, *Jour. Soil and Water Conserv.*, **30**, 215–219.

Woodruff, N. P., 1956. Wind-blown soil abrasive injuries to winter wheat plants, *Agron. Jour.*, **48**, 499–504.

Woodruff, N. P., and Armbrust, D. V., 1968. A monthly climatic factor for the wind erosion equation, *Jour. Soil and Water Conserv.*, **23**, 103–104.

Woodruff, N. P.; Lyles, L.; Dickerson, J. D.; and Armbrust, D. V., 1974. Using cattle feedlot manure to control wind erosion, *Jour. Soil and Water Conserv.*, **29**, 127–129.

Woodruff, N. P.; Lyles, L.; Siddoway, F. H.; and Fryrear, D. W., 1972. How to control wind erosion, *USDA, Agric. Inf. Bull. No. 354.*, 22p.

Woodruff, N. P., and Siddoway, F. H., 1965. A wind erosion equation, *Soil Sci. Soc. America Proc.*, **29**, 602–608.

Zingg, Z. W., 1951*a*. Evaluation of erodibility of field surfaces with a portable wind tunnel, *Soil Sci. Soc. America Proc.*, **15**, 11–17.

Zingg, A. W., 1951*b*. A portable wind tunnel and dust collector developed to evaluate the erodibility of field surfaces, *Agron. Jour.*, **43**, 189–191.

Zingg, A. W., 1953*a*. Wind-tunnel studies of the movement of sedimentary materials, *Proc. 5th Hydraulic Conference, Iowa Institute of Hydraulics Res. Bull. 34*, 111–135.

Zingg, A. W., 1953*b*. *Some Characteristics of Aeolian Sand Movement by Saltation Process.* Paris: Editions du Centre National de la Recherche Scientifique 13, Quai Anatole, France, 197–208.

Zingg, A. W., 1954. The wind erosion problem in the Great Plains, *Trans. Am. Geophys. Union*, **35**, 252–258.

Zingg, A. W., and Woodruff, N. P., 1951. Calibration of a portable wind tunnel for the simple determination of roughness and drag on field surfaces, *Agron. Jour.*, **43**, 191–193.

Cross-references: *Conservation; Management of Soils; Wind Erosion. Vol. III: Deflation; Eolian Transport; Erosion; Wind Action. Vol. IVA: Soil Erosion. Vol. XVI: Erosion Hazards.*

Z

ZERO-TILLAGE

Zero-tillage is a system of seedbed preparation without primary or secondary cultivation where mechanical soil disturbance and manipulation is limited to seeding and fertilizer placement. A continuous cover of mulch (q.v.) of crop residues or dead weeds is an essential component.

Relevance to Shifting Agriculture

Shifting agriculture is a system of land management under which food is produced by reliance on nature or simple technology rather than on modern technology. A small tract is cultivated until the yield can no longer sustain the farm family. The tract is then abandoned temporarily and another piece of land is cleared for cultivation. Approximately 5 to 6 billion ha in the tropics are cultivated under this system of subsistence farming. With the increasing pressure of population and desire to improve standards of living, this system has to be replaced by land management that will provide greater production but on a sustained basis.

The essential components of soil management in the tropics include: (1) minimum disturbance of the soil surface, (2) continuous ground cover of dead crop residue, (3) mechanisms for replenishing nitrogen by leguminous cover crops, and (4) maintenance of soil structure through natural agencies such as earth worms and termites (Greenland, 1975). The zero-tillage system of soil management appears to meet most of these requirements.

Form of Zero-tillage

Although the possibilities of zero-tillage are promising, its adaptation to specific soil conditions depends on soil types, nature of the crop to be grown, and financial resources of the farmer. Under shifting cultivations, where farm sizes are less than 2 ha, only chemical or manual weed control and planting directly through mulch are required. For commercial, large-scale farms, however, the zero-tillage concept may be applied with one of the following modifications, as described by Wittmus et al (1973):

1. *Zero-till* or *no-till* is a system whereby a crop is planted directly into a seedbed untilled since the harvest of a previous crop. This is also referred to as *sod plant, slot plant,* or *slit plant.*
2. *Till-planting* is a strip-tillage system in which a small strip is cut through the stubble ahead of the surface planters in one operation.
3. *Minimum tillage* is the minimum soil manipulation necessary for crop production under given soil and climatic conditions.
4. *Stubble mulching* or *mulch tillage* involves preparing soil in such a way that plant residues and other materials are left to cover the soil surface before and after crop establishment.

In this entry *conventional tillage* (q.v.) means a system of seedbed preparation including both primary and secondary tillage operations such as plowing, harrowing, and disking.

Historical Background

Zero-tillage techniques have been used in the last decade in the United States, United Kingdom, and Europe. Reviews by Kuipers (1970) and Baeumer and Bakermans (1973) and a symposium organized by the Soil Conservation Society of America (1973) outline the development of zero-tillage techniques. Early experiments conducted at Rothamsted provided the necessary guidelines for the minimum-tillage concept. From these trials and later work, Russell (1941, 1953) concluded that the major advantage of tillage was weed control and that the nitrogen deficiency symptoms on shallow-tilled plots were primarily due to weed competition.

Experiments in Ohio showed that maize yields decreased over a period of years with shallow cultivation and that weed problems were serious (Page, 1946). Page, Willard, and McCune (1946) found that tillage could be detrimental to soil structure and concluded that "good soil structure—tilth—is not made and may easily be destroyed by gasoline."

Zero-tillage became feasible with the availability of herbicides in the 1950s. Using herbicides, Davidson and Barrons (1954) and Skogley (1957) devised a reduced-tillage system for corn production in the United States. A till-plant system was developed in the 1960s for row crops to prevent soil erosion and to minimize energy input. Herbicides became an essential input in this system. Because of the serious weed problems with reduced tillage, a satisfactory herbicide should kill existing vegetation and prevent weed germination and establishment without harmful effects on crops.

As a result, the total acreage cropped with various forms of zero-tillage methods in the United States in 1972 reached nearly one million hectares (Bagley, 1973). The acreage of cereals under direct drilling in the United Kingdom in 1971 was about 4000 ha (Cannell and Ellis, 1970). The future of zero-tillage and its problems for different crops in the United States has been reviewed by Van Doren and Triplett (1973).

Even though the zero-tillage concept was developed in temperate regions, those techniques have not been widely adopted because of delayed soil warming under mulch in the spring. This delay is an advantage in the tropics where soil temperatures can be too high in the early stages of crop growth. The maintenance of porosity and structure favorable for root growth poses special problems in the tropics where soils are not subject to periodic freezing and thawing. Adverse effects of tillage are diffcult to overcome. Zero-tillage thus holds considerable promise in the tropics.

Soil Conditions under Zero and Conventional Tillage

Soil Structure. Both soil structure (q.v.) and pore-size distribution (q.v.) with their effects on infiltration (q.v.) and water content and retention (q.v.) can be damaged by manipulation of many soils in the tropics. One important effect is that of reducing earthworm activity. In contrast, the use of mulches stimulates the activity of earthworms, which in turn helps keep the surface soil porous and in good physical condition. On an Alfisol in Nigeria, for example, the quantities of earthworm casts produced per hectare per month were at the rate of 19 tons on mulched and 1 ton on plowed plots (Table 1). Earthworm casts are aggregates of soil particles that are resistant to raindrop impacts. Worm channels also facilitate root penetration into the deeper layers of soils. The activities of earthworms and other invertebrates can maintain a soil under zero-tillage in a highly porous condition with a lower bulk density. The bulk density (q.v.) of surface soil 40 days after seeding was 1.3 g/cc under zero-tillage as compared to 1.4 g/cc under conventional tillage.

TABLE 1. Maximum Worm Activity Under Zero-till and Conventional Plowing.

Crop	Worm Casts/m^2/week Zero-till	Plowing
Maise	2600	0
Cowpeas	1800	70
Mean	1700	35

Infiltration Rate. As a result of good soil structure, high porosity, and minimal crusting, infiltration can be maintained at a very high rate under zero-tillage. For example, four years after forest clearing in Nigeria, the infiltration rates of soil with zero-tillage declined from 3.5 to 1.1 cm min^{-1} and from 3.5 to less than 0.1 cm min^{-1} for a plowed bare ground surface (Lal, 1976). Zero-tillage without adequate crop residues will also reduce infiltration rates. The infiltration rate of zero-tillage plots without mulch was 0.5 cm min^{-1} as compared to 0.7 cm min^{-1} with mulch.

Soil Erosion. Because zero-tillage prevents soil erosion, conventional control measures are usually not required. The high infiltration rate and better soil structure and pore-space distribution minimizes runoff and soil loss. Mechanical cultivation of soil reduces the aggregate stability and promotes puddling (q.v.) and sealing of the surface, thereby lessening the infiltration capacity and increasing runoff. Table 2 shows the runoff and erosion loss from conventionally plowed and zero-till soybeans on various slopes. Soil loss from zero-tillage was negligible even on a 15% slope. Mean soil loss from plowed plots was 2000 times greater than that from zero-tilled plots. The runoff from plowed plots was nine times higher than zero-till plots.

The zero-till technique thus promises to prevent erosion in highly erodable soils in the tropics.

Soil Moisture Regime. Crops on zero-till plots are less susceptible to drought than on

TABLE 2. Soil and Water Loss from Zero-till and Plowed Soybeans. (Rainfall = 42 mm)

Slope (%)	Runoff (% of Rainfall) Zero-till	Plowed	Soil Loss (kg/ha) Zero-till	Plowed
1	0.9	8.8	0.2	40.0
5	2.1	24.5	5.0	620.0
10	2.4	13.6	1.0	7620.0
15	2.4	21.4	1.0	6910.0
Mean	1.9	17.1	1.8	3800.0

(Lal, 1975. Data presented at the 67th Annual Meeting of the ASA, Knoxville, August 24-30, 1975.)

conventionally plowed land. Soil moisture storage is greater in zero-till plots because of minimal water runoff, changes in the moisture-release characteristics, and reduced evaporation losses due to the insulating effect of a mulch layer. Plants grown with a zero-till system can, in general, withstand short dry spells much better than those on conventionally plowed unmulched plots. The lesser drought susceptibility of crops grown on zero-till plots may be attributed to differences in rooting patterns and changes in pore-size distribution and moisture status of soils.

Soil Temperature. Supraoptimal soil temperature conditions in some tropical soils can seriously affect plant growth with an unmulched, conventionally plowed system of soil management. The optimum soil temperature for maize and soybeans is 25 to 30°C. Soil temperature higher than 35°C can be injurious to crop yield (Lal, 1975*a*, *b*). The maximum soil temperature measured at 5 cm depth after conventional plowing has been as high as 45, 43, and 41°C, respectively, under maize, soybeans, and cowpeas. The maximum soil temperature under similar conditions with a zero-till system was 34°C under all these crops. The soil temperature difference even at 20 cm depth can be as much as 4 to 5°C between two tillage systems. In addition to the differences in the absolute maximum soil temperature, there are also phase differences in the propagation of temperature waves. The maxima in surface layers can be two to three hours later in zero-tilled than in conventionally tilled plots.

Nutrient Status and Availability. The organic matter content of the surface soil is maintained at a higher level with zero-tillage techniques than with conventional plowing. For the surface 0–10 cm layer, the organic carbon content and total nitrogen in the zero-tillage plots five years after forest clearing were 1.30% and 0.103%, respectively. The organic carbon and total nitrogen in the plowed plots were 0.86% and 0.07%, respectively. Because of the differences in organic matter content due to tillage systems, there can be significant differences in the cation exchange capacity of the surface soil. There is also evidence that phosphorus availability is enhanced in the mulched layer of zero-tilled soils. However, there may be greater nitrogen requirements for a zero-till system, particularly in the first few seasons of crop growth.

Crop Response to Zero-Tillage

Mulch farming techniques applied to Alfisols in Sri Lanka suggest that cotton yields can be improved by 32%; the plants were, however, less vigorous in the early stages of growth during wet years. (Abeyratne, 1960). Thick mulch during heavy rains acted as a sponge, and cotton seedlings were adversely affected. In Ghana, Kannegieter (1969) reported that zero-till techniques using *Puerarie phaseoloides* as sod crops subsequently increased the yield of maize. Nigerian experience indicates that maize grown with zero-tillage can produce grain yields equivalent to those with conventional plowing in good seasons and better yields in seasons with frequent drought stress (Lal, 1974, 1975*a*).

Germination and Seedling Establishments. Under tropical conditions with well-drained soils, germination and seedling establishment with a zero-tillage system may be better than with conventional plowing, provided seeding is done with equipment capable of planting through crop residues or cover crops. Soybean emergence in zero-till plots at the International Institute of Tropical Agriculture (IITA) was, for example, significantly better than with conventional plowing, probably because of better soil moisture and temperature regime and lack of surface crust formation. The time required for soybean emergence in conventionally plowed plots was longer. No difficulties have been experienced with seedling establishment for maize, soybean, cowpeas, pigeonpeas, and cassava with zero-till techniques.

Root Growth. Restricted root growth in zero-till plots during the seedling stage could be related to the compacted surface layer, the inadequate amounts of mulch, or to excessive wetness. Experiments conducted on Alfisols in Nigeria indicate that although root growth and development is restricted in the seedling stage, root proliferation in zero-tillage plots is superior in terms of depth of penetration at later stages of growth. Thus the superior yield potential of a zero-tillage system during drought stress is attributed to a deeper root system.

Plant Growth and Yield. Crop response to tillage systems depends on various factors such as soil type and climatic conditions. If the initial soil conditions are favorable, zero-tillage can produce a yield equal to that with conventional tillage; under adverse conditions such as drought, zero-tillage may actually out-yield conventional plowing.

Even though the initial growth of a crop under zero-tillage is sometimes suppressed, it usually improves during the later stages of growth. The annual grain yield of various crop rotations with conventional plowing and zero-tillage techniques has been reported by Lal (1975*a*). The annual grain yields obtained with zero-tillage at IITA in 1973 were, for example, 27% higher for maize-maize, 59% higher for maize-cowpea, 3% higher for pigeonpeas-maize, 22% lower for soybeans-soybeans, and 27% higher for cowpea-cowpea rotation than with conventional plowing. The yield of maize with zero-tillage has been higher in three out of four years.

Weed Competition. Weeds can be a serious problem with zero-tillage, particularly for the first year or two. However, if carefully controlled in the initial stages of land development, weeds under zero-tillage may be less of a problem than with conventional plowing. Thick mulch layers in zero-tillage helps suppress weed growth. Rhizomatus weeds such as *Imperata cylindrica* and *Talinum triangulare* pose a more serious problem with zero-tillage than with plowed plots. Systemic herbicides can help eradicate such weeds. Weeds of the graminae family have been observed to be more prevalent in conventionally plowed land. With time, some weeds resistant to herbicides, may dominate the zero-till plots. Adequate crop rotation may help solve this problem

Insects, Pests, and Diseases. The incidence of insects, pests, and diseases depends on many factors. Limited experiments conducted at IITA indicate that the population of parasitic nematodes in maize is five times greater in plowed than in zero-tillage plots. Similarly, the insect infestation, i.e., stalk borer damage in maize, has been shown to be much higher with conventional plowing.

Desired Improvements

For soil and water conservation, zero-tillage is a desirable approach. Under many circumstances, zero-tillage may be the best approach. Presently, however, the zero-tillage technique is not perfected as it is still soil and crop specific. Its present specificity calls for improvements to make it more versatile.

Conclusions

Zero-tillage techniques are synonymous with conservation tillage, zonal tillage, no-tillage, slit or slot tillage, and minimum tillage. The essential component is the minimal disturbance of soil surface, use of crop residues as mulch and weed control by alternate means other than plowing. The advantages of zero-tillage for soil and water conservation, improvement of soil structure and maintenance of organic matter, and nutrient-retention capacity of the soil are large. The advantages open avenues for zero-tillage as a substitute for forest fallow in the areas of shifting cultivation in the humid tropics.

Grain yields of crops such as maize, cowpeas, pigeonpeas, and soybeans obtained with zero-tillage have been encouraging because they compare favorably with conventional tillage. During drought stress, zero-tillage has proved to be better than conventional plowing.

Problems associated with zero-tillage that should be investigated include the development of (1) equipment for planting and fertilizer placement, (2) alternate systems of weed control, and (3) suitable crop rotations that include cover crops.

R. LAL

References

Abeyratne, E. L. F., 1960. Dry land farming in Ceylon, *Trop. Agric. (Ceylon)*, **112**(3), 191–229.

Baeumer, K., and Bakermans, W. A. P., 1973. Zero-tillage, *Adv. Agron.*, **25**, 77–120.

Bagley, G. R., 1973. Conservation tillage doesn't cost –It will pay, *in Conservation Tillage, The Proceedings of a National Conference of the Soil Conservation Society of America 1973,* 1–4.

Cannell, R. Q., and Ellis, F. B., 1973. *Reduced cultivation for cereal crops, Annual Report 1973.* Berkshire: Letcombe Laboratory, Wantage, 43–49.

Chaminad, R., 1970. Agricultural research work carried out by I.A.R.T., *L'Agron. Trop.*, **25**(2), 2–16.

Charreau, C., and Fauck, R., 1970. Mise au point sue l'utilisation agricole des sols de la sablo argileux de la zone tropicale seche Ouest-Africaine et ses incidences agronomiques, *L'Agron. Trop.*, **26**(9), 903–78.

Davidson, J. H., and Barrons, K. C., 1954. Chemical seedbed preparation–New approach to soil conservation, *Down to Earth* (Winter).

Greenland, D. J., 1975. Bringing the green revolution to the shifting cultivator, *Science,* **190**, 841–844.

Haines, W. B., and Keen, B. A., 1925*a*. Studies on soil cultivation, II, *Jour. Agric. Sci.*, **15**, 387–394.

Haines, W. B., and Keen, B. A., 1925*b*. Studies on soil cultivation, III, *Jour. Agric. Sci.*, **15**, 395–406.

Haines, W. B., and Keen, B. A., 1928. Studies on soil cultivation, IV, *Jour. Agric. Sci.*, **18**, 724–733.

Kannegieter, A., 1969. Zero cultivation and other methods of reclaiming pueraria fallowed land for food crop cultivation in the forest zone of Ghana, *Trop. Agric. (Ceylon)*, **123**, 1, 2.

Keen, B. A., 1930. Studies on soil cultivation, V, *Jour. Agric. Sci.,* **20**, 364–389.

Keen, B. A., and Haines, W. B., 1925. Studies on soil cultivation, I, *Jour. Agric. Sci.*, **15**, 375–386.

Keen, B. A., and Cashen, G. H., 1932. Studies on cultivation, VI, *Jour. Agric. Sci.*, **22**, 126–134.

Kuipers, H., 1970. Historical notes on the zero-tillage concept, *Neth. Jour. Agric. Sci.*, **18**, 219–224.

Lal, R., 1973. Soil erosion in shifting agriculture, *FAO Soils Bull. 24,* 48–71.

Lal, R., 1974. No-tillage effects on soil conditions and maize production in Western Nigeria, *Plant and Soil,* **40**, 321–331.

Lal, R., 1975*a*. Role of mulching technique in tropical soil and water management, *IITA–Technical Bull. No. 1,* 38p.

Lal, R., 1975*b*. No-tillage effects on soil properties under different crops in Western Nigeria, *Soil Sci. Soc. America Proc.*, **40**(5), 762–768.

Lal, R., 1976. Soil erosion problems on an Alfisol in western Nigeria and their control, *IITA Monograph 1*, 117–124.

Page, J. B.; Willard, C. J.; and McCune, G. W., 1946. Progress report on tillage methods in preparing land for corn, *Soil Sci. Soc. America Proc.*, **11**, 70–80.

Russell, E. W., 1953. The effects of methods of cultivation on crop yield, *Landbouwk. Tijdschr.*, **65**, 169–184.

Russell, E. W., and Keen, B. A., 1938. Studies on soil cultivation, VII, *Jour. Agric. Sci.*, **28**, 212–223.

Russell, E. W., and Keen, B. A., 1941. Studies on soil cultivation, X, *Jour. Agric. Sci.*, **31**, 326–347.

Skogley, C. R., 1957. Seedbed preparation for corn after sod. Ph.D. Thesis, Rutgers University, New Brunswick, New Jersey.

Stibbe, E., and Ariel, D., 1970. No-tillage as compared to tillage practices in dry land farming of a semi-arid climate. *Neth. Jour. Agric. Sci.*, **18**, 293–307.

Van Doren, D. M., Jr., and Triplett, G. B., Jr., 1973. *The Future of No-tillage in the U.S.A., 6th Internat. Conf. Soil Tillage, Wageningen, Sept. 24–29, 1973*, 21–23.

Wittmus, H. D.; Triplett, G. B., Jr.; and Greb, B. W., 1973. Concept of conservation tillage systems using surface mulches, *Conservation Tillage–Proceedings of a National Conference of the Soil Conservation Society of America*, March 28–30, 1973, in Des Moines, Iowa, 5–12.

Cross-references: *Conservation; Management of Soils; Moisture Management; Mulch; Tillage.*

SUBJECT INDEX

Boldface entries represent article titles; boldface numbers indicate page of main articles.

AUTHOR CITATION INDEX